W9-ACX-207

THE CORROSION AND OXIDATION OF METALS: SCIENTIFIC PRINCIPLES AND PRACTICAL APPLICATIONS

BY

ULICK R. EVANS, Sc.D., F.R.S., F.I.M.

HONORARY FELLOW, KING'S COLLEGE, CAMBRIDGE
EMERITUS READER IN THE SCIENCE OF METALLIC CORROSION,
CAMBRIDGE UNIVERSITY

LONDON
EDWARD ARNOLD (PUBLISHERS) LTD.

First published in 1960
Reprinted 1961

Printed in Great Britain by
Butler & Tanner Ltd., Frome and London

PREFACE

Of the Author's last two books on corrosion, the short " Introduction to Metallic Corrosion " was intended for students, and appears to be fulfilling its purpose; it is, however, too elementary to satisfy those engaged in pure-science research, and too condensed to afford safe guidance to engineers or industrial chemists in practical problems. The long " Metallic Corrosion, Passivity and Protection ", published in 1937, received an encouraging welcome; a drastic revision—almost a new book—appeared in 1946. To produce a third edition would be difficult for the author and probably useless to the reader. Owing to the vast out-pouring of corrosion papers in recent years, any book which endeavoured to summarize even a selection of them would be either long and expensive, or so condensed as to be misleading; neither form would give the reader any clear picture of the subject.

I have decided to write a new book based on a new plan. The present volume represents my personal views on corrosion, with a summary of the experimental work which has influenced me in reaching those views: this, of course, includes much work carried on outside Cambridge. The opinions reached by other investigators have not been neglected; the Author Index contains about 3000 names. However, a different policy has been adopted for the introduction of divergent views regarding scientific mechanism from that suited to the sections dealing with engineering or industrial applications.

In the scientific sections, it has seemed best to let the different schools of thought speak for themselves. Accordingly, at suitable places (generally at the ends of paragraphs, sections or chapters), references are provided to papers and books expressing views, not necessarily inconsistent with those put forward in the present book, but approaching the subject from different angles. The reader is strongly urged to consult some of these in libraries. To help his choice, a few words are often added to indicate the subject matter of each paper quoted and the class of reader likely to be helped by it; but no attempt is made to summarize in a few lines a paper occupying many pages.

In the industrial sections, it has generally been found possible to summarize the facts established by the observations of competent observers, and then give references where fuller details can be found.

I have, of course, been influenced, consciously or otherwise, by other papers too numerous to be cited; a writer who fails to find his name in the index is asked to accept my apologies, with the assurance that its omission does not necessarily mean that he has not contributed to the conclusions reached in the book.

The placing of references at the end of paragraphs is necessary for the objects explained above. It is believed that this will not unduly interrupt the thread of the argument. A reader will be well advised to have pencil and paper at hand; when he reaches references, he will make a note of any which seem to concern him; otherwise he will pass on.*

The choice of title has presented some difficulty; the word " corrosion " lays too much stress on the unlovely aspects of those reactions in which metal passes from the elementary to the combined state; yet those reactions should have special attractions for the pure scientist, since they conform to rational laws and principles, so that the reaction-rate can sometimes be calculated from physical measurements, with good agreement between calculated and observed values. There appears, however, to be no suitable alternative to the term " corrosion "; the word has been used throughout the book to include dry oxidation, but, to avoid misapprehension, it has seemed best to add the word " oxidation " in the title.

The first chapter headed " The Approaches to Corrosion ", contains much which might well be placed in the preface—but for the desire to keep this preface short. I explain therein that the subject of corrosion possesses potential interest to diverse classes of persons, including pure scientists, applied scientists and engineers; that there is also an important economic aspect and a relationship with public health. Corrosion is in fact a vast edifice which appears entirely different according to the angle from which it is approached. Clearly a book designed to help various classes of readers demands some tolerance from all. I am not convinced that it is necessary, or even advisable, to write a separate book for each class; it is indeed good discipline for one man to be compelled to look through the eyes of another.

The needs of engineers require some discussion here. I have lectured to engineers for thirty years, and have never adopted the plan (advocated in some quarters) of giving them a statement more over-simplified than would be provided, say, for metallurgists. Except in one respect, the engineer can understand the subject as easily as anyone—more easily, indeed, than those persons who are allergic to mathematics. It has to be recognized, however, that the engineer today is taught very little chemistry—a matter which ought to be rectified for a number of reasons. To help him, I have provided an appendix presenting—admittedly in highly condensed form—those principles of chemistry needed for the understanding of corrosion, choosing examples from compounds mentioned in the body of the book.

A second appendix—devoted to physical chemistry, especially electrochemistry—may make a wider appeal; many pure scientists today receive

* There are, of course, five places in any book where references can be inserted:—(1) Within paragraphs, (2) at the ends of paragraphs, (3) in foot-notes below each page, (4) in short lists at end of chapters, (5) in one long list at the end of the volume. Most of these plans have been tried and rejected. The end of paragraphs has proved the only suitable position. The plan of foot-note references, used in my earlier books, was abandoned for the present one, since it too often means that, unless there is much repetition of names, the man who did the main experimental work remains unmentioned in the text; this partly results from the development of team-work, leading to papers with four to eight authors.

little instruction in electrochemistry—and that mainly from the thermodynamic angle, with neglect of kinetic aspects. Both appendices are highly condensed. If they fail to provide what is wanted, text-books on chemistry or electrochemistry must be consulted.

To help those who have forgotten the mathematics which they learnt at school, or who react unfavourably to symbolism, the first eighteen chapters have been kept almost free from equations (except in foot-notes); thereafter, the treatment becomes quantitative. I have included in the quantitative section matter which, although not mathematical, appears inappropriate for the earlier, more elementary, section.

It is hardly to be expected that the average reader will read the whole book. Possibly, however, he will read with some care the chapter discussing the part of the subject with which he is most concerned, will skim the rest, and afterwards treat the book as a work of reference. In some cases, however, it may be justifiable to hope for something more than this. Where-ever possible, a chapter has been made self-contained. Someone already possessing a general knowledge of corrosion may perhaps care to treat individual chapters as isolated essays, reading one of them when he finds time and inclination—so that in due course the greater part of the book will have been calmly read and pondered. Such a reader may be helped by the first section of each chapter, which for lack of a better word has been called the "synopsis"; besides summarizing the contents of the chapter, it shows the relation to others, including earlier chapters where the elements of the knowledge have been introduced, and later chapters where further development may be sought. The endeavour to make the chapters self-contained has involved, of course, some repetition—which seems, in the circumstances, justifiable.

In some branches of industry and technological developments, a reader may have access to confidential files containing extensive data unknown to myself—or, if known, not available for printing. Such branches receive little direct treatment in this book, but the general principles laid down may prove useful to a privileged reader in applying the special data to which he has access.

I have made no attempt to treat the subject historically although a few "anecdotes" have been included. Those interested in the history of science will find a "Historical Note" in my "Introduction to Metallic Corrosion".

Nearly all the references quoted have been consulted in the original papers; where only abstracts have been accessible, these abstracts are quoted in the text. Russian papers have in many cases been consulted in translations, and the references to these translations are quoted; unless otherwise stated, the translations are those prepared by the Consultants' Bureau, New York. Patents are only quoted in cases where there is no adequate information in papers published in journals; the fact that a patent may not be mentioned in respect of certain inventions or processes should not be taken to imply that no patent protection exists.

The book would have been impossible but for the kind help of many friends who have provided information on matters where they possess expert knowledge. I am particularly grateful to those who have kindly acted as "scrutineers" by reading one or more chapters. These have all pointed out numerous errors present in the drafts sent them and have made many valuable suggestions, the great majority of which I have adopted or utilized indirectly. I wish, however, to make it perfectly clear that the appearance of a scrutineer's name does not mean that he has made himself even remotely responsible for the correctness of the facts stated in the chapter or that he necessarily agrees with the interpretations offered. I am also asked to state that in no case has the scrutineer been the author of the chapter indicated, although in a few cases a scrutineer has very kindly added a section, or rewritten a few pages. A list of scrutineers appears opposite.

A list of those who have helped in other ways is provided on p. viii; I would express my sincere thanks to all of these.

I would also take the opportunity to acknowledge my gratitude to others who have helped Corrosion Research at Cambridge University during the past fifty years. Two of these are no longer alive—Col. C. T. Heycock and Dr. W. H. Hatfield. The others have in turn been Heads of the University Department of Metallurgy; they are Professor R. S. Hutton, Professor G. Wesley Austin and Professor A. H. Cottrell.

U. R. EVANS

19 Manor Court,
Grange Road, Cambridge.
1959.

* * *

At the last moment I have introduced (p. 1042) an Addendum, designed to correlate Chapters V and VII which deal with Inhibition and Anodic Passivation, and to suggest why some solutions render iron passive on simple immersion whilst others require application of anodic current. By selecting a simple example (iron in copper nitrate solution), a simple explanation is provided for the formation of an oxide-film by a liquid which might well be expected to produce a soluble salt. This may prove a more acceptable treatment of the subject than that offered in Chapters V and VII, and readers may care to study the Addendum along with those chapters.

U. R. E.

ACKNOWLEDGMENTS

LIST OF SCRUTINEERS WHO HAVE READ THROUGH CHAPTERS IN DRAFT FORM (*see also statement opposite*)

Chap. I	F. P. Dunn
Chap. II	J. N. Wanklyn
Chap. III	J. N. Wanklyn; parts by W. H. J. Vernon
Chap. IV	F. Wormwell
Chap. V	R. S. Thornhill
Chap. VI	(Mrs.) V. E. Rance
Chap. VII	T. P. Hoar
Chap. VIII	F. Wormwell
Chap. IX	J. P. Chilton; with list of references by R. S. Thornhill
Chap. X	C. Edeleanu; also A. J. Gould
Chap. XI	H. G. Cole; also A. H. Cottrell
Chap. XII	E. C. Potter; parts by H. S. Campbell
Chap. XIII	W. H. J. Vernon; parts by J. C. Hudson, E. Ll. Evans and F. E. Jones
Chap. XIV	J. E. O. Mayne
Chap. XV	S. C. Britton
Chap. XVI	H. K. Farmery; also A. J. Gould
Chap. XVII	A. J. Gould; also H. K. Farmery
Chap. XVIII	R. B. Waterhouse; parts by G. T. Callis and W. H. Wheeler
Chap. XIX	D. Eurof Davies; parts by J. C. Hudson
Chap. XX	J. N. Wanklyn
Chap. XXI	R. S. Thornhill
Chap. XXII	J. N. Agar; also G. F. Peaker
Appendix I	A. J. Berry; parts by J. E. O. Mayne and J. P. Chilton
Appendix II	J. N. Agar

LIST OF OTHERS WHO HAVE PROVIDED INFORMATION AND ADVICE OR GIVEN PERMISSION TO USE MATTER OR DIAGRAMS

J. A. Allen, W. E. Ballard, W. Beck, G. Bianchi, E. Billig, C. E. Bird, Sir L. Bragg, (Miss) D. M. Brasher, W. R. Braithwaite, L. de Brouckère, R. S. Brown, A. Bukowiecki, G. H. Cartledge, J. V. Cathcart, F. A. Champion, A. W. Chapman, J. C. Chaudhari, W. D. Clark, S. G. Clarke, M. Cohen, N. Collari, W. J. Copenhagen, J. B. Cotton, L. S. Darken, M. Darrin, R. L. Davies, E. H. Dix, C. Domb, G. G. Eldredge, C. Evans, A. J. Fenner, J. E. Field, F. W. Fink, P. J. E. Forsyth, U. F. Franck, F. C. Frank, B. Fullman, H. Gerischer, P. T. Gilbert, P. Gooding, H. Grubitsch, P. E. Halstead, P. Hamer, R. A. F. Hammond, K. Hauffe, T. M. Herbert, P. Hersch, P. W. Heselgrave, A. Hickling, W. F. Higgins, W. Hirst, A. M. Horsfield, B. Ilschner, V. D. Johnson, A. Keller, J. N. Kenyon, W. T. King, K. Kölbl, P. Lacombe, J. Lewis, R. A. Lowe, G. M. W. Mann, C. W. Marler, R. May, T. Mills, J. W. Mitchell, L. Mullins, R. Olivier, R. N. Parkins, L. G. Patrick, F. B. Pickering, R. Piontelli, R. C. Plumb, J. R. Postgate, M. Pourbaix, M. Pražák, M. J. Pryor, N. Rayner, R. Ll. Rees, A. Rius, B. G. Robbins, M. W. Roberts, W. D. Robertson, J. P. Saville, G. Schikorr, W. J. Schwerdfeger, B. A. Scott, H. T. Shirley, S. C. Shome, L. L. Shreir, H. Silman, M. T. Simnad, G. C. Smith, F. N. Speller, K. A. Spencer, H. A. Streicher, N. Stuart, H. H. Svensson, W. H. Tait, C. A. J. Taylor, D. Taylor, (Mrs.) C. F. Tipper, T. H. Turner, H. H. Uhlig, C. Wagner, C. J. Walton, S. Wernick, L. S. Whitby, D. Whitwham, A. B. Winterbottom, W. A. Wood, S. R. Woolcock, K. H. R. Wright, W. F. K. Wynne-Jones.

Added later: W. Bullough, A. H. Goodger

Other Helpers

(Mrs.) E. Johnson (typing of text)
(Mrs.) E. Hassall (typing of correspondence)
E. L. Billington and C. A. J. Taylor (preparation of new diagrams)

The diagrams include some which have appeared in the Author's earlier books, but the majority are new—drawn in the same style. Some of these are original sketches, but many closely follow diagrams included in published papers—with kind consent of the authors of those papers, to whom acknowledgment is made in the legends. Acknowledgment is here made to the journals in which the papers appeared; these are:

Annali di Chimica applicata
Bulletin of the Institute of Metal Finishing
Bulletin de la Societé chimique Belge
Carnegie Scholarship Memoirs (Iron and Steel Institute)
Corrosion
Discussions of the Faraday Society
Industrial and Engineering Chemistry

Journal of the American Chemical Society
Journal of Applied Chemistry
Journal of the Chemical Society
Journal (previously Transactions) of the Electrochemical Society
Journal of the Institute of Metals
Journal of the Iron and Steel Institute
Korrosion und Metallschutz
Metallurgia Italiana
Monatshefte für Chemie
Nature
Neues Jahrbuch für Mineralogie, Geologie und Palaeantologie
Proceedings of the American Society for Testing Materials
Proceedings of the Institution of Mechanical Engineers
Proceedings of the Royal Society
Reçueil des Travaux chimique des Pays-Bas et de la Belgique
Transactions of the Faraday Society
Transactions of the Institute of Metal Finishing
Zeitschrift für Anorganische Chemie
Zeitschrift für Elektrochemie

CONTENTS

QUALITATIVE SECTION

CHAPTER I

THE APPROACHES TO CORROSION

SYNOPSIS

The chapter opens with an explanation of such terms as Corrosion, Erosion and Oxidation. It is shown that the subject of Corrosion in its widest sense possesses potential interest to numerous classes of persons, including Pure Scientists, Engineers, Architects, Economists and Health Authorities. Attention is then turned to various manners of regarding the subject, which presents a very different appearance according to the direction from which it is approached. Corrosion is regarded by different classes of persons (1) as the kinetics of chemical reactions involving metals and non-metals, (2) as the outcome of the varying electron affinities of different elements, (3) as the electrochemistry of short-circuited cells, (4) as the demolition of crystal-structure, (5) as a smelting process in reverse, (6) as a branch of chemical thermodynamics, or (7) as a disease afflicting metals, which it behoves us to cure. These different ways of regarding Corrosion will appeal to different types of mind, but there is benefit to be gained by each from an endeavour to understand alternative aspects of the subject, and thus gain sympathy with those whose approach is a different one.

The Terminology of Corrosion

General. Much work has been devoted to the compiling of Glossaries which provide formal definitions of corrosion terms; beyond question these have their uses. It may be doubted, however, whether the average man, coming up against a technical term with which he is unfamiliar, receives much enlightenment on reading a formal definition. A better way of helping him may be to describe a situation showing the need for a term to cover a group of phenomena, and then to introduce the name which has been agreed upon to fill that need.

Corrosion and Erosion. It has long been known that metals are liable to undergo changes in which they lose their characteristic properties; in most cases, the metal ceases to be an element, and becomes a compound. Iron exposed to damp air is changed to a brownish hydrated oxide, known as rust. Iron placed in salt water with access of air also produces rust, and the change is known to be connected with electric currents flowing through the liquid. If heated in air, iron becomes covered with a dark scale, consisting, usually, of more than one oxide. If placed in dilute sulphuric acid, it passes into solution with evolution of hydrogen gas, and the solution on evaporation and cooling deposits crystals of ferrous sulphate.

The name *Corrosion* is a convenient one to describe these and similar chemical and electrochemical changes, in which a metal passes from the elementary to the combined state. However the term is today sometimes employed where a solid metal passes into solution in another (liquid or molten) metal—although this is not generally regarded as a chemical change; thus a steel vessel used for containing molten zinc slowly loses thickness, the iron being dissolved by the zinc, and the vessel is then said to be suffering corrosion. Those who desire a formal definition may say that Corrosion is the destruction of metal or alloy by Chemical Change, Electrochemical Change or Physical Dissolution. This excludes Mechanical Processes which destroy the value of a metallic article by grinding it away as metallic dust. Such purely mechanical damage, in the absence of chemical change, is called *Erosion*.

Conjoint Action. Often Corrosion and Erosion operate simultaneously, and the damage produced is frequently far greater than the damage caused when they proceed one at a time. This disastrous combination of destructive agents is generally called *Conjoint Action*. Certain types of Conjoint Action deserves special mention here. *Stress-Corrosion Cracking* denotes the cracking caused when a steady tensile stress acts on a metal in a corrosive environment, whereas cracking caused by alternating or cyclic stresses in a corrosive environment is called *Corrosion Fatigue*. Sometimes the cracking, whether due to Stress Corrosion or Corrosion Fatigue, passes along Grain-Boundaries, and is then known as *Intergranular* or *Intercrystalline* Cracking; sometimes it passes across the grains, and is then called *Transgranular* (*Intragranular*) or *Transcrystalline* Cracking.

Oxidation. The conversion of the surface portions of metal to oxide on heating in air or oxygen has been mentioned above. It is generally termed Oxidation. In the present book, Oxidation is regarded as a type of Corrosion. However, since some authorities would not include oxidation in the subject of corrosion, it has seemed advisable—in the interests of clarity—to introduce the word " Oxidation " into the title.

Other Special Types of Corrosion. Generally corrosion starts on the surface of a metallic specimen. If a large portion of the surface is affected, it is said to be *General Corrosion*; if only small areas, it is called *Localized Corrosion*; if confined to small points, so that definite holes are produced in an otherwise almost unattacked surface, we speak of *Pitting*. A special form of pitting, often undercutting the surface layers of the metal, is sometimes met with where bubbles in a rapid water-stream have impinged on the surface, preventing the maintenance of a protective layer; this is called *Impingement Attack*. If attack extends inwards, following grain-boundaries and leaving the interior of the grains unattacked, the corrosion is said to be *intergranular* or *intercrystalline*. If it produces grooves following grain-boundaries, or perhaps (on a much larger scale) following the water-line on a partly immersed metallic plate, we may speak of *Grooving*; grooves may also be developed along the line where two dissimilar metals meet, or in zones running parallel to a weld. If in rolled or extruded metals, corrosion extends along certain planes parallel to the surface, it is called *Layer Corrosion*; under atmospheric conditions, the voluminous corrosion-product formed along these planes may lever the intervening layers apart, so that the material divides into flakes, recalling pastry; this is known as *Foliation*.

Etching. The metallographist uses certain corrosive reagents to develop the structure of polished micro-sections. The reagent may show up the grains by producing crystal-facets, so that some grains appear dark and some bright when viewed under the microscope; or it may attack the grain-boundaries, developing them as dark lines; or again it may attack and colour certain constituents of an alloy, which can thus be identified. Although these are all examples of corrosion, the metallographist rarely uses that word; he speaks, generally, of *etching* his sections.

Other scientists are actively engaged in corrosion work without employing the term. Metal-physicists treat metallic surfaces with certain acid mixtures, and count the pits which appear along grain-boundaries or on slip-planes, believing, rightly or wrongly, that they reveal to him the number of dislocations present. Other physicists study oxidation or film-growth, using the theory of lattice defects to interpret their results. Each of these scientists is studying a corrosion process, and his work may be contributing much to our knowledge of the subject; but some of them would be highly indignant to be described as " Corrosionists ".

Many-sided Character of Corrosion. The subject of Corrosion—in the wide sense indicated above—may be likened to a large building with many wings, an edifice which different classes of visitor will approach from different points of the compass and will get entirely different impressions according to the angle of view. If the structure is to be understood, it will be well—before entering—to walk around and inspect it from all sides. Such a preliminary inspection cannot be expected to give full understanding; that will come later. One cannot hope to see all the features of a building by looking through the windows, but at least one can gain some knowledge as to how the various rooms are related to the fields and gardens outside.

Classes of Persons potentially interested in Corrosion

The Pure Scientist. Even in these utilitarian days there survives a type of man who studies the workings of Nature for their own sake; he finds them interesting—perhaps beautiful. Since he shares the outcome of his studies with others, who also find them interesting and sometimes beautiful, he should not be regarded as selfish or anti-social; the justification for encouraging, and indeed supporting, the pure scientist is the same as the justification for supporting the artist, sculptor or composer, who also produce what is interesting or beautiful, sharing the fruits of their labour with others. However, although the pure scientist, as such, has no economic aims, the results of his studies, which enable us to understand the mechanism of natural phenomena and to predict effects in untried cases, may have economic or political results which can be literally shattering. Whether —over the whole range of Science—the growth of understanding is good or bad for Humanity, is a matter about which many views are held today. In the particular case of Corrosion, however, a clear understanding of the mechanism is necessary, if undesired wastage of metal is to be avoided; here almost everyone would agree that advance in such understanding is beneficial to Mankind.

Despite the aversion widely felt to a singularly ugly name, corrosion reactions should make a real appeal to any pure scientist who can bring himself to study them. Human interest is aroused by phenomena which can be represented or explained by simple laws and principles. Subjects where the facts are a confused, inexplicable jumble, requiring to be memorized, possess no fascination. Corrosion reactions, whatever was the situation half a century ago, can today be presented on a logical scheme and explained on a quantitative basis. They would today appeal strongly to the young chemical student—but for the fact that no place is found for them in his curriculum.

Oxidation, however, is today making an appeal to the physical investigator, because oxidation processes depend upon the action of such things as lattice defects in which certain schools of physicists are keenly interested. Modern knowledge of the combination of metals with oxygen—surely a chemical process—is largely due to the work of physicists or the physical wing of physical chemistry; the average chemical student is told nothing about it and, if he wishes to learn, he must do so in his own time. It is hoped that for some of these, the present book may be helpful.

It should be pointed out that Oxidation and Corrosion Reactions—considered as Pure Science—have reached a more satisfactory stage than many other branches of kinetics. The absolute values of oxidation or corrosion velocities can be calculated from purely electrical measurements and agree well with the values obtained by direct gravimetric or chemical observation; this agreement is obtained without any arbitrary assumption regarding the value of constants—such as often has to be made in obtaining agreement between observed and calculated values in chemical kinetics. The accord between theory and practice, brought out in Chapters XX and XXI, constitutes valid claims for Oxidation and Corrosion to be regarded seriously as branches of Pure Science.

The Engineer and Chemical Engineer. For over a century, progress in engineering has been controlled largely by the development of materials. In the early days of steam engines, it was realized that very high pressures and temperatures would bring increased efficiency, but, until strong materials were available, limits were set to what was attainable. Both in structural and mechanical engineering, the designer has long been able to make an approximate calculation of the dimensions needed to withstand his mechanical stresses, but, in the early days, he wisely left a considerable margin of safety; it is likely that this margin of safety really took care of weakening due to chemical action—for which, in general, engineers have not consciously made allowance. Recent developments have, however, increased the peril of break-down due to corrosion. Improved mathematical technique tends to make the engineer-designer feel justified in reducing the margin of safety; he would be right in doing so, if the mechanical factors were the only ones, but in cases where formerly the margin of safety has been providing against corrosion, the reduction must increase the hazards.

Again, the very success of the metallurgist in providing mechanically stronger material, which the engineer will probably use at the maximal stresses considered to be safe on solely mechanical reasoning, greatly increases risk of break-down owing to chemical causes, unless the resistance to corrosion is improved to the same extent as the mechanical strength—which is rarely the case.

Another feature of the situation requires attention. Engineers have long known that in some situations the safe performance of material is decided not by its strength under steady tensile stress, but by the much lower " fatigue strength " which it will exhibit under alternating or fluctuating stress; if this is exceeded, the material will fracture, not immediately, but after a period during which a fatigue crack is slowly advancing across the stressed member. Recent events have made the majority of engineers acutely aware of the perils of fatigue; what is not sufficiently appreciated is that resistance to fatigue is enormously reduced by chemical influences, and that, in a corrosive environment, members can fracture under what would normally be regarded as a " safe " range of stress. If corrosive influences are overlooked, accurate calculations on the mechanical side may lead to dangerous delusions.

The position is made worse by the fact that very little chemistry and probably no electrochemistry is included in the teaching of young engineers and architects; yet, without some basic knowledge of those subjects, a sound policy for avoiding the dangers of corrosion in engineering and architectural design is impossible. There still appear to be engineering and architectural courses in which the students hear little of chemistry and nothing of corrosion. The situation seems to be improving, but even today many young engineers and architects must start to get up their corrosion knowledge for themselves after their education is supposed to be complete. In case they turn to this book, the two appendices—devoted to chemistry and electrochemistry respectively—may prove helpful.

It is commonly stated at Corrosion Meetings that anti-corrosive measures should start at the drawing-board. Most corrosion specialists must have experienced a feeling of hopelessness at being called in at the eleventh hour to provide corrosion-resistance in some finished plant or structure built of the wrong materials brought together in the wrong way. Too often the designer has not merely been ignorant of corrosion, but has lacked interest in it. Now interest is not something which can be created at will, but it is a fallacy to imagine that something which makes no appeal to the interest can safely be neglected. If the engineer's present rate of progress in overcoming purely mechanical break-down continues and if the remarkable success of the metallurgist in providing strong materials also continues, then chemical break-down may become a more serious menace than mechanical break-down; if this occurs, the designer who is not interested in corrosion will be a man who is not interested in engineering safety.

The argument has been heard that it is unnecessary to trouble the engineer with the theoretical aspects of Corrosion; give him the facts, it is urged, without troubling him about causes; provide him with tables showing the rate of wastage of different materials in different environments, and he will be able to choose a material which will meet his requirements without risk of dangerous thinning or weakening. This idea that empirical data could be conveniently accumulated and presented in tables neglects the number of factors affecting corrosion that are capable of independent variation. The average supply water contains perhaps six constituents requiring consideration; the flow velocity and the temperature are other factors affecting results. Thus even for a single material, if we wish to consider in our tables ten values of each factor, we require 10^8 or 100,000,000 experiments to cover all possible combinations. Doubtless some combinations could be omitted, but having regard to all the materials, and the important influence of geometrical factors, it is clear that there are not enough trained investigators in the world to obtain the empirical information sufficient to cover all combinations of conditions likely to arise.

The engineer will have in practice to apply measurements of corrosion-rates obtained in one situation in making forecasts of behaviour in other situations. If he has no knowledge of the scientific mechanism, this will be pure guess-work. If he has such a knowledge, the prospect of inferring from information available some idea of behaviour under new conditions is not an entirely hopeless one. Quite apart from the intellectual satisfaction of knowing the reason of events, the engineer needs an understanding of corrosion mechanism if he is to face his technical problems.

In new branches of engineering and applied chemistry such as the raising of power from nuclear reactors or the remarkable new industry known as Petroleum Chemistry (Petrochemistry), the facts contained in the present book may be of no direct value. Presumably, however, the corrosion staffs within these branches are accumulating the pertinent facts which are duly being stored in files. If so, the principles set forth in the present book may be indirectly useful in helping the corrosion specialist within these branches to apply the laboratory measurements to practical cases. At least the book may provide some warnings about factors which might otherwise be neglected—for instance the risk that irradiation may cause invisible films which would normally be regarded as protective to become much less reliable.*

Numerous specifications and codes of practice exist to help the engineer in his struggle against corrosion. These refer not only to the composition of materials, but to the carrying out of protective processes. Gratitude

* Other cases where radio-active material may accelerate corrosion are discussed by G. H. Cartledge, *Werkst. u. Korrosion* 1958, **9**, 473, esp. Fig. 6 on p. 501.

is due to members of committees which have devoted time and trouble to the difficult but unexciting task of framing these documents, but most of those who have served on such committees will be the first to recognize that it is impossible to design a form of words which will ensure success; in most cases it would be possible to carry out work in such a way that there is no infringement of the relevant specification and yet fail to obtain satisfactory results. This should always be borne in mind in deciding between estimates from competing contractors; it may not always be cheapest to accept the lowest, even though they all purport to give the same result—namely work carried out in accordance with a certain specification. Information should be sought as to which, if any, of the contractors have detailed experience of providing protection under conditions similar to those likely to exist in the case under consideration, which of them have built up a reputation for reliability in such work, which of them possess an intelligent and imaginative technical staff to plan the work and which will co-operate in methods of inspection designed to ensure that the precautions set out on paper will really be observed in practice. If these matters receive no attention, and the lowest tender is accepted without such enquiry, there may be a heavy price to pay for a " penny-wise pound-foolish policy ". This applies with particular force if the work is to be carried out for a government department where those who have to decide these matters, with the excellent intention of saving the tax-payer money, tend to accept the lowest estimate—a policy which, in the end, can easily prove very costly.

The Architect. The neglect of corrosion precautions by the architect is today as serious a national problem as their neglect by the engineer and presents a new difficulty. The engineer, although not—it may be hoped—without appreciation of aesthetics, aims primarily at producing a structure or machine which will fulfil its functions, and for that reason is more than willing to consider corrosion-precautions when once he is convinced that their neglect will endanger proper functioning. The architect is primarily and rightly proud of his profession as a branch of Fine Art; he is well aware that a house has to be lived in as well as admired from the outside, but his training contains too little chemistry and metallurgy for the assessment of the claims of salesmen (themselves uninformed rather than deliberately dishonest) who represent their several materials as needing no precautions against Corrosion. This book will give the architect no direct help in the choice between proprietary materials—trade names are avoided wherever possible—but the general principles set forth in Chapters VI, XIII and XIV may perhaps provide some assistance in forming a judgment of the various claims, and should enable the architect to avoid mistakes of planning which have been made in the past. Some years ago it was not uncommon to find dissimilar metals placed in direct contact without regard to the corrosion-couples set up; experience has taught the architect that this is wrong, and it is likely that today combinations of metals are being avoided which would really be safe. However, cases are still met with where water is allowed to run off copper roofing material into aluminium pipes, with complete disregard for the fact that traces of metallic copper will be redeposited on the aluminium, thus setting up vast numbers of little corrosion-couples. Moreover the possibility of crevice-corrosion at contact between metals and materials which are non-conductors of electricity, the corrosion dangers connected with water-traps and those due to the condensation of acidic moisture in rooms where town-gas is burnt without a proper flue are not sufficiently appreciated by many architects today. Nor does he always seem to make proper use of the organizations established in the interest of the user which would give unbiassed information in difficult cases, such as the Building Research Station, Watford, and the National Chemical Laboratory, Teddington.

The Economist. Many estimates have been made of the economic cost of Corrosion. In 1949 Uhlig gave 5500 million dollars as the annual tax exacted directly by the ravages of corrosion in the United States. Recently Vernon has calculated that the total expenditure within the United Kingdom, including the cost of preventive measures and of metal losses due to Corrosion, must be of the order of 600 million pounds annually (H. H. Uhlig, *Proceedings of the United Nations Scientific Conference on the Conservation and Utilization of Reserves*, 1950, Vol. II, p. 213. W. H. J. Vernon, " Conservation of Natural Resources ", *Instn. civ. Engrs.* 1956–57, p. 105, esp. p. 130. For Australian figures, see H. K. Worner, Corrosion Symposium, Melbourne University, 1955–56, p. 1).

It is clear that a modest sum spent in Corrosion Research, even if it merely succeeded in reducing the annual toll by a small fraction of its present amount, would be remunerative. In the U.S.A. a Corrosion Research Council has been formed to sponsor long-term research projects; leading industrial undertakings are to be asked to make a contribution of 0·01% of their annual profit. If devoted to the scientific study of corrosion processes, this should pay a handsome dividend (*J. electrochem. Soc.* 1955, **102,** 75C).

The possibility of economy through Corrosion Research and the consequent conservation of materials is perhaps better realized in America than in Europe; a directory published in 1950 by the (American) National Association of Corrosion Engineers seemed to suggest that there were then more corrosion specialists in Houston (Texas) or Tulsa (Oklahoma) than in the whole of the United Kingdom. Some British industrial concerns which could probably achieve great savings by careful study of maintenance problems have no corrosion specialist on their staff. Even a better use of existing knowledge (notably that concerning the proper preparation of steel surfaces before painting) should carry the possibilities of great economies which might do something to off-set the ever increasing wage-bill, and thus reduce the danger of a price spiral. The possibility of avoiding trouble by careful attention to corrosion problems at the design stage has been emphasized by many competent authorities, but it is doubtful whether their advice has been widely heeded.

The true cost of Corrosion cannot be reckoned in a money sum representing replacements and maintenance. We have to visualize cases where some plant or machine which has been working perfectly is suddenly brought to a stand-still by a corrosion break-down. This may be due to leakage consequent on perforation, and here the localized character of some types of corrosion has to be borne in mind. Localization of attack can cause perforation even when the total destruction of material at other parts of the tank or tube is negligible. Again Corrosion may lead to partial blockage of tubes, due to the fact that the products of Corrosion usually occupy a larger volume than the metal destroyed whilst they are being formed; this may cause a slower supply of liquid where it is running under gravity, or higher fuel costs when it is being pumped through the pipes. Rarer but more serious results of Corrosion include explosions in closed vessels or the failure of stressed members; these have led to loss of life.

It is doubtful whether all cases where Corrosion has really been the initial cause of a disaster are attributed to it at the " enquiry "; practically every fracture is, in its final stage, genuinely due to mechanical forces, although it is possible that that final stage would never have been reached but for stress-intensification due to corrosion-pitting or corrosion-trenching in the early stages, or to stress-corrosion cracking, probably caused by faulty heat-treatment.

Shortages of certain materials have occasioned public anxiety on many occasions. The remedy usually prescribed is increased production, but much could be done by the reduction of wastage and an increase in the useful life of articles; this would entail a study of Corrosion. It is, however, a strange fact that Shortage and Surfeit often go together. Some years ago there was a shortage of sulphur—required for sulphuric acid—and arrangements were made to mine anhydrite (calcium sulphate), thus providing a fresh source of supply. Yet all the time, vast quantities of sulphur oxides were being thrown into the air owing to the burning of coal containing pyrites; this was causing much destruction to stone-work and metal-work; also (as was discovered during smog periods) it inflicted damage on human health. The possibilities of turning this waste sulphur into the much-needed sulphuric acid or ammonium sulphate had from time to time been discussed, but the decision always seems to turn upon the question as to whether such a method of obtaining sulphuric acid (or some other saleable sulphur compound) was cheaper than methods starting from brimstone, pyrites or anhydrite. Clearly, in view of the damage done by the liberated gases, the question of subsidizing a process which was not in itself directly remunerative, deserved more attention than it was then receiving (see also p. 503).

About the same time, nickel was in short supply, and severe measures were taken to restrict the amount of nickel available for the plating of steel. Now thin nickel plating contains, or develops, more pores than thick plating, and will protect for a shorter period. Any temptation to apply inadequate coats will tend to give the plated articles short lives, so that they will quickly pass to the scrap heap, carrying with them whatever nickel has been applied, besides wasting the steel. Since thinly plated parts will quickly require replacement, it is at least arguable that the policy of restricting nickel for plating—far from economizing the metal—may have accelerated consumption. These are matters worthy of careful investigation.

The importance of Conservation of Materials is becoming better appreciated today. The discussions published by the United Nations in the Proceedings of the Conference on the Conservation of Resources (1949) deserve study. Other economic aspects of corrosion are presented by A. Keynes, *Corros. Tech.* 1956, **3,** 226; *Chem. and Ind.* (*Lond.*) 1958, p. 398; 1959, p. 666. Sulphur and nickel shortages in relation to corrosion receive consideration from U. R. Evans, *Metal Ind.* (*Lond.*) 1951, **78,** 366; 1951, **79,** 500; see also editorial comments (*Metal Ind.* (*Lond.*) 1951, **78,** 357).

The Health Authority. The provision of a good water-supply involves many corrosion problems, several of which are discussed briefly in this book. The necessity to avoid the leakage which might arise from the internal or external attack upon pipes, and the clogging with voluminous corrosion product, has already been mentioned. Acute problems arise when the pipes or containers consist of a poisonous metal. Lead—a cumulative poison—is an important case, and the conditions favouring *plumbo-solvency* —as it is called—deserve attention.

Containers made of Terne plate (steel coated with lead-tin alloy) are used for petrol or other stores, and in primitive countries have provided natives with a supply of cooking utensils obtainable free of charge from the scrap heap; naturally this has caused outbreaks of lead-poisoning.

Other cases of poisoning have occasionally been reported when galvanized iron (zinc-coated) vessels have been used for preparing lemonade or vegetables. The results have generally been unpleasant but temporary; zinc is not a cumulative poison.

Some scares about the danger of metal-poisoning arising from corrosion have proved on investigation to be without foundation; it may be useful to give an example. For over 20 years, there has been a deeply rooted idea that aluminium cooking vessels are dangerous to health. In the early days—before experience was gained—aluminium vessels occasionally came on the market which developed pits and were difficult to keep clean; it is possible that such vessels may have caused illness through decayed food remaining in the pits, but the question has little more than historical importance. Also in the early days, it was customary to clean aluminium with sodium carbonate or powders containing it, and this caused marked attack; washing powders are now obtainable which do not attack aluminium; generally these either contain sodium silicate or are based on synthetic detergents.

The idea that aluminium compounds are poisonous is a mistake; articles in popular journals which have been responsible for this idea prove on examination to have been written by people whose words betray amazing ignorance of scientific matters, but they have clearly made an impression on credulous readers. Endeavours by qualified persons to present the matter fairly have met with limited success; there is a type of mind which is less influenced by a balanced statement than by the irresponsible utterances of the semi-charlatan.

Definite evidence exists that aluminium—taken by the mouth—is non-poisonous. Baking powders containing aluminium salts have been used in some countries and must have introduced far more aluminium into the human system than could possibly be introduced accidentally by corrosion of a cooking-vessel. Also a scientific experiment was carried out about 1929 at Yale University, undertaken to discover the fate of aluminium introduced by the mouth; the object was not to discover whether it was poisonous, since the organizers had no misgivings on that matter. A group of twenty-seven subjects—mostly medical students—were fed with considerable quantities of aluminium salts each day, and suffered no ill effects. One of them, writing to the Author seven years later, stated definitely that he had never experienced any symptoms or sensations which he could attribute to the experiments.

On the other hand, fallacious arguments have occasionally been advanced in support of the idea that some particular metal is harmless. It is not sound reasoning to argue that " traces of metal X, introduced by the corrosion of vessels or containers made of X, can be regarded as without danger to health, because traces of X are found in natural foodstuffs ". Before such an argument can be accepted, it is necessary to ask about the state of combination. If metal X is locked up in foodstuffs as an organic complex, it might be harmless, whereas the same amount of X present as a salt or hydroxide in the corrosion-product might be harmful.

In cases where corrosion would cause traces of poisonous metal to enter food or water, the maximum permissible content is usually laid down either by regulation, or in books by recognized authorities. The Author is not qualified to specify limits of safety, but the principles developed in this book may help responsible authorities to keep below the limits recorded elsewhere. As regards water, valuable information will be found in the book of J. C. Thresh, J. F. Beale and E. V. Suckling, " Examination of Waters and Water Supplies, " 6th edition, edited by E. W. Taylor (Churchill), esp. p. 550, where the maxima contents permitted in the U.S.A. are discussed; the values (0·1 ppm. for lead, 3 ppm. for copper and 15 ppm. for zinc) are regarded as reasonable, but the copper value, although below the toxic limit, may allow attack on galvanized iron (see p. 205 of the present book). A useful discussion of materials for the food industries is provided by R. Falconer, *Chem. and Ind.* (*Lond.*) 1954, p. 1058.

Other medical aspects of corrosion science deserve brief mention. The blunting of surgical and ophthalmic instruments—as well as razor blades —is largely due to corrosion of the sharp edges on a microscopic scale; clearly sterilizing liquids should be non-corrosive. In orthopaedic surgery, it is believed that corrosion of the metal parts (plates, bolts and screws) inserted in the body can set up inflammation, and the correct choice of alloys is all-important; not only the parts which are to be left in the body, but the tools (screw-drivers, spanners and forceps) used for fixing them require to be of the correct composition, since there is reason to believe that transfer of metal from tool to appliance can set up corrosion couples. Here again, no technical advice can be given, but references may be useful; C. S. Venable and W. G. Stuck, *J. Bone Jt. Surg.* 1948, **30** (A), 247; E. G. C. Clarke and J. Hickman, *ibid.* 1953, **35** (B), 467; F. P. Bowden, J. B. P.

Williamson and P. G. Laing, *ibid.* 1955, **37** (B), 676; also *Lancet* 1957, p. 1081, and the first chapter of "Surgical Materials" (editor, L. Gillis, publisher, Butterworth). Cf. C. G. Fink and J. S. Smatko, *Trans. electrochem. Soc.* 1948, **94**, 271, 396.

Manners of Regarding Corrosion

(1) **Corrosion regarded as a Branch of Chemical Kinetics.** Reactions involving various non-metallic elements receive detailed study from both organic and inorganic chemists, along with the properties of the products resulting therefrom. Alloying of metallic elements receives corresponding attention from metallurgists. Reactions involving combination of metallic with non-metallic elements, and the properties of the products—in other words, corrosion reactions and corrosion-products—would seem to have equal claims.

However, in many cases the rate at which a corrosion reaction proceeds is not controlled by chemical considerations. Oxidation and analogous changes lead to the production of films on the metal, and the rate of combination is governed by the maximum rate at which material can pass through the film of oxide or other compound; often the reaction starts rapidly but becomes slower and slower as the film becomes thicker and thicker. As already stated, such phenomena are more likely to interest the physicist.

In cases where no film is formed, e.g. attack upon metal by acid containing an oxidizing agent, the reaction may be capable of proceeding so rapidly that the rate of change is limited by the rate of replenishment of one of the reagents. Such cases certainly conform to the laws found in the text books of physical chemistry, but are of no interest in connection with the mechanism of the chemical reactions. Even where the controlling process is some reaction proceeding at the metallic surface, this is generally an electrochemical reaction.

It is, perhaps, understandable that the general chemist, and even the physical chemist who has not specialized in electrochemistry, is somewhat disinclined to concern himself with corrosion reactions. However, there is another approach in which these reactions can be made to illustrate the Periodic Arrangement of the Elements; such an approach, which should be more acceptable to the general chemist, will now receive consideration.

(2) **Corrosion regarded as the Outcome of Electron Affinities of Metals and Non-metals.** In the Periodic Table,* the non-metals (oxygen, sulphur, chlorine, iodine, etc.) stand before the inert gases, and their atoms are electron acceptors. If an oxygen atom can capture two extra electrons, it acquires something of the structure and something of the stability of an inert gas atom, although differing from the latter in carrying a negative charge. Metals, however, follow the inert gases and are essentially electron donors. It is not surprising that if an oxygen atom and an atom of a bivalent metal come together, the oxygen captures two electrons from the

* A form of the Periodic Table, designed to assist corrosion studies, is provided on p. 954.

metal, and union of the O^{--} anion with the metallic cation is attended by a drop in free energy, so that the reaction is one which can occur spontaneously.

As already stated, the fact that the oxide produces a film over the metal surface, isolating metal from oxygen, causes the oxidation to slow down as the film thickens; direct union of metal and oxygen is only important at high temperatures and largely depends on lattice defects in the oxide structure, which defects permit the passage either of cations outwards or of anions inwards. However, at low temperatures and in the presence of a suitable aqueous solution (e.g. sodium chloride), the capture of electrons by oxygen and the loss of electrons by the metal can occur in such a way that the destruction of the metal can continue indefinitely without building up a film of a kind which would interfere with attack. For instance, if a plate of zinc is immersed vertically in salt solution, with air above the liquid, oxygen can take up electrons from the zinc at the water-line, producing OH^- ions.* The constant removal of electrons allows zinc ions, Zn^{++}, to enter the liquid lower down the plate, without causing an accumulation of electric charge anywhere. The Zn^{++} and OH^- ions will meet, but, instead of forming a protective film, they will form either a flocculent precipitate of zinc hydroxide, $Zn(OH)_2$, or a membranous wall growing out at right angles to the metal. Thus the rate of attack on the metal, instead of slowing down with time, continues almost unchanged.

The reason why this (electrochemical) attack can continue indefinitely (even at temperatures where direct oxidation would stifle itself by film-formation) is that both primary products are freely soluble bodies; since the main ions in the solution are Na^+ and Cl^-, we can regard the " cathodic product " formed at the water-line as NaOH and the " anodic product " formed lower down as $ZnCl_2$. If the solution had been one capable of producing a sparingly soluble film either at the " cathodic area " or the " anodic area ", the electrochemical corrosion would have stifled itself—and this is an important factor in the inhibition of corrosion by suitable additions to the liquid.

The electrochemical corrosion of partly immersed zinc is really a type of oxidation, but one in which the oxygen is consumed at one place (the water-line), the metal consumed at a second place (lower down), and the oxide (in hydrated form, as hydroxide) precipitated at a third place. It is this spatial separation of the various parts of the process which avoids stifling by protective films and thus allows corrosion to proceed unchecked.

The approach to the electrochemical mechanism just indicated is the one most likely to appeal to the general chemist; the electrochemical specialist will probably regard the matter rather differently—as suggested below.

(3) **Corrosion regarded as the operation of short-circuited electrochemical cells.** Electrochemists often study the effect of current provided

* Oxygen ions, O^{--} are unstable in water. If formed, they would soon disappear by the reaction $O^{--} + H_2O = 2OH^-$, but it is doubtful whether they are formed at all; the reactions by which O_2 is reduced to OH^- are suggested on p. 102.

from an external source when it passes through a cell consisting of two metal electrodes separated by an aqueous solution. They also study primary cells which themselves provide a current and can operate some contrivance placed in the external circuit. Corrosion-cells represent the intermediate, and perhaps simpler, case where there is no current applied from an external source, and no current supplied to an external circuit. The cell may be set up by electrodes of two different metals in metallic contact and joined by solution covering them both; or the electrodes may consist of the same metal surrounded by solutions of different concentrations; or they may be placed in the same solution with better accessibility of oxygen to one electrode than to the other. In nearly all cases, the cell is a short-circuited one, and the resistance of the metallic path joining the two areas (which constitute cathode and anode respectively) is negligibly small. If the liquid connecting the two areas is a concentrated salt solution, the liquid path may also have a negligible resistance; in such cases, the factor controlling the current is polarization, usually caused by the limited rate of replenishment of oxygen to the cathode. To understand these cases, some familiarity with electrochemistry is necessary. It is remarkable that most electrochemical text-books which discuss in some detail electrolytic (current-consuming) cells and primary batteries (current-producing cells) do not apply the same principles to the interesting, intermediate, case of corrosion-cells, where an external E.M.F. is neither applied nor generated; an exception is provided by the admirable book on " Electrochemistry " by E. C. Potter (Cleaver-Hume Press).

(4) **Corrosion regarded as the Demolition of Crystal Structure.** Metals are in general crystalline in structure, and the destruction of metals has points in common with the destruction of crystalline matter by volatilization or dissolution. Volatile metals like cadmium can be grown as beautiful little crystals by cooling of the vapour; on heating these crystals, they disappear again. The same is true of non-metallic crystals, such as iodine —on which much classical investigation has been performed. The main conclusion from modern work on crystal growth is that the growth of a perfect crystal would be a very slow affair, but that, given certain defects in the structure,—especially the spiral dislocations studied by Frank—growth proceeds readily. Conversely, defects in the structure should facilitate the removal of material, if the conditions are such as to favour removal rather than deposition. It is not surprising to find that metallic corrosion tends to start at points of structural disarray—even though it often spreads out. The matter is complicated and the local cracking of films often plays a part (p. 108). But enough has been said to explain why the modern development of crystal physics is of great importance to the Corrosionist, and why the Crystal Physicist often uses what is really a corrosion process to show up the sites of certain defects (dislocations) by the production of pits.

It also explains the use of corrosion (etching) processes by metallographists. A freshly polished section of metal will be covered with a layer of disarrayed material which is easily attacked by reagents, and the attack penetrates downwards until slowed up by layers of accurately arrayed atoms; thus facets are produced, representing those crystal planes along which the atoms are densely packed. Since the inclination of these planes will differ from one crystal-grain to another, some grains will reflect light up the microscope tube and appear bright, whilst others appear dark, so that the grain-structure is shown up. In other cases, the grain-boundaries are themselves preferentially attacked, and the grain-structure is revealed by a net-work of dark lines.

(5) **Corrosion regarded as Smelting in Reverse.** Except for the noble metals, such as gold, metals occur in the earth's crust as certain stable compounds, usually oxides, hydrated oxides or sulphides; sometimes basic sulphates, basic chlorides or carbonates are met with. In reducing the " ore " to the metallic state, energy must be expended to overcome the affinity between the metal and non-metal; sometimes, as in the case of aluminium, this energy is applied as an electric current, but more often, as in the reduction of iron, the oxide is heated with carbon which, at high temperatures, has an even greater affinity for oxygen than has the metal.

The metal, thus produced, represents an energy-rich state, and if, as usually happens in service, it is exposed to oxygen and/or water, or to sulphur compounds, the return to the low-energy state in which they originally occurred in the earth (oxide, hydrated oxide or sulphide) is a reaction involving drop of free energy—i.e. an operation which will occur spontaneously. It is not surprising to find that iron heated in air acquires a scale of oxide, that iron exposed to air and water produces rust (hydrated oxide) or that copper exposed to an atmosphere containing a trace of certain sulphur compounds develops " tarnish films " containing copper in combination with sulphur (and often oxygen also). Since we live surrounded by oxygen and water vapour, it would be sensible, instead of asking the question " Why do metals corrode ? ", to enquire why—under many circumstances—metals manage to escape corrosion. It is hoped that this book will provide something of an answer to that question.

(6) **Corrosion regarded as a Branch of Chemical Thermodynamics.** The approach to corrosion just indicated, which may appeal to the Process Metallurgist, reminds us that Corrosion will only occur where it results in a drop of free energy. The recent development of Chemical Thermodynamics has made it possible to construct diagrams defining those conditions where corrosion is impossible (in absence of a supply of energy from outside), and those conditions where corrosion is possible, and does, in fact, generally occur. The diagrams also show the conditions under which the formation of protective films becomes possible—such as may often interfere with attack. This graphical method of defining the thermodynamical regions is entirely due to the genius of Pourbaix, whose work has provided a novel and attractive approach to the subject. His method is discussed in Chapter XXI, but the reader should also study his book (M. Pourbaix, " Thermodynamics of Dilute Aqueous Solutions ", translated by J. N. Agar (Arnold)).

(7) **Corrosion regarded as a Disease of Metals.** Many practical men accept corrosion as something which—at least on many cheap materials—must be expected, unless something is done to prevent it. Most of the Corrosion Research in progress today is concerned with preventive methods.

Sometimes it is possible to add an " inhibitor " to water which would otherwise be corrosive, and in recent years water-treatment has furnished results which are most impressive. One objection which must be mentioned, however, is that, in the case of many inhibitors, an insufficient addition will make matters worse; sometimes an insufficient addition actually increases the total attack on the metal, but even if the total attack is diminished the corroded area is diminished to a still greater degree, so that the intensity of attack (corrosion per unit area) may be increased; such inhibitors must be regarded as dangerous.

Inhibitors for adding to water are discussed in Chapter V, whilst volatile inhibitors intended to prevent atmospheric corrosion receive consideration in Chapter XIII.

Electrochemical principles indicate that the application of an external E.M.F. to metal may either increase or decrease the corrosion-rate. If the article is made an anode, corrosion is greatly stimulated, unless the conditions are such as to establish " passivity " by producing a protective film. If it is made the cathode, corrosion can often be prevented; this is the basis of the important method known as *Cathodic Protection*, usually employed to supplement protection by some suitable coating; it is discussed in Chapter VIII.

In general, protective coatings are adopted in the effort to avoid corrosion. These may consist of another (more resistant) metal; or they may consist of an organic layer, generally carrying an inorganic pigment, and then called paint; or an inorganic layer such as vitreous enamel or cement may be used. Protective Coatings are discussed in Chapters XIV and XV.

Final Remarks. If at this point some readers feel that they have not fully understood all that has been written, they should not be unduly discouraged. The introductory chapter may have served to break the ground, and doubtful points will become cleared up later. It may be thought fit to read this chapter again after the perusal of the rest of the book—or of such parts of it as concern the reader. It is designed to provide a Retrospect as well as a Prospect.

Other References

It is convenient here to mention some general books on Corrosion, such as those of J. C. Hudson, " Corrosion of Iron and Steel " (Chapman & Hall), F. N. Speller, " Corrosion: Causes and Prevention " (McGraw-Hill), H. H. Uhlig (Editor), " Corrosion Handbook " (Wiley; Chapman & Hall). The last-named work includes a glossary providing formal definitions; another glossary presented to the Inter-Society Corrosion Committee in 1958 will be found in *Corrosion* 1958, **14,** 319t. A table of equivalent German, French and English terms appears in a *Beiblatt* to DIN 50900 (1958). A welcome arrival is the new Anti-corrosion Manual (1958), issued by the publishers of the journal *Corrosion, Prevention and Control*; this contains a theoretical section by W. H. J. Vernon, whose " Spring Lecture " to the Society of Chemical Industry Corrosion Group should receive study (*Chem. and Ind.* (*Lond.*) 1958, p. 1381).

The works mentioned above deal mainly, but not exclusively, with the practical aspects of Corrosion. Authors who approach the subject from the scientific angle without losing sight of practical issues include G. Schikorr, " Die Zersetzungserscheinungen der Metalle " (Barth, Leipzig); E. Jimeno, " El Problema de la Corrosion Metalica " (Ministero de Marina, Madrid); and Ir H. van der Veen, " Corrosie " (Waltman, Delft). Russian work deserves study, especially the book by N. P. Zhuk on " Corrosion and Protection of Metals " (Mashgiz). The views of G. W. Akimov based on electrode potentials may be read in English in *Corrosion* 1955, **11,** 477t, 515t, or in French in " Theorie et Méthodes d'Essai de la Corrosion des Métaux " (Dunod) published in 1957—which is a translation of an earlier Russian book. The views of N. D. Tomashov—particularly on methods of increasing corrosion resistance by alloying—will be found (in English) in *Corrosion* 1958, **14,** 229t.

Books dealing with particular parts of the subject (e.g. Oxidation) are mentioned at the ends of the appropriate chapters.

CHAPTER II

SIMPLE OXIDATION OF SINGLE METALS

Synopsis

The chapter opens with a re-statement and extension of certain arguments from the previous chapter. Oxidation and wet corrosion are presented as alternative electron-exchange processes; it is shown that simple oxidation is important at high temperatures, wet corrosion at low ones. The interference colours produced by oxide-films within a certain range of thickness are then briefly discussed; tables showing the thicknesses associated with these colours are presented at the end of the chapter, whilst discussion of the causes of the colour and methods for measuring films in this range of thickness is deferred to Chapter XIX. The special features of films on copper and iron respectively receive a brief discussion.

The three main classes of curves connecting film-thickness and time are then presented, although the mathematical derivation of the various equations is deferred until Chapter XX. The various laws which are important at low temperatures where oxidation soon comes almost to a stand-still are grouped together for purposes of the present chapter as the *Log Laws*. At higher temperatures, it is pointed out that oxidation may sometimes be rectilinear at first and become parabolic later; or it may be parabolic at first and become rectilinear later. Emphasis is laid on the part played by lattice defects in film-thickening. It is pointed out that growth may occur by metal moving outwards or oxygen inwards; there is evidence that both movements can sometimes occur simultaneously.

Then metals are classified according to their oxidation behaviour. After a brief reference to certain noble metals which cannot suffer oxidation within that range of temperatures where the oxide would decompose, the Pilling-Bedworth Principle of 1923 is introduced in a modified form. The ultra-light metals, which form *porous* films and burn in air, are first discussed; then the metals which build *compact* films by *inward movement of oxygen* and sometimes develop *high stresses* in the film; finally, those building *compact* films by *outward movement of metal*—a process which sometimes leaves *cavities* at the base of the film. Sometimes both stresses and cavities appear to be absent—possibly owing to simultaneous movement in both directions. The chapter ends with a discussion of oxidation at relatively low temperatures; this includes the invisible films formed at atmospheric temperature, and a further discussion of the interference colours formed at higher temperatures.

General Character of Films on Metal

Importance of Physical Processes in Oxidation. The combination of a metal with oxygen would seem at first sight to provide a simple example of chemical change. However, metallic oxidation possesses two characteristics not generally met with in reactions commonly investigated by chemists. The oxide-film, as it grows, shuts off the metal more and more from the oxygen, so that the rate of growth is often controlled, not by a chemical reaction, but by the passage of either metal (outwards) or of oxygen (inwards) through the solid film—that is by a physical process; this passage through the film depends essentially on lattice defects.

Oxidation as Electron-Exchange. The production of an oxide-film should not be regarded simply as a union of metallic atoms with oxygen atoms, but rather as an exchange of electrons—a point of view suggested in the previous chapter. A cuprous oxide crystal consists, not of copper atoms and oxygen atoms arranged on a lattice, but of cuprous ions (copper atoms with one electron missing) and half the number of oxygen ions (oxygen atoms with two extra electrons each)—thus preserving electrical neutrality. The exchange of electrons which occurs when copper and oxygen unite represents a passage to a more stable state—and thus one which can occur spontaneously, without external provision of energy. Oxygen stands in the Periodic Table (p. 954) two places before the inert gas, neon; by accepting two extra electrons, an oxygen atom becomes an oxygen ion (O^{--}) and acquires something of the structure and something of the stability of the inert gases.

Furthermore, the arrangement of ions in oxides is generally such that oppositely charged ions are placed closer together than like-charged ions. The lower oxides of iron, cobalt and nickel (FeO, CoO and NiO), like lead sulphide (PbS) and sodium chloride ($NaCl$), are built up of cations and anions arranged at alternate corners of cubes, an arrangement clearly conducive to electrical stability.

It is not surprising that, speaking generally, the oxidation of metal exposed to air proceeds readily, at least at high temperatures, where the particles concerned are relatively mobile. At room temperatures, however, the film of oxide soon isolates the metal from the air, and the rate of oxidation becomes, in dry, pure air, negligibly slow before ever the oxide has become visible to the eye. This may not be true if the films contain sulphur, but in reasonably pure air direct oxidation is, at room temperature, unimportant as a method of destruction. Unfortunately another method of exchanging electrons, known as electrochemical corrosion, which requires the presence of water, and often leads to hydroxide rather than anhydrous oxide, assumes importance at low temperatures; in such a type of corrosion, the product is often formed, not as a protective film, which would stifle further attack, but as a loose precipitate, or possibly a membrane formed at a distance from the metal; if so, corrosion, once started, continues unchecked. This is called *wet corrosion*; it is discussed in Chapter IV. The present chapter is devoted to direct oxidation, which is important at elevated temperatures.

The engineer desires to use material on which oxidation will stifle itself before there is serious loss of thickness. This would generally happen, even at fairly elevated temperatures, if he was prepared to wait until the oxide-scale had become sufficiently thick, and was confident that it would remain undamaged. Unfortunately, a thick scale is more liable to damage than a thin film, and indeed where the coefficient of expansion of oxide and metal differ, detachment is liable to occur spontaneously on cooling if the film has come to exceed a certain critical thickness.* Thus it is advisable, at least for the highest temperatures, to choose materials on which the oxidation-rate becomes slow when the scale (or film) is still thin.

Colour effects produced by Oxidation. If a strip of nickel, previously cleaned by abrasion, is heated in a flame at one end, a " wedge-shaped film " of oxide is produced which becomes thicker as the heated end is approached (fig. 1). At that end, the colour after heating is grey,

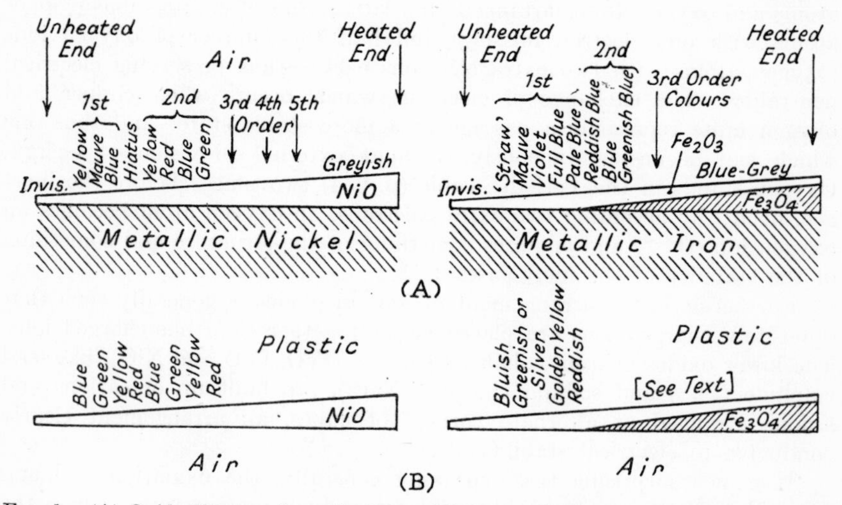

FIG. 1.—(A) Oxide-films on " gradient-tinted " Iron and Nickel. (B) The same films transferred to plastic, the metal being dissolved away.

but the central portion displays a series of beautiful colours due to interference between the light reflected from the outer and inner surfaces of the oxide-film; several " orders " of colours will be noticed, red occurring five times if the oxidation is carefully carried out. Towards the unheated end, the appearance of the metal remains unchanged. That, however, is no proof that oxide is absent; on the contrary, practically all metals which

* If oxide-covered metal is cooled, the total strain-energy per unit area arising from unequal contraction between metal and oxide is likely, in the simplest case, to be proportional to the thickness y, say ky; for a smooth interface between metal and oxide the work of detachment per unit area W_A should be nearly independent of y, and detachment should be possible without performance of external work when $ky > W_A$ or $y > W_A/k$. This will not be true if there is interlocking of metal and scale.

have been exposed to air, even at ordinary temperature, carry oxide—often invisible.

A strip of copper or iron, similarly heated, develops practically the same sequence of colours, but on iron the colours produced by the thicker films may be weak or even absent; iron oxide is less transparent than nickel oxide, so that the reflection due to the inner surface becomes weak when the film becomes thick. However, the fact that the colour-sequence is nearly the same on all metals, shows that the colour depends on the film-thickness, and is not a specific property of the oxide. The recurrence of the same colour several times in the sequence is easily explained. If the film-thickness is such that the paths travelled by the light reflected from the inner and outer surfaces of the film respectively differ by a distance equal to half the wave-length of green light, there will be partial extinction of green rays, and the specimen, viewed in daylight, will appear red-mauve, since red-mauve is complementary to green. A rather similar colour will, however, be produced when the film has thickened sufficiently to produce a path difference of $1\frac{1}{2}$, $2\frac{1}{2}$ or $3\frac{1}{2}$ green wave-lengths, and thus a reddish tint re-appears several times in the sequence, although its character is not quite the same at its various re-appearances. The situation is, however, rather complicated; not only the character, but also the sequence, of the second-order colours, differ slightly from those of the first-order colour, for reasons explained on p. 789. Tables showing the thickness of the various colour-films are given on pp. 55, 56.

At the more strongly heated end of the strip, colour-effects due to interference cease, and the appearance is decided by the specific colour of the oxide. Viewed on the metal, the film is usually bluish-grey or black, but in the case of the films on a copper strip, it is sometimes possible by bending the strip sharply to detach the film as tiny flakes; it will then be seen that, although the outer layer is blackish, generally consisting of cupric oxide (CuO), the inner layer is red, consisting of cuprous oxide (Cu_2O). Nickel develops only a single layer (NiO), whilst iron can sometimes develop three layers (see below).

The films of the interference-colour region cannot, in general, be removed as flakes by merely bending the metal, but it is possible to transfer the film to a sheet of transparent plastic. The oxidized metal is cemented to the sheet (with the oxide-film next to the plastic), and the metal is then removed by anodic corrosion by a method described on p. 785, leaving the film adhering to the plastic. It will then be found that the film still shows colours—best seen when viewed from the side where the metal had originally existed. In the case of the film on nickel, which consists of a single layer, the colour at any point is found to have become roughly complementary to that observed at the same point when the film was still on the metal; thus the place which was originally yellow becomes blue; that which was originally green becomes red, and *vice versa*. Optical principles would lead us to expect this complementary relationship.

A similar complementary relation can be seen in the case of iron oxide films under circumstances where only a single layer is present. Thus an iron strip heated for a short time at one end gives colours extending over three orders. The colours are due to a ferric oxide-film (Fe_2O_3) which, in the first-order region, rests directly on the metal; the films, transferred to plastic, show weak colours roughly complementary, at each point, to the colours originally exhibited at the same point when the film was backed by the metal; once more, the colours are best seen by viewing the specimen from the side where the metal had been, but they can also be discerned on looking through the plastic. However, in the second-order region, there is a layer of nearly opaque magnetite (Fe_3O_4) between the Fe_2O_3 film and the metal, and this makes it difficult to see the colours from the inner side after the removal of the metal; when the specimen is viewed through the plastic, colours can still be seen, but they are now identical with those which existed before the removal of the metal; nor is this surprising, since the reflecting surface at the inner face of the ferric oxide-film, which was magnetite when the metal was present, remains magnetite after metal has been removed (fig. 1, p. 20).

The observation of colours is assisted if glass is used instead of plastic as the support. Eurof Davies, after heating an iron specimen in oxygen for a chosen time at a chosen temperature, thus producing a uniform tint, cemented the specimen to glass with epoxy-resin, and removed the metal by anodic action. In cases where the film consisted of two layers, Fe_2O_3 and Fe_3O_4, the latter being sufficiently thin to be transparent, four different colours could be obtained from a given film:—

(1) the colour observed when the film was still on the metal and was viewed by reflected light.
(2) the colour observed after removal of the metal, when the film was viewed by transmitted light.
(3) the colour observed after removal of the metal, when the film was viewed by reflected light through the glass.
(4) the colour observed when the film was viewed by reflected light from the " inner " side (where the metal had been).

Specimens consisting of a single (Fe_2O_3) layer showed the same colour whether viewed through the glass or from the " inner " side (D. Eurof Davies, U. R. Evans and J. N. Agar, *Proc. roy. Soc.* (*A*) 1954, **225**, 443, esp. p. 452).

Films on Copper. When gently heated in air, copper develops a cuprous oxide-film, which, as it thickens, exhibits a complete succession of interference tints up to the fourth order; the colours appear in the same sequence as on nickel, although the character of the earlier colours is slightly modified by the specific colour of the metallic background. However, copper differs from nickel in having two oxides stable under these conditions. As the cuprous oxide-film thickens, tiny black spots of cupric oxide appear outside it, so that the colours are marred by a slightly sooty appearance.*

* In the early years of corrosion research, during the preparation of specimens designed to exhibit the colour sequence, this disfigurement was avoided by

As the thickness increases beyond the interference-colour range, the black cupric oxide spreads laterally over the whole surface so that the appearance viewed from the outside is black, although particles which have been chipped off are red on their under surface (with some varieties of copper, saucer-shaped particles fly off spontaneously on rapid cooling, the curvature being caused by the unequal contraction of the two layers).

The fact that cupric oxide does not appear rapidly at all points even at such temperatures and pressures as would render it a stable phase is an example of a low " nucleation rate "—such as is met with frequently in corrosion and accounts for many unexpected phenomena. It would seem that only on rare occasions do a sufficient number of copper and oxygen atoms chance to come together in the correct positions to form a nucleus of CuO, on which others can array themselves to make it grow into a cupric oxide grain. Thus under conditions of oxygen pressure where cupric oxide ought to be a stable phase, the oxide-film may consist wholly of cuprous oxide; even when cupric oxide is formed, it appears only as isolated grains, producing the sooty appearance already mentioned. Doubtless small changes in surface state will affect the chance of forming a nucleus, so that different experimenters working under different conditions have occasionally obtained results which at first sight appear contradictory. The reader may care to compare the findings of H. A. Miley, *J. Amer. chem. Soc.* 1937, **59,** 2626; A. L. Dighton and H. A. Miley, *Trans. electrochem. Soc.* 1942, **81,** 321; R. F. Tylecote, *Metallurgia* 1956, **53,** 191; K. R. Dixit and V. V. Agashe, *Z. Naturf.* 1955, **10,** 152; D. W. Bridges, J. P. Baur, G. S. Baur and W. M. Fassell, *J. electrochem. Soc.* 1956, **103,** 475; J. Paidassi, *ibid.* 1957, **104,** 749; *Acta Met.* 1958, **6,** 561.

Electron diffraction studies at Poona suggest that the oxide formed on the dodecahedral faces of a copper single crystal is at the out-set always cuprous oxide. The precise character depends on the conditions. It may be amorphous or crystalline and in the latter case it may exhibit *epitaxy* —which means that some definite relationship exists between the orientation of the atomic layers in the oxide with those in the metallic basis.* In the Indian experiments, the oxide formed at 125°C. was first amorphous Cu_2O but later became crystalline, exhibiting epitaxy with the metal below;

heating in a feebly oxidizing atmosphere, e.g. the upper part of the flame from a spirit lamp, which provided conditions unfavourable to the formation of the higher oxide.

* Epitaxy is a phenomenon not confined to oxide-films; it can occur when one metal is electrodeposited on another. In most cases, if the deposited material is placed with its crystal planes oriented in a certain way, the distance between atoms on one side of the plane of separation is *nearly* the same as that on the other side, so that with a *small* amount of strain on each side (the atoms being brought unnaturally close together on one side and unnaturally far apart on the other) a smooth passage from one phase to the other is attained. This is not necessarily produced by placing corresponding crystal-planes parallel. If both substances are cubic, but have different lattice parameter, considerable strain would result if the cubic planes were brought parallel; probably some pair of planes can be found which will give better fit between the inter-atomic distances, but in general there will always be *some* strain.

heating for 12 hours produced no further change and there was no indication of CuO—even superficially. On the other hand at 250° and 300°, the Cu_2O films were soon covered with crystalline CuO (A. Goswami and Y. N. Trehan, *Trans. Faraday Soc.* 1956, **52,** 358. Cf. B. Lustman and R. F. Mehl, *Trans. Amer. Inst. min. met. Engrs.* 1941, **143,** 246).

Films on Iron. At high temperatures, iron develops three layers, which correspond roughly to ferrous oxide (FeO), magnetite (Fe_3O_4) and ferric oxide (Fe_2O_3), although, as first emphasized by Pfeil, the composition of each layer changes continuously on passing from one surface towards the other; the metal-content is highest at the interface closest to the metallic base. Thus the formulae can, at best, only represent the exact composition at one particular level. In the case of the innermost layer, commonly known as *Wüstite*, Pfeil considered that, even at its inner interface, it contained less metal (more oxygen) than the formula FeO would suggest. This is confirmed by the more recent equilibrium diagram of the iron–oxygen system due to Darken and Gurry (fig. 2); Bénard, however, thinks that at high temperatures, the composition of the ferrous oxide phase at points next to the metal should, under equilibrium conditions, approach FeO. Fig. 2 shows that FeO decomposes to Fe + Fe_3O_4 below about 570°C. (L. B. Pfeil, *J. Iron St. Inst.* 1929, **119,** 501; 1931, **123,** 237. L. S. Darken and R. W. Gurry, *J. Amer. chem. Soc.* 1946, **68,** 798. J. Bénard, *Bull. Soc. chim. Fr.* 1949, p. D117. J. Aubry and F. Marion, *C.R.* 1955, **240,** 1770. G. Chaudron, *ibid.* 1955, **240,** 1771. H. J. Engell, *Z. Elektrochem.* 1956, **60,** 905; *Arch. Eisenhuttenw.* 1957, **25,** 109).

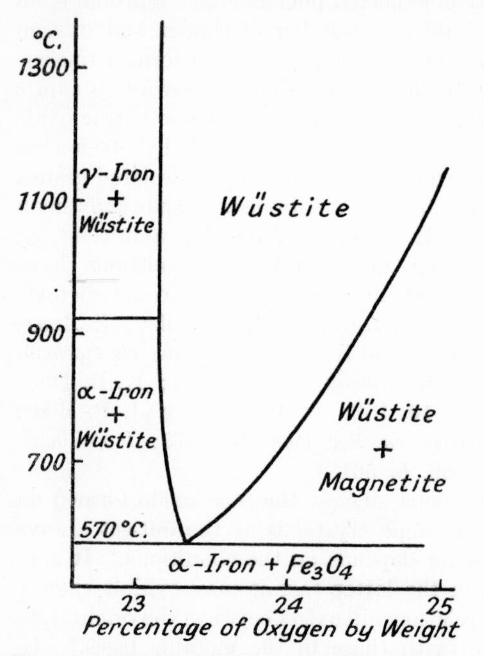

FIG. 2.—Diagram showing decomposition of Wüstite (FeO with deficiency of iron) below 570°. True FeO would contain 22·3% oxygen (diagram based on L. S. Darken and R. W. Gurry).

Ferrous oxide, magnetite and also γ-ferric oxide possess cubic crystal structures; Fe_3O_4 can be regarded as derived from FeO by removal of one quarter of the iron (some of the ferrous ions must become ferric to preserve electrical neutrality). It is possible to remove further iron and obtain a series of cubic solid solutions of which γ-Fe_2O_3 is the end member. The γ-oxide, however, is generally believed to be unstable at high temperatures;

the stable form (α-Fe_2O_3), which is the essential constituent of the film responsible for interference tints, possesses a rhombohedral structure, unrelated to that of the cubic oxides just mentioned. Recent electron-diffraction observations in Cohen's laboratory at Ottawa on iron heated at 320°C. for 1740 min. in oxygen at 20 mm. pressure point to the presence of small amounts of γ-Fe_2O_3 between the magnetite and α-Fe_2O_3 layers and suggest the manner in which magnetite changes to α-Fe_2O_3; apparently the magnetite first loses iron so that the composition becomes Fe_2O_3 without loss of cubic structure, and then a rearrangement of atoms to rhombohedral structure follows. Readers with crystallographic interests should study the papers of H. J. Goldschmidt, *J. Iron St. Inst.* 1942, **146,** 157P; E. J. W. Verwey, *Z. Kristallogr.* 1935, **91,** 65; K. H. Buob, A. F. Beck and M. Cohen, *J. electrochem. Soc.* 1958, **105,** 74.

There is, however, some conflict of views. Finch and Sinha regard γ-Fe_2O_3 as the stable high-temperature form, while David and Welch think that γ-Fe_2O_3 only exists at room temperature if stabilized by the presence of combined water. Cohen, however, quotes magnetic experiments by Selwood, who finds that γ-Fe_2O_3 is changed to α-Fe_2O_3 on heating; he also believes that γ-Fe_2O_3 is stable at low temperatures in the absence of occluded water or OH-radicals. The Author is inclined to accept Cohen's views (M. Cohen, Priv. Comm., March 19, 1958. G. I. Finch and K. P. Sinha, *Proc. roy. Soc.* (*A*) 1957, **241,** 1. I. David and A. J. E. Welch, *Trans. Faraday Soc.* 1956, **52,** 1642, esp. p. 1649).

Knowledge of the situation has been retarded by the fact that the structures of γ-Fe_2O_3 and Fe_3O_4 are closely related. At one time it seemed impossible to distinguish them with confidence by X-rays or electron diffraction. There are, however, slight differences in the patterns furnished by the two oxides. Certain lines occur in the pattern of γ-Fe_2O_3 but not in magnetite; Eurof Davies, heating magnetite at 400°C., found that oxygen was taken up, and when about 80% of the ferrous iron had been oxidized, the new lines appeared in the X-ray patterns. In some cases he obtained information by detaching the film from the metal and observing it chemically. Earlier workers had found a variation of lattice parameter with composition on passing along the series, and this variation has been used by Vernon and associates at Teddington in estimating the composition of oxide formed under different conditions; they examined the films by electron diffraction after removal from the metal—obtaining somewhat different results from those workers who had examined the films still attached to the metal.

By such methods, considerable information has been obtained. The Teddington group, studying abraded iron oxidized around 200°C., found the film to be duplex with rhombohedral α-Fe_2O_3 overlying cubic oxide. The cubic material formed at 225°C. initially has a composition relatively near to γ-Fe_2O_3, but rather rapidly changes to Fe_3O_4; at 180°C., however, the change is slow, and even after 453 hours, the composition Fe_3O_4 is not reached. Moreau and Bardolle consider the oxide present below the α-Fe_2O_3 layer to be magnetite if formed at 300°C. but γ-Fe_2O_3 if formed at 250°C. The Cambridge work (on iron prepared by heating in hydrogen) indicated the invisible film formed at low temperatures to have a composition near to γ-Fe_2O_3. At 175° and 225°C., the film was α-Fe_2O_3 without cubic oxide; at 250°C. the film was solely α-Fe_2O_3 for 8 hours, but then developed an inner layer of cubic oxide; at 300°C. the film was α-Fe_2O_3 overlying cubic oxide from the start. The cubic oxide formed at these temperatures was not true γ-Fe_2O_3 but contained ferrous iron. On abraded (as opposed to hydrogen-prepared) iron, cubic oxide as well as α-Fe_2O_3 was formed as low as 175°C.—an observation which removed an apparent discrepancy with the Teddington results.

Despite some uncertainties, the results from different laboratories agree in showing that the oxygen-content in the cubic phase (where present) tends to be higher at low temperatures than at high ones; for rough purposes, one may provisionally call the invisible oxide formed at room temperatures γ-Fe_2O_3 and the visible oxide formed above 200°C. magnetite.

The situation, however, is not simple, and the reader should study the original papers by W. H. J. Vernon, E. A. Calnan, C. J. B. Clews and T. J. Nurse, *Proc. roy. Soc.* (*A*) 1952, **216,** 375; D. Eurof Davies and U. R. Evans, *J. chem. Soc.* 1956, p. 4373; J. Moreau and J. Bardolle, *C.R.* 1955, **240,** 524. Information regarding X-rays and electron diffraction patterns is provided by E. J. W. Verwey, *Z. Kristallogr.* 1945, **91,** 65; *Bull. Soc. chim. Fr.* 1940, p. D123; G. Hägg, *Z. phys. Chem.* (*B*) 1935, **29,** 95; H. P. Rooksby, Chapter X of C. W. Brindley's book, " X-ray Identification and Crystal Structures of Clay Minerals ", 1951, esp. fig. X-3 opposite p. 257 (Mineralogical Society).

Above 570°C. all three layers are normally present on iron; the inner (FeO) layer may often be the thickest of the three. It should, however, be absent if heating is conducted below 570°C., since FeO only becomes a stable phase above that temperature (fig. 2, p. 24). If iron is heated strongly and cooled, the FeO layer decomposes to an intimate mixture of $Fe + Fe_3O_4$ about 570°C., unless the cooling is very rapid (G. Chaudron and R. Collongues, *Rev. Métallurg.* 1951, **48,** 917; *C.R.* 1952, **234,** 728). For structure of scale on wire or strip, where the decomposed Wüstite is sometimes sandwiched between two magnetite layers, see S. Garber, *Nature* 1959, **183,** 1387).

Magnetite is a stable phase even at low temperatures, and it might be expected that in the interference-colour range the ferric oxide films (which contribute the colours, being fairly transparent) would be separated from the metal by relatively opaque magnetite. For films showing colours of the second and higher orders, this appears to be the case, and it is probably true of first-order films if time is allowed. A given colour can be obtained by (1) long exposure at a low temperature or (2) a shorter exposure at a higher temperature; if the second method is adopted, the ferric oxide colour-film usually rests directly on the metallic surface.*

* The absence of the magnetite below rapidly produced first-order colour-films is probably another case of low nucleation number; but it is a little surprising that the invisible cubic oxide-film (essentially γ-Fe_2O_3) is produced so rapidly on air-exposure at an ordinary temperature.

The Process of Oxidation

Three Main Classes of Oxidation Curves. Work at Bristol University on oxidation at low pressures has shown that at − 183°C. copper rapidly takes up a very small amount of oxygen, which is considered to represent a monolayer of adsorbed oxygen; there is no further uptake (fig. 3). At room temperature (20°C.) there is rapid oxidation at first, but after about 2 hours the thickness of the oxide-film has become almost constant at about 24 Å; at 74°C., the oxidation is more rapid and reaches a greater thickness (J. A. Allen and J. W. Mitchell, *Disc. Faraday Soc.* 1950, **8,** 309; J. A. Allen, *Trans. Faraday Soc.* 1952, **48,** 273).

At pressures of the order of one atmosphere the oxidation of many

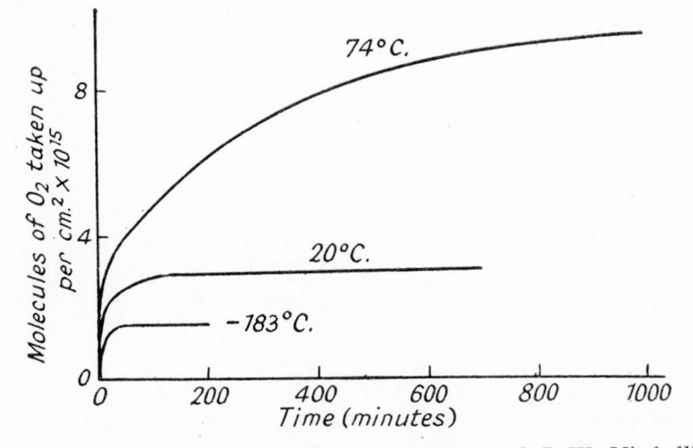

FIG. 3.—Oxygen-uptake by Copper (J. A. Allen and J. W. Mitchell).

metals obeys different laws at different ranges of temperature; if the temperature only slightly exceeds room temperature, the thickening starts rapidly, but soon becomes so slow that for practical purposes oxidation can be considered to have ceased. Hart, for instance, finds that aluminium exposed to dry oxygen quickly builds a film about 30 Å in thickness, beyond which oxidation comes practically to an end (R. K. Hart, *Proc. roy. Soc.* (*A*) 1956, **236,** 68).

Most of the equations put forward to explain this transient growth at relatively low temperatures involve logarithms or exponentials. They can be grouped together for present purposes as the *log laws*; their mathematical form and theoretical derivation are discussed in Chapter XX.* However,

* It may, however, be convenient to state the various equations for the three log laws which are met with at temperatures too low for normal thermal movement of ions. Since oxidation is not uniform it may be best to consider the weight-increment per unit area W rather than the thickness y obtained in time t. The inverse logarithmic law $1/W = K_1 - K_2 \log (at + 1)$ should be obeyed if the ions are pulled through the film by the strong electrical field present in very thin films. The asymptotic law $W = K_3(1 - e^{-K_4 t})$ should be

above a certain temperature, instead of one of the log laws, the thickening obeys a *parabolic law*; this implies a greater risk of wastage over long periods, since thickening does not in general become negligibly slow until the film has become relatively thick (involving, as already explained, a greater risk of cracking or flaking). The simplest type of parabolic law can be expressed by the equation

$$dy/dt = k/y \quad \text{or} \quad \tfrac{1}{2}y^2 = kt + k'$$

where y is the thickness at time t, whilst k and k' are constants depending on temperature but independent of time. This is, of course, the equation

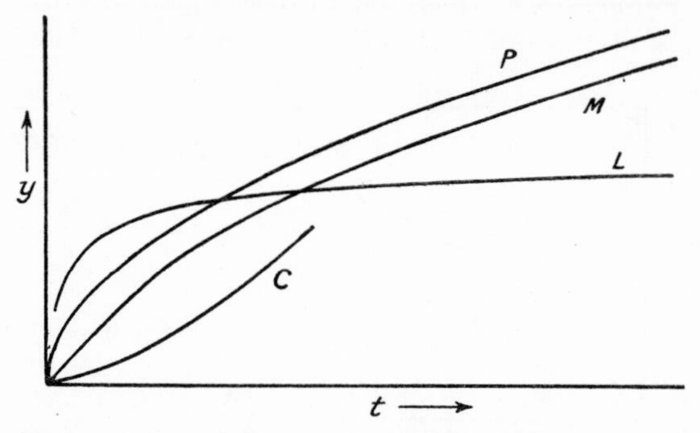

FIG. 4.—Main Types of Growth Laws. *P*, Simple Parabola. *M*, Mixed Parabola. *L*, " Log-law " *C*, " Concave-upwards ".

of a parabola. Fig. 4 shows a curve (*P*) conforming to the parabolic law, and for comparison another curve (*L*) typical of a log law. However, in most cases, parabolic thickening is slightly less simple, since, under conditions favouring parabolic growth, a metal is first attacked at an almost constant rate—which gradually declines until the curve becomes a parabola; the curve obtained is known as the " mixed parabola " (curve *M*). Bénard

obeyed if there is diffusion through pores, any one of which is liable to become blocked by material formed within it. The direct logarithmic law

$$W = K_5 + K_6 \log (at + 1)$$

should be obeyed if there is diffusion through pores, which tend to block their neighbours by the compressional force developed. In certain circumstances, it may be difficult to distinguish experimentally between the three relationships; moreover in practice all these things may occur together, giving intermediate relationships. Hence it is convenient to bracket them together as the " log laws ". A parabolic law should be obeyed if the temperature is high enough to permit normal thermal movement in the absence of a potential gradient, so that the function of the gradient (whether a chemical-potential gradient or an electrical-potential gradient) is *directive* rather than *causative*. This may be a simple parabola $W^2 = K_7 + K_8t$, if the growth-rate is controlled simply by the rate at which material can pass through the film, but a mixed parabola

$$K_9W^2 + K_{10}W = K_{11} + K_{12}t,$$

if there is partial control by one of the interface reactions. In extreme cases where the first term can be neglected, this passes into the rectilinear law. All these matters are more fully discussed in Chapter XX.

found that iron heated at 1000°C. yielded a curve which was straight at first and then became roughly parabolic (J. Bénard, *Bull. Soc. chim. Fr.* 1949, p. D89).

Provided that the value of k is reasonably small (which is true of most common metals when the temperature is not too high), a material which thickens by the parabolic law is often accepted by the engineer as reasonably resistant. Nevertheless its employment may be unsatisfactory—as can be shown by an example. Suppose that, after one hour, the oxidation-rate of some metallic surface is suffering wastage at a rate about ten times that which the engineer would regard as a tolerable maximum; the rate will, of course, decrease with time, and it is only necessary to wait 100 hours to reach the tolerable value. He might be prepared to accept such a situation—were it not for the fact that the film will then be ten times as thick, with increased risk of flaking or cracking; obviously if flaking occurs, rapid oxidation will set in again. In contrast, one of the log laws will generally bring oxidation almost to a stand-still at a small thickness. The practical problem attending choice of materials for use at relatively high temperatures is to find something which thickens according to a favourable growth-law; it will be seen later that certain alloying materials, introduced into iron, come near to achieving this.

At still higher temperatures, especially in atmospheres containing only small amounts of oxygen or only weakly oxidizing substances, the oxidation-rate is sometimes found to become nearly constant; the extreme case, where the rate is independent of time, is represented by the rectilinear law

$$dy/dt = k \quad \text{or} \quad y = kt + k'$$

where k and k' are constants, depending on temperature.

Obedience to the rectilinear law may merely mean that the replenishment of oxygen (or some compound furnishing oxygen) cannot keep up with the rate at which an oxide-film could thicken, so that the oxygen-replenishment rate assumes control of the whole process. This may easily happen in atmospheres where the content of oxygen (or oxidizing agent) is low. However, it may occur in air, or a gas-mixture containing plenty of oxygen, at the outset of the process, when the films are very thin, and therefore capable of transmitting material at a rate higher than even a good oxygen-supply can sustain. Bénard's experiment on iron at 1000°C. —just described—affords an example.*

However, rectilinear growth may arise in another way. If the oxide-film is formed in a strained condition (which, as explained later, often occurs when it is formed by oxygen moving inwards—not metal outwards), it may, at high temperatures, keep breaking down or flaking off as quickly as it is formed. This spontaneous film break-down, without rough treatment from any external sources, is to be expected when a certain critical

* The oxidation of iron in CO_2/CO mixtures (which is parabolic at 850°C.) follows a rectilinear law at 900°C.; the rate-determining factor is here considered to be the chemisorption of the oxygen formed by the decomposition of the carbon dioxide (K. Hauffe and H. Pfeiffer, *Z. Metallk.* 1953, **44**, 27).

thickness of strained oxide has been formed—for reasons analogous to those given on p. 20 (esp. foot-note). Thus at roughly equal intervals of time, we may expect the film to become non-protective and oxidation to be speeded up; as a result the oxidation-rate is for rough purposes constant (see also fig. 120(C), p. 732).

At somewhat lower temperatures, the mechanical break-down may be avoided, at least for a considerable time until a critical thickness is reached. Thus we may obtain parabolic thickening over an intermediate range of temperature, but rectilinear thickening at high temperatures. Metals which show approximately parabolic growth at intermediate temperatures and approximately rectilinear growth at higher ones include aluminium, thorium, niobium and uranium; other metals, such as chromium and nickel, show a sudden speeding up of the oxidation-rate at a certain thickness—which probably arises in a similar way. On chromium, in the range 950–1050°C. the spurt in oxidation occurs when the thickness reaches 4800 Å; above 1050°C., a second spurt occurs about 42000 Å. On nickel, the films crack away from the metal at 1000°C. (E. A. Gulbransen and K. F. Andrew, *J. electrochem. Soc.* 1957, **104,** 334, 451).*

Tungsten often shows parabolic growth at first and later rectilinear growth; here there are two oxides, a dark blue dense layer of W_4O_{10} overlain with yellow powdery WO_3; the rate of oxidation is inversely proportional to the thickness of the lower denser layer—explaining the parabolic law; but after a time the thickening of that layer almost ceases, since W_4O_{10} becomes transformed into WO_3 as quickly as it is produced, so that the oxidation-rate becomes constant (W. W. Webb, J. T. Norton and C. Wagner, *J. electrochem. Soc.* 1956, **103,** 107. Cf. J. P. Baur, D. W. Bridges and W. M. Fassell, *ibid.* 1956, **103,** 266).

On niobium, the oxidation-rate at 375°C. falls off steadily with time, but at 400° it falls off until a certain thickness (corresponding to 60–80 μg./cm.^{2}) is reached, and then increases, the curve becoming linear. The oxide which appears smooth and coherent under the electron microscope during the period of decreasing oxidation-rate, begins to show changes when the linear period is approached. The greater part of the surface still carries a film which is continuous and probably protective, exhibiting interference colours when viewed optically, but at places there are blister-like cracks; in the linear range itself the whole surface is covered with rough, white oxide.

* J. N. Wanklyn adds the comment, "One can imagine two cases of film-cracking changing 'parabolic' behaviour to 'rectilinear'. In the first, as the film approaches a certain critical thickness, it begins to crack; oxide added is therefore unprotective and the thickness of the protective part of the film from then onwards is approximately equal to the critical thickness. But another possibility is that, when the protective oxide has built up 'parabolically' to a critical value, the cracks which then form reduce the 'protective thickness' to *much less* than the 'critical value'. This could occur if the cracks were hard to *nucleate* but easy to *propagate*, as is compatible with stress-raising effects at the sharp tips of cracks. This would give a linear portion which is steeper than the tangent to the parabola." He suggests that this may occur in the "break-away" or "transition" effects found on zirconium alloys (p. 46).

It is evident that the loss of protectivity is due to mechanical break-down of the film under its internal stresses, which are only able to cause this break-down after a certain thickness has been reached for reasons analogous to those explained in footnote* of p. 20. The break-down occurs at somewhat lower thicknesses at higher temperatures. At 375 and 425°C. the early part of the curves, after initial irregularities, fit the parabolic law (E. A. Gulbransen and K. F. Andrew, *J. electrochem. Soc.* 1958, **105,** 4. J. V. Cathcart, J. J. Campbell and G. P. Smith, *ibid.* 1958, **105,** 442).

On zirconium, the film is compact and fairly protective up to about 400–500°C., above which the protective character declines; the failure of the film to withstand the stresses is attributed to the softening of the metallic base, which yields to the stresses—as indicated by J. Hérenguel, D. Whitwham and J. Boghen, *C.R.* 1956, **243,** 2060; *Rev. d'Alumin.* 1957, **244,** 611.

Most investigators consider that films on zirconium thicken by the movement of oxygen inwards, but some evidence in favour of zirconium ions moving outwards is reported by E. A. Gulbransen and K. F. Andrew, *J. Metals* 1957, **9,** 394. The state of the oxygen present in solid solution in the metal below the oxide-film is discussed by J. P. Pemsler, *J. electrochem. Soc.* 1958, **105,** 315.

On hafnium there seems little doubt that the oxygen moves inwards; radioactive markers have been found to remain at the oxide–oxygen interface. Moreover, hafnium foil oxidized at 1200°C. from *both faces* until the oxidation meets in the centre develops two oxide-scales which bend in different directions and part company from one another—showing clearly the internal stresses developed during oxidation. This work deserves study (W. W. Smeltzer and M. T. Simnad, *Acta Met.* 1957, **5,** 328, esp. pp. 331, 332).

The case of uranium is more complicated, as there are several oxides; the films have been studied by electron diffraction, and measurements have also been made of the oxidation of lower to higher oxides. Readers should consult the papers of D. Cubicciotti, *J. Amer. chem. Soc.* 1952, **74,** 1079; Y. Adda, *C.R.* 1956, **242,** 126; R. K. Hart, *Trans. Faraday Soc.* 1953, **49,** 299; J. Loriers, *C.R.* 1952, **234,** 91; J. S. Anderson, L. E. J. Roberts and E. A. Harper, *J. chem. Soc.* 1955, pp. 3939, 3946; also for uranium-zirconium alloys, S. Barnartt, R. G. Charles and E. A. Gulbransen, *J. electrochem. Soc.* 1957, **104,** 218.

The temperature below which the parabolic law is replaced by a log law varies with the material and with conditions. Teddington work on steel prepared by abrasion showed the transition to occur at 200°C.; Cambridge work on iron freed from oxide by heating in hydrogen in a vessel which was then evacuated and brought to the experimental temperature before oxygen was admitted, placed it somewhat above 300°C. There is no necessary discrepancy.

Several graphical methods are available for deciding the law which is being obeyed. In cases where a series of specimens have been heated for different times, and the weight increase W measured, a simple test is to plot W^2 against t, when a straight line indicates parabolic thickening (fig. 5);

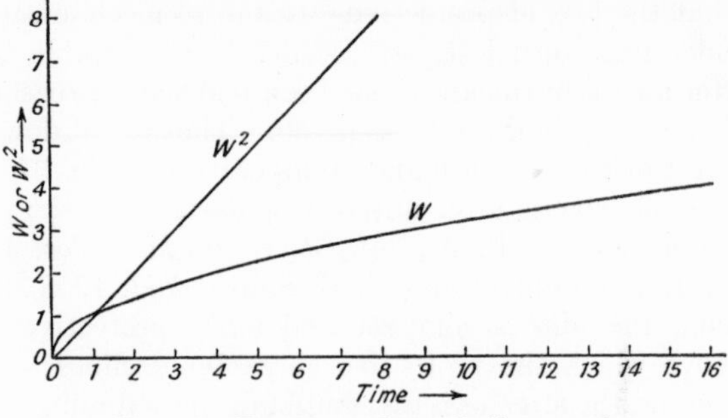

FIG. 5.—Features of parabolic growth.

if the rectilinear law is being obeyed, a straight line will be obtained on plotting W against t. Testing for the log laws is discussed on p. 839.

Importance of lattice-defects. At very low temperatures such as − 183°C. (p. 27), the affinity of metal for oxygen is apparently satisfied by a single layer of " chemi-adsorbed " oxygen attached to the metal surface; probably electrons pass from the outermost layer of metal atoms (which become cations) to the oxygen atoms (which become anions), so that the two layers together can be regarded as a two-dimensional oxide-film (not necessarily identical with any oxide known in the massive three-dimensional state).* An ordinary three-dimensional oxide-film is only likely to be formed if the temperature is sufficiently high to render the particles appreciably mobile, so that either metal passes outwards through the film to meet the oxygen, or oxygen inwards to meet the metal.

At really high temperatures such movement becomes a frequent occurrence, owing to the presence of lattice defects. Cuprous oxide, generally written Cu_2O, contains rather less copper than that formula would indicate; it is believed that some of the cation sites where Cu^+ ought to exist, are really vacant, neutrality being maintained by the fact that some of the other sites are occupied by Cu^{++}, instead of Cu^+. Many lower oxides, such as NiO and FeO, have similar vacant cation sites. Other oxides, such as ZnO, Fe_2O_3 and ZrO_2, appear to contain more metal, or less oxygen, than the formula would predict (see p. 33); in some cases (e.g. ZrO_2) there seem to be vacant anion sites, whilst in others (e.g. ZnO) the excess metal may be present as interstitial cations along with quasi-free electrons.†

* K. Hauffe (private communication) considers that the chemisorbed oxygen atoms each capture only one electron, forming O^-, not O^{--}. The factors favouring chemiadsorption, which seems to be absent on gold, are discussed by B. M. W. Trapnell, *Proc. roy. Soc.* (*A*) 1953, **218,** 566; M. A. H. Lanyon and B. M. W. Trapnell, *ibid.* (*A*) 1955, **227,** 387; T. B. Grimley and B. M. W. Trapnell, *ibid.* 1956, **234,** 405; T. B. Grimley, Chap. 14 of W. E. Garner's book, " Chemistry of the Solid State ", 1955 (Butterworth). See also W. E. Garner, *Chem. and Ind.* (*Lond.*) 1951, p. 1010, who discusses the adsorption of oxygen on an oxide surface and the dissociation of oxygen molecules to provide adsorbed oxygen atoms, which take up electrons to produce oxygen ions.

† The excess zinc in zinc oxide produces traces of hydrogen when the oxide is dissolved in acid—a method used for determining the excess (H. J. Allsopp, *Analyst* 1957, **82,** 474).

At high temperatures vacancies can move about, since an adjacent ion can move into a vacancy, thus creating a new vacant position, which a fresh ion can enter from the far side—and so on. Thus the existence of lattice-defects allows passage of material through an oxide film; metal can pass outwards if these are cation vacancies, or oxygen inwards if there are anion vacancies. The defects also allow passage of electrons; a Cu^{++} ion can accept an electron from a neighbouring Cu^{+} ion which becomes a Cu^{++} ion and is thus able to accept an electron from another Cu^{+} ion on the far side. Evidently the electronic conductivity will increase as the composition diverges from that represented by the formula. The effect of oxygen-pressure on conductivity has been used to distinguish between oxides which have metal deficiency and those which have metal excess (or oxygen deficiency); in the former group (such as Cr_2O_3) the conductivity increases as the pressure of oxygen in the atmosphere with which they are in contact increases, whereas in the latter group (which includes ZnO and Fe_2O_3) it decreases (J. M. Bevan, J. P. Shelton and J. S. Anderson, *J. chem. Soc.* 1948, p. 1729).

Solid-state physicists call the group of oxides in which there is metal deficiency, *p-conductors*; these possess vacant cation sites and conduct electricity at high temperature by the movement of *electron holes* (places where an electron is missing); in the case of cuprous oxide the electron holes are the places where a Cu^{+} ion has lost an electron, being converted into a Cu^{++} ion. The other group where metal is in excess are called *n-conductors*; their conductivity is due to the presence of practically free electrons; there are also extra cations in interstitial positions or alternatively vacant anion sites.

If we imagine a block of so-called cuprous oxide at 1000°C., it is likely that the vacancies are moving about at random. If, however, we pass to consider a film of cuprous oxide attached to a copper surface on its inner side (with oxygen gas on the outer side), the movement will no longer be at random; the vacancies will now tend to move towards the metal, which is another way of saying that the cations will move, on the whole, outwards towards the oxygen.

There are two reasons why movement in one particular direction should prevail over movement in other directions. First, when oxygen from the gas phase becomes adsorbed as atoms on the outer surface of the oxide, they will acquire electrons from the metal, thus becoming oxygen ions (negatively charged), and since the metal is left positively charged, the electric field set up will clearly encourage cations to move outwards. Secondly, there will be a concentration gradient, since the film will become more rich in oxygen near the outer surface. For both these reasons, copper cations will move, on the whole, outwards, and from time to time sites between the outermost layer of oxygen ions will become occupied by copper cations, thus providing a new layer of cuprous oxide; on this layer of oxide, fresh oxygen can be adsorbed, leading to another layer of oxide and so on. Thus oxidation proceeds, fresh oxide being continually deposited on the *exterior* of the film.

Whether growth is due to metal moving outwards or oxygen inwards, the parabolic law is normally to be expected, provided that the directive force remains constant. For instance, if we assume, for simplicity, that a constant E.M.F. operates across the film, the movement of ions will be proportional to $1/y$, by Ohm's Law. Since the rate of growth is determined by the rate of arrival of ions (either cations outwards, or anions inwards) it follows that $dy/dt = k/y$—the parabolic equation. The same conclusion it reached if the transport consists of diffusion under a concentration gradient. It is right to add, however, that this statement of the situation is something of an over-simplification. Fuller treatment, with references to the papers of Wagner and of Mott, will be found in Chapter XX.

There is fairly definite evidence that the growth of Cu_2O, NiO and FeO films occurs by the outward passage of cations, and that the oxidation of titanium and probably zirconium occurs mainly by the inward passage of anions. It seems that Fe_3O_4 grows by outward movement, and Fe_2O_3 by inward movement. The best evidence depends on the use of " markers " placed upon the original surface of the film, which are found after oxidation, either buried at the base of the film in cases where cations move outwards, or clinging to the exterior surface of the film in cases where anions move inwards. Pfeil, in his classical work, first used markers to demonstrate the outward passage of material. He applied a " distemper " of chromic oxide particles to an iron surface, and after oxidation of the iron, found the chromium at the base of the film. Later, radioactive substances were used as markers, notably in Mehl's laboratory at Pittsburgh, since the exact position in a cross-section of the film is then easy to locate (L. B. Pfeil, *J. Iron St. Inst.* 1929, **119,** 501; 1931, **123,** 237. M. H. Davies, M. T. Simnad and C. E. Birchenall, *Trans. Amer. Inst. min. met. Engrs.* 1951, **191,** 889; 1953, **197,** 1250. L. Himmel, R. F. Mehl and C. E. Birchenall, *J. Metals,* 1953, **7,** 827; M. Simnad and A. Spilners, *ibid.* 1955, **7,** 1011. Other radioactive methods are described by J. Bardeen, W. H. Brattain and W. Shockley, *J. chem. Physics* 1946, **14,** 714).

On cobalt also there is evidence of movement in both directions, giving two layers of different physical character, the outer one hard and coherent and the inner one friable; both possess (at least above 950°C.) the composition CoO. The dividing line between the two layers is coincident with the original metallic surface and it is fairly certain that the porous oxide is formed by oxygen moving inwards and the coherent oxide by metal moving outwards (A. Preece and G. Lucas, *J. Inst. Met.* 1952–53, **81,** 219; confirmed by C. A. Phalnikar, E. B. Evans and W. M. Baldwin, junr., *J. electrochem. Soc.* 1956, **103,** 429, esp. p. 433).

The case of nickel is described by Ilschner and Pfeiffer, who arranged a platinum wire in contact with the surface of nickel foil; after oxidation at 1000°C. for 48 hours, the wire was found to be located within the film, at the junction of two layers, which had different appearances when viewed under dark-field illumination; presumably one had grown by outward movement, and the other by inward movement (B. Ilschner and H. Pfeiffer, *Naturwissenschaften* 1953, **40,** 603).

The case of iron is analogous, but slightly more complicated owing to the presence of three oxide layers. It was studied in detail by Pfeil (p. 34) and recently by Sachs, who confirms Pfeil's picture of counter-current diffusion in all essential respects, and has given the following picture of the film-thickening process. Since iron ions are considerably smaller than those of oxygen, it might be expected that most of the scale-formation would depend on iron diffusing outwards. However, this outward movement produces cavities at the scale-base and thus reduces the effective cross-section available for the outward diffusion of iron, whilst providing extra facilities for inward movement of oxygen, by dissociation of the oxide and gaseous diffusion, as well as by ionic diffusion through the lattice. A steady state should ultimately be established at which iron and oxygen are moving in opposite directions in equal amounts, and the markers should define the position where this is occurring.

Nevertheless, Sachs shows that some caution is necessary in the use of wire markers, especially in the case where the wire consists of a metal which can interact with the oxide or with oxygen, or where it is so thick as to screen the metal. His paper should be studied (K. Sachs, *Metallurgia* 1956, **54,** 11, 109. Cf. A. de S. Brasunas and N. J. Grant, *Trans. Amer. Soc. Metals* 1952, **44,** 1117. R. A. Meussner and C. E. Birchenall, *Corrosion* 1957, **13,** 677t, esp. p. 681t).

Some observations by Pfeiffer and Ilschner are instructive in suggesting that various changes must be occurring simultaneously. Iron wire heated at 700–1000°C. became a *tube* with an internal cavity—which clearly points to *outward* movement; but this can hardly have been the only phenomenon, since a uniform outward movement would soon have produced an annular cavity between the oxide-scale and the metallic core—which cavity would put an end to oxidation. Similar wire heated at 500–600°C. was oxidized completely to the centre, without any cavity being produced, which clearly points to *inward* movement. It is difficult to escape the idea that movement occurs in both directions, accompanied by plastic flow, which—it has been suggested—can be the result of simultaneous ionic movement in two directions, with the consequent disappearance of any rigid frame-work (H. Pfeiffer and B. Ilschner, *Z. Elektrochem.* 1956, **60,** 424).

Mackenzie and Birchenall find that iron parallelepipeds heated in oxygen at 800–850° develop a central cavity with sharp edges and corners—the dimension of the cavity corresponding precisely with those of the original iron block; however, on heating iron in a mixture of hydrogen and water vapour adjusted to give only wüstite, without higher oxide, no cavity appears. The difference is ascribed to the fact that magnetite and ferric oxide possess a nonplastic " cage " which is absent in wüstite; thus subsidence is possible when wüstite is the only phase, whereas the cavity remains when the two other oxides are present. This idea is supported by creep experiments on the three oxides, which showed wüstite to be more plastic than the other two oxides (J. D. Mackenzie and C. E. Birchenall, *Corrosion* 1957, **13,** 783t; see also C. E. Birchenall, *J. electrochem. Soc.* 1956, **103,** 619; H. Engell and F. Wever, *Acta Met.* 1957, **5,** 695).

Whether growth depends on cations moving outwards or anions inwards, a parabolic law would be expected at temperatures high enough to permit of the vacancies moving about without help from external agencies (which, if present, merely serve to give direction to the predominant movement). At lower temperatures, where particles are less mobile, and the unassisted movement of a vacancy would be a very rare occurrence, parabolic thickening will not occur. Under such conditions, we may still have thickening whilst the film is very thin, since the strong electric field may force a movement of vacancies (or of ions in the opposite direction) from one position to the next; as the film thickens, the field diminishes, and at a certain range of thickness, the field falls to such a low value that movement practically ceases, thus accounting for the general character of the log law curves. Explanation is impossible without mathematics, but in Chapter XX a general equation is easily arrived at—of which the two limiting cases are (1) the parabolic equation and (2) one of the logarithmic equations.

Qualitatively we can summarize the situation thus:—

Very Low Temperatures. Movement through a film is *impossible*; only a chemi-adsorbed oxygen film is produced.

Low Temperatures (including room temperature). Movement through a film is possible, but *only with help* from an electric field; this leads to a film which almost ceases to thicken after it has reached a certain range of thickness.

Higher Temperatures. Movement through the film is possible *without help*, but in absence of a field or gradient it would proceed at random; the field or gradient (which arises from the affinity of metal for oxygen) directs it in one direction, so that thickening occurs according to the parabolic law.

Still Higher Temperatures. Oxidation occurs, in some cases, at a constant rate (rectilinear thickening). Sometimes, however, the constant oxidation-rate at the highest temperatures may actually be smaller than at somewhat lower temperature, as a result of sintering of the film-substance.

Under certain conditions oxidation may start slowly and actually become more rapid with time. This has most often been observed in researches carried out at low pressures, where (after the immediate formation of a very thin uniform film, possibly representing two-dimensional oxide), further oxidation must await the formation of three-dimensional nuclei, which, when oxygen is scarce, will only arise on rare occasions; we then get curves which are concave upwards (fig. 4, curve *C*, p. 28). Curves of such a shape have been obtained by Wagner and others during the oxidation of copper at low pressures; the concavity was absent from the curves obtained at higher pressures (C. Wagner and K. Grünewald, *Z. phys. Chem.* (*B*) 1938, **40**, 455. See also D. H. Bangham, *J. sci. Instrum.* 1945, **22**, 230, quoting early work by V. Bloomer).

The fact that at low pressure nuclei appear only infrequently, followed by lateral spread over the surface, has been demonstrated in some very beautiful photo-micrographs; the embryo crystals appear in forms related to the crystal structure of the metal, so that their orientation varies from grain to grain (J. Bardolle and J. Bénard, *C.R.* 1951, **232,** 231; 1954, **239,** 706; *Rev. Métall.* 1952, **49,** 613. E. Gulbransen, W. R. McMillan and K. F. Andrew, *J. Metals* 1954, **6,** 1027).

Work in Finch's laboratory has shown that films formed on metal at low temperatures may be amorphous; Hart, for instance, found that tin foil exposed to air below 130°C. develops an amorphous film, whereas at higher temperatures the film is crystalline (R. K. Hart, *Proc. phys. Soc.* (*B*) 1952, **65,** 955).

It is possible that crystalline films may be less protective than amorphous ones, since movement can occur most easily along the crystal-boundaries, where the atoms are more losely packed than in the grain-interiors (see fig. 79, p. 377). Mills has explained a sudden spurt observed in the oxidation of copper at a certain stage by assuming that the first-formed film is *pseudomorphic*, continuing the structure of the underlying metal, until suddenly it *recrystallizes*, becoming less protective (T. Mills and U. R. Evans, *J. Chem. Soc.* 1956, p. 2182).

Classification of Metals according to Oxidation Behaviour

Noble Metals. We have seen that on a single metal, the growth of the oxide-film follows different laws according to the temperature of exposure to oxygen; different metals, however, exposed to oxygen at comparable temperatures, behave quite differently. On base metals the behaviour is decided more by the nature of the oxide-film produced than by the affinity of the metal for oxygen. Obviously, however, where the metal has no affinity for oxygen (i.e. where the oxide would decompose to metal and oxygen at the temperature under consideration), no oxidation can be expected. Immunity from oxidation due to such a cause is only obtained with noble metals;* the baser metals, heated even under very low oxygen pressures, become oxidized, although sometimes where the oxide is volatile, the surface remains clean and bright; this can occur in the case of molybdenum (for details of the oxidation of molybdenum, see L. Northcott, " Molybdenum ", Chapter 8 (Butterworth); also E. S. Jones, J. F. Mosher, R. Speiser and J. W. Spretnak, *Corrosion* 1958, **14,** 2t).

Optical studies by Tronstad suggest that silver becomes slowly oxidized if heated in air below 180°C., and that the oxide decomposes when the temperature is raised above 180°C. This is probably due to the fact that the pressure of oxygen corresponding to the equilibrium

$$2Ag_2O \rightleftharpoons 4Ag + O_2$$

becomes $\frac{1}{5}$ atmosphere at about 180°C., which accords reasonably well with thermodynamic data provided by Ellingham. However, the reverse changes are not produced on cooling, and Tronstad's observations have not been

* Free energy changes attending oxidation processes are conveniently summarized in diagrams prepared by H. J. T. Ellingham, *J. Soc. chem. Ind.* 1944, **63,** 125. Fuller diagrams are provided by A. Glassner, U.S.A.E.C. Report No. ANL 5107.

fully explained. The fact that three-dimensional oxide decomposes at about 180°C. would not preclude the existence of a two-dimensional film existing at much higher temperatures, since the decomposition pressure of such a film—essential chemi-sorbed oxygen attached to the surface—should be much lower. It seems that such a film is formed at high temperatures, since Eurof Davies has detected a film about 5 Å thick on silver heated in oxygen at 800°C.; it is absent on silver heated in hydrogen at 450°C. There is probably no inconsistency between the various published data, but the reader should study and compare the papers (L. Tronstad and T. Höverstad, *Trans. Faraday Soc.* 1934, **30,** 1114. H. J. T. Ellingham, *J. Soc. chem. Ind.* 1944, **63,** 125, esp. fig. 1, p. 127; D. Eurof Davies, *Nature* 1957, **179,** 1293. See also W. E. Garner and L. W. Reeves, *Trans. Faraday Soc.* 1954, **50,** 254; G. H. Twigg, *ibid.* 1946, **42,** 657; E. C. Williams, *Bull. Inst. Met.* 1954, **2,** 183, who brings out the effect of copper in the silver).

If silver is melted in air, oxygen dissolves in it, since a *dilute* solution of oxygen or oxide in silver is capable of standing in equilibrium with oxygen at $\frac{1}{5}$ atm.; on cooling below the melting-point, the solid silver crystals deposited are nearly free from oxygen, so that the concentration in the liquid portion increases, and there is a sudden eruption of gas, known as *spitting*. If the solidification is carried out under high pressure, a eutectic of silver and silver oxide will be found between the grains—as demonstrated by N. P. Allen, *J. Inst. Met.* 1932, **49,** 317 (see especially his photo-micrograph facing p. 320).

Gold does not seem to develop an oxide-layer under ordinary circumstances. Wilson's measurements of the electrical resistance at the contact made where a cone with rounded tip slides on a moving plate (both made of gold) indicate that there is probably some adsorption of oxygen, which appreciably affects the resistance at very light loads; he found no evidence of the presence of an oxide-film in its ordinary sense (R. W. Wilson, *Proc. phys. Soc.* (*B*) 1955, **68,** 625).*

Wilson found that platinum out-gassed *in vacuo* behaved similarly to gold, but after the ordinary methods of preparation the resistance was higher than that calculated on the assumption that the metal is bare; whether this is due to a layer of oxide or of a solid solution of oxygen in platinum is uncertain. Certain colour-changes observed by early investigators on heating platinum are today generally ascribed to impurities, but the effect recorded in the classical work of Rideal and Wansbrough-Jones—who found that a platinum wire heated at 1670–2170°C. in a bulb surrounded by liquid air loses more weight if the bulb contains oxygen than if it is evacuated—is regarded as genuine; the loss may be partly due to oxidation on the metal surface but partly to volatilization of metal followed by union with oxygen in the gas phase. Hunt, however, states

* Cf. L. G. Carpenter and W. N. Mair, *Nature* 1957, **179,** 212, whose results suggest that gold oxide is stable in the gas phase at high temperatures. See also W. J. Müller and E. Löw, *Ber. dtsch. chem. Ges.* 1935, **68,** 989, who conclude, from microscopic observations and measurements of passivation-times, that gold exposed to air at room temperatures develops an oxide-film.

that platinum and iridium are essentially free from oxide-films at all temperatures up to their melting-points ("J.C.C.", *Platinum Metals Rev.* 1957, **1**, 55; L. B. Hunt, *ibid.* 1957, **1**, 74. E. K. Rideal and O. H. Wansbrough-Jones, *Proc. roy. Soc.* (*A*) 1929, **123**, 202. Cf. G. C. Fryburg, *J. chem. Phys.* 1956, **24**, 175 for attack by activated oxygen containing free atoms).

The Pilling–Bedworth Principle. In their classical paper of 1923, Pilling and Bedworth divided the metals which develop oxide-films into two main classes:

(1) The ultra-light metals which when oxidized yield oxide occupying a *smaller* volume than the metal destroyed in producing it.

(2) Other metals where the oxide, if uncompressed, would occupy a *larger* volume than the metal destroyed.

Simple calculation will show that the first class includes sodium, potassium, calcium and magnesium, whilst the second class includes all the heavier metals (iron, copper, lead, zinc) and also some fairly light metals (aluminium, beryllium and titanium).

Pilling and Bedworth argued that the oxide-film formed on the first group must be porous and non-protective, whilst that on the second group should be compact and protective. In arriving at this conclusion they assumed that the oxide-film thickens by the movement of oxygen *inwards* to reach the metal (N. B. Pilling and R. E. Bedworth, *J. Inst. Met.* 1923, **29**, 529).

Later, Pfeil (p. 34) showed that sometimes metal moves *outwards* to meet the oxygen. For that reason, some revision of the Pilling–Bedworth principle is necessary, but even today the relation between metal-volume and oxide-volume—to which Pilling and Bedworth attached importance—remains of great significance. Indeed, whilst it is correct to state that the original Pilling–Bedworth views are not applicable to cases where the oxide grows by outward movement of metal, it seems possible that the volume relationship may help to decide the all-important question as to whether oxygen shall move inwards or metal outwards. On a metal where a porous oxide would be predicted, the oxygen can travel inwards as *molecules* through the pores up to a point close to the unchanged metal, so that inward movement is favoured, whilst outward movement will be appreciably hindered by the porosity of the film. Other things being equal, inward passage of oxygen is more likely for a member of the first class than the second.

Ultra-light Metals. Pilling and Bedworth studied the oxidation of calcium wire, by measuring the increase of electrical resistance which occurs as the outer layers are converted to oxide. At 300°C. (fig. 6) the curve is straight and the oxidation-rate constant—indicating that the porous film produced has no protective value, as indeed their principle would predict. At 400 and 500°C. the curves are straight at first, but, on reaching a certain film-thickness, suddenly rise, probably because the heat produced by the oxidation failed to escape as quickly as it was generated through a layer of that thickness, so that the temperature no longer remained constant. At 550°C. calcium wire ignited at once and burnt violently, leaving a powdery, partly fused mass of oxide.

Pilling and Bedworth used calcium which was far from pure, and oxygen containing 0·07% water. Gregg and Jepson, using purer calcium, found that in dry oxygen there was very little oxidation below 475°C., but that it became measurable at 500°C.; their 600°C. graph lay *below* that for 500°C. Water vapour (3%) accelerated the reaction, which could be detected at room temperature and was easily measurable at 50°C. (S. J. Gregg, Priv. Comm., Oct. 31, 1958; also S. J. Gregg and W. B. Jepson, *J. Chem. Soc.*, 1960, p. 712.

When Pilling and Bedworth published their results it was generally thought that the gradient of the straight lines obtained at 300°, 400° and 500°C. represented the " velocity of chemical reaction " between calcium and oxygen; the pores in the oxide-film were believed to reach the unchanged

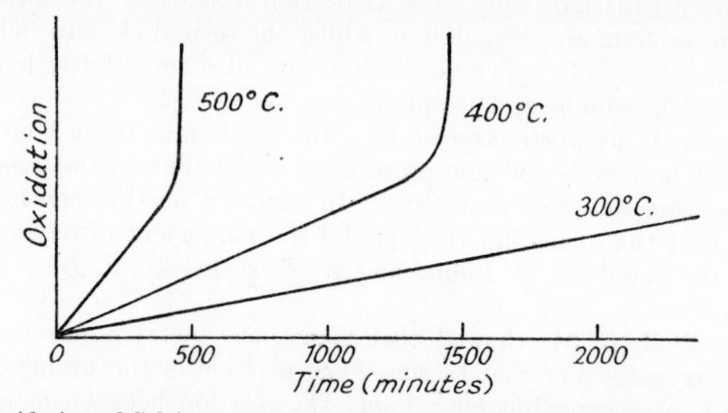

FIG. 6.—Oxidation of Calcium, measured by loss of electrical resistance of a wire; the ordinates are proportional to oxygen taken up (N. B. Pilling and R. E. Bedworth). *Note:* in dry oxygen the behaviour of calcium is different (see text).

metal, so that there was no physical restriction to the oxygen-supply. Dunn and Wilkins, however, argued that, if the oxidation-rate was really controlled by the rate of a chemical reaction, the increase with temperature should be far more rapid. They suggested that the lowest layer of oxide produced next the metal is continuous and pseudomorphic with the metal; clearly owing to its unnatural volume it is under strain and breaks down to yield a porous layer as soon as it exceeds a certain thickness. After that point, there will still be a very thin continuous layer of constant thickness next the metal; this layer is in a state of tension, but, being braced by attachment to the metallic basis, does not break, provided that the critical thickness is not exceeded; outside this comes a thicker layer of porous oxide, throughout which the formation of the pores has relieved the tension. The oxidation-rate is controlled by the passage of oxygen through the inner layer of constant thickness, and thus remains constant as long as the temperature is prevented from rising (J. S. Dunn and F. J. Wilkins, " Review of Oxidation and Scaling of Heated Solid Metals " 1935, p. 67 (H.M. Stationery Office)).

There is an independent reason for believing that, even where the bulk of the oxide is porous and non-protective in appearance, the rate is controlled by movement through the solid oxide. If this were not the case, it would be hopeless to try to retard the oxidation of, say, magnesium by alloying additions, unless they were carried out on a scale so large as to bring the volume of metal destroyed down to that of the oxide produced. Now Huddle, by modest additions to magnesium, has succeeded in achieving a marked reduction in its inflammability—which is difficult to explain unless movement through solid oxide, and not merely through pores, is admitted. He found that unalloyed magnesium was resistant up to 400°C.; above this cracking, flaking and powdering of the film occurred; ignition was observed about 500°C. The purest metal showed the highest resistance, and most additions promoted film break-down; small quantities of calcium and beryllium, however, raised the temperature at which oxidation becomes appreciable to above 600°C. (R. A. U. Huddle, *Amer. Inst. chem. Eng. Preprint* **108** (1955); *J. Inst. Met.* 1955–56, **84,** 479; cf. W. E. Dennis, p. 477; A. B. McIntosh, p. 480; J. G. Ball, p. 247).

Recent work at Oak Ridge shows clearly that the oxidation of sodium in dry oxygen commences rapidly but, at least at low temperatures, ultimately becomes extremely slow—showing that the film has some protective properties. The thickness at which growth practically ceases is 50 Å at − 79°C., but at 48°C. the thickness reached 1500 Å—very much higher than the limiting thickness reached on copper at similar temperatures. Moreover at 48°C. thickening does not become negligibly slow; indeed there seems to be an increase of oxidation-rate after about 25,000 min. The results seem to fit in with the idea of Dunn and Wilkins that the film—although distended—does not break up under the internal stresses until a certain thickness has been reached. The fact that even in this range of thickness, oxidation proceeds more quickly on sodium than on copper seems consistent with Huddle's squeeze theory—according to which the movement of matter through an unbroken film is controlled by the energy needed by the ions to squeeze their way from one site to the next and proceeds most quickly if the ions are spaced well apart. Whatever is thought about this, the curves should receive study (J. V. Cathcart, L. L. Hall and G. P. Smith, *Acta Met.* 1957, **5**, 245. Cf. W. H. Howland and L. F. Epstein, *Industr. engng. Chem.* 1957, **49**, 1931, who find that the purest sodium oxidizes more slowly than the less pure variety, and R. N. Bloomer, *Nature* 1957, **179**, 493, who has studied the oxidation of barium).

Burning of Metals. Since, in absence of complicating factors, the ultra-light metals oxidize at constant rate so long as the heat evolved in the reaction is removed, it is clear that any failure to remove the whole of the heat, so that the temperature rises, will cause the reaction to proceed faster and faster. Thus metals like sodium, potassium, magnesium and calcium, if first heated in air from an external source, may start to become hotter as a result of the oxidation, finally reaching a temperature at which the oxide is incandescent; in other words, metals of this class " burn in air ".

In contrast, members of the class forming compact oxides do not normally burn in air. Here the oxidation-rate, at constant temperature, will fall off as the film thickens; even if, in the opening stages, the whole of the heat evolved is not removed, the slight rise in temperature will not compensate for the effect of film-thickening. Thus, in general, there is no burning of these metals—assuming that they are in compact form. Even " heavy " metals, however, can burn in air when present as a sponge, a fine powder or very thin wire, since the high surface/volume ratio may then allow rapid oxidation, generating a large amount of heat in proportion to the mass to be heated, and the rise in temperature will be self-accelerating; thus oxidation may be speeded up, despite the film-thickening. In such a case, the heavy metal may be said to " catch fire ". Burning of this type is more probable if the temperature to which the specimen is brought by means of an outside source of heat is elevated—or if the atmosphere consists of oxygen instead of air. In the case of wire, the smaller the diameter, the lower is the temperature needed to " set it on fire " (G. Tammann and W. Boehme, *Z. anorg. Chem.* 1934, **217,** 225).*

In Eurof Davies' work (p. 22), iron powder (80-mesh) was heated in hydrogen at 360°C. to reduce the oxide present on each powder particle, then cooled to 40°C. and exposed to oxygen; as soon as oxygen entered the vessel, the powder started to glow, and subsequent X-ray study revealed magnetite, ferrous oxide and metallic iron—the survival of which showed that oxidation had not penetrated to the centre of the particles.

Metals forming compact oxide-film by inward movement of oxygen. If an oxide-film is formed on a metal of the second class by oxygen moving inwards, it is likely that the oxide will be in a state of strain; in the direction parallel to the surface, the atoms will be closer together than in a lump of unstrained oxide, and doubtless abnormally far apart in the direction normal to the surface. Furthermore if the metal was itself in a state of internal stress before the oxidation, these stresses will be inherited by the oxide; they may then serve either to increase or to decrease the strain arising from the volume-relation of oxide and metal. In general, metal which has undergone any sort of surface treatment will either have compressional stress in the surface layer balanced by tensional stress lower down, or *vice versa*; in both cases, the film will have a stress-distribution tending to make it curl up when it is loosened from the metallic basis;

* J. N. Wanklyn adds the comments:—(1) With some metals, including magnesium, the high vapour pressure of the metal is an additional factor influencing oxidation and ignition. (2) Ignition does not always occur when the heat evolved by the reaction exceeds the heat removed by cooling, for the rate of heat-loss generally increases as the metal becomes hotter than its surroundings (conduction, convection and radiation all increase). Thus a stable temperature above that of the surroundings is possible. Only if the rate of heat-production rises so steeply as to outstrip the rate of heat removal can the temperature rise (in principle) indefinitely. In practice it rises only until the reaction-rate is limited by the oxygen-supply to the reacting surface. This change in " controlling factor " could be incorporated in a definition of " ignition ". (3) When metals of the second class do burn in air, behaviour depends greatly on previous history. A long exposure at a relatively low temperature will produce a thick protective film, causing a slower oxidation-rate in the higher range of temperature, and consequently a higher ignition temperature. See W. R. de Hollander, U.S.A.E.C. Report HW 44989.

if there are no inherited stresses, but only the strain due to the volume-relation, the film will not, in general, curl but wrinkle.*

The reality of these internal stresses is shown by experiments on films from heat-tinted nickel, which seems to develop films partly by oxygen moving inwards and partly by metal moving outward (p. 34). The films from superficially oxidized nickel have been transferred to glass carrying a film of petroleum jelly by the method described on p. 785. When the jelly is softened by warming, the films assume a shape representing minimum constraint; the thinner ones curl up and the thicker ones wrinkle (U. R. Evans, " Symposium on Internal Stresses in Metals and Alloys ", 1947–48, p. 291 (Inst. Met.)).

Huddle thinks that compression will not occur even after inward movement of oxygen if the metal is smooth on an atomic scale; with ordinary surfaces oxide-formation in micro-cracks causes a volume increase and compressional stress; this point of view deserves careful consideration (R. A. U. Huddle, Nuclear Engineering and Science Congress, Amer. Inst. min. met. Engrs., 1955).†

The fact that the compact films formed on the heavier metals are in a state of compressional stress parallel to the surface might perhaps be expected to make them more protective than if they were without strain; almost certainly any fissure which exists in the film will have its walls firmly pushed together. The strain cannot cause a collapse of a very thin film; the wrinkling of the film, which occurs after transfer to soft petroleum jelly, cannot occur as long as the film is firmly attached to the metal. Wrinkling can only occur after detachment, and for a thin film the work needed for detachment would exceed the decrease in strain energy which occurs when the film is able to assume a shape of minimum constraint. Thus the *thin* films formed by inward penetration of oxygen into metals of this group should be highly protective.

But at a certain thickness, the argument will cease to apply, since the strain energy which can be released may now be sufficient to supply energy for the detachment (p. 20, footnote). Thus we should expect the films to break down spontaneously when a certain thickness is reached.

This expectation is realized. The oxidation of titanium is largely a process of oxygen passing inwards through the film, probably by oxygen anions taking advantage of vacant anion sites in the lattice. When it reaches the metal-oxide interface, part of it enters the metal in solid solution,

* If the film consists of two layers, it may curl. Thus duplex films on iron consisting of α-ferric oxide upon magnetite curl up into tight rolls (U. R. Evans, *Nature* 1949, **164,** 909).

† J. N. Wanklyn adds the comments:—(1) This reasoning is sound, provided that all the volume-increase is taken up by outward movement of the outer oxide surface in a direction *perpendicular* to the general level. If (e.g. for crystallographic reasons) this condition is not fulfilled, and there is generally some expansion *parallel* to the surface, stress will develop. (2) With inward growth, no real healing of cracks is possible, as they are " left behind " at the outer interface; in some cases they might be squeezed together by compressive forces.

and part of it is used to form fresh oxide and thus increase the thickness of the film. The film is in a state of strain, and when a certain thickness is reached, it breaks down and becomes non-protective. The matter has been carefully studied by Jenkins, who writes, " Observation on the structure of scales formed on titanium indicate that the thin dense slate-grey scale formed at low temperatures is replaced at high temperatures by a thick porous, yellow-brown scale. This scale is largely composed of layers of oxide which have been twisted and shattered like natural rock strata. The transformation in scaling behaviour is believed to occur when the thin, dense scale grows beyond a certain maximum thickness; when this has been attained, growth stresses established within the scale could partially shatter the outer layers. These stresses result from the growth of fresh oxide at the interface between scale and [metallic] core " (A. E. Jenkins, *J. Inst. Met.* 1953–54, **82,** 213, esp. p. 219).

Jenkins showed that at low temperatures, oxidation is parabolic, being controlled by diffusion of oxygen through a compact scale; at intermediate temperatures it is parabolic until the compact scale reaches the critical thickness, when it suddenly breaks down, giving a porous scale so that further thickening follows a linear law. At still higher temperatures, we once more get parabolic thickening; probably in this temperature range, the stresses which would cause the break-down are relieved by annealing and sintering, so that the film remains relatively protective.

The composition of the films formed at high temperatures is complicated; below the TiO_2 layer comes usually a layer of Ti_2O_3 and below that a layer of variable composition, probably TiO with vacant anion sites. There is evidence, based on marker experiments, that at high temperatures titanium moves outwards as well as oxygen inwards (W. Kinna and W. Knorr, *Z. Metallk.* 1956, **47,** 594; also *Tech. Mitt. Krupp* 1956, **14,** 99. Cf. A. L. G. Rees, " Chemistry of the Defect Solid State " 1954, p. 9 (Methuen); D. and M. K. McQuillan, " Titanium " 1956, p. 420 (Butterworth)).

Thus the films on titanium are protective at low temperatures, but not at high ones. The astonishingly protective power of the film on titanium in preventing attack by acid and chloride solutions and the fact that protection does not break down at surface defects—as often happens on metals which develop oxide-films by outward passage—are discussed later (p. 342). The protective films, generally formed at ordinary temperatures, are thin and invisible. In contrast, titanium shows poor resistance to hot oxidizing gases.

Valuable information has been provided by Kofstad and Hauffe, whose observations with radioactive tracers provide proof of the inward movement of oxygen. They point out that oxygen is more soluble in α-titanium than in β-titanium owing to the larger internal spaces in the former. The three oxides in the films, nominally TiO_2, Ti_2O_3 and TiO, all have compositions varying over perceptible ranges; TiO, which has the rock-salt crystal-structure, possesses the widest range, stretching from $TiO_{0\cdot6}$ to $TiO_{1\cdot25}$. The growth curve varies with the temperature and duration of the oxidation, being in different cases (1) logarithmic, (2) cubic, (3) parabolic, (4) linear,

(5) broken, due to cracking, with linear periods (P. Kofstad and K. Hauffe, *Werkst. u. Korrosion* 1956, **7**, 642).

Zirconium, which stands in the same group of the Periodic Table as titanium, presents analogies. Here the evidence for the inward movement of oxygen is based on careful marker experiments; the marker is found on the outside of the film after oxidation, whereas on copper it is found within or below the film. It also shows another feature characteristic of inward growth, the formation of longitudinal grooves along the edges of specimens after oxidation—producing what the Germans call *kreuzformig* (cruciform) structure. The same phenomenon has been noticed during the oxidation of tungsten, and also on titanium placed in molten aluminium; Scheil states that it is produced by rectilinear as opposed to parabolic growth (fig. 7), but this view is not universally accepted (E. Scheil, *Z. Metallk.* 1937, **29**, 209; R. Kieffer and K. Kölbl, *Z. anorg. Chem.* 1950, **262**, 229; J. Mackowiac and L. L. Shreir, *Acta Met.* 1956, **4**, 556. G.

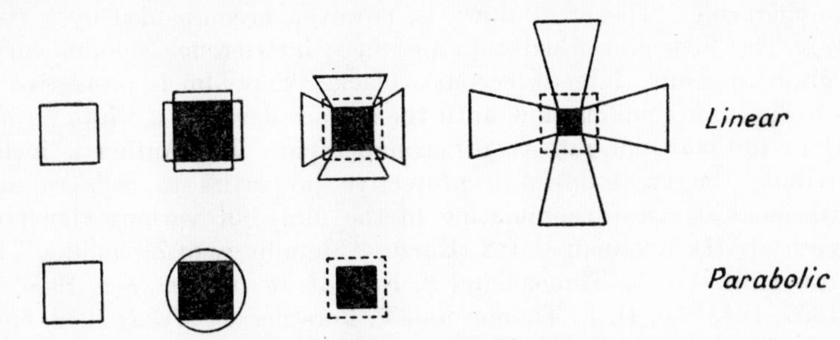

Fig. 7.—" Cruciform " Oxidation connected with rectilinear—as opposed to parabolic —growth (R. Kieffer and F. Kölbl).

Schikorr, " Die Zersetzungserscheinungen der Metalle " 1943, p. 55 (Barth, Leipzig)).

If the protective character of zirconium at low temperatures is connected with lateral compression in the oxide-film, this clearly involves lateral tension in the metal below, and protection should be absent if the metal is insufficiently strong to withstand such tension. Recent work in Hérenguel's laboratory emphasizes the importance of this factor. Zirconium heated in air below 500°C. developed a compact, adherent and protective oxide-film, which remained thin; at higher temperatures the scale became powdery at the surface and had a complex internal structure. It is believed that above 500°C., the metallic basis is too weak to resist deformation, and the self-protection against oxidation declines. Alloying additions designed to confer immunity from serious oxidation at high temperatures should be selected to confer resistance to flow in the metallic phase (J. Hérenguel, D. Whitwham and J. Boghen, *C.R.* 1956, **243**, 2060).

Much of the experimental work on zirconium refers to oxidation by pure water heated under pressure. Strictly, this is not a case of simple oxidation, but the behaviour shows an analogy to the sudden film breakdowns already mentioned, and may conveniently be mentioned here. When zirconium is exposed to hot water, the film thickens slowly at a rate which declines as the thickness increases, until suddenly, at a certain point, known as the *breakaway*, the film spalls, leaving the metal almost unprotected. There are special alloys containing tin and other elements which avoid spalling; the best known is Zircaloy 2,* but even this alloy exhibits a sudden increase in reaction-rate at a certain point, which is known as the *transition*. Thus in all cases there appears to be a deterioration in protective properties, although it may take various forms and perhaps arise from more than one factor. It is possible that break-away and transition are two separate phenomena, due to different causes. Thomas and Kass consider that break-away is associated with hydrogen pick-up by the metal, but that transition is definitely not associated with hydrogen. Investigations at the Battelle Institute in high-temperature water (600°F.) show that crystallographic changes from tetragonal to monoclinic zirconia occur in the film, but there is no proof that this is a direct cause of the break-down of protective character. The break-down is, however, accompanied by a visible change. The first film when thin produces interference colours varying from grain to grain; later it becomes black. This film is protective and the corrosion-rate remains slow until the surface develops a white powdery material; the corrosion-rate then becomes rapid. The Author is inclined to attribute the break-down of protective properties to collapse under compressional stresses accumulating in the film, but various view-points deserve study (B. Lustman and K. Kerze, " Metallurgy of Zirconium " 1955 (McGraw-Hill). D. E. Thomas and S. Kass, *J. electrochem. Soc.* 1956, **103,** 478; 1957, **104,** 261; D. E. Thomas and F. Forscher, *J. Metals* 1956, **8,** 640. J. N. Wanklyn, *Nature* 1956, **177,** 849. C. M. Schwartz, D. A. Vaughan and G. G. Cocks, Battelle Memorial Institute, Report No. 793, Dec. 17, 1952 (declassified 1955). E. A. Gulbransen and K. F. Andrew, *Corrosion* 1958, **14,** 32t; *Trans. met. Soc. A.I.M.E.* 1958, **212,** 281).

Zirconium, like titanium, shows remarkable resistance to acids and alkalis, but a resistance which depends on a film in strong compression has its disadvantages. Even in the absence of corrosive reagents, there are risks of explosions or fires if the film should break down, exposing the highly reactive metal. Fatal accidents have occurred when stores of zirconium scrap, after lying quietly without incident for long periods, have been disturbed, generally when slightly wetted with water. The film, although too thin to crack spontaneously, can do so if a small amount of energy is supplied from outside by impact or friction. If the film then spalls or shatters, exposing the metal, a violent reaction may occur with large generation of heat; the rapid evolution of hydrogen may then lead to a hydrogen-air explosion, and the pressure wave sent out may fracture the film on zirconium in other parts of the store, and cause many secondary explosions. Soon the whole store may be ablaze and water-jets directed on it will only make matters worse. In transporting zirconium scrap, great care should

* The composition is 1·5% Sn, 0·12% Fe, 0·10% Cr, 0·05% Ni, and the rest zirconium. Zircaloy 3 with 0·25% Fe and 0·25% Sn is said to possess superior resistance.

be taken to avoid vibration; vermiculite has been used as a precaution against shock; personnel should wear fire-suits.

Metals forming compact oxide-films by outward movement of metal. In cases where film-thickening occurs through metallic cations moving outward through the film, and taking up positions between oxygen ions on the external face to form fresh layers of oxide, there would seem to be no reason for the development of any large strain in the film, since the ions on arrival are free to take up positions of minimal energy. Actually, for reasons of epitaxy (p. 23, foot-note), some strain may be introduced, but it is less serious than that arising from the differences in volume of metal and oxide, or from the inheritance of internal stresses present in metal which has suffered deformation before oxidation. On the other hand, the movement of atoms from the metal into the film as ions will leave vacancies at the interface and these may coalesce to form cavities, so that the film may locally cease to be anchored to the metal. Furthermore, although the film will not inherit stresses from the metal, the removal of metal from the outer layer into the film may upset the equilibrium between compressional and tensional internal stresses, so that a resultant stress arises which may rupture the film at places where it is unsupported. Such considerations, which are more fully developed on p. 110, may explain why the films on iron and zinc are so much less protective against, say, chloride solutions than those on titanium, and why the break-down occurs at surface defects.

The amount of strain which can be produced in films built by outward movement and exhibiting epitaxy with the metallic basis is a matter about which different views are held. If an inter-atomic spacing in the oxide-lattice can be found which is nearly the same as some inter-atomic spacing in the metal, then it would be expected that there will be an attempt to continue such a spacing in the film. That this actually happens is indicated by the X-ray work of R. F. Mehl, E. L. McCandless and F. N. Rhines, *Nature* 1934, **134,** 1009; 1936, **137,** 702; R. F. Mehl and E. L. McCandless, *Trans. Amer. Inst. min. met. Eng.* 1937, **125,** 531. Cf. J. H. van der Merwe, *Disc. Faraday Soc.* 1949, **5,** 201.

In cases where the fit is inexact, slight strains must be introduced into the oxide to reconcile the discrepancy. In the case of polycrystalline metal, the strain will vary from crystal to crystal. Thus the distance between ions (in the direction parallel to the surface) will be different in the oxide covering one grain from that in the oxide covering the next grain, and the energy needed for an ion to squeeze its way from site to site should vary also. This may be expected to affect the rate of movement through the film, even though the oxide crystallizes in the cubic system.* If the film growth is obeying the parabolic law, the value of k should vary with the crystal direction—explaining a very beautiful phenomena observed by

* According to some authorities, the rate of movement through a *cubic* crystal should be independent of direction. The Author feels doubt regarding this argument even for unstrained cubic crystals; it is certainly untrue when the crystals are strained.

many experimenters. When the section of a polycrystalline metal is heated in oxygen to give films of interference-colour thickness, each grain develops a different colour—indicating that the film-thickness covering the different grains is not the same.

The variation of oxidation-rate with crystal direction has been studied in some detail by A. T. Gwathmey and A. F. Benton, *J. phys. Chem.* 1942, **46,** 969; W. W. Harris, F. L. Ball and A. T. Gwathmey, *Acta Met.* 1957, **5,** 574; B. Lustman, *Trans. electrochem. Soc.* 1942, **81,** 359, esp. p. 368; J. Bénard and J. Talbot, *C.R.* 1947, **225,** 411; J. Bénard and J. Bardolle, *ibid.* 1951, **232,** 231; 1954, **239,** 706; J. Moreau and J. Bardolle, *ibid.* 1955, **240,** 524; E. A. Gulbransen and R. Ruka, *J. electrochem. Soc.* 1953, **99,** 360; T. N. Rhodin, *J. Amer. chem. Soc.* 1951, **73,** 3143. Cf. work on growth on different crystal faces of aluminium by S. J. Basinska, J. J. Polling and A. Charlesby, *Acta Met.* 1954, **2,** 313; K. M. Carlsen, *ibid.* 1957, **5,** 58; also that on germanium, by J. T. Law and P. S. Meigs, *J. electrochem. Soc.* 1957, **104,** 154.

Cavities at the base of the Film. More important than the strains produced by the outward movement is the possibility of the formation of cavities at the base of the film. Wherever an atom has passed into the oxide-film as an ion, a vacancy is left behind. Now vacancies can diffuse through metal as well as through oxide, and can join one another; thus, instead of a vast number of vacancies of atomic size dispersed over the metal-oxide interface, we may sometimes obtain a small number of holes large enough to be seen under the microscope. Such holes can only be explained by the diffusion of vacancies.

This cavity-formation may be better understood if we recall that when a number of crystals of a salt are placed in a saturated solution, the large crystals often grow and the small ones vanish—a change which reduces the interfacial energy of the system. Similarly if, at one point on the metal-oxide interface, a fairly large cavity has appeared, it will continue to grow at the expense of vacancies of atomic size present at points around it.

An admirable example has been provided from Fontana's laboratory on thin iron sheet heated at 1090°C. It was observed that the scale on one face became loose, being separated from the metal by a narrow but extensive cavity, whilst the scale on the other face was entirely adherent; apparently a cavity on one side had absorbed the vacancies from the opposite side of the metal, so that the crevice between oxide and metal was confined to a single face.

The same research included a study of wire. If a wire (fig. 8(A)) were to develop oxide by inward movement of oxygen, it should be converted into a slightly larger wire of solid oxide. If the mechanism were outward movement, we might expect to find the metal wire converted into a " tube " of oxide. Actually, this result was obtained, but the tube was highly asymmetric. First a cave developed on one side (fig. 8(B)), covered with a loose scale, which in due course started to leak (fig. 8(C)), causing, on the floor of the first cave, a fresh layer of oxide; this in due course came to constitute the roof of a second cave (fig. 8(D)). Later other caves appeared and the tube ultimately obtained contained many separate channels (B. W. Dunnington, F. H. Beck and M. G. Fontana, *Corrosion* 1952, **8**, 2. Cf. A. de S. Brasunas, *Met. Progr.*, 1952, **62**, Dec. p. 88).

Where the oxide is sufficiently plastic, it will subside into the caves, filling them completely, as indicated by certain experiments on copper spheres and tubes at 1000°C. carried out by W. J. Moore, *J. chem. Phys.* 1953, **21**, 1117.

This plastic flow, which often prevents the formation of cavities under conditions where they might be expected, may be really the result of simultaneous movement of cations and anions in the two directions as already suggested. At fairly high temperatures, where the oxide is plastic, any important cavity which may form is likely to be filled by the subsidence of the roof, which re-establishes contact with the floor (in that way interfacial energy will be reduced). Thus oxidation will proceed, on the whole, smoothly, and, provided that conditions are otherwise favourable to parabolic growth, the curves will show only slight deviations from that law.

If, however, the oxidation is occurring at a temperature where the

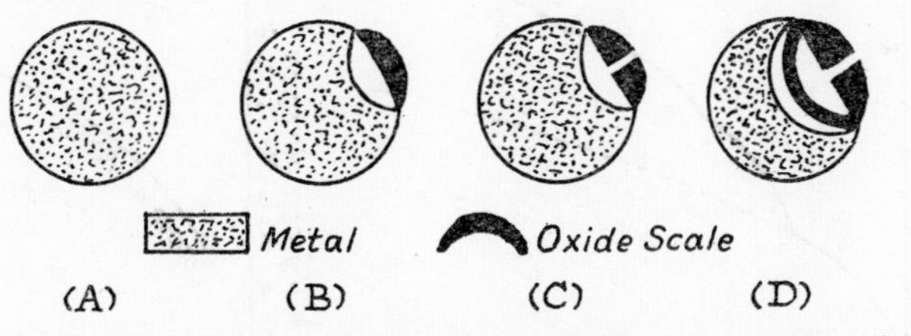

FIG. 8.—Formation of caves and multiple scales on Iron wire (schematic), based on B. W. Dunnington, F. H. Beck and M. G. Fontana.

oxide is not plastic, the roof may still hold up, even when the cavity has extended over a large fraction of the specimen area. Assuming that there is no leakage of oxygen through the roof, the cavity will extend and oxidation should ultimately become very slow. If on the other hand, the scale periodically breaks, the oxidation, after it has become comparatively slow, should suddenly become rapid at each break. This was shown in early work by Pilling and Bedworth (p. 39). Copper gave smooth continuous curves at 800°C., where the scale is plastic, but discontinuous curves at 500° C., where the scale is brittle (fig. 9).

The physical character of the film, however, depends very much on minor constituents of the metal. As shown by Tylecote, pure cuprous oxide films are remarkably ductile at elevated temperature, but films produced on copper containing phosphorus are far less ductile, owing to the presence of cuprous phosphate; such films exfoliate on cooling, since they cannot withstand the strain imposed by differences between the thermal contractions of scale and metal; on phosphorus-free copper the strains imposed by cooling can be withstood, and the film adheres. However, even under conditions of constant temperature the films may become leaky after they have reached a certain thickness. Tylecote used a recording balance, which enabled the weight-changes on specimens to be followed over periods of several days; his results show clearly sudden breaks in the oxidation-time curves (R. F. Tylecote, *J. Inst. Met.* 1950–51, **78,** 301, 327; 1952–53, **81,** 681; *Metallurgia* 1956, **53,** 191).

Caplan and Cohen have endeavoured to correlate the number of breaks in the oxidation curves of iron-chromium alloys with the number of scale-layers found on the specimen (representing, presumably, the roofs of successive cavities); exact agreement can hardly be expected, since some of the cavities will fail to extend all over the entire surface of the specimen, but in a number of cases there is satisfactory correspondence—which supports the idea that the sudden increase in oxidation-rate occurs when a cavity-roof has become leaky and a new scale is starting to form on the cavity-floor (D. Caplan and M. Cohen, *J. Metals*, 1952, **4,** 1057; *Corrosion*, 1959, **15,** 141t).

Many other examples of discontinuous curves could be quoted. During

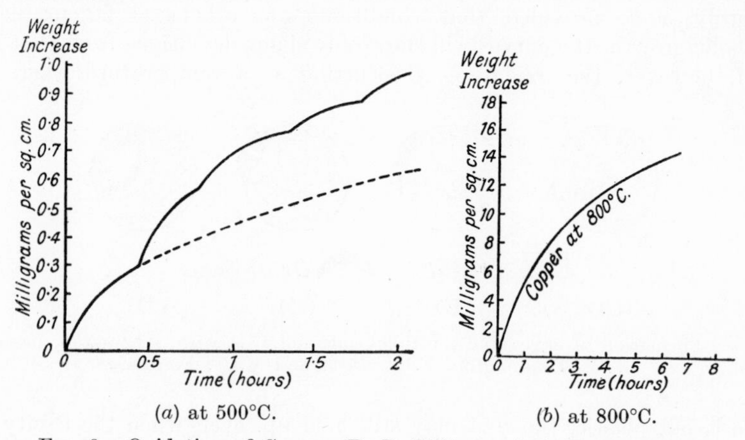

(*a*) at 500°C. (*b*) at 800°C.

Fig. 9.—Oxidation of Copper (R. B. Pilling and R. E. Bedworth).

his early work Vernon observed sharp breaks in the curve representing the increase of weight of aluminium exposed indoors to London air over some months; several times oxidation appeared to have come to an end, and then suddenly there was a fresh increase in weight (W. H. J. Vernon, *Trans. Faraday Soc.* 1927, **23,** 152).

Such considerations may serve to indicate why laboratory experiments do not always give correct predictions regarding service behaviour, which involves longer periods and rougher usage than those commonly introduced into scientific studies. For technical assessment of materials, there is an increasing use of tests involving fluctuating temperatures and operations designed to cause deliberate damage to the film.

In a case where the vacancies do not coalesce to form larger cavities, they nevertheless diminish the area over which metal atoms can move into the scale as ions, and thus the effective area available for the oxidation process is reduced. This should cause the oxidation-rate to fall off more rapidly than the parabolic law would predict, and calculations (p. 836) show that, under idealized conditions, a log law will govern the situation. It seems possible that certain instances of obedience to a log law under conditions where parabolic growth might have been expected can be explained in that way.

On the other hand cavity formation may cause faulty adhesion, and break-away of a film which would otherwise be protected. Gulbransen and Andrew, studying high-purity nickel, record that the film is protective below 900°C., but that at higher temperatures the oxidation-rate does not fall off as quickly as the parabolic law would predict; above 1000°, the oxide cracks away from the metal. Whether this is due to cavity formation or other cause cannot be stated at present (E. A. Gulbransen and K. F. Andrew, *J. electrochem. Soc.* 1957, **104,** 69C).

The formation of cavities below scales has received detailed attention from several investigators; special attention should be given to the papers of A. Dravnieks and H. J. McDonald, *Trans. electrochem. Soc.* 1948, **94,** 139; C. E. Birchenall, *J. electrochem. Soc.* 1956, **103,** 619; H. Pfeiffer and B. Ilschner, *Z. Elektrochem.* 1956, **60,** 424; A. de S. Brasunas, *Met. Progr.* 1952, **62,** Dec., p. 88; D. W. Juenker, R. A. Meussner and C. E. Birchenall, *Corrosion* 1958, **14,** 39t.

Outgrowths from Films. The inner surface of a film often tends to interlock with the metal, as a result of preferential growth inwards along grain-boundaries or other paths of weakness—a matter discussed on p. 69. The outer surface is often smooth to the naked eye, although cracks are sometimes visible; in certain circumstances blisters appear, associated with caves (p. 80), or arising, as on copper, from the presence of two oxide layers undergoing different thermal expansion (p. 23). Even when the film appears smooth to the eye, the electron microscope may detect out-growths, probably springing from pores in the film proper. Thus Pfefferkorn reports needles at right angles to the surface, whilst Halliday and Hirst describe eruptions giving rise to blisters or mounds. On aluminium heated for 30 min. at 500°C., the smaller hillocks may have a height of 60 Å and diameter 0·5 μ; nickel oxidized for 15 min. at 500°C. forms a yellow NiO film with eruptions up to 0·5 μ high and 5 μ across the base (G. Pfefferkorn, *Naturwissenschaften* 1953, **40,** 551. J. S. Halliday and W. Hirst, *Proc. phys. Soc.* (*B*) 1955, **68,** 178).

Possibility of Simultaneous Movement in Both Directions. It has been shown that movement of anions inwards tends to produce compressional stress (except for ultra-light metals), whilst movement of cations outwards tends to produce cavities. However, in many cases where either stress or cavities might reasonably be expected, they do not appear; instead there is a plastic yielding of the film. This result, as has been suggested by Hauffe and others, may be connected with movement in both directions, and there is no reason why such simultaneous movement should not occur. If cations move out sufficiently to avoid the development of stress and anions move in sufficiently to avoid the formation of cavities, a highly resistant film should develop. On nickel there is evidence for movement in both directions and this may be one reason for the good resistance of nickel and many nickel alloys, both to high-temperature and low-temperature attack by many reagents. Iron is a more complicated case (p. 34).

Oxidation at Relatively Low Temperatures

Formation of Invisible Films at Room Temperature. Since there still appears to be some scepticism regarding the formation of oxide on metal exposed cold to air, in cases where the bright surface undergoes no visible change, it is advisable to state the evidence for the existence of invisible films. The question is important, since exposure of, say, iron to air alters its chemical behaviour. Although iron, even after long exposure to cold air, is still attacked by strongly corrosive liquids, and although, even when freshly scraped, it remains unattacked by passivating liquids, yet its behaviour to liquids adjusted to stand on the border-line between corrosion and passivity is altered by previous air-exposure. In his statistical work, Mears used 0·07M sodium bicarbonate as a border-line liquid, and found that it caused rusting at a scratch-line with decreasing frequency as the period of air-exposure between the engraving of the scratch and application of the liquid was lengthened; the proportion of scratches developing rust were 84%, 65%, 58%, 48% and 27% after exposures of 0·25, 2·0, 16·0, 128 and 1024 minutes respectively (R. B. Mears and U. R. Evans, *Trans. Faraday Soc.*, 1935, **31**, 532).

An earlier example comes from the work of Vernon who found that iron, after exposure for 11 months to dust-free air, although undergoing no visible change, acquired resistance to rusting under dew-point conditions; the same conditions caused rapid rusting on specimens which had not received the prolonged exposure to dust-free air. He obtained somewhat analogous results on lead and copper (W. H. J. Vernon, *Trans. Faraday Soc.* 1927, **23**, 159).

Many experiments point to the up-take of oxygen by metal exposed cold to air. Measurements based on weight-increase are rejected by those who disbelieve in invisible oxide-films on the ground that adsorbed oxygen could cause a weight-increase; but it is difficult, for instance, to believe that the weight-increases of the magnitude observed on zinc and iron by accurate workers can be accounted for by adsorbed oxygen. (See, for instance, W. H. J. Vernon, E. I. Akeroyd and E. G. Stroud, *J. Inst. Met.* 1939, **65**, 301.)

However this may be, there are methods which serve to distinguish oxide-films from oxygen-films. The polarized light method (p. 790) gives an idea of the refractive index of the film substance, and shows that it is oxide, not oxygen. Winterbottom states that the films produced on iron mirrors exposed to air at room temperature for 0·1, 1, 10 and 100 hours have thicknesses of 20 Å, 24 Å, 31 Å and 35 Å respectively. He remarks that these numbers agree well with Sloman's vacuum-extraction determinations on cold-rolled iron, and summarizes the matter by saying that "the oxide-films on these metals, Cu, Fe and Al, grow rapidly at room temperature to a thickness of the order of 20–40 Å" (A. B. Winterbottom, *Trans. electrochem. Soc.* 1939, **76**, 326).

Some measurements made by Winterbottom of the invisible films produced on copper at room temperatures are quoted in Table I; these are based on the assumption that the optical constants of the metallic copper substrate are the same in all cases—an assumption which he considers to be unjustified; however this may be, they show the order of film-thickness reached and also indicate how the thickness is influenced by the surface condition of the metal (A. B. Winterbottom, Priv. Comm., Feb. 6, 1958).

TABLE I

THICKNESS OF FILMS ON COPPER EXPOSED TO LABORATORY AIR

(A. B. Winterbottom)

Preparation	Thickness
Annealed, mechanically polished, bright-annealed in hydrogen	10 to 15 Å after 0·1 hr. 15 to 20 Å after 1 hr. 30 Å after 300 hr.
Annealed, mechanically polished	20 to 30 Å after 0·1 hr. 30 to 35 Å after 2·5 hr. 50 to 60 Å after 264 hr.
Cold-rolled, mechanically polished	15 to 20 Å after 0·1 hr. 15 to 25 Å after 2·5 hr. 25 to 35 Å after 264 hr.
Annealed, ground (incident light parallel to scratches)	5 Å after 0·1 hr. 7 Å after 2 hr. 20 Å after 145 hr.
Annealed, ground (incident light perpendicular to scratches)	12 Å after 2 hr. 17 Å after 38 hr.

Other convincing evidence comes from electron diffraction studies of the metallic surface. Mayne and Pryor prepared iron specimens by fine abrasion followed by etching in N/10 hydrochloric acid and examined them after different exposures to dry air. After 30 minutes exposure the pattern of metallic iron was still visible, but there were two broad oxide haloes. A specimen exposed to dry air for 48 hours showed a series of sharp rings characteristic of cubic oxide, whilst the iron rings could no longer be detected. Rather similar results had been obtained by Nelson on films of pure iron prepared by evaporation. The freshly prepared films showed the rings of metallic iron, but after exposure to air for some hours, they became fainter and two new rings appeared indicating cubic oxide (J. E. O. Mayne and M. J. Pryor, *J. chem. Soc.* 1948, p. 1831, esp. p. 1832. H. R. Nelson, *J. chem. Phys.* 1937, **5**, 252. Also Hancock and Mayne, p. 776 of this book).

Allen, using a gas-absorption process (p. 27), studied the up-take of oxygen by copper at room temperature; his measurements corresponded to an oxide-film 15 Å thick after ½ hour and 20 to 25 Å thick after 17 hours; the oxygen-uptake was too great to be attributed to adsorption.

There appears to be no reason to doubt the real existence of oxide-films on most metals after exposure, unheated, to air. In many cases, the films have been isolated by dissolving away the metal from below, and become visible after removal from the bright surface.

In recent papers the thicknesses assigned to the invisible film produced on a metallic surface at room temperature are generally lower than those assigned 25 or 30 years ago, and it is often concluded that the early measurements were incorrect. Although some of the technique used in pioneer work today appears open to criticism, there is no reason to think that the early assessments of thickness involved serious errors. It seems more likely that, owing to the limited experience of surface preparation available to the earlier investigators, the films really *were* thicker than the films formed on the better prepared surfaces used today. It is known (p. 176) that, on an abraded surface, the oxide-film keeps cracking under the internal stresses present and then healing itself by the formation of more oxide. If a surface has been coarsely abraded, it will have developed a thicker film by the time that oxidation becomes slow than if it has been carefully prepared. The fact that the crack-heal process continues longer in the first case is seen by observing the time taken for a pure iron surface to acquire immunity from the deposition of copper when tested with copper nitrate solution. Under conditions where a finely abraded surface becomes passive to copper nitrate after a few minutes exposure to air, it requires several hours if the abrasion has been coarse (U. R. Evans, *J. chem. Soc.* 1927, p. 1020, esp. p. 1030. Z. *Elektrochem.* 1958, **62**, 619. See also p. 1042 of this book).

Interlocking of oxide and metal on abraded surfaces. Early work on film-stripping at Cambridge and more detailed studies at Teddington indicated that on the thinner films on iron and copper (including those giving the earlier interference tints), there was considerable penetration of oxidation into the metal, so that below the homogeneous oxide-film there was a mixed zone of metal and oxide. Recent work on electron emission at Glasgow confirms this and provides information on the surprisingly deep extension of the mixed zone. The unstable, deformed oxide formed below the surface of abraded metal is capable of emitting electrons when illuminated with certain wave-lengths under conditions where the relatively stable oxide-film on the surface fails to do so. Aluminium abraded in air with medium-grade carborundum paper and kept in air for two days (after removal of loose debris) was found by Grunberg and Wright to contain an oxide-phase containing oxygen-ion vacancies extending to a depth of about 10^{-3} cm. below the surface (U. R. Evans and J. Stockdale, *J. chem. Soc.* 1929, p. 2651; W. H. J. Vernon, E. A. Calnan, C. J. B. Clews and T. J. Nurse, *Proc. roy. Soc.* (*A*) 1953, **216**, 375; L. Grunberg and K. H. R. Wright, Conference on Lubrication and Wear, Instn. mech. Engrs. 1957, paper **67**; L. Grunberg, *Brit. J. appl. Phys.* 1958, **9**, 85).

Thickness of Oxide Films producing Interference Tints. It will be convenient at this point to give some figures for the " thickness " of the oxide present on metal which has been tinted by heating in air. The corresponding numbers for iodide-films are given in Chapter III, whilst measuring methods, along with a brief discussion of the causes of the colours, will be found in Chapter XIX.

When a rough metallic surface is heated in air, the oxide may penetrate unequally at different points, and in extreme cases a " mixed zone " of oxide and metal may be left below the true oxide-film—as already stated. It is clearly impossible to express by a single " thickness number " all the features of a complicated situation. Two parameters, however, are often used to give some idea of the thickness. One is the *Mean General Intercept*, the average length of the intercepts which would be made by a vast number of lines drawn perpendicular to the general plane of the surface; if we know the amount of oxide per unit area, the mean general intercept is easily obtained on dividing by the density of the oxide. The other is the *Mean Local Intercept*, the average length of the intercepts made by lines drawn perpendicular to the local plane at different points on a rough surface; it is obtained on dividing the mean general intercept by the *roughness factor*—the true surface area (in $cm.^2$) of a face measuring 1×1 cm. The roughness factor may perhaps lie between 1 and 10 according to the nature of the surface, but for a given surface it may vary with the method used to determine it. It is clearly an arbitrary quantity for surfaces with " cracks " running into the metal; the figure obtained will depend on the " breadth " below which a crack ceases to be regarded as a crack and becomes a mere " lattice-distortion "—just as the length of the coast-line of England will vary according to the width of an estuary below which " coast " is considered to become " river-bank ". Thus the mean general intercept appears the more unambiguous measure of film-thickness.

For the films obtained by heating metals in air, some numbers are shown in Tables II to V; these represent Mean General Intercepts. Measurements on iron by Vernon and Wormwell gave 350 Å for the first-order straw and 590 Å for the first-order blue, these also being Mean General Intercepts; by adopting a roughness factor of 2·5, based on Erbacher's work, they arrived at 140 Å and 240 Å respectively for the Mean Local Intercepts.

Comparing the thicknesses based on Constable's pioneer optical researches (Table II) and Miley's early electrometric investigations (Table III) with the later numbers obtained for copper by Campbell and Thomas (Table IV), and for iron by Eurof Davies (Table V), we may gain the impression that the early numbers for both metals were too high. The factors involved may be indicated. As regards the optical numbers, the discrepancy is

TABLE II

THICKNESS OF OXIDE-FILMS ON COPPER AND NICKEL (F. H. Constable)

Copper		Nickel	
Dark brown	380 Å	Pale brown	490Å
Red brown	420	Dark brown	540
Very dark purple	450	Purple	570
Very dark violet	480	Very dark violet	600
Dark blue	500	Very dark blue	760
Pale blue green	830	——	
Pale silvery green	880	Silvery green	1120
Yellowish green	970	Yellow green	1200
Full yellow	980	Yellow	1260
Old gold	1100	Straw	1350
Orange	1200	Yellow brown	1660
Red	1260	Dark brown	1720

probably due to neglect of the phase-jump at the metal-oxide interface. When this is allowed for, the numbers calculated for the reds of the first and second order from optical data agree much better with the results of Campbell and Thomas (A. B. Winterbottom, *Trans. electrochem. Soc.* 1939, **76**, 326).

TABLE III

THICKNESS OF OXIDE-FILMS ON COPPER AND IRON (H. A. Miley)

Copper		Iron	
Dark brown	370 Å	Straw	440 Å
Red brown	410	Reddish yellow	530
Purple	460	Red brown	560
Violet	485	Purple	625
Blue	520	Violet	695
Yellow (2nd order)	940	Blue	725
Orange	1170		
Red	1240		

TABLE IV

THICKNESS OF FILMS (Cu_2O) ON COPPER (W. E. Campbell and U. B. Thomas)

	Electrometric	Weight-increase
"Darkened"	190 Å	180
Light rose	260	270
Deep rose	330	300
Light blue	410	390 and 400
Yellow II	760	730
Orange II	980	990

TABLE V

THICKNESS OF FILMS (Fe_2O_3 layer only) ON IRON (D. Eurof Davies, U. R. Evans and J. N. Agar)

Pale straw	(15 min. at 225°C.)	77Å
Golden brown	(2 hr. at 225°)	103
Reddish brown	(4 hr. at 225°)	132
Red	(8 hr. at 225°)	152
Deep red	(3 hr. at 250°)	159
Purple	(6 hr. at 250°)	184
Violet	(9 hr. at 250°)	210
Blue	(12 hr. at 250°)	224
Greenish blue	(2 hr. at 300°)	244
Greenish silver	(5 hr. at 300°)	303
Greenish straw	(8 hr. at 300°)	350
Straw	(11 hr. at 300°)	380
Greyish brown	(8 hr. at 325°)	521
Greyish pink	(10 hr. at 325°)	606

As regards the early electrometric results, it is customary to attribute the high values obtained by Miley to his use of an open cell, but it is equally possible that the apparent discrepancy is due to the fact that he used specimens manually abraded under somewhat heavy pressure, and that his values include not only the film responsible for the colours, but also the oxide present in the mixed zone of metal threaded by oxide, the existence of which was noted in early work by Stockdale (p. 58).*

The fact is that an observation of colour does not by itself provide a measurement of the amount of oxide on (or just below) the metallic surface. Vernon and his colleagues found that, whereas above 200°C. (where abraded iron oxidizes uniformly by the parabolic law) the oxygen present in the film was equal to the total oxygen-uptake, this was not true below 200°C. (where the oxidation penetrates downwards into the metal, the log law being now obeyed); in this lower temperature-range, the oxygen in the strippable film was much less than the total oxygen-uptake, and interference colours were absent (or restricted to a faint straw) even when the total oxygen-uptake, as judged by weight-increase, would have been sufficient to produce colour on the assumption that a uniform film had been formed.†

A careful study of the literature may perhaps lead the reader to the conclusion that there is no necessary discrepancy between the measurements obtained by different investigators. The references are: F. H. Constable,

* It is clearly better to use a closed cell, and the methods preferred at Cambridge are briefly described in Chapter XIX (see especially figs. 127, 129 and 131, pp. 773, 778, 780). However, without wishing to reopen old discussions it may be stated that, in general, the error introduced by the use of an open cell should become small if the current density employed is high compared to the *Reststrom* corresponding to the movement of oxygen through the *Diffusionschicht*—which is normally small under stagnant conditions, when once a steady state has been established (this argument becomes invalid for very thin films, since possibly the steady state may only be reached at a late stage in the reduction). The main objection to the use of a high current-density is that, unless an oscillograph or equivalent instrument is available, an accurate potential-time curve cannot be obtained; nevertheless the completion of the reduction may be accurately determined by setting the potentiometer at a reading corresponding to the potential representing the end-point and observing on a stop-watch the exact instant at which the needle of the null-point galvanometer crosses the zero-mark.

† A film formed on an uneven surface which itself suffers oxidation unevenly cannot be defined by a single thickness number; a distribution curve (p. 913) is needed for an adequate description. The colour produced depends mainly on the *modal* value of the curve whereas most of the experimental methods (micro-gravimetric, gas-absorption and electrometric) provide *mean* values. Since two distributions corresponding to the same means may have their modes at entirely different places, a given colour may represent two quite different values for the "oxygen per unit area". If the distribution curve has two modes, highly anomalous results are to be expected. The mean thicknesses needed to produce a colour may appear either surprisingly high or surprisingly low. In Eurof Davies' work on hydrogen-reduced surfaces (p. 58), oxidation was believed to take place by oxygen entering at pores in the film and causing oxidation to spread sideways below the film, producing a thick layer around each pore; when this thick layer reached the value appropriate for interference of some wave-length the complementary colour would be produced—although the mean thickness would have been quite insufficient for the interference of that wave-length; thus the numbers appeared abnormally small. The numbers obtained by Vernon and his colleagues (p. 58) on abraded iron oxidized at temperatures below 200°C. show the opposite effect; here much of the oxidation penetrated inwards into the metal, and only early colours were obtained even when the oxygen-uptake was quite high.

Proc. roy. Soc. (*A*) 1927, **115,** 570; 1927–28, **117,** 376; 1929, **125,** 630. U. R. Evans and H. A. Miley, *Nature* 1937, **139,** 283; H. A. Miley, *Carnegie Schol. Mem.* 1936, **25,** 197 (see criticism by G. D. Bengough and F. Wormwell, *J. Iron St. Inst.* 1937, **135,** 412P); W. E. Campbell and U. B. Thomas, *Trans. electrochem. Soc.* 1936, **76,** 303, esp. p. 312. D. Eurof Davies, U. R. Evans and J. N. Agar, *Proc. roy. Soc.* (*A*) 1954, **225,** 443, esp. p. 454. U. R. Evans and J. Stockdale, *J. chem. Soc.* 1929, p. 2651. W. H. J. Vernon, E. A. Calnan, C. J. B. Clews and T. J. Nurse, *Proc. roy. Soc.* (*A*) 1953, **216,** 375, esp. p. 378. See also p. 775 of this book.

Other References

Two authoritative text-books on oxidation are now available: K. Hauffe, " Oxydation von Metallen und Metallegierungen " (Springer); and O. Kubaschewski and B. E. Hopkins, " Oxidation of Metals and Alloys " (Butterworth). Much information about oxidation will be found in another book by K. Hauffe, " Reaktionen in und an festen Stoffen " (Springer). Hauffe has contributed an article in the English language in " Progress in Metal Physics " Vol. 4, p. 71. A résumé article by Bénard deserves study, particularly as it provides a useful collection of his own curves. (J. Bénard, *Métaux et Corros.* 1950, **25,** 241; see also J. Bénard and O. Coquelle, *C.R.* 1946, **222,** 796, 884; *Rev. Métall.* 1947, p. 113. J. Bénard, *Bull. Soc. Chim. Fr.* 1949, D89, D117.) Special attention should be given to the classical papers of C. Wagner (*Z. phys. Chem.* (*B*) 1933, **21,** 25; also " Atom Movements ", 1951, p. 153 (Amer. Soc. Metals); *J. electrochem. Soc.* 1952, **99,** 369) and of N. Cabrera and N. F. Mott (*Rep. Progr. Phys.* 1948–49, **12,** 163). Another treatment deserving study is that of W. J. Moore, *J. electrochem. Soc.* 1953, **100,** 302; a more elementary survey is provided by W. W. Smeltzer, *Corrosion* 1955, **11,** 366t, whilst useful résumé articles are due to M. T. Simnad, *Industr. engng. Chem.* 1956, **48,** 586; 1957, **49,** 617. Special attention is directed to the extensive work of Gulbransen, largely based on electron diffraction; this includes studies of aluminium, zinc, iron, nickel, copper, cobalt, chromium, molybdenum, tungsten, titanium, zirconium, columbium, tantalum as well as many alloys. (E. A. Gulbransen with J. W. Hickman and W. S. Wysong, *Metals Technology* 1947, **14,** 3, 6; *Amer. Inst. min. met. engrs. Tech. Pub.* **2144** (1947), **2224** (1947), **2226** (1947); *J. phys. coll. Chem.* 1947, **51,** 1087; 1949, **53,** 698; *Anal. Chem.* 1948, **20,** 158; *Rev. Métallurg.* 1948, **45,** 181, 287; *J. electrochem. Soc.* 1949, **96,** 364; 1951, **98,** 241; 1952, **99,** 360, 393, 402; *Industr. engng. Chem.* 1951, **43,** 697; *J. Metals* 1954, p. 1027; *Annals New York Acad. Sci.* 1954, **58,** 830.)

The quantitative results of Eurof Davies on iron, and of Hart on aluminium receive further discussion in Chapter XX, where the theory of Mott and Cabrera also receives notice; the work of Mills on copper is described in Chapter III.

Attention should be given to a recent paper on magnesium by S. J. Gregg and W. B. Jepson, *J. Inst. Met.* 1959, **87,** 187, which, with the discussion, if printed, will help the understanding of the oxidation of other metals, notably titanium and zirconium.

CHAPTER III

EFFECT OF OTHER ELEMENTS ON OXIDATION AND FILM-GROWTH

Synopsis

Chapter II was concerned with the action of oxygen on unalloyed metals. The present chapter takes account of a second constituent in one or other phase. This may be present in the metal (either as an unwanted impurity or as an intentionally introduced alloying constituent); or in the attacking phase, which may itself be a gas or a liquid.

First, we consider a second constituent in the metallic phase. This may behave during oxidation in two ways. It may enter the main oxide-film, either increasing or diminishing lattice-defects, so that the oxidation-rate may be speeded up or retarded; here Hauffe's Valency Rule has been found helpful in many cases. Alternatively, the second constituent may accumulate as a new phase at the base of the main oxide-film; under favourable conditions, this may obstruct passage between metal and film-substance and consequently retard oxidation, despite the fact that the composition of the main portion of the film remains almost unchanged; the form of the curve may cease to be parabolic and approach a log-law shape. On the whole, this alteration of growth law to a more favourable type seems a better method of preventing oxidation-wastage than the other method, which accepts the parabolic law and merely seeks to diminish the value of the rate-constant.

Attention is then turned to the formation of sub-scales in alloys, due to preferential attack on one alloying constituent, so that oxidation penetrates inwards along grain-boundaries or other favoured paths. It becomes clear that internal attack, if allowed to proceed unchecked, may cause increased damage; for the grains, becoming completely surrounded by oxidized matter, may be loosened and finally dislodged. However, a modicum of attack along grain-boundaries or favoured paths can, if suitably controlled, improve the adhesion of the main scale by " pegging in " ; this prevents spalling and builds up oxidation-resistance.

We then consider a second constituent in the external basis. Here sulphur deserves notice, since even small amounts may stimulate attack in two ways; the number of lattice-defects in the film-substance may be increased, and its melting- or softening-point may be lowered. Since the rate of attack is always greatly increased when a film becomes molten or partially molten, the effect of sulphur can be a serious one.

Next the catastrophic effects sometimes produced by vanadium and molybdenum demand attention. These metals are in common use as alloying constituents—often without serious results. However, the presence of vanadium in the scale—often derived from the ash of oil-fuel—may have devastating effects. The main cause is again the formation of liquid constituents in the scale, although the increase of lattice-defects may in some cases be important.

Finally, we discuss the growth of iodide-films, which has interest for the pure scientist. They can be produced by the action of iodine vapour, or, more conveniently, by means of iodine dissolved in an organic solvent; beautiful interference colours are here observed. A table relating colour to film-thickness is provided.

The quantitative aspect of film-growth is deferred to Chapter XX.

Influence of a Second Element in the Metallic Phase

Two effects of Alloying Constituents. A minor constituent in the metal phase can influence oxidation in two ways. The choice between them depends largely on whether it can distribute itself through the main oxide phase, forming an oxide in which some of the cations of the main metal are replaced by the minor constituent. If that is the case, the velocity may be accelerated or retarded, but generally the growth law will remain the same; assuming it to be parabolic in the original metal (at the temperature under consideration), it is likely to remain parabolic, but the value of the constant, k, will alter. If, however, the minor constituent introduces a new phase at the base of the oxide-layer, the changes may be more drastic; for instance, the parabolic law may give place to a log law—which, from the engineer's stand-point, is usually desirable.

Cases where the minor constituent enters the main oxide phase. If the two constituent metals have the same valency, replacement of cations of the major metal by cations of the minor metal will cause no radical change in the scale, but if the valencies are different, the number of vacant sites may be seriously altered.

Consider the replacement of nickel ions by ions of lithium, which has a lower valency; if electrical neutrality is to be preserved, one Ni^{++} must be replaced by *two* Li^{+} ions. Now nickel oxide normally contains vacant cation sites (like cuprous oxide); of the two Li^{+} ions, only one can occupy the position of the Ni^{++} ion displaced, so that the other is likely to settle into a site formerly vacant. Thus the number of vacancies will be diminished, reducing the rate at which ions can cross the film; the oxidation constant, k, will fall, and, during a fixed oxidation-period at a fixed temperature, the total oxidation will be less than if lithium had not been introduced.

Evidently in an oxide containing *cation vacancies* the introduction of a minor constituent of *lower* valency than the main metal will *diminish* the oxidation-rate, but by the same argument the introduction of a minor constituent of *greater* valency will *increase* the oxidation-rate. Conversely, in an oxide with metal excess (which may involve *anion vacancies*), a minor constituent with *lower* valency will *increase* the oxidation-rate, whilst one with *higher* valency will diminish it. This important " Valency Rule" is due to Hauffe, and examples of its applicability are found in the papers of his school, and also those of Wagner and Kubaschewski. See especially K. Hauffe, *Werkst. u. Korrosion* 1951, **2,** 131, 221, 243, also " Progress in Metal Physics ", Vol. **4,** p. 71; C. Wagner and K. E. Zimens, *Acta chem. Scand.* 1947, **1,** 547; O. Kubaschewski and O. von Goldbeck, *Z. Metallk.* 1948, **39,** 158.

A few of these examples may be quoted. The oxidation of nickel (a divalent metal with a deficiency of metal in its oxide—nominally NiO) is retarded by the presence of oxide of lithium (Li_2O) in the gas phase, but accelerated by the presence of oxide of molybdenum (MoO_3) in the gas phase. It is also accelerated by traces of chromium or manganese in the metallic phase, since either of them will enter the oxide in the trivalent state. Conversely, the oxidation of zinc (a divalent metal with excess of metal in its oxide) is accelerated by monovalent lithium and retarded by trivalent aluminium in the metallic phase.*

Whilst Hauffe's rule has often proved useful in explaining the effect of a minor constituent, it does not always give a correct prediction; lithium oxide vapour does not influence the oxidation-rate of iron at 550°C. and 620°C., and only does so at 850°C. after certain preliminary treatments (E. Brauns and A. Rahmel, *Werkst. u. Korrosion* 1956, **7,** 448).

Clearly other factors can influence the situation and sometimes these may mask the effect of valency change. Huddle considers that the magnitude of the parabolic rate-constant is decided largely by the energy needed for an ion to squeeze its way from one site to an adjacent site; the replacement of a few cations of the predominant metal by cations of a different size will certainly alter the " squeeze-energy " involved, and Huddle has found this principle useful in planning additions to enhance the resistance

* It should be noted that an addition may affect ionic conductivity (which determines oxidation-rate) and electronic conductivity in opposite ways. Zinc oxide contains interstitial zinc ions and also electrons; let the concentrations be C_i and C_e respectively, such that $C_i C_e = K$, where K is constant at a given temperature. If alumina is added to the zinc oxide Al^{+++} ions will replace Zn^{++} in the lattice, and their higher ionic charge will lead to vacant cationic sites which will cause zinc ions to leave interstitial positions and fill the vacant sites. This means that C_i will decline and C_e must therefore rise to maintain K as constant. Thus although the presence of aluminium in zinc decreases the oxidation-rate, the presence of alumina in zinc oxide will increase the electronic conductivity; in such oxides, the conductivity, as normally measured, is predominantly electronic. For oxides containing vacant cation sites, additions of lower ionic charge are needed if the conductivity is to be raised; thus the presence of lithium increases the conductivity of nickel oxide (K. Hauffe and A. L. Vierk, *Z. phys. Chem.* 1950, **196,** 160; K. Hauffe and J. Block, *ibid.* 1950, **196,** 438 K. Hauffe, *Metall.* 1950, **4,** 462. E. J. W. Verwey, *Bull. Soc. chim. Fr.* 1949 p. D93. C. Wagner, *J. electrochem. Soc.* 1952, **99,** 346C, esp. p. 351C).

of magnesium (R. A. U. Huddle, lecture at Nuclear Engineering and Science Congress, sponsored by Amer. Inst. min. met. Engrs., 1955).

Cases where a new oxide phase is introduced. Consider a metal like iron or copper, which produces a film growing by outward passage of cations, and insufficiently protective to justify extensive use of the unalloyed metal at a really high temperature. Let us now add a minor constituent, possessing a greater affinity for oxygen, which forms an oxide almost immiscible with that of the major metal. Particles of that oxide are likely to accumulate at or near the base of the main scale; in due course, they may come to form a layer shutting off the alloy from the main scale, and thus largely put a stop to oxidation.

Let us call the main metal A and the minor constituent B. Since the solubility of the oxide of B in that of A is small, cations of B will rarely pass into the scale, but, since the affinity of B for oxygen is greater than that of A, the B atoms in the metal phase next to the interface will tend

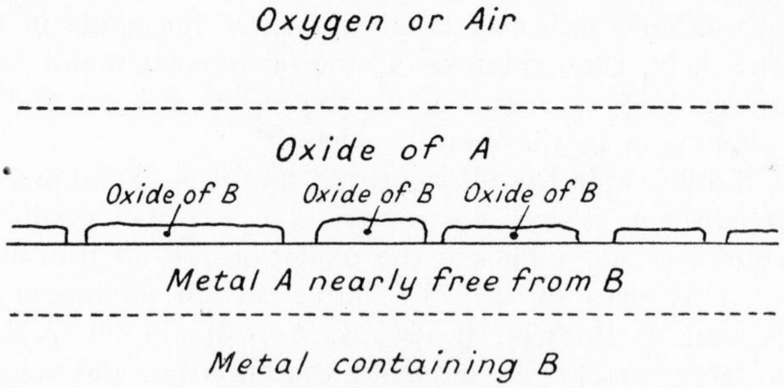

Fig. 10.—Formation of thin protective layer at base of main oxide-scale during the oxidation of certain alloys.

to reduce to the metallic condition the A which is present in the oxide phase at the base of the film. Thus the metal phase at the interface will become impoverished in B, and nearly all the B in that region will be precipitated as oxide (fig. 10). When the layer of oxide at B has become continuous, oxidation will become slow, since atoms of A will only slowly pass through this intervening layer. If the two oxides are practically insoluble in one another, and if all the oxide of B is precipitated exactly at the interface, mathematical arguments suggest (p. 837) that the parabolic law should be replaced by a log law—a desirable change.

The most favourable case occurs when B is a metal of fixed valency, forming an oxide nearly free from lattice defects, and thus offering no facilities for cations to pass through it. Beryllium (which has a fixed valency) is added as an alloying constituent to copper to increase resistance to oxidation. Some instructive experiments at Brussels have demonstrated the diminution of the oxidation, and have indicated that this is due to a beryllia layer formed at the *base* of the cuprous oxide scale; in the experiments carried out at 500°C., the up-take of oxygen continued after attack on the alloy had ceased, but it was now employed in converting cuprous oxide to cupric oxide; when the cuprous oxide had completely disappeared, oxidation came to an end (fig. 11) (L. Hubrecht and L. de Brouckère, *Bull. Soc. chim. Belg.* 1951, **60,** 311; 1952, **61,** 101, 205; a different interpretation is offered by E. A. Gulbransen, *Corrosion* 1956, **12,** 637t. The effect of other elements in copper is described by J. P. Dennison and A. Preece, *J. Inst. Met.* 1952, **81,** 229).

In the case of iron, additions of aluminium, silicon and chromium have been used to control oxidation. Each of these three elements can be regarded as having a fixed valency; although in some chemical compounds chromium exerts valencies of 2 and 6, only a valency of 3 is likely to

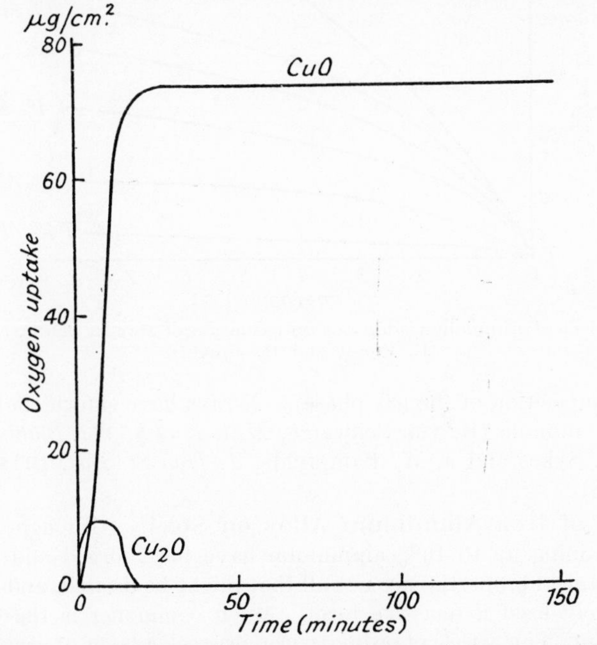

FIG. 11.—Effect of Beryllium addition on oxidation of Copper at 500°C. (L. de Brouckère and L. Hubrecht).

be stable under the conditions prevailing in an oxide-scale. The foreign element tends to accumulate as oxide near the base of the main oxide-scale (which consists of the usual three layers, essentially FeO, Fe_3O_4 and Fe_2O_3); even if the foreign element is present in an amount insufficient to form a continuous layer, it evidently interferes considerably with the passage of iron from the metal into the ferrous oxide (" wüstite ") phase, since the addition produces a marked change in the shape of the oxidation–time curve; on unalloyed iron the curve conforms approximately to a mixed parabola, but as the content of (say) aluminium is increased, there is a departure from parabolic form, and the shape of the curve (fig. 12) becomes characteristic of a protective film (A. M. Portevin, E. Prétet and H. Jolivet, *J. Iron St. Inst.* 1934, **130,** 219; cf. L. B. Pfeil, p. 34 of this book).

Although the alloys with less than 6% aluminium develop black scales similar to those on unalloyed iron, white areas appear on the scales above 8%, whilst above about 14% the scale is often entirely white (the composition needed for an all-white scale will doubtless depend on temperature

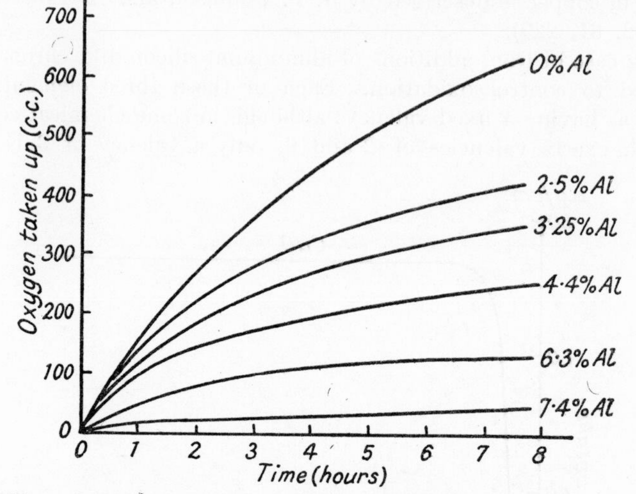

Fig. 12.—Effect of Aluminium additions on oxidation of Iron at 900°C. (A. Portevin, E. Prétet and H. Jolivet).

and the composition of the gas phase). X-rays have shown that the white material is alumina (H. von Schwarze, *Mitt. Forsch. Ver. Stahlwerke* 1932, **2,** 263. C. Sykes and J. W. Bampfylde, *J. Iron St. Inst.* 1934, **130,** 389, esp. p. 408).

Layers of Iron-Aluminium Alloy on Steel. Although low-carbon steels containing up to 16% aluminium have been forged and rolled hot, their mechanical properties are not all that might be desired, and aluminium steel is rarely used in massive form. Much commoner is the practice of producing, upon an article of ordinary iron or steel, a layer of iron-aluminium alloy. Such a plan is widely favoured for furnace fire-bars and annealing boxes; it makes use of the excellent corrosion-resistance of the alloy combined along with the acceptable mechanical properties and cheapness of the unalloyed material. Another application of steel coated with iron-aluminium alloy is in resisting hydrogen sulphide at high temperature (E. R. Backensto, R. C. Drew and C. C. Stapleford, *Corrosion* 1956, **12,** 6t. G. Sorell and W. B. Hoyt, *ibid.* 1956, **12,** 213t).

One process (*Calorizing*) consists in embedding the articles to be coated in a powder mixture of aluminium, alumina and ammonium chloride (proportions 49/49/2) and heating at about 850–950°C.; this is well above the melting-point of aluminium (659°C.) and the particles of metal would probably coalesce if the alumina was omitted; the ammonium chloride is presumably added as a flux to remove the thin oxide layer which surrounds each particle of metal, possibly converting it in part to volatile aluminium chloride. In another process, aluminium is sprayed from a metallizing pistol (p. 601) on to the surface, which is afterwards heated at about 850–950°C. to allow alloying between aluminium and iron; a layer of bitumen is applied before the heating, to prevent undue oxidation in the opening stages before alloying has had time to occur.

Steels containing Chromium. When chromium is added to steel, the oxidation-rate falls off as the chromium-content rises, but considerable amounts are required to make oxidation really slow; Yearian's curves suggest that 20% chromium may be needed for resistance at 925°C. and over 25% for resistance at 1000°C. If sufficient chromium is added, the protective scale is in general essentially chromic oxide, although, on steels containing much manganese, the spinel $MnCr_2O_4$ may be present. When the chromium addition has been insufficient to give protection at the temperature to be faced, the scale is more like that on unalloyed steel; there is α-ferric oxide on the outer surface, covering a spinel layer which may be Fe_3O_4 at its outer limit, but contains increasing chromium as the metal is approached; it is best written $Fe(Fe_{2-x})Cr_xO_4$. A wüstite phase may also be present, but this usually contains chromium; it is best written (Fe,Cr)O (H. J. Yearian, E. C. Randell and T. A. Longo, *Corrosion* 1956, **12,** 515t; J. Moreau, *C.R.* 1953, **236,** 85; J. Moreau and J. Bénard, *ibid.* 1953, **237,** 1417; there is not complete agreement between the two groups of investigators).

In practice, chromium, aluminium and silicon are all added to steel intended for high-temperature use, whilst some of the materials designed for the severest conditions (e.g. gas-turbine blading) are essentially nickel-chromium alloys with other constituents present. Sometimes iron as well as nickel is present, and titanium may be added to prevent the removal of the chromium as carbide (the titanium fixes the carbon as a stable carbide or as a compound containing titanium, carbon and nitrogen).

If mechanical considerations could be disregarded, an 18/8 chromium/nickel steel, similar to the familiar austenitic stainless steel, might prove sufficiently oxidation-resistant for some purposes. In practice, however, it may be necessary to add small amounts of other materials selected to improve the mechanical character, such as silicon and tungsten, and often to increase the chromium and nickel contents. It is not difficult to find materials capable of withstanding oxidation at constant temperature, but in service the material is alternately cooled and heated—so that the protective scale has every chance of cracking off. Shirley considers that ordinary 18/8 stainless steel should not normally be used above 800°C.; a steel with 23% chromium, 11·5% nickel and 2·75% tungsten can be used up to 1050°C., whilst one with 23% chromium and 21% nickel is serviceable up to 1100°C. Among nickel-free materials, steel with 13% chromium can stand 750°C., that with 21% chromium 900°C., and that with 29% chromium 1100 to 1150°C.; if 3·4% silicon is present, 8% chromium is sufficient to give a permissible range of 800–900°C. (H. T. Shirley, Priv. Comm., Nov. 29, 1957; further information is provided by O. Edstrom, *J. Iron St. Inst.* 1957, **185,** 450).

The oxidation of nickel-chromium steels has been studied by Yearian and his colleagues, who have obtained samples of the scale by flaking, scraping or controlled abrasion and examined them by X-rays. Here again the principal constituent on the protective type of alloy seems to be Cr_2O_3; a solid solution, $(Cr,Fe)_2O_3$, is present, but the iron content is only found to be high in cases where there has been rapid attack. Alloys containing manganese develop considerable amounts of the spinel $MnCr_2O_4$, and this is not inconsistent with good resistance to corrosion; but, where spinels containing nickel, chromium and iron are present, the attack is more rapid. Evidently the spinel is less protective than Cr_2O_3; it probably contains more vacant sites, and thus allows the passage of cations outwards. On those alloys which suffer serious attack, no Cr_2O_3 is found, but spinels (mainly in the inner regions) and $(Cr,Fe)_2O_3$ solid solution (mainly in the outer regions) are present (H. J. Yearian, with H. E. Boren, R. E. Warr, W. D. Derbyshire and J. F. Radavich, *Corrosion* 1956, **12,** 561t; 1957, **13,** 597t).

Alloys containing Cobalt. Many alloys which have come into favour for high-temperature resistance contain cobalt. When heated strongly, unalloyed cobalt forms two main layers, both essentially CoO; experiments with a wire marker suggest that the inner layer is produced by movement of oxygen inwards and the outer layer by movement of cobalt outwards—in analogy with the behaviour of nickel (p. 34). Below a certain temperature (900°C. according to Phalnikar, but 950°C. according to Preece), there is an outer layer consisting essentially of Co_3O_4. Small amounts of chromium greatly *increase* the oxidation-rate—as Hauffe's Valency Rule would predict; however, beyond a certain composition, the scale, instead of containing both metals, consists solely of Cr_2O_3, and is very protective. Preece and Phalnikar agree in placing this sudden falling off of oxidation-rate at about 25% chromium, and also in stating that the simple oxide Cr_2O_3 is more protective than the spinels found in some complex scales; there is, however, incomplete agreement on some other points, and the reader should study the original papers (A. Preece and G. Lucas, *J. Inst. Met.* 1952, **81,** 219, with discussion on p. 727. C. A. Phalnikar, E. B. Evans and W. M. Baldwin, *J. electrochem. Soc.* 1956, **103,** 429).

Phalnikar observes that, although cobalt alloys with more than 20% chromium resist scaling well during continuous heating, the scales formed spall badly on cooling to room temperatures. This would seem to preclude the use of such alloys in service unless the spalling-resistance can be improved by addition of small additions of such elements as silicon which peg the external scale into the basis metal; this " pegging in " plan has worked well for the nickel-chromium alloys used in electric heating elements (p. 71).

Other troubles besides oxidation-resistance have to be considered by the designer of high-temperature plant; often the " creep " of the hot material when subjected to tensile stress is his main preoccupation. The two problems, however, cannot be considered separately; if creep occurs in the alloy, there is danger that the oxide-scale may rupture, which will accelerate oxidation; conversely, reduction of cross-section of the metallic portion by oxidation will increase the stress (load per unit area) and thus accelerate creep. Any composition selected must provide resistance to *both* troubles. Fortunately many alloying constituents favoured for improvement of mechanical properties tend to improve oxidation-resistance—or at least do not worsen it; two possible exceptions, molybdenum and vanadium, are considered later.

Deposition of minor constituent as metal. When the minor constituent possesses a smaller affinity for oxygen than the major one, it will usually be found near the base of the main scale in the metallic state and not as oxide. Thus, when an iron-nickel alloy is heated in air, both metals may start to pass outwards into the (wüstite) layer as (ferrous) cations, but any nickel oxide present in solid solution in the layer of scale next to the metal will quickly interact with metallic iron to yield iron oxide and metallic nickel; wüstite itself is capable of reducing nickel oxide if its iron content exceeds 72%—as indicated by Sachs. In his early work, Pfeil (p. 34), heating steels containing nickel, found particles of metallic nickel embedded in the inner layer of oxide. Copper in steel may also accumulate at that level. The presence of metallic copper below the scale is not, however, an advantage; it tends to diffuse along grain-boundaries into the metal and renders the steel brittle (W. Püngel, *Arch. Eisenhuttenw.* 1951, **22,** 143, esp. p. 146).

Nickel is not harmful in that way, and an accumulation of nickel at the right place in the scale may hinder oxidation. Sachs has studied the distribution of nickel in the thick scale formed on heating steel containing 1·5 to 3% nickel for 24 hours at 1200–1250°C. The main accumulation of metallic particles (largely nickel) occurs some way from the final position of the interface between the oxide-scale and the "core" of unoxidized metal, but in most experiments no nickel was found in the part of the scale situated outside the original surface of the specimen. The nickel is sometimes found as groups of isolated particles and sometimes as a filigree pattern; the two types are probably formed by different mechanisms—described in Sachs' second paper (K. Sachs, *J. Iron St. Inst.* 1957, **185,** 348; 1957, **187,** 93. Cf. R. T. Foley, C. J. Guare and H. R. Schmidt, *J. electrochem. Soc.* 1957, **104,** 413; M. J. Brabers and C. E. Birchenall, *Corrosion* 1958, **14,** 179t).

Many heat-resisting alloys contain nickel, and although its main function may be to improve mechanical properties, or perhaps to stabilize an austenite phase, its effect in controlling the oxidation-rate should not be overlooked. Under reducing conditions in presence of sulphur compounds, the possible formation of low-melting eutectics containing nickel sulphide requires to be borne in mind. Occasionally alloys containing nickel have proved disappointing in reducing atmospheres containing sulphur, although in other situations their performance has exceeded expectations. The subject is complicated, and in choosing a material to withstand drastic conditions, empirical experience of the behaviour of the various alloys in similar service may be more valuable than prediction from scientific theory. The best advice comes from those who combine practical experience with scientific knowledge.

However, oxidation of an alloy containing a relatively noble metal as minor constituent does not necessarily lead to increased resistance. Theoretical arguments developed by Wagner lead us to the conclusion that a film of uniform thickness will only be stable if diffusion in the metallic phase is rapid compared to diffusion in the oxide of the less noble metal. Otherwise the alloy-oxide interface is likely to become rugged. In extreme cases we may get protruding sections of the oxide of the less noble metal interspersed with slender trunks of an alloy rich in the more noble metal. Wagner has calculated that, until the noble metal content reaches 50% (atomic), the oxidation-rate will remain of the same order of magnitude as that of the less noble metal containing no alloying constituent; in other words, no marked advantage can be expected from alloying (C. Wagner, *J. electrochem. Soc.* 1956, **103,** 571).

Effect of Carbon on Oxidation. So many common materials contain carbon that it is well to consider the possibility of an otherwise protective film being ruptured by the evolution of carbon monoxide below it. This is most likely to occur with metals having a relatively low affinity for oxygen; when the metal has a high affinity for oxygen, the carbon may be retained in the alloy. Wagner and his colleagues found that nickel containing carbon, and also tungsten carbide, reacted with oxygen more rapidly than the corresponding carbon-free metal (W. W. Webb, J. T. Norton and C. Wagner, *J. electrochem. Soc.* 1956, **103,** 112).

Sub-Scales. Another possibility occurs when the minor constituent of an alloy possesses a much higher affinity for oxygen than the major one. Copper alloys containing small amounts of silicon or manganese have been studied by Rhines. Even in unalloyed copper which is undergoing oxidation there will be a small amount of oxygen present in solid solution in the metallic phase, and presumably at the metal-oxide interface approximate equilibrium will exist between the dissolved oxygen (in the metal) and the combined oxygen (in the scale). Clearly, the oxygen in the metal can never become supersaturated with respect to the oxide, and there is no reason to expect the separation of copper oxide particles in the interior. However, since silicon or manganese have a greater affinity for oxygen, it is reasonable to anticipate that particles of silica or manganese oxide may separate. Rhines found that, in these and analogous cases, a " sub-scale " is formed, consisting of particles of silica or other oxide dispersed in a metallic matrix. The lower boundary of the sub-scale gradually advances with time inwards into the metal, and at high temperatures, where the diffusion of dissolved oxygen through solid copper is possible, the particles are as frequent in the grain-interiors as along the grain-boundaries. Rhines heated a copper alloy containing 0·1% silicon in air at about 1000°C., and then etched a section in ammonia containing hydrogen peroxide; he found the surface to darken over the zone in which silica had been precipitated. The boundary of this darkened zone ran accurately parallel to the surface, and its thickness (Y) increased with time in rough obedience to a parabolic law, $dY/dt = k/Y$, as would be expected. At lower temperatures (e.g. 600°C.), where normal lattice-diffusion becomes a slow process, the silica is preferentially deposited along the grain-boundaries, since the atomic disarray at such boundaries favours diffusion. The different behaviour at high and lower temperatures is shown schematically in fig. 13. For further details, see papers by F. N. Rhines, with W. A. Johnson, W. A. Anderson, A. H. Grobe and B. J. Nelson, *Trans. Amer. Inst. min. met. Engrs. Tech. Pub.* **1162, 1368, 1439** and **1617** (1940–43). The formation of external films

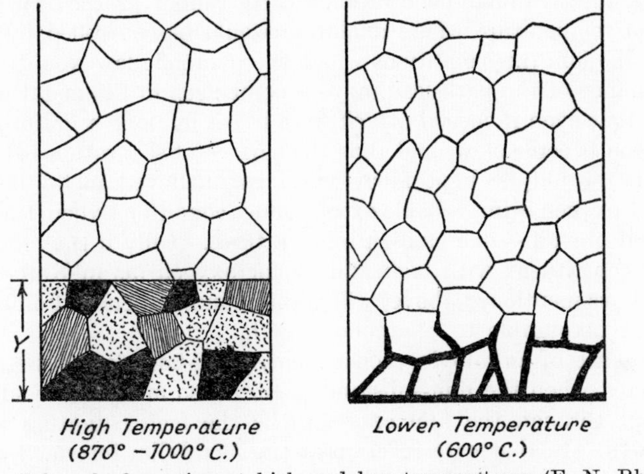

FIG. 13.—Sub-scale formation at high and low temperatures (F. N. Rhines).

on copper-manganese alloys has been studied by F. Bouillon and M. Jardinier, *J. Chim. phys.* 1956, p. 817.

The precipitation of small oxide particles in a metallic matrix is capable of modifying mechanical properties, favourably or otherwise. The possibility of attaining improvement has received study from J. W. Martin and G. C. Smith, *J. Inst. Met.* 1955, **83,** 417; E. Gregory and G. C. Smith, *ibid.* 1956–57, **85,** 81.

Intergranular Penetration into the Metallic Phase. The formation of a sub-scale affords proof that oxygen can penetrate inwards into an alloy, and Rhines' work shows that such penetration often proceeds most readily along grain-boundaries. Even in unalloyed metal, grain-boundary penetration is met with, and this includes cases where the growth of the external scale is associated with cations moving outwards. It would seem that oxygen diffuses inwards along grain-boundaries, producing oxide which, near the surface, may form a continuous network but at deeper levels may appear in a section as isolated spots (fig. 14). An example of internal oxide formed along the grain-boundaries of iron heated in oxygen is provided in an early paper by J. E. Stead, *J. Iron St. Inst.* 1921, **103,** 271.

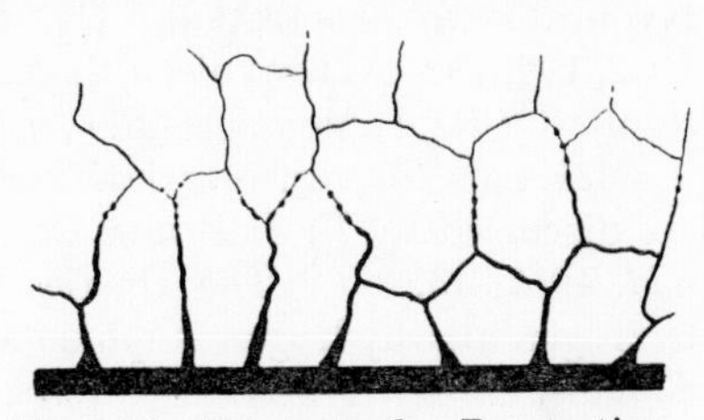

FIG. 14.—Intergranular Penetration.

Clearly the distribution of oxide shown in fig. 14 cannot be explained unless we assume movement of oxygen inwards as well as metal outwards. It would seem to provide yet another example of anions and cations moving simultaneously in opposite directions. Such simultaneous movement may avoid the cavity-production characteristic of purely cationic movement and likewise the strain-production characteristic of purely anionic movement. However, intergranular penetration clearly cannot proceed far under conditions of temperature where oxidation would be controlled by movement through the film (i.e. conditions where the parabolic law would be obeyed); the resistance of the path leading to a point deep in the metal will be very high, so that even if special attack were to set in along a grain-boundary it would soon become slow, and thus the surface of separation between oxide and meta]would be kept fairly flat. In unalloyed metal intergranular attack by oxygen is improbable except under conditions where the rectilinear law would prevail—such as high temperatures. Only if the metal contains a minor constituent with high affinity for oxygen (as in Rhines' work on copper alloys mentioned above) will grain-boundary penetration occur under conditions favourable to parabolic growth.

In absence of an alloying constituent, grain-boundary penetration may be expected when the film-substance contains so many lattice-defects that movement through the film ceases to be the factor limiting the rate of film-growth. In that case reaction at the metal-film interface will assume control, and such a reaction is likely to occur most readily where the atoms are in disarray—i.e. at a grain-boundary. The necessary conditions exist in sulphide-films, which are full of defects, as indicated later. A picture of nickel exposed to sulphur at a high temperature, which has developed a complete network of sulphide round the grains, is provided by E. N. Skinner, "Corrosion Handbook" 1948, p. 1109 (Editor, H. H. Uhlig; publishers, Wiley; Chapman & Hall).

The fact that this intergranular penetration of sulphur proceeds more readily into nickel than into iron can be explained on the principles just developed. The combination of iron with sulphur follows the parabolic law—which will be unfavourable to deep penetration; in contrast, the combination of nickel with sulphur follows a rectilinear law—which will be favourable to penetration. (K. Hauffe and A. Rahmel, *Z. phys. Chem.* 1952, **199,** 152; A. Dravnieks, *Industr. engng. Chem.* 1951, **43,** 2897. Cf. later results by R. A. Meussner and C. E. Birchenall, *Corrosion* 1957, **13,** 677t.)

Materials for resisting hot gas mixtures containing hydrogen sulphide are discussed by F. J. Bruns, *Corrosion* 1957, **13,** 27t; see also C. Phillips, *ibid.* 1957, **13,** 37t. The kinetics of the reaction of steel with mixtures of hydrogen and hydrogen sulphide are discussed by A. Dravnieks and C. H. Samans, *J. electrochem. Soc.* 1958, **185,** 183.

The Pegging-in Principle. A limited amount of penetration along grain-boundaries (and perhaps also along sub-grain boundaries, slip-planes or similar favoured paths) can anchor the film and reduce the danger of spalling, thus improving the protective character of a scale, especially under conditions of fluctuating temperature or periodical bending. On the other hand, excessive penetration may weaken the material, so that under slight stress, metallic grains, surrounded completely by an oxide or sulphide envelope, become detached; in that case, the indirect damage involved in the loss of metallic grains will vastly exceed the direct damage represented by the conversion of metal into oxide or sulphide. If " pegging in " is to be used to improve film-adhesion, it must be carefully controlled, and there may be advantages in adding an alloying constituent which will improve the mechanical strength of the metal phase, so that a grain which is nearly surrounded with an oxide or sulphide envelope but is still joined to its neighbours by tenuous metallic bridges, will resist detachment. Thus the addition of the pegging-in constituent should be regulated to provide sufficient, but not excessive, anchorage; a second element should be added to strengthen the metallic phase, and perhaps a third to provide a thin impenetrable layer of a foreign oxide at the basis of the main scale; the function of that layer would be to slow down the main oxidation-rate by the same mechanism as is operative in copper-beryllium alloys (p. 62). The conditions for success are complicated, and it is not surprising that much of the progress made has been achieved by empirical studies of the situation, although possibly some industrial investigators have been guided by scientific principles more than would appear from what they have been allowed to publish.

Nickel-chromium Alloys. Examples of the successful application of the pegging-in principle are provided by the improvement achieved in the alloys used as wire or ribbon for electrical resistance furnaces and cookers, as well as for heaters in dwelling-rooms. Various conditions must be fulfilled. The material must be ductile but must resist creep and distortion at high temperatures; it must not interact with the refractories with which they may come into contact (these refractories should be kept free from chlorides, sulphides and the like); it must develop a protective oxide-film, containing at least one layer through which ionic movement is slow; and that film must be sufficiently pegged-in to prevent spalling.

For most purposes, alloys based on nickel-chromium (usually in a ratio of about 80/20) are still used. The " Nimonic " series are essentially nickel-chromium alloys with additions designed to improve mechanical properties—such as are required for gas-turbine blades and components likely to be highly stressed whilst hot. Nimonic 80 is an 80/20 alloy containing titanium and aluminium with high creep-resisting properties. Nimonic C is a similar alloy, without guaranteed creep-resistance, but with good resistance to crazy cracking (intergranular oxidation accelerated by temperature fluctuation). Nimonic 75 is also an 80/20 alloy stiffened with carefully controlled amounts of titanium carbide. Nimonic DS, with 37% Ni, 18% Cr, 2% Si and the rest iron, has less resistance to ordinary oxidation but shows satisfactory resistance to green rot (p. 72). Particulars of properties at different temperatures will be found in a brochure " The Nimonic Alloys " (Henry Wiggin & Co.).

In the absence of additions and complicating factors, the oxidation of nickel-chromium alloys develops three layers:—

(i) Within the metallic phase, Cr_2O_3 granules are dispersed in the metal.

(ii) Next, there is a two-phase structure with the cubic spinel ($NiCr_2O_4$) dispersed in a NiO matrix.

(iii) Outside this comes pure NiO (J. Moreau and J. Bénard, *C.R.* 1953, **237**, 1417).

In practice, various elements are added; these include silicon, calcium, zirconium, thorium and cerium in different cases. The function of each individual constituent is not always clear, but probably the effect of the correct composition is to secure correct conditions at the base of the scale. The oxidation-rate as determined by experiments conducted at constant temperature is of limited value in predicting the relative utility of different alloys under service conditions of frequent heating and cooling. The "A.S.T.M. useful life test" alternately heats the wire for 2 minutes at 1175°C., then cools it to room temperature, heats it again and so on—thus giving the scale every chance to break off if adhesion is poor; the time in hours needed to change the electrical resistance by 10% is regarded as a measure of useful life.

Gulbransen and Andrew find that 2% of manganese, which changes the composition of the outer scale, has practically no effect on the life as measured in the test; conversely silicon additions, which have no appreciable effect on composition or structure, nevertheless do affect the life. They believe that silicon, calcium and other additions, which do not, by reason of their ionic size and/or charge, enter the main part of the oxide-film, become concentrated at the basis as SiO_2, CaO or a silicate, and play a determining part in resisting detachment. They have introduced a new test in which oxidation is carried out until the film reaches a certain thickness; then the alloy is cooled, stretched to give a certain percentage increase in length, and once more heated. The stretching naturally causes some short-term damage, but this heals during the re-heating, and the long-term effect may be either favourable or the reverse. A low-silicon alloy shows an improvement after 1% strain but deterioration after 2, 3 or 4%; a high-silicon alloy shows a long-term improvement even after 1, 2, 3 or 4% strain. The film-thickness at the time of straining is a decisive factor; even low-silicon alloys show an improvement if this is small. The high-silicon alloys, although less damaged by strain than low-silicon alloys, show a greater high-temperature oxidation-rate—as would be expected from Hauffe's valency rule; replacement of Cr^{+++} by Si^{++++} must increase the number of vacant sites. Readers are recommended to study the original reports (E. A. Gulbransen and K. F. Andrew, *Amer. Soc. Test. Mat. Spec. Tech. Pub.* No. **171** (1955); *J. electrochem. Soc.* 1959, **106**, 294; E. A. Gulbransen and W. R. McMillan, *Industr. engng. Chem.* 1953, **45**, 1734).

An objectionable form of intergranular oxidation is met with in some nickel-chromium alloys; it is characterized by general swelling and a breakage with green fracture, and has received the name *green rot*. It occurs when the alloy is exposed to a carburizing atmosphere, such as occurs when electric heaters are used in a furnace handling oily components. Carburization of the chromium leads to swelling and the precipitated carbide becomes oxidized; since the precipitation of carbide occurs throughout the depth of the metal, the damage is really serious. Details are provided by J. C. Wright, *Bull. Inst. Met.* 1952, **1,** 112; E. H. Bucknall and L. E. Price, *Rev. Métallurg.* 1948, **45,** 129; N. Spooner, J. M. Thomas and L. Thomassen, *J. Metals* 1953, **5,** 844; L. B. Pfeil, *Chem.* and *Ind.* (*Lond.*) 1955, p. 208, esp. p. 214. H. R. Copson and F. S. Lang, *Corrosion* 1959, **15,** 194t.

For heating elements required to withstand high temperatures, consideration may be given to alloys containing cobalt such as *kanthal* (70% iron, 1·5–5% aluminium, 20–30% chromium and 2·5% cobalt), which is resistant to about 1500°C. The nickel-chromium types, as produced today, can be used up to 1150°C. or even, if the element is not too slender, to about 1250°C.; this has been achieved by the regulated additions of other elements as described above.

Materials are available for still higher temperatures, but in general only in furnaces designed to avoid contact with ceramics and to keep the heating elements in a protective atmosphere or *in vacuo*; under such conditions, it is stated that platinum can be used up to 1600°, niobium to 2230°, tantalum to 2400° and tungsten to 2560°C. (R. Kieffer and F. Benesovsky, *Metallurgia* 1958, **58,** 119).

The Growth of Cast Iron. Another example of internal oxidation—discovered at a much earlier period of metallurgical history—is provided by the volume-increase of cast iron produced by alternate heating and cooling through the $\alpha \rightleftharpoons \gamma$ transformation-temperature in an oxidizing atmosphere. Any iron heated through this temperature undergoes a contraction, but under ideal circumstances a homogeneous sample of metal should return to its original volume on slow cooling; if, however, as may happen in some cast irons, there is internal bursting around the graphite lamellae, this is unlikely to be repaired and some permanent effect must be expected. However, on high-quality iron, little harm occurs if there is no oxidation; Pearson shows that alternate heating and cooling in a non-oxidizing atmosphere usually produces little growth, and, although growth has sometimes been reported after heating *in vacuo*, this may have been due to gases occluded within the material. If, however, heating is conducted in air, the alternate entry and expulsion of gas at each cooling and heating leads to internal oxidation and a permanent change of volume. There is, however, a complex interplay of factors, so that many apparent anomalies are on record; tests carried out under certain conditions give no guidance to behaviour under different conditions; of a number of cast irons tested by Honegger, the one showing the best behaviour in steam at 500°C. was the worst in air at 650°C.* (E. Honegger, " Das Gusseisen " 1928, p. 25 (Eidg. Materialprüfungsanstalt)).

Two main factors are needed for deterioration, (1) access of oxygen or an oxidizing agent, and (2) passage through the $\alpha \rightleftharpoons \gamma$ temperature; the importance of passage through that temperature was emphasized in the

* J. N. Wanklyn adds the comment that in these cases the $\alpha \rightleftharpoons \gamma$ change cannot have been involved, the temperature being too low.

classical work of Benedicks and Löfquist. To eliminate the first factor will often be impracticable, but the second can be avoided in two ways—

(*a*) by raising the $\alpha \leftrightharpoons \gamma$ change above the working-range of temperature;
(*b*) by using material which normally exists in the γ-iron state at room temperature.

The raising of the transformation temperature can be accomplished by large additions of silicon. Curiously enough, small amounts of silicon had been shown in early work by Rugan and Carpenter to promote growth, and indeed much of the low-temperature growth in ordinary cast iron is really due to the oxidation of the silicon present, although at higher temperatures the oxidation of the larger graphite flakes and the iron around them seems to be the important factor. The presence of 4–10% silicon in an iron containing its graphite in a finely divided state gives a material in which growth of the type connected with passage through the $\alpha \leftrightharpoons \gamma$ point is materially diminished (C. Benedicks and H. Löfquist, *J. Iron St. Inst.* 1927, **115,** 603. Cf. R. F. Rugan and H. C. H. Carpenter, *ibid.* 1909, **80,** 29; 1911, **83,** 196).

High-silicon irons, however, are brittle and sometimes crack if heated and cooled suddenly. It may be better to use irons which are normally austenitic. Nickel is generally used as austenite-stabilizer, whilst chromium and silicon are generally present to improve the resistance to oxidation; some of these irons contain copper—which may improve resistance to atmospheric corrosion at lower temperatures.

The behaviour of a given cast iron depends greatly on the atmosphere to which it is exposed, and sometimes a weakly oxidizing atmosphere produces more damage than a strongly oxidizing atmosphere. A catastrophic oxidation of a cast iron containing flake graphite has been reported in carbon dioxide containing a little carbon monoxide; when carbon monoxide was absent, exposure to pure carbon dioxide was found to produce no serious damage. The reason suggested is that in the less strongly oxidizing atmosphere the external oxide-film is thin, and the undamaged graphite flakes provide paths through it; with carbon dioxide free from monoxide, the outer oxide is thicker, and the graphite flakes within it disintegrate, so that no such paths exist (S. R. Billington and B. C. Woodfine, *Metallurgia* 1957, **55,** 213).

The growth of cast iron—and also the green rot of nickel-chromium alloys—can be regarded as examples of " conjoint action ", similar to the phenomena described in Chapter XVIII where mechanical and chemical influences, acting together, produce more damage than if they acted separately; in the two cases mentioned, the mechanical action arises from volume-changes set up within the material.

Influence of a Second Element in Gas, Liquid or Ash

Effect of Sulphur introduced into Scale. Sulphur may affect film-growth in two ways. (1) It may lower the melting-point, or at least the temperature at which the film becomes pasty through the formation of a liquid eutectic within a mainly solid mass. (2) It may introduce additional lattice-defects.

It is arguable that these changes are merely diverse manifestations of a single effect. According to some physicists, a liquid only differs from a crystalline solid in the fact that defects, which are rare occurrences in a solid, are so numerous in a liquid that it can change its shape under weak forces such as gravity.* This picture of a liquid as an extremely defective solid suggests that compounds rich in defects should generally have low melting-points, and also explains why a solid film becomes suddenly less protective when it melts.

Sulphides are more rich in defects than oxides. The mineral pyrrhotite is assigned, in old-fashioned books on mineralogy, the formula $Fe_{11}S_{12}$; it is now usually regarded as ferrous sulphide (FeS) in which $\frac{1}{12}$ of the cations sites are vacant and another $\frac{1}{6}$ carry Fe^{+++} ions instead of Fe^{++}, thus maintaining neutrality. However, vacancies among sulphides can occur even when there is no departure from stoichiometric composition. Cuprous sulphide (Cu_2S) has a different lattice arrangement from that of cuprous oxide (Cu_2O) and its Cu^+ ions are in a highly disordered state; thus the diffusivity of the cations is high and the electrical conductivity of cuprous sulphide at 1000°C. is far superior to that of cuprous oxide, despite the fact that the composition departs very little from that represented by the formula Cu_2S—as shown by Wagner. See H. Braune and O. Kahn, *Z. phys. Chem.* 1924, **112,** 270; C. Tubandt, H. Reinhold and W. Jost, *Z. anorg. Chem.* 1928–29, **177,** 253; P. Rahlfs, *Z. phys. Chem.* (*B*) 1935–36, **31,** 157; C. Wagner, *J. chem. Phys.* 1957, **26,** 1602; also Priv. Comm., Oct. 28, 1957.

The greater frequency of defects in sulphides may explain why copper, which in pure air must be heated if we desire to obtain interference colours (due to cuprous oxide), can develop precisely the same colour-sequence at room temperature in air containing hydrogen sulphide; if these are due to a sulphide-film, containing lattice-defects as an essential feature of its structure, the possibility of ionic movement and film-thickening at room temperature is easily understood.

It may serve also to explain some early observations. The author at one time used to demonstrate interference tints by means of specimens made by applying the extreme tip of a blowpipe flame to a plate of copper of suitable thickness; a complete sequence of colours appeared as expanding rings with the most advanced colours at the centre. Such specimens, once prepared, could be preserved in a desiccator and underwent no further change. Similar coloured rings were obtained without heating by holding a drop of ammonia sulphide held just above a copper plate, but in that case the specimens could not be preserved unchanged in a desiccator, since

* A rapid change of shape is only possible if there exist plenty of vacancies, into which atoms or molecules can move, thereby creating other vacancies. D. K. C. MacDonald (*Nature* 1956, **177,** 23) states that there is evidence that a crystal melts when the holes reach a concentration of the order of 10^{-3}. See also discussion on melting in *Trans. Faraday Soc.* 1956, **52,** 882–885; J. O'M. Bockris and N. E. Richards, *Proc. roy. Soc.* (*A*) 1957, **241,** 44, esp. p. 56.

the colours continued to alter with time in a sense which indicated that all the films were thickening, even though the air of the desiccator contained no sulphur compound. Apparently copper cations could diffuse outwards through the pervious sulphide film and form oxide on the exterior, thus increasing the film-thickness and changing the colour.

In his classical work, Vernon showed that very small amounts of hydrogen sulphide in air produce colour changes (tarnishing) at ordinary temperature, and followed gravimetrically the film-growth responsible for this change. He found the parabolic law, $dy/dt = k/y$, to be obeyed—thus providing support for the idea that the changes produced by sulphur compounds at room temperature are similar to those obtained in pure air at elevated temperature, but differ from those responsible for the invisible film formed during exposure to pure air. Nevertheless the films were not pure sulphide films, but oxide containing a certain small proportion of sulphur (10·0 to 15·7% of the oxygen). Vernon showed that the value of the constant, k, was proportional to the amount of sulphur in the atmosphere as measured by means of alkaline lead acetate; if the number of defects is proportional to the rate at which hydrogen sulphide molecules reach the surface, this instructive result is easily understood (W. H. J. Vernon, *Trans. Faraday Soc.* 1927, **23**, 122).

It has been mentioned that copper, first exposed to air containing hydrogen sulphide, can undergo further film-growth, if subsequently exposed, unheated, to pure air. If, however, the exposure to pure air occurs first, film-growth during subsequent exposure to air containing sulphide is largely prevented. Vernon found that copper exposed for some months to a relatively pure ("summer") atmosphere in London suffered very little increase in weight when the pollution due to winter arrived, although specimens first exposed in winter were attacked rapidly.

The occurrence of tarnishing on articles fashioned from silver, copper or copper alloy during exposure to air containing sulphur compounds greatly diminishes the attractiveness of these materials. A uniform interference tint produced on a chemically clean surface by air containing hydrogen sulphide possesses a beauty of its own, but the surface of the articles under consideration may not be clean, and the colours produced are far from uniform; they may appear as smears or even "thumbographs". Moreover, possibly because the sulphur in the atmosphere occurs largely as sulphur dioxide, the effect is often not a bright colour but an unattractive dirty brown.

Prevention of Oxidation and Tarnishing by Selective Oxidation of Alloys. The rational method of preventing tarnishing is to introduce into the silver or copper an alloying constituent chosen to produce a film which becomes protective whilst it is still too thin to be visible. Aluminium and beryllium, which are effective in reducing high-temperature oxidation, suggest themselves. Price and Thomas tried the introduction of 1% aluminium into standard silver, thus producing an alloy containing 6·5% copper and 1% aluminium, but the results were disappointing. They explained the failure to prevent tarnishing in this way by supposing that the film formed on such alloys was not pure alumina but alumina containing copper or silver, where the foreign atoms would introduce lattice-defects. Accordingly they developed a treatment designed to produce pure alumina; they heated the alloy in hydrogen containing a small regulated quantity of water vapour (equivalent to 0·1 mm. mercury pressure). Such a treatment can convert aluminium to alumina, since aluminium exerts so great an affinity for oxygen that it can displace hydrogen from water vapour; no corresponding oxidation of copper or silver can occur, owing to their lower affinity for oxygen. The treatment caused no alteration in appearance, but an invisible protective film had evidently been formed, since the surface, when exposed to an atmosphere which rapidly tarnished ordinary silver, suffered no visible change at the end of 75 days (L. E. Price and G. J. Thomas, *J. Inst. Met.* 1938, **63,** 21, 29; 1939, **65,** 247).

The same principle, known as *Selective Oxidation*, was applied to prevent tarnishing changes of the copper-aluminium alloys sometimes used for cheap ornamental articles owing to their resemblance to gold; it was extremely effective. The process was also found to confer resistance to high-temperature oxidation. Price and Thomas calculated, on the basis of Wagner's equation (p. 848), that, if they could obtain an entirely copper-free alumina film, the value of k in the parabolic equation $dy/dt = k/y$ should be reduced to less than 1/80,000 that obtained on unalloyed copper. Their experiments showed that k was actually reduced to about 1/200,000 of the value for unalloyed copper—a remarkable confirmation of their predictions from theory. There is no doubt that the improvement was due to the treatment in damp hydrogen, since alloying alone—without damp hydrogen treatment—leads to far less satisfactory results. Fröhlich, studying similar alloys, had obtained a much smaller improvement over unalloyed copper (K. W. Fröhlich, *Z. Metallk.* 1936, **28,** 368).

Effect of Sulphur Dioxide on Oxidation of Copper. The idea that the action of sulphur is due to the introduction of lattice-defects into the oxide-film is supported by the observations of Mills on the oxidation of copper in the range 88–172°C. Statements made about the oxidation of copper in this range by various authors have sometimes been contradictory, probably due to varying surface conditions; some experimenters found parabolic growth under circumstances where others claimed a log law. Mills developed a method of cleaning his surfaces which gave reproducible results. He used anodic etching in 10% nitric acid, followed by washing in water and then in acetone; after wiping off a small amount of loose black matter with filter-paper, he subjected his specimens to cathodic treatment in sodium dihydrogen phosphate, and finally reduced any oxide-film still present by heating the specimens in hydrogen at 400°C., replacing it with nitrogen, reducing the temperature to the value chosen for the oxidation and then admitting oxygen or the gas mixture to be used in the experiment.

In absence of sulphur dioxide, the oxidation process at 88° consisted of two stages; film-growth started rapidly but gradually became slow—as usually occurs; then suddenly the rate of oxidation increased once more, gradually slowing off again. This was probably due to the fact that the first oxide-film was "pseudomorphic"—continuing the crystal-structure of the copper base; when a thickness of about 300 Å was reached, recrystallization set in, rendering the film suddenly pervious again. At 120°C., there was a similar break at about 300 Å, and both stages seemed to obey a log law. At 172°C., the curve showed no break; possibly the pseudomorphic film, if it ever existed, recrystallized at once. At this temperature, the curve was parabolic at first, but later oxidation became less rapid than the parabolic law would demand—probably owing to formation of cavities below the film. (A derivation of the new log law which was found to be obeyed is suggested on p. 836.)

The presence of sulphur dioxide modified the situation. At 88°C. it allowed the "break" (attributed to recrystallization) to occur earlier, probably because vacancies in the film-substance favour atomic rearrangement. At 172°C. the sulphur greatly accelerated oxidation, increasing the parabolic constant (k), and deferred the time at which the oxidation-rate dropped below the value calculated from the parabolic law; both effects are attributable to an increase in the number of vacancies, which would favour ionic movement through the film, and also its subsidence into cavities. Another effect of sulphur was the formation of a film of cuprous sulphate outside the oxide film; but that did not affect the oxidation-rate (T. Mills and U. R. Evans, *J. chem. Soc.* 1956, p. 2182).

Oxidation by Combustion Products. Oxidation of steel at high temperatures may be greatly accelerated if the oxidizing gas mixture contains sulphur dioxide. Hatfield found that the introduction of sulphur dioxide and moisture simultaneously into air greatly increased the oxidation-rate of steel at 900°C., although sulphur dioxide alone had little effect (W. H. Hatfield, *J. Iron St. Inst.* 1927, **115**, 483, esp. p. 486).

The hot gaseous mixture produced by the burning of fuel often causes rapid oxidation of steel. Naturally the oxidation-rate is greatest if free oxygen is present (i.e. if excess of air has been used for the combustion); oxidation may be slow or absent if the gaseous mixture is definitely reducing, although if sulphur compounds are present, hydrogen sulphide in a reducing mixture may then cause serious destruction to some materials—notably alloys rich in nickel.

Nevertheless—as shown by work in Cobb's laboratory—serious oxidation is possible even in mixtures containing no free oxygen or sulphur compound. If burnt in deficiency of air, coke or coal will yield a mixture of carbon dioxide and carbon monoxide diluted with much nitrogen; there will be some water vapour, but the amount of water is far greater if the fuel has been oil or town gas, which consists largely of hydrocarbons and hydrogen.

It is well known in blast-furnace reactions that a mixture of carbon monoxide and dioxide will only reduce oxide-ore to metallic iron if the CO_2/CO ratio is kept below a certain value, which depends on temperature; conversely, above that value, iron will be oxidized. The normal burning of coal will produce a gas mixture containing enough carbon dioxide to oxidize iron; this could be prevented by restricting the air-supply, but such restriction is to be condemned in the interests of fuel-economy and smoke-prevention. If small amounts of sulphur compounds are present, the oxidation due to combustion-products becomes more rapid. If coal is washed to remove the pyrites usually present before being burnt in a furnace, the wastage of furnace parts is reduced; it is, however, not eliminated, since some of the sulphur present in coal occurs in a form not capable of removal by washing. Sulphur in combustion-products can also accelerate corrosion of another kind where water condenses on relatively cold metal (p. 468).

Cobb's measurements of iron-wastage by gas mixtures representing the combustion-products of three types of fuel, with and without added sulphur dioxide, are shown in Table VI. It will be seen that the scaling-rate increases with the amount of water-vapour in the mixture, and is further increased by small additions of sulphur dioxide (H. C. Millett and J. W. Cobb, *Trans. Instn. Gas Engrs.* 1935–36, **85,** 610; 1936–37, **86,** 388; also J. W. Cobb, C. B. Marson and H. T. Angus, *J. Soc. chem. Ind.* 1927, **46,** 61T, 68T).

TABLE VI

WASTAGE OF IRON BY GAS MIXTURES (J. W. Cobb and H. C. Millett)

Mixture representing Products of Combustion from:—	Composition of Mixture (Percentage by volume)			Scaling Rate at 1000°C. (mg./cm.^{2} in 1 hour)			
	H_2O	CO_2	N_2	No SO_2	0·05% SO_2	0·10% SO_2	0·20% SO_2
Dry, high-temperature coke	2	18	80	4·4	11·2	14·6	18·6
Coal, oil or producer gas	10	10	80	7·5	17·5	22·5	26·5
Coal gas	20	10	70	12·0	21·5	27·8	32·0

The blistering of scale on steel during the reheating which precedes rolling into sheet form may affect the quality of the final product. Many years ago, Griffiths found that, under conditions where the heating of strip in pure oxygen at 850–1000°C. gave a non-adherent scale which flaked off as a sheet leaving only a thin oxide-film on the metal, heating in air often gave an adherent scale with local blisters; oxygen containing nitrogen exceeding 30% or carbon dioxide exceeding 50% also gave blisters. Air containing steam developed a smooth dark scale which could be easily stripped from the metal, leaving a silvery surface (R. Griffiths, *J. Iron St. Inst.* 1934, **103,** 377).

The causes of Griffith's observations deserve investigation both on account of their scientific interest and the practical advantage of producing a self-stripping scale. The action of steam may possibly be due to the production of hydrogen below the scale. The effect of nitrogen in air in altering the type of scale is not merely due to dilution of the oxygen, since pure oxygen at 1/5 atmosphere pressure produced no blistering. Presumably if once a local cavity below the scale is formed by the mechanism described on page 48, it will continue to grow, developing a scale-blister, unless the roof cracks or subsides. If only oxygen is present, the difference of pressure which must arise (since any oxygen leaking into the cavity will be used up in oxidizing the floor) will cause cracking or subsidence of the roof; fresh cavities will then be formed elsewhere, until the scale is largely undermined and capable of flaking off as a sheet. If the atmosphere is air, any leakage into the cavity will introduce nitrogen, which is not taken up in oxidizing the floor, and a large difference of pressure is avoided, leading to formation of a blister. A research designed to test this explanation might be well worth carrying out.

Oxidation of metal placed in the oxidizing part of a flame introduces factors not operative in oxidation by combustion-products coming from such a flame, since at the high temperature of a flame molecules and free radicles can exist, of a kind not met with at low temperatures; these include C_2, CH, OH, H and O. They are not in chemical equilibrium with one another, and in some flames they are not even in thermal equilibrium. All sorts of reactions with metals would here appear possible, and research on the subject would be welcome (D. T. A. Townend, *J. Iron St. Inst.* 1939, **139,** 553P; A. G. Gaydon and H. G. Wolfhard, " Flames " (Chapman & Hall); G. Porter, *Endeavour,* Oct. 1957, p. 224, esp. p. 228).

Protective Atmospheres. The mixture produced by burning coal gas or oil in air will oxidize iron if the H_2O/H_2 ratio exceeds a certain value. If this mixture is used as a source of heat in annealing or heat-treatment, either the amount of unburnt hydrogen must be increased—which is uneconomical—or the water must be removed. " Protective atmospheres " for use in these processes can be made by burning gas with a limited supply of air, so that hydrogen remains in excess, and then removing the water vapour; an alternative method is to burn the gas with the correct amount of air and later to add unburnt gas. The subject of " protective atmospheres " has been authoritatively discussed by Ivor Jenkins, " Controlled Atmospheres for the Heat Treatment of Metals " 1946 (Chapman & Hall); see also H. A. Fells, *Trans. Instn. Gas Engrs.* 1933–34, **83,** 598.

The oxidation of iron by the combustion products of gas is greatly increased by the presence of sulphur dioxide—as shown by the figures of Cobb (p. 79), but sulphur compounds become particularly undesirable in the annealing of copper and nickel alloys where it is desired to avoid tarnishing to bright surfaces. An atmosphere suitable for bright annealing of nickel must be free from sulphur; if town gas is used, the organic sulphur compounds present must be removed; processes are described by H. J. Hartley and E. J. Bradbury, *J. Inst. Met.* 1951–52, **80,** 297.

However, it is better to start with a fuel free from sulphur. In regions where a supply of butane or pentane is commercially available, these hydrocarbons constitute a suitable sulphur-free source. More general use is made of the mixture of nitrogen and hydrogen obtained by *cracking* ammonia, i.e. decomposing it into its elements by passing over a catalyst (such as spongy iron ore containing oxides of the rare earths as promoters); alternatively a mixture more rich in nitrogen can be obtained by *burning* ammonia with a suitable quantity of air. Water should be removed as far as possible from these atmospheres, silica gel being sometimes employed. Burnt ammonia is cheaper than cracked ammonia, since most of its nitrogen comes from the air; it can be used as a protective atmosphere for steel in situations where cracked ammonia would normally be too expensive.*

Atmospheres rich in hydrogen may be inadvisable for the treatment of copper of the grades which contain oxide between the grains, as the internal generation of steam can lead to cracking. This matter has been studied by W. E. Ruder, *Trans. electrochem. Soc.* 1916, **29,** 515; N. B. Pilling, *J. Franklin Inst.* 1918, **186,** 373; C. E. Ransley, *J. Inst. Met.* 1939, **65,** 147; 1940, **66,** 175. E. Mattsson and F. Schückher, *ibid.* 1959, **87,** 241.

Carbon monoxide should also be kept low where oxide-containing copper is to be treated, since the reduction of copper oxide to porous copper is undesirable and may leave paths for the entry of hydrogen (W. C. F. Hessenberg and E. C. Mantle, *Metal Ind.* (*Lond.*) 1953, **83,** 279, 301, 323, 363, 377).

Catastrophic Oxidation. Two elements, molybdenum and vanadium, can, in certain circumstances, cause sensational destruction. Both are common constituents of certain alloys much used by engineers on account of their good mechanical properties; molybdenum, in particular, is valuable in conferring creep-resistance. Steels containing 0·5% molybdenum, with or without chromium, are widely used for steam-pipes and super-heaters. Steels containing more molybdenum are available, but the additions must be regulated; contents beyond 2–3% become dangerous at temperatures where MoO_3 would be volatile and disrupt the scale—despite the fact that higher amounts of molybdenum would improve creep resistance further (H. T. Shirley, *Chem. and Ind.* (*Lond.*) 1954, p. 425).

Vanadium can be introduced from an external source, since its compounds are important constituents of the ashes of certain oil-fuels. If these ashes come into contact with hot steel, catastrophic destruction may occur, especially if sodium sulphate (and perhaps chloride) are present.

Such effects are commonly attributed to the " catalytic " action of vanadium (or molybdenum), but the use of the word " catalytic " provides no explanation and may even leave the wrong impression. " Catalysis " means the acceleration of a *chemical* reaction produced by a small amount of some substance which remains unchanged. In situations where oxidation is controlled by a *physical* process (e.g. movement through the film, whether of cations outwards or anions inwards), a catalyst cannot directly speed up the destruction, since the quasi-chemical boundary reactions can keep up with the movement through the film even when no catalyst is present. In such cases we seem to be dealing with an " opening agent "—a substance which renders the film more pervious. Only when the movement through

* J. N. Wanklyn points out that in some circumstances ammonia, at least when " cracked ", can " nitride " steel, and should not therefore be regarded as a completely inert atmosphere.

the film has been facilitated, either by the provision of pores filled with gas or liquid, or by the increased number of lattice-defects in a solid film-substance, can any catalytic acceleration of the boundary reaction come into play. Given an open film, the catalytic effect of vanadium in speeding up the oxidation of sulphur dioxide to trioxide may be genuinely important, since sulphur trioxide is itself corrosive; but it is doubtful whether catalysis plays a part in the combination of the metal with oxygen.

The main destructive effect of the two elements would seem to be connected with the formation of liquid or gaseous compounds. At furnace temperatures, the most stable oxide of molybdenum (MoO_3) is volatile; the most stable oxide of vanadium (V_2O_5) and several other vanadium compounds are liquid at comparatively low temperatures; various low-melting eutectics are formed by the compounds of both metals.

Rathenau and Meijering have shown that, if a furnace atmosphere comes to contain MoO_3 (perhaps owing to the presence of molybdenum wire), the oxidation of aluminium-copper alloy (8% copper) is appreciably enhanced at 470°C.—despite the low vapour pressure of MoO_3; it suddenly becomes very marked about 530°C.—close to the temperature at which a MoO_3–Cu_2O eutectic appears (a mixture of MoO_2, MoO_3 and Cu_2O can develop a liquid phase at 470°C.). The formation of liquid is the cause of corrosion, which tends to penetrate along grain-boundaries (G. W. Rathenau and J. L. Meijering, *Metallurgia* 1950, **42,** 167; *Nature* 1950, **165,** 240).

It is stated that the dangers due to molybdenum can be mitigated by the presence of silicon and chromium in the alloy. The two elements, present together, give far better effects than either used separately, and the same is true of aluminium and chromium. The reason may, however, merely be that the combination favours better forging (H. T. Shirley and L. B. Pfeil, *Chem. and Ind.* (*Lond.*) 1954, p. 425).

The facts regarding vanadium have been established by a number of investigators. Useful references include P. Schläpfer, P. Amgwerd and H. Preis, *Schweiz. Arch. angew. Wiss.* 1949, **15,** 291; A. de S. Brasunas and N. J. Grant, *Trans. Amer. Soc. Met.* 1952, **44,** 1117; G. W. Cunningham and A. de S. Brasunas, *Corrosion* 1956, **12,** 389t; W. E. Young and A. E. Hershey, *ibid.* 1957, **13,** 725t; F. C. Monkman and N. J. Grant, *ibid.* 1953, **9,** 460; W. R. Foster, M. H. Leipole and T. S. Shevlin, *ibid.* 1956, **12,** 539t; G. Lucas, M. Weddle and A. Preece, *J. Iron St. Inst.* 1955, **179,** 342; S. H. Frederick and T. F. Eden, *Proc. Instn. mech. Engrs.* 1954, **168,** 125; L. B. Pfeil, *Trans. Inst. Marine Engrs.* 1954, **66,** 169; W. Betteridge, K. Sachs and H. Lewis, *J. Inst. Petroleum* 1955, **41,** 170; E. Fitzer and S. Schwab, *Berg-Hüttenmann Monatshefte* 1953, **98,** No. 1, p. 7; G. H. Tandy, *J. appl. Chem.* 1956, **6,** 68; S. K. Coburn, *Corrosion* 1956, **12,** 122t; also general discussion, *J. Iron St. Inst.* 1956, **182,** 195–199.

Foster and his colleagues state that vanadium pentoxide (V_2O_3) is rarely found, as such, in oil ashes, but in cases where sodium sulphate is present, this interacts with V_2O_5 to give a number of complex vanadates, some of which appear to be more corrosive than simple $NaVO_3$ or V_2O_5. Of the mixtures examined, that with 12% sodium sulphate and 88% vanadium pentoxide is found to be the most corrosive; on heating, SO_3 is driven off and a corrosive complex, probably $Na_2O.V_2O_4.5V_2O_5$ is left; the vanadate $Na_2O.3V_2O_5$ also appears to be very corrosive, whilst attack by the SO_3 expelled is also possible. The product formed by interaction between V_2O_5 and Na_2SO_4 is liquid well below the melting-point (680–690°C.) of pure V_2O_5.

In a certain gas-turbine where the inlet temperature was 650°C., the product formed by interaction of V_2O_5 and Na_2SO_4 was a liquid, and severe corrosion from the ash-particles resulted. Below the melting- or softening-point of the ash, there is little trouble on steels in such situations. That is not true, however, of copper-base alloys (such as aluminium bronze, with about 9% aluminium); these suffer at lower temperatures—which suggests that other influences may be at work. It has been shown by researches in several laboratories that vanadium compounds can sometimes increase the oxidation of various alloys at temperatures too low to permit of a liquid phase. Possibly the presence of vanadium in the scale increases the number of lattice-defects, in accordance with Hauffe's rule (p. 61); vanadium and molybdenum will normally exert valencies higher than that of the main constituent of the scale.

If the only cause of the trouble were the presence of liquid constituents, a remedy would suggest itself, namely the addition of substances calculated to raise the melting-point, and much experimentation has been carried out on these lines. There has been a modicum of success, but opinions differ as to which additive is best. Some regard zinc oxide as the most promising; others prefer lime or barium oxide. Thermodynamic calculations by Young and Hershey (p. 82) show that magnesia would be good in absence of sulphur, owing to the high melting-point produced. In practice, the fuels almost always contain sulphur, and the magnesia would be converted to sulphate at the temperatures and pressures involved in gas-turbine service. Moreover there is reason to believe that more than one cause is operative; if so, a remedy is only to be found if *all* causes are provided for.

Preece (p. 82) considered that no heat-resisting alloy depending on chromium for its resistance can withstand oxidation in presence of vanadium pentoxide. Others regard this view as unduly pessimistic and report fairly good service results with steel containing 26% chromium or even less. (See H. T. Shirley, G. Burns and others, also reply by A. Preece, *J. Iron St. Inst.* 1956, **182,** 195–199.)

There is little doubt, however, that stainless steel of ordinary chromium content does fare badly, especially if the vanadium oxide is mixed with sodium sulphate. Cunningham and Brasunas (p. 82) picture the situation on such alloys in the manner suggested by fig. 15. The scale is porous, and has a layer of molten oxide below it. The Na_2SO_4–V_2O_5 mixture can absorb oxygen, and pass it on to the metal; the absorptive power is greatest in mixtures containing about 16–20% sodium sulphate—and those are the most corrosive. These authorities consider the oxides of calcium, strontium or barium to be the best additives.

Austrian researches by Fitzer and Schwab (p. 82) suggest that an answer may be found in alloys containing silicon; these seem, however, to be difficult to work. Hopeful results from Nimonic alloys (p. 71) have been reported by K. Sachs, *Metallurgia* 1958, **57,** 167. Attempts to prevent oil-ash oxidation of stainless steel by coatings of aluminium produced by hot-dipping have given disappointing results (J. E. Srawley, *Corrosion* 1958, **14,** 37t). An assessment of the present situation is provided by a Task Group of the National Association of Corrosion Engineers (*Corrosion* 1958, **14,** 369t).

Apart from a metallurgical solution of the problem, there remains the possibility of removing vanadium from the fuel oil, but this is unlikely to be easy, as the vanadium occurs as a stable " porphyrin ", belonging to that group of compounds of which haemoglobin and chlorophyll are the best known examples.

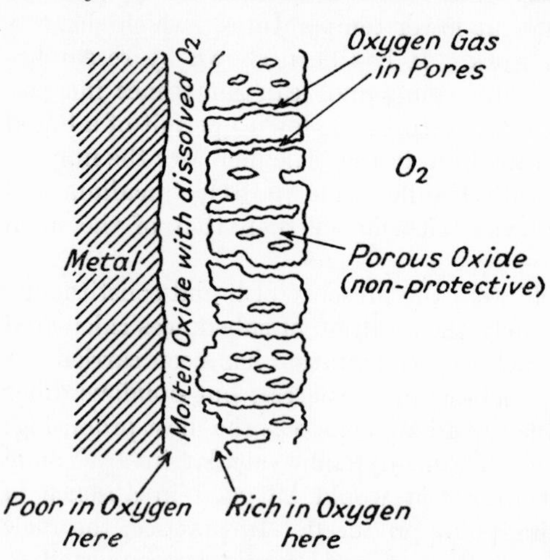

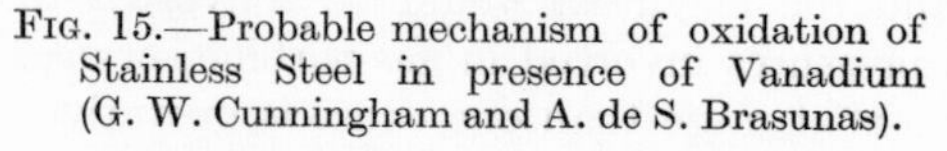

Fig. 15.—Probable mechanism of oxidation of Stainless Steel in presence of Vanadium (G. W. Cunningham and A. de S. Brasunas).

Probably some cases of corrosion by ash have been ascribed to vanadium which are really due to other causes. Shirley points out that sodium or calcium sulphate, although almost harmless as ash-constituents by themselves, produce damage if a small amount of chloride is present. Probably the sulphate is reduced to sulphide. Air-heater tubes and stator blades can thus suffer. A Nimonic alloy with 74% nickel, 20% chromium and 2% titanium develops small spheres of a eutectic containing Ni, NiS and NiO; no vanadium is necessary for this form of attack (H. T. Shirley, *J. Iron St. Inst.* 1956, **182,** 144).

An excellent summary of the parts played by molybdenum and vanadium in catastrophic oxidation is provided by L. L. Shreir, *Rep. Progress App. Chem.* 1955, **40,** 149.

Growth of Iodide-films. In his classical work, Tammann studied the development of colours on silver exposed to iodine vapour at room temperature, and reached the conclusion that the thickening of the silver-iodide film obeyed the parabolic law; his method of obtaining the thickness from the colour was not very accurate, but the general conclusions were probably correct. Later Bannister, working at Cambridge, studied the growth of silver-iodide films on silver immersed in a solution of iodine in chloroform or another organic solvent. Like Tammann, he used colour to indicate thickness, but before starting his measurements, he calibrated the colour-scale—which extended over five " orders "—against measurements of thickness obtained gravimetrically; since the gravimetric method was itself found to agree with measurements obtained by two other methods,* confidence can be placed in his results. He found that, with a chloroform solution of any given concentration, the thickening did indeed follow the parabolic law, but that the value of the constant, k, varied with the concentration, and that the use of a different solvent affected the curve. His results are best explained on the assumption that iodine moves inwards through the film—not silver outwards. Bircumshaw and Everdell, who made a detailed study of the behaviour of copper in aqueous solutions of iodine, also formed the opinion that iodine was passing inwards (G. Tammann, Z. *anorg. Chem.* 1920, **111,** 78; G. Tammann and W. Köster, *ibid.* 1922, **123,** 196. U. R. Evans and L. C. Bannister, *Proc. roy. Soc.* (*A*) 1929, **125,** 370. L. L. Bircumshaw and M. H. Everdell, *J. chem. Soc.* 1942, p. 598).

Russian work on the action of chlorine gas, wet and dry, on metals is reported in *Corros. Prev. Control* 1957, **4,** April, p. 49, esp. p. 50. The action of iodine on a tantalum wire partly covered with silver, which provides evidence of local cell action, is described by C. Ilschner-Gensch and C. Wagner, *J. electrochem. Soc.* 1958, **105,** 198.

Thickness of Films responsible for Interference Colours. The thicknesses of silver-iodide films extending over five orders of colours as determined by Bannister, are shown in Table VII. The surface was not smooth and numbers represent the *Mean General Intercepts*, i.e. the mean of the intercepts which would be made by the film on a vast number of lines drawn perpendicular to the plane of the surface as a whole (p. 55); no account has been taken of the roughness factor. It should be noted that each colour appears over a considerable *range* of thickness; the figures given in the Table represent, as nearly as possible, the centre of the range.

TABLE VII

THICKNESS OF IODIDE-FILMS ON SILVER (U. R. Evans and L. C. Bannister)

Colour	Thickness
Yellow I	200 Å
Red I	430
Blue I	550
Silvery Hiatus	800
Yellow II	1150
Red II	1650
Blue II	1950
Green II	2250
Yellow III	2450
Red III	2900
Green III	3400
Red IV	4100
Green IV	4750
Red V	5600

* These methods were the measurement of the quantity of electricity needed for the cathodic reduction of the silver iodide to silver, and the nephelometric measurement of the iodine liberated into the solution; further details are given on pp. 788, 975.

Other References

The text-books mentioned on p. 58 include the oxidation of alloys, but reference should also be made to the experimental papers of two of their authors; O. Kubaschewski (with A. Schneider and O. von Goldbeck), *J. Inst. Met.* 1948–49, **75,** 403; 1949, **76,** 255; *Metalloberfläche* 1953, **7,** A113; 1954, **8,** A33; K. Hauffe, *Werkst. u. Korrosion* 1951, **2,** 131, 221, 243; *Metall.* 1950, **4,** 462; *Metalloberfläche* 1954, **8,** A97; also (with H. J. Engell and H. Pfeiffer), *Z. Elektrochem.* 1952, **56,** 366, 390; *Z. Metallk.* 1952, **43,** 364.

Useful résumé papers regarding materials suited to withstand the drastic conditions prevailing in gas turbines and similar situations are provided by E. W. Colbeck and J. R. Rait, *Iron St. Inst. Spec. Rep.* 43 (1951) and by L. B. Pfeil, *Chem. and Ind.* (*Lond.*) 1954, p. 425; 1955, p. 208. A survey of the high-temperature behaviour of chromium-rich alloys has been provided by A. H. Sully and E. A. Brandes, *J. Inst. Met.* 1952–53, **81,** 578. The creep-resistance and general properties of the Nimonic group of alloys (nickel-chromium with other elements), much used in gas-turbines, are discussed by D. C. Herbert and D. J. Armstrong, *Engineering* 1953, **175,** 605, and L. B. Pfeil, *ibid.* 1954, **177,** 620. Materials designed to resist attack on steel by high-pressure hydrogen (which attacks mild steel at 200°C. at 350 atm. and at room temperature at 4000 atm.), are discussed by N. P. Inglis, " Materials of Construction in Chemical Industry " (Soc. Chem. Ind.) April 1950, p. 86. Further information on the subject is due to J. D. Hobson and C. Sykes, *J. Iron St. Inst.* 1951, **169,** 209, and to G. A. Nelson, *Trans. Amer. Soc. mech. Engrs.* 1951, **73,** 205.

Many of the researches of E. A. Gulbransen, with J. W. Hickman, K. F. Andrew and W. R. McMillan (*Rev. Métallurg.* 1948, **45,** 181, 287; *Trans. Amer. Inst. min. met. Engrs.* 1949, **180,** 534; *Industr. engng. Chem.* 1953, **45,** 1734; *J. electrochem. Soc.* 1954, **101,** 163) concern alloy systems, particularly nickel-chromium—as already mentioned on p. 72. Other electron-diffraction examinations of alloys containing iron, nickel and chromium are due to I. Iitaka, I. Nakayama and K. Sekiguchi, *J. Sci. Res. Inst.* (*Japan*) 1951, **45,** 57. The work of Preece deserves study both in its scientific and industrial aspects (see especially A. Preece and G. Lucas, *J. Inst. Met.* 1952–53, **81,** 219; J. P. Dennison and A. Preece, *ibid.* 1952–53, **81,** 229; J. P. Dennison, *ibid.* 1953–54, **82,** 117).

Summaries of five researches on the oxidation of the alloy systems U–Zr, Co–Cr, Ni–Mn, Fe–Ni and of chromium will be found in *J. electrochem. Soc.* 1956, **103,** 203C. Micro-balance experiments, relating the oxidation-resistance of alloys with the equilibrium diagrams, are described by P. Spinedi, *Metallurg. ital.* 1957, **40,** 363. Recent papers on iodide films include J. L. Weininger, *J. electrochem. Soc.* 1958, **105,** 577; D. M. Smyth and M. Cutler, *ibid.* 1959, **106,** 107. For sulphide films on copper, see J. Oudar and J. Bénard, *Acta. Met.* 1959, **7,** 295. For behaviour of magnesium in carbon dioxide, see M. L. Boussion, R. Darras and D. Leclerq, *Rev. Met.* 1959, **56,** 61.

CHAPTER IV

ELECTROCHEMICAL CORROSION

Synopsis

It is first explained that electrochemical corrosion in a salt solution with access of oxygen can produce relatively rapid destruction of metal even at ordinary temperatures, where simple oxidation would build a film and therefore stifle itself. This occurs because the electrochemical mechanism, where the anodic and cathodic part of the matter are spatially separated, provides a way of converting metal to oxide, hydroxide or basic salt without the production of an obstructive film. Then a description is given of the behaviour of various metals immersed partially or completely in various salt solutions; it is emphasized that attack often starts from surface defects and that the " corrosion-pattern " produced depends greatly on surface condition. It is shown how hard natural waters develop on steel a clinging, partially protective film of " chalky rust ". Corrosion by distilled water containing oxygen is next discussed; here the growth of protective films is often slow and incomplete, so that pitting is sometimes met with—e.g. on zinc. Reference is made to the fact that certain soft natural waters take up sufficient lead from lead pipes to become dangerous as drinking supplies.

The last part of the chapter is devoted to four influences which are important in deciding the corrosion-pattern. These are (1) the action of internal stresses left after scratching, and surface defects generally, in providing starting-places for attack; (2) the mutual protective effect which causes attack in progress at one part to depress the probability of its initiation at points around. (3) The various factors which decide between spreading, healing or continued local attack (pitting); and (4) differential aeration, a paradoxical effect of unequal oxygen distribution; it is observed in certain cases that oxygen is needed for corrosion, but that attack is largely directed on those regions which are unfavourably situated for oxygen replenishment.

Qualitative evidence is presented for the part played by electric currents in wet corrosion, but the measurement of those currents —and the demonstration that they are equivalent to the corrosion-rate—must be deferred to the quantitative section (Chapter XXI).

Corrosion by Salt Solutions with Access of Oxygen

Distinction between Simple Oxidation and Electrochemical Corrosion. In Chapter II it was shown that oxidation is really an exchange of electrons between oxygen and metal. The "simple oxidation", which occurs on exposure to oxygen at elevated temperatures leads to a film, and usually the attack—fast at the outset—becomes slower as the film thickens. At low temperatures, the rate of simple oxidation becomes negligible before the oxide even becomes visible. However, in the presence of an aqueous salt solution, a different (electrochemical) mechanism is possible, which allows oxygen to take up electrons at one part of the surface, whilst the metal gives them up at another, thus avoiding any accumulation of electric charge; in many salt solutions, no film is formed, so that the attack proceeds unretarded. The "electrochemical" type of corrosion is important at low temperatures, just as the film-forming type is important at high temperatures. This may be made clear by some simple examples of the behaviour of various metals partly or wholly immersed in various liquids.

Zinc partly immersed in Potassium Chloride Solution. If a rectangular zinc plate is partly immersed in potassium chloride solution, either clamped vertically (fig. 16(A)) or placed sloping in a narrow beaker

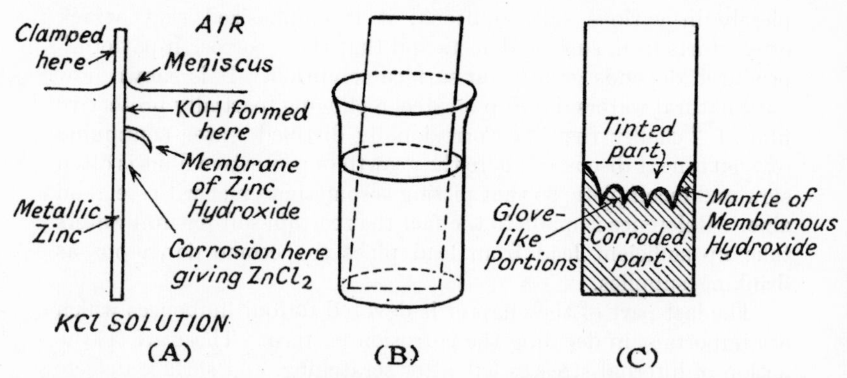

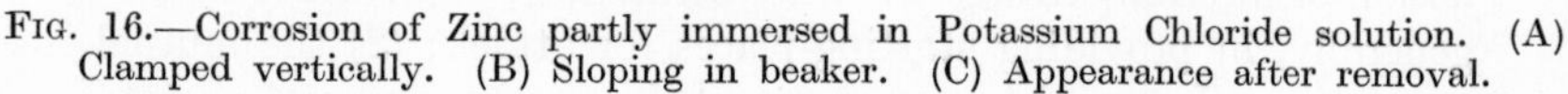
Fig. 16.—Corrosion of Zinc partly immersed in Potassium Chloride solution. (A) Clamped vertically. (B) Sloping in beaker. (C) Appearance after removal.

(fig. 16(B)), the lower portion soon becomes attacked; there is usually a zone just below the water-line which remains unattacked, although after some time it develops weak interference tints; these can best be seen after the specimen has been taken out, washed and dried (fig. 16(C)).

The explanation is simple. Just below the water-line oxygen from the air uses up electrons in a "cathodic reaction", which proceeds in steps but leads to a final result most simply expressed thus:—

$$O + H_2O + 2e = 2OH^- \quad \text{(Cathodic Reaction)}$$

The cathodic reaction uses up electrons, and thus makes possible the entry of metal into the liquid as cations (atoms deficient in electrons)—lower down the specimen; this may be written

$$Zn = Zn^{++} + 2e \quad \text{(Anodic Reaction)}$$

If the cathodic (electron-consuming) and anodic (electron-providing) reactions occur at the same rate, no electric charge will accumulate. Since the main ions present in the solution are K^+ and Cl^-, we can regard the cathodic and anodic products as potassium hydroxide (KOH) and zinc chloride ($ZnCl_2$) respectively. Both are freely soluble, and no film will be formed over the lower part of the specimen, so that the corrosion, once started, will continue at almost constant rate. The constancy of the rate is well shown by the results of Borgmann, reproduced in fig. 17; it will be observed that alloying constituents in the zinc produce only small changes in the corrosion-rate—in contrast to the attack by sulphuric acid, discussed on p. 310, where minor constituents have a marked effect (for details, see C. W. Borgmann and U. R. Evans, *Trans. electrochem. Soc.* 1934, **65**, 249; cf. G. Bianchi, *Corrosion* 1958, **14**, 245t).

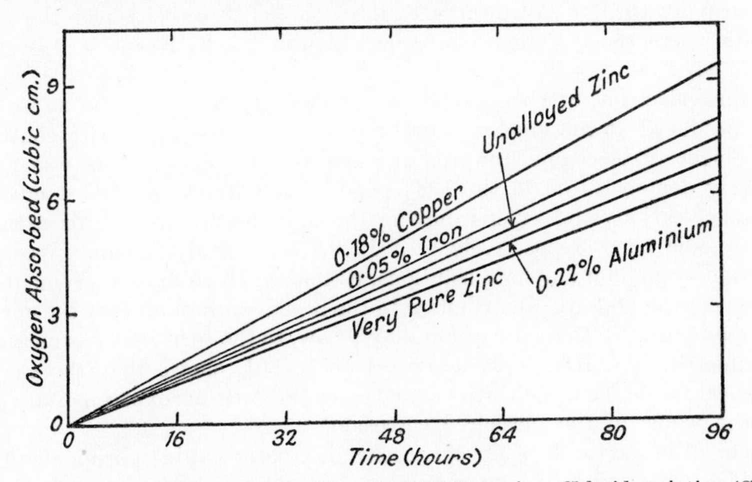

FIG. 17.—Corrosion of Zinc and its Alloys in 0·1N Potassium Chloride solution (C. W. Borgmann and U. R. Evans).

At the places where the OH^- and Zn^{++} meet, sparingly soluble bodies, essentially zinc hydroxide or basic chloride, appear;* owing to the position of formation, they will not interfere with attack. Only at the zone just below the water-line, where OH^- is in excess, will a solid film (oxide or hydroxide) be formed in contact with the metal, accounting for the slowly developing interference colours. This film will hinder its own growth, and does not represent any appreciable destruction of metal; its formation, however, is important, since the cathodic reaction has to take place, not on a clean metallic surface, but on an oxidized surface, and this must affect the rate at which the corrosion process can proceed. Probably there are only

* Under certain conditions, a basic chloride or even zinc oxide may be formed. See W. Feitknecht (with H. Weidmann and E. Häberli), *Helv. chim. Acta*, 1943, **26**, 1911; 1949, **32**, 2294; 1950, **33**, 922. Also Dissertations (Berne) by his collaborators, R. Petermann (1945) and F. M. Aebi (1946). Czech work suggests that the product formed on zinc in KCl varies between $Zn(OH)_2$ and $4Zn(OH)_2.ZnCl_2$, the $ZnCl_2$ content increasing as the KCl concentration rises (I. Sekerka and K. Smrček, *Collect. Czech. Chem. Comm.* 1957, **22**, 712). In the present section the word " hydroxide " is used to include basic salt.

relatively few places on an oxide surface having the catalytic character suited to the cathodic process, and, notwithstanding the proximity of an ample oxygen supply, the corrosion process is limited by the rate of replenishment of oxygen at those points.

Many of these statements are based on direct experiment. The formation of alkali near the water-line and of zinc chloride lower down have been demonstrated by chemical tests, and the formation of zinc hydroxide (or basic salt) where they meet is directly visible. Part of this hydroxide is formed at a distance from the metal and falls to the bottom of the containing vessel as a white flocculent precipitate, but close to the metal it builds a membraneous wall or " mantle " starting out at right angles to the metal, and usually turning downwards to form glove-like screens over the peak-points where the corroded area comes highest (U. R. Evans, *J. Inst. Met.* 1923, **30**, 239).

Since electrons are being taken up at the cathodic zone (the water-line) and produced at the anodic zone (lower down), there must be a continual movement of electrons upwards through the metal, and this was qualitatively demonstrated in the 1923 paper. Simultaneously there must be trains of cations and anions moving through the liquid towards cathodic and anodic zones respectively. In 1939, Agar, studying zinc in sodium chloride or sodium sulphate solutions, estimated these ionic movements by observing the potential distribution in the liquid, and found that the current was equivalent, in the sense of Faraday's Law, to the corrosion-loss measured gravimetrically. His work, described on p. 861, shows that the electrochemical mechanism, indicated above, accounts, within experimental error, for the whole of the measured corrosion.

The ultimate result of the electrochemical corrosion is to bring about the combination of zinc, oxygen and water, to form zinc hydroxide. It is thus a form of oxidation, but differs from simple oxidation in that oxygen is consumed at one place (the water-line), the zinc at another place (the anodic zone, lower down), whilst the zinc hydroxide appears at a third place. It is this spatial separation of the three parts of the process that allows electrochemical attack to continue unchecked even at room temperatures, where simple oxidation, with its production of an obstructive film, would stifle itself.

Corrosion currents can be brought about in various ways. If a specimen consists of two different metals, one will be anode and the other cathode; this is known as *galvanic corrosion*. Where the specimen is chemically and physically uniform, currents can be generated if differences are maintained in the composition of the liquid covering different areas; in the example given above, there is better replenishment of oxygen in the meniscus zone than elsewhere, and the currents thus generated are known as *differential aeration* currents (p. 127).

The absence of stifling in electrochemical corrosion is conditional on the anodic and cathodic products being freely soluble; in cases where either product is sparingly soluble, corrosion is generally slow and frequently almost absent—as shown in the next chapter.

Iron partly immersed in Potassium Chloride Solution. The case of iron differs from that of zinc in that ferrous hydroxide ($Fe(OH)_2$ or $FeO.H_2O$), which might be expected to appear as the secondary corrosion-product, is relatively soluble and highly unstable, readily taking up oxygen. If the experiment is carried out in a broad vessel, the solid precipitate is $FeO(OH)$ or $Fe_2O_3.H_2O$, the familiar yellow-brown " rust ". If the beaker is narrow, or the oxygen-content of the atmosphere low, much of the solid corrosion-product will be black magnetite (Fe_3O_4). Sometimes green bodies appear—which may be intermediate hydroxides or basic salts; in either case they contain both ferrous and ferric iron; these ferroso-ferric compounds have been studied by W. Feitnecht and G. Keller, *Z. anorg. Chem.* 1950, **262**, 61.

In general, partly immersed iron specimens develop a well-defined mantle of membraneous rust separating the attacked from unattacked zones. The interference tints seen on the unattacked zone, after the specimen has been taken out and dried, are brighter on iron than on zinc; often they are brighter than the tints obtained on iron by heating in air.

The quantitative establishment of the electrochemical mechanism in the case of iron was accomplished by Hoar in 1932. His method was different from that afterwards used by Agar on zinc. Measurements of potential drop at the surface allowed the electric current to be estimated, and, as in the case of zinc, its strength was found to be equivalent to the corrosion-velocity; details will be found on p. 863, or in the original paper (U. R. Evans and T. P. Hoar, *Proc. roy. Soc.* (*A*) 1932, **137**, 343).

Hoar found that, even on a steel specimen where a considerable fraction of the area remained immune from attack, most of the cathodic reaction was occurring in the meniscus zone, where a good supply of oxygen was available to replenish those spots catalytically suitable to the cathodic reaction. The cathodic reaction seemed to control the situation; the corrosion-rate was found by Borgmann to be much faster when oxygen, instead of air, was the gas phase.

Conversely, if oxygen is absent altogether from the gas phase, there is very little attack. The only possible cathodic reaction is then the production of hydrogen by a reaction

$$2H^+ + 2e = H_2$$

which proceeds sluggishly in neutral solutions (it can be rapid in acid solutions where hydrogen ions are plentiful). In contrast to corrosion in presence of oxygen, attack by a salt solution extends right up to the water-line when the gas above the liquid is nitrogen or hydrogen (U. R. Evans and C. W. Borgmann, *Z. phys. Chem.* (*A*) 1934, **146**, 153, esp. p. 162).

Factors determining the Corrosion-pattern on Iron and Zinc. The distribution of attacked and unattacked areas varies considerably with different samples of iron and zinc, and particularly with the surface condition. The physical structure of the metal appears more important than the chemical composition. Electrolytic iron, if abraded, develops patterns similar to some samples of abraded steel; corrosion occurs over streaks or arch-shaped areas which descend from a limited number of isolated spots. However, the behaviour varies with the previous treatment. Sand-blasted steel gives a systematic array of little rust-tufts; a mantle is formed, separating a highly corroded zone below it from an only slightly corroded zone above it; the number of rust-tufts per unit area in the zone above the mantle is roughly the same as on the zone below it, but the rust-tufts are of microscopic dimensions, whereas those below the mantle are so large that they form an almost continuous blanket of rust.

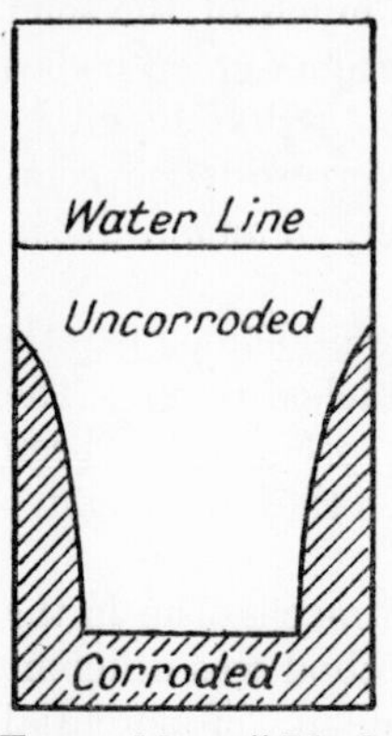

Fig. 18. — " Ideal Distribution " of corrosion on Steel (met with only under rather exceptional conditions).

On abraded steel, the pattern is different. One high-quality steel, specially prepared by W. H. Hatfield for the Cambridge researches, was found, if abraded and exposed to dry air before its partial immersion in N/2 potassium chloride, to be attacked only along the cut edges. This corrosion-pattern was named the *Ideal Distribution* (fig. 18), but it is not usually met with on ordinary steel; even on that specially prepared steel it appeared only at fairly high salt concentrations.

Both on iron and zinc it is rare to find corrosion confined to the cut edges. It is interesting to observe the behaviour of sheet zinc tested in the as-rolled condition in chloride or sulphate solution. An early research showed four types of attack (fig. 19), (1) *regional corrosion* covering an area bounded usually by a straight, horizontal line, (2) *point corrosion* starting at points, and streaming downwards as streaks (rolled zinc usually has one side brighter than the other, and point corrosion develops more quickly on the brighter side), (3) *edge-point corrosion* streaming down from starting-points situated on the cut edges (these were more closely spaced than the starting-

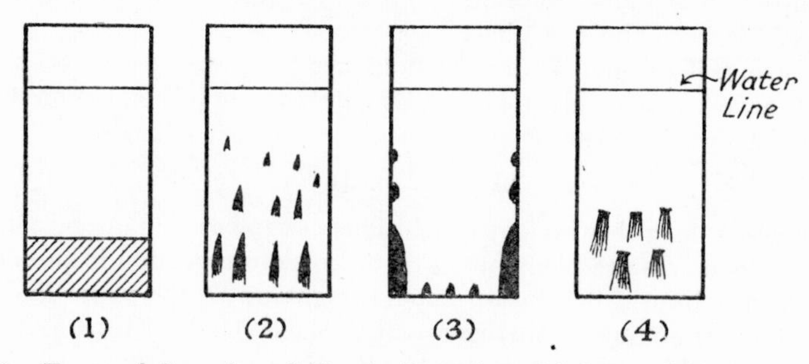

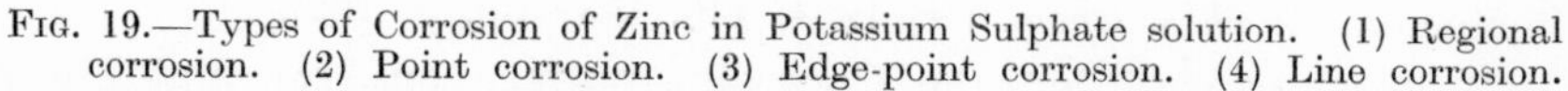

Fig. 19.—Types of Corrosion of Zinc in Potassium Sulphate solution. (1) Regional corrosion. (2) Point corrosion. (3) Edge-point corrosion. (4) Line corrosion.

points on the face), (4) *line corrosion* descending as streams from short lines —evidently scratch-marks; the scratches were usually visible, but only a few of the visible scratches present on the specimen seemed capable of acting as founts for line corrosion. Scratch-lines traced intentionally on the surface served as starting-places of corrosion, but the streaming descended, not from the whole length of the scratch-line, but merely from isolated points situated upon it (U. R. Evans, *J. Soc. chem. Ind.* 1926, **45**, 37T).

In these and other experiments, it was noticed that sulphate solutions produced a more adherent corrosion-product than chloride; on rolled zinc, the corrosion descending from a starting-point gives a broad arch-shaped area in sodium chloride, a narrower stream in potassium sulphate and a mere wisp in magnesium sulphate. Potassium sulphate solution is convenient for studying the starting-points of attack on zinc, since the corroding areas are left permanently whitened even after washing and rubbing, whereas the corrosion-product formed in chloride solution is loose and non-adherent. The different action of the two ions seems to be due to the better flocculating power of the divalent SO_4^{--} compared to the monovalent Cl^-; experiments on the titration of zinc salt solution with alkali showed that the amount of alkali needed to produce a definite precipitate was smaller when SO_4^{--} was the main anion instead of Cl^-. Thus when a plate of metallic zinc is placed in a sulphate solution containing a little dissolved oxygen, the small amount of alkali quickly formed by cathodic reaction over the entire face will precipitate some of the zinc ions emerging at weak points in physical contact with the metal, giving adherent white matter. However, the surface condition affects the character of the product, which is more diffuse after coarse than fine grinding; the behaviour of finely ground zinc in chloride resembles that of coarsely ground zinc in sulphate (U. R. Evans, *J. chem. Soc.* 1929, p. 111, esp. p. 115).

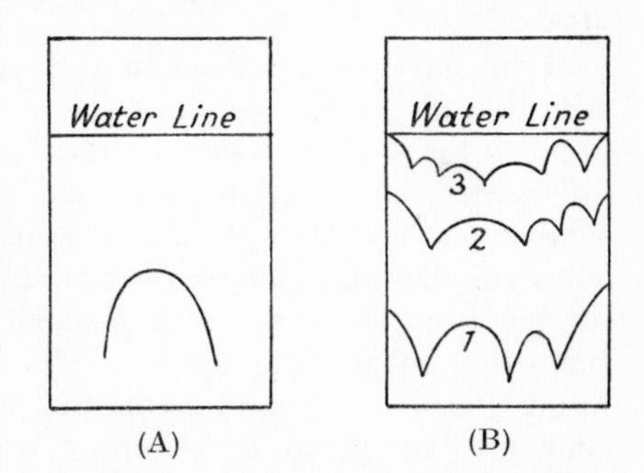

FIG. 20.—(A) Single arch-shaped area of attack. (B) Multiple mantles formed at progressively more advanced stages.

In a chloride solution, corrosion starting at a point usually spreads downwards and outwards, forming an arch-shaped area (fig. 20(A)). But, although corrosion *from any particular starting-point* extends *downwards*, attack *as a whole* can be said to extend *upwards* as time passes, since starting-points in the upper regions only become active after a considerable period. Probably this slow upward extension results from the progressive exhaustion of the orignal oxygen throughout the liquid; in the end, the cathodic reaction may be confined to the meniscus, where replenishment of oxygen from the air can continue. If the arch-shaped areas in the lower part join together to form one large corroding region, the upper edge of this represents the dividing line between the regions where alkali and metallic salts remain respectively in excess; along this indented line, the membraneous mantle of white zinc hydroxide or yellow rust grows out roughly at right angles to the metal. Sometimes a mantle will be formed low down in the early stage. followed later by a second and then a third mantle at successively higher levels; the formation of multiple mantles is illustrated in fig. 20(B) (U. R. Evans, *Industr. engng. Chem.* 1925, **17**, 363).

The highest mantle often touches the water-line at the two sheared edges of the plate and occasionally at one or more points in the centre; there is an element of chance here, since clearly a suitable starting-point may or may not exist very close to the water-line itself. If an arch does touch the water-line, corrosion at its apex is likely to be rapid, owing to the proximity of the source of oxygen for the cathodic reaction.*

A protected zone at the water-line usually persists even after many weeks, provided that the air (or oxygen) above the liquid is free from acidic substances, and provided that precautions are taken against evaporation; if carbon dioxide or sulphur dioxide is present to destroy the cathodically formed alkali, corrosion may soon extend up to the water-line at all points; if evaporation occurs, corrosion may *appear* to move up to the water-line, but what is really occurring is that the water-line falls to meet the corroding area.

Iron partly immersed in a Magnesium or Calcium Salt Solution. Iron in a magnesium sulphate solution containing oxygen starts to corrode in much the same way as in a potassium or sodium salt solution, but certain differences soon arise, due to the fact that magnesium hydroxide is not very soluble; on the outside of the arch-shaped areas, the cathodic reaction quickly starts to deposit solid magnesium hydroxide, which apparently excludes oxygen from the parts covered, and helps to set up an anodic reaction; the white magnesium hydroxide is then acted upon by the anodically formed ferrous sulphate, giving first a bright green compound (apparently similar to the green ferroso-ferric compounds already mentioned, except that some of the ferrous iron is replaced by magnesium); later a pale clinging rust, also containing magnesium, is formed by absorption of oxygen. At this stage, further magnesium hydroxide is being deposited at a higher level, and in turn sets up anodic attack, so that ultimately corrosion will extend almost to the water-line; in the final stage, a narrow band within the meniscus zone is covered with a white magnesium compound (probably hydroxide or basic carbonate), whilst the remaining area may be covered with clinging pale rust, differing from ordinary dark rust in possessing a slightly protective character. The corrosion-rate is found to be slow, much slower than that caused by potassium sulphate solution, and usually somewhat slower than that caused by the distilled water used to make the magnesium sulphate solution—evidently because the magnesium hydroxide hinders the replenishment of oxygen at points favourable to the cathodic reaction. Whatever attack does occur is, even in the early stages, fairly well spread out (U. R. Evans, *J. Soc. chem. Ind.* 1928, **47,** 55T, esp. p. 57T).

Thus at low temperatures magnesium salts can in some circumstances

* In the zone where oxygen can be replenished easily the "probability" of corrosion is lower than in areas remote from the oxygen-supply, but the intensity of attack, when it does occur, is greater. The analogy to the statistical experiments on drops exposed to different oxygen–nitrogen mixture (p. 138) will be evident.

be regarded as inhibitors of corrosion. However, under boiler conditions the acid formed by hydrolysis of magnesium salts is dangerous; magnesium chloride is an especially undesirable constituent of boiler water since the hydrogen chloride is volatile and can attack the relatively dry parts of the system. Moreover, an industrial case has been reported where a vessel containing cold magnesium salt solution suffered severe general attack, probably because acid fumes in the air prevented the formation of the white band which normally interferes with the cathodic reaction.

Fig. 21 reproduces some early curves which show that partly immersed steel corrodes more slowly in magnesium sulphate than in potassium sulphate, where the cathodic product is freely soluble; the rate is practically the same in potassium chloride as in potassium sulphate, and the particular steel used in the research of 1929 is corroded no faster than pure electrolytic iron (more recent work, however, showed that some steels corrode faster than others). Zinc definitely corrodes faster than steel (U. R. Evans, *J. chem. Soc.* 1929, p. 111, esp. p. 124).

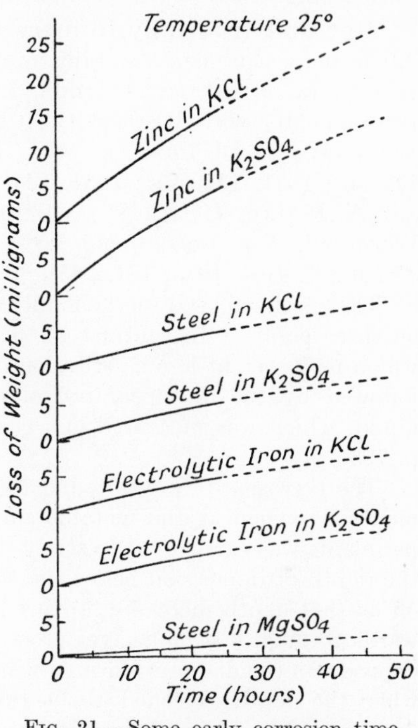

FIG. 21.—Some early corrosion–time curves (U. R. Evans).

Natural waters containing calcium bicarbonate and oxygen also produce a clinging rust—provided that they contain only just sufficient carbonic acid to keep the calcium carbonate in solution; in such a water, the small rise in the pH value produced by the cathodic reaction suffices to throw down adherent calcium carbonate, which then interacts with anodically formed ferrous products and oxygen to form a pale clinging deposit, containing calcium as well as iron; ultimately only the narrow meniscus zone, little more than 1 mm. broad, remains covered with white calcium carbonate. In such cases, the corrosion-rate may be much lower than in a sodium salt solution and, being well distributed, the attack is relatively harmless. At one time, the "chalky rust" was regarded as a mixture, but Haase's researches suggest that a double calcium iron carbonate is present (L. W. Haase, "Werkstoffzerstörung und Schutzschichtbildung im Wasserfach" 1950 (Verlag Chemie)).

The presence of calcium bicarbonate in many hard natural waters often causes them to corrode steel more slowly than the average soft water, but the rapid attack associated with soft waters is also due to the fact that soft waters are so often acid waters. The presence of calcium and magnesium in sea-water is one reason why sea-water is less corrosive than sodium chloride solution, as suggested on p. 165.

Many natural waters contain both calcium and magnesium. In hard fresh waters from a limestone source, the calcium-content will exceed the magnesium-content, whereas in sea-water the magnesium exceeds the calcium. It is not always certain whether it is the calcium carbonate present in the " chalky rust " or the magnesium hydroxide which is slowing down the cathodic reaction, and a research to decide this point in different types of waters might have useful results (see also p. 165).

Zinc and Iron fully immersed in Potassium Chloride Solution. Accurate measurements on fully immersed specimens extending over several years were carried out at Teddington by Bengough, first with Stuart and Lee and later with Wormwell (G. D. Bengough, J. M. Stuart, A. R. Lee and F. Wormwell, *Proc. roy. Soc.* (*A*) 1927, **116**, 425; 1928, **121**, 88; 1930, **127**, 42; 1931, **131**, 494; 1931, **134**, 308; 1933, **140**, 399; G. D. Bengough and A. R. Lee, *J. Iron St. Inst.* 1932, **125**, 285; G. D. Bengough and F. Wormwell, *Rep. Corr. Comm. Iron St. Inst.* 1935, **3**, 123; 1936, **4**, 213. *J. Iron St. Inst.* 1935, **131**, 285).

The specimen used was generally a machined disc, supported horizontally on three points. In contrast to the Cambridge experiments quoted above which used weight-loss as criterion of corrosion, the progress of attack was followed by the disappearance of oxygen from the gas-space above the liquid, which was measured in a gas burette. The method is described on p. 795.

The corrosion-rate was found to be controlled by the rate of replenishment of oxygen at the metallic surface. A very interesting series of experiments was carried out to study the effect of depth of immersion. When the depth of immersion below the surface was small, the corrosion-rate fell off as that depth increased, apparently because transport of oxygen downwards through a *shallow* layer proceeds by *diffusion*, and the rate of supply depends on the distance through which the oxygen must diffuse. However, when the depth exceeded about 1·5 cm., the rate became independent of depth, the transport being attributed to *convection*. The origin of the convection currents was a subject of some discussion twenty years ago, and probably a number of factors play their part. Researches at Cambridge seemed to indicate that the falling of corrosion-product over the sides of the specimens may have contributed, since different curves were obtained when the apparatus was designed in such a way as to prevent this falling over the edge (U. R. Evans and R. B. Mears, *Proc. roy. Soc.* (*A*) 1934, **146**, 153, esp. p. 159).

An important point emerging from Bengough's work was that the cross-section of the experimental vessel affects the total corrosion far more than the area of the specimen. It is known that gaseous molecules do not easily pass from a gas-phase into a liquid under *stagnant* conditions.* By doubling

* The rapid increase in the *solution rate* of oxygen when the water surface is visibly disturbed is shown by the measurements of A. L. Downing and G. A.

the area of the liquid surface we may double the rate of passage, and thus greatly increase the rate at which oxygen can be replenished at the surface of a totally immersed metal specimen.

The use of oxygen instead of air was found to enhance the rate of attack, whilst increase of the oxygen-pressure enhanced it further, provided that the specimen was placed in the liquid before the pressure was increased. If, however, the specimen—still dry—was exposed to the high-pressure oxygen before the liquid was introduced, then very high pressures tended to reduce the corrosion-rate (G. D. Bengough and F. Wormwell, quoted by U. R. Evans, "Metallic Corrosion, Passivity and Protection," 1946 edition, p. 297 (Arnold)).

The starting-points of attack were distributed sporadically and were often too numerous to be counted; they occurred chiefly on edges and burrs (points of disarray) or places where the specimen rested on the glass support-points (probably because the points interfered with oxygen-replenishment and rendered the shielded metal anodic). The corrosion, starting locally, spread out, and in the more concentrated solutions ultimately affected the whole surface. This at one time seemed to suggest that the corrosion was not electrochemical, since—it was argued—anodic attack over an entire surface would leave no room for a cathodic area. Later Noordhof, using a special dielectrode (p. 865), found that on a horizontal, completely immersed surface of zinc the anodic area kept moving about, so that, after a long period, each spot on the surface had suffered corrosion at one time or another, although at any given instant, part of the surface was anodic and part cathodic. In the early stages when only part of the specimen was seen to be suffering attack, Noordhof found that the electrical instrument always showed the part suffering visible attack to be anodic; the total corrosion calculated from the electrical observations agreed fairly well with that obtained directly. Thus once again the electrochemical mechanism appeared to be established. The wandering about of the anodic area is perhaps not surprising since the whole of a horizontal surface (apart from the edges) would be equally well situated for oxygen replenishment, and anodic attack will probably fall on areas of atomic disarray; when the "loosest" areas are used up, the attack moves elsewhere (G. S. Noordhof, reported by U. R. Evans, *Proc. International Congress of Pure and Applied Chemistry* (*London*) 1947, **5**, 743; see also U. R. Evans, *Corrosion* 1951, **7**, 238).

The fact that the total corrosion-rate under conditions of total immersion is controlled by the rate of oxygen-supply has sometimes been regarded as evidence that the mechanism is not electrochemical. This line of argument is open to doubt, since if there is a limited supply of oxygen, the strength of the current connected with the cathodic reduction of oxygen will itself be limited by the rate at which oxygen can diffuse up to the cathodic

Truesdale, *J. appl. Chem.* 1955, **5**, 570. The solubility of oxygen in pure water and sea-water is discussed by A. L. H. Gameson and K. G. Robertson, *ibid.* 1955, **5**, 502. See also E. Schaschl and G. A. Marsh, *Corrosion* 1957, **13**, 243t, who find that agitation greatly increases the corrosion-rate of steel in buffered salt solution, provided that the oxygen-concentration is above 2 ppm.

areas. Such oxygen-diffusion currents have been extensively discussed. See, for instance, U. R. Evans, L. C. Bannister and S. C. Britton, *Proc. roy. Soc.* (*A*) 1931, **131,** 355, esp. p. 369; I. L. Rosenfel'd and K. A. Khigalova, *Doklady Akad. Nauk C.C.C.P.* 1955, **104,** 876.

Acceleration of Corrosion by Contact with a second Metal. The corrosion-rate of iron partly immersed in sodium chloride is largely determined by the rate of oxygen uptake at the water-line. If the iron plate is electrically connected to a copper plate, also partly immersed, the oxygen taken up at the water-line of the copper will suffer cathodic reduction and will produce attack on the iron, which constitutes the anode of the cell Fe | NaCl | Cu, so that its corrosion will be accelerated (the copper receives *cathodic protection*). The anodic and cathodic products ($FeCl_2$ and NaOH) will meet and produce rust or magnetite—according to the nature of the oxygen-supply—at a distance from either metal, so that there will be no stifling of the attack by corrosion-product. This acceleration of attack by a second metal—which may cause intense corrosion if the ratio of cathodic to anodic area is high—is discussed in Chapter VI.

Acceleration of Corrosion by application of an external E.M.F. A still greater degree of acceleration can be obtained by providing a current from an external source, as in fig. 22; the cathode may consist either of the same metal as the anode (here iron) or a different one. If the whole of the current is employed in anodic attack, the corrosion-rate—which can be calculated by applying Faraday's Law—may be made very high, but often at high current densities, the metal becomes passive—a matter discussed in Chapter VII.

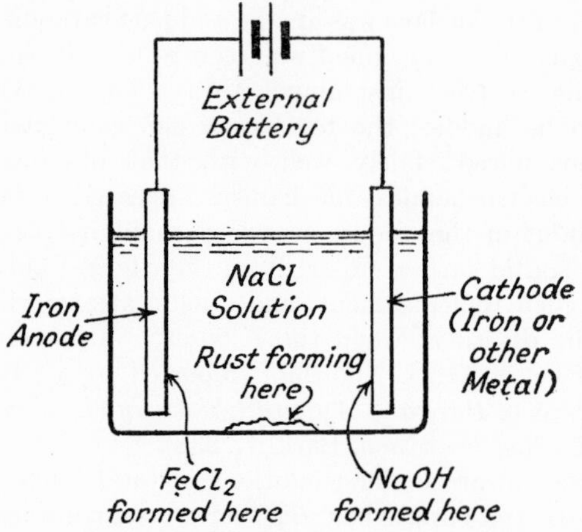

FIG. 22.—Acceleration of corrosion by external E.M.F.

Corrosion under an external E.M.F. has been found useful in the laboratory for many purposes. It has been employed to show up the position of the sensitive points which form the starting-places of corrosion (p. 108); here the current should be continued for the shortest possible time so as to avoid "spreading". Another application is in the transfer of films from a metal plate to a glass or plastic support by attaching the plate with adhesive (or even with petroleum jelly) to the glass or plastic, and dissolving away the metal completely from the other side by anodic action, so that the film is left on the support; the precautions needed to avoid secondary effects are discussed on p. 786. In all these cases, any access of cathodically formed alkali to the anode might cause disturbance. The risk can be reduced by interposing a diaphragm, but it is well to adopt a second precaution by choosing a liquid which does not yield alkali as cathodic product. If to the sodium chloride solution normally used in such work, zinc sulphate is added, the main cathodic product will be metallic zinc, and even if some alkali is formed, it will be consumed immediately in precipitating zinc hydroxide. The sodium chloride/zinc sulphate mixture used at Cambridge (p. 786) is known as " stripping solution ", but it has been used for other purposes besides stripping.

Corrosion by Distilled Water containing Oxygen

Introduction. It might reasonably be expected that distilled water would cause less corrosion than a salt solution; it might even seem that, when once the water had become saturated with oxide or hydroxide, corrosion ought to cease altogether—since the anodic reaction should then start to build a protective film. However, at these low solubilities considerations of nucleation become important. When a hot solution of, say, potassium chloride is concentrated by evaporation and cooled down to a temperature at which it becomes definitely super-saturated, there is a good chance of a sufficient number of potassium and chlorine ions coming together in positions suitable to constitute a stable nucleus; in a large body of liquid such nuclei are almost certain to appear *somewhere*, and solid matter will crystallize upon them until equilibrium is reached. But in a saturated—or even super-saturated—solution of a metallic oxide or hydroxide—the concentration is very low, so that the ions are normally far apart; the chance of them coming together in sufficient numbers to form a stable nucleus is a relatively rare event. If a metal is placed in water saturated with its oxide or hydroxide—whichever is the least soluble—a film should slowly be built up, but much time must elapse before it covers the entire surface; when the greater part of the surface is protected and a small part remains film-free, there is a risk of pitting, since cathodic reduction of oxygen on the protected part will itself promote anodic entry of metallic ions into the water at the anodic part; this attack—concentrated on a small area—may become very intense. If a small part of the surface is shielded from replenishment of oxygen, points within that part are likely to remain film-free longest and to suffer pitting; pieces of film-substance, broken off from some other part, and lodging on the surface, may promote just the oxygen-screening needed to produce pits.

Zinc. In potassium chloride solution, zinc suffers attack which starts locally and soon spreads out; it is suggested on p. 116 that this spreading is due to the destruction of the oxide film by zinc chloride, the anodic product. In distilled water, no similar destruction is possible, and the attack should either remain localized or cease. Under stagnant conditions, it remains localized, and produces deep pits, quickly causing perforation of thin sheet. This is not due to any impurity in the water or metal. Davies found that very carefully prepared water produced pits on zinc; if anything, they were more sharply cut than those formed in water of lower quality; moreover zinc of different degrees of purity gave much the same corrosion-pattern.

Solid corrosion-product was formed around the pits which were also loosely filled with white material; both oxide and hydroxide were present. Zinc also passed into the liquid in colloidal solution; the water acquired the power to produce a Tyndall cone, and electrophoretical experiments indicated a negative charge on the colloidal particles (U. R. Evans and D. Eurof Davies, *J. chem. Soc.* 1951, p. 2607).

Apparently a protective film is constantly being built up over a zinc surface in distilled water, but it takes time to complete itself, and tends to break down. Bengough's curves for zinc totally immersed in distilled water or very dilute potassium chloride solution are asymptotic and indicate the gradual cessation of corrosion—in marked contrast to his straight corrosion–time graphs in relatively concentrated potassium or sodium chloride solution. With increasing concentration, there is a gradual passage from

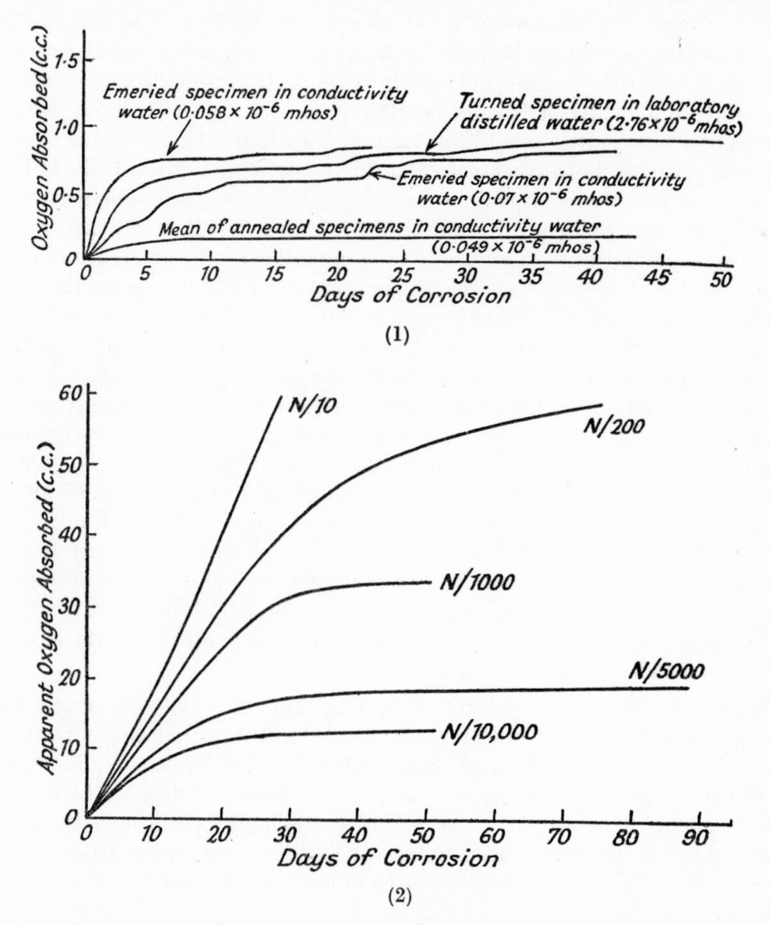

(1)

(2)

Fig. 23.—Corrosion of Zinc (1) in Distilled Water after various treatments; (2) in Potassium Chloride solution (annealed specimens) (G. D. Bengough, J. M. Stuart and A. R. Lee).

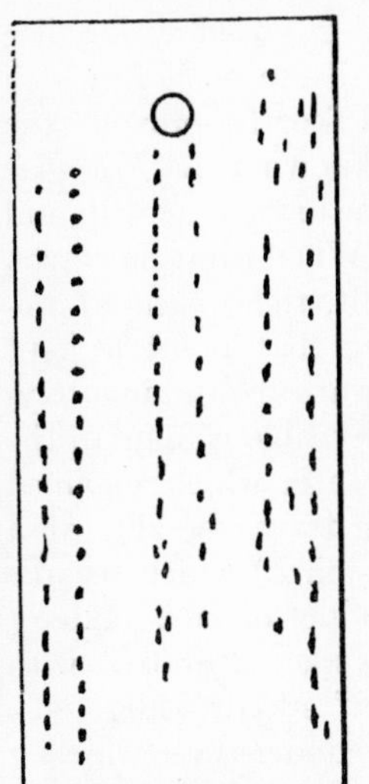

Fig. 24.—Vertical arrangement of pits in Zinc exposed to Distilled Water (G. D. Bengough and O. F. Hudson).

the asymptotic to the rectilinear type (fig. 23).* However, on un-annealed specimens placed in distilled water, the attack, after coming to rest, keeps starting again, in a manner recalling the oxidation of copper or aluminium (p. 50); it is probably attributable to periodic collapse of the film. Another curious observation made by Bengough (in his early work with O. F. Hudson) was the tendency of pits on a vertical zinc surface in distilled water to arrange themselves in roughly vertical lines (fig. 24) (G. D. Bengough and O. F. Hudson, *J. Inst. Met.* 1919, **21**, 59. G. D. Bengough, J. M. Stuart and A. R. Lee, *Proc. roy. Soc.* (*A*) 1927, **116**, 425; 1928, **121**, 88).

Eurof Davies (see above) confirmed the vertical arrangement of pits on several varieties of zinc and found it to be independent of the rolling or abrasion direction; he attributed it to oxygen-screening by particles of solid corrosion-product originating at an active pit and sinking by gravity until they chance to lodge at points arranged vertically below the original pit. If the small areas shielded happen to contain points which, although not sufficiently sensitive to develop attack with the small supply of oxygen which is normally available, are sufficiently sensitive to develop attack when the supply is further reduced by the screening, pits are likely to be formed; thus the vertical arrangement explains itself. Davies obtained lines of closely arranged pits by stretching a polythene thread round a surface immersed in distilled water; the additional screening at the two crevices produced on either side of the thread (fig. 25) gave rise to two lines of pits or trenches spaced much more closely than the pits developed at points distributed sporadically over the rest of the surface.

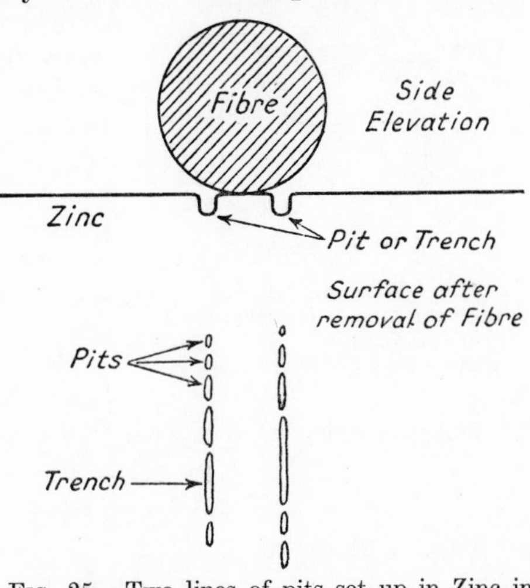

Fig. 25.—Two lines of pits set up in Zinc in Distilled Water by contact with polythene thread (D. Eurof Davies and U. R. Evans).

If oxygen-replenishment can be made perfect over the

* Bengough's curve for the corrosion of zinc in N/10,000 potassium chloride solution obeys the same asymptotic equation as that developed for film-growth conditions (p. 834).

whole surface, pitting should be avoided. Davies (fig. 26) used a zinc disc specimen slipped over a vertical rod, joined eccentrically to a shaft which could be driven from a motor, and furnished with a foot to prevent the specimen from slipping off. The type of apparatus is known as an *Eccentric Whirler*. The main vessel can be filled with nitrogen or oxygen at will, and distilled water introduced directly from a quartz still. When rotation starts, the zinc disc mounts up the rod; owing to the loose fit, there is no appreciable oxygen-screening at the point of contact, since that point is kept continually changing. In experiments with distilled water containing oxygen, all pitting and all visible change to the zinc were avoided; the water, which remained clear, came to contain traces of zinc at a concentration (10^{-5} g.-ion/l.) which accords with the solubility-product of zinc oxide or α-hydroxide; it was super-saturated with respect to the " stable " ε-hydroxide.

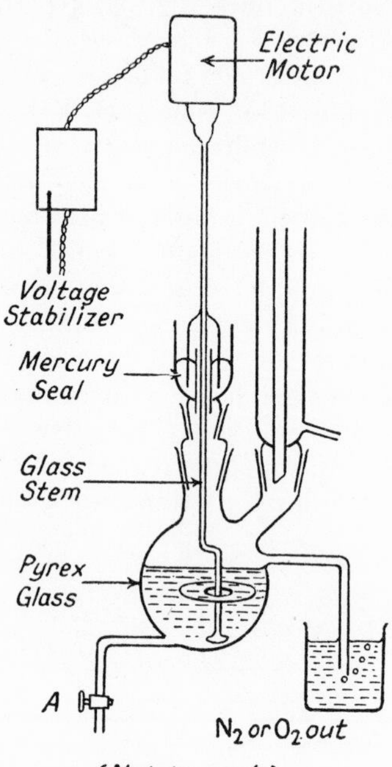

FIG. 26.—Whirling apparatus as used for Zinc in Distilled Water —diagrammatic (D. Eurof Davies and U. R. Evans).

It seems likely that, under these whirling conditions, a film is built up by a mechanism not unlike that operative in dry oxidation, but where one part of the surface is screened, whether by a polythene thread or corrosion-product, this part becomes anodic to the cathodic area upon which oxygen can be reduced, using up electrons.*

The reduction of oxygen in distilled water (and doubtless in aqueous solutions) appears to occur in at least two steps; the first main step leads to hydrogen peroxide, which is then reduced to hydroxyl. Hydrogen peroxide is found among the corrosion-products of zinc, aluminium and other metals (J. R. Churchill, *Trans. electrochem. Soc.* 1939, **76**, 341; G. Bianchi, *Corrosion et Anticorrosion* 1957, **5**, 146).

It is possible that the steps should themselves be sub-divided. However for our present purposes we can write the two stages

$$(1) \quad 2O + 2H^+ + 2e = H_2O_2$$
$$(2) \quad H_2O_2 + 2e = 2OH^-$$

These add up to

$$2O + 2H^+ + 4e = 2OH^-$$

or

$$O + H^+ + 2e = OH^-$$

The loss of H^+ and the gain of OH^- is equivalent to the net gain of $2OH^-$,

* It had previously been questioned whether distilled water, with its poor conductivity, *could* carry differential aeration currents, but Davies' experiments gave direct demonstration of the existence of such currents in distilled water.

so that, since H_2O is present in great excess of H^+ and OH^-, the result is for most practical purposes the same as that expressed by the equation put forward on p. 88.

$$O + H_2O + 2e = 2OH^-$$

Iron. An " Eccentric Whirler " had been used in earlier Cambridge work on iron in distilled water. Pure electrolytic iron whirled in distilled water often remained bright, although some specimens developed a few tiny spots of rust. Specimens partly immersed in the same water under stagnant conditions, rusted freely; probably the relatively soluble ferrous hydroxide was formed at the more deeply immersed part, and was converted to the less soluble rust near the water-line; the rust, formed at a distance from the point of attack, did not protect. The non-protective character of ordinary rust must be ascribed mainly to the fact that iron has two sets of compounds, the lower (hydrated) oxide being more soluble than the higher one.

Steel discs, exposed to distilled water and oxygen in the eccentric whirler, developed appreciable adherent rust, although the liquid remained clear; possibly, although differential aeration was successfully avoided by the whirling, other sources of corrosion current would be set up on steel; for instance the cell iron | cementite would remain operative (U. R. Evans, *Industr. engng. Chem.* 1925, **17**, 363, esp. p. 370; " Corrosion of Metals " 1926 edition, p. 108 (Arnold)).

Lead. Particularly important is the case of lead, on account of the danger of poisoning which arises if traces of lead find their way into soft drinking water. Curious inconsistencies have from time to time been reported in the behaviour of lead in distilled water, according to the previous history of the metal and the carbon dioxide content of the water. Although lead oxide is generally considered a sparingly soluble substance, it possesses a low nucleation number and for that reason generally fails to build a protective film. A super-saturated solution tends to deposit solid upon existing crystals rather than to start new ones. This was shown clearly in the work of Mayne who obtained a bright lead surface by casting a specimen against polished stainless steel and then placed it in distilled water. After some hours, the metal became etched, and subsequently crystals were observed under the microscope; after some days, the surface had become covered with an open net-work of beautiful little crystals, but there was no continuous film, and the persistence of corrosion was not surprising (J. E. O. Mayne, reported by U. R. Evans, " Metallic Corrosion, Passivity and Protection " 1946 edition, p. 353 (Arnold)).

In some of Mayne's experiments, the red form of the oxide was the one to appear, but the yellow form seemed to be more frequently produced. It is likely that anodic and cathodic reactions, producing Pb^{++} and OH^- respectively, occurred at almost adjacent points (perhaps at the same points), rendering the liquid super-saturated with respect to the hydroxide as well as to both forms of oxide; in such cases, presumably, chance atomic movements decide whether a nucleus of yellow or red oxide first appears.

If carbon dioxide is present in the water, a protective film may sometimes be formed, covering the whole surface. Liddiard and Bankes found that lead placed in stationary water containing oxygen and a small amount of carbon dioxide was at first strongly attacked, and some white precipitate was formed; however, the rate of attack fell off as a film of lead carbonate was formed over the surface. In a water containing more carbon dioxide, the water may remain clear, perhaps suggesting that attack has been avoided; any such suggestion, however, is deceptive, since a solution of carbon dioxide has a solvent action for lead carbonate, which passes into the liquid as bicarbonate. It is known that treatment of a soft natural water with carbon dioxide often prevents turbidity, but dangerous amounts of lead can pass into solution. The attack on lead by many waters is increased if chlorides are present, although the traces of chloride left after chlorination of a public supply will usually have no appreciable effect (E. A. G. Liddiard and P. E. Bankes, *J. Soc. chem. Ind.* 1944, **63**, 39. Cf. F. J. Liverseege and A. W. Knapp, *ibid.* 1920, **39**, 27T).

The previous history of the lead affects behaviour. Lead placed in a vessel containing distilled water below air produces on some occasions lead carbonate as a scum floating on the water–air surface, but at other times builds a protective film on the metal; in the first case, lead continues to pass into the liquid. German work has shown that pre-exposure to damp air containing normal amounts of carbon dioxide, by producing lead carbonate nuclei, favours the second (more desirable) alternative; the use of a vessel of broader cross-section, which facilitates passage of carbon dioxide into the water, has a similar effect (O. Bauer and E. Wetzel, *Mitt. Mat. Prüf. Amt.* 1916, **34**, 347; O. Bauer and G. Schikorr, *Mitt. Mat. Prüf. Anstalten, Sonderheft* 1936, **28**, 69, G. Schikorr, *Korros. Metallsch.* 1940, **16**, 181).

Liddiard and Bankes showed that the addition of 16 ppm. of calcium bicarbonate to pure water can prevent the up-take of lead, but the presence of sodium chloride has an unfavourable effect. Many natural soft waters, and also most rain-waters, are highly *plumbo-solvent*, i.e. capable of taking lead into solution; water from peat-bogs sometimes contains acids derived from the roots of bilberry, heather or other plants, which prevent the formation of protective films on lead. It is dangerous to pass such waters through lead pipes if they are to be used for drinking; it should be remembered that lead is a cumulative poison, and its effects may not immediately be apparent. Treatment with milk of lime or whiting, or percolation through a limestone bed, is to be preferred to mere neutralization of the acid by sodium carbonate, since calcium bicarbonate is formed, which in many cases may eliminate danger; but apparent exceptions are on record. The situation is complicated, and organic matter other than the acids mentioned can increase the corrosion of lead (G. Miles, *J. Soc. chem. Ind.* 1948, **67**, 10).

Whatever the treatment favoured—it is necessary to test the treated water before deciding that it can safely be passed through lead pipes. Methods of estimating lead traces in water are described in the Liddiard–Bankes paper, whilst a useful survey is provided by H. Ingleson, *Water*

Pollution Res. Tech. Paper No. 4 (1934); *Analyst* 1938, **63**, 546; 1940, **65**, 403; and by J. C. Thresh, J. F. Beale and E. V. Suckling, " Examination of Waters and Water Supplies " 6th edition, edited by E. W. Taylor (Churchill); esp. p. 244; also pp. 550, 551 for permissible maxima of concentration.

A graphical presentation of conditions favouring the corrosion of lead by waters with or without carbon dioxide is offered by P. Delahay, M. Pourbaix and P. van Rysselberghe, *J. electrochem. Soc.* 1951, **98**, 57.

Aluminium. The behaviour of aluminium in distilled water varies both with the temperature of the water and its purity. Hart, studying high-purity aluminium, found that below a critical temperature (between 60 and 70°C.), an amorphous film is formed at first, which then becomes boehmite (γ–AlO.OH, orthorhombic); finally bayerite (β–$Al_2O_3.3H_2O$, monoclinic) is produced. At 20°C. the velocity of attack increases with time during the period of boehmite formation but falls off during the bayerite period. Above the critical temperature (say at 80°C.) only boehmite films are formed on the top of the initial amorphous film, and the rate of attack falls off steadily with time (R. K. Hart, *Trans. Faraday Soc.* 1957, **53**, 1020; cf. H. Raether, *C.R.* 1948, **227**, 1247; J. D. Keller and F. Edwards, *Metal Progr.* 1948, **54**, 35, 195).

The boehmite film formed by hot water can vary considerably in its protective properties. Altenpohl finds boiling for 4 hours in water of very high purity produces a colourless film conferring high resistance to subsequent attack. After such treatment, aluminium of 99·5% purity is found to resist N/20 hydrochloric acid for 14 days, although the untreated aluminium is attacked by that acid within an hour. However, small traces of certain impurities in the distilled water lead to a dark film which is much less resistant; as little as 0·001% silica, or 0·0004% of citric acid, causes the film to be much thinner and less protective. At first sight this seems to conflict with the fact that the addition of silicate to many supply waters reduces the attack, but Altenpohl's film-thickening curves suggest that silica also inhibits the thickening of the protective film. Thus whereas prolonged treatment in very pure water at 100°C. leads to a film too thick to be crossed by ions at, say, 25°C., the necessary thickness is not reached if silica is present. Sulphuric acid is also unfavourable to the growth of a protective film, but rather larger amounts are needed (0·005%). Ammonia, on the other hand, is a favourable constituent, and it was possible to obtain highly protective films by boiling with the supply water of Singen (S. Germany) if ammonia (0·1N) was present; after such treatment, aluminium resisted cold supply water for 1½ years, and resisted N/20 hydrochloric acid for 4 days. After-treatment of the film for 30 min. in 2% water-glass, although reducing the protective properties of the film formed in pure water, improved that of the film formed in Singen supply water containing ammonia; the films thus treated resisted N/20 hydrochloric acid for 13 days. The paper contains much surprising information and deserves study. (D. Altenpohl, *Aluminium* 1957, **33** 78).

Altenpohl also produced protective films by treatment in superheated steam. The crystalline boehmite film produced by treatment in boiling water or steam is much thicker than the natural oxide-film. He recommends similar treatments for preventing the darkening which is produced by some hard supply waters on vessels and cooking utensils (D. Altenpohl, *Aluminium* 1955, **31**, 10, 62; 1957, **33**, 78; *Metalloberfläche* 1955, **9**, A118; *Metall* 1955, **9**, 164).

The darkening noted on cooking utensils has been explained in various ways. Some authorities regard it as the start of an interference-colour series, whilst others attribute it to iron. Others again postulate the formation of small particles of metallic aluminium embedded in oxide or hydroxide; if, for instance, corrosion penetrates along grain-boundaries, which are converted to oxide, leaving the grain-interiors metallic, any light striking the surface may become lost by repeated reflection between the metallic particles, causing the surface to appear dark for the same optical reason as is responsible for the darkness of platinum black. Possibly all these factors may operate in different circumstances.

In apparent contrast with Altenpohl's discovery that silica in otherwise pure water prevents the formation of a film capable of providing protection against impure water or certain chemical solutions, stands Bryan's experience that silicon in the metallic phase is helpful in building up a protective film on aluminium exposed to boiling water. He found that aluminium containing silicon evolves hydrogen readily when placed in boiling water, but after 8 hours the specimen practically ceases to gain weight, showing that the film has become protective. In contrast, pure aluminium continues to gain weight at a rate which becomes nearly constant. If iron is present, more silicon is needed to provide an adherent protective film than if it is absent; apparently the iron combines with the silicon and renders it ineffective unless silicon is present—as rather suggested by the early results of Bailey (J. M. Bryan, *J. Soc. chem. Ind.* 1950, **69**, 169; G. H. Bailey, *J. Inst. Met.* 1913, **9**, 79).

Some Principles governing the distribution of Wet Corrosion in nearly Neutral Solutions

General. The information already provided has indicated that wet corrosion is connected with electric currents flowing between anodic and cathodic areas—sometimes well separated. The quantitative treatment of Chapter XXI shows that in all cases hitherto studied the corrosion-rate is equivalent to that current in the sense of Faraday's Law, and often the strength of the current is controlled by the supply of oxygen to points where the cathodic reaction can proceed, although in dilute solutions the conductivity becomes an important factor.

However, for practical purposes, it is not the total destruction of metal but the intensity of attack (corrosion-rate per unit area) which matters, and for this purpose we need to know something of the factors which decide the corrosion-pattern. Obviously it is advantageous to have the corrosion uniformly spread out, and it is noteworthy that reduction in the oxygen-supply not only diminishes the total corrosion, but generally causes such corrosion as still occurs to be well distributed.

Unfortunately the corrosion-pattern is far less " reproducible " between duplicate experiments than the corrosion-velocity. Two plates of the same metal intended to be identical, if carefully prepared and partly immersed in the same salt solution for a given time, will usually suffer the same total corrosion; but the distribution of the attack will probably vary seriously between one specimen and another, and the proportion of the surface which remains unattacked will also vary; it follows that the loss of thickness at the part where corrosion is most intense will not be the same on the so-called " duplicate " specimens. It may seem surprising that the rate of total attack is so nearly the same whereas the corrosion-pattern varies from one specimen to another. The reproducibility of the corrosion-rate probably arises from the fact that the strength of the corrosion-current depends largely on the rate of oxygen-replenishment at points suited for the cathodic reaction and, since these are believed to be closely spaced and numerous, at least on iron and zinc, Bernoulli's principle (the law of averages) will ensure that the rate of destruction of metal is nearly the same for duplicate specimens; for aluminium, where the cathodic points are less numerous and spaced more widely apart, the reproducibility is less good. On the other hand, the corrosion-pattern depends on the situation of points where the attack starts, and these are spaced well apart even on iron and zinc—as is obvious to the eyes; thus the pattern which will appear on any single specimen cannot be predicted.

The corrosion-velocity is a matter which can be discussed in terms of elementary mathematics, and this is done in Chapter XXI. The maximal loss of thickness—or the depth of the deepest pit, if corrosion takes the form of pitting—is a question involving probability, and is therefore discussed in Chapter XXII, which deals with Statistical Considerations.

Each of the factors which influence the corrosion-pattern is relatively simple and clear-cut, but the fact that a number of factors may operate simultaneously renders the situation somewhat complicated. It may be well to summarize the four important factors and then to consider them in greater detail:—

(1) The points where corrosion starts are generally determined by *Surface Defects*, which may represent places where the metallic atoms are disarrayed and therefore escape most easily into the liquid, or places where the invisible film has recently been damaged; more often they seem to represent places where internal stress exists in the metal.

(2) The situation is modified by the *Mutual Protective Effect*; rapid corrosion developing at one point greatly reduces the probability of corrosion developing at neighbouring points.

(3) The decision between corrosion (*a*) *ceasing* at a point where it has started, (*b*) *remaining localized*, producing a *pit* or (*c*) *spreading out* to give relatively general corrosion, depends largely on the corrosion-product and its possible action on the substance of the invisible film.

(4) For many metals, the distribution of attack in neutral salt solutions is greatly influenced by the considerations of *oxygen replenishment*; in some liquids, the most " aerated " parts remain free from attack. The

corrosion-velocity is often determined by the rate of oxygen-uptake at the aerated (cathodic) parts, even though the corrosion may be occurring at the unaerated (anodic) parts; as already stated, currents set up by inequalities in the oxygen supply are known as *Differential Aeration Currents.* The distribution mentioned, with protection at the aerated parts, is characteristic of sodium and potassium salt solutions; in acid, corrosion may be more general, and often attack is most severe in the parts where oxygen replenishment is best.

These four matters now deserve more detailed consideration:—

1) Surface Defects

The Starting-points of Attack. Early work at Cambridge on zinc and iron plates partly immersed in potassium or sodium chloride solution showed that corrosion usually started at cut edges or surface defects. Since, however, it was observed that under such conditions the unequal oxygen distribution was also influencing the corrosion-pattern, experiments were designed to avoid complications due to the presence of two super-imposed factors. These were carried out on fully immersed specimens exposed to oxygen-free liquid, the specimen being subjected to anodic attack by application of an external E.M.F. The metals studied were again generally zinc and iron, but there were a few experiments on aluminium. The general conclusions reached were the same as those reached in the experiments on partly immersed specimens without applied E.M.F., and in many other experiments carried out about the same time under different conditions. The position may be summarized as follows:—

(1) On sheared plates, corrosion usually starts at points on or very near the edge of a plate.

(2) On plates having imperfect surfaces, corrosion also starts at points where defects (usually elongated cavities) can be detected under the microscope.

(3) On rolled specimens, it also starts at certain points usually situated on lines parallel to the original rolling direction.

(4) On abraded specimens, it also starts at certain points situated on some (but not all) of the more conspicuous abrasion lines.

(5) On bent sheet-specimens, it often starts at points situated along the bending axis, either on the convex or the concave side.

(For details, see U. R. Evans, *J. chem. Soc.* 1929, pp. 92, 111 (2 papers).)

Stress distribution at a scratch-line. In the experiments just quoted the attack starting at surface defects may in a few cases have been due to a recent damaging of the air-formed film. Some of the surface defects, however, where attack was observed to start, had probably been made at the rolling-mill at least a year earlier; it was suspected that the real cause was internal stress in the metal, which would be expected to exist at the sort of places where corrosion did, in fact, start. It is known that the presence of tensile stress in metal may render it anodic towards unstressed metal. If, therefore, small areas of tensile stress exist at surface defects, this would explain why corrosion starts at such places. It appears advisable to consider the distribution of stresses near the surface of metal which has received local mechanical treatment. It is obvious that on any specimen both tensional and compressional stresses must be present in equilibrium, otherwise the metal would change shape until such an equilibrium was established. It is also known that different sorts of mechanical treatment may leave

either (1) a layer in tension overlying a layer in compression,
or (2) a layer in compression overlying a layer in tension.

The first of these combinations is probably the most dangerous.

Consider a scratch-line made by a weighted needle drawn over a surface under a load sufficient to produce permanent deformation (i.e. to engrave a trench) on a metal of the class which exhibits a yield-point. Whilst the

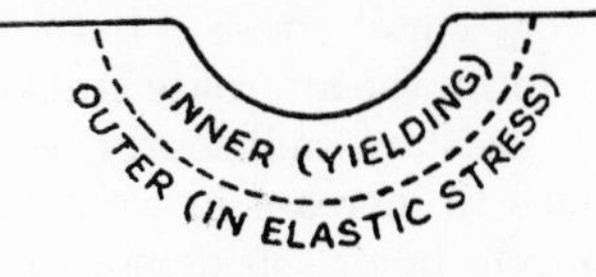

(a) During passage of weighted needle

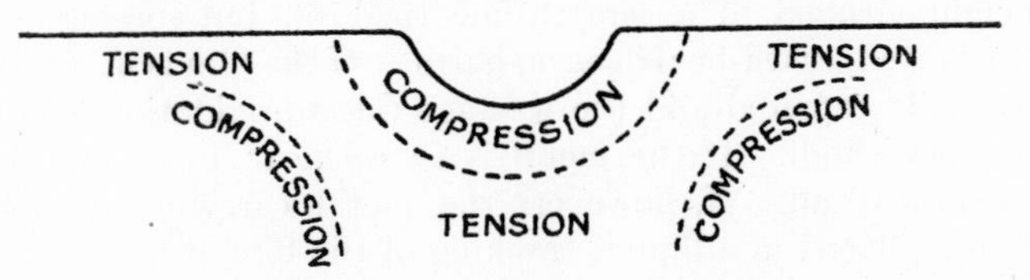

(b) After the needle has passed on

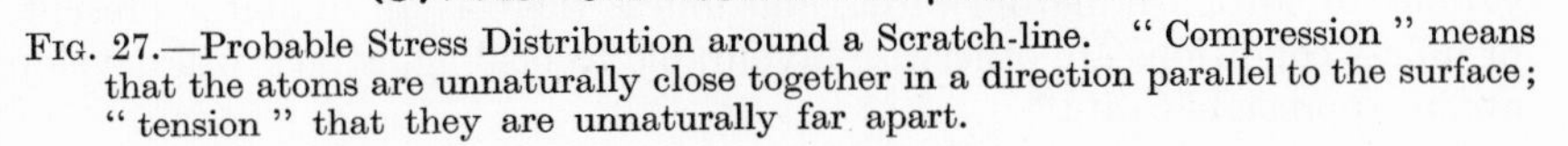

Fig. 27.—Probable Stress Distribution around a Scratch-line. " Compression " means that the atoms are unnaturally close together in a direction parallel to the surface; " tension " that they are unnaturally far apart.

weighted needle is passing over a point (fig. 27(*a*)), the surface layers must yield, relieving the stress over a certain region, but further from the needle there will be a region where the strain will remain in the elastic range, the atoms being pressed unnaturally close together in a direction normal to the surface, which means that they will be unnaturally far apart in a direction parallel to the surface. When the needle has passed on, the atoms in the region under elastic strain will tend to spring back towards the spacings characteristic of unstrained metal, but this automatically brings the atoms in the yielded region unnaturally close together; equilibrium will be established when compressional stresses in the second region come into equilibrium with tensional stresses in the first; below the scratch-line compression overlies tension, but at each side we may expect to find a region where tension overlies compression (fig. 27(*b*)).

Effect of Stresses on a film in air. We must now consider how the distribution of internal stresses is altered when an invisible film is produced by exposure to dry air. It will be assumed that we are dealing with a heavy metal on which the film thickens by material passing from the metal outwards through the films as cations. This removal of material from the outermost layer of the metal into the film during oxidation must upset equilibrium between the tension and compressional layers. In the regions on either side of the scratch-line where tension overlies compression, the tensional layer will be very slightly reduced in thickness, and the equilibrium will only be restored if the atoms in the compressional layer move rather further apart in a direction parallel to the surface (approaching more nearly their natural spacing). This must put a tension on the film. The tensional force will be a weak one, but the film is very thin. The tension would do no harm if the film was firmly anchored, but the outward movement leaves vacancies at the interface, which may sometimes segregate to form definite cavities (p. 48). Where cavities exist, the film will be left unsupported and will then crack under a very small stress. If the metal is still exposed to air, a new film will at once form, the alternate crack-heal-crack-heal sequence will continue, until the stresses are sufficiently used up. This may be one reason why metal continues to increase in weight—at a rate which is just detectable—for several months in air—as demonstrated by W. H. J. Vernon, *Trans. Faraday Soc.* 1935, **31**, 1668, esp. p. 1675.

There is experimental evidence that crack-heal does really take place in the neighbourhood of a scratch-line (p. 176), but so long as the metal remains unwetted, crack-heal does no harm. If the film is in air, the damage is automatically repaired and the wastage due to oxidation even over long periods is very small. If the metal is *in vacuo* or in argon, no chemical change occurs at all. If, however, the metal carrying an invisible air-formed film is placed in a liquid, cracking of the film at a scratch-line, even through an occurrence rare in time and space, will sooner or later allow cations to start leaking outwards, and unless there is an inhibitor present, or unless conditions are otherwise favourable to healing, attack will develop at an appreciable rate.*

Experiments on Corrosion set up by Scratch-line. The part played by local stresses in corrosion is confirmed by the Author's work on thin rolled nickel sheet which, under anodic attack in stripping solution (p. 786), was found to undergo intense localized corrosion at certain points, leading to perforation; except at these sensitive points the original surface remained quite uncorroded. Some of the perforations (but not all) were

* The argument does not necessarily involve the assumption that cracking occurs more frequently after immersion but merely that conditions for healing, which are present in air, are absent in water. However, wetting does alter the effective strength of materials; a crack cannot develop spontaneously unless the reduction of strain energy brought about by such cracking exceeds the greatest increase of energy involved at some stage in forming the two new walls of the crack; this depends on the medium in which the material is placed. The stresses still remaining in a film which has ceased to crack in air may be sufficient to cause resumption of cracking in water. The effect of wetting on the effective strength of materials has been studied by C. Benedicks and R. Härden, *Arkiv för Fysik* 1951–52, **3**, 407; *Rev. Métall.* 1948, **45**, 9.

observed to lie on barely visible scratch-lines, some of which had probably been produced accidentally long before the experiment.

It was found possible to produce the effect at will by tracing scratch-lines with a weighted needle before the anodic action was started. The perforations produced on these scratches and elsewhere were counted. Exact numbers cannot be stated, since a " scratch-area " has no sharp demarcation, but the perforations per unit area on the scratch-line were probably about 1000 times as numerous as elsewhere. A surprising observation was that the load on the needle within wide limits hardly affected the number of perforations *per unit length,* although the breadth of the scratch made under a high load was much greater than at that of a scratch made at a low load. If the attack had started at the central region, where compression overlies tension, the facts would be difficult to explain; but, if it starts at points situated close to the two margins of the scratch, in the regions where tension overlies compression, then the width of the scratch, as such, ceases to be important, and the facts receive an easy explanation. This points to local tensional stress as the factor deciding the starting-point of attack.

Two series of experiments were then carried out on heat-tinted nickel. In the first series a scratch-line was engraved and the metal heated to produce a film showing interference colours; in the second, the metal was first heat-tinted and then a scratch-line engraved, so that presumably a gap was present in the film when the specimen entered the liquid for anodic treatment whereas in the first series the oxide-film should have been continuous. It might be expected that there would be more attack on specimens where scratching followed tinting, but for the thinner films (corresponding to early interference tints) the reverse effect was encountered; specimens scratched before tinting suffered perforation under conditions which those scratched after tinting survived. This suggests that gaps in the film have little influence in determining the seat of attack and that stress in the metal may be more important. A further observation suggests that it is tensile stress which is important—as would be expected. Nickel scratched before being tinted to a first-order colour suffered anodic attack, the perforation line taking a zig-zag course along the scratch (fig. 28(*c*)); this suggests that the attack had started at individual points situated on either side of it, later extending into the centre and joining up to give the zig-zag perforation. If the attack is following the tensional regions, as shown in fig. 27(*b*), that is exactly what we should expect.

In the thicker films (corresponding to late colours), the scratch-lines made after tinting produced far more damage than those made before tinting, and in this case there were often two rows of perforations on either side of the scratch (fig. 28(*a*)); here gaps were evidently more important than stresses, which may have been removed (by annealing) during the heating. Probably these thicker films break at the margins during scratching and are pushed down inwards, as suggested in fig. 28(*b*).

Further information was obtained by Faerden, who studied specimens oxidized at various temperatures for 30 min. in an electric tube furnace, a 2·5 cm. scratch being engraved either *before* or *after* the tinting. Both

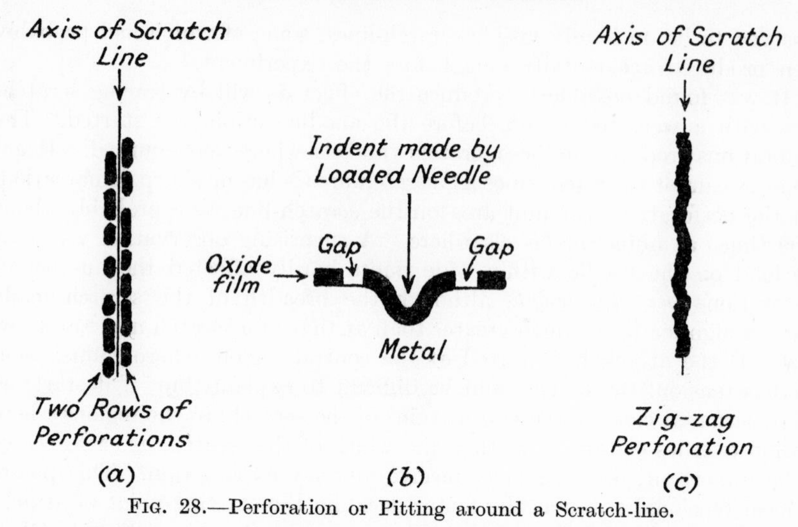

FIG. 28.—Perforation or Pitting around a Scratch-line.

as-rolled and abraded nickel was used, and gave similar results; the observations recorded in Table VIII refer to scratching at a 500-grams load, but results obtained at 200 grams were not very different. Each specimen carried two scratches, one made before and one after the oxidation, and the anodic treatment continued for 10 min at 50 ma, the area exposed to anodic action being 1 cm.^{2}.

TABLE VIII

ATTACK AT SCRATCH-LINES ON NICKEL (A. Faerden and U. R. Evans)

Temperature of Oxidation	Incidence of Attack
150°, 200° and 250°C.	Both scratch-lines attacked
300°, 350° and 400°	Preferential attack on scratch made *before* oxidation
450° and 500°	Results variable
550° and 600°	Preferential attack on scratch made *after* oxidation

This seemed to confirm the conclusion reached previously that at the lower range of temperature stresses are more important than gaps. It remained necessary, however, to ascertain whether gaps really did exist in the films on specimens scratched after oxidation and were absent on the others. To investigate this point, Faerden made another set of specimens and transferred the films, by a method similar to that described on p. 785, to glass covered with polystyrene paint pigmented with white lead; after the nickel had been dissolved away anodically, leaving the film clinging to the white lead, the specimen was placed in dilute ammonium sulphide solution, which showed up any gaps existing in the film by the blackening of the white lead. It was found that gaps really did exist when scratching followed oxidation, but were absent when it had preceded oxidation.

If tensile stress in the metal rather than gaps in the film represents the factor determining the starting-point of attack, it is easy to see why heat-tinting after scratching fails to prevent attack at the scratch-line, unless the temperature has been sufficiently high for stress-relief. A little further consideration will show why heating often makes the situation worse. Oxidation at a place where a tensional zone on the surface overlies a compressional zone, will upset the equilibrium, by destroying the extreme outermost layer of the tensional zone; since now the tensional zone is thinner at the outset, equilibrium is only restored when the tensional stress (load per unit area) has become greater than before, so that the probability of attack on the metal will be increased.

This seems to complete the line of argument and to show the importance of internal stresses in the metal; the reader may care to study the original papers, which, however, offer a slightly different interpretation (U. R. Evans, Symposium on Internal Stresses in Metals and Alloys (Inst. Met.) 1947–48, p. 291; also Symposium on Properties of Metallic Surfaces (Inst. Met.) 1952–53, p. 253; A Faerden, *ibid.* p. 275. For latest views see U. R. Evans, Hothersall Memorial Lecture, *Trans. Inst. Met. Finishing.* 1957, **31**, 139).

(2) The Mutual Protective Effect

Early stages of the Corrosion-pattern. Fig. 29 shows the distribution of corrosion on steel plates partly immersed in sodium chloride solution. It is evident that intense corrosion at a given point tends to protect the adjoining area. The patterns were obtained on a steel specially prepared to provide specimens free from serious surface defects, so that attack starts at the sheared edges rather than elsewhere (on specimens cut from an ordinary steel, corrosion would start also at points on the face, extending downwards

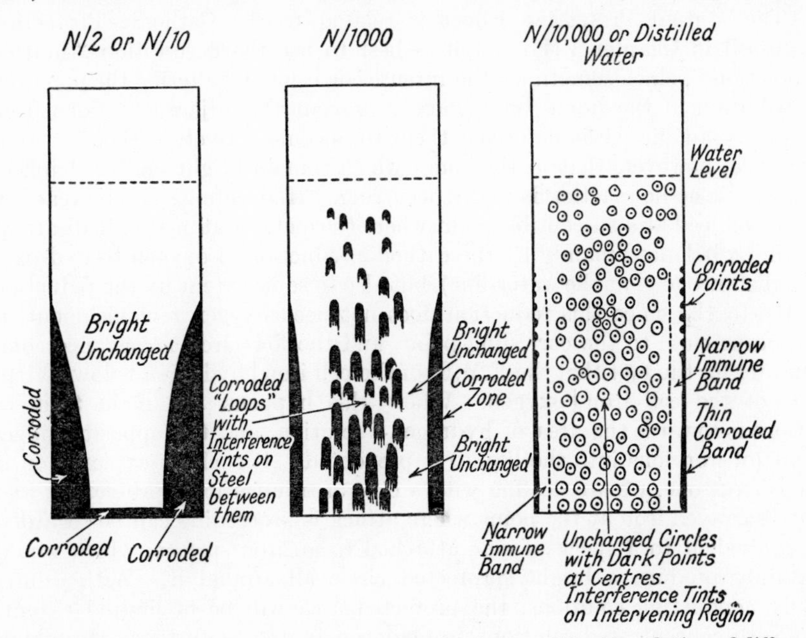

FIG. 29.—Early stages of attack on stee plate by Sodium Chloride solution of different concentrations.

as arch-shaped areas). In concentrated (N/2 or N/10) chloride solution, the intense corrosion occurring along the cut edges of the special steel serves to protect the whole of the rest of the face. The fact that the absence of attack over the main part of the face is due to protection from the edges was shown by Britton's experiments on specimens with edges protected by means of baking varnish; in these circumstances, attack did indeed start from points on the face, producing the usual arch-shaped areas. Corrosion at the *most* susceptible points (on the cut edges) will prevent attack at slightly less susceptible points (on the face); but when this edge-attack is prevented by baking varnish, the attack will then start at the points on the face (U. R. Evans, *Trans. electrochem. Soc.* 1930, **57**, 407, esp. p. 414; S. C. Britton and U. R. Evans, *ibid.* 1932, **61**, 441, esp. p. 454).

In N/1000 solution, where the conductivity is low, the protection can extend only a limited distance from the cut edges (fig. 29(*b*)). In N/10,000 solution, the protected zone becomes very narrow; the main central portion shows change, but in the early stages, the attack starting at various points on the face itself protects the area immediately around it, so that each dark point of corrosion is surrounded by a bright, unchanged circle. Outside these circles the main part of the face is covered with interference colours, suggesting that here the protection is insufficient to prevent iron cations moving outwards through the film, but that they move outwards only slowly and are precipitated by alkali and oxygen on arrival; thus a film is built up in contact with the metal, sufficiently thick to produce interference colours, and loose rust is, for a time, avoided on this area.

The Mutual Protective Effect is related to the Cathodic Protection discussed in Chapter VIII, but it is best to use the description "mutual protection", since sometimes the intense corrosion, by altering the potential distribution in the liquid, may merely prevent the adjacent regions from becoming anodic without forcing them to become actively cathodic. It is probable, however, that in the zones which remain bright and unchanged, a slight cathodic reaction is really occurring. If an intense anodic reaction has developed at a susceptible point where the metallic atoms are in disarray, it will be balanced mainly by the cathodic reduction of oxygen to hydroxyl at a distance (e.g. at the water-line), but also to some extent by the reduction of H^+ to H_2; since this reduction does not need oxygen replenishment, it can occur close to the corroding point, and the low circuit-resistance compensates for the fact that the E.M.F. of the cell iron | hydrogen is lower than that of the cell iron | oxygen. Where this happens, the main train of cations is towards the area of hydrogen-liberation, and the opposite movement (of iron ions into the liquid) is prevented. If such an explanation is the correct one, we are dealing with a case of true cathodic protection, and the disarrayed iron at the point where attack is proceeding can be regarded as equivalent to a speck of zinc attached to an iron surface, which would certainly produce a bright, protected circle all around it. With dilute, badly conducting solutions, the protected area will be of limited extent, but with concentrated solution, protection may extend further—stretching right across a specimen 2·5 cm. broad in the case of an N/10 solution

Good examples of the mutual protective effect are furnished by the crack-heal experiment described on p. 176, but perhaps the most striking is that of concentric corrosion rings.

Concentric Corrosion Rings. If, for any reason, intense corrosion has developed at a centre, most points within a certain distance are protected, giving a circle much more immune from attack than the surface as a whole. This in turn makes the probability of attack on an annular area outside the protected circle abnormally great, and a ring of corrosion here develops, surrounded, in its turn, by a second protected circle, outside which comes a second corroded ring, and so on. The phenomenon can occur on zinc and on iron—and has been observed at Teddington, Cambridge and elsewhere; one of the best examples was provided by Homer, working at Birmingham, on iron placed in a solution containing sodium chloride (a corrosive salt) and sodium carbonate (an inhibitor) in proportions adjusted to the border-line between corrosion and passivity; Homer's picture shows four concentric rust-rings (C. E. Homer, *Rep. Iron St. Inst. Corr. Comm.* 1934, **2**, 225; esp. fig. 88 (plate XXIV). Cf. G. Chaudron, *Helv. chim. Acta* 1948, **31**, 1553, esp. fig. 5, p. 1562).

(3) Spreading, Healing or Pitting

General. An important feature of corrosion is that the attack which starts at sensitive spots, sometimes *spreads out*, apparently when the corrosion-product can dissolve the invisible protective skin around the points of origin; sometimes on the other hand, it ceases to develop, the weak place in the film being apparently *healed*; more rarely it remains localized, producing intense attack or *pitting*.

On most metals, anodic action may, according to the pH, lead either to (1) active corrosion, with formation of a soluble salt, or (2) a solid oxide, hydroxide or basic salt which will thicken the original film and possibly heal the weak places. In a definitely acid liquid the first would generally be the " expected reaction "—the reaction which leads to the lowest energy state; for any oxide or hydroxide momentarily formed would be unstable, and capable of dissolving in the acid liquid with further decrease in energy. In a weakly alkaline solution (insufficiently concentrated to redissolve the hydroxide) the second would be the " expected reaction", since any soluble salt momentarily formed would be unstable and capable of throwing down oxide or hydroxide. It is true that the " expected reaction "—the one which would lead to the state of lowest energy—is not always the reaction which is met with. Nevertheless, the principle suggested above is a fairly safe guide to behaviour; experience shows that more often than not, soluble products are obtained in slightly acid liquids and solid products in weakly alkaline liquids.*

For most heavy metals it is true to say that a solution of an ordinary salt

* Where corrosion is occurring very slowly and reversible conditions can be assumed, only the " expected reaction " is possible and the principle is infallible. Where corrosion is speeded up by application of an external E.M.F., as in anodizing, activation energy enters into the decision between alternative reactions, and film-formation may occur from an acid bath. See p. 1042.

(chloride or sulphate) tends to deposit hydroxide or basic salt at pH 7; a clear solution is only obtained if a small amount of acid is added. Let us consider a specimen of such a metal—first exposed to air and then placed in a strictly neutral solution of a sodium or potassium salt; if the " expected reaction " occurs, this will tend to heal the sensitive spots in the air-formed film, or at least thicken that film, rather than yield a soluble corrosion-product. However, the formation of metallic oxide or hydroxide requires the consumption of OH^- ions, and thus lowers the pH value; provided that the acid formed accumulates on the spot, and does not (in the time available) expend itself in secondary reactions, we shall sooner or later reach a stage where active attack can set in, producing a soluble salt.

When once a soluble salt starts to be formed at the sensitive spots, there is the possibility that it may affect the protective film on the surrounding area. On vertical or sloping specimens of zinc or iron in chloride solution, the attack starting at isolated points on the face is observed to extend downwards and to some extent sideways over *arch-shaped areas*—evidently the areas covered by the heavy anodic product sinking under gravity; on horizontal surfaces, the spreading usually leads to roughly circular areas if conditions are stagnant. In running water, the attack often spreads in streaks parallel to the flow-direction.

The reasons for this extension—and for the healing which occurs in other cases—now demand consideration; it will be convenient to consider the behaviour of the various metals one by one.

Zinc. There is some evidence that when a zinc plate is partially immersed in potassium or sodium chloride solution, thickening of the film takes place before active corrosion sets in. Certainly adherent white matter (producing rude interference colours when the specimen is held at a suitable angle to the light) occurs on the lower part of the specimen, around the points which subsequently develop active corrosion. It is a little uncertain whether this adherent matter results from thickening by anodic production of solid matter, or whether it is due to soluble anodic and cathodic products (zinc salt and alkali) being formed at contiguous points, and interacting so close to the metal as to give adherent oxide or hydroxide. This second explanation was preferred when the phenomenon was first reported (U. R. Evans, *J. chem. Soc.* 1929, p. 111, esp. p. 112).

When once active corrosion, with the production of soluble zinc chloride, has set in at these sensitive points, it starts to extend outwards and downwards as arch-shaped areas. Nor is this surprising. It has long been known that zinc chloride interacts with zinc oxide—this being the basis of the so-called " oxy-chloride cements " made by stirring zinc oxide powder with concentrated zinc chloride solution and moulding the mixture into shape; the oxide particles dissolve, but the solution produced is super-saturated in relation to a hydrated basic chloride, which separates out as a new solid phase, takes up in its formation all the water present, so that the whole mixture sets into a solid mass.* On our specimen of metallic zinc immersed

* There are two important basic chlorides (hydroxychlorides), which have been studied by W. Feitknecht and his collaborators. F. M. Aebi, Dissertation,

in chloride solution, the zinc chloride formed at the sensitive points will certainly be capable of dissolving the oxide-film at points below it, but in view of the large volume of liquid, it is doubtful whether any solid oxy-chloride will be deposited; certainly the arch-shaped area will not come to be covered completely with a continuous film of oxy-chloride, and the development of active attack over this area is to be expected.

Iron. Corrosion on iron starts at isolated points, as in the case of zinc, and from some of these, the attack extends as circles, streaks or arch-shaped areas. Not all the points where corrosion is observed to start, however, suffer continued attack. In an early research, plates of pure electrolytic iron partly immersed in chloride or sulphate solution showed corrosion-products streaming down from surface defects within an hour of immersion, but later the attack at most of these points ceased to develop; after 21 hours corrosion was confined to a region at the bottom bounded by a straight horizontal line. Healing at points where attack has started was observed more commonly on pure iron than on steel (U. R. Evans, *J. chem. Soc.* 1929, p. 111, esp. p. 117).

On sloping specimens of ordinary iron or steel, corrosion starts at points and extends outwards and downwards as arch-shaped areas. These cannot be due to direct attack upon ferric oxide by ferrous chloride (or by the acid resulting from its hydrolysis); the solubility product of ferric oxide is very low, and even hydrochloric acid, which should ultimately dissolve it, acts only very slowly. If, however, metallic iron is also present "reductive dissolution" occurs quite readily (as explained on p. 225); the cell

Iron | Acid liquid | Ferric oxide

in which the ferric oxide is the cathode, brings the oxide into solution, not as Fe^{+++} but as Fe^{++} ions. Such a cell can be set up at any weak spot on the film, and, as the film around the original weak spot is destroyed, the cell resistance will decline. Thus reductive dissolution should rapidly destroy the film on the arch-shaped area over which the anodically produced ferrous chloride is flowing; even weakly acid solutions are capable of supporting reductive dissolution and ferrous chloride solution would be expected to do so, although quite incapable of causing direct dissolution.

To ascertain whether the liquid descending from the starting-point of attack is really capable of destroying the oxide-film, special experiments were carried out with specimens of heat-tinted iron on which a scratch-line had been engraved, before being partly immersed (with the scratch-line vertical) in sodium chloride solution. Corrosion soon commenced at the scratch-line, and the corrosion-products, sinking downwards and outwards, were observed to destroy the colours over an arch-shaped area. To put the matter beyond doubt, it was arranged for the corrosion-products from

Berne, 1946, p. 10, gives the following solubility products (the figures in brackets are the numbers adjusted for one equivalent of zinc):—

Hydroxychloride II . .	$(ZnCl_2.4Zn(OH)_2)$,	$8{\cdot}2 \times 10^{-76}$	(1×10^{-15})
Hydroxychloride III . .	$(ZnCl_2.6Zn(OH)_2)$,	1×10^{-110}	$(1{\cdot}8 \times 10^{-16})$
Oxide, in its most active form	(ZnO) ,	7×10^{-17}	(7×10^{-17})

a scratch-line on a half-immersed specimen to flow off on to a second specimen, placed lower down and out of electrical contact with the first—so that complications due to differential aeration were avoided; here again the colours disappeared on the area over which the corrosion-products were flowing, but not elsewhere. If the corrosion-products can destroy the relatively thick films produced by heat-tinting, they can certainly destroy the invisible films present on unheated specimens.

Healing occurs regularly on iron and steel in regions where cathodically formed alkali comes to predominate; one example is afforded by the water-line zones on a partly immersed specimen which develops interference tints but generally escapes active corrosion (p. 89). The principles involved are, however, better illustrated by the behaviour of drops of sodium or potassium chloride solution placed on a horizontal steel surface. The distribution of anodic and cathodic points is conveniently shown up by the addition of the so-called *ferroxyl indicator* (phenolphthalein plus potassium ferricyanide);

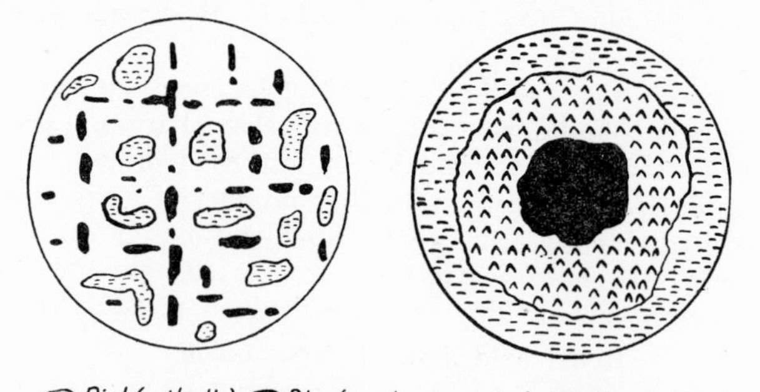

(a) (b)

Fig. 30.—Effect of drop of Potassium Chloride solution containing Ferroxyl Indicator: (a) primary distribution, (b) secondary distribution.

these two reagents are indicators for alkali and ferrous salts respectively, and thus render visible the cathodic and anodic points as pink and blue spots.*

If a drop of sodium chloride solution containing a little ferroxyl indicator and saturated with air is placed on an abraded steel plate, small pink and blue spots soon appear over the whole drop area; the blue (anodic) points generally tend to lie along abrasion lines. This is the *primary distribution* (fig. 30(*a*)); it persists until the oxygen originally dissolved in the liquid begins to be

* The amount of each reagent must be kept as low as possible. If large amounts of ferricyanide is added, the distribution of cathodic and anodic spots will not be the same as in the absence of indicator, since the reduction of ferricyanide can serve to support the cathodic reaction at places which oxygen cannot reach, whilst the precipitate of ferrous ferricyanide formed at points where anodic action has started, may act as an oxygen screen and thus perpetuate anodic attack at a place where, in absence of indicator, it would die away; a coloration rather than a precipitate should be aimed at. A suitable solution is produced by adding 0·5 c.c. of 1% alcoholic phenolphthalein and 3 c.c. of 1% potassium ferricyanide (freshly prepared) to 100 c.c. of N/10 chloride (U.R. Evans, *Metal Ind.* (*Lond.*) 1926, **29**, 481).

exhausted; then gradually, in the central portion, where, owing to the greater depth of liquid, oxygen-replenishment from the air is slowest, alkali formation almost ceases, and the pink spots may be seen to disappear. If the production of ferrous chloride is now predominating at the centre, that of alkali must prevail at the margin, since in the drop as a whole the anodic and cathodic products are being formed in equivalent quantities. Thus at this stage, any iron ions formed at weak spots in the marginal zone (where alkali is in excess) will be deposited as solid hydroxide or oxide at the surface, healing the weak places, so that anodic attack ceases, and the blue spots disappear; we have then a blue interior and a pink margin separated by a brown rust-ring, due to interaction between anodic and cathodic products along with oxygen. This is the *secondary distribution* (fig. 30(*b*)). If the original liquid has been freed from oxygen (by boiling, with subsequent cooling to room temperature in nitrogen before the two indicators are added) the secondary distribution is observed from the outset; if the liquid has been supersaturated with oxygen (by bubbling in oxygen at about 0°C., and then warming to room temperature without disturbing the liquid surface), the primary distribution persists for an abnormally long period.

Tuberculation and Pitting in Steel Pipes. The early stages of the corrosion of steel by supply water are probably connected with electric currents passing from point-anodes to the main surface as cathode; the intensity of attack is due to the large cathode/anode ratio, and the rate of penetration determined by the rate of supply of oxygen to the surface outside the pit. On the inside of a steel pipe which carries mill-scale when put into service, the anodic points at the out-set may be tiny (possibly invisible) gaps in the scale. The causation of the original points of attack is not very important, however, as the corrosion-pattern often changes with time. We will consider what happens when the corrosion-pattern has become stable, in a case where the anodic attack is still at that stage concentrated upon small points.

In the case of iron in a water containing, say, sodium sulphate in addition to calcium bicarbonate, we may picture each anodic point where ferrous sulphate is being formed, covered with a blister of membraneous ferric hydroxide, following a roughly hemispherical surface and separating the region where ferrous sulphate is in excess from the region which has become alkaline, owing to the cathodic reaction, and contains oxygen. Probably calcium carbonate is also formed along the same surface of separation between the alkaline region and the slightly acid region, and will provide a supporting structure for the membrane—a view held by T. E. Larson and R. V. Skold, *Corrosion* 1958, **14**, 285t.

Already at this stage a shallow pit will have begun to form below the blister, but a new change now becomes possible which will greatly increase the intensity of pitting. If a small amount of oxygen passes in through the membrane, a small concentration of Fe^{+++} ions may be produced, and soon the solubility product of magnetite, which is very low, and of various green ferroso-ferric compounds (hydroxides or basic salts), which are also very sparingly soluble, will be surpassed. Any of these compounds may separate as a solid; the blisters present on corroding steel surfaces generally contain magnetite or green ferroso-ferric compounds. If any of these compounds separate, acid will be liberated, and will continue to accumulate until equilibrium is established—represented, for instance, by

$$Fe^{++} + 2Fe^{+++} + 4H_2O \rightleftharpoons Fe_3O_4 + 8H^+$$

Baylis found that the interior of *tubercles* (the large, usually elongated blisters found inside water-pipes) showed a pH value of about 6, irrespective of the pH value in the main body of the water flowing through the pipe; he also showed that the concentration of sulphates or chlorides was higher within the tubercles than outside; in general each tubercle overlies a pit (J. R. Baylis, *Met. Chem. Eng.* 1925, **32**, 874).

The pH value of 6·0 recorded by Baylis doubtless represents a compromise value at which the production of acidity in the outer part of the blister (just within the membrane) which oxygen will reach exactly balances the consumption of acidity in the pit through attack upon the metal. An alarming situation may be envisaged in which acid is continually acting on the metal, producing ferrous salts, which are continually regenerating fresh acid by oxidation and hydrolysis, so that further attack can then occur. At first sight, it might seem that, so soon as a small amount of ferrous salt has been produced within the membrane, corrosion would continue indefinitely—even if the original electrochemical action were to cease (e.g. by stifling of the cathodic reaction on the main surface outside the blister with a " chalky " film). Fortunately this does not occur, since hydrolysis is never complete, and provided that fresh SO_4^{--} ions are not pumped through the membrane by the potential gradient, the pitting should gradually become slow and cease. The matter requires experimental investigation, and the electrochemical situation is far from simple, but there would seem to be a reasonable prospect of retarding, and perhaps arresting pitting, in such cases, by shutting off the supply of dissolved oxygen.*

The attack of the acid on the steel within the pit may well be of the hydrogen evolution type, but it is unlikely that much hydrogen is evolved in bubbles; probably the hydrogen atoms, formed in the pits by the cathodic reaction $H^+ + e = H$ (an alternative reaction to the oxygen-reduction proceeding on the cathode area outside) will diffuse into the metal, as explained in Chapter XI; if to a small extent they join to form H_2 molecules, these

* If the original amount of ferrous sulphate is Q and the fraction which liberates acid by oxidation and hydrolysis is α, the total destruction of metal including that destroyed in forming the original amount will be in absence of replenishment $Q(1 + \alpha + \alpha^2 + \alpha^3 + \ldots) = Q/(1 - \alpha)$. It might perhaps be argued that when acid liberated by hydrolysis is consumed by attack on metal, a further quantity should be liberated, so that ultimately hydrolysis should proceed to completion; this would mean that $\alpha = 1$, and a finite quantity of ferrous salt should produce an infinite quantity of corrosion. In practice, however, SO_4^{--} is probably removed as a green ferroso-ferric basic salt, or by adsorption on the membrane of ferric hydroxide, or again by anchorage to micelles of colloidal ferric hydroxide which are equivalent to basic salts of variable composition. Thus in practice α is likely to be less than unity, making the total corrosion produced finite.

can diffuse outwards dissolved in the water. However some hydrogen may be formed in bubbles; this might be the cause of the occasional bursting of the tubercles noted by Baylis, although other explanations (such as the voluminous character of the corrosion-products) suggest themselves.

Sulphate-reducing bacteria undoubtedly play a part in the internal corrosion of some pipe-lines. The matter is discussed on p. 297, but reference may here be made to K. R. Butlin, M. E. Adams and M. Thomas, *Nature* 1949, **163**, 26.

Aluminium. It is fairly certain that on both aluminium and tin the first anodic effect is a film-thickening; serious destruction of the metal, with formation of soluble corrosion-products, comes later. With aluminium, the existence of various ions, Al^{+++}, $Al(OH)^{++}$, $Al(OH)_2^{+}$, leads to a variety of possible anodic reactions.

$$
\begin{aligned}
Al &= Al^{+++} && + 3e \\
Al + H_2O &= Al(OH)^{++} + H^+ && + 3e \\
Al + 2H_2O &= Al(OH)_2^{-} + 2H^+ && + 3e \\
Al + 3H_2O &= Al(OH)_3 + 3H^+ && + 3e
\end{aligned}
$$

The last reaction will give a solid corrosion-product; whether $Al(OH)_3$ is actually obtained—or $AlO(OH)$ or Al_2O_3 by loss of water—does not affect the argument.

Possibly we should picture these changes as occurring simultaneously, but certainly the upper ones will be favoured by acid conditions and the lower ones in nearly neutral conditions. Under ideal circumstances, where the reactions occur in such amounts as to keep the various ions in equilibrium with one another, the last equation should predominate about pH 7, whilst the first probably only becomes important below pH 6. At first sight, it might seem that in a neutral liquid, film-repair would prevail, and corrosion would be avoided indefinitely. This overlooks the fact that the last (film-building) reaction automatically generates H^+. Provided that the acidity formed at the anode is not dispersed into the body of the liquid by stirring or destroyed by cathodically formed alkali, the pH at an anodic point will sink, and conditions soon become increasingly favourable to destructive corrosion with the formation of soluble products. Edeleanu has calculated that, provided dispersal of acid is avoided, the pH will have sunk into the range favourable to the formation of soluble products before hydroxide (or oxide) has been formed in sufficient amount to cover the entire surface with a monomolecular layer. Only when the pH is sufficiently low for the first reaction to predominate will the formation of fresh acidity cease.

This production of acidity at an anode has been definitely established; Edeleanu followed the changes of pH close to an aluminium specimen made an anode by means of an external current. Fig. 31 shows the drop of pH in chloride solution with time; the drop becomes slow at about pH 5·3; in sulphate solution the descent continues to pH 4·7 (C. Edeleanu and U. R. Evans, *Trans. Faraday Soc.* 1951, **47**, 1121).

In the case of a specimen to which no external current is applied, the small amount of acidity produced by local anodic action at a susceptible point may either accumulate on the spot, or it may be dissipated by stirring or convection; alternatively, if the solution is a sodium or potassium salt solution, the small amount of acidity formed at the anodic points may be neutralized by the larger amount of alkali produced at the cathodic area around, and if this occurs, we may expect film-building to continue. Early experiments on aluminium plates partly immersed in potassium chloride or sulphate showed no loose corrosion products in the opening stages. On abraded specimens in N/10 sulphate, the liquid was only very slightly cloudy after 4 days and there was only a limited amount of settled corrosion-product

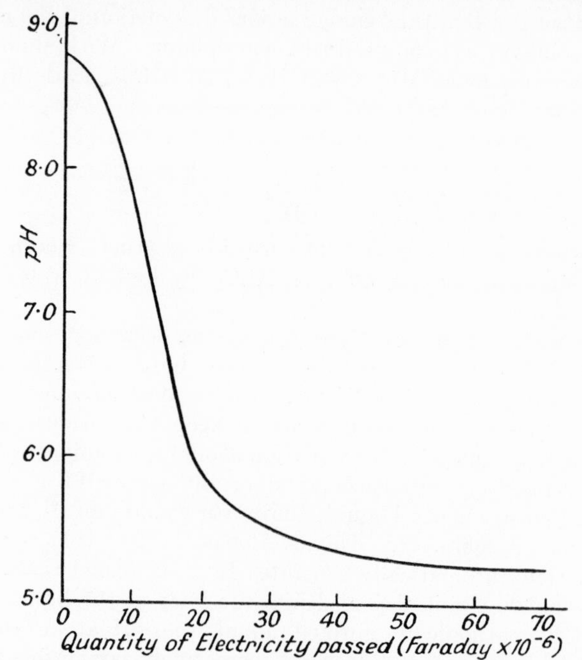

FIG. 31.—Sinking of *p*H value as Corrosion proceeds on an Aluminium Anode in Chloride solution (C. Edeleanu and U. R. Evans).

after 15 days; on iron there is measurable production of loose corrosion-product within an hour (U. R. Evans, *J. chem. Soc.* 1929, p. 111, esp. p. 115).

On " as-rolled " aluminium the corrosion-product appeared as an adherent white deposit at points along blemish marks, also at the edges and along the water-line. Specimens taken out after appropriate intervals, dried and examined under suitable illuminations, provided evidence of film-thickening. Around the points where attack had started, rings of interference colours were observed, the advanced tints being in the centre; the inner (thicker) rings showed white cloudiness by scattered light, with interference colours complementary to those obtained from the same rings by regularly reflected light—a phenomenon capable of optical explanation.* It is not quite

* See, for instance, U. R. Evans, *Chem and Ind.* (*Lond.*) 1926, p. 211, esp. p. 213.

certain whether the films producing the colours were due to the thickening of the original air-formed film, or to aluminium ions leaking out into a liquid where cathodically formed alkali was in excess.

One specimen taken out from sulphate solution after 4 days displayed a first-order yellow colour over the main surfacc—the yellow being so faint that it might easily have escaped observation; the second- and third-order colours were often very clear and beautiful.* Another specimen taken out after 7 days displayed on different parts the entire sequence of interference tints, ranging from the first-order yellow to the fifth-order red.

It may be asked, why, in cases where film-building is prevailing, corrosion ever becomes serious; it seems likely that the alkali formed by the cathodic reaction slowly dissolves the film over considerable regions, thus exposing the metal to attack—which may become fairly general. If, however, the liquid is one in which the cathodically produced alkali cannot accumulate, and if conditions are such that the anodically produced acid can accumulate at the original starting-points of attack, then we may expect pitting, since the low pH will prevent film-forming reactions, and the attack, concentrated on a limited number of microscopic areas, may become rather severe. The danger of pitting is greatly increased, however, if traces of copper are present in the liquid; the copper may slowly be deposited by the cathodic reaction and then provide a cathodic surface far more efficient than an alumina film. The accumulation of alkali is prevented if the liquid contains calcium bicarbonate, which reacts with the alkali. It has been found that supply waters often produce pitting on aluminium if they contain (1) calcium bicarbonate, (2) chloride, (3) oxygen and (4) copper traces. The necessity of these four factors for pitting by waters of the type met with in South England was established in an important research by F. C. Porter and S. E. Hadden, *J. appl. Chem.* 1953, **3,** 385. (Cf. O. Sverepa, *Werkst. u. Korrosion* 1958, **9,** 533.)

The deposition of relatively noble metals on the cathodic areas around the pits has been established by Aziz using tracer methods, based on the radioactive isotopes of cobalt and lead; he has also measured the corrosion currents flowing around pits on aluminium placed in various fruit juices (P. M. Aziz, *J. electrochem. Soc.* 1954, **101,** 120; *Corrosion* 1957, **13**, 536t.)

The spreading of corrosion as arch-shaped areas—a characteristic feature on sloping zinc or iron specimens—was not seen on aluminium in the Author's 1929 research, although the circles of initial attack became elongated downwards. Schikorr noticed that both the anodic and cathodic products (aluminium chloride and sodium hydroxide) can attack aluminium with evolution of hydrogen, thus leading to more general corrosion—as suggested above (U. R. Evans, *J. chem. Soc.* 1929, p. 115; G. Schikorr, *Mitt. Mat. Prüfungsanstalten* 1933, Sonderheft **22,** p. 22. Cf. New Zealand work quoted on p. 132. Unpublished work by K. F. Lorking and J. E. O. Mayne should receive study when available).

Tin. An instructive case of localized corrosion, explained by Hoar on

* The reason for the later colours being brighter than the early ones in cases where the film-substance is very transparent and the metal highly reflective is given in references indicated on p. 790.

principles rather similar to those afterwards used by Edeleanu during his study of aluminium, concerns the black spots which at one time used to cause much distress to users of tinned dairy equipment. The salts of tin are strongly hydrolysed, and it may be expected that in a neutral solution anodic attack at a weak spot on the air-formed film covering tin will not cause tin cations to pass into the liquid, but rather cause an increase in the film-thickness by deposition of oxide and hydroxide. Hoar showed that local thickening did occur, causing first-order interference tints over the whole surface after 72 hours in N/10 potassium chloride, with colours of a higher order (indicating greater film-thicknesses) immediately around the points where the attack was developing. After a certain time the accumulated acidity at these points apparently became sufficient for the formation of soluble stannous ions, undermining the film, so that break-down occurred, with the formation of the black spots. For details, and the interpretation of the potential changes which enabled Hoar to follow the build-up and breakdown of the film, the original paper should be consulted (T. P. Hoar, *Trans. Faraday Soc.* 1937, **33**, 1152).

The blackness of the spots is probably due to the absence of reflection from the locally roughened surface; there is no need to postulate a black tin compound (S. C. Britton, Priv. Comm., June 22, 1952).

Copper. In contrast to iron, where the ferric compounds are generally less soluble than the ferrous compounds, the lower (cuprous) series of copper compounds are less soluble than the higher (cupric) series. Cuprous chloride, for instance, is sparingly soluble, and it might be expected that a copper immersed in, say, potassium chloride solution would escape corrosion, since any weak spots in the air-formed film would be healed with cuprous chloride. Actually a copper specimen placed in potassium chloride solution shows rapid change, although for some days this takes the harmless form of surface coloration (best seen after removal and drying of the specimens), the liquid remaining free from precipitate. The early stages of copper corrosion in potassium chloride and sulphate solutions were found at Cambridge to vary greatly with the previous abrasive treatment of the metal; the facts are complicated, but the following summary will serve to convey typical phenomena.

The changes start, as on other metals, at isolated spots, which are most plentiful near the cut edges; but instead of producing loose matter, as on iron or zinc, they lead to oval areas of bright interference tints on the finely ground specimens, or rather dirty brownish patches on coarsely ground specimens. It is believed that solid cuprous chloride formed by the anodic reaction, being only very slightly soluble, is converted *in situ* into cuprous oxide by the cathodically formed alkali creeping along the surface; the oxide-film thus formed is responsible for the coloration. Meanwhile the rest of the surface develops lighter colours, believed to be formed by cuprous ions leaking through the original film into a liquid where cathodically formed alkali already predominates; if so, they should be regarded as analogous to the colours formed near the water-line on zinc or iron.

During this early period, the liquid remains clear; the corrosion-products are entirely adherent. However, after about 3 days, green basic cupric chloride appears, relatively loose, probably formed by the action of oxygen on the small amount of cuprous chloride in solution, and the liquid becomes cloudy (U. R. Evans, *J. chem. Soc.* 1929, p. 111, esp. p. 118; see also W. Feitknecht and W. Schütz, *Rev. Métallurg.* 1955, **52**, 327).

No pitting was observed in the research at Cambridge just mentioned. The pitting of copper is a relatively exceptional phenomenon; it has been investigated by May, whose paper deserves study. He pictures the copper surface covered with a uniform scale of oxide which locally becomes faulty, either through the existence of crevices in the metal, the settlement of small particles on the surface during film-formation, mechanical scratching or some similar cause. Sometimes the fault may heal itself by the anodic formation of solid cuprous chloride, but sometimes the cuprous chloride, which is appreciably soluble, will diffuse out into the liquid as soon as it is formed, and be converted either to the basic cupric chloride by oxidation and hydrolysis, or perhaps to cuprous oxide by the action of cathodically formed alkali. If this occurs, the repair of the film by plugging with cuprous chloride will not occur, and the cuprous chloride (a reducing agent) will prevent fresh oxygen from reaching the metal, so that repair by oxide-formation will now be impossible; thus attack, once started, will continue* (R. May, *J. Inst. Met.* 1953–54, **82**, 65).

Especially in fresh waters, pitting is most likely to occur when the main surface is one on which the cathodic reaction (normally reduction of oxygen) can proceed readily, since this will increase the current at the small anode. Unfortunately copper pipes sometimes carry on their interior surface carbonaceous films (often almost invisible) derived from lubricant used in the drawing process; such a film can provide an efficient cathode. Alternatively, they sometimes carry a special type of oxide-film which also acts as an efficient cathode. Every endeavour is now made at the tube works to control the furnace atmosphere and thus avoid the production of either carbonaceous or oxide layer; but if either layer should exist on the pipes when put into service with certain waters, pitting is liable to develop.

Fortunately only a few waters produce pitting even if the carbonaceous layer or dangerous type of oxide is present. Campbell has discovered that the majority of natural waters contain an organic substance which suppresses the trouble.† The *suppressor* has not been isolated, but electrochemical tests are available for its detection. It appears to be a colloidal substance, with particles carrying negative charges. It is removed by treatment with alum, the negative particles being flocculated by the positive Al^{+++} ions.

Copper pipes, even if free from carbonaceous and dangerous oxide films, may suffer pitting if used in hot-water systems carrying certain soft

* The fact that sometimes the healing of the fault with solid cuprous chloride fails to occur suggests that the nucleation-energy is high; research into the nucleation characteristics of different corrosion-products would be welcome.

† The suppressor is believed to be adsorbed on the growing edges of cuprous oxide crystals, and hinder their growth, so that, instead of a few large crystals, a mass of small ones is formed—which is far more likely to provide a continuous protective layer.

moorland waters containing manganese. The oxides of manganese are deposited cathodically around the sensitive spot where anodic attack is starting, and can take up additional oxygen from the water to produce a higher oxide, which will then stimulate the cathodic reaction. If the suppressor is present, pitting will not occur, but unfortunately the alum treatment often applied to such waters for other purposes will remove the suppressor (H. S. Campbell, *J. Inst. Met.* 1950, **77**, 345; *Trans. Faraday Soc.* 1954, **50**, 1351; *J. appl. Chem.* 1954, **4**, 633; *Proc. Soc. Water Treatment and Examination* 1954, **3**, 100; see also remarks by P. T. Gilbert, p. 111).

Location of Corrosion Spots and Pits. The question is often asked whether the points at which local corrosion develops are decided by some peculiarity in the metal or by some external factor—such as contact with solid particles. If the cause first suggested is operative, the manufacturer should endeavour to eliminate the peculiarity and thus avoid the danger of spots or pitting; in the second case, the remedy must be left to the user.

Doubtless in some cases internal factors do play a part in deciding the sites of localized corrosion—provided that external conditions are such as to keep it localized. Homer's work on the corrosion of steel in chloride-carbonate solution shows that local corrosion tends to develop at certain sulphide inclusions (p. 115), but it rarely remains so intense as to be described as " pitting "; often the attack either spreads out or ceases altogether. Probably most of the rust spots seen on steel exposed indoors to air are due to settlement of certain kinds of dust particles—as indicated by Vernon (p. 493).

Zinc and aluminium sheet, exposed to certain waters or salt solutions, show point corrosion, sometimes developing into pitting. The points are often more numerous on one face of the sheet than the other. This may be due to dust-particles, probably consisting of iron, copper or carbonaceous material, settling on the upper surface at the rolling-mills, and being pushed into the metal; thus the spots are more frequent on the upper surface. It is not certain, however, whether the pitting or localized corrosion could be prevented altogether by avoiding such settlement of dust.

It is fairly certain that in some cases pits occur at random, their sites being independent of any inclusion or other local peculiarity in the metal. Consider copper or a copper alloy placed in water containing sodium chloride and oxygen. A cuprous oxide film develops on the surface, probably through the electrochemical reactions suggested above. Doubtless from time to time this film becomes leaky at certain places, but they will soon be repaired, and although stresses, internal or applied, may be important in deciding the regions favourable to this crack-heal phenomenon, they will not, in general, determine the sites of pitting. However, in addition to Cu^+ ions a small amount of Cu^{++} may be formed. The cathodic reaction

$$O + H_2O + 2e = 2OH^-$$

is balanced mainly by

$$Cu = Cu^+ + e$$

but partly by

$$Cu^+ = Cu^{++} + 2e$$

Energy considerations suggest that it is possible to reach a concentration of Cu^{++} such that the solubility product of the basic chloride $CuCl_2.3Cu(OH)_2$ (*atacamite*) is exceeded. But the complicated formula of atacamite suggests that only rarely will the necessary number of Cu^{++}, Cl^- and OH^- ions chance to come together in such a manner as to constitute a stable crystal nucleus. This exceptional occurrence probably depends on the random motion of ions in the liquid, and green atacamite may not be noticed for some days. When once a crystal nucleus has been formed, however, it is likely to grow, and will keep low the local concentration of Cu^{++} ions (and also that of Cu^+ ions, which must be oxidized to form fresh Cu^{++} ions). Thus the point where the nucleus appears is likely to be anodic towards the area around and attack, once started, will continue and produce a pit.

In the argument suggested it has been assumed that the crystal nucleus will be formed close to a crack or leakage point in the film without actually blocking it. If there is no immediate blockage, the pit is unlikely to be blocked at a later stage, since the precipitation of atacamite requires OH^- ions as well as Cu^{++} and Cl^-, and is unlikely to take place at the bottom of a pit.

(4) Differential Aeration and Moto-electric Currents

Differential Aeration Currents on partly immersed plates and in drops on horizontal plates. It has already been stated that the corrosion of partly immersed plates is connected with currents flowing between the aerated cathodic area at the water-line and an anodic area lower down. The correctness of this view is indicated by the fact that these currents have been measured, by several independent methods, and are found to correspond to the corrosion-rate in the sense of Faraday's Law; the methods are described in Chapter XXI.

The currents in drops have been mentioned in connection with the ferroxyl indicator, but some additional experiments—which can be carried out without the use of that indicator—may now be described. A drop of potassium (or sodium) chloride (or sulphate) placed on a horizontal steel sheet, develops corrosion over the central (anodic) zone, surrounded by a cathodic ring at the periphery, which remains immune; they are separated by a ring of rust, which soon spreads along the liquid surface over the whole central part of the drop forming a membrane separating liquid and air. The fact that this " corrosion-pattern " is really due to the better access of oxygen to the metal at the periphery than at the centre, where the " depth " of the liquid is greater, was shown by special experiments in which it was arranged for the periphery of the drop to be surrounded with nitrogen, whilst an oxygen current was blown down from a jet placed somewhat above the centre of the drop; the pattern is then reversed (fig. 32), an immune

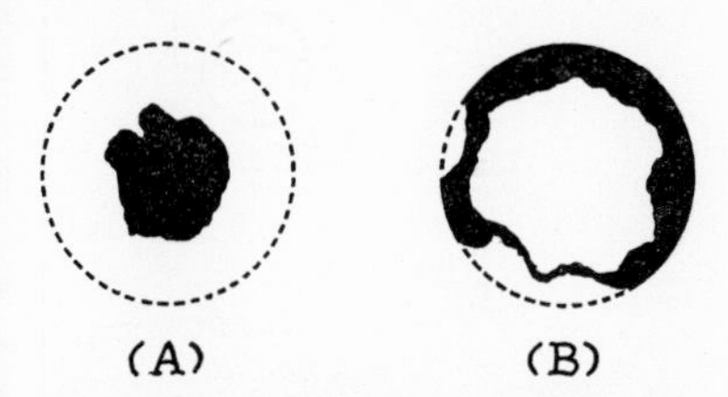

FIG. 32.—Distribution of Corrosion produced on Steel by a drop of Potassium Chloride solution, (A) surrounded by air, (B) with nitrogen around the flanks of drop but oxygen blown on to centre (corroded areas shown black).

centre being surrounded by a ring of corrosion at the periphery (U. R. Evans, *Korros. Metallsch.* 1930, **6**, 173).

A beautiful demonstration of the electric currents flowing between the centre and peripheral part of a drop has been provided by Blaha. He placed an iron specimen carrying the drop in a magnetic field, and observed that the drop started to rotate, in the same way as a wheel with current flowing along its spokes would rotate in a magnetic field; on reversal of the magnetic field, the drop rotated in the opposite direction (F. Blaha, *Nature* 1950, **166**, 607).

An earlier demonstration of electric currents flowing in drops was provided by E. Baisch and M. Werner, *Bericht über Korrosionstagung* 1931, **1**, 84, 87.

Differential Aeration Currents set up by a divided cell. Early Cambridge work was helped by a study of differential aeration currents between two separate electrodes of the same metal placed in two compartments of a cell, oxygen being bubbled over one of them. It cannot be claimed that such cells provide faithful " models " of the natural corrosion of (say) half-immersed plates, since the anodic and cathodic electrodes are not contiguous or co-planar, the external resistance will certainly be unnaturally

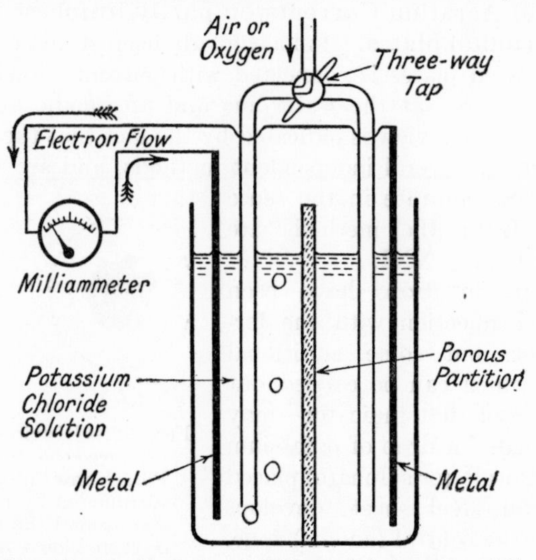

FIG. 33.—Early form of Differential Aeration Cell.

high, and the internal resistance will probably be unnaturally high also. In 1932, Hoar succeeded in measuring the differential currents on a half-immersed plate without cutting the metal or introducing other unnatural features; since that time, the study of divided cells has become less important. However, the simple apparatus used in 1923 is still instructive in showing qualitatively how easily these currents can be set up. The cell (shown diagrammatically in fig. 33) is divided into two compartments by a porous diaphragm of low resistance; both cells are filled with the same liquid (say, potassium chloride solution), and contain two electrodes of the same material (say, zinc) joined to a milli-ammeter, the scale of which has a central zero-point. The three-way tap allows a stream of oxygen or air to be directed into one or other compartment at will. A current is registered and it is found that the electrode in the compartment which is thus aerated is the cathode; when the gas-stream is diverted to the other compartment, the current quickly dies away and reverses its direction—often within a few seconds.

In the early Cambridge experiments, carried out in apparatus designed to keep the resistance low, the unaerated anode usually lost more weight than the aerated cathode; in the form of cell used by Bannister, its loss agreed closely, in the sense of Faraday's law, with the current generated (U. R. Evans, *J. Inst. Met.* 1923, **30**, 239 esp. pp. 263, 267; U. R. Evans, L. C. Bannister and S. C. Britton, *Proc. roy. Soc.* (*A*) 1931, **131**, 355, esp. p. 360).

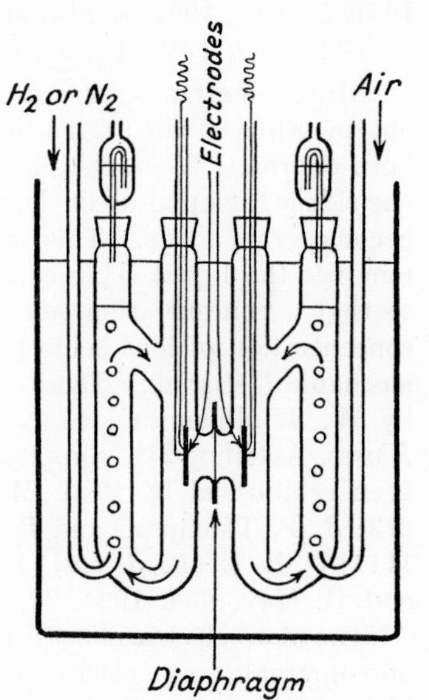

FIG. 34.—Improved Differential Aeration Cell (H. Grubitsch).

In general, there will be competition between the main anode situated in the unaerated compartment and any local anodes which may develop on the aerated electrode. If the resistance of the cell, or of the ammeter, is high, these local cells set up at different points on the so-called cathode will claim a large proportion of the total current flowing. Indeed, certain experimenters who from time to time have decided to repeat the work, but who do not seem to have appreciated the importance of keeping the resistance low, have found the aerated electrode to lose more weight than the unaerated electrode. Behaviour under these conditions varies considerably with surface preparation. The use of roughly abraded surfaces for the (aerated) cathode is more favourable to local cells than smooth ones; this may be because the bottoms of the grooves left by the abrasion act as (unaerated) anodic points—a useful suggestion due to M. Werner, *Werkst. u. Korrosion* 1952, **3**, 347, esp. p. 349.

The situation has been cleared up neatly by Grubitsch—using the elegant form of differential aeration cell shown in fig. 34, where two electrodes are placed close to (but on opposite sides of) a cellophane diaphragm; they are washed with N/10 sodium chloride carried round the apparatus by means of two gas lifts, the gas being oxygen on one side and either nitrogen or hydrogen on the other. As in the early Cambridge work, the weight-loss of the anode corresponded approximately to the total coulombs recorded by the electrical apparatus. The total corrosion (combined weight-loss of anode and cathode) was fairly constant in all experiments, being doubtless determined by the rate of supply of oxygen to specifically cathodic points on the aerated electrode. However, the part of that total corrosion which fell upon the unaerated electrode naturally varied with the conditions, and thus the current recorded by the electrical apparatus varied with the conditions also; for instance, it was depressed when the external resistance was raised, or when a smooth cathode was replaced by a roughened cathode. The work, which deserves study, is described by H. Grubitsch, *Monatshefte für Chemie* 1955, **86**, 752. Grubitsch discusses the results of K. Wickert, *Metalloberfläche* 1950, **4**, A181; 1951, **5**, A94, A96; also (with H. Wiehr and E. Jaap) *Werkst. u. Korrosion* 1950, **1**, 299; 1957, **8**, 6. Views diverging from those of Wickert have been expressed by G. Schikorr, *Metalloberfläche* 1951, **5**, A93, A96; see also M. Werner, *Werkst. u. Korrosion* 1952, **3**, 347; F. Tödt, *ibid.* 1950, **1**, 244; G. Masing, *ibid.* 1950, **1**, 433.

Moto-electric Currents. Differential aeration currents are easily obtained in a divided cell when both the electrodes are composed of zinc or both of iron. When they are both copper, the effect of bubbling air into one side is the opposite of that obtained with zinc; the electrode on that side becomes the anode. This is because the bubbling causes stirring, and removes the liquid layer containing cuprous ions from the metal surface, so that a concentration cell is set up, the copper surrounded by the low concentration of ions being the anode; the same results can be caused by mechanical stirring. These " moto-electric " currents have been studied by W. J. Müller nd K. Konopicky, *Sitzungsberichte Akad. Wien Math. Klasse*, Abteiling IIb, Supplement, 1929, p. 79. The general principle had been established by R. J. McKay, *Trans. electrochem. Soc.* 1922, **41**, 201, 213; J. F. Thompson and R. J. McKay, *Industr. engng. Chem.* 1923, **15**, 1114; U. R. Evans, *J. Inst. Met.* 1923, **30**, 239, esp. p. 270; G. D. Bengough and R. May, *ibid.* 1924, **32**, 81, esp. p. 138.

The moto-electric effect is useful in explaining cases of localized corrosion on copper alloys at places where the relative velocity of liquid and metal is highest. An example is described on p. 753.

Probably every metal really exhibits both the differential aeration and moto-electric effects, but since they act in the opposite direction, one is usually masked. On zinc, iron and usually cadmium, differential aeration prevails; on copper and sometimes lead, the moto-electric effect prevails.*

* The reason why stirring generally moves the potential of zinc in the positive direction and that of copper in the negative direction will best be understood by the reader if he applies the graphic constructions of Chap. XXI. It may, however, be helpful to suggest a simple explanation at this point. Where a copper is immersed in, say, potassium chloride solution containing oxygen, the anodic reaction (passage of copper into the ionic state) must at first be balanced by the cathodic reduction of oxygen, but under nearly stagnant conditions where copper ions can accumulate, the cathodic reaction may in the later stage be largely the deposition of copper ions. If the liquid is in motion, the concentration of copper ions present in the layer next the metal when a steady state has been established will be smaller, and the potential at which cathodic and anodic reactions occur at the same rates will become more negative as a result of the motion. It is true that the stirring would favour oxygen-replenishment, but with a noble metal like copper the corrosion-rate is so slow

LaQue has described interesting results obtained by whirling flat circular discs in sea-water, so that points near the centre move with a small velocity and those near the periphery with a high velocity. In the case of iron discs, the outer, rapidly moving region becomes passive owing to the ready renewal of oxygen, whilst the part near the centre suffers corrosion. In contrast, on Admiralty brass, the parts near the periphery are strongly attacked, owing to the carrying away of copper ions from those parts, whilst the centre is relatively immune. Rather similar experiments had been described by earlier workers, but were often wrongly interpreted; it had been claimed that the linear speed prevailing at the circular line separating the attacked and unattacked regions represented a critical velocity needed to prevent attack. As LaQue points out, this speed has no absolute significance. There will be a cell set up between the rapidly moving and slowly moving regions—a differential aeration cell in the case of iron, and a moto-electric cell in the case of brass; the rapid attack on the anodic area will help to protect the cathodic area. For details of the results, see F. L. LaQue, *Corrosion* 1957, **13**, 303t.

Other References

Important studies of the cathodic reduction of oxygen have been carried out by P. Delahay, *J. electrochem. Soc.* 1950, **97**, 198, 205; *Corrosion* 1951, **7**, 146; P. van Rysselberghe, J. M. McGee, A. M. Gropp, R. D. Williams and P. Delahay, *Corrosion* 1950, **6**, 105; G. Bianchi and B. Rivolta, *La Chimica e l'Industria*, 1954, **36**, 358.

Various studies of rust and other corrosion products should receive attention. These include papers by W. Feitknecht and G. Keller, *Z. anorg. Chem.* 1950, **261**, 61; J. E. O. Mayne, *J. chem. Soc.* 1953, p. 129; A. Fricke, Th. Schoon and W. Schröder, *Z. phys. Chem.* (*B*) 1941, **50**, 13. The importance of crystal structure of corrosion-products in determining corrosion and the importance of metastable compounds is emphasized by W. Feitknecht, *Schweiz. Arch. angew. Wiss.* 1952, **18**, 368, also " Korrosion VIII " (1954–55), p. 35 (Verlag Chemie). The influence of light in accelerating

and the corresponding oxygen consumption so small that, unless conditions are exceptionally stagnant, we may picture the oxygen-concentration as being that existing in the body of the solution, and thus unaffected by the rate of stirring. Consequently the effect of stirring is to move the potential in a negative direction. With a reactive metal like zinc, the oxygen consumption will be considerable, and in the absence of stirring oxygen may become exhausted in the layer next the metal. Thus stirring, which replenishes oxygen, raises the potential at which the sum of all possible cathodic reactions balances the sum of all possible anodic reactions. The fact that zinc ions are swept away by the stirring makes little difference. Unless the zinc ion concentration becomes high, the exact value of that concentration hardly affects the balance potential (for reasons analogous to those explained in the case of cadmium on p. 1018); if oxygen is present in excess, the reduction of oxygen would take place more easily than the redeposition of zinc, whilst if oxygen is absent, the liberation of hydrogen should (after allowing for over-potential) occur at a potential not very different from that at which zinc is redeposited. Thus the net effect of stirring is to shift the potential of zinc in a positive direction.

corrosion processes—directly or indirectly—was studied many years ago by L. W. Haase, *Z. Elektrochem.* 1930, **36**, 456, and by C. O. Bannister and R. Rigby, *J. Inst. Met.* 1936, **58**, 227. A recent research is contributed by B. Lovreček and E. Korkut, *Werkst. u. Korrosion* 1957, **8**, 277, who used a fluorescent light-source—which may reduce the possibility that the effect is really due to convection set up by heat-absorption.

An interesting graphical representation of differential aeration is published by M. Pourbaix, " Revue de l'École polytechnique de l'Université libre de Bruxelles " 1950, Nos. 5, 6.

The early history of differential aeration currents and a useful account of their mechanism is provided by W. Lynes, *J. electrochem. Soc.* 1951, **98**, 3C; 1956, **103**, 467. The currents were studied by several early scientists, most of whom were not very interested in their connection with metallic corrosion. One of the first was E. Marianini, *Ann. Chim. Phys.* 1830, **45**, 28, who was mainly concerned with the unsatisfactory state of the understanding of electrochemical cells existing in his time. A beautiful demonstration of differential aeration currents was provided by R. Adie, *Edinburgh New Phil. J.* 1845, **38**, 99; *Phil. Mag.* 1847, **31**, 350, whilst differential aeration cells were investigated—but, it seems, wrongly interpreted—by E. Warburg, *Wiedemanns Annalen der Physik u. Chemie* 1889, **38**, 321. Probably the first investigator to apply the principle of the cell to the practical problems of corrosion was J. Aston, *Trans. Amer. electrochem. Soc.* 1916, **29**, 449.

Reference may be made to an attractive semi-popular account of the subject, profusely illustrated in colours, entitled " Corrosion in Action " by F. L. LaQue, T. P. May and H. H. Uhlig (International Nickel Company); this is based on an instructive cinematograph film.

Interesting research in New Zealand on aluminium corrosion in chloride solution depends on drilling a hole in the aluminium specimen and immersing it vertically, connected to a graphite counter-electrode. The heavy anodic product formed in the hole descends over the face, and the area suffering corrosion extends downwards. The pH value within the pit is found to become lower than elsewhere, but the spreading of the corrosion is attributed to the power of concentrated aluminium chloride to dissolve oxide or hydroxide. As an explanation of *spreading*, this seems acceptable; it is analogous to that offered for spreading on zinc on p. 116. As presented, it affords no explanation of *pitting*; whether cases where the attack remains localized can be explained without involving the explanation of p. 121, future work must decide. The paper deserves study (T. Hagyard and J. R. Santhiapillai, *J. appl. Chem.* 1959, **9**, 323).

CHAPTER V

SOLUBLE INHIBITORS

SYNOPSIS

The argument starts by re-stating the essential condition for unimpeded electrochemical attack—namely that both anodic and cathodic products shall be freely soluble. A solution containing a salt chosen to produce an insoluble substance by either the anodic or the cathodic reaction is likely to inhibit corrosion; such salts may be called *anodic* or *cathodic inhibitors*. However, inhibition is sometimes met with in a solution where soluble electrodic products would seem to be predicted, and it is not always easy to know beforehand whether a substance will cause corrosion or passivity.

The warning is then sounded, that, even when the inhibitive properties of a substance have been established, it may do more harm than good if added in insufficient amount; the corrosion-rate may then sometimes exceed that which would be maintained if the inhibitor were completely absent. In such cases the attack, being concentrated on a limited area, may become really intense; the attack at a water-line can be particularly serious.

Certain special inhibitors now receive consideration, such as chromates and nitrites; more than one opinion is held about the manner in which these substances suppress corrosion.

Calcium bicarbonate—an inhibitor which is present in many natural waters and often enables them to be run (untreated) through iron or steel pipes without fear of corrosion troubles—is next considered. The treatment of soft, acid waters to produce calcium bicarbonate receives attention, followed by a discussion of silicate treatment. Consideration is given to the corrosive action of sea-water, which is often less than that of plain sodium chloride solution. Attention is then turned to condensed phosphates, and to various inhibitive mixtures suitable for preventing attack in cooling systems.

Cases where a substance can stimulate or inhibit corrosion—according to its concentration—next claim consideration. Even oxygen, under ordinary conditions a powerful accelerator of corrosion, can act as an inhibitor if present at high concentrations, or is constantly replenished by rapid water-movement. Several other inhibitors are only efficient if oxygen is present.

Next, experiments are described illustrating the *Crack-Heal* phenomenon shown by films, and its importance in explaining why the ordinary air-formed film on metals is not protective except in very mild conditions, and why in general it is necessary to maintain a small concentration of inhibitor in waters even after a fairly complete film has been built up over the metallic surface.

Other factors involved in inhibition are considered—particularly the adsorption of certain organic inhibitors at places where corrosion would otherwise start. It is suggested that the coverage of part of the surface may increase the rate of oxygen-replenishment at the parts left uncovered, so that oxygen acts as an inhibitor in circumstances where it might accelerate attack in absence of the " adsorption inhibitor "; this may explain why the protective film found on the metal is often essentially oxide, even though the presence of some organic compound is needed for successful inhibition. Alternative explanations for the inhibitive action of these substances are offered.

General Principles governing Inhibition

Action of Drops of Different Salt Solutions on Iron. In Chapter IV it was explained that electrochemical corrosion is often serious in those cases where both anodic and cathodic products are freely soluble. The same argument suggests that, if one or other product should be sparingly soluble, there is likely to be less attack. The validity of the prediction was shown in early studies of the attack produced by drops placed on a horizontal surface of abraded steel. Despite the simple character of this work, the results are sufficiently instructive to justify a brief summary here.

Each of the numbers recorded in Table IX represents the average of measurements made on four (or more) drops; the individual measurements for the shortest time (5·5 hours) show excellent agreement, but there is considerable scatter for the experiments lasting longer than 24 hours, which have, therefore, been omitted; the scatter was apparently due to the spreading of the drop caused by the alkali formed by the cathodic reaction at the margins—a spreading which varied from drop to drop. For details, see U. R. Evans, *J. Soc. chem. Ind.* 1924, **43,** 315T; the results are also reviewed in *J. electrochem. Soc.* 1956, **103,** 73, esp. p. 77. The scatter of the individual results is shown in Table XXXIII (p. 917) of this book.

TABLE IX

CORROSION OF STEEL BY SINGLE DROPS of diameter about 1 cm., at 29°C., expressed in mg.

Liquid	Time 5·5	18·5	24·0 hours
Distilled water	0·26	0·74	0·99
N/10 KCl	0·65	1·98	—
N/10 K_2SO_4	0·74	2·10	—
N/10 KNO_3	0·78	2·35	—
N/10 (M/30) Na_2HPO_4	—	—	0·005*
N/10 $ZnSO_4$	—	—	0·18
N/10 $NiSO_4$	—	—	0·68

(The sign — indicates " no experiment ")

* Average of five experiments. A sixth drop produced 0·65 mg. and the result is omitted (see text).

It will be noticed that the three potassium salts (chloride, sulphate and nitrate), which give soluble anodic and cathodic products, caused more corrosion than the distilled water from which the solutions were made. However, zinc sulphate caused less corrosion than the distilled water; here we would expect a sparingly soluble cathodic product—whether zinc hydroxide or oxide—and actually the entire area (particularly the peripheral zone) was observed to become covered with an adherent grey-buff deposit containing iron as well as zinc. This was probably formed by partial interaction of the zinc hydroxide with the iron salts resulting from the slow anodic reaction; it would produce a surface on which the cathodic reaction could not easily take place, since the thick zinc hydroxide layer would itself be a bad conductor for electrons, and would screen from oxygen any potentially cathodic spots on the metal below it. Nickel sulphate allowed considerably more corrosion than zinc sulphate, possibly because the cathodic deposit contained metallic nickel as well as hydroxide or oxide; metallic nickel would constitute an efficient cathode. Sodium phosphate, which would be expected to produce a sparingly soluble anodic product (an iron phosphate) caused, in five cases out of six, no visible change to the metallic surface and a barely detectable amount of iron in the liquid, which remained clear; but in the sixth case a considerable amount of corrosion was produced, presumably because the drop chanced to rest on an area containing an exceptionally sensitive point. (The inhibitive action of phosphates was, however, later found to be somewhat less simple than has just been suggested; it is discussed on p. 136.)

Two Classes of Inhibitor. Zinc sulphate and sodium phosphate can both be regarded as " inhibitors " of corrosion on iron. It is natural to distinguish *cathodic inhibitors*, like zinc sulphate, which cause interference with the cathodic reaction, from *anodic inhibitors*, like sodium phosphate, which stifle the anodic reaction. At one time, great importance was attached to this distinction, since it was believed that cathodic inhibitors, although less efficient than anodic ones, were " safe ", in that they did not localize and thus intensify such attack as occurred. Consequently methods were worked out for deciding whether some particular chemical, known to diminish corrosion by water, was a cathodic or anodic inhibitor. The situation is now known to be less simple (p. 165)—which should be borne in mind by those reading the earlier papers on the subject (U. R. Evans, *Trans. electrochem. Soc.* 1936, **69,** 213; E. Chyżewski and U. R. Evans, *ibid.* 1939, **76,** 215). For further views on the performance of zinc salts as inhibitors, see H. B. Jonassen, *Corrosion* 1958, **15,** 375t.*

* The opinion has been heard that there is no such thing as the inhibition of the cathodic reaction and that the so-called cathodic inhibitors merely serve to increase the resistance of the path joining anodes to cathodes. In case the reader should be of this way of thinking, it may be well to point out three ways in which a discontinuous deposit on the part of the surface which would otherwise be the cathode can interfere with the cathodic reaction. (1) A slimy, porous deposit, incapable of preventing anodic attack, may nevertheless shut off the supply of oxygen and thus interfere with the cathodic reaction. Experimental support is provided by the generation of current by scraping one of

Nature of the protective layer. In the case of a cathodic inhibitor, the obstructive material is usually visible, and can be scraped off and identified. It is not necessarily a single compound, since often the original cathodic product interacts with an anodic product. Thus a zinc sulphate solution produces, not white zinc hydroxide, but the buff-coloured deposit mentioned above. Similarly magnesium sulphate or calcium bicarbonate undoubtedly obstructs the cathodic reaction by depositing magnesium hydroxide or calcium carbonate, but these are soon converted to green deposits containing iron which later become pale brown (p. 94).

Anodic inhibitors, on the other hand, usually produce thin films which are invisible whilst clinging to the metal, although they become visible when isolated from the metallic surface (e.g. by dissolving the metal below them with iodine). In the case of films produced by sodium phosphate on iron, it was at one time expected that these would prove to consist of an iron phosphate. However, Mayne and Menter have examined the film obtained from iron immersed for two days in sodium phosphate solution containing oxygen, after detaching it by Vernon's alcoholic iodine method (p. 784). Where the solution used had been N/10 Na_3PO_4, electron diffraction showed the film to consist almost entirely of " cubic oxide " (i.e. either Fe_3O_4, or γ-Fe_2O_3, or a member of the series of solid solutions of which these are the two end-members); in the case of N/10 Na_2HPO_4, cubic oxide was again the main constituent, but relatively large particles of $FePO_4.2H_2O$ were found embedded in the oxide matrix. The invisible films formed on iron which had been immersed in sodium borate or sodium carbonate in presence of air were also found to consist of cubic oxide.

In these experiments, the original air-formed film was first removed by treatment with hydrochloric acid in a closed vessel, the acid being replaced by the inhibitor solution without introduction of air, although oxygen was present in solution in the inhibitive liquid, and there was access of air during the immersion. The presence of oxygen is, indeed, necessary if

a pair of identical electrodes covered with a visible slimy deposit; the scraped electrode becomes the cathode, although in cases where the pair of electrodes carries an invisible and truly protective film, the scraped electrode becomes the anode (U. R. Evans, *J. Inst. Met.* 1925, **33**, 27, esp. p. 31). (2) Cathodic reactions often tend to occur preferentially at certain catalytically favourable points, as is suggested by the fact that in acid corrosion the hydrogen bubbles are usually seen to rise from a limited number of points, whilst in corrosion by neutral salt solutions the cathodic polarization curves obtained on plotting current against potential are often straight, suggesting that the cathodic polarization measured is largely an *IR* drop over the approaches to the catalytically active points (if such an *IR* drop were absent, straight lines would be obtained only on plotting log (current) against potential). If now the catalytically active points become covered with a substance which has no electronic conductivity, the cathodic reaction must be diverted to other points where the activation energy will be greater, and the effect will be to increase the polarization and therefore reduce the strength of the current flowing. (3) Even if no points exist peculiarly favourable to the cathodic reaction, the covering up of part of the surface reduces the area available for the cathodic reaction, so that a given current represents a greater current density and an increased polarization; the effect will be to reduce the corrosion-rate.

corrosion is to be suppressed completely in a sodium phosphate solution) Pryor and Cohen found that, above pH 7·25, sodium phosphate (M/10; produced no corrosion in the presence of air, but perceptible corrosion (with hydrogen-evolution, presumably, as the cathodic reaction) when the liquid was carefully deaerated.* For details see J. E. O. Mayne and J. W. Menter, *J. chem. Soc.* 1954, p. 103; M. J. Pryor and M. Cohen, *J. electrochem. Soc.* 1951, **98**, 263.

Iron can be kept in sodium hydroxide solution in the presence of oxygen without visible change, and here also the invisible film which is formed on the surface and which prevents further attack has been studied by electron diffraction: the main constituent is cubic oxide. It would seem that the formation of a cubic oxide is an essential feature of a protective film. That is not surprising since, whether formed by direct oxidation or otherwise,† this oxide can continue the crystal structure of the metal, and the film is less likely to leak or break down than one composed of other compounds (whether α-Fe_2O_3 or a hydroxide) where discontinuity of structure is likely. The suggestion that films of cubic oxide continue the structure of the underlying metal is supported by the study of films obtained from rolled iron which had been immersed in potassium chromate; these films, after stripping, show " preferred orientation ", probably inherited from the underlying metal (p. 47); moreover the patterns of the stripped films are clearer than those obtained, by the reflection method, from the same films when still on the metal, suggesting that, when in contact with the metal, the structure of the film had to be slightly distorted in continuing the structure of the metal basis (J. E. O. Mayne with M. J. Pryor and J. W. Menter, *J. chem. Soc.* 1949, p. 1831; 1950, p. 3229; 1954, p. 99).

Mechanism of Inhibition in presence of Oxygen. It has been shown (p. 103) that iron whirled in water containing oxygen escapes sensible corrosion, such as would occur under stagnant conditions. The invisible protective film is probably γ-ferric oxide or perhaps magnetite. The behaviour of a specimen depends primarily on happenings at the most sensitive spot (possibly a place where the atoms are unnaturally far apart). If the rate of supply of oxygen to that spot is such that the loose matter can be converted (by any mechanism) to oxide more quickly than ions could pass into the liquid, then corrosion is likely to be prevented; if, however, the oxygen arrives too slowly to prevent the emergence of iron as cations into the liquid, the supply of oxygen to the area around will make matters worse, since it will consume electrons by the cathodic reaction and accelerate the emergence of metallic ions at the anodic point. Thus oxygen, which may play an inhibitive role if the experimental area contains no highly sensitive spot, can stimulate attack if such a spot is present.

* There was at one time some unprofitable discussion as to whether phosphate or oxygen should be regarded as the true inhibitor, but clearly a combination of the two is needed; neither substance acting alone prevents corrosion.

† Mayne, Menter and Pryor suggest that possibly in cases where an air-formed film is present before immersion in sodium hydroxide, ferrous hydroxide is formed at weak spots in the film by anodic action and is then converted *in situ* into cubic oxide.

This is brought out by Mears' statistical studies of drops of pure water or very dilute potassium chloride solution placed on iron and surrounded by gas mixtures containing oxygen and nitrogen in different proportions. The results are summarized in fig. 35. The proportion of drops developing rust diminishes as the oxygen-content of the gas-mixture is raised, whilst the corrosion per drop producing attack increases (R. B. Mears and U. R. Evans, *Proc. roy. Soc.* (*A*) 1934, **146,** 153, esp. p. 164; *Trans. Faraday Soc.* 1935, **31,** 527, esp. p. 530).

It will easily be understood, therefore, why a specimen which remains immune in distilled water under whirling conditions favourable to oxygen replenishment will suffer rusting under stagnant conditions. If the water contains sodium chloride, this leads to a more soluble anodic product. The salt also increases the conductivity, enabling the current to travel long distances, so that oxygen arriving at regions remote from a sensitive spot

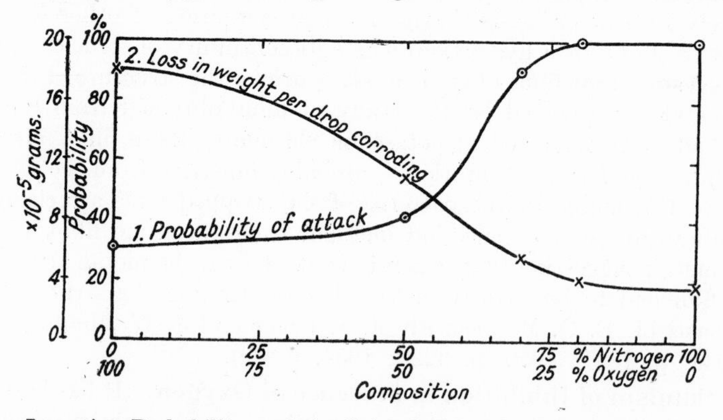

FIG. 35.—Inception Probability and Conditional Velocity of Corrosion of Iron below drops of Distilled Water in Oxygen–Nitrogen mixtures (R. B. Mears and U. R. Evans).

can play an effective part in the cathodic reaction, and thus enhance the anodic attack at the sensitive spot. Both factors will make inhibition more difficult to bring about by rapid oxygen replenishment; nevertheless, work at Teddington at extremely high water-velocities has provided evidence that oxygen can function as an inhibitor even in the presence of chloride (F. Wormwell and H. C. K. Ison, *Chem. and Ind.* (*Lond.*) 1951, p. 293).

If, instead of sodium chloride, we use a solution of sodium phosphate (Na_2HPO_4), it becomes easy to inhibit solutions even under stagnant conditions; for the potential anodic product, an iron phosphate, will tend to repair gaps in the film of cubic oxide that is quickly built up over the major part of the surface. A supply of oxygen will now help the protective mechanism not only by causing the phosphate ions to migrate towards the sensitive spots (the anodes), but also by the formation of cubic oxide over the minor sensitive spots, which would otherwise have to be protected by phosphate. Thus phosphate and oxygen, acting together, prevent corrosion; the film is found to be largely cubic oxide, but contains plugs of hydrated ferric phosphate, presumably produced at just those points where corrosion would have started in absence of phosphate. The fact that the iron phosphate is hydrated and that ferric (rather than ferrous) phosphate is produced suggests that the substance is formed at the outside of gaps in the film, where the ions are on the point of escape.*

The amount of phosphate in the film has been studied at Ottawa by means of radioactive tracers. Phosphate was detected on iron immersed in a solution of either disodium or trisodium phosphate (Na_2HPO_4 or Na_3PO_4), but the former salt solution produced the larger amount of iron phosphates, which was found to be $FePO_4.2H_2O$; the amount was greater if the original air-formed film had been removed by a 15-second dip in 0·2N hydrochloric acid than if there had been no acid dip; this is easily understood. The large amount of phosphate present has been taken to mean that the hydrated ferric phosphate forms thicker deposits than the oxide, suggesting that it is fundamentally less protective than oxide and must reach a greater thickness before it checks its own growth. Similar conclusions have been reached by Pryor when studying films formed in sodium hydroxide solutions; he considers that the film of cubic oxide contain inclusions of hydrated oxide, representing plugs formed at discontinuities. Further arguments, some based on potential measurements, should be studied in the original papers (M. J. Pryor, M. Cohen and F. Brown, *J. electrochem. Soc.* 1952, **99**, 542; M. J. Pryor, *ibid.* 1955, **102**, 163, 199C).

The oxygen-concentration needed to produce inhibition will depend on the OH^- concentration. In researches at Washington, water was brought to a given pH value by means of sodium hydroxide and/or carbon dioxide and tested at different oxygen concentrations. At any fixed pH an increase in oxygen concentration caused an increase in the corrosion-rate up to a maximum, and then diminished it. The maximum occurred at a smaller oxygen concentration when the pH was high than when it was low (E. C. Groesbeck and L. J. Waldron, *Proc. Amer. Soc. Test. Mater.* 1931, **31**, II, 285; cf. H. H. Uhlig, D. N. Triadis and M. Stern, *J. electrochem. Soc.* 1955, **102**, 59; esp. fig. 2, p. 60).

Influence of the Anion in Inhibition. Although the protective film produced by most inhibitors consists largely of oxide, the main anion has an overwhelming influence in deciding between inhibition and corrosion; it is generally difficult to produce inhibition in presence of chlorides, and generally easy to produce it in presence of anions forming sparingly soluble salts with the metal. Solubility, however, cannot be the only factor involved; aluminium sulphate is freely soluble and yet an oxide-film is formed on aluminium exposed to anodic attack in a bath containing sulphuric acid (p. 241).

Imagine a specimen of metal immersed in a dilute aqueous solution of salt. The majority of the particles are water molecules carrying no charge as a whole, and moving almost at random even when a corrosion current is passing between anodic and cathodic areas; however, each molecule possesses a positive and negative portion, and if a molecule chances in its

* The importance of this observation appears on p. 151 (footnote).

kinetic wandering to come near the anodic area, it will generally be kept by the potential gradient in such an orientation that the oxygen part is nearest to the metal and the hydrogen part furthest from it. Thus, whilst there is free movement throughout the liquid, there will be a tendency for the water molecules next the anodic area to arrange themselves as an imperfect oriented phalanx, which may be represented thus:—

```
        |  O H
        |    H
        |
        |  O H
Metal   |    H
        |  O H
        |    H
        |  O H
        |    H
```

This water phalanx is most nearly complete in very dilute solutions. When anions are present in any notable concentration, they must move towards the anodic area, rather than away from it, and, since each has a negative charge *as a whole*, they will tend to displace the water molecules which carry no net charge. The displacement can occur most readily if the anion consists of a single atom with an additional electron—e.g. Cl^-; but it can also occur if the anion is less simple, such as OH^- or SO_4^{--} or CrO_4^{--}. If we call the ion " X ", we now get a situation which we can picture thus:—

```
        |  O H
        |    H
        |  O H
        |    H
        |
Metal   |  X
        |  O H
        |    H
        |  O H
        |    H
```

With any ordinary metal, an anion X which has reached the metallic surface at an anodic point can move no further. The transfer of current across this surface must therefore consist of metal cations moving in the opposite direction, but here there are two possibilities:—

(1) The metal may pass into the solution as hydrated cations, giving in effect a solution of a soluble salt; once started, this process can continue, so that the metal suffers corrosion.

(2) Cations of the metal may move outwards into positions between the oxygen atoms of the phalanx, displacing from the water molecules the appropriate number of hydrogen ions; this can produce the same electrical transfer, but leaves solid oxide or hydroxide on the metal. When a complete layer of oxide or hydroxide has been built up, further water molecules will attach themselves and a second layer of oxide or hydroxide will be formed. The process then repeats itself until the film is thick enough and sound enough to prevent further cations moving outwards.

At any place where the structure of the film is faulty, it will grow to a greater thickness before growth stops; spots and rings of colour on a specimen immersed in a nearly inhibitive salt solution represent places where the film is weak (possibly owing to structural defects in the basic metal); at such places the film reaches interference-colour thickness whilst elsewhere it remains invisible. Coloured rings have been noticed on several metals (on aluminium in absence of inhibitor, as explained on p. 122). In the case of iron, the material composing these colour-rings was at one time regarded as hydroxide, ($Fe(OH)_3$) or ($FeO(OH)$), but it is probably γ-Fe_2O_3. In cases where the colour-rings are set up through an insufficient addition of nitrite to water containing a trace of chloride, Cohen found γ-Fe_2O_3 by means of X-rays; the same oxide is probably produced when other inhibitors are used (M. Cohen, Priv. Comm., Oct. 17, 1957).

If the cation pushed into the liquid comes from the oriented water molecules, we can obtain either hydroxide or oxide. Suppose the metal is divalent, then every M^{++} ion which passes out of the film into a position between the O atoms belonging to the water molecules must displace two hydrogen ions; if these both come from the same water molecule, we build up a film of oxide; if they come from different water molecules, we may expect hydroxide. Experiment suggests that anhydrous oxide is in fact generally formed; presumably it thickens by the cations moving outward through the film, as indicated by the arrows, to sites between the oxygen ions.

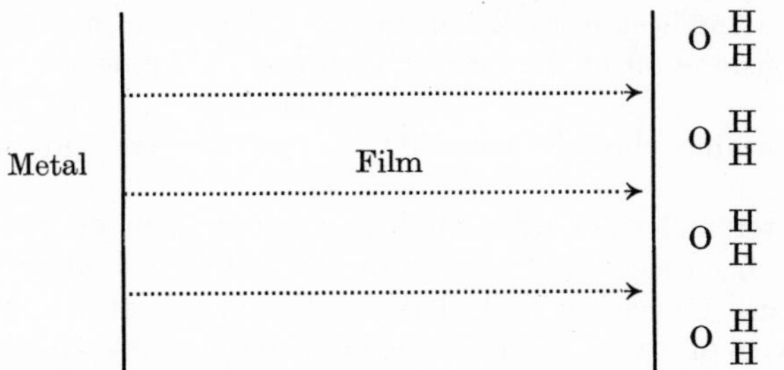

If the solution is alkaline, the phalanx of OH^- ions is likely to be more complete than the phalanx of H_2O molecules present in a nearly neutral liquid, on account of the negative charge. In that case the removal of hydrogen from the OH^- ions next the metal leaves O^{--} (which, with the M^{++} cations at intervening points, give a new layer of oxide) whilst each H^+ ion pushed off joins a OH^- further out to form water. This building up of an oxide-film will continue until it becomes sufficiently protective to prevent further outward movement of cations.

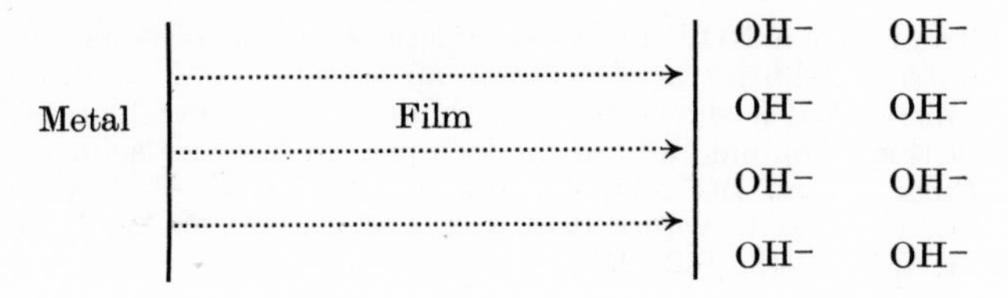

It will be noticed that whilst the pushing off of H^+ from a phalanx of water molecules might produce *either* an oxide- *or* hydroxide-film, the pushing off of H^+ from a phalanx of OH^- ions can *only* lead to an oxide-film. Since oxide-films are generally more protective than hydroxide-films, this may be one reason why alkaline conditions favour inhibition, although the depressed solubility of the film-substance in an alkaline liquid is also an important factor.

Salts containing oxygen introduce fresh possibilities, since here there can exist, not only stable anions but under some conditions cationic groups. This is best exemplified by the acids. Nitric acid which in dilute and moderate solutions ionizes into H^+ and NO_3^- and has all the properties of a strong acid, sometimes behaves, at high concentrations, as though it were the base $[NO_2]OH$; the cationic group $[NO_2]^+$ has a real existence in the liquid—especially if sulphuric acid is also present.* Similar cationic groups appear to exist in sulphur and chromium compounds and form the basis of such bodies as sulphuryl and chromyl chlorides (SO_2Cl_2 and CrO_2Cl_2), which are reasonably stable in absence of water and require an appreciable time to react with water giving HCl and either sulphuric or chromic acid. It would seem that small quantities of $[SO_2]^{++}$ or $[CrO_2]^{++}$ may have a momentary existence in neutral sulphate or chromate solutions where the main ions containing S or Cr are SO_4^{--} or CrO_4^{--}.

Consider now iron placed in a chromate solution; the potential gradient will drive CrO_4^{--} ions up against the anodic areas—presumably with their positive parts furthest from the metal. In that case, a possible reaction will be the pushing off of the CrO_2^{++} group on to the nearest pair of water molecules, forming chromic acid $CrO_2\begin{matrix}\diagup OH \\ \diagdown OH\end{matrix}$. The two protons displaced from the water molecules will switch to the next pair of water molecules further away from the metal; such a proton switch can continue up to the cathodic area. Oxygen will be left on the metal at the anodic areas, giving a layer of oxide. The cathodic reaction may be the reduction of oxygen by the arriving H^+ ions, or, if oxygen is absent, the reduction of chromate. In the latter case, an oxide of chromium will be found in the film; work at Teddington (p. 784) has shown that there is more chromium in the film when oxygen is absent than when it is present. The mechanism of passivation by chromates is further discussed on p. 245.

The passage of metallic cations into the liquid to give a solution of the metallic salt is only likely when the cations can co-exist with the anions present—as in a sodium chloride solution. Where the metallic cations is incompatible with an anion present, the places where corrosion would

* The nitronium ion NO_2^+ probably functions as the nitrating agent when "mixed" (nitric-sulphuric) acid acts on an organic compound. The basic character of nitric acid is shown by the formation of the perchlorate NO_2ClO_4. For details of this compound and of the ions present in nitric acid, see J. Weiss, *Ann. Rep. Progr. Chem.* 1947, **44,** 60, esp. pp. 74, 75; D. R. Goddard, E. D. Hughes and C. K. Ingold, *Nature* 1946, **158,** 480; E. G. Cox, G. A. Jeffrey and M. R. Truter, *ibid.* 1948, **162,** 259.

otherwise start are likely to be plugged up with some solid compound, and attack will be stifled; thus—as already stated—iron placed in common sodium phosphate (Na_2HPO_4) solution containing oxygen develops an oxide-film studded with plugs of iron phosphate (p. 136). Similarly iron placed in potassium chromate solution develops an oxide-film containing small amounts of chromium, since Fe^{++} is incompatible in solution with CrO_4^{--}; Hoar showed that potassium chromate added to a ferrous sulphate solution throws down a precipitate containing trivalent iron and trivalent chromium (T. P. Hoar and U. R. Evans, *J. chem. Soc.* 1932, p. 2476, esp. p. 2478).

Decision between Corrosion and Passivity. In the cases considered above, the question as to whether corrosion or passivity will develop depends on which one of the two possible reactions will prevail; whether metallic ions will move into the solution giving a soluble salt (active corrosion), or whether film-formation, leading to passivity, will predominate. Under conditions close to equilibrium, one of these two reactions may be possible and the other impossible—the situation being conveniently shown by the Pourbaix diagrams (p. 900). Most metals will suffer corrosion, yielding soluble salts, in slightly acid solutions, but will become passive by development of an oxide-film in mildly alkaline solutions; in neutral solutions, oxide-growth may occur at first, but the consumption of OH^- ions itself renders the liquid acid, and later the conditions may become favourable for corrosion. At those points where the metal is in disarray (the atoms being spaced unnaturally far apart), the oxide-film will thicken most rapidly (rings of interference colours are often observed, but these can arise in more than one way); considerable local acidity is thus produced and conditions become favourable to the formation of soluble salt; the development of pits or spots of corrosion-product is the result—as shown by Hoar on tin and Edeleanu on aluminium (pp. 121, 124).

If, however, the metal is made the anode by means of an E.M.F. from an external battery, there is sufficient energy supplied to render *either* of the two reactions a possible one, and the decision between anodic corrosion and film-production will depend, not on considerations of the free-energy change, but on the activation energy needed; thus reactions occur which would not be expected close to equilibrium conditions. As already stated, an aluminium anode in a bath containing sulphuric acid may develop a film of alumina, although the formation of soluble aluminium sulphate might well have been expected. Such cases are considered in Chapter VII.

We can now see why it is difficult to form a protective oxide-film in a solution containing chloride ions in appreciable amounts; no assumptions need be made regarding adsorption. It is reasonable to expect that chloride ions moving under the potential gradient towards places where anodic attack would normally start will, by virtue of their negative charge, tend to dispel the phalanx of oriented water molecules, which carry no charge as a whole, and thus alter the conditions in a sense unfavourable to film-formation. However, there is good reason to think that adsorptive forces also come into play. Piontelli has established the fact that chloride and other anions affect cathodic as well as anodic reactions; those potent in promoting cathodic reactions are often potent in promoting anodic reactions; he relates their effect to deformability (p. 907).

Pryor, discussing aluminium, considers the effect of ionic size and ionic charge in pulling cations through an oxide-film. Chloride ions, being small and negatively charged, set up intense localized electrical fields at places where they are attached, and thus draw the aluminium ions through the film; they also create defects of the type needed for the movement. The ions of lead soaps, although also carrying charge, are very large, and the field is much less intense; thus—it is argued—lead soaps are inhibitors, unless chlorides are present; as little as 5 ppm. chloride is enough to break down passivity (M. J. Pryor, *Z. Elektrochem.* 1958, **62**, 782).

Cases where Complex-anion Formation is possible. Hitherto we have mainly considered that the various anions are being pushed up against the anodic points of the surface by the prevailing potential gradient, but free to leave as a result of thermal movement. Certain anions, however, after reaching a surface will be firmly held there by specific adsorptive forces. CN^- ions, for instance, are well known to attach themselves firmly to a surface—the fact that they " poison " certain reactions is evidence of this; it is unlikely that a CN^- ion, having reached, for instance, a silver surface, will break away from the silver, and if CN^- is to leave the surface it will bring silver with it as $[Ag(CN)_2]^-$. Thus a new anodic reaction

$$Ag + 2(CN)^- + e = [Ag(CN)_2]^-$$

becomes possible which involves the same electrical transfer as the simple reaction

$$Ag + e = Ag^+$$

There is, however, a difference between the two types of anodic reaction; the first involves *negative* ions moving *against* the electric potential gradient, and the second *positive* ions moving *with* the gradient; as a result, for a given departure from the equilibrium potential, the first will occur less frequently than the second; reactions involving complex anions are more " polarized " than those involving simple cations. Nevertheless it is clear that strong adsorption does not necessarily lead to inhibition and may sometimes favour corrosion.

Another interesting case is that of the hydroxy-acids such as tartaric acid

$$COOH.CHOH.CHOH.COOH$$

or citric acid

$$COOH.CH_2.C(OH)(COOH).CH_2.COOH,$$

where the hydroxyl groups may help to anchor the anions, which, once they reach the surface under the electrical gradient, are likely to be held fast. There are two possibilities—the removal of the organic anion with metal attached, forming a complex anion, or the removal of the radicle, leaving oxygen behind, as in the case of chromate. As to which of these occurs probably depends on the relative strengths of the chemical bonds which have to be broken if either we are to obtain complex-anion formation (leading to corrosion) or oxide-formation (leading to passivity). Without knowledge of bond-strength prophecy is difficult, but cases are known where these acids, or their salts, can either promote or prevent corrosion.

Adverse Effects caused by an Insufficient Quantity of Inhibitor

Stimulation and Intensification of Attack. The curves of Pryor and Cohen (fig. 36), which show the effect of solutions of different inhibitor concentrations, make it clear that in some cases the addition of an insufficient quantity actually increases the total attack, but that the corrosion-rate drops suddenly when a critical value is passed. A reason can be suggested for the increased attack. The corrosion-rate depends mainly on the oxygen-supply to the cathodic area, and if a large part of the specimen remains

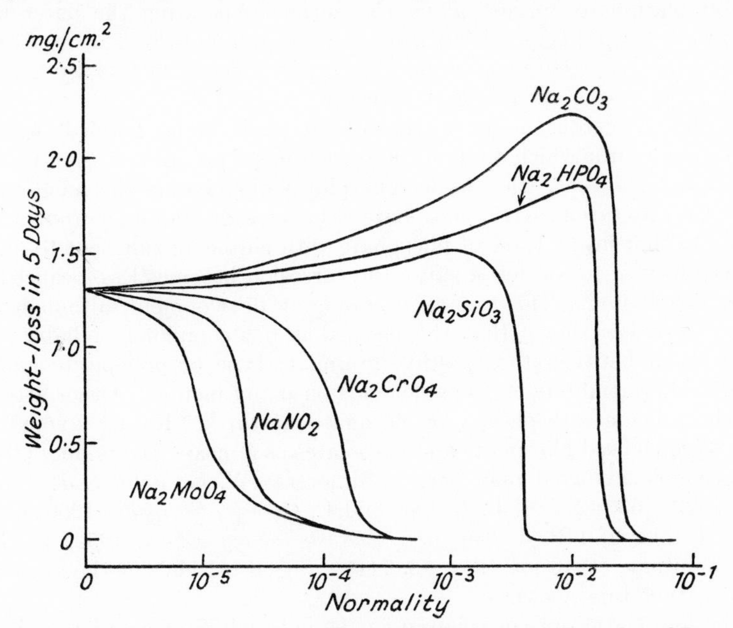

FIG. 36.—Effect of Concentration of various Inhibitors on the Corrosion of Iron (M. J. Pryor and M. Cohen).

uncorroded (as when the inhibitor is present in amounts slightly less than the critical amount), this area is available for the cathodic reaction, so that the anodic attack will become more rapid. Since furthermore the corrosion is now concentrated on the *small* anodic area instead of being spread over a large area (as in the absence of inhibitor), the *intensity* of corrosion (corrosion per unit area of the part affected) is increased even more than the total corrosion. The intensification of attack after insufficient addition of inhibitor had been shown in the Cambridge researches of 1925 and 1927, whilst some of Mears' measurements of 1935 had indicated that in certain cases even the *total* corrosion can be increased (p. 138). This was then regarded as something exceptional. The Pryor–Cohen curves of 1953 showed that the increase in the total corrosion is a more common phenomenon than had been supposed (U. R. Evans, *J. Soc. chem. Ind.* 1925, **44,** 163T; 1927, **46,** 347T; M. J. Pryor and M. Cohen, *J. electrochem. Soc.* 1953, **100,** 203).

The situation is the more serious owing to the fact that the attack, having once started upon a single limited area, continues there in preference to other parts, thus causing rapid penetration into the metal. This is because a blister, forming over the site of attack, prevents the inhibitor from approaching the metal at the place where it is needed if the corrosion is to be arrested. If, for instance, iron is immersed in a mixture of sodium phosphate and chloride, with the phosphate-content too low to prevent attack completely, so that corrosion develops at one sensitive spot, the phosphate soon becomes exhausted around this part, and a membrane of iron phosphate appears along the surface separating the outer liquid (essentially $Na_2HPO_4 + NaCl$) from the inner liquid ($FeCl_2 + NaCl$ without phosphate); the membrane, forming a blister, effectually excludes inhibitor from the corroding surface.*

If this dangerous intensification of attack is to be avoided, we must select an inhibitor which does not produce a membranous corrosion-product. Palmer noticed that, whereas ferrous phosphate favours the membranous type of precipitation, ferric phosphate assumes a crystalline, granular form. He found that in mixtures of phosphate with chromate sufficient to oxidize the ferrous iron to the ferric state, the corrosion becomes less localized and better distributed. This was an encouraging development, although more recent work has shown that the method is not completely reliable, since the decision between the healthy (granular) type of precipitate and the undesirable (membranous) type depends on many factors. Other attempts have been made with phosphate–chromate mixtures. Kahler favours mixtures of condensed phosphate and chromates as a means of avoiding pitting; the principle involved may here be different (W. G. Palmer, *J. Iron St. Inst.* 1949, **163,** 421; M. L. Kahler and C. George, *Corrosion* 1950, **6,** 331; H. L. Kahler and P. J. Gaughan, *Industr. engng. Chem.* 1952, **44,** 1770; F. L. Whitney, *Corrosion* 1957, **13,** 711t. See also Ride's method discussed on p. 174 of this book).

The localization of corrosion at inadequate inhibitor contents is largely avoided when certain organic inhibitors are used. Sodium benzoate, developed as an inhibitor at Teddington under Vernon, possesses the valuable characteristic that any attack which occurs is not permanently concentrated on a few limited regions. Wormwell states that, if benzoate is added in insufficient quantity, the attack which occurs is wide-spread and uniform.

* R. S. Thornhill points out that when the amount of inhibitor is limited it is gradually consumed, so that the corrosion, at first localized, gradually becomes general; this lessens the danger of pitting attack. Blisters are sometimes filled with gelatinous or spongy material which acts as an ionic sieve. He suggests that a membrane will only be dangerous if it transmits Cl^- and not $(PO_4)^{---}$. However, even if the PO_4^{---} can penetrate a diaphragm, it will speedily be precipitated by the Fe^{++} present, thus thickening the diaphragm and preventing the PO_4^{---} from reaching the metal at the corroding point; even diaphragms pervious to PO_4^{---} would seem to be potentially dangerous, if they prevent mixing of the excess of Fe^{++} with the body of the liquid.

The sodium salts of several other organic acids also render water non-corrosive; sodium cinnamate and *m*-nitrocinnamate in 1% solution serve to protect a machined surface of grey cast iron, and also galvanized iron—neither of which is protected by 1% benzoate. Steel is protected by 0·5% benzoate; inhibition is aided by water movement, but breaks down if the pH value is reduced below 6. Chlorides and sulphates seem to militate against protection (this is true of most inhibitors). In some cases the necessary concentrations of organic inhibitors is high, but certain bodies hold out real promise; *o*-nitrohydrocinnamate gives protection at 21 ppm. —which suggests that it is more effective than other substituted benzoates, or even than chromate (F. Wormwell and A. D. Mercer, *J. appl. Chem.* 1952, **2**, 150; F. Wormwell, *Chem. and Ind.* (*Lond.*) 1953, p. 556. See also "Chemistry Research" 1957, p. 14 (H.M. Stationery Office)).

Organic inhibitors which—as emphasized by Pryor and Cohen (p. 146) require the presence of oxygen—seem generally to build up a film of cubic oxide; radioactive tracer measurements by Brasher and Stove suggest that there is a little carbon in the film, but the amount is small compared to the chromium found in the film left on iron after immersion in chromate solution (D. M. Brasher and E. R. Stove, *Chem. and Ind.* (*Lond.*) 1952, p. 171).

It is not certain how these organic inhibitors function. One possibility worthy of consideration is that the organic ions or molecules become adsorbed loosely on the metal surface, attaching themselves to certain points and then leaving them, so that there is always a certain proportion of the surface covered; at any moment only the uncovered fraction needs to be converted to oxide by the oxygen present, so that small concentrations of oxygen under stagnant conditions might have the same effect as high concentrations or as rapid whirling (see p. 183, footnote).

Another possibility is that the inhibitors serve to increase the conductivity of the liquid and thus allow the cathodic reduction of oxygen at a distance from a weak point to play a part in raising the anodic current density. If chlorides or similar ions were present, this would merely increase the corrosion-rate, but if the salt is one which increases the conductivity without catalysing the anodic reaction, the potential may be raised to the level where passivity becomes possible, a film being formed by the alternative reaction already mentioned; organic sodium salts might easily possess the required properties. Such an explanation has been put forward by Hancock and Mayne. The notion is consistent with the more general statement put forward by Stern who regards inhibitors as acting by producing local-cell currents which anodically polarize the metal to the potential region appropriate to passivity (P. Hancock and J. E. O. Mayne, *J. appl. Chem.* (in the press); M. Stern, *J. electrochem. Soc.* 1958, **105**, 156C).

Other organic substances which act as restrainers in acid pickling baths are considered on p. 410.

Water-line Attack in presence of Inhibitors. A solution containing a corrosive salt (e.g. sodium chloride), with an inhibitor (e.g. sodium phosphate or carbonate) added in insufficient amount to prevent attack altogether, produces intense localized corrosion at certain places on a plate specimen, e.g. surface defects or points on or near the cut edges of the specimen. If the total corrosion is under cathodic control, the combination of large cathodic area and small anodic area lead to intense attack and rapid loss of thickness. Particularly, however, is there a tendency for rapid attack along the water-line of a specimen partly immersed in unstirred liquid, or at the margin of a drop placed on a horizontal surface.

The effect on the corrosion of half-immersed plates produced by increasing additions of sodium carbonate to a sodium chloride solution is suggested in fig. 37. Small additions merely diminish the attacked area, rather larger ones restrict attack to a few points, from which spreading is only slow, and still larger amounts, which might be expected to eliminate attack altogether, cause it to set in along the water-line; thin sheet specimens may suffer perforation at the water-line in a period which would produce no serious loss of thickness in sodium chloride free from carbonate. Given sufficient concentration of inhibitor, attack can be stopped altogether, but the amount

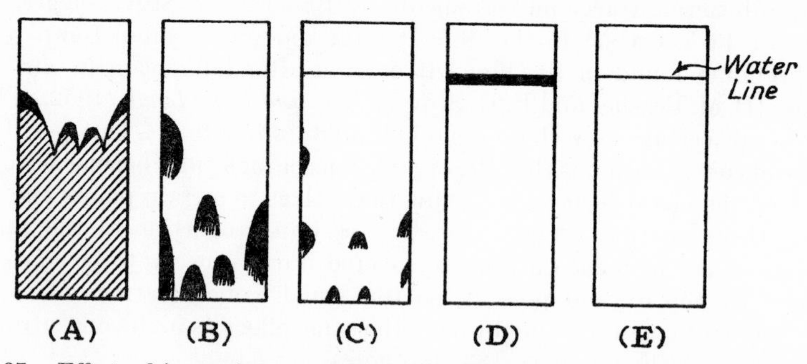

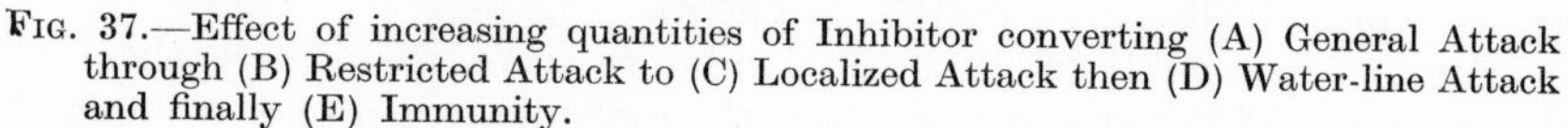

FIG. 37.—Effect of increasing quantities of Inhibitor converting (A) General Attack through (B) Restricted Attack to (C) Localized Attack then (D) Water-line Attack and finally (E) Immunity.

necessary is considerable. Around the critical concentration there is a lack of reproducibility; some specimens will escape attack entirely, whilst others which chance to carry a highly sensitive spot at the level of the meniscus will develop an intense attack, starting at the spot in question, but usually spreading to form a trench along the whole length of the water-line. A large excess of inhibitor is needed to prevent attack; Peers, using 0·017M sodium chloride, found that inhibitor additions exceeding 0·08M were insufficient to prevent water-line attack on the majority of the specimens; this same result was observed—whether the inhibitor chosen was sodium carbonate, sodium phosphate (Na_2HPO_4), sodium hydroxide or potassium chromate (A. M. Peers and U. R. Evans, *J. chem. Soc.* 1953, p. 1093).

The attack at the water-line in a chloride–carbonate solution takes the form of a triangular box (fig. 38(A)), the upper curved surface of which is a shiny dark brown membranous cover, following the water-surface, whilst the lower side is less compact; the box is full of pasty matter, and the metal forming the third side of the box is usually found to have lost appreciable thickness. At places where ferrous salts from within the box have broken out and sunk downwards, narrow vertical streaks are produced (fig. 38(B)); sometimes these form detached trunks or tubes, bounded by strong membranous hydroxide formed along the surface separating the two regions where Fe^{++} and CO_3^{--} ions are respectively in excess (fig. 38(C)).* When once a water-line box has been produced, attack at this level is likely to continue, since, even if inhibitor should find its way into the box, it will at once be precipitated by the excess of iron salts present. If, however, the inhibitor is in considerable excess, a varnish-like deposit (fig. 38(D)) is sometimes formed in close contact with the metal surface, and below such an adherent coating there may be little or no attack; this varnish-like deposit is best seen when silicate, not carbonate, is used as inhibitor (U. R. Evans, *J. Soc. chem. Ind.* 1927, **46**, 347T; for more detailed information regarding the contents of blisters, see M. J. Pryor, *Corrosion* 1953, **9**, 467, esp. p. 470).

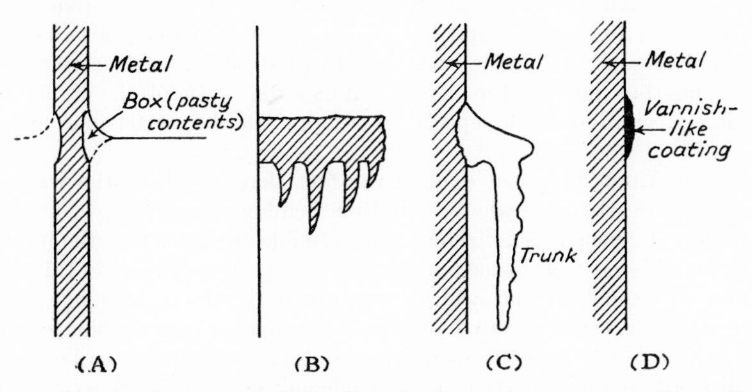

FIG. 38.—Intense Corrosion at Water-line (A, C and D are cross-sections; B is a front view).

Other inhibitors (phosphates, hydroxide or chromates) show similar effects in presence of chloride. Sodium silicate was once thought to be rather less apt to cause localized attack than the others—at least at low chloride concentrations; but the difference is not great.

Sodium vanadate, an inhibitor which would never be used in service, has furnished some interesting laboratory results. At concentrations insufficient to prevent corrosion completely a large number of little round

* Although not a water-line effect, the fascinating "hollow whiskers" produced at Teddington may conveniently be mentioned here. Recent studies of the behaviour of iron in rapidly flowing warm water (55°C) containing calcium bicarbonate in an amount insufficient for complete inhibition reveal the formation of hollow whiskers consisting of Fe_3O_4 and/or α-FeO.OH; the diameter is about 0·15 mm., the wall-thickness 0·02 mm. whilst the length sometimes reaches 9 cm. Apparently the ferrous salt produced at a point-anode is carried away by the flow as a slender thread and reacts with the bulk liquid around, which has higher oxygen-concentration and higher pH-value, to produce one of these beautiful tube-lets (G. Butler and H. C. K. Ison, *Nature* 1958, **182**, 1229).

pits appear. Additions to 0·1M potassium chloride were found by Thornhill to produce the following results:—

0 to 0·0025M sodium vanadate . . .	General corrosion
0·0085M ,, ,, . . .	Attack more localized
0·025M ,, ,, . . .	Perforation at water-line and pitting
0·050M ,, ,, . . .	Pitting
0·50M ,, ,, . . .	Nearly complete immunity

(R. S. Thornhill, *J. Iron St. Inst.* 1942, **146**, 73P; esp. p. 94P.)

Causes of Water-line Attack. On a partly immersed specimen, the presence of inhibitor may cause intense attack just below the water-line—the region which, in a sodium chloride solution free from added inhibitor, would be immune. The immunity close to the water-surface in plain sodium chloride solution is clearly due to the formation of a cathodic product (sodium hydroxide) which is itself an inhibitor. At any sensitive spot on the metal situated in the water-line region, the iron, instead of passing into the liquid as soluble chloride, will be converted on the surface to oxide or hydroxide, thus healing the defect. When the body of the liquid itself contains an inhibitor, no special immunity at the water-line need be expected.*

This does not, however, explain the fact that attack at the water-line is *more* intense than elsewhere. In 1938 Schikorr suggested, in a private letter, that the meniscus acted as an inaccessible crevice where the inhibitor will not readily be replenished; minor sensitive spots occurring in this region will fail to be healed and will develop into corrosion, although similar minor sensitive spots situated on the lower part of the specimen (where there will be better replenishment of inhibitor) would easily be healed. A careful study of the matter by Peers has confirmed Schikorr's explanation; the probability of break-down was found to be greatest when the meniscus crevice was narrowest (its shape can be varied by sloping the specimen). Peers also removed certain objections to Schikorr's theory which were once held to be serious.† He thinks, however, that in some cases, the accumula-

* In such a case, the total corrosion will be small, and the oxygen-uptake slow, so that there will be a sufficient supply of oxygen to the immersed parts of the surface.

† Corrosion will occur if the maximal replenishment-rate of inhibitor is less than the rate of metal-ion formation by anodic action in the absence of inhibitor. Now the constriction of approach to a sensitive spot at the tip of the meniscus crevice will undoubtedly reduce the replenishment-rate of inhibitor, but to exactly the same extent will it reduce the rate of ion migration, since any constriction, whatever its geometrical form, will increase diffusion-resistance and electrical resistance by the same factor. Thus the probability of break-down should be precisely the same at a point situated in the meniscus as elsewhere. Peers finds, however, that the special susceptibility of points in the meniscus zone is only met with if the metal carries a film (e.g. heat-tinted iron) or the liquid contains a film-forming substance (e.g. an inhibitor). If a film is present, the ion must emerge through a pore in that film, so that the additional resistance imposed by the meniscus is small compared to that imposed by the pore, which is in series with it, and the former can be neglected. Thus the

tion of chloride ions which migrate towards any incipient anode, and will accumulate in the meniscus crevice more easily than elsewhere, also plays an important part (A. M. Peers and U. R. Evans, *J. chem. Soc.* 1953, p. 1093; A. M. Peers, *Trans. Faraday Soc.* 1955, **51,** 1748).

Peers' work shows that attack at the meniscus by liquid containing inadequate amounts of inhibitor is analogous to other types of crevice corrosion, discussed on pp. 207–12. Hersch is examining the function of the zone above the meniscus proper, which often carries a very tenuous film of moisture and which seems to play a part in the cathodic reaction (P. Hersch, *Nature* 1957, **180,** 1407; also Priv. Comm., Feb. 17, 1958).

Stagnant water-line conditions are sometimes set up in industrial plants during shut-down periods, since, even where a vessel is nominally kept "full" of water, unsuspected air-pockets may be present. However, the water-line may not always occur at the same level during successive shut-down periods, and if the metal is thick, the attack may extend sideways before there has been serious loss of thickness—especially where the material is a steel having a laminated texture. Thus in service, troubles due to water-line attack by water containing an inadequate quantity of inhibitor occur less frequently than might be expected from laboratory experiments on thin sheet.

Ascertainment of an inhibitive or corrosive character in a treated liquid. Hancock and Mayne have developed an interesting method of testing a salt solution to which inhibitor has been added. They subject iron to anodic polarization at a very low current density such as $10\mu A/cm.^2$. If the liquid contains sufficient inhibitor to prevent corrosion in long-time tests (without applied current), the potential soon rises to about 0·8 or 0·9 volt, on the hydrogen scale; if the amount has been insufficient and the liquid is still corrosive, the potential descends to negative values; near the critical concentration, the potential may first rise and then fall. Higher or lower current densities give similar effects, but the changes occur more quickly or more slowly. In adopting such a method to predict behaviour of a liquid in large-scale plant, it must be recollected that where the probability of attack is small on a small area it may be large on a large area (see p. 914). Thus for practical purposes the liquid should receive considerably more inhibitor than the concentration which produces these fluctuations (P. Hancock and J. E. O. Mayne, *J. appl. Chem.* 1957, **7,** 700).

Chromates and Nitrites

Inhibitors containing Oxygen. If the inhibitive power of oxygen under conditions of whirling is due to replenishment of oxygen at sensitive

meniscus constriction does not affect the rate of ion production. However, the inhibitor need only reach the mouth of the pore; the fact that the film built on iron in sodium phosphate contains hydrated ferric phosphate (p. 139) shows this. Consequently the constriction will seriously reduce the replenishment-rate. If, therefore, at a spot in the meniscus, the replenishment-rate is reduced whilst the replenishment requirements remain the same, the probability of corrosion will be enhanced.

spots, it would be reasonable to expect more soluble oxidizing agents to be effective. This is, in fact, found to be the case under slightly alkaline and sometimes under neutral conditions. It is not generally true of strongly acid liquids, where most oxidizing agents increase attack by providing a new cathodic reaction; instead of the evolution of hydrogen, the oxidizing agent is reduced—which produces a higher working E.M.F.

Chromates and nitrites are much used to inhibit corrosion. These are oxidizing agents, but it seems fairly certain that the inhibitive properties do not directly depend on the oxidizing power. Pertechnetates (such as $KTcO_4$) have been found by Cartledge to be exceptionally effective at concentrations as low as 5 to 10 ppm. although—judging from their redox potentials—they are less strong oxidizing agents than chromates; permanganates which are strong oxidizing agents, and perrhenates (analogous in formulae to pertechnetates) are not good inhibitors; indeed the former often stimulate attack. Thus other factors than oxidizing power seem to be involved. Molybdates and tungstates act as inhibitors, as shown by Robertson; but here air must be present, as shown by Pryor and Cohen. Chromates and nitrites inhibit whether air is admitted or not, but the concentration of nitrite needed for inhibition is lowered if oxygen is present (G. H. Cartledge, *J. phys. Chem.* 1955, **59,** 979; 1956, **60,** 28, 32, 1037; 1957, **61,** 973 (with R. F. Sympson); *ibid.* 1956, **60,** 1571, discussing the passivity produced by osmic acid; also *J. Amer. chem. Soc.* 1955, **77,** 2658; *Corrosion* 1955, **11,** 336t; W. D. Robertson, *J. electrochem. Soc.* 1951, **98,** 94; M. J. Pryor and M. Cohen, *ibid.* 1953, **100,** 203).

Iron remains bright and unaffected in water containing chromate or nitrite, but if chlorides are also present the amount of inhibitor must be increased; insufficient additions of chromate to a chloride solution often produces intense localized attack, particularly at the cut edges or at the water-line, and the total attack may be increased (see below).

Composition of the Films produced by Chromate. The main constituent of the invisible film found on iron after immersion in chromate is cubic oxide (J. E. O. Mayne and M. J. Pryor, *J. chem. Soc.* 1949, p. 1831). The presence of some chromium in the film has been established by chemical study of the stripped film (T. P. Hoar and U. R. Evans, *J. chem. Soc.* 1932, p. 2476), and by the radioactive tracer method (M. Simnad, *J. Metals* 1950, **188,** 1220; D. M. Brasher and E. R. Stove, *Chem. and Ind.* (*Lond.*) 1952, p. 171; N. Hackerman and R. A. Powers, *J. phys. Chem.* 1953, **57,** 139).

The measurements of different investigators seem at first sight to present discrepancies, and surface condition may be important. The amount of chromium taken up undoubtedly depends on the time of pre-exposure to air, being least if there has been long exposure to air, as was noticed in Hoar's early work. Recent research at the Chemical Research Laboratory, Teddington, shows that the amount increases with the period of immersion in the chromate solution. The Teddington investigators consider that the growth in air is continued on immersion in chromate solution, the same growth-law being followed but with a different rate-constant. They find that the amount of chromate in the film is greater if oxygen is present in solution than if it is excluded. An explanation is suggested on p. 142 (D. M. Brasher, A. H. Kingsbury, A. D. Mercer and C. P. De, *Nature* 1957, **180,** 27, 28 (2 letters); D. M. Brasher and A. H. Kingsbury, *Trans. Faraday Soc.* 1958, **54,** 1214; also various views expressed at the Heiligenberg Symposium, *Z. Elektrochem.* 1958, **62,** 619–827).

The invisible protective film produced by nitrites on iron appears to be cubic iron oxide. Cohen, who has studied both chemical and magnetic properties, finds it to be γ-ferric oxide at high nitrite concentrations and magnetite at low concentrations (100 ppm.) in absence of oxygen. In presence of chloride, inclusions of γ-$Fe_2O_3.H_2O$ (*lepidocrocite*) appear; these are detectable under the electron microscope and their appearance is accompanied by reduced protection, as suggested by the fact that the potential becomes less noble and less steady as chloride is added (G. W. Mellors, M. Cohen and A. F. Beck, *J. electrochem. Soc.* 1958, **105,** 332; also M. Cohen, *Priv. Comm.*, Oct. 17, 1957).

There are many analogies between nitrites and chromates in their " inhibitive " powers, but one difference should be noticed. Nitrite yields ammonia as cathodic reduction product, a soluble substance which itself may exert inhibitive properties if it is not quickly carried away by a water stream; in the case of chromate, no soluble inhibitor is produced by the reduction, although some investigators attach value to the solid chromium compounds that sometimes result from the cathodic reaction.

Causes of inhibition by chromates. A careful study of the behaviour of iron in chromate solution containing a radioactive isotope carried out by Cohen and Beck, accompanied by measurement of the chromium lost from the solution and of the chromium appearing in the film, suggests that film formation proceeds in two steps; first the formation of relatively soluble ferrous hydroxide, which will fail to protect directly, but if converted to γ-Fe_2O_3 in physical contact with the metal may yield a protective film.

Cohen thinks that iron placed in water (whether chromate be present or absent) produces ferrous hydroxide with evolution of hydrogen; if chromate is absent, the ferrous hydroxide moves out into the liquid and will often be converted to non-protective rust by dissolved oxygen at a distance from the metal. If chromate is present, the conversion to protective ferric oxide will occur at the metallic surface. Cohen's measurements of the chromium and iron present in the film support the view that the main function of chromate is to oxidize ferrous oxide or hydroxide to ferric oxide, although they do not exclude the possibility that other reactions may proceed simultaneously. The passivity is attributed to a film at least 100 Å thick containing about 25% Cr_2O_3, the remainder being γ-Fe_2O_3 (M. Cohen and A. F. Beck, *Z. Elektrochem.* 1958, **62,** 696).

If oxygen is present, this will help in oxidizing the ferrous hydroxide, explaining the measurements of Miss Brasher and her group at Teddington who find (also by radioactive tracer methods) less chromium in the film if oxygen is present in the liquid. If the oxygen replenishment-rate is so high as to form a protective ferric oxide film in absence of chromate (as must occur in whirling experiments), no chromium need be expected in that film.

If the iron has been exposed to air before immersion in chromate, less chromium-uptake will be required to repair weak places in the air-formed film and to thicken it to the point where thickening becomes slow than if the specimen is comparatively free from an air-formed film at the outset —explaining Hoar's early results as well as the Teddington measurements.

Gerischer, commenting on the results of Cohen and Beck, remarks that the film formed on iron by pure chromate solution approximates in composition to the passive film formed on chromium steels; in both cases he attributes the good corrosion-resistance to the extremely low solubility of the mixed oxide-layer (H. Gerischer, *Angew. Chem.* 1958, **70,** 285).

A slightly different mechanism may be worthy of consideration. As suggested in Chapter VII, the anodizing of aluminium in chromic acid by means of a high external E.M.F. may be explained by the pulling away of a $(CrO_2)^{++}$ group from each $(CrO_4)^{--}$ group (kept pressed against the metal by the potential gradient), thus leaving oxygen ions behind to form oxide with the aluminium ions moving outwards through the film. It is reasonable to ask whether on iron immersed without applied E.M.F., the oxide-film formed is not merely the anodic product of a local cell mechanism, with the cathodic reaction proceeding at the outer surface of the same growing film. This cathodic reduction will be the reduction of chromate (probably to Cr_2O_3) if insufficient oxygen is present—explaining the chromium-content of the film. If oxygen is present, it should (on grounds of free-energy drop) be reduced in preference to chromate—thus explaining the Teddington observation that the chromium-content of the film diminishes as the oxygen-content of the liquid increases. It also explains why some earlier investigators working in open vessels detected no chromium in the film at all.

Such a theory also suggests why the salts which act as inhibitors by building up an oxide-film rather than a film of some insoluble compound, are substances which contain oxygen in the anion, along with a cationic radicle (chromyl, sulphuryl, benzoyl, acetyl, etc.). These radicles would seem to have sufficient stability to be transferred unchanged to the nearest water molecules, leaving their oxygen on the metal; compounds such as chromyl chloride, sulphuryl chloride, benzoyl chloride, acetyl chloride exist; they react with water to give the corresponding acids. There is little doubt that the radicles are sufficiently stable for the switch indicated above to take place.

Any reader wishing to consider this alternative explanation in the case of chromates may be glad to find the thermodynamical and electrochemical data for the various reactions involved (such as the cathodic reduction of chromate and oxygen) conveniently collected by E. Deltombe, N. de Zoubov and M. Pourbaix, *Cebelcor Rapport Technique* No. **41** (1956), p. 19. The reasons for the strongly contrasting behaviour of pertechnetates (which are stronger inhibitors than chromates), perrhenates and permanganates (both poor inhibitors and sometimes stimulators of attack) are clearly brought out by M. Pourbaix and N. de Zoubov in *Cebelcor Rapports Techniques* Nos. **49, 50** and **51**; the basic principles involved in Pourbaix's work are discussed briefly on p. 900 of this book. See also M. Pourbaix, *Z. Elektrochem.* 1958, **62,** 670, but compare B. V. Ptitsyn and V. F. Petrov, *J. gen. Chem. U.S.S.R.* 1956, p. 3233 (English translation, p. 3601).

If iron is placed in a chromate–chloride mixture, the iron may pass rapidly into solution at certain susceptible spots where CrO_4^{--} will soon become exhausted; along the surface where $FeCl_2$ and K_2CrO_4 meet, a visible blister of basic matter containing iron and chromium compounds in membranous form will appear. This will prevent CrO_4^{--} ions from reaching the point where they are needed if corrosion is to be stopped; the membrane may not be completely impervious to CrO_4^{--} ions, but any CrO_4^{--} ion passing through it will at once be destroyed by interaction with Fe^{++}. The combination of small anode and large cathode* will cause intense corrosion, as is frequently observed. In such situations, the solid matter is, in the words of King, " only a hindrance to formation of a really effective film of iron oxide "; in the Author's opinion, these words are true of matter formed at a perceptible distance from the metal, but not of matter deposited in contact with the metallic surface. The reader, however, should form his own opinion after studying the original paper (C. V. King, E. Goldschmidt and N. Meyer, *J. electrochem. Soc.* 1952, **99,** 423).†

It is clear that under border-line conditions the decision between solid-film formation and soluble-salt formation will be made in the very early stages when the only chromium attached to the metal is that present in a layer of CrO_4 groups pressed up against the surface by the potential gradient (or perhaps by adsorptive forces). If once solid-film formation has started to prevail, it will interfere with soluble-salt formation; if once soluble-salt formation has established itself, it will undermine any films which previously existed. In this limited sense, it can be said that passivity is " caused " by the protective action of an adsorbed (or quasi-adsorbed) layer—as claimed by some American research groups. But in no practical sense can a monomolecular adsorbed layer be called " protective "; it does not protect the metal against further action by the chromate solution, otherwise the film would cease to thicken. Nor does it protect against ordinary supply waters; Darrin states that, when water is treated with chromate to prevent rusting, it is safe gradually to reduce the chromate concentration, but that this reduction should be spread over some months; in other words, protection only becomes important when a three-dimensional film of appreciable

* This may occur whether the cathodic reaction is the reduction of O_2, or of CrO_4^{--}, or even the evolution of hydrogen.

† In the Cambridge work with chromate-chloride solutions, coloured rings showing the usual sequence of interference tints and indicating the greatest film-thickness nearest to the centre were often found, around points where localized corrosion was in progress, and sometimes around other points—perhaps those which were just about to start corroding. Interesting observations by Miss D. M. Brasher in chromate-sulphate solution deserve notice, although it is not certain that they are the same phenomenon; fairly uniform interference colours developed over the whole protected (cathodic) areas of mild steel specimens undergoing slight local attack in these solutions (F. Wormwell, Priv. Comm., Jan. 2, 1959).

thickness has been attained, and indeed some of Darrin's specimens showed interference colours (M. Darrin, *Industr. engng. Chem.* 1946, **38,** 368).

Nevertheless, detailed knowledge of the early adsorption stages is very desirable, and Hackerman's studies of adsorption in solutions containing chromium present in different forms (e.g. $HCrO_4^-$, CrO_4^{--} or $Cr_2O_7^{--}$) deserve close attention (N. Hackerman and R. A. Powers, *J. phys. Chem.* 1953, **57,** 139; cf. H. H. Uhlig and A. Geary, *J. electrochem. Soc.* 1954, **101,** 215).

It has been noticed that some American investigators consider that the " cause " of chromate-protection is a layer of adsorbed CrO_4, whilst most European investigators ascribe it to a three-dimensional oxide-film. In part, this difference can be resolved by saying that the word " cause " has been used in two different senses; but there is probably a genuine divergence of opinion, and a research conducted at Hokkaido University, Japan—designed to clear up the matter—possesses considerable interest (G. O. Okamato, N. Nagayoma and Y. Mitami, *J. electrochem. Soc. Japan* (English Version) 1956, **24,** 69).

The Japanese investigators found a threshold concentration of chromate (6 to 7 $\times$ 10^{-4} mol./l. in the case of steel wire) above which the corrosion-rate suddenly fell off, whilst the potential suddenly rose. Considerable time, however, was taken for the potential to reach its final value. On electrolytic iron pickled in acid, the potential shown immediately after immersion in chromate solution was almost the same below the threshold concentration as above it; after 2 hours, however, it was + 0·1 volt in solutions containing chromate in excess of the threshold concentration and − 0·6 volt in solutions with chromate below the threshold. Passivity, once produced by immersion above the threshold concentration, survived after transfer to solutions below it. For instance, iron placed for 30 hours in 1 $\times$ 10^{-3} mol./l chromate continued to show a potential of + 0·1 volt even when transferred to 2 $\times$ 10^{-4} chromate—retaining the value unchanged after 30 hours in the more dilute solution; in contrast, iron placed for 30 hours in 2 $\times$ 10^{-4} chromate (too dilute to produce passivity) showed − 0·6 volt after transfer to 1 $\times$ 10^{-3} chromate, and the potential rose only gradually, reaching − 0·35 volt after about 5 hours; it did not rise further, and the iron never became passive. The facts agree with Darrin's experience of practical water-treatment which has shown that if, at the outset, a high chromate concentration is used to produce passivity, it may be reduced subsequently without corrosion setting in.

The Japanese investigators also measured the surface resistance and condenser capacity of iron wire placed in 0·1 mol./l. chromate, noticing that the resistance increased with time and that the capacity fell; the potential meanwhile was rising slowly. All three quantities moved quickly during the first 50 hours, the rate of movement steadily declining with time, so that nearly constant values were reached after 300 hours. This points strongly to the formation of a film, thickening fast at the outset, and more slowly as the film increasingly hinders its own growth. The slow movement and the fact that passivity, if once produced in strong solutions, survives in solutions too weak to initiate it, is considered by the Japanese scientists to support the film theory, and to be difficult to reconcile with an adsorption theory.* They calculate from their numbers film-thicknesses between 46 and 55 Å.

After-effect of Chromate Treatment. The possibility of gradually diminishing the chromate-concentration during chromate treatment of industrial waters shows that the film built up has some effect in rendering the iron passive, but it is doubtful whether the chromate-content can ever be reduced to zero without risk of rusting. Some improvement in the behaviour of iron in chromate-free water is possible, however, if the treatment is so adjusted as to destroy preferentially the most susceptible matter (U. R. Evans, *Nature* 1938, **142,** 160).

Concentrations of Chromate or Nitrite needed for Inhibition. Advice regarding quantities of inhibitor required to render water non-corrosive (e.g. a cooling-water circulating in an industrial plant) is extremely difficult to provide, and there are apparent discrepancies between laboratory results and practical experience. Thornhill has shown that in the laboratory $7{\cdot}5 \times 10^{-3}$M potassium chromate added to Cambridge supply water (which contains calcium and sodium bicarbonates as main constituents) prevents attack on abraded mild steel under stagnant conditions; there is attack at first, since some hours are needed to build up the film, but thereafter the corrosion-rate becomes negligible. Under moving-water conditions much lower concentrations suffice; data obtained at Teddington for mild steel (turned surface) exposed to conductivity water moving at 35·6 ft./min. at 25°C., showed that 2×10^{-5}M potassium chromate suffices for inhibition (alternatively 10^{-5}M sodium nitrite or 5×10^{-3} sodium benzoate). Passing, however, to industry, we find van Brunt recommending $1\frac{1}{2}$% of sodium dichromate for the cooling system of power rectifiers, with further increases if chloride is to be present; later information indicates that 0·5% dichromate with sufficient alkali to convert it to normal chromate has prevented trouble for 11 years, but even this is about $1{\cdot}7 \times 10^{-2}$M—nearly one thousand times the Teddington figure. For the cooling-jackets of internal combustion engines, Darrin recommends starting at 500–1000 ppm. and gradually reducing to 200–400 ppm. (R. S. Thornhill, *Research* 1952, **5,** 324; C. van Brunt and E. J. Remsheid, *Gen. Elect. Rev.* 1936, **39,** 128; *Elect. Times* 1941, **99,** 373; M. Darrin, *Corr. and Materials Protection,* May, June 1947, p. 6; L. Cavallaro and A. Indelli, *Bull. Centre d'Étude et de Documentation des Eaux* 1951, No. 11, p. 41; Y. D. Yale, *Corrosion* 1946, **2,** 85).

Doubtless the apparent discrepancy between laboratory and service results is partly due to higher temperature and the presence of other salts in industrial plants. Kroenig states that a water which requires 0·05% dichromate at 20°C. may need 0·2% at 80–90°C., and perhaps 1% if chlorides are present; presence of dissimilar metals in the system will also demand

* They mention difficulties in following the arguments used by advocates of the adsorption theory—notably the applicability of electrostatic principles in calculating the potential of a system where corrosion must be proceeding. The same difficulties have been felt by European critics.

an increase of inhibitor concentration (W. O. Kroenig and S. E. Pawlow, *Korros. Metallsch.* 1933, **9**, 268).

Another difference between industrial conditions and laboratory experiments is probably the fact that the ratio of area to volume is not the same. An even more important difference is that in an industrial plant there are so often places where replenishment of inhibitor is very slow; these may be re-entrant corners or crevices which a good designer should, so far as possible, avoid; but they may also be places where foreign matter, such as sand, cotton-waste used for cleaning, metal filings, scale-particles or rust from another part of the plant, has settled and screened the surface locally from inhibitor. Such conditions produce a small anode surrounded by a large cathode, with consequent intense attack. Probably chromates are rather specially liable to set up attack at places which they cannot reach directly. On the other hand they have characteristics of great practical advantage such as the power to prevent corrosion by sulphur suspensions—as shown by work at Teddington ("Chemistry Research" 1957, p. 15 (H.M. Stationery Office)).

Nitrites are largely used in the oil industry—to deal with corrosion due to water entrained in the oil. Used in connection with mechanical scrapers they can assist in obtaining a clean, rust-free pipe with high flow factor. They find application in other branches of chemical industry. In considering their use for cooling-systems, it should be recollected that nitrite sometimes suffers bacterial oxidation to nitrate, which is not an inhibitor. Sherwood states that nitrites can prevent corrosion even when chlorides are present, provided that sufficient is added. However, the amount needed increases with the chloride content and the use of nitrite may become expensive when the salinity is high (P. W. Sherwood, *Corros. Tech.* 1954, **1**, 113). See also A. Wachter (with S. S. Smith), *Industr. engng. Chem.* 1943, **35**, 358; 1945, **37**, 749, who provides curves indicating the amount of nitrite needed at different chloride concentrations, and a useful summary by T. P. Hoar, *Corrosion* 1958, **14**, 103t.

Risks of Pitting. There is yet another reason why laboratory experiments, however accurate, cannot easily be translated into service practice. In presence of chlorides, corrosion, if it starts at all in water treated with chromate, will take the form of pitting, and the risk of this happening must be regarded from the statistical standpoint, as suggested on p. 914. If at a certain chromate-content the average frequency of pit development is one pit per square foot, then on a laboratory specimen measuring one square inch, the odds are over 100 to 1 *against* a pit being obtained (assuming that the effect of sheared edges can be neglected); in contrast, within an industrial vessel with internal surface measuring one square yard, the odds are about 8000 to 1 in *favour* of a pit being obtained.* Discrepancy is seen to be inevitable. Thornhill has provided data regarding the rapid perforation of steel when chromates and chlorides are present in unfortunate proportions (R. S. Thornhill, *Research* 1952, **5**, 324).

* The "expected number" in the sense of Poisson's Theorem is 9. The probability of obtaining 0 when the expected number is 9 is only $1{\cdot}2341 \times 10^{-4}$.

Pitting also occurs when nitrites are added in insufficient amount. The subject is discussed by Hoar, Cohen, Bowrey and others (T. P. Hoar, *J. Soc. chem. Ind.* 1950, **69**, 356; see also photographs by M. Cohen, R. Pyke and P. Marier, *Trans. electrochem. Soc.* 1949, **96**, 260; also S. E. Bowrey, *Trans. Inst. Marine Eng.* 1949, **61**, 57, who discusses the use of nitrites for preventing attack on turbine journals).

Even in situations where pitting on part of a steel surface would not have serious results (which may be the case if the metal is thick at those points where the risk of corrosion is greatest), it is essential to add these inhibitors in ample amount. If any corrosion-products (containing ferrous compounds) should arise, they will interact and destroy the inhibitor. However, when once inhibition has fairly been established, it may be safe to reduce the amount of inhibitor maintained in solution. In the case of chromate—as already stated—Darrin recommends that the reduction should be gradual—spread over a period of one to three months. In many other types of inhibitor, notably condensed phosphates, a gradual reduction in the concentration has been found possible.

If at the outset the amount of chromate added is too low to prevent corrosion, there will be no ultimate economy, since that which has been added will rapidly be destroyed—probably in interacting with the ferrous compounds formed. If a liberal addition of chromate is made when treatment starts, there will be a saving in the long run, since the destruction of chromate should come to an end after a few weeks—as shown in fig. 39 (M. Darrin, *Industr. engng. Chem.* 1946, **38**, 368).

One disadvantage of chromate or nitrite is the need for chemical estimations at frequent intervals to ensure that the necessary concentration is being maintained; this is especially important when nitrites are used, since here the inhibitor can be destroyed, not only by chemical interaction with ferrous corrosion-products, but also, it would seem, by bacterial action.

However, bacteria can sometimes be helpful. At one chemical works, it was found that the cooling water, which contained traces of ammonia, was not producing the corrosion suffered at other works belonging to the same concern; investigation showed that the inhibition was caused, not directly by the ammonia, but by nitrite formed from it by bacterial oxidation. The authorities, therefore, tried the addition of nitrite to cooling water at the other works where corrosion was being experienced, but found that—however much nitrite was added only a trace could be detected in the cooling water; the rest was converted, probably by bacterial oxidation, to nitrate. It was then found that if a sufficient quantity of unionized ammonia (10 ppm. or more) was present, the nitrite could be maintained in the water; if the concentration dropped below 10 ppm., the oxidation of nitrite occurred. Possibly the unionized ammonia kills the bacteria or prevents them from functioning, but this is uncertain. Whatever the explanation, the treatment of cooling water by this process has greatly reduced corrosion—without, however, stopping it. It will be mainly useful in plants where there is a source of ammonia available too weak to be worth concentrating (W.D. Clark, Imperial Chemical Industries, Ltd., Priv. Comm., Feb. 28, 1958).

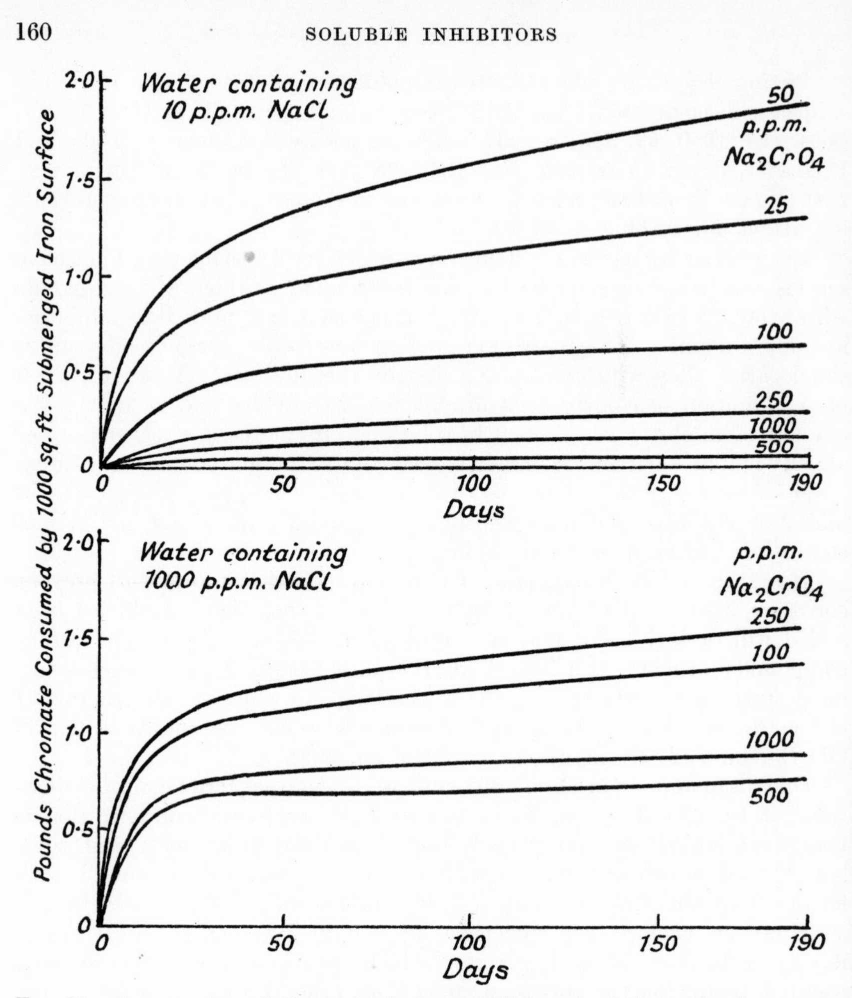

FIG. 39.—Consumption of Chromate during establishment of Protection in presence of Chloride (M. Darrin).

Calcium Bicarbonate as Inhibitor

The State of Carbon Dioxide in Hard Waters. The cheapest—and possibly the most useful—inhibitor for water-supplies is calcium bicarbonate, which is already present in most hard waters, and can be obtained in acid soft waters by passing them over calcium carbonate in suitable form. However, protective films of chalky rust are only formed from bicarbonate waters if the carbon dioxide content is limited to the amount needed to stabilize the bicarbonate, so that the smallest rise of pH suffices to render the liquid next the metal super-saturated with calcium carbonate. In general, protection is only obtained if sufficient oxygen is present to convert ferrous compounds to the less soluble ferric compounds at a point very close to the metal; solid material precipitated at a distance cannot protect.

A calcium-bicarbonate water may contain four kinds of carbon dioxide:—

(1) the CO_2 which goes to make up $CaCO_3$

(2) the CO_2 needed to convert $CaCO_3$ to $Ca(HCO_3)_2$

(3) the CO_2 needed to stabilize the $Ca(HCO_3)_2$ (a plain solution of $Ca(HCO_3)_2$ would decompose, depositing solid $CaCO_3$, until sufficient stabilizing CO_2 had been liberated to stop further decomposition).*

(4) any other CO_2 present. The fourth kind is the *aggressive carbon dioxide* or *aggressive carbonic acid,* which, if present, will cause the water to dissolve any $CaCO_3$ film already present and will interfere with the formation of a protective $CaCO_3$ film on bare metal.

Treatment of Aggressive Waters. Even waters from a chalky source may contain aggressive carbonic acid—especially at certain seasons of the year, determined by the state of the vegetation in the collecting area. Such water could, theoretically, be rendered suitable for passage through steel pipes by percolation through a bed of broken limestone, but in practice it is more convenient to cascade it; passage down a series of waterfalls arranged to keep breaking the liquid surface and facilitating passage of gas through the interface allows most of the carbon dioxide of the fourth kind to escape, and leaves the water close to the composition at which it would stand in equilibrium with solid calcium carbonate.† Cascading has been used with success at certain British water-works for overcoming corrosion problems (E. L. Streatfield, *J. Soc. chem. Ind.* 1939, **58,** 313; J. C. Thresh, J. F. Beale and E. V. Suckling, "Examination of Waters and Water Supplies" (Churchill)).

The cascading also introduces oxygen, which will aid the conversion of ferrous to ferric iron close to the metallic surface and thus favour the production of a protective film. Haase, in a private letter to the Author, has expressed the view that, for German waters, the main function of cascading is to introduce oxygen, since the carbon dioxide content is in such cases not perceptibly reduced. However, a British case known to the Author concerned the attack on the zinc of galvanized pipes, and here the success must have been due to the removal of carbon dioxide, not to the up-take of oxygen, since the corrosion of zinc is accelerated both by the presence of carbon dioxide and that of oxygen (see determinations by G. D. Bengough and O. F. Hudson, *J. Inst. Met.* 1919, **21,** 37, esp. p. 75).

* The concentration of CO_2 needed for stabilizing the solution should, in the absence of other salts, be proportional to the third power of the $(HCO_3)^-$ content—according to calculations by G. Bodländer, *Z. phys. Chem.* 1900, **35,** 23, which agrees with the early experimental results of T. Schloesing, *C.R.* 1872, **74,** 1552; 1872, **75,** 70. The well-known data of J. Tillmans and O. Heublein, *Gesundheits-Inginieur* 1912, **35,** 669 also refer to the simple case where no other salts are present, but a more general treatment, with curves showing the relation between the free carbon dioxide and the semi-combined carbon dioxide, has been provided by M. Pourbaix, Soc. Roy. Belge des Ingenieurs et des Industriels, 1952, No. 4, p. 17; also *Cebelcor Rapport Technique* No. **40** (1956).

† R. S. Thornhill remarks that common practice is to pass water up a vertical pipe and allow it to spill over a system of collars fitting round the pipe. Another method, used in conjunction with base exchange, is to blow air into a tower down which water is spilling.

Although it is generally assumed that a water containing less carbon dioxide than would correspond to equilibrium is non-corrosive, this may not always be the case. A water which deposits calcium carbonate spontaneously as a loose slime (through dissociation of unstable calcium bicarbonate) and not as a coating on the metal (through the cathodic reaction) is by no means non-corrosive; the slime—far from preventing attack—may actually intensify it at the places on which settlement takes place, owing to differential aeration.* Such possibilities are discussed by L. W. Haase, *Arch. Metallk.* 1949, **3,** 114. Cf. R. S. Thornhill, *Chem. and Ind.* (*Lond.*) 1956, p. 403.

Haase has also commented on the distribution of corrosion in a pipe net-work, and its tendency to occur at dead-ends where there is no renewal of oxygen. He has emphasized the need to remove organic matter if the oxygen-content of the water is to be kept at a level suitable for the maintenance of a protective film. If the organic matter is removed by flocculation with sodium aluminate, oxygen can remain at a fairly constant level (5 mg./l.) even after long standing in pipes; he gives examples from experience at Solingen, where iron, manganese and organic matter have been separated by filtration and chlorination, after which the oxygen-content in the remote sections of the pipe net-work was found to be the same as in the near sections (L. W. Haase, *Werkst. u. Korrosion* 1950, **1,** 4; *Vom Wasser* 1955, **22,** 420; 'Korrosion VIII" (1954–55), p. 94 (Verlag Chemie)).

A method developed by Baylis about 1926 for rendering harmless those waters which originally contained aggressive carbon dioxide has received extensive use in American cities. The water is treated with sufficient alkali to be just super-saturated with respect to calcium carbonate, and in this form it lays down a chalky coating on the pipe-walls. At a suitable stage (sometimes described as the attainment of "egg-shell thickness"), the alkali addition is reduced to such a point that the water ceases to deposit further chalky matter, but does not redissolve what has been deposited already. It would seem advisable not to overdo the super-saturation in the first stage, since calcium carbonate thrown down by chemical precipitation in the water, rather than by the cathodic reaction, is likely to be un-protective (J. R. Baylis, *Industr. engng. Chem.* 1926, **18,** 370; 1927, **19,** 777; 1928, **20,** 1191; *J. New Eng. Water-Works Assoc.* 1953, **67,** 38).

Baylis' process has been unsuccessful in waters rich in sodium chloride, such as occur in Western Australia. Doubtless if there is more sodium than calcium present, the latter will become exhausted at the cathodic areas, so that sodium hydroxide will come to be the main cathodic product, precipitating calcium carbonate at a distance where it can serve no useful purpose. The failure is analogous to the failure of sodium phosphate as an inhibitor in presence of a chloride, except that a cathodic process—not an anodic one—is involved.

Soft acid waters of moorland regions, which contain organic acids,

* The crystalline character of the calcium carbonate produced by the thermal decomposition of calcium bicarbonate in a heated water differs from that thrown down by the cathodic reaction, e.g. at a water-line. See J. N. Wanklyn and U. R. Evans, *J. Soc. chem. Ind.* 1949, **68,** 171.

besides carbonic, and often relatively little calcium, will not be rendered non-corrosive by cascading. They are better treated with calcium carbonate. One method is to pass them through beds of broken limestone, but if the pieces are large the rate of reaction is slow, whilst if they are small there is danger of clogging. More satisfactory is the use of dolomite calcined under conditions chosen to decompose the magnesium carbonate, giving magnesia, but to leave calcium carbonate unchanged; both magnesia and calcium carbonate play a part in the water treatment. The correct physical form as well as the chemical character of the product is important, and the conditions of burning must be carefully regulated. The subject has received authoritative discussion by L. W. Haase (" Werkstoffzerstörung und Schutzschichtbildung im Wasserfach " 1951 (Verlag Chemie)), whilst the chemistry of the decomposition of dolomite is discussed by R. A. W. Haul and J. Markus, *J. appl. Chem.* 1952, **2,** 298, and by H. T. S. Britton, S. J. Gregg and G. W. Winsor, *Trans. Faraday Soc.* 1952, **48,** 63, 70.

Dutch, French and American views on water-treatment will be found in *Bull. Centre Belge d'Étude et de Documentation des Eaux* 1951, No. 11 presented by W. F. J. M. Kral (p. 3), G. Chaudron (p. 8) and E. S. Hopkins (p. 35) respectively, and much American experience is recorded in *Industr. engng. Chem.* 1952, **44,** 1736–1760.

It is, of course, always possible to destroy acidity in water by direct addition of alkali, and probably milk of lime is the cheapest form; sodium hydroxide is sometimes more convenient. Sodium silicate, the solution of which shows an alkaline reaction, deserves consideration, and is regarded by some authorities as the most economical method of avoiding corrosion. Any corrosive acid in the water is neutralized with the simultaneous production of colloidal particles of silicic acid. Even in waters free from aggressive acid colloidal silicic acid is usually formed by hydrolysis. The colloid is normally stable, but it seems that, as soon as corrosion starts, silica is adsorbed on the corrosion-product and is thought to play a part in providing a protective coating. Lehrman and Shuldener have examined the films formed on brass or galvanized iron hot-water pipes which had been carrying New York City water; the water had been treated with silicate, receiving 8 to 12 ppm. of silica, and the films contained silica along with the oxides of iron, copper and zinc; zinc and copper silicates were not detected (L. Lehrman and H. L. Shuldener, *J. Amer. Water Works Assoc.* 1951, **43,** 175; J. W. Wood, J. S. Beecher and P. S. Laurence, *Corrosion* 1957, **13,** 719t).

The researches quoted seem to suggest that, before silica can form a protective film, a layer of iron oxide must be formed. The mechanism is not very clear; it seems, however, difficult to picture how the addition of silicate can arrest attack when once definite tuberculation has started; presumably it should be added before tuberculation has developed.*

Silicate treatment of water has been extensively used in the United States. The results have been summarized by Speller who states that small

* R. S. Thornhill, however, states that silicate is one of the few inhibitors that can be used on a rusty surface.

additions retard, but do not completely inhibit, attack on iron, lead, copper and brass pipes—the effects being usually more pronounced for hot-water than for cold-water systems. He recommends at the out-set additions sufficient to increase the silica-content of the water by at least 8 ppm. (many natural waters contain some silica before treatment); when a film has formed throughout the system, the amount can be reduced—perhaps to 5 ppm. If chlorides are present, much larger additions of silicate are called for, whilst the presence of moderate amounts of calcium or magnesium salts also interfere with the production of a uniform silica film; Speller states that the addition of polyphosphate (2 ppm.) with the silicate will overcome this trouble. Of the various grades of silica, he recommends the variety described as " Na_2O, $3{\cdot}3SiO_2$ " for waters with pH above 6, but Na_2O, $2SiO_2$ for more acid waters. Hot waters may be run through a receptacle containing glassy silicate of the Na_2O, $3{\cdot}3SiO_2$ type; this dissolves slowly, leaving a skeleton of silica, which should be removed before recharging.* Calcium is believed to be beneficial; some American writers add silica up to the concentration where the liquid would be supersaturated with calcium silicate; polyphosphate is useful in avoiding actual precipitation. Silicate has been effective in correcting the "red-water plague" so prevalent when some soft waters are run through iron pipes (F. N. Spelle "Corrosion: Causes and Prevention" 1951, pp. 390–395 (McGraw-Hill); also Priv. Comm., March 22, 1958).

Like other alkaline inhibitors, silicate localizes and intensifies attack when added to saline waters in insufficient amounts; it seems unwise to use it for waters containing much chloride without detailed investigation.

In American writings frequent reference is made to the *Langelier Index*, the difference between the actual pH value of a water and the pH value reached when the water has been brought into equilibrium with calcium carbonate; if the first is the lower, the water cannot build a chalky film, and would be capable of dissolving such a film if one should already exist on the pipes. Possibly the significance of the index has been exaggerated, but it is important to enquire whether a water is in the right state, and a convenient method is to pass it over limestone lumps enclosed in a wash bottle, and ascertain whether the pH is changed in the process. Alternatively, the " alkalinity ", as obtained by titration with acid, may be determined before and after the treatment; if the water is in equilibrium, the two determinations should be identical (C. P. Hoover, *J. Amer. Water Works Assoc.* 1938, **30,** 1802; 1942, **34,** 1425. See also the handbook on " Water Quality and Treatment " 1951, pp. 293–299, published by the American Water Works Association).

Behaviour of Sea-Water. The existence of calcium salts in sea-water is probably one of several reasons why under many conditions it is far less corrosive than a plain sodium chloride solution; the presence of magnesium salts may contribute to the same result, and possibly the saponin and other organic compounds may act as inhibitors—although their influence

* R. S. Thornhill considers the addition of one of the commercial silicate liquors more convenient.

does not seem to have been studied. N/10 sodium chloride has sometimes been used as a substitute for sea-water in laboratory corrosion tests, but the two liquids behave quite differently. In tests carried out by C. A. J. Taylor at Cambridge on partly immersed specimens of thin steel sheet, it was found that N/10 sodium chloride always produced a greater weight-loss than sea-water (from the English Channel), and the difference increased as the tests were made longer. Thus in 2-day tests the sodium chloride produced 1·9 times the weight-loss obtained in sea-water; in 8-day tests 2·3 times; in 32-day tests 3·0 times; and in 128-day tests 3·3 times. Indeed after 128 days the sea-water produced no more weight-loss than a hard supply water from a chalk source (containing calcium bicarbonate as the main constituent) and considerably less than water from the same chalk source which had been softened by base exchange (replacing much of the calcium bicarbonate by sodium bicarbonate). Moreover, whereas the attack by the sea-water was sufficiently well distributed to cause no perforation, both the fresh waters mentioned caused perforation at certain points near the water-line—which shows that it is not always possible to avoid intense localized corrosion by adopting a cathodic inhibitor, as was at one time hoped. The measurements also showed that the replacement of calcium bicarbonate by sodium bicarbonate increases appreciably the corrosive power—as would be expected on theoretical grounds (U. R. Evans, *Chem. and Ind.* (*Lond.*) 1941, p. 867).

It is generally assumed that the relative slowness of corrosion by sea-water is due to a " chalky film " deposited on the cathodic areas, but some authorities consider that magnesia rather than calcium carbonate plays the major role in inhibiting the cathodic reaction; the importance of magnesium salts was emphasized in an early paper by Cazaud. Later, Schikorr, who found that *zinc* was much less corroded by a sea-water spray than by a spray of sodium chloride solution, ascribed this to the formation of a layer of magnesium hydroxide; by adding magnesium chloride to sodium chloride solution, he greatly reduced the corrosion-rate, which continued to fall off with time (R. Cazaud, *Rev. Métallurg.* 1934, **31,** 560; G. Schikorr, *Z. Metallk.* 1940, **32,** 314, esp. fig. 1 and table 1, p. 315).

The varying corrosive character of different sea-water samples to copper alloys, possibly due to sewage contamination, is discussed by T. H. Rogers, *J. Inst. Met.* 1949–50, **76,** 597.

Condensed Phosphates

Character of Condensed Phosphates. The condensed phosphates are a large group of compounds containing chains or rings of phosphorus and oxygen atoms joined alternately, thus —P—O—P—O—P—. Useful reviews have been provided by B. Topley, *Chem. Soc. Quarterly Review* 1949, **3,** No. 4, p. 345, and by T. Pitance, *Bull. Centre Belge d'Étude et de Documentation des Eaux* 1950, No. 8, p. 471.

Many of these substances are glassy, and the solubility varies greatly; some are described as " insoluble " and others as " hygroscopic and easily soluble, without definite limit, in water ". Several of them are efficient inhibitors and the addition of a few parts per million to a natural water may render it non-corrosive when in movement. Their main use is the prevention of the unwanted separation of calcium carbonate in solid form from water, but their corrosion-inhibitive properties are also important.

Practical uses of condensed Phosphates. Several commercial products consisting of condensed phosphates are on the market, the best known being *Calgon*, sometimes described as a hexametaphosphate, $(NaPO_3)_6$, but better expressed as $(NaPO_3)_n$, since the state of polymerization usually exceeds 6; one authority has mentioned 20 as the value of n. The material is stated to contain about 67% of P_2O_5. Another commercial product, known as *Micromet*, is a glassy polymer of sodium calcium metaphosphate which dissolves only slowly; this may be a convenience, since the water requiring treatment can be made to percolate through a bed consisting of micromet in pieces of suitable size; in favourable cases, it becomes relatively non-corrosive, taking up both hexametaphosphate and also calcium, which is needed for inhibition of corrosion. Other manufactured products contain zinc, and sometimes manganese, metaphosphates. A mixture of a quickly dissolving zinc metaphosphate and a slowly dissolving zinc calcium metaphosphate is stated to have given promising results for protecting galvanized tanks against very soft water in Australia (*Chem. and Ind.* (*Lond.*) 1952, p. 825).

In general, the presence of condensed phosphates in water only prevents corrosion if the water is in motion. This should be borne in mind when designing treatment-systems for water passing through a plant; the designer should make most particular enquiries whether there are likely to be regions of stagnancy, or idle periods when the flow is stopped. Given a reasonable rate of flow, considerable reduction of corrosion may be hoped for—as shown by Hatch and Rice. In fresh water—such as is used for town supply—a few parts per million may sometimes serve to prevent appreciable attack, but the amount needed varies greatly, and has to be ascertained by trial. If the amount added is too small, movement of the water (or aeration) may increase attack instead of diminishing it. It has been claimed that in one large city 1 ppm. proved adequate, but generally much larger quantities are needed, and at least 20 ppm. in systems with poor circulation; generally the level of concentration maintained can be reduced as a protective film is built up. For industrial cooling systems, still higher concentrations may be needed (see Thornhill, p. 168). The dependence of successful inhibition upon working conditions represents the main practical drawback to the use of condensed phosphates.

Calcium must be present for proper protection, and at some works lime, as well as calgon, is added, either to increase the calcium content of the mixture or to neutralize the acidity which will be liberated when the metaphosphate is converted by heat into orthophosphate. For some waters, however, lime will be unnecessary.

Parham and Tod state that the ratio of phosphate to calcium carbonate must be less than 2; most natural waters contain enough calcium to fulfil this condition, since even "soft" waters often carry 10 to 20 ppm. of calcium hardness and could thus tolerate 20 to 40 ppm. of phosphate.

Further information will be found in papers by G. B. Hatch and O. Rice, *Industr. engng. Chem.* 1945, **37,** 752; O. Rice, *J. Amer. Water Works Assoc.* 1947, **39,** 552; P. N. Parham and C. W. Tod, *Chem. and Ind.* (*Lond.*) 1953, p. 628.

Eliassen (p. 168) finds that the presence of iron is also advisable for the building up of protective films. One reason for the need of calcium and iron suggests itself on reading Seelmeyer's work on inhibition by orthophosphates; the protection is ascribed to a layer of calcium iron double phosphate, the solubility of which is stated to be below 0·1 mg./l. at pH 5 to 9 (G. Seelmeyer, *Werkst. u. Korrosion* 1951, **2,** 17).

Risk of Pitting. One of the first questions to be asked about any inhibitor is the effect of under-treatment. Cohen has found serious pitting by Ottawa water after under-dosing with condensed phosphate; the view that under-dosing causes pitting has been challenged in some quarters, but Parham and Tod (see above) agree that localization of attack can occur if the water has a high pH value; they state that it is absent if the pH is kept at or below 6. The Ottawa tap water (which had received a lime-treatment) showed a pH value of 8·6, and about 50 to 60 ppm. of calgon were needed for the best inhibition under moving-water conditions (M. Cohen, *Trans. electrochem. Soc.* 1946, **89,** 105; G. B. Hatch, *Industr. engng. Chem.* 1952, **44,** 1775, 1780).

In general, it seems best to use a low pH value, and thus minimize the risk of pitting; pH 5 or 5·5 appears suitable for an all-iron system, but it should be raised to 6 or even 7 if copper or lead are present along with iron. The fact that condensed phosphates can be used for relatively acid waters and in some cases at remarkably low concentrations are arguments in their favour, but the need for water movement is a disadvantage. The absence of pitting under weakly acid conditions may be due to the fact that an oxide-film would normally suffer reductive dissolution, so that the combination of large cathode and small anode is not set up. This reductive dissolution may not occur, however, if the oxide present is a thick mill-scale, and engineers should avoid applying the results of laboratory experiments conducted on scale-free specimens to large-scale systems where the steel will carry mill-scale.

Many of these remarks probably apply to other condensed phosphates, but reliable information appears to be wanting in some cases. Pyrophosphates and triphosphate are less hygroscopic and more stable than calgon, which changes into orthophosphate if stored in a damp condition. Information about heptaphosphate ($Na_9P_7O_{22}$) is provided by R. Eliassen (*Water and Sewage Works* 1946, **93,** R163), and about tripolyphosphate ($Na_5P_3O_{10}$) by G. Ammer (*Vom Wasser* 1949, **17,** 128).

Thornhill sums up the practical uses of condensed phosphates thus:

" The optimum dose for water treatment seems to vary widely. In the writer's experience using a very soft lake water at a pH of about 6·7, 70 ppm. of calgon and 25 ppm. of micromet protected steel under conditions of turbulent flow in the cold. . . . Hot solutions of calgon hydrolyse and are converted to sodium dihydrogen phosphate with a fall in pH, and if calcium ions are present, calcium phosphate is precipitated. Calcium phosphate scale may prove troublesome with cooling waters treated with calgon, if the temperature rise is appreciable.

" In the large-scale industrial use of calgon there are two methods of approach. One is to adjust the pH of the water by alkali treatment so as to bring the water on to the verge of scaling, and to use 2 to 5 ppm. calgon to prevent this becoming troublesome. The other method is to maintain the pH at about 7, if necessary by addition of acid, and then to use calgon doses of 100 ppm. which can be dropped to about 20 ppm. after a time. The first method is applicable to not very corrosive waters, but may give rise to calcium phosphate scales on hot surfaces. The latter method, which is being used successfully in recirculating systems, is applicable to pitting waters where it is of advantage to spread out the corrosion by depressing the pH. Calgon is best used in neutral or slightly acid solution, with a high initial shock dose to lay down a protective film " (R. S. Thornhill, *Research* 1952, **5**, 324, esp. p. 331).

Theory of Inhibition. Condensed phosphates are believed to act by depositing a film on the cathodic part of the surface. Experiments described by Raistrick on the current generated by cells of the type Zn | Fe, Fe | Cu, Zn | Pt, Fe | Pt containing water treated with calgon, show that, if the cathode was taken out, cleaned with acid, and replaced, the current suddenly increased, although no such increase was obtained on cleaning the anode. Such experiments suggest that calgon is a cathodic inhibitor. However, Shome finds it to act as an anodic inhibitor, whilst Lamb and Eliassen believe it to be either anodic or cathodic according to circumstances, although they think that it should be " classed as a cathodic inhibitor under practical operating conditions " (B. Raistrick, *Chem. and Ind.* (*Lond.*) 1952, p. 408; S. C. Shome, *Bull. Central Electrochem. Res. Inst. Karaikudi* 1954, **1**, No. 4, p. 64; J. C. Lamb and R. Eliassen, *J. Amer. Water Works Assoc.* 1954, **46**, 445; also J. C. Lamb, Priv. Comm., Jan. 18, 1955. Cf. J. L. Mansa and W. Szybalski, *Corrosion* 1952, **8**, 381).

Lamb and Eliassen, using a radioactive phosphorus isotope, have demonstrated the production of a film on the cathode, and it is found that this is built up more readily if traces of iron are present in the water, and that the current furnished by the cell falls more quickly under such conditions. The liquid was found to contain colloidal particles, presumably containing calcium, iron and phosphate, since the film deposited on the cathode carried all these ingredients.* The desirability of iron for the building up of a protective film may suggest a new explanation for the need of movement in the liquid and the advisability of relatively acid conditions, since iron from the anodes must reach the cathode without being precipitated as a membrane over the anodic points.

Experiments on the addition of calgon to a differential aeration cell show that the film formed on the cathode reduces the supply of oxygen

* The results suggest that the colloid particles containing iron carry a positive charge; according to Pitance (p. 165 of this book), the particles formed by adsorption of condensed phosphates on $CaCO_3$ nuclei carry a negative charge.

to it, and in service this may increase the amount of oxygen available for the anodic points, thus reducing the E.M.F. in two separate ways (E. Olsen and W. Szybalski, *Corrosion* 1950, **6,** 414; J. L. Mansa and W. Szybalski, *ibid.* 1952, **8,** 381).

It has been demonstrated that calgon can alter the crystal habit, or even the crystal system, of calcium carbonate grown from a super-saturated solution; indeed the state of hydration can be changed (R. F. Reitemeier and T. F. Buehrer, *J. phys. Chem.* 1940, **44,** 535, 552; R. Brooks, L. M. Clark and E. F. Thurston, *Phil. Trans. roy. Soc.* (*A*) 1950–51, **243,** 145; M. Verbestel, *Travaux du Centre Belge d'Étude et de Documentation des Eaux* 1945, p. 256).

The alteration of crystal-habit may have important practical results. The cathodic film was supposed by Raistrick to be essentially calcium carbonate deposited in a crystalline form of a type suited to protect the greatest possible area. He suggests that since the inter-atomic distance of the —O—P—O—P—O— chain corresponds closely to the inter-atomic distances in the basal (0001) plane of calcium carbonate, the phosphate attaches itself to such planes, preventing their growth; the result of this is the crystals develop in layers, so that the small amount of calcium carbonate protects a far greater surface than if the crystal grew out in other forms.

The fact that calcium carbonate is an essential ingredient of the film is indicated by one of Raistrick's experiments with the cell

Zinc | Water containing $CaSO_4$ and metaphosphate | Iron

The current fell off with time if carbon dioxide was present, but not in any marked degree if it was absent. The formation of the film in relatively acid water is not inconsistent with the idea, since the layer of liquid next a cathode is always less acid than the body of the liquid; furthermore the adsorbed phosphate may reduce the rate of dissolution. However, Lamb and Eliassen's demonstration of the importance of iron in building up protective films may necessitate some modification of Raistrick's theory.

Inhibitors for Cooling Systems

Oil Inhibitors. Anti-corrosive fluids are on the market which, when added to water in small amounts, impart a milkiness and render the water non-rusting. These fluids are allied to cutting oils but are made with non-corrosive ingredients; actually they are really water-in-oil emulsions* which " invert " when added to excess of water, forming oil-in-water emulsions; the dispersed particles are small before the inversion, so that the original fluids appear clear, but large after the inversion, explaining the milkiness. The oil particles carry negative charges, and thus tend to be deposited at the anodic points of the incipient corrosion-process, or to be precipitated by the emergent iron ions; this tends to prevent the corrosion

* A soap, along with a fatty acid or alcohol, is present to stabilize the colloidal system. See T. P. Hoar and J. H. Schulman, *Nature* 1943, **152,** 102; J. W. McBain and K. E. Johnson, *J. Amer. chem. Soc.* 1944, **66,** 9.

process from developing. Evidently the oil is likely to be deposited preferentially at just those places where it is needed to prevent attack, but a general oily film seems to be produced over a steel surface and if this is excessive, there would seem to be danger of interference with heat-transfer in cases where these oils are used for treating cooling waters. They have been successfully used, however, for cooling systems in Diesel engines, and for automobile cooling systems. Other applications include addition to the wash-water used in removing the mud from agricultural implements and machines. On a very different scale, they may help to solve certain problems concerning the sterilization of ophthalmic instruments, where corrosion is a factor in reducing the sharpness of a cutting-blade.

Patterson and Jones established two important facts when they proved (1) that oil inhibitors can suppress corrosion on steel (and in some cases cast iron) even when appreciable amounts of chloride are present, (2) oil inhibitors can halt attack which has already started. Even when corrosion had been proceeding in dilute sodium chloride solution for 7 days, the addition of oil inhibitor arrested it; this effect was obtained at 72°C. in a brackish mixture containing 2% by volume of sea-water (W. S. Patterson and A. W. Jones, *J. appl. Chem.* 1952, **2,** 273).

The possibility of using anti-corrosive oils in water-cooling circuits at chemical works has been considered by Hamer, Powell and Colbeck, who have studied their properties in the laboratory. A favourable feature is the after-effect; after standing in water containing 0·5% of the oil, a surface was afterwards brought into contact with oil-free water, but no rust appeared. However, when string had been tied around a specimen before immersion in water containing the oil, intensified attack was produced below the string, doubtless through the combination of large cathode and small anode. This behaviour, which seems to indicate a potential danger having regard to the geometry of many cooling circuits, along with a tendency to form scums due to calcium soaps in hard water, and a reduced stability of the emulsions at high temperatures, has led to doubts about their suitability for industrial cooling systems (P. Hamer, L. Powell and E. W. Colbeck, *J. Iron St. Inst.* 1945, **151,** 109P. See also A. Ferri, *Metallurg. ital.* 1950, **42,** 261, who assigns the oils to the anodic and " dangerous " class of inhibitors).

Blended Tannins. The tannins represent a large family of natural products—varying in chemical behaviour. Some can combine readily with dissolved oxygen, others produce on metallic surfaces films which in certain cases possess protective character, whilst others prevent calcium carbonate from being precipitated in objectionable form. By suitable blending of tannins from different sources, mixtures are obtained which have given valuable results in averting corrosion and scaling in the water-cooling systems of engines—particularly those of the Diesel type. Hancock has described their use on 450 Diesel and Diesel-electric locomotives over 7 years; hard waters were softened, whilst medium and soft waters were treated with soda ash and tannin; no difficulties were encountered (see E. L. Streatfield, Paper to Diesel Engine Users Association, Feb. 17, 1955, with discussion by J. S. Hancock and B. G. Houseman, who provides a useful list of treatments suited for the various classes of engines; also B. G. Houseman, *Chem. and Ind.* (*Lond.*) 1956, p. 59).

Tannin has long been used in steam locomotives, largely to avoid attack upon the copper, which is apt to suffer by the moto-electric effect (p. 130). On one railway in South America, trouble had been caused with soft or very alkaline water; it was overcome by adding solid tannin to the tenders (100 grams per cubic metre for the more corrosive waters but smaller quantities for others). The paste produced dissolved slowly, and served to prevent corrosion during the round trip. Experience over many years has proved such methods to be effective—provided that they are adjusted to suit local conditions (R. J. Barham, Priv. Comm., Dec. 23, 1948).

Inhibitors for Motor Cooling Systems. In summer, supply water, with its high specific heat and good conductivity, is commonly used as cooling agent for motor-engines, but in winter the addition of an antifrigerant (generally glycol) is necessary; sometimes methanol is used. Since a circulating system often comprises several metals in contact, serious corrosion occurs unless a suitable inhibitor is present. Some disagreement has been evident between the views held, and importance may be attached to an objective examination of the various inhibitors carried out at Zürich. They have been tested, as additions either to supply water alone or to supply water containing antifrigerant; both single metals and combinations were tried; the combination most studied comprised iron, copper, soft solder and aluminium (or an aluminium alloy containing copper). On the whole, a mixture of triethanolamine phosphate and sodium mercapto-benzthiazole (TEP + NaMBT) behaved best, but good results were obtained from borax, and also from oil inhibitors; it was recognized that in the latter case there might be danger of de-emulsification in service (P. Schläpfer and A. Bukowiecki, Mitteilungen über Kühl-und Frostschutz-mittel für den Motorfahrzeugbetrieb Schweiz. Ges. für das Studium der Motorbrennstoffe, Zürich, 1949).

Most of the other inhibitors tried at Zürich were found to involve difficulties. Chromates, for instance, were found to react with alcohols. Best and Roche, however, disagree on this point, stating that chromates are sufficiently stable, and suggest the use of lithium chromate which is freely soluble even in absolute methanol; they admit, however, that chromates are undesirable when small areas of aluminium are present in contact with large areas of other metals. They suggest that borax may be useful in preventing a rise in pH to levels dangerous to aluminium* (G. E. Best and E. A. Roche, *Corrosion* 1954, **10,** 217, 223; for other views upon inhibitors for coolants in contact with aluminium, see G. G. Eldredge and R. B. Mears, *Industr. engng. Chem.* 1945, **37,** 736, esp. p. 740):

Borax, which inhibits the corrosion of zinc by glycol at high temperatures, may accelerate it at lower temperatures, apparently through the formation of a complex (D. Caplan and M. Cohen, *Corrosion* 1953, **9,** 284).

The combination of sodium benzoate (1·5%) and sodium nitrite (0·1%)

* R. S. Thornhill states that sodium silicate is effective in glycol–water mixtures.

as additions to the 20% glycol antifreeze is the result of distinguished work at Teddington. In absence of benzoate, the nitrite protects cast iron, but increases the attack on soldered joints; if benzoate is present, steel, cast iron and solder are all protected, provided that a short initial period of heating is given. The combination does not always prevent the corrosion of aluminium alloys—for reasons suggested below. The laboratory investigations, and extensive tests on working vehicles, are described by F. Wormwell, A. D. Mercer and H. C. K. Ison, *J. appl. Chem.* 1953, **3,** 22, 133; W H. J. Vernon, F. Wormwell, E. G. Stroud and H. C. K. Ison, *M.I.R.A. Bulletin* 1948 (4th Quarter), p. 16; 1949 (4th Quarter), p. 19. See also W. H. J. Vernon, *J. Soc. chem. Ind.* 1947, **66,** 137, esp. p. 138; F. Wormwell, *Chem. and Ind.* (*Lond.*) 1953, p. 556.

The TEP–NaMBT mixture has been studied in detail by Squires. TEP alone gives good protection to combinations of metals which do not include copper or a copper alloy. If copper is present, TEP stimulates attack upon it, giving copper compounds which proceed to stimulate attack on aluminium alloys. The addition of NaMBT prevents attack on copper and thus the admirable inhibitive properties of TEP on the other materials are realized. The combination does not prevent attack on nickel, and it is best adapted for systems where corrosion has not been allowed to start. The corrosion-product of solder may remove TEP from the liquid, and cases have been reported where the NaMBT has been removed by old rust present in the system—either by reaction or adsorption. It would seem advisable to clean a system thoroughly before this combination of liquids is applied in an old vehicle (A. T. B. P. Squires, " Corrosion Inhibitors for Ethylene Glycol Water Coolants for Piston Engines " (Rolls-Royce); also Diesel Engine Users Association, Discussion, Feb. 17, 1955).

Triethanolamine is a weak base, and when the TEP–NaMBT combination is used, the pH value does not rise to levels where aluminium would be attacked, as may happen with some other inhibitive combinations which are in many ways attractive. For instance the mixture of sodium benzoate and sodium nitrite, which protect steel and cast iron respectively, has given admirable results in most conditions, is not invariably satisfactory in combinations including aluminium, on account of the pH rise (F. Wormwell, *Chem. and Ind.* (*Lond.*) 1953, p. 556; cf. P. F. Thompson and K. F. Lorking, *Corrosion* 1957, **13,** 531t; L. C. Rowe, *ibid.* 1957, **13,** 750t).

In many situations, either the benzoate–nitrite or the TEP–NaMBT combination would prove satisfactory, but cases are known where each of them has shown behaviour which is less than perfect; further work on inhibitors for such systems would seem advisable. The subject is discussed in detail in papers prepared for the Symposium on the Protection of Motor Vehicles from Corrosion, organized in March 1958 by the Soc. Chem. Ind. Corrosion group; see papers by A. T. P. B. Squires, and by A. D. Mercer and F. Wormwell. Inhibitors for railway diesel-engines cooling systems are discussed by J. I. Bregman and D. B. Boies, *Corrosion* 1958, **14,** 275t.

The action of water–glycol and water–*iso*propylalcohol mixtures on aluminium has recently been studied in the laboratory; it was found that soluble oil is an effective inhibitor—especially if a buffer is present to keep the pH value near 7·0. This is true both of static conditions or conditions of rapid movement, but with water free from antifrigerant, although there is complete inhibition under conditions of rapid movement, the oil causes acceleration under static conditions. Moreover the inhibition is less effective when there is contact with brass (S. B. Twiss and J. D. Guttenplan, *Corrosion* 1956, **12,** 311t).

Today glycol is obtainable with the appropriate mixture of inhibitors already added; some simple tests by Turner suggest that the commercial mixture is effective in preventing corrosion, evaporation and freezing. It is probably advantageous to leave the mixture in the system during the summer, instead of replacing it with plain water (T. H. Turner, *Corros. Prev. Control,* 1956, **3,** July, p. 29).

Use of Chromate as Inhibitor in Freezing Brine. It has already been mentioned that the chromates are useful inhibitors for water when the chloride concentration is very low, but cause localized attack when it is higher. With very high chloride concentration, such as occurs in the calcium chloride solutions used as freezing brine, the addition of chromate can reduce the corrosion without localizing it, and such additions are largely used in the refrigerating industry. The American Society of Refrigerating Engineers recommend that to every 1000 cubic feet of calcium chloride brine 100 lb. of $Na_2Cr_2O_7.2H_2O$ should be added together with enough sodium hydroxide to convert it to Na_2CrO_4; for mixed $CaCl_2$–$MgCl_2$ brines or for sodium chloride brine 200 lbs. of dichromate are required, but for the latter liquid 100 lb. of $Na_2HPO_4.12H_2O$ can be substituted. It may arouse surprise that inhibitors are effective in these very strong chloride liquors, but it is likely that a situation arises in which the whole surface becomes slowly attacked, and there is no unattacked portion left to act as cathode; thus although attack is not stopped, the intense pitting characteristic of the combination of big cathode and small anode is avoided (see also G. E. Best and J. W. McGraw, *Corrosion* 1956, **12,** 286t, for information regarding the effect of different chromate additions to calcium chloride and sodium chloride brines).

Industrial Cooling Systems with Circulating Water. In many large works and power stations the cooling water, after absorbing heat during its passage through the coolers, is itself cooled by passing down open towers, or by being sprayed into ponds; although some of the heat is given up directly to the air, much of the cooling is due to the latent heat of evaporation, which means that the concentration of chlorides—or other salts unfavourable to inhibition—rapidly increases. Also in industrial districts large amounts of acid are taken up from the air, and unless suitable steps are taken, corrosion of the cooling pipes or jackets is likely; at Sheffield the pH of certain circuits falls to pH4. It is, of course, possible to add alkali, but the amount needed may be considerable. In some circumstances, the use of condensed phosphate is worth considering, since it can be effective in a water showing an acid reaction; it would generally be useless, however, in plants with long idle periods; moreover, in calculating the amount needed, it must be remembered that the metaphosphate will not only be used up in maintaining the film (probably a small matter after the opening stages), but that $(NaPO_3)_n$ may be converted into NaH_2PO_4 by the heat, so that acidity is developed; some phosphate will be lost by being blown away with the spray from the towers (but this occurs—whatever inhibitor is used). The use of zinc compounds along with condensed phosphates to build a film quickly may deserve consideration.

Many factors, besides the mineral content shown in the analysis, can cause troubles in cooling water. Hurst has emphasized the dangers of silt, which prevents oxygen from reaching certain places and sets up attack of the differential-aeration type. Slime-producing bacteria build adherent masses which plug equipment and restrict flow, besides setting up differential-aeration attack. Moreover, on steel they produce conditions favourable to the activities of sulphate-reducing bacteria (E. H. Hurst, *Corrosion* 1957, **13,** 696t).

When the atmosphere is not acid, the pH may rise by loss of carbon dioxide in the towers, and the water may deposit calcium carbonate in the cooling jackets or pipes, unless this is prevented by the addition of suitably blended tannins, lignins, condensed phosphates or other dispersing agents (C. W. Drane, *Chem. and Ind.* (*Lond.*) 1956, pp. 403, 1367).

A promising method is being developed at Melbourne to deal with a water which, although originally containing only 15 ppm. of chloride, rapidly increases in chlorine-content on repeated passage round the circulating system. The water is first treated with the *acid* phosphate (KH_2PO_4) which serves to remove the original oxide-film from steel (presumably by reductive dissolution), and then with small doses of dichromates ($K_2Cr_2O_7$). When these relatively oxidizing conditions have been established, reductive dissolution becomes impossible and oxide is once more a stable phase; a new and more uniform film is therefore built up which confers remarkable freedom from attack. When once the film has been established, the necessary further addition of each chemical is small (R. N. Ride, Symposium on Corrosion (Melbourne University), 1955–56, p. 267; *J. appl. Chem.* 1958, **8,** 175. See also p. 186 of this book).

The idea of starting with a film-free surface when building up the protective film appears to be a sound one, and the protection should be better than if the chromate serves only to patch up discontinuities in an air-formed film. Thus the method holds out promise, but it must not be too readily assumed that it would be satisfactory for waters differing greatly in composition from that of Melbourne. Moreover, in assessing the results of any experiments which may be carried out with a phosphate inhibitor, it must be remembered that ferrous phosphate is white—so that the absence of rust colour may not prove the absence of corrosion. It should also be noted that the *dihydrogen* phosphate must be used in Ride's process—a matter about which there was some misconception when reports of the process first reached this country.

An inhibitive process which possibly involves somewhat similar principles has been in use for some years by a large chemical concern in the United

Kingdom, although not in systems involving cooling towers or spray ponds —for which chromates, with their toxic properties, are considered to be normally unsuitable. In applying the process to a system which is severely corroded at the out-set, calgon (about 1000 ppm.) is added to penetrate and remove corrosion-products. The system is then drained and chromate added to the water, so that a film is built up on a relatively clean surface. In the steady state, chromate reckoned as Na_2CrO_4 is controlled between 1500 and 3000 ppm., no further phosphate being added. Chloride is kept below 30 ppm. whilst a pH value between 6 and 7 is generally recommended. Control tests are normally carried out once a month (but in one plant three times a week). In general, potassium chromate is preferred to the sodium salt, as a suitable commercial grade is available.

The concern is now using the process in six types of cooling circuits, namely (1) a ship's diesel engine, (2) an air compressor, (3) a refrigerator circuit, (4) a land diesel engine, (5) locomotive diesel engines, (6) turbo-alternator oil coolers. The combination of metals involved includes cast iron, steel (sometimes galvanized), copper and its alloys and in one case stainless steel. The first case (the ship's engine) was started in 1953, so that there is now four years' experience of the method, which appears to be proving satisfactory (P. Hamer, Priv. Comm., July 30, 1957).

Substances which are Corrodents or Inhibitors according to their Concentration

Oxygen. Of the numerous substances which stimulate attack at low concentrations but repress it at high ones, the most interesting is oxygen; its behaviour has already been discussed in connection with Mears' studies of drops in oxygen–nitrogen mixtures (p. 138). For practical purposes, however, the solubility of oxygen is so low that, except at high pH values or in the presence of another inhibitor, it must be regarded as a stimulator of corrosion. In hard waters containing calcium bicarbonate, however, the presence of oxygen may assist the building up of a protective film (p. 162), whilst many recognized inhibitors, notably sodium phosphate, are only effective in presence of oxygen (p. 137).

Hancock and Mayne suggest that the ineffectiveness or unreliability of oxygen as an inhibitor in pure water is due to the low conductivity of the water. If a salt can be added which increases the conductivity without stimulating the anodic reaction, then the cathodic reduction of oxygen at a distance from a weak point in the film will serve to increase the anodic current density at that weak point; it may even reach the value at which the potential is raised sufficiently to permit the formation of cubic oxide, and the weak point—whatever its exact character—will be healed. In that way they explain the inhibitive power, in presence of oxygen, of benzoate, acetate and borate (P. Hancock and J. E. O. Mayne, unpublished work).

Such a suggestion appears helpful. Certainly if oxygen at a distance from the weak point becomes effective as a cathodic stimulator, it will greatly increase requirements for metal atoms in the correct energy state to enter the liquid as ions which must be supplied if anodic attack is to continue;

if they cannot be supplied, the alternative reaction (the formation of solid oxide, and consequent healing of any gap or point of loose structure in the film) may be expected. Clearly, however, the salt must not be one providing ions which themselves lower the energy barrier between metal and liquid, and thus various salts, such as chloride, do not serve to promote passivity; in fact they have the opposite effect.

Fluorides. Chapman found that a dilute potassium fluoride solution corrodes iron about as quickly as the corresponding potassium chloride solution, but above 0·8N the attack suddenly ceases. Apparently the solubility of ferrous fluoride is depressed by the presence of excess of fluoride ions, so that the substance, which can in dilute solution be regarded as "soluble", must in strong potassium fluoride solutions be classed with the film-forming compounds which inhibit attack (A. W. Chapman, *J. chem. Soc.* 1930, p. 1546).

Carbonates and Bicarbonates. Probably for analogous reasons, either sodium bicarbonate or sodium carbonate can cause rapid attack on iron at low concentrations, whereas iron can be kept unchanged in a concentrated solution of either salt—provided that chlorides and sulphates are absent. At the border-line concentration, there is lack of reproducibility: of two specimens, or two drop-areas, one will rust and the other remain bright, and behaviour comes to depend on past history; prolonged pre-exposure to air depresses the "probability" of attack (see p. 52). The concentration representing "border-line conditions" varies with the nature of the iron or steel used and other circumstances, but in Mears' scratch-line experiments (p. 52), it fell at 0·07M sodium bicarbonate. For many purposes, carbonate–bicarbonate buffer mixtures have been convenient, as in the experiments on the crack-heal phenomena described below.

Demonstration of the Crack-heal Phenomenon. Rectangles of steel, abraded, degreased, tinted by heating in air for 40 min. at 200°, were engraved with a series of parallel scratch-lines with a blunt gramophone needle pressed down under a 200-gram load, and placed in a sloping position in a buffer solution containing sodium carbonate and sodium bicarbonate, each at 0·011M concentration. The scratch-lines were entirely immersed but the upper part of each specimen, used as a handle, remained dry; this part carried no scratch-lines. After 5 min. the specimen was withdrawn into air in such a way that, as it emerged, it came into contact with filter-paper soaked with the same solution, and was laid down horizontally for another 5 min., with the wet filter-paper clinging to the surface by capillarity. The filter-paper developed rust-spots at the points on the scratch-lines where corrosion had started. The specimen was then returned to the liquid, and the filter-paper, which detached itself automatically on immersion, was preserved as a record of the first pattern of rust-spots. After 5 min. the specimen was again withdrawn, being brought into contact with a fresh filter-paper on which subsequently the second pattern was obtained, and so on. The whole work was carried out at 25°C. in a thermostated room.

In one experiment two specimens were subjected to different cycles of treatment. One of these (fig. 40(I)) underwent during sucessive 5-min.-periods:—

(1) immersion in the liquid (shown as " *L* " on fig. 40)
(2) development of rust-spots on clinging wet paper, being then dried with filter-paper
(3) exposure, dry, to air (shown as " *A* ")
(4) development of rust-spots on clinging wet paper, followed by (1) (2) (3) (4) (1) (2) (3) (4) (1) . . . etc.

The other specimen (fig. 40(II)) underwent no *A*-treatment; the cycle consisted of 5-min. *L*-periods separated by 5-min. periods for the development of rust-spots on the clinging wet paper.

A brief study of the patterns will show that when the surface never came into *direct* contact with the air (oxygen could, of course, diffuse

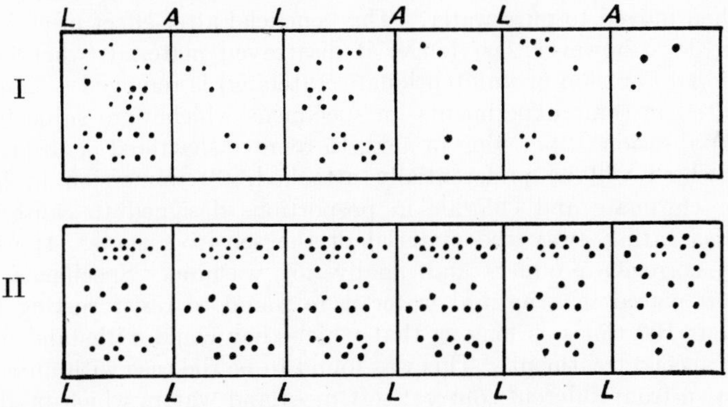

Fig. 40.—(I) Pattern of rust-spots during successive 5-minute periods separated by 5-minute periods alternately in liquid (*L*) and in air (*A*). (II) Pattern during successive periods in liquid.

inwards through the sodden paper during the " development " of the rust-spots) the pattern remained almost unchanged. In contrast, when the specimen during alternate periods was exposed dry to air, the old spots were found to disappear after each air-exposure; on subsequent immersion, fresh spots appeared, but at new points. These results—which were confirmed by other experiments—show that air-exposure does really produce healing at points where active corrosion had been taking place, but that, after such healing, fresh points of attack can then develop elsewhere. It would seem that, from time to time, discontinuities keep appearing spontaneously in the film, but that if there is already active corrosion occurring at other points in the vicinity, they do not develop into active corrosion-sites, since they receive cathodic protection. It is only when these other points of attack have been healed by air-exposure, that the fresh discontinuities arising on occasions at points distributed along the scratch-line can really develop into fresh active corrosion sites (U. R. Evans, *Nature* 1946, **157**, 732; *J. electrochem. Soc.* 1956, **103**, 73, esp. p. 79).

The argument developed on p. 110 would lead us to expect that a film would periodically break down at the sites of scratch-lines, surface grazes or other local defects of a kind involving an assortment of tensional and compressional stresses in equilibrium; thus the phenomenon of crack-heal should cause no surprise. Its experimental demonstration makes it clear why an air-formed film is powerless to prevent corrosion under severe conditions, although it may determine behaviour if conditions are only slightly corrosive; thus exposure to air may increase the power to resist liquids standing on the border-line between the corrosive and inhibitive classes (p. 52) and also to resist certain types of atmosphere (p. 484).

The demonstration of crack-heal also enables us to understand the necessity for maintaining a small concentration of inhibitor in a water even after a period sufficient for the formation of a continuous film over the entire surface. Iron which has been immersed for some time in potassium chromate solution without suffering any visible change, may gradually start to rust when moved to pure water. The beneficial after-effect of chromate immersion does, however, last longer if disarrayed material (whether left by intentional abrasion or unintentional scratching) is removed. This was demonstrated in some experiments on specimens which were subjected in turn to (1) abrasion, (2) etching in acid, to remove shattered or disarrayed material, which will be preferentially attacked, (3) immersion in liquid containing chromate and chloride in proportions designed to cause local attack, thus eating away the material of the sensitive spots, (4) wiping away the corrosion-product, and finally (5) washing. Specimens thus treated and placed in a natural water were found to resist rusting for a period about 100 times as long as that which they could withstand in the absence of special treatment. This was found to be the case with numerous waters drawn from different sources; soft moorland waters which produced rust in about an hour on untreated iron required some days after the treatment to develop rusting; hard water from a chalk source acquired a rusty tint in about a day with untreated metal but required some months after the treatment. The procedure has probably no practical importance, but serves to show the general correctness of the argument developed (U. R. Evans, *Nature* 1938, **142,** 160).

Other Factors in Inhibition

Inhibition in the absence of insoluble Anodic or Cathodic Products. In most of the cases so far considered, it has been possible to refer inhibition to the formation of a protective or obstructive substance by either the anodic or the cathodic reaction. In general, where corrosion is brought to a standstill, there appears to be a film of oxide (ferric oxide when the metal is iron) usually carrying plugs of some other compound at certain isolated points, probably representing the places where, in absence of an inhibitor, corrosion would have started. There are, however, cases where other factors seem to be playing a part. They will be considered in the next section.

Adsorption. The addition of certain organic compounds (generally amines or substituted ureas) to acid greatly diminishes the velocity of attack, without stopping it altogether. These compounds, which will b considered in Chapter XI, are best known as " restrainers ", and the decrease in the corrosion-rate is attributed to adsorption of the organic compounds. Even in a nearly neutral liquid, organic colloids greatly retard attack—as shown especially by the classical work of Friend. He found, for instance, that the action of lead acetate or copper sulphate on zinc was greatly retarded by agar and other colloids; also that the presence of agar in distilled water reduced its attack on iron to 2·7% of its former value, that on lead to 1·5%, whilst the action on zinc practically ceased; egg albumin and gum tragacanth were found to posses similar inhibitive properties. These cases are generally attributed to adsorption of the colloid on the metal (J. A. N. Friend and J. S. Tidmus, *J. Inst. Met.* 1925, **33,** 19; J. A. N. Friend and R. H. Vallance, *J. chem. Soc.* 1922, **121,** 466).

In some cases, there is definite evidence that the colloid really does attach itself to the metal surface. The removal of gelatine from the liquid has been directly demonstrated by Morris, and the proportion of the surface covered has been calculated by Machu—whose views on the effect of adsorption deserve study (T. N. Morris, Report of Director of Food Investigation 1931, p. 192; W. Machu, *Korros. Metallsch.* 1934, **10,** 277, esp. p. 284; 1937, **13,** 1, 20; 1938, **14,** 324; *Trans. electrochem. Soc.* 1937, **72,** 333).

Probably the surface never carries an uninterrupted phalanx of adsorbed molecules, but the attachment may produce an effect out of all proportion to the fraction of the area covered, since the molecules will attach themselves at the points where the greatest free force-field exists to attract them, and these are generally just the places where corrosion would tend to start. In neutral salt solutions, where an oxide-film will generally be present, the points may be gaps or weak points in the film; in acid, where the metal will often be film-free, they will be places of disarray (particularly areas where the atoms are abnormally far apart, so that less energy is needed for them to pass, as cations, into the liquid); or again they may be ledges produced at the edges of incomplete atomic layers and particularly at the corners produced at the ends of incomplete atomic rows.

The attachment may, in different cases, occur by the physical forces such as exist between molecules in a liquid, or by the chemical forces such as exist between atoms in a molecule. In the first case (physical adsorption,) the molecules may keep attaching and detaching themselves at different points, whereas in the second case (chemi-adsorption or chemi-sorption) they are more likely to remain attached. The distinction may be of practical importance, since in the second case any corrosion which does take place is likely to be concentrated on the spots which remain uncovered, and thus the intensity is likely to be greater than if the same amount of attack occurs first at one bare point and then at another. On the other hand, chemi-adsorption, which may exhaust the chemical activity of the surface metallic atoms at sensitive spots, is likely, other things being equal, to reduce the total attack more than physical adsorption.

In some cases, the two types of adsorption may occur together. This may be the explanation of the observations of Hackerman and Cook who washed steel powder with a benzene solution of an aliphatic acid, amine or ester, and afterwards with fresh solvent; part of the material could be washed away but part remained attached. The reactivity of the powder (as measured by the time needed to evolve a given volume of hydrogen when placed in 4N hydrochloric acid) fell off with the amount of matter taken up—giving a direct proof of the efficacy of adsorption in reducing (but not preventing) attack. They reached the conclusion that acids and amines attached themselves at different places, and successive treatment with both types of reagents was more effective than one alone. Chains of about ten carbon atoms produced most inhibition; longer ones were less effective, especially after the "reversibly adsorbed" material had been washed away (N. Hackerman and E. L. Cook, *J. electrochem. Soc.* 1950, **97,** 1).

There has long been discussion whether, in the case of acid attack, the restrainers (e.g. amines and thioureas) act by retarding the anodic or the cathodic reaction, or whether perhaps they clog the surface generally and prevent replenishment of acid. An extensive investigation by Hoar and Holliday has shown that there is interference with both anodic and cathodic reactions.* The quinolines are primarily anodic inhibitors, but at high concentrations they limit the cathodic reaction also; the thioureas inhibit both reactions—the cathodic reaction mainly at low concentrations and the anodic reaction mainly at high concentrations. The paper is not easy to summarize but should certainly be studied in the original (T. P. Hoar and R. D. Holliday, *J. appl. Chem.* 1953, **3**, 502; cf. N. Hackerman and J. D. Sudbury, *J. electrochem. Soc.* 1950, **97,** 109; B. L. Cross and N. Hackerman, *Corrosion* 1954, **10,** 407).

Much simpler carbon compounds can retard acid attack in some cases. Uhlig has shown that carbon monoxide greatly retards the attack of hydrochloric acid on stainless steel, and this is generally attributed to adsorption (H. H. Uhlig, *Industr. engng. Chem.* 1940, **32,** 1490; C. V. King and E. Rau, *J. electrochem. Soc.* 1956, **103,** 331, esp. p. 335).

Electron Donation. Although in a general way it is probably true to say that the best inhibitors are the strongly adsorbed substances, there are many exceptions, and other factors seem to come into play. In the case of chemi-adsorption, donation of electrons from the adsorbed body to the metal may interfere with the anodic reaction, since the latter process itself leaves electrons in the metal. Such ideas have been developed by Hackerman to explain the special efficacy of organic compounds containing sulphur, which he considers will be better donors of electrons than compounds containing oxygen or nitrogen (N. Hackerman and A. C. Makrides, *Industr. engng. Chem.* 1954, **46,** 523; *J. phys. Chem.* 1955, **59,** 707).

Variable Action of Surface-active Agents. The extensive work of

* They obtained the "true anodic current" on a specimen to which an anodic or cathodic current was applied, from the corrosion-rate, and deduced the true cathodic current by subtracting the applied anodic current from, or adding the applied cathodic current to, the true anodic current.

Holness, Ross and Langstaff on copper, iron, tin, nickel and aluminium show that in nearly neutral liquids, molecules which become concentrated in the liquid next the metal may affect corrosion in different ways. Some increase attack, some reduce it and some have no effect. Similarly Piatti found that some wetting agents can prevent pitting of steel by supply water but not others. Of two derivatives of the same mono-acetylated diamine, the hydrochloride stops attack almost entirely, whilst the methyl-ammonium-sulpho-methylate increases it. Often the explanation is uncertain. Possibly the attachment to an oxide-film of a molecule which has positive and negative charges unequally distributed along a chain of atoms may alter the electric field available for pulling metallic ions through the film at weak places, and thus either help or hinder the inception of attack.* Whatever the explanation, the matter has practical importance in view of the use of surface-active substances in detergents, and the papers should receive study (H. Holness and T. K. Ross, *J. appl. Chem.* 1951, **1**, 158; 1952, **2**, 520, 526; H. Holness and R. D. Langstaff, *ibid.* 1956, **6**, 115, 140; T. K. Ross, *Metal Treatm.* 1953, **20**, 183. L. Piatti, *Werkst. u. Korrosion* 1952, **3**, 186).

Electrocapillarity and Inhibition. Much interest has been aroused by the application to corrosion of ideas developed out of the study of the dropping mercury electrode and the capillary electrometer. Of these, the first has practical utility in the detection of traces of inhibitors. Largely due to the extensive work of Heyrovský, a dropping electrode in which mercury is allowed to descend as drops from a jet is used to provide a constantly renewed surface on which cathodic reactions can take place. If the potential is gradually altered in a direction increasingly favourable for the reduction of substances present in the aqueous liquid into which the mercury drops are falling, the current will increase suddenly when a potential value is reached appropriate to the reduction of any particular substance, and the height of the step in the curve of current plotted against time will indicate the concentration of the substance present. If the liquid contains oxygen, we first obtain a step corresponding to the reduction of oxygen to H_2O_2, and then a distinct step interpreted as reduction to OH^-; as the potential becomes further depressed, we notice steps corresponding to the deposition of any metals present in the liquid, first the noble metals, then the more reactive ones and finally the alkali metals, since under these conditions the formation of sodium or potassium amalgam is perfectly possible. As an analytical tool for estimating small amounts of metal in an aqueous solution, the dropping electrode is of great value; the step-height corresponding to various concentrations is first determined by using solutions containing the metals in known amount, and the apparatus can

* Another possible explanation of the varying behaviour of different substances may be found in their effects on the interfacial energy between oxide and water. For reason explained on p. 110, a surface which has been exposed to air so long that film-cracking has practically ceased, may resume film-cracking on being wetted with a liquid containing a substance producing a low interfacial energy; a substance which increases the interfacial energy might be expected to prevent even the infrequent cracking which would occur during exposure to air. There is limited evidence of this, and further research would be welcome.

then be used to estimate unknown quantities in another liquid, since the step-height is normally proportional to concentration.

Now it has been found that sometimes when the potential is first raised to the value appropriate for the reduction of (say) oxygen, there is a temporary rise in the current value to an abnormally high value after which it descends to the normal value. The theory of this " maximum " is still a matter for discussion, but it is certainly connected with an adsorption effect favourable to the provision of reducible material, since if a " capillary-active " substance is added to the liquid, the maximum is diminished and finally suppressed. This is presumably due to the fact that the capillary-active substance is adsorbed in preference to the substance taking part in the reduction, and the degree to which the maximum is diminished affords a measure of the amount of substances present in the liquid which are adsorbed with particular ease. Now it is these substances which—in some cases—act as inhibitors, and the suppression of a maximum affords a useful method of detecting and roughly estimating inhibitors present in small amounts. The investigators at Ferrara have demonstrated the suppression of the maximum by thiourea and other pickling restrainers. Campbell has found the dropping electrode valuable for detecting and estimating the unknown inhibitor present in many natural waters which prevents attack on copper pipes (A. Indelli and G. Pancaldi, *Gazz. chim. ital.* 1953, **83,** 555; H. S. Campbell, *J. appl. Chem.* 1954, **4,** 633. The dropping electrode as used for estimating metal traces is briefly described by E. C. Potter " Electrochemistry " (1956 edition), p. 203 (Cleaver-Hume). See also J. Heyrovský, *Trans. Faraday Soc.* 1924, **19,** 692; D. Ilkovič, *Coll. Trav. chim. Tchécosl.* 1936, **8,** 13).

The application of other ideas of electrocapillarity are only slowly beginning to take shape. Those interested may care to study papers by A. Indelli and G. Pancaldi, *Gazz. chim. ital.* 1953, **83,** 555; C. P. De, *Nature* 1957, **180,** 803; H. C. Gatos, *ibid.* 1958, **181,** 1060.

Practical Use of substituted ammonias. The use of ammonia in boilers is mentioned in Chapter XII. Being volatile, it will reach remote parts of the system where moisture may condense. Substituted ammonias are also used both in steam heating-systems and in boiler plants. These will serve to destroy acidity due to carbon dioxide, but some of them, such as octadecylamine, are very sparingly soluble, and can probably produce no appreciable alkaline reaction; it seems probable that they act as adsorption inhibitors. Cyclohexylamine has been widely used in steam heating-systems.

Amines are also used in the oil industry to prevent the corrosion caused by the water accompanying the oil. In most American oil wells the inhibitors used are soluble in the oil phase but dispersible in the aqueous phase; some concerns use inhibitors soluble in water but insoluble in oil (see *Corrosion* 1957, **13,** 743t, 744t). The composition of proprietary inhibitors is not always disclosed, but the formula for one oil-soluble compound with inhibitive properties is stated to be

$$CH_3(CH_2)_7CH{=}CH(CH_2)_7{-}CO{-}\underset{\displaystyle CH_3}{\underset{|}{N}}{-}CH_2{-}COOH$$

(R. M. Pines and J. D. Spivack, *Corrosion* 1957, **13,** 690t).

An aliphatic diamine of high molecular weight has given satisfactory results in reducing the corrosion of the internal casing in sour oil wells, according to J. A. Caldwell and M. L. Lytle, *Corrosion* 1956, **12,** 67t; see also D. A. Shock and J. D. Sudbury, *ibid.* 1954, **10,** 289; *Corros. Tech.* 1955, **2,** 190.

Ammonia is much used to prevent corrosion in oil distillation plants at places where trouble would otherwise arise from hydrogen sulphide, or from hydrogen chloride produced by hydrolysis of magnesium chloride in the watery liquid accompanying the oil. It is said that methyl-diethanolamine has been successful in combating hydrogen sulphide and carbon dioxide. Here also numerous proprietary inhibitors of undisclosed composition are available.

Mixtures of sulphonic acids and amines have been found to provide excellent inhibition against corrosion when oil is present—being better than either substance acting by itself; possibly the amine sulphonate is the true inhibitor. In absence of oil they are ineffective, and their function seems to be to make oil particles adhere to the metal surface. Probably the polar end of the molecules attaches itself to the metal, the oily ends enter the oil particle which thus becomes anchored, and the whole surface becomes covered with a barrier-film of oil (A. H. Roebuck, P. L. Gant, O. L. Riggs and J. D. Sudbury, *Corrosion* 1957, **13,** 733t).

Effect of Partial Coverage in increasing the Oxygen Replenishment-rate. It was explained in Chapter IV that, in the absence of an inhibitor, an increase in the rate of oxygen-replenishment—as in the whirling apparatus—could greatly diminish or even prevent corrosion. If we consider a sensitive spot and imagine the rate at which metallic atoms would there pass into the liquid as ions in the absence of oxygen, then if we supply oxygen at a much greater rate than this, so that there is always some dissolved oxygen on the spot, we may hope to prevent the passage into liquid, by keeping the metallic cations on the surface as oxide. Any oxygen-supply less than that equivalent to the ion production must fail to arrest the passage into the liquid, and will indeed stimulate attack by taking part in the cathodic reaction on the area around the sensitive spot.

In the absence of other inhibitor, oxygen will only serve to prevent or diminish corrosion under quite exceptional circumstances—such as exist in the whirling apparatus. If, however, the greater part of the surface is covered up, either with loosely attached molecules or with a three-dimensional film, the oxygen-supply at any gap in the covering layer will be increased, since now, even under stagnant condition, oxygen can approach the gap by diffusion from all directions.* Thus under conditions where

* Imagine a small circular bare patch of radius r and area $A(=\pi r^2)$ surrounded by a region covered with the film. It is clear that the number of atoms acquiring sufficient energy to pass into the liquid within a given time

otherwise corrosion would occur, an oxide-film may be built up. It is possible that this may explain the inhibitive function of such bodies as sodium benzoate, where the protective film produced consists mainly of oxide.

The idea may be applied to explain the facts of inhibition by condensed phosphate. If this forms a film on the cathodic points, oxygen becomes available to prevent emergence of metal as ions at the points which would otherwise be anodic. Even so, movement in the liquid is needed to provide oxygen-replenishment, but a rate of movement which would be inadequate in the absence of condensed phosphate, will now suffice, even at relatively low pH value; if the concentration of condensed phosphate is too low, motion in liquid increases corrosion (see p. 166).

These explanations are put forward tentatively, and may prove to be wrong. In the case of inhibition by benzoates, acetates and similar organic salts, the mechanism based on the switch of the benzoyl, acetyl or other radicle, leaving oxygen on the metal, may appear more probable. The explanation suggested by Hancock and Mayne (p. 147) also deserves consideration.

Inhibition by Substances which prevent loose Corrosion Product. If, when a plate of iron is first placed in water or aqueous liquid, a small amount of loose rust is formed over a sensitive spot, this may hinder the access of oxygen (or inhibitor, if present) and attack will continue. If the loose rust (or loose corrosion-product, in the case of another metal) is brushed away, the chance of avoiding attack is improved. If the liquid contains some compound which will prevent the formation of the loose rust—or alter its physical character—so that there is no hindrance to the replenish-

is proportional to A, whereas it is shown below that the replenishment-rate of oxygen is proportional to $\sqrt{A}$; so that, although at large values of A the actual replenishment-rate may be less than the replenishment-rate needed for passivity, the reverse will become true if A is sufficiently small. Thus provided that the gaps in the layer are extremely small, concentrations of oxygen in the liquid will suffice for passivity which would not suffice when the gaps are large.

The actual rate of replenishment under critical conditions when the oxygen is consumed as it reaches the surface is given by C/R where C is the oxygen-concentration in the body of the liquid, and R is the diffusion-resistance. The diffusion paths will approach the gap radially until they come to a hemisphere of radius r, after which they will approach the metal normally; there will really be a gradual transition from the radial to the normal portions, but for the present purposes we may assume an abrupt passage, and regard R as $(R_1 + R_2)$, where R_1 and R_2 respectively represent the resistances of the radial and normal parts of the paths. Of these R_1 is the sum of the resistances of numerous hemispherical shells of area $2\pi\rho^2$ and thickness $d\rho$, so that

$$R_1 = \int_{\rho=\infty}^{\rho=r} \frac{D}{2\pi\rho^2} d\rho = \frac{D}{2\pi} r^{-1}$$

where D is the diffusivity coefficient. Clearly R_2 must also be proportional to r^{-1} since if it were proportional to any other power of r, the value of $R_1 + R_2$ would depend on the unit of length adopted. Thus $R = (R_1 + R_2)$ is proportional to r^{-1} and the rate of arrival of oxygen, for any fixed value of C, is proportional to $\sqrt{A}$.

ment of oxygen (or other film-forming agent), then corrosion may be greatly reduced. Polyhydric alcohols have been found to act in this way; beyond a certain concentration they peptize the corrosion-product of iron (i.e. bring it into colloidal solution); the total corrosion-rate falls off, the iron produced being found in solution. Their effectiveness increases with the number of OH groups present, so that sorbitol (with six groups) is more efficient than erythritol (with four), which is more efficient than glycerol with three), whilst glycol (with two) is the least efficient of all (W. S. Patterson, *J. Soc. chem. Ind.* 1934, **53,** 298T; W. S. Patterson and R. C. A. Culbert, *ibid.* 1935, **54,** 327T).

Perhaps the good protection obtained from mixtures containing chromate and condensed phosphate should be explained in the same way (p. 146). Other examples are described by King. He found that the efficacy of chromate in preventing attack in a liquid containing potassium nitrate and hydrochloric acid could often be increased if a complexing or chelating agent were added to prevent a loose precipitate. Apparently the protection of iron is increased by the addition of the sequestering agent EDTA (ethyldiamine-tetracetic acid) to the chromate, although ultimately pitting occurs.* The protection of zinc was improved by addition of fluoride—although in the latter case more than one explanation has been offered for the efficacy of the chromate–fluoride mixture (C. V. King, E. Goldschmidt and N. Mayer, *J. electrochem. Soc.* 1952, **99,** 423; C. V. King and E. Hillner, *ibid.* 1954, **101,** 79; C. V. King and E. Rau, *ibid.* 1956, **103,** 331).

Uhlig considers that condensed phosphates act in a similar manner, removing hydrous oxide films, and permitting higher surface concentrations of oxygen for a given concentration in the liquid (H. H. Uhlig, D. N. Triadis and M. Stern, *J. electrochem. Soc.* 1955, **102,** 59).

Cause of dependence on previous exposure to Oxygen. Finally, it can now be seen why, at critical concentrations of inhibitor, behaviour is affected by past history, and particularly on length of exposure to air. Let us recall Mears' work (p. 52) in which he traced scratch-lines on iron and exposed them to air for different times before placing on the surface drops of 0·07M sodium bicarbonate; whilst the proportion developing rust was 84% after 0·25 min. exposure, it gradually decreased with the exposure-time and became only 27% after 1024 min. exposure.

The scratching process will produce, on each side of the scratch-line, zones where metal in tensional stress will overlie metal in compressional stress (p. 109). If there were no oxidation the stresses would be in equilibrium, but the conversion of part of the outer tensional material to oxide, will leave compression in excess of tension, and the atoms in the compressional part will start to move further apart, transmitting a tension to the film. Now if the film has been produced by outward movement, there is a possibility of sufficient vacancies at the base to give bad anchorage, and at some places there may be a cavity between film and metal extending for an appreciable distance; if so, there is a likelihood that the film will crack

* R. S. Thornhill remarks that the sodium salt of EDTA is alkaline and can attack copper.

at such a place. Fresh film will then be formed, and alternate cracking and healing will occur, explaining the phenomena presented on p. 1042. However, gradually the stresses in the compressional zone will be used up and cracking will become less frequent. The longer the exposure to air, the smaller is the chance of a gap being found, when the drop is placed in position, sufficiently large to make the actual replenishment-rate less than the replenishment-rate needed for passivity; thus the proportion of scratch-lines which suffer rusting will fall off steadily with the time of air-exposure.

Other References

Technical advice regarding inhibitive water treatment will be found in the books of L. W. Haase, " Werkstoffzerstörung und Schutzschichtbildung im Wasserfach " 1951 (Verlag Chemie); F. N. Speller, " Corrosion: Causes and Prevention " (McGraw-Hill); J. C. Thresh, J. F. Beale and E. V. Suckling " Examination of Waters and Water Supplies " (Churchill).

For the theoretical aspects, readers should study R. B. Mears (" A Unified Mechanism of Passivity and Inhibition "), *Trans. electrochem. Soc.* 1949, **95,** 1; L. Cavallaro, *Chimie et Industrie* 1952, **68,** 511; L. Cavallaro and G. P. Bolognesi, *Atti Accad., sci. di Ferrara* 1946–47, **24,** 59, 235; L. Cavallaro and A. Indelli, *ibid.* 1946–47, **24,** 261; T. P. Hoar (" The Mechanism of Anodic Inhibitors of Corrosion: a General Treatment "), Association Belge pour l'étude, l'essai et l'emploi des matériaux (1954); H. Fischer, *Werkst. u. Korrosion* 1955, **6,** 26; C. C. Nathan, *Corrosion* 1956, **12,** 161t; H. C. Gatos, *Corrosion* 1956, **12,** 23t. For means of distinguishing physical from chemical adsorption, see J. J. Kipling and P. B. Peakall, *J. chem. Soc.* 1957, p. 834. See also B. M. W. Trapnell, " Chemisorption " 1955 (Butterworth).

Special attention should be given to the various views on inhibition presented to the Internationales Kolloquium über die Passivität der Metalle, published in the *Zeitschift für Elektrochemie* (1958), **62,** 619 to 827. The importance of adsorption is emphasized in papers by H. H. Uhlig and N. Hackermann; the importance of film-formation is discussed by U. R. Evans. The papers by G. H. Cartledge on pertechnetates, M. Cohen and A. F. Beck on chromates, W. R. Buck and H. Leidheiser on cation effects and M. J. Pryor on the effect of different compounds on the corrosion of aluminium, all deserve careful study. The Author's views on inhibition are presented in simplified form on p. 1042 of this book.

Readers should also study the latest information regarding inhibitive processes. Experimental treatment in the auxiliary (2000 gallon) cooling circuit at Calder Hall power station has supported Ride's work (W. T. Edwards, J. E. LeSurf and Mrs. P. A. Hayes, U.K.A.E.A. (I.G.) Report 52 (RD/C), 1959). Recent reports show that borate, which is already in wide use in the U.S.A., is performing well for internal combustion engines manufactured by a large British company (R. J. Brown, Priv. Comm. April 3, 1959).

CHAPTER VI

BIMETALLIC CONTACTS AND CREVICE CORROSION

SYNOPSIS

The chapter opens with a discussion of the *galvanic corrosion* set up when two dissimilar metals are in direct contact (or connected by an electronic conductor) and are also joined by a bridge of electrolytically conducting liquid. The metal which is anodic often suffers special corrosion, whilst attack on the cathodic metal is sometimes diminished. It is shown that the idea that the magnitude of the danger introduced by the use of dissimilar metals in contact can be assessed directly by the magnitude of the difference between the potentials of the two metals, as recorded in the tables, involves a fallacy; in particular the " quarter-volt " criterion, still apparently adopted in some quarters, should be regarded as completely unreliable.

The acceleration of the attack upon the anodic metal produced by connection to a cathodic metal is usually most serious when the cathodic metal is kept free from film. In general, the intensity of attack becomes most serious when the ratio of cathodie to anodic area is high; gaps in a coating of a cathodic metal may cause intense attack on the (anodic) basis metal; the internal pitting of copper pipes sometimes set up at gaps in a carbonaceous coating provides another example of the same principle.

The use of dissimilar metals may lead to corrosion trouble even when they are out of contact; thus water flowing through copper pipes and then through a galvanized tank may first dissolve a trace of copper, which is then deposited as metal on the galvanized surface, setting up corrosion-cells on a microscopic scale.

We then pass to *crevice attack*. Even when only one metal is present, severe corrosion may be set up in crevices, produced at chinks between metal plates or at places where sediment or other matter has settled. This is often due to poor replenishment of oxygen or of inhibitor at such places. Crevice attack in atmospheric corrosion, due to retention of moisture after the more exposed surfaces have become dry, is discussed in Chapter XIII.

Finally attention is given to corrosion at soldered and welded joints; electrical contact between dissimilar materials, metallographic changes set up by the heating, stresses introduced by the joining operation and residues of flux may all cause trouble. In some cases, it may be advisable, where possible, to submit an article to heat-treatment after welding.

Galvanic Corrosion

General. Intense corrosion—exceeding that normally experienced where only one metal is present—often occurs where a " base " (reactive) metal is in contact with a more " noble " (unreactive) metal; the effect of the combination is frequently to accelerate the corrosion of the base metal and to reduce, or even prevent altogether, the attack on the noble metal. Such behaviour, which has long been known, is generally attributed to a galvanic cell set up with the base metal as anode. Essentially, this explanation is correct; a *galvanic cell* (with the electrodes consisting of two *different* metals) will usually generate a much stronger current than, say, a *differential aeration cell* (with both electrodes of *the same metal*). However, the situation is not simple. It has been possible to point to instances where contact between dissimilar metals in service has caused no special damage, and others where crevices between metal plates of the same kind, or even contact with non-conducting substances (stones, sand, cotton-waste and the like), has caused intense local attack; such a phenomenon, known as *crevice corrosion*, is discussed later in this chapter.

Bimetallic Couples. The early measurements of corrosion proceeding on combinations between iron and a second metal, reproduced in Tab X are due to O. Bauer and O. Vogel, *Mitt. Mat. Prüf. Amt. Berl.* 1918, **36,** 114

TABLE X

BIMETALLIC CORROSION OF IRON AND A SECOND METAL IN 1% SODIUM CHLORIDE (O. Bauer and O. Vogel)

Corrosion of Iron (mg.)	Second Metal	Corrosion of Second Metal (mg.)
183·1	Copper	0·0
181·1	Nickel	0·2
171·1	Tin	2·5
183·2	Lead	3·6
176·0	Tungsten	5·2
153·1	Antimony	13·8
9·8	Aluminium	105·9
0·4	Cadmium	307·9
0·4	Zinc	688·0
0·0	Magnesium	3104·3

The numbers show that iron is protected by contact with a baser metal, whereas it suffers serious corrosion in contact with a relatively noble metal; but the amount of corrosion is much the same for five of the noble metals studied, namely tungsten, lead, tin, nickel and copper. The attack on the second metal varies enormously. If we disregard aluminium, which carries a highly protective film, it seems to increase as we pass towards the base end of the Table of Normal Electrode Potentials (p. 312).

Some prevailing fallacies. Such considerations were probably responsible for the belief that it is possible to decide which combinations are the most dangerous by consulting the Table of Normal Potentials. It is known that the E.M.F. of a Daniell-type cell, formed by two metals placed respectively in solutions containing their salts at equivalent concentrations, is obtained approximately by subtracting the normal potential of the baser metal from that of the nobler metal—with due allowance for signs. Under such conditions, the further apart the two metals stand in the Table, the larger will be the E.M.F. of the combination. In absence of polarization, the current which such a cell can send through a given resistance, and thus the corrosion-rate of the baser metal forming the anode, will increase as the difference between the two Normal Electrode Potentials becomes greater.

However, as a practical guide to combinations which will produce dangerous stimulation to the baser metal and provide useful protection to the nobler metal under service conditions, the Table of Normal Potentials has proved of limited value. Nor is this surprising, since each number printed in that table refers to film-free metal placed in a solution containing its ions at normal activity. In service we rarely meet with film-free metals and are rarely concerned with liquids containing normal activity of ions; thus discrepancies between the table and practical experience are to be expected. Even qualitatively, predictions conflict with service behaviour. The Table would seem to suggest that aluminium ought to protect zinc. Actually zinc protects aluminium and its alloys; Akimow has described how the whole length of a 4-metre bar of duralumin placed in sea-water received protection through contact with zinc at one end; potential measurements showed that in point of fact the zinc did function as the anode of the cell—in apparent contradiction to the Table, which of course refers to quite different conditions (G. W. Akimow, *Korros. Metallsch.* 1930, **6,** 84).

Results like this prompted the practical man to reject the Table of Normal Electrode Potentials and adopt a different table which showed empirical potentials obtained by measurements on the various metals placed in sea-water; the choice of sea-water as standard liquid, despite its variable character, was doubtless prompted by the fact that the most serious cases of galvanic corrosion were being met with under marine conditions. Accordingly, suggestions were made that, if the potential measurements obtained empirically on two metals in sea-water differed by more than a certain amount, the combination must be regarded as a dangerous one. In certain quarters 0·25 volts has been adopted as the limit of safety for severe conditions of service. This " quarter-volt criterion " eliminated a few of the worst absurdities which had accompanied the use of the Table of Normal Potentials, but the results were still unsatisfactory. Once again, it is easy to see the reasons. The procedure is based on the assumption that the working E.M.F. of a bimetallic combination can be obtained by subtracting the potential provided by one half-cell from that provided by the other —a procedure *legitimate for reversible potentials* (such as are given in Table of Normal Electrode Potentials) but *wrong for irreversible potentials*, such as are obtained in sea-water. In any case, the E.M.F. alone does not determine the current generated by the cell; the current is controlled mainly by polarization considerations, and the corrosion-rate, which is proportional to the current, will not necessarily be highest with those combinations which give the highest E.M.F. The quarter-volt criterion must be condemned as being completely unreliable, and its pseudo-scientific flavour only makes it the more dangerous.

The falsity of the predictions obtained by this sort of procedure is shown by an example quoted by Goddard. The potential of stainless steel measured in sea-water is much the same as that of copper, so that if the argument indicated above were right, the cell

Aluminium | Stainless Steel

should give much the same E.M.F. as the cell

Aluminium | Copper

Yet contact of aluminium with stainless steel produces relatively little stimulation of attack except under special conditions, whereas contact with copper commonly has disastrous results (H. P. Goddard, *Corrosion* 1955, **11,** 542t, esp. p. 546t).

Effect of Films on the Cathodic Member. The reason for the apparent discrepancy just mentioned is that copper provides an efficient *metallic* cathode, whereas stainless steel provides an inefficient *oxide* cathode. The slowness with which aluminium itself corrodes arises largely from the fact that the cathodic reaction has to proceed on an inefficient cathodic surface. It is generally believed that the main invisible oxide film on the aluminium plays no part in the corrosion, either because it is a bad conductor of electricity, or because it is a poor catalyst. In the picture presented in fig. 41(1), based largely on the ideas of Masing and Altenpohl, the places

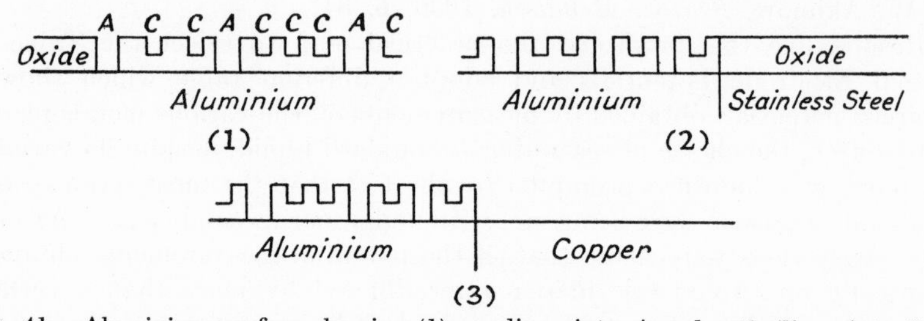

FIG. 41.—Aluminium surface showing (1) anodic points *A* and cathodic points *C*, (2) Effect of junction with Stainless Steel, (3) Effect of junction with Copper.

where the film has broken down to form definite gaps are the anodes (*A*), whilst the cathodic points (*C*) are the places where the film is thin, so that electrons can pass across it. Whether or not this provides a complete picture of the situation, Pryor and Keir have provided evidence that the cathodic reaction on aluminium occurs only at isolated points, and that an aluminium cathode polarizes rapidly; so soon as even a very weak current begins to flow, the potential shifts in such a direction as will oppose the flow of current. If now we join a plate of aluminium to stainless steel (fig. 41(2)) we extend the cathodic area available, but the extension still consists of an oxide-covered surface, which will form an inefficient cathode. If, on the other hand, we join aluminium to copper (fig. 41(3)) there is a possibility that the cathodic reaction will soon reduce any oxide originally present on the copper surface, and thereafter keep that surface clean; if so, instead of an oxide, which is unsuited as a site for the cathodic reaction, we obtain an efficient metallic cathode; stimulation may be expected and is, indeed, obtained (G. Masing and D. Altenpohl, *Z. Metallk.* 1952, **43,** 404, 433; M. J. Pryor and D. S. Keir, *J. electrochem. Soc.* 1955, **102,** 605).

It is possible by relatively simple calculation * to decide whether copper is likely to be kept free from oxide by contact with aluminium; the result answers the question in the affirmative—thus explaining why contact with copper is dangerous, whereas contact with stainless steel is relatively harmless.

It is interesting to ask the same question in respect of a contact with iron.† Here the answer is somewhat less definite. It is true that in certain

* The calculation is *not* based on the normal electrode potential of copper in contact with cupric ions as recorded in the Table of Normal Electrode Potentials (p. 312), since that value represents the equilibrium

$$Cu \rightleftharpoons Cu^{++} + 2e \quad (+\ 0{\cdot}345 \text{ volt})$$

We are here concerned with the equilibrium

$$Cu \rightleftharpoons Cu^{+} + e \quad (+\ 0{\cdot}522 \text{ volt})$$

The reason is that the equilibrium between copper and its two ions

$$Cu + Cu^{++} \rightleftharpoons 2Cu^{+}$$

is represented by the relationship

$$\frac{[Cu^+]^2}{[Cu^{++}]} = \text{constant } (K) \quad \text{or} \quad [Cu^+]^2 = K \times [Cu^{++}]$$

where K is of the order 10^{-6}. It is clear that, although cupric ions may predominate at high concentrations, cuprous ions must come to predominate at sufficiently low ones. Since cuprous oxide is sparingly soluble, with a solubility product $[Cu^+][OH^-]$ equal to $1{\cdot}2 \times 10^{-15}$, it is the cuprous ions which will fix the potential. In water buffered at neutrality (pH = pOH = 7), the cuprous ion concentration will be $1{\cdot}2 \times 10^{-8}$. The potential of Cu in $1{\cdot}2 \times 10^{-8}$N Cu^+ is $+\ 0{\cdot}522 + (0{\cdot}059 \times \log_{10}(1{\cdot}2 \times 10^{-8}))$ that is $+\ 0{\cdot}522 - 0{\cdot}467 = +\ 0{\cdot}055$ on the hydrogen scale. Slightly different values would be found at other pH values, but they will range around the zero point of the hydrogen scale. Now the potential of the aluminium electrode will be a compromise value depending on the state of the oxide, but in practice it is always found to be strongly negative. Thus the cell

$$Al \mid Water \mid Cu_2O \mid Cu$$

will be capable of generating an E.M.F. capable of reducing Cu_2O and keeping areas of the copper free from oxide.

† Here the oxide will be Fe_2O_3 which is even less soluble than Cu_2O, and the cathodic reaction best capable of keeping the oxide clean is not a reduction to metal, but to ferrous ions, which pass into the solution ("the reductive dissolution" discussed on p. 225). The equilibrium

$$Fe_2O_3 + 6H^+ + 2e \rightleftharpoons 2Fe^{++} + 3H_2O$$

depends on the pH value and the ferrous ions accumulating at the electrode surface. E. Deltombe and M. Pourbaix (Int. Comm. Electrochem. Thermodynamics and Kinetics, 6th Meeting (1954), p. 120) give it as

$$+\ 0{\cdot}728 - 0{\cdot}177(\text{pH}) - 0{\cdot}059 \log_{10} [Fe^{++}]$$

If we assume a neutral liquid under conditions sufficiently stagnant to allow situations such as are liable to occur in a marine environment, reductive dissolution of the invisible ferric oxide through contact with aluminium is very likely to occur, and iron may then form an efficient cathode, so that the combination of steel and aluminium in electrical connection is a dangerous one. Experience shows that in the superstructures of ships, serious corrosion does sometimes occur unless precautions are taken to insulate the two materials from one another; various ingenious designs, described by Whiteford, allow this insulation to be achieved; Whiteford's paper, with other pronouncements on the subject, deserves careful study. On the other hand, it is possible to envisage conditions where the iron will not function as a " metallic " cathode; thus there is no reason to distrust reports of the successful use of aluminium and steel in contact in certain situations; but those who hope to avoid trouble without careful, consideration of the electrochemical factors cannot expect that their luck will always hold. For discussion of the subject, with special regard to ships, see J. M. Whiteford, *Shipp. World*, Sept. 2, 1953; D. C. G. Lees, *Chem. and Ind.* (*Lond.*) 1954, p. 949; S. F. Dorey, *J. Inst. Met.* 1953–54, **82**, 497, esp. p. 499.

The calculations needed to answer the questions as to whether contacts of aluminium with copper or iron must be considered dangerous, involve electrochemical principles with which some readers may not be familiar, and have therefore been placed in footnotes. But all readers should at least recognize that the problem is far from simple. Those who expect to obtain an answer by performing a subtraction sum on two potential values culled from a table are sadly deceiving themselves.

It is particularly difficult to predict behaviour where highly protective films exist on metals. Contact of ordinary steel with stainless steel, or contact of ordinary steel with aluminium, has been known in some cases

the ferrous ions to accumulate to a concentration (or strictly, activity) of N/100, this becomes

$$+ 0{\cdot}728 - (7 \times 0{\cdot}177) - (0{\cdot}059 \times (-2)),$$

which is $-$ 0·393 volt on the hydrogen scale. Now the potential of aluminium in a chloride solution is usually much more negative than this value, and consequently contact between aluminium and iron in presence of, say, sea-water represents a dangerous situation—as is known to be the case. If the number of + 1·035 volts, adopted by some authors for the standard value, is taken instead of + 0·728 volt, the conclusion is the same. However, Pryor points out that the pH sensitivity of the reductive dissolution of ferric oxide is three times that of the reduction of dissolved oxygen, and thus in sea-water containing a good supply of oxygen the film might survive and the effect of contact with aluminium would be less destructive, especially as this good oxygen-supply will reduce the danger of the destruction of the film on the aluminium (p. 518). The question is a complicated one. Different predominant anions, different pH values, different flow conditions (involving different values of [Fe^{++}]), and different rates of oxygen-replenishment, would all affect the decision. There is no reason to be surprised when confronted with cases where aluminium and iron have been used in contact without disastrous results. The influence of oxygen is pointed out by M. J. Pryor, Priv. Comm., April 15, 1958. The reductive dissolution of ferric oxide films by contact with aluminium—and particularly the influence of the pH value—has been studied experimentally by M. J. Pryor and D. S. Keir, *J. electrochem. Soc.* 1958, **105**, 629.

to stimulate corrosion of *both* materials. Probably the ordinary steel is first anodic, but the corrosion products interfere with the oxygen supply needed to maintain the protective film on the other material, which then starts to corrode (see p. 340).

The Catchment Area Principle. Bimetallic contacts may increase corrosion troubles in two ways. (1) The total attack on the anodic metal may remain the same, but is concentrated upon a smaller area, so that the intensity increases; (2) the total attack is increased.

The first effect, demonstrated in a classical paper by Whitman and Russell, may best be understood by considering a rectangle of steel fully immersed in flowing water containing dissolved oxygen (fig. 42(A)). If the flow is not too rapid, the rate of corrosion may be controlled by the rate of arrival of oxygen, which is completely reduced to hydroxyl by the cathodic reaction and allows the anodic reaction (representing corrosion) to take

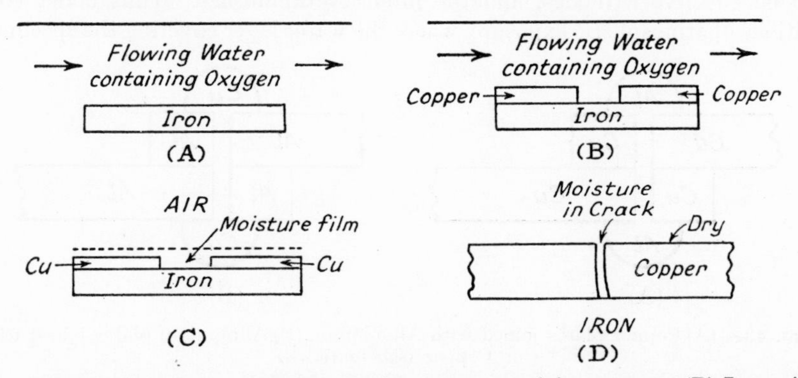

FIG. 42.—(A) Bare Iron exposed to flowing water containing oxygen, (B) Iron with three-quarter area covered with Copper and exposed to flowing water containing oxygen, (C) as B, but carrying only moisture-film, (D) crack in Copper-coating containing moisture; outer face dry (schematic).

place at an equivalent rate. Now suppose that three-quarters of the steel has been plated with copper (fig. 42(B)); if alternative cathodic reactions (e.g. hydrogen liberation) are excluded, the total attack will not be increased, since the whole of the oxygen was used up when the steel was bare. However, the copper will largely escape attack, and any oxygen reaching the copper will suffer reduction to hydroxyl as before, but the corresponding anodic reaction will be concentrated on the bare quarter. The amount of steel destroyed will remain the same, but the *intensity* of attack (the corrosion per unit area of the part affected) will be multiplied by four. The copper is being used as a *catchment area* for oxygen, the effects being passed on to the steel; the convenient expression " catchment area principle " came to be adopted some years after Whitman and Russell published their results, but their paper deserves study (W. G. Whitman and R. P. Russell, *Industr. engng. Chem.* 1924, **16,** 276).

The catchment area principle is applicable to cases where the corrosion-rate is fixed essentially by the rate of oxygen-replenishment, that is, where corrosion is under " cathodic control ". In such cases it possesses great practical importance. Clearly if the cathodic area is made large, the total amount of oxygen reaching it will be large and the current large; then if the anodic area is made small, the current density becomes high and the corrosion-intensity high. Up to a point the ratio of cathodic area to anodic area fixes the intensity of attack. At first sight, it would seem possible to obtain an enormous corrosion-intensity by simply making the anodic area small compared with the cathodic area, and indeed rather severe pitting can be obtained at tiny pin-holes in a coating of a metal cathodic to iron. Clearly, however, there are limits. If the liquid is a good conductor (e.g. sea-water) and the specimen is deeply immersed, the distant parts of the cathodic coating can play an effective part in the cathodic reaction; but if the liquid is a supply water with poor conductivity, or if there is only a shallow layer of water, the distant parts of the coating cease to be effective cathodes, and the intensity diminishes. Thus under conditions of atmospheric exposure, where the water layer covering the specimen

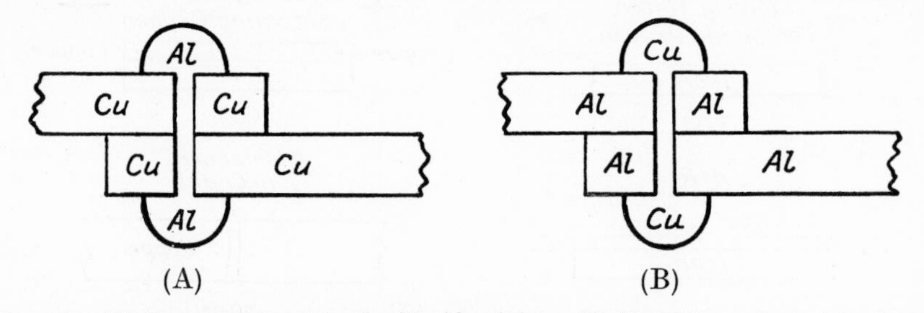

Fig. 43.—(A) Copper plates joined with Aluminium, (B) Aluminium plates joined with Copper (schematic).

may be a mere moisture film, the attack at breaks in a cathodic coating is not generally intense (fig. 42(C)); only the parts of the coating very near the exposed steel function as cathode. Under nearly dry conditions, it may be that only the walls of the crack in the coating, which may carry a small amount of water held by capillarity, function as cathode (fig. 42(D)).

Evidently the ratio of the areas of the two materials forming a bimetallic combination can be important under immersed conditions, but less important under conditions of atmospheric exposure, unless the design is such that rain water can be entrapped around the joint. Consider, however, conditions of immersion, and compare an assemblage formed of copper plates fastened together with aluminium bolts or rivets (fig. 43(A)) with one formed of aluminium plates fastened together with copper bolts or rivets (fig. 43(B)). In the first case, the total attack, determined by the oxygen reaching the large copper area, will be wholly concentrated on the small aluminium fasteners, which will be intensely attacked, soon become loose, and finally disappear. In the second case, the attack will probably be slower, and will be well distributed over a considerable area of the aluminium plates, so that the damage will be less. Both arrangements, however, represent bad practice, and it is far better never to use aluminium and copper in contact.

The area ratio may not always be so important. In the experiments of Borgmann and Koenig, carried out under fully immersed conditions, the corrosion-rate of the couples Fe | Cu, Fe | Ni and Fe | Cr was independent of the area ratio; the same was true of Zn | Cu, but on Zn | Ni, Zn | Fe and Zn | Cr, hydrogen was evolved, and the area ratio became important. Probably in cases where the cathodic reaction is the reduction of oxygen, the passage of oxygen through the water-surface fixed the total attack (C. W. Borgmann and R. T. Koenig, *J. electrochem. Soc.* 1950, **97,** 87C. See also M. J. Pryor and D. S. Keir, *ibid.* 1957, **104,** 270, who studied the effect of stirring as well as area-ratio).

Cases where Contact increases the total attack. If we pass to a situation favourable to oxygen-replenishment, such as occurs at a water-line, the presence of a second metal may not merely redistribute the corrosion but increase it. Early measurements by the author, on vertical rectangles of steel sheet partly immersed in N/10 potassium chloride,

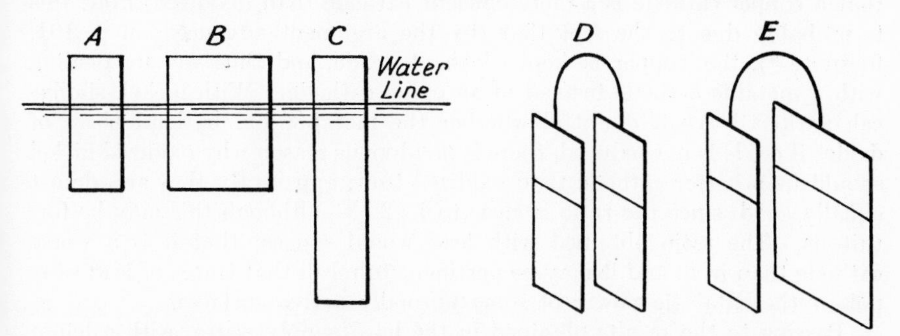

FIG. 44.—(*A*) Partly immersed Steel rectangle. (*B*) as *A* but breadth doubled. (*C*) as *A* but length doubled. (*D*) as *A* but joined to cathodic metal of same breadth. (*E*) as *A* but joined to cathodic metal of doubled breadth.

showed that when the breadth of the specimen was increased (as in passing from Case (*A*) to (*B*) of fig. 44) the corrosion was greatly increased, since the zone where oxygen could be replenished was increased; when, with the breadth kept constant, the vertical length of the immersed portion was increased (as in passing from Case (*A*) to (*C*)), there was only a small increase in the corrosion-rate and doubtless the intensity of attack declined; the fact that there was any increase at all indicates that the control of corrosion was not purely cathodic (U. R. Evans, *J. chem. Soc.* 1929, p. 111, esp. p. 122).

Now let us consider what will happen if the steel is joined electrically to a strip composed of a metal cathodic to steel, having identical dimensions (Case *D*) or double the breadth (Case *E*). If the catchment principle governed the situation, we should expect the attack on the steel to be doubled in Case (*D*) or trebled in Case (*E*). This experiment was tried, three different metals being used as cathodic contact pieces. The results are shown in Table XI (U. R. Evans, *J. Soc. chem. Ind.* (*Lond.*) 1928, **47,** 73T).

Had the Catchment Principle been operative, the ratio should have been

TABLE XI

RATIO OF THE CORROSION PRODUCED IN 14 DAYS ON SAND-BLASTED STEEL STRIPS WITH IMMERSED PORTION MEASURING, ON EACH SIDE, 2·5 CM. BROAD AND 4·0 CM. DEEP

(*a*) without contact piece, (*b*) with contact piece of single size, and (*c*) with contact piece of double size. (Each number represents the mean of two experiments)

Contact Metal	N/10 Sodium Chloride (*a*) (*b*) (*c*)	Cambridge Supply Water (*a*) (*b*) (*c*)
Copper	1 : 2·80 : 4·17	1 : 1·25 : 1·58
Nickel	1 : 1·91 : 2·86	1 : 0·72 : 0·83
Lead	1 : 1·24 : 1·47	1 : 0·93 : 0·93

1 : 2 : 3 in all cases. It is apparent that, in sodium chloride, copper produces a greater acceleration than that principle would predict—in other words, that a copper cathode is a more efficient cathode than (oxidized) iron; this is probably due to the fact that (by the argument advanced on p. 191, footnote *), the copper is kept clean of oxide, and thus we are dealing with a metallic cathode instead of an oxide cathode. With nickel, similar calculations leave it doubtful whether the metal would be kept clean of oxide; if oxide is not reduced, there is no obvious reason why oxidized nickel should be a better cathode than oxidized iron; apparently they are almost equally good, since the ratio is close to 1 : 2 : 3—although this may be fortuitous. The ratio obtained with lead would suggest that it is a worse cathode than iron, and it may be pertinent to recall that traces of lead salts poison the catalytic power of some normally active surfaces.

Passing to the results obtained in the hard supply water, with calcium bicarbonate as main constituent, we notice that only with copper as contact piece was any acceleration produced at all. The different behaviour of the hard water may be due to various causes. The electrical resistance of the liquid path between the two metals will be much higher, and the precipitation of calcium carbonate on the second metal cathode will interfere with the cathodic reaction (on iron a similar precipitation occurs, but the conversion into clinging rust causes a complication). Why connection with nickel or lead should have diminished the corrosion (apparently significantly) is uncertain; perhaps the presence of a second electrode parallel to the surface interfered with the liquid circulation set up by the falling corrosion-products—which circulation may play a part in replenishing oxygen at active points near the water-line on the steel specimen.

A study of the zinc/steel and aluminium/steel couples in normal sodium chloride by Pryor and Keir deserves study by those interested in the electrochemistry of bimetallic systems. Under a wide range of conditions, both aluminium and zinc are anodic to mild steel and protect it completely; in both cases the current flowing and the weight-loss of the anode metal are proportional to the area of the steel cathode and are much less dependent on the area of the anode. The galvanic corrosion-rates are controlled by oxygen-supply and are thus markedly influenced by stirring. Under stagnant conditions, the zinc/steel couple delivers more current than the aluminium/steel couple, but the situation is reversed at high stirring rates, which have the rather unexpected effect of causing uniform attack on the aluminium, but deep pitting of the zinc (M. J. Pryor and D. S. Keir, *J. electrochem. Soc.* 1957, **104,** 270; M. J. Pryor, *Corrosion* 1958, **14,** 1t).

Empirical Experience of Bimetallic Contacts. The arguments of pp. 188–192 make it clear that many factors affect the decision as to whether the corrosion of a given metal will or will not be increased by connection with a second metal. In attempting to help the engineer-designer, it seems better, instead of developing complicated electrochemical arguments (which, in some cases, must involve uncertainties), to provide facts based on practical experience, even though this is necessarily incomplete. For service designers this information has been collected by a British Service Committee, and the results have been made available to the public. They are arranged as a Table, which shows at a glance whether the corrosion of one material is increased by contact with a second material. No attempt is made to indicate the degree of corrosion suffered by the first material in the absence of the second; the Table is merely concerned with the danger of attack being *accelerated* when the second material is connected to it.

The Table represents the combined knowledge of a group of experienced people, and thus has authority behind it; but its compilers do not claim that all the statements are correct, and indeed it must always be impossible to predict behaviour regarding new situations which may arise in the future; a combination which has shown itself safe in past employment may prove unsatisfactory in some fresh use which cannot at present be foreseen. An appeal is made in the Table, asking any designer or user who finds the information misleading to report his experiences to certain officials (who are indicated), so that corrections or amplification can be made in future editions. Thus it may be hoped that in a few years' time a really reliable statement of the situation will become available. The document, which has an introduction by U. R. Evans and Vera E. Rance, is entitled "Corrosion and its Prevention at Bimetallic Contacts" (H.M. Stationery Office).

Some of the features of the document may be summarized. The behaviour of the noble metals, like gold, platinum, rhodium and silver is not adversely affected by contact with a baser metal (although, as explained later, the second metal, which is the anode of the combination, will usually suffer). Likewise certain "pseudo-noble" materials, such as 18/8 stainless steel, chromium and titanium, which owe their resistance to protective films, are not adversely affected by contact with a baser metal; but 13% Cr stainless steel suffers increased corrosion in contact with many noble or pseudo-noble materials, including nickel and copper alloys, as well as with 18/8 stainless steel. The behaviour of monel metal, inconel and nickel-molybdenum alloys is slightly worsened by contact with gold or platinum, whilst that of nickel and many copper alloys is definitely worsened by contact with those two noble metals, and to some extent by contact with monel, chromium or 18/8 stainless steel. The behaviour of lead, tin and solder deteriorates in contact with the (nobler) metals which appear above lead in the Table of Normal Electrode Potentials—and also with stainless steel. Mild steel should not in any case be used in the uncoated condition where water or condensed moisture is to be present, but attack upon it is stimulated by contact with practically all metals standing above iron in the Table of Normal Potentials; also by stainless steel, chromium and titanium. Cadmium and zinc suffer acceleration of attack in contact with most materials (except perhaps magnesium and its alloys), but they are extremely useful for the coating of other metals which are to be involved in bimetallic combinations—as explained later. Magnesium and its alloys should not normally be used uncoated, except in very dry atmospheres, but contact with practically any other material accelerates the attack; cadmium and zinc are the least dangerous.

There are exceptions to many of these statements, and the document quoted should be studied. Thus liquids containing substances like cyanides, citrates and tartrates, which form complex ions, may upset predictions. Among non-metals, contact with carbon, graphite and even coke can stimulate corrosion on most metals; graphite in packing has caused trouble, and where possible mica is to be preferred.

The document lays stress on the fact that the condition of the metal may affect behaviour. Particularly in aluminium-copper alloys of the H.15 type, heat-treatment alters potential, the naturally aged (W) alloy being more noble than the artificially aged (WP) alloy. The noble character of the stainless steels is dependent upon the protective film on the metal, and the maintenance of this film is, in turn, dependent on the continuous presence of oxygen. Even within a given class of alloy, composition may affect potential; cases are known where the corrosion of one brass has been accelerated by contact with another brass of higher copper-content.

It may be interesting to compare this British document with a somewhat simpler American scheme put forward in connection with naval aircraft. Materials are here divided into four groups, and combinations are regarded as " dissimilar " only if the two components come from different groups; a joint between two materials from the same group can be treated—it is thought—in the same way as a joint between two parts composed of the same material. The groups consist of:—

I Magnesium and its Alloys
II Cadmium, Zinc, Aluminium and their Alloys
III Iron, Lead and Tin and their Alloys (other than stainless steel)
IV Copper, Chromium, Nickel, Silver, Gold, Platinum, Titanium, Cobalt, Rhodium and their Alloys. Stainless steels. Graphite.

Some of these materials such as magnesium, iron and some aluminium alloys require, of course, very careful protection by suitable coverings, even when no second material is present. However, special measures are called for if they are to be joined to a material from another group. It is recommended that bolts used to attach plates together should be dipped into a priming paint immediately before installation; the exposed surface should be touched up after assemblage. In the case of magnesium alloys, the

American authors advise caulking of the joint with zinc chromate paste or rubber cement to form a water-tight connection; or the parts should be definitely insulated from one another by means of vinyl tape (N. E. Promisel and G. S. Mustin, *Corrosion* 1951, **7**, 339).

At first sight, it might appear that the requirements of these American authors (whose views do not necessarily represent the official requirements of the American Navy Department) are less stringent than the British requirements, since a pair of materials which are regarded as " dissimilar " in British practice are treated in the American scheme as though they were identical. However, the treatment of surfaces composed of a single material is carefully prescribed and is probably sufficiently elaborate to take care of cases where two materials, not identical but from the same group, are present.

Methods of avoiding trouble at joints by the use of jointing compounds —with special reference to aluminium in contact with other materials— are discussed on p. 201. The problem of bimetallic junctions in electrical equipment, in connection with the choice of finish, is discussed by E. C. J. Marsh, *Electroplating* 1954, **7**, 88.

Contacts involving Light Metals. Teeple has described atmospheric exposure tests involving magnesium alloys bolted in electrical contact with other metals. Each specimen consisted of four discs differing in size, bolted together with a stainless steel bolt insulated from the discs by means of bakelite washers and bushes. The top (smallest) and third disc consisted of magnesium alloy, whilst the second and fourth were of the contact material. Aluminium contributed less to the weight-loss of magnesium than the other materials tested, but zinc or zinc-plated steel were also relatively harmless. Nickel or stainless steel, red brass, monel and ordinary carbon steel all increased the attack; carbon steel (despite its lower position in the table of potentials) stimulated attack more than the other materials, possibly because the film on it was destroyed by reductive dissolution. These relationships were valid, on the whole, for all the exposure stations and for both the magnesium alloys tested (H. O. Teeple, *Amer. Soc. Test. Mater. tech. Pub.* **175** (1955), p. 89).

If contact between magnesium and steel is unavoidable, danger can be lessened by plating the latter with zinc or cadmium, which may then with advantage be phosphated and lacquered; covering with a zinc chromate primer has been widely recommended (L. F. LeBrocq and H. G. Cole, *Métaux et Corros.* 1949, **24**, 211; N. H. Simpson, *Corrosion* 1950, **6**, 51; see also *Mater. and Meth.* Feb. 1950, **31**, 49).

The exposure of magnesium alloy in an unprotected condition (like that of steel) is normally inadvisable even if no bimetallic contact is present. A chromating treatment (p. 589) increases the resistance, but even in the chromated condition, a magnesium alloy may suffer at a junction with another metal if a volume of water (as opposed to a mere film of moisture) can collect at the joint, and in all such cases insulation of the two materials is advisable. In any case, nuts, bolts, washers, screws and studs made of steel, copper or brass, should be galvanized, cadmium-plated or otherwise protected; but such coverings do not absolve the designer from the need to avoid accumulation of water; in presence of sea-water, magnesium has been known to suffer along a contact with cadmium-plated steel. On the other hand, if jointing compound (D. T. D. 369A) is used, it has sometimes been possible to dispense with insulation, since good design, if it achieves the prevention of any collection of water at the joint, may sometimes compensate for a combination which appears to involve electrochemical risks. Advice should, however, be sought before adopting any hazardous contact. Jointing compound should be squeezed out to give a fillet at least $\frac{1}{4}$ inch broad, and tape, if used, should also extend beyond the edge of the faying surfaces; everything should be done to avoid a film of moisture connecting the two dissimilar materials.

American opinion appears somewhat less optimistic than British opinion regarding the behaviour of magnesium in contact with other metals. It is, however, difficult to lay down hard and fast rules. Much depends on geometry. As pointed out by Higgins, bimetallic effects can be greatly diminished by attention to details of design and assembly. He recommends a particular form of nut, the " Galvanut ", incorporating a plastic washer, which increases the length of the electrolytic path provided by any liquid connecting the two metals. He also points out that the introduction of iron shot-blasting or abrading with emery can greatly diminish resistance to chloride solutions—which resistance can largely be restored by high-voltage anodic treatment in ammonium acid fluoride solution (W. F. Higgins, *Chem. and Ind.* (*Lond.*) 1958, pp. 218, 1604).

Edwards has commented on the galvanic corrosion set up on magnesium tubes in water if they carry a discontinuous coating of embedded carbon arising from the lubricant (W. T. Edwards, *Chem. and Ind.* (*Lond.*) 1958, p. 219).

Where magnesium is in contact with aluminium in sea-water or other saline liquid (such as sodium nitrate, sometimes used on aircraft as a " defroster "), the alkali formed at the aluminium cathode may attack the aluminium (see " Corrosion and its Prevention at Bimetallic Contacts " 1956, notes (*a*) (*b*) (*c*) and (*h*) (H. M. Stationery Office)).

The behaviour of aluminium has already been mentioned. Contact between aluminium and steel structures has been known to stimulate the attack on both metals, although there is variability of behaviour. When, as in aircraft, contact between strong alloy steel and strong aluminium alloys is unavoidable, cadmium-plating of the steel is usually relied on, although some designers distrust cadmium-plating in the belief that it causes hydrogen brittleness (see Chapter XI).

Raine, referring apparently to the attachment of aluminium-sheathed cables with steel cleats, states that the aluminium-spraying of the steel avoids trouble. Pool has discussed the joining of structures involving aluminium alloy components with steel bolts, gussets and splice plates; to avoid corrosion, the steel plates may be zinc-sprayed, the bolts cadmium-plated and the contact surfaces brought together with a wet coat of primer; finally the whole assembly should receive a three-coat paint system (P. A. Raine, *Chem. and Ind.* (*Lond.*) 1954, p. 958; E. E. Pool, p. 957).

Britton's exposure trials in various atmospheres (also others in salt-spray and with intermittent immersion in the sea) deserve study; steel nuts and screws coated with zinc, cadmium or tin-zinc alloy were exposed in contact with five different aluminium alloys. Tin-zinc alloy coatings proved better than zinc in marine atmospheres and better than cadmium in inland urban conditions (S. C. Britton and R. W. de Vere Stacpoole, *Metallurgia* 1955, **52**, 64).

Very serious results may occur when suitable precautions are not taken. Aluminium rain-water piping placed in electrical contact with copper has suffered rapid perforation; but in this case insulation will not provide a remedy if rain first flows over the copper and then on to the aluminium, carrying traces of copper salts. Failure has occurred on an aluminium drain-pipe fixed with a cast iron bracket (*Metal finish. J.* 1955, **1**, 85).

It must be remembered that aluminium alloys vary greatly among themselves, although it is common in many circles to call them all "aluminium". An aluminium-magnesium alloy will generally be anodic towards steel, whereas an aluminium-copper may sometimes be cathodic. This, however, depends on composition and particularly heat-treatment; von Zeerleder has described experiments on an alloy which was distinctly cathodic to a steel when "naturally aged" and anodic when "artificially aged" (A. von Zeerleder, *J. Inst. Met.* 1931, **46**, 169; esp. discussion, p. 180).

In the case of aluminium alloys containing copper, troubles at junctions are generally reduced where the material has been clad with unalloyed aluminium, which is anodic towards alloys containing copper; sometimes benefit may be obtained by spraying the whole assembly with unalloyed aluminium. In other cases, trouble can often be avoided by the use of jointing compounds, which can be used to fill the crevice between the metals and are sometimes advantageously spread over the face outside; one of these consists of a mixture of barium chromate and kaolin in a slow-drying oil-varnish, but there is a feeling in some quarters that a more soluble chromate (perhaps strontium chromate) would be better. Several proprietary jointing compounds are on the market, some designed to harden quickly and others to remain soft for considerable periods. After assemblage, the whole should be painted, preferably with a chromate paint. Wray's tests on aluminium-iron couples indicate alkyd resin to be a better medium than phenolic resins for the purpose; good results were obtained from an alkyd-base primer pigmented with calcium chromate (R. I. Wray, *Corrosion* 1954, **10**, 50, esp. p. 55). Extensive Dutch exposure tests on compound specimens joined with rivets or bolts are described in Metaalinstituut T.N.O., 3e Aflevering, Feb. 1956. See also F. Pearlstein, *Metal Finishing* 1956, **54**, No. 4, p. 52. Other exposure tests on aluminium in contact with other metals are described by E. Zurbrugg, *Rev. Alumin.* 1957, **244**, 647.

The corrosion of steel-cored aluminium used in overhead electrical conductors may involve three metals (since the steel is usually zinc-coated) and in certain situations crevice corrosion (p. 207) plays a part. The matter has been examined in detail by J. S. Forrest and J. M. Ward, *J. Instn. elect. Eng.* 1954, **101**, Part II, p. 271, with discussion on p. 283.

Cases where contact with stainless steel has caused severe corrosion to aluminium (which ceased when insulation was provided, or when the stainless steel parts were replaced by aluminium parts) are described by A. A. Brouwer, *Industr. engng. Chem.* 1958, **50**, April, p. 73A.

The behaviour of titanium in couples has naturally aroused much interest. Normally titanium resembles stainless steel in its electrochemical behaviour. Inglis states that the titanium itself is generally unaffected by contact with a second material, but that the attack on other material may be greatly increased, especially if it is mild steel or gun-metal. This accords with general experience, but certain statements from other sources may appear contradictory. The opinion has been expressed that the corrosion behaviour of titanium alloys may be regarded as equivalent to that of unalloyed titanium. Since titanium is really a highly reactive material which becomes passive owing to a film in a state of high compression, small changes in composition or surface condition would be expected to alter behaviour, although alloying does not necessarily increase efficiency as a cathode; the figures of Paige and Ketcham seem to show that in normal sodium chloride solution, coupling with a titanium alloy containing 1·8% chromium and 0·9% iron causes less acceleration to most base metals than coupling with unalloyed titanium or stainless steel. Contact with any of these three materials was found to increase the attack on cadmium nine times—a matter deserving notice in view of the reliance placed on cadmium-plating for preventing corrosion in aircraft. It appears to be established that in marine situations the corrosion of steel or aluminium alloys can be seriously increased by coupling to titanium (which does not itself normally suffer); even such relatively noble materials as monel metal and copper suffer an appreciable increase of corrosion (N. P. Inglis, *Chem. and Ind.* (*Lond.*) 1957, p. 180, esp. p. 188; J. B. Cotton, *ibid.* 1958, p. 640; H. Paige and S. J. Ketcham, *Corrosion* 1952, **8,** 413; D. Schlain and C. B. Kenahan, *ibid.* 1958, **14,** 405t; H. B. Bamberger, P. J. Cambourdis and G. E. Hutchinson, *Trans. electrochem. Soc.* 1954, **101,** 442; D. Schlain, C. B. Kenahan and D. V. Steele, *ibid.* 1955, **102,** 102).

It should seem that painting of such contacts is advisable, and readers should study future papers for authoritative recommendations.

Avoidance of Bimetallic Corrosion. Adverse effects at contacts may frequently be avoided by insulating the two metals from one another. This may involve special design, with insulating bushes and washers where plates are to be bolted together. Complete insulation may sometimes be essential where the combination is to be immersed in water, but if it is to be exposed to the atmosphere and covered merely with a film of condensed moisture, it may be permissible to allow electrical contact between the two metals and break up the liquid path, which would otherwise connect them, by means of a large insulating washer. All insulating materials should be impervious to water; felt is unsuitable.

Occasionally insulation is impossible for functional reasons, and often it is inconvenient. In such cases, bad effects may sometimes be avoided by coating the first metal with the second, or with a metal chosen for compatibility; or both components may be coated with a single suitable metal. Coating may be achieved by electro-deposition, hot-dipping or metal spraying. In air-frame assemblies for British service aircraft it is mandatory that all bolts, screws, pins, bushes, etc. made from non-stainless steels shall be cadmium-plated where such parts are to be in contact with light alloys (whether based on aluminium or magnesium). Cadmium-plating is also recommended for such items made from stainless steel. These precautions are effective only whilst the coating remains intact; a break in a coating on the anodic metal may have serious consequences; it may be well to protect the coating by passivation or painting.

In many cases it is well to fill crevices between dissimilar plates with a jointing compound, and this may also serve to prevent crevice corrosion; they are most effective when used as an impregnation on a rot-proof fabric placed between the faying surfaces. The outer surfaces around joints should receive very careful paintings, and where one of the metals is non-ferrous, a chromate paint will usually be more suitable than the red lead type. As stated above, there is some doubt about the best chromate pigment; clearly it should be sufficiently soluble to render the water non-corrosive, but not so soluble as to be washed away quickly; reference should be made to a paper by H. G. Cole (*J. appl. Chem.* 1955, **5**, 197) and discussion thereof (*Chem. and Ind.* (*Lond.*) 1956, p. 152).

In some cases, non-metallic coating materials may be used, whilst in others it may be effective to subject the whole joint to cathodic protection.

Distribution of Attack near a Boundary between Metals. When a plate consisting of two metals joined together in a single plane is fully immersed in water of low conductivity, the attack on the anodic metal will be confined to a narrow zone close to the junction-line, for reasons of conductivity; the total attack will probably be slow but its localization may cause corrosion to be relatively intense (fig. 45(A)). If, through the

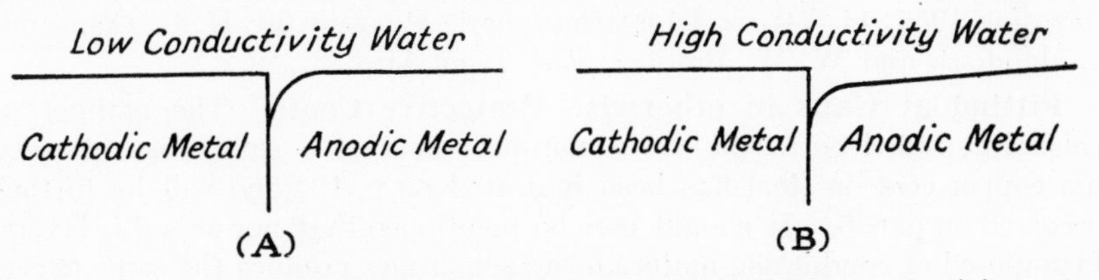

Fig. 45.—Distribution of Attack near junction of metals in (A) low-conductivity water (B) high-conductivity water (schematic).

presence of salts in the water, the conductivity rises, the attack is better spread out (fig. 45(B)) and the total corrosion will become greater; however, an increase in water-conductivity will not in itself increase the intensity along the marginal strip; there is some current density which can never be exceeded for reasons of polarization (p. 869), and this should be reached at the junction-line, whether the conductivity be good or bad. If the salt reduces polarization or prevents the formation of protective skins, it may increase the intensity of attack at the junction-line. The addition of chloride to water increases the intensity of anodic attack on magnesium.

If, instead of being totally immersed, the plates are merely covered with a thin film of moisture, the attack is likely to be confined to the neighbourhood of the boundary-line, even if the liquid is a good conductor. Both total corrosion and intensity will generally be greatest if the air contains acid fumes which dissolve in the moisture film, as in industrial regions, or salt, as near the sea; clearly intensity will be increased by traces of chloride flux—as mentioned later in connection with soldering and welding. The distribution of attack near a bimetallic junction has been studied mathematically by Waber; for references, see p. 217.

Under immersed conditions in the sea, a couple can cause corrosion at great distances from the cathodic component. Many marine engineers have long believed that bronze propellers on ships promote pitting on the hull. Others have contested this idea on the grounds that there is as much pitting near the bow as at the stern; but the criticism is without weight, since the steel is only exposed at small blemishes in the paint, and the bottle-neck approaches to the exposed anodic points interpose resistances large compared to the broad path joining the propeller to the neighbourhood of the ship's bow; thus it is to be expected that there will be as much pitting near the bow as at the stern. At both places the attack may become quite intense, owing to the small size of the exposed anodic areas which leads to a large cathode/anode ratio, and the rapid movement of the propeller which favours the cathodic reaction by assisting oxygen-replenishment. Zinc protector blocks around the propeller shaft may counteract this effect, but are only effective if the contact is good and the zinc surface kept clean; efficacy also depends on the quality of the zinc (p. 288).

Much attention has been given to the design of laboratory apparatus for testing bimetallic corrosion. Arrangements for complete immersion are described by J. F. Willging, J. P. Hirth, F. H. Beck and M. G. Fontana, *Corrosion* 1955, **11**, 71t, and for atmospheric exposure by H. G. Compton, A. Mendizza and W. W. Bradley, *ibid.* 1955, **11**, 383t.

Pitting at Gaps in otherwise Protective Coats. The manner in which intensified corrosion can sometimes occur at a small discontinuity in a copper coat on steel has been indicated on p. 194 and will be further discussed on p. 636. It should here be mentioned that non-metallic layers, if composed of conducting material, can sometimes produce the same effect. Mill-scale on steel, consisting largely of magnetite, may act as a cathode, and since in practice the scale, owing to rough treatment or rapid cooling, generally contains visible or invisible discontinuities, steel covered with mill-scale often becomes pitted. The scale may not increase the total corrosion—sometimes it actually diminishes it; however, the attack tends to be concentrated on a number of small points, and the local loss of thickness will often be greater than if the mill-scale has been removed. Speller describes the effect of mill-scale in localizing corrosion. In some tests at Pittsburgh, certain portions of the interior of a steel pipe were machined free from scale, whilst other portions were left with the scale adhering;

the latter were perforated by pitting before the machined section had lost one fifth of its thickness (F. N. Speller, "Corrosion: causes and prevention" 1951, p. 44 (McGraw-Hill), quoting J. O. Handy and W. P. Wood).

It should be noticed, however, that different types of scale, and different waters, behave differently. In preparing demonstration specimens in the laboratory, designed to show intense corrosion at gaps made (intentionally) in the oxide-film present on steel sheet previously heated in air, the Author found that, on some materials, plain sodium chloride did not produce the intense attack which he desired to demonstrate; the corrosion did start at the gaps, but then turned sideways, undermining the oxide, so that corrosion, although localized at the outset, later became relatively general; it was found that a small addition of sodium carbonate (or carbonate-bicarbonate buffer) to the chloride solution served to keep the corrosion localized and thus intense.

An example of intensified corrosion at gaps in a cathodic coat is provided by the pitting of copper pipes carrying a carbonaceous film, which serves as cathode towards copper (p. 125). Another is the pitting of condenser tubes by polluted waters containing certain sulphur compounds; these produce on the brass surface a film capable of protecting the brass below, except where it is kept broken by the impingement of bubbles or otherwise, in which case the large cathode/anode ratio causes severe pitting (p. 476).

Effect of Dissimilar Metals out of contact. Dissimilar metals may sometimes cause trouble even when they are not in electrical contact. If water containing free carbonic acid flows through a copper pipe and then through a galvanized tank or cistern, serious corrosion may be set up; such water can take up traces of copper (as copper bicarbonate) and later deposit metallic copper on the tank or cistern.

$$Cu^{++} + Zn = Cu + Zn^{++}$$

The micro-cells zinc | copper may then cause rapid attack on the zinc coating, and when the zinc has disappeared the micro-cells iron | copper may set up attack on the steel, finally causing perforation; the damage to the copper pipe in such cases will usually be trivial. The micro-cells may be more dangerous than the macro-cell set up when the metals are in electrical contact (the copper pipe being then a large cathode and the tank or cistern a large anode). It has even been claimed that deliberate electrical connection between copper and galvanized iron can improve the situation, since the zinc then affords cathodic protection to the copper, so that the micro-cells do not develop; but it is doubtful whether the trouble can be completely prevented by bonding the two materials.

Kenworthy has studied the incidence of trouble on housing estates. On one estate, where copper pipes were installed with galvanized tanks, *half* the installations failed in four years, although on another estate using the same water where both pipes and tanks were of galvanized iron there were *no failures*. On two other estates, both fitted with copper pipes and galvanized tanks, but using different waters, *every* installation failed in four years on one estate (where the water carried 4·1 ppm. free carbonic acid, and took up 0·32 ppm. of copper) whereas on the other estate (where the water had only 1·1 ppm. of free carbonic acid and took up only 0·03 ppm. of copper), *no failures* were reported. He suggests lime addition as a remedy, or aeration, which would remove much of the free carbonic acid (L. Kenworthy, *J. Inst. Met.* 1943, **69**, 67).

Traces of copper taken up from copper pipes may cause pitting to aluminium cooking vessels; 1 part of copper in 50 million suffices if calcium bicarbonate, a chloride and oxygen are present (F. C. Porter and S. E. Hadden, *J. appl. Chem.* 1953, **3**, 385).

Steel and aluminium in electrical contact may be dangerous, as already mentioned, but trouble is not necessarily avoided by insulating the two materials from one another, if the corrosion-products from the steel are allowed to reach the aluminium. There are various possibilities. If the cathodic and anodic products of the steel arrive at the aluminium without mixing, they may each be expected to set up corrosion; the cathodically formed alkali is likely to attack the aluminium, forming an aluminate, whereas the anodically formed iron salts may deposit iron or magnetite, with the possibility of a corrosion-couple; alternatively, if the two products meet close to an aluminium surface, interacting and forming ferrous hydroxide, which then absorbs oxygen to yield rust, the oxygen-removal may set up attack of the differential-aeration type, which may be intense if the anodic area is small.

Bird has studied these various possibilities. Small amounts of ferrous salt (10 or 100 ppm.) added to N/2 sodium chloride were found actually to diminish the total amount of corrosion on aluminium and two of its alloys; however, on the alloy HS15, the attack was localized, so that deep pitting was produced; a soluble ferrous salt also caused pitting to commercial aluminium when oxygen was present, but not when it was absent. Probably a coating of metallic iron was deposited, which protected the aluminium or alloy surface at most places, but, by providing a large cathode caused intense anodic attack at gaps in the coating. If instead of using an iron salt in solution, rust (prepared in a separate vessel) was added so as to cover the lower part of the specimen, serious intensification of attack occurred, although the total corrosion was diminished in most cases. This localized attack was probably due to oxygen screening, although, since magnetite was formed, it is possible that the couple aluminium | magnetite may have played a part.

It would seem that, whether or not there is electrical contact, the proximity of iron or steel to aluminium and its alloys may—in some circumstances—cause intensification, even though the total amount of aluminium destroyed is actually diminished. This is not necessarily a reason for avoiding the use of the two materials in conjunction, but it does suggest that the designer should exercise some ingenuity in avoiding a geometrical arrangement likely to cause the iron corrosion-products to sweep or settle on the light materials (C. E. Bird and U. R. Evans, *Corros. Tech.* 1956, **3**, 279).

Variable Polarity. Noble metals like silver and copper are cathodic to iron in almost all liquids, whilst magnesium or zinc (at least freshly abraded zinc) are almost always anodic. Metals like tin and lead, which stand close to iron in the Potential Series, display a variable polarity. Tin, for instance, is cathodic to iron in hot distilled water, but anodic in many organic acids—a matter of some importance in connection with fruit canning; the reason is that the formation of complex ions with the organic acids keeps low the concentration of simple tin cations. In other cases, vigorous aeration of the normally anodic member of a couple may cause it to act as cathode.

Although freshly abraded zinc is, at ordinary temperatures, strongly anodic towards iron in all ordinary waters or solutions, the polarity gradually reverses at high temperatures. The reversal, which has long been known, has been studied in some detail by Kenworthy and Smith, who find that, whereas a plate of galvanized iron suffers no rusting in cold water even though the steel basis is exposed at the cut edges, there is rusting in hot water; the iron receives cathodic protection at relatively low but not at high temperatures. This should be remembered in planning hot-water systems; it is essential that the steel basis should nowhere be exposed. Nitrates and bicarbonates seem to favour the reversal of polarity, whilst sulphates and chlorides decrease the risk of its occurring. The cause of the reversal of polarity is uncertain, but it is likely that zinc hydroxide formed on the zinc surface, changes to zinc oxide when the temperature is raised; zinc oxide has an appreciable electronic conductivity which increases with rise of temperature, so that the cathodic reaction (reduction of oxygen or liberation of hydrogen) can proceed appreciably on the hot zinc oxide surface, the iron functioning as anode. At lower temperatures, the zinc oxide is too poor a conductor to allow of the cathodic reaction, whilst the hydroxide, if it exists in slimy form, may act as oxygen-screen and help to keep the zinc anodic. The change of polarity is shown graphically in fig. 101, p. 640 (L. Kenworthy and M. D. Smith, *J. Inst. Met.* 1944, **70**, 463; P. T. Gilbert, *J. electrochem. Soc.* 1952, **99**, 18; R. B. Hoxeng and C. F. Prutton, *Corrosion* 1949, **5**, 330; H. L. Shuldener and L. Lehrman, *ibid.* 1958, **14**, 545t; J. L. Mansa, *Corros. Tech.* 1954, **1**, 102; G. Masing, " Lehrbuch der allgemeinen Metallkunde " 1950, p. 564 (Springer)).

Crevice Corrosion

General. Corrosion may give special troubles at crevices for several reasons. In a metallic structure exposed to the atmosphere, crannies between plates remain wet after a rain-storm for a considerable period, whereas the external faces dry quickly; corrosion will, therefore, continue longer in the crannies. Moreover, in such cases the voluminous nature of the corrosion-product may tend to lever the plates apart, even breaking rivets if these have been spaced at excessive distances from one another. Other cases of crevice trouble in atmospheric attack are discussed in Chapter XIII.

The special corrosion seen at crevices under conditions of complete immersion must, however, arise from other causes; the main one is the slow replenishment of oxygen—or some inhibitive chemical—in the recesses of a crevice. This not only refers to crevices or inaccessible corners arising from bad design, but also the crevices formed where foreign matter has settled or become wedged against the walls of a pipe or vessel through which water is required to flow (e.g. as part of a cooling system). In the early days when the corrosion of brass condenser tubes was a serious problem, those studying the subject were puzzled by the fact that corrosion was found, not only in cases where the foreign body was a carbonaceous conductor (capable of acting as a cathode), but also when it was a non-conductor such as stone, sand, shell or cotton-waste (see p 473.).

Oxygen Exhaustion in Crevices. It is now generally agreed that many cases of corrosion trouble are due to the fact that, in a crevice, oxygen-replenishment is so slow that quite minor sensitive spots, which would not

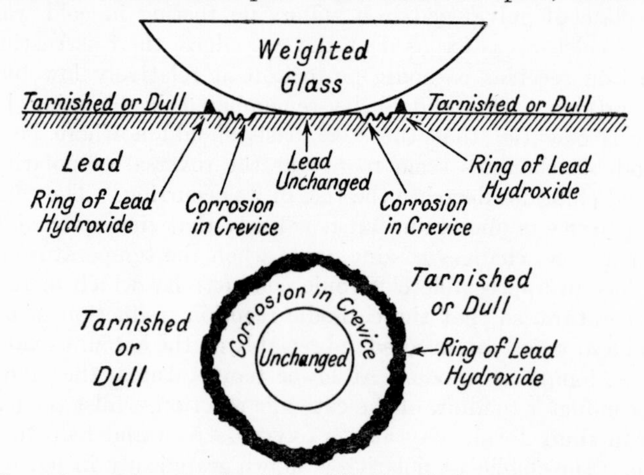

FIG. 46.— Corrosion produced on Lead in a Chloride solution at contact with a lenticular glass surface.

become sites of active corrosion if situated elsewhere, can, in their sheltered situation, develop attack. Crevice corrosion can easily be produced in the laboratory. If a piece of lead sheet is laid in a dish containing potassium chloride solution, with a lenticular glass placed on it, a ring of intense corrosion gradually develops in the annular crevice; the same result can be obtained by placing a specimen of lead of lenticular form on a flat porcelain surface—the whole being immersed in N/2 potassium chloride solution. After a week, the appearance will be somewhat as shown in fig. 46. The main part of the surface, accessible to oxygen and therefore cathodic, has become dull and often shows iridescent colours—analogous to those seen close to the water-line on iron or zinc (p. 88). Around the point of contact, where there has been a narrow annular crevice, there is marked etching of the lead—exposing crystal faces; the specimen, after cleaning and drying, shows points which flash out suddenly when the specimen is rotated in the light; this etching is clearly the result of anodic attack. The etched (anodic) ring is separated from the dulled (cathodic) area by a ring of white matter (lead hydroxide or a basic salt) produced by interaction between the anodic and cathodic products (lead chloride and potassium hydroxide); lead chloride, although sometimes regarded as " insoluble ", is in fact too soluble to produce a film over the anodic area—such as might occur in a sulphate or phosphate solution (U. R. Evans, *J. Inst. Met.* 1923, **30**, 267).

Clearly, however, crevice corrosion will not occur under all circumstances. Had the experiment just described been tried with iron instead of lead, there would probably have been *less* corrosion in the crevice than elsewhere, since on iron, sensitive spots are sufficiently numerous to start attack even where there is no special shielding from oxygen. If, for example, anodic attack started at two points, one inside the crevice and one outside it, it would develop more quickly at the point outside, owing to the greater proximity to the places where the cathodic reaction (which needs oxygen-replenishment) is occurring. If, however, the liquid contains enough inhibitor to prevent corrosion except at points exceptionally well situated for its development, intense corrosion may be expected in the crevices, and little outside. Thus crevice corrosion on iron and steel is mainly met with in water containing an inhibitor in inadequate amount.

Experiments designed to decide between different causes of crevice corrosion. Mears carried out a series of tests in a liquid (N/20 sodium carbonate) chosen as being on the border-line between corrosion and inhibition for the steel used (which contained 1·16% carbon); he studied the behaviour of steel rods upon which other rods (glass or steel) were placed at right angles (fig. 47). Corrosion took place mainly at the crevices existing at the contacts between the upper and lower rods, but did not occur at all of these crevices. Statistical experiments were carried out to ascertain the effect of conditions on the frequency of attack at the contacts. It was found that this frequency

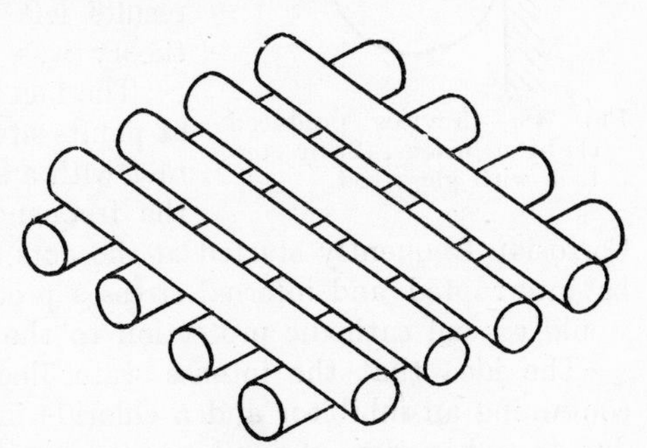

FIG. 47.—Corrosion at crevices between rods.

(1) increased with the size of the upper rods
(2) was greater for steel-steel contacts than steel-glass contacts
(3) was smaller under conditions of agitation than under conditions of stagnancy
(4) was reduced when cuts were deliberately made into the steel near the contact
(5) was approximately the same after ten minutes as after two days (R. B. Mears and U. R. Evans, *Trans. Faraday Soc.* 1934, **30**, 417).

The main object of Mears' research was to ascertain the mechanism of crevice corrosion. Two theories seemed at that time to deserve consideration. The one attributed corrosion to exhaustion of oxygen or inhibitor in the crevice, where replenishment would be slow. The other accounted for the special attack on steel at contact with glass on the supposition that the protective substance which elsewhere would be deposited on the steel surface might at the contacts be deposited preferentially on the glass, leaving the steel bare. The fact, however, that attack was developed more frequently at steel-steel than at steel-glass contacts seemed difficult to reconcile with the second theory; that potentially protective matter might adhere to glass in preference to steel was possible, but there was no obvious reason why it should desert one steel rod for another. To make the comparison between steel-glass and steel-steel a fair one, a special series of experiments was carried out in which the glass rods consisted of sealed tubes containing lead shot, so as to bring their weights up to those of the steel rods; this did not alter the results described above. In case silicate, dissolved from the glass, had been inhibiting the corrosion in some cases, a series of experiments was carried out with sodium silicate of appropriate concentration as the liquid chosen; the conclusions, however, remained unaffected. The fact that large rods, which by their smaller curvation would bring the two surfaces close together over a larger area, gave a higher frequency than small rods was consistent with either theory. However the exhaustion theory was favoured by the fact that stirring of the liquid decreased the frequency of attack. Viewed as a whole, the results left little doubt that the exhaustion theory was the correct one.

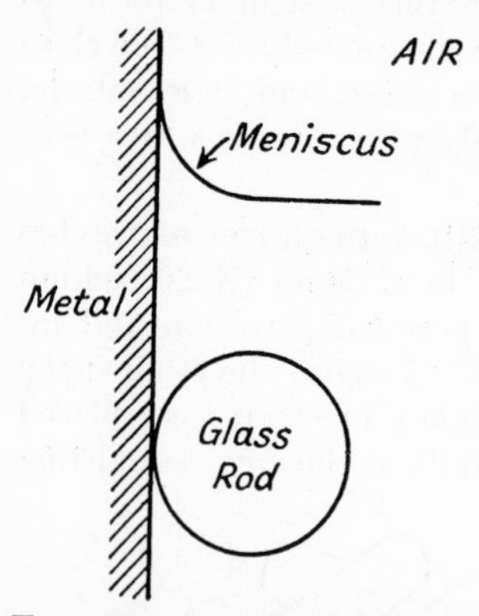

FIG. 48.—Crevices produced (1) by meniscus, (2) by contact with glass rod.

The fact that small cuts, engraved in a rod at points situated close to the place where contact with a second rod would occur, depressed the frequency of attack is easily understood. Corrosion frequently started at the cuts, where the protective films would be interrupted and internal stresses produced; such corrosion at the cuts would extend cathodic protection to the near-by contact areas.

The idea that the intense water-line attack produced by a solution containing an inhibitor and a chloride in unfortunate proportions is itself due to exhaustion of inhibitor (which is used up in film-repair whereas chloride is not), has been discussed on p. 150 in connection with Peers' work. Peers compared the effect of the crevice produced by the meniscus (fig. 48), with that of glass rods of different sizes; when the diameter of the rod chosen was such as to give a curvature similar to that of the meniscus, there was a satisfactory similarity in the results.

Although crevice corrosion can be set up by liquid containing an inhibitor alone (e.g. dilute sodium carbonate), it is more serious when a chloride is present, with corresponding larger proportion of the inhibitor (carbonate or phosphate); the total corrosion produced in a given time is greater for the more concentrated mixture, and if the distribution is equally unfavourable, the intensity of attack will be greater.

Crevice Corrosion on materials developing highly protective films. The danger of crevice corrosion must be borne in mind when employing materials which—given sufficiency of oxygen—develop highly protective films; the fact that such materials are commonly regarded as resistant to corrosion makes the initiation of attack, at places which are normally out of sight, a serious matter. Stainless steel is prone to crevice attack in situations where the ordinary user would not expect it, particularly crannies and stagnant corners within industrial vessels through which some liquid normally non-corrosive to stainless steel is to be run; the experience of manufacturers may be helpful in avoiding such trouble. Crevice corrosion has been reported below bakelite washers on austenitic stainless steel. It has also appeared at gaskets in chemical plants, where the presence of asbestos and fibrous material seems disadvantageous (E. V. Kunkel, *Corrosion* 1954, **10**, 260; J. A. Collins, *ibid.* 1955, **11**, 11t; O. B. Ellis and F. L. LaQue, *ibid.* 1951, **7**, 362).

Even titanium, which is normally free from any tendency to crevice corrosion, may fail unexpectedly at crevices in a strongly acid environment. Titanium plates tested in aerated 1·6–1·75N hydrochloric acid, which were completely resistant when well spaced apart, generally suffered attack when placed face to face in the acid, with a narrow cranny between a pair of plates. In some circumstances, the exposed face also became affected, and it has been suggested that in such cases the film was destroyed by cathodic action (D. Schlain and C. B. Kenahan, *Corrosion* 1956, **12**, 422t).

Aluminium, often a highly resistant metal, is apt to fail at places where oxygen-replenishment is too slow to keep the protective film in repair. Doubtless the complete exclusion of oxygen needed for such an effect will rarely occur under service conditions, but the phenomenon can be produced in the laboratory. R. L. Davies has described how aluminium placed in oxygen-free distilled water sealed from oxygen underwent no change for some weeks—after which hydrogen evolution started and rapid corrosion set in. This occurred after 4 weeks at 40°C., but the phenomenon has also been observed at ordinary temperature; it is prevented by small amounts of sodium silicate (R. L. Davies, *Iron St. Institute Spec. Rep.* 1952, **45**, 131).

Inglis has described a test for crevice corrosion. A tubular specimen is provided with a tightly fitting polythene inset which " shields " it from oxygen; the whole is exposed to sea-water. In his experiments pitting and grooving were produced within 100 hours on chromium-nickel austenitic stainless steel and to a small extent on certain copper alloys; none had occurred on titanium even after 5000 hours (N. P. Inglis, *Chem. and ind.* (*Lond.*) 1957, p. 180, esp. p. 187).

Design in avoiding Crevice Corrosion. In a special discussion on " corrosion and design ", members of the Society of Chemical Industry Corrosion Group laid emphasis on the importance of avoiding, in structures, any crevices or pockets where water can collect; even small horizontal ledges, on which water can rest, are dangerous. Turner suggested that the use of dome-shaped members, in places where flat members are used today, might assist drainage. Where crevices are inevitable, adequate protective coatings must be applied; for steel windows, zinc coating is considered satisfactory—provided it is sufficiently thick (W. E. Ballard, V. E. Stanbridge, J. R. Dewhurst and T. H. Turner, *Chem. and Ind.* (*Lond.*) 1954, pp. 957–958).

Another authoritative account of crevice corrosion with practical suggestions for avoiding it has been provided by E. H. Wyche, L. R. Voigt and F. L. LaQue, *Trans. electrochem. Soc.* 1946, **89**, 149. Crevice corrosion in aircraft is discussed by O. E. Kirchner and F. M. Morris, *Corrosion* 1951, **7**, 161, and in diesel engines by F. N. Speller and F. L. LaQue, *ibid.* 1950, **6**, 209, esp. p. 215.

Joints

Soldered Joints. Since at a soldered joint, the " outcrop " of the solder represents an area small compared with the area of the pieces joined, it is better, in principle, to choose a solder cathodic towards the plate, at least if the joint is to be exposed to liquid and not merely to the atmosphere. Thus a union involving steel joined with silver solder is likely to suffer less serious deterioration than one joined with a zinc-rich solder. Where, for technical reasons, a solder anodic to the plate must be chosen, the joint should be so designed that the solder area exposed is not too small.

The polarity of solders may vary with conditions. Tin-lead solder may be expected to be cathodic to steel in many supply waters, but will probably be anodic in some fruit juices, since the organic acids form complex ions containing tin, thus keeping low the concentration of tin cations and shifting the potential in the negative direction; since any attack upon the solder may bring poisonous lead into the liquid, the geometrical design of cans used for food deserves attention (H. Cheftel, " La corrosion du blanc fer " 1935, p. 18 (J. J. Carnaud, Paris)).

Soft solder will usually be anodic to copper, and trouble has been experienced at joints on copper ball valves in water cisterns; L. Kenworthy (*J. Inst. Heat Ventilating Engrs.* 1940, **8**, 24), has suggested modification of design, whilst C. E. Homer (*Tin and its Uses* 1942, March, No. 12, p. 1) advocates a special solder with 85% tin and 15% lead.

Despite its small area, the action of the solder in accelerating attack on the main material may sometimes be important; the corrosion of 14% chromium stainless steel may sometimes be increased by contact with silver solder. Where the soldered joints are unlikely to become wetted with liquid water, the polarity of the solder may become unimportant, although it must itself be resistant to atmospheric attack. When the soldering of aluminium is required for electrical apparatus where there will be no corrosive liquids, the addition of zinc to lead-tin solders improves the corrosive-resistance of the joint; for dry situations solders consisting mainly of tin and zinc have proved satisfactory (R. Chadwick, Priv. Comm., March 23, 1955).

Most solders are liable to be attacked by alkali. Interest, therefore, attaches to a solder containing 25% indium (with 37·5% Pb, 37·5% Sn) giving joints which have resisted 25% potassium hydroxide at 50°C. for a month without loss of strength; there was indeed a slight gain in strength, probably due to age-hardening (S. M. Grymko and R. I. Jaffee, *Mater. and Meth.* 1950, **31**, 59).

The main danger of corrosion introduced by soldering arises from corrosive and hygroscopic residues from the flux—particularly chlorides. If a chloride flux has been used, a thorough washing under the tap is commonly recommended. Certain fluxes containing borates leave glassy masses which are less easily removed. Even in the absence of borates, residues consisting of zinc oxychloride, which possess cement-like properties (p. 617), may resist tap-water, but in this case a rinse with acidified water will usually overcome the difficulty (S. C. Britton, *Chem. and Ind.* (*Lond.*) 1956, p. 657).

Wherever possible, the use of non-corrosive flux is advisable; some, at least, of the "cored solders" on the market (which contain their own flux) are made with a non-corrosive variety. Resin is a safe flux but not, by itself, sufficiently vigorous. An "activator" must be added; useful ingredients of such fluxes include aniline hydrochloride or hydroxylamine hydrochloride. The advantage of these amine hydrochlorides is that they decompose at the soldering temperature and leave no ionizable chloride. Hydrazine hydrochloride is considered to be a good addition to resin-cored solder for use on copper which has become oxidized (S. C. Britton "Corrosion Resistance of Tin and Tin Alloys" 1952, p. 34 (Tin Research Inst.); G. L. J. Bailey and H. C. Watkins, *Metal Ind.* (*Lond.*) 1949, **75**, 551. For composition of resin and activated resin fluxes, see W. R. Lewis, "Notes on Soldering" 1948, pp. 13–15 (Tin Research Institute). See also *Tin and its Uses* May 1954, p. 9).

Welded Joints. In welded joints, the electrochemical contrast between weld metal and parent metal is considerably smaller than that between solder and the plates joined by it. At first sight it would seem that the electrode chosen should be such as to give weld metal cathodic to the plates, but often other considerations prevail. In welding aluminium intended for service with nitric acid, there is a danger that the acid will pick out closed-in blowholes along the weld-line; this is least likely to happen if pure aluminium, which develops a very protective film, is used as weld-metal—despite the fact that pure aluminium is, in many liquids, anodic towards commercial aluminium. In most materials, the attack on the weld-line will be least when the porosity is least. On steel this calls for an absence of gas evolution by reaction between carbon and iron oxide; in other words, welds should be made with exclusion of air.

The danger of flux residues, mentioned in connection with soldering, exists also in welding. It may be necessary in some cases to use mechanical methods, such as impact tools, to remove residues; Liebman recommends sand-blasting. Corrosive constituents may be fluorides, chlorides and, for some metals, alkali; even where the metal is resistant to alkali, alkaline residues may cause deterioration to a paint-coat, if subsequently applied; here vinyl coatings—or other alkali-resisting types—may prove useful (A. J. Liebman, *Corrosion* 1954, **10**, 147).

Many years ago Meunier conducted some tests lasting a year in which welded steel was dipped every 3 minutes into 3% sodium chloride; pitting was produced on welds containing oxygen but absent in cases where precautions had been taken against oxygen-uptake during welding—a procedure which also gave the best mechanical properties. Rather later, Meryon described tests continued for 30 months on steel specimens welded with four types of electrodes. They were exposed (*a*) to sea air, (*b*) in the sea at the wind-and-water level, (*c*) in the sea completely immersed, (*d*) in the sea in contact with brass. The exposure did not seriously lower the breaking strength, and in tests (*a*) and (*b*) there was little difference between the corrosion of the weld and that of the parent metal. Meryon concluded that welding introduces no fresh corrosion problems in marine service (F. Meunier, *C.R.* 1933, **196**, 271; E. O. Meryon, *Iron and Steel Institute Corrosion Committee Reports* 1935, **3**, 101. Cf. J. P. Moore, *Corros. Tech.* 1954, **1**, 92, who discusses the relative corrosion-resistance of different zones).

On some other materials, the situation is less satisfactory. Marshall states that nickel alloys are often specially attacked along the weld-line, and likewise aluminium exposed to nitric acid, but not to industrial waters (W. K. B. Marshall, *Soc. Chem. Ind. chem. Engng. Group*, March 14, 1944).

However, the situation at the centre of the weld-line is not the only one needing consideration. The parent metal on both sides undergoes heating and subsequent cooling, and among the infinite variety of heat-treatments thus unintentionally carried out at points situated at different distances from the weld, it would not be surprising that some will adversely affect the corrosion-resistance. The two materials where this unintended heat-treatment leads to most deterioration are the austenitic stainless steels and the temper-hardening aluminium alloys.

If 18/8 Cr/Ni stainless steel is heated, it may become susceptible to intergranular corrosion. The change can be brought about by a short heating at about 700° or a much longer heating at a lower temperature; the relationship between time and temperature needed to cause trouble for one particular steel is shown in curves of fig. 49 (E. C. Rollason, *J. Iron St. Inst.* 1933, **127**, 391; 1934, **129**, 311).

Now at a weld there must always be two zones—one situated on each side of the weld-line—which have reached the dangerous temperature-range during the operation; inevitably, in the absence of special precautions, intergranular corrosion along these lines will occur if the welded article is afterwards exposed to a corrosive environment; in the severest conditions, the corrosion penetrates downwards along the grain-boundaries, so that grains may even fall out. The deterioration is usually determined by the precipitation of chromium carbide, leaving a zone depleted of chromium, the element responsible for the protective film on stainless steel; at these moderate temperatures the necessary atomic rearrangement can only occur at grain-boundaries (places of loose structure), susceptible zones will appear around the grains and the attack will be intergranular. The three alternative ways of avoiding trouble depend respectively on (1) the use of stainless steel containing either titanium or niobium (columbium), elements which have a stronger affinity for carbon than has chromium; (titanium is satisfactory for the parent metal, but tends to oxidize if introduced into the electrode, where niobium is better); (2) the use of a stainless steel practically free from carbon; (3) reheating the whole welded assembly to above 1100°C., which should bring all carbides into solution, followed by fairly rapid cooling; for large articles, this calls for a large furnace, and special care in arranging support for them during the operation. The danger of quenching-stresses produced during cooling must not be overlooked.

Ordinary mild steel sometimes suffers grooving around welds, notably in equipment for handling sulphuric acid; the trouble is mainly noticed under mildly corrosive conditions, since when corrosion is severe the general thinning swamps any local effects due to couples. Under milder conditions, the differing potentials of zones heated to different ranges may set up corrosion cells. Nelson reports that spheroidized steel, although less attacked than annealed steel when tested as an isolated specimen, is anodic to it,

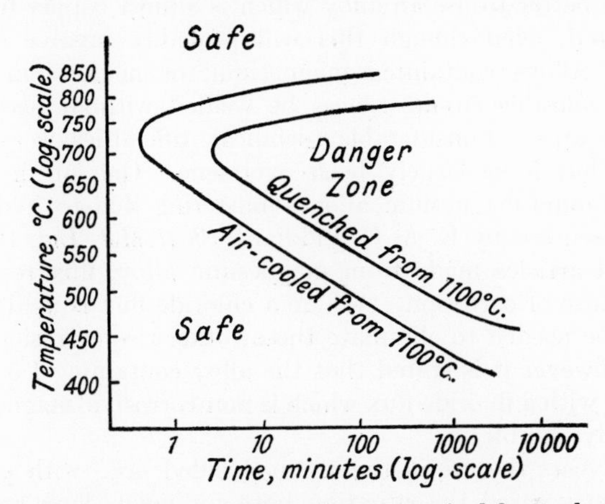

FIG. 49.—Relation between Time and Temperature required for production of susceptibility to intergranular attack on one particular Stainless Steel (E. C. Rollason).

and the grooves tend to follow zones of partial spheroidization. If the article can be normalized after welding, trouble is often avoided. If it is too big for a furnace, special welding techniques are suggested (G. A. Nelson, *Corrosion* 1958, **14**, 145t).

The temper-hardening class of aluminium alloys (which acquire strength by a carefully regulated heat-treatment) present special difficulties in welding. On unwelded articles it is possible to obtain these alloys in a condition combining high strength and reasonable resistance to corrosion; but treatment at certain combinations of time and temperature produces susceptibility to stress corrosion (p. 673), and it is fairly certain that around the weld-line there will be places where unfortunate combinations have existed, so that there will be zones of low resistance to corrosion cracking. The matter is discussed by J. H. Reimers, " Aluminium " (Tanum, Oslo), 1947 edition, p. 255.

In the absence of special precautions, welds in some aluminium alloys containing copper may suffer rapid destruction if exposed to sea-water.

Here again, if the article is not too large, the troubles may be avoided by suitable furnace treatment after the welding.

If heat-treatable alloys containing copper are used, it may be best to join them by means of rivets—a plan which avoids welding troubles, although obviously precautions against crevice attack are needed, and the right rivet material must be chosen. Steel rivets are only suitable if the joint is to be painted, and copper-bearing aluminium alloy rivets are not to be recommended; magnesium-bearing aluminium alloys are generally preferred, but the magnesium must not be so high as to produce exfoliation (p. 679). Useful information is provided by M. Cook, *Light Metals* 1951, **14**, 327. The opinions of certain other authorities on the effect of welding on corrosion-resistance are summarized by U. R. Evans, *Corros. Tech.* 1951, **1**, 136, esp. p. 138.

It may be better to use an alloy which is almost copper-free, if welding is contemplated, even though this will probably involve some sacrifice of strength. Alloys containing magnesium, or magnesium with silicon, can under favourable circumstances be welded without necessary loss of corrosion-resistance. Considerable technical difficulties were at one time encountered, but have largely been overcome. One of the problems in welding aluminium-magnesium alloys—blistering due to hydrogen-uptake—has been discussed by E. A. G. Liddiard, *Sheet Met. Ind.* 1947, **24**, 1857.

On welded articles made from magnesium alloy, flux residues are the most likely cause of corrosion. Where a chloride flux is used, scrubbing or pickling will be needed to eliminate these; otherwise corrosion troubles are probable. However it is stated that the alloy containing 1·5% manganese can be welded with a fluoride flux which is non-corrosive, magnesium fluoride being sparingly soluble.

In recent years the use of the argon-shielded arc—with elimination of flux—has greatly eased the situation both for magnesium and aluminium alloys. The argon should be substantially free from nitrogen; the gas cylinders should not be completely emptied, but should be refilled when there is still a positive pressure, so that any leakage is outwards. Helium is also used in welding, but the two gases have different electrical properties; for some purposes argon is better, for others helium. Corrosion tests of multi-arc-welded high-strength aluminium alloys are described by L. W. Smith, *Trans. electrochem. Soc.* 1946, **89**, 83.

Internal Stresses near Welds. The presence of internal stresses is known to increase the tendency to corrosion and corrosion-cracking in several materials. The stress-corrosion-cracking of welded joints in steel has been discussed by Parkins (see below). Since locked-up stresses usually occur near welds, stress relief must be considered. Experience shows that heat-treatment designed to afford relief of stress does greatly diminish or remove the liability to corrosion-cracking, in most cases where it would represent a serious trouble. Boyd, however, considers that the benefit does not arise from the disappearance of the stress, but from an improvement of metallurgical structure (G. M. Boyd, *British Welding J.* 1954, **1**, 560).

The stress relief is best carried out at high temperatures—usually above 600°C. in the case of steel. The recommended temperatures for different materials* are conveniently tabulated by Parkins (R. N. Parkins, *British Welding J.* 1955, **2**, 495, esp. p. 499).

Processes of low-temperature relief (working about 200°) are available, but opinion varies greatly regarding their efficacy. On the whole, the feeling of those of those who have studied the matter is that, although the process, applied with understanding and precision, is capable of improving the stress situation, the degree of control needed is such as will not always be available in industry; moreover, the process is applicable only to welded objects of simple geometrical design. Nevertheless, engineers will do well to consider the method and will find much useful information regarding service experience—mainly in Germany—provided by W. Liebig, *Schweissen u. Schneiden* 1955, **7**, 355; K. Wellinger, F. Eichhorn and F. Löffler, *ibid.* 1955, **7**, 7; H. G. Kunz, *ibid.* 1955, **7**, 291. R. Gunnert, *British Welding J.* 1955, **2**, 200.

Some years ago, Hamer and Colbeck urged that welded boilers should receive a stress-relieving treatment, and such precautions are now generally enforced in this country. Chaudron likewise recommended annealing at 600°C. for steel tanks intended for transport of mixed nitric–sulphuric acid (P. Hamer and E. W. Colbeck, *Chem. and Ind.* (*Lond.*) 1944, p. 163; G. Chaudron, *Métaux et Corros.* 1939, **14**, 71).

The welded tanks used for the removal of vitreous enamel by means of alkali call for careful stress relief, preferably at 650°C. It has been recommended that " no tank should be put into service with 50% caustic soda without having been so treated " (N. L. Evans, *Found. Tr. J.* 1940, **62**, 102).

Stress-corrosion-cracking experienced near welds at gas-works, where moisture containing ammonia and ammonium sulphide condenses, has been largely overcome by stress relief at 450–500°C., or by peening (*Instn. Gas. Engrs. Lond. Communications* No. **398** (1951), **517** (1957)).

Hydrogen at Welds. The introduction of hydrogen into steel from the welding electrode has received much study. It may arise from loosely held water (which can be expelled by heating the electrode before use) or from organic material, silicates, clay and other components. The effect of such hydrogen in increasing liability to cracking has been well summarized by C. L. M. Cottrell, *British Welding J.* 1954, **1**, 167, who recommends low-hydrogen electrodes.

Other References

The distribution of attack along a junction is a complicated subject. It was qualitatively discussed on p. 203. Those interested in the quantitative aspect should study Waber's mathematical papers (J. T. Waber, *J. electrochem. Soc.* 1954, **101**, 271; 1955, **102**, 344 (with M. Rosenbluth); 1955, **102**, 420; 1956, **103**, 64 (with B. Fagan); 1956, **103**, 138 (with J. Morissey and J. Ruth)). Some caution is needed in applying mathematical

* On aluminium alloys, the possibility that the stress-relief treatment may introduce undesired structural changes requires to be borne in mind.

equations to service problems since, however elaborate the mathematical treatment may appear to be, it probably leaves out of account factors operative in practice. Comparison between experimental determinations of distribution and theoretical predictions are provided by W. J. Schwerdtfeger and I. A. Denison, *Bur. Stand. J. Res.* 1955, **54,** 61. The effect of the ratio of cathodic to anodic areas on the corrosion-current flowing, which is shown to be greatest when the ratio is 1 : 1 (although of course the intensity of corrosion is not greatest at this value), is discussed by M. Stern, *Corrosion* 1958, **14,** 329t.

The effect of contact with four noble metals (Rh, Pd, Ir and Pt) on the acid corrosion of base metals (Al, Cd, Co, Fe, Pb, Mn, Ni, Sn, Ti and Zn) with results difficult to explain on the basis of overpotential alone—is described by W. R. Buck and H. Leidheiser, *Nature* 1958, **181,** 1681. Their work on the effect on acid corrosion of small amounts of the salt of a second metal also deserves study (*Z. Elektrochem.* 1958, **62,** 690).

Five Russian papers on crevice corrosion by Rosenfel'd and his colleagues should command attention (*Zhur. fiz. Khim.* 1956, **30**, 2724; 1957, **31**, 72, 328, 2328; 1958, **32**, 66). The first concerns steel (0·3% C, 0·25% Cr and 0·25% Ni) placed in NaCl or NaCl + Na_2SO_4 solution with crevices of different widths; the second concerns steel in NaCl + Na_2SO_4 containing an inhibitor ($NaNO_2$, $K_2Cr_2O_7$, Na_2HPO_4 or $ZnSO_4$), the corrosion being greatest with slits of intermediate width. The third deals with bimetallic corrosion below thin layers of electrolyte, the total corrosion being less, but the local attack at junctions greater, than with deep immersion. The fourth concerns crevice corrosion of aluminium and its alloys, including clad materials; it is shown that the increased corrosion in narrow slits is due to hindrance of oxygen access and to a change of pH in the slit, which reaches values where Al is no longer resistant; if the metal in the slit is in contact with metal reached by the bulk of the solid, the pH value may sink to pH 3·2–3·6, although if the entire surface exposed is within the slit, it may be 8·0–8·2. The fifth paper concerns low-chromium steel and stainless steel with 13–17% chromium.

Troubles around welds in nickel-molybdenum alloys exposed to acid, somewhat similar to those experienced in stainless steel, have recently been studied; the addition of vanadium (or niobium) largely prevents intergranular attack, if the contents of molybdenum or silicon are made high and that of iron low (G. N. Flint, *J. Inst. Met.* 1959, **87,** 303).

CHAPTER VII

ANODIC CORROSION AND PASSIVATION

Synopsis

The argument starts by recalling that, in the types of wet corrosion hitherto considered, the corrosion-rate has usually been limited by the rate of oxygen-replenishment; it is pointed out that, if an external E.M.F. is applied, no such limit will be imposed. At first sight, any corrosion-rate, however large, might seem to be obtainable, but in practice passivity supervenes; passivation is favoured by stagnant conditions. It is shown that a film—normally invisible—is formed on an anode which becomes passive. The objections raised to the view that passivity produced in acid on an iron anode is caused by an oxide-film are examined and found to be untenable; films which in absence of an anodic current would be destroyed by acid through " reductive dissolution " are stable under anodic conditions. The mode of formation of a film on an anode is discussed, and reasons are suggested for the influence of anions—especially chlorine ions, which are unfavourable to passivity. Cases where anodic corrosion has been deliberately applied to some useful purpose, e.g. in the making of a pigment, receive brief notice.

Attention is then turned to the anodic production of relatively thick films on aluminium and other metals belonging to the A-groups of the Periodic Table; it is explained that in some baths the thickness reached is proportional to the E.M.F. applied. It is shown that certain methods of " anodizing " produce a compact protective barrier film next the metal, covered by a porous outer layer, which is capable of taking up dye-stuffs; this layer may be rendered non-porous and protective by subsequent sealing treatment. It is convenient to include in the present chapter processes for producing protective films on aluminium similar to those obtained by anodizing, but requiring no external E.M.F.

Finally, electropolishing is discussed, the difference between the surface condition obtained by mechanical and electrical polishing methods respectively being emphasized. The theory of electropolishing receives consideration, and some views are quoted regarding the industrial possibilities of electropolishing and chemical polishing.

Some of the phenomena discussed qualitatively in this chapter receive quantitative treatment in Chapter XXI.

Conditions determining the decision between Corrosion and Passivity

Factors limiting the Corrosion-Rate. In the types of corrosion hitherto considered, the rate of attack has in most cases been controlled by the rate of oxygen-replenishment and is normally slow. A small piece of metal can, however, suffer severe corrosion if it is connected to a larger piece of a cathodic metal which will act as catchment-area for oxygen; intensification of attack, due to localization upon a small area, is also sometimes obtained when a small inert body (e.g. a stone) rests against a large metallic surface, and screens the small area from oxygen or inhibitor.

The present chapter is devoted to a type of corrosion which needs no oxygen-supply, so that there would seem, at first sight, to be no limit set to the corrosion-rate obtainable; this is the corrosion produced by an external E.M.F., which, if sufficiently high, can force huge currents through a cell of the type

Metal A (anode) | Liquid | Metal C (cathode)

(where C can, if desired, be the same metal as A). If the whole of this current is devoted to corrosion of the anode, there is clearly no limit to the corrosion-rate; fortunately the phenomenon of passivity may come in, so that above a certain current density, the current is mainly employed, not on the dissolution of metal, but on the evolution of oxygen. Under stagnant conditions, and particularly on horizontal surfaces, passivity sets in after corrosion has proceeded for a short time, even at relatively low current densities; it is favoured by the presence of an inhibitor, and, in the case of iron, by alkaline conditions; on most metals, unfortunately, the presence of chlorides is unfavourable to passivity.

Conditions favouring Anodic Corrosion. The reactions produced by an external E.M.F. are similar to those already met with. If (fig. 22, p. 98) the cell

Fe sheet | NaCl solution | Cu sheet

is connected through a resistance to an external source of E.M.F. (e.g. an accumulator) with the copper joined to the negative pole, iron will pass into solution as soluble ferrous chloride, whilst alkali will be formed at the copper cathode, where the reaction may be the reduction of oxygen to OH^- if the current density is *very low* (e.g. when the external resistance is very high). If the current density is higher, the oxygen-replenishment may fail to keep pace with the current, and the only cathodic reaction possible will be the evolution of hydrogen; since this reaction destroys hydrogen ions

$$2H^+ + 2e = H_2$$

it will render an originally neutral solution alkaline—just as does the reduction of oxygen. The anodic and cathodic products will interact, together with dissolved oxygen, to form loose flocculent rust ($Fe_2O_3 \cdot H_2O$ or $FeO(OH)$), which, being precipitated well away from the iron, will not obstruct attack; if the oxygen-supply is limited, black magnetite or green ferroso-ferric compounds (p. 91) may be thrown down.

If the liquid is, say, dilute sulphuric acid, it will remain clear. If the main part of the iron surface is protected by a suitable varnish film, leaving only a small area bare, the intensity of attack on that small area is greatly increased, and the iron sheet anode may soon become perforated.

Conditions favouring Passivity. If the iron surface is arranged horizontally, and particularly if it is protected from convection currents by means of a rubber hood (fig. 50(*a*)), the onset of passivity is greatly favoured. This device was much used by W. J. Müller, whose extensive work placed the study of passivity on a mathematical basis. Müller found that an anode of iron, zinc, cadmium or copper ultimately becomes passive, even when the liquid used is dilute sulphuric acid; passivity appears quickly

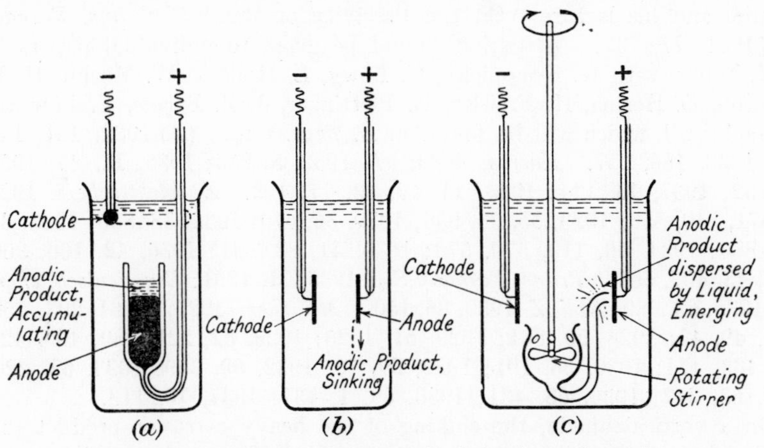

FIG. 50.—Cells for studying anodic behaviour of metals.

(*a*) With horizontal anode protected from chance stirring—conditions favouring passivity (W. J. Müller).
(*b*) With vertical anode, allowing heavy anodic products to sink—less favourable to passivity.
(*c*) With dispersal of anodic products with mechanical stirrer—preventing passivity unless current density is high (W. J. Shutt and A. Walton).

when the current applied is strong, but, even if it is feeble, the metal will usually become passive after a sufficient period of waiting. The time is needed for the accumulation of a dense solution of metallic sulphate to accumulate on the anode surface; ultimately this becomes super-saturated and a layer of tiny doubly refracting crystals appears on the surface—which in several cases Müller identified as the hydrated sulphate by means of the polarizing microscope. These crystals shield the greater part of the surface, and although the total current passing (under Müller's condition of working) falls off, the current density on the part still bare increases. An important result of the shielding is that the potential at the bare parts becomes more positive, so that other reactions become possible, such as the formation of oxide; when once oxide has been formed, the attack on the metal is obstructed and most of the current is then devoted to the production of oxygen. For our present purposes the reaction may be conveniently written

$$2H_2O = O_2 + 4H^+ + 4e$$

and it will be noticed that it will increase the acidity.* Attack does not cease altogether; a very slow passage of iron into solution (as ferric, not ferrous ions) continues, as was shown 38 years ago by C. A. Lobry de Bruyn, *Rec. Trav. chim. Pays-Bas* 1921, **40**, 30. A possible mechanism is suggested by K. J. Vetter, *Z. Elektrochem.* 1955, **59**, 67.

W. J. Müller's monumental researches were carried out with numerous collaborators, of whom Konopicky and Machu deserve special mention. The earlier parts are described in his book "Die Bedeckungstheorie der Passivität der Metalle und ihre experimentelle Begründung" 1934 (Verlag Chemie) and his lecture "On the Passivity of Metals" (*Trans. Faraday Soc.* 1931, **27**, 737). Attention should be given to individual papers by W. J. Müller with K. Konopicky, O. Löwy, L. Holleck, W. Machu, H. K. Cameron, O. Hering, H. Freisler, E. Plettinger, J. Z. Briggs, E. Löw and E. Nachtigall, which will be found in *Z. phys. Chem.* (*A*) 1932, **161**, 147, 411; 1933, **166**, 357. *Korros. Metallsch.* 1932, **8**, 253; 1935, **11**, 25; 1936, **12**, 132; 1937, **13**, 144; 1938, **14**, 49, 63, 77, 198. *Z. Elektrochem.* 1928, **34**, 571, 840, 850; 1929, **35**, 93, 656; 1930, **36**, 679; 1932, **38**, 850; 1933, **39**, 872, 880; 1934, **40**, 119, 570, 578; 1935, **41**, 83, 641; 1936, **42**, 166, 366; 1937, **43**, 407, 561. *Trans. Faraday Soc.* 1935, **31**, 1291; *Ber. deutsch. chem. Ges.* 1935, **68**, 989; *Koll. Z.* 1939, **86**, 150. *Mh. chem.* 1927, **48**, 61, 293, 559; 1928, **49**, 47; 1928, **50**, 861; 1929, **51**, 1025; 1929, **52**, 221, 289, 409, 425, 442, 463, 474; 1930, **56**, 191; 1932, **59**, 73; 1932, **60**, 359; 1933, **62**, 220; 1933, **63**, 347; 1936, **68**, 431; 1936, **69**, 1, 437; 1937, **70**, 113.

On a vertical anode, the sinking of the heavy corrosion-product (fig. 50(*b*)) delays, or possibly prevents, the arrival of passivity, whilst if the liquid is deliberately stirred, as in the arrangement shown in fig. 50(*c*), used by Shutt and Walton chiefly in the study of gold and chromium, passivation becomes still more difficult; below a certain current density, it will never occur at all (W. J. Shutt and A. Walton, *Trans. Faraday Soc.* 1934, **30**, 914; R. H. Roberts and W. J. Shutt, *ibid.* 1938, **34**, 1455).

Demonstration of the Film. On iron, the deposition of ferrous sulphate (or possibly basic sulphate in some circumstances) during the process of passivation causes the surface to become dull, as was noticed by Hedges (p. 232) and Olivier (p. 229); but when once the evolution of oxygen has fairly set in, the sulphate crystals are pushed away by the oxygen bubbles and generally redissolve; when fully passive, the iron surface appears bright and very similar to its original condition. However, a film is really present, as was shown in the Author's apparatus with a sloping anode (fig. 51)—an arrangement chosen to allow the surface to be kept under observation through a microscope (U. R. Evans, *Nature* 1930, **126**, 130).

The material was cold-rolled electrolytic iron, abraded before the experiment, and the liquid was N sulphuric acid. An E.M.F. of 6 volts was

* An originally neutral liquid would become acid.

applied. At first the anode suffered corrosion, but the high current density soon led to passivity so that the evolution of oxygen commenced. If subsequently the current was momentarily interrupted (by operating the left-hand key) the anode was found still to be passive when the circuit was restored; but if the interruption was too long, the anode was found to have become active again, and a fresh period of activity (corrosion) was necessary before passivity (and oxygen-evolution) once more set in. A little experience made it possible to predict the instant at which the circuit must be restored so as to catch the electrode " in the act " of changing over, being still passive in most parts but already active at a few points. If the right moment was seized, attack started at the active points and gradually extended under the film present on the passive parts, so that the film, thus undermined, peeled off; when once out of contact with the metal, the film was clearly visible, and was found to be reasonably stable in acid. Success depended

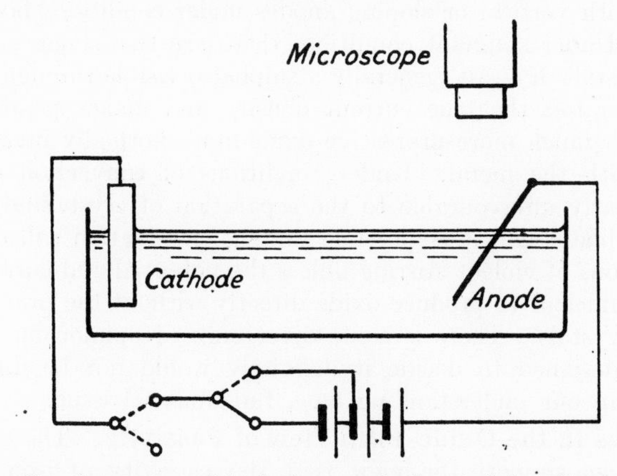

FIG. 51.—Apparatus with sloping anode for studying film removal.

on choosing the right moment for switching on and off the current and it was found useful to vary the E.M.F., between 4 and 6 volts, by means of the right-hand key, according to the requirements of the situation, which was watched through the microscope; some slight experience was needed, but after this had been gained, it was found easy, by suitable operation of the two keys, to isolate the film. Three principles required to be borne in mind in carrying out the isolation:—

(1) the film when in contact with metal is destroyed by acid if no current (or only a weak current) is passing, owing to reductive dissolution (a phenomenon explained later);

(2) the film, when in contact with metal, resists destruction by acid when a strong current is passing;

(3) the film, once it is out of contact with metal, resists rapid destruction by acid, whether current is passing or not.

The isolated film was less easy to preserve in acid than that previously obtained from iron rendered passive in potassium chromate solution (p. 152), owing to internal stresses; in some cases the film as it detached itself from the anode rolled up neatly " like a carpet ", forming long tight rolls which, under low-power magnification, resembled fibres. In other cases the fragments twisted themselves together into forms recalling cobwebs. However, although mechanically flimsy, the films were chemically stable, and transparent specimens could survive an hour in normal acid; some samples carried opaque inclusions, evidently metallic iron, and these dissolved quickly (through the reductive dissolution process to be described); the presence of the iron in the film is a sign of the interlocking of metal and oxide at an abraded surface, for which there is other evidence (U. R. Evans and J. Stockdale, *J. chem. Soc.* 1929, p. 2651; W. H. J. Vernon, E. A. Calnan, C. J. B. Clews and T. J. Nurse, *Proc. roy. Soc.* (*A*) 1953, **216**, 375).

A distinction should be drawn between the passivation of a horizontal anode under stagnant conditions—as in Müller's work—and passivation carried out with vertical or sloping anodes under conditions favourable to convection. Under stagnant condition, there are two stages: (1) the formation of crystals of a salt (generally a sulphate) visible through the microscope, which raises the true current density and makes possible (2) the formation of a much more protective oxide-film—normally invisible whilst in contact with the metal. Under conditions of convection or stirring, which are clearly unfavourable to the separation of crystalline salt, there may be only one stage; it is possible that no passivation will occur at all under conditions of violent stirring unless the potential and current density applied are sufficient to produce oxide directly without the prior formation of sulphate crystals. There is some divergence of opinion on this point; experiments designed to decide it definitely would not be difficult, and would clear up our understanding on a fundamental issue.

Difficulties in the Oxide-Film View of Passivity. The experiments described above support the view that the passivity of iron anodically polarized in acid is due to an invisible oxide-film, which becomes visible when it ceases to be in contact with the bright reflecting metal. Some early objections to such a view require, however, to be discussed. Fifty years ago it used to be argued that any oxide-film would at once dissolve in the acid liquid. Such an argument had two weak points. First, the liquid next the anode surface may, at the moment of passivation, no longer be dilute sulphuric acid but essentially concentrated ferrous sulphate. Secondly, although oxide-films are normally dissolved by acid whilst in contact with the metal, they resist acid under anodic conditions, and also resist it when they have been removed from the metallic base. It may be helpful to recall some early experiments on iron carrying visible films, which directly demonstrate these points (U. R. Evans, *J. chem. Soc.* 1930, p. 478, esp. p. 480, supplemented by later experiments).

Specimens of iron were "gradient-tinted", i.e. heated along one edge to give a range of interference tints. When such a specimen was dipped for one minute in 0·01N sulphuric acid, the colours suffered a change, disappearing completely on the less strongly heated portions where the films were thin, whilst the more strongly heated parts showed colour changes indicative of loss of thickness (fig. 52(A)). In some cases the thicker films were undermined and peeled off; after detachment they could remain in acid for long periods without serious change.

It appeared desirable to ascertain whether this immunity extended to the thinnest films.* Accordingly (at a later date) the film from a gradient-tested metal was transferred to plastic (by the method described on p. 785). The transferred film was dipped up to a certain level in 0·01N sulphuric acid for 1 minute and was now found to suffer no change; it was impossible to decide, after the specimen had been taken out and dried, the level up to which the specimen had been immersed in acid. Evidently, *the film is only attacked by the acid if it is in contact with the metal.*

It was also found that if a gradient-tinted specimen was dipped into 0·01N sulphuric acid containing an excess of chromic acid, it suffered no change; here again it was impossible afterwards to decide where the upper

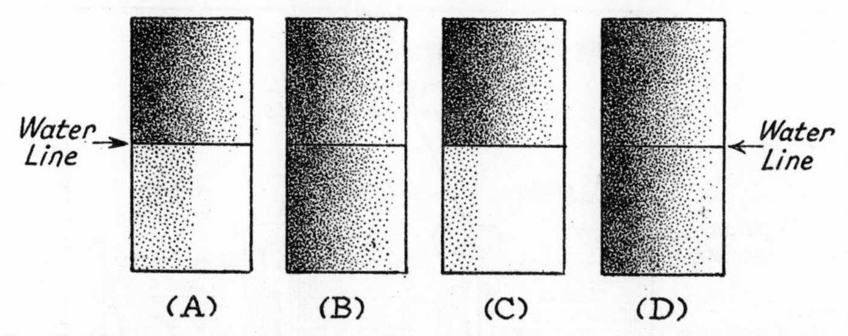

FIG. 52.—Gradient-tinted steel specimens. (A) dipped in 0·01N sulphuric acid, (B) dipped in 0·01N sulphuric acid containing chromic acid, (C) cathodically treated in 0·01N sulphuric acid, (D) anodically treated in 0·01N sulphuric acid.

limit of the wetted area had been (fig. 52(B)). If a pair of specimens connected to an external battery were treated in 0·01N sulphuric acid (free from chromic acid), the cathodically treated specimen (fig. 52(C)) suffered a destruction of the colours which proceeded further than on a specimen immersed without current (fig. 52(A)), whereas the anodically treated specimen (fig. 52(D)) underwent no change; once more it was impossible afterwards to decide where the water-line had been.

Reductive Dissolution. The facts just stated show clearly that the *direct solution* produced by acid acting on ferric oxide films is *very slow*, although *under cathodic conditions* destruction of the film is *rapid*. The rather rapid attack in the absence of applied current is attributed to local cells set up at the discontinuities which appear spontaneously in films,

Iron (Anode) | Acid | Ferric Oxide Film (Cathode)

The cathodic action reduces the ferric oxide to the ferrous condition so that it passes directly into the liquid; this is known as *Reductive Dissolution.* If the tinted specimen as a whole is made an anode by means of an external

* The films responsible for first-order colours consisted of ferric oxide alone; those responsible for high-order colours were duplex, with ferric oxide overlying magnetite (p. 22).

E.M.F., then local cathodic action is prevented, and the film remains unharmed; the same occurs on simple immersion if the liquid contains chromic acid, a strong oxidizing agent which will be reduced in preference to ferric oxide, thus acting as " whipping-boy " and so enabling the film to escape damage.

The suggestion that the failure of ferric oxide films to protect under certain circumstances was due to reductive dissolution, and not to direct dissolution, was confirmed by Pryor, who not only studied oxide-films on iron, but also ferric oxide in powder form; he prepared it by precipitating ferric ammonium sulphate with ammonia, carefully washing, and igniting the precipitate at a chosen temperature. Pryor found that the direct action of acid was slow, and the acid extract contained much ferrous iron, although the oxide was nominally Fe_2O_3; this accords with the work of

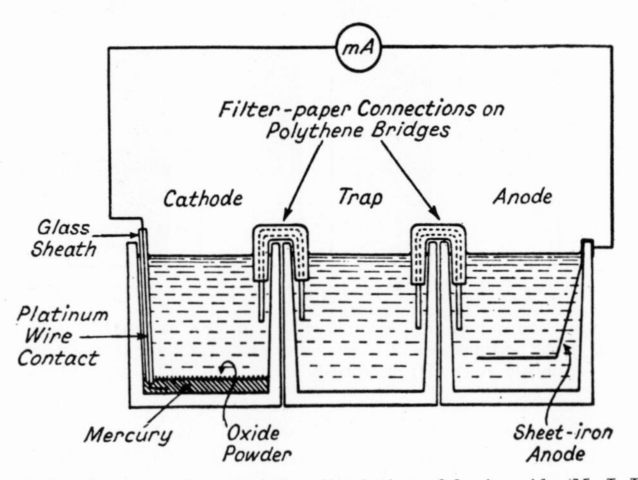

FIG. 53.—Cell for demonstrating reductive dissolution of ferric oxide (M. J. Pryor and U. R. Evans).

Pfeil (p. 24) who found that, in the so-called ferric oxide, there is a deficiency of oxygen, presumably through the presence of ferrous ions associated with lattice-defects; if dissolution occurs preferentially at faults in the structure, the presence of ferrous iron in the liquid is explained (M. J. Pryor and U. R. Evans, *J. Chem. Soc.* 1949, p. 3330; 1950, pp. 1259, 1266, 1274).

When Pryor floated the same ferric oxide powder on mercury (fig. 53) and connected the vessel by means of filter-paper bridges, through an intermediate vessel, to a final vessel containing acid in which was immersed a piece of iron joined electrically to the mercury, the attack on the oxide became vigorous; the iron in the right-hand vessel acted as anode of the short-circuited cell, and the mercury with the iron oxide on it as cathode; the cathodic reaction may be regarded as manufacturing lattice-defects in the ferric oxide, so that dissolution becomes easy. In such an experiment, ferrous ions appeared in the left-hand compartment by cathodic dissolution of the ferric oxide and also in the right-hand compartment by anodic attack on the iron. The cell shown in fig. 53 really represents a model of the local cell set up at each discontinuity of a ferric oxide film on heat-tinted iron (fig. 54). The " discontinuity " is shown on the diagram as a definite gap in the film, but in practice it may be merely a place where the oxide structure contains sufficient defects to allow rapid passage of iron cations outwards into the acid.

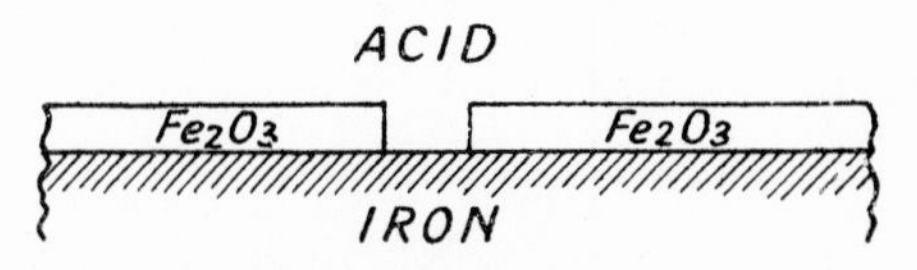

FIG. 54.—Reductive dissolution of ferric oxide on iron exposed to acid.

By insertion of a milliammeter, Pryor was able to measure the current passing between anode and cathode, and compare it with the rate of reductive dissolution of the ferric oxide. If oxygen was excluded, the dissolution efficiency—in the sense of Faraday's Law—was over 90% at the outset, declining later; in presence of oxygen it was much lower, evidently because parts of the current was then employed in the reduction of oxygen. Thus dissolved oxygen in a liquid should reduce the risk of destruction of a passivating film on metal—and this is found to be the case.*

Disappearance of Passivity after shutting off the Anodic Current. If an iron anode is rendered passive by anodic treatment at a sufficiently high current density in an acid liquid, and if the external current is then shut off, the passivity will disappear, as already explained, owing to reductive dissolution. In definitely acid liquid, the loss of passivity (shown by a sudden tumble of potential) may occur quickly—probably within a few seconds. In feebly acid liquid, passivity may survive longer, and in neutral solutions longer still; in a definitely alkaline solution, it may remain indefinitely. Even in a liquid which does not preserve passivity if current is shut off altogether, passivity may survive when the current density has been reduced to a level too low to produce passivity in a similar electrode which is initially active. It is easier to preserve passivity on an anode which is already passive than to produce it on an anode which is active at the outset.

This is shown in a beautiful experiment due to Pourbaix (fig. 55). Two cells, each containing two similar iron wire electrodes, dipping into N/10 sodium bicarbonate, are joined in series to an external battery providing about 3 volts (two Leclanché cells). Since the cells are joined in series and since the immersed area is the same for all the wires, the anodic current density must be equal in both cells. The only difference between them is that the left-hand cell is furnished with a short-circuiting key, *C*. Shortly after the start of the experiment, this key is pressed down for a few seconds, and the cessation of current through the left-hand cell causes its anode to become active, whilst the anode in the right-hand cell remains passive.

* In the particular case of stainless steel in dilute sulphuric acid, where there is attack in absence of oxygen and passivity in its presence, it may be, not the oxygen, but the Fe^{+++} ions formed when oxygen is present, which suffer reduction at points of temporary leakage, thus preventing reductive dissolution of the film. See Chapter IX.

The pressure on the key is then discontinued, so that the current again flows through both cells, but the anode in the first cell remains active, and

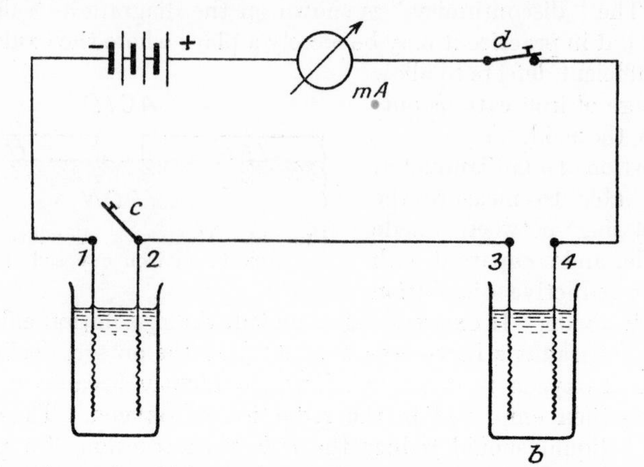

FIG. 55.—Experiment in which the same current passing through two cells maintains one anode passive and causes anodic dissolution on the other (M. Pourbaix).

that in the second cell passive—notwithstanding the equality of the current density in both (M. Pourbaix, *Cebelcor, Rapport Technique* No. **20** (1954) p. 3; also *Corrosion* VIII (1954–55) p. 9, esp. p. 12 (Verlag Chemie)).

Anodic Polarization Curves. When an E.M.F. is applied to a cell fitted with an iron anode, the variation of current with the potential drop at the anode presents some interesting but rather complicated features, which have been studied by several experimenters. The pioneer researches on the subject, including even the admirable work of W. J. Müller, were not entirely satisfactory, since the electric circuit employed was not suited to fix the potential at some chosen value (V), maintaining it constant until the current had taken up the corresponding value (I). More recently *potentiostatic* methods have become available for fixing V at a value which will not be affected by happenings within the cell, and satisfactory curves relating I and V can now be traced. Although many present-day potentiostatic methods depend in the use of electronic apparatus, it is possible to use an ordinary potential-divider, provided that the resistance of the bridge is sufficiently low.

Galvanostatic methods may also be used. Here the circuit is such as to provide a current which will remain constant—whatever may happen within the cell. The use of a high E.M.F. (large compared to the polarization changes at the electrodes) applied through a high resistance adjusted to fix the current at the required value, is the simplest method.

Under suitable conditions, the same curve showing the relation of current to potential should be obtained whether (1) we fix the potential successively at a series of different values and allow time for the current to take up a steady value in each case, or (2) fix the current at a series of values and allow the potential to become steady. The two curves obtained by U. F. Franck for an iron anode in sulphuric acid by galvanostatic and potentiostatic methods respectively lie close together.

Potential-current curves for an iron anode in sulphuric acid have been obtained under satisfactory conditions by Bartlett, Franck and Olivier, working in the U.S.A., Germany and Holland respectively. The three sets of results are mainly in good accord. It is convenient here to choose Olivier's work for description, since, starting from iron, he passed to the alloys of iron and chromium; some of Franck's results are discussed in Chapter XXI. Olivier acknowledges that his circuit was based on that of Bartlett, and that some of his experiments were essentially repetitions of those of Franck. All three researches deserve study in detail, and it is unfortunate that in two cases the full results exist only in thesis form. The references are: J. H. Bartlett, *Trans. electrochem. Soc.* 1945, **87**, 521; J. H. Bartlett and L. Stephenson, *J. electrochem. Soc.* 1952, **99**, 504. U. F. Franck,

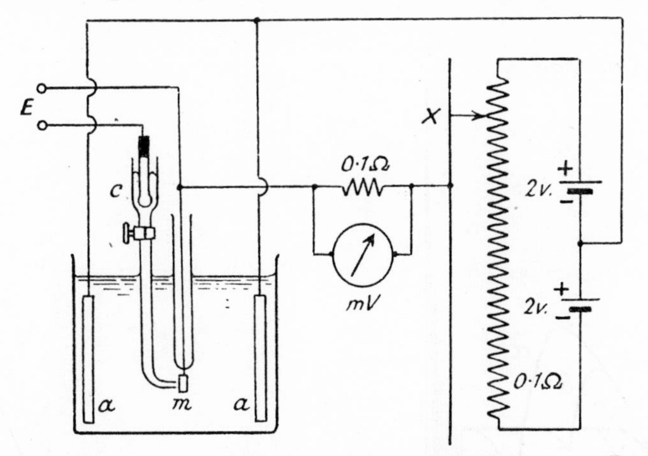

Fig. 56.—Potentiostatic circuit for determining polarization curves (R. Olivier).

Thesis, Göttingen, 1954. R. Olivier, Thesis, Leyden, 1955. Partial accounts of their results will be found in papers by R. Olivier, *Int. Comm. electrochem. Thermodynamics and Kinetics*, 1955, 6th Meeting, and by U. F. Franck, *Z. Elektrochem.* 1958, **62**, 649.

Fig. 56 shows the circuit used by Olivier. The potential falling across the cell is decided by the position of the sliding contact (X) on the resistance arm, which consists not of a wire but of a tube of stainless steel 3·5 metres long; its resistance is only 0·1 ohm. The battery is connected at the mid-way point between the two accumulators, so that current can be passed through the cell in either direction at will. Provided that the tube-resistance is low enough, happenings in the cell will not affect the potential applied. Since 4 volts imposed over 0·1 ohm provides a current of 40 amperes, water-cooling is needed, to prevent the resistance material becoming oxidized. Accumulator plates, a (lead carrying lead peroxide), are used as non-polarising cathodes. The potential drop at the surface of the anode, m, relative to a saturated calomel electrode, is measured on a potentiometer connected at

E. By moving *X* steadily along the resistance tube, measurements of *I* corresponding to different values of *V* can be obtained, giving results such as those shown in fig. 57, which refer to the cell

Iron (anode) | 10% H_2SO_4 | PbO_2 | Pb
(accumulator plate
as cathode)

Interpretation of the Polarization Curves. In the absence of any applied current (the state represented by point *O* on fig. 57) the iron is attacked by the acid, the local anodic reaction

$$Fe = Fe^{++} + 2e$$

being balanced by the evolution of hydrogen on local cathodes

$$2H^{+} + 2e = H_2$$

This balance occurs at about— 0·25 volt on the hydrogen scale.* If now

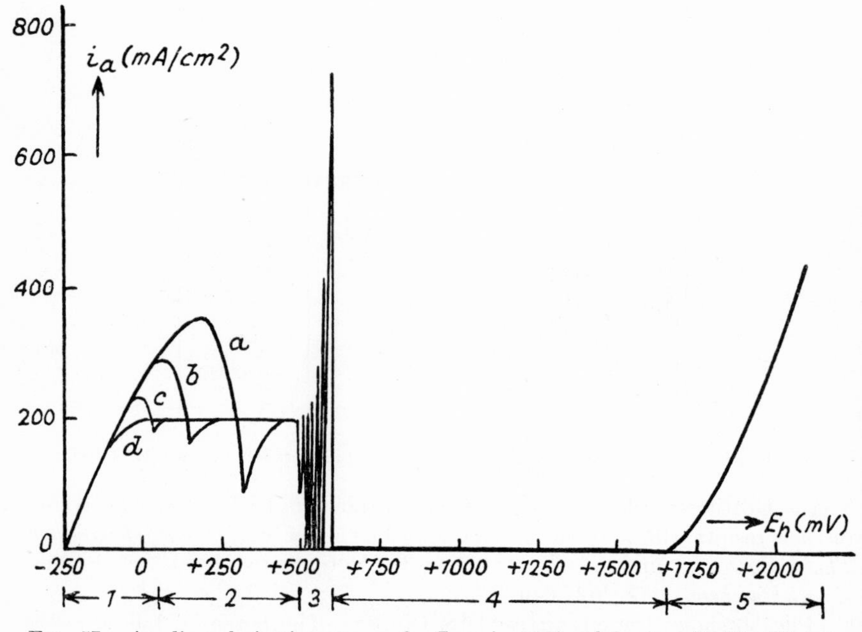

FIG. 57.—Anodic polarization curves for Iron in 10% sulphuric acid (R. Olivier).

the potential is moved in the positive direction, so that the specimen as a whole becomes an anode, the anodic reaction is stimulated whilst the hydrogen evolution slackens and soon disappears. When it has ceased, the current flowing becomes the equivalent of the corrosion-rate, in the sense of Faraday's Law. The current continues to increase as the potential is moved in the positive direction (Region 1) until suddenly it reaches a

* The apparatus shown in fig. 56 gives measurements relative to the saturated calomel electrode; they can be converted to the hydrogen scale by adding 0·241 volts.

maximum, and starts to fall off. This occurs when the liquid next to the layer has become so super-saturated with ferrous sulphate that a crystalline layer appears over the surface, causing a visible dullness to the surface; the crystalline matter can be detected through bi-refringence in the polarizing microscope. Owing to considerations of nucleation energy, there is always a lag in the appearance of the crystal layer, and the shape of this part of the curve depends on the rate at which the potential is being increased; the more rapid the increase, the higher will be the maximum current density attained before the inevitable tumble, which, however, is more severe after a high maximum, since, the super-saturation being greater, the amount of ferrous sulphate crystallizing out will be more considerable. Ultimately the current density will rise again to a steady state at which the production of ferrous sulphate by the anodic attack exactly balances dissolution by the liquid; the rate of dissolution depends on convection and diffusion but not on potential, so that in the steady state (Region 2), the curve becomes horizontal; the current in this horizontal region corresponds to anodic corrosion of iron exposed at the end of the channels which thread the crystalline layer.

The horizontal curve is replaced at + 0·45 volt by a state of violent potential oscillation (Region 3).* This strange behaviour is certainly due to the fact that the formation of solid oxide-film on the metal exposed at ends of the channels threading the crystals has become possible. Calculations based on early measurements by Flade (concerned actually with the *loss* of passivity) appear to indicate that in 10% sulphuric acid passivations would need a potential of + 0·580 volt.† However, in the channels between the crystals, the acidity is far lower; apparently the pH value is about 2·2—which would allow the formation of oxide at + 0·450 volt. If, however, the iron here becomes covered with oxide, anodic attack is practically prevented, and the current now falls sharply to a negligible value. In absence of current the dissolution of the ferrous sulphate crystals can proceed without any compensating formation of fresh crystalline matter, so that the iron soon becomes exposed to acid of nearly full (10%) strength; when this acid reaches the metal, the oxide already formed will almost immediately disappear by reductive dissolution, which at that potential is a possible reaction. Consequently, iron is, for the moment, bare, and a large current can again flow, with rapid attack leading to a fresh crust of crystalline sulphate, and then to the formation of fresh oxide at the end

* The oscillations take place even if the potential is held constant; it should not be supposed that certain potentials within Region 3 favour high currents and others low ones—as the drawing of the curve, if taken too literally, might perhaps suggest.

† Flade found that the passivity of an originally passivity anode disappears when the potential is reduced below 580 − (59 × pH) mv. If the pH of 10% sulphuric acid is about zero, we may expect passivity at + 0·580 volt in such acid, but at about + 0·450 volt at a pH of about 2·2 (F. Flade, *Z. phys. Chem.* 1911, **76,** 513). Various views on the Flade potential are expressed by K. E. Heusler, K. G. Weil and K. F. Bonhoeffer, *Z. phys. chem.* (*Bonhoeffer Gedenkband*) 1958, **15,** 149; H. Göhr and E. Lange, *Z. Elektrochem.* 1957, **61,** 1291; H. H. Uhlig and P. F. King, *J. electrochem. Soc.* 1959, **106,** 1.

of the channels, producing a fresh drop of current. This accounts for the violent oscillations of current.

The oscillations continue up to a potential of about + 0·580 volt, and then cease, since above this level the oxide-film becomes stable, even in the presence of the 10% acid—as already stated. Over the new potential-range (Region 4), the current flowing is very low, being only just sufficient to replenish the direct dissolution of the oxide by the acid; it will be recalled (p. 226) that direct dissolution (in contrast with reductive dissolution) is extremely slow. For many practical purposes, it may be said that, over this range, the " passive " electrode is behaving like platinum or other noble metal. The almost current-less state continues until about 1·66 volts, when the evolution of oxygen becomes possible (Region 5). The magnitudes of the current and corrosion-rate in the passive ranges have been discussed by K. J. Vetter, *Z. Elektrochem.* 1955, **59**, 67. See also K. J. Vetter, " Passivierende Filme und Deckschichten " 1955–56, p. 72 (Editors, H. Fischer, K. Hauffe and W. Wiederholt: publisher, Springer).

The oscillations of current observed and interpreted by Olivier have long been known. They received detailed study in early work by E. S. Hedges and J. E. Myers, *J. chem. Soc.* 1925, pp. 445, 1013. E. S. Hedges, *ibid.* 1926, pp. 1533, 2580, 2878; 1927, p. 2710; 1928, p. 969; 1929, p. 1028. See also U. F. Franck, Dissertation, Göttingen, 1954.

Franck's work, which presents parallelism to that of Olivier, deserves special study. A useful summary (in English) of anodic passivation was provided by K. F. Bonhoeffer, *Corrosion* 1955, **11**, 304t.

Anodic Behaviour of Iron-Chromium Alloys. Olivier has also studied the polarization curves of the alloys of iron and chromium; fig. 58 shows them plotted with a logarithmic scale for the current density so as to clarify the representation of the very small currents passing in the potential-range corresponding to passivity. It will be noticed that the current becomes smaller as the chromium-content rises. The vertical broken lines represent the limits of the region of oscillations. The alloy with 2·8% Cr gives a curve similar to that of pure iron, and develops a crystalline sulphate layer, which is absent at higher chromium contents.

It will be noticed from fig. 58 that anodic corrosion sets in again at about + 1·2 volts; this is known as the *transpassive region*, and it has been studied for a series of iron-chromium alloys in N sulphuric acid by Pražák. He finds that a breakdown of passivity occurs above + 1·2 volts on alloys in which the chromium-content exceeds about 16%; the chromium enters solution in the hexavalent state, giving chromic acid, which, being an energy-rich compound, can only be formed at a high potential. Alloys containing between 18% and 30% chromium, again become passive at about + 1·8 volts but alloys containing 35% chromium or more are continuously attacked and do not show this " secondary passivity ". Pražák explains his results neatly by applying Tammann's principle of parting limits (p. 351); for this purpose, he regards the film-substance as a spinel, $Fe_{II}(Fe_{III}, Cr_{III})_2O_4$, in which all the divalent sites are occupied by ferrous ions, whereas the trivalent sites are occupied partly by ferric and partly by chromic ions—according to the composition. Of the trivalent ions, only Cr^{+++} is removable in the soluble condition at high potentials, and it can be shown that when the chromium-content reaches 16 to 18%, continuous paths of chromium ions into the interior become possible. The secondary passivity is attributed to the blocking of narrow channels by oxygen (M. Pražák and V. Čihál, *Z. Elektrochem.* 1958, **62,** 739).

It is shown in Chapter IX that in absence of applied current stainless steel resists acids which attack unalloyed iron, since the chromium in the

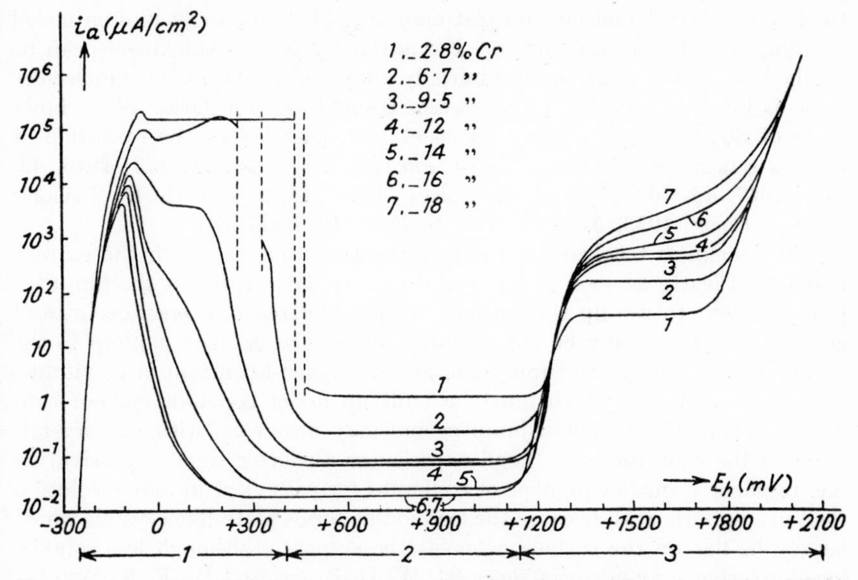

FIG. 58.—Anodic polarization curves for Iron–Chromium Alloys in 10% sulphuric acid (R. Olivier).

film prevents reductive dissolution, divalent chromium being unstable. On application of a feeble *cathodic* current from an external source, so as to depress the potentials, the stainless steel may be attacked, the iron passing into solution in the divalent state and chromium in the trivalent state. On application of an *anodic* current from an external source, there may be some anodic attack around a potential of about 0·0 volt but passivity sets in between about + 0·2 and + 1·2 volts; above this level certain iron-chromium alloys suffere anodic attack—as already explained—the chromium entering solution as soluble chromic acid.

Thickness of anodically formed films. Olivier has obtained a measure of the thickness of the film on a passive anode by using a constant-current circuit, obtaining from the time of passivation the number of coulombs needed for passivity. The constant current is obtained by the plan adopted in several laboratories, of using an E.M.F. (270 volts in this case) large compared to any potential changes likely to occur in the cell, in series with a large resistance. Having measured the time of passivation at several values of the current,* he calculates the thickness, after adopting a roughness factor of 7 (obtained by the B.E.T. method†); the thickness arrived at was 6 atom layers for alloys with 9·5% or 12% chromium, 3 atom layers for those of 14, 16 and 18% chromium, whilst on unalloyed chromium the film was regarded as a unimolecular layer.

It may be of interest here to mention the thickness obtained by other experimenters using other methods and studying different metals for the anodes. Hickling, using a constant current method, finds that the potential jumps to a value at which the current is used in oxygen evolution, when the thickness is only 1 molecule on platinum or gold, 1 to 2 molecules on nickel or about 4 molecules on copper. For antimony greater thicknesses can be reached before the jump occurs; initially a layer of Sb_2O_3 about 9 molecules thick is formed, and this still increases through the formation of an inner layer of Sb_2O_5, which reaches about 17 molecules before the evolution of oxygen sets in (A. Hickling, *Trans. Faraday Soc.* 1945, **41**, 333; 1946, **42**, 518; 1947, **43**, 762 (with J. E. Spice); 1948, **44**, 262 (with D. Taylor); *J. phys. Chem.* 1953, **57**, 203 (with S. E. J. El Wakked)).

The thickness reached must depend on the value at which the current is fixed. Clearly as soon as the metal cannot move through the film at a rate sufficient to use up this current, some other reaction *must* occur, and generally that can only be the evolution of oxygen, so that a jump to the potential high enough to supply energy for oxygen-liberation is inevitable. The amount of film which must be built up under constant current conditions before the potential-jump occurs may also vary with the crystal-habit of the solid formed. Lead experiences a shorter time of passivation (i.e. requires a smaller number of coulombs for passivity) in iodide solution than in sulphuric acid, because the lead iodide forms plate-like crystals which largely lie flat on the surface, whereas the sulphate, although less soluble, forms relatively rounded particles (G. W. D. Briggs and W. F. K. Wynne-Jones, *J. chem. Soc.* 1956, p. 2966. Cf. W. Feitknecht, *Z. Elektrochem.* 1958, **62**, 795. See also work on cadmium in alkaline solutions by K. Huber, *J. electrochem. Soc.* 1953, **100**, 376, esp. p. 379, and that on zinc in alkali by R. Landsberg, *Z. phys. Chem.* 1957, **206**, 291).

Excessive importance should not be attached to these thickness numbers. They represent the thickness at which the fixed current can no longer be employed exclusively in adding to the film-thickness, so that oxygen-

* Below a certain current density (i_0) passivity is never obtained at all; presumably this represents the rate of removal of material by diffusion and connection; at current densities (i) exceeding i_0, the time of passivation t_p is found to be expressed by $(i - i_0)t_p = Q$, where Q, measured in coulombs per cm.², permits a calculation of the thickness. The equation had been established by Shutt and Walton (p. 222) and by U. F. Franck, *Z. Naturforschung* 1949, **4A**, 378. Under some circumstances, however, the relationship may be more complicated; see R. Landsberg (with H. Bartelt and G. Just), *Z. Elektrochem.* 1957, **61**, 1162; *Z. phys. Chem.* 1958, **209**, 124.

† This method of measuring the true surface area of a metallic sample is based on the adsorption of an inert gas at a very low temperature. Olivier used krypton at − 196°C., and adopted 18·5 Å as the area of an adsorbed krypton atom (R. Olivier, Priv. Comm., Jan. 6, 1956).

evolution must set in. It does not follow that the film-thickening ceases after the potential-jump has occurred. The classical work of Tronstad, carried out by the polarized light method (p. 790), suggests that the film still continues to thicken after passivity has set in (L. Tronstad, *Z. phys. Chem.* (*A*) 1932, **158**, 369; *Trans. Faraday Soc.* 1933, **29**, 502).

The thickness corresponding to the start of oxygen-evolution will also vary with the current density. If this is high, a quicker movement through the film is needed to keep up with it; furthermore the polarization of the uncovered portion is greater. In Hickling's work on antimony (p. 234) the number of coulombs which passed before the occurrence of the potential-jump was greater at 0·01 amps/cm.^{2} than at 0·05 amps/cm.^{2}. Presumably at very high current densities, the jump might occur when the average film-thickness was less than one molecule, although such passivity (if indeed the term were appropriate) would probably disappear as soon as the current was discontinued.

Under potentiostatic conditions, the factors defining the thickness of a passive film are more easily defined. Presumably, after the current has been flowing for some time, the film-thickness takes up a fairly constant value at which the rate of dissolution by the acid or other liquid exactly balances its rate of regeneration by the current. It is interesting to remember that lattice-defects, which are favourable to the movement of ions through the film, also favour dissolution of the film-substance; of these two factors the first tends to produce a thick film and the second a thin film.

Tronstad's work, mentioned above, is particularly instructive, since it not only shows the growth of film-thickness during anodic treatment, but the decrease of thickness during cathodic treatment. Nickel, for instance, was studied at constant current density in N/10 sulphuric acid, alternately as anode and as cathode during successive 15-minute periods. The optical changes indicated a building up of a film when the nickel mirror was the anode, and its partial dissolution during the cathodic periods; however, cathodic treatment did not completely destroy the film, so that the next anodic period brought the thickness to a level not previously attained. After four cathodic and four anodic periods, the thickness had become sufficient to produce interference colours.

The Alternative Anodic Reactions. In Chapter V it was pointed out that when a metal was immersed in a salt solution, alternative reactions had to be considered. If for the moment we exclude cases like lead in a sulphate solution or silver in a chloride solution where films of sparingly soluble salts might be built up, the two possibilities are (1) the passage of the cations into the liquid to form a salt solution, and (2) the building up of an oxide-film. In absence of an applied E.M.F. the decision depends mainly on the relative drops of free energy, and in general acid conditions favour the production of a soluble salt whilst feebly alkaline conditions favour film-formation; on some metals neutral solutions may cause film-formation at first, but the acidity developed at points of atomic disarray where the anodic change proceeds most readily establish in the end conditions favourable for the local formation of soluble salt.

However, when a current is furnished from an external source of E.M.F., under approximately galvanostatic conditions (the current strength being roughly fixed by the external circuit), then the criterion will not necessarily be the magnitude of the free energy changes. If the current imposed be high, the first change can only proceed if the necessary number of atoms in the right energy condition to enter the liquid as cations is provided; the number available may be limited. Thus, whilst at low values of the applied current, the choice between the two alternative reactions may still be decided on considerations of free energy, a new criterion is met with at high current densities—namely the activation energy needed for each reaction. If this is less for a film-forming reaction than for the formation of soluble salt, we may obtain film-formation, followed by oxygen-evolution, even under acid conditions—simply because that is the only way in which the imposed current be used up. Thus it comes about that iron, which dissolves in dilute sulphuric acid to give soluble ferrous sulphate in the absence of an applied current, may become passive when at a high anodic current density is applied, the large imposed current being expended on the evolution of oxygen.

Conversely a material like stainless steel which at a low current density builds up a film on the anode, may start to behave differently if the current density is made so high that the anodic potential is raised above a certain " break-down potential ". Most of the cases studied have been carried out under conditions which were neither potentiostatic nor galvanostatic. Brennert used a cell fitted with a stainless steel anode, applying a gradually increasing E.M.F. and measuring at each stage the current flowing and also the anode potential. The anode potential rose steadily until at a certain value it suddenly collapsed, the current simultaneously increasing. This value was the *break-down potential*, and its measurement for different types of stainless steel has been regarded by some authorities as a guide to resisting power—a criterion which not all corrosionists would today accept. Brennert obtained somewhat analogous results with a tin anode in a chloride solution (S. Brennert, *Jernkontorets Annaler* 1935, p. 281; *Int. Tin Dev. Council Tech. Pub.* D2 (1935); *Korros. Metallsch.* 1936, **12**, 46; N. A. Nielsen and T. N. Rhodin, *Z. Elektrochem.* 1958, **62**, 707).

Similar break-down potentials have been observed on aluminium; if the potential is raised step-wise, very little current passes until the break-down potential is approached when conditions become erratic; above the break-down potential, a large increase in current occurs; pitting occurs above the break-down potential, but not below it. The level of the break-down potential is raised by the presence of CrO_4^{--} or HPO_4^{--} ions, but brought low by Cl^- ions. The manner in which the break-down potential depends on the relative concentrations of inhibitor ions and Cl^- ions has been quantitatively studied by P. J. Anderson and (Miss) M. E. Hocking, *J. appl. Chem.* 1958, **8**, 352.

At relatively low anodic current densities, passivity is favoured by conditions which keep low the solubility of the film-forming substance. A metal like iron, forming oxides which possess basic properties but no marked acidic properties, quickly becomes passive when exposed to anodic action in an alkaline solution, and generally remains passive in such solutions after the current has ceased to flow; a film is easily formed with the large concentration of OH^- ions present, and, once formed, does not easily dissolve. In an acid solution, as already stated, iron requires to be raised above a certain critical potential if passivity is to be produced, and an appreciable time must elapse before passivation is achieved; passivity disappears rapidly if the current is shut off.

In contrast, metals like molybdenum forming oxides which possess acidic properties and no pronounced basic properties, show the converse behaviour, dissolving anodically in an alkaline solution (to give molybdates) at high current efficiency, but often becoming passive in an acid solution, although here various factors are involved. The resistance of molybdenum to hydrochloric acid may be connected with a film of relatively insoluble chloride (E. Deltombe, N. de Zoubov and M. Pourbaix, *Cebelcor*, *Rapport Technique* No. **35** (1956)).

Although an iron anode polarized to a high potential in a nearly neutral solution may behave as an inert electrode, the current being devoted mainly to the evolution of oxygen, a chromium anode, polarized to a high potential, dissolves freely as yellow chromate; this is simply because the hexavalent compounds of chromium are highly soluble and stable in neutral (or even acid) solution. Iron dissolves in *concentrated* alkali at high anodic potentials to form similar, but less stable, hexavalent compounds, the ferrates. Antimony becomes passive when made the anode in dilute acid or dilute alkali, in which a film can be built up; it does not become passive in concentrated acid or alkali, which would dissolve the oxide (S. E. S. El Wakkad and A. Hickling, *J. phys. Chem.* 1953, **57,** 203).

The anodic behaviour of lead is of practical importance in connection with accumulators, and possesses also much theoretical interest. It has been studied in sulphuric acid solution by W. Feitknecht and A. Gaumann, *J. Chim. phys.* 1952, **49,** C135. H. R. Thirsk and W. F. K. Wynne-Jones, *Trans. Inst. Met. Finishing* 1952–53, **29,** 260. J. Burbank with A. C. Simon, *J. electrochem. Soc.* 1953, **100,** 11; 1956, **103,** 87; 1957, **104,** 693; 1959, **106,** 369; J. J. Lander, *ibid.* 1951, **98,** 213, 220; 1956, **103,** 1; in halide solutions by G. W. D. Briggs and W. F. K. Wynne-Jones, *J. chem. Soc.* 1956, p. 2966; and in alkaline solutions by P. Jones, H. R. Thirsk and W. F. K. Wynne-Jones, *Trans. Faraday Soc.* 1956, **52,** 1003.

Those seeking detailed information regarding the anodic behaviour of individual metals should study the following papers:—

D. R. Turner, *J. electrochem. Soc.* 1951, **98,** 434 (discussing nickel), with comments by T. P. Hoar, *ibid.* 1952, **99,** 275; R. Landsberg and M. Hollnagel, *ibid.* 1954, **58,** 680 (discussing nickel); J. H. Bartlett with L. Stephenson and R. S. Cooper, *ibid.* 1954, **101,** 571; 1958, **105,** 109 (discussing copper); T. P. Dirkse, *ibid.* 1955, **102,** 497 (discussing zinc); K. Huber, *ibid.* 1953, **100,** 376 (discussing magnesium, zinc and cadmium); P. E. Lake and E. J. Casey, *ibid.* 1958, **105,** 52 (discussing cadmium); N. Hackerman and C. D. Hall, *ibid.* 1954, **101,** 321 (discussing titanium); D. S. McKinney and J. C.

Warner, *ibid.* 1950, **97,** 86C (discussing titanium); D. R. Turner, *ibid.* 1956, **103,** 252 (discussing germanium); A. L. Ferguson and D. R. Turner, *ibid.* 1954, **101,** 382 (discussing silver); H. Göhr and E. Lange, *Z. phys. Chem.* 1958, **17,** 100 (discussing silver); H. Lal, H. R. Thirsk and W. F. K. Wynne-Jones, *Trans. Faraday Soc.* 1951, **47,** 999 (discussing silver); P. Jones and H. R. Thirsk, *ibid.* 1954, **50,** 732 (discussing silver); J. S. Halliday, *ibid.* 1954, **50,** 171 (discussing copper); H. Lal and H. R. Thirsk, *J. chem. Soc.* 1953, p. 2638 (discussing copper); R. Landsberg and M. Hollnagel, *Z. Elektrochem.* 1946, **60,** 1098 (discussing nickel); M. L. Levin, *Trans. Faraday Soc.* 1958, **54,** 935 (discussing beryllium); G. R. Hoey and M. Cohen, *J. electrochem. Soc.* 1958, **105,** 245 (discussing magnesium); T. L. Boswell, *ibid.* 1958, **105,** 239 (discussing indium); S. E. S. El Wakkad and S. H. Emara, *J. chem. Soc.* 1952, pp. 461 (discussing platinum at very low current density); S. E. S. El Wakkad, A. M. Shams El Din and H. Kotb, *J. electrochem. Soc.* 1958, **105,** 47 (discussing zinc at very low current density). Attention should be paid to the Spanish work on the anodic and cathodic behaviour of nickel, copper, silver and platinum; see A. Rius, J. Llopis with I. M. Tordesillas, M. Serra and F. Colom, *Anales real Soc. espan. fis. y quim* 1952, **48B,** 23, 35, 719, 861; 1955, **51B,** 11, 21, 379.

Influence of anions in deciding between corrosion and passivation. It is difficult to produce, or even to maintain, passivity in presence of chlorides. An iron electrode, treated anodically in dilute sodium hydroxide solution (free from salts), soon becomes passive; if similarly treated in a liquid containing sodium hydroxide and excess of chloride, it remains active, passing into solution as ferrous chloride, and being precipitated as hydroxide (rust, if the oxygen-supply is sufficient) at a perceptible distance from the metal, so that the solid is not protective.

The fact that metals are usually more difficult to render passive in chloride than in sulphate solutions is sometimes ascribed to the greater solubilities of chlorides, and there is no doubt that solubility is an important factor in passivation. Silver, where the solubility of the chloride is low, readily becomes passive when subjected to anodic action in a chloride solution; lead which has a sparingly soluble sulphate becomes passive when subjected to anodic action in a sulphate solution very much more quickly than iron, although the variation of current with time obeys the same law —as demonstrated by W. J. Muller and W. Machu, *Mh. Chem.* 1933, **63,** 347.

However, the unfavourable effect of chlorides upon passivation is too general to be ascribed to solubility differences, and should probably be explained by the argument suggested on p. 140. In a very dilute solution the potential gradient will cause the water molecules near the anode surface to orient themselves with the oxygen portion nearest the metal, providing an easy mechanism for film-formation; the cations from the metal, instead of passing out into the liquid, take up places between the oxygen ions, and in their stead hydrogen ions from the water molecules move out into the liquid. If, however, the liquid contains an appreciable concentration of anions carrying a negative charge, the same potential gradient will cause them to displace from the metal surface the water molecules which carry no net charge. If the anion in question is OH^-, this will provide a mechanism for oxide-film formation, better than that provided by H_2O molecules. If the anion is CrO_4^{--} it may provide a film-forming mechanism in virtue of the fact that the anion itself contains a cation $(CrO_2)^{++}$; the same is true of SO_4^{--}, but becomes important only at high current densities, where activation energy rather than free energy change becomes the criterion deciding between corrosion and passivation; the behaviour of metals—especially aluminium—in baths containing CrO_4^{--} and SO_4^{--} is discussed on p. 245. Chloride ions, however, will displace water molecules and break up the water phalanx without providing any new mechanism for passivation, and the fact that their presence is unfavourable to passivity need cause no surprise.

Piontelli has found that chlorides diminish the polarization, not only of anodic processes but also of cathodic processes; in the latter case the solubility of the chlorides cannot affect the situation. He also believes that in many solutions the metal–liquid interface is covered with a phalanx of adsorbed, oriented water molecules, which form a barrier opposing passage of ions *in either direction*, except in the case of those possessing unusual energy. Thus in order to produce cathodic deposition at any appreciable rate, it is necessary to depress the potential well *below* the equilibrium value, and in order to produce anodic attack at any appreciable rate, it is necessary to raise it well *above* the equilibrium value. If, however, the liquid contains chlorides, some of the chloride ions will secure positions on the metal–liquid interface, and the barrier becomes less formidable. It is now possible to obtain appreciable cathodic deposition or anodic dissolution at potentials relatively close to the equilibrium value; the " catalytic " action of chlorides is attributed to the " deformability " of the chloride ion* (R. Piontelli, *Z. Elektrochem.* 1951, **55**, 128; see also p. 907 of this book).

If the passage of a certain anodic current density requires that the potential shall be raised to a greater distance above the equilibrium value in a sulphate than in a chloride solution, passivity will be easier to produce in the first case. Suppose that the solution is (or has become) sufficiently acid to make the oxide or hydroxide an unstable phase, it will be impossible to produce a film unless the potential is raised a certain distance above the equilibrium value. (This is conveniently shown in the Pourbaix diagram for iron, fig. 174, p. 903.) At lower potentials, the formation of an oxide or hydroxide film will be impossible—on grounds of energy. Now if, in chloride solution, the necessary elevation of potential is only attained at a very high current density, whilst in sulphate solution it occurs at a lower one, it becomes easy to understand why it is difficult to render metals passive in chloride solution, but easier in sulphate solution. Cartledge attributes the effect of Cl^- ions (and also S^{--} ions) to the influence of the electrostatic field set up by the negative charge at the site of adsorption, which, he considers, will influence the activation energy for electrodic processes and facilitate both the formation and dissolution of metallic cations (G. H. Cartledge, *J. phys. chem.* 1956, **60**, 32, esp. p. 36).

* T. P. Hoar remarks that HS^- ions may act similarly.

There may, however, be another reason for the manner in which chlorides oppose passivity. Cl^- ions may be introduced into the lattice of the film, occupying some of the sites normally occupied by O^{--}. Clearly, if electrical neutrality is to be preserved, their presence will require that some of the cation sites become vacant—so that ions can move more easily through the film, which becomes non-protective; if an air-formed film is present on a specimen before it is made the anode in a cell, the attack on the metal will undermine the film which will finally become detached. If chlorides can enter the oxide, it will clearly be difficult to maintain—much less to build—a protective film in presence of chlorides—except on metals like silver, which have a sparingly soluble chloride, or cadmium, which forms a stable basic chloride.*

Production of Pigments by Anodic Corrosion. In the next chapter, we shall meet many examples where anodic corrosion in service is undesired and indeed disastrous; it may, therefore, be appropriate at this point to provide a few examples of the deliberate employment of anodic corrosion for useful purposes in industry. One of the examples chosen is also instructive in emphasizing the importance of the geometrical factor in corrosion.

If we wish to manufacture the pigment lead chromate, starting from metallic lead, it would seem at first sight reasonable to make the lead an anode in a potassium chromate solution. When, however, this is tried, it is found that the lead chromate, being the direct anodic product, forms an adherent deposit, interfering with the corrosion of the metal, so that the current is partly employed in oxidizing the lead chromate to the peroxide. After a layer of pure peroxide (PbO_2) has been formed over the surface, the current will be employed mainly on the evolution of oxygen, and no further lead chromate will be obtained. If, however, we use a solution containing potassium chromate with excess of sodium chlorate, the lead will dissolve as soluble lead chlorate, and will be precipitated as lead chromate at a slight distance from the metal, so that no stifling of attack will occur. This was shown in early work by M. le Blanc and E. Bindschedler, *Z. Elektrochem.* 1902, **8,** 255. Cf. C. Wagner, *J. electrochem. Soc.* 1954, **101,** 60.

More recent applications of anodic attack include the electromachining methods used in industry for shaping metals without introducing cold-work, and also for producing thin discs; in pure science, anodic attack has been used for the cutting of single crystals along any desired plane without distorting the structure. See R. Piontelli, R. Rivolta and G. Sternheim, *Rev. sci. Instrum.* 1955, **26,** 1206.

* This view is partly due to E. J. W. Verwey (priv. comm.). An interpretation of the activating influence of chlorides on chromium, based on the stability of soluble cations like $Cr(OH)^{++}$ and $Cr(OH)_2^+$ when the anion is Cl^-, is put forward by E. Deltombe, N. de Zoubov and M. Pourbaix, *Cebelcor Rapport Technique* No. **41** (1956). Cf. G. H. Cartledge, *J. phys. chem.* 1956, **60,** 32; *Z. Elektrochem.* 1958, **62,** 670.

Production of Thick Films by Anodizing

Anodic Behaviour of Aluminium. Whereas the film produced on iron by anodic treatment in dilute sulphuric acid disappears—perhaps within a few seconds—when the current is switched off, the state of affairs is different on aluminium—a metal of fixed valency where reductive dissolution is impossible, and direct dissolution of the oxide slow. Here anodic treatment in a sulphuric acid bath is the basis of a method for producing thick oxide-films which appreciably enhance resistance to corrosion, and also serve as a suitable basis for paint coats.

Numerous anodizing processes have been worked out based on chromic, sulphuric, oxalic and phosphoric acids. Their industrial usefulness is discussed on p. 249. For the moment the scientific interest of the film-formation will claim our attention.

Difference between anodized films and natural (air-formed) films. As already stated, aluminium exposed to dry air develops a film which possesses some protective properties, but does not entirely resist damp conditions. This is indeed shown by the growth-curves. Hart, confirming earlier workers, showed that in dry oxygen the film, after rapid thickening at the outset, practically ceases to grow when a certain value is reached; in damp oxygen, however, growth does not come to an end, but continues at a slow rate (R. K. Hart, *Proc. roy. Soc.* (*A*) 1956, **236,** 68).

It is possible to improve the protective value of the air-formed film by exposure to hot water or steam (p. 250), but the more usual method of obtaining increased resistance is by anodic treatment. This gives a much thicker film; in the formation of the air-formed film, the cell operative is Al | O_2 (1/5 atmosphere), which gives an E.M.F. of about 2·7 volts; the potential gradient falling over the film diminishes as the film thickens and soon becomes too small to move ions across the thickness, so that thickening becomes negligibly slow. In anodizing, 50 or 100 volts can be applied, and, if the mechanism remains the same, much greater thicknesses should be obtainable—as is found to be the case.

Structure of Films. Much skilful research has been carried out on the micro-structure of the films formed in anodizing, but at one time there was some difficulty in reconciling the results obtained in the U.S.A. with those obtained in France. Today a fairly definite picture of the facts can be presented; there is still some disagreement regarding their interpretation.

Boric acid, borate and nitrate baths confer thin, non-porous layers only, with the thickness proportional to the E.M.F. used. The sulphuric acid or chromic acid baths also give a thin, hard, barrier layer next the metal, but if time be permitted, a thicker, porous layer is produced outside the barrier layer (fig. 59). The structure of this part is fibrous (Clarke has compared it to a layer of closely packed bristles)—the fibrosity being well brought out in the photographs of Lacombe and Beaujard. The fibres consist of hexagonal prisms with spherical ends, each with a centrally located pore, sometimes described as star-shaped in cross-section. In most cases the central pore does not extend to the metal, but it is possible that a few of the pores penetrate the barrier zone, reaching the metal. The cellular structure has been closely studied by Keller; the dimensions vary with conditions of formation, but there may be billions of cells per square inch. The thickness of both layers also vary with the conditions of formation. Dunbar and Stephens suggest 0·1 to 1 μ for the barrier layer, and anything up to 200 μ (sometimes even 500 μ) for the outer porous film. The original papers deserve study (P. Lacombe and L. Beaujard, *Journées État Surface* 1945, p. 44; P. Lacombe, *Trans. Inst. Met. Finishing* 1954, **31,** 1; F. Keller, M. S. Hunter and D. L. Robinson, *J. electrochem. Soc.* 1953, **100,** 411; A. F. Dunbar and H. A. Stephens, Corrosion Symposium, Melbourne University 1955–56, p. 17 esp. p. 47; W. N. Bradshaw and S. G. Clarke, *J. electrodep. tech. Soc.* 1949, **24,** 147; R. L. Burwell and T. P. May,

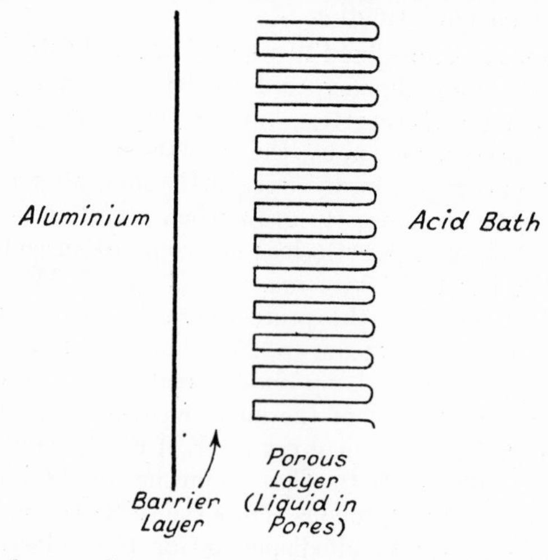

FIG. 59.—Barrier layer and porous layer produced by anodizing. *Not to scale.* The porous layer is often *much* thicker than the barrier layer. Some authorities show the pores as rounded at the bases.

J. electrochem. Soc. 1948, **94,** 195; H. W. L. Phillips, Symposium on Properties of Metal Surfaces 1952–53 (Inst. Met.), p. 237; R. B. Mason, *J. electrochem. Soc.* 1955, **102,** 671).

The arrival of the pores is interesting. Machu states that at first the film is structure-less, even under a magnification of $\times$ 40,000. Later round or ring-formed thickenings appear, which are often lens-shaped; these increase in number as the film thickens, until finally the whole surface appears to be granular. The centres of the rings represent pores. In the case of a film of thickness 50–100 Å produced in a borate buffer bath, the diameter of the rings may be 200–300 Å and that of the pores 50–100 Å (W. Machu "Nichtmetallische anorganische Überzüge" 1952, p. 7 (Springer), quoting H. Mahl).

Gerischer describes the structure of the film as removed by Pryor using the iodine method as very thin γ-Al_2O_3 layer overlaid with thick porous

β-$Al_2O_3.H_2O$. Dekker and van Geel state that whilst the films formed in sulphuric, phosphoric or oxalic acid baths are amorphous and porous (with pore-diameter 10^{-5} cm.), those obtained in borate, boric acid, succinate or citrate baths are crystalline (H. Gerischer, *Angew. Chem.* 1958, **70**, 296; A. J. Dekker and W. Ch. van Geel, *Phillips Res. Rep.* 1947, **2**, 213).

It seems that films on aluminium anodized in an oxalic acid bath, besides containing many micro-pores, also contain a few macro-pores, visible under low magnification or even to the naked eye; these occur at corners and rough parts of the surface. Both types can absorb impregnating material or dyestuffs, but only the macro-pores permit the initiation of attack on the metal (O. Kubaschewski and A. von Krusenstjern, *Metalloberfläche* 1952, **6**, A97).

Mechanism of the formation of the Barrier Layer. It is important to ask the reasons for the production of oxide-films by anodic treatment with external E.M.F. As already stated aluminium behaves differently in different baths. If anodized in a neutral nitrate solution or in a weakly acid borate bath, it produces a single thin barrier film, which appears to be non-porous and protective; in the more strongly acid anodizing baths, containing chromic, sulphuric, phosphoric or oxalic acid, a similar non-porous barrier film is formed at the base, covered with a much thicker porous film, which does not add much to the protection, unless subsequently sealed.

The formation of an oxide-film (rather than a soluble salt) in a neutral or weakly acid bath is not surprising, since this will represent the final, stable end-product.* The thickening of the film will practically cease when the potential gradient has become so small that ions cannot easily be moved from one site to the next. Since most of the E.M.F. of the bath will be used in producing a potential drop across the oxide-film, the thickness attained when current has almost ceased to flow and thickening has practically stopped, will be proportional to the E.M.F. applied, being 14·5 Å per volt for aluminium† (16·0 Å per volt for tantalum).

This attainment of a limiting thickness proportional to the E.M.F. would be expected whether the film is built by movement of cations outwards or of anions inwards. Actually the work of Plumb, based on the use of radioactive tracers, has indicated that the growth is due to cations moving outwards. Plumb coated a specimen first with a thick non-porous layer from an ammonium tartrate bath and then with a thinner layer (too thin for porosity) from a sulphuric acid bath containing a radioactive isotope of sulphur. The specimen was radioactive after the treatment, but soon lost its activity when the outer portion (only) of the anodized film was dissolved away in a phosphoric–chromic acid solution, showing that the

* In a neutral or faintly acid bath, oxide is a stable phase, provided a very small concentration of aluminium ions is present, whereas any large concentration of aluminium ions would be unstable and would tend to deposit a solid hydroxide or oxide.

† Some authorities give rather lower values for aluminium—such as 13·8 Å per volt.

second layer had been built up outside the first one. If the thin (radioactive) layer from the sulphuric acid bath was first deposited and then the thick layer from the tartrate bath, dissolution of the outer part of the film caused no serious drop of radioactivity. These observations show that the oxide is formed at the oxide–electrolyte interface and it may be concluded that the film grows by aluminium ions moving outwards, not by oxygen ions moving inwards. Plumb also showed that the films contain excess of aluminium, and that oxide layers formed from a phosphate bath contained phosphorus as an essential constituent. If it is assumed that P^{+++} ions replace Al^{+++} in the lattice, the approximate composition of the film may be written 3·5% P_2O_3 and 96·5% Al_2O_3 (R. C. Plumb, Priv. Comm.; see also R. C. Plumb (with J. E. Lewis), *J. electrochem. Soc.* 1958, **105,** 154C, 155C. Cf. H. W. McCune, *ibid.* 1959, **106,** 63).

The thickness of the barrier layer will vary from metal to metal, since the energy needed for an ion to pass from one stable position over an energy hump to the next stable position varies from one oxide to another. The earlier numbers appearing in the literature for the thickness per volt show discrepancies, but a useful table has been provided by van Rysselberghe and his colleagues (P. van Rysselberghe, with G. B. Adams, M. Maraghini and H. A. Johansen, *J. electrochem. Soc.* 1955, **102,** 502; 1957, **104,** 339, esp. Table V, p. 345. Cf. A. Güntherschulze and H. Betz, *Z. Phys.* 1934, **91,** 70; 1934, **92,** 367; E. J. W. Verwey, *Physica* 1935, **2,** 1059. K. J. Vetter, *Z. Elektrochem.* 1954, **58,** 230; 1955, **59,** 711 (with K. G. Weil). For tantalum, careful measurements are provided by A. Charlesby and J. J. Polling, *Proc. roy. Soc.* (*A*) 1954–55, **227,** 434. See also D. A. Vermilyea, *J. electrochem. Soc.* 1957, **104,** 140, and work on niobium by L. Young, *Trans. Faraday Soc.* 1955, **51,** 1250).

Where baths based on sulphuric, phosphoric or oxalic acids have been used, the total film-thicknesses may greatly exceed the value calculated from the voltage.* The thin compact barrier film, which never exceeds 14·5 Å per volt and may well be less, is overlaid by an outer film threaded by pores carrying highly conducting liquid up to the outside of the barrier film. The potential drop over the barrier film is sufficient to account for its thickness.

In discussing the formation of the barrier layer, the real difficulty is to show why in a strongly acid sulphuric acid bath, where the expected anodic product would be soluble aluminium sulphate, solid alumina is ever obtained. The fact that, once obtained, the oxide remains largely undissolved is not so surprising; the rate of acid-attack upon alumina, especially

* On metals where there is appreciable dissolution of the oxide, the thickness may be much *less* than the theoretical value. For the anodic film on iron, reductive dissolution should usually be avoided owing to the high potential involved, but a slow direct dissolution may be sufficient to cause the thickness to settle down at a thickness where fresh building of the film-substance (due to ionic movement through the film) is compensated by dissolution in the acid. The steady-state thickness will depend on pH value of the liquid. See also K. J. Vetter in "Passivierende Filme und Deckschichten" (H. Fischer, K. Hauffe and W. Wiederholt) 1956, p. 72, esp. p. 84 (Springer).

if pure, is extremely slow (p. 321); moreover, the liquid next to the oxide layer may be less acid than the body of the bath. The difficulty, however, disappears on considering the two alternative anodic reactions mentioned on p. 140. Let us assume that the acid used is H_2XO_4, where X may be Cr or S, and that it dissociates to give $(XO_4)^{--}$ ions (or possibly $(HXO_4)^{-}$ or $(X_2O_7)^{--}$ ions).

Most of the current between the aluminium to be anodized and the cathode conveying the current from an external source is carried through the liquid by the proton-switch mechanism (p. 1012), but a small amount is carried by $(XO_4)^{--}$ ions moving in the opposite direction (i.e. towards the anode). When an $(XO_4)^{--}$ ion reaches the oxide-coated aluminium, it can move no further, and is pressed against the surface, presumably with at least two oxygen atoms in contact with the alumina.

$$\begin{array}{r|l|l} \text{Al} & Al_2O_3 & \begin{array}{l} O \\ \quad > XO_2 \\ O \end{array} \quad \begin{array}{lllll} HO.H & HO.H & HO.H & HO.H \ldots & HO.H \\ \\ HO.H & HO.H & HO.H & HO.H & HO.H \end{array} \;\Big|\; \text{Cathode} \\ \text{Base} & \text{Film} & \end{array}$$

(If $(HXO_4)^{-}$ atoms predominate, the situation will be

$$\begin{array}{r|l|l} \text{Al} & Al_2O_3 & O\text{—}XO_3H \quad HO.H \quad HO.H \quad HO.H \quad HO.H \ldots HO.H \;|\; \text{Cathode} \\ \text{Base} & \text{Film} & \end{array}$$

and the wording of the argument requires slight modification—which can be left to the reader).

Two anodic reactions are now possible:

(1) Al^{+++} ions may pass into the liquid producing what is effectively an aluminium salt solution; these ions are extracted from the outer part of the oxide-film but the sites are filled by fresh Al^{+++} moving outwards and coming originally from the metal.

(2) $(XO_2)^{++}$ ions may be pulled off, joining the nearest H_2O molecules, displacing protons and regenerating H_2XO_4; the displaced protons are switched on to the next set of H_2O molecules, which then lose protons on the far side, and the proton switch will then continue right up to the cathode surface. This reaction produces a new layer of oxide, consisting of the oxygen ions left behind by the $(XO_2)^{++}$ ions, and the newly-arriving Al^{+++} ions which take up appropriate positions between them.

Now of these alternatives, reaction (1) undoubtedly provides the greatest drop in free energy, and may be expected under conditions close to reversibility, i.e. at low current densities; there is indeed a range of potentials over which reaction (1) is possible and reaction (2) cannot occur at all, for reasons of energy. However reaction (1) probably requires far more activation energy than (2) since it involves breaking the bond between aluminium and O, which is apparently strong in sesquioxides (probably it is higher for Al than Fe; corundum is harder than haematite). It is true that the ions are probably extracted from points where the structure is faulty, so that the work involved for each extraction is less than would be calculated from the work needed to break a sapphire, but it is probably a good deal higher than that involved in severing the bond between $(XO_2)^{++}$ and O^{--}.

At high current densities where reaction (1) would demand a large number of aluminium cations in the right energy state to detach themselves from the film and pass into the liquid, we may expect reaction (2) to predominate, although (1) will also occur. This agrees with the observed facts. The main product of anodizing is solid alumina, but plenty of aluminium sulphate is found in the sulphuric acid—more than could easily be explained by the attack of acid on the alumina already formed.

Clearly for successful anodization the bath must have a suitable composition. Water is needed for the proton switch mechanism, but the amount of free water should be kept at a minimum; for free water will help reaction (1) which required the provision of a sheath of water molecules for each Al ion entering the liquid; again free water will hinder reaction (2) by providing a phalanx of water molecules over the outer surface of the growing oxide-film, in competition with the XO_4^{--} ions. A bath which contains a large " reservoir " of water combined with the acid molecules but only a small concentration of free water molecules may be expected to give best results.

The solutions found suitable for anodizing do, in fact, contain much less free water than total water. In sulphuric acid, for instance, much of the water is probably present as acid hydrate (as suggested by the freezing-point curve, which shows a maximum at a composition corresponding to $H_2SO_4.H_2O$); thus the water activity lies far below the water concentration. The presence of glycerol or glycol, which possess affinity for water, may reduce the free-water content still further, and these substances are often added to the anodizing bath. Phosphoric and chromic acids also possess a high affinity for water, readily absorbing it from ordinary air—as does sulphuric acid. The case of oxalic acid is less simple. The solid is generally written $COOH.COOH,2H_2O$, but there is some reason to believe that the two water molecules do not represent water of crystallization, but water of constitution; indeed the crystal-structure seems to indicate that they may be present not as H_2O but as $[H_3O]^+$ ions (A. F. Wells, " Structural Inorganic Chemistry " (1945), p. 376 (Clarendon Press)).

Thus the facts are consistent with the idea that the formation of a protective oxide-film in an acid bath needs free water, but that the amount must be kept low. In Chapter IX it will be seen how the same principle helps in explaining the behaviour of aluminium towards alcohols, acids and phenols in absence of an applied E.M.F.

Mechanism of the Formation of the Porous Layer. In the anodization of aluminium in an acid bath, it seems likely that both oxide and soluble salt are formed simultaneously, but that in the early stages a continuous, compact oxide-film will be formed possessing a nearly constant thickness, since if at any point the film is momentarily thinner than elsewhere, current will be concentrated on that point and the thickness will be built up. However, as the limiting thickness (14·5 Å per volt) is approached, the growth of the compact film must become slow; there is, however, nothing to prevent the attainment of great thicknesses by the formation of an outer porous film, through simultaneous formation of aluminium oxide and aluminium sulphate (fig. 59, p. 242). Supposing that the thickness of the compact barrier layer is rather less than the limiting thickness, it is clear that the outward movement of Al^{+++} ions through the film will continue, since the liquid is a good conductor, and most of the E.M.F. will fall over the barrier layer. Of the aluminium cations moving outwards, part will enter the liquid, producing what is virtually a solution of aluminium sulphate, and thus keeping the film porous, whilst the other part will form fresh oxide at the base of the solid part of the porous film, pushing outwards the solid matter already present—so that the porous film continues to thicken indefinitely.*

The liquid in the pores is likely to become less acid than that in the body of the bath, but acidity will not disappear altogether. The ratio Al^{+++}/H^{+} should become stabilized at a certain value for the following reasons. If the only anodic product was solid oxide, the pH value would steadily drop, since six OH^{-} ions are used in producing each Al_2O_3 molecule; if the only product was Al^{+++}, the pH would rise, since mobile H^{+} ions will

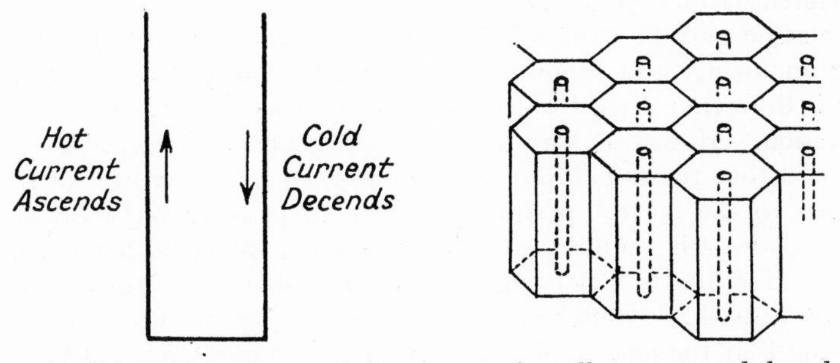

FIG. 60.—Types of thermal convection current in tall narrow and broad vessels (hexagonal type), possibly presenting an analogy to the hexagonal structure of anodized films.

carry more current than the Al^{+++} and SO_4^{--} ions, causing the region at the bottom of the pores to become partially denuded of H^{+} ions. Any drop in the pH will be unfavourable to the production of solid oxide, and any rise will be unfavourable to the production of Al^{+++} ions. Thus a steady state must ultimately be established at which the pH neither rises nor falls, and this will serve to standardize the porosity of the film substance also.

The structure of the porous outer layer has still to be explained; it can best be described as an array of hexagonal pillars each containing a central pore filled with liquid (fig. 60); fresh alumina is deposited at the base of the pillars which are pushed steadily outwards; in effect, therefore, oxygen is

* This may be the interpretation of an interesting experiment described by Champion. The porous oxide-film present on anodized aluminium was dyed and the specimen was then submitted to further anodization; the new, colourless film appeared below the dyed film suggesting that the porous layer is pushed outwards by the new matter deposited at its base (F. A. Champion, letter of July 16, 1955, suggesting, however, a somewhat different interpretation. See also N. D. Pullen and B. A. Scott, *Trans. Inst. Met. Finishing* 1956, **33,** 163 Cf. the experiments of T. Rummel, *Z. Physik* 1936, **99,** 518, esp. p. 537).

moving inwards combined with hydrogen (as water) along the central channels, and then moving outwards in the pillars combined with aluminium; the position of the hydrogen is relatively stationary since the inward movement of hydrogen (combined as water) is roughly balanced by the outward movement of the ionic hydrogen.

This circulation of oxygen is analogous to the circulation of water by thermal convection in a vessel heated from below. In a tall, narrow vessel, the hot water will generally rise at one side and sink at the opposite side, the direction of circulation being determined by some chance irregularity of geometry, or an asymmetry in the initial heat-application; once established, circulation may continue, even if the original asymmetry is removed. But in a broad, narrow vessel, heated uniformly below, the circulation may in some circumstances consist of hexagonal cells in which there is downward movement in the centre and upward movement near the boundaries of the hexagons (cf. O. G. Sutton, *Proc. roy. Soc.* (*A*) 1950, **204,** 297 and references quoted by him).

The same may be expected if oxygen is circulating in the growing film during anodizing, and should not be altered by the fact that the motive power is different; if so, the hexagonal cell structure receives a reasonable explanation. However this view is not generally accepted, and it is recognized that the radial movement of oxygen from the base of the pores outwards to the correct position in order to produce oxide at the base of the pillars still requires some elucidation.* Most authorities consider that an oxide-film is first formed and then is slowly and locally dissolved by the acid bath. Hunter and Fowle account for the dissolution-rate by supposing that the acid liquid at the base of the pores reaches the boiling-point and contains about 50% sulphuric acid (M. S. Hunter and P. Fowle, *J. electrochem. Soc.* 1954, **101,** 517, esp. p. 518).

The anodization of other metals. Several other metals of the " A " group of the Periodic Table develop films under anodic treatment. Their behaviour presents analogy to that of aluminium, but there has been divergence of opinion regarding mechanism. Papers deserving study include those of A. Charlesby and J. J. Polling, *Proc. roy. Soc.* (*A*) 1955, **227,** 434 (discussing tantalum); D. A. Vermilyea, *Acta Met.* 1953, **1,** 282; *J. electrochem. Soc.* 1954, **101,** 389; 1955, **102,** 207, 655; 1956, **103,** 690 (discussing tantalum); L. Young, *Trans. Faraday Soc.* 1956, **52,** 502, 515 (two papers discussing niobium); also *ibid.* 1957, **53,** 841 (discussing tantalum); R. D. Misch and E. S. Fischer *J. electrochem. Soc.* 1956, **103,** 153 (discussing hafnium); N. P. Inglis and M. K. McQuillan, *Endeavour* 1958, **17,** 77 (discussing titanium); I. S. Kerr and H. Wilman, *J. Inst. Met.* 1955–56, **84,** 379 (discussing beryllium); M. Hardouin, *Metal. Ind.* (*Lond.*) 1955, **87,** 385, 408 (discussing magnesium); W. McNeill and R. Wick, *J. electrochem. Soc.* 1956, **103,** 204C (discussing magnesium). Cf. J. H. Greenblatt, *J. electrochem. Soc.* 1956, **103,** 539 (discussing magnesium in chloride solution, where

* A research designed to clear up this point is described by T. P. Hoar and N. F. Mott, *J. Phys. Chem. Solids* 1959, **9,** 97.

soluble and insoluble material is produced); H. A. Johansen, G. B. Adams and P. van Rysselberghe, *ibid.* 1957, **104,** 339 (discussing several metals).

Of the papers mentioned, that of Charlesby and Polling requires special notice. They use the bright interference tints, which extend over several orders, in assessing the thickness and show clearly that the thickness ultimately attained is a linear function of the E.M.F. applied. Thickening increases until the potential gradient has fallen to 0·06 volt per Å, and beyond this it comes practically to rest. Any weaker field is insufficient to force the ions through the energy barriers, which—if we accept a line of reasoning due to Vermilyea—are not localized at the metal–oxide interface —as was once thought—but occur within the film itself.

Industrial Anodizing Processes. The original process of Bengough and Stuart based on a chromic acid bath is still used. It is described by G. D. Bengough and J. M. Stuart, British Patents 223,994 and 223,995 (1923); " The Anodic Oxidation of Aluminium and its Alloys as a Protection against Corrosion " 1926 (H.M. Stationery Office). See also G. D. Bengough and H. Sutton, *Engineering* 1926, **122,** 274. A new chromic acid process designed to give ductile films, very protective and suitable for dyeing, is described by A. W. Brace and R. Peek, *Trans. Inst. Met. Finishing* 1957, **34,** 232; see also *Metal finish. J.* 1958, **4,** 179. The anodizing of aluminium alloys containing copper, which presents difficulties, is discussed by J. Hérenguel and P. Lelong, *ibid.* 1958, **4,** 20.

In the first recommendations of Bengough and Stuart, the article was to be scrupulously cleaned and made the anode in 3% chromic acid at 40°C., the E.M.F. being gradually raised to 50 volts. Sutton and his colleagues preferred to increase the E.M.F. gradually to 40 volts in 15 min., maintain it there for 35 min., raise it in 5 min. to 50 volts and maintain it there for a further 5 min.; then the article must be washed and dried. Steel may be used as cathode material. The chromic acid should be free from sulphate; a trace of chloride in the bath—or left on the metal as flux residue—can cause serious trouble (H. Sutton and A. J. Sidery, *J. Inst. Met.* 1927, **38,** 241; H. Sutton, *J. Electrodep. tech. Soc.* 1929–30, **5,** 1; 1938–39, **15,** 77. J. W. Willstrop and H. Sutton, *ibid.* 1938–39, **15,** 53).

The non-conducting properties of the film confer good " throwing power " (i.e. power to deposit on remote parts of an article of complicated shape); for instance, the whole internal surface of a tube can be covered effectively by the use of a cathode placed outside it; the parts near the end are first coated and thus insulated, so that current is automatically diverted to the remoter parts. The yellow colour of the chromic acid helps in the detection of fissures.

The chromic acid process gives good results for aluminium and its lower alloys, but difficulties arise if the content of heavy metals exceeds about 4 to 5%. Moreover the necessity of gradually raising the voltage in the chromic acid process is a disadvantage; the sulphuric acid process employs a voltage which is constant and usually lower (15–30 volts), although rigorous control of the conditions is needed. Sometimes A.C. is superimposed on the D.C.

The sulphuric acid bath has attained special importance owing to the variety of film properties obtained by suitable choice of acid concentration, temperature, current density and time of treatment; the baths usually contain glycol or glycerine (E. Hermann, *Schweiz. tech. Zeitsch.* 1933, **8,** 285; S. Wernick, *J. Electrodep. tech. Soc.* 1933–4, **9,** 153; see also W. Campbell, *ibid.* 1951–52, **28,** 273, who discusses the production of hard films).

As regards corrosion-resistance, films obtained in the chromic acid bath are probably better than the others; chromate residues are clearly more desirable from the corrosion stand-point than sulphate residues, although they may produce stains. Objections to the sulphuric acid bath, however, largely disappear if the coating is subsequently sealed. In absence of sealing, the porosity of films formed in sulphuric acid baths appears to exceed that of films formed in chromic acid or oxalic acid (L. A. Cosgrove, *J. phys. Chem.* 1956, **60,** 385).

The oxalic acid process can be carried out with D.C., A.C. ora combination; the bath is usually 7 to 10% oxalic acid; sometimes 0·1% chromic acid is added. In one form of the process a short A.C. treatment is followed by a long D.C. treatment. Technical details of the various processes are conveniently provided by W. Machu "Nichtmetallische anorganische Überzüge" 1952, pp. 1–99 (Springer). For performance of different types, see F. Flusin, *Rev. Alumin.* 1957, **243,** 525.

The choice of anodizing process will depend on the purpose which the film is expected to perform in service. Films which are helpful in resisting corrosion are those which carry both the thin compact barrier layer and the thicker porous layer, which will serve for keying purposes if an outer protective layer (whether lanoline or paint) is to be applied, or which can be sealed by treatment with steam or hot water. Borate baths, which confer the barrier films only, without the outer porous layer, are valuable for certain purposes—for instance, in rectifiers and capacitors—but are not often used in connection with corrosion.

The rectifying effect is due to the fact that an oxide-coated aluminium electrode passes electrons more easily when it functions as cathode than as anode; similar valve-action is shown by tantalum and zirconium.

Sealing of Anodized Coatings. Where the anodized or treated aluminium is to be varnished or painted, pores may not be fatal, and the larger ones may even provide keying and improve adhesion; certainly porosity is welcome when the surface is to be dyed. However it seems best for most purposes to seal the pores, and many chemical processes are known. Solutions of cobalt or nickel acetate have their advocates, but Whitby states that sodium silicate is at least as good, and dichromate better. For many purposes boiling water gives admirable results, and it is found that demineralized water is better than supply water, which introduces dissolved salts (L. Whitby, *Metal. Ind.* (*Lond.*) 1948, **72,** 400; A. E. Bratt, *Trans. Inst. Met. Finishing* 1952–53, **29,** 336, esp. p. 346; A. E. Durkin, *Iron Age* 1941, **167,** May 10, p. 96).

Hart has used electron diffraction in studying the nature of the substance produced by boiling water on aluminium, whether in the mechanically polished, electropolished or anodized state; his anodized specimens included aluminium treated in baths of both classes. He found that the pattern produced corresponded to boehmite (γ-AlOOH) (R. K. Hart, *Trans. Faraday Soc.* 1954, **50,** 269).

In Hart's paper the boehmite formation is attributed to aluminium ions moving outwards through pores in the film and interacting with hydroxyl formed at the outer surface. The three reactions involved may be written:—

(1) Anodic attack on the metal below the oxide

$$Al = Al^{+++} + 3e$$

(2) Cathodic production of hydrogen and hydroxyl on the outside of the films

$$2H_2O + 2e = H_2 + 2OH^-$$

(3) Precipitation of boehmite where the anodic and cathodic products meet

$$Al^{+++} + 3OH^- = AlO(OH) + H_2O$$

This may be the true mechanism where the metal carries only a barrier film, but where, as on aluminium anodized in a sulphuric acid bath, an outer, porous film is present, it seems that the sealing of this film is due to hydration of the porous alumina. Spooner found that when such a film, after being detached from the metal, is treated in pure, boiling water, it increases in weight, the weight-gain being nearly the same as that observed on specimens of metal still carrying their anodized film. In both cases, the weight-gain was clearly connected with the conversion of alumina to boehmite, since electron diffraction gave good boehmite patterns; before the " sealing " in boiling water, the films (whether examined on the metal or after detachment) yielded no ring-pattern (R. C. Spooner, *Nature* 1956, **178,** 1113; see also D. Altenpohl, *Z. Metallk.* 1957, **48,** 306).

Bradshaw and Clarke observed that the stress in the anodic film changes from tension to compression on sealing in hot water, indicating a strong positive plugging action (W. N. Bradshaw and S. G. Clarke, *J. Electrodep. tech. Soc.* 1949, **24,** 147, esp. p. 161).

Chemical Processes. Processes designed to build up a protective film on aluminium without external E.M.F. are naturally welcome on grounds of economy and convenience. Many such processes have been developed and, although the older ones at least were probably inferior to anodizing, they are still much used—generally with success. The *M.B.V. process,* of German origin, is perhaps the best known; the articles are dipped in a hot solution containing sodium carbonate (5%) and chromate (1·5%); 3 to 5 minutes in the boiling bath, or 30 minutes at 30–40°C. in a similar mixture containing sodium hydroxide, usually suffice (W. Helling and H. Neunzig, *Aluminium* 1938, **20,** 536; P. Mabb, *Metallurgia,* 1936, **13,** 109).

The *Alrok* process involves immersion in a hot solution of sodium carbonate and potassium dichromate for 10 to 20 minutes, followed by sealing in potassium dichromate. The *Alodine* process (also called *Alodizing* or *Alochrome*) uses a solution containing chromic and phosphoric acids along with a fluoride accelerator; the pH should be controlled. The method has the advantages of speed and ease of application. The film—stated to consist of amorphous phosphates and chromates—is highly adherent; the metal can be sharply bent without detachment of the film and without undue loss of protective power. Immersion for 2 minutes or spraying for 30 seconds suffices. It should be followed by a plain water rinse, and also an acidulated rinse, if it is to be used as a basis for paint; but if no subsequent painting is contemplated, the rinse should be omitted and the article dried by moderate heating; the thickness of the coat is stated to be 0·6 to 3·2 μ. Pollack considers that the process, which is cheaper than anodizing, is equally reliable. This, and other processes which dispense with an external E.M.F., are much used in the aircraft and motor industries (A. Pollack, *Werkst. u. Korrosion* 1952, **3,** 352; D. Armstrong, *Metal Progr.* 1953, **63,** No. 6. p. 104; W. F. Castell, *Plating* 1954, **41,** 1409. R. Stricklen, *Mater. and Meth.* 1952, **35,** No. 2, p. 91. S. Heslop and A. Faulkner, Soc. Chem. Ind. Symposium on Protection of Motor Vehicles from Corrosion 1958, p. 45, esp. p. 51).

Several variants of the Alodine process exist. One gives a coat consisting of amorphous phosphates of aluminium and chromium, whilst another gives a coat containing oxides and chromates; both are said to confer considerable corrosion-resistance even on unpainted surfaces, but Stockbower states that the second type of film is the best. In both cases, good paint-bonding power results, but often a film of crystalline zinc phosphate, similar to that obtained in the phosphating of steel, is used for paint-bonding purposes. Details regarding procedures are furnished by E. A. Stockbower, *Metal finish. J.* 1956, **2,** 459.

Chemical processes for producing protective films without external E.M.F. are used on other metals and are described elsewhere in this book; they considerably improve the resistance of magnesium (p. 589), of zinc p. 586), cadmium (p. 651) and tin (p. 647).

Electropolishing

Different Methods of Polishing. Another important application of an anodic process is the electropolishing process introduced by Jacquet. The ordinary mechanical polishing method renders rough surfaces smooth and lustrous by causing material from the prominences to flow, filling or covering up the depressions; the whole surface becomes covered with a layer of flowed material, generally known as the *Beilby layer,* which is usually a mixture of metal, oxide and often particles of polishing powder. The surface colour of a mechanically polished surface varies with polishing agent used (which may be fine ferric oxide, chromic oxide or alumina), whilst the appearance varies with the technique of the polisher, especially when the operation is conducted by hand. In contrast, electropolishing depends on the dissolving away of the prominences at a rate greater than the depressions, so that the surface, as attack proceeds, becomes increasingly level, without the production of a Beilby layer. The result is far more reproducible, and the human factor is largely eliminated. Reflectivity measurements show that a number of electropolished specimens of very pure aluminium all gave nearly the same values, whereas mechanically polished specimens gave values which varied on different parts of the same specimen and, still more, between one specimen and another (P. A. Jacquet, *Nature* 1935, **135,** 1076; *Trans. electrochem. Soc.* 1936, **69,** 629; *C.R.* 1935, **201,** 1473; 1937, **205,** 1232; 1939, **208,** 1012 (with A. Rocquet); 1951, **232,** 71; *Rev. Métallurg.* 1940, **37,** 210; 1945, **42,** 133; P. Jacquet and L. Capdecomme, *Nature* 1938, **141,** 752. Also P. A. Jacquet's book, " Le Polissage electrolytique des Surfaces métalliques et ses Applications " 1948 (Éditions Métaux)).

When a rough surface is subjected to anodic action under conditions favourable to corrosion (as opposed to passivity), the current density will, from considerations of resistance, be highest at the prominences, so that these will tend to be eaten away more quickly than the depressions, making the surface somewhat smoother. This will happen in almost any bath, and is not peculiar to the liquids recommended by Jacquet and others for polishing purposes. Moreover, even in a bath chosen for its electropolishing properties, the ratio of the current density at prominences to that at depressions does not change significantly when we pass from conditions unsuited to polishing to those where good polishing is obtained—a point established by Edwards in researches based on the grooved copper sheets used by the makers of " micro-groove " gramophone records (J. Edwards, *J. electrochem. Soc.* 1953, **100,** 189C, 223C; *J. Electrodep. tech. Soc.* 1953, **28,** 133).

The projections and depressions on such surfaces are of course gross compared to the irregularities which interfere with specular reflection of smooth but not bright surfaces. Whilst almost any bath will serve for anodic smoothing, only certain baths will serve for anodic polishing—and even these are effective only under certain conditions of potential and current density.

Hoar and Mowat believe that true polishing occurs only if and when a compact solid film is present on the anodic surface. Impressive support for such a view was provided by Hoar and Farthing. They found that when mercury is brought into contact with a copper specimen which is being subjected to anodic attack in phosphoric acid solution, it " wets " the surface under etching conditions, but refuses to wet it under polishing conditions—which would seem to establish the existence of a non-metallic layer. This view, however, has not gained universal acceptance. Rowland electropolished palladium in molten sodium or potassium chloride free from oxygen—where, he argued, a film must be absent. Other authorities believe that the film operative in electropolishing is not solid but liquid —perhaps a viscous layer. The various views should be studied in the original papers (T. P. Hoar and J. A. S. Mowat, *Nature* 1950, **165,** 64; T. P. Hoar and T. W. Farthing, *ibid.* 1952, **169,** 324. P. R. Rowland, *Nature* 1953, **171,** 931. J. Edwards, see above. K. F. Lorking, see p. 255).

In the Author's opinion, Hoar and his colleagues are probably correct in saying that a solid film helps electropolishing. The cause may be that responsible for a phenomenon noticed during Miley's work on the reductive dissolution of oxide-films present on heat-tinted iron (p. 58); the metallic surface left behind after the film had been removed was smoother than the original surface before tinting. The same smoothing after oxidation and film-removal was noticed by other Cambridge investigators, and it was found that, starting from an abraded surface, it was possible, by alternately oxidizing and removing the film by reductive dissolution a few times, to develop a bright surface, recalling—although inferior to—the surface obtained in an approved electropolishing bath. (The method was studied by Eurof Davies with a view to its adoption as a method of preparing the surface of specimens before an oxidation experiment, but in the end hydrogen-reduction was preferred.) There seems little doubt that the principles involved are the same as those involved in anodic polishing under conditions which produce a film on the anode.

It is easy to see why the surface separating metal from oxide should become progressively flatter as the oxide film thickens. Consider a sub-microscopic circular area on a flat portion of the metallic surface, and suppose that the film is thickening by metal cations moving outwards under the potential gradient, that is at right angles to the surface. The region available for their movement from a flat surface will be a *cylinder*. If now the area is situated on a prominence (i.e. on a convex surface), the region will be a *cone*, the cross-section of the available path *increasing* as we pass outwards; if, however, it is situated in a depression (i.e. on a concave surface), it will be a *cone* with the cross-section decreasing as we pass outwards. It is clear that under a constant E.M.F. there will be far more rapid passage from the prominences than from the depressions, and we shall get smoothing on a sub-microscopic scale—that is, brightening.

This explains the brightening obtained by alternate dry oxidation and film destruction. The same brightening will occur on an anode covered with a solid film, provided that the specific resistance of the solid film-substance is high compared to that of the liquid bath, so that both the surfaces of the film can be regarded as equipotential surfaces; under these circumstances the movement must be at right angles to those surfaces. This might also happen with a liquid layer possessing quasi-solid properties and a specific resistance high compared to that of the body of the liquid; possibly a viscous liquid layer would serve. It becomes easy to see that almost any liquid—and any electrical condition—will suffice for smoothing on a gross scale, but only a few liquids can eliminate the final sub-microscopic irregularities and achieve true brightness; even these can do so only under electrical conditions suitable for the production of a solid, or quasi-solid layer* (cf. H. F. Walton, *J. electrochem. Soc.* 1950, **97**, 219; C. Wagner, *ibid.*

* It is less easy to accept explanations based on the argument that on a film-covered surface the removal of metal is conditioned by random movement of vacancies in the film substance and is independent on the crystalline structure of the metal, so that crystallographic etching (which might cause micro-roughening) is avoided. It is unlikely that the movement of vacancies will be uniform near a rough surface, since the potential gradient will not be uniform; the idea that oxidation is independent of the crystalline orientation of the metal is

1954, **101,** 225; J. F. Nicholas and W. J. McG. Tegart, *ibid.* 1955, **102,** 93C).

The idea of a viscous layer as the necessary factor in polishing may have arisen from the fact that some mixtures used as polishing baths do possess high viscosity. This is not, however, true of all. Good polishing on aluminium has been obtained at Cambridge in a bath containing anhydrous aluminium chloride (10 grams) and anhydrous zinc chloride (45 grams) dissolved in ethyl alcohol (144 c.c.) containing water (32 c.c.) and normal butyl alcohol (16 c.c.)—a mixture which is not viscous. The existence of this bath also disposes of an idea—once prevalent—that polishing cannot be carried out in presence of a chloride. Although for many purposes not equal in polishing efficiency to the baths based on perchloric acid or anhydride, the Cambridge chloride bath has been used in France by Lacombe for polishing stainless steel (p. 664), in an Australian laboratory for the polishing of zinc and recently at Birmingham in a research on the dynamics of slip and twinning (U. R. Evans and D. Whitwham, *Trans. Electrodep. tech. Soc.* 1947, **22,** 24; D. S. Kemsley and W. J. McG. Tegart, *C.S.I.R. (Australia) Div. of Tribophysics, Phys. Met. Rep.* No. **7** (1948). R. L. Bell and R. W. Cahn, *Proc. roy. Soc.* (*A*) 1957, **239,** 494, esp. p. 497. Cf. M. Halfawy, *Experientia* 1951, **7,** 175, whose theory is based on viscosity).

Lorking, studying the polishing of copper in phosphoric acid, suggests that the chains of molecules are arranged in a manner similar to the condensed molecules of polyphosphate. The density of packing is thought to be greatest over the concave parts of the surface, explaining the preferential removal of the convex parts. Such an arrangement is equivalent to the semi-solid film. The idea deserves serious consideration (K. F. Lorking, *Australian Dept. of Supply Res. and Div. Branch, Aeronautical Research Laboratories, Report Met.* **18,** Sept. 1956, p. 15).

Edwards (p. 253) has put forward an *acceptor theory*—of which the Hoar–Mowat theory can be regarded as a special case. In explaining the polishing of copper in a phosphoric acid bath, Edwards considers that the rate of dissolution of copper is determined solely by the rate at which certain "acceptors" reach the surface (by diffusion and convection, not by electric migration); they are stated to be "phosphate ions of some sort". They will reach the surface most readily at the prominences, where the concentration-gradient is steeper, thus accounting for the removal of such prominences. The importance of concentration-gradients had been emphasized by Elmore, who, however, believed the controlling factor to be the concentration-gradient of metal ions moving outwards, not that of an "acceptor" inwards (W. G. Elmore, *J. appl. Phys.* 1939, **10,** 724; 1940, **11,** 797).

Some of the theories of electropolishing are based on the study of curves connecting potential and current density in polishing solutions. Many

difficult to reconcile with the observation that different grains of a specimen develop different tints. T. P. Hoar adds the comment, "I think *both* effects may well be important. A high-resistance solid film (high cation energy barriers for movement *through* the film, and *few* and *random* places to move) may assist random cation movement from metal $\rightarrow$ film."

authors have produced such curves, but there has not been complete agreement about the form. Readers may care to compare the results of A. Hickling and J. K. Higgins, *Trans. Inst. Met. Finishing* 1952–53, **29**, 274; I. Épelboin and C. Chalin, *C.R.* 1948, **226**, 324; I. Épelboin, *Métaux, Corrosion, Industries* 1956–57, **376**, 378; *Rev. Métallurg.* 1952, **49**, 863; 1958, **55**, 260 (with M. Froment and G. Normanski); H. F. Walton, *Trans. electrochem. Soc.* 1950, **97**, 219; H. T. Francis and W. H. Colner, *ibid.* 1950, **97**, 237.

Hickling and Higgins used a potentiostat, which enabled them to decide with certainty the current density corresponding to each potential value, and obtained curves of the form shown in fig. 61, which represents copper in phosphoric acid. The little peak at the left-hand end of the plateau

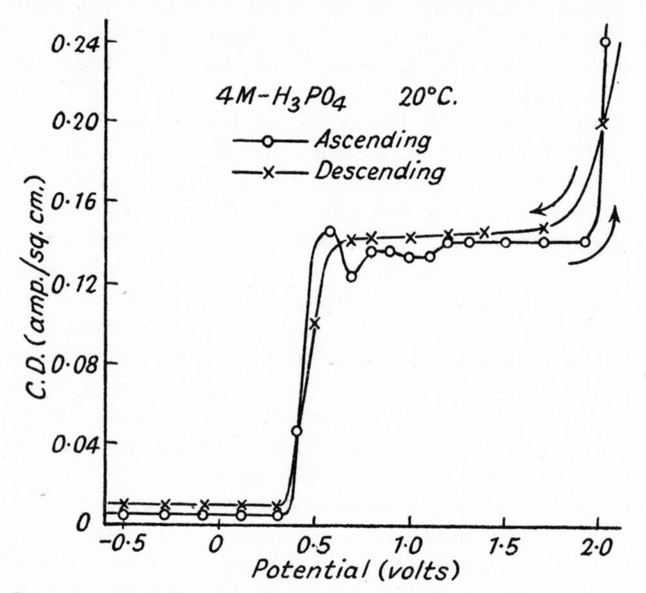

FIG. 61.—Current–potential curves for a Copper anode in 4M phosphoric acid (A. Hickling and J. K. Higgins).

is seen on the curve for increasing potential, but does not occur on the return curve obtained when the potential is reduced. The curves should be compared with those obtained for iron in sulphuric acid by Olivier (fig. 57, p. 230), a system which produces passivity instead of polishing. The main difference is that on the plateau of Hickling's curves, where current is independent of potential, the current flowing is far stronger than on the flat part (Region 4) of Olivier's curves. On Hickling's curves, the plateau corresponds to polishing conditions.* The current strength on the plateau is made lower if the phosphoric acid contains copper phosphate; indeed the polishing current can become extremely small if the bath is saturated with copper phosphate; at any copper phosphate concentration, however,

* In a phosphoric acid bath, but not in a cyanide bath, polishing also occurs on the rising portion on the right, where oxygen is being evolved.

the current is increased if the liquid is stirred. All these facts support the idea that the anode, which on the left-hand rising portion of the curve is film-free and dissolves freely, is covered with a copper phosphate film in the plateau region; the current passing will then adjust itself to a value, independent of potential, which will exactly serve to replenish the amount of copper phosphate lost by passage into the solution—an amount increased by stirring but decreased by the presence of copper phosphate in the acid; thus the facts are explained. Hickling considers that there is no fundamental difference between polishing and passivity. If the film substance is reasonably soluble, we get polishing; if it is sparingly soluble, we get passivity.

In discussing the Hickling–Higgins paper, Hoar has quoted views put forward many years ago by Vernon and Stroud. Brightening and smoothing, he points out, are two different effects; when, as often happens, they occur together, the result is polishing (T. P. Hoar, *Trans. Inst. Met. Finishing* 1953, **29**, 292; W. H. J. Vernon and E. G. Stroud, *Nature* 1938, **142**, 477, 1161).

Composition of Electropolishing Baths. The liquids found to polish metal are of many different kinds. Some are extremely simple, like the 25% potassium hydroxide used in the polishing of zinc by Vernon and Stroud. Others are similar to plating baths, such as the cyanide solution used for silver and cadmium (B. Shuttleworth, R. King and B. Chalmers, *Metal Treatm.* 1947, **14**, 161; A. W. Hothersall, *Rep. Progress appl. Chem.* 1947, **32**, 112; J. Liger, *Bull. Soc. chim. Fr.* 1944, **11**, 568). Lead has been polished in an acid acetate bath (E. Jones and H. R. Thirsk, *Nature* 1953, **171**, 843). Some relatively cheap baths for steel, based on nitric acid, are put forward by H. J. Merchant, *J. Iron St. Inst.* 1947, **155**, 179.

Two types of baths, based on phosphoric and perchloric acids respectively, deserve special attention, being capable of polishing a wide range of materials.

Phosphoric Acid Baths are used considerably for the polishing of copper (where 50/50 vol./vol is usual) and also for its alloys; Perryman adds sulphuric acid in polishing tin bronze. Lorking recommends similar baths for chromium and its alloys (E. C. W. Perryman, *Metallurgia* 1942, **46**, 55; K. F. Lorking, *Trans. Inst. Met. Finishing* 1955, **32**, 451).

A mixture of phosphoric acid (37%) with water (7%) and glycerine (50%)—the latter probably serving to increase the viscosity—has been put forward for stainless steel and nickel silver by H. Evans and E. H. Lloyd, *J. Electrodep. tech. Soc.* 1946–47, **22**, 73; another bath for stainless steel (P. A. Charlesworth, *ibid.* 1950, **26**, 43) contains phosphoric acid (22%), sulphuric acid (55%) and water (23%) with aniline (about 2%) added; a Russian bath ("Protection of metals from Corrosion" 1950 (Mashgiz), contains phosphoric acid, sulphuric acid, glycerine, chromic acid and water. P. A. Jacquet (Compagnie générale de Constructions Téléphoniques, Notice VII) recommends that zinc and magnesium should be polished in alcoholic phosphoric acid, whilst J. Hérenguel and F. Santini (*Métaux et Corros.* 1946, **21**, 131) put forward a bath with phosphoric acid (15%), sulphuric acid

(70%) and water (15%) for aluminium alloys. See also a set of papers in *Trans. Inst. Met. Finishing* 1956, **33,** which include chemical brightening processes as well as electropolishing, esp. N. D. Pullen and B. A. Scott (p. 163), A. W. Brace and T. S. de Gromoboy (p. 177), G. E. Gardam and R. Peek (p. 198), F. Baumann and H. Neunzig (p. 211), K. Sachs and M. Odgers (p. 245).

Perchloric Acid Baths. The original bath introduced by Jacquet contained acetic anhydride; according to the instructions given in his book, the bath is made by mixing 215 c.c. of perchloric acid of density 1·479 at 20°C. with 785 c.c. of pure acetic anhydride (98%). The acid of density 1·479 can be prepared by adding 114 c.c. of distilled water to 445 c.c. of commercial concentrated acid of density 1·61 (55°Be.). The mixing must be carried out by introducing the acetic anhydride gently and in small portions into the perchloric acid which is kept cool with a rapid flow of water. For avoiding overheating he recommends an apparatus due to Goodyear. The book should be consulted for details (P. A. Jacquet, " Le polissage electrolytique des surfaces métalliques et ses applications " 1948, p. 11 (Éditions Métaux); J. H. Goodyear, *Metal Progr.* 1944, **46,** 106).

Jacquet has recommended similar mixtures for polishing many materials, including steel, tin and titanium.

The use of this acetic anhydride bath has led to serious explosions, probably caused by departure from instructions. Later Jacquet introduced a safer bath which contains only 50 c.c. of 60–65% perchloric acid (density 1·59–1·61) along with 1000 c.c. of glacial acetic *acid* (not anhydride); in contrast with the older bath, practically no heat is evolved when the ingredients are brought together. Jacquet states that with this bath good electropolishing is obtained at an anodic current density of 12 amps/dm.^{2} or even 30 amps/dm.^{2}. The E.M.F. required is 30 to 50 volts, and in practice polishing is controlled by regulating the voltage. The method is intended for steel, but is also suitable for polishing chromium-plate (P. A. Jacquet, *C.R.* 1948, **227,** 556; *Rev. Métallurg.* 1949, **46,** 214; L. Médard, P. A. Jacquet and R. Sartorius, *ibid.* 1949, **46,** 549).

The original bath with acetic anhydride probably gives the best results for aluminium alloys. Care should be taken not only during mixing, which causes much evolution of heat, but also when disposing of waste solution. During use, care should be taken to switch off the current when introducing or withdrawing the articles to be polished, so as to avoid arcing, and to avoid the use of certain plastics in the fittings. A triangular diagram constructed by Jacquet shows which mixtures of perchloric, acetic anhydride and water are explosive, and which merely inflammable; the recommended mixtures lie well away from the explosive region, but clearly the proportions should be strictly adhered to. The diagram is printed in Jacquet's paper with Médard and Sartorius (see above) and also in his useful survey of electrolytic and chemical polishing (P. A. Jacquet, *Metallurg. Reviews* 1956, **1,** 157, esp. fig. 7, p. 189).

A bath consisting of alcohol with a small amount of perchloric acid (which evolves relatively little heat on mixing) is recommended as almost free from dangers of explosion by A. de Sy and H. Haemers, *Metal Progr.* 1948, **53,** 368.

Use of Electropolishing in Scientific Work. Electropolishing enables a flat surface, free from any Beilby layer and work-hardening, to be obtained in preparing sections intended for microscopic examination. After etching, structural details are brought out with unusual clarity; much information has been gained by the use of electropolished sections which would otherwise have been missed (see, for instance, P. A. Jacquet, *C.R.* 1953, **237,** 1248, 1332).

At one time, it was thought that an electropolished surface of aluminium, as it left the bath, was free from any film; the unusually low (negative) potentials measured on such surfaces were thus interpreted. It is probable, however, that a film does exist on the freshly electropolished surface, and increases on exposure to air; indeed, Hoar's theory of electropolishing demands the presence of a film. In the case of copper polished in phosphoric acid, there is no doubt of contamination with a film containing phosphorus, since this has been detected in tracer research by Simpson and Hackerman. Jacquet states that the contamination on copper and zinc polished in phosphoric acid can be detected analytically, and considers that the thickness must lie between 30 and 120 Å—in fair accord with an estimate based on electron-diffraction studies by Raether, who observed that the film was thick enough to obscure the underlying metal. Allen, who found the contamination (probably basic copper phosphate) objectionable for his oxidation work, removed it by washing in 10% phosphoric acid (P. A. Jacquet and M. Jean, *C.R.* 1950, **230,** 1862. H. Raether, *Mikroskopie* 1950, **5,** 101. J. A. Allen, *Trans. Faraday Soc.* 1952, **48,** 273. N. H. Simpson and N. Hackerman, *J. electrochem. Soc.* 1955, **102,** 660. E. C. Williams and M. A. Barrett, *ibid.* 1956, **103,** 363).

Use of Electropolishing in Industry. Many opinions have been expressed as to how far electropolishing will displace mechanical polishing in industrial processes. The result of electropolishing is sometimes less pleasing to the eye than mechanical polishing, since it does not always completely obscure surface defects, and sometimes tends to reveal them; this may provide a valuable warning in cases where such defects would be likely to cause breakage in service, but is a disadvantage in the preparation of an attractive finish on cheap objects prepared from material of indifferent quality. Doubtless by careful selection of steel or brass sheet, a brilliant finish could be produced without any mechanical polishing, but that may not always be economic. The various aspects of the question have been discussed by A. W. Hothersall, *J. Electrodep. tech. Soc.* 1946, **21,** 287; P. A. Jacquet, *Métaux et Corros.* 1950, **25,** 237; *Atomes*, Feb. and June 1953; C. L. Faust, *J. Electrodep. tech. Soc.* 1946, **21,** 181; H. E. Zentler-Gordon, *ibid.* 1950, **26,** 55; P. Michel, *Sheet Met. Ind.* 1949, **26,** 2169; 1950, **27,** 165, 267; R. Mondon, *Metal. finish. J.* 1955, **1,** 513.

Today electropolishing is being employed regularly for stainless steel—especially for small articles; on this material mechanical polishing is relatively costly and there is an economic advantage in using the electrical method. For aluminium, electropolishing and also chemical polishing are being increasingly used, especially on superpurity material (S. Wernick, Priv. Comm., April 15, 1958).

Lacombe has referred to the " virtual abandonment of mechanical polishing ", in France, " in favour of the chemical or electrolytic method, wherever bright finishing, i.e. films which do not alter the reflecting power of the underlying metal, are wanted " (P. Lacombe, *Trans. Inst. Met. Finishing* 1954, **31,** 1, esp. p. 9).

Electropolishing occasionally reduces resistance to fatigue, despite the apparent advantage of removing roughnesses which might act as stress-raisers. This is a matter of importance in connection with aircraft; the question of the restoration of fatigue resistance by subsequent heat-treatment is considered by A. T. Steer, J. K. Wilson and O. Wright, *Aircraft Production* 1953, **15,** 242.

Recent developments of anodizing and electrobrightening in the motor industry are described by F. Baumann and H. Neunzig, *Metal finish. J.* 1956, **2,** 131; alkaline electrobrightening and anodizing of aluminium is described by N. D. Pullen and B. A. Scott, *ibid.* 1956, **2,** 127 (see also below); bright anodizing in a bath containing ammonium bifluoride and nitric acid is discussed by G. E. Gardam and R. Peek, *ibid.* 1956, **2,** 130.

Brightening followed by Anodizing. There are many purposes where a high reflectivity for light is desirable on a surface, accompanied by wear-resistance; aluminium reflectors provide an example. The film produced by anodizing in a chromic acid bath is too opaque, and even that obtained in an oxalic or sulphuric baths reduces the reflectivity; by using sodium bisulphate, however, the reflectivity is not reduced and may even be increased.

In the *Brytal* process, described by Pullen, the articles are treated in a solution containing sodium carbonate and trisodium phosphate without current until the natural oxide skin has been dissolved in the alkali; it is convenient to attach the articles to the anode bar during this first (current-less) stage which need only last about 20 seconds. When a vigorous etching action has developed, the current is switched on at about 10 volts; the etching soon ceases, and after about 5 minutes, the reflectivity is far superior to that of the original milled surface; the article is then washed and transferred to the bisulphate bath where it is anodized at about 10 volts at 35°C. Technical details are provided by N. D. Pullen, *J. Inst. Met.* 1936, **59,** 151; N. D. Pullen and B. A. Scott, *Trans. Inst. Met. Finishing* 1955–56, **33,** 163; A. Ensor, *J. Electrodep. tech. Soc.* 1942, **18,** 13; A. W. Brace, *Metal finish. J.* 1955, **1,** 253.

There has been lively discussion as to whether bright anodized finishes will replace the much favoured combination of nickel with chromium plating on motor-cars and in its many indoor and domestic uses; if this came about, it might ease the nickel supply problem. Super-pure aluminium brightened and anodized has a total reflectivity superior to a nickel-chromium finished surface, and only slightly inferior to lacquered silver plate; the specular reflectivity is superior to lacquered silver, but inferior to the nickel-chromium finish. An interesting debate on the relative merits and demerits of the two rival finishes will be found in *Metal finish. J.* 1956, **2,** 357.

Chemical Polishing. A process which would produce some of the results of electropolishing without demanding an external E.M.F. would have obvious advantage. For some materials, such methods are available. Various processes for the chemical polishing of aluminium are compared by H. Fischer, *Metall,* 1952, **6,** 491; H. Silman, *Product Finishing* 1956, **9,** 52; A. W. Brace, *Metal Ind.* (*Lond.*) 1957, **90,** 147, 153.

Perhaps the most interesting development is the Marshall smoothing process for steel, based on a dip into a bath containing oxalic acid, sulphuric acid and hydrogen peroxide. This can be used to restore a sufficiently smooth surface to steel articles which have become rough through rusting and subsequent rust-removal. Hammond, who points out that polishing costs are a major item in producing a decorative finish on steel, states that the process leaves the surface in a suitable condition for plating in some, but not all, plating baths (W. A. Marshall, *J. Electrodep. tech. Soc.* 1951–52, **28,** 27; R. A. F. Hammond, *ibid.* 1951–52, **28,** 61).

The process leaves a film about 60 Å thick on the metal. Presumably it is produced by an anodic reaction similar to that operating in electropolishing with an external E.M.F., but maintained in this case by a local cathodic reaction consisting of the reduction of hydrogen peroxide to water. The mechanism of the thickening has been examined by Hickling and his colleagues (A. Hickling, with W. A. Marshall, E. R. Buckle and A. J. Rostron, *Trans. Inst. Met. Finishing* 1951–52, **28,** 47; 1954–55, **32,** 229).

For metallographic purposes, a chemical bright dip seems equal or better than anodic polishing. The American bath known as R5, originally developed for industrial purposes, reveals sub-grain boundary structures on micro-sections under conditions when they are not easily shown up by other processes; it contains phosphoric acid (73–83%), nitric acid (2–5%) and water (14–33%) and is used in the metallurgical laboratory at about 85°C. (W. C. Cochran, U.S. Patent 2,650,157; British Patent 659,747; R. C. Plumb, Priv. Comm., April 1, 1958).

A review of chemical and electrolytic polishing of non-ferrous metals is provided by P. Grivel, *Metal finish. J.* 1956, **2,** 173.

Other References

Authoritative information about electropolishing will be found in works by P. A. Jacquet, " Le Polissage Electrolytique des surfaces métalliques et ses applications " 1948 (Édition Métaux); S. Wernick, " Electrolytic Polishing and Bright Plating of Metals " 1948 (Redman); D. S. Kemsley and M. J. McC. Tegart, " Electrolytic Polishing of Metals in Research and Industry " 1948 (Australian Council of Scientific and Industrial Research); W. J. McC. Tegart, " Electrolytic and Chemical Polishing of Metals in Research and Industry " (Pergamon Press). Anodizing and chemical oxidation are discussed by W. Machu, " Nichtmetallische anorganische Überzüge " 1952 (Springer).

Several of the papers presented to the Internationales Kolloquium über die Passivität der Metalle present valuable ideas about anodic attack; they appear in *Zeitschrift für Elektrochemie* 1958, **62**, 619–827. Of these the papers of K. G. Weil, K. J. Vetter and U. F. Franck deal with the general mechanism of anodic passivation. B. Le Boucher describes a new technique for anodic polarization, whilst B. N. Kabanov puts forward a method based on the measurement of electro-impedance at high frequency. R. Ch. Burstein discusses the effect of small traces of oxygen on the behaviour of metal electrodes. Among the papers devoted to the behaviour of different materials may be mentioned those of Th. Heumann (chromium); P. Charlon and E. Darmois, and M. J. Pryor (aluminium); O. Rüdiger and W. R. Fischer (titanium); M. Pražák and C. Carius (stainless steel); K. Schwabe, R. Piontelli and colleagues, G. Okamoto and colleagues (nickel); W. Feitknecht (lead); and J. H. Bartlett (copper). Y. M. Kolotyrkin includes chromium, nickel and stainless steel in his researches. The structure of the films responsible for passivity is described in some detail by K. Huber. I. Épelboin presents an interesting theory of electropolishing.

A chapter on anodic processes by T. P. Hoar in " Modern Aspects of Electrochemistry ", Vol. II, edited by J. O'M Bockris should receive careful study (to be published in 1959).

Some of the papers in " Passivierende Films und Deckschichten " (edited by H. Fischer, K. Hauffe and W. Wiederholt, published by Springer) deal with matters mentioned in this chapter, notably those of K. J. Vetter, Th. Heumann, H. Spähn and J. Heyes. See also section on Anodic Processes by H. R. Thirsk, *Ann. Rep. chem. Soc.* 1957, **54,** 18.

Recent papers on the anodic behaviour of aluminium include P. J. Anderson and M. E. Hocking, *J. appl. Chem.* 1958, **8,** 352 (discussion *Chem. and Ind.* (*Lond.*) 1959, p. 355), and R. C. Plumb (with J. E. Lewis), *J. electrochem. Soc.* 1958, **105,** 496, 498, 502 (3 papers); D. J. Stirland and R. W. Bicknall, *ibid.* 1959, **106,** 481; film-formation on zirconium is discussed by G. B. Adams, T. S. Lee, S. M. Draganov and P. van Rysselberghe, *ibid.* 1958, **105,** 660. A discussion of the Flade potential—stressing the importance of excess oxygen—is provided by M. J. Pryor, *ibid.* 1959 (in the press).

Important. The Addendum (p. 1042) correlates this chapter with Chapter V, explaining why an applied anodic current is needed for passivity in some solutions and not in others. Some readers may find the discussion of oxide-film formation there provided simpler than the treatment given in either chapter.

CHAPTER VIII

BURIED AND IMMERSED METAL-WORK

Synopsis

The chapter opens with a discussion of damage to pipes and cable sheaths caused by electric currents straying from tramways; such currents can produce destruction much more rapidly than "natural" corrosion. Methods of avoiding trouble by suitable electrical connection or insulation are reviewed.

Attention is then turned to the natural corrosion of pipe-lines due to electrochemical action, including long-line currents produced when a pipe passes through different soils, and differential aeration currents produced by unequal distribution of air-pockets or similar causes. Micro-biological corrosion, especially that due to sulphate-reducing bacteria, receives special consideration.

The general subject of the protection of pipes by means of coatings is next discussed. It is pointed out that even quite elaborate coating systems often fail to give complete protection, and that the provision of cathodic protection is frequently advisable. This may depend on sacrificial anodes (magnesium alloy, zinc or aluminium alloy), or on power-impressed schemes involving an external source of E.M.F.; in the latter case, the anodes may consist of carbon, platinum or a lead alloy, and should then enjoy a long life, or of scrap iron, which requires periodical renewal. Attention is called to the difficult problem of "interference"—accidental damage to neighbouring buried structures which have not been included in the cathodic protection scheme.

The protection of pipes against internal attack, especially by means of cement linings, next receives consideration. The dangers arising from micro-biological action again demands attention; corrosion may lead to the thinning of the walls or the reduction of the channel available for water flow.

It is found convenient to include at the end of the present chapter two subjects which at first glance may appear unrelated, but which involve the principles developed in the earlier part. The first concerns ship hulls, where stray currents sometimes cause damage, and cathodic protection is being increasingly used; the second concerns the corrosion of steel embedded in concrete structures. Ships' paints are discussed in Chapter XIV.

Stray Electric Currents

Attack on Pipes by Stray Currents. The last chapter provided three examples of anodic action put to useful purposes—Anodizing, Electro-polishing and Pigment Manufacture. The opposite case—where anodic action is disastrous—is met with only too frequently. One of these—happily becoming less common, owing to the disappearance of tramways—is corrosion by stray currents.

The old-fashioned tram-car received its current from an overhead conductor and returned it to the rails; D.C. from transformer stations was supplied to the overhead conductor at certain " positive feed-points " and was removed from the rails at " negative feed-points ". Given good electrical connection between the individual lengths of rails, most of the return current did in fact follow the intended path; but if the connections were bad, the road ill-paved, the ground damp and particularly if salt had been thrown down to prevent freezing up of the " points ", a fraction of

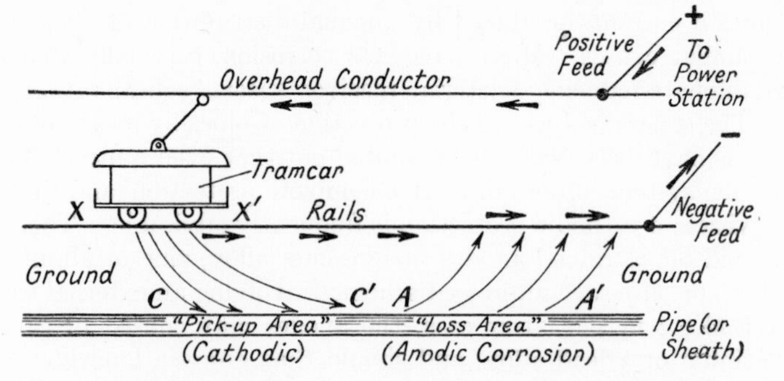

FIG. 62.—Corrosion of buried pipes by stray electric currents (the arrows show the direction of movement of " positive " electricity).

the current passed into the ground, joined a buried pipe or sheath (below the rails) and re-joined the rails closer to the negative feed-point; sometimes, if the pipes or sheaths reached by the stray current did not run parallel to the tram-line, that current might ultimately return to a different set of rails running along a different street.

In all such cases two cells in series are set up (fig. 62), namely:—

Cell I (on the left). Rails (anode) | Soil | Pipe (cathode)
Cell II (on the right). Pipe (anode) | Soil | Rail (cathode)

The first cell only causes corrosion to the rails, which are accessible and need periodical renewal for other reasons; here corrosion is not disastrous. The second cell may cause a sudden leakage of a pipe, or failure of a bunch of telephone lines—which requires the digging up of the roads, with consequent traffic-obstruction and heavy cost for labour and renewal. In the old days, break-downs used to occur at points decided by causes sometimes difficult to ascertain; in one case, it was noticed that, among a series of identical branch pipes entering a row of houses, the pipe entering one particular house failed at frequent intervals, whilst the others remained unharmed; this was ultimately traced to the habits of an ice-cream merchant who at the end of his day invariably disposed of his saline freezing-mixture in the gutter opposite the house in question.

Attack on Cable-Sheathing by Stray Currents. The lead alloy sheaths which surround telephone cables suffer in the same way. The cables are usually laid in ducts of stone-ware or timber, but these often become leaky. Here the attack on lead is not confined to the region of Cell II where the lead is anode; in the region of Cell I, where an iron pipe would be immune, a lead sheath may be attacked if the ground-water is salt; sodium hydroxide, produced by the cathodic reaction, attacks the lead giving soluble sodium plumbite, which in due course hydrolyses depositing lead monoxide in red tetragonal crystals, the stable form of PbO (the yellow or green forms are metastable); this red product constitutes a danger-signal that attack of the cathodic type is occurring. Salt, if used to prevent the freezing of tramway points, greatly increases the alkaline attack on cables nearby (R. M. Burns, *Trans. electrochem. Soc.* 1941, **79**, 307, esp. p. 309; A. G. Andrews, *Corrosion* 1948, **4**, 93; R. H. Pope, *ibid.* 1954, **10**, 324).

At the anodic sections, the attack is more severe, and tends to penetrate preferentially along grain-boundaries, so that the rate of penetration into the sheath greatly exceeds that which Faraday's Law would predict on the assumption of uniform corrosion. This intergranular type of attack, when found on cable sheaths, is generally regarded as evidence that stray currents have been at work. Likewise the presence of lead chloride in the corrosion-product, in amounts exceeding about 5%, is generally held to prove that stray currents have been responsible for the damage—unless indeed the soil is itself exceptionally saline; migration of chlorine ions towards anodic points causes them to accumulate there, even when their concentration in the body of the soil is quite low. In the absence of stray currents the accumulation of chlorine ions coming from distant parts of the soil will not usually be met with, nor will there be special intergranular attack; in " natural " corrosion the anodic and cathodic products, being formed at points very close together, will interact to form basic lead carbonate which will tend to obstruct any attack along grain-boundaries, whereas in stray-current attack the anodic and cathodic areas may be hundreds of yards apart, so that obstruction cannot occur. It should, however, be recognized that long-line currents set up on a line passing from one soil to another (as explained on p. 268) might cause an accumulation of lead chloride on the anodic portion (H. S. Campbell, *Chem. and Ind.* (*Lond.*) 1956, p. 1207. See also W. G. Radley, *Nature* 1945, **156,** 437; J. Gerrard and J. R. Walters, *Chem. and Ind.* (*Lond.*) 1956, p. 1060).

The presence of the puce-coloured peroxide (PbO_2) on corroded lead sheaths is often regarded as a proof of stray-current attack; where large amounts are observed, it constitutes valid evidence, since on grounds of energy, this strong oxidizing agent could not be formed except when a high E.M.F. is operating. However, most of the chemical tests used to detect peroxide start with the acidification of the sample of corrosion-product.

In such a case, a weak positive reaction is no proof of stray-current corrosion, at least where the lead has been in contact with mortar or other alkaline material. The action of lime and oxygen on lead can produce red lead (Pb_3O_4) even in absence of an external E.M.F., and on acidification this gives peroxide and a lead salt.

A.C. Corrosion. There has been some difference of opinion as to whether A.C. current can produce corrosion on lead and iron. It seems likely that A.C. can enhance natural corrosion, under conditions where it would in any case start but would otherwise become slow as a protective layer is built up. It may be that the A.C., by alternate reduction and oxidation of such layers, can keep them porous and allow the attack to proceed; but there has been little systematic study of the matter. Scarpa, however, has found that A.C. promotes corrosion of steel in concrete containing chlorides (p. 303).

Several observers have reported cases of corrosion to lead cable sheaths which seem to be attributable to A.C. currents. According to one view,

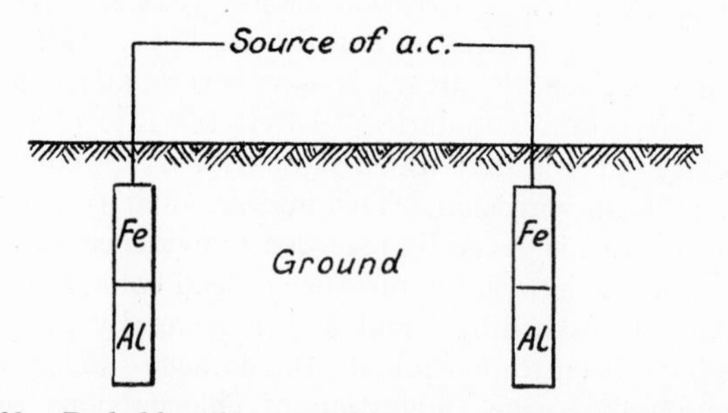

Fig. 63.—Probable effect of A.C. on an Iron–Aluminium couple.

the currents are first rectified, and then produce corrosion, but this is uncertain; any rectification of current which is observed may be the result of the corrosion rather than its cause. Nevertheless it does seem that the effect of A.C. is most serious on metals prone to act as rectifiers. Raine (see below) states that lead is not seriously affected by A.C. up to 12 volts, whereas aluminium suffers at small E.M.F.s unless the current density is very low.

This is not surprising. Consider two compound pieces each made of iron and aluminium in contact buried in the earth, with an A.C. applied to them (fig. 63). During the half-cycle when the left-hand piece is anodic, the iron will bear the brunt of the anodic attack, since aluminium does not readily function as an anode; during the next half-cycle the aluminium will play a more important role as cathode than it had previously played as anode, so that the cathodic reaction will not deposit so much iron as had been dissolved during the previous half-period. It is true that aluminium and iron ought not to be used in contact even if A.C. currents are absent, but probably a far greater rate of attack would become possible in presence of A.C. than in its absence. Moreover, the same state of affairs might ultimately be set up even if no piece of iron had been deliberately clamped on to the aluminium, since the dissolution and re-deposition of any iron present as an impurity in the aluminium would probably, in the long run, produce an aluminium-iron couple.

A detailed investigation of such matters is over-due, but some useful information is provided by P. A. Raine, *Chem. and Ind.* (*Lond.*) 1956, p. 1102; R. L. Davies, *ibid.* 1956, p. 1196; cf. H. F. Schwenkhagen, " Korrosion VIII ", p. 59 (Verlag Chemie).

Prevention of Stray Currents on Pipes and Sheaths. Although the disappearance of the tram-line is probably mainly responsible for the diminution of stray-current trouble during recent years, the steps which were being taken to combat it some decades ago are sufficiently instructive to deserve attention. To some extent, they are still in use, and depend on three different principles:—

(1) At the area where the pipe (or sheath) would otherwise be anodic to the rails, it is bonded electrically to the latter, thus short-circuiting Cell II (p. 264), and making the pipe, as a whole, cathodic; this is known as *drainage.* If the current drained from the pipe by simple bonding is still insufficient to make it cathodic at all points, a " booster " may be introduced to increase the drainage. If the drainage is excessive, then a resistance must be interposed at the bond. It is undesirable to use any more drainage than is really necessary, since if (say) a water-pipe which is being drained becomes strongly cathodic towards (say) a gas-pipe, which has not been bonded to the rails, the effect will be to set up anodic attack on the gas-pipe. It is usually advisable, when a drainage system is being planned, that a Committee should be formed representing all interests—water, gas, telephone, electric supply and traction—to arrange a scheme which provides maximum protection and minimum disturbance. An example of such co-operation in an American city is described by R. M. Lawall, *Corrosion* 1949, **5**, 79 (other examples are cited in the discussion of Lawall's paper on p. 84).

Much of the danger of over-drainage can be avoided by automatic devices. Pike describes an apparatus for mitigating trouble to cables by stray currents from D.C. railways. It automatically closes a drainage bond at times when stray currents are being picked up sufficiently to render the sheath anodic, opening it again when drainage is no longer required (V. B. Pike, *Corrosion* 1952, **8**, 311).

As a means of avoiding corrosion due to stray current, drainage seems to have given satisfaction in some parts of the world, but it is probably not well suited for the complex conditions existing underground in a large British city; the problems introduced have much in common with the interference problems which confront those who design cathodic protection schemes (p. 293).

(2) The system successfully adopted by the Telephone Department of the British Post Office for combating stray-current corrosion to cable-sheaths is the opposite of drainage; the conducting paths provided by the sheaths are interrupted by insulating gaps. Similar insulating gaps are possible

on pipes and power-cables. The gap-system *diminishes* the current flowing in the sheath (or pipe) whereas drainage usually *increases* that current. Clearly the gaps must be sufficiently numerous to make the current very small, since any current which does continue to flow will produce corrosion on the anodic side of each gap; in water-pipes this corrosion will occur largely on the interior, in cable-sheaths on the exterior. The design of gaps for cable-sheaths is described by W. G. Radley and C. E. Richards, *J. Inst. elect. Eng.* 1939, **85**, 685, esp. p. 699, and that for pipes by W. Beck, *Corrosion* 1949, **5**, 175; see also E. T. Pearson, *ibid.* 1955, **11**, 535t.

(3) Another system was used for protecting a gas-pipe system in a district of Great Britain where stray currents had been causing damage. The pipe lengths were electrically bonded, and coated as perfectly as possible. Then at the anodic section only (Cell II on p. 264), long earth bars, bonded to the pipes, were sunk into the soil, or in some places buried in special beds of wet coke-breeze; the anodic attack, which would otherwise have fallen on the pipes, was concentrated on the bars, which lost about 1 inch of length each year, so that they would have needed replenishment from time to time. The method seemed to be fulfilling its purpose, but the removal of the offending tramways prevented a convincing demonstration of its efficacy (E. E. Jeavons and H. T. Pinnock, *Gas. J.* 1930, **191**, 203, 255; supplemented by private communication from S. C. Waddington (General Manager, S. Staffs. Mond Gas Co.), Nov. 5, 1934).

Methods of measuring stray currents flowing along cable-sheaths are described by Radley and Richards (see above); at times of the day when few tram-cars are running, the sudden increase in the current flowing along the sheath, associated with the starting of an individual car after a halt, is easily detected.

Stray-current damage to ships and buildings is described on pp. 299, 303.

Other Forms of Electrochemical Corrosion to Pipes and Cable-Sheaths

Long-line Currents. Where a pipe-line passes from one soil (A) to another soil (B), a cell of the type

Steel | Soil A | Soil B | Steel

may cause serious attack on the steel in the section where it is anodic. The two soils may differ in pH value or in texture; oxygen-replenishment is better in an open soil than in a compact soil, so that the steel in the latter tends to become anodic. If one soil contains sulphides, organic acids of peaty origin or sewage contamination, considerable E.M.F. may be set up. These *long-line currents* may flow over considerable distances along the pipe. Cases are known where the current strength has reached 5 amperes and the length of flow has exceeded a mile. Ewing has described their prevention by means of insulating couplings. This plan is a sound one if there is a sharp passage from Soil A to Soil B, but if the geology is such that the pipe passes from one to the other several times, the situation is less simple.

If soils A and B really represent the same soil with different water-contents (and hence different air-contents), the position of the junction may vary with the weather. For a discussion of long-line currents see E. R. Shepard, *Industr. engng. Chem.* 1934, **26**, 729; L. W. Ewing, *Corrosion* 1954, **10**, 315.

Clearly, if a pipe is buried at the junction of two horizontal strata, so that its upper part is in contact with Soil A and the lower half with Soil B, the result may be disastrous, since the resistance will be far lower than in the case of true long-line currents.

Differential Aeration. Where a pipe is buried in a uniform soil or formation, an E.M.F. can still be set up, if, for instance, it approaches or reaches the surface at certain places; it will usually be cathodic at those places whilst the deeply buried parts will be anodic. Shepard (see below) quotes E.M.F.s of over 0·5 volt between areas exposed to the wetted and drier parts of the same soil. In other cases, the anodic attack may be localized to a large extent at salt pockets in the soil, or places where an underground water-course is cut. Even assuming the absence of such factors, corrosion-cells may be set up where there are air pockets in an otherwise moist soil; in general, the metal at the air pockets will be cathodic, the corrosion being concentrated at the points shielded from oxygen-replenishment. Such a corrosion-pattern due to oxygen-distribution may occur where the character and moisture-content of the soil is such as to favour air pockets; but even a soil which would be free from air pockets when consolidated may contain them, if the back-fill is applied in such a way as to entrap air between the spadefuls. The earlier American investigators laid stress on the fact that it is the physical, rather than the chemical character of soil which determines the distribution of attack. Some British authorities consider that certain cases, which in the early days of research into soil-corrosion the American investigators were inclined to attribute to differential aeration currents, may really have been due to sulphate-reducing bacteria. It is, however, possible that the microbiological type of corrosion was less important at the sites studied by the earlier American investigators than at some locations in the Old World. Today both factors are recognized as being serious causes of corrosion.

On some other metals, corrosion is particularly liable to occur at crannies inaccessible to air. On lead, the corrosion-rate is dependent on the size of the soil particles, as shown by Burns and Salley, who buried lead specimens for periods up to 5 months in moist inert sands—maintaining the temperature at 40°C. They found that, within certain limits, the corrosion-rate increased as the particle size increased; the attack was attributed to "oxygen concentration cells" set up as a result of partial or complete exclusion of oxygen from the points of contact of metal and soil. In the Author's opinion, the exclusion of carbon dioxide from these crannies may possibly have played an important role, since basic lead carbonate is more protective than lead oxide. However this may be, the paper deserves study (R. M. Burns and D. J. Salley, *Industr. engng. Chem.* 1930, **22**, 293).

Marsh and Schaschl state that on buried steel pipes differential aeration is important at the water-table; but in soil saturated with stagnant liquid, it has no effect even when the oxygen-concentration in the bulk of the liquid at different points is very different (G. A. Marsh and E. Schaschl, *Corrosion* 1957, **13**, 695t).

This matter requires consideration. Possibly the rate of diffusion of dissolved oxygen through the water between the soil particles (where the molecules may perhaps be oriented and form a formidable barrier to diffusion) may be too slow to cause trouble, and it is only those places which seasonally come into contact with gaseous oxygen that can act as cathodes to the areas lower down.

Mill-Scale and Corrosion. Undoubtedly, even in the absence of bacteria, numerous factors can influence the corrosion-pattern. If the pipe carries mill-scale, attack may in the early stages be localized at breaks in that scale, but Shepard states that this influence tends to die away with time, giving way to a new pattern determined by oxygen-distribution. Probably the influence of mill-scale in concentrating corrosion at discontinuities is more important on the inside of pipes (p. 204) than on the outside; it is undoubtedly important in water-tanks and on ships. Much depends on the nature of the scale; mill-scale on steel is often undermined, allowing the corrosion to spread before it has become too intense; in contrast, foundry scale on cast iron pipes is stated by Kuhn to have been the cause of pitting, the attack being concentrated at breaks in the scale, which covers perhaps 92% of the surface. It is perhaps unlikely that such scale will act as an efficient cathode, but in the anodic sections of a pipe-line suffering from long-line currents, the effect of localization of the anodic attack on the gaps may well be serious (R. J. Kuhn, *Industr. engng. Chem.* 1930, **22**, 335; E. R. Shepard, *ibid.* 1934, **26**, 723, esp. p. 727).

Corrosion around leakage points. The mechanism of the formation of tubercles covering pits on the interior of pipes was discussed on p. 119, but there is reason to think that a similar mechanism can explain the external corrosion outside cast iron water-pipes noticed where water is leaking out into the soil—possibly at places where the cast iron has suffered graphitic softening extending through the pipe wall. In such cases the membranes of the new external tubercles are of necessity at some distance from the pipe walls, so that the tubercles include soil particles and small stones; when the first tubercles burst new membranes are formed at a greater distance, and this process is repeated, until finally there is a large adherent mass of soil particles cemented together by iron oxides; this agglomeration of soil on to the pipe may occur even in soils which in their original state are very loose.*

Metal-ion concentration cells. Local concentration-differences can play a part in determining the corrosion-pattern of some metals. This is more important on lead than iron. The intensified attack sometimes met with on lead pipes buried in chalky soils may be connected with the removal of lead ions as basic carbonate at places where lumps of chalk press against

* I am indebted to W. R. Braithwaite and W. T. King for information and suggestions on these points.

the lead; these places would become anodic to the rest, owing to the concentration cell set up (p. 1017).

Prediction of corrosive properties of soils. Many attempts have been made to correlate the corrosive value of different soils with their pH value or conductivity; it would be extremely convenient to be able to predict, by measuring one of these quantities, whether a given soil is likely to cause serious attack on buried pipes. In exceptionally acid soils, corrosion can occur in absence of oxygen, since hydrogen can be liberated (it is not necessarily evolved in bubbles, but may sometimes merely diffuse away in solution). In such cases, conductivity is unimportant, since the cathodic points where hydrogen is formed will be close to the anodic points where corrosion occurs. However, a low pH value of a sample is not necessarily an indication of serious corrosive power, since, unless the soil possesses a strong buffer action, the pH will start to rise as soon as corrosion starts. Probably the total acidity, as obtained by titration, is a better indication of power to cause serious attack. This view was put forward by I. A. Denison and R. B. Hobbs, *Bur. Stand. J. Res., Wash.* 1934, **13**, 125.

Where the soil is insufficiently acid to liberate hydrogen, and where microbiological corrosion is not expected, soil conductivity becomes important, since the current may have to flow long distances from the cathodic parts (where there is a supply of oxygen) to the anodes. Thus in alkaline soils there is a certain correlation between soil conductivity and corrosive power, but it is far from perfect. Shepard states that American soils with resistances below 500 ohm/cm. usually cause severe corrosion; above 1000 ohm/cm. it is difficult to forecast corrosive power from resistance. Waters describes a practical means of measuring soil-resistance, and classifies non-acid soils as (*a*) very severely corrosive (0–900 ohm/cm.), (*b*) severely corrosive (900–2300), (*c*) moderately corrosive (2300–5000), (*d*) mildly corrosive (5000–10,000) and (*e*) very mildly corrosive (above 10,000) (E. R. Shepard, *Bur. Stand. J. Res., Wash.* 1931, **6**, 683; F. O. Waters, *Corrosion* 1952, **8**, 407).

Interest attaches to a form of probe designed in Australia by Lorking which can be pushed into any soil and provides measurements enabling the corrosive character of the soil to be assessed. It is a development of an American probe known as the Columbia rod, which consisted of a rod with a steel tip and copper stem, separated from one another by insulating material, but joined through flexible wire to the terminals of a current measurer. The deflection on the scale (calibrated on soils of known corrosive power) is taken as a measure of the corrosive power of the (unknown) soil into which the probe is inserted. Lorking has improved the Columbia rod by embedding a platinum disc in the insulating material which can be joined to one terminal of a potentiometer, the other one being joined to a calomel electrode inserted elsewhere in the soil. This gives the redox potential (p. 1033) of the soil and by application of the principles of the Pourbaix diagram (p. 900), an idea of the corrosive character of the soil is obtainable —provided that sulphate-reducing bacteria are absent (R. K. Lorking, Corrosion Symposium, Melbourne University, 1955–56, p. 106. For the use of redox potential in indicating aggressiveness of a soil due to activity of sulphate-reducing bacteria, see R. L. Starkey and K. M. Wight, " Anaerobic corrosion of iron in soils " 1945 (Amer. Gas. Ass. Monograph). Other information regarding the redox probe technique is provided by F. E. Costanzo and R. E. McVey, *Corrosion* 1958, **14**, 268t).

Prevention of Corrosion by means of a Surround. Where a soil is believed to be specially corrosive, the trench in which the pipe is to be laid may be filled with porous brick or other substance designed to isolate the pipe from the corrosive soil, and also to allow the rain to wash away any salt which may be present. Gravel has been advocated as a surround, but recent tests show that corrosion was not reduced by a gravel surround on a short length of pipe running through an aggressive clay. A 3-inch surround of Portland cement gave better results, but aluminous cement was not effective (" Chemistry Research ", 1952, p. 14 (H. M. Stationery Office)).

Some success appears to have been achieved from the use of chalk as a surround for pipes which have to pass through a clay formation liable to produce microbiological corrosion. A sand surround is useful, and may also serve to prevent mechanical damage to protective coatings in rocky districts, but it must be free from salt and clayey matter. Different samples of sand have given different performances; the Dutch Corrosion Committee suggest that those furnishing the best results may have contained chalk (E. F. Reid, *Instn. civ. Engrs.*, *Selected Papers* 1934, **154**, 44; *Stichting voor Materiaalonderzoek Comm. Med.* 1935, **10**, 28).

Speller describes the use of dense concrete for protecting pipes where they cross stretches of corrosive ground. The old way was to surround the pipe with a wooden box and pour in the mixture of Portland cement and sand, wet enough to flow round the pipes; a more economical and convenient method is to use a set of portable steel formers. A pipe-line protected by a surround of cement and sand where it runs through brackish marshes and soil carrying acid drainage has remained free from corrosion for 40 years (F. N. Speller, " Corrosion: causes and prevention " 1951, pp. 596–99 (McGraw-Hill)).

Influence of Character of Iron and Steel on Corrosion by Soil. Extensive comparative tests on different ferrous materials have been carried out at the U.S. Bureau of Standards by K. H. Logan, *Trans electrochem. Soc.* 1933, **64,** 118; *Bur. Stand. J. Res. Wash.* 1939, **22,** 109; 1942, **28,** 57, 379; I. A. Denison and M. Romanoff, *ibid.* 1952, **49,** 299, 315. A useful summary is provided by M. Romanoff, *Nat. Bur. Stand. Circ.* 579 (1957). Other tests for a British organization* have been carried out by J. C. Hudson and G. P. Acock (" Corrosion of Buried Metals ", *Iron St. Inst. Spec. Rep.* **45**, p. 1, 1952).

The most conspicuous results of both tests is that—whilst the rate of penetration varies greatly from one soil to another—there is little difference

* A research Sub-Committee of the Institution of Civil Engineers working in collaboration with the British Non-ferrous Metals Research Association and the Corrosion Committee of the Iron and Steel Institute (now a Committee of the British Iron and Steel Research Association).

between the behaviour of one ferrous material and another, unless expensive stainless steels are used. Small amounts of chromium, copper or nickel produce no marked effect on corrosion-rates. In the British tests, there was some difference between cast iron tubes manufactured by different casting methods, but this could be ascribed largely to their different thicknesses. Steel and cast iron seem to corrode at about the same rate; if the thickness were the same, the steel would probably outlive the iron; in practice, the iron tube is generally thicker and tends to outlive the steel.

Corrosion of Lead Cable-Sheaths not connected with Stray Currents. The methods described on pp. 267–268 have largely removed the menace of stray currents, even in localities where tramways or electric railways using D.C. still operate. However, other types of attack can clearly continue; the formation of the red corrosion products mentioned above can occur, even when stray currents are absent, in alkaline conditions caused by lime-mortar or concrete. Wolf and Bonilla state that the most serious cases occur at the points of contact with copper bonding straps under fire-proofing materials, where the bimetallic combination causes local anodic attack on the lead (E. F. Wolf and C. F. Bonilla, *J. electrochem. Soc.* 1941, **79**, 307).

A frequent type of failure has come to be known as " phenol corrosion ", since it was formerly attributed to phenol in the tar used for impregnating the hessian or other protective bandages which are usually wrapped around the lead; Davies and Coles, however, cite cases where this type of corrosion has occurred with tars almost free from phenol and has failed to occur with tars containing much phenol; they demonstrate experimentally that phenol at low concentrations in water diminishes rather than increases the corrosion of lead. Their results suggest that the so-called phenol corrosion may be largely due to acetic and similar acids produced by decomposition of the hessian wrappings, probably through bacterial action—which would indicate that by using a material proof against such action the trouble can be avoided. If in any case phenol or cresol has helped attack, its action has probably been indirect; the polar group in the molecule may have aided absorption of water into the wrapping (E. L. Coles and R. L. Davies, *Chem. and Ind.* (*Lond.*) 1956, p. 1030; with discussion by U. R. Evans, p. 1198, B. Lunn, p. 1199, H. L. Pedersen, p. 1199, D. C. Hewitt, p. 1200, D. W. Glover, p. 1201, T. W. Farrer, p. 1201 and F. Wormwell, p. 1202. See also E. de Famo, *Telegraphen u. Fernsprech Technik* 1932, p. 267; W. Hess and R. Dubuis, *Technische Mitteilungen P.T.T.* (*Switzerland*) 1956, p. 172; W. Hess, *Werkst. u. Korrosion* 1956, **7**, 649. General confirmation of the ideas of Davies and Coles is provided by the findings of J. C. Senez and F. Pichinoty, *Corros. et Anticorros.* 1957, **5**, 203. The corrosion of lead by organic acids is further described by E. L. Coles, J. G. Gibson and R. M. Hinde, *J. appl. Chem.* 1958, **8**, 341).

In view of the corrosion set up by hessian and similar wrappings, new forms of protective coatings are being increasingly used. These include polyethylene sheaths extruded over the reinforcement, with a special compound inserted below it to prevent longitudinal penetration of water. Polyvinyl chloride is also used, but the possible formation of hydrogen chloride requires to be borne in mind. High-voltage power cables of the pipe type are often protected with glass silk impregnated with a bituminous compound (R. Tellier, *Chem. and Ind.* (*Lond.*) 1956, p. 1202, esp. p. 1203).

In the U.S.A. cables are often laid in ducts of timber, which when new may give off acetic acid fumes, causing serious attack on the lead or alloy forming the sheath. This trouble can be largely avoided if ammonia vapour is passed through new wooden ducts.

When stone-ware ducts are used and moisture finds its way into them, bare lead sheaths will probably suffer crevice corrosion at the places where the lead rests on the stone-ware. This may be particularly intense if wet silt exists between lead and duct-material; in one case severe attack was attributed to sulphide in the silt; the advice given in such cases is to clean out the ducts (J. Gerrard and J. R. Walters, *Chem. and Ind.* (*Lond.*) 1956, p. 1060; " Chemistry Research " 1955, p. 11 (H.M. Stationary Office); see also *Corrosion* 1956, **12**, 257t.)

Lead-sheathed power cables laid in concrete tunnels suffer attack from the alkali liberated by the cement as it hardens. The concrete should be thoroughly cured before the lead is introduced; a procedure is described by R. I. Perry, *Corrosion* 1956, **12**, 207t.

Tests upon specimens of lead, copper and aluminium buried in the soil at five sites for five years are described by P. T. Gilbert and F. C. Porter, *Iron St. Inst. Spec. Report* **45**, p. 55 (1952).

Cinders in the ground can cause serious corrosion to buried lead sheaths. Experiments by Robson and Taylor showed that specimens buried wholly in clay or wholly in cinders underwent no change, but if two specimens, one buried in clay and the other in cinders, were joined, a current was found to be flowing, and the (anodic) specimen in clay underwent serious attack which closely corresponded to the current measured; the whole body of the cinders acted as cathode. Field tests showed analogous results; lead pipes buried wholly in clay or wholly in cinders escaped serious attack, but where two pipes laid respectively in clay and in cinders were connected, the pipe in the clay lost weight and became pitted. Coating with polyvinyl chloride or polyethylene gave protection (W. W. Robson and A. R. Taylor, *Chem. and Ind.* (*Lond.*) 1956, p. 1111).

The corrosion of aluminium cable-sheaths is curiously spasmodic; some lengths may suffer no damage whilst others develop pitting. Raine considers differential aeration at crevices to be the main cause. Protection is undoubtedly needed, but here again the results are variable. Jute hessian and paper have behaved well in some situations and given no protection in others. The pattern of minute (anodic) pits surrounded by dull (cathodic) areas with a pH below 5 in the pits is correlated with the winding; the deepest pits lie below the intersection of the woven hessian bands and at overlapping points (points of greatest pressure), despite the fact that the hessian is separated from the aluminium by paper. There is today a tendency to employ polyvinyl chloride (as tape or extruded sheath) in the place of cellulose-based reinforcements for bitumen (P. A. Raine, *Chem. and Ind.* (*Lond.*) 1956, p. 1102).

Microbiological Corrosion

Limits to Corrosion in absence of Bacteria. In absence of sulphate-reducing organisms, corrosion is usually serious only in soils which hold moisture but also contain air pockets. Sandy soils which normally keep dry will cause a little corrosion at first, but the ferric product formed in close contact tends to be protective. Clay soils, if water-logged so that they permit no ready replenishment of oxygen, will produce a little attack of the hydrogen-evolution at first, but so soon as hydrogen and ferrous ions have sufficiently accumulated for the potential of the half-cell $H | H^+$ to balance that of $Fe | Fe^{++}$, the attack will become slow. Such attack as continues largely depends on the slow arrival of oxygen (to destroy the hydrogen) or the slow removal of ferrous ions (by diffusion outwards).*

Sulphate-reducing bacteria. In clays where certain organisms flourish, the situation is entirely changed. Sulphate-reducing bacteria grow under anaerobic conditions, but require the presence of sulphate and sufficient pollution to provide food for their growth; where such conditions prevail, they can remove hydrogen and prevent it accumulating by the cathodic process so that the corrosion process can continue indefinitely, instead of being brought to an end by the "polarization" process described in the previous paragraph. Since the hydrogen reduces the SO_4^{--} to S^{--}, the final effect is that the oxygen present in the SO_4^{--} is made available for the cathodic process.

The existence of the bacteria was discovered by the Dutch investigator, M. W. Beijerinck, in 1895, although their importance in corrosion was only established by his compatriot, von Wolzogen Kuhr, in 1922. The most recent Dutch views are summarized by C. A. H. von Wolzogen Kuhr, in *Metaal-instituut T. N. O. Publikatie* **48** (1956), p. 6 (see references to British workers on pp. 276, 279).

* In such a case the oxygen may be said to "depolarize" the cell, removing the hydrogen which has "polarized" it. Some authorities believe that even in situations where oxygen is readily replenished, e.g. at the water-line of a vertical iron specimen half-immersed in sodium or potassium chloride, the first reaction is the production of hydrogen, which is then "depolarized" by means of oyxgen. However, under the alkaline conditions known to exist at this position, the cathodic deposition of hydrogen atoms

$$H^+ + e = H$$

would be a rare event, and in an exceptional case where it did occur the reverse reaction

$$H = H^+ + e$$

would follow without delay—quite irrespective of the action of oxygen. The potential—which has been measured—lies far away from the equilibrium

$$2H^+ + 2e \rightleftharpoons H_2$$

In such a situation the production of OH^- ions is not due to the action of oxygen on hydrogen already deposited, but to some such reaction as

$$O + H_2O + 2e = 2OH^-$$

In the Author's opinion, the use of the word "depolarization" is here inappropriate.

The organism has been given many names, but *Desulphovibrio desulphuricans* now seems to be favoured by microbiologists. Lately its study has been assisted by the electron microscope; its form is a slightly bent cylinder only about 2×10^{-3} mm. long with a flagellum about 100 Å thick (pictures are provided in "Chemistry Research" 1948, pp. 20, 21 (H.M. Stationery Office)).

The organisms, which can live far away from iron, require hydrogen or some reducing substance for their life process. Growth only occurs if there is some such reaction as

$$SO_4^{--} + 8H = S^{--} + 4H_2O$$

At an iron or steel pipe buried in clay, which has become polarized through the formation of hydrogen (probably atomic) at the cathodic places (and of Fe^{++} ions at the anodic ones), the bacteria can obtain their necessary hydrogen, so that the cathodic process is "depolarized"; the S^{--} ions formed combine with the accumulated Fe^{++}, giving black ferrous sulphide, so that the anodic process is depolarized also. Thus the attack, instead of being brought to rest, continues indefinitely (cf. the views of W. H. J. Vernon, *J. roy. Soc. Arts* 1949, **97**, 592; F. Wormwell and T. W. Farrer, *Chem. and Ind.* (*Lond.*) 1952, p. 108; 1954, p. 1444; J. N. Wanklyn and C. J. P. Spruit, *Nature* 1952, **169**, 928).

The corrosion-product of this type of attack, instead of being rust-coloured, is black, owing to the presence of ferrous sulphide; it may be pyrophoric—a danger in the oil industry. It should not be too readily assumed that wherever a black corrosion is found on iron or steel, microbiological action is responsible; magnetite is also black, and may be formed instead of rust when the oxygen-supply is deficient. However, if the corrosion-product is non-magnetic and produces the smell of hydrogen sulphide when acidified, the responsibility for the corrosion can generally be attributed to sulphate-reducing bacteria.

A somewhat different view of this type of corrosion has, however, been put forward. The bacteria can develop hydrogen sulphide quite independently of metallic iron; walkers venturing into marsh-lands near the sea-coast sometimes become unpleasantly aware of the fact. It has been supposed, therefore, that the bacteria merely serve to generate hydrogen sulphide which then attacks the iron. Such a simplified view has its attractions, but hardly explains the facts. Butlin and Vernon have pointed out that oxygen which would stimulate attack by hydrogen sulphide, must be excluded at least intermittently if the microbiological attack is to take place (K. R. Butlin and W. H. J. Vernon, *Proc. chem. engng. Group, Soc. chem. Ind.* 1949, **31**, 65; K. R. Butlin, *Research* 1953, **6**, 184; H. J. Bunker, *J. Soc. chem. Ind.* 1939, **58**, 93. Cf. F. E. Kulman, *Corrosion* 1953, **9**, 11).

Another somewhat unorthodox view is sometimes expressed today, which requires more respectful consideration. Whilst it is likely that thirty years ago cases of attack which were really due to sulphate-reducing bacteria received other explanations, it is possible that the pendulum has now swung too far in the microbiological direction. The opinion has lately been heard that in certain places where bacteria are found below the corrosion-product, they are not the *cause* of the attack, but rather the *result* of corrosion originating from non-biological factors; the anaerobic conditions set up below the rust have—according to this view—proved favourable to the development of the bacteria. In judging any particular outbreak of corrosion, such a possibility should always be borne in mind.

Damage due to Sulphate-reducing Bacteria. Steel and cast iron can both suffer through microbiological corrosion. Bunker has mentioned the puncturing of a steel main 1 cm. thick after 9 years. In the case of cast iron the attack is less obvious than on steel, since the graphite net-work holds the corrosion-product in place, so that the iron, although greatly softened, may appear unchanged until probed with a knife; the soft " graphitized " product often contains unchanged metallic iron as well as graphite and various iron compounds. The attack on cast iron generally proceeds more slowly than on steel, and cast iron pipes are usually thicker than steel pipes. Nevertheless perforation of cast iron pipes can occur after some years burial in clay containing active bacteria; bursting may occur before perforation is complete.

The damage is not confined to buried pipes. Sulphate-reducing bacteria can be a serious cause of corrosion in cooling systems and heat-exchangers, since some strains can withstand relatively high temperatures. Bacterial action is today suspected as the cause of certain leaks in oil wells previously ascribed to other factors (W. C. Koger, *Corrosion* 1956, **12**, 507t).

Bacterial attack occurs in marine mud, which can provide the necessary anaerobic conditions. Copenhagen states that at Cape Town piling 0·6 inch thick has been perforated in 12 years, with the usual black ferrous sulphide accompaniment (W. J. Copenhagen, *Trans. royal Soc. S. Africa* 1934, **22**, 103. *S. African Industrial Chemist, Feb.* 1954).

Patterson has described damage produced on a ship which had rested several hours a day on infected mud; the rivets were most affected. He recommends in such cases a paint capable of withstanding the abrasion arising from movement of the ship on the mud and also, of course, resistant to hydrogen sulphide (W. S. Patterson, *Trans. N.E. Cst. Instn. Engrs. Shipb.* 1951, **68**, 93).

Sulphate-reducing bacteria have caused corrosion to steel jetties in the Lower Thames (" Chemistry Research " 1955, p. 17 (H.M. Stationery Office)).

Prevention of Microbiological Corrosion. Sulphate-reducing bacteria cause most damage in nearly neutral soils; the most favourable pH value for their action is usually about 7·5 and conditions become adverse to their development above pH 9. Bacterial corrosion of the casing at oil wells has been effectually prevented by making the drilling mud alkaline —as described by K. Doig and A. Wachter, *Corrosion* 1951, **7**, 212.

This may provide the reason why microbiological attack can usually be prevented by cathodic protection, which raises the pH value around the cathode. Other explanations are possible, and there has been some disagreement as to whether cathodic protection is always effective; it has, however, been confidently recommended by A. Weiler, *Chem. and Ind. (Lond.)* 1954, p. 716; see also W. C. R. Whalley, *ibid.* 1954, p. 687.

Much work has been carried out at Teddington in searching for an inhibitor to stop this type of attack. Some soils which might be expected to cause corrosion due to sulphate-reducing bacteria do not in fact do so, and metal has remained buried in them for many centuries without serious corrosion. Examination by the Teddington workers has revealed that certain soils can inhibit the activities of the bacteria. In an interesting case at York it was concluded that the inhibitive substance was probably a tannin derived from an early leather factory on the site (" Chemistry Research " 1952, p. 13; 1954, p. 15; 1955, pp. 16, 17 (H.M. Stationery Office); T. W. Farrer, L. Biek and F. Wormwell, *J. appl. Chem.* 1953, **3**, 80).

Recent work at Teddington, still unpublished, indicates that when tannins inhibit sulphate-reducing bacteria they do so merely by virtue of their acidity—depressing the pH to a level unfavourable to the bacteria. Further work is in progress, and the results of practical tests must be awaited before the efficacy of tannins in protecting buried structures can be assessed (F. Wormwell, Priv. Comm., July 22, 1958).

Other processes involving bacteria. The reduction of sulphate to sulphide is only one of a number of reductions capable of being carried out by bacteria with the aid of elementary hydrogen; besides SO_4^{--}, there can be reduction of SO_3^{--}, $S_2O_3^{--}$, $S_4O_6^{--}$, $S_2O_5^{--}$, $S_2O_4^{--}$ and colloidal sulphur (" Chemistry Research " 1949, pp. 24–33 (H.M. Stationery Office)).

Bacterial reductions of CO_3^{--} and NO_3^{-} are also known, carbonates being reduced to methane, whilst nitrates may yield ammonia or nitrogen. Von Wolzogen Kuhr (p. 275) writes the cathodic reductions in the three cases thus:—

$$H_2SO_4 + 8H = H_2S + 4H_2O$$
$$CO_2 + 8H = CH_4 + 2H_2O$$
$$HNO_3 + 8H = NH_3 + 3H_2O$$

The complete corrosion changes (including the anodic reaction) can be expressed:—

$$4Fe + H_2SO_4 + 2H_2O = 3Fe(OH)_2 + FeS$$
$$4Fe + CO_2 + 6H_2O = 4Fe(OH)_2 + CH_4$$
$$4Fe + HNO_3 + 5H_2O = 4Fe(OH)_2 + NH_3$$

These are based on the theoretical equations of von Wolzogen Kuhr above. In practice, examples of corrosion attributed to the agency of nitrate-reducing bacteria have been reported by D. H. Caldwell and J. B. Ackerman (*J. Amer. Wat. Wks. Ass.* 1946, **38**, 61), but no cases of corrosion caused solely by the activity of the methane bacteria appear to be on record.

Various strains of sulphate-reducing bacteria differ from one another in their biochemical make-up and hence in their biochemical behaviour, but in general ferrous salts promote growth, whilst reducing agents, such as sulphide, cysteine and thioglycollates also often have stimulatory action. The function of the pigments present in the organisms is interesting but lies outside the scope of this book. For details see J. P. Grossman and J. R.

Postgate, *Nature* 1953, **171,** 600; 1959, **183,** 481; *J. gen. Microbiol.* 1951, **5,** 714, 725; 1952, **6,** 128; 1953, **9,** 440; 1955, **12,** 429; 1956, **14,** 545; 1956, **15,** 186; 1957, **17,** 387 (with D. Littlewood); 1959, **20,** 252 (with M. E. Adams); M. E. Adams and T. W. Farrer, *J. appl. Chem.* 1953, **3,** 117.

An indirect danger from hydrogen sulphide produced by bacterial action is that at a water-surface it can be oxidized to sulphur. Farrer and Wormwell have shown that elementary sulphur dispersed in a 5% suspension of bentonite clay can cause uniform etching on steel at a rate faster than attack by 3% hydrochloric acid; loose ferrous sulphide and magnetite is formed; probably the sulphur stimulates the cathodic reaction in the same way as oxygen, but since it is on the spot in solid form, there is no limitation of the attack by the replenishment-rate (T. W. Farrer and F. Wormwell, *Chem. and Ind.* (*Lond.*) 1953, p. 106).

Bacteria also exist which are capable of converting sulphur to sulphuric acid, so that pH values as low as 1·0 have been met with; such organisms have aided the corrosion of cast iron and also the destruction of cement (T. H. Rogers, *Métaux et Corros.* 1948, **23**, 177; C. B. Taylor and C. H. Hutchinson, *J. Soc. chem. Ind.* 1947, **66**, 54).

Protection of Pipes

Tests on Coatings for Pipes. The tests mentioned on p. 272 supply valuable information regarding the efficacy of different protective coatings; elaborate tests on the subject have been carried out in other countries—notably in Holland. A little caution is, however, needed in using the results to calculate the probable lives of protected pipe-lines. All the tests mentioned have been carried out with great care, but short lengths of pipes often last better than long lengths, possibly because the short specimens are not affected by long-line currents. A working pipe-line does not corrode uniformly over the whole length; some sections act as anodes, and the attack upon them is stimulated by the cathodic reaction on other sections, which remain almost untouched. This factor cannot easily be introduced into tests carried out on short lengths. Caution in using the results of such tests has been recommended by W. F. Higgins, *Iron and St. Inst. Spec. Rep.* **45,** p. 111 (1952); see also R. W. Bailey, *Chem. and Ind.* (*Lond.*) 1956, p. 1206; J. C. Hudson, p. 1207.

Nevertheless tests such as these mentioned have provided most instructive information. In Logan's American tests, steel pipes coated with lead seemed to be giving promise during the first years, but specimens buried for 16 years showed most unsatisfactory results; this was presumably because lead is cathodic to iron in most soils, so that, when once the coating had been eaten through at a point, attack on the exposed steel became accelerated. Coatings of zinc, which is normally anodic to iron, were somewhat more hopeful, except in acid soils, for which they are clearly unsuited, as the zinc itself is rapidly attacked. Except in one soil, a coating of 2·8 oz./ft.^{2} prevented measurable rusting for 10 years, the nature of the ferrous basis being without importance. In the British tests, Hudson and Acock found that a zinc coating of 2 oz./ft.^{2} protected steel for 5 years, except when the specimens were buried in railway ashes.

Coatings based on Tar and Bitumen. Most of the coatings applied today contain organic materials, and their constituents first demand discussion. Often they are based on black products derived from coal tar (which is rich in ring-hydrocarbons) or from natural sources, such as oil-fields and asphalt-fields. Coal tar, on distillation, provides a number of fractions boiling at different temperature ranges, and leaves a black residue (pitch) in the retort. A different type of pitch is also left in the retort during the distillation of petroleum products; the end-products of petroleum-distillation are generally classed with the constituents of natural asphalt as bitumen. It is possible by dissolving or softening a pitch with a selected portion of the distillate to obtain mixtures suitable for protective coatings on pipes. These may contain a volatile solvent, so that a thin layer applied by brushing or spraying dries in air at ordinary temperature, leaving a coating which (if undamaged) will serve to protect against mild conditions, or will provide a basis to which the heavier type of coating (obtained by hot application, as explained below) will adhere better than to the untreated pipe. Pearson states that the adhesion and continuity of so-called "bituminous paints" (really solutions of coal-tar pitch in solvent naphtha) can be greatly improved by adding a few per cent of tung oil varnish (A. R. Pearson, *Chem. and Ind. (Lond.)* 1953, p. 639; tentative specifications for coal-tar coatings are given in *Corrosion* 1956, **12**, Jan. p. 75).

Alternatively, the mixture can be adjusted to be solid at ordinary temperatures, but liquid when heated, so that a pipe supported by a chain passing through it can be lowered into the molten mixture and kept there until it has attained the temperature of the bath; it is then withdrawn, whereupon the coating quickly solidifies. It is thus possible to coat exterior and interior simultaneously. Various mixtures have been used for hot-dipping, and an old favourite consisting of coal tar containing 6 to 8% of raw linseed oil, applied by dipping at 150°C., is known as Angus Smith solution—although that name is used somewhat loosely.* It is generally believed that *tar distillates* from *vertical* retorts are inferior to those from *horizontal* retorts, although C. O. Thomas, who made a careful comparison of materials, considered the *pitch* left behind in both types of retort to be about equally good (C. O. Thomas, *J. Instn. Engrs. Australia*, 1934, **6**, 337; *Commonwealth Engineer* 1937, **24**, 293).

Neither the coatings which dry by evaporation of a solvent nor those obtained by hot-dipping are sufficiently thick or tough to withstand severe conditions; it is doubtful whether they suffice even for mildly corrosive conditions, owing to the risk of mechanical damage during transport or burial. Two methods are available for producing coatings which are thicker and more robust.

Mixtures containing Fillers. The first method depends on the

* Apparently the original Angus Smith solution contained coal tar, tallow, linseed oil, slaked lime, resin and naphtha.

addition to the coal tar or bitumen of about 30% by weight (perhaps 20 to 30% by volume) of a suitable "filler", which may be powdered talc, mica, pumice, asbestos, kieselguhr or slate; C. O. Thomas (see above) recommended limestone.

There has been difference of opinion regarding the relative merits for this purpose of coal tar and natural bitumen (asphalt). Thomas found that mixtures based on asphalt undergo a gradual disintegration and that residual oil-bitumen fails to wet the metal. In the early years of the Coolgardie pipe-system (Western Australia) where the soil was saline and very corrosive, a mixture of coal tar and bitumen perished more quickly than a mixture of Trinidad bitumen and "maltha", a tar-like distillation-product of petroleum; but both coatings seem to have perished rapidly in the salt soil (P. V. O'Brien and J. Parr, *Proc. Instn. civ. Engrs.* 1917–18, **205**, Part 1, p. 310, esp. p. 327).

Bitumen can be greatly improved by blowing air (or, for some purposes, steam) through it, and blown bitumen is now in wide favour. The blowing, which causes partial oxidation, raises the softening temperature without loss of hardness—a point emphasized by Spencer and Footner, who state, however, that coal-tar products, although more brittle than bitumen, show greater resistance to oil and adhere better (K. A. Spencer and H. B. Footner, *Chem. and Ind.* (*Lond.*) 1953, p. 448).

Mixtures containing fillers are variously described as coal tar enamels or bituminous mastics. Application in the field is described by S. D. Day, *Corrosion* 1949, **5**, 221.

Rather similar mixtures can be used for producing thick internal coatings within pipes, and, by centrifugal application, a very smooth surface can be obtained, which will reduce resistance to water-flow; the main objection to thick internal linings of this type is that adhesion is not always good, and, at least in large pipes, vibration may cause detachment and sagging of the lining; endeavours have been made to overcome this by suitable grooving of the interior surface.

An American Committee has laid down rules for "mastic systems" for the external protection of buried pipes. A priming coat of specified composition is followed by a mixture of asphalt, mineral aggregate and mineral filler (which may include asbestos fibre) and the finished coating should be painted with whitewash. The aggregate should be sand of specified gradation—either a loose product resulting from the natural disintegration of rock, or artificially crushed material; the filler may be limestone dust or other mineral dust "approved by the engineer" (N.A.C.E. Committee T-2H, *Corrosion* 1957, **13**, 347t).

Wrapping of Pipes. The second method is to wrap the pipe with a continuous bandage which passes through a trough containing the hot black mixture just before it is wound round the pipe; the wrapping can be carried out mechanically. At one time hessian was largely used as bandage, but like all such fabrics it is liable to rotting in presence of certain cellulose-decomposing bacteria; the acids formed in this way (acetic, propionic and the like) are definitely corrosive, and other products arising from the action of the cellulose-decomposing bacteria provide nutrient for sulphate-reducing bacteria, so that two different types of corrosion can be set up. The possibility of impregnating the hessian with a substance unfavourable to the development of the cellulose-decomposing bacteria has been considered, but some of these anti-rot materials are copper compounds and may introduce fresh corrosion problems.

It is thus better to wrap the pipes with an inorganic material, suitably impregnated with bituminous or similar mixture. Asbestos products are used, but their mechanical properties are not ideal. Woven glass cloth is admirable for hand application but expensive. For long pipe-lines Spencer and Footner recommend fibre-glass felt consisting of glass fibres bonded together with a synthetic resin and applied mechanically. They attach importance to the strength and hardness of the wrap—especially the outer layer, which must resist damage resulting from stones and the like pressed against the surface during burial; the weight of the soil above the pipe and seasonal volume-changes may also cause damage. Doubtless this emphasis on hardness represents sound advice—at least for warm climates—but at low temperatures hard coatings tend to crack under vibrational stresses, and in tests at − 17°C. described by Beck, the best results were obtained by a bandage impregnated with grease, but covered with an asphalt layer. Arguments in favour of non-rigid, greasy materials have also been put forward by J. S. G. Thomas, but Spencer and Footner reply that " hot applied enamel coatings find the greatest popularity among large pipeline operators today, as is shown by the mileage of pipes coated with this type of material ". The various views are expressed by W. Beck, *Corrosion* 1949, **5,** 405; J. S. G. Thomas, *Chem. and Ind.* (*Lond.*) 1953, pp. 568, 848; K. A. Spencer and H. B. Footner, p. 568; A. J. Banks, *ibid.* 1954, p. 717; P. Clark, p. 718.

The wrapping of pipes with fibre-glass has become a common method today, to avoid the rotting characteristic of cellulose materials. Small pipes are wrapped at the mill and large ones on the site. Details are provided by A. M. Robertson, *Chem. and Ind.* (*Lond.*) 1950, p. 643.

The authorities at Teddington consider that, to resist bacterial attack, pipes should receive a coating at least $\frac{3}{8}$ inch thick, consisting of coal tar or bitumen reinforced with a neutral material such as glass fibre or asbestos. Hessian should not be used, for reasons already indicated (F. Wormwell and G. Butler, *J. Inst. Heating and Ventilating Engineers*, 1956, **23,** 461; K. R. Butlin and W. H. J. Vernon, *J. Instn. Wat. Engrs.* 1949, **3,** 627).

The use of plastic bandages is becoming increasingly common. The subject is reviewed by R. Rohm, " Korrosion VIII ", 1954–55, p. 17 (Verlag Chemie).

Combination of Methods. In practice these different modes of application are combined, and schemes of graduated complexity are available for withstanding conditions of varying severity. As an example may be quoted the following Dutch specification. For the severest conditions, five layers are prescribed:

(1) A ground coat of blown asphaltic bitumen, not exceeding 0·5 mm.

(2) A coat of blown asphaltic bitumen containing filler (the combined thickness of (1) and (2) must exceed 6 mm.).
(3) A sandwich of woollen cloth impregnated with blown asphaltic bitumen 2 mm. thick between two layers of asbestos cloth, felt, glass-cloth or rot-proof jute.
(4) A shield of blown asphaltic bitumen containing filler.
(5) External cording with twisted straw fibre, which serves to prevent damage during transport and is removed when the pipes are laid in the soil.

For some conditions, (5) can be omitted; moderate conditions demand only (1) and (2), whilst for mild conditions (1) will suffice, but its thickness should then reach 1·5 mm. (*Centraal Instituut voor Materiaalonderzoek Mededeling* No. **13** (1949)).

Many other schemes have been published, and probably adherence to almost any specification coming from a responsible source would give good results. However, even today it is still possible to see pipes laid out for burial carrying, apparently, a single coating broken up by rust patches, so that—if there are stray or long-line currents—the attack on the anodic section, concentrated upon the breaks in the coat, may be more intense than if no coating had been applied at all.

The need for robustness in coatings can hardly be over-estimated. The danger of damage is not confined to the periods of transport and burial, but remains serious in all soils where seasonal volume-changes produce stress. The damage due to soil stress in certain clays which periodically crack on drying is considerable, and may even produce a net-work of cracks on the *interior* lining of the pipes—as shown in a picture published by P. W. Lewis, *Corrosion* 1957, **13,** 489t.

Cathodic Protection

Need for Cathodic Protection. It might reasonably be expected that with the elaborate coating systems available today, the spectre of soil corrosion would have vanished. However, whether owing to faults of manufacture, damage incurred during transport or laying, or deterioration after burial, corrosion is still a common trouble. It has been stated that 27% of the burstings of pipes in 1947 must be attributed to corrosion—which could probably have been avoided in most cases, had there been less callousness among those responsible. The annual cost of replacing corroded pipes in Great Britain has been estimated as £5,000,000 (W. H. J. Vernon, *Chem. and Ind.* (*Lond.*) 1952, p. 404).

It is generally considered that the most economical way of prolonging the lives of pipes today is by combining a suitable coating with the application of a cathodic current provided from an external anode. The prevention of corrosion on an entirely uncovered pipe by cathodic protection is not impossible, but would involve a large expenditure of electric power, and the plan is only adopted in special circumstances. Likewise, it would be possible to provide coatings so perfect as to dispense with cathodic protection, but this would involve a high expense for materials, manufacture and inspection. It seems agreed that in most situations a combination of the two protective methods will minimize the total cost, but doubts are often felt in deciding whether it is best to use (1) a rather poor covering combined with high electric power costs, or (2) a more expensive covering with lower power costs. As to which of these policies will prove the cheaper may vary from one situation to another. Those seeking to reach such a decision and to convince themselves that cathodic protection can really save money, should study the papers of D. H. Lewis and O. C. Mudd, *Chem. and Ind.* (*Lond.*) 1954, p. 93; R. M. Wainwright, *Corrosion* 1953, **9,** 51; 1955, **11,** 146a; L. G. Sharpe, *ibid.* 1955, **11,** 227t; see also *Chem. and Ind.* (*Lond.*) 1955, p. 622. The economics of cathodic protection are ably reviewed by W. H. J. Vernon, "Conservation of Natural Resources" (1957), pp. 116–119 (Instn. civ. Engrs.), whilst A. Keynes, *Chem. and Ind.* (*Lond.*) 1958, p. 398, discusses from the economic angle the optimum combination of covering and cathodic protection.

Principles involved in Cathodic Protection. Since an anodic current stimulates corrosion, it might be expected that a cathodic current would prevent it, but electro-chemical theory suggests that corrosion will only cease when a certain concentration of metal ions has accumulated at the surface. If the ferrous ion concentration has reached, say, N/100, and we then depress the potential artificially to − 0·5 volt (hydrogen scale),* no further iron can enter solution (if we move the potential to more negative values, we may even hope to redeposit some of the iron as metal). Clearly if there is N/100 concentration of iron ions next the pipe and none in the body of the soil, there will be a slow diffusion outwards, and if the potential is controlled exactly at − 0·5 volt, we shall continue corrosion at just such a rate as will replenish the iron ions lost by diffusion. This may not be negligible, but it will be much smaller than the free corrosion-rate in absence of applied cathodic polarization. Now − 0·5 volt on the hydrogen scale is not far off the value which practical men find must be applied to a pipeline to keep corrosion reasonably slow, and some authorities explain this on the mechanism suggested.

It seems likely, however, that the mechanism in many practical cases is a different one. Calcium bicarbonate is present in many soil-waters and indeed in most natural waters. If the steel pipe as a whole becomes cathodic, then calcium carbonate will be deposited upon it, and any ferrous ions which would otherwise pass into the liquid will be precipitated as ferrous carbonate; if the pipe is mainly coated with a protective covering, the function of the current may be to block any gaps in the coating with a chalky plug.

Most cathodic-protection specialists seem unwilling to express an opinion as to which of these two mechanisms is the one generally operative in practice.† The question is not merely of academic interest, since if the

* The normal electrode potential of iron is about − 0·440 volt, so that the equilibrium potential of Fe | N/100 Fe^{++} is about $-0{\cdot}440-(2\times0{\cdot}030)=-0{\cdot}500$ volt, assuming that the activity can be regarded as equal to the concentration.

† The graphical explanation generally put forward, wherein corrosion is

second mechanism is operative then two practically important conclusions emerge: (1) cathodic protection, as we know it, requires the presence of calcium (or magnesium) in the soil water; (2) temporary stoppages of the current do not matter, since the chalky plugs deposited at breaks in the coating whilst the current is flowing will remain and prevent iron from entering the liquid until current is started again. Practical decisions would be helped if it were decided whether the object is (*a*) to maintain a potential at which the passage of iron is slow, being equal to that lost by diffusion, or (*b*) to maintain a chalky filter which will arrest the iron as it emerges. Carefully planned research to distinguish between the two mechanisms is over-due.

Electrical Conditions needed for Prevention of Attack. Experience has shown that, if corrosion is to be stopped, the potential must normally be kept at about − 0·85 volt relative to the half-cell:

$$\mathrm{Cu} \mid \text{saturated } \mathrm{CuSO_4.5H_2O}$$

which is about − 0·85 + 0·32 or − 0·53 volt on the hydrogen scale. In presence of sulphate-reducing bacteria, it must be kept at a still lower level; Wormwell and Farrer advise − 0·95 volt. The object may here be to maintain a high pH unfavourable to the bacteria, or perhaps to block the surface with a film of chalk and thus prevent the bacteria coming into

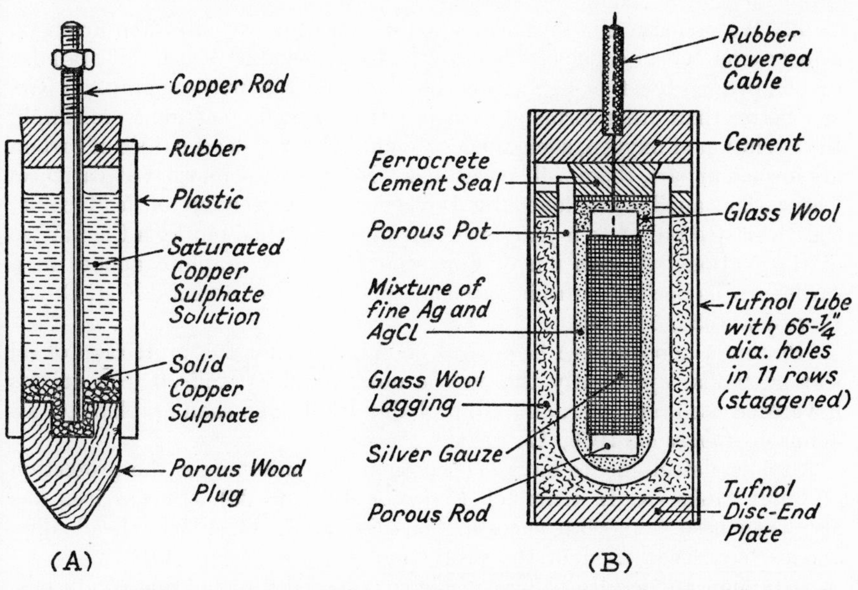

FIG. 64.—(A) Essential features of copper/copper sulphate electrode as used in cathodic protection. (B) Silver/silver chloride electrode (J. T. Crennell).

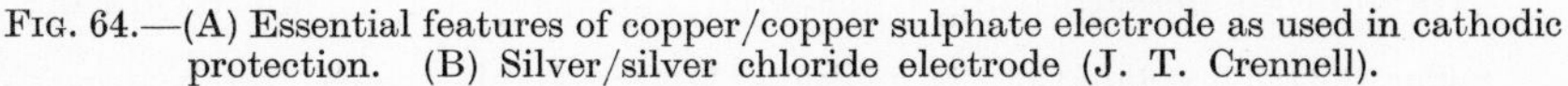

supposed to cease when the potential is brought down to the open-circuit potential of the cathode, really assumes the first mechanism, since the existence of the open-circuit potential at (say) − 0·5 volt is only possible if ferrous ions have accumulated to a concentration of (say) N/100. See p. 894.

contact with the hydrogen-charged metal* (T. W. Farrer, *Chem. and Ind.* (*Lond.*) 1954, p. 683).

The essential features of a "saturated copper sulphate" electrode, suitable for determining the potential drop on the surface of a buried pipe-line are shown in fig. 64(A) whilst fig. 64(B) shows a form of the electrode:

Ag | Sea-water saturated with AgCl

which is much used in the cathodic protection of ships or marine structures. By a coincidence, the potential of these two electrodes is nearly the same, the silver electrode being 0·05 volt less noble than the copper one.† Further particulars regarding the copper sulphate electrode are provided by G. N. Scott, *Corrosion* 1958, **14,** 136t.

Methods of Applying Cathodic Protection. In order to make a pipe-line cathodic at a potential of − 0·85 volt (or − 0·95 volt), we need an anode. There are three possibilities:—

(1) We can use a *sacrificial* anode of magnesium alloy, aluminium alloy or zinc and connect it by a wire to the pipe-line. No external source of E.M.F. is then needed, but the magnesium alloy (or other material) is "sacrificed" in generating the current, so that the anodes need periodical replacement—a source of inconvenience and expense; moreover, power raised by consumption of magnesium or zinc is more expensive than power supplied from a generating station.

(2) We can use non-sacrificial anodes (generally carbon) often arranged as a ground-bed of considerable size—perhaps including a mass of granular carbon pressed together between carbon bars; for this purpose graphite bars last better than other forms of carbon. If the anodic reaction were simply the evolution of oxygen, such anodes should last for ever; if oxides of carbon are formed in part, there will be wastage. Results are somewhat at variance. The carbon anodes employed by Doyle for the protection of lead cable-sheaths lasted only 4 to 5 years, whereas in tests carried out by a Committee of the National Association of Corrosion Engineers graphite anodes in a carbonaceous back-fill were in excellent condition after 4 years (E. J. Doyle, *Corrosion* 1955, **11,** 57t).

Platinum anodes have been used on ships and for the protection of oil storage tanks; apparently the cost is not excessive. Silicon iron anodes have been used on a pier, and are probably less fragile than graphite (C. R. Johnson, *Corrosion* 1956, **12,** 157t).

Stainless steel anodes have been pronounced suitable for the protection of water tanks (A. L. Kummel, *Corrosion* 1950, **6, 35).** Silicon iron has also been used here with success (*Corrosion* 1957, **13,** 103t). Lead alloy anodes have been used in the protection of ships (see p. 301).

Titanium anodes are being considered, especially for marine conditions.

* The latter explanation is perhaps less probable, since the hydrogen-production is increased by the cathodic polarization.

† The *normal* electrode potential of silver is, of course, much more noble than that of copper, but since copper sulphate is freely soluble and silver chloride very sparingly soluble, the concentration correction (p. 1017) brings the potentials of the two electrodes close together.

Used alone they suffer rapid anodic corrosion in sea-water; but if a relatively small piece of platinum wire is welded to the titanium, or if, alternatively, a thin, porous film of platinum (even as little as 0·000005 inch thick) is deposited on it, the composite anode has been found by Cotton to resist well; the platinum serves to maintain the potential of the combination at a level appropriate to the maintenance of passivity. The artifice is analogous to that used by Tomashow, who prevented the attack on stainless steel by acids normally able to attack it, by means of a thin layer of platinum, the principle being essentially that employed in Edeleanu's potentiostat researches (p. 336). Titanium anodes carrying electrodeposited platinum coatings about 0·0001 inch thick are on trial under service conditions (J. B. Cotton, *Platinum Metals Review* 1958, **2**, 45; cf. N. D. Tomashow, *Z. Elektrochem.* 1958, **62**, 717).

Where non-sacrificial anodes are used, the E.M.F. must be supplied from an external source. The electrical circuit is rather more complicated, but the cost of current less, than when sacrificial anodes are used. If there is A.C. power on the spot, a rectifier will furnish the D.C. current. If not, an oil or petrol engine must generally be installed to drive a dynamo; windmills have been used for this purpose in the U.S.A. and also in Russia. Non-sacrificial anodes, if buried in the ground, should be vented for the escape of the gas liberated by the anodic reaction (W. R. Schneider, *J. Amer, Water Wks. Assoc.* 1952, **44**, 413).

(3) We can use a scrap-iron anode (often an old rail or hawser). This will not last for ever but the cost of renewal—although not negligible—will be far less than with a magnesium or zinc system. Scrap steel anodes, whether hawsers or old rails, are now often surrounded by powdered coke breeze.

The use of scrap-iron anodes does not dispense with the need for an external E.M.F. and the complexity of the system is not reduced. However, the power costs are lower than with a carbon anode, since the anodic reaction, instead of being the evolution of oxygen, which needs a high positive potential, will now be anodic attack on the iron, which proceeds at a negative potential.

Systems requiring an external source of E.M.F. are known as *power-impressed* schemes.

Choice of Sacrificial Anode Material. The composition of a sacrificial anode requires careful consideration. The so-called magnesium anode is a magnesium alloy containing 6% aluminium, 3% zinc and 0·2% manganese; it is sometimes buried in a special back-fill of gypsum and bentonite. Variations in the aluminium and zinc contents are considered relatively unimportant, but copper, nickel, iron, lead and tin are detrimental impurities; the manganese added counteracts the bad effect of iron—perhaps also that of copper.

The shape of the magnesium anodes is important; it must be chosen to give maximum output of current and longest life; the copper connection to the wire which joins the anode to the pipe-line must be rightly placed —otherwise copper may enter the magnesium, with disastrous effect (O.

Osbourne (with H. A. Robinson), *Corrosion* 1951, **7,** 2; 1952, **8,** 114; see also P. M. Aiken and G. L. Christie, *ibid.* 1951, **7,** 406).

The usefulness of bentonite (p. 970) in preparing ground-beds for magnesium anodes (and still more for zinc anodes) depends on the form of its particles. These consist of flat sub-microscopic ribbons, which link up with one another at their corners to form a structure highly porous but nevertheless possessing a modicum of rigidity. Thus the gypsum (or other addition) is held in position, whilst good conductivity of the mass is attained. Details of the *gelation structure* of bentonite are provided by M. B. M'Ewen and M. I. Pratt, *Trans. Faraday Soc.* 1957, **53,** 535.

Unalloyed aluminium would be useless as an anode, owing to film-formation, but an alloy containing 5% zinc is used to some extent. Russian investigators report favourable results with aluminium-zinc-calcium alloys (p. 300).

Early attempts to use zinc as an anode gave disappointing results. It is now recognized that the zinc must either be extremely pure, or contain additions designed to counteract the effect of iron, the most objectionable impurity. Zinc anodes of suitable composition may prove effective and economical—provided that high currents are not required; there are, however, situations with which they cannot cope, and it may sometimes be advisable to use different materials at different points on the same pipe net-work. Thus a protective scheme for gas mains at New Orleans contains 1200 galvanic anodes of which 1000 are zinc (some are magnesium), along with 40 ground-beds supplied with current from rectifiers (S. E. Trouard, *Corrosion* 1957, **13,** 151t).

Zinc anodes should be buried in gypsum and bentonite, mixed in equal proportions. The open structure reduces the risk of blockage of the current paths with corrosion-product, but whatever backfill is used, it is impossible to avoid the formation of obstructive films on ordinary zinc anodes containing impurities. It is only since the time when zinc anodes of special composition became available that the use of zinc has received serious consideration for cathodic protection. Zinc anodes containing iron usually cease to supply current after a short life. Probably the iron impurity provides local cathodes, so that anodic and cathodic products formed side by side interact to give a zinc hydroxide or a cement-like basic salt; the surface becomes insulated and current ceases to flow. Information about the effect of iron in zinc anodes is provided by R. B. Teel and D. B. Anderson, *Corrosion* 1956, **12,** 343t. See also J. Kruger, O. R. Gates and M. C. Bloom, *J. electrochem. Soc.* 1955, **102,** 223C.

In Great Britain attempts to improve the situation by keeping the iron content of the zinc very low have met with limited success, but much progress has been made in providing antidotes for iron—especially by the use of aluminium along with silicon. Crennell's results suggest that small aluminium additions can produce anodes which remain active even when the iron content is 200 ppm.; the alloys containing aluminium and silicon have a small grain-size—a further advantage over pure zinc, which is said to develop larger pits (J. T. Crennell and W. C. G. Wheeler, *J. appl. Chem.*

1956, **6**, 415; 1958, **8**, 571; see also *Zinc Bulletin*, Spring Issue 1957). American experience with an anode containing aluminium and cadmium is reported by E. C. Reichard and T. J. Lennox, *Corrosion* 1957, **13**, 410t.

The current efficiency of zinc is higher than that of magnesium, which may furnish only about half the total current predicted by direct application of Faraday's Law, owing to local cells evolving hydrogen at local cathodes. Thus for mildly corrosive soils of low resistance, or for new pipe-lines where there are only very small faults in the coating, zinc should receive serious consideration (F. Ritter, *Werkst. u. Korrosion* 1955, **6**, 523, esp. p. 525).

It must, however, be remembered that the effective E.M.F. provided by the cell zinc | iron is much smaller than that of the cell magnesium | iron. Thus in soils of high resistance, zinc may be useless as an anode, since, when provision has been made for the IR drop, it may be impossible to polarize the pipe down to the necessary potential; this is particularly the case if the pipe is not well covered, with large gaps in the protective coat. In such situations, magnesium anodes are preferable, but even these have their limitations, and there are situations where only a power-impressed scheme can provide the necessary current.

In some soils (and in the sea) there may be an advantage in supplying ample current in the early stage so as to form the chalky film—after which current requirements may be reduced. There is a patented process for depositing a chalky film by applying a high current for a short time. Thus after applying 50 ma./ft.^{2} for 5 days, the subsequent protection can be carried out with 3 ma./ft.^{2} instead of the usual 6 to 10 ma./ft.^{2}. Such a method is recommended by E. P. Doremus and G. L. Doremus, *Corrosion* 1950, **6**, 216; also by S. P. Ewing, *ibid.* 1951, **7**, 410 and by H. S. Preiser and B. L. Silverstein, *J. Amer. Soc. naval Engrs.* 1950, **62**, 881. See G. C. Cox, U.S. Patent 2,200,469 (1940) and 2,417,064 (1947); British Patent 540,487 (1941).

In contrast, where the conductivity is high—as in salt water—an anode providing a small E.M.F. may be more economical, and here zinc or the aluminium alloy containing 5% zinc require serious consideration. The latter is being used considerably in U.S.A. for protection in the sea, and is said to avoid the damage to paint-work due to cathodically formed alkali which sometimes occurs with magnesium or power-impressed systems, unless precautions are taken. Other aluminium alloys are being developed for the same purpose. The aluminium alloy with 5% zinc gives a potential almost as low as that of unalloyed zinc when no current is being taken from it, but it polarizes more readily and the current obtainable is probably lower (T. P. May, G. S. Gordon and S. Schuldiner, Cathodic Protection Symposium 1949, p. 158 (Electrochem. Soc. and N.A.C.E.)).

Information about the relative merits of magnesium and zinc as anodes has been conveniently collected by J. van Muylder and M. Pourbaix, *Cebelcor*, *Rapport Technique* **34** (1956). Arguments for zinc anodes are put forward by R. W. Bailey, *Chem. and Ind.* (*Lond.*) 1954, p. 714, and for magnesium anodes by K. N. Barnard, *ibid.* 1954, p. 715; also by W. F. Higgins, *Civil Eng. and Pub. Works Rev.* 1949, **44**, 712; 1950, **45**, 43.

Methods of ascertaining in the field whether sufficient current is being applied to give protection. As already stated, it is generally assumed that if the pipe potential is everywhere at least 0·85 volt on the negative side of the saturated copper sulphate electrode (0·95 volt if sulphate-reducing bacteria are present), corrosion is avoided. To ascertain whether this potential has been obtained it is usual to place the tip of the copper sulphate electrode on the ground just over the buried pipe-line, join it to one terminal of a potentiometer or high-resistance voltmeter, the other terminal of which is joined by a wire to the pipe at some point where it comes to the surface (or near enough to be conveniently tapped). If, when the protecting current is applied the potentiometer registers a value exceeding 0·85 (or 0·95) volt, the pipe is considered to be adequately protected. If the depth of burial is small compared to the distance between the pipe

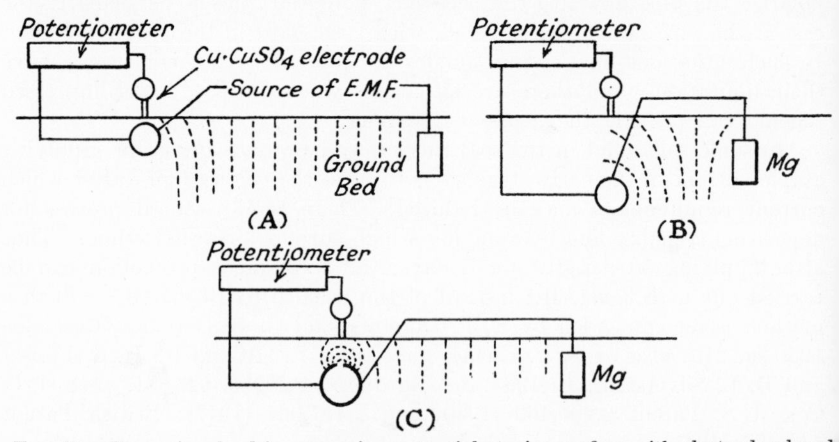

FIG. 65.—Errors involved in measuring potential at pipe surface with electrode placed on ground. In (A) the error is negligible, but in (B) and (C) several equipotential surfaces fall between the pipe and the point of measurement—involving serious inaccuracy.

and the external anode (fig. 65(A)), this method may often be sufficiently accurate; if the depth of burial is considerable, or if the external anode is placed close to the pipe (as may often happen when sacrificial anodes are used instead of a power-impressed scheme), there will be an error due to the fact that the equipotential surfaces are not perpendicular to the ground (fig. 65(B)). Even with shallow burial, an error is likely to be introduced if the pipe is well coated with protective covering except at one or two small gaps; the resistances of the bottle-neck approaches to these gaps will cause the equipotential surfaces to come close together (fig. 65(C)), and any measurement taken on the ground surface is rendered inaccurate. In all such cases, a margin of safety must be allowed—preferably based on practical experience of similar situations. Proposals for the calculation of the marginal allowance from theoretical principles have not always been sound.*

* If the shape of the equipotentials in the ground is known with certainty, calculation is possible. But shapes based on small-scale laboratory work carried

The criterion mentioned above (— 0·85 or — 0·95 volt, to the copper sulphate electrode) is not universally favoured, and other criteria for sufficiency of applied current have been proposed. Some authorities, for instance, state that the potential of the pipe-lines should be depressed by 0·3 volt below the value observed when no protecting current is flowing. These various criteria are discussed by J. W. Boon, *Metaalinstituut T.N.O. Pub.* **42** (1956); G. Bianchi " Il Problema della Protezione dei Metanodotti dal Punto di Vista della Corrosione Elettrolitica " (Stazione sperimentale per i combustibili) 1955;* W. J. Schwerdtfeger (with O. N. McDorman), *Bur. Stand. J. Res., Wash.* 1951, **47,** 104; 1957, **58,** 145. See also R. P. Howell, *Corrosion* 1952, **8,** 300; L. P. Sudrabin, *ibid.* 1956, **12,** 60t; 1957, **13,** 351t (with F. W. Ringer); discussion, *ibid.* 1957, **13,** 835t; M. C. Miller, *Petroleum Engineer* 1946, p. 55; E. Schaschl and G. A. Marsh, *Corrosion* 1957, **13,** 243t. See also p. 894 of this book. Values for the protective potentials for copper and aluminium are suggested in " Chemistry Research " 1956, p. 17 (H.M. Stationery Office) and by P. W. Heselgrave, *Chem. and Ind. (Lond.)* 1957, p. 556.

One of the criteria requires special discussion. It appears to have arisen from some pure-science work carried out about 1931 at Cambridge University, designed to provide understanding of the mechanism of corrosion. It was observed that the curves connecting applied cathodic current and potential of a metal specimen immersed in liquid containing oxygen showed a change in direction at a value of the current sufficient to give complete cathodic protection. Reasons were given for the break in the curve at this point. Reference to the original paper will make it clear to the reader that this was a pure-science investigation, and that there was no aim at working out a protective process; it was also pointed out that the case of iron is less simple than that of other metals. Nevertheless practical men interested in cathodic protection started to use the change of direction which they found in the I–V curves obtained on working pipe-lines or structures as an indication of the protective current; sometimes good results were obtained; sometimes the criterion proved unreliable.

There are several reasons why the method in its original form is unsuitable for working pipes and structures. One important reason is that in practice different parts of the structure will be placed at different distances

out without regard to the requirements of dimensional analysis may be misleading. The form of equipotentials and current-lines in two cells of geometrically similar form, one N times as big as the other, will *not* be the same if the two cells are filled with the same liquid. To obtain the same form, the specific conductivity must be changed also, by a factor of N. (J. N. Agar and T. P. Hoar, *Disc. Faraday Soc.* 1947, **1,** 158).

* Bianchi's masterly report may be recommended to those interested in the scientific basis of protection. Unfortunately his criterion, although eminently sound, would be difficult to apply in soil. He showed that if the tip of a probe is moved along a straight line joining a local anode on the specimen to be protected to the external anode, the potential will steadily fall if protection is complete. If the current passing is too weak for complete protection, the potential will first rise (owing to the influence of the local cathode) and then fall (on approaching the external anode). See p. 896.

from the anode. We may regard these different parts as the separate cathodes of a number of cells of different internal resistance, joined in parallel to the source of E.M.F. On plotting the current (I) against the applied E.M.F. (V), each cell will provide a different I–V curve. It may be that each of the individual curves would show a sharp break, but they will occur at different points on different curves; thus the master-curve obtained by adding the current of the various cells together and plotting against V, will show a change of direction which is not sharp but rounded. The point where the change of direction is most marked will represent the part where the *greater* part of the structure is becoming protected, but the engineer wishes to know the current at which *all* parts are receiving protection. Recent laboratory work at Emeryville suggests that this is indicated by the point where the master-curve becomes straight again after the rounded elbow has been passed—thus suggesting a means of determining the value of current sufficient to give protection. The reader should study the original papers and form his opinion. The original pure-science work was described by U. R. Evans, L. C. Bannister and S. C. Britton, *Proc. roy. Soc.* (*A*) 1931, **131**, 355, esp. p. 367. Endeavours to show the limitations of the criterion for practical problems were made by S. C. Britton, *Corrosion* 1951, **7**, 403 and later by U. R. Evans, *ibid.* 1957, **13**, 833t, whilst the more hopeful criterion now put forward from Emeryville will be found in a paper by E. W. Haycock, *ibid.* 1957, **13**, 767t. See also W. J. Schwerdtfeger, *Bur. Stand. J. Res., Wash.* 1958, **60**, 153. The effect of temperature on cathodic protection is discussed by G. R. Hoey and M. Cohen, *Corrosion* 1958, **14**, 200t.

Siting of Anodes or Ground-beds. Where a power-impressed scheme is used, a few large ground-beds may suffice, set some way from the pipeline so as to spread out the current fairly uniformly over the length. So long as there are no other metal pipe-lines or structures in the neighbourhood, this plan has obvious advantages, and it should be possible to explore the pipe-line with the copper | copper sulphate electrode, to ensure that at all points it has been polarized below the required level. The wire from the copper pole of the electrode can be attached to one terminal of a potentiometer or high-resistance voltmeter, the other terminal of which is joined by a long flexible lead to the pipe at any accessible position. The electrode is then taken to various points along the pipe-line, for measurement of the potential, and the current flowing between ground-bed and pipe is adjusted until the potential at all parts of the pipe has been depressed to the appropriate level (— 0·85 or — 0·95 volt relative to the saturated copper sulphate electrode). Allowance must be made for the source of error mentioned above; if part of the potential drop falling over the resistance between pipe and ground-bed is included in the potential measured, the value obtained will be wrong; this is most likely if the resistance of bottle-neck approaches to bare spots on the pipe constitutes an appreciable part of the resistance between the pipe and the distant ground-bed.

When magnesium (or zinc) blocks or rods are used as anodes, they have to be spaced at closer distances along the pipe, although sometimes about ten per mile may suffice to depress the potential sufficiently. The distance between anode and pipe may be smaller than in power-impressed schemes, and where, for any reason, a magnesium anode is close to the pipe, some discretion has to be exercised in taking the potential measurements to avoid the error just mentioned. Another question to be answered is the siting of the magnesium anodes. It may be best to conduct a preliminary survey to discover the places where in absence of applied (protective) current the anodic and cathodic areas would lie, and particularly to discover whether the cathodic reaction or the anodic reaction tends to be localized. If the pipe comes close to the surface at one place, the cathodic reaction may be concentrated there; in that case it may be best to place a magnesium anode rather close to that spot—so that the current which would otherwise flow between the small cathode and the numerous anodic spots spread out along the length of the pipe, shall be deflected to the magnesium anode, which is nearer at hand and provides a higher E.M.F. If, on the other hand, the anodic reaction tends to be localized, say at a salt nest in the soil, the magnesium should be placed close at hand, so that the current which would otherwise be flowing to this spot from the numerous well-distributed cathodes shall be directed on to the magnesium.

Cathodic Protection of Lead. The methods described above are not only applicable to steel, but have proved effective for lead cable-sheathing. In the United Kingdom, 100 sections of telephone cable are now protected by magnesium anodes, and there are 72 power-impressed schemes (J. Gerrard and J. R. Walters, *Chem. and Ind.* (*Lond.*) 1956, p. 1060, esp. p. 1065; L. M. Plym, *Corrosion* 1956, **12,** 331t).

To ensure protection on lead, the potential needed is different from that on iron. Compton considered that it should be made about 0·1 volt more negative than the potential assumed by lead in the soil in the absence of current; in most soils this means about – 0·7 volt relative to the Cu | sat. $CuSO_4$ electrode (K. G. Compton, *Corrosion* 1956, **12,** 553t).

Interference. Imagine (fig. 66) a power-impressed system intended to protect the pipe *PQR*, with a large ground-bed *G* placed well away from it, so as the distribute the current over bare spots scattered along the length of the pipe. If now there happens to be another pipe (or sheath) *XYZ* running at an angle to *PQR*, it is clear that part of the current may run through the ground between *G* and *X*, then along *XYZ* to *Y*, finally passing through the ground to join *PQR* at *Q*. This will tend to give protection at *X*, but to increase corrosion at *Y*; if premature perforation occurs at *Y*, the owners of *XYZ* will have a just grievance against the owners of *PQR*. Moreover, if there should be a poor electrical contact at a joint at, say, *J*, so that part of the current passes through the ground at this point, there will be damage at the anodic side of the joint.

Two remedies suggest themselves. If the owners of *XYZ* would agree, the two pipes could be bonded at *YQ*, thus preventing anodic attack at *Y*. However, there are two possible objections. The actual current passing along *XYZ* will be greatly increased by this bonding, and the damage on the anodic side of *J* will be increased still further. Furthermore, the power costs to the owners of *PQR* will be increased, and the owners of *XYZ* are unlikely to be willing to pay a share.

Alternatively, the ground-bed might be placed at *G′* instead of *G*. On a two-dimensional plan, this suggestion looks sensible. However, in some situations the advantage may prove slight. Supposing that the upper part of the ground in which *G* and *XYZ* lie is relatively dry, with a water-rich region deeper down, the main resistance of the current path will be a nearly vertical passage of current at *G* and at points on *XYZ*; where the current passes through the wet region, it can spread out in all directions, and the resistance is small; the increase in resistance of the path between the ground bed and bare spots on *XYZ* may not in that case be greatly altered by increasing the horizontal distance between the two.

A form of test often recommended for discovering whether the cathodic

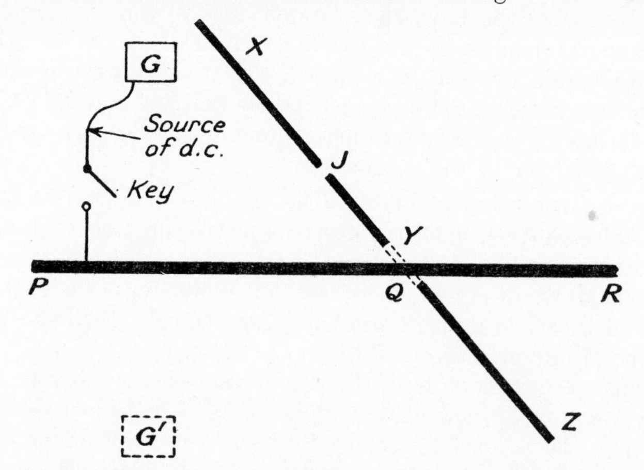

FIG. 66.—" Interference ". Cathodic protection of pipe *PQR* may cause anodic attack to pipe *XYZ*.

protection of *PQR* will cause trouble to *XYZ*, depends on ascertaining whether the potential at points on *XYZ* between pipe and ground undergoes serious change at moments when the protective current is switched on or off by means of a key. The result of such tests, however, have sometimes proved puzzling, even to experts. In cases where the situation is geographically simple, experience of similar cases, applied with understanding of the electrochemical principles involved, may lead to a reasonable solution of the difficulties; but the problem in a country where there are intricate net-works of conductors serving different purposes is a very different one from that of a single long pipe-line crossing a desert.

Clearly the use of a large number of magnesium anodes placed close to the pipe will cause less risk of interference than a small number of power-impressed ground-beds placed at a distance. On paper, the ideal scheme would seem to be a continuous magnesium anode running close beside the pipe. In practice the use of magnesium ribbon—at first sight so attractive—may be found inconvenient, owing to the trouble and expense involved in maintenance and renewal. For cases to which the method is suitable, a ribbon form of anode is today available (H. L. Davis, *Corrosion* 1955, **11**, 295t).

If interference troubles are to be avoided, close co-operation between the various interests involved (Power, Light, Gas, Water, Telephone and Transport) is essential; a joint regional committee on which all are represented is advisable. Where co-operation has been good, the difficulties have often been overcome to the satisfaction of all. A case is described by M. C. Miller, *Corrosion* 1956, **12**, 247t.

A British Joint Committee for the Co-ordination of the Cathodic Protection of Buried Structures was formed in 1953. Its activities are indicated in a pamphlet entitled " Cathodic Protection of Buried Structures " edited by W. T. J. Atkins (Central Electricity Generating Board, London).

The resident engineer or chemist attached to some undertaking will not generally be asked to design his own cathodic protection scheme, but he may be called upon for an opinion as to whether cathodic protection should be used, or whether, as an alternative, a serious attempt should be made to obtain a fault-free protective covering—despite the cost; fault-detectors are available which may help this alternative solution of the problem. If cathodic protection is decided upon—and often this will be the correct choice—it is highly advisable that the engineer on the spot should understand at least the basic principles involved, and be able to hold his own in discussions with the specialist called in to design the scheme. He ought to know enough electrochemistry to enable him to distinguish the real expert who combines a wide practical experience of cathodic protection with adequate electrochemical understanding of the principles involved, from the semi-charlatan who uses grand scientific language to impress rather than to inform; moreover, he ought to know what information derived from his own special knowledge of the structures and local topography is worth bringing to the attention of the expert.

Cathodic Protection of Tanks. Pipe interiors rarely receive cathodic protection, but it has proved effective for the protection of water-tanks, either as sole protection or as supplement to painting, in which case the paint must be alkali-resistant, at least if the water contains sodium salts. Since it is undesirable to introduce heavy metals into the water, graphite is preferable to steel in power-impressed systems, and magnesium to zinc in sacrificial anodes; it is doubtful whether zinc would supply the necessary protection in some waters, and in hot-water systems it might actually become cathodic to steel (p. 207). The danger of damage from ice-pressure may render it advisable to withdraw fragile graphite anodes from outdoor tanks in severe winter weather.

The use of magnesium anodes in protecting a crude oil tank in which water collected at the bottom, is described by J. H. Graves, *Corrosion* 1956, **12**, 254t.

Cathodic Protection in Special Situations. The cathodic protection of ships is discussed on p. 300 and that of cable-sheaths on p. 293. Authoritative advice for its use in various other situations is provided by K. A.

Spencer, *Chem. and Ind.* (*Lond.*) 1954, p. 2, discussing pipes, tank exteriors, jetties, bridges, pontoons, buoys and ships; B. H. Tytell and H. S. Preiser, *Corrosion* 1957, **13**, 515t, discussing ships; M. G. Duff and I. D. G. Graham, *Corros. Tech.* 1957, **4**, 9, discussing small ships; E. E. Nelson, *ibid.* 1957, **13**, 315t, discussing a sea-wall; J. K. Ballou and F. W. Schremp, *Corrosion* 1957, **13**, 507t, discussing oil-well casings; W. C. R. Whalley, *Chem. and Ind.* (*Lond.*) 1954, p. 140, discussing oil pipe-lines and tanks in the Middle East; H. M. Powell, *Corros. Prev. Control* 1954, **1**, 166, discussing water-box interiors in heat exchangers; G. L. Olson and H. V. Beezley, *Corrosion* 1949, **5**, 249, discussing gas transmission lines; J. P. H. Zutphen, *ibid.* 1955, **11**, 37t, discussing open-box coolers; J. B. Prime, *ibid.* 1954, **10**, 165, discussing problems at the cooling-water intake of a power station; F. E. Jones, *Chem. and Ind.* (*Lond.*) 1957, p. 1405, mentioning magnesium anodes in water-tanks as an auxiliary protection; W. S. Merrithew, *Corrosion* 1952, **8**, 90, discussing a softening-plant problem; A. P. Farr, *ibid.* 1953, **9**, 108, S. C. Moore, *ibid.* 1953, **9**, 112 (two papers) discussing drill-stems in oil-fields; J. F. Hirshfeld and L. P. Schaefer, *ibid.* 1952, **8**, 140, discussing radiant heating systems.

Internal Corrosion of Pipes

General. The treatment given to water before entering pipe-lines has been described on p. 161 and certain internal coatings have been mentioned on p. 280. Corrosion of the internal pipe-walls is likely where water treatment has been neglected or where the protective coatings are thin and discontinuous. Some raw waters cause, not only corrosion of iron and steel pipes, but serious blockage with rust and those elongated rust-coloured membranous blisters filled with black material, known as "tubercles" or "tuberculations". In a typical case, described by Baylis, a pipe had become two-thirds full of rust in 50 years, and the corrosion had eaten half-way through the walls. In general, each important tubercle is found to cover a pit.

Although tuberculations are often associated with microbiological action, they can be formed in a purely electrochemical manner, as described on p. 119. In the opening stages, attack may start at small anodic points (possibly gaps in the mill-scale if this has not been removed), and membranous blisters are set up, elongated in the direction of water-flow, representing the surfaces separating the regions within which iron salts from the anodes and alkaline liquid from the cathodes respectively predominate. If, however, this electrochemical action represented the sole mechanism, we should get empty blisters, whereas in fact they are often found to be full of magnetite or other ferroso-ferric compounds; moreover, the electrochemical action should slow down if a layer of chalky rust is produced over the cathodic surface outside the blisters. It is probable that a small amount of oxygen moves inwards through the membrane, oxidizing a small part of the ferrous ions to ferric ions so that the solubility product of magnetite is soon surpassed; consequently either magnetite or another ferroso-ferric compound is thrown down, regenerating acid (sulphuric acid if SO_4^- is the main anion) which is able to act on further metal, producing fresh ferrous salts; then further acid is produced, and although, for reasons explained on p. 120, regeneration does not continue indefinitely, fresh acid can be formed so long as the water contains oxygen. Although oxygen may be a desirable component of water containing calcium bicarbonate as the main constituent, since it will throw down chalky rust in contact with the metal over the whole surface, it is generally an undesirable constituent of water containing sodium salts.

Thornhill, in a laboratory study of tuberculation by running water, found the greatest amount of blister-formation with a synthetic water made to contain 20 ppm. of sodium bicarbonate and 16 ppm. calcium sulphate. Waters containing calcium or magnesium sulphate alone gave general rusting but no tuberculation. The general distribution of attack in liquids containing calcium and sodium remained much the same if the SO_4^{--} was replaced by Cl^- (R. S. Thornhill, *Chem. and Ind.* (*Lond.*) 1951, p. 1201. Further details regarding tuberculation are furnished by J. R. Baylis, *Met. Chem. Eng.* 1925, **32,** 874; *J. Amer. Water Wks. Ass.* 1926, **15,** 598, esp. p. 606; *J. New England Water Wks. Ass.* 1953, **67,** 38; T. E. Larson and R. V. Skold, *J. Amer. Water Wks. Ass.* 1957, **49,** 1294).

Microbiological Corrosion and Blockage. Some of the worst cases of pipe blockages are connected with " iron bacteria ", particularly *Gallionella Ferruginea*, an organism which derives the energy needed for its life process from the oxidation of ferrous to ferric iron; soluble ferrous salts are oxidized and the matter deposited on the sheaths of the organism is hydrated ferric oxide—effectively rust. The masses of rusty material which obstruct pipes may appear wholly unorganized at their external portions, which represent the remains of dead organisms, but the inner parts nearer to the metal, will be seen, when examined under the microscope, to show a structure characteristic of living organisms. At the base next the metal walls, which are often seriously pitted or furrowed, sulphides can sometimes be detected, and it is likely that sulphate-reducing bacteria are here at work.

Two main theories have been put forward. Olsen and Szybalski consider that when a small amount of ferrous salt has been formed by anodic attack at a susceptible spot (probably some point shielded from oxygen-replenishment), the organisms convert it to hydrated ferric oxide, and this shields the surface still more effectively from oxygen, so that the attack, once started, soon becomes vigorous, as a result of differential aeration cells; aeration of the water favours the attack, and, although inoculation with the bacteria assists matters, the same result can be produced in a purely inorganic way (E. Olsen and W. Szybalski *Corrosion* 1950, **6,** 405).

Butlin considers that, however the corrosion starts, the conversion of ferrous salts to hydrated ferric hydroxide provides anaerobic conditions at the basis of the mass which are favourable to the development of sulphate-reducing bacteria. Thus at the stage when both destruction of the walls and blockage of the channel are proceeding rapidly, the sulphate-reducing bacteria are responsible for the corrosion, and the iron bacteria for the production of the voluminous matter which obstructs water-flow (K. R.

Butlin, private communications to L. Minder and the author, Jan. 23, 1947 and Feb. 3, 1949).

Such a view is confirmed by the examination of pipes which have failed. The rust on these contains much sulphide and many sulphate-reducing bacteria in its inner portions, but there is no sulphide in the outer portions. Some difficulty has been experienced in establishing the state of affairs on pipes which have failed accidentally, since the examination usually takes place a considerable time after the pipe has ceased to perform its usual function. However, the examination of a freshly opened pipe by the Teddington investigators has provided definite evidence of microbiological corrosion (" Chemistry Research " 1948, p. 23 (H.M. Stationery Office)).

Butlin's theory, which appears very probable, provides a further instance of bacteria " hunting in couples "; another example (p. 282) was the decomposition of hessian by cellulose-decomposing bacteria providing nutriment for development of the more dangerous sulphate-reducing type.

The provision of thick internal coatings for pipes, adequate to prevent the production of ferrous salts, should effectually prevent blockage by the ferric compound, provided always that the water, as it enters the pipe, contains no iron. Ordinary coatings are of limited value. An early paper by Brown described how water containing iron held in solution by organic acids can develop a tangled mass of sheaths and threads due to bacterial action even on pitch-coated pipes. Neutralization of the acid followed by filtration was regarded as the best way to prevent this trouble. Parker has described troubles on heating coils at a sewage treatment plant where bacterial activity was found to remove hydrocarbon constituents of a coal-tar paint; a vinylite paint pigmented with aluminium was found to be satisfactory (J. C. Brown, *Min. Proc. Instn. civ. Engrs.* 1903–4, **156,** 1; C. D. Parker, Corrosion Symposium (Melbourne University) 1955–56, p. 186).

Use of Cement in Pipes. Cement linings may be produced by centrifugal action or by the use of mandrels. Large conduits, like the Catskill Aqueduct, are lined after installation; Speller's book contains a picture of six men inside the aqueduct engaged in the task of lining it with a mixture of sand and Portland cement (1 : 2) to a thickness of 2 inches; the linings were found to be in good condition after 18 years, the metal being well protected. A cement-lined water main at Norfolk, Virginia, was found to be as good as new after 20 years, whereas an uncoated cast iron main, carrying the same water, had developed so much rust that the useful diameter was reduced by 25% (and the carrying power, presumably, by 44%*), whilst pits extended half-way through the walls. Speller adds that these and other experiences have been so favourable that the use of cement lining for cast iron mains has become quite general for corrosive water conditions. He describes the jointing of pipes purchased ready-lined at the mill (F. N. Speller, " Corrosion: causes and prevention " 1951, pp. 370–76 (McGraw–Hill)).

Cement-lined conduits suffer little loss of flow capacity with time, so that the inside-diameter allowed in the design can be smaller than in the

* $1 - (3/4)^2 = 0{\cdot}44$.

case of pipes carrying bituminous coatings. The protection of the metal may be partly connected with the alkalinity due to lime set free during the setting process; any increase of hardness to the water is small and probably temporary. Some iron is precipitated as ferric hydroxide on the walls, but this does not impair the protective qualities of the coating.

Internal coatings based on plastics are discussed by J. C. Watts, *Corrosion* 1955, **11,** 210t. Troubles in copper pipes due to carbonaceous layers have been mentioned on p. 125.

Corrosion and Protection of Other Types of Buried, Immersed or Embedded Structures

Corrosion by Sea-water. Steel work exposed to sea-water is often protected by painting. Anti-corrosive paints and anti-fouling paints for ships are discussed in Chapter XIV. Certain materials are commonly used in sea-water in the unpainted condition, and some recent tests in Port Hueneme Harbour (California) have produced rather unexpected results. Of the materials tested, lead showed greatest resistance; monel metal developed pits below marine growth, whilst the aluminium alloys tested also developed pitting, which in one case tended to occur along scratch-lines and reached a maximal depth of 70 mils (and an average depth of 52 mils) in 30 months. There was some pitting on stainless steel, but none on copper. Magnesium alloy specimens became perforated by corrosion (C. V. Brouillette, *Corrosion* 1958, **14,** 352t). Corrosion-resisting steels for use in sea-water are discussed by C. P. Larrabee, *ibid.* 1958, **14,** 501t.

Stray-current Damage to Ships. Both on ships and landing structures serious anodic corrosion of steel due to stray currents has been met with, and cathodic protection schemes have been applied; since the principles are similar to those involved on pipes, it is convenient to discuss such matters in the present chapter.

During the fitting-out period of a newly launched ship, the vessel is usually brought to the edge of a basin (wet dock), electric cables are run out to the ship from sub-stations or generating plants on land for purposes of welding (and also lighting, since the ship's dynamos are not available). For welding, one cable is attached to the hull, whilst the other is joined to the electrodes. If only a single ship is on a circuit, no harm generally results. If two are on the same circuit, the two hulls will not, except as a coincidence, be at the same potential, and relatively large currents will pass between the two ships, causing corrosion to whichever of them is the anode. If both ships were bare of paint and mill-scale, the attack would be well distributed and not dangerous; in practice they will have been painted, and attack will be concentrated at places of accidental damage to the paint, including scratch-lines made during the launching operations, or at bare places where the plates had been in contact with the poppets and launching cradle. The corrosion digs rapidly into the metal producing trenches. The aspect of the attack is characteristic, and was once aptly described by a ship-yard foreman who remarked, " looks like someone trying sabotage with a cold chisel ".

It might be thought that connection of the two ships with a stout copper cable would eliminate the trouble, but although copper is a better conductor than sea-water, the cross-section of the water path is so enormously greater that an appreciable fraction of the current may still adopt the water route, with disastrous results. The only way of escaping trouble is to arrange the electric circuits in a manner designed to avoid the cell Ship | Water | Ship; this should, in general, not be difficult.

Cases of this sort are known to the author, and rather similar ones are described by J. C. Hudson, " Corrosion of Iron and Steel " 1940, p. 173 (Chapman & Hall). See also R. A. Ponfret and L. M. Mosher, *Corrosion* 1948, **4**, 227; H. W. Mahlqui, *ibid.* 1953, **9**, 1).

Cathodic Protection on Ships. Cathodic protection is now being used in several navies, especially for reserve (inactive) ships; for active ships, there is difference of opinion as to whether it is worth while. The systems in use in the Royal Canadian Navy were made public some years ago by K. N. Barnard with G. L. Christie, *Corrosion* 1950, **6**, 232; 1951, **7**, 114; *Research* 1952, **5**, 117.

Methods adopted in the British Navy have been published by J. T. Crennell, *Chem. and Ind.* (*Lond.*) 1954, p. 204. I. D. Gessow, *Corrosion* 1956, **12**, 100t, has described cathodic protection in the U.S.A. Navy. See also W. A. Bowen, *Corrosion* 1956, **12**, 317t; E. E. Nelson, *ibid.* 1957, **13**, 122t; H. S. Preiser and F. E. Cook, *ibid.* 1957, **13**, 125t. C. F. Schrieber, *ibid.* 1958, **14**, 126t.

Numerous types of anodes have been tried, and apparently the magnesium alloy anodes used for protecting pipes (p. 287) are generally considered the most convenient when the anodes have to be attached to the vessel, at least for small ships; for the design of the attachments, the original papers should be consulted. Magnesium is not generally suitable for fresh water.

Russian work suggests that aluminium-zinc-calcium alloys, in which the calcium occurs as the compound Al_3Ca, offer possibilities; apparently calcium hydroxide, an alkaline anodic product, prevents passivity. Tests under conditions where magnesium protectors of standard size were completely consumed in 25 to 30 days produced only 8 to 10% corrosion of anodes of the new alloy after twice that period.* Tests in the Black Sea confirm the conclusions from laboratory work (O. F. Zhurakhofsky, I. N. Franzevich and E. L. Pechentkovsky, *UkSSR Reports* 1957, No. 6, pp. 569, 575).

For large vessels, or groups of ships, there is much to be said for impressed power systems which are not only cheaper, but more versatile and easier to control, thus minimizing damage to paint. Graphite anodes—despite their fragility—can be recessed into the hull or bilge keel and thus avoid breakage, whilst presenting no obstruction to water flow. Platinum or platinum-clad silver has been used with success; platinum-coated titanium anodes are under trial (p. 287). Steel anodes, despite the need for replacing

* It seems doubtful, however, whether the types of magnesium alloy or zinc, used for comparison with the new alloy, possessed the composition which would have been chosen for practical use in Great Britain (regarding magnesium alloys, this opinion is confirmed by W. F. Higgins, Priv. Comm., Jan. 13, 1959).

them periodically, may sometimes prove more convenient than the brittle graphite. A promising anode material is a lead alloy which can be extruded into convenient form, so that the large areas can be protected with simple electrical connections—a considerable convenience in installation; the alloy used contains 1% silver and 6% antimony. Details of the system, which comprises automatic adjustment of potential, are provided in *Corrosion, Prevention and Control* 1958, **7**, July, p. 53, supplemented by R. A. Lowe, Priv. Comm., Aug. 6, 1958.

It is generally recommended that the potential of the hull should be kept at − 0·80 to − 0·85 volt relative to the Ag | AgCl electrode shown in fig. 64(B) (p. 285). The electrochemical principles involved are the same as in the protection of pipes. There may be some advantage in applying a high current at the outset, so as to deposit a chalky film, which will allow protection at a low current later; sea-water contains sufficient calcium bicarbonate to produce such a film through cathodic action. The magnesium salts, present in much larger amount, may also play a part by producing a magnesium hydroxide film, which gives an adherent protective rust (p. 94).* Schikorr has shown that the addition of magnesium salts to a sodium chloride solution greatly reduces the attack especially during long periods. Nevertheless it seems that calcium carbonate occurs in the chalky film produced on cathodically protected surfaces in greater amount than any magnesium compound—as shown by the analyses of I. B. Ulanovsky, *J. appl. Chem. U.S.S.R.* 1956, **29**, 1056 (English translation p. 1143).

The main objection raised against cathodic protection when it was first tried on ships was the manner in which the cathodically formed alkali either softened or loosened the paints then commonly used for marine purposes. It was soon found necessary, where a magnesium anode was to be attached to a ship, to cover the area around the attachment with a layer of some alkali-resistant material; ordinary paint in this region is quickly stripped off. It is, however, gradually becoming recognized that —whatever the source of current—the whole of the ship's paint should be of an alkali-resistant character; many of the classes of paints based on the synthetic resins (vinyl, polystyrene, epoxy-resins, etc.) do resist alkali, and doubtless others will be introduced (R. P. Devoluy, *Corrosion* 1953, **9**, 2; L. P. Sudrabin, F. J. LeFebvre, D. L. Hawke and A. J. Eickhoff, *ibid.* 1952, **8**, 109; cf. H. W. van der Hoeven, *J. Oil Colour Chem. Ass.* 1957, **40**, 667).

It is pointed out in Chapter XIV that there is a great advantage in using alkali-resistant paints in marine situations even if no cathodic protection is contemplated. Many years ago the author showed that drops

* In cathodic protection in pure sodium chloride solution free from calcium compounds, the sodium hydroxide formed as the cathodic-product plays the same sort of part as the chalky film, provided that it can be held in place. In laboratory experiments, the author once found that the protection of a vertical steel cylinder could be maintained at a lower current when a screw thread had been cut into the surface, despite the increase of area—since the recesses served to retain the sodium hydroxide solution (U. R. Evans, *Metals and Alloys* 1931, **2**, 62).

of salt water placed on a painted steel surface quickly loosened the peripheral cathodic zone, unless either (1) alkali-formation was prevented or (2) the vehicle was alkali-proof. Hence, whether cathodic protection is used or not, alkali-sensitive paints are undesirable for salt water, and alkali-resistant paints should be chosen (U. R. Evans, *Trans. Amer. electrochem. Soc.* 1929, **55**, 243).

Corrosion of Steel embedded in Concrete

General. The use of concrete or other mixtures for the internal or external protection of pipes has been described on pp. 272, 298, but numerous types of structural design involving steel embedded in concrete still require discussion. These include steel-frame buildings and concrete reinforced with steel rods or wire. Since the effective tensile strength of concrete is poor, there is danger of cracks developing owing to temperature changes, volume changes, service stresses and the like. From the purely mechanical standpoint there is an obvious advantage if tensile stress can be applied to the steel in such a way that the stresses, transmitted to the concrete, keep it in compression. Thus the so-called pre-stressed concrete is gaining popularity, for example in the walls of reservoirs and in the manufacture of pipes. The procedure should be further sub-divided into pre-tensioning and post-tensioning practices, according as the stress is applied to the steel before or after the concrete has hardened. These introduce different methods of transmitting the stress from the steel to the concrete so as to put the latter into compression. When post-tensioning is used, properly designed anchorage plates, arranged at right angles to the wires, can be inserted in such a way that when the wires are put into tension, the concrete is *ipso facto* put into compression; the wires usually pass through ducts in the concrete and the space between the two materials is afterwards filled with cement grout; some stress-transmission may later develop between wire and grout. When pre-tensioning is used the transmission depends at least in part on the chance roughness of the steel surface. Copenhagen's tests have shown that the mechanical bond is less satisfactory for a bright surface than a rusty surface, and even less satisfactory for a surface carrying mill-scale. Certain specifications disallow the use of rusty or scale-covered steel for reinforcements, but it is doubtful whether such injunctions are often observed, or indeed whether rust *could* be avoided under conditions commonly prevailing at the site of erection; it is understood that some engineers actually favour a rusty surface, after removing loose rust, for improving the bond, but the plan, whatever its mechanical advantages, introduces chemical dangers. However, the bond does not depend solely on surface roughness. As the tension in the steel becomes relaxed and the length of the wire diminishes the diameter must slightly increase, and the compressive stress in a radial direction will improve the bond. The formation of fresh rust would probably also improve the bond, owing to the volume-increase, but this is not a healthy manner of obtaining stress transmission.

Inhibitive Character of Cement. Clean steel buried in Portland cement concrete in the absence of added salts usually remains bright and unattacked, as has been shown by experiments in the laboratory and also by observations on buildings dismantled after many years of service. The immunity is generally attributed to the alkali liberated during the setting of cement, but it is not impossible that other constituents of cement contribute to inhibition. The passivity of the steel persists (though not in the presence of chlorides) even if the steel is subjected to anodic action from stray currents. The passivity of steel produced by cement may be relied on if the surface is bright and clean so that the cement can come into close contact with the steel everywhere. If there are rust-patches containing ferrous salts, which would precipitate the alkali or other inhibitor diffusing inwards through the rust towards the metal, the possibility of small anodic and large cathodic areas appears to introduce a risk. It is doubtful whether this has often been the cause of structural failure, but the matter deserves investigation in the laboratory and the field. The danger will be greatest if the rust contains sulphates or chlorides.

Effect of Chlorides. If through design or accident chlorides are present in the concrete, passivity may break down, and the voluminous character of the product formed where the ferrous chloride formed by anodic attack interacts with the alkali in the concrete can develop stresses high enough to crack the concrete (A. A. Knudson, *Trans. Amer. Inst. elect. Engrs.* 1907, **26,** 231).

Early in the century many examples were recorded of damage due to stray currents acting on steel in concrete; in one electrochemical factory where brine was being used, the floor beams lost nearly all their strength, the concrete being shattered by internal expansion and the steel largely corroded away. At that period, when stray currents were known to be rampant in the ground in many American cities (in 1903 one 6-inch pipe at Brooklyn was found to be carrying 70 amperes), fears were entertained regarding the safety of tall buildings, and special attention was paid to the state of the steel and concrete, whenever a building had to be demolished. A survey carried out in 1913 showed corrosion to have occurred only in cases where considerable chloride could be detected in the cement; it is uncertain whether, in most of these cases, stray currents had played any major part (E. B. Rosa, B. McCollum and O. S. Peters, *U.S. Bur. Stand Tech. Paper* **18,** (1913)).

Scarpa found that iron buried in cement–sand mixtures almost free from chlorides suffered practically no attack, but that where 1% sodium chloride was present, there was much corrosion, although only where oxygen and atmospheric moisture gained access; he concluded that there is danger of localized attack at cracks or where the cement mixture is porous; iron partly buried in the cement-mixture showed the usual differential-aeration corrosion-pattern. When an A.C. was applied to iron in concrete, there was serious corrosion when chlorides were present, but not otherwise; the attack took the form of pitting, covered with rather voluminous oxide (O. Scarpa, *Rendiconti dell' Academia Naz. dei Lincei* 1951, 10, Fasc. 3, 4 (2 papers)).

Magee, studying steel rods embedded in concrete and buried in the ground, found no effect when A.C. was applied; but D.C. at 25 volts caused corrosion and cracked the concrete if the soil contained sulphate; he recommended interposing asphalt either between ground and concrete or between concrete and steel (G. M. Magee, *Corrosion* 1949, **5**, 378).

Most of the earlier cases of corrosion due to chlorides were attributable to the presence of sodium chloride, which was sometimes intentionally added to the cement or concrete to prevent freezing. In more recent years, however, the addition of calcium chloride to accelerate the development of strength has found favour in many quarters, although it is not now recommended for use in connection with pre-tensioned construction, or for the cement grout used to fill the ducts through which the wires pass in post-tensioned construction. This change of opinion has largely arisen from a serious failure of pre-stressed pipes at Regina (Canada)—a case examined by R. H. Evans, *Proc. Instn. civ. Engrs.* 1955, **4**, Part III, p. 725, esp. p. 748.

In his useful survey of corrosion in contact with concrete, Halstead includes a short section on pre-stressed concrete. He considers that " it seems desirable to avoid the use of calcium chloride in concrete to be stressed by thin wires ". With regard to the use of steam-curing in connection with calcium chloride—which has been criticized—he says " no doubt the elevated temperature can increase the rate of attack of steel, but there is no particular reason to think that the initial corrosion is progressive " (P. E. Halstead, *Chem. and Ind.* (*Lond.*) 1957, 1132, esp. p. 1136).

Much useful information has been collected by R. H. Evans (see above); S. B. Hamilton, *Nat. Bdg. Res. Studies*, **24** and **25**; F. E. Jones, *Chem. and Ind.* (*Lond.*) 1957, p. 1409; M. P. Pirotte, p. 1410. However, further knowledge about prestressed concrete containing calcium chloride is badly needed, and researches into the question are in progress at H.M. Building Research Station, Watford, and also at the Eidg. Material-prüfungs- u. Versuchsanstalt, Zürich. The reports should be studied when available.

Whether calcium chloride should be allowed in concrete surrounding " unstressed " steel is a matter on which opinions differ. It must here be borne in mind that usually the steel will be subject to service stresses even when there is no pre-tensioning or post-tensioning. Experience of refrigerating-brines suggests that calcium chloride is less corrosive than sodium chloride, probably because the cathodic product (calcium hydroxide or in some cases calcium carbonate) is less soluble. It has been thought that in cement mixtures containing calcium chloride the Cl^- ion activity is reduced to a safe level owing to the formation of complex ions, but no physical-chemical study of the system appears to have been made. In other systems of complex formation which have been studied in the laboratory, the complex has generally been found to be more or less dissociated. Thus arguments based on complex-formation should, for the moment, be regarded with reserve. Such observations as have been recorded are not all in agreement. In the laboratory Bukowiecki found that cement containing calcium chloride produced rusting under conditions where the same cement without chloride produced none. However in Muller's tests on concrete blocks containing embedded bars exposed in the open air or at half-tide level in the River Tees, the behaviour was said to be the same whether calcium chloride was present in the cement or not. It is hoped that the tests at the Building Research Station—mentioned above—will settle this apparent discrepancy. Meanwhile the reader may like to study the two papers: A. Bukowiecki, *Schweiz. Arch. angew. Wiss.* 1953, **23**, 97, esp. p. 104; P. P. Muller, *Magazine of Concrete Research* 1954, **6**, No. 6, p. 37.

Even where no chloride is deliberately added to the concrete, it may reach the steel from the outside. This occurs, for instance, near a coast—particularly if the concrete is porous and the cover thin. In discussing the corrosion of steel reinforcements in concrete in wharfs in New South Wales, Moore has expressed the view that today concrete is generally porous, because, in order to obtain workable mixes, it is necessary to add more water than is required for the hydration of the cement. If the amount of moisture present in the voids under service conditions was the same at all parts of the steel, corrosion was found to be slight, even when the cover was thin. If the conditions were not uniform, corrosion currents were set up; steel in fairly dry concrete was cathodic to steel in similar concrete fully saturated; the E.M.F. sometimes reached 0·2 volt. Heavy covering was not in itself sufficient to prevent corrosion; indeed blocks with a $1\frac{1}{2}$-inch cover could be made anodic to those with a $\frac{1}{4}$-inch cover by suitably varying the degree of saturation. On partly immersed piling the fully immersed portion was anodic relatively to the dry portion, but corrosion occurred principally at the junction of the two. These observations can probably be accounted for on the basis of differential aeration, but Moore does not commit himself to an opinion regarding the cause of the potential variation (D. D. Moore, *Maritime Services Board of N.S.W., Research Report* No. **20** (1947)).

Similar trouble has occurred on bridges near the coast in South Africa, and has been studied by Copenhagen, who also criticizes the common statement that a sufficient depth of covering necessarily ensures safety; he points out that $\frac{1}{2}$ inch of impervious concrete may protect better than 2 or 3 inches of a pervious mix. He considers the various micro-cells and macro-cells which may be set up and cause attack at the anodic portions. Where the steel carries broken mill-scale, the high cathode–anode ratio at the cells established at gaps in the scale may be an important cause of trouble. Where the concrete is locally cracked, locally thin, or locally pervious, both carbon dioxide and oxygen will reach certain small areas of the steel; of these the carbon dioxide (and sulphur dioxide, if present) will neutralize the alkalinity produced during the setting of the concrete, and thus tend to produce a local anode, whereas the oxygen will favour cathodic polarity; thus two of the constituents of air act in opposing senses, and laboratory experiments by Bird suggest that the effect of acid will generally prevail, setting up a small anode surrounded by a large cathode—a particularly dangerous combination. Copenhagen considers that the application of modern principles of concrete technology could give a relatively impermeable concrete which would " extend the life of most structures by a large factor ". He adds that " no matter how well prepared a specification may be, it requires most careful supervision and site-control for it to be complied with in practice " (W. J. Copenhagen, *S. African Builder* 1953, **31**, Dec. p. 33;

S. African Industrial Chemist 1957, **11**, No. 10 (with D. A. Lewis, reporting also work by C. E. Bird)).

American experience on the spalling of pervious concrete from an American reinforced bridge in a salt-water environment is described by S. F. Stratfull, *Corrosion* 1957, **13**, 173t.

Types of Corrosion Trouble in Concrete. Assuming that chloride or other corrosive influence reaches the steel, events may take different courses according to circumstances. If there is sufficient oxygen to produce voluminous rust by interaction with the anodic and cathodic products (e.g. ferrous chloride and sodium hydroxide) compressive forces must develop, and if at certain points these are sufficient to force the concrete away from the steel, cracking and spalling will develop, leaving the steel unprotected, so that the situation will deteriorate further. The actual mechanical breakdown will probably occur only after an interval needed for the compressive stress to develop and may be somewhat sudden, but symptoms of trouble should be noticed earlier by rust-stains on the outside of the concrete and these will develop gradually. On the other hand, if the depth of cover is ample, the concrete well made and the adhesion good, there seems no reason why rust-stifling (p. 762) should not develop—in which case the deterioration will be avoided. This is probably the main reason why the provision of adequate cover can often be effective in preventing trouble. If the function of a thick layer of concrete was merely to slow down the inward diffusion of salt, oxygen or other substance, increasing the thickness of the concrete would merely postpone the day on which the trouble would commence. Research into the possibility of rust-stifling below concrete, the thickness needed and the type of design most favourable to this welcome occurrence is desirable, as knowledge of the matter is almost non-existent today. The same research might embrace a study of steel specimens covered by different thicknesses of concrete in different parts, and the measurement of the corrosion currents generated by such an arrangement.

When the steel is strongly in tension (and that state of affairs is not confined to pre-tensioned or post-tensioned steel) the possibility of stress-corrosion cracking must be considered. The structure of the steel commonly used for wires in concrete is roughly sorbitic, with carbide particles in a ferritic matrix; this is the structural condition which Parkins (p. 683) found to be associated with stress-corrosion cracking in nitrate solution. The steel contains 0·6 to 0·7% carbon and is used in two conditions, " as drawn " and " heat-treated "; the heat-treatment involves quenching and tempering, and is considered to increase susceptibility to cracking (R. N. Parkins, Priv. Comm., July 2, 1958, quoting W. O. Everling, " World Conference on Pre-stressed Concrete ", San Francisco, 1957).

A clear distinction should be drawn between the gradual thinning of the reinforcement due to fairly uniform corrosion, which will produce a slow but steady loss of strength, and the entirely different phenomenon of stress-corrosion cracking; the latter may only start to develop after some years, and then cause an alarmingly rapid drop in strength. In the case at Regina studied by R. H. Evans (p. 304) stress-corrosion cracking does not seem to have been involved, and this is not surprising, since calcium chloride was apparently present in considerable amount—which would favour general corrosion; stress-corrosion cracking is more likely to occur under conditions on the border-line between corrosion and passivity.

Stress corrosion is considered in Chapter XVI, but the application of the principles there discussed to pre-stressed concrete has hardly started. Further research, particularly involving study of the effect of heat-treatment, is urgently called for. The argument that few cases of stress-corrosion cracking have occurred hitherto does not completely dispel anxiety lest cases may arise in years to come. Pre-stressed concrete has now been in use for many years and the only serious mishaps reported seem to have occurred in special cases where there has been some departure from recommended practice. It must, however, be remembered that in many types of stress-corrosion cracking (not connected with concrete), the trouble only shows itself after an induction period.

Prevention of Corrosion of Steel in Concrete. Various methods of preventing corrosion have been explored. One proposal is to add a soluble inhibitor to the concrete, or to the grout used in the post-tensioning method. A beneficial effect from inhibitor additions might reasonably be expected if the steel surface is clean and chlorides absent—but under such conditions there is unlikely to be serious trouble even without added inhibitor. Where rust-patches occur, there is a risk that the rust, preventing the inhibitor from reaching the metal below it, may establish the combination of small anodes and large cathodes—making matters worse. Until such possibilities have been investigated in detail, caution should be observed in recommending the addition of inhibitors.

From the purely chemical stand-point, it would seem rational to apply a protective paint to the surface of the steel. Engineers reject such proposals on the ground that a paint-coat would prejudice the bond, and it is possible that with the paints obtainable today this may be true. Possibly detailed research might lead to a paint sufficiently adherent to the metal, and providing a sufficiently rough surface to provide a bond at least as good as that obtainable at present. Presumably a bond to a rough painted surface would be more reproducible than a bond dependent on chance roughness of the steel, which must vary from one wire to another according to the amount of scale, rust and surface irregularities present. If once the right rough covering layer could be found, the situation might be improved even from the purely mechanical standpoint. It appears desirable that investigations into rational protective methods should be carried out. The difficulty of finding a paint sufficiently robust to avoid being scraped away locally may not be so serious as would at first sight appear—since paints are already known which protect at gaps (p. 615).* If the roughness of the

* F. E. Jones adds the comment that, if it is necessary to choose a paint which protects at gaps, this would restrict the choice of paint available. He considers that at first sight it is optimistic to hope for a paint with the necessary properties; the bond cannot be better than the shear-strength of the paint-film itself—which, in a thick coating or a two-coat system, may be poor (F. E. Jones, Priv. Comm.).

paint-coat cannot be made great enough to provide the required bond when applied to straight wires, some improvement might be obtained by altering the geometry; twisted square rods, or round rods with knobs or indentations, have been used for special purposes, and might be applied to this one, although it is feared that such devices would increase the expense.

Where piping coated with concrete, or steel bars embedded in concrete, is joined to similar steel coated with bitumen or enamel, a cell is set up, the steel in the alkaline surround being the cathode; serious corrosion may result, and insulated flange-joints should be applied to pipes at such situations (D. Hendrickson, *Corrosion* 1952, **8**, 212).

It has been suggested that similar cells may be set up where two different types of cement have been used. This is highly probable if the types are sufficiently dissimilar to produce different pH values; thus aluminous cement, which provides less alkali than Portland cement, might produce such a cell in the rather unlikely event of the two being used together. In one particular case where a two-cement cell was suspected, however, examination by R. H. Evans (p. 304) showed that no such cell was operative,

Cement mixtures may themselves be attacked and disintegrated by water containing sulphate. This is generally attributed to the formation of a very voluminous sulpho-aluminate, but other factors may be involved.

Other References

The external and internal coatings of pipes receive authoritative discussion from F. N. Speller (" Corrosion: causes and prevention " (McGraw-Hill)) and in the " Report of the Ministry of Health Deterioration Committee on Cast Iron and Spun Iron Pipes " 1950 (H. M. Stationery Office). Cathodic Protection receives discussion from many angles in " Cathodic Protection: Symposium and Discussion " 1953 (Society of Chemical Industry): " Kathodische Bescherming van ondergrondre Buisleidingen " 1955 (Metaalinstituut T.N.O., Delft): " Cathodic Protection: a Symposium " 1947 (Electrochemical Society and National Association of Corrosion Engineers). See also section by K. A. Spencer and D. A. Lewis in the " Anti-corrosion Manual " 1958, pp. 35–54 (Corrosion Prevention and Control). A Russian book on " Cathodic Protection of Pipe-lines and Storage Tanks " by V. A. Pritula has been translated and published by H.M. Stationery Office (1953).

Recent publications include a discussion of the corrosivity of soils by N. D. Tomashov and Y. N. Mikhailovsky, *Corrosion* 1959, **15**, 77t; current-potential relations in cathodic protection by W. J. Schwerdtfeger, *ibid.* 1958, **14**, 446t; the use of lead-platinum couples as anodes by L. L. Shreir, *Platinum Metals Rev.* 1959, **3**, 44; and the cathodic protection of bronze propellors against cavitation by H. G. Duff, *Corros. Tech.* 1958, **5**, 250.

CHAPTER IX

HYDROGEN EVOLUTION AND ACID CORROSION

Synopsis

The chapter opens by recalling that most of the wet corrosion processes so far considered have depended on the reduction of oxygen for the cathodic reaction. Certain alternative cathodic reactions now claim our attention.

The main cathodic reaction in the attack of non-oxidizing acids upon the baser metals is the evolution of hydrogen. This occasionally occurs on the nobler metals, but only in exceptional circumstances—which are first discussed. Then the attack of acids on the baser metals—especially iron, zinc and aluminium—is considered; it is pointed out that on zinc and aluminium there is an " induction period " before attack becomes vigorous, but that this phenomenon arises from different causes in the two cases. The different degrees to which oxide-films on different metals provide protection from acid is noted, and shown to depend on more than one factor. Evolution of hydrogen from alkaline and neutral salt solutions then receives consideration; the latter is important in the case of magnesium.

Next, attack by oxidizing acids (or mixtures of acids and oxidizing agents) claims attention; here the cathodic reaction is the reduction of the oxidizing acid (or oxidizing agent). Nitric acid is considered in some detail and it is shown that on some metals the cathodic reaction is autocatalytic—explaining why attack is more violent at inaccessible crannies and why the velocity of corrosion is diminished by movement in the liquid. The passivity of iron in nitric acid then comes up for notice.

After that, the various materials which are used in chemical industry to resist acids and alkalis are considered in turn. The theory of the resistance of stainless steel is developed; the special uses of aluminium, the remarkable properties of titanium and zirconium, the resistance conferred on metals by various alloying additions and the behaviour of noble metals, including the phenomenon of parting limits, all receive attention.

General

Alternative Types of Cathodic Reactions. In the discussion of corrosion by nearly neutral liquids in Chapter IV, the main cathodic reaction met with was the reduction of oxygen. Even from neutral solutions, however, a small amount of hydrogen can be evolved by iron and zinc, whilst on magnesium, hydrogen evolution constitutes the main cathodic reaction; the presence of oxygen is not needed for the rapid attack of sodium chloride solution upon magnesium. The alkali metals and alkaline earth metals decompose water violently, even in absence of salt; the heat evolution is such that the hydrogen sometimes catches fire.

Evidently we have two important alternative cathodic reactions; these proceed in steps (pp. 102, 877) but the final results can be written:—

$$2H^+ + 2e = H_2 \quad \text{(hydrogen-evolution type)}$$

and

$$O_2 + 4e + 2H_2O = 4OH^- \quad \text{(oxygen-reduction type)}$$

The first occurs at a potential lower than the second, that is closer to the potential at which the anodic reaction

$$M = M^{n(+)} + n(e)$$

can proceed. Thus the E.M.F. driving the hydrogen-evolution type of attack is much smaller than that driving the oxygen-reduction type. If oxygen could be replenished at great speed, its reduction would be the most probable cathodic reaction, but in acid solutions the hydrogen-evolution type of corrosion can also occur with considerable speed, and since the oxygen-replenishment will rarely proceed with sufficient rapidity to compete with this, the hydrogen-evolution type of corrosion usually predominates.

Evolution of Hydrogen as a Cathodic Reaction. There are three situations where the liberation of hydrogen becomes important, namely:—

(*a*) where the metal has a very negative normal electrode potential, as has *magnesium*;

(*b*) where the liquid has a very high hydrogen ion concentration, i.e. is *acid*;

(*c*) where the liquid keeps the concentration of normal cations continuously low through the formation of *complex ions*; for many metals this occurs in alkaline liquids or in cyanide solutions.

Impurities in the metal affect the rate of attack far more when the cathodic reaction is the evolution of hydrogen than when it is the reduction of oxygen, since there are some metals (said to possess a low overpotential) on which hydrogen is evolved at a potential relatively close to the reversible value, whilst on others (metals of high overpotential) it only occurs at potentials far removed from the reversible value. Certain materials, like impure zinc, are attacked slowly at first, but the corrosion-rate increases with time as a residue of impurities accumulates, from which the hydrogan can be evolved. Fig. 67 shows how greatly the presence of a second constituent in zinc may alter the rate of attack during total immersion in N/2 sulphuric acid; these curves should be compared with fig. 17 (page 89) which shows how little the rate of attack on zinc during partial immersion in N/10 potassium chloride is affected by an alloying constituent. Comparison between these two figures will bring out the increase of corrosion-rate with time in the case of acid attack, in contrast with its remarkable constancy in the case of attack by the chloride solution (R. Vondráček and J. Izák-

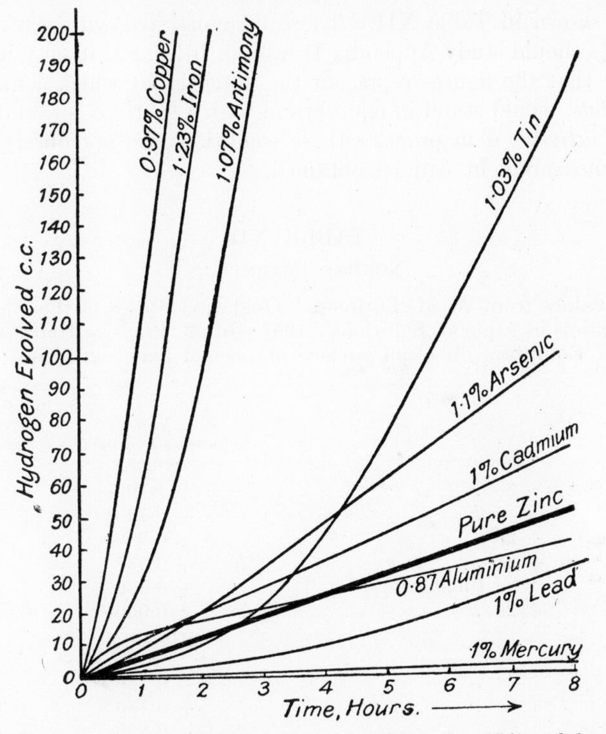

FIG. 67.—Effect of minor constituents on corrosion by N/2 sulphuric acid (R. Vondráček and J. Izák-Križko).

Križko, *Rec. Trav. chim. Pays-Bas* 1925, **44**, 376; R. Vondráček, *Coll. Czech. chem. Comm.* 1929, **1**, 627. Cf. C. W. Borgmann and U. R. Evans, *Trans. electrochem. Soc.* 1934, **65**, 249, for the effect of alloying in neutral solution).

Reduction of an Oxidizing Agent as a Cathodic Reaction. A different type of corrosion occurs when the acid is itself an oxidizing agent (e.g. nitric acid or hot concentrated sulphuric acid), or when an oxidizing agent is added to a non-oxidizing acid; a mixture of dilute sulphuric acid and potassium chlorate will attack a metal like copper, which is sufficiently noble to withstand dilute sulphuric acid in absence of oxygen or an oxidizing agent. In these cases, the working E.M.F. is higher than that available when the cathodic reaction is oxygen-reduction, and the attack, if it occurs at all, may become violent; with nitric acid, the action is often autocatalytic and the rate increases with time—for reasons suggested later. However, nitric acid and concentrated sulphuric acid, being strong oxidizing agents, may also produce films on the metal surface; when once a film of sesquioxide (p. 225) has appeared, it dissolves only very slowly in the acid, and the attack on the metal dies down; thus iron, which is vigorously attacked by dilute nitric acid, becomes passive in concentrated acid.

Table of Normal Potentials. It will be convenient to tabulate, at this point, the values of the normal potentials of the principal metals.

They are shown in Table XII. Those unacquainted with the meaning of such values should study Appendix II (esp. p. 1018), but it may be useful to state here that the figures represent the potential at which a metal, *in the film-free state*, would stand in equilibrium with a solution containing its ions *at normal activity*; if in practice these conditions are not observed, a very different measurement can be obtained.

TABLE XII

NORMAL POTENTIALS

Based on values from W. M. Latimer, " Oxidation States of the Elements and their Potentials in Aqueous Solutions " 1952 edition (Prentice-Hall). (Metal-Ion Equilibria on clean surface at normal ionic activities)

Equilibrium	Volt (normal hydrogen scale)	Correction for ten-fold change of activity (effective concentration), volts
$Au \rightleftharpoons Au^{+++} + 3e$	+ 1·50	0·0197
$Ag \rightleftharpoons Ag^{+} + e$	+ 0·7991	0·0591
$Hg \rightleftharpoons (Hg_2)^{++} + 2e$	+ 0·789	0·0295*
$Cu \rightleftharpoons Cu^{++} + 2e$	+ 0·337	0·0295
[$Cu \rightleftharpoons Cu^{+} + e$ (unstable)†	+ 0·522	0·0591]
$H_2 \rightleftharpoons 2H^{+} + 2e$	0·000 (arbitrary zero)	0·0591
$Pb \rightleftharpoons Pb^{++} + 2e$	− 0·126	0·0295
$Sn \rightleftharpoons Sn^{++} + 2e$	− 0·136	0·0295
$Ni \rightleftharpoons Ni^{++} + 2e$	− 0·250	0·0295
$Cd \rightleftharpoons Cd^{++} + 2e$	− 0·403	0·0295
$Fe \rightleftharpoons Fe^{++} + 2e$	− 0·440	0·0295
$Zn \rightleftharpoons Zn^{++} + 2e$	− 0·763	0·0295
$Al \rightleftharpoons Al^{+++} + 3e$	− 1·66‡	—§
$Mg \rightleftharpoons Mg^{++} + 2e$	− 2·37‡	—§

* The fact that the potential shifts by 0·0295 volt and not by 0·0591 volt for every ten-fold concentration change provides evidence that the mercurous ion is $[Hg_2]^{++}$ and not Hg^{+}.

† If it were possible to obtain momentarily a normal concentration of Cu^{+} ions, it would at once decompose when brought into contact with copper, by the reaction $2Cu^{+} = Cu + Cu^{++}$ which runs nearly to completeness at high concentrations, although at low ones it comes to equilibrium when the concentration of Cu^{++} is very low. This explains why we have to use the potential for $Cu \rightleftharpoons Cu^{+} + e$, and not $Cu \rightleftharpoons Cu^{++} + 2e$ as the starting point of our calculation on p. 191.

‡ These equilibria are not normally realized in aqueous solution owing to film-formation. The numbers in such cases are usually calculated from thermodynamic data.

§ In practice the concentration of cation added hardly affects the potentials of the metals as normally met with. Plumb's measurements for Al, however, depend both on the Al^{+++} and H^{-} concentrations (R. C. Plumb, Priv. Comm., April 1, 1958). See also p. 1018 of this book.

Hydrogen-evolution Type of Attack

Conditions under which Noble Metals can liberate Hydrogen. It is sometimes stated that the noble metals standing above hydrogen in the Table of Normal Potentials, such as gold, platinum, silver, copper and mercury, are incapable of liberating hydrogen gas when placed in acid, whereas base metals, standing below it, can do so. This statement—as it stands—is not quite accurate, and there are apparent exceptions which will be considered shortly. To make it accurate, the statement should be reworded as follows:—

"If a metal is placed in acid of normal activity containing cations of that metal at normal activity, free from oxygen but saturated with hydrogen at one atmosphere pressure, then corrosion is *impossible* for a metal standing *above* hydrogen in the table, and *possible* for a metal *below* it if the surface is free from film; in the last case, however, the corrosion need not always be rapid."

Pure acid does not contain metallic cations. Thus, if copper is placed in pure dilute sulphuric acid free from oxygen and hydrogen, it should displace a little hydrogen at first; but as soon as the molecular hydrogen and the ionic copper in the liquid have reached such a level that the potential established for the equilibrium

$$H_2 \rightleftharpoons 2H^+ + 2e$$

is equal to that established for the equilibrium

$$2Cu^+ + 2e \rightleftharpoons 2Cu$$

the E.M.F. vanishes and corrosion ceases. This must happen long before the liquid becomes saturated with hydrogen so that no hydrogen bubbles need be expected.* For all practical purposes, it can be said that copper remains unattacked by dilute sulphuric acid, provided that air is excluded.

In 1934, the late Prof. Oliver Watts sealed up in a tube containing N/2 sulphuric acid a specimen of bright copper (part of which had been platinum-

* The equilibrium potential of H_2 (sat. at 1 atm.) | H^+ (normal) is of course $\pm$ 0·000 volt and rises by 0·030 volt for every ten-fold reduction of hydrogen pressure; the solubility of hydrogen at 25°C. is about 0·000015 g./l. or about 7×10^{-6}M. The equilibrium potential of Cu | Cu^+ (normal) is + 0·522 volt and descends by about 0·059 volt for every ten-fold reduction of the cuprous ion activity (the cupric ions can be neglected for reasons explained below Table XII). When the acid starts to corrode the copper, Cu^+ ions and H_2 molecules are formed. Thus the potential Cu | Cu^+ rises with time, whilst the potential H_2 | H^+ falls with time; when they become equal, the driving E.M.F. becomes zero, and corrosion must stop, unless the hydrogen can escape continuously from the liquid into the gas phase.

If the cuprous ion concentration were to reach 10^{-7}M, the potential Cu | Cu^+ would have risen to + 0·522 − (7 × 0·059) volt, i.e. about 0·11 volt. Now the production of 10^{-7}M of (Cu^+] would involve the production of 5×10^{-8}M of [H_2] which is rather less than 10^{-2} times the saturation value, so that the potential H_2 | H^+ would have fallen almost to 2 × 0·030 volt, i.e. to 0·06 volt (unless hydrogen were to pass out of the liquid continuously—as might happen if a vacuum were maintained above the liquid). Under ordinary conditions, the driving E.M.F. would have disappeared *before* the copper ion concentration reached 10^{-7}M, since the Cu | Cu^+ potential would have risen to the level to which the H | H^+ potential had fallen. Thus in absence of an applied E.M.F. hydrogen gas cannot be evolved (except under vacuum) nor will copper be detected by ordinary tests in the liquid when equilibrium is set up and action ceases. In other words, there will be no appreciable corrosion of copper by a non-oxidizing acid in absence of oxygen.

In the calculation given, only rough figures are indicated; the laws of electrochemistry will not be obeyed accurately at these very low concentrations, so that great exactitude would serve no useful purpose.

plated, so as to provide every facility for the evolution of hydrogen), and presented it to the Author; the specimen has been in his possession for 25 years; the copper remains bright and uncorroded, and the acid shows no sign of a blue colour. In contrast, a strip of sheet copper partly immersed by Mr. C. Taylor in N/2 sulphuric acid with air above the liquid was eaten right through at the water-line within 3 weeks, the liquid becoming blue.*

Conditions under which Noble Metals can liberate Hydrogen. The attack of a non-oxidizing acid on copper in absence of oxygen will only continue if the molecular hydrogen and cationic copper are both removed as quickly as they are formed, so that equilibrium is never attained. This happens when copper is placed in boiling concentrated hydrochloric acid. In that liquid, simple copper cations will never accumulate, since the copper passes into the state of complex anions $[CuCl_2]^-$; the equilibrium

$$[CuCl_2]^- \rightleftharpoons Cu^+ + 2Cl^-$$

is established when the complex ions vastly exceed the simple ones. Meanwhile, the hydrogen is continually removed in the bubbles of hydrogen chloride and water which are being boiled off; if the distillate is condensed (giving hydrochloric acid) a small amount of gas is left uncondensed, and this has been identified as hydrogen. Thus copper is attacked by boiling concentrated hydrochloric acid even in absence of oxygen. (U. R. Evans, *J. Inst. Met.* 1923, **30**, 239, esp. p. 256; confirming W. A. Tilden, *J. Soc. chem. Ind.* 1886, **5**, 85).

Similarly, copper will displace hydrogen from potassium cyanide solution even in the cold, passing into the liquid as complex $[Cu(CN)_2]^-$ ions; this occurs despite the low concentration of hydrogen ions (the cyanide solution is slightly alkaline, owing to hydrolysis, since HCN is a weak acid). No oxygen is needed for the attack of cyanide on copper. In contrast, gold, which has a normal potential considerably more positive than that of copper, requires oxygen for dissolution in cyanide—a fact well known in the gold industry. Gold resists most acids, but dissolves in aqua regia (p. 962).

The dissolution of copper in ammonia (with or without ammonium salts) is also made possible through the formation of complex ions, but in this case they are cations; here oxygen is needed (p. 967). The reaction, which is diffusion-controlled in certain circumstances, has been studied by S. Uchida and I Nakayama, *J. Soc. chem. Ind. Japan* (English version) 1933, **36**, 416B, 635B under pipe-flow conditions, and also by B. C. Y. Lu and W. F. Graydon, *J. Amer. chem. Soc.* 1955, **77**, 6136.

Conditions under which Base Metals can liberate Hydrogen. Metals standing below hydrogen in the table of potentials are capable, under normal conditions, of liberating hydrogen gas from non-oxidizing

* The attack on a vertical copper sheet partly immersed and wetted above the water-line appears to be connected with electrical currents flowing between a cathodic zone on the wetted area above the water-line and an anodic zone at the meniscus; for evidence, see U. R. Evans, *J. Inst. Met.* 1923, **30**, 239, esp. p. 272. Slightly different views are expressed by A. I. Kinevsky, *J. appl. Chem. U.S.S.R.* 1955, **28**, 1088 (English translation, p. 1043).

acids, i.e. dilute hydrochloric or sulphuric acid. Electrochemical theory suggests that this should only fail to occur

(1) if the hydrogen ion concentration is very low (such solutions, however, can hardly be termed " acid ");

or (2) if the metal ion concentration is extremely high (the solubility of the salt, however, sets a limit to the concentration);

or (3) if the hydrogen gas pressure is very high (this will rarely occur, however, under working conditions).

There are certain other conditions where liberation of hydrogen, although a " possible reaction " in the sense that it can proceed with a decrease of free energy, occurs *so slowly as to be unimportant*. These cases, which are in part connected with overpotential, will first be considered.

Experiment shows that lead, tin and nickel, which stand immediately below hydrogen in the table, liberate hydrogen from dilute hydrochloric acid only very slowly (in contrast, iron, zinc and aluminium generally liberate it rapidly, although in the last two cases the attack is usually slow at the outset, as shown later).

There is more than one cause for the slowness of the attack on lead, tin and nickel. Lead resists dilute sulphuric acid, even in presence of oxygen, owing to the low solubility of lead sulphate. However, the slow attack by hydrochloric acid is due to a different cause, since the solubility of lead chloride is about 1% at 20°C. Here it is the high overpotential value of hydrogen evolution on a lead surface which is responsible for the sluggish reaction (p. 1026). The same factor explains the feeble reaction of tin with dilute hydrochloric acid; even a very slow evolution of hydrogen will depress the potential of

$$2H^+ + 2e = H_2$$

almost down to the level of

$$Sn = Sn^{++} + 2e$$

so that the E.M.F. has almost vanished before the rate of attack becomes serious. In both cases contact with platinum black—a surface of low overpotential—accelerates the evolution of hydrogen.

Nickel has a lower hydrogen overpotential than lead or tin, but the anodic reaction is subject to a high overpotential, which causes the potential of

$$Ni = Ni^{++} + 2e$$

to rise and reduce the E.M.F. to a very small value before the attack becomes serious. In presence of oxygen, the attack is much greater since, although the anodic reaction is not affected, the basic E.M.F. is much higher, and a greater corrosion-rate is reached before the E.M.F. becomes too small to force the corrosion-current over the small resistance of the corrosion-circuit.

Hydrogen Evolution by Iron. When iron is placed in dilute acid, hydrogen is evolved quite freely, despite a high anodic potential, since the basic E.M.F. is here much higher than on nickel (the normal potential of iron is $-$ 0·44 volt, as opposed to $-$ 0·25 volt for nickel). The anodic overpotential is greatly reduced by the presence of hydrogen sulphide (see

p. 878). Thus the addition of hydrogen sulphide to the acid, or the presence in the metallic phase of iron or manganese sulphide which liberates hydrogen sulphide in acid, greatly stimulates the attack. If, however, the iron (or steel) contains copper in sufficient quantity to remove hydrogen sulphide from the liquid by precipitation as stable copper sulphide, then the attack is much slower. The action of citric acid on steels containing Cu and S in a ratio above 1·38 is slower than that on steels in which the ratio stands below that limit—an important fact established by T. P. Hoar and D. Havenhand, *J. Iron St. Inst.* 1936, **133**, 239P; see also T. P. Hoar's later work with T. N. Morris and W. B. Adam, *ibid.* 1939, **140**, 55P; 1941, **144**, 133P.

Although the overpotential of hydrogen evolution on iron is lower than that on tin, it is much higher than on platinum. Thus the addition of a trace of platinum salt to normal hydrochloric acid in contact with iron (which at once precipitates metallic platinum) greatly increases the rate of attack, the platinum acting as a cathodic surface on which hydrogen can readily be eliminated; we have now the two reactions occurring on different areas

anodic reaction involving attack on the iron $Fe = Fe^{++} + 2e$
cathodic reaction involving gas evolution from the platinum $2H^+ + 2e = H_2$

In the case of pure iron, both reactions occur on the iron surface, and quite likely at contiguous points— or even at the same point at different moments. Some types of iron, however, already contain a second phase favouring hydrogen evolution. For instance, grey cast iron, containing graphite flakes, evolves hydrogen more easily than relatively pure iron, and in that case the addition of platinum salt produces no acceleration (W. D. Richardson, *Trans. Amer. electrochem. Soc.* 1920, **38**, 265).*

The behaviour of cast iron in acids is, however, not simple. Some samples, which are attacked by acid more quickly than steel at the out-set, are less attacked in experiments lasting over some days—probably owing to the accumulation of a protective skin of silica (J. R. Rylands and J. R. Jenkinson, *Proc. Instn. mech. Engrs.* 1948, **158**, 405, esp. fig. 3, p. 406).

When iron is first placed in acid, appreciable quantities of iron ions appear in the solution within the first few seconds. This rapid initial attack, observed by Tödt, has been shown to depend greatly on the previous history of the specimen by the experiments of Grubitsch, who attributes it to the

* Recent work by Buck and Leidheiser suggest that other factors besides overpotential play a part in determining the effect of traces of foreign metals in acid corrosion; they studied the effect of various cations at 10^{-7} to 10^{-5}M concentration on the corrosion of Al, Cd, Mg and Zn in citric acid (also of Cu, Ag, Ti, Mo and W in HCl). Ruthenium, nickel, palladium and platinum caused acceleration in most cases, but their order of efficacy varied with the metal under attack, and parallelism with overpotential was far from perfect. Lead inhibited corrosion of iron and aluminium, whilst mercury inhibited that of nickel, tin and zinc—as also did lead and tin to some extent, the order varying with the metal under attack; perhaps poisoning of spots which would normally catalyse the cathodic reaction may be a cause. For experimental results see W. R. Buck and H. Leidheiser, *Corrosion* 1958, **14**, 308t; *Z. Elektrochem.* 1958, **62**, 690.

reductive dissolution of an oxide-film (F. Tödt, R. Freier and W. Schwarz, *Z. Elektrochem.* 1949, **53**, 132, esp. figs. 7 and 8, pp. 140, 141; see also " Korrosion VIII ", 1954–55, p. 6 (Verlag Chemie). H. Grubitsch, *Z. Metallk.* 1956, **47**, 188; see also p. 129 of this book).

The attack of non-oxidizing acids on carbon steel is greatly affected by previous heat-treatment; the effect of the temperature of annealing before quenching on the corrosion by 1% sulphuric acid is very marked—as shown in the curves of H. Endo, *Sci. Rep. Tohoku Univ.* 1928, **17**, 1245, 1265, esp. p. 1261.

The blistering, embrittlement and cracking caused when atomic hydrogen penetrates into the interior, instead of being harmlessly evolved as molecular hydrogen on the external surface, may sometimes be more dangerous than the destruction of metal; these troubles are discussed in Chapter XI.

Hydrogen Evolution by Zinc. When we pass from iron to zinc—where the basic E.M.F. is greater but the overpotential of hydrogen evolution higher—the effect of impurities becomes far more important. In fig. 67 (p. 311) Vondráček's straight-line graph for unalloyed zinc shows a slow evolution of hydrogen, but no tendency to increase with time; on such a material, contiguous points must act as cathodes and anodes;* or perhaps a given point can act as cathode at one moment and suffer anodic attack the next. However, where a second metal is present, the attack is usually stimulated and becomes increasingly fast as the second metal accumulates as a dark sponge on the zinc surface—so that the curves are no longer straight; here the second metal acts as cathode and the zinc as anode.† The metals which cause most stimulation to the attack upon zinc by N/2 sulphuric acid (copper and iron) are those which are known by independent electrochemical experiments, to possess relatively low overpotential values.‡ Certain other metals, however, retard attack, notably aluminium, which readily forms obstructive films resistant to acid (see p. 319). Lead depresses attack in the early stages, but after 7 hours the corrosion of lead-containing zinc has become about at least as fast as that of lead-free zinc. Tin also depresses attack in the early stages, but after a few hours the attack is quicker than in the case of tin-free zinc. The causes are complicated, but it may be said that any relatively noble metal, whilst in solid solution in

* The special reason for the spatial separation of anodic and cathodic areas is absent. If on zinc in sodium chloride, the anodic and cathodic products appeared at contiguous points, an obstructive film of zinc hydroxide would be formed; in acids, film formation is unlikely.

† If the second metal forms a second phase in the original zinc, it will remain as an undissolved residue. If it exists in solid solution, it will be dissolved and then reprecipitated as a sponge.

‡ The complaint has been heard that the " explanation " of the effect of impurities on the acid attack upon zinc based on " overpotential " is merely a case of giving a learned name to the phenomenon, which still remains unexplained. This is not entirely just. Overpotential is a quantity which can be measured by electrochemical experiments quite independently of the attack of acids upon metals. If once its existence be accepted, the fact that certain minor constituents in zinc stimulate acid attack receive an explanation, without further *ad hoc* assumptions.

zinc, is likely to slow down the anodic reaction; later, when redeposited as a sponge, it is likely to stimulate the cathodic reaction even if the overpotential is high, since the surface area available for the cathodic reaction becomes large. This probably explains those cases where initial depression of attack is followed by stimulation.

Commercial zinc, placed in dilute acid, starts to corrode slowly, but as the impurities are reprecipitated as a black scum, the gas evolution becomes much more rapid. Different grades of zinc behave differently, but the Author's early experiments with zinc produced by distillation may be quoted. In one variety, the impurities were largely present at the grain-boundaries, and in the early stages microscopic observation showed the bubbles being evolved mainly from these boundaries. In another experiment, at the stage when the whole surface had become covered with a black sponge, gas evolution being already general, part of the surface was wiped free from the scum, and it was then noticed that hydrogen practically ceased to be evolved from the wiped area, although it continued elsewhere.

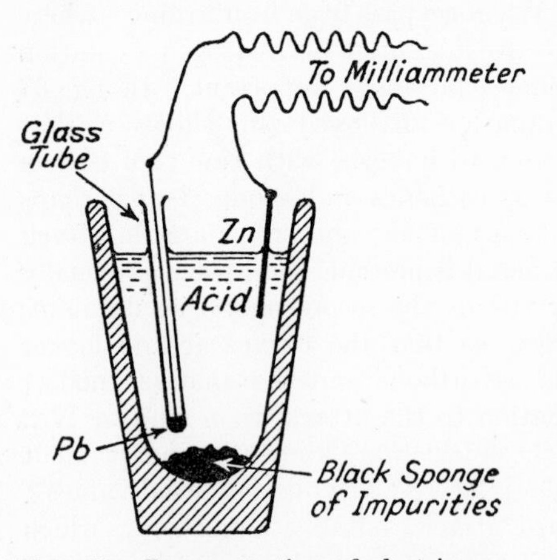

FIG. 68.—Demonstration of electric currents passing between Zinc as anode and black residue of impurities as cathode.

In another case, a quantity of the black residue was collected, and placed in a vessel with rounded bottom containing acid; into the same acid was dipped a strip of clean sheet zinc connected to one terminal of a milliammeter (fig. 68); a lead rod sheathed by a glass tube (the lead being rounded at the bottom by previous heating) was connected by a flexible wire to the other terminal. When the lead rod was pressed down into contact with the black spongy mass, a current was indicated on the meter showing the sponge to be the cathode, and simultaneously gas was observed rising from the sponge; when the lead contact was raised, the gas evolution ceased and the current fell to a low value (although not quite to zero, since the lead surface itself continued to act as an inefficient cathode). These and other observations leave no doubt regarding the manner in which the precipitated impurities provide cathodic surfaces from which hydrogen can be liberated (U. R. Evans, *J. Inst. Met.* 1923, **30**, 239, esp. p. 254; 1925, **33**, 27, esp. p. 37).

The delay observed between the introduction of zinc into acid and the time when gas evolution becomes conspicuous is often called the *Induction Period*, but it has no sharp termination; the author's early curves (fig. 69) showed hydrogen evolution developing gradually; curves II and III reveal a slight unevenness, possibly due to fluctuations in the electrical contact between the precipitated impurities and the zinc.

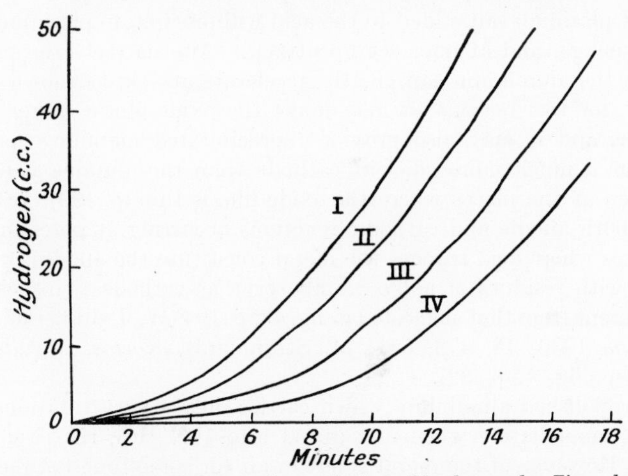

FIG. 69.—Early curves showing how the evolution of hydrogen by Zinc placed in acid starts slowly and increases with time; the *irregularities* (especially noticeable on curves II and III) are *genuine*.

Hydrogen Evolution by Aluminium. An induction period is noticed when aluminium is introduced into acid, but arises from a cause different to that discussed in connection with zinc; it is due to the slow rate at which acid dissolves the air-formed film of alumina, which must be removed before rapid attack on the metal can start.

The period of induction is much shorter on impure aluminium than on the purest variety, and much shorter in concentrated than in dilute hydrochloric acid; a short preliminary dip in concentrated acid will permit immediate attack when the aluminium is subsequently introduced into relatively dilute acid, which would otherwise show an induction period. The reason why the induction period—measurable perhaps in minutes on impure metal—may last several days on the purest varieties, is easily understood. The pure oxide formed on the pure aluminium contains few lattice-defects, so that its rate of dissolution by the acid is slow; moreover, even when it has been eaten through at some points, exposing the metal, the area where pure oxide still remains cannot act as an efficient cathode, owing to the poor conductivity for electrons. The varying behaviour of aluminium of different degrees of purity in acid are described by W. Palmaer, " Corrosion of Metals " 1931, Vol. **2**, p. 106 (Svenska Bokhandelscentralen A.B.); see also H. Moore and E. A. G. Liddiard, *Chem. and Ind.* (*Lond.*) 1935, p. 786, esp. p. 789.

The fact that the invisible film on aluminium resists acids for a considerable time whereas the film on iron disappears in a few minutes—or even seconds—is due to the fixed valency of aluminium, which prevents the reductive dissolution responsible for the rapid destruction of films on iron (p. 225). However, the presence of ions of a second (noble) metal in the liquid will allow the deposition of that metal on the surface of the aluminium and this may hasten the on-set of corrosion. Although pure aluminium placed in dilute acid evolves practically no hydrogen for hours or even days, a trace of platinum salt added to the acid will precipitate platinum metal as a black sponge, and at once set up attack. Alternatively the presence of copper in the aluminium can greatly accelerate attack, as shown by Müller and Löw, for two reasons; it may make the oxide-film a better electronic conductor, and it may also provide reprecipitated metallic copper which will act as a much more efficient cathode than the alumina is ever likely to be, even at the places where the oxide-film is thin (p. 190). Straumanis pictures both anodic and cathodic reactions occurring at pores in the film; those pores where acid reaches the metal constitute the anodes, while those filled up with residues of noble metals serve as cathodes; this picture is a little different from that suggested in fig. 43, p. 194 (W. J. Müller and E. Löw, *Aluminium* 1936, **18**, 478, 541; M. Straumanis, *Korros. Metallsch.* 1938, **14**, 81, esp. fig. 1, p. 82).

Electropolished aluminium which carries only a very thin film liberates hydrogen immediately when introduced into 10N HCl, the reaction being violent. However, if the metal is left in air for some time between electropolishing and introduction into the acid, it behaves at first as a relatively inert material, and the attack only becomes rapid after an induction period —presumably representing the time needed to dissolve the oxide-film (P. A. Jacquet, Assoc. Ital. Metallurgica, 2nd Congress, 1948, p. 31. G. Chaudron, *Helv. chim. Acta* 1948, **31**, 1553).

The fact that the induction periods observed on zinc and aluminium are due to two distinct causes is illustrated by the different behaviour of the two metals towards an oxidizing acid solution, in which the cathodic reaction will not require the liberation of hydrogen. Nitric acid attacks pure aluminium more slowly than impure aluminium, presumably because the same delay occurs in dissolving the pure oxide as is noticed in hydrochloric acid. In contrast, spectroscopically pure zinc is attacked almost as quickly as ordinary zinc by an acid liquid containing hydrochloric acid and potassium nitrate (which can be regarded as containing nitric acid, since both H^+ and NO_3^- are present). Details are provided by C. V. King and M. M. Braverman, *J. Amer. chem. Soc.* 1932, **54**, 1744, esp. p. 1756.

Factors governing the Rate of Destruction of Oxide Films by Acids. There is no reason to doubt that the contrast just presented is connected with the quicker dissolution of zinc oxide in acid than that of alumina. The rate of dissolution of an oxide cannot be decided by casual experiments on the powdered oxides taken from the laboratory bottles since the grain-size of the powders will not, in general, be the same, and thus the surfaces exposed will differ. However, in the case of massive oxide as found in the earth, zinc oxide dissolves far more rapidly than alumina, which the mineralogical text-books describes as " insoluble " in acids—meaning thereby " slowly soluble ". However, this leaves a question still to be answered—why does zinc oxide dissolve so much more quickly than alumina? An experimental study of the dissolution-rates of the different oxides and their relation to crystal-structure and defect-structure is long overdue.

Perhaps the most probable explanation is that based on lattice-defects.

Zinc oxide contains metal in excess of the formula ZnO. If the excess metal is interstitial, it might easily enter the liquid, and the process having started and the structure being thereby loosened, dissolution would probably continue. Nickel oxide, which has a similar formula, contains less metal than corresponds to NiO, and is resistant to acids; heat-tinted nickel can remain some time in acid without appreciable destruction of the colours. Pure alumina, which is only very slowly attacked by acids, is believed to have a composition fairly close to Al_2O_3, but presence of iron or copper in alumina should increase the ratio of metal to oxygen ions, since trivalent ions will probably be replaced by divalent ones; it appears that a film of impure alumina containing such metals does dissolve more quickly than a film of pure alumina, the induction period being far shorter on impure than on pure metal—as already stated.

The Difference Effect. Much interest has been shown in the effect of bringing a piece of platinum into contact with metal which is evolving hydrogen from acid. When the metal is zinc, the attack is essentially controlled by the cathodic reaction, being fixed by the rate at which hydrogen can be evolved on the impurities; if platinum is introduced, connected electrically to the zinc, this provides an efficient cathodic surface, and the total hydrogen production is increased, but the part of the hydrogen coming from the zinc surface is diminished. The phenomenon—first observed by Thiel—is known as the *positive difference effect*. If the experiment is repeated with aluminium where the rate of attack is controlled by the anodic reaction, being fixed by the state of the film, the effect of contact with platinum is often to increase, not only the total corrosion-rate, but also the amount of hydrogen coming from the aluminium, since the increased corrosion extends the damage to the film—as indeed is confirmed by visual observation; this is the *negative difference effect*. A negative effect is found on magnesium placed in sodium chloride, which confirms Whitby's belief that the oxide-film on magnesium controls the corrosion-rate. Aluminium placed in alkali (which rapidly destroys the film, quite apart from platinum contact) or amalgamated aluminium in acid (where the amalgamation prevents the film from being protective, as is seen from the increased reactivity of the amalgamated metal in many liquids), give the positive effect. Aluminium and titanium exhibit a positive effect in hydrofluoric acid, which attacks these metals more easily than most other acids.

Further information will be found in the papers of A. Thiel and J. Eckell, *Z. Elektrochem.* 1927, **33,** 370; A. Thiel and W. Ernst, *Korros. Metallsch.* 1930, **6,** 97; W. O. Kroenig and W. N. Uspenskaja, *ibid.* 1935, **11,** 10; 1936, **12,** 123; L. Whitby, *Trans. Faraday Soc.* 1933, **29,** 1318; M. Straumanis, *Z. phys. Chem.* (*A*) 1930, **148,** 349; *Korros. Metallsch.* 1938, **14,** 67, 81; *J. electrochem. Soc.* 1955, **102,** 304 (with Y. N. Wang); 1958, **105,** 284; W. J. Müller, *Korros. Metallsch.* 1938, **14,** 49, 77, 83; *Trans. electrochem. Soc.* 1939, **76,** 171; V. Čupr, *Korros. Metallsch.* 1939, **15,** 256.

Evolution of Hydrogen in Alkali. Whilst most metals which form simple cations are attacked by acids, the attack frequently decreases as the pH rises, and then finally rises again in the alkaline range. Thus zinc (fig. 70) is readily attacked in the acid range, forming soluble zinc salts, shows a minimum attack about pH 11, and then dissolves in definitely alkaline solutions; zinc in strong sodium hydroxide yields sodium zincate Na_2ZnO_2, ionizing as $2Na^+$ and $(ZnO_2)^{--}$. The minimum at pH 11 is believed to correspond to the conditions under which the solubility of the hydroxide is minimal The curve reproduced in fig. 70 is due to B. E. Roetheli, G. L. Cox and W. B. Littreal, *Met. Alloys*. 1932, **3,** 73.

This state of affairs is not peculiar to zinc. Most metals show a rise of corrosion-rate at both ends of the pH scale, although the position of the minimum varies in different cases. On aluminium the minimum occurs on the acid side of neutrality and the rise at the alkaline end is very marked. The rapid attack of aluminium by washing soda giving soluble sodium aluminate was a serious matter in the early days of the aluminium saucepan (see p. 10). Other metals showing a rise at the alkaline end of the pH range include tin and lead, which form stannites and plumbites respectively; copper forms cuprites, but there should be little attack at either acid or alkaline end if dissolved oxygen is absent. The minimal attack is said to occur at about pH 8 on lead, and between 8 and 9 on copper. Cadmium shows no rise of corrosion at the alkaline end of the pH range (M. Pourbaix, C.I.T.C.E. Meeting 1954, p. 137, quoting Russian measurements by A. V. Shavelow).

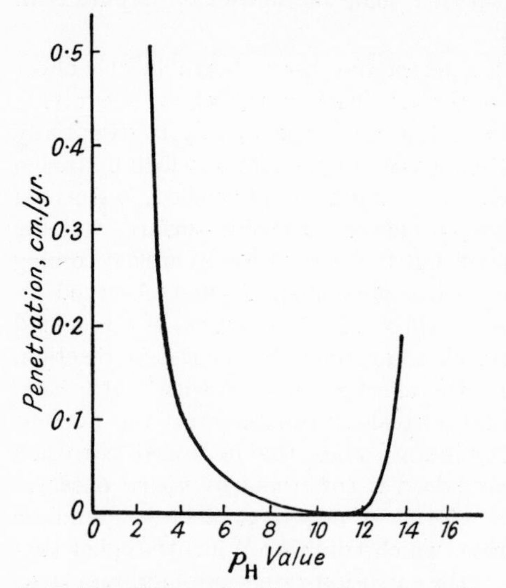

FIG. 70.—Effect of pH on Corrosion of Zinc (B. E. Roetheli, G. L. Cox and W. B. Littreal).

Few metals are completely proof against attack by alkali. Silver, nickel and certain stainless steels are the best.* Iron is remarkably resistant to cold, dilute sodium hydroxide, but dissolves in hot concentrated alkali, producing soluble sodium ferroate† (Na_2FeO_2), which is analogous to sodium zincate, but much less stable; unlike the zincate, the ferroate can only exist in presence of concentrated alkali, and, as will be seen in connection with boiler corrosion (p. 456), the solution is apt to deposit magnetite.

It may appear strange that zinc and aluminium can liberate hydrogen from alkaline solution as easily as from acid solution, seeing that the concentration of hydrogen ions in alkali is far lower. However, the basic

* Nickel is appreciably attacked by sodium hydroxide, in absence of added water, in the range 600–1000°C.; see D. M. Mathews and R. F. Kruh, *Industr. engng. Chem.* 1957, **49,** 55.

† *Hypoferrite* would be a better name.

E.M.F. of zinc placed in alkali is not very different from that operating when it is placed in acid; although a lowering of the pH concentration produces a marked shift of the potential $H_2 | H^+$ in the negative direction, the formation of complex ions such as $[ZnO_2]^{--}$ shifts that of $Zn | Zn^{++}$ to a comparable extent, since the concentration of simple Zn^{++} is kept very low in an alkaline solution. Thus the E.M.F. is not greatly altered. The production of hydrogen by the cathodic reaction is probably not due to the discharge of the few hydrogen ions present even an alkaline solution

$$H^+ + e = H$$

but rather to the dragging away of the OH portion of the adsorbed water molecules, leaving H atoms.

$$H_2O + e = OH^- + H$$

The rate of dissolution of aluminium in sodium hydroxide depends on many factors, being stimulated by impurities of low overpotential in the metal, and dependent, in rather complicated fashion, on the alkali concentration. For details see M. A. Streicher, *Trans. electrochem. Soc.* 1948, **93,** 285; 1949, **96,** 170, esp. p. 180.

Evolution of Hydrogen from pure Water. It is convenient here to mention in passing corrosion of the hydrogen-evolution type in the absence of acids or alkalis. Some metals can liberate hydrogen from salt-free water. This occurs rapidly where the hydroxide is soluble—familiar examples being provided by sodium and potassium. Thallium, a more noble metal, possesses a soluble hydroxide, and is appreciably attacked by warm water in the absence of oxygen—although oxygen greatly stimulates the attack (J. T. Waber and G. E. Sturdy, *J. electrochem. Soc.* 1954, **101,** 583).

Many other metals liberate hydrogen from hot water, especially from water heated under pressure so that temperatures far exceeding 100°C. are obtained. Special importance attaches to the case of iron where the lower (ferrous) hydroxide is appreciably soluble. These matters are discussed in Chapter XII.

Evolution of Hydrogen by Magnesium. The hydroxide of magnesium is sufficiently soluble to confer an alkaline reaction on water, but not so high as to prevent the formation of a protective film when magnesium is placed in cold water. Thus the attack is slow, and is little increased by the presence of such salts as nitrate or sulphate; it is, however, greatly stimulated by chloride. The attack of many salt solutions tends to fall off with time if the pH value is maintained about 10·2—at which the solubility of magnesia hydroxide is low (W. F. Higgins, *Chem. and Ind.* (*Lond.*) 1958, 1604, esp. p. 1605).

The effect of impurities of low overpotential in causing attack on magnesium by chloride solution is very marked. If the impurity content is below a certain " tolerance limit " (0·017% for iron, and 0·0005% for nickel), attack soon becomes very slow, apparently because the hydroxide formed by interaction between anodically produced magnesium chloride and the sodium hydroxide formed on cathodic specks of impurity covers up these specks, and attack then ceases. Above the limit, serious attack continues; it has been suggested that if the impurity specks are sufficiently numerous (and hence close together), the corrosion caused by one speck will have exposed a second speck before the first speck has been detached by undermining or covered up. This seems to be confirmed by Beck's observation that, although the addition of 8% aluminium to pure magnesium greatly alters the tolerance limit for iron, the limit occurs at the same distance of separation between the iron-bearing particles which are supposed to act as cathodes. When the distance becomes so large that one particle is dislodged before the next is reached, the rapid attack does not develop (W. Beck, Priv. Comm., Dec. 22, 1957).

The fact that sulphates cause much less attack than chlorides may be due to the fact that they precipitate magnesium hydroxide, which would otherwise exist in colloidal solution close to the metal; but alternative explanations are available. The low tolerance limit for nickel has been attributed to the low overpotential of the compound Mg_2Ni. The bad effect of iron and nickel is largely neutralized if manganese is present; it has been suggested that manganese surrounds the iron particles, and prevents them from acting as cathodes. Alloys containing manganese are much used where there is to be risk of exposure to sea-water; they tend to develop golden-yellow protective films.

The facts regarding tolerance limits are presented, and various interpretations discussed, in papers by R. E. McNulty and J. D. Hanawalt, *Trans. electrochem. Soc.* 1942, **81,** 423; J. D. Hanawalt, C. E. Nelson and J. A. Peloubet, *Trans. Amer. Inst. Min. Eng.* 1942, **147,** 273; F. A. Fox and C. J. Bushrod, *J. Inst. Met.* 1944, **70,** 325; 1945, **71,** 255; 1946, **72,** 51; F. A. Fox and J. K. Davies, *ibid.* 1947, **73,** 553; C. J. Bushrod, *ibid.* 1947, **73,** 567; F. W. Fink, *Metal. finish. J.* 1955, **1,** 324; A. Beerwald, *Z. Metallk.* 1941, **33,** 28. For the electrochemistry of magnesium corrosion, see G. E. Coates, *J. chem. Soc.* 1945, pp. 478, 480, 484 (3 papers); *J. Inst. Met.* 1945, **71,** 457. For reaction in hydrochloric acid, see R. Roald and W. Beck, *J. electrochem. Soc.* 1951, **98,** 277.

The effect of iron in stimulating the corrosion of magnesium is brought out in statistical studies by E. R. W. Jones and M. K. Petch, *J. Inst. Met.* 1947, **73,** 129.

Attack by Oxidizing Acids

Attack of Nitric Acid on Base Metals. Those metals like magnesium, aluminium and zinc which liberate hydrogen from hydrochloric acid are readily attacked by nitric acid, but the hydrogen atoms, instead of joining to form molecular hydrogen, react with nitric acid to form ammonia (or rather ammonium nitrate); in some cases a salt of hydroxylamine (NH_2OH) is formed. The cathodic reaction probably starts by the discharge of hydrogen ions

$$H^+ + e = H$$

followed by some such reaction as

$$8H + HNO_3 = NH_3 + 3H_2O$$

which may itself occur in steps. The continuous removal of atomic hydrogen as it is formed, serves to maintain a higher working E.M.F. than if the final cathodic product were H_2. Thus the attack on magnesium and zinc may be somewhat violent; in the case of magnesium, where the basic E.M.F. is sufficient to produce atomic hydrogen very quickly, some of this may escape oxidation, so that the gases evolved may contain molecular hydrogen, provided that the nitric acid is dilute (oxidizing properties decline with dilution more than acidic properties). On aluminium, the oxidizing properties of nitric acid serve to build a protective oxide-film, which, like other sesqui-oxides, is only slowly dissolved by acid. Indeed in the concentrated acid the attack is so slow that aluminium vessels can be used for storage and transport; the diluted acid permits more attack, but this is much slower than that produced by acid of the same concentration acting on such noble metals as copper and silver.

Attack of Nitric Acid on Noble Metals. Metals like silver, copper or mercury which cannot liberate hydrogen from hydrochloric or sulphuric acid, are unlikely to produce hydrogen or hydrogen-rich products, such as ammonia. However, the removal of atomic hydrogen by interaction with nitric acid to give oxides of nitrogen (NO_2 or NO) should provide a reaction involving a drop of free energy, and therefore capable of proceeding spontaneously. The attack upon these metals displays certain "autocatalytic" features, the reaction being accelerated by the corrosion-products. It occurs most readily at stagnant corners where the metal specimen rests against the walls of the containing vessel, so that the products of reaction accumulate next to the metallic surface; conversely attack is slow if the liquid is stirred, so that the reaction-products are continually removed from the metal. Thus whilst a rapid rotation of the specimen is favourable to corrosion of a base metal in nitric acid—by replenishing the reagent—it almost prevents the attack upon a noble metal like copper. Conversely, the addition of urea to the nitric acid, which is said to favour the attack upon base metals, greatly retards that upon copper; the urea destroys nitrous acid, and breaks the autocatalytic cycle. Other substances which destroy nitrous acid inhibit the attack of nitric acid on copper. Russian work has shown sodium thiosulphate to be efficient (S. A. Balizin and G. S. Parfenov, *J. Appl. Chem. U.S.S.R.* 1953, **26**, 795; p. 723 in English translation).

In considering this abnormal behaviour, it is convenient to divide the reaction into its anodic and cathodic components, even if they proceed at the same points. The anodic reaction such as

$$Cu = Cu^{++} + 2e$$

is doubtless the same as in attack by sulphuric acid in presence of oxygen, and calls for no comment. The abnormality is clearly connected with the cathodic reaction, since it is found on studying cells of the type Pt | HNO_3 | Pt, operated by an external E.M.F.; here also the cathodic reduction of nitric acid to nitrous acid presents autocatalytic features, being stimulated by the addition of nitrous acid (a reduction product), and restrained by stirring, which removes nitrous acid from the electrode surface as it is formed; it is also restrained by adding urea, which destroys nitrous acid (H. J. T. Ellingham, *J. chem. Soc.* 1932, p. 1565).

Now let us consider copper placed in nitric acid. Presumably the cathodic reaction starts as usual

$$H^+ + e = H$$

If the atomic hydrogen accumulates, the E.M.F. soon vanishes; if it is removed as fast as formed by some such reaction as

$$H + HNO_3 = H_2O + NO_2$$

the attack can continue; this reaction requires two particles coming together in the right energy state, and may not proceed very rapidly. However, when once NO_2 molecules have accumulated at the electrode surface, opportunities arise for a rapid, self-stimulating change; such opportunities are present from the moment of immersion, if, as sometimes happens, the sample of acid used already contains a small amount of nitrogen peroxide. For an adsorbed NO_2 molecule can take up an electron without waiting for a collision with a second particle; the reaction

$$NO_2 + e = NO_2^-$$

is the type of simple reaction which may be expected to proceed readily. Now NO_2^- is the ion of the nitrous acid, and equilibria such as

$$H^+ + NO_2^- \rightleftharpoons HNO_2$$

are known to adjust themselves very rapidly; thus the NO_2^- will spend part of its life as HNO_2, and may be expected to react with nitric acid molecule giving *two* NO_2 molecules

$$HNO_2 + HNO_3 = 2NO_2 + H_2O$$

These will give 2 NO_2^- ions, and thus 4 NO_2 molecules; then 8, 16, 32 . . . molecules appear and so on. We see, why, after a brief period of induction, the action becomes self-stimulating and rapid, provided that the chain of events is not broken by stirring (which removes the nitrous acid from the metal surface) or by the presence of urea (which destroys it).*

It might appear that the acceleration should proceed without limit until an explosion occurs, but actually when sufficient nitrous acid has accumulated, other reactions become possible, such as

$$H + HNO_2 = NO + H_2O$$

or, alternatively,

$$2HNO_2 = NO_2 + NO + H_2O$$

Either of these will bring about the production of one molecule of NO and the *net* destruction of one molecule of HNO_2 (since in the second case the

* Various equations have been suggested in the literature for the reaction between nitrous acid and urea, but if we regard urea $(NH_2)_2CO$, as the anhydride of ammonium carbonate, $(NH_4)_2CO_3$, we may expect the liberation of CO_2 and the formation of ammonium nitrite (NH_4NO_2), which, as is well known, easily decomposes to nitrogen and water. In practice, other products are formed simultaneously, but the essential equation can probably be expressed

$$CO(NH_2)_2 + 2HNO_2 = CO_2 + 2N_2 + 3H_2O$$

NO_2 will regenerate one fresh molecule of HNO_2). Let us assume the first method of destruction to occur. When we have reached a *steady state* at which the concentration of nitrous acid becomes constant, its *destruction* by the sequence

$$H^+ + e = H$$

followed by

$$H + HNO_2 = NO + H_2O$$

must exactly balance its *production* by the sequence

$$\begin{aligned} H^+ &+ e &&= H \\ H &+ HNO_3 &&= NO_2 + H_2O \\ NO_2 &+ e &&= NO_2^- \\ NO_2^- &+ H^+ &&= HNO_2 \end{aligned}$$

Thus, in the steady state, we can obtain the complete cathodic reaction by adding all six equations together, which gives

$$3H^+ + HNO_3 + 3e = NO + 2H_2O$$

This must balance the anodic reaction

$$Cu = Cu^{++} + 2e$$

In order to obtain electrical balance the cathodic reaction must be doubled and the anodic reaction trebled; the complete reaction obtained by adding them together is

$$6H^+ + 2HNO_3 + 3Cu = 3Cu^{++} + 2NO + 4H_2O$$

This involves no accumulation of electrons, but, when thus written, contains unbalanced cations; $6NO_3^-$ should, therefore, be added to each side, and the result, re-written in molecular form, is:—

$$8HNO_3 + 3Cu = 3Cu(NO_3)_2 + 2NO + 4H_2O$$

which is the equation presented, in the elementary text-books, for the action of nitric acid on copper. Evidently it represents the change which establishes itself when a *steady state* has been reached, at which the nitrous acid concentration ceases to increase.

The same conclusion is reached if it is assumed that the destruction of nitrous acid depends on the alternative equation

$$2HNO_3 = NO_2 + NO + H_2O$$

Various views regarding the mechanism of the cathodic reaction in corrosion by nitric acid—reasonably consistent with one another—will be found in papers by U. R. Evans, *Trans. Faraday Soc.* 1944, **40,** 120; C. N. Hinshelwood, *J. chem. Soc.* 1947, p. 694; K. J. Vetter, *Z. phys. Chem.* 1950, **194,** 199, 284. The attack on silver is described by E. Raub, *Metalloberfläche* 1953, **5,** B17.

Action of Nitric Acid on Iron. We have seen that most of the baser metals produce with nitric acid compounds rich in hydrogen (such as the salts of ammonia or hydroxylamine), whereas the noble metals produce compounds rich in oxygen (such as NO_2, HNO_2 or NO). Iron shares the properties of both groups, and yields both ammonium salts and nitrous acid, besides the gases NO_2 and NO; since ammonium nitrite and ammonium nitrate when heated yield nitrogen and nitrous oxide respectively

$$NH_4NO_2 = N_2 + 2H_2O$$
$$NH_4NO_3 = N_2O + 2H_2O$$

it is not surprising to find among the gases evolved a certain amount of nitrogen and nitrous oxide; we thus meet with the complete series, NO_2, NO, N_2O and N_2.

The action of dilute nitric acid on iron tends to start slowly and become violent, but its autocatalytic cycle may be more complicated than that associated with the corrosion of copper. For the nitric oxide (NO) which in the case of copper escapes from the liquid (and is oxidized to red NO_2 only when it mixed with air), can, in the case of iron, be retained by combining with the ferrous salts to form a brown soluble nitroso-compound, such as $Fe(NO_3)_2.NO$. The NO can then be reoxidized by nitric acid, and thus re-enter the cycle. The brown liquid seen clinging to a piece of iron which has been immersed in nitric acid until the attack becomes violent and is then withdrawn, contains a nitroso-compound.

Whilst rather dilute nitric acid (density 1·2) acts violently on iron with evolution of red fumes, concentrated acid (density 1·4) produces at first a little attack which soon dies away, the metal becoming passive. Different varieties of iron behave somewhat differently, but the vacuum-melted electrolytic iron sheet used in the Author's early experiments evolved a few gas bubbles at first—after which action ceased within about 2 minutes. However, when a strip of iron, held at the top, had its lower portion pushed into concentrated acid, then taken out and allowed to drain, a violent reaction would start quite suddenly at some point—usually at the top of the wetted region—and spread rapidly downwards; this reaction soon died down, and if the specimen was then again dipped into acid there was no action. If, whilst the violent reaction was still proceeding, the specimen was plunged into acid with sufficient motion to dissipate the corrosion-products, the reaction ceased; but if the specimen was introduced gently and held partly immersed, reaction ceased over the main surface, but continued violently at the water-line, so that the strip was soon eaten through at that level; the lower part then fell to the bottom of the vessel, where it usually remained unattacked (U. R. Evans, *J. chem. Soc.* 1927, p. 1020, esp. p. 1036; cf. E. S. Hedges, *ibid.* 1928, p. 969).

The special attack noted at the water-line by the Author (also, under somewhat different circumstances, by Hedges) is probably due to the fact that any slight initial reaction uses up HNO_3 molecules and produces H_2O molecules, thus diminishing the ratio N_2O_5/H_2O; if there is no stirring, the ratio, which at the outset is high enough to ensure passivity, may locally become low enough for violent attack. The meniscus zone favours the development of attack since it is a corner less accessible to replenishment by concentrated acid than the main part of the surface, and also a place from which the oxides of nitrogen can readily escape into the air.

In some of the Author's experiments, it was noticed that the unwetted areas above the water-line developed interference tints whilst the gases were being violently evolved, suggesting that perhaps NO_2, and not nitric acid, is the oxidizing agent responsible for the production of the oxide-film on the metal. This is far from certain, but it would explain why passivity is not immediately established on iron when plunged into concentrated nitric acid, since time is needed to form the NO_2; the opening attack allows NO_2 and HNO_2 to accumulate by the autocatalytic cycle already discussed in connection with copper, and when the NO_2 has reached a certain concentration, the oxidizing power becomes sufficient to form the oxide-film on the metal. It is possible, however, that HNO_2 is the oxidizing agent which serves to build the oxide-film.*

Different opinions have been expressed regarding the nature of the film produced. At one time, there was a feeling that none of the known oxides of iron possessed the required properties. Since, however, a strip of heat-tinted iron can be dipped into concentrated nitric acid without loss of colour, there seems to be no need to postulate a special oxide. Doubtless a film produced by nitric acid or similar strong oxidizing agent will contain more oxygen than that produced on heating in air, and this will mean that it will contain fewer lattice-defects; the ferric oxide-film produced by heating iron in air has *less* oxygen than corresponds to the formula Fe_2O_3 (see p. 24)—showing that lattice-defects are present; if they are eliminated, the material should be more protective.

The thickness of the film produced by nitric acid was found by Tronstad's optical method to be 25 to 30 Å on iron and 100 Å on steel; the film was, however, thicker at some points than others; several of the specimens showed interference colours when viewed at grazing incidence. Chromic acid produced films 30 to 40 Å thick. The optical method probably tells us the thickness existing at parts of the surface running parallel to the general plane, and will not be invalidated by oxide formed in internal cracks; thus it gives a better idea of the thickness for our purpose than some other methods. The original paper deserves study (L. Tronstad and C. W. Borgmann, *Trans. Faraday Soc.* 1934, **30,** 349).

For detailed electrochemical discussion of the formation of the film, the reader is referred to K. J. Vetter, *Z. Elektrochem.* 1951, **55,** 675; 1952, **56,** 16 (with H. J. Booss); 1952, **56,** 106. Other views are expressed by H. C. Gatos and H. H. Uhlig, *J. electrochem. Soc.* 1952, **99,** 250; H. H. Uhlig and T. L. O'Connor, *ibid.* 1955, **102,** 562; cf. U. R. Evans, *Z. Elektrochem.* 1958, **62,** 619, esp. p. 623.

The passivity produced on iron or steel by nitric acid is not reliable for practical purposes; occasionally it breaks down and violent attack develops. In the Author's experiments (p. 328), most of the specimens

* Recent observations in Fontana's laboratory that stainless steel is passivated by sulphuric acid containing NO, but this is prevented by the presence of urea, could be similarly interpreted. The urea would destroy HNO_2. The facts are given by W. P. McKinnell, L. F. Lockwood, R. Speiser, F. H. Beck and M. G. Fontana, *Corrosion* 1958, **14,** 9t.

which had been rendered passive in concentrated nitric acid and then quickly washed were found to be deprived of the power to precipitate metallic copper when dipped into M/20 copper nitrate solution,* but occasionally a specimen quickly became coppered; this anomalous behaviour—which may be due to some surface defect—is exceptional on a small specimen, but the probability of break-down would evidently be much greater if the area involved was large, i.e. a vessel at a chemical factory. If once break-down of passivity starts, the decrease of the ratio N_2O_5/H_2O in the liquid by the reaction starting at the point in question is likely to cause it to spread. Thus it is not safe to use steel for the storage or conveyance of nitric acid. However, for mixtures of nitric and sulphuric acid, steel is often used in industry, since any water produced will become closely bound to the sulphuric acid molecules; thus any incipient break-down is unlikely to spread or continue (see p. 331).

In many respects, the behaviour of iron in nitric acid shows the opposite features to that in hydrochloric or dilute sulphuric acid. A high anodic current favours the on-set of passivity (p. 236), and thus under border-line conditions contact with a noble metal may produce passivity on active metal. For instance, although acid of density 1·4 almost always renders iron passive and that of density 1·2 almost always causes corrosion, the effect of acid of density 1·3 is variable, and may depend on previous history. If a specimen of iron has become passive, contact with zinc will generally render it active; if a specimen is active, contact with platinum may render it passive; to some extent behaviour varies from specimen to specimen. It will be recalled that in dilute hydrochloric acid contact with platinum would stimulate, and contact with zinc diminish, corrosion. Many interesting effects of contact are brought out in the early correspondence between C. T. Schonbein and M. Faraday, *Phil. Mag.* (iii) 1836, **9**, 53, 57. See also H. L. Heathcote, *J. Soc. chem. Ind.* 1907, **26**, 899 for an admirable survey and many interesting experiments.

Action of Sulphuric Acid on Metals. Whilst dilute sulphuric acid is usually regarded as a non-oxidizing acid, concentrated sulphuric acid possesses oxidizing properties, being reduced to sulphur dioxide; it renders iron passive. Steel or cast iron vessels are commonly used at works for the accommodation of fairly concentrated acid, but accidental dilution below the safe limit (generally about 68% but varying with the material) may have serious consequences; the possibility of water absorption from the air must not be forgotten. Nor must it be assumed that, because a given sample of steel resists the acid at the out-set, this will continue. Sometimes a vessel of Thomas steel may possess an outer layer which resists sulphuric or nitric acid well, but when once slow corrosion has

* Later experiments on heat-tinted iron showed that the concentration and pH value of a copper salt solution affected behaviour as well as the nature of the anion. Specimens which would deposit copper from one solution would fail to do so from another; a statement that a specimen is "active" or "passive" according to the copper salt test has no meaning unless the solution is accurately defined (see U. R. Evans, *J. Iron St. Inst.* 1940, **141**, 219P; esp. p. 225P).

penetrated to the segregate, attack becomes rapid (M. Werner, *Z. Metallk.* 1953, **44**, 37).

The difference between dilute and concentrated sulphuric acid is shown by the fact that the attack by dilute acid is greatly stimulated by the presence of oxygen (as is also the attack by hydrochloric acid), whereas the attack by concentrated sulphuric acid, itself an oxidizing agent, is hardly affected by oxygen (as is also the case with nitric acid). These relations were brought out by the early measurements of W. G. Whitman and R. P. Russell, *Industr. engng. Chem.* 1925, **17**, 348. Some interesting curves showing how, in both nitric and sulphuric acid, the corrosion-rate first rises with concentration and then falls are reproduced by G. Schikorr, " Die Zersetzungserscheinigungen der Metalle " 1934, p. 130 (Barth).

The attack upon iron by 75% sulphuric acid is greatly diminished if ferrous sulphate is present in the acid and movement of the acid past the metal surface causes greater attack than stagnant conditions where ferrous sulphate can accumulate—as is easily understood. Readers interested should study the curves provided by H. W. van der Hoeven, " Korrosion VIII ", 1954–55, p. 53 (Verlag Chemie).

Action of Mixed Acids on Iron and Steel. It has been explained why the passivity of iron in nitric acid cannot be relied on in industry; if at any susceptible point, attack on the metal has started, the evolution of oxides of nitric acid lowers the ratio N_2O_5/H_2O and brings the concentration out of the passivation range into the corrosion range; many reactions using nitric acid in chemical industry produce water as one of the products, so that the acid may become locally diluted. If, however, there is also present sulphuric acid, which has an affinity for water, the activity of *free* water in such cases may be kept low. This is doubtless the reason why mild steel and cast iron vessels can be used in industry for organic nitration reactions—despite the production of water. However, the composition must be carefully controlled; spent acid must be dehydrated and replenished with fresh nitric acid. Caney and Henshall state that, provided the water content does not exceed 32% and the sulphuric acid content is not less than 12%, " mixed acid " (i.e. a mixture of nitric and sulphuric) can be handled in mild steel or cast iron—although in the latter case, close attention must be paid to the structure; the graphite must be primary (deposited from the melt) and not secondary (arising from decomposition of solid cementite). A " weeping casting " is unacceptable, since the concentrated acid penetrating outwards through pores will absorb moisture from the atmosphere, and then start to corrode the iron from the outside (R. J. T. Caney and B. Henshall, Corrosion Symposium (Melbourne University) 1955–56, p. 416).

Triangular diagrams representing the behaviour of mixtures of HNO_3–H_2SO_4–H_2O are frequently employed at chemical works, and show at a glance the proportions which are safe or dangerous for different purposes. Much useful information has been collected by E. Rabald, *Werkst. u. Korrosion* 1956, **7**, 652. See also C. P. Dillon, *Corrosion* 1956, **12**, 623t, who considers also corrosion by the acid vapours.

Safe ranges of acid composition will, however, vary from one material to another. It has been recommended that steels for mixed acids should be low in sulphur and high in carbon (J. Eddy and F. A. Rohrman, *Industr. engng. Chem.* 1936, **28,** 30).

Another mixture of acids possessing industrial interest is that produced in the chamber process for making sulphuric acid. This contains oxides of nitrogen in a quantity which may be about 10% when expressed as HNO_3, the major constituent being sulphuric acid. The mixture can act upon iron, but the rate of attack quickly falls off, producing a protective layer, which Russian research has shown to be ferric sulphate. It may be about 3 μ thick. The larger the amount of oxides of nitrogen present, the faster is the attack at the out-set, and the quicker does that attack come to an end. The curves deserve study (E. I. Litvinova, *J. appl. Chem. U.S.S.R.* 1956, **29,** 1521; English translation p. 1641).

Attack by mixtures of Non-oxidizing Acids with Oxidizing Agents. It has been stated (p. 314) that copper, which is not appreciably attacked by dilute sulphuric acid in the absence of oxygen, is quickly attacked when oxygen is present. A more soluble oxidizing agent may render attack even more vigorous; in presence of potassium dichromate or chlorate, dilute sulphuric acid attacks copper rapidly (O. P. Watts and N. D. Whipple, *Trans. Amer. electrochem. Soc.* 1917, **32,** 257, esp. p. 268).

Iron, which can liberate hydrogen from non-oxidizing acids, is attacked more rapidly if potassium nitrate, ferric chloride or other oxidizing agent is present; in such cases, no hydrogen is formed. The rate of attack permitted by the electrochemical reactions is so rapid that under ordinary conditions the corrosion velocity is controlled by the rate of replenishment of a reagent, i.e. by the replenishment of oxidizing agent if acid is in excess, or the replenishment of acid if oxidizing agent is in excess. This is brought out in the extensive experiments of King, who studied pure iron in various acids containing potassium nitrate or ferric chloride at various stirring speeds and temperatures. In hydrochloric acid containing ferric chloride, the corrosion-rate at low stirring speeds was controlled solely by diffusion, the rate increasing about 1·35 times for every 10°C. rise of temperature; at higher stirring speeds, the chemical or electrochemical reaction at the metallic surface will no longer be able to keep up with the rate of supply of reagents, and the reaction is no longer diffusion-controlled; the rate now increases 2·0 to 2·4 times every 10°C. Clearly in cases where the corrosion-rate is controlled entirely by diffusion across the almost motion-less diffusion-layer clinging to the metal (which Bowden and Agar considered to be 0·001 to 0·005 cm. thick in stirred aqueous solution, becoming thinner as the stirring becomes more rapid), it should be independent of the metal under attack. Early work by van Name and Hill showed that the velocity of attack by strongly acid ferric chloride solution was almost the same for a number of metals; the same is true of attack by iodine dissolved in an iodine solution (C. V. King, *J. Amer. chem. Soc.* 1939, **61,** 2290 (with M. B. Abramson); C. V. King, *J. electrochem. Soc.* 1950, **97,** 191 (with M. Hochberg); 1950, **97,** 290 (with H. Salzberg); 1952, **99,** 295 (with F. S. Lang). A. C. Makrides and N. Hackerman, *ibid.* 1958, **105,** 156. F. P. Bowden and

J. N. Agar, *Chem. Soc. Ann. Rep.* 1938, **35,** 93; R. G. van Name and D. U. Hill, *Amer. J. Sci.* 1916, **42,** 301; cf. E. Brunner *Z. phys. Chem.* 1904, **47,** 56, esp. p. 65). The behaviour of zinc, cadmium and iron in acidified chromic chloride solution has been studied by C. V. King (with N. Mayer, E. Hillner and R. Skomoroski), *J. electrochem. Soc.* 1953, **100,** 473; 1956, **103,** 261; 1957, **104,** 417.

In some cases, the oxidizing agent can establish passivity, at least if films are already present on the metal as it enters the acid. Heat-tinted iron loses its colours in M/100 sulphuric acid within about a minute, but if the liquid contains chromic acid in relatively large amounts (M/10) in addition to sulphuric acid (M/100), the colours are unchanged after 24 hours. If the sulphuric acid is in excess, however, the results are different. The attack by M/20 sulphuric acid upon untinted iron is increased steadily by additions of chromic acid up to M/50 (U. R. Evans, *J. chem. Soc.* 1930, p. 478).

In normal sulphuric acid, Gatos found that the dissolution-rate of iron first increased with the addition of ferric ions up to 0·47 gram-ion/litre, and then decreased again until at 4·0 gram-ion/litre of Fe^{+++} it was again as slow as in the absence of ferric ions; this may have been due to anodic polarization. Hydrogen was evolved when the Fe^{+++} concentration was low, but became less important when plenty of Fe^{+++} was present; over part of the range studied, the corrosion-rate was controlled by the Fe^{+++} supply (H. C. Gatos, *J. electrochem. Soc.* 1956, **103,** 286).

Resistant Materials for Special Situations

Causes of Acid Resistance of Stainless Steel. If chromic acid can prevent the destruction of a ferric oxide film on iron or low-carbon steel by dilute sulphuric acid, and likewise prevent the attack on the metal below the film—this suggests that even oxygen, if present in large amounts, might protect both film and metal. It has been stated (p. 223) that, after iron has been rendered passive by anodic action in acid, the current can be switched off momentarily without loss of passivity; apparently oxygen, produced by the anodic action, and possibly present in a specially active form, can, whilst it remains, prevent the reduction of the ferric oxide, being itself preferentially reduced. However, if the interruption of current is continued too long, the residual oxygen is used up, the iron becomes active and must be passivated afresh. Experiments by Berwick on heat-tinted iron showed that here a similar (but longer) interruption of current was possible without loss of passivity; the time which could be tolerated varied greatly from specimen to specimen, and tended to increase with the length of the period during which the current had been flowing before the interruption; on M/10 sulphuric acid, it sometimes exceeded 1 min. at 24°C. and approached 5 min. at 6°C.; the variability may be attributed to the sporadic manner in which discontinuities in films appear. However, there appears to be an upper limit to the length of the interruption which can be tolerated, and this may be due to the limited solubility of oxygen, since clearly when the oxygen concentration is reduced to a certain value, the reduction of the oxide will become the easier reaction; such an explanation seems reasonable—whether the residual oxygen is considered to be (1) dissolved in the liquid or (2) present in the solid oxide-film, perhaps as an unstable higher oxide, as many authorities suppose, or perhaps (3) present in the solid oxide, eliminating the lattice-defects connected with ferrous ions, and thus conferring stability (U. R. Evans and I. D. G. Berwick, *J. chem. Soc.* 1952, p. 3432).

The results suggest that effective resistance to acid attack cannot easily be produced on unalloyed iron by a charge of oxygen, but that if the oxide-film was composed of material less easily reducible to the divalent condition, the small concentration of dissolved oxygen normally present in dilute acid might suffice to prevent reductive dissolution of oxide around a discontinuity; this would provide a chance of repairing the discontinuity itself by oxidation of the exposed metal. Clearly what is needed is an alloying constituent which is unstable in the divalent state, so that the reduction of oxygen even at great dilution will take precedence over the reduction of oxide. Chromium is just such an element; the lack of stability of its divalent state is shown by the fact that solutions of chromous salts readily liberate hydrogen, giving chromic salts.

This line of reasoning suggests that the resistance of iron-chromium-nickel alloys towards acids, which has long been known, should be ascribed to the presence of oxygen or other oxidizing agent in the liquid—which in turn explains the fact that the resistance sometimes breaks down at inaccessible crevices. Berwick has shown that 18/8 stainless steel, exposed in the active condition to M/2 sulphuric acid, corrodes at first, but, if sufficient oxygen is present in the acid, becomes passive, the passivation being indicated by a sudden rise in the potential.* The time needed for active stainless steel to become passive was found to increase as the amount of oxygen introduced into the vessel containing the acid was diminished; below a certain oxygen-concentration, the stainless steel never became passive at all (I. D. G. Berwick and U. R. Evans, *J. appl. Chem.* 1952, **2,** 576).

These observations confirm the idea that the function of chromium in stainless steel is to provide a film which can resist reductive dissolution, provided always a trace of oxygen or oxidizing agent is present. Resistance is greatly improved if the acid contains a substance which will maintain the potential at a level at which reductive dissolution becomes inherently improbable. Edeleanu has explained the action of cupric or other weakly oxidizing ions in preventing attack by acid by their power to control the potential at a safe level, by virtue of a redox system such as $Cu^{++} | Cu^{+}$ (C. Edeleanu, *Nature* 1954, **173,** 739; *J. Iron St. Inst.* 1958, **188,** 122).

It is, however, possible that the metallic copper, deposited on the surface, provides an area on which the cathodic reduction of oxygen or oxidizing agent can occur more readily, thus avoiding the reductive dissolution of

* It was shown by special experiments that the corrosion ceased when the potential rose, so that in this particular case the rise of potential could be adopted as a reliable criterion of the cessation of corrosion.

the film. The nickel of the alloy, which does not enter the oxide and must be enriched in the metal phase just below the film, may act in the same way; this would explain why a stainless steel containing nickel as an " austenite stabilizer " is more resistant to corrosion than one in which manganese is introduced for the purpose.

The same line of argument enables us to understand Tomashow's interesting observation that the deposition of a discontinuous layer of platinum or copper on stainless steel favours passage into the passive state and to some extent increases resistance to sulphuric acid, although enhancing corrosion in hydrochloric acid where passivity is not to be hoped for; 0·05 to 0·1 μ of platinum or 0·8 to 2·0 μ of copper suffices (N. D. Tomashow, *Z. Elektrochem.* 1958, **62**, 717. See, however, limitations of this method indicated in *Platinum Metals Review*, 1958, **2**, 117).

Chilton's work suggests that in Berwick's experiments the passivity brought about by adding oxygen to the sulphuric acid was largely due to the redox system $Fe^{+++} | Fe^{++}$ set up by the attack on the metal before passivity is established. If a sufficient concentration of ferric ions had been produced in the liquid, the cathodic reduction of Fe^{+++} to Fe^{++} in the liquid around leakage-points in the film would proceed in preference to the reductive dissolution of the oxide-film. The existence of oxide on the metal at the moment of introduction into the liquid also influences behaviour. Hatwell dipped stainless steel specimens into hydrochloric acid, then exposed them to air and finally dipped them into sulphuric acid. He found passivity if the concentration of sulphuric acid exceeded 9%; if the air-exposure was omitted, the specimen remained active at all concentrations up to 33% (C. Edeleanu, *Nature* 1954, **173**, 739; J. P. Chilton, quoted by R. A. U. Huddle in unclassified paper on " Fundamental Aspects of Metallic Corrosion in Water-cooled Reactors " 1957, p. 20 (A.E.R.E.); H. Hatwell, Thesis, Paris, 1954).

Edeleanu has shown that at a low (negative) range of potential stainless steel is active in acid, at an intermediate range passive, whilst at a high range it again becomes strongly attacked. The safe intermediate region probably represents conditions favouring a valency of 3; the lower range permits partial assumption of a valency of 2 (i.e. reductive dissolution of the film), whilst the high range permits production of a soluble compound (chromic acid) in which chromium exerts a valency of 6. Small variations in the composition of the steel affect the upper and lower potential limits defining the safe intermediate range; thus the composition of the steel best suited to withstand non-oxidizing acids (which would tend to dissolve the metal in a divalent state) is different from that most suitable to withstand oxidizing acids, which would tend to produce hexavalent chromium (C. Edeleanu, *Nature* 1954, **173**, 739; *Metallurgia* 1954, **50**, 113; *Corros. Tech.* 1955, **2**, 204; *J. Iron St. Inst.* 1957, **185**, 482; 1958, **188**, 122. Cf. G. H. Cartledge, *Nature* 1956, **177**, 181; *J. electrochem. Soc.* 1957, **104**, 420).

Behaviour of Stainless Steel towards Acids. If the failure of dilute nitric acid to preserve unalloyed iron in the passive condition is attributable to the reductive dissolution of the film, it is natural to turn to stainless steel, in which the presence of chromium, a metal unstable in the divalent conditions, may render the film stable. It has just been shown that dilute sulphuric acid may render 18/8 Cr/Ni stainless steel passive if oxygen is present, but it would be risky to try and make use of this fact in an industrial situation; the chance of break-down at some surface defect or in a corner where oxygen would not be replenished, would be very real. However, towards nitric acid, even relatively dilute, stainless steel behaves well and is extensively used. For dilute sulphuric acid, stainless steel containing molybdenum is generally satisfactory (p. 346). However, the behaviour of the stainless steels to acid depends not only on composition, but also on thermal treatment and cold work (see C. Carius, *Cebelcor, Rapport Technique* No. **66,** (1958) p. 13). The presence of chloride also affects behaviour; the behaviour of five stainless steels to 8N nitric acid containing 0·2N chloride (a liquid of importance in atomic energy schemes) has been described by I. I. Tingley, *Corrosion* 1958, **14,** 273t. Behaviour in 30% nitric acid is discussed by J. M. West, *J. appl. Chem.* 1959, **9,** 1.

Anodic Protection. The use of stainless steel in contact with an acid which would under normal conditions attack it may become possible if the potential is adjusted to such a level as to prevent reductive dissolution due to local cells. This method is sometimes called *Anodic Protection.* To attain safe conditions, it is necessary to raise the potential, instead of depressing it as in Cathodic Protection (Chapter VIII). It is still too early to assess the industrial importance of anodic protection, but a demonstration arranged by Edeleanu (p. 335) at a conversazione organized by the Society of Chemical Industry Corrosion Group in 1954 made a notable impression. A model plant was arranged in which 50% sulphuric acid, brought to its boiling-point, was pumped continuously round a circuit constructed of an alloy which would normally be attacked violently by that acid; a potentiostatic arrangement for maintaining the potential within the safe correct range prevented any attack during the whole period of the session (C. Edeleanu, *Metallurgia* 1954, **50,** 113).

Character of the Film on Stainless Steel. Much research has been carried out on the isolation and study of the invisible films from stainless steel; usually the metal is dissolved from below by means of a solution of iodine or bromine in an organic solvent—a method developed at Teddington. The film is generally found to contain the metals in a proportion differing from that in which they exist in the alloy, but the nature of the enrichment depends on the treatment to which the material has been subjected. It would seem that enrichment in certain elements—notably those which are not liable to reductive dissolution—improves the corrosion-resistance of the alloy; chromium-enrichment is regarded as valuable by Vernon, and silicon-enrichment by Rhodin. Some of the statements made by reliable authorities can only be reconciled by assuming that an element which confers resistance to one reagent is less beneficial in regard to a second. Rhodin's results suggest that silicon confers resistance to ferric chloride, but Uhlig states that for some purposes 18/8 stainless steel prepared from pure materials is equally resistant to pitting, perhaps more resistant, than a commercial alloy containing silicon.* The reader should read and compare the results of different investigators (W. H. J. Vernon, F. Wormwell and T. J. Nurse, *J. Iron St. Inst.* 1944, **150,** 81; E. M. Mahla and N. A. Nielsen, *Trans. electrochem. Soc.* 1948, **93,** 1; T. N. Rhodin, *Corrosion* 1956, **12,** 123t, 465t; N. A. Nielsen and T. N. Rhodin, *Z. Elektrochem.* 1958, **62,** 707; W. P. McKinell, R. Speiser, M. G. Fontana and F. H. Beck, *ibid.* 1958, **62,** 733; H. H. Uhlig, *Corrosion* 1956, **12,** 135t; T. Tokumitu, *Nature* 1940, **145,** 580).

The attack on stainless steel is most objectionable if it takes the form of pitting or stress-corrosion cracking; the latter is discussed on p. 680. Streicher has studied the formation of pits on various samples of stainless steel—mainly during anodic treatment in M/10 sodium chloride; under these conditions, and also during simple immersion, grain-boundaries rather than non-metallic inclusions constitute the sites of pit initiation (M. A. Streicher, *J. electrochem. Soc.* 1956, **103,** 375; a second paper, discussing causes of grain-boundary attack, will appear in *ibid.* 1959, **106,** 161).

Choice of Stainless Steels for Industrial Purposes. In using stainless steel in a chemical plant the possibility of the breakdown of passivity at weld-lines should be kept in mind (p. 214); either low-carbon material or alternatively material stabilized with niobium (columbium) or titanium must be adopted. Under oxidizing conditions (e.g. in nitric acid) niobium is a better stabilizer than titanium, but the opposite may be true under reducing conditions (e.g. in sulphuric acid), although there is not a great difference (C. Edeleanu, Priv. Comm., Oct. 12, 1956; cf. N. P. Inglis, *Chem. and Ind.* (*Lond.*) 1957, p. 180).

Many different grades of stainless steel, with varying chromium and nickel contents (some containing molybdenum also) are available today, and are suited to withstand different conditions. Some are entirely austenitic, others are martensitic or ferritic in different degrees. In general, the purely austenitic types are the most resistant. Some information useful in making a choice is provided by L. Guitton, *Métaux et Corros.* 1950, **25,** 91, 164; K. Bungardt and H. J. Rocha, *Werkst. u. Korrosion* 1952, **3,** 209. E. Smith, *Chem. and Ind.* (*Lond.*) 1953, p. S9; N. P. Inglis, *ibid.* 1957, p. 180; as well as the books of Monypenny and Keating (pp. 340, 341). See also Symposium on " Evaluation Tests for Stainless Steels ", A.S.T.M. Spec. Tech. Pub. **93** (1949).

It is important that the choice should be made aright. For resisting sour oil, ferritic stainless steel is said to be worse than ordinary carbon steel, but austenitic stainless steel is more satisfactory (C. P. Larrabee and W. F. Rogers, *Corrosion* 1951, **7,** 276). Behaviour in hot citrus fruit juice is discussed by J. M. Bryan and J. W. Selby, *J. Sci. Food Agriculture* 1951, No. 8, p. 359; behaviour in organic acids by C. F. Poe and E. M. van Vleet, *Industr. engng. Chem.* 1949, **41,** 208, and in ferric chloride by H. A. Liebhafsky

* In Monypenny's tests on steels containing 8 or 14% chromium without nickel, an increase of silicon considerably reduced the rate of attack by nitric and sulphuric acid if the acid concentration was not too high; the silicon contents used were 1·2 to 3·5% (J. H. G. Monypenny, " Stainless Iron and Steel " 1931, pp. 302–307 (Chapman & Hall)).

and A. E. Newkirk, *Corrosion* 1956, **12,** 92t. The choice of stainless steel for different purposes in chemical industry, where sometimes a nickel-free type may prove satisfactory, is discussed by H. Braun, *Werkst. u. Korrosion* 1952, **3,** 93.

The influence on corrosion of the so-called *sigma phase,* which appears in highly alloyed irons, deserves attention. Conditions for its appearance in stainless steel, with and without molybdenum, are defined by W. P. Rees, B. D. Burns and A. J. Cook, *J. Iron St. Inst.* 1949, **162,** 325; L. Smith and K. W. J. Bowen, *ibid.* 1948, **158,** 295; H. W. Kirby and J. I. Morley, *ibid.* 1948, **158,** 289; M. G. Fontana, *Industr. engng. Chem.* 1951, **43,** July, p. 91A; T. P. Hoar and K. W. J. Bowen, *Trans. Amer. Soc. Met.,* 1953, **45,** 443; A. H. Sully, *J. Inst. Met.* 1951–52, **80,** 173; R. O. Williams and H. W. Paxton, *J. Iron St. Inst.* 1957, **185,** 358. See p. 664 of this book.

The behaviour of stainless steel is greatly affected by mechanical treatment, and cases are known where deformation has either increased or decreased the susceptibility to attack. The situation is complicated by the fact that some types of austenitic steel develop martensite when deformed —thus greatly increasing the strength. This might be expected to provide starting-places for attack—at least under stress conditions (p. 681). However, often the reverse is the case, and it is stated that certain types of stainless steel, hardened at very low (sub-zero) temperatures, become particularly resistant to acids. (See *Corrosion,* May 1955, Vol. 11, p, 75; cf. however, E. P. Hedley, *J. Assoc. S. Africa Mech. Elect. Eng.* 1934, **8,** 106, for a case where deformation reduced the resistance of 13% chromium steel.)

It is not unusual to try to ascertain the relative values of different samples of stainless steel by studying their behaviour in boiling 65% nitric acid. The test has probably little value except where the material is intended for use in hot nitric acid, and the results are apt to be variable according to the conditions of carrying out the test. Nitric acid, acting on stainless steel, produces chromic acid which catalyses thc attack, so that if the corrosion-products are allowed to accumulate, the corrosion-rate increases during the test; if the testing apparatus is so designed that they do not accumulate, the rate is constant. Apart from this, traces of fluoride greatly stimulate attack. Other tests (especially for use in connection with weld-troubles) are suggested on pp. 693–694, but those who decide to use the nitric acid test should first consult papers by H. T. Shirley and J. E. Truman, *J. Iron St. Inst.* 1952, **171,** 354; J. E. Truman, *J. appl. Chem.* 1954, **4,** 273; also J. M. West, *J. appl. Chem.* 1959, **9,** 1.

The autocatalytic acceleration of the attack which starts when once chromic acid has been formed has caused much anxiety in industry—notably in " reboilers "; the attack is often initiated at weld craters which usually contain hair-line cracks. A research at Oak Ridge suggests that the only way of avoiding the trouble is to keep the temperature below the level at which nitric acid oxidizes chromium to the hexavalent state (J. L. English *Corrosion* June 1956, **12,** p. 65).

In atomic energy plants, the possibility that fission products—or the direct effect of radiation—may affect the behaviour of stainless steel requires to be borne in mind. The fission products may affect the redox potential maintained in the plant, and thus alter the conditions under which the attack of stainless steel by oxidizing acids can take place; sometimes the effect is beneficial—as pointed out by McIntosh. For non-oxidizing acids, Simnad's demonstration that structural defects introduced into sesqui-oxides by irradiation can accelerate dissolution may come to have practical importance in the future, although probably his irradiation was more drastic than would be met with in the type of atomic power stations favoured at present. The atomic energy specialist should study the information published in connection with any other information to which he may have access (A. B. McIntosh, *Chem. and Ind.* (*Lond.*) 1957, p. 687; M. Simnad and R. Smoluchowski, *J. chem. Phys.* 1955, **23,** 1961; A. H. Cottrell, *Metallurg. Reviews* 1958, **1,** 479, esp. p. 519).

In the analogous case of aluminium, German work has shown that corrosion is undoubtedly speeded up by exposure to radioactive substances; this may be due to the formation of hydrogen peroxide, or to damaging of the protective skins or to changes in the metal at spots where α-particles impinge, giving local cells; the mechanical strength is also reduced (K. Lintner, E. Nachtigall and E. Schmid, *Metall* 1957, **11,** 31).

The fact that radioactivity can cause deterioration to the mechanical protection of metallic substances irrespective of corrosion may be held to reduce the responsibilities of the corrosionist, since there is no advantage in diminishing the rate of deterioration due to corrosion to a level small compared to the rate due to other causes. However he ought to bear in mind the possibility of conjoint action analogous to that discussed in Chapter XVIII, where chemical agencies and mechanical agencies acting together produce more damage than either acting separately.

Pre-treatment of Stainless Steel. Much attention has been paid to processes designed to improve the passivity of a stainless steel vessel by pre-treating the surface with nitric acid before it goes into service. Some regard this as a process for conferring a protective oxide-film, but probably the main benefit is the destruction of unreliable material on the surface, such as specks of ordinary steel which have accidentally been rolled or pressed in. Hydrochloric acid, which certainly does not passivate, is stated to confer much the same effect (H. H. Uhlig and E. M. Wallace, *Trans. electrochem. Soc.* 1942, **81,** 511; H. H. Uhlig, *Corrosion* 1951, **7,** 401).

Guitton has cited a case where 18/8 stainless steel was attacked when subjected to nitric acid, but became passive if subjected first to sulphuric acid then to nitric. This also suggests that the function of the pre-treatment is to remove sensitive material produced by cold-work or otherwise, leaving a surface favourable to the building up of a protective film. Other experiments described by Guitton are explicable on similar lines (L. Guitton, *Métaux* 1945, **20,** 68; 1947, **22,** 47).

This pre-treatment of stainless steel may perhaps be compared to the pre-treatment of unalloyed iron in chromate–chloride solutions which is designed to destroy the susceptible material (p. 157). However, a definite passivating treatment can beyond question be useful in some cases. At a certain American dye-works, stainless steel vessels were required to hold in turn (1) hot detergent, (2) fresh water for rinsing, (3) hot dye-bath free from chloride, (4) hot dye-bath containing chloride. The chloride in the fourth stage of the cycle was found to produce pitting. The addition of sufficient inhibitor at that stage to overcome the effect of the chloride was not a practical proposition, but the addition of inhibitor (such as sodium chromate) to the rinsing water and salt-free dye-bath used in the second and third stages built up a film sufficiently resistant to survive the fourth stage (F. N. Speller, *Corrosion* 1955, **11,** 303t. Cf. M. A. Streicher, *J. electrochem. Soc.* 1956, **103,** 375, who has studied the effect of passivating treatment on the number of pits).

The fact that oxygen, which stimulates acid attack on most metals, may be needed for the passivity of stainless steel, has produced embarrassment in some situations. At a certain chemical works, a product had to pass, first through silver equipment and then through stainless steel vessels. When oxygen was present, sufficient silver was dissolved to contaminate the product; when (to avoid this happening) oxygen was scrupulously removed from the liquid before it entered the silver section, serious attack on the stainless steel developed. In the end it was found necessary to maintain oxygen-free conditions in the silver section, and inject oxygen at the entry to the stainless steel section (F. H. Keating, Priv. Comm.).

Although stainless steel resists ordinary concentrated nitric acid, it is appreciably attacked by the fuming acid, especially at elevated temperatures. Indeed according to Sands, the attack becomes appreciable below 100%. He states that drums filled with 93% acid exposed three months in sunshine developed no pressure and caused no pitting, whereas with 99% both pressure and pitting were noticed (G. A. Sands, *Industr. engng. Chem.* 1948, **40,** 1937). The action of both white and red fuming acids is discussed by W. Kaplan and R. J. Andrews, *ibid.* 1948, **40,** 1946; also by J. D. Clark and M. A. Walsh, *Trans. N.Y. Acad. Sci.* 1955, **17,** 279).

Information regarding the behaviour of stainless steels and nickel-chromium-iron alloys generally to anhydrous ammonia is provided in *Corrosion* Oct. 1956, **12,** p. 93; and to sulphuric acid in a series of diagrams by J. Bünger, *Werkst. u. Korrosion* 1956, **7,** 322. The behaviour of stainless steel in different mixtures of nitric and sulphuric acid, with information about the frequency of pits, is discussed by P. E. Krystow and M. Balicki, *Corrosion* 1956, **12,** 449t.

The function of nickel in stainless steel is mainly that of an austenite stabilizer, although it may serve other purposes—perhaps as a catalyst in the reduction of oxygen, which is necessary if the reductive dissolution of the film is to be avoided (p. 227). In view of nickel shortage, efforts have been made to replace it by other stabilizers, such as manganese or even nitrogen. High-manganese stainless steel is considered by some authorities to be comparable to 13% chromium, but definitely inferior to 18/8 chromium-nickel steel, in respect to corrosion-resistance.

Further information affecting the choice of stainless steel for industrial processes is provided by J. H. G. Monypenny " Stainless Iron and Steel "

(Chapman & Hall); F. H. Keating " Chromium-Nickel Austenitic Steels " (Butterworth); L. Columbier, " Aciers inoxydables, aciers réfractaires " (Dunod); H. Braun, *Werkst. u. Korrosion* 1952, **3,** 93; L. Guitton, *Métaux et Corros.* 1950, **25,** 164; E. Smith, *Chem. and Ind.* (*Lond.*) 1953, p. S9. N. P. Inglis, *ibid.* 1948, p. 202; 1957, p. 180, discusses stainless steel and other resistant alloys, including those based on copper, aluminium and titanium.

Behaviour of Aluminium towards Acids. The importance of avoiding reductive dissolution in acids suggests the use of aluminium, a metal of invariable valency. Towards non-oxidizing acids, the film on aluminium may only delay the attack (p. 219), but where the acid is an oxidizing agent, the resistance is effective and permanent, since film-repair becomes possible. Concentrated nitric acid is often stored and transported in aluminium vessels, provided that chlorides—always fatal to passivity—are absent; the pure grades are the most resistant and should be preferred, at least for weld-lines, crannies and places where attack would be dangerous. Care should be taken that, after pure aluminium has been paid for, the advantage is not lost through a pick-up of iron from tools, or perhaps through iron, copper or carbonaceous dust particles settling from the air and being pressed into the soft metal during fabrication. Dilute nitric acid naturally produces more attack, and stainless steel is generally preferred for handling it.

Papers discussing the uses of aluminium for the handling and storage of nitric acid include those of G. A. Sands, *Industr. engng. Chem.* 1948, **40,** 1937; N. P. Inglis, *Chem. and Ind.* (*Lond.*) 1957, p. 180. The behaviour towards fuming acid, of the white and red varieties, is discussed in *Corrosion* Oct. 1956, **12,** p. 77; see also W. W. Binger, *Corrosion* Jan. 1952, Topic Section, p. 1.

Behaviour of Aluminium towards Organic Acids and Alcohols. In Chapter VII, it was explained that aluminium anodically treated in certain baths containing little free water but a considerable reservoir of combined water develops an oxide-film instead of soluble compounds. Rather similar considerations may explain the behaviour of aluminium placed in organic acids, phenols or alcohols, without external E.M.F.—the movement of ions being in that case maintained by local cells. A certain amount of water is needed to provide a protective film, but it should be kept low. 1% acetic acid attacks aluminium rather readily, but the corrosion-rate diminishes as the concentration rises, and the 99% acid is almost without action; however, the removal of the last 0·05% of water increases the corrosion-rate 100 times. Similarly in the case of propionic and butyric acids, ethyl, butyl and amyl alcohols, as well as phenol, the anhydrous substances attack aluminium, but a small amount of water is protective (R. Seligman and P. Williams, *J. Soc. chem. Ind.* 1916, **35,** 88; 1918, **37,** 159T; G. Schikorr, " Die Zersetzungserscheinungen der Metalle " 1943, p. 75 (Barth); H. O. Teeple, *Corrosion* 1952, **8,** 14; A. B. McKee and W. W. Binger, *ibid.* 1957, **13,** 786t. Useful information is collected by J. M. Bryan, " Aluminium and Aluminium Alloys in the Food Industry " (H.M. Stationery Office); see also " Aluminium in the Chemical and Food Industries " (British Aluminium Co.)).

Whenever aluminium plant is to be used in industry, it should be recollected that small traces of mercury may greatly diminish its power to resist attack. They may pass through the plant in solution, or as vapour (R. C. Plumb, M. H. Brown and J. E. Lewis, *Corrosion* 1956, **12,** 277t).

Behaviour of Titanium, Zirconium or Thorium. The corrosion resistance of the tetravalent metals of Group IVA of the periodic table (p. 954) is apparently due to the character of the oxide-films formed, since the metals in the film-free state possess a high affinity for non-metals. Titanium, which is outstanding in its resistance to sea-water, also resists hydrochloric and sulphuric acids at low concentrations and nitric acid at all ordinary concentrations; its behaviour in fuming acid is discussed later. An excellent account of "The New Metal: Titanium" is provided by M. Cook, *J. Inst. Met.* 1953–54, **82,** 93. For resistance to nitric acid, see N. P. Inglis, *Chem. and Ind.* (*Lond.*) 1957, p. 180, esp. p. 186. For general corrosion-resisting properties see J. B. Cotton, *ibid.* 1958, p. 68; for resistance to sea-water, see J. B. Cotton and B. P. Downing, *Trans. Inst. Marine Engrs.* 1957, **69,** 311. For comparison of alloys, see D. Schlain and C. B. Kenahan, *Corrosion* 1958, **14,** 405t.

Cotton has found that anodic protection of the type worked out by Edeleanu for stainless steel is particularly well suited for titanium, which, if the potential is controlled in the range favourable to passivity, becomes almost completely resistant to hydrochloric acid. Whilst the resistance of titanium is certainly due to a film, it is not always the original air-formed film which confers the best protection. When titanium is placed in 2 % hydrochloric acid at 40°C., the potential first declines, indicating the destruction of the air-formed film, but then rises again, apparently owing to the formation of a secondary film, more resistant than its predecessor; after this the material will withstand 2% hydrochloric acid up to 100°C. (O. Rüdiger, R. W. Fischer and W. Knorr, *Tech. Mitt. Krupp* 1956, **14,** 82).

The general use of titanium in chemical industry is restricted by its high price. There are good prospects of a cheapening, and it has been argued that, even if titanium remains very much more expensive than stainless steel when purchased in sheet form, the discrepancy between the prices of a finished plant should be much less pronounced—owing to the cost of fabrication which is probably not so very different for the two materials. The life of a titanium plant should, in most cases, be much longer (G. T. Fraser, W. L. Finlay and A. G. Caterson, *Mater. and Meth.* 1956, **43,** May, p. 112; N. P. Inglis and M. K. McQuillan, *Endeavour* 1958, **17,** 77).

However, titanium does not resist all reagents. It is not completely resistant to sulphuric acid and is readily dissolved by hydrofluoric acid; moreover traces of fluoride can cause attack by acids normally resisted. It is slowly acted on by aqua regia, and is not completely resistant to boiling hydrochloric acid. Information about its behaviour in hydrofluoric and other acids is provided by M. E. Straumanis, *J. electrochem. Soc.* 1951, **98,** 234 (with P. C. Chen); 1954, **101,** 10 (with C. B. Gill). Behaviour to various reagents is surveyed by J. B. Cotton and H. Bradley, *Chem. and Ind.* (*Lond.*) 1958, p. 640.

Zirconium has many potential uses in chemical industry, being resistant to hydrochloric acid at comparatively high concentrations, and also to alkali. It can now be obtained as tubes and in most useful forms; welding has presented difficulties, but it can now be carried out in argon. Zirconium-clad steel, if it can be produced cheaply, may prove to have many applications. Titanium-cladding is discussed by R. C. Bertossa, *Iron Age* 1957, **180,** Oct. 31, p. 59. Titanium-plated steel, not yet in the production stage, has promise; fused baths appear most suitable (F. W. Fink, *Industr. engng. Chem.* 1958, **50,** Jan. p. 129A).

Both titanium and zirconium are subject to hydrogen embrittlement—a matter which should receive serious consideration before deciding on their use (H. M. Burte and others, *Metal Progr.* 1955, **67,** No. 5, p. 115; G. T. Muehlenkamp and A. D. Schwope, *Corrosion* 1955, **11,** 182a; L. B. Golden, I. R. Lane and W. L. Acherman, *Industr. engng. Chem.* 1953, **45,** 782; H. R. Toler, *Bull. Amer. Ceramic Soc.* 1955, **34,** 4).

The uses of titanium and/or zirconium in industry are discussed by L. B. Golden, I. R. Lane and W. L. Acherman, *Industr. engng. Chem.* 1952, **44,** 1930; 1953, **45,** 782; L. Rotherham, *Chem. and Ind.* (*Lond.*) 1954, p. 1164; M. E. Straumanis and J. I. Ballass, *Z., anorg. Chem.* 1955, **278,** 33; G. L. Miller, *Corros. Prev. Control* 1954, **1,** 23.

Zirconium has been proposed for use in surgery, where it might prove free from some of the disadvantages of stainless steel (J. Cauchoix and J. Lavard, *Métaux et Corros.*, 1950, **25,** 115).

Thorium may be mentioned at this point, although it does not seem to be used appreciably in the chemical industry. It dissolves only very slowly in hydrochloric and sulphuric acids and even shows resistance to hydrofluoric acid—being in that respect superior to titanium, zirconium and tantalum (M. E. Straumanis and J. I. Ballass, *Z. anorg. Chem.* 1955, **278,** 33).

Explosions and Fires due to Titanium and Zirconium. The highly protective character of the oxide-films on titanium and zirconium is almost certainly due to the fact that they are formed by oxygen penetrating inwards, so that the oxide (which, if uncompressed, would occupy a larger natural volume than the metal from which it is formed) contains compressional stresses parallel to the surface. These tend to close up any fissures, and the thin films are highly protective. If a certain thickness is exceeded, however, the strain energy per unit volume becomes greater than the work involved in detaching the metal, and spalling or other failure occurs; thus the thick films are non-protective. The matter has been discussed on pp. 43–46, where evidence that the break-down is due to stresses in the film is provided.

The sudden break-down of resistance on both metals has been the cause of fatal accidents and destructive fires. If a quantity of zirconium scrap in store is allowed to become damp, the film may approach the thickness at which spontaneous break-down would occur, and a relatively small amount of disturbance (e.g. slight impact, vibration or perhaps even stirring) may supply sufficient additional force to break the skin at a few points, causing a sudden evolution of hydrogen, along with much heat which may cause local explosion of the hydrogen–air mixture present in the interstices of the scrap. The wave sent out from this first small explosion may combine with the internal stresses in causing the films to break down throughout the mass; violent explosions, possibly killing men close at hand and injuring others at a distance, will quickly occur. The surface-rich condition of the scrap contributes to the violence of the reaction. The actual explosion is usually that of a hydrogen–oxygen mixture, and some water is needed for the reaction, but the danger is greatest if the scrap is merely damp; if the whole is fully immersed in water, the danger of starting an explosion is less, since the heat-capacity of the mixture is greater. However, when once a fire has started, the presence of much water will not prevent the reaction; in fact both water and carbon dioxide, which would normally be used to extinguish fires, here make matters worse. Once the fires have started, it is said to be almost impossible to extinguish them, but control is possible by means of dry powder; a special type of powdered sodium chloride (having the grains coated with stearine to prevent clogging) is favoured in some quarters. The greatest care should be taken in dealing with zirconium scrap or powder; it should be handled in the smallest possible quantities.

Violent titanium explosions have occurred during the testing of specimens in red fuming nitric acid. This reagent usually produces little attack on titanium, but occasionally some slight movement or impact has caused a disastrous explosion leading, in one case, to the loss of a valuable life. The explosions generally seem to be preceded by an attack on the metal leading to *pyrophoric material*. This material is produced on many titanium alloys (and probably also on titanium not intentionally alloyed but containing certain impurities) when the grain-boundary material is dissolved away, leaving a powder, each particle of which represents a single crystal-grain, protected presumably by a film. Although the powder apparently represents the least active part of the original material, the large surface area presented is favourable to the starting of a violent reaction, if an impact or shock causes the film to break down, first on a few particles, and then, as a result of a compression-wave, everywhere; since the strain-energy per unit area in a film increases as the film thickens, the danger is likely to increase with the period of exposure to acid.

Bomberger has found that the reaction of the red acid with the titanium, leading to pyrophoric material, is much more violent in experiments carried out in sealed flasks. The lessened danger associated with open flasks has been attributed to the access of oxygen or up-take of moisture—which would certainly tend to slow down the reaction. On the other hand, the difference may really be due to the fact that the use of closed flasks prevents the escape of nitrogen peroxide, which is possibly needed for the setting up of an autocatalytic cycle, as in the attack of ordinary nitric acid on copper. It is known that the danger of red nitric acid increases with the concentration of NO_2 in the original reagent. Gilbert and Funk have published a diagram showing the combinations of NO_2 and H_2O contents which will lead to violent reaction. Relative safety is obtained at the higher H_2O contents and the lower NO_2 contents, but there is no abrupt passage from absolute safety to certainty of violent reaction; over a range of intermediate concentrations behaviour is irreproducible; the probability of disaster diminishes as the NO_2 content diminishes and as the H_2O content increases. Experimental observations require to be carried out statistically; too much reliance has hitherto been placed on single observations. One investigator has stated that no danger exists when the water content exceeds 0·7%, but other authorities have observed explosions at twice this water content; indeed the disastrous explosion at College Park seems to have occurred with acid containing 1·5 to 2·0% water. That explosion is described in detail in *Corrosion*, Aug. 1955, Gen. News Section p. 86. See also L. L. Gilbert and C. W. Funk, *Metal Progr.* 1956, **70,** Nov. p. 93. J. B. Rittenhouse and C. A. Papp, *Corrosion* 1958, **14,** 283t; H. B. Bomberger, *ibid.* 1957, **13,** 287t.

The fact that these explosions occur on only rare occasions suggests that they depend on some unusual combination of conditions. It seems that danger is increased if there are internal stresses in the metal of a type which will keep opening out cracks and thus exposing film-free surfaces more quickly than films can be formed by the oxidizing action of the acid; however, the need for stress is denied by some authorities. The danger is also increased if the rapidly evolved oxides of nitrogen can form a gas space around the metal, so that it becomes dry, then starts to burn, reaching temperatures exceeding the melting-point; a spattering of metal, which has clearly become molten at the time of the accident, is often found after these explosions—a proof of dry oxidation, since the metal, if continually wetted, could not reach a temperature appreciably exceeding the boiling-point of the acid. Combustion in oxides of nitrogen will cause more heat than combustion in air. If the attack of nitric acid on titanium has the same autocatalytic character as that on copper (p. 325), an inaccessible crevice will constitute a dangerous situation.

It is possible, therefore, that an explosion will only occur when three or four different conditions are spontaneously fulfilled. Supposing that only three conditions need to be fulfilled, and that each of these is fulfilled only once in ten days. Then, if the fulfilments are "uncorrelated" in the statistical sense, we shall get the dangerous combination only once in 1000 days—explaining the relative rarity of these disasters and also the difficulty in establishing the exact conditions needed to produce them. If only a single factor was at work, the responsibility would have been settled long ago. Three possible conditions might be (1) internal stress, (2) crevices favourable to an autocatalytic cycle, (3) geometry suitable for the formation of a gas pocket next the metal. These three conditions are merely cited as examples; others could be suggested. Some friction between the particles when material has reached the sensitive condition may be necessary. Whatever the factors, the general principle that the cause of these disasters is to be found in a combination of rather rare conditions seems very probable. Further investigation conducted on a statistical basis is desirable.

Acid Resistance conferred by alloying elements forming oxides without basic properties. Many of the metals so far considered possess well-developed sulphates and nitrates, and their resistance to sulphuric and nitric acid depends largely on the slow dissolution-rates of the oxides. Some metals, however, do not yield such salts—notably those of groups VA and VIA. Tantalum and molybdenum form well-developed tantalates and molybdenates, but normal sulphates are not met with; even their normal chlorides do not seem to be stable in aqueous solution. Tantalum chloride ($TaCl_5$), for instance, can be prepared by passing chlorine over a heated mixture of the pentoxide (Ta_2O_5) and carbon, but is at once hydrolysed by water. Thus it is not surprising that acids, other than hydrofluoric, fail to attack the metal. *Tantalum* is useful in building plants designed for operations involving chlorine, aqua regia, or nitrous acid containing halogen acids; it is, unfortunately, expensive. Other uses of tantalum are discussed by C. W. Balke, *Chem. and Ind.* (*Lond.*) 1948, p. 82; F. G. Cox, *Corr. Prevention and Control* March, 1958, **5**, p. 44.

Niobium resists most acids, except hydrofluoric, but is rapidly attacked by caustic alkali. It offers no great advantage over tantalum, except that it is much lighter (the specific gravity being 8·57 against 16·6), so that in certain circumstances its employment might reduce costs (A. B. McIntosh, *J. Inst. Met.* 1956–57, **85**, 367. See also " Technology of Columbium (Niobium) "—papers presented at a Symposium held in May 1958 (Electrochemical Society)).

Molybdenum is stated by one authority to be passive in 20% hydrochloric acid, and to be rendered active in potassium hydroxide—an interesting contrast to iron, which tends to be passive in alkali and active in acid. The reason for the contrasting behaviours is clear; the oxides of iron have basic properties and those of molybdenum acidic properties. Some of the figures published in regard to the resistance of molybdenum to hot hydrochloric and other acids suggests many potential uses for the material, but the claims may be optimistic, and different grades seem to vary in their behaviour; difficulties in welding molybdenum may have limited its use in the past, but today welding in argon or helium with exclusion of oxygen should overcome these. There is, however, considerable discrepancy between the published corrosion-rates, especially in sulphuric acid. Those interested should compare the following references:—C. A. Hampel, Rare Metals Handbook 1954, p. 283 (Reinhold); Kirk-Othren, " Encyclopaedia of Chemical Technology " 1952, **9**, 193 (Interscience Publishers); M. G. Fontana, *Industr. engng. Chem.* July 1952, p. 71A; W. Rohn, *Z. Metallk.* 1926, **18**, 387. L. Rotherham (Priv. Comm., Jan. 8, 1955) thinks that Rohn's numbers are on the low side.

Molybdenum is today mainly used as a component of alloys. The 18/8 stainless steel, which cannot be relied upon to resist dilute sulphuric or hydrochloric acids, acquires considerable resistance when molybdenum (2 to 4%) is added; such alloys are useful in the paper industry for operations involving sulphurous acid, and in many branches of chemical industry. Although valuable as an addition to stainless steel, the addition of small amounts of molybdenum to ordinary mild steel increases the corrosion by dilute sulphuric acid and sodium chloride solution, doubtless because the molybdenum is present as a separate phase—presumably carbide (L. Aitchison, *J. chem. Soc.* 1915, p. 1531; *J. Iron St. Inst.* 1916, **93,** 77, esp. p. 87).

Extremely high resistance to acids is obtained from alloys containing nickel and molybdenum as main constituents, with relatively small amounts of iron. The *Hastelloy* series are extremely useful for severe conditions but are by no means cheap. Hastelloy A (58% Ni, 20% Mo, 20% Fe, 2% Mn) resists hydrochloric acid in absence of oxygen. Hastelloy C (now known in Great Britain as Langalloy 5R), which has less iron (58% Ni, 17% Mo, 6% Fe, 14% Cr, 5% W), resists it in presence of oxygen; it is a casting alloy, not available as sheet or lining.

Some useful information about the chemical resistance of the (British) Langalloys and (American) Hastelloys—along with that of other materials, including such organic materials as teflon—is provided by E. Franke, *Werkst. u. Korrosion* 1951, **19,** 40.

Another application of nickel-molybdenum alloys is the handling of hot concentrated alkali; information regarding these and other materials for the purpose is provided by A. Guitton, *Métaux et Corros.* 1950, **25,** 84.

Since acid resistance can be obtained by introducing, as alloying constituents, metals which possess no marked base-forming properties, the use of non-metals would seem rational. Silicon is the one mainly used for this purpose. It was suggested (p. 316) that the improved behaviour of cast iron towards acids in long-continued tests was due to the gradual building up of a silica film, which will be almost insoluble in acid. If large amounts of silicon are added, the materials become resistant from the out-set; about 14% is needed to provide a material proof against sulphuric acid and about 17% for hydrochloric acid. Unfortunately the mechanical properties of the high-silicon cast irons are as bad as the chemical properties are good; the brittleness increases seriously when the silicon is raised from 14% to 17%. The alloys can be cast, but not rolled, and the castings are extremely fragile; experience, however, in design and casting-procedure has taught how to avoid internal cavities of the sort most liable to endanger the service life; today acid pumps, valves and the like, composed of silicon iron, are extensively used. To some extent, risk of breakage is reduced by a stress-relieving anneal. The iron with 14 to 16% silicon has established itself at the sulphuric acid works; to attain the extra resistance needed for hot hydrochloric acid, it is not unusual to add 3–4% molybdenum instead of raising the silicon content, thus avoiding the extreme fragility characteristic of the 17% silicon alloy.

The use of high-silicon iron in chemical industry is discussed by J. E. Hurst and R. V. Riley, *J. Iron St. Inst.* 1947, **155,** 172; A. Parker, *J. Instn. mech. Engrs.* 1948, **158,** 414; J. Dodd, *Corros. Tech.* 1955, **2,** 37; and their metallography by K. M. Guggenheimer and H. Heitler, *Trans. Faraday Soc.* 1949, **45,** 137 and by R. Lefébure, *Métaux et Corros.* 1950, **25,** 9, 44, 67.

Uses of unalloyed metals in Chemical Industry. Although most of the materials used to resist acids and other aggressive liquids in industrial processes are alloys (often of the stainless steel class), it is interesting to note certain situations where single metals are useful. Of these, platinum and copper are examples of metals chosen for their intrinsic inert character; the others (sometimes even silver) owe their utility to the formation of films, and will fail to resist if the film is destroyed by abrasion, repeated bending, thermal shock or chemical action. The following list, based on a table provided by Werner, is instructive:—

Ag is used for HCl, Cl_2, molten NaOH, photographic materials, pharmaceutic products.

Cu (or alloys) for acetic acid, acetic anhydride, other weak acids, organic reactions at high temperatures.

Pb for dilute H_2SO_4, H_2SO_3, roaster gases, chromates.

Ni for alkalis, sometimes for dilute mineral acids or organic acids.

Fe for concentrated H_2SO_4, "mixed acid", alkalis, some salt solutions, organic products.

Al for nearly concentrated HNO_3, nearly concentrated acetic and other organic acids.

Mg (sometimes) for HF and fluorides (M. Werner, *Z. Metallk.* 1953, **44**, 37).

Resistance due to an insoluble anodic product. In most of the cases just cited, the metals resist acids because the salt formed with the anion is sparingly soluble. The use of silver to resist hydrochloric acid and of lead to resist sulphuric acid are connected with the low solubilities of silver chloride and lead sulphate. Magnesium, which is so violently attacked by the majority of acids, has occasionally been used for the storage of hydrofluoric acid, a reagent which attacks many otherwise highly resistant materials like tantalum; this is ascribed to the low solubility of magnesium fluoride. Apparently it has stood up to the conditions, although certain plastics are usually preferred.

The employment of lead to resist sulphuric acid is extensive, but owing to its weakness, it is generally used as a lining for steel or wood; attempts to increase mechanical strength by alloying generally diminish the chemical resistance. The effect of minor constituents on the behaviour of lead is a large subject, and the various grades of lead behave differently in different parts of a single sulphuric acid plant. In some parts, cells of the type

$$Pb \,|\, H_2SO_4 \,|\, H_2SO_4 + HNO_3 \,|\, Pb$$

are set up at discontinuities in the sheathing, and contribute to attack.

In general, most kinds of lead remain unattacked by cold dilute sulphuric acids owing to the formation of a lead sulphate film, but for hot concentrated acid the film on some varieties breaks down. Two main types of lead are used in the chemical industry. One is a very pure grade showing an analysis better than 99·99% and sometimes reaching 99·998%, whilst the other has a small content of a protective element deliberately added; this may be silver or copper (usually less than 0·1%); other impurities must be excluded—especially bismuth which greatly reduces the corrosion-resistance. Werner recommends that the silver added as the protective element should consist of filings or " precipitated silver ", which is rolled into the surface. The relatively coarse silver particles then form the cathodes of the cell Pb | H_2SO_4 | Ag and the layer of lead sulphate formed over the anodic surface is more uniform and protective than if the distribution of anodic and cathodic areas had been left to chance variations in the composition of the metal or the liquid. There is some difference of opinion, however, whether the very pure grade or the grade containing silver (or copper) is the best.

The whole subject is complicated. Reference should be made to the writings of M. Werner, *Z. Metallk.* 1930, **22,** 342; 1932, **24,** 85; *Korrosion u. Metallsch.* 1930, **6,** 134; A. Schünemann, *ibid.* 1933, **9,** 325; O. Kröhnke and G. Masing, " Die Korrosion metallischer Werkstoffe " Band II (1938), p. 496 (Hirzel); O. Heckler and H. Hanemann, *Z. Metallk.* 1938, **30,** 410; S. W. Shepard, *Corrosion* 1951, **7,** 279; K. H. Roll, *ibid.* 1951, **7,** 454. Attack by other acids is discussed by G. Schikorr, " Zersetzungserscheinungen der Metalle " 1943, pp. 83, 84.

Behaviour of Noble Metals and Alloys. Many of the metals already mentioned have a simulated nobility due to film-formation; if the film breaks down, corrosion may be serious. Metals standing at the head of the Table of Normal Potentials are intrinsically noble, but unfortunately very expensive. Platinum plant is used for special purposes in the chemical industry; the financial aspect is connected with the interest on the capital locked up, since the platinum itself can be recovered almost intact when the plant is dismantled.

Silver is much cheaper, but less resistant; its special uses in presence of chlorides have been mentioned, but it also finds extensive employment in the manufacture of pharmaceutical and photographic chemicals. It resists caustic alkali, provided that nitrates are absent, but is attacked by many oxidizing salts, such as ferric sulphate solution, where the attack is sufficiently rapid for the corrosion to be controlled by the rate of supply of ferric ions to the surface (H. Salzberg and C. V. King, *J. electrochem. Soc.* 1950, **97,** 290).

Silver has uses in handling hydrofluoric acid—a reagent which attacks many normally resistant materials such as tantalum. The corrosion-rate increases when dissolved oxygen has access to the liquid, or when sulphur compounds are present, but under average conditions, silver is considered to withstand the acid up to 40% concentration, even at the boiling-point. Further information regarding choice of materials to resist fluorine compounds is provided by R. Landau, *Corrosion* 1952, **8,** 283; see also A. Guitton, *Métaux et Corros.* 1950, **25,** 34; E. Lingnau, *Werkst. u. Korrosion* 1957, **8,** 216.

Surveys of the industrial use of precious metals—platinum (whether unalloyed or hardened with 6% rhodium), gold, silver, palladium—have been contributed by J. M. Pirie, *Chem. and Ind.* (*Lond.*) 1947, p. 91, 134; E. M. Wise, *Métaux et Corros.* 1949, **24,** 87; C. Engelhard, *ibid.* 1950, **25,** 170.

Copper has usefulness in resisting oxygen-free acids—especially hot acetic acid—but in most cases the corrosion-rate increases when oxygen is present, and there may be special attack at points of high relative velocity between liquid and metal—due to the motoelectric effect (p. 130). Introduction of silicon or beryllium improves resistance to acids, and alloys with aluminium possess considerable resistance to sea-water and to polluted atmospheres.

Copper has sometimes been recommended for hydrofluoric acid, but the attack increases with aeration and movement in the liquid. Some of the nickel-copper alloys which are serviceable for that acid under fully immersed conditions, suffer severely at a water-line, if one is present.

A copper alloy containing 1·9% beryllium is considerably used for springs, scraper-blades, seals and the like in chemical processing equipment; details of its behaviour are provided by J. T. Richards, *Corrosion* 1953, **9**, 359.

Monel metal, a nickel-copper alloy, has achieved extensive use in connection with acids; it finds, for instance, application in pickling equipment. This was originally a " natural " alloy made directly from the ore of a certain mine which contained the two metals in the desired proportions (67% nickel and 30% copper) although carefully controlled quantities of other metals were always added, such as manganese (1·25%), iron (1·25%), and smaller quantities of carbon and silicon. Synthetic alloys were, of course, tried but apparently gave some disappointment at first, as the carbon was not always in solid solution, and the graphite flakes promoted attack. Monel metal, and nickel-copper alloys generally, display considerable resistance to dilute sulphuric and even hydrochloric acid, but the rate of attack is greatly increased by the presence of oxygen.

Parting Limits. In using an alloy based on a noble metal to resist corrosion, it is naturally desirable to add as much of the cheaper constituent as is possible without loss of corrosion-resistance. In general, the corrosion-resistance declines—sometimes suddenly—when the content of the baser material exceeds a certain value. This has, indeed, long been known through practical experience in a branch of industry *where corrosion is actually desired*—namely the *parting* of alloys. In separating gold from silver, it is customary to subject the gold-silver alloy to corrosive conditions designed to dissolve the silver and leave the gold as a porous skeleton or sludge. This can be accomplished by simple immersion in an oxidizing acid, such as nitric or hot concentrated sulphuric acid (which is cheaper), or by making the alloy an anode with an external E.M.F. *Electrolytic parting* of a gold-silver alloy is sometimes carried out in two stages, first in a silver nitrate solution which leaves an anode sponge of gold still containing some silver; this is melted down and used as an anode in an acid solution of gold chloride.

It has long been known that parting is only successful if the less noble metal is present in sufficient quantity. Consequently in the purification of gold containing only small amounts of baser metal, it is necessary to melt it with a sufficient proportion of silver, after which the parting takes place smoothly. The limit occurs when a definite fraction of the atoms consists of the attackable species. Thus Tammann and Brauns, studying annealed gold-silver alloys in hot sulphuric acid, found that at 150°C. the alloys are almost unattacked if more than half the atoms are of gold, but that when only 49% consist of the noble metal, attack becomes perceptible. Tammann established similar *parting limits* for other alloy systems such as gold-copper and gold-palladium, but the limit depends on the reagent and the conditions observed; sometimes parting requires three-quarters of the atoms to be of the attackable species. He also found that generally a sharp limit is only obtainable after annealing, which brings the two species into a definite orderly arrangement (the atoms of the minor constituent being arranged on a " superlattice "); in unannealed alloy, although the atomic sites as a whole constitute a regular lattice, the assignment of the two kinds of atoms to the sites is entirely sporadic, so that it may chance that, even in alloys which contain plenty of the attackable species, clusters of " attackable " atoms may chance to be surrounded by an envelope of the nobler species and thus escape attack. When the arrangement is " ordered " by annealing, we suddenly pass, at a certain composition, from (1) a state in which only those atoms of the attackable species situated close to the surface are accessible to the reagent to (2) a state in which continuous paths of attackable atoms extend from the surface into the interior, so that a clean separation of the two metals becomes possible.

Since Tammann carried out his work, the existence of ordered arrangements of the species, and of super-lattices, has been established by X-ray work, thus confirming, in principle, Tammann's interpretation of his early chemical investigations of parting limits. His papers—if not wholly free from errors—deserve study as examples of original thinking of a kind not too common today. Those worth consulting include G. Tammann, *Z. anorg. Chem.* 1919, **107,** 1; 1931, **200,** 209 (with E. Brauns); *Ann. Physik* 1929, **1,** 309, 321; U. Dehlinger and R. Glocker, *ibid.* 1933, **16,** 100, esp. p. 108; L. Graf, *Korrosion u. Metallsch.* 1935, **11,** 34, esp. p. 40; *Z. Metallk.* 1949, **40,** 275. Cf. work on the ε-brasses, which behave differently according to the reagent, by C. W. Stillwell and E. T. Turnipseed, *Industr. engng. Chem.* 1934, **26,** 740. The behaviour of the copper-gold alloys in ammonia is discussed by J. I. Fisher and J. Halpern, *J. electrochem. Soc.* 1956, **103,** 282.

Other References

In seeking a material to withstand an acid, alkali or other chemical in an industrial process, tables must be consulted, but great caution is needed in applying the figures printed in them to a practical case.

The behaviour of a material will vary greatly with the flow conditions, presence or absence of water-lines, presence of inaccessible crannies in the plant, *débris* on the floors of vessels, applied or locked-up stresses (especially in materials susceptible to stress corrosion), movement in the water, and particularly impingement of air-bubbles, vacuum cavities and suspended sand-particles. Furthermore, it must be recollected that minor constituents of liquids greatly alter behaviour. Sometimes they diminish attack; lemonade and vinegar behave differently from citric and acetic acids of corresponding strength, just as sea-water behaves differently from sodium chloride solution. Furthermore, as explained on p. 771, " ipy " values calculated on the assumption of uniform corrosion, are dangerously misleading if there is localization of attack.

After these warnings, it is right to pay a tribute to those who have devoted so much care to collecting figures for the help of designers and users. The best-known tables are found in the " Corrosion Handbook ", edited by H. H. Uhlig (Chapman & Hall); O. Bauer, O. Kröhnke and G. Masing, " Die Korrosion metallischer Werkstoffe " (Hirzel); E. Rabald, section of " Ullmanns Encyklopädie der technischen Chemie " 1951, **1,** 935 (Urban u. Schwarzenberg); E. Rabald, " Corrosion Guide " (Elsevier); Dechema-Werkstoffe Tabelle (Verlag Chemie); M. Werner, " Werkstoffe in der chemischen Technik " 1953 (Hanser); R. J. McKay and R. Worthington, " Corrosion Resistance of Metals " (Reinhold).

Much of the information presented in brochures published by commercial organizations is extremely useful, if not always uniformly objective.

There are many articles devoted to materials suitable for resisting certain classes of reagents in chemical industry. Materials for dealing with alkali are discussed by A. Guitton, *Métaux et Corros.* 1950, **25,** 84; the action of molten alkali is dealt with by D. D. Williams, J. A. Grand and R. R. Miller, *J. Amer. chem. Soc.* 1956, **78,** 5150 (see also p. 356 of this book). Materials for sulphuric acid are discussed by E. Rabald, *Werkst. u. Korrosion* 1956, **7,** 652; J. Bünger, *ibid.* 1956, **7,** 322; G. C. Lowrison and F. Heppenstall, *Corros. Tech.* 1956, **3,** 174; those for the phosphoric acid industry, by J. C. Barber, *Corrosion* 1958, **14,** 357t; those for hydrochloric acid by M. G. Fontana, *Industr. engng. Chem.* 1950, **42,** Sept., p. A111; those for hydrofluoric acid, by A. Guitton, *Métaux et Corros.* 1950, **25,** 34; M. Schussler, *Industr. engng. Chem.* 1955, **57,** 133; R. Laurdau, *Corrosion* 1952, **8,** 283 (dealing also with fluorine). Materials suitable for the oil industry are dealt with by E. Smith, *Chem. and Ind.* (*Rev.*) 1953, p. S9; and by J. F. Mason, *Corrosion* 1956, **12,** 199t; those suitable for pickling baskets are discussed by E. E. Halls, *Sheet Met. Ind.* 1949, **26,** 2127. Materials capable of withstanding various photographic solutions are given in early papers by J. I. Crabtree, G. E. Matthews and H. A. Hartt, *Industr. engng. Chem.* 1923, **15,** 666; 1924, **16,** 13, 671, and in Kodak Data sheet PR-6.

Materials useful in paper-making plant are discussed by R. P. Whitney, *Corrosion* 1949, **5,** 435 and by N. S. Mott, *Paper Trade J.* 1951, **132,** Jan. 26, p. 31; those for ion-exchange systems by J. F. Wilkes, *Corrosion* June 1952, Topic Section.

Discussions of the behaviour of materials in chemical industry generally are provided by N. P. Inglis, *Chem. and Ind.* (*Rev.*) 1948, p. 202; Sir C. Hinton, *J. Inst. Met.* 1952–53, **81,** 465; M. Werner and W. Ruttmann, *Z. Ver. dtsch. Ing.* 1952, **94,** 1113; J. B. Cotton, *Corros. Tech.* 1957, **4,** 345. Fuller details will be found in the Conference organized at Birmingham in 1950 by the Society of Chemical Industry on " Materials of Construction in Chemical Industry ". A meeting dealing with similar subjects was held in London in 1956 by the Corrosion Group of that Society.

Many of the publications mentioned above contain information about organic materials which have to be considered seriously as alternatives to metals for the conveyance and storage of corrosive substances, and even for the carrying out of chemical processes. At one time many of the organic materials which show satisfactory chemical resistance were mechanically weak or at least softened when warmed. In this respect, some more recently introduced materials show great improvement. New types of polyvinyl chloride, for instance, possess sufficient strength, hardness and impact resistance to replace metals for some purposes; an American Committee has recommended rigid PVC for resisting concentrated hydrochloric acid, sodium hydroxide, chromic acid and many other chemicals (*Corrosion* 1956, **12**, 183t).

Special interest attaches to organic compounds containing chlorine and fluorine. Chlorosulphonated polyethylene is stated to withstand many oxidizing agents, and should be useful for tank-linings, hoses and valve diaphragms; it is stated to withstand nitric acid, chromic acid, chlorine and sodium hypochlorites (R. McFarland, *Corrosion* 1956, **12**, 197t).

Particularly valuable are the properties of compounds containing fluorine, which are less polar than those containing chlorine, and often surprisingly inert to reactive substances, besides possessing in many cases water-repellent and even oil-repellent properties. The low friction makes them useful in connection with frettage—as discussed in Chapter XVIII. The most important of these, polytetrafluorethylene (" teflon "), is inert to sodium hydroxide and most reagents, insoluble in nearly all solvents and only starts to soften at about 300°C. By replacing one of the fluorine atoms in C_2F_4 by chlorine, giving $(C_2F_3Cl)_n$, a limited solubility and workability is introduced without much loss of inertness. A good account of these compounds is provided by H. R. Leech, *Research* 1952, **5**, 450.

Apart from these relatively new products, old-established materials like glass, ebonite and rubber are much used, whilst carbon (often as graphite) and numerous kinds of acid-resisting and chemical-resisting bricks, find employment in the chemical industry. An article on non-metals in chemical plant construction has been supplied by H. W. Cremer and G. Brearley, *Chem. and Ind.* (*Lond.*) 1957, p. 374.

General accounts of plastic construction materials are provided by R. B. Seymour, *Corrosion* 1953, **9**, 152, and by J. L. Huscher, *ibid.* 1953, **9**, 272; plastic piping by R. W. Flournoy, *ibid.* Aug. 1953, Topic Section; castings in epoxy-resin in Aero Research Tech. Notes, Bull, **138** (1954); polyethylene foil in plant construction by F. R. Himsworth, *Chem. and Ind.* (*Lond.*) 1950, p. 555; glass plastic laminates by R. S. Treseder, *Corrosion* July 1954, Tech. Comm. News, p. 18, and also S. W. Shepard, *ibid.* 1954, **10**, 215. See also book on " Plastics for Corrosion-resisting Applications " by R. B. Seymour and R. H. Steiner (Reinhold).

Several papers presented at the Internationales Kolloquium über die Passivität der Metalle concern corrosion-resisting materials; they appear in a special number of the *Zeitschift für Elektrochemie*, 1958. In particular, those of N. A. Neilsen and T. N. Rhodin, M. Pražák, C. Carius, N. D. Tomaschow on stainless steel deserve study. H. H. Uhlig has conveniently collected his views regarding the explanation of the resistance of alloys, based on the presence of an unfilled D-band in certain atoms; some remarks on the possibilities and limitations of such lines of thought appear at the end of the paper by U. R. Evans.

The subject of attack of alkali has been only briefly discussed in this chapter, but the following list of papers, kindly prepared by R. S. Thornhill, may prove useful for reference.

Ammonium Salts, Ammonia and Amines

H. Rohrig and J. Roch	Action of cold aqueous ammonia on aluminium sheet of normal and super purity	*Aluminium* 1939, **21**, 128
H. J. Mcdonald and M. Feller	Corrosion of iron by ammonia	*Chem. Met. Engng.* 1943, **50**, (8), 111
J. Bell	By-product ammonia	*Coke and Gas*, 1950, **12**, 206
R. S. Treseder and R. F. Miller	Corrosion in the ammonolysis of aliphatic chlorides	*Corrosion* 1951, **7**, 225
C. F. Pogacar and E. A. Tice	Corrosion problems in the modern by-product coke plant	*Corrosion* 1951, **7**, 76; 128
H. E. von Steinwehr and R. Weber	Formation of γ Fe_2O_3 at low temperatures in a strongly hygroscopic medium	*Naturwiss.* 1952, **39**, 140
J. N. Gordon	Corrosion of stainless steels in liquid ammonia cleaning	U.S. Atomic Energy Comm. N.P.–6432, 1953
W. G. Lloyd and F. C. Taylor	Corrosion by, and deterioration of, glycol solutions and glycol-amines	*Industr. engng. Chem.*, 1954, **46**, 2407
A. W. Bamforth	Corrosion problems of ammonium sulphate manufacture	*Corros. Tech.* 1954, **1**, (2), 31
W. W. Binger and H. W. Fritts	Aluminium alloy heat exchangers in the process industries	*Corrosion* 1954, **10**, 425
T. L. Keelen and R. C. Anderson	Surface interaction between metallic nickel and ethylene diamine solutions	*J. phys. Chem.* 1955, **59**, 881
W. Fuchs and J. Brandes	Investigations into the problem of copper corrosion in steam power stations. Part I. The dissolution of copper in very dilute aqueous (ammonia) solutions	*Mitt. Verein Grosskesselb.* 1955, **37**, 670
S. C. Britton and D. G. Michael	Corrosion of tin by aqueous solutions of ammonia	*J. appl. Chem.* 1955, **5**, I

G. Roberti, F. Gianni and G. Bombara	Protection of ammonium sulphate crystallizers by means of inhibitors	*Metallurg. Ital.* 1956, **48,** 281
N. Hackermann, R. M. Hurd and E. S. Snavely	Corrosion rates of mild steel in NH_4NO_3–NH_3–H_2O solutions	*Corrosion* 1958, **14,** 203t

Caustic Soda Solutions—Aluminium

C. E. Ransley and H. Neufeld	Absorption of hydrogen by aluminium attacked by caustic soda solution	*Nature* 1947, **159,** 709
M. A. Streicher	Dissolution of aluminium in sodium hydroxide solutions	*J. electrochem. Soc.* 1948, **93,** 285; 1949, **96,** 170
ibid.	Dissolution of aluminium in sodium hydroxide solutions—effect of gelatin and potassium permanganate	*Industr. engng. Chem.* 1949, **41,** 818
M. E. Straumanis and N. Brakšs	Rate of solution of highest purity aluminium in sodium hydroxide solutions	*J. electrochem. Soc.* 1949, **95,** 98
ibid.	Effect of minor alloying elements on the rate of dissolution of aluminium in bases	*J. electrochem. Soc.* 1949, **96,** 310
W. Machu and M. K. Hussein	Action of alkaline solutions on aluminium and aluminium alloys	*Werkst. u. Korrosion* 1954, **5** (8/9), 295
T. H. Orem	Influence of crystallographic orientation on the corrosion-rate of aluminium in acids and alkalies	*Bur. Stand. J. Res.* 1957, **58,** (3), 157
H. Makram	Corrosion of aluminium in an alkaline medium	*C.R.* 1957, **244,** 3153; **245,** 1060
S. Heath and G. Tolley	Dissolution of aluminium in alkaline solutions	*Chem. and Ind.* 1957, **77,** 367

Caustic Soda Solutions—Iron

G. Grube and H. Gmelin	Influence of superimposed alternating current on anodic ferrate formation	*Z. Elektrochem.* 1920, **26,** 153
ibid.	Electrolytic formation of the alkali salts of ferrous oxide and ferric oxide	*Z. Elektrochem.* 1920, **26,** 459
K. Taussig	Investigations of the action of alkalies and various salts on iron	*Arch. Warmewirt.* 1927, **8,** 337
E. Berl and F. von Taack	Protective action of sodium sulphate on the attack of ingot iron by alkalies and salts under high pressure.	*Arch. Warmewirt.* 1928, **9,** 165

V. V. Stender and B. P. Artamonov	Electrochemical protection of iron from corrosion in alkalies	*Trans. electrochem. Soc.* 1937, **72**, 389
R. Scholder	Anionic iron	*Angew. Chem.* 1936, **49**, 255
G. Nilsson	Promotion of the corrosion of iron in alkaline solutions by zinc and aluminium	*Nature* 1943, **152**, 189
G. Nilsson	Behaviour of activated iron in sodium hydroxide solutions	*Nature* 1946, **157**, 587
M. Pourbaix	Corrosion of iron by solutions of caustic soda	*Bull. Tech. Assn. Ing. Brussels* 1946–47, (1) 67; (3), 109
I. A. Ammar and S. A. Awad	Effect of some corrosion inhibitors and activators on the hydrogen overpotential at iron cathodes in NaOH solutions	*J. phys. Chem.* 1956, **60**, 871
S. L. Levina	Effect of inhibitors on the dissolution of iron in alkaline solutions	*J. appl. Chem. U.S.S.R.* (Engl. transl.) 1956, **29**, 1457

Caustic Soda Solutions—Nickel

Int. Nickel Co.	Resistance of nickel and its alloys to corrosion by caustic alkalies	Dev. and Res. Div. Tech. Bull T6, 1945
M. Volchkova, L. G. Antonova and A. I. Krasil'skchikov.	Anodic behaviour of nickel in alkaline solutions	*J. phys. Chem.* (*U.S.S.R.*) 1949, **23**, 441; 714
Anon.	Resistance of nickel and its alloys to caustic alkali	*Chimica e Industria* 1956, **38**, 804
Anon.	Pure nickel in the soda industry	*Rev. Nickel* 1956, **1**, 17

Fused alkalies

U. Perret	Corrosion of pots used in the fusion of caustic soda	*Chimica e Industria* 1938, **20**, 133
C. M. Craighead, L. A. Smith and R. I. Jaffee	Screening tests on metals and alloys in contact with sodium hydroxide at 1000 and 1500°F.	U.S. Atomic Energy Comm. BMI-706, 1951
C. M. Craighead, L. A. Smith, E. C. Phillips and R. I. Jaffee	Corrosion by fused caustic	U.S. Atomic Energy Comm. AECD-3704, 1952
R. A. Lad and S. L. Simon	Study of corrosion and mass transfer of nickel by molten NaOH	*Corrosion*, 1954, **10**, 435t
R. S. Young, D. A. Benfield and K. G. A. Strachan	Alkali fusion pots	*Chem. Tr. J.* 1954, **135**, 486
A. F. Forestieri	Effect of additives on corrosion and mass transfer in caustic soda–nickel systems under free-convection conditions	U.S. Natl. Adv. Comm. Aero. Res. Memo E54E19, 1954

L. D. Dyer, B. S. Borie and G. P. Smith	Corrosion products of nickel by fused alkali hydroxide under oxidizing conditions	U.S. Atomic Energy Comm. ORNL-1667, 1954
R. S. Peoples, P. D. Miller and H. D. Hannan	Reaction of nickel in molten sodium hydroxide	U.S. Atomic Energy Comm. BMI–1041, 1955
H. Lux and T. Niedermaier	Transformations and equilibria in alkali hydroxide melts. I. Action of alkali hydroxide melts on gold, silver and platinum	*Z. anorg. Chem.* 1955, **282,** 196
D. D. Williams, J. A. Grand and R. R. Miller	Reactions of molten sodium hydroxide with various metals	*J. Amer. chem. Soc.* 1956, **78,** 5150
J. N. Gregory, N. Hodge and J. V. G. Iredale	Static corrosion of nickel and other materials in molten caustic soda	Atomic Energy Res. Est. C/M-272, 1956
ibid.	Corrosion and erosion of nickel by molten caustic soda and sodium uranate suspensions under dynamic conditions	Atomic Energy Res. Est. C/M-273, 1956
E. M. Simons, N. E. Miller, J. H. Stang and C. V. Weaver	Corrosion and components studies on systems containing fused sodium hydroxide	U.S. Atomic Energy Comm. BMI-1118, 1956
G. P. Smith, M. E. Steidlitz and E. E. Hoffman	Two phase product formed in the reaction between sodium hydroxide and Inconel	U.S. Atomic Energy Comm. ORNL-2129, 1957
ibid.	Action of sodium hydroxide melts on alloys of nickel, molybdenum and iron at 815°C.	U.S. Atomic Energy Comm. ORNL-2131, 1957
ibid.	Corrosion and metal transport in fused sodium hydroxide. Part I. Experimental procedure	*Corrosion* 1957, **13,** 561t
G. P. Smith and E. E. Hoffman	Corrosion and metal transport in fused sodium hydroxide. Part 2. Corrosion of nickel-molybdenum-iron alloys	*Corrosion* 1957, **13,** 627t
D. M. Matthews and R. F. Kruh	Reactions in the sodium hydroxide–nickel system	*Industr. engng. Chem.* 1957, **49,** 55
G. P. Smith, M. E. Steidlitz and E. E. Hoffman	Corrosion and metal transport in fused sodium hydroxide. Part 3. Formation of composite scales on Inconel	*Corrosion* 1958, **14,** 47t

Materials of Construction

Author	Title	Reference
H. G. Haase	Cast iron resistant to acids and alkalies	*Stahl u. Eisen.* 1927, **47**, 2112
M. V. Pershké and L. Popova	Corrosion researches and factors governing the selection of alloys in the construction of equipment for the manufacture of caustic soda and caustic potash	*Chimie et Industrie* 1930, Spec. No. March, 232
J. L. Everhart	Metals that resist alkali corrosion	*Chem. Met. Engng.* 1932, **39**, 88
Various authors	Caustic soda versus construction materials	*Chem. Met. Engng.* 1949, **56**, (12), 213; 1950, **57**, (1), 213; (2), 215
W. Krannich	Corrosion protection as a factor in economics	*Chemische Industrie* 1950, **2**, 323
Anon.	An integrated alkali-chlorine plant	*Industrial Chemist,* 1951, **27**, 115
M. G. Fontana	Corrosion by caustic soda	*Industr. engng. Chem.* 1952, **44**, (6), 81A
H. O. Teeple	Use of austenitic nickel cast iron in caustic manufacture	*Iron and Steel* 1952, **25**, 38
P. J. Gegner	Corrosion in caustic of nickel-iron welds obtained in the fabrication of nickel-clad vessels	*Corrosion* 1956, **12**, 2616
P. A. Hembold	Corrosion of evaporators. Attack on nickel by sodium hydroxide	*Corros. Tech.* 1956, **3**, 384; *Corr. et Anticorr.*, 1956, **4**, (8), 271
Anon.	Cast iron kettles for chemical purposes	*Giessereitechnik* 1956, **2**, 211
J. L. Weis	Nickel-lined steel vessels for caustic service	*Industr. engng. Chem.* 1957, **49**, (6), 69A

Miscellaneous Alkalies

Author	Title	Reference
G. Assarsson	Action of lime solutions on aluminium	*Z. anorg. Chem.* 1930, **191**, 333
E. I. Levitina	Behaviour of magnesium in alkaline solutions	*J. gen Chem.* U.S.S.R. 1954, **24**, 216
W. Z. Friend and F. L. LaQue	Some case histories of corrosion problems in chemical process equipment	*Trans. Amer. Inst. Chem. Engrs.* 1946, **42**, 849
B. B. Morton	Metallurgical methods for combating corrosion and abrasion in the petroleum industry	*J. Inst. Petroleum* 1948, **34**, 1
W. Z. Friend and J. F. Mason	Corrosion tests in the processing of soaps and fatty acids	*Corrosion* 1949, **5**, 355
R. P. Whitney and S. T. Han	A bibliography of alkaline digester corrosion 1954–1955 supplement	*TAPPI*, 1956, **39**, (4), 182A

W. A. Wright	Cleansers for aluminium dairy equipment	*Dairy Engng.* 1956, **73,** 167
W. A. Mueller	Corrosion studies of carbon steel in alkaline pumping liquors by the potential-time and polarization curve methods. Part I. Theory, methods and selected results	*Can. J. Tech.* 1956, **34** 162
	Part II. Mixture of white with oxidized or non-oxidized black liquor	*TAPPI* 1957, **40,** 129
H. W. McCune	Corrosion of aluminium by alkaline sequestering solutions	*Industr. engng. Chem.* 1958, **50,** 67

Recent publications include a Swiss study of the behaviour of aluminium in alkalis and amines (A. Bukowiecki, *Werkst. u. Korrosion* 1959, **10,** 91, with comparative information about heavy metals on p. 105).

CHAPTER X

CRYSTALLOGRAPHIC CORROSION

Synopsis

The aim of the present chapter is to supply certain information regarding the crystalline structure of metallic and non-metallic material, and to apply this to the understanding of certain features of corrosion. Many readers will already possess a full understanding of the matter discussed in the opening sections, and may indeed be disposed to criticize the over-simplified presentation; they must use their own judgment about " skipping " certain paragraphs.

It is explained why crystals of salts grown from solution usually develop geometrical shapes (e.g. cubes) whilst the grains in cast metals are usually *allotriomorphic*, their shapes being decided by the surfaces along which crystals starting at different points have run up against one another. The production of metallic monocrystals is briefly mentioned, and the development of crystal-facets by etching a section of polycrystalline metal is discussed.

It is emphasized that the corrosion behaviour of a metal depends less on the particular type of crystal-structure on which its atoms are arranged than on the defects in that structure; the various types of defects, and the manner in which impurities collect in or around them, demand attention.

First, *screw dislocations*, which permit crystals to grow without any " starting of fresh atomic layers ", are discussed; the " decoration " of crystals which produces a spiral pattern, not only shows that the screw arrangement—envisaged by the theorist before being produced in the laboratory—has a real existence, but also illustrates the fact that atoms situated at the corners of a shelf are chemically more reactive than atoms forming part of a level surface.

Next consideration is given to *edge dislocations* which permit gliding to occur along planes of dense atomic packing at stresses far below the theoretical strength of a perfect crystal, and to perform this operation without the two gliding portions becoming detached; this type of dislocation also has been rendered visible by recent research. The dislocations which occur at *grain-boundaries* or *sub-grain boundaries* are also discussed. The accumulation of small amounts of impurities at dislocations within the grains and also at grain-boundaries is important as a factor in localized corrosion; this is met with even when the impurity is present in amounts far too small for the production of a second phase. The tiny etch-pits which certain acid reagents produce at the sites of dislocations are discussed, along with the possibility of using them to count the dislocations; and the question as to whether the pitting is due to the dislocations as such or to the impurities associated with them receives attention. The interesting phenomenon known as *tunnelling* is briefly mentioned.

Then attention is turned to the different rates of deposition of crystalline matter in different crystallographic directions, which leads us to the converse phenomenon, the removal of material by corrosion at different rates in different directions.

Next, consideration is given to the special attack on metals at places of strain, e.g. bends on a deformed strip; this may sometimes be due to the breaking of an invisible film and sometimes to the distortion of the metallic structure. Finally, the variation of electrode potential with stress and with crystal direction receives a brief discussion.

Some of the subjects opened in this chapter are pursued further in chapters XVI (intergranular and stress corrosion) and XVII (corrosion fatigue).

The Crystalline State

Crystalline Structure of Metals. It is well known that the atoms of metals are arranged, like those of salts, on definite architectural plans; however, whereas a crystal of a salt grown from solution, usually displays an external shape (a cube in the case of sodium chloride) which at once conveys the idea of atoms being arrayed on a " crystal lattice " (or three-dimensional pattern), a metallic article, obtained by casting from molten metal, takes the shape of the mould. Nevertheless, the atoms are arrayed on lattices, and it is this internal arraying of the atoms, not the external form, which is the real characteristic of the crystalline state.

Most metals conform to three relatively simple crystal structures.* A few metals, like β-manganese, present complicated lattices; rather intricate types also occur in alloys.†

External Form of Crystals grown from Solution. That a crystal growing from a slightly super-saturated solution will tend to assume a geometrical form related to the atomic arrangement is readily understo d. Even if the shape at the outset is irregular, the deposition of matter will occur at such places as will convert the irregular into a regular geometrical

* The three simpler types are the body-centred cubic, the face-centred cubic and the close-packed hexagonal. Details will be found in books on crystal-physics. A useful elementary statement is provided by F. Seitz, " Physics of Metals " (McGraw-Hill). See also G. E. Doan and E. M. Mahla, " Principles of Physical Metallurgy " (McGraw-Hill).

† Other arrangements are met with in non-metals, an interesting example being provided by graphite, in which the atoms are packed closely together on parallel layers, with a relatively great distance between the layers, so that the layers can glide upon one another—explaining the lubricating properties of graphite; compounds like molybdenum sulphide, which are built on a similar layer-like structure, are also lubricants.

outline (fig. 71). For consider a rugged part of the surface (*AB*) which is not parallel to the close-packed layers of atoms, so that incomplete layers meet the surface, and consider the deposition of a single additional atom. It might attach itself at a position such as *X*, where it would be starting a new layer, but such a position will be unstable since there is attachment on only one side, and the atom is soon likely to leave the surface again; on the other hand at *Y*, there is attachment to more than one other atom and the atom there deposited is more likely to remain attached. On the whole, the atoms are likely to be deposited in such places as will help to complete the incomplete layers, and convert the rugged surface into a smooth one bounded by a plane face; thus our irregular shape will grow into a regular shape. When all the incomplete layers have been completed (bringing the surface of the crystal to the level *CD*), the problem of starting a new layer is a more difficult one, and theory suggests that on a perfectly

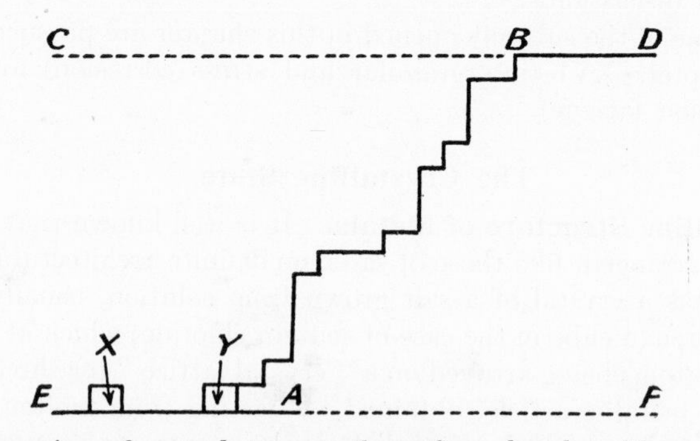

FIG. 71.—Conversion of rugged to smooth surfaces by deposition or dissolution (schematic).

arranged crystal it will happen spontaneously only on rare occasions unless the degree of super-saturation is made very great. The fact that crystallization can proceed from solutions which are only slightly super-saturated is due to a form of structural imperfection (the screw dislocation) which prevents a final flat surface from being obtained—however much fresh matter is deposited; this form of growth is discussed on p. 368.

We may now pass to the reverse process, and suppose that an irregular mass is placed in a slightly under-saturated liquid; we shall now obtain dissolution instead of crystallization, but once again there will be a tendency to pass from an irregular to a regular geometrical shape. An atom situated at the end of an incomplete layer requires less energy to detach itself from its neighbours than an atom in a completed layer, the removal of which would leave a hole. Thus dissolution will first remove any completely disarrayed atoms, then remove the incomplete rows and thus leave a flat —or nearly flat—surface (such as *EF*). The facets exposed by " etching " are planes in which the crystallographic arrangement provides a dense atomic packing (i.e. a large number of atomic centres per unit area situated on a single plane).

The argument just developed explains why deposition and dissolution of crystalline tend to produce flat crystal-faces, but there are other factors which favour less simple forms. The growth of a crystal from a solution can only continue if the concentration over the surface is kept slightly above the saturation value; continuous replenishment of the material deposited is needed, and, whether this occurs by diffusion or convection, it will take place most rapidly at the corners, so that growth is likely to prolong the corners, producing long spines (fig. 72); finally this leads to *dendritic* (tree-like) forms. If a crystal is being produced by the cooling of molten material below its melting-point, crystallization can only occur if the heat of solidification is disposed of, and this also can occur preferentially at the corners, again favouring dendritic growth.

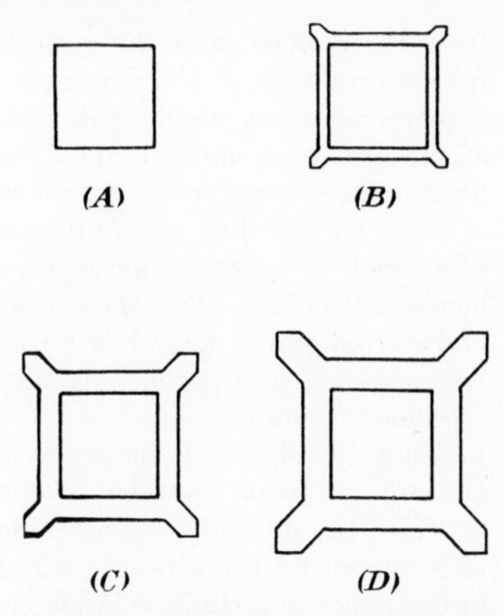

FIG. 72.—Spines, leading to dendritic growth, produced by preferential deposition at corners. (*A*), (*B*), (*C*), and (*D*) suggest, schematically, the successive stages.

Form of Crystal-grains in Metal cast from the Fused State. Where numerous crystals are being deposited close together, considerations of geometry will prevent them from assuming simple shapes. Consider molten metal poured into a mould. If heat is abstracted by the walls, crystal nuclei will appear close to the walls, and the crystals will grow out from them in all directions; in the direction parallel to the walls, they will soon run into one another, and growth in this direction will cease; however, it can continue in the direction at right angles to the walls, producing long narrow " columnar " crystals.* If, on the other hand, the cooling is slow and the temperature throughout the molten mass is fairly constant, crystal-nuclei will appear at points sporadically dispersed throughout the whole mass; the crystals will then grow outwards until they meet one another, producing an *equi-axed* structure, i.e. a structure in which there is no one direction in which the grains, as a whole, are longer than in other directions. Here the form of each crystal-grain will not be decided by the atomic arrangement; the grain-boundaries will represent the surfaces along which the crystals growing from different nuclei happen to have met one another. The growing crystals will often have adopted a dendritic (tree-like) form, mainly because the heat evolved in crystallization is most easily dissipated

* There is often a preferred orientation in this columnar zone, the direction being influenced by minor constituents. In very pure lead, the direction is $\langle 111 \rangle$; in definitely impure lead and alloys it is $\langle 100 \rangle$. A theoretical explanation is presented by W. A. Tiller, *J. Metals* 1957, **9**, 847; A. Rosenberg and W. A. Tiller, *Acta Met.* 1957, **5**, 565.

from the corners, and the grain-boundaries in cast metal are usually far from straight, since the dendrites interlock. On *annealing* (i.e. heating at a temperature at which, although it is well below the melting-point, the atoms are appreciably mobile), there will be a tendency for the grain-boundaries to straighten, since a short straight boundary represents a state of lower energy than a long indented boundary. If the metal is deformed (" worked ")—an operation which distorts the structure and thus causes an increase of energy—and then heated, new nuclei may be formed, and new undistorted grains may grow out, until the whole is recrystallized; the new " secondary " grains generally possess straighter outlines than the original " primary " grains.

Thus the shape of the grains in *polycrystalline* metal (metal consisting of many grains or crystals), is *allotriomorphic* (determined by the positions at which the mutually approaching crystals have met) and not *idiomorphic* (determined by the arrangement of the atoms throughout the material, as in the cubes of sodium chloride crystallizing from solution). If by chance a grain happens to exhibit, on a micro-section of a metallic casting, 4 or 6 sides, that is no evidence that the metal is cubic or hexagonal. Occasionally, on sections of alloys, idiomorphic crystals are seen—for instance, the cubes of the compound SbSn in the tin-antimony-copper alloy used in bearings; the cubes are formed whilst the main mass of the alloy is still liquid, so that there is nothing to prevent the compound from assuming the shape characteristic of the internal arrangement of the atoms; but such cases are unusual.

Single Crystals. By suitable treatment, the lattice throughout a mass of metal can be brought into the same orientation, so that the whole becomes a single crystal; the fact that the mass may be a round bar does not prevent it from being a crystal, provided that the layers of atoms are parallel to one another throughout the whole mass.* Today cylindrical single crystals are most often produced by melting metal in a tube of appropriate form, and lowering the temperature to a point just below the freezing-point, where the chance of the spontaneous formation of a stable crystal-nucleus is very small; if then one end is " seeded " with a small crystal, this will grow along the tube until the whole mass is one large solid crystal. Clearly the risk of spontaneous crystallization from a second point, which would spoil the result, can be diminished by an appropriate device. In one method, largely due to Bridgman, a tube containing metal originally above the melting-

* The Author remembers Sir Harold Carpenter telling him of the occasion when he and Miss Elam (now Mrs. Tipper) first exhibited single crystals to the Royal Society. He has endeavoured to obtain confirmation, or correction, of the following account from those present—but without success. The events—as the Author remembers them—are roughly as follows. The two experimenters had felt slightly apprehensive as to whether the distinguished audience would accept a round bar as a crystal—so deeply rooted was the idea that a crystal was something possessing a simple geometrical form bounded by flat faces, such as a cube or octahedron. However, any nervousness was rapidly dispelled when, at the opening of the discussion, Sir William Bragg grasped the round bar, and announced with confidence "this is a single crystal!" Nobody ventured to contradict him.

point is slowly moved through a tube furnace arranged to produce a temperature gradient; when a given point in the tube reaches a sufficiently low temperature, the metal will solidify, continuing the orientation already adopted by the portion which had reached the cool region earlier; thus a single crystal is produced. Alternatively, the furnace can be moved, the metal remaining stationary. A method worked out by Piontelli avoids movement altogether; the current conveyed to the coils is gradually altered in such a way as to displace the temperature gradient—an arrangement which increases the chance of success (R. Piontelli, *Rendiconti Istituto Lombardi di Scienza e Lettere* 1957, **91**, 347).

The earlier method developed by Carpenter and Elam in their classical work is a little different. The starting-point is solid metal which has been slightly deformed, rendering the structure unstable; it is then heated cautiously to a temperature where there is an appreciable chance that at some point the atoms will rearrange themselves to form a stable, undeformed nucleus, which will then grow out until the atoms have adopted a single orientation throughout the whole mass; in other words the whole mass of metal becomes a single crystal (*mono-crystal*). If, through mischance, two nuclei appear spontaneously, a *bicrystal* is formed. By suitable use of temperature gradients, the risk of producing a bicrystal, instead of the desired monocrystal, can be minimized.

Monocrystalline metal is softer than polycrystalline metal (consisting of many grains), in which the grain-boundaries interfere with the gliding, as explained later.

Information provided by Etching regarding the orientation of grains. Although the shapes of the grains in cast metal usually tell us nothing about the arrangement of the atoms, we can obtain that information by etching. If a section of a polycrystalline block of metal is cut, ground, polished (mechanically, or by the electrical method described in Chapter VII), then etched with a suitable acid mixture, etch pits appear at scattered points in the grains, bounded by flat facets; sometimes these are conveniently described as " negative crystals", since the shape is that which would be produced by pushing a tiny cube or tiny octahedron into a soft wax surface. The form arises from the fact that the acid starts to attack the metal at points which are physically and/or chemically different from the rest, and leads to the production of flat faces along planes in which the atomic centres are most densely packed, such as the cubic and octahedral planes; as already explained, when once one atom of a layer has been removed, it is easier to remove the remaining atoms of that layer than to start removing atoms from the next layer, and consequently at any moment the front line developed by the attack will represent a crystallographic plane. If the metal crystallizes in the cubic system, the etch-figures may have facets parallel to the cube planes or to the octahedral planes, according to the reagent used.

Clearly if the orientation of two adjacent grains is different, the facets will be inclined at different angles in different grains, and the shape of the pits themselves will appear different (fig. 73). This method of demonstrating and determining the orientations existing in different grains has been greatly developed by Wyon in Lacombe's laboratory, where the following mixture has been found convenient for aluminium:—

Fuming nitric acid ($d = 1{\cdot}33$)	150 c.c.
Hydrochloric acid ($d = 1{\cdot}19$)	80 c.c.
Butyl cellosolve	60 c.c.
Hydrofluoric acid ($d = 1{\cdot}14$)	5 c.c.

The reagents should be mixed in the order named in a container at 0°C. and aged 1 hour at 15°C. before use. All reagents should be absolutely free from impurities, especially silicon, iron and copper. Glass vessels should be avoided; teflon or polyethylene vessels are suitable (P. Lacombe, Priv. Comm., Jan. 15, 1957; March 17, 1958).

Other mixtures, which confer convenient distributions of pits, have been arrived at by suitable blending of the various ingredients, which had been found to confer different properties; thus hydrochloric acid was found to favour localized attack and hydrofluoric acid general attack; water, which had been included in certain mixtures used by earlier workers, increased the number of the pits, whilst diminishing their size.

FIG. 73.—Etch Figures on different grains.

Those wishing to use these methods of studying orientation should consult the paper of P. Lacombe and L. Beaujard, *J. Inst. Met.* 1948, **74**, 1, and also—especially for the determination of crystal-direction from angular measurements on the etch-figures—that of G. E. G. Tucker and P. C. Murphy, *ibid.* 1952–53, **81**, 235. Anodic attack has also proved useful in developing etch-figures on different grains; see J. Bardolle and J. Moreau, *C.R.* 1954, **238**, 1416; R. Jacquesson and M. Menenc, *ibid.* 1950, **230**, 959.

Etch-figures occur during natural corrosion processes more often than is sometimes supposed. They have been observed on the under-side of an aluminium roof corroded by condensed moisture; the reasons why they had previously been overlooked, is the nature of the liquid commonly used to " clean " the corroded surfaces (often a mixture of nitric, hydrochloric and hydrofluoric acids); this rounds off the facets and obliterates the structure (P. Brenner, F. E. Faller and E. Höffler, *Aluminium* 1956, **32**, 6, 64, esp. p. 65).

The facets developed by etchants enable the metallographist to distinguish one grain from another, since when a micro-section is viewed through a microscope, some grains which have developed facets inclined at such an angle as to reflect the light up the tube will appear bright, whilst others will appear dark. It is instructive to view such a section under oblique illumination through a microscope fitted with a rotating stage, since then a grain which has appeared bright in one position becomes dark when the stage is rotated, whilst others which were formerly dark may become bright.

The grain-structure may also be shown up by a reagent chosen to produce special attack along the grain-boundaries. A trench here produced can often be seen as a dark line, or a pair of dark lines close together. However, lines developed at the boundaries may not always represent trenches; if one grain is attacked more than another, the difference in level can produce a line in some circumstances. There are also cases where the grain-boundary is less attacked than the grain-interior and stands up as a ridge; tin of 99·99% purity—or its alloy with 1% copper—is stated to resist carbonate solutions better along the boundaries than elsewhere (G. Derge, *Trans. electrochem. Soc.* 1939, **75**, 449, esp. p. 459).

Aluminium containing a trace of iron also sometimes leaves a honey-comb structure due to grain-boundaries in relief (G. Chaudron, address at Ferrara, Nov. 7, 1955).

A simple but ingenious method of bringing out the grain-boundaries of iron in the austenitic state (stable, in many alloys, only at a high temperature) has been worked out in Japan. The specimen is exposed to air at the appropriate high temperature for a short time—perhaps only a few seconds —and quickly plunged into borax which removes the oxide, leaving thin grooves along the grain-boundaries, where oxidation had proceeded rather more quickly than elsewhere (Y. Imai and H. Hirotani, *Sci. Rep. Res. Inst. Tohoku Univ.* (*A*) 1957, **9**, 467).

In studying alloys, the metallographist uses reagents which darken or colour one phase and leave others unchanged. This method of distinguishing phases has been developed empirically. Sometimes, when the reagent contains a copper or silver compound, the coloration or darkening depends on the deposition of the noble metal; in alloys where one phase consists of metal containing carbon in solid solution, the removal of metal may leave a dark carbonaceous residue. On steels, sodium picrate darkens the cementite, cupric ammonium chloride darkens areas rich in phosphorus, whilst nitric acid in methyl alcohol is sometimes used to develop the grain-boundaries. A scientific study of such techniques in the light of our new understanding of corrosion and passivity might yield results of value both to metallographist and corrosionist. Edeleanu's method depending on anodic attack at a potential controlled to allow attack on certain phases and to leave others unchanged has many practical possibilities; a similar principle has been developed in Czecho-slovakia by Pražák (C. Edeleanu, *J. Iron St. Inst.* 1957, **185**, 482; M. Pražák and V. Čihal, *ibid.* 1957, **187**, 48).

Defects in Crystals

Influence of defects on properties of metals. If metals consisted of perfect crystals, their properties would be vastly different from those of metals as we know them today. Theoretical calculations, for instance, have indicated that defect-free crystals ought to possess a tensile strength greatly superior to that of every-day metal, and recently this prediction has been verified, since " metallic whiskers ", perhaps 1 μ in diameter, can be produced with a strength up to 400 tons/in.^{2}; these appear to be single crystals free from defects (H. K. Hardy, *Research* 1955, **8**, 57; G. W. Sears, *Acta Met.* 1955, **3**, 361, 367. Whiskers produced on heating aluminium-magnesium alloys *in vacuo* are described by P. J. E. Forsyth, P. G. Partridge and D. A. Ryder, *Royal Aircraft Est. Tech. Note* **286** (1958)—unclassified. For conference on " Whiskers ", see *Nature* 1958, **182**, 296).

Much of the research work leading up to our knowledge of defect-structure has been carried out on non-metals. It is often convenient to consider cases where crystals are formed by deposition from vapour on a cool surface, and destroyed by volatilization on heating. Very beautiful little crystals of iodine—or of a volatile metal such as cadmium—can be produced by condensation from the vapour phase.

Screw Dislocations. When a crystal grows from a slightly super-saturated solution, atoms will be deposited easily on uncompleted layers, but when these have been completed, a starting of a new layer will be difficult; for the first atom to be attached is in a very unstable position, being in contact with the mass on one side only (fig. 74 (A)); it is likely to leave the surface again before being joined by others. Only if conditions are very far from those of equilibrium (as happens when a crystal is in

FIG. 74.—Instability of (A) First atom starting a new plane, (B) Single hole.

contact with a strongly super-saturated solution) are additional atoms likely to be added so quickly as to produce a stable group. Conversely in dissolution, the formation from a slightly under-saturated solution, of the first vacancy in a complete layer, represents an unstable situation (fig. 74 (B)); it is likely to be filled up again, before a sufficient number of neighbouring atoms can leave to produce a hole of stable size. Only if the solution is extremely undersaturated will attack develop on a complete layer in a perfect crystal. Calculations based on such arguments suggest that under conditions close to equilibrium, the starting of new layers, or the removal of the first atom from an existing complete layer, should be a rare event, and that perfect crystals should only grow, and only shrink, at an appreciable rate when conditions become far removed from equilibrium. Yet actually both growth and shrinkage occur at appreciable rates close to equilibrium conditions. An iodine crystal stands in equilibrium with iodine vapour at a certain pressure; if the pressure is raised slightly above the equilibrium value, deposition occurs and the crystal grows; if it is depressed slightly below that value, vaporization occurs and the crystal shrinks. How are these facts, established by observation, to be reconciled with predictions from theory?

The answer, largely due to Frank, is that in practice we are not dealing with perfect crystals, but with crystals containing screw dislocations, an arrangement wherein the atoms are arranged, not on absolutely smooth planes, but as cork-screw inclines of very gentle gradient, terminating in the manner suggested in fig. 75; when we trace a layer of atoms round the central axis, we arrive at a level of one (or more) atoms above the former position. With atoms thus arranged, there is bound to be a ledge representing the end of an incomplete atomic layer; in such a case, even a very slight departure from equilibrium conditions should make it possible for fresh atoms to be added in the corner provided by the ledge. However many extra atoms are added, we never complete the layer, and never destroy the ledge; all that happens is that its position moves around the axis, and we continue to build up the endless staircase. In other words, a mechanism is provided permitting deposition to continue indefinitely and easily on a crystal very close to equilibrium conditions, i.e. on a crystal exposed to a vapour only slightly above the equilibrium vapour pressure, or to a solution

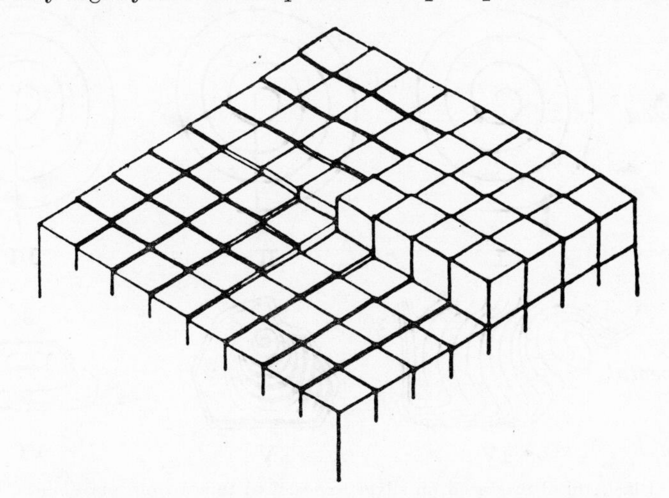

FIG. 75.—The end of a screw dislocation (F. C. Frank).

only slightly super-saturated, or to a melt held only slightly below the melting-point.

It might perhaps be expected that the ledge would rotate around the axis during growth like the spoke of a wheel; but at a position situated far from the axis, this would require a very rapid supply of atoms in the correct energy state, since the linear rate of growth would become steadily greater with the distance from the axis. In practice, the "spoke" will lag, and we obtain a catherine wheel or spiral, the central parts being raised to a slightly higher level than the outer parts, and deposition occurring at the corners provided by the various ledges. Actually the ledge will itself have have kinks, the atoms being deposited preferentially in the kinks; such matters, however, cannot be considered here. For detailed treatment, see F. C. Frank, *Disc. Faraday Soc.* 1949, **5**, 48. Cf. experimental studies of growth and evaporation of crystals by S. A. Kitchener and R. F. Strickland-Constable, *Proc. roy. Soc.* (*A*) 1958, **245**, 93.

Gould has aptly described the situation as a helix of low pitch, growing in irregular manner both radially and in thickness; the pitch may be as small as the inter-atomic distance (A. J. Gould, Priv. Comm., Feb. 19, 1957).

When first the conception of these screw dislocations arose, it was considered unlikely that they would ever be directly observed. However, it is today possible to view and photograph the spirals with ordinary microscopic equipment. The fact that this can be done is due to two rather unexpected developments. In some cases the height of the ledge (i.e. the vertical distance between the floors of the staircase), instead of being equal to the inter-atomic distance, has been found to be much greater, and the ledges become visible without special treatment. However, even where the ledge-height is only about 2Å—too small for direct observation—

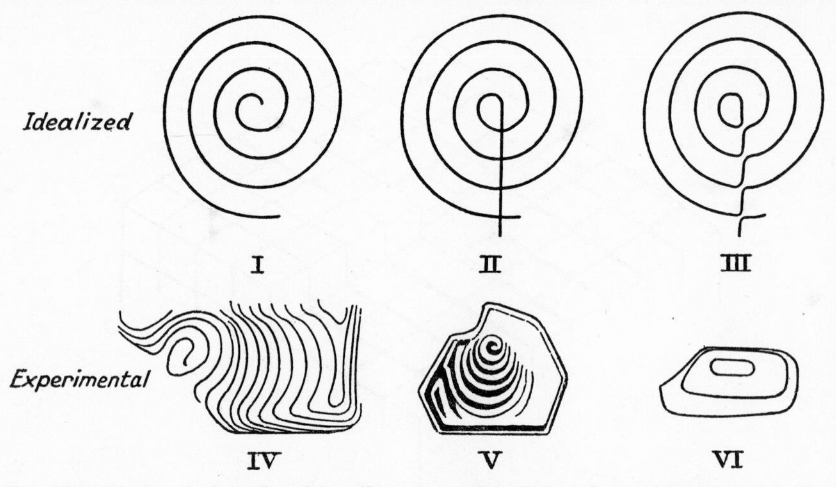

Fig. 76.—Black spiral produced on silver exposed to fumes from plasticine. I is the idealized form predicted by theory; movement of a dislocation may change it to II and rearrangement to remove sharp corners may then give III. The real forms observed in practice (IV, V and VI) are less perfect; the closed loops of VI may possibly arise by the movement of two dislocations of opposite signs (A. J. Forty and F. C. Frank).

the outline of the spiral becomes clearly visible, in some cases, owing to the fact that the atoms at the ledge are, as would be expected, more reactive than those on the flat face; if a very small trace of tarnishing agent is present, and is replenished very slowly, the metal will be darkened along the spiral line and nowhere else. Silver crystals, which have been developed by exposure to the vapours given off by plasticine, show a black spiral line, probably due to silver sulphide (fig. 76); the spiral can be photographed at a magnification of only $\times$ 3000. This method is instructive, since it provides direct evidence of the spiral mechanism of crystal growth, and also of the superior reactivity of atoms situated at corners compared to that of atoms situated at flat surfaces. Attention should be given to the photographs of A. J. Forty and F. C. Frank, *Proc. roy. Soc.* (*A*) 1955, **217**, 262; see also

F. C. Frank, *J. Inst. Met.* 1956–57, **85**, 581. For growth spirals in electro-deposited copper, see H. J. Pick, *Nature* 1955, **176**, 693.

It might be expected that when a crystal showing spiral growth is exposed to corrosive conditions, there would be steady removal of atoms along the ledge, and that instead of building up the staircase—as in crystal growth—we should dismantle it. Information in regard to metals is limited, but the corrosion of spiral crystals of silicon carbide has received much study. When place in fused alkali or borax, these crystals develop a deep polygonal pit at the axis of the spiral, and sometimes saucer-shaped pits at points situated along the spiral line itself. This behaviour certainly shows that atoms are most easily removed from places where they stand in positions irregular with respect to the layers of atoms around them—which is particularly the case at the axial line of the spiral. However, the localization of the attack at certain points on the spiral demands explanation. One possibility is that a protective film of silica is present over the main surface, and that, when this has once broken down at a point, attack will continue there rather than elsewhere. Another explanation is that activation energy is needed for the removal of atoms—even those situated at the corners provided by the ledge; only an exceptional atom which chances to have acquired abnormal energy can escape from the solid phase into the liquid, but when once it has passed through a position of high energy where it is still under attraction from neighbouring atoms on the solid surface, the final effect is a drop of energy, with local development of heat, which may serve to provide activation energy for other neighbouring atoms.

These two explanations are put forward tentatively; the facts, however, are not in doubt, and are described by F. H. Horn *Phil. Mag.* 1952, **43**, 1210, and by G. Gevers, S. Amelinckx and W. Dekeyser, *Naturwissenschaften* 1952, **39**, 448.

The fact that the removal of atoms from metal can follow a mechanism which is apparently the exact reverse of that involved in the deposition of atoms, is indicated by the observation of spiral pits, produced when material is removed by evaporation (V. G. Bhide, *Nature* 1958, **181**, 1006).

Behaviour of Metals under Stress. Before edge dislocations can be understood, the behaviour of metal under stress must be sketched (fig. 77). When a metallic rod is subjected to a small tensile force, it undergoes slight *elastic elongation* (the inter-atomic distance increasing in the stress direction, to an extent roughly proportional to the stress*) but returns to its original dimensions when the applied force is discontinued. At higher stresses, it is found that a slight *permanent elongation* remains and the stresses at which this amounts to 0·1 and 0·2% are called the 0·1 and 0·2% *Proof Stresses*, respectively. In most non-ferrous materials, there is a gradual passage from the *Elastic Range*, where the metal returns to its original length when the stress is released, to the *Plastic Range*, where elongation is permanent. On some materials (including steel, after certain

* The stress where this proportionality ceases is known as the *Limit of Proportionality*.

heat-treatments followed by ageing) a sudden inelastic elongation occurs at a certain stress, known as the *Yield Point*.*

Since the elongation must cause the rod to become thinner, it follows that, for a constant load, the stress (force per unit area) will *increase*, as elongation proceeds, but at first the fact that deformed material is stronger than undeformed (due to the work-hardening or work-strengthening discussed below) more than compensates for this. If, however, the load applied is increased sufficiently, the local narrowing of cross-section may at some point increase the stress more than can be met by increased strength, and the rod starts to neck off and then breaks. The percentage elongation of the rod, and the percentage loss of area at the point of fracture, are usually recorded after a tensile test, since they provide measures of ductility. The stress at which fracture occurred, the ultimate tensile strength (U.T.S.), is also recorded; the value is usually obtained by dividing the maximum value of the load sustained before fracture by the *original* area, although it is possible to allow for the reduction of area.

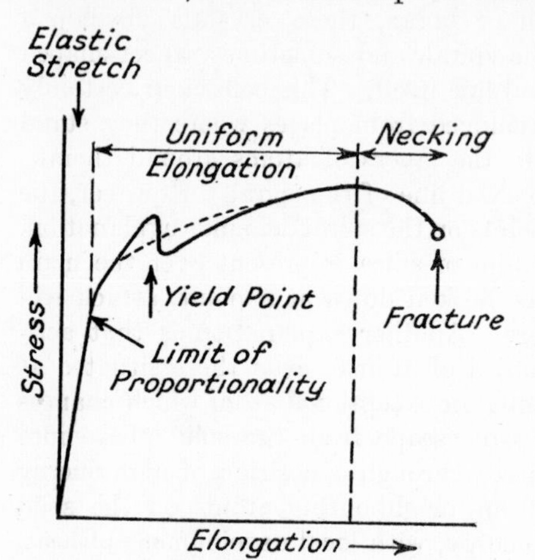

FIG. 77.—Elongation under stress (schematic). (*Note.* Only certain materials show the complications at the yield point; others show smooth curves as suggested by the broken line.)

It is established by direct observation that the plastic deformation of metal—whether under tension or compression—involves the gliding of slices of metal over one another. Such gliding occurs along crystal planes corresponding to dense atomic

* The Yield Point is believed to be caused by the presence of atoms of carbon or nitrogen which can occupy interstitial positions in the iron lattice. Given time, the carbon or nitrogen atoms tend to diffuse to dislocations and diminish the energy associated with them by minimizing the discontinuity of structure—as suggested by Cottrell. This in effect anchors the dislocation, making it difficult to move them to a new region where there are no impurities, unless work is performed to provide the increase of energy involved in the removal. When once, however, a dislocation has moved it can continue to do so easily. Thus the presence of carbon or nitrogen strengthens iron, increasing the value of the stress which must be applied before deformation can occur. However, when once this has been applied, the material continues to deform under decreasing stress (A. H. Cottrell, " Dislocations and Plastic Flow in Crystals " 1953 edition, pp. 134–145 (Clarendon Press)). The fact that hydrogen also can cause a yield under conditions where the yield points associated with carbon and nitrogen have been eliminated is indicated by H. C. Rogers, *Acta Met.* 1954, **2,** 167; 1956, **4,** 114. Curiously enough, hydrogen eliminates the yield point due to carbon. (See also W. D. Robertson, *Corrosion* 1957, **13,** 437t, esp. fig. 8, p. 442t.) For rate at which impurities flow to an edge dislocation, see R. Bullough and R. C. Newman, *Proc. roy. soc.* (*A*) 1959, **249,** 427.

packing, and is most easily produced in those grains where the gliding plane makes an angle of about 45° with the axis of the rod; at high values of the load, however, other planes, less favourably oriented, will also provide gliding. When the rod consists of a single crystal, the gliding along planes will produce a series of steps—often about 10^{-4} cm. apart—at the edges of the crystal, which can be observed under the microscope. Such gliding will render the specimen longer but thinner. Similar steps can be seen when a polished section of polycrystalline metal is strained; dark lines appear on the polished surface, varying in direction from one grain to another and clearly indicating the direction of gliding, although it seems likely that each line represents a number of short glides along neighbouring planes rather than one long glide on a single plane (p. 705).

Within the interior of polycrystalline metal, a simple gliding process would cause each grain to assume a stepped outline, provided that its neighbours "made way" for the steps developed; in practice, however, the neighbours would also be producing steps, and—except by coincidence—the steps formed by adjacent grains would not fit into one another. Thus at a stress which would suffice to produce gliding in each grain if it existed in isolation, a simple gliding process becomes impossible, except perhaps in the interior of large grains—and some other method of deformation, demanding a greater application of energy, must be introduced. That is probably one reason why polycrystalline metal is stronger than a single crystal; the strength usually increases as the grain-size becomes smaller. One alternative method of deformation involves the breaking up of the original grains into fragments, which are swung round into a favourable direction; in this way the material acquires preferred orientation, since certain crystal planes of the various fragments tend to orient themselves parallel to the direction of the applied force. Preferred orientation is particularly marked on rolled metal, which shows different physical properties (and different corrosion behaviour), according as it is tested in different directions.

After the break-down of the original structure, it is clear that gliding becomes increasingly difficult and that the resistance to random sets of applied forces, which would easily cause change of shape in the original material, will have become considerable; this is one possible cause of work-hardening, but other factors contribute to that phenomena.*

The deformed material is naturally in a state of internal stress since the fragments will have their inter-atomic spacings altered within the elastic range; of course, throughout an article as a whole, when it is not subjected to an external force, the internal stresses must balance one another (otherwise the article would spontaneously alter its shape until this balance was

* For instance, the early stages of deformation cause the dislocations present originally in the metal to "pile up" near grain-boundaries, so that they cease to be available for further gliding. Fresh dislocations can be formed, but this needs energy. Thus the work involved tends to increase as the deformation increases. Work-hardening received detailed discussion at a special meeting of the Royal Society in 1957 (*Proc. roy. Soc.* (*A*) 1957, **242**, 145–227).

obtained). However, of the tensile and compressive forces, one type may predominate near the surface; thus in a cold-drawn bar, the outer zones are usually in tension, and the interior in compression. (These facts are important in corrosion, being responsible for the " season cracking " of brass pressings, tubes, cartridge-cases, and the like—as indicated in Chapter XVI.)

Annealing at a low temperature may largely remove the objectionable internal stresses, by allowing the atoms to resume their equilibrium distances, without complete loss of the valuable work-hardening. Annealing at a higher temperature (or perhaps even for longer times at the lower temperature) may cause recrystallization, with the formation of new grains, which may be larger or smaller than the first set according to the degree of deformation and the temperature employed; the work-hardness is lost, but if conditions are chosen to refine the grain, the change may be of considerable practical value.

Generalization about annealing is dangerous, but the typical stages through which a deformed metal passes when heated may often be something like these:—

(1) *Recovery:* a partial return to the original physical properties which had been altered by deformation.

(2) *Polygonization:* the formation of blocks (sub-grains) with the atomic rows set at very slight angles to one another; this greatly reduces the strain energy present, for instance in metal which has been bent.

(3) *Recrystallization,* due to the migration of the existing interfaces, so that some grains invade their neighbours.

A useful statement of present views, with discussion of the differences between polygonization and recrystallization is provided by A. G. Quarrell, *J. Inst. Met.* 1957–58, **86**, 475; see also M. Deighton, *Bull. Inst. Met.* 1956, **3**, 93. Cf. N. Thorley, *J. Inst. Met.* 1950, **77**, 141; A. J. Kennedy, *Metallurgia* 1955, **52**, 265.

Edge Dislocations. When metal is subjected to mechanical force, it undergoes a change of shape by means of *slip* within the grains; whole layers of atoms glide relatively to one another. In fig. 78(A) the block of metal is considered to be perfect in structure, and when a force is applied, one part is pulled in one direction, and the other part in the opposite direction; after passing through the state shown in fig. 78(B), we arrive at the state of fig. 78(C). Such slip would, however, require a very strong force in metal where the crystalline structure is perfect. When the atoms are half-way (fig. 78(B)), the system will be very energy-rich, and since every atom in the plane reaches this energy-rich situation simultaneously, the stress required for gliding in perfect metal must be enormous.

If, however, the metal contains certain structural defects (dislocations) the amount of work needed for slip is greatly reduced; this is most easily explained in the case of the defect known as an *edge dislocation* and shown in fig. 78(D). Here it is noticed that there is one less atom in the row placed below XY than in the row above it; the atoms minimize the discontinuity by graduating their positions as shown; atoms in other planes parallel to the paper assume similar distributions, so that the " edge " runs at right angles to the paper. If now we apply force such as would tend to pull the upper group to the right and the lower group to the left, it is clear that at any one moment during the gliding only a few atoms are in the energy-rich state. Thus the energy-increment involved at any stage in the shift becomes small, and the force required for formation becomes small also. The position of the edge dislocation moves as gliding proceeds.*

The existence of dislocations undoubtedly reduces the strength of metal, allowing deformation to occur at a stress far below what would be needed on perfect material; but the structure described is also responsible for the fact that when one portion is gliding relatively to the other, the two

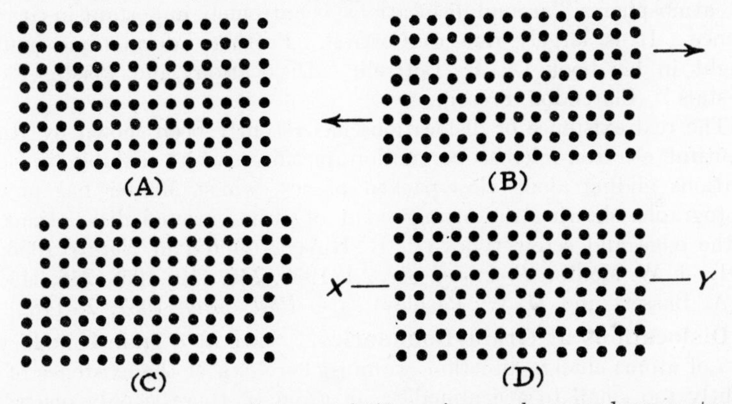

FIG. 78.—Deformation of a perfect crystal (A) requires much energy because structure must go through an energy-rich intermediate stage (B) before reaching (C). If the crystal is imperfect containing a dislocation, this energy-rich condition at the half-way position occurs at different times for different groups of atoms and much less work is required; the dislocation itself moves, as the upper half glides relatively to the lower half.

portions do not part company. For, except at the part of the gliding planes close to a dislocation, the atoms on both sides are in almost perfect continuity and the two portions continue to form a single structure. In other words, metals subjected to stress will deform without cracking—a property largely responsible for the value of metals. Some people are apt to take this valuable property for granted, and when, as in fatigue, they allow gliding to take place backwards and forwards, they confidently expect the two portions to continue to hold together indefinitely; if, after a few million cycles of travel to and fro, the expectation is not realized, and cracking here and there occurs instead of non-destructive gliding, they feel aggrieved,

* It is thus possible to explain what is said on p. 822, about a dislocation acting as the source or sink of vacancies—a matter of some importance in connection with oxidation, where vacancies may be created at the base of an oxide-film. A vacancy diffusing to the dislocation of fig. 78(D) would not alter the structure except that the position of the first plane deficient by one atom would be shifted by one unit; a shift of one unit in the other direction would create one vacancy in the plane of the paper and presumably in other planes also.

and formulate complicated theories to account for the failure. To the author, the most surprising thing about fatigue failures is that they do not occur sooner and more frequently. Fatigue, and corrosion fatigue, are discussed in Chapter XVII.

Since the distance between atoms is irregular near a dislocation, it is probable that atoms of an impurity, differing in size from those of the main metal, will tend to settle at or around such places—in such a way as to minimize the discontinuity of structure. If now a stress is applied, gliding is only possible if the dislocation is detached from its " atmosphere " of foreign atoms—involving extra work; thus a second substance in solid solution often hardens and strengthens the metal. Its atoms are sometimes said to " anchor " the dislocation. This idea of foreign atoms collecting as " atmospheres " around dislocations, is extremely important in corrosion-science. It is largely due to Cottrell. Further information should be sought in his book (A. H. Cottrell, " Dislocations and Plastic Flow in Crystals " (Clarendon Press)).

The real existence of dislocations has recently been shown by Hirsch's beautiful electron-microscope motion pictures, which exhibit single dislocations gliding along close-packed planes, whilst Menter has produced photographs showing the arrangement of atoms around dislocations both of the edge and screw types (P. B. Hirsch, quoted in *Nature* 1956, **178**, 1091. J. W. Menter, *Proc. roy. Soc.* (*A*) 1956, **236**, 119; 1958, **246**, 345 (with G. A. Bassett and D. W. Pashley); also *Phil. Mag.* 1957, **2**, 1482).

Dislocations at Grain-Boundaries. Since at a grain-boundary, the rows of atoms change direction, we must here expect the existence of holes slightly too small to accommodate an atom (if there is only one type of atom present). The sort of arrangement likely to occur at the boundaries is shown in Bragg and Nye's " rafts " of bubbles on a liquid surface, where the rows of bubbles differently oriented represent different grains, and gaps inevitably occur at the boundaries where these meet (fig. 79). It is not surprising that diffusion can occur along grain-boundaries at temperatures too low for it to occur through the lattice forming the grain-bodies. The intergranular penetration of oxygen and sulphur—discussed in Chapter III —becomes intelligible.

It might be expected, perhaps, that the grain-boundaries would be surfaces of weakness, but, at relatively low temperatures, where deformation occurs by gliding of slices within the grains, this is not the case. At very high temperatures, it becomes easier to pull the grains apart than to cause gliding within each of them. Many years ago, Jeffries suggested the name *equi-cohesive temperature* for that temperature at which the strength of the grains and their boundaries became equal; above the equi-cohesive temperature, the fracture obtained on breaking a specimen in tension was intergranular; below that temperature it crossed the grains (Z. Jeffries, *J. Amer. Inst. Met.* 1917–18, **11**, 300).

However, it is known today that the rate of pulling is important. The slow extension of metals subjected to tension at rather high temperatures, known as *Creep*, is partly connected with gliding of slices within the grains over one another, and partly with the grains sliding over one another at their boundaries; recent work has shown an almost constant proportionality

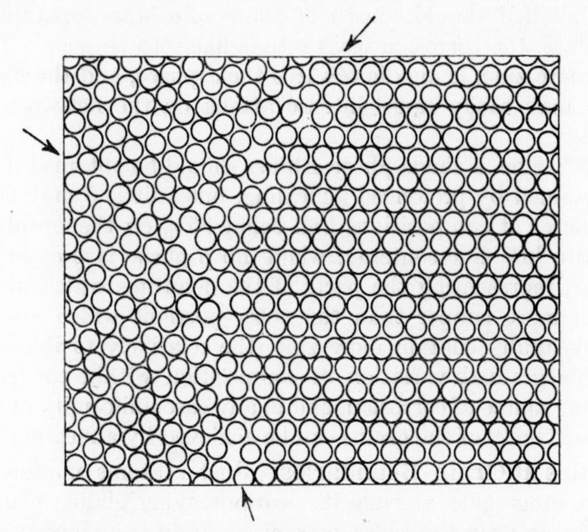

FIG. 79.—Bubble-raft model with grain boundaries indicated by arrows (Sir L. Bragg and J. F. Nye, *Proc. roy. Soc.* (*A*) 1947, **190**, 474).

between the two types of movement (D. McLean and M. H. Farmer, *J. Inst. Met.* 1956–57, **85**, 41).

At boundaries of sub-grains, where the angle between the atomic rows on either side is a very small one, the holes may be spaced well apart, and the distances between them can be calculated by trigonometry. The importance of this will appear shortly.

Introduction of a Second Metal

Structure of Alloys. When an alloying constituent is introduced in a limited amount, a solid solution is formed, the two types of atoms occupying sites on a common lattice. In some cases, complete mutual solubility is possible at all compositions, but in most alloy systems, when the content of the second constituent exceeds a certain amount, a second phase appears. In the gold-magnesium system (fig. 192, p. 977) alloys in which not more than 18% of the atoms are magnesium will, in a state of equilibrium, consist of a single phase (essentially gold with magnesium in solid solution); beyond this other phases appear, believed to represent compounds such as AuMg, $AuMg_2$, etc., with excess of one or other metal in solid solution; in this case there are maxima in the curve relating melting-point and composition at the points representing " pure " AuMg, $AuMg_2$ and so on.

In some cases, the compound from which the various solid solutions are supposed to be derived may be " meta-stable " in the " pure " state and there are no maxima. This is likely to occur in the case where the melting-points of one metal is much higher than the other, as in the copper-zinc system (fig. 193, p. 978). Here, *at equilibrium*, brasses containing up to

35% zinc should consist of a single phase (α), but beyond this a second phase (β) appears. The structure will depend on heat-treatment and mechanical treatment, but if it should consist of grains of α-brass separated by a net-work of β-brass, the corrosion-resistance is likely to be poor. The β-phase, containing more zinc, is anodic towards the α-phase, and the high cathode–anode ratio may easily cause intense attack on the net-work of β-brass (see p. 472).

Below 35% zinc, a single-phase α-brass should be obtained if equilibrium has been obtained by prolonged annealing, but a little β may be present at grain-boundaries at compositions well below 35% on the freshly cast alloy. Even at quite low zinc-contents, where no β-phase is present, the centre of the grains, representing the parts of the dendrites which first solidified, may contain more copper than the outer portions which are the last to freeze and naturally contain more zinc (the relatively fusible constituent). Probably this gradual variation of composition within the grains is not often a serious source of corrosion, and in any case uniformity of composition over the whole body of the grains can be achieved by annealing.

Impurities at Grain-Boundaries. Just as atoms of a foreign element will tend to congregate at edge dislocations along gliding-planes, so they congregate in or near the holes present at grain-boundaries; small atoms may settle in the holes, filling them up; large ones, occupying positions near the holes, will reduce the distortion of the structure, and minimize strain energy.

In cases where the composition along grain-boundaries differs from that representative of grain-interiors (whether this be due to a separate phase forming an intergranular net-work, or to foreign atoms settling around dislocations), the occurrence of intergranular corrosion should cause no surprise. The phenomenon is discussed in greater detail in Chapter XVI, although a few examples are introduced into the present chapter. Grain-boundary attack may occur if the net-work is anodic to the main part of the grains, or if the intergranular material can catalyse a cathodic reaction —e.g. the evolution of hydrogen.

Annealing does not prevent the tendency of the traces of the minor constituent present, even in very nearly pure metal, from settling at dislocations, either along grain-boundaries or within the grains. If anything it favours such settlement, which may give rise to micro-pits when a section is treated with a reagent capable of selective attack. This method has been used by some physicists to determine the sites of dislocations. There has been some difference of opinion whether dislocations alone can cause these micro-pits, or whether impurities settling at dislocations are a necessary condition; it may be convenient to present the facts and offer a possible interpretation.

Grain-Boundary Phenomena. The difference in the corrosion behaviour between grain-boundaries and grain-interiors depends on both physical and chemical factors. Since in different grains the atoms are arranged in layers possessing different orientations, it is likely that an atom at the boundary will be less firmly held than at an atom within the grain;

for if it is placed at a point which would represent its accurate position as a continuation of an atomic row in one grain, it will not be in an accurate position with respect to its neighbours in the other grain. It would not be surprising to find that, even in pure metals, atoms were more easily removed from grain-boundaries than from grain-interiors, but this will probably be true only of atoms very close to a grain-boundary. On the other hand, if a trace of a second constituent is present in the metal, it will often tend to accumulate at or near the grain-boundaries, and a zone of appreciable width following the boundaries may possess electrochemical properties different from those of the grain-interiors. The electrochemical contrast is best brought out if the amount of impurity is very small—just enough to allow the boundary-zones to take up all the impurity that they are capable of holding and leave the grain-interiors in a pure condition. If the grain-boundaries are anodic to the grain-interiors, they may suffer preferential attack; if they are cathodic, grain-boundaries may stand out in relief after exposure to a corrosive liquid; examples of each of these phenomena have been given (p. 367).

It would seem that in the majority of cases, the presence of a minor constituent is needed for preferential attack at the grain-boundaries. In the Teddington researches on zinc in potassium chloride, the attack upon zinc of exceptional purity produced in the grain-interiors geometrical pits, either six-sided or otherwise displaying hexagonal symmetry, whereas on zinc of slightly lower purity, these geometrical figures were usually absent, but there was attack upon the grain-boundaries (G. D. Bengough, A. R. Lee and F. Wormwell, *Proc. roy. Soc.* (*A*) 1931, **131**, 494, esp. p. 510).

However, crystallographic considerations as well as purity affect the chance of attack at grain-boundaries—especially in acid solutions. Lacombe and Yannaquis showed that the attack on very pure aluminium by 10% hydrochloric acid occurs at those boundaries where there is a large misfit between the grains. The situation is complicated, and the angle which the plane of the boundary makes with the rows of closely packed atoms, as well as the relative orientation of these rows in the two crystals—affects the question as to whether a particular boundary will be attacked or not; thus sometimes one part of a boundary will suffer whilst another part of the same boundary remains immune; on one specimen the straight portion of a boundary separating two twin-crystals remained unchanged, whilst the curved portion was attacked; since the atomic rows in twin-crystals are in mirror-imagine positions, the discontinuity of structure favourable for attack would at a straight boundary be absent (P. Lacombe and N. Yannaquis, *Rev. Métallurg.* 1948, **45**, 68; *C.R.* 1947, **224**, 921; 1948, **226**, 498; G. Wyon and J. M. Marchin, *Phil. Mag.* 1955, **46**, 1119).

Erdmann-Jesnitzer has studied the manner in which the orientation difference between two adjacent grains affects intergranular attack on an aluminium alloy containing 7% magnesium, made from the purest materials; he carried out experiments in multiple, so as to overcome the effect of scatter. Up to an angular difference of 4° there is practically no attack; then it suddenly increases and becomes nearly independent of angular difference between 15° and 20°, after which there is a further slow increase (F. Erdmann-Jesnitzer, *Werkst. u. Korrosion* 1958, **9**, 7).

Since the aluminium used in Lacombe's work was extremely pure, his experiments were at one time considered to prove that the attack was determined solely by crystallographic considerations, and not by impurities. However, Thomas and Chalmers, using a lead alloy containing 5% bismuth to which only 1 part in 10^{10} of polonium had been added, showed that the polonium segregates at grain-boundaries to an extent which depends on the difference in orientation of the grains, being four times as great when the angular difference was 25° as when it was below 15°; this suggests an alternative explanation of Lacombe's results. Metzger and Intrater have studied the action of aluminium of different purities in hydrochloric acid, and consider that iron tends to segregate at the grain-boundaries, and affects the attack there; they also find that acid containing traces of copper produces attack at grain-boundaries where pure acid produces none. Thus whilst Lacombe's work had shown up attack only at high angle boundaries, they obtained deep and selective attack on boundaries between sub-grains (where the difference of orientation is very small) by using 7% hydrochloric acid containing 0·5 p.p.m. of cupric chloride; the attack was so deep and selective that, after some days, entire sub-grains became undermined and fell out. If the acid was free from copper, it was necessary to increase the concentration to 20% to obtain comparable rates of attack.

Perryman has also studied aluminium of different iron content. His results, quoted in greater detail on p. 661, show that on furnace-cooled specimens, the grain-boundary attack first increases as the iron content is increased and then declines. He is in some disagreement with Metzger and Intrater—more in regard to interpretation than experimental facts. He considers that at very low iron contents, the iron will be concentrated only at grain-boundaries, and thus produces a lower hydrogen overvoltage at these points, explaining the rapid boundary attack and local evolution of hydrogen from the crevices which is observed. With higher iron contents there will be iron in the grains; the over-voltage will be the same everywhere —making the attack more uniform.

The very low concentration of noble metal (whether in the solid or liquid phase) needed to direct the attack on to grain-boundaries is not really surprising. To saturate the grain-boundaries to an extent which will minimize strain-energy should require only a very small amount of noble constituent; a second phase is not needed to explain the results. If the noble impurity is present in solid solution in the metal, it will pass into the acid in the initial attack, and will then be reprecipitated; given a large amount, it will be reprecipitated over the whole surface, but if only traces are present, the concentration produced in the acid will be very small, and nucleation of the nobler element is only likely to occur at places of loose structure, such as at the grain-boundaries; it will there be precipitated at a separate phase, explaining the local evolution of hydrogen and the intense corrosion. The same explanation appears suitable if the noble metal is introduced as a trace of salt intentionally added to the acid.

That is the explanation which the Author is inclined to favour. Readers, however, should compare the facts and interpretations advanced by different investigators, and form their own opinions. The references to Lacombe papers are given on p. 379. The others are:—W. R. Thomas and B. Chalmers, *Acta Met.* 1955, **3,** 17; M. Metzger and J. Intrater, *Nature* 1954, **174,** 547; *J. Metals* 1953, **5,** 821; 1954, **6,** 661; E. C. W. Perryman, *ibid.* 1953, **5,** 911; B. A. Bilby and A. E. Entwisle, *Acta Met.* 1956, **4,** 257; A. J. Forty and F. C. Frank, *J. Phys. Soc. Japan* 1955, **10,** 656; F. Montariol, *C.R.* 1957, **244,** 2163; *Corrosion et Anti-corrosion* 1958, **6,** 101.

The Spacing of Dislocations. Very interesting results are obtained on germanium where the attack along grain-boundaries often produces, not a continuous trench, but rows of tiny pits spaced out at uniform intervals. If the boundary separates two grains with the layers arranged at only slightly different angles to one another, geometrical considerations predict that holes must occur spaced out at definite intervals along the grain-boundaries. This can be demonstrated by means of bubble raft similar to that of fig. 79, p. 377, where the atoms are represented by bubbles of the same size floating on the surface of a liquid, and closely packed together. Areas where the rows of bubbles have different inclination represent different grains, and the lines separating the different areas represent the grain boundary. It is inevitable that along the boundary there must be holes just too small to take an extra bubble of the same size; the discontinuity at such places would be reduced if a bubble of different size (representing an impurity atom) is introduced there, whereas its presence elsewhere would increase the distortion of the structure. The bubble raft presents a picture of the state of affairs in a single plane; but in three-dimensional metal, we will have similar holes in other planes parallel to those of the raft, so that there exist lines of holes, or dislocation-lines.

If the angular difference between the atomic layers in two adjacent grains is known, the theoretical spacing of the dislocations can be calculated. On germanium, etched in a mixture of nitric, hydrofluoric and acetic acids, with a small content of bromine, Vogel found that the pits were spaced out at a distance which agreed well with the calculated distance between the dislocations. This impressive work certainly suggests that the pits in such a case are formed at the sites of the dislocations, although it remains uncertain whether it is the dislocation itself or the presence of foreign atoms (perhaps iron) associated with the dislocation which is responsible for the pitting. Whichever is the case, the phenomenon deserves study, as a means of making evident the effect of crystallographical factors on corrosion (F. L. Vogel, *Acta Met.* 1955, **3**, 245; Sir L. Bragg, *Proc. roy. Inst. Gt. Britain* 1955, **35**, 844, esp. p. 849).

Although in these circumstances the pits do seem to mark the sites of dislocations at grain-boundaries, it is not possible, in all circumstances, to obtain the number of dislocations by merely counting the pits. If that were a sound method, the numbers of pits would be found to be independent of the etching agent, whereas, at least in the case of pits in grain-interiors, the number developing varies according to the etching agent adopted; the number, for instance, increases with the amount of water in the acid mixture —as shown by Lacombe and Beaujard (p. 366). Clearly the various counts cannot all represent correct estimates of the dislocations present.

A suggestion that, on plastically strained zirconium, pitting by an etching agent is due, not directly to the dislocation but to the preferential segregation of impurities at such defects, in the manner suggested by Cottrell, is made by Billig. In the case of aluminium, it is fairly certain that the pits present in the interior regions of grains are due to associations of dislocation with iron as the impurity; this is brought out in a skilful research by Wyon and Lacombe, who show that the number of etch-pits become fewer as the iron-content of the metal becomes smaller; if as a result of heat-treatment, the impurity is precipitated as a separate phase, there is no formation of etch-pits. In studying extremely pure aluminium obtained by zone-melting, Lacombe is now using reagents containing a few parts per million of copper or other impurity, so as to bring the association of dislocations and impurities which are otherwise more difficult to show up.

The situation is complicated and the reader should study the facts and views recorded by G. Wyon and P. Lacombe, Phys. Soc. Report of Conference on Defects in Crystalline Solids 1954, p. 187. Cf. A. H. Cottrell, Phys. Soc. Report of Conference on the Strength of Solids 1948, p. 30; E. Billig, *Proc. roy. Soc.* (*A*) 1956, **235**, 37, esp. p. 48; B. A. Bilby and A. R. Entwisle, *Acta Met.* 1956, **4**, 257; T. H. Schofield and A. E. Bacon, *ibid.* (in the press).

Observations on non-metallic crystals seem to support the view that impurities are necessary for pitting in some cases, but not in all; lithium fluoride behaves differently according to the etching agent used; some reagents attack all types of dislocation, whilst others attack only those which have previously absorbed impurity during annealing (J. J. Gilman and W. G. Johnston, quoted in *Nature* 1956, **178**, 1091).

Pitting at emergence points of gliding-planes. Rows of pits can occur, not only along grain-boundaries, but also along the small cliffs produced on an originally smooth surface as a result of deformation. Forsyth

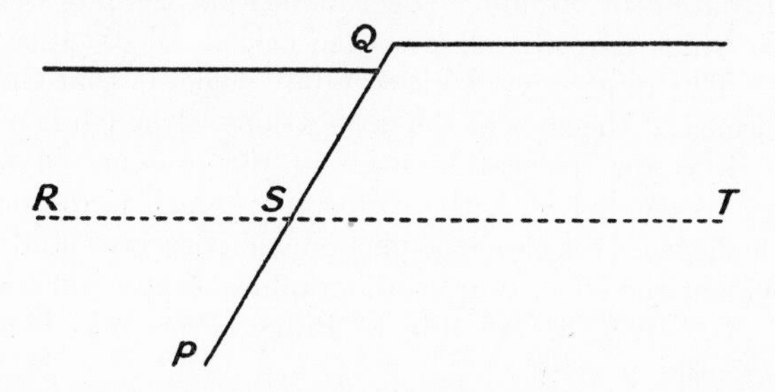

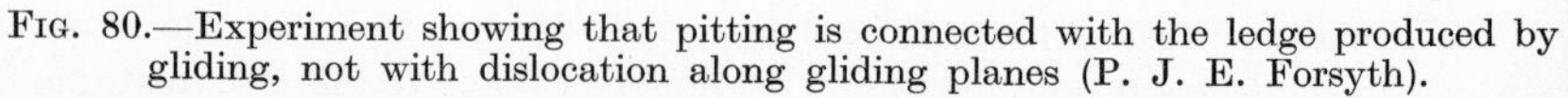
FIG. 80.—Experiment showing that pitting is connected with the ledge produced by gliding, not with dislocation along gliding planes (P. J. E. Forsyth).

electropolished a surface of pure aluminium, and subjected it to a slight strain which caused gliding along a single plane (PQ) producing a small ledge at Q (fig. 80); a very short dip in a mixture of hydrochloric, nitric and hydrofluoric acid developed a line of pits along the line of the ledge. When, however, the surface was again made flat by electropolishing and exposed to the acid, pitting was not observed at S where the gliding-plane cut the new surface, RT. This indicates that, in the case under study, the special attack was due to the greater ease of removing atoms from the edge of a ledge (as explained on p. 370) than from a flat surface; it was not due to disorganization along the plane of gliding. Readers may like to compare Forsyth's views with those of Amelinckx, who had made somewhat similar observations, but interpreted them differently (P. J. E. Forsyth, *Phil. Mag.* 1954, **45**, 344; S. Amelinckx, *ibid.* 1953, **44**, 1048).

Tunnelling Pits. Anodic attack in acid sodium chloride upon aluminium of 99·8% purity produces a remarkable tunnelling effect—shown up by Burger, Tull and Harris, using the electron microscope. The attack gives rise to tunnels of cross-section varying from 0·07 to 1·2 μ, running straight into the metal and suddenly turning at right angles. All the tunnels follow three mutually perpendicular directions, presumably related to the cubic crystal structure. The explanation of such observations is still uncertain, but it is possibly due to impurity atoms tending to associate themselves with edge dislocations (p. 376) (F. J. Burger, V. F. G. Tull and P. H. Harris, *Bull. Inst. Met.* Sept. 1955, **3**, 6; cf. E. C. Pearson, H. J. Huff and R. H. Hay, *Canadian J. Tech.* 1952, **30**, 311).

Variation of Corrosion Velocity with Crystal Direction

Production of Crystal-faces on Spheres. Extremely instructive work has been carried out on the growth or dissolution of single crystals of spherical shape. Before discussing metals, it may be well to describe some early work by Nacken on the crystallization of the organic compound, salol (p. 993). Nacken grew his crystals on a copper hemisphere C (fig. 81(1)) immersed in liquid salol (L) which was kept accurately at the melting-point. The hemisphere was attached to the copper rod R, which was surrounded with a current of cooling-water; by adjusting the temperature of the cooling-water, heat could be abstracted from C at any desired rate, and salol was deposited on C at a rate determined by the removal of the heat of crystallization. If the salol crystal was grown very slowly, its form (fig. 81(2)) was roughly hemispherical (strictly, it was ellipsoidal). If, however, the speed of crystallization was slightly increased, a smooth face appeared, indicating that in one direction the rate of arrangement of the atoms was unable to keep up with the requirements and thus in this direction the outgrowth was proceeding more slowly; the crystal remained rounded in some parts but flat in others (fig. 81(3)). On further speeding up of the rate of heat-removal, additional flat faces were produced (fig. 81(4)) and at the highest rates the crystal was wholly bounded by flat faces (fig. 81(5)) (R. Nacken, *Neues Jahrbuch Min. Geol. Pal.* 1915, **11**, 133).

The work of Gwathmey on the dissolution of copper spheres represents the converse of Nacken's experiment. A copper single crystal was ground to spherical form, and subjected to anodic attack in phosphoric acid, the cathode being a larger sphere, placed, concentrically, around it. Clearly if the corrosion velocity had been controlled by some factor other than crystallographic, the anode would have remained a sphere, becoming gradually smaller. In fact, however, the control proved to be crystallographic and crystal faces appeared. It was found that the corrosion-rate perpendicular to the cubic faces was only 70% of that perpendicular to the octahedral faces.

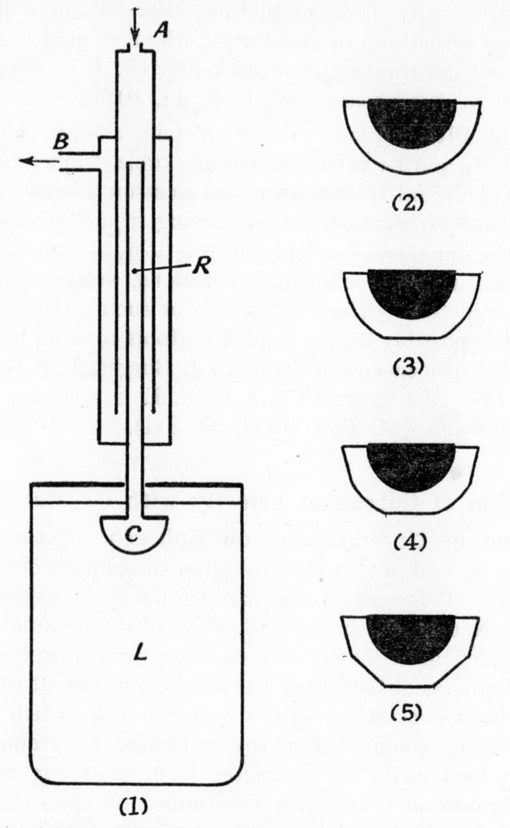

FIG. 81.—Growth of Salol crystals, with production of curved or flat surfaces, according to rate of heat removal (R. Nacken).

In another research the spherical single crystal of copper was suspended in acid copper sulphate with access of air; there was no attack in the direction perpendicular to the cubic faces, but rapid attack perpendicular to the octahedral faces. When the attack was speeded up by making the sphere an anode with an external E.M.F., the sphere was converted to a rough octahedron (A. T. Gwathmey and A. F. Benton, *Trans. electrochem. Soc.* 1940, **77**, 211; H. Leidheiser and A. T. Gwathmey, *ibid.* 1947, **91**, 95; W. R. Buck and H. Leidheiser, *ibid.* 1957, **104**, 474).

Effect of reagent on the crystal-plane developed by etching. It has been mentioned that the production of etch-pits depends on the reagent used. Aluminium yields pits bounded by cubic faces in aqueous or alcoholic hydrochloric acid, but by octahedral faces in dry hydrogen chloride. Zinc, which crystallizes in the hexagonal system, develops six-sided pits on etching; these are bounded by sloping sides, representing pyramidal faces, when the etching agent is sodium hydroxide, but by prism faces when it is hydrochloric acid (M. Mahl and L. Stranski, *Z. phys. Chem.* 1942, **52**, 257; also R. E. Richardson, quoted by C. H. Desch "Chemistry of Solids" 1934, p. 74 (Cornell Univ. Press)).

Doubtless the development of different faces is connected with the fact that corrosion-rate varies with direction in a different manner according to the reagent. This was shown by Glauner's work on copper crystals subjected to 0·3N acid in presence of 0·1N hydrogen peroxide; octahedral faces dissolved faster than cubic faces when the acid was hydrochloric or acetic, but slower when it was propionic (G. Glauner, *Ber. Korrosionstagung* 1933, **3**, 39. Cf. results on aluminium recorded by T. H. Orem, *Bur. Stand. J. Res.* 1957, **58**, 157).

As to why one reagent should favour attack in one direction, and a second reagent attack in a second direction, cannot be stated with certainty. Presumably the anodic reaction (passage of copper into the ionic state) and also the cathodic reaction (reduction of hydrogen peroxide) are the same in all acids. If propionic ions or molecules are adsorbed on the octahedral faces better than on the cubic where the inter-atomic distance would be different, and if such adsorption obstructs one or other electrodic reaction, the behaviour of propionic acid would be explained; but this is speculation.

Interesting information on crystal-orientation as a factor in the attack of acid on iron is provided by Engell. Under anodic attack at low current densities, the rate is independent of crystal-orientation, whereas at high current density, attack is quicker on octahedral than on cubic planes. Probably the removal of iron proceeds in two steps: (1) the removal of atoms from corners and shelves sideways to positions on the flat part of the face, and (2) their passage thence into the liquid. At low current density, the second stage presents difficulty and controls the rate of change, so that crystal orientation has no effect; with the high electrical fields needed to give high current density, the passage outwards proceeds readily and the first stage assumes control; in that case, crystal-structure becomes important. On simple immersion of iron in nitric acid, which produces high current densities in the local cells set up, differently oriented crystals act as anodes and cathodes, and special attack occurs at the grain-boundaries (H. J. Engell, *Arch. Eisenhuttenw.* 1955, **26**, 393).

The variation of corrosion-rate with crystal direction is not confined to cases where the corrosion product is soluble, but is met with even where a solid film is formed. A section of polycrystalline metal heated in air or oxygen develops a different colour on each grain—as was mentioned on p. 48; this shows that the thickness of the oxide-film produced varies from grain to grain.

Corrosion of Strained Metal

Cases where a Film is absent. Several experimenters have observed that the deformation of metal generally increases its attack by dilute acid. This may be due to a number of causes. The disarray in the atomic structure is likely to facilitate the detachment of an atom from the metallic phase and its passage into the liquid; the energy needed for detachment will surely be less than in the case of an atom occupying a stable position relative to its neighbours. However, many mechanical operations, such as rolling, break up the original grains and the pieces tend to rotate until certain crystal directions come to lie parallel to the rolling direction (p. 678); the structure now presents preferred orientation, and since certain crystal faces must predominate on the surface of the rolled metal, there is a chance that these will be faces liable to suffer rapid attack. In the case of steel, there is also a chance that straining may bring to the surface a segregate, containing say, sulphide, which will accelerate the attack; this effect is met with on iron strained in torsion, and differs from the other effects mentioned in that annealing does not restore the original resistance (B. Garre, *Korrosion u. Metallsch.* 1927, **3**, 1).

The increase in the rate of attack of dilute sulphuric acid on iron after strain has been shown in experiments on cold-drawn wire, also on iron strained in a tensile machine, and iron strained under torsion. Endo found that the increase in the corrosion-rate produced by tension or torsion is small on low-carbon iron, but increases with the carbon-content. Details will be found in the papers of E. Heyn and O. Bauer, *J. Iron St. Inst.* 1909, **79**, 109, esp. p. 159; P. Goerens, *Carnegie Schol. Mem.* 1911, **3**, 320, esp. pp. 374, 394; J. A. N. Friend, *ibid.* 1922, **11**, 103; H. Endo, *Sci. Rep. Tohoku Univ.* 1928, **17**, 1265.

Apparent exceptions are found, on examination, to fit well into the scheme of things. For instance, whereas the corrosion of steel and aluminium by dilute acid is increased by cold-work and, on the whole, diminished by annealing, the reverse is true of pure tin, where rolling reduces the subsequent attack by hydrochloric acid. This is probably due to the fact that tin, after serious deformation, recrystallizes at ordinary temperature—whereas other metals require heating to bring it about. Thus rolled tin may easily present a more perfect crystalline structure than cast tin (B. Garre, *Korrosion u. Metallsch.* 1930, **6**, 200).

Again Friend, who had found that straining increased the rate of attack by dilute sulphuric, later found that it produced no serious effect on corrosion by stagnant sea-water. However, in such a liquid, the corrosion-rate is controlled essentially by the rate of arrival of oxygen, probably passing inwards through a rust layer, to the metallic surface; if the unstrained iron is corroding as fast as the oxygen-supply will permit, there is no reason why oxygen should be supplied any faster to the strained iron, and the latter will corrode at the same rate (the two papers are: J. A. N. Friend, *Carnegie Schol. Mem.* 1922, **11**, 103, dealing with acid attack; and *J. Iron St. Inst.* 1928, **117**, 639, dealing with sea-water).

Müller and Buchholtz compared the corrosion of steel (*a*) unstrained, (*b*) strained to 90% of the yield point, (*c*) stretched to give 5% permanent elongation; the specimens were apparently kept under stress throughout the testing periods, which lasted 3, 6 and 9 months. The tests in *running* water (where the oxygen supply would probably be sufficient to keep up with the increased corrosion-rate of the stressed iron) showed that condition (*c*) produced more attack than (*a*); the reverse was sometimes observed on steel exposed to the atmosphere, where the stress may have caused the rust to break off more easily, so that the unstressed iron remained covered with damp rust after the stressed iron had dried (E. W. Müller and H. Buchholtz, *Arch. Eisenhuttenw.* 1935–36, **9**, 41).

Very interesting patterns are produced by acid cupric chloride (*Fry's Reagent*) acting on iron which has been strained locally and heated at 200°C. Two formulae are given by Jevons; one, intended for macro-etching, contains cupric chloride and hydrochloric acid in water; the other, intended for micro-etching, also contains alcohol. Some changes in the material can usually be seen without etching. The so-called *Hartmann* or *Lüders' lines*, which appear at an angle of about 50° to the deformation direction and are really wedges of stressed material, are often visible without any special chemical treatment. They are best seen on scale-covered iron because the scale tends to chip off along a series of roughly parallel but distinct zones which have undergone mass deformation, the intervening region being relatively undeformed. On atmospheric exposure, the rusting along these lines shows them up even more clearly. In absence of scale or rust, the lines can be developed by heating half an hour at 200°C, followed by grinding, polishing and etching with Fry's reagent. The patterns of intersecting stressed zones produced around places where an indent has been made into the metal, or at other regions of localized deformation, have received much study from T. H. Turner and J. D. Jevons, *J. Iron St. Inst.* 1925, **111**, 169; J. D. Jevons, *ibid.* 1925, **111**, 191; E. W. Fell, *Carnegie Schol. Mem.* 1927, **16**, 101,; 1937, **26**, 123; *J. Iron St. Inst.* 1935, **132**, **75**.

Fry's reagent is often used by the police in identifying stolen property by developing identification numbers which have been obliterated by a thief using a file; the strain left in the metal below the punch marks is sufficient to promote the action of the reagent, so that the numbers reappear.

The patterns produced on polycrystalline aggregates by Fry's reagent do not seem to have much relation to crystallographic direction, but in other cases etching patterns produced on strained material are closely connected with certain crystal planes. Compression of the rhombo-dodecahedral faces of iron single crystals, followed by etching, produces lines parallel to the principal slip-plane, and is attributed to severe damage to the crystal resulting from movement during slipping (L. B. Pfeil, *Carnegie Schol. Mem.* 1927, **16**, 153, esp. p. 156).

An interesting example of the effect of strain on corrosion is provided by the behaviour of iron towards nitric acid. It has been explained (p. 325) that reactive metals like magnesium, zinc and aluminium, when attacked by nitric acid, produce hydrogen-rich bodies like ammonia, whereas the nobler metals like copper produce oxides of nitrogen. Iron, standing between the two groups, show both types of reaction product, together with nitrogen, probably formed from the decomposition of ammonium nitrite. However, the proportion of the decomposition products is altered by strain in the metal; iron, when strained, becomes closer to zinc and less close to copper in its behaviour—as is easily understood. The composition of the reaction products can in fact be used for the detection of strain due to cold working or other causes, and a study of the reaction products from pearlitic steel has led to the view that the ferrite present in pearlite is in a state of strain. This may be due to the differential contraction of cementite and ferrite on cooling; the strains found in steel after a long sub-critical anneal have been similarly explained by Parkins (p. 683). The information obtained from the behaviour of iron in nitric acid deserves study (J. H. Whiteley and A. F. Hallimond, *Carnegie Schol. Mem.* 1918, **9**, 1).

The attack upon aluminium in hydrofluoric acid was found by Straumanis and Wang to be accelerated if the aluminium had been hammered, which may produce cracks, but not by ordinary cold-work (M. E. Straumanis and Y. N. Wang, *Corrosion* 1956, **12**, 177t).

Cases where a Film is present. In most of the cases cited above, the presence of acid liquid would be inconsistent with existence of an oxide-film on the metal. Where a film can exist, local strain (especially bending) may promote corrosion by breaking it, since oxide is less ductile than metal. This effect has been noticed on iron, steel and zinc sheet bent locally and tested in M/10 potassium sulphate or chloride, and on copper tested in silver nitrate. In potassium salt solutions, attack was noticed on both sides at the bend—especially on the convex side, where the film would suffer most. It might reasonably be thought that the special attack at the bend was due to strain in the metal, and to distinguish between this and the effects of film-damage, a set of three parallel experiments was performed by the Author.

(1) Copper was finely abraded, bent locally and at once tested in M/25 silver nitrate; it showed general blackening due to deposition of silver (copper and silver nitrate give metallic silver and copper nitrate). There was no special phenomenon at the bend.

(2) Copper was abraded, kept 9 days in a desiccator, then bent and placed in silver nitrate; it showed marked deposition of silver at the bend and practically none elsewhere, except at isolated spots. This suggests that the air-formed film produced during the nine days was protective against silver nitrate, but failed to protect where damaged by the bending.

(3) Copper was abraded, bent and then exposed 9 days to dry air in a desiccator. It was then placed in silver nitrate but showed no special phenomenon at the bend, since the air-exposure had healed the damage.

Evidently, *in this particular case*, damage to the film and not disarray of the crystal structure, was the cause of local attack (U. R. Evans, *J. chem. Soc.* 1929, p. 92, esp. p. 101).

Potential Changes brought about by Straining. There is no doubt that straining can greatly increase corrosion velocity; the strained metal acts as a more reactive material than the annealed metal. Some commentators have expressed surprise that the electrode potential, as commonly measured, does not always show a shift in the negative (reactive) direction; the potential change is often small, sometimes irregular, and may even occur in the positive direction. Other commentators have argued that the strain energy left in the metal after deformation is so small that no serious change would be expected. However, measurements of energy stored in iron wire which has been twisted show it to be appreciable (G. I. Taylor and H. Quinney, *Proc. roy. Soc.* (*A*) 1934, **143**, 307).

In order to decide whether the energy left in strained metal is capable of causing a serious change in the potential, it is necessary to consider its distribution. Only strain energy near the surface will affect the potential. If the distortion is localized, a small strain might cause a large local energy increase—explaining an unexpectedly large potential change. Such an exaggerated effect is most likely where the main part of the surface is covered with a film nearly impervious both to ions and electrons—as is generally believed to exist on aluminium. In other cases the potential change may be unexpectedly small, since the strain at the surface may facilitate both the cathodic and anodic reactions. If it facilitated the cathodic reaction alone, the potential would rise; if it facilitated the anodic reaction alone, it would fall; if it facilitated both, there might be little or no change. Simnad has studied the electrochemistry of cold-rolled and annealed iron in acid; details of his research are given on p. 908, but it may be here stated that the stimulation of the two reactions by cold work roughly cancelled one another, as regards the effect on the potential, although the velocity of the attack was much greater for the cold-rolled metal than for the annealed metal (M. T. Simnad and U. R. Evans, *Trans. Faraday Soc.* 1950, **46**, 175).

In some cases an important shift of potential in the negative direction occurs, even when the stretching is confined to the elastic range. This is particularly true of aluminium alloys which have received a heat treatment such as produces material along grain-boundaries anodic to the grain-interiors. Measurements of the " true surface area " (p. 598) show that even an apparently smooth surface must have fissures running into the interior and generally these will follow the grain-boundaries. If so, stretching of the material will tend to open wider these tiny inter-granular fissures—thus exposing the anodic material better to the liquid, so that the measured potential (a " compromise " between the potentials at the cathodic and anodic points) is shifted in the negative direction. Farmery has constructed a *differential stress cell* consisting of two specimens of an aluminium alloy containing 7% magnesium which had received a heat-treatment chosen to produce the intergranular change mentioned; one of these was unstressed and the other was stressed, the liquid being sodium chloride solution containing bicarbonate. A current was generated, the stressed electrode being the anode; the strength of the current increased as the applied stress was increased (H. K. Farmery and U. R. Evans, *J. Inst. Met.* 1955–56, **84**, 413, esp. p. 418).

In many materials, the grain-boundaries are probably anodic relative to the grain-interiors and in such cases a stretching within the elastic range tends to shift the potential in an anodic direction when the liquid is, say, a sodium or potassium salt solution, free from ions of the metal under study. Simnad measured the potential of a given point on an iron wire bent into the form of a bow (fig. 82); the point to be studied (D) was in contact with a strip of filter paper (F) kept wet with N/10 potassium chloride, and this led to a calomel electrode. The two ends of the wire were held in bearings (B, B') and it was possible to rotate it through 180°, when desired. Clearly when the point under study was on the convex side of the bow, it was in tension but, when it came round to the concave side, it passed into compression; in that way, the potential *at the same point* could be measured first in tension and then in compression, thus eliminating irregularities due to variation of potential from one point to another along the wire. It was

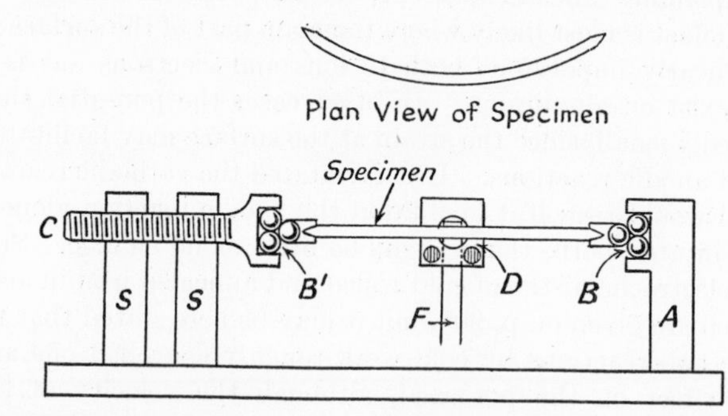

Fig. 82.—Measurement of potential at a given point on a bent wire when that point is alternatively under tension and compression—curvature exaggerated (M. T. Simnad and U. R. Evans).

found that tension on the whole shifts the potential in an anodic direction, and compression in a cathodic direction (U. R. Evans and M. T. Simnad, *Proc. roy. Soc.* (*A*) 1947, **188**, 327, esp. p. 386).

When, however, the solution contains ions of the metal in question, the contrary effect is produced—at least with noble metals. A sharp distinction must be drawn between (1) the *irreversible* potential of, say, iron in potassium chloride, where the cathodic reaction (reduction of oxygen) is not the same as the anodic reaction (the passage of iron ions into the liquid), and (2) the *reversible* potential of, say, silver in silver nitrate, where the cathodic and anodic reactions are equal and opposite. In the latter case, there is a possibility of metal passing into solution at one place and being redeposited at another, in such a way as to reduce the total surface area. The stable state will generally be reached when the surface area has become minimal, and this can be attained by the anodic dissolution on the face and cathodic deposition in the fissures. In such a case, therefore, a tensile stress which opens the fissure should render the potential cathodic and a compressional stress anodic (fig. 83). This is the effect found for silver and gold tested in solutions of their respective salts by R. E. Fryxell and N. H. Nachtrieb, *J. electrochem. Soc.* 1952, **99**, 495.*

The question as to whether different crystal faces should or should not show different potentials under equilibrium conditions has been the subject of much confused thinking. It is not always remembered that only a few faces can continue to exist in stable equilibrium with the solution of the metallic salt.† As already stated, a copper crystal dissolves more quickly in a direction perpendicular to the octahedral faces than in other directions, so that a copper crystal bounded by other faces and subjected to anodic attack tends to become an octahedron. If we assume the octahedral face to be the stable form, and if a copper crystal bounded by smooth cubic faces is placed (without applied current) in a copper salt solution, it may be expected that the atoms will dissolve at some points and be deposited at others, in such a way as to produce a number of microscopic octahedral facets on the originally smooth cubic faces. This means that at a potential at which copper can be deposited from a given solution on an octahedral

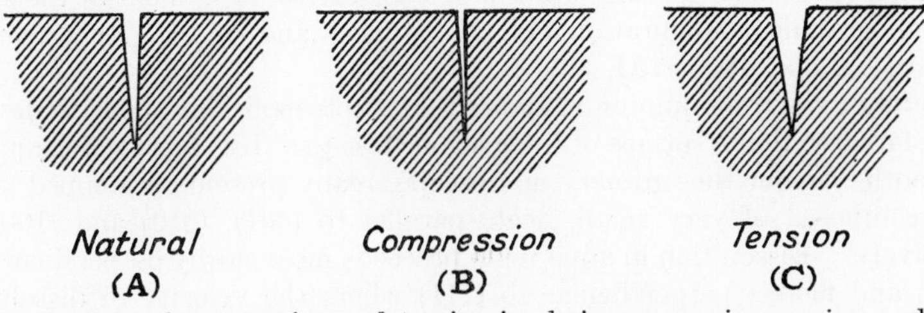

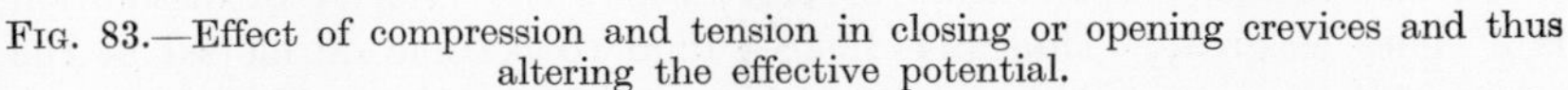

FIG. 83.—Effect of compression and tension in closing or opening crevices and thus altering the effective potential.

face, it can simultaneously be dissolved from a cubic face; evidently, therefore, the equilibrium value of the potential (at which deposition and dissolution balance one another) must be different for the original smooth cubic face from that for an octahedral face, but that as the minute octahedral facets develop (by dissolution and redeposition) on the originally cubic face, the latter should approach a potential similar to that of the face of an octahedral crystal (perhaps it will not reach quite the same value, owing to the extra energy of atoms situated at the corners of the facets).

* It is well to point out, however, that the spontaneous dissolution of metal at one place and deposition at another does not always lead to a decrease in area. Thus the autogenous lead tree sometimes formed when a freshly cut surface of lead is immersed in a lead nitrate solution involves an increase in area. It appears to be a concentration cell effect, caused by the fact that at certain places the nitrate is reduced to nitrite—a complexing anion—so that the Pb^{++} concentration will there be kept low; anodic attack will occur at those points, and cathodic deposition of dendritic lead elsewhere (A. Thiel, *Ber. dtsch. chem. Ges.* 1920, **53**, 1066; 1921, **54**, 2755).

† In some cases, even where, say, a cubic face possesses less energy per unit area than any other, the most stable form of a crystal of given mass might be a cube with corners cut off by small octahedral faces, since that would give a smaller total area than a simple cube. Thus there may be more than one stable face. Cf. I. N. Stranski, *Disc. Faraday Soc.* 1949, **5**, 13.

This matter has been studied in W. D. Robertson's laboratory. Single crystals of copper were cut so as to expose faces corresponding to different crystal orientations, and, after grinding to correct slight errors in the angle of cutting, were electropolished; the potential of the cell

$$\text{Cu (chosen face)} \mid CuSO_4 \mid Hg_2SO_4 \mid Hg$$

was then measured. (In most cases, the "chosen face" was purposely amalgamated with mercury, to avoid errors caused by the accidental access of mercury traces to the copper surface; this slight complication does not affect the general argument.) The results show that only the octahedral plane provided a stable value of the reversible potential

$$Cu \rightleftharpoons Cu^{++} + 2e$$

which remained constant from the first reading. All the other crystal planes gave at the outset different values, which after some hours started to move towards the value for the octahedral plane—(this was not reached, however, within 17 hours). Polycrystalline electrodes approached the stable state, provided that sufficient time was allowed for the establishment of the stable crystallographic configuration (W. E. Tragert and W. D. Robertson, *J. electrochem. Soc.* 1955, **102**, 86).

Bussy's work on aluminium shows that electropolishing tends to develop cubic faces, so that a surface of inclination close to a (101) face, which appears "smooth" under the ordinary microscope, really presents a stepped structure composed of very small facets parallel to (001), (010) and (100) respectively. Dissolution in aqua regia proceeds most slowly perpendicular to (001), and fastest perpendicular to (111) where the velocity of dissolution is 3 times as great. Thus {111} faces tend to become developed at the expense of the slow {011} and {001} faces, which gradually disappear. A mono-crystal cast in spherical form and placed in aqua regia develops {111} faces after 5 hours (P. Bussy, Thesis (Paris) 1954).

Other References

A full understanding of crystallographic corrosion demands a study of crystal growth and structural defects. Reference should be made to the discussion on Crystal Growth held at Bristol in 1949 (*Disc. Faraday Soc.* 1949, **5**), to the Symposium held in 1950 at Pocono Manor, Pennsylvania on "Imperfections in nearly perfect crystals" (report edited by W. Shockley, J. H. Hollomon, R. Maurer and F. Seitz (Wiley, New York; Chapman & Hall, London)) and to the Faraday Society discussion on Crystal Imperfections (1959). See also W. J. Sinnott, "Solid State for Engineers" (Wiley, New York; Chapman & Hall, London).

Information on etching agents for metallographic sections is provided by R. H. Greaves and H. Wrighton, "Practical Microscopical Metallography" (Chapman & Hall), and by the A.S.M.E. Handbook.

Recent Hungarian work on slip-sources in sodium chloride as revealed by etch-pits may come to be useful in the study of metals (Z. Morlin, *Nature* 1959, **183,** 1319).

CHAPTER XI

HYDROGEN CRACKING AND BLISTERING

Synopsis

It is first recalled that under circumstances where hydrogen overpotential becomes high, the destruction of metal by an acid is often much less rapid than would otherwise be the case; if, however, the overpotential arises from the sluggishness of the second stage of hydrogen liberation (the production of gaseous molecular hydrogen from atomic hydrogen), the actual destruction of metal may be replaced by a far more serious menace—the physical deterioration of the part of the metal which escapes chemical attack. Especially on iron, the hydrogen atoms, unable to escape as gas from the metallic surface, build up a concentration gradient and diffuse into the interior; if conditions are favourable in the interior for the atoms to join together to form molecules, huge pressures of hydrogen may be built up in blow-holes or other cavities, and this can cause hydrogen blistering in soft steels or hydrogen cracking in strong steels. Experiments are described demonstrating the passage of hydrogen through thin iron sheet, the hydrogen being produced either with or without an external E.M.F. It is shown that catalytic poisons, which prevent the formation of molecular hydrogen on the external surface, allow the building up of the concentration gradient required to push the hydrogen through the sheet.

The evil influence of sulphur in steel then receives consideration. It can exert an unfavourable effect both at anode and cathode. The catalytic acceleration of the anodic reaction by hydrogen sulphide can largely be corrected if the steel contains sufficient copper to precipitate it as copper sulphide. The worst trouble occurs through its influence at the cathodic areas where hydrogen sulphide may poison the second stage of the cathodic reaction, and thus lead to concentration gradients of atomic hydrogen. In practical cases, the hydrogen sulphide is sometimes an original constituent of the watery phase but may also be produced by the action of acid on sulphides in the steel.

Next, attention is turned to practical troubles arising in the pickling of steel—carried out in industry for the removal of mill-scale or rust. It is shown that pickling by simple immersion in acid can cause six different troubles; of these the first three can be mitigated by cathodic pickling, which however tends to worsen the last three; conversely, anodic pickling tends to rectify the last three but certainly worsens the first two (perhaps the third also).

The use of *restrainers* is then discussed both from the scientific and practical stand-points. The elimination of hydrogen after pickling receives consideration, and the question of the introduction of hydrogen during electroplating comes up for discussion. Attention is paid to pickling in molten baths.

The various types of hydrogen trouble met with in service are then discussed in greater detail. First, blistering, with its relation to blow-holes and inclusions, receives attention, and then cracking, which is so serious in strong non-stainless alloy steels. Hydrogen in steel may originate in the furnace atmosphere used during manufacture and can be introduced during welding, but probably the most usual modes of introduction are associated with pickling and plating; these matters are briefly discussed. Some of the special troubles of metal in the " sour oil-fields " due to the presence of hydrogen sulphide are probably due to hydrogen—although often called " sulphide cracking " and attributed by some authorities to sulphide net-works formed within the material and causing failure under stress. The chapter ends with a discussion of hydrogen troubles in non-ferrous materials. The question as to whether hydrogen plays a part in caustic cracking, nitrate cracking and similar cases of undoubted stress-corrosion is deferred to Chapter XVI.

General

Types of Damage arising from Corrosion of the Hydrogen-Evolution Type. The corrosion produced when acids or alkalis attack metal with evolution of hydrogen has been discussed in Chapter IX. In practice, this is not a frequent cause of direct wastage, since the material chosen for handling acids or alkali will be one which resists such reagents, through film-formation or otherwise. Even where metals have to be pickled in acid for the removal of scale or rust, the period of immersion is so short that the destruction of metal is small and can be further diminished by the use of a partial inhibitor or *restrainer*. Nevertheless if hydrogen is left in the interior, damage can be produced of a kind far more serious than the actual destruction of metal.

Hydrogen Overpotential. The cathodic production of hydrogen often consists of two main stages:—

(1) The *discharge* of hydrogen ions, generally represented by

$$H^+ + e = H$$

(perhaps more accurately written as

$$H_3O^+ + e = H + H_2O)$$

(2) The formation of *molecular from atomic hydrogen*,* *either* by

$$H + H = H_2$$

or by

$$H + H^+ + e = H_2$$

* Hoar and Holliday consider that during the attack on iron by dilute sulphuric acid, $H + H = H_2$ predominates over $H + H^+ + e = H_2$. They regard

(the last reaction is perhaps more accurately written

$$H + H_3O^+ + e = H_2 + H_2O)$$

On one material, namely black platinum, both of these stages proceed smoothly, so that rapid evolution of gaseous hydrogen is observed at potentials only slightly below the theoretical potential of the equilibrium

$$2H^+ + 2e \rightleftharpoons H_2$$

for the liquid in question. On other cathodic surfaces, the ions require help from a potential gradient if they are to be discharged, as explained on pp. 1024–1026; if we wish to obtain a liberation of gaseous hydrogen at any appreciable rate, the potential must be depressed considerably below the equilibrium value. This extra depression is known as *overpotential*; it can be considerable on certain metals, such as zinc, lead and mercury, even for a slow hydrogen evolution-rate. On that account, as explained on p. 315, corrosion of the hydrogen-evolution type is often very slow in cases where rapid attack might otherwise be expected.

Rate-determining step. A high hydrogen overpotential can greatly reduce the total destruction of metal. It may, however, arise from sluggishness either in the first or the second stage. As regards the actual destruction of metal, it matters little which stage is sluggish (or rate-determining) since clearly the sluggish stage must set the pace for the other.* However, in determining the damage due to hydrogen in the metal, it becomes important to decide which is the sluggish stage; if the first proceeds readily and the second sluggishly, a concentration of atomic hydrogen will be built up at the metallic surface, and (since it is not changing quickly into molecular hydrogen) it may diffuse into the interior of the metal.

Probably such cases, where the second stage is " rate-determining ", are in natural corrosion the more common. Nevertheless the situation varies from metal to metal. Gerischer, studying reactions at a cathode where an E.M.F. is applied from an external source, considers the rate-controlling reaction on mercury and silver to be

$$H^+_{solution} + e \longrightarrow H_{adsorbed}$$

but on copper

$$H^+_{solution} + H_{adsorbed} + e \longrightarrow H_{2adsorbed}$$

(H. Gerischer and W. Mehl, *Z. Elektrochem.* 1955, **59**, 1049).

the anodic spots as places where there are incomplete rows or layers of atoms, or places where adsorbed H_2S distorts the lattice and produces surface forces which reduce the activation energy needed for the escape of iron cations (T. P. Hoar and R. D. Holliday, *J. appl. Chem.* 1953, **3**, 502, esp. p. 510). For the effect of control by the first and second stages of the cathodic reaction, see J. O'M. Bockris and B. E. Conway, " Modern Aspects of Electrochemistry " 1956, p. 187 (Butterworth). Also J. O'M. Brockris and E. C. Potter, *J. electrochem. Soc.* 1952, **99**, 169.

* The rate-determining stage is sometimes described as the " slow stage " or " slow step ", but in a steady state both stages proceed *at the same rate*, although this is determined mainly by the one which (if its natural rate were allowed full play by an infinitely fast replenishment of material and an infinitely fast removal of products) would proceed least fast. It seems better to use the word " sluggish ".

The question of the rate-determining step becomes specially important in connection with the choice of a restrainer addition to the acid used in the pickling of steel. If the only object is to avoid destruction of metal or waste of acid in the attack, it matters little whether the restrainer selected is one which retards the first stage or one which retards the second stage; but clearly the risks of hydrogen moving into the interior of the metal are greatly enhanced if the restrainer allows the first stage to proceed smoothly, but interferes with the second stage. One substance, hydrogen sulphide, which may be formed by the action of pickling acid on sulphide inclusions in the metal, poisons the second stage of the cathodic reaction, but actually—in the case of iron—accelerates the anodic reaction

$$Fe = Fe^{++} + 2e$$

The anodic acceleration increases the rate of production of atomic hydrogen at the cathode and since its conversion into molecular hydrogen is prevented, the evolution of high-pressure hydrogen gas at points in the interior of the steel, where no poison exists, becomes a serious menace; it may take the form of blistering or cracking, as shown below.

Passage of Cathodically Produced Hydrogen into Metal

Passage of Hydrogen through Iron under Applied E.M.F. Some classical work in Aten's laboratory serves to suggest the mechanism of the passage of hydrogen into iron. An electrolytic cell (fig. 84) containing

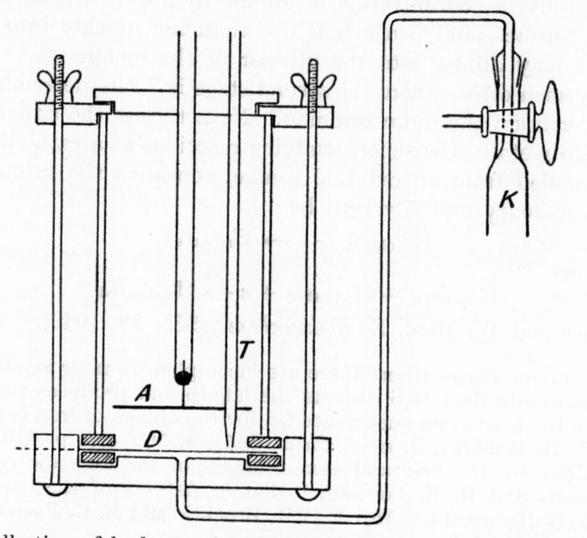

Fig. 84.—Collection of hydrogen forced through Iron cathode (A. H. W. Aten and M. Zieren).

N/10 sulphuric acid or N/10 sodium hydroxide was provided with a platinum anode (*A*), whilst the cathode was a thin horizontal iron sheet (*D*) at the bottom, the space below it being connected to a gas-burette (*K*), which indicated the rate of passage of gas through the iron; a tubulus (T) leading to a hydrogen electrode served to measure the potential at the cathode, which becomes depressed if hydrogen accumulates in the metal. When an E.M.F. was applied, hydrogen was evolved on the cathode, and in pure dilute sulphuric acid rose as bubbles from the upper surface; very little passed through the iron. When, however, a " poison ", such as arsenious oxide or mercuric chloride was added, the hydrogen began to pass through and was collected in the gas-burette; the movement of potential provided evidence that a concentration of atomic hydrogen was being built up on the upper surface in contact with the electrolyte. This is easily understood, since experience of catalysis shows that " poisons " occupy sites where reactions involving adsorbed atoms would be likely to occur and thus prevent their occurrence. If the hydrogen atoms are prevented from forming hydrogen molecules, the concentration gradient of atomic hydrogen will cause diffusion through the metal; on reaching the further side, where no poison exists, the atoms can combine readily to form molecules and thus emerge as gas. It is easy to see that if the cathode had been a thick metal plate containing small cavities (instead of a thin sheet with a gas space beyond it), hydrogen gas would be formed in these cavities, where considerable pressures might be built up. The effect of such pressure may vary with the material, and with the position of the cavities; if they are holes near the surface of a relatively soft steel, the steel may be forced up in blisters; if, however, they are situated in the interior of a strong steel and particularly if they are micro-fissures rather than rounded holes, then the effect may be to start internal cracking.

Further details of the experiments mentioned will be found in the original papers of A. H. M. Aten, *Rec. Trav. chim.* 1930, **49**, 641 (with M. Zieren); 1931, **50**, 943 (with P. C. Blokker).

State of Hydrogen in Metals. There is little doubt that the hydrogen travels through iron as atoms, and that molecular hydrogen cannot pass unless definite pores are present. Only on the assumption of passage as atoms is it possible to explain the fact—established by Bodenstein in an early quantitative research—that the rate of passage through walls of a given thickness is proportional to the square root of the current density* (M. Bodenstein, *Z. Elektrochem.* 1922, **28**, 517).

In the same early paper, which still deserves study, Bodenstein showed that hydrogen arriving at the far side of the cathode could combine with nitrogen to form ammonia, or with oxygen to form water—which could hardly happen if the hydrogen arriving were molecular. It is, however,

* If C_F is the concentration of hydrogen atoms at the front of the cathode (next the liquid) and C_B that at the back (next the gas-space) the diffusion rate will be proportional to $(C_F - C_B)$, or, if $C_B \ll C_F$, approximately to C_F. On the other hand, the rate of evolution of hydrogen as bubbles of gas by combination of two H atoms to give H_2 is proportional to C_F^2. Now the total production of atoms by the electrochemical reaction is proportional to the current I. If the main part of this escapes as gas bubbles, then approximately $C_F^2 \propto I$, and since the small portion diffusing through the iron is proportional to C_F, it must be proportional to $\sqrt{I}$.

likely that not all the hydrogen is in the same form; some seems to be more mobile than the rest. In the case of hydrogen entering palladium it is generally considered that the first part dissociates into protons and electrons, which serve to fill up the (incomplete) D-band of the palladium atoms; the protons, being doubtless held by electrostatic forces near the sites of the electrons, cannot easily move. When the ratio H/Pd has reached 0·6, the D-band becomes full, and any additional hydrogen entering the metal will do so as atoms, which occupy octahedral holes in the face-centred cubic lattice; these atoms, being uncharged, are not held by any electrostatic forces and can diffuse under a concentration gradient (N. F. Mott and H. Jones, "Theory of the Properties of Metals and Alloys" 1936, p. 200 (Clarendon Press); J. R. Lacher, *Proc. roy. Soc.* (*A*) 1937, **161,** 525; D. P. Smith, "Hydrogen in Metals" 1948, p. 104 (Univ. Chicago Press); B. Svensson, *Ann. Phys.* 1933, **18,** 299; C. Wagner, *Z. phys. Chem.* 1944, **193,** 407; J. P. Hoare and S. Schuldiner, *J. electrochem. Soc.* 1955, **102,** 485; 1956, **103,** 178, 237).*

It is possible that on iron also we must distinguish between mobile and fixed hydrogen. It seems unlikely that molecular hydrogen formed by union of atoms at cavities will be capable of diffusion out through the lattice, although, owing to its high pressure, it will quickly escape if there are pores leading to the surface. Hydrogen introduced by heating iron in a hydrogen atmosphere at 800°C. is mobile at 300–600°C., since it is expelled by heating the iron *in vacuo*; the rate of expulsion is controlled by diffusion within the metal; phase-boundary phenomena do not exercise any control (T. M. Stross and F. C. Tompkins, *J. chem. Soc.* 1956, p. 230; cf. P. L. Chang and W. D. G. Bennett, *J. Iron St. Inst.* 1952, **170,** 205).

It may be unwise, however, to assume that the situation in iron is the same as in palladium. Smialowski points out that the dissolution of hydrogen in palladium is exothermic, and in iron endothermic. He inclines to the view that hydrogen enters iron during cathodic treatment in acid as protons (M. Smialowski, *Neue Hütte* 1957, **2,** 621, esp. p. 625).

Fischer and Heiling, depositing hydrogen by cathodic action on the outside of a beaker-shaped iron cylinder, found that the potential on the inside surface (where there was no cathodic current) became depressed by hydrogen diffusing through the metal; at both surfaces the potential was a linear function of the logarithm of the current density—in accordance with Tafel's equation (p. 1024) (H. Fischer and H. Heiling, *Z. Elektrochem.* 1950, **54,** 184).

It is interesting to note that hydrogen-charged iron behaves essentially as a hydrogen electrode in definitely acid solution, the potential varying

* A. H. Cottrell adds the following comment: "What I think happens is that the electron of the hydrogen atom does join the D band, but the electrostatic field of the proton attracts around this a charge of one electron from the conduction electron band in the metal. This attraction also leads to the withdrawal of one quantum state from the band, so that the dissolved hydrogen does cause the band to become filled up, although not in the sense that was thought some years ago." He refers to a discussion of the matter by J. Friedel (*Phil. Mag.* 1952, **43,** 153).

with the pH value, according to the usual law (p. 1029), although the value is always less noble than that of the reversible hydrogen electrode. At a rather higher pH value, it behaves as an iron electrode, the potential then varying with the Fe^{++} ion concentration (J. D'Ans and W. Breckheimer, *Naturwissenschaften* 1951, **38**, 282).

Action of other Catalytic Poisons in Hydrogen Transmission. Various other elements—all catalytic poisons—act in the same way as arsenic or mercury. Some early measurements by Baukloh and Zimmermann, although not wholly free from inconsistencies, served to bring out the important effect of selenium and tellurium in favouring the passage of hydrogen through iron (presumably by preventing its evolution as gas). Phosphorus, antimony and bismuth are now known to act in the same way (W. Baukloh and G. Zimmermann, *Arch. Eisenhuttenw.* 1935–36, **9**, 459; Y. Lindblom and S. E. Dahlgren, *Metal finish. J.* 1955, **1**, 517).

Passage of Hydrogen through Iron without applied E.M.F. In the researches hitherto described, the total rate of hydrogen production was decided by the current imposed from an external source. Different

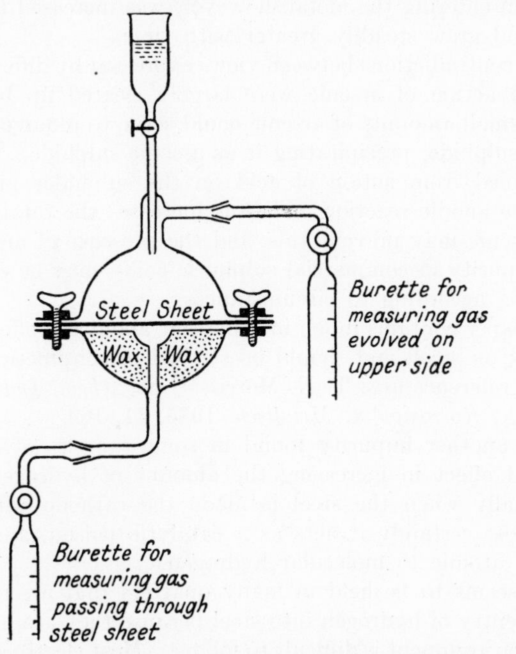

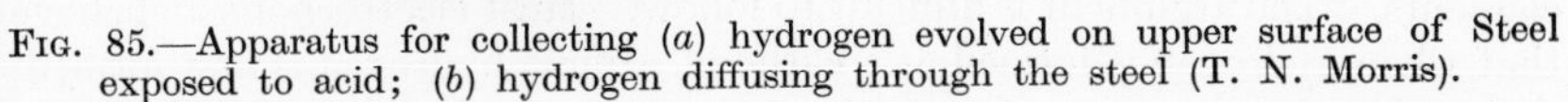
FIG. 85.—Apparatus for collecting (*a*) hydrogen evolved on upper surface of Steel exposed to acid; (*b*) hydrogen diffusing through the steel (T. N. Morris).

conditions were studied by Morris, who exposed steel to acid without any applied E.M.F. He used an apparatus (fig. 85) designed to measure separately (1) the hydrogen " evolved " on the near face of the iron sheet in contact with the acid, and (2) the hydrogen which " diffused " through the metal and appeared as gas on the far side. Although the experimental situation is simpler than in Aten's research, the electrochemical factors involved are more complicated, since the current flowing is no longer decided by the experimenter, but arises from the action of local cells. In general, the presence in the liquid of a substance which stimulates the anodic reaction will increase the total hydrogen production at the cathode; conversely an anodic inhibitor will reduce the total hydrogen produced.* But unless the control is solely anodic, a cathodic inhibitor will also reduce the total of hydrogen produced; if it is a poison that interferes only with the second stage of the cathodic reaction, the *proportion* of that hydrogen which enters the metal may be increased; on the other hand, if the inhibitor is one which interferes with the first stage, the concentration of atomic hydrogen in the steady state will be reduced, and the amount entering the metal should be diminished also. Morris found in fact that gelatine, beet sugar and tin citrate reduced the total production of hydrogen by citric acid, and that they also decreased the *proportion* of this which diffused into the metal. Arsenic appeared slightly to stimulate the total hydrogen evolution in the early stages, but after 50 hours there was little change; the proportion entering the metal, however, was increased by the presence of arsenic, and grew steadily greater with time.

Apparent contradictions between views expressed by different authorities regarding the action of arsenic were largely cleared up by Bablik, who showed that small amounts of arsenic could serve to remove small amounts of hydrogen sulphide, precipitating it as arsenic sulphide. Now hydrogen sulphide, formed from action of acid on the sulphides present in steel, accelerates the anodic reaction and thus increases the total hydrogen production. Arsenic may prevent this, and the presence of arsenic—often an accidental impurity in commercial sulphuric acid—may be either beneficial or the reverse, according to circumstances.

Bablik's paper contains much information about the effect of additions in acid attack on steel, and should be studied in conjunction with that of Morris. The reference are: T. N. Morris, *J. Soc. chem. Ind.* 1935, **54,** 7T; and H. Bablik, *Korrosion u. Metallsch.* 1935, **11,** 169.

Selenium, another impurity found in some samples of sulphuric acid, has a marked effect in increasing the amount of hydrogen entering the metal—especially when the steel is made the cathode with an external E.M.F. Almost certainly it acts as a catalytic poison, and prevents the conversion of atomic to molecular hydrogen.

The idea seems to be held in many quarters that arsenic or selenium facilitate the entry of hydrogen into steel because they are hydride-forming elements. The argument is difficult to follow. Most electrochemists believe that when hydrogen is formed at a cathode, atomic hydrogen is first formed, which then becomes transformed into molecular hydrogen—provided that there is nothing to interfere with this second stage; arsenic or selenium, if present, prevent the second stage—in which case the only alternative is

* This assumes "mixed control", partly anodic, partly cathodic (p. 875); if the control were purely cathodic, the addition of an anodic stimulator would not affect the total hydrogen production.

for atomic hydrogen to sink into the metal. There is no difficulty in imagining the movement of atomic hydrogen through the metal, and strong evidence exists that it does move in this form. There would be, however, considerable difficulty in picturing the passage of AsH_3 or SeH_2 molecules; nor is the entry of the hydrogen into the metal likely to be easier if AsH_3 or SeH_2 is initially formed on the surface. Moreover, some elements which favour the entry of hydrogen into iron (like zinc, cadmium and mercury) do not form gaseous hydrides. In the case of chromium, discussed on p. 415, the hydrogen-charged metal can itself be regarded as a solid hydride, but it differs from gaseous hydrides such as AsH_3 and SeH_2 in having a composition not determined by the ordinary valency laws.

The up-take of hydrogen by steel placed in contact with acid is stated to increase linearly with the square root of the time until a saturation value is reached, after which it appears to cease. This saturation value is highest in solutions of low pH, but depends also on the state of the metal—being much greater for cold-worked than for hot-rolled steel—possibly because hydrogen is trapped in defects left by cold-work and at sub-grain boundaries (L. S. Darken and R. P. Smith, *Corrosion* 1949, **5**, 1).

Rough surfaces are less liable to the entry of hydrogen into the metal than smooth ones, doubtless because they provide points favourable to the conversion of atomic to molecular hydrogen; this beneficial influence of roughness has been observed both in cathodic pickling and plain acid pickling (J. Duflot, *Rev. Métallurg.* 1952, **49**, 35, esp. p. 39; Y. Lindblom and S. E. Dahlgren, *Metal finish. J.* 1955, **1**, 517).

The fact that residual stresses present in fabricated articles made of high-tensile steels tend to favour hydrogen embrittlement is a serious problem in the aircraft industry, since stress-relief at an adequate temperature may be precluded by the low tempering temperature required in some types of steels for the development of high strength (T. F. Kearns, *J. electrochem. Soc.* 1955, **102**, 223C).

It must be recollected that external (applied) stress, internal (residual) stress due to fabrication and stresses set up by evolution of high-pressure hydrogen within the metal may be additive. Failure due to a combination of factors may occur during pickling or plating, or later, in service, when an external load comes to be applied.

Stainless steels are less suceptible to hydrogen-deterioration than other steels—a point which deserves note, since stainless steels of high strength are now being developed for structural purposes; publications on this matter should receive attention as they appear.

Bastien has studied the effect of hydrogen on the mechanical properties of (ordinary) steel, by measuring the number of bends, applied to and fro, which hydrogen-charged wire can withstand before fracture occurs; in general, the number withstood decreases as the hydrogen-content increases. He also studied the liberation of hydrogen after removal from the acid and finds that a trace of sulphide in the acid, besides favouring the up-take of hydrogen in the metal, also interferes with its liberation. In his experiments, exposure to air at 100–300°C.—a treatment which rather rapidly removed hydrogen from annealed iron and restored its power to withstand the bending tests—failed to restore the mechanical properties of cold-worked metal into which hydrogen had apparently penetrated more deeply. Bastien concludes that the imperfections and disarray introduced by deformation assist the movement of hydrogen through the material. This conclusion may seem to clash with the results of Andrew and his colleagues, who found that hydrogen deposited cathodically on the outside of a cylinder passed inwards less rapidly if the iron had suffered much cold-deformation. The method of testing, however, was different in the two researches; Andrew applied an external E.M.F., whereas Bastien dipped wire-specimens into hydrochloric acid, apparently containing a small amount of sulphide, without applied E.M.F. Probably also the type of cold-deformation was different, and in Andrews' material produced the paths of diffusion running parallel to the surface—which might explain the apparent contradiction (P. Bastien, *Métaux et Corros.* 1939, **14,** 43; 1950, **25,** 248, esp. p. 260; *C.R.* 1939, **208,** 188; 1941, **212,** 706, 788; 1942, **214,** 354; 1949, **228,** 1651 (with P. Azou); *Rev. Métallurg.* 1948, **45,** 301. J. H. Andrew, U. V. Bhat and H. K. Lloyd, *J. Iron. St. Inst.* 1950, **105,** 382).*

Importance of the Copper/Sulphur Ratio in Steel. One curious feature emerging from Morris' work (p. 648) was the fact that certain steels were corroded much faster than others. The " fast " steels evolved hydrogen more rapidly than the " slow " ones, but the contrast between the hydrogen diffusing through the metal was even more pronounced; there was practically no diffusion through the slow steels in the opening stages, although it developed later; a fast steel showed a considerable hydrogen diffusion at the outset but it increased little with time.

It would seem reasonable to explain the difference by assuming that the fast steels produced more hydrogen sulphide than the slow ones. Yet the analytical figures published by Morris assign to one of the fast steels practically the same sulphur content as to one of the slow ones. A possible cause of the discrepancy is suggested by the work of Hoar and Havenhand (p. 948) who showed that it is not the absolute sulphur content of the steel, but the ratio of sulphur to copper which is important. A steel containing appreciable sulphur but also plenty of copper will generate hydrogen sulphide which, however, will at once he removed as copper sulphide; a second steel with the same sulphur content but no copper will leave the hydrogen sulphide in the liquid—accelerating the anodic attack on the metal. As the cathodic hydrogen evolution must keep pace with anodic attack (in the absence of dissolved oxygen), this will also be greater in the case of the second steel, and the proportion of that hydrogen diffusing into the metal—instead of being evolved as gas—will be greatly enhanced. Morris's analytical figures

* A. H. Cottrell suggests the following explanation of the apparent discrepancy between Bastien and Andrew. He thinks that the cold-working in Andrew's experiment led to so much irregularity at the surface of the iron that the accretion from atomic to molecular hydrogen took place to a far greater extent than in unworked iron. On the other hand, when a poison was present as in Bastien's experiments, the effect of having a large number of recombination sites on the surface would no longer be significant.

do not indicate the contents of copper (the significance of that element was not appreciated at the time when Morris worked); but it is at least probable that the difference between the fast and slow steels is connected with their contents of elements capable of precipitating sulphur.

Effect of hydrogen sulphide on Hydrogen Diffusion. The influence of hydrogen sulphide on hydrogen diffusion is illustrated by some delightfully simple experiments described by Marsh. His method depended on the use of stout rods made from the steel to be tested, bored down the axis to produce a cylindrical hole into which a more slender rod fitted loosely, leaving a narrow annular space; a dial pressure gauge at the top communicated with this annular space. To conduct an experiment, it was only necessary to insert the outer rod into the chosen liquid, and watch the movement of the dial pointer. When high-quality steel was immersed in dilute hydrochloric acid, there was a vigorous attack, but practically the whole of the hydrogen gas was evolved on the external surface and no build-up of pressure was observed. When a solution of hydrogen sulphide (a much weaker acid) was used, the attack was much slower, but (since hydrogen sulphide acts as a poison for the second stage of the cathodic process) the hydrogen formed, instead of being evolved on the external surface as gas, diffused through the steel, so that the dial pointer showed a build-up of pressure. If steel of poor quality was used in with hydrochloric acid, there was again a build-up of pressure, probably due to hydrogen sulphide generated by the acid acting on sulphides present in the metal; in one experiment, Marsh observed the pressure rising to 60 atmospheres in 9 days, but at that point a leak occurred and the pressure dropped (G. A. Marsh, *Corrosion* 1954, **10**, 101).

Pressure obtainable by internal production of Hydrogen. The pressures attained experimentally are small compared to those theoretically possible. In an experiment in which hydrogen is being generated by means of an external E.M.F. on an iron cathode, each 30 mv. of overvoltage (the depression of the potential below the equilibrium value of the potential at 1 atmosphere hydrogen) represents a 10-fold increase of pressure—neglecting various sources of error which may become serious at high values of the overpotential. In practice, an overpotential of at least 200 mv. may be expected if hydrogen is being evolved in bubbles—which at first sight suggests that, if the atomic hydrogen diffusing inwards is liberated as molecular hydrogen at cavities in the interior, raising blisters, the pressure might be of the order of 10^7 atmosphere; this, however, assumes that the electrochemical laws continue valid at these levels and that there is no leakage—conditions unlikely to be fulfilled. Before such an astronomical pressure can be reached, some leaks will have been developed, capable of allowing the escape of molecular hydrogen, and the pressure will then cease to rise. Undoubtedly, however, it requires a high pressure to " blow bubbles " in solid steel, and the blisters actually observed in such cases provide ocular evidence that high pressures are really produced. Experiments with hollow cathodes have produced pressures of 200–300 atmospheres, but clearly these are not equilibrium values.

More refined calculations by Ubbelohde suggest that pressures of about $6{\cdot}6 \times 10^{18}$ atmospheres ought theoretically to be obtained. The rate of emergence into vacuum is higher than into oxygen, which apparently blocks by adsorption some of the sites where emergence would occur; at very high current density, however, the effect of oxygen is reversed, and oxygen helps emergence by "extracting" the hydrogen. Those interested in the mechanism should consult the original papers: F. W. Thompson and A. R. Ubbelohde, *J. appl. Chem.* 1953, **3,** 27; V. C. Ewing and A. R. Ubbelohde, *Proc. roy. Soc.* (*A*) 1955, **230,** 301.

Elementary electrochemical considerations (p. 1017) show that the equilibrium hydrogen pressure at which the potential $H_2 \mid H^+$ exactly balances $Fe \mid Fe^{++}$ must diminish as we pass from definitely acid to nearly neutral solutions—in absence of complexing substances, assuming the Fe^{++} concentration to remain constant; when the pH value reaches the point when $Fe(OH)_2$ is precipitated as a solid phase, the decline of pressure should cease. In the presence of cyanide, which will keep low the concentration of Fe^{++}, it may become very high even in alkaline liquid.

Smialowski and his colleagues have used a method for assessing the pressure based on the expansion of iron when charged cathodically with hydrogen from N sulphuric acid containing 0·004 g./l. of As_2O_3. This process produces two sorts of cavities, some being open on to the surface, whilst others lie beneath the surface so that the hydrogen is entrapped and develops pressures which are the cause of the expansion; these internal bubbles are believed to be formed at lattice defects, micro-pores and non-metallic inclusions. The magnitude of the expansion of the specimens suggests that the pressures developed probably exceed 10^5 atmospheres, possibly reaching 10^7 atmospheres.

It is, however, not the cavities, as such, which are the cause of brittleness in the metal. The Polish investigators studied wires made from two steels; one was an unsound steel containing much sulphur and oxygen so that the cavities developed were numerous and the increase of length accompanying hydrogen-up-take considerable; the other was a steel killed with silicon, aluminium, titanium and vanadium, which showed few cavities and little length-increase. Nevertheless, the two materials showed almost the same brittleness, as measured by the number of twists which the wires could withstand before breaking. The embrittlement appears to be connected with the atomic hydrogen, which doubtless diffuses into the cracks as they extend—as suggested on p. 418 (A. Krupkowski and M. Smialowski, *Proceedings of the Electrochemical Conference*, Warsaw, 1955 (published 1957), p. 295; J. Foryst and M. Smialowski, p. 319).

Pickling and Pickling Restrainers

Acid Pickling of Steel. As explained on pp. 559, 623 the removal of mill-scale and rust from steel is necessary before the application of a metallic coat and advantageous before the application of paint. This can be accomplished by mechanical means (grit-blasting or grinding), but is often carried out by pickling in acid. Either heated sulphuric acid or unheated hydrochloric acid (which generally becomes warm owing to the heat evolved in the reaction) is generally used; suitable conditions are 5–20% sulphuric acid at 65–98°C. or 14% hydrochloric acid at 30–50°C. Sometimes mixtures of the two acids are used. Practical information is provided by A. B. Winterbottom and J. P. Reed, *J. Iron St. Inst.* 1932, **126,** 159; E. E. Halls, *Electroplating* 1952, **5,** 143; A. G. Gardner, *Metal finish. J.* 1955, **1,** 117; E. L. Streatfield, *Chem. and Ind.* (*Lond.*) 1955, p. 855; W. Bullough, *Metallurg. Reviews* 1957, **2,** 391. E. D. Martin, *Blast Furnace and Steel Plant* 1948, **36,** 825, 929, 1089.

Of the two acids, hydrochloric is sometimes preferred as it is more effective in removing the carbonaceous " smut " which tends to be left on the surface by sulphuric acid. However, recent work throws doubt upon the general advantage of using hydrochloric instead of the (cheaper) sulphuric acid unless the temperature can be kept at 50°C. 10% sulphuric acid at 70°C. is equivalent to 20% hydrochloric acid at 50°C. in removing either annealing or welding scale from mild steel; in each case, a temperature rise of 10°C. doubles the rate of removal. An accumulation of iron salt in the bath increases the time needed for descaling (D. Jackson, A. J. Stedman and R. V. Riley, *Metal finish. J.* 1955, **1,** 435).

A freshly pickled steel surface exposed to the atmosphere quickly becomes yellow owing to small amounts of rust; the yellow staining is particularly marked after hydrochloric acid pickling, and is probably due to traces of hygroscopic ferrous chloride hidden in microscopic cavities, which undergoes oxidation and hydrolysis. Such a surface cannot be regarded as an ideal one for the application of a coating, and the practice of following the ordinary pickling by a short dip in hot phosphoric acid has much to commend it. In Footner's *Duplex Process*—extensively used today before the painting of large steel plates intended for the oil industry—the plate, suspended by chains from cranes, is dipped in turn into three narrow vertical tanks containing respectively

(1) 5% sulphuric acid at 60°C. (the plate remains there for about 15 to 20 minutes or for such time as is necessary to ensure complete removal of mill-scale; it may be withdrawn at intervals for inspection);

(2) wash-water at 60°C. (two short dips);

(3) 2% phosphoric acid at 85°C. for 3 to 5 minutes.

After this the plate is withdrawn, and rapidly becomes dry, since the heat in the steel causes the water to evaporate; the paint (sometimes red lead containing carbon) is applied to the plate whilst it is still hot—thus avoiding condensation of moisture.

The plates after the phosphoric acid treatment carry a greyish film of an iron phosphate, which prevents the rapid rusting mentioned above, and serves as a layer into which the paint can " key ", ensuring good adhesion. If painting has to be delayed, the phosphate coat confers appreciable protection, although inferior to the iron, zinc or manganese phosphate coats produced in the proprietory " phosphating " processes (pp. 562–568).

The details of this important method of pickling deserve study and will be found in papers by H. B. Footner, *Rep. Corr. Comm. Iron Steel Inst.* 1938, **5**, 369; see also *The Engineer* 1937, **164**, 665.

Mechanism of de-scaling in acid. It is still commonly believed that the removal of mill-scale by acid is a direct attack, the oxides dissolving to give the corresponding chlorides or sulphates. This is certainly untrue. Much of the scale is undermined, and will be found, undissolved, as flaky particles at the bottom of the pickling tank. The true mechanism of scale-removal depends, however, somewhat on the thermal history of the steel.

As explained on pp. 24–26, iron heated in air above about 570°C. carries three layers of oxide, corresponding roughly to the Fe_2O_3, Fe_3O_4 and FeO; but below 570°C. the FeO layer splits into Fe and Fe_3O_4, and thus in cases where the rolling of steel has been finished below 570°C., the structure of the scale is duplex, whereas if it is finished above 570°C., it is triplex at that temperature; however, the ferrous oxide phase will largely decompose during slow cooling into an intimate mixture of metallic iron and magnetite—a mixture which has quite different properties from the true magnetite layer above it.

The direct action of an acid on ferric oxide to give a soluble ferric salt is slow (p. 226), but reductive dissolution of a ferric oxide scale in contact with metal to give the ferrous salt proceeds more readily, as explained on p. 225 (however, the discussion there presented refers to much thinner films produced at lower temperatures; the ferric oxide of mill-scale will certainly react less rapidly). Magnetite is more resistant to reductive dissolution than ferric oxide. If, therefore, the steel has been finished below 570°C. and carries a duplex scale, slow attack starts at any place where the scale is cracked, and penetrates along the steel/magnetite interface, undermining the scale, so that, in course of time, it becomes undermined and peels off. This undermining may be partly due to the attack of acid on the metal, with evolution of hydrogen; the bubbles, formed under the scale, probably help to dislodge it. However, it seems possible that slow reductive dissolution of the magnetite, by operation of the cell

Iron | Acid | Magnetite

plays its part, and that the lower part of magnetite layer, which lies in contact with the metal, passes into solution as a soluble ferrous salt (direct attack would give a mixture of ferrous and ferric salts). Probably the existence of cracks in the scale—which would admit the acid to the interface between metal and oxide—is important for rapid pickling—a factor emphasized by F. Wever and H. J. Engell, *Arch. Eisenhuttenw.* 1956, **27**, 475; see also W. Lueg, W. Dahl and H. J. Engell, *Z. Stahl u. Eisen* 1956, **76**, 1678, and, especially, S. Garber, *J. Iron Steel Inst.* 1959, **192**, 153.

At the best, de-scaling of this type of steel is a slow operation, and whilst the operator is waiting for the detachment of the scale from regions which are proving stubborn, the acid is destroying metal at those areas where the steel has already become bare of scale.

The triplex scale present on steel finished above 570°C. and then cooled is more quickly removed, since acid can work its way rapidly along the lowest layer; the intimate mixture of magnetite and iron may favour more rapid functioning of the cell already mentioned, owing to the large iron/magnetite surface; if so both phases will pass into the liquid as soluble ferrous salts; any undecomposed ferrous oxide will suffer slow direct dissolution, and, according to Engell and Garber (see above), rapid electrochemical corrosion; Garber's views relate specially to scale on steel strip.

Troubles arising in Acid Pickling. Whether the removal of scale proceeds smoothly or sluggishly, there is always some reaction between metal and acid, and this may produce six bad effects:—

(1) valuable metal is destroyed;

(2) the surface becomes roughened (this is particularly serious if a polished surface, which has become slightly rusty in storage, has to be pickled to remove the rust-traces, since the operation of restoring the original smooth surface is an expensive one);

(3) acid is unnecessarily consumed;

(4) hydrogen-evolution fills the air of the pickling-shop with acid spray, which is unpleasant to those operating the process and probably prejudicial to their health;

(5) hydrogen blisters may be produced in the metal;

(6) hydrogen cracking may occur, either during the pickling or afterwards when the part is under load.

Advantages of Cathodic Pickling. Of the six effects just mentioned, the first three are connected with the attack on the metal, and the last three with the production of hydrogen. If instead of merely dipping the steel in acid, it is made a cathode, by means of an E.M.F. from an external source, the first three troubles are mitigated (since attack at local anodes is hindered if the surface as a whole is cathodic), but the last three troubles are likely to be enhanced; the presence of arsenic or selenium especially promotes the last two troubles, and great care should be taken regarding the purity of the acid if " cathodic pickling "—as it is called—is employed.

However, by adding a trace of tin salt to the acid, the advantages of cathodic pickling can be obtained without the disadvantages, since a very thin layer of metallic tin is deposited on any part of the metal which becomes bare of scale; such a layer not only hinders evolution of hydrogen in bubbles, owing to the high over-potential of tin, but also imposes a barrier through which atomic hydrogen cannot easily pass, thus preventing atomic hydrogen from entering the metal and avoiding the last two troubles. Moreover, since tin has a high hydrogen overpotential, the current which would fall on the prominent parts of an intricate surface is deflected to the recessed parts as soon as the former are coated with tin, so that de-scaling of the recesses occurs. This is the basis of the Ballard–Dunn process, which has proved useful in the automobile industry. The main anodes may be of silicon iron, but a few tin anodes are also used to maintain the tin-content of the bath. The process is described by C. G. Fink and T. H. Wilber, *Trans. electrochem. Soc.* 1934, **66,** 381; R. Springer, *J. Electrodep. tech. Soc.* 1937, **13,** Paper 5, p. 6; J. Kronsbein, *ibid.* 1940, **16,** 55. See also *Product Finishing* June 1949, p. 74.

At one time, it was proposed to use lead instead of tin for this purpose, but that would probably be undesirable, since traces of lead in a cathodic pickling bath may cause embrittlement, as shown by U. Trägardh, *Corros. Tech.* 1954, **1,** 164.

Advantages of Anodic Pickling. Whilst cathodic pickling tends to stop the first three troubles and may (in the absence of tin) enhance the last three, the effect of making the steel the anode will have the reverse effect, since it should effectually prevent the evolution of hydrogen and its entry into the metal; on the other hand, some forms of anodic pickling increase the destruction of metal and roughening, and possibly also the consumption of acid. If, however, the steel quickly becomes passive, destruction of metal may be minimized. The Specification DTD 901C recommends a current density of 100 amps/ft.^{2} in sulphuric acid diluted with twice its volume of water; this quickly produces passivity—as shown by copious oxygen evolution; it is understood that, on clean steel, the loss of metal is very small.

If the object is to remove mill-scale, a considerable destruction of metal would seem to be necessary before the scale can be undermined by anodic attack. Anodic pickling is perhaps more often used for material from which the scale has already been removed by grit-blasting or other mechanical process, the object being to produce a surface suitable for plating. The procedure still enjoys considerable favour for alloy steels of fairly high strength, but there is some doubt about its suitability for steels of the very highest strength; some fear exists that it may reduce fatigue resistance (probably through the destruction of surface layers carrying beneficial compressive stresses), whilst the need to remove the " smutty " residue left after anodic treatment (perhaps with wire wool) is a practical disadvantage. In some cases a method of bombarding with a jet of water containing very fine alumina in suspension (a process which has received the misleading name of " vapour honing ") is being used instead of anodic pickling for obtaining a surface suited to plating.

At one time, anodic pickling in baths containing chromic acid was favoured for steel articles before nickel-plating. In Laban's process, the pieces to be pickled are made anodic in 30% sulphuric acid (by volume) containing about 3 ounces per gallon of potassium bichromate (about 2% by weight); the current density is started at 100 amps./ft.^{2}, but is allowed to fall off as the steel becomes passive. The steel leaves the bath with a dull satiny finish, which provides a good " key " for the subsequent nickel deposit. The oxygen evolved when the work becomes passive exerts a scouring action. Presumably hydrogen evolution at the cathode is avoided (N. R. Laban, *J. electrodep. tech. Soc.* 1929–30, **5,** 127, esp. p. 128; H. Sutton, *ibid.* 1935–36, **11,** 117, esp. p. 127).

Possibilities of Oxidizing Acids. It may seem remarkable that a simple dip in an oxidizing acid bath which would pickle a surface without introducing hydrogen is not more widely adopted as a simple means of avoiding hydrogen troubles without demanding an external E.M.F. and without the worry of joining the metallic articles to an electric circuit. But practically every oxidizing acid or acidic mixture introduces some new objection. Nitric acid, for instance, is unpleasant owing to the fumes evolved and the action on the skin. It has, however, been used in one situation where spring clips were found to become embrittled on pickling in hydrochloric or sulphuric acid followed by electro-tinning; the problem was solved by using nitric acid for pickling and then tinning by hot-dipping (C. E. Homer, Tin and its Uses, Oct. 1941, No. 11, p. 14).

Where nitric acid is regarded as objectionable, sulphuric acid containing an oxidizing agent might be considered. Hydrochloric acid containing potassium dichromate has been effective in Japan for high-chromium steels, although the danger of dichromate, which is a cathodic stimulator as well as an anodic inhibitor, is recognized (H. Endo and S. Isihara, *Sci. Rep. Tohoku Univ., Series A*, 1950, **2**, 209).

Use of Restrainers. The most usual method of avoiding the six troubles mentioned above is by adding to a sulphuric or hydrochloric acid pickling bath a substance chosen to restrain the attack on the metal, whilst still permitting the destruction or undermining of the oxide-scale. Such substances are commonly referred to as " inhibitors ", but the corrosion-rate is not brought down to so low a level as to justify the term, and is only tolerated because the time of immersion in the pickling bath is comparatively short. The term " restrainer " is perhaps preferable. Restrainers are commonly used in ordinary acid pickling baths without the application of current, but Machu advocates a combination of restrainer and cathodic current as the most economical procedure; the current required is much lower than would be needed in absence of restrainer, whilst the restrainer concentration falls far below that needed in absence of current. The combination avoids the main disadvantage of each method, the fear of hydrogen brittleness associated with cathodic treatment at high current densities, and the loss of time associated with the presence of restrainers, some of which appreciably retard the removal of the scale. The current density can be reduced to 1/100 that needed in absence of restrainer, and the restrainer concentration to 1/10 that needed in absence of current (W. Machu and O. Ungersboch, *Arch. Eisenhuttenw.* 1942, **15**, 301).

If the only object of a restrainer was to cut down the destruction of metal and the consumption of acid, any substance could be used which obstructed one of the essential steps of the corrosion process (it would not greatly matter whether the step obstructed was the anodic reaction or one stage of the cathodic reaction). There are in fact a large number of substances which do appreciably diminish the corrosion-rate, including many cheap and familiar products such as glue, size, gelatine and molasses. In practice, however, additional properties are demanded of a restrainer. It is not sufficient simply to cut down the total corrosion, that is the total hydrogen production; it is particularly necessary to diminish that fraction of the hydrogen which enters the metal and causes either blistering or cracking. A restrainer which poisoned the second stage of the cathodic reaction without interfering with the first stage of the cathodic reaction or with the anodic reaction, would, for many purposes, do more harm than good; it would cause a concentration of atomic hydrogen to be built up, and thus promote a diffusion of hydrogen into the metal.

Morris (p. 400) studying the effect of various additions has measured separately the " hydrogen evolved " as gas and the " hydrogen diffusing " through the metal; he showed that the two amounts are not always affected in the same way by an addition to the acid. Other authors report similar results. Glue additions greatly reduce the amount of hydrogen gas evolved from the pickling bath but produce only a negligible effect on the risk of cracking (V. A. Wardell, *Chem. Age Met. Sect.* 1935, **33,** 9).

At one time, there were many commercial restrainers on the market which failed to answer requirements. In 1948, Zappfe and Haslem tested fourteen so-called " inhibitors ", including some of the most popular commercial preparations of the day, and found that on stainless steel, eleven of them increased embrittlement and none of them greatly reduced it; on mild steel, however, some of the reagents tested did produce an appreciable improvement (C. A. Zappfe and M. E. Haslem, *Wire and Wire Products* 1948, **23,** 933, 1048, 1120).

Several of the substances used in practice today as restrainers are organic bodies containing nitrogen and sulphur; these polar atoms probably serve for the attachment of the molecules to metal at points where otherwise some essential step in the corrosion process would occur. It seems reasonable to suppose that the presence of two polar groups ensures better attachment than a single one, but this view is not universally held. In any case, it might be asked whether two nitrogen groups would not be as efficient as one nitrogen and one sulphur group; perhaps the answer is that the two different elements serve to provide attachment at different types of site. The proprietary restrainers now on the market contain such compounds as di-ortho-tolyl-thio-urea or dihydro-thio-ortho-toluidine. Substances with nitrogen in the ring, like quinoline and its derivatives, are also effective.

Interest is felt in restrainers which are very sparingly soluble so that the effectiveness is, beyond a certain minimal addition needed to provide complete coverage, almost independent of the amount added to the bath. They have been studied by Jenny, who has also examined the effect of adding an emulsifying agent along with the restrainer; this improves the inhibiting action by aldehydes but reduces the efficiency of mercaptans (R. Jenny, *Corrosion et Anti-corrosion* 1955, **3,** 189).

Restrainers are also added to the acid used in removing rust, chalky scale and other matters from pipe-systems, boiler-tubes and the like. After the derusting of a water-supply pipe system, the " inhibited " acid is sometimes followed by a slurry of cement which leaves an excellent protective lining. For cleaning steel, zinc and galvanized steel, quinoline has sometimes been chosen as restrainer (F. N. Speller, E. L. Chappell and R. P. Russell, *Trans. Amer. Inst. Chem.* 1927, **19,** 165. M. J. van der Wal, *Korrosion u. Metallsch.* 1937, **13,** 41. W. Wiederholt, " Korrosion VIII ", 1954–55, p. 125 (Verlag Chemie)).

Theory of Restrainers. Considerable discussion has arisen regarding the manner in which restrainers act. That they attach themselves to the surface—probably by the nitrogen or sulphur atom—is generally accepted. According to one view, the attached molecules, extending outwards from the metallic surface like the bristles of a brush, prevent the ready replenishment of acid at the metallic surface—so that the acid becomes exhausted and attack becomes slow. According to another view, the molecules which are most efficient as restrainers are those which, when attached through the polar atom or atoms, lie flat, so that each molecule covers up a considerable element of surface; tests on aniline and its alkyl derivatives show that, as the side chain is made longer, the restrainer-efficiency increases. Mann and his colleagues obtained excellent restrainer action with straight-chain di-amines and triamines, the results improving with the length and number of the chains attached to the nitrogen (up to three). Nevertheless some very small molecules can act as restrainers—notably formaldehyde, as pointed out by Batta.

Some authorities consider that the restrainer molecule attaches itself at the points where a cathodic or anodic reaction would otherwise occur. The earlier investigators often favoured the idea that the restrainer interfered with the cathodic reaction, and it is not improbable that in the early days when more importance was attached to preventing the destruction of metal than to avoiding the introduction of hydrogen, some of the substances in use did act in this way. Most of the restrainers favoured today seem mainly to interfere with the anodic reaction, although there is generally some retardation of the cathodic reaction also. This was the conclusion reached by Hoar and Holliday after an exhaustive electrochemical study of the compounds which are effective as restrainers. Hoar's principal method of distinguishing anodic from cathodic inhibitors depends on observation of the potential movement produced when the restrainer is added to the acid surrounding the iron specimen; movement in the positive direction denotes an anodic inhibitor, whilst movement in the negative direction indicates a cathodic inhibitor; for reasons see fig. 156, p. 869. The paper deserves study (T. P. Hoar and R. D. Holliday, *J. appl. Chem.* 1953, **3,** 502; see also T. P. Hoar, Corrosion Symposium (Melbourne University), 1955–56, p. 124).

Other important papers, some expressing divergent views, include those of E. L. Chappell, B. E. Roetheli and B. Y. McCarthy, *Industr. engng. Chem.* 1928, **20,** 582; W. Beck and F. von Hessert, *Z. Electrochem.* 1931, **37,** 11; C. A. Mann, B. E. Lauer and C. T. Hultin, *Industr. engng. Chem.* 1936, **28,** 159; C. A. Mann, *Trans. electrochem. Soc.* 1936, **69,** 115; H. Pirak and W. Wenzel, *Korrosion u. Metallsch.* 1934, **10,** 29; E. G. R. Ardagh, R. M. B. Roome and H. W. Owens, *Industr. engng. Chem.* 1933, **25,** 1116; G. Batta, L. Scheepers and L. Bousmanne, *Rev. Métallurg.* 1951, **48,** 105; W. Machu, *Korrosion u. Metallsch.* 1934, **10,** 277; 1937, **13,** 1, 20; 1938, **14,** 324; *Trans. electrochem. Soc.* 1937, **72,** 333; L. Cavallaro, A. Indelli and G. Bolognesi (with G. Pancaldi and L. Felloni), *Gazz. chim. Italiana* 1953, **83,** 540, 555, 563, 573, 583, 596; *Annali di Chimica* 1955, **45,** 554, esp. p. 556; H. Fischer, *Werkst. u. Korrosion* 1955, **6,** 26; H. C. Gatos, *Corrosion* 1956, **12,** 23t; R. N. Ride, *J. electrochem. Soc.* 1956, **103,** 98; J. E. Swearingen and A. F. Schram, *J. Phys. Coll. Chem.* 1951, **55,** 180.

At first sight, theory would seem to predict that the ideal restrainer would be one that would interfere with the first stage of the cathodic reaction since this would impose no hindrance on the dissolution of the oxide-scale (direct or reductive), whilst definitely diminishing attack on the metal, and also the total production of hydrogen, whether " evolved " or " diffused ". This requirement may, however, be unobtainable.* In any case, good results would be expected from a restrainer which exerts a reasonable control upon the anodic reaction, permitting it to proceed at such a rate as to balance the *desired* cathodic reaction (the reductive dissolution of magnetite and/or ferric oxide), but not at such a rate as to balance an *undesired* cathodic reaction (the formation of hydrogen); it would seem that this is how the restrainers used today do their work.

Italian work has brought out the important fact that some restrainers, besides repressing attack on the metal, also accelerate the attack on the oxide. This effect ought to be taken into account when assessing the relative merits of restrainers. Instructive information regarding the behaviour of the various thioureas, amines and isothiocyanates both on metal and oxide has been provided by L. Cavallaro and L. Felloni, *Metallurg. ital.* 1952, **44,** 366.

Some commercial restrainers contain foaming agents; the foam, acting as a blanket, reduces the emission of acid spray into the air, besides collecting the scum and particles of detached scale; if, however, it is desired to view the work during the process, a non-foaming preparation is preferable.

Many authorities recommend the addition of wetting agents as well as restrainers to the pickling bath; some wetting agents themselves possess inhibitive properties, such as the sodium salt of a hydrocarbon-sulphamido-acetic acid, $CH_3(CH_2)_{11-17}SO_2NH.CH_2.COONa$. This also serves as a rust preventative when the articles are being raised and dried in air. The use of a wetting agent sometimes tends to spread out such attack as does occur, thus preventing pitting. The practice is recommended by K. F. Hager and M. Rosenthal, *Corrosion* 1950, **6,** 225; see also L. Cavallaro, address at Ferrara, May 7, 1955.

It is suggested on p. 648 that tin salts retard the attack by acids on steel by precipitating the hydrogen sulphide which can be formed from sulphide in the steel and which would act as an anodic stimulator. The use of stannous chloride as a pickling restrainer has been examined in the laboratory. The best results were obtained when 0·05% $SnCl_2.2H_2O$, 0·025% of gelatine and 0·05% of cresol-sulphuric acid (which prevents

* If we adopt the view that, during the cathodic production of hydrogen in absence of restrainer, it is the second stage which is rate-determining, and that this second stage is not a union of pairs of atoms to give molecules but a discharge of hydrogen ions on a surface largely covered with hydrogen atoms

$$H + H^+ + e = H_2$$

then any restrainer which interferes with the first stage (also a discharge process) is likely also to interfere with the second stage. Cf. P. J. Hillson and E. K. Rideal, *Proc. roy. Soc.* (*A*) 1949, **199,** 295; P. J. Hillson, *Trans. Faraday Soc.* 1952, **48,** 462.

oxidation) were present in sulphuric acid (7% by weight) at 80°C. (T. P. Hoar and S. Baier, *Sheet. Met. Ind.* 1940, **14,** 947).

A useful list of pickling restrainers, indicating the amounts needed, will be found in the Corrosion Handbook 1948, pp. 910–912 (Editor—H. H. Uhlig. Publishers—Wiley: Chapman & Hall).

Removal of Hydrogen after Pickling. If it has been decided to use acid pickling, and if the precautions indicated above have not been successful in preventing up-take of hydrogen, much of the hydrogen will escape from the articles on storage; if they can be exposed to warm air, the hydrogen will be expelled more quickly. Immersion in hot water, however, is more effective in releasing the hydrogen than exposure to hot air. This is not difficult to understand. Hydrogen is believed to be held interstitially between the iron atoms, and can presumably diffuse to any point on the surface, but only certain points will be favourable to the combination of hydrogen atoms in pairs into molecules; if some of the points where this would normally occur have been blocked by the adsorption of poisons, such as arsenic or selenium, the escape of hydrogen as gas is greatly restricted. If, however, the surface is immersed in water, the hydrogen atoms can pass into the liquid anywhere as ions, by the anodic reaction

$$H = H^+ + e$$

the electron being used up at points catalytically favourable for the cathodic reactions

$$H^+ + e = H$$

followed by $$H + H = H_2$$

or by $$H^+ + H + e = H_2$$

In hot water molecular hydrogen will be produced at the few favourable points, and this should allow hydrogen atoms to stream out as ions over the main part of the surface. The superiority of water at 100°C. to air at 100°C., as a means of eliminating hydrogen, has been emphasized by I. G. Slater, *J. Iron St. Inst.* 1933, **128,** 237.

If an air-bath is used for eliminating hydrogen, the temperature should be well above 100°C. However, it is much better, instead of trying to expel hydrogen after it has entered, to prevent that entry altogether, since clearly expulsion of hydrogen cannot repair mechanical damage already sustained. For instance, in the case of high-tensile steel, cracking may sometimes start in the pickling bath itself, and it is certain that, when once a crack has appeared, no subsequent expulsion of hydrogen can put matters right. It may be best to avoid plain dipping into non-oxidizing acids for high-tensile material, and either to use anodic pickling (with due precautions to prevent hydrogen absorption when the steel is withdrawn for washing) or the use of an oxidizing acid. These matters are discussed, with special reference to the pickling of spring steel, by J. S. Jackson, *J. Electrodep. tech. Soc.* 1952, **28,** 89.

Possibility of Hydrogen Absorption during Plating. There is a wide-spread belief that the electroplating of steel increases the risk of hydrogen cracking, especially in non-stainless alloy steels of high tensile strength

Particularly is it felt that cadmium-plating constitutes a peril, but it is by no means certain that all the hydrogen present in cadmium-plated articles has been introduced during the plating operation. There is some evidence that if a small amount of cadmium salt finds its way into a cathodic pickling bath, it will increase the amount of hydrogen taken up. Similarly, embrittlement after zinc-plating is liable to occur if a trace of zinc has gained access to the alkaline bath used for the preliminary cleaning of the articles; it seems that zinc—by poisoning the places where atomic hydrogen would turn in to molecular hydrogen—can promote the entry of hydrogen into the steel. When this arises, the plating process may be wrongly blamed; after the true cause has been established, it should be possible to avoid the contamination.

When the pickled article reaches the cadmium-plating bath, there may be at the out-set a certain up-take of hydrogen by the steel; but when once a thin coating of cadmium has been deposited over the whole surface, it is likely that this will hinder the passage of hydrogen atoms inwards and retard entry into the steel. If, however, through a badly designed pickling procedure, the steel already contains hydrogen when it enters the plating bath, the cadmium coating will interfere with the escape of that hydrogen, and to that extent plating does increase a risk of hydrogen trouble. This, however, probably only occurs when the cadmium becomes relatively thick, and in any case the right course would seem to be, not to avoid plating but to improve pickling—or even to avoid it altogether, by the use of " vapour-honing " (p. 408) followed by anodic etching and electropolishing. Several variants on these procedures are in industrial use, mainly for strong non-stainless alloy steels, and appear to be generally, but not invariably, effective. Where the preparation of the surface consists in abrasion with hard powder suspended in water, it must be borne in mind that the formation of hydrogen by water acting on the oxide-free surface is a possibility that cannot be ruled out.

For mild steel, or even non-stainless alloy steels of medium strength, it would seem that cadmium-plating is not dangerous—provided that pickling conditions are correct. However, cases have been reported where bolts of low-alloy steel, heat-treated to produce high strength, and then cadmium-plated, have failed in service. It is not uncommon after the cadmium-plating of high-carbon steel to apply a heat-treatment designed to expel the hydrogen, but this is not invariably effective, as indicated below. Fears have been expressed that baking may cause blistering of the coating, but it seems that if the adhesion is good blistering does not occur. Bad adhesion has been caused by certain proprietary lubricants used in stamping and pressing, which are difficult to remove (*Bull. Inst. Met. Finishing* 1956, **6**, No. 2, p. 84).

It is generally agreed that some cadmium baths are safer than others —at least for the plating of high-tensile alloy steels; for mild steel, there is probably little to choose. Zappfe and Haslem coated mild steel in three different cadmium-plating baths and five different zinc-plating baths. In all cases, mild steel after being plated showed satisfactory bending properties; plating caused less embrittlement than cathodic pickling or plain acid pickling. Stainless steel, however, was seriously embrittled by plating, in some cases more than by pickling; acid zinc-plating baths caused less trouble than cyanide baths* (C. A. Zappfe and M. E. Haslem, *Plating* 1950, **37**, 366).

Some interesting results regarding the behaviour of different cadmium plating baths have been made public in the U.S.A. It was found that certain baths when used at high current density gave deposits which did not interfere with the expulsion of hydrogen by baking, but the deposits obtained at low current density presented an obstacle to escape so that the plated steel remained brittle even after baking. Investigation showed that the deposits obtained at high current density had an abnormally low density, indicating porosity; presumably it was the porosity which allowed the hydrogen to escape. It is stated, that on adding nitrate to a standard cyanide cadmium-plating baths, embrittlement is prevented, and provided that " a suitable organic agent " is present to maintain throwing and covering power, excellent results can be obtained without undue difficulty (W. F. Hamilton, M. Levine and R. C. Mauer, *Soc. Automotive Eng.* Sept. 30 to Oct. 5, 1957).

Another American research has given encouraging results which have been disclosed in greater detail, so that judgment is easier. Cadmium deposited from ammoniacal baths containing certain aliphatic amines such as glycine and β-alanine (amino-propionic acid) showed improved behaviour under a sustained load test; another promising bath contained ammonium amino-butyrate (P. N. Vlannes, S. W. Strauss and B. F. Brown, *Electroplating and Metal Finishing*, 1958, **11**, p. 85).

In a third American process, a flash coating of cadmium is applied, sufficiently porous to allow the escape of hydrogen when the plated metal is baked—after which thick cadmium is deposited (H. H. Johnson, E. J. Schneider and A. R. Troiano, *Iron Age* July 31, 1958, **182**, p. 47).

The situation in regard to chromium-plating is different from that of cadmium-plating. It is impossible to deposit chromium on a cathode without introducing much hydrogen, and it is generally believed that the deposit first deposited on the cathode is not chromium, but an unstable hydride—as indicated by the work of Snavely. Gradually, however, some of the hydrogen migrates into the steel, leaving the chromium atoms at greater distances from one another than would be the case in stable chromium; in other words, the deposit develops a tensile stress. Snavely calculates that a contraction of 15% would be necessary to produce stress-free chromium. Nor is it easy to rectify the situation by procedures which would be effective for unplated steel. Levi found it necessary to heat chromium-plated steel above 400°C. in order to regain the original " bend value " of unplated steel. The reason why heating of chromium-plated steel is ineffective in eliminating hydrogen at a range of temperature suitable for driving hydrogen out of pickled steel is that in the former case much of the hydrogen, instead of escaping into the air, moves down into the steel.

* Perhaps the cyanide acts as a poison for the reaction $H \longrightarrow H_2$.

It is necessary to heat at relatively high temperatures (perhaps 490°C.) to elimimate this hydrogen from the steel basis. Levi's experiments with the mass spectrograph show that up to about 350°C., heating merely causes the hydrogen to leave the chromium (not the steel)—which presumably makes matters worse. His observations may serve to explain the results of Hammond and Williams (see p. 724); these investigators found it necessary to heat chromium-plating steel above 440°C., to eliminate tensile stresses and thus secure good fatigue resistance (C. A. Snavely, *J. electrochem. Soc.* 1947, **92,** 537; C. A. Snavely and C. L. Faust, *ibid.* 1950, **97,** 466; C. Levi and G. A. Consolazio, *ibid.* 1956, **103,** 624).

De-scaling in molten Sodium Hydroxide. Several methods for removing scale from steel depend on reduction to the metallic state in molten caustic soda. In one method, due to Tainton, the steel is made the cathode with an external E.M.F. It is possible, however, to carry out the process in an easier manner, which dispenses with electrical connections, by the use of sodium hydride. The articles to be de-scaled are simply dipped, without external current, into molten sodium hydroxide at 350–370°C. containing about 2% sodium hydride; the hydride can be prepared on the spot in an ancillary plant by the direct union of metallic sodium and hydrogen; in practice it may be convenient to use cracked ammonia (p. 80) in place of hydrogen. The adherent scale on the steel is reduced to metal which takes the form of loose powder or non-adherent flakes (if the steel contains chromium, the reduction-product may contain a lower chromium oxide). The reduction proceeds rapidly, being sometimes complete within a minute of the time when the article has reached the temperature of the bath. At the appropriate moment, the article is removed, allowed to drain and then quenched in cold water, so that the steam evolved pushes off most of the loose reduction-product, although it may be necessary to use jets of high-pressure water to complete the removal. It is not uncommon to follow the process with a dip in phosphoric acid. Safety precautions (screens, goggles and protective clothing) are, of course, essential. It appears that at the temperatures used no hydrogen is taken up, and the steel is not cracked or embrittled. This may excite surprise to those familiar with caustic cracking (p. 455), but it should be pointed out that caustic cracking does not occur in anhydrous sodium hydroxide, although it is a serious menace in solutions of the range 35–70%. Information about hydride de-scaling is provided by N. L. Evans, *J. Electrodep. tech. Soc.* 1948–49, **24,** 9, 15; *Metalloberfläche,* 1949, **1,** B53; *Corros. Tech.* 1954, **1,** 273 and in the "Finishing Handbook" 1951, p. 29.

A molten-bath process designed mainly for nickel-chromium steels (heat-resistant or stainless) depends on an oxidizing mixture containing sodium hydroxide along with either sodium nitrate or potassium permanganate (H. C. Smith, *Wire and Wire Products* 1950, **25,** 1050, 1085).

Types of Hydrogen Troubles

Hydrogen Blistering. In relatively soft steels, the atomic hydrogen sinks into the metal, and becomes converted to molecular hydrogen at points below the surface; the high pressures thus developed raise up the surface layers of the steel, producing blisters. This sometimes happens during pickling, and in such a case it is useless to try to repair the damage by expelling the hydrogen afterwards. More often, it occurs during some subsequent treatment to which the steel is subjected; this may be an annealing process. If the steel is to be galvanized or lead-coated, the blisters may be produced whilst the steel is in the molten metal; if the steel is to be enamelled, the hydrogen may cause faults in the coating. Worst of all —blistering may occur in service.

In an early paper, Edwards identified the points of evolution of high-pressure hydrogen with inclusions of oxide or possibly sulphide, and this has been confirmed by many subsequent observers. Blow-holes also can provide collecting places for the gas. Sometimes a line of inclusions may occur on a sheet, joined by a narrow crack, and blisters will then tend to be produced along such a line. In general, only inclusions or cavities near the surface are likely to start blisters.

There is some doubt as to whether it is the inclusions as such, or cavities associated with them, which form the collecting places for gas. Non-metallic inclusions may tend to develop cavities when the steel is rolled, since they are less malleable than the metal and may fail to flatten during rolling sufficiently to maintain contact with the steel at all points; they may even shatter, producing crevices between the pieces formed. Again, blow-holes may be formed in casting but some of these weld up on rolling and can be disregarded; holes which are lined with non-metallic matter may refuse to weld up, and it is these which are most likely to form collecting-places. In all such cases, the inclusion forms the starting-point of the blister mainly because of associated cavities.

Theory suggests hat an inclusion composed of a substance having no adhesion to the metal could form the starting-point of a gas-bubble even if there were no cavity associated with it, since in such cases no work would be needed to separate the steel from the inclusion—apart from that involved in the deformation. Thus it cannot be stated with certainty the fact that the gas-blisters start at inclusions is due to pre-existing cavities. It is, however, fairly certain that the material forming the inclusion affects the question, and further work to distinguish between " safe " and " dangerous " inclusions would be welcome. Meanwhile the reader should read the papers of C. A. Edwards, *J. Iron St. Inst.* 1924, **110,** 9; A. Hayes, *Trans. Amer. Soc. Steel Treating* 1930, **17,** 527; T. P. Hoar and S. Baier, *Sheet Met. Ind.* 1940, **14,** 947; T. Skei, A. Wachter, W. A. Bonner and H. D. Burnham, *Corrosion* 1953, **9,** 163. The deformation of inclusions on mechanical rolling has been studied by F. B. Pickering, *J. Iron St. Inst.* 1958, **189,** 148.*

* Pickering points out that cavity-formation at the ends of inclusions during rolling depends on the ability of metal to flow into the holes caused (momentarily at least) when the inclusions break up; it is more likely to occur at low rolling-temperatures than at higher ones and more at brittle inclusions than at plastic ones. He agrees that it is not necessary to postulate cavities of appreciable size in explaining hydrogen troubles, and points out that dislocation pile-ups are really cavities on an atomic scale, whilst the interface between an inclusion

Hydrogen Cracking. As already stated, articles composed of steel of high tensile strength (either high-carbon steels or non-stainless alloy steels) sometimes develop appreciable cracks in the pickling bath, and clearly no heat-treatment to expel hydrogen can thereafter restore their original conditions. In cases where there is no cracking when the articles are put into service, the presence of hydrogen may render them brittle, so that they fracture instead of deforming when subjected to stress. If once a crack has started, the diffusion of hydrogen into it may maintain a high pressure and cause the extension of the crack to continue; this is easily understood, but it is not equally certain what determines the starting-points of the cracks. It is possible that harmless micro-cracks often exist unsuspected in many materials—particularly those of high tensile strength—as a result of stresses produced by fabrication, quenching or welding. In absence of hydrogen, a crack will only extend until a point is reached beyond which the activation energy necessary for any further extension would exceed the decrease of strain energy produced by such an extension; at this point, cracking should cease. If, however, a high pressure of hydrogen is generated within a micro-crack, extension may continue indefinitely and the cracking will become dangerous; the direction of the cracking will then be decided by the direction of the stresses, internal and applied, but the force needed to produce the rupture in the metal is mainly derived from the hydrogen; in the neighbourhood of a weld-line, the welding-stresses determine the direction of the cracking.

The crack will only continue to extend if the hydrogen diffuses from the metal around into the cavity, and this requires time, especially at low temperatures. Thus the effect of hydrogen on the strength of metal as determined by a mechanical test depends greatly on the character of that test; the effect of hydrogen is marked in tests at low strain-rates, but practically absent in impact tests, where the time is far too short for the required diffusion of hydrogen to take place. For the same reason hydrogen has no effect in tests carried out at low temperatures (e.g. − 110°C.), where diffusion is negligible, even when the rate of straining is slow. Hydrogen hardly affects the mechanical properties of steel at very fast deformation rates but greatly increases fragility at lower rates; the elastic deformation is unaffected but the damping capacity is increased (P. Bastien and P. Azou, *Soc. française Met.* Oct. 1955; cf. H. G. Vaughan and M. E. de Morton *J. Iron St. Inst.* 1956, **182**, 389; J. T. Brown and W. M. Baldwin, *J. Metals* 1954, **6**, 248).*

Interesting studies of true stress-strain curves have been carried out on

and the surrounding metal can be regarded as a big thin cavity, since there is no evidence that the interface is coherent. Slag particles (especially basic slag) may contain more hydrogen than the surrounding steel, and may constitute a source of hydrogen conveniently close to the cavities (F. B. Pickering, Priv. Comm., June 14, 1958).

* A. H. Cottrell refers to the interesting discussion of hydrogen cracking by Stroh, who has shown how the development of cracking can be traced systematically to adsorption of hydrogen on the face of the crack (A. N. Stroh, *Phil. Mag.* 1957, **46**, 968).

steel charged with hydrogen from an acid bath to which a few drops of a phosphorus solution in carbon disulphide had been added to poison the second stage of the cathodic reaction. At only 5–6 ppm. of hydrogen the ductility (as measured by the elongation) fell to a minimum, but greater contents of hydrogen produced no further effect. The embrittlement was greatest at low strain-rates and decreased as the rate of straining was increased; as already stated, an impact-test furnished no evidence of embrittlement (J. B. Seabrook, N. J. Grant and D. Carney, *J. Metals* 1950, **188,** 1317).

It would seem that the transition from atomic hydrogen (accommodated at interstitial places in the lattice) into molecular hydrogen (at the cracks) probably largely occurs after the deforming force is applied, so that a bending force, instead of providing a smoothly curved article, produces destructive cracking. However some authorities attribute hydrogen embrittlement to the pre-existing existence of hydrogen at high pressure in cracks or perhaps at the junction of the " mosaic blocks " or sub-grains; in some of these theories, molecular hydrogen at such points seems to be postulated. Views deserving study include those of C. A. Zappfe and C. E. Sims, *Trans. Amer. Inst. Min. Met. Eng.* 1941, **145,** 225; F. de Kazinczy, *J. Iron St. Inst.* 1954, **177,** 85; *Jernkontorets Annaler* 1956, **140,** 347; J. D. Hobson and J. Hewitt, *J. Iron St. Inst.* 1953, **173,** 131; J. H. Andrew and H. Lee, Symposium on Internal Stresses in Metals and Alloys 1947–48, p. 265 (Inst. Met.); J. E. Draley and W. E. Ruther, *J. electrochem. Soc.* 1957, **104,** 329; I. Class, *Werkst. u. Korrosion* 1955, **6,** 237; " Korrosion VIII ", 1954–55, p. 115 (Verlag Chemie); J. T. Brown and W. M. Baldwin, *J. Metals* 1954, **6,** 298; N. J. Petch and P. Stables, *Nature* 1952, **169,** 842.

The fact that a period must elapse before hydrogen can diffuse into a cavity and initiate crack-propagation has been demonstrated by measuring the electrical resistance of specimens charged with hydrogen, cadmium-plated, baked (to distribute the hydrogen) and then stressed; the propagation of the crack was shown by an increase of the resistance (H. H. Johnson and A. R. Troiano, *Nature* 1957, **179,** 777).

A recent paper by Troiano and his colleagues deserves quoting. " The strain-rate dependence of hydrogen embrittlement reflects variations in the time available for hydrogen to diffuse into the highly stressed regions. In a fast strain-rate test there is insufficient time for a damaging quantity of hydrogen to diffuse into the region of maximum triaxility and embrittlement is minimized. However, with decreasing strain-rate, more hydrogen can diffuse into the highly stressed region and embrittlement increases." They consider, however, that " hydrogen embrittlement is but one facet of the general problem of fracture " (J. G. Morlet, H. H. Johnson and A. R. Troiano, *J. Iron St. Inst.* 1958, **189,** 37, esp. p. 44).

Among the excellent French papers on this subject, these of Bastien's have already received notice (p. 402). Others deserving attention include J. Duflot (see below), G. Chaudron, A. Portevin and L. Moreau, *C.R.* 1938, **207,** 235; L. Moreau *ibid.* 1944, **218,** 353.

Duflot has shown that the cathodic charging of steel with hydrogen develops cracks or holes which may be either intercrystalline or intracrystalline, and which soon become easy to detect—sometimes visible to the naked eye; occasionally cracks follow the rolling direction. He believes, however, that the union of hydrogen atoms to form molecules occurs only at imperfections in the structure which were present before the charging started. He points out that when coarse-grained steel is placed in acid, the grains are sometimes caused to fall apart; this does not happen with steel cathodically charged with hydrogen. Presumably in the first case, anodic attack as well as development of hydrogen pressure play a part in the disintegration (J. Duflot, *Rev. Métallurg.* 1952, **49**, 35).

Special interest attaches to the papers of Zappfe, since he courageously championed the view that hydrogen was the cause of many troubles at a time when such a view was unpopular; perhaps he may have gone too far at times, but the reader should decide this for himself. Zappfe's views on the detrimental effect of hydrogen in enamelling are mentioned on p. 592. The chief general papers are C. A. Zappfe and C. E. Sims, *Metals and Alloys* 1941, **13**, 444, 584; 1942, **14**, 56; C. A. Zappfe and M. E. Haslem, *Trans. Amer. Soc. Met.* 1947, **39**, 213.

An interesting type of breakage is described by Bell and Sully. It occurred in certain high-tensile steel bolts, which had been pickled and cadmium-plated; there was no immediate failure, but it took place within a few minutes of being tightened, and only on bolts which had undergone pickling and cadmium-plating. Tests on circlip samples, held in stress by means on screws which forced the ends apart, showed that cathodic pickling for $\frac{1}{2}$ hour in 2·5N sulphuric acid greatly reduced the deflection needed to produce breakage; but storage for 115 hours at room temperature brought about considerable recovery. The experimenters attribute the cracking to the gradual building up of pressures inside discontinuities in the lattice, owing to the atomic hydrogen gradually diffusing to these points and forming molecular hydrogen on arrival; but they think that the atomic hydrogen present interstitially in the lattice also affects cohesion directly. The papers, which deserve study, are W. A. Bell and A. H. Sully, *J. Iron St. Inst.* 1954, **178**, 15; W. A. Bell, G. J. Metcalfe and A. H. Sully, *J. Electrodep. tech. Soc.* 1950–51, **27**, 91.

The stamping properties of steel sheet are not affected after cathodic treatment in pure sulphuric acid, but if selenium or arsenic are present they suffer severely; hydrogen sulphide has much less effect (V. Baukloh and K. Gehlen, *Métaux et Corros.* 1938, **13**, 78).

Other sources of Hydrogen in steel. Pickling is not the only operation which may introduce hydrogen into steel. Some can enter from the furnace atmosphere at the steel-works, whilst an important up-take can occur during welding, especially where moisture is present in the coating of the welding electrode, or where these coatings contain cellulose or clay; heating the electrode before use can eliminate the loosely held water, but not the combined water. Hydrogen is probably responsible for the cavities known as "fish-eyes"—one of the welder's most serious troubles. Zappfe considers that "fish-eyes" in welding are the equivalent of "white-spots" in castings and " flakes " in forgings; in all three cases he holds hydrogen responsible (C. A. Zappfe and C. E. Sims, *Welding J.* 1940, **19,** 377S; C. A. Zappfe, *Trans. Amer. Soc. mech. Engrs.* 1944, **66,** 89).

Modern views on the responsibility of hydrogen for welding faults and the advantages of low-hydrogen electrodes are discussed by C. L. M. Cottrell, *British Welding J.* 1954, **1,** 167.

One view widely held is that the adverse effect of hydrogen near welds is not due to the building up of huge gas-pressures in cavities but to the anchoring of dislocations by hydrogen atmospheres, so that ductile deformation becomes impossible. Atomic hydrogen arising from different sources will produce the same effect, so that the result of the three modes of introduction (in the furnace, during welding and during pickling) may be additive —a matter sometimes overlooked. It is probably necessary, however, to distinguish between mobile and immobile hydrogen (p. 398). Thus the total hydrogen-content, as obtained by analysis, may not be a reliable measure of liability to hydrogen troubles.

It is probable that the importance of hydrogen-introduction during steel manufacture as a factor in subsequent cracking has been exaggerated in some quarters, and attention may be directed to a reasoned paper by C. Sykes, H. H. Burton, C. C. Gegg, *J. Iron. St. Inst.* 1947, **156,** 155. See also J. D. Hobson and C. Sykes, *ibid.* 1951, **169,** 209.

Nevertheless the connection between hydrogen-content and the production of the so-called " hair-line cracks " or " flakes " cannot be disregarded. Many views have been expressed, but the concise interpretation offered by Huddle in 1942 when discussing a paper by Andrew and others still deserves consideration. He pointed out that hydrogen, like manganese, is a carbide stabilizer and, whilst it exists in any considerable quantity in the metal, will help the retention of austenite on cooling down steel at a rate which (in absence of such a stabilizer) would have produced a complete $\gamma \rightarrow \alpha$ transformation at the usual range of temperature; for this stabilization evidence had been provided by Andrew. Huddle also emphasized the fact that the solubility of hydrogen in γ-iron is much greater than in α-iron. If then a steel containing hydrogen has been cooled down from furnace temperature, so that it carries retained austenite, the hydrogen needed to stabilize that austenite may pass away—either slowly by diffusion outwards during storage, or more suddenly during subsequent heating. At some point the austenite, being no longer stabilized, will switch over into martensite, which, being unable to retain the greater part of the hydrogen in solid solution, will evolve it catastrophically, and the stresses produced, added possibly to welding-stresses already present, may cause cracking. Applied stress may also cause austenite to change into martensite, with consequent hydrogen evolution and cracking. Since 1942, cases have been brought up where hair-line cracks and the like have been produced in steel after treatment which has not carried it through the $\gamma \rightleftharpoons \alpha$ point, but the most recent work by Kumar and Quarrell does confirm the fact that hydrogen can have a marked effect on martensite formation, depressing the temperature at which this occurs by 50°C. in some steels, although, when at last it does take place, it occurs with a " pronounced burst "—so that it is complete sooner than in nitrogen-treated control specimens. The reader may like to study the evidence and compare the views (J. H. Andrew, A. K. Bose, G. A. Greach, H. Lee and A. G. Quarrell, *J. Iron St. Inst.* 1942, **146,** 193P; A. U. Huddle, *ibid.* 1942, **146,** 262P; C. L. M. Cottrell, *ibid.* **154,** 176, 273; R. Kumar and A. G. Quarrell, *ibid.* 1957, **187,** 195. A. Barker and A. Wainwright, *Nature* 1956, **177,** 1136).

Hobson considers that the practice commonly adopted at steel-works of cooling down alloy steel billets slowly in pits so as to allow the hydrogen time to escape is a sound one, and does much to prevent hair-line cracks. He points out, however, that hydrogen diffusion becomes abnormally slow below about 150°C. and suggests that the continuance of the slow cooling below 100°C. is of little practical use (J. D. Hobson, *J. Iron St. Inst.* 1958, **189,** 315).

Hydrogen Troubles in the Oil-Fields. In the so-called " Sour Oil-Fields " where the water accompanying the oil usually contains hydrogen sulphide, it is not uncommon to find blisters blown in the outer surface of oil pipes, containers and other equipment by the hydrogen which has been produced at the inner surface by the action of hydrogen sulphide. Whilst on the softer steel, the trouble takes the place of blistering, serious embrittlement may be produced on stronger materials, such as alloy steels; the cracking of springs may be a serious problem. The situation is discussed in papers from Emeryville by R. T. Effinger, M. L. Renquist, A. Wachter and J. G. Wilson (Amer. Petroleum Inst. Div. of Refining, May 1, 1951); J. P. Fraser and R. S. Treseder, *Corrosion* 1952, **8,** 342; T. Skei, A. Wachter, W. A. Bonner and H. D. Burnham, *ibid.* 1953, **9,** 163; M. H. Bartz and C. E. Rawlins, *ibid.* 1948, **4,** 187. The symposium reported in *Corrosion* 1952, **8,** 326–360, reveals several different points of view regarding the causes of the troubles. See especially W. D. Robertson, *ibid.* 1952, **8,** 359; 1957, **13,** 437t (with A. E. Schuetz).

The laboratory experiments described by the Emeryville authors show that the passage of hydrogen through steel diminishes as the pH is raised but does not entirely cease even when the liquid is appreciably alkaline with ammonia. Moreover if cyanides are present, the hydrogen pressure which can be reached is greatly increased in alkaline solutions—either because the removal of Fe^{++} ions as complex anions increases the driving E.M.F., or, as Skei and his colleagues suggest, because cyanides prevent the formation of a protective sulphide-scale.

Several carbon steels were examined but none of them resisted transmission of hydrogen. On the other hand no transmission and no blistering occurred on monel metal, nickel or stainless steel either in acid sulphide solutions or in alkaline solutions containing both sulphide and cyanide.

After a discussion of hydrogen troubles in a hydrogen-sulphide absorption plant, Bradley and Dunne make recommendations for its mitigation recommending stainless steel and monel metal for various pump, valve and other parts, with aluminium conduits and instrument tubing in certain places to avoid atmospheric attack (B. W. Bradley and N. R. Dunne

Corrosion 1957, **13,** 238t; L. Cauchois, J. Didier and E. Hergog, *ibid.* 1957, **13,** 263t; B. W. Neumaier and C. M. Schillmoller, *Corros. Tech.* 1956, **3,** 357).

Hydrogen sulphide can cause troubles of another kind by producing an intergranular network of the metallic sulphide—especially in certain alloy steels—and such intergranular weakening may perhaps be mistaken for hydrogen embrittlement arising from hydrogen sulphide. Probably there are situations in the oil industry where both troubles occur together. At high temperatures, as in catalyst units, true intergranular sulphide-penetration on low-alloy steels may be important (J. P. Fraser and R. S. Treseder, *Trans. Amer. Soc. mech. Engrs.* 1955, **77,** 817; L. W. Vollmer, C. N. Bowers and W. J. McGuire, *ibid.* 1955, **77,** 823; see also *Corrosion* Oct. 1956, **12,** p. 73).

The trouble in the sour oil-fields and other cases of cracking in presence of hydrogen sulphide has led to much discussion as to whether hydrogen embrittlement or stress-corrosion cracking is mainly responsible; translated into electrochemical language, the question to be settled is whether atomic hydrogen developed by the cathodic reaction diffuses inwards and develops dangerous pressures in internal cavities, or whether the anodic reaction, by converting the metal to sulphide along grain-boundaries or other paths of weakness, permits failure at stresses normally too low for breakage. Some experiments carried out in the American Oil Industry a few years ago seemed to point to sulphide-formation, but a recent extensive research in W. D. Robertson's laboratory has shown why this evidence is inconclusive, and has supplied convincing arguments in favour of hydrogen being the essential cause. It was shown, for instance, that the various phenomena observed (a loss of bend-strength, followed by recovery on standing in air) could be produced equally well by cathodic charging of the iron (which involves no corrosion) or by immersion in hydrogen sulphide solution (which involves corrosion); such differences as were observed are consistent with the hydrogen-mechanism. The disappearance of the yield phenomena and their reappearance on standing in air is also consistent with the fact that hydrogen is the operative cause. It is difficult to see how recovery of bending properties on storage, or of yield phenomena, could occur if conversion of metal to sulphide were the cause. Incidentally, the use of phosphorus dissolved in carbon disulphide as a poison for cathodic charging (a practice favoured by some investigators) was shown by Robertson to be unsuitable; carbon disulphide without phosphorus produced a large up-take of hydrogen, and the addition of phosphorus caused no increase in that up-take. The research included a study of nickel steels in various states (ferritic, martensitic and austenitic) as well as unalloyed metal (A. E. Schuetz and W. D. Robertson, *Corrosion* 1957, **13,** 437t; for influence of hydrogen on yield-point, see H. C. Rogers, *Acta Met.* 1954, **2,** 167; 1956, **4,** 114).

Role of Hydrogen in Stress Corrosion. Whilst cases of stress corrosion may sometimes have been wrongly attributed to hydrogen embrittlement, it is possible that internal hydrogen pressures play an essential role in certain types of breakages which are properly called " stress corrosion ", i.e. breakages which occur where stress and corrosion operate together, but which would be absent if either the stress or the corrosion was eliminated. Hydrogen may, for instance, play an important part in the cracking near welds in gas-works plant and perhaps even in caustic cracking of boilers; these matters are discussed on pp. 455, 685. The fact that those steels which have proved best in withstanding nitrate cracking are similar to those which have proved suitable for use with a natural gas containing hydrogen sulphide would seem to support the view that nitrate cracking is connected with hydrogen. The steels contain chromium and aluminium (G. Chaudron, quoting E. Herzog, lecture to Fédération des Industries chimiques de Belgique, Dec. 13, 1957).

Hydrogen Troubles in Non-ferrous Materials. Hydrogen can cause trouble in aluminium and its alloys as well as in steel, and here also it can originate in different ways. At one time it was largely introduced in manufacturing the alloys by the remelting of scrap carrying corrosion products, largely aluminium hydroxide or hydrated oxide; the water combined as hydroxide acted on the metal, liberating hydrogen; this to a large extent dissolved in the liquid metal, but was liberated at the moment of casting, causing numerous cavities. Such a source of porosity can today be avoided by suitable precautions—including the washing of the molten metal with gases, such as chlorine, nitrogen or a mixture of the two. Hydrogen-porosity in aluminium, and recommendations for its avoidance, are discussed by D. Hanson and I. G. Slater, *J. Inst. Met.* 1931, **46,** 187, 216; 1935, **56,** 103. For solubility of hydrogen on molten aluminium, see L. L. Bircumshaw, *Phil. Mag.* 1926, **1,** 510; C. E. Ransley and N. Neufeld, *J. Inst. Met.* 1948, **74,** 599. For methods of determining the hydrogen content of aluminium and its alloys, see C. E. Ransley and D. E. J. Talbot, *J. Inst. Met.* 1955–56, **84,** 445. The alterations in the elastic properties of aluminium produced by charging with hydrogen are demonstrated by L. Moreau and G. Chaudron, *C.R.* 1944, **219,** 554.

Hydrogen may also enter aluminium or its alloys during welding, especially if the plates or electrodes carry hydrated corrosion-products.

The up-take of hydrogen by aluminium can occur during pickling, but it remains near the surface. Pickling sometimes affects the fatigue strength, but this may often be due to removal of beneficial surface layers, rather than to up-take of hydrogen. In Sutton's experiments on mechanical specimens, fatigue resistance dropped by 31% after pickling; by a special procedure based on a dilute mixture of sulphuric and hydrofluoric acids followed by 50% nitric acid, the loss was brought down to 6%, and could be made still lower if pickling was followed by immersion in boiling water. Details will be found in papers by H. Sutton and W. J. Taylor, *J. Inst. Met.* 1934, **55,** 149; I. G. Slater, *ibid.* 1934, **55,** 160; H. Sutton and T. J. Peake, *ibid.* 1936, **59,** 59.

The use of sodium hydroxide for pickling aluminium alloys is considered inadvisable in some quarters, although opinion varies on this question. Australian investigators recommend that articles should be immersed in boiling water as soon as they emerge from a caustic soda pickling bath; chromic acid anodizing should then follow at once (H. E. Arblaster and

P. F. F. Thompson, Corrosion Symposium (Melbourne University) 1955–56, p. 379, esp. p. 380).

On copper and its alloys, hydrogen troubles take a different form. If cuprous oxide is present between the grains, heating of the material in a hydrogen-rich atmosphere causes an internal evolution of steam, and thus sets up cracking. References are given on p. 81.

Lead treated cathodically at above 10–50 ma./cm.^{2} in certain solutions may suffer disintegration, giving colloidal lead. One view is that lead hydride is formed and decomposes. The work of Angerstein seems to show that disintegration only occurs if a salt of an alkali metal or some other metal forming an unstable alloy with lead is present; the lead-sodium or other alloy reacts with water producing hydrogen and breaks up into small particles. This seems to explain the experimental results, whereas the proposed hydride mechanism fails to explain them. Catalytic poisons, which would prevent the change of atomic into molecular hydrogen and should thus encourage penetration of hydrogen into lead, do not promote the disintegration; arsenic indeed prevents it. Dibenzyl sulphoxide, which had been found in Polish work to prevent hydrogen penetration on other metals, did not affect the disintegration of lead (H. Angerstein, *Proceedings of Electrochemical Conference at Warsaw* (1955–57); English version on p. 451. Cf. H. W. Salzberg, *J. electrochem. Soc.* 1953, **100,** 146; 1957, **104,** 701 (with L. W. Gastwirt); also *Cebelcor*, *Rapport technique* No. 64 (1958), p. 12).

A type of intergranular corrosion of magnesium produced by cathodic polarization in sodium chloride has been attributed to hydrogen penetration by G. R. Hoey and M. Cohen, *J. electrochem. Soc.* 1958, **105,** 245.

The power of different metals to transmit hydrogen is very different. Palladium and iron transmit it readily; nickel, niobium, tantalum and molybdenum allow only slight transmission, whilst copper gives no detectable transmission (H. B. Wahlin, *J. appl. Phys.* 1951, **22,** 1503). The serious effect of hydrogen on titanium and zirconium was mentioned on p. 343; the diffusion of hydrogen and deuterium in zirconium has been studied by E. A. Gulbransen and K. F. Andrew, *J. electrochem. Soc.* 1954, **101,** 560.

Prevention of Attack on Non-ferrous Metals. Many attempts have been made to find substances which, added to acid, would inhibit or at least restrain attack on aluminium and zinc. Some investigators have measured the effect upon hydrogen evolution, but a retardation of the production of hydrogen gas does not necessarily indicate the end of trouble; it may in some cases indicate that the hydrogen-evolution type of attack is being replaced by the oxygen-reduction type, or—what is worse—that the hydrogen, instead of escaping as gas, is passing into the metal, possibly causing blistering, disintegration or embrittlement. Work before the second world war suggested that acridine or hexamethylenetetramine might be valuable in preventing attack on aluminium—but the published results on acridine have not been confirmed by later workers. Recently Heimann has found that mixtures of cyclic aldehydes and bases are valuable in preventing acid attack on aluminium, zinc and tin (notably for aluminium in 10%

hydrochloric acid)—being much more effective than either substance alone. The aldehyde may be benzaldehyde or phenyl-acetaldehyde; the base may be nicotine or urea (not thiourea). Further results will be awaited with interest (H. Heimann, U. K. Patent Applications 36,290 and 38,575, Nov. 27 and Dec. 18, 1956).

Other References

A book, now in preparation, which should receive careful study when available, is M. Smialowski's " Hydrogen in Steel " (Pergamon Press). Meanwhile attention should be given to his lecture and the discussion which follows (*Chem. and Ind.* (*Lond.*) 1958, p. 1460).

The bulk of research-work carried out on hydrogen in steel has been very large. Some workers have been interested in hydrogen taken up from furnace atmospheres, others in hydrogen taken up during welding, others again in hydrogen taken up during pickling or plating; each group of workers appears strangely unaware of—or indifferent to—the work of the other groups, although a combination of the results obtained in the different researches would aid interpretation and would avoid the putting forward of unacceptable mechanisms. Another feature noticed among investigators in English-speaking countries is their unawareness of the carefully planned quantitative work carried out in Germany and Holland a quarter of a century ago, which in some respects will bear comparison with anything that is being produced today. Many investigators would benefit also by a study of results achieved by those who have entered the subject from another angle. These will be found in references scattered throughout the chapter.

The relation of hydrogen-cracking with other forms of brittle fracture makes advisable close attention to that important subject. An authoritative review is available from (Mrs.) C. F. Tipper, *Metallurg. Reviews* 1957, **2,** 195; the book by W. D. Biggs, " Brittle Fracture of Steel " (Macdonald & Evans) (in the press) should receive study when it appears.

Recent papers include a study of the delayed fracture of high-tensile steel by A. R. Troiano, *Corrosion* 1959, **15,** 207t; on the removal of furnace-absorbed hydrogen from large forgings by J. D. Hobson, *J. Iron Steel Inst.* 1959, **191,** 342; on the rôle of hydrogen in the corrosion of zirconium by high-temperature water by J. N. Wanklyn and B. E. Hopkinson, *J. appl. Chem.* 1958, **8,** 796; and on hydrogen-cracking of steel in moist hydrogen sulphide by P. G. Bastien, *Pittsburgh Conference on Stress-Corrosion Fracture,* 1959.

CHAPTER XII

BOILERS AND CONDENSERS

Synopsis

After some introductory remarks, a short sketch is given of the types of boilers used today—which many readers can clearly omit. It is then pointed out that one object of boiler-water treatment is to avoid the deposition of scale, which would cause risk of overheating and blown tubes; the principles of water-softening and conditioning are briefly indicated. However, corrosion is an important menace, especially in high-pressure boilers, and for this reason it is necessary, except in boilers working at very low pressures, to remove oxygen; much can be expelled mechanically or thermally, but the last traces are generally removed by means of sulphite or hydrazine; the former is unsuitable for the highest pressures, and hydrazine, which introduces no solid matter, is being increasingly employed, although opinions vary regarding its reliability. Most boiler-chemists recommend that the water should be kept slightly alkaline; for the highest pressures, where the introduction of solid matter is undesirable, ammonia or an amine is now sometimes used.

Troubles due to alkali are next considered. Although an alkaline reaction may be desirable in the ordinary type of boiler, there is a potential danger in adding this as sodium hydroxide. If sodium hydroxide should accumulate unduly at crevices (e.g. in the seams of a riveted boiler) it may set up " caustic cracking ", whilst if it accumulates below deposits of solid salts, it can attack the steel, and cause deterioration of the metal, probably associated with the entry of hydrogen. Both types of troubles are to some extent avoided if the alkalinity is controlled by means of phosphate—although opinions differ regarding the efficiency of this method. Nitrates are also used to prevent caustic cracking and appear to be effective in locomotive boilers; the so-called sulphate-ratio method is today under suspicion.

For boilers working at the highest pressures, there is a school of thought which prefers an ultra-pure water to which no chemical additions have been made at all; such a policy, carefully supervised, has worked well in places, but appears risky if used without skilled chemical control. Recent work in connection with atomic-power production has made it clear that high-temperature water—even though pure—can be highly corrosive to many materials which resist relatively cool water; the ranges of conditions within which it is safe to use different materials are briefly indicated.

Next, consideration is given to the corrosion problems of economizers and air-heaters, due to condensation of acid moisture; methods are discussed for reducing the sulphur trioxide content of the gas phase, which is the main cause of the trouble.

We then pass on to condensers. The materials which were used for the tubes in earlier times are briefly reviewed. The deposit attack produced where inert materials settle, acting as oxygen-screens, is described, along with methods of prevention based on keeping clean the interior of the tubes. Dezincification is also discussed, with methods of avoiding it by means of arsenical brass. Next impingement attack, due to air-bubbles, claims attention; some relatively new alloys, which have largely solved this problem, are mentioned. The end of the chapter includes a discussion of tube corrosion by polluted water—a problem which unfortunately has not yet been fully overcome.

Boilers

Introductory Remarks. The majority of boiler mishaps cannot be attributed directly to corrosion. Blown tubes, for instance, arise from overheating, itself usually caused by deposits either on the water-side or on the fire-side, which interfere with heat-transmission; they may also be caused by bad disposition of the burning fuel. However, decarburization of the steel through the action of hydrogen can render the tubes weaker; thus chemical deterioration may often be a contributory cause of mechanical break-down. The various phenomena are so much interwoven that any account of corrosion in boilers must also take notice of deposition of solid matter (scaling).

Probably no living writer could, in a single chapter, tell the engineer how to avoid corrosion in all the various types of boilers operating today. Certainly the present author is unqualified to do so. What can be attempted is to provide a scientific basis which may enable the engineer to take better advantage of the empirical information already accessible to him. Much of that information possesses real value, being based on practical experience, but some of the interpretation provided for corrosion phenomena in boilers has been inaccurate and is useless for deciding rules to be observed in service.

For example, boiler corrosion is sometimes " explained " by writing down the equation

$$3Fe + 4H_2O = Fe_3O_4 + 4H_2$$

When the person who has provided this explanation is asked why the majority of boilers escape serious harm, he replies that a protective film of magnetite is formed; when asked about the reaction for the film-formation, he writes down the same equation over again. If now the same equation has to do duty to explain both disease and cure, it is clearly unhelpful in predicting the conditions likely to confer immunity—even though it may represent the summation of the changes occurring in both cases. If, however, the equation is split up into its simpler components, and if geometrical factors receive due consideration, it becomes easier to understand why sometimes there is corrosion and sometimes protection.

Types of Plant. Innumerable patterns of power boilers and condensers are described in the technical literature, and in useful brochures issued by manufacturers. For some readers, however, it may be useful to recall the main classes to mind before proceeding further. These are:—

(1) *Fire-tube boilers*, where the hot gases from burning fuel pass through tubes surrounded by the water which is to be turned into steam; these include locomotive boilers and Scotch marine boilers.

(2) *Water-tube boilers of the drum type*, where water circulates, generally by convection, around paths formed by banks of tubes joining the drums, the tubes being arranged in parallel, and surrounded by hot gases coming either from solid fuel burning on a grate—or from pulverized fuel, oil or natural gas injected into the system. The steam drawn off from an upper drum at a temperature which is determined by the pressure, is usually further heated by being passed through *superheater* tubes. The feed-water usually receives a preliminary warming by passage through *economizer* tubes and *heat-exchangers*, whilst the air for the furnace is generally passed through a *pre-heater*; in both cases part of the heat of the exit gases is transferred to the in-coming water or air.

(3) *Boilers of the " Once Through " type*, where water is forced through externally heated tubes, and emerges as highly superheated steam—above the critical pressure and temperature, i.e. in a region where steam and water do not exist as distinct phases. The Benson boiler is of this type.

The forms of construction are numerous and varied, but distinction must be drawn between:—

(*a*) **Riveted Boilers** which involve seams—with consequent risk of leakage; the steel near the rivets sometimes carries high internal stress, possibly due to bad alignment of rivet holes or to over-caulking carried out in an endeavour to avoid leakage—a situation which sometimes leads to caustic cracking, as explained later.

(*b*) **Welded Boilers** which, if properly stress-relieved, should be free from such troubles.

Boilers are also classified as low-pressure, medium-pressure and high-pressure, but there is no sharp distinction; moreover, the high pressure of today may become the medium pressure of tomorrow. Actually, it is not so much high pressures, as the high temperature which accompany them, that increase the corrosion-risks.*

* E. C. Potter has suggested that " numerical information relating to pressures, temperatures, percentage make-up steam output " would " put the reader in the picture with regard to the modern scale of things ", adding—what is very true—that " it is the huge scale of steam production by a modern boiler which produces extraordinary problems in chemical analysis ". He regards pressures below 300 lb./in.^{2} as " low ", those in the range 300–1000 as " medium ", 1000–2000 being " high ", 2000–3200 " very high ", whilst pressures exceeding 3200 are " super-critical ". He states that the highest in the world is 4500 (U.S.A.), whilst the highest in the U.K. will shortly be 2700. As regards steam output, he regards anything below 5×10^4 lb./hr. as " small ", 5×10^4 to

For boiler plants, ferrous materials are largely used, and in the less hot portions, high-quality mild steel is suitable, care being taken that segregates —particularly sulphide segregates—do not lie at the surface. The position of the segregate relatively to the surface of a tube depends on the character of the steel (whether killed, semi-killed or rimming) but also the method of tube-making. For information see G. R. Woodvine and A. L. Roberts, *J. Iron St. Inst.* 1926, **113,** 219; see discussion by H. O. Hibbard, p. 223; also F. A. Ruddock, *Engineering* 1930, **129,** 632.

For superheaters, alloy steels are generally necessary and those containing 0·5% molybdenum with or without 1% chromium are much used; for more elevated temperatures steels with 1% molybdenum and 2·25% chromium have been introduced. For the most modern power stations with superheat above 565°C. (1050°F.) stainless steel (18/12/1 Cr/Ni/Nb) is used (E. C. Potter, Priv. Comm., Feb. 28, 1957. For materials used at German Benson Stations see M. Werner and W. Ruttman, *Z. Ver. dtsch. Ing.* 1953, **95,** 811, 827; see also papers by H. Tietz, p. 802, 825; H. Buckholtz, p. 809; and, for turbine materials, R. Schinn, p. 818. See further K. Dangl, *Mitteilungen der VGB* 1954, **31,** 265).

After performing its duty in the engines (probably turbines), the steam, now greatly reduced in pressure and temperature, passes to the condensers, which are usually of the water-tube form. The roughly horizontal tubes are fixed at their ends into two vertical tube-plates, which separate two water-boxes from the central steam space. Cooling water (often salt and frequently polluted) passes through the tubes, which are usually made from a copper alloy; the steam passes around their exterior and is condensed to water, which is extracted by a pump and passes to a well, whence it is returned to the boiler. An efficient plant uses almost entirely condensate for feeding the boilers; except at these works where some of the steam is required for chemical processes, district heating or other purpose which does not allow it to be recovered, only a small proportion of make-up water should be needed; this make-up water, however, requires careful treatment.

Object of Water-Treatment. The condensate is usually very slightly acid or alkaline and contaminated with small amounts of copper and other metals; it should, however, be free from salts and oxygen, and the plant-designer should bear in mind the advantage of avoiding fresh up-take of oxygen before the re-entry of the water into the boiler. The raw water used to make up for what is not recovered will certainly contain salts, and may come from a polluted source; it usually requires treatment before it is fed into the boiler. The main object of such treatment is to prevent the formation of adherent scale on the interior surface of the tubes. Such

3×10^5 as " medium ", 3×10^5 to 10^6 as " large ", whilst anything about 10^6 is " very large ". A plant on order for U.K. with 2×10^6 will represent the biggest in the world. Readers desiring further data will find it conveniently provided in " Central Station Boilers " (Babcock & Wilcox), but they should remember that such information quickly becomes out of date, whereas the chemical and electrochemical mechanism of the reactions involved remains—one must hope—independent of time.

scale would interfere with heat transfer and cause overheating, with weakening of the material and risk of sudden tube failure; a non-adherent sludge, which in many types of boilers can be expelled by means of a blow-down, is less dangerous. The overheating may also lead to reduced thermal efficiency, increased oxidation on the fire-side and often accelerated corrosion on the water-side.

For all these reasons, some treatment of the make-up water is needed. This should include (1) *softening* (whether by chemical treatment, distillation or demineralization) to remove scale-forming and sludge-forming matter as far as possible, (2) *conditioning*, to ensure that any matter deposited is of sludge-forming rather than scale-forming character.

Some general remarks on scale-formation are here necessary, although discussion of the individual constituents of scale lie outside the scope of this book; reference should be made to the papers of L. M. Clark and C. W. Bunn, *J. Soc. chem. Ind.* 1940, **59,** 155, for identification of some of the most objectionable constituents; see also K. V. Aubrey, *Nature* 1954, **174,** 81.

Most hard waters contain calcium bicarbonate and calcium sulphate; the former is generally regarded as a sludge-former (decomposing to calcium carbonate and carbon dioxide when heated), and the latter as a scale-former; calcium phosphate and magnesium hydroxide generally form sludges, but both calcium carbonate and magnesium hydroxide can " bake on " to the boiler tubes, forming an adherent mass, in certain circumstances—e.g. if the feed-water contains oil (P. Hamer, *J. Inst. Heating and Ventilating Engrs.* 1956, **23,** 476).

Certain double silicates form adherent scales prejudicial to heat transfer and difficult to detach; they may cause the temperature to reach the softening point of steel, so that the tube bulges and perhaps bursts. Silica is a most undesirable constituent of feed-water, and several ways are known for removing it, of which passage through a suitable resin column is probably the easiest. It is objectionable for another reason, since silica is volatile in steam at pressures above about 200 atm. and can produce deposits on the turbine blades; these do not necessarily consist of pure silica; if alkali is also carried over the deposits may consist of sodium silicate and are then relatively easy to wash off. The estimation of silica in boiler-water is not easy to carry out, partly because of the very low concentrations at which its action becomes important.* Analytical methods, with sources of error, together with a description of a continuous silica recorder, are provided by H. F. J. Scrase, *Chem. and Ind.* (*Lond.*) 1956, p. 919.

There is still doubt regarding the factors which decide between the formation of scale and of sludge. It was once considered that any compound of which the solubility declines with rising temperature is deposited on the hot walls, producing scale, whereas a compound having a solubility rising

* E. C. Potter makes some pertinent comments about the difficulties of estimating very low concentrations of dissolved gases and metals (e.g. at parts per 10^9 or lower), stating that he knows of no other field where concentration as low as 1 part in 10^{10} can be relevant.

with temperature is thrown down as sludge. However, exceptions to this rule are known, and the decision between scale and sludge probably depends on the relative ease of nucleation at different places. Certain substances, if present in a state of super-saturation, most readily develop nuclei on the walls of the containing vessel, and thus form adherent scales—possibly owing to some simple relationship between the inter-atomic distances characteristic of the oxide covering the metallic surface and of the substance to be deposited. In other cases, nuclei first appear in the liquid at points situated at an appreciable distance from the metal and a sludge is formed. In the case of calcium bicarbonate, decomposition sometimes seems to occur along the surfaces of steam bubbles into which the carbon dioxide can pass; thus the calcium carbonate forms microscopic particles clinging to the bubbles and rises to the surface with them, forming a scum.

The removal of oil from feed-water is most important, as it interferes with heat transfer, especially when present in conjunction with silicate. The minute oil droplets usually carry negative charges and the presence of polyvalent cations tends to discharge them, permitting coalescence to bigger globules which rise more quickly. Aluminium salts are often used for the removal of oil, but the process probably depends in part on entrapment of the oil in a floc of aluminium hydroxide. One method is to add sodium aluminate, and then alum or crude aluminium sulphate; the aluminate anions and aluminium cations interact to precipitate aluminium hydroxide which removes oil, along with suspended solid particles and colloidal matter (T. Fordyce, *Proc. Instn. mech. Engrs.* 1935, **129,** 40; G. McNeill, *ibid.* 1943, **150,** 107; W. Gregson, *ibid.* 1943, **150,** 95. The determination of oil in water is described by G. P. Pringle, *J. Soc. chem. Ind.* 1941, **60,** 173).

Feed-water Purification or Softening. Distillation is generally regarded as one of the most reliable methods of obtaining a pure water. On ships it is generally used as the sole method of purification, but on land it has largely been displaced by demineralization—or the two are being used together. Distillation is important in cases where the raw water-supply is very hard or impure, or where the amount of make-up is large; that will occur if much of the steam generated is being used for process purposes, or for heating under conditions where there is no recovery of the condensed water, so that the amount of water extracted from the main condensers attached to the boilers is far less than the water evaporated to produce the steam. In a typical power-station, about 99% of the water evaporated may return as condensate, and the treatment of the 1% of make-up water to render it suitable for boiler purposes can generally be carried out by demineralization.

Soda-lime softening by addition of sodium carbonate and hydroxide, precipitating calcium as carbonate and magnesium as hydroxide, usually leaves the water alkaline if the reagents are added in excess; since an alkaline reaction is usually desired (p. 451), that may be no disadvantage. Magnesium hydroxide, however, tends to remain suspended as positively charged colloidal particles, and sodium aluminate is often employed to flocculate it; its hydrolysis produces colloidal particles of aluminium hydroxide carrying negative charges, and the two oppositely charged sets of particles then precipitate one another. The subject is discussed by C. W. Bunn and L. M. Clark, *J. Soc. chem. Ind.* 1938, **57**, 399; L. M. Clark and P. Hamer, *ibid.* 1935, **54**, 25T; R. B. Beal and S. Stevens, *ibid.* 1931, **50**, 307T.

Ion-exchange softening methods are being increasingly employed. In the original form, the hard water was passed through a bed of zeolite, in which the calcium salts were replaced by sodium salts; a water originally containing calcium bicarbonate emerged with sodium bicarbonate. Sodium bicarbonate (which is corrosive at low concentrations) loses carbon dioxide when heated, producing sodium carbonate and then (at temperatures prevailing in a modern boiler) sodium hydroxide; the latter is useful as an inhibitor up to a certain point, but may produce caustic cracking if it becomes concentrated (pp. 453–461). When a zeolite bed becomes exhausted of sodium it can be regenerated by running through it a solution of sodium chloride, which brings about the replacement of calcium by sodium.

In the modern form of ion-exchange (*demineralization*), resins are used —a procedure which makes it possible to remove all the objectionable ions. Synthetic resins containing acidic groupings can replace Ca^{++} or Mg^{++} cations by H^+ ions; a resin containing alkaline groups can replace SO_4^{--}, Cl^- and other anions by OH^-. A water which has passed in turn through columns containing both acid and alkaline resins, should emerge relatively free from all foreign ions; when exhausted, the two resins can be re-activated by washing with acid and alkali respectively, so as to re-introduce the H^+ and OH^- ions into the resins. In one variant of the process, the feed-water is run through a column consisting of a mixture of particles of the two resins; when the time comes for re-activation, the two resins must be separated, by taking advantage of their different gravities (T. R. E. Kressman, *Research*, 1952, **5**, 212; *Chem. and Ind.* (*Lond.*) 1957, p. 1473; S. R. M. Ellis, *Chem. and Process Eng.* 1955, **36**, 79; see also E. C. Potter and J. F. Moresby, " Ion Exchange and its industrial application " 1955 (London; Soc. Chem. Ind.)).

Water deprived of all ions other than H^+ or OH^- is termed " deionized " or " demineralized " water. One school of thought regards it as ideal for boiler purposes, but there are others who regard it as more corrosive than water containing traces of ions; the argument—which the Author is inclined to question—is that pure water is " hungry water ", which " having been robbed of most of its ions is very anxious to acquire more " (W. Z. Friend, *Corrosion* 1957, **13**, 81t, esp. p. 84t).

An interesting electrical method of removing salts from water depends on passage through the central compartment of a three-compartment cell, of which the anode and cathode are placed in the two outer compartments. The two diaphragms consist of membranes permeable to anions and cations respectively, so that all the current passing through the first is carried by anions and all that passing through the second by cations; thus both ions are removed from the water in the central compartment. Doubts have been expressed regarding the suitability of the method for boiler-water treatment with the membranes available at present, but the principle would seem capable of development and the paper deserves study (C. H. de Whalley, *Chem. and Ind.* (*Lond.*) 1958, p. 8. Cf. C. B. Amplett, *Metallurg. Reviews* 1956, **1**, 419, esp. p. 466).

One of the advantages of dimineralizated water is its remarkable freedom from silica; sometimes the silica-content is below the limit of analytical detection. This is also true of water prepared by distillation since silica is not volatile in steam at 100°C. Silica, however, can be introduced from the air, and from men's boots, when a boiler is opened up—also in a new boiler from concrete brick.

Apart from the oxygen-content, which requires special determination (p. 449), the quality of condensate is largely assessed on its electrical conductivity. The Central Electricity Generating Board specify an upper limit of 0·3 micro-ohms/cm.^{2} at 20°C. (in absence of ammonia or volatile bases) for plants operating at 1500 and 2350 p.s.i.

Boiler-water Conditioning. The processes already mentioned may remove nearly all dissolved substances from water, but small traces will be found even in the most carefully prepared make-up water.* When most of the water in a boiler has been turned into steam, the remaining liquid may ultimately become super-saturated with some substance; deposition of solid usually first occurs below the flanks of steam bubbles forming on the metal walls. It is important to know whether this will be a substance which tends to cling to the metal (leading to a scale) or to the bubbles (leading to a scum, and ultimately a sludge) or to neither (leading directly to a sludge).† It was the American, Hall, who first insisted that the Solubility Product Principle (p. 1011) is capable in favourable circumstances of predicting whether scale or sludge will be produced; he uses it to decide how much phosphate or carbonate should be added to a water containing, say, calcium sulphate to ensure that evaporation will render it super-saturated with respect to calcium phosphate or carbonate (sludge-formers) before it becomes super-saturated with respect to calcium sulphate (a scale-former). Given a knowledge of the solubility products of calcium phosphate, carbonate and sulphate and other necessary factors, the calculation is relatively simple.‡

If phosphate is chosen as conditioning agent, the cheapest form is common sodium phosphate ($Na_2HPO_4.12H_2O$), provided that the calcium

* E. C. Potter adds that the concentrations involved are around 1 part in 10^8.

† E. C. Potter points out that the rates of re-dissolution of solids deposited by evaporation constitute an important factor; in general these are much slower than the rate of deposition. He suggests that the phenomenon of " hide-out " is partly connected with this fact.

‡ The Solubility Products of $CaCO_3$ and $CaSO_4$ at the temperature of a 150 lb./in.^{2} boiler have been given as 0·085 and 0·7 respectively (R. E. Hall, G. W. Smith and H. A. Jackson, Report on " Treatment of Feed Water ", National Electric Light Association, 1923–24, p. 11). In simple theory, the ratio CO_3^{--}/SO_4^{--} expressed in equivalents should exceed 0·085/0·7 or 0·12 to ensure a sludge-former being first thrown down. Owing to considerations of nucleation, presence of magnesium salts, activity factors and other complications, these ratios should be exceeded by an ample margin.

content of the water is sufficiently low to ensure that no calcium phosphate will be deposited in the feed-pipes. If there is danger of this, a condensed phosphate, such as Calgon $(NaPO_3)_x$, can be used (p. 166); a water containing calcium gives no precipitate with Calgon, so long as only metaphosphate is present; thus the feed-pipes remain clean. When the water reaches the boiler, the higher temperature allows the combination of metaphosphate with water to form orthophosphate

$$NaPO_3 + H_2O = NaH_2PO_4$$

and calcium is precipitated as a sludge. It will be noted that the *acid* orthophosphate is formed—a fact which must be borne in mind when calculating the amount of alkali needed for addition to the boiler-water.

However, there is another school of thought which demands that, instead of adding solid chemicals to the water, which must leave some sort of solid residue, the water should be made as nearly as possible free from solids, and that the only chemicals added shall be volatile bodies; thus the alkaline reaction should be obtained by means of ammonia (or perhaps an amine) rather than sodium hydroxide or phosphate, whilst hydrazine should be used as oxygen-scavenger instead of sodium sulphite. Some go even further, and insist that the water shall be ultra-pure and that no chemicals shall be added to it at all. Such policies have been adopted in Germany for boilers of the Benson type. The matter is considered later (p. 464).*

Oxygen in Boiler-water. It has long been known to engineers that, if boiler-water contains oxygen, it is liable to produce pitting; also that alkali to some extent counteracts the bad effect of oxygen. It is now known that, at least at the high temperatures reached in power boilers, serious corrosion trouble can occur even if the oxygen is reduced to negligible amounts, but this should not blind us to the importance of oxygen in many types of boiler plant. It is well to start by asking why oxygen produces pitting.

Some laboratory experiments carried out by Gould are helpful in explaining the situation; they were carried out at about 100°C., but some of the principles established probably apply to temperatures prevailing in working boilers, although undoubtedly around 300°C. other phenomena may come in, which cannot be detected at 100°C. The experiments were carried out in glass apparatus, but it is unlikely that the alkali or silica pick-up from the glass affected the conclusions (A. J. Gould and U. R. Evans, *J. Iron St. Inst.* 1947, **155,** 195; U. R. Evans, *Engineering* 1953, **175,** 602; A. Thiel and H. Luckmann, *Korrosion u. Metallsch.* 1928, **4,** 169).

Gould studied the behaviour of steel partly immersed in boiling water over periods up to 76 days, under two conditions: (1) in sealed tubes, with nothing except steam above the water-line, oxygen being excluded (2) in open tubes, with air above the water (fig. 86). When the liquid used was

* E. C. Potter after stating that he knows of only *two* adherents of the " Pure Water School " refers to one undeniable drawback; " if there is a mechanical failure leading to leakage, then it is always aggressive contamination which enters.

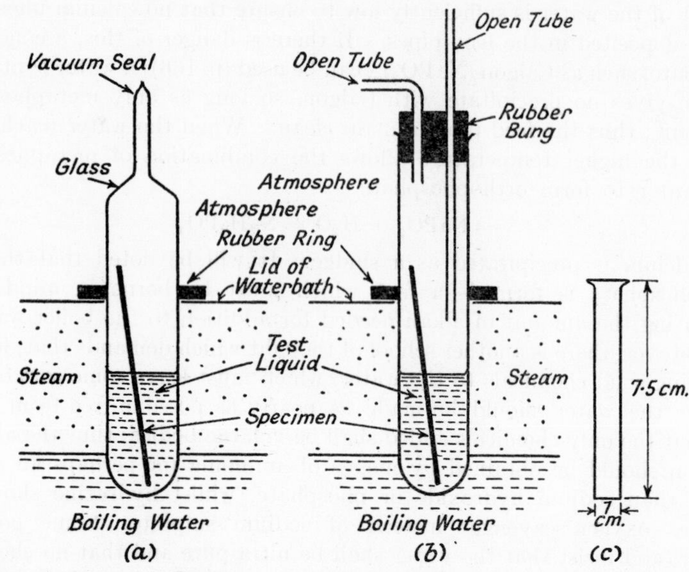

Fig. 86.—Apparatus for studying the behaviour of Steel in boiling water under (*a*) anaerobic, (*b*) aerobic conditions, using specimen of form shown in (*c*) (A. J. Gould and U. R. Evans).

distilled water, the sealed-tube experiments produced a little hydrogen at first, but its evolution soon became slow, and the metallic surface became covered with a visible scale (fig. 87(*a*)). After the experiment the scale

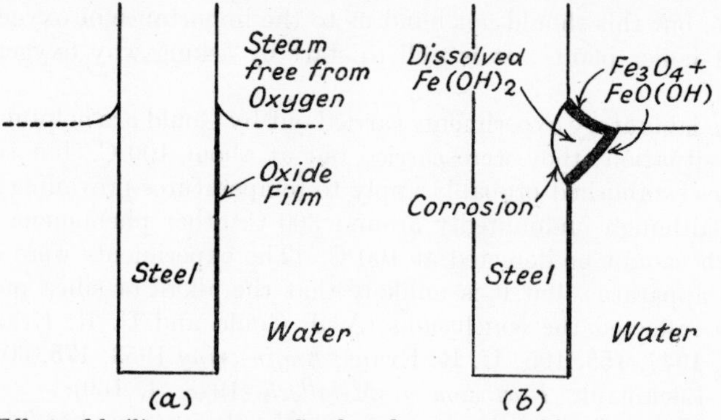

Fig. 87.—Effect of boiling water on Steel under (*a*) anaerobic, (*b*) aerobic conditions (A. J. Gould and U. R. Evans).

was transferred to a transparent support by a method similar to that of p. 785, and was identified as magnetite.* It is reasonable to believe that

* It is unlikely that the composition of the scale was altered during transfer, although this possibility cannot be excluded. The much thicker scale produced in working boilers is generally regarded as largely magnetite, although a more accurate study of its composition would be desirable.

it was the formation of the magnetite film which slowed down both the attack upon the metal and the evolution of gaseous hydrogen. The manner in which the rate of hydrogen evolution falls off as the scale develops is shown in the curves of fig. 88, due to Thiel and Luckmann.

A similar falling off of hydrogen-production has been noted at higher temperature both in the laboratory experiments of Fellows and in the Ulrich's measurements of hydrogen in the steam generated by working boilers; Ulrich's analysis of Fellows' results and his own accord with the idea of parabolic film-thickening in most cases, although the value of the constant k in the equation $w^2 = kt$ departs in some circumstances from that predicted from theory. These papers deserve study (C. H. Fellows, *J. Amer. Water Wks. Ass.* 1929, **21**, 1373; E. Ulrich, article in " Passivierende Filme

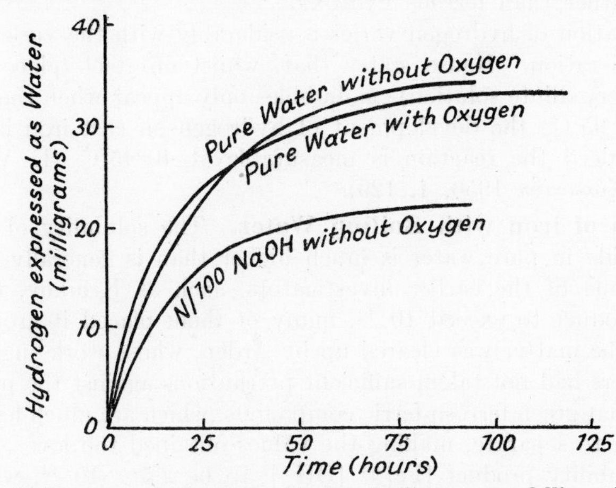

FIG. 88.—Hydrogen evolution from Steel in boiling water, falling off with time (A. Thiel and H. Luckmann).

und Deckschichten " 1955–56, p. 308, esp. pp. 323–327 (editors, H. Fischer, K. Hauffe and W. Wiederholt: publisher, Springer)).*

In his experiments in tubes which were open to the air, Gould found a large amount of magnetite and rust, formed especially at the water-line; the products were either loose or attached to the metal at certain points only (fig. 87(*b*)). At the water-line the product tended to be formed along the air–water interface and clung to it; when the solids were removed, the metal was found to have suffered serious corrosion, leaving a groove parallel to the water-line. The observations all suggest that the metal passed into the solution as Fe^{++} ions; ferrous hydroxide, which, as shown below, is appreciably soluble in water, presumably diffused out from the metal, and

* E. C. Potter sounds a word of warning against the adoption of hydrogen measurements as a measure of boiler-corrosion; some of the hydrogen may circulate round the system and be counted over and over again, whilst other hydrogen may pass right through the steel tubes into the furnace and be lost to measurement. In any case, it is not the total corrosion which concerns us but its distribution.

at the air–water interface was converted by oxygen to the less soluble compounds, magnetite and rust, which, being formed (at least along some parts of the water-line) at a distance from the metal, provided no protection; at the same time, the transformation of ferrous hydroxide to the less soluble bodies prevented the water from becoming super-saturated with ferrous oxide or hydroxide; thus the action continued indefinitely, instead of being stifled, as happened in the tubes from which air was excluded.

The addition of alkali in the presence of oxygen reduces the solubility of all the hydroxides and oxides of iron, owing to the " Common Ion Effect " (p. 1011). Thus—given the right amount of alkali—very little ferrous hydroxide can exist in the liquid, and precipitation at a distance becomes impossible. For reasons suggested later, the film-substance is likely to be magnetite rather than ferrous hydroxide.

The liberation of hydrogen varies considerably with the variety of iron under consideration. Haase states that, whilst on steel (placed in pure water or a very dilute solution) gas-bubbles only appear when the temperature reaches 95°C., the development of hydrogen on cast iron is rapid at 60–80°C.; indeed the reaction is measurable at 40–45°C. (L. W. Haase, *Werkst. u. Korrosion* 1950, **1,** 129).

Reaction of iron with Boiling Water. The solubility of pure ferrous hydroxide in pure water is much higher than is generally supposed. Although some of the earlier investigators, such as Kriukov, found the solubility product to exceed 10^{-14}, many of them placed it around 10^{-15} or lower. The matter was cleared up by Arden, whose work suggests that earlier workers had not taken sufficient precautions against the presence of oxygen, so that green ferroso-ferric compounds, which are much less soluble, were tending to separate, making the values obtained too low. He found the true solubility product $[Fe]^{++}$ $[OH^-]^2$ to be $2{\cdot}5 \times 10^{-14}$, whilst that of the green compound was $6{\cdot}4 \times 10^{-18}$ (T. V. Arden, *J. chem. Soc.* 1950, p. 882. P. A. Kriukov and G. P. Awsejewitsch, *Z. Elektrochem.* 1933, **39,** 884. L. H. N. Cooper, *Proc. roy. Soc.* (*B*) 1935, **118,** 419; 1937, **124,** 299. R. Fricke and S. Rihl, *Z. anorg. Chem.* 1943, **251,** 414. D. L. Leussing and I. M. Kolthoff, *J. Amer. chem. Soc.* 1953, **75,** 247).

It follows from Arden's numbers, that if water is kept free from oxygen and is maintained at the neutral point ($p_{OH} = p_H = 7$) by buffering salts, the saturated solution will contain $\dfrac{2{\cdot}4 \times 10^{-14}}{(10^{-7})^2} = 2{\cdot}4$M of ferrous ions, whilst in acid water the solubility will be higher still. Conversely, addition of alkali by raising the value of $[OH^-]$ will depress the equilibrium concentration of $[Fe^{++}]$; in presence of sodium hydroxide, ferrous hydroxide can be regarded for many purposes as a sparingly soluble compound.*

* At a certain temperature, apparently not known, ferrous hydroxide would lose water, giving ferrous oxide; below the temperature in question anhydrous ferrous oxide will have a higher solubility than ferrous hydroxide. However, ferrous oxide is metastable and should decompose to iron and magnetite at temperatures below 570°C., although at temperatures met with in most boilers the reaction is likely to be slow (see p. 26).

The solubilities of magnetite, ferric oxide and ferric hydroxide are much lower than that of ferrous hydroxide.* Thus even if a very small concentration of ferric ions is introduced into water containing ferrous ions but unsaturated with respect to solid ferrous hydroxide, it will immediately become highly super-saturated with respect to magnetite or hydrated magnetite; this explains why early determinations of the solubility of ferrous hydroxide invariably gave values far too low, since the precautions for excluding traces of oxygen were inadequate.

It is now easy to understand why the presence of oxygen in boiler-water, unless uniformly distributed, is likely to give rise to pitting, since on any areas where oxygen fails to reach the metallic surface, soluble ferrous hydroxide, formed at the surface, will be converted to loose magnetite (or perhaps, in extreme cases, to rust) at a distance from the site of attack; on such areas, attack will continue indefinitely.

Evidence of a spontaneous change from the ferrous to ferric state is provided by the work of Schikorr, who showed that ferrous hydroxide prepared by precipitation of ferrous salts with alkali sometimes decomposes into magnetite and hydrogen. Later, Wanklyn found that the decomposition does not occur with pure ferrous hydroxide, which remains unchanged in contact with water even at 100°C.; hydrogen was evolved at room temperature, however, in presence of excess ferrous sulphate if certain additions were made to the solution before precipitation with alkali; these additions included platinum chloride, colloidal platinum, nickel sulphate, nickel powder and copper powder. Wanklyn's results have been generally confirmed and extended by Shipko and Douglas, who find that above 178°C. the decomposition of ferrous hydroxide also produces metallic iron; if we consider that the first effect of heating ferrous hydroxide is a loss of water, the facts can easily be understood, since at any temperature below 570° ferrous oxide should decompose at once to iron and magnetite (p. 24). For studies of the behaviour of ferrous hydroxide, see G. Schikorr, *Z. Elektrochem.* 1929, **35,** 65; U. R. Evans and J. N. Wanklyn, *Nature* 1948, **162,** 27; F. J. Shipko and D. L. Douglas, *J. phys. Chem.* 1956, **60,** 1519.

Wanklyn's observation that the decomposition of ferrous hydroxide took place readily in presence of a metallic phase, such as platinum or copper, suggests that the transformation is electrochemical. Presumably an anodic reaction

$$2Fe^{++} = 2Fe^{+++} + 2e$$

is balanced by its cathodic counterpart

$$2H^{+} + 2e = H_2$$

—the latter doubtless proceeding in steps.

If this view of the Schikorr reaction is accepted, the production of a

* W. Feitknecht, *Chimia* 1952, **6,** 3, esp. p. 9 gives the solubility products of Fe_2O_3 and FeOOH as 10^{-42} and 10^{-39} respectively. The value for magnetite must be extremely low since gelatinous precipitates of ferrous and ferric hydroxides when mixed at once give a black product which after drying is found to be magnetic (H. O. Forrest, B. E. Roetheli and R. H. Brown, *Industr. engng. Chem.* 1931, **23,** 650, esp. p. 651).

magnetite film on a steel surface exposed to hot water is represented by a combination of the simple changes:—

Anodic	Fe	$= Fe^{++} + 2e$	(1)
followed to a small extent by	Fe^{++}	$= Fe^{+++} + e$	(2)
Cathodic	$H^{+} + e$	$= H$	(3)
followed by	$H + H$	$= H_2$	(4*a*)
or by	$H + H^{+} + e$	$= H_2$	(4b)

Owing to the very low solubility of magnetite, it should become the stable solid phase as soon as the Fe^{+++} concentration exceeds a certain very low value; thus the films formed on iron are magnetite, rather than ferrous oxide or hydroxide, which are relatively soluble. Moreover, since reaction (2) requires a metallic surface, it is to be expected that the magnetite will appear as a film on close union with the metal, and thus should be protective, provided that no oxygen is present in the water. In presence of oxygen, however, the dissolved ferrous hydroxide may be converted to magnetite at a distance from the surface, where it cannot be protective, so that attack continues.

The work of Linnenbom on the reaction between pure iron and water in absence of oxygen should here be noted, although the iron used by him was either in powder form or was foil which had been subjected to alternate oxidation and reduction so as to reach a surface-rich state. At room temperature, the surface of the iron remained " bright and silvery in appearance even after 42 days contact with water. The solution after only a few hours showed a strong Tyndall effect." Evidently colloidal matter was being formed, and after four days the pH was 9·3 and the content of soluble iron 0·4 ppm., gradually diminishing to 0·08 ppm. after about 30 days. Linnenbom considered that the reaction product at room temperature was ferrous hydroxide. Visible black magnetite was only observed in a few cases where air had leaked in, or after oxygen had intentionally been admitted. At 60°C. the formation of black magnetite proceeded with fair rapidity even in absence of oxygen, and the iron content was only about 0·008 ppm. after 10 days—suggesting that a less soluble product was being formed. Experiments in an autoclave at 300°C. gave much black magnetite; the iron-content was higher, although it diminished with time, tending to approach 0·05 ppm. as a lower limit.

These experiments would suggest that on pure iron in oxygen-free water reaction (2) is not important at room temperature, and that the essential product is a colloidal solution of ferrous hydroxide, whereas at 60°C. or higher temperatures reaction (2) proceeds with sufficient rapidity to lead to magnetite as the main product, with consequent diminution in the soluble iron-content. They do not exclude the possibility of an invisible magnetite film at room temperature, but such a question is not important in considering boiler reactions. Linnenbom expresses his conclusions somewhat differently, but the reader may feel that there is no fundamental difference of opinion; the paper deserves study (V. J. Linnenbom, *J. electrochem. Soc.* 1958, **105,** 322).

Returning to reactions at higher temperatures, we must admit the possibility that at first the cathodic and anodic reactions may start at distinct points situated on the metallic surface; but, as soon as a uniform oxide-film has been produced over the whole area, reaction (1) must occur at the inner surface of that film, the ferrous ions moving outwards across the film, as in high temperature oxidation; probably reaction (2) then occurs at the outer (scale–water) interface. Reactions (3) and (4) also occur at the outer surface; thus hydrogen is produced and the film thickens. This is essentially the mechanism demonstrated by Wagner and others (Chapter II) for the oxidation of metals heated in dry air. However the oxygen ions needed to produce, with the metallic ions arriving through the film, fresh layers of oxide on the outer face of the film, are supplied, not from the oxygen molecules of the air, but from adsorbed water molecules which have lost their hydrogen in the cathodic reaction.

Effect of Scale Breakage. In boilers operating at low temperatures, the attack on the steel becomes slow when the scale is fairly thin. At higher temperatures, however, the scale soon becomes thick, and there is danger that it will become broken by stresses set up by differences in thermal expansion if the working temperature fluctuates rapidly. Small breaks in the scale will produce the usual dangerous combination of large cathode and small anode, leading to intense localized attack. Clearly care must be taken not to cool down boilers too rapidly from a high temperature, since if once the scale becomes cracked, intense attack may set in.* Even if a boiler could work at constant temperature, abrasion due to impinging particles (or perhaps erosion by a high-velocity water-stream) would probably damage the scale locally, with disastrous results (see A. A. Berk, *Trans. Amer. Soc. mech. Engrs.* 1955, **77**, 441; cf. S. T. Powell and L. G. von Lossberg, *Corrosion* 1949, **5**, 71).

Hydrogen degeneration of the steel. It was explained on p. 19, that, during oxidation in air at relatively low temperatures, film-growth is controlled by the (physical) movement of ions across the film, so that film-growth, rapid at first, becomes slow as the film becomes thick. In boiling water also, the evolution of hydrogen is likely to become slow as the film thickens (fig. 88, p. 437). However, in dry oxidation at still higher temperatures, it sometimes happens that one of the boundary reactions cannot keep pace with the maximum possible movement through the film, and in such cases the film continues to thicken with constant velocity. The same control by a boundary reaction might occur at very high temperatures in a boiler. Suppose, for instance, reaction (4*a*) or (4*b*) were to become the controlling factor, then atomic hydrogen will accumulate, and the rate of attack will not fall off in any marked degree as the thickness of the oxide-scale increases. This may lead to a new type of trouble. Suppose that

* E. C. Potter adds the useful comment that " The important point about cooling down boilers (even temporarily) is that parts of the system normally under pressure may come under vacuum. These parts are unlikely to be vacuum-tight, and atmospheric oxygen and carbon dioxide are sucked in to the deteriment of the boiler surfaces."

the atomic hydrogen is unable to escape with the steam as molecular hydrogen with sufficient rapidity, it will build up a concentration gradient, and diffuse into the metal, producing high-pressure molecular hydrogen in internal cavities; under such circumstances embrittlement or cracking may be expected. Potter reports that in some British high-pressure boilers the steel is altered and embrittled far below the surface, the pearlite being converted to ferrite (iron), apparently through removal of carbon by hydrogen. He thinks that the transformation of carbide to ferrite is by itself sufficient to account for the observed deterioration of mechanical properties (E. C. Potter, *Research* 1955, **8,** 450; *Centre Belge. d'Étude et de Documentation des Eaux,* 1956, Bull. No. **32,** p. 73; also Priv. Comm., Feb. 3, 1956).

Calculations from densities support Potter's idea that decarburization would itself cause the deterioration without assuming any disruption by high-pressure gas. When the hydrogen converts iron carbide to iron, the mass is likely to become porous, since (assuming the densities of Fe_3C and Fe to be 7·40 and 7·86 respectively) a gram-molecule of Fe_3C should occupy about 24 c.c., whilst the equivalent volume of 3Fe is only about 21 c.c.* If water has access to the pores left where pearlite originally existed, it can almost certainly transform the remaining iron to magnetite, as the individual grains of iron will be too small to protect themselves effectively with oxide sheaths. Thus the whole decarburized zone is likely to become converted to oxide, provided that the original pearlite regions are continuous or sufficiently close together. It would seem that the extent of the oxidation of the decarburized layer may depend on the metallographic structure of the original steel.

The idea that hydrogen is the main cause of cracking in boilers has long been held by Zappfe; but the mechanism put forward by him is rather different, since a gas pressure is supposed to be built up until the cohesive strength of the material is passed and disruption occurs. Arguing against that view, Schroeder and Berk produced pictures of boiler cracks which showed no signs of decarburization. Such evidence, however, may not in itself be fatal to the belief in the responsibility of high gas pressures, since the cracking might be caused by molecular hydrogen; indeed a given quantity of atomic hydrogen will generate twice as many molecules of H_2 as of CH_4, and thus generate double the pressure. Against this must be set the fact that H_2 might escape under circumstances where CH_4 would fail to do so.† Probably, however, Schroeder and Berk are right in thinking that hydrogen cracking is only important at high temperatures, where various experimenters, including Potter (above), have produced convincing evidence of decarburization. The reader may care to compare the two opposing views

* These numbers are based on densities at room temperature; the basis for a calculation at the operating temperature is not available, but the same conclusion would probably be reached.

† C. H. Desch, quoted by E. C. Potter, recovered CH_4 from steel which had undergone caustic cracking by drilling it under mercury; similar steel was "de-embrittled" by temporary heating to about 900°C.

(C. A. Zappfe, *Trans. Amer. Soc. mech. Engrs.* 1944, **66,** 81; W. C. Schroeder and A. A. Berk, *ibid.* 1944, **66,** 117).

Apparently, this type of deterioration rarely occurs at the power stations situated in certain countries of continental Europe. It occurs rather frequently in Great Britain, and to a considerable extent in the U.S.A. and Russia. It is possible that in British stations some minor impurity acts as a poison, preventing equation (4*a*) or (4*b*) which would allow the hydrogen to escape harmlessly with the steam; if so, an accumulation of atomic hydrogen might be built up, with consequent embrittlement. The poison might be arsenic from the condenser tube alloys often used in British condensers, or hydrogen sulphide formed from particles of metallic sulphides existing in the steel at the surface of the tubes. There is at present no definite information about this matter, but an investigation into such possibilities is in progress.

So long as the factor controlling the thickening of the scale is the passage of ions through the scale, variations in composition of the steel between one part of the surface and another should not affect the rate of oxidation, and the scale should be of nearly uniform thickness, the steel being attacked evenly, as is found to be the case in low-pressure boilers. So soon as boundary reactions start to exercise control, the factors favouring uniform scale thickness cease to operate, and in fact the boundary below the thick scale formed at high temperatures tend to be irregular, producing numerous saucer-shaped cavities in the metal.

It is pertinent to ask whether solid matter deposited from the water might assist hydrogen-formation, since such deposits may obstruct heat transfer, causing the temperature to become abnormally high; different opinions are held about this. Deposits may also act in a different way, since, if macroscopically porous, they permit the concentration of certain dissolved compounds to become concentrated at places screened by the deposit. Iron at high temperatures, like zinc at ordinary temperatures, can liberate hydrogen readily both at low and high pH values, although there is an intermediate range where the attack is slight. If the water contains rather too much alkali, the alkali concentration may reach high levels below the deposit with liberation of hydrogen; embrittlement may then be produced.* Very much the same effect, however, can be produced if the water contains too little alkali. In one case reported from America, tubes broke in a position near the drum where they were only slightly inclined to the horizontal, and examination showed removal of the carbide from the steel, along with intergranular cracking—phenomena described as " characteristic of other cases where hydrogen embrittlement has been found ". In this particular instance, the alkali-content of the water was too low. Arrangements were made to add alkali and also sulphite, and during the next three years there was no further trouble. It is unwise, of course, to generalize from a single instance, but the paper contains

* E. C. Potter considers that there may be a delicate balance depending upon temperature between the rate of hydrogen-diffusion through steel and the rate of reaction of hydrogen with Fe_3C.

information deserving study (C. E. Kaufman, W. H. Trautman and W. R. Schnarrenberger, *Trans. Amer. Soc. mech. Engrs.* 1955, **77**, 423, esp. p. 427).

Where the water is too alkaline, excessive concentrations of alkali may be built up below the deposits. This is an example of the phenomena known as *alkali hide-out*, which apparently sometimes occurs in the absence of deposits; it can lead to hydrogen deterioration of the type described. Alkaline hide-out is related to the caustic cracking met with in riveted boilers and described later, but in that case the concentration occurs in crevices between plates or around rivets, and not in the channels below deposits. Some of the sodium compounds intentionally added for water-conditioning or oxygen-removal (sodium phosphate, sulphate or sulphite) are very liable to cause hide-out; Leicester considers the deposition of sodium ferrous phosphate to be specially dangerous. Hall has recommended the use of potassium salts, which are more soluble than sodium salts, in their place. In softening water by zeolite treatment, it would be possible to use potassium chloride instead of sodium chloride to regenerate the mass (R. E. Hall, *Trans. Amer. Soc. mech. Engrs.* 1944, **66**, 457; J. Leicester, *Trans. Inst. mar. Engrs.* 1957, **69**, 129).

Use of Volatile Alkalis. Clearly alkalinity maintained by means of sodium (or potassium) hydroxide is only effective on the parts of the surface reached by the boiler-water. To prevent corrosion at other parts of the circuit, a volatile alkali is needed and ammonia has been used—mainly in Germany. The conditions of equilibrium between ammonia and carbon dioxide are naturally important, and information, useful to those contemplating the adoption of this method, will be found in the charts and tables provided by C. D. Weir, *J. appl. Chem.* 1957, **7**, 505.

Various organic bases are now much used for the same purpose. These are discussed on p. 182 (E. W. F. Gillham, *Trans. Inst. mar. Engrs.* 1957, **69**, 140).

Copper in Boiler-waters. As to whether the presence of copper in boiler-water increases the corrosion of the tube is a question on which different opinions are held. Copper is found in most boiler-water owing to slow attack in the condenser or at other points in the external circuit where copper alloys are used. If the water is sufficiently alkaline and otherwise suitable, copper should be expelled with the sludge as copper oxide or a basic salt. If the water becomes polluted with ammonia or amines derived from sewage, the attack on the copper alloys may be increased, assuming oxygen to be present, and there is a risk that copper will be deposited on the tubes as metal; there may be other factors—besides a high ammonia-content—which would lead to the deposition of metallic copper. In view of the fact that metallic copper is known to be a catalyst for the Schikorr reaction, it becomes at least a theoretical possibility that magnetite will be built up to greater thicknesses at the places where the metallic copper exists. This might mean that at other places the magnetite film will remain relatively thin. If we consider the location of the equations presented on p. 440 on a surface carrying copper in some parts and no copper on others, then equation (1) (formation of ferrous cations) can proceed unhindered on the copper-free area, whilst the changes (2), (3) and (4) leading to magnetite will proceed on the copper, ultimately burying it and stopping the action until fresh metallic copper is deposited; on the copper-free area the attack might continue unhindered (the formation of magnetite being relatively slight) and this has been considered as a possible explanation of pitting—although it is by no means certain that it is the correct one (U. R. Evans, *Engineering* 1953, **175,** 602).

During the war, partly owing to enemy action, the water used at certain London power-stations became polluted, its ammonia-content reaching abnormal levels; metallic copper often appeared in the magnetite scale formed in the tubes, especially near the expanded ends. The thickness of the magnetite formed was often phenomenal; barnacle-like growths sometimes reached $1\frac{1}{2}$ inch diameter; these possessed a laminated structure and sections often revealed metallic copper between the laminations. A similar barnacle-type of oxidation has been met with in an American plant (R. Ll. Rees and E. A. Howes, *Combustion* Feb. 1949, **20,** p. 49; L. E. Hankison and M. D. Baker, *Trans. Amer. Soc. mech. Engrs.* 1947, **69,** 479).

It seemed natural at that time to attribute the oxidation to the presence of metallic copper, and if this is the case, the explanation suggested above seems the most reasonable one. Many authorities, however, think that the copper was not the cause of the unusual oxidation, although probably in the case quoted above, the presence of copper and the abnormal corrosion of the boiler had a common cause—namely the pollution of the water. Gillham has stated that copper has been detected in two-thirds of the boilers in Great Britain by the methods available to the power-station chemist (possibly more sensitive methods would reveal its presence in the others); about half have suffered corrosion and the others not. The numbers, which refer only to boilers working above 350 lb./in.^{2}, are shown in tabular form below:—

	Copper present	No copper found or seen
Corrosion found	133	24
No corrosion found	292	185

(E. W. F. Gillham, *Trans. Inst. mar. Engrs.* 1957, **69,** 140; E. C. Potter, Priv. Comm., July 31, 1958).

Copper can be determined in boiler-water by means of sodium di-ethyl-thiocarbamate. In many power-stations, copper at concentrations below 1 part in 2×10^8 is determined by using ion-exchangers as concentrators (E. C. Potter and J. F. Moresby " Ion Exchange and its Industrial Application " 1955, p. 92 (published by *Soc. chem. Ind.*). For other methods see D. Parkhouse, *Chem. and Ind.* (*Lond.*) 1957, p. 224).

The question of using steel containing copper in boiler tubes is a different one. Careful experiments at Teddington on a model boiler of the Scotch marine type suggest that it reduces pitting in untreated brackish water, but confers no advantage in similar water treated with alkali, where the pitting is slight in any case (G. Butler and H. C. K. Ison, *Trans. Inst. mar.*

Engrs. 1957, **69**, 121; " Chemistry Research " 1956, p. 13 (H.M. Stationery Office)).

Débris acting as Oxygen-Screens. In old-fashioned boiler plant operating at a low pressure, the pitting produced by boiler-water containing oxygen is often related to the settling of foreign particles; these may be flakes of loosened oxide-scale or rust—perhaps even stony scale—brought from another part of the boiler or from the external feed system. It is generally considered that these act as local oxygen-screens. Some authorities go so far as to suggest that the screened area constitutes a small anode, with the area around it, where oxygen can be replenished, as large cathode; the pitting is thus attributed to differential aeration currents. Such a mechanism is possible where salts have become concentrated near the heating surface. However, where the water remains relatively pure, it is more likely that the attack below the *débris* is of the hydrogen-evolution type, the hydrogen gas being liberated on the spot; the ferrous hydroxide then moves outwards and is converted by dissolved oxygen to magnetite just outside the screened area; such action can continue indefinitely, since the liquid will be prevented from becoming super-saturated with ferrous hydroxide or oxide at the place where corrosion is occurring.

Permissible Limits for Oxygen. In boilers working at pressures too low to produce caustic cracking or decarburization, reasonable freedom from corrosion can be obtained by keeping the oxygen-content of the water low, and the water slightly alkaline. Oxygen and other gases can to a large extent be expelled thermally, for instance by spraying the pre-heated water into an evacuated chamber; carbon dioxide is largely expelled at the same time, but acidity due to mineral acids remains.

In another form of de-aerator, the water previously heated almost to its boiling-point by passage through tubes surrounded by hot vapour, enters the de-aeration chamber and is there made to flow down a rack of trays, where it is brought into direct contact with an ascending current of air-free steam, which carries off the liberated gases (H. Chambers, *Engng. and Boiler House Rev.* 1952, **67**, 222).

Without the installation of any special de-aeration plant, a considerable reduction of oxygen-content can be achieved by introducing the make-up water at the condenser where a low pressure already prevails; Potter considers that if the make-up is only about 1 to 2%, oxygen-contents can thus be reduced to 0·02 ppm.

Hamer considers that for pressures up to about 450 lb./in.^{2}, oxygen should not exceed 0·05 c.c./litre, whilst for higher pressures the maximum is 0·02 c.c./litre;* this, he thinks, could be obtained by mechanical de-aeration or by means of sodium sulphite. Johnson thinks that in a closed-feed system, with the surge tank properly blanketed with steam at a slight positive pressure, it should be possible to keep oxygen well below 0·03

* In quoting figures, it has seemed best to retain the units used in the original papers, but there is a tendency today to prefer parts per million (ppm.) to ml./l., which is dependent on pressure and temperature; 1·00 ml./l. at N.T.P. is 1·43 ppm. by weight.

c.c./litre under steady conditions; it will rise higher when conditions fluctuate.* In large power-plants, Hewson regards 0·01 c.c./litre as not difficult to attain by a well-designed closed-feed system (P. Hamer, *Chem. and Ind.* (*Lond.*) 1940, p. 64; G. W. Hewson, *ibid.* 1941, p. 764; W. J. Johnson, Corrosion Symposium (Melbourne University) 1955–56, p. 539).

Chemical De-aeration. Where thermal de-aeration fails to reduce the oxygen-concentration sufficiently, lower contents can be obtained by the addition of chemicals. At power-stations these are usually regarded as second-line defences and in some cases they could probably be dispensed with; as designs improve, this statement will become true for an increasing number of power-stations.

Of the chemicals useful in removing oxygen, sodium sulphite is still often employed, although hydrazine is needed at the higher range of pressures. Sulphite combines with oxygen, being converted to sulphate, but in some waters the reaction is slow and this should be borne in mind when deciding the plane of injection. It would seem that certain constituents present in moorland water inhibit the removal of oxygen by sulphite. In contrast, metallic ions such as cobalt and copper accelerate it; copper is often present in the water, having been taken up from condenser or other tubes; cobalt salts are sometimes added intentionally for the purpose, being more efficient than copper.†

Sulphite treatment tends to produce an acid condensate, especially in high-pressure boilers, probably because traces of sulphur dioxide are released by hydrolysis and come over with the steam. For very high-pressure boilers, sulphite is definitely unsuited and indeed can cause severe corrosion; about 250°C., it is reduced by iron or hydrogen—presumably to sulphide, a most undesirable constituent (W. C. Schroeder, A. A. Berk and E. P. Partridge, *Proc. Amer. Soc. test. Mater.* 1936, **36,** II, 721, esp. p. 746).

Hydrazine is used today at a number of high-pressure plants, including those fitted with drum boilers as well as the Benson type. A considerable excess over the amount theoretically necessary for the destruction of the oxygen present must be used. Some recommend 150–200% excess, but 50% is found sufficient at one British power-station. A divergence of opinion exists as to the general suitability of hydrazine, although it appears to have given satisfaction in certain large plants, including Leverkusen (p. 465). Much of the uncertainty is due to the fact that in the presence of hydrazine the ordinary methods for estimating oxygen fail, and an analyst may report the absence of oxygen in a water where in truth it is present. Potter has been developing methods for estimating oxygen in presence of hydrazine (see p. 451).

* E. C. Potter remarks that this depends on the design of the boiler; modern design aims at eliminating these variations as far as possible.

† A. J. Berry draws attention to the use in analysis of alcohols (primary or secondary) in preventing the deterioration of sulphite solutions due to conversion to sulphate by air; apparently the alcohols break the chain of autocatalytic reactions responsible for the oxidation; possibly moorland waters contain similar chain-breakers. Traces of copper ions favour the chain-reaction.

The combination of oxygen and hydrazine can be speeded up by passing the water, previously warmed, through a carbon catalyst bed. But the fact that small amounts of hydrazine and oxygen can co-exist in a water shows that reactions such as

$$N_2H_4 + O_2 = N_2 + 2H_2O$$

proceeds only very slowly in the body of the water. It is generally agreed that the elimination of oxygen, if it occurs, is a wall-reaction; whether it consists of the direct interaction between adsorbed oxygen and absorbed hydrazine molecules to give nitrogen and water, or whether (as Stones suggests) the oxygen converts magnetite to ferric oxide which the hydrazine reduces again to magnetite, thus destroying both oxygen and hydrazine, is a matter of opinion. Copper appears to act as a catalyst, and manganese salts certainly speed up the reaction.

The Author is inclined to attribute the benefits of hydrazine in boilers, not to the destruction of oxygen which proceeds slowly and uncertainly, but to the provision of an alternative anodic reaction; it acts, according to this view, in a manner analogous to a sacrificial anode of magnesium or zinc (p. 286). In absence of hydrazine, the cathodic reduction of oxygen

$$2H_2O + O_2 + 4e = 4OH^-$$

is balanced by the harmful anodic corrosion process ($Fe = Fe^{++} + 2e$). In presence of hydrazine it is balanced by a harmless anodic reaction such as

$$N_2H_4 + 4OH^- = N_2 + 4H_2O + 4e$$

so that the corrosion of iron does not take place. According to this view, corrosion is prevented and oxygen destroyed simultaneously since the two reactions add up to

$$N_2H_4 + O_2 = N_2 + 2H_2O$$

Such reactions would probably occur more easily on copper than on oxide-scale.

This idea—put forward tentatively—is in line with German beaker experiments in which air (freed from carbon dioxide by bubbling through alkali) was passed over an iron sheet immersed in condensate; this naturally caused corrosion and the amount of iron compounds suspended in the water steadily increased with time, reaching 24·5 mg. after 8 hours. If the water contained 0·6 g./l. of hydrazine, the iron content after 8 hours was only 0·02 mg./l.—slightly lower than in the original condensate! It is certain that the hydrazine was not removing the oxygen as it bubbled through the water; unless it is regarded as an adsorption inhibitor, it must have been acting as sacrificial anode (E. A. Woodward, *Power* (*N.Y.*) 1953, **97**, No. 11, p. 91; esp. German work quoted on pp. 93, 212).

Although the experiments quoted were conducted at room temperatures and at hydrazine concentrations vastly exceeding those used in boiler practice, they may perhaps serve as some sort of pointer to the probable function of small concentrations of hydrazine at high temperatures, and suggest that the fact that hydrazine and oxygen can co-exist for comparatively long periods does not necessarily mean that hydrazine is ineffective in stopping corrosion.

Practical experience of the use of hydrazine has varied from one power-station to another, but in most cases favourable results have been reported. The case of one British power-station, where sulphite had proved unsatisfactory and hydrazine has performed well, is described by W. F. Stones, *Chem. and Ind.* (*Lond.*) 1957, p. 120. See also M. A. Pearson, *ibid.* 1956, p. 1383. General information about British stations has been furnished by R. Ll. Rees, Priv. Comm., March 8, 1957. American experience is provided by J. Leicester, *Trans. Amer. Soc. mech. Engrs.* 1956, **78**, 273; M. D. Baker and M. Marcy, *ibid.* 1956, **78**, 299; N. L. Dickinson, D. W. Felgar and E. A. Pirsh, World Power Conference, 1958 (in the press). General information is conveniently collected in the Proceedings of the International Conference on "Hydrazine and Water Treatment" held at Bournemouth 1957, published by Whiffen, Wigmore Street, London. See also C. Moreland, Dissertation, Birmingham; S. R. M. Ellis and C. Moreland, *Chem. and Process Engng.* 1955, **36**, 79; 1956, **37**, 47. For German experience see p. 465 of this book. For other views, see F. G. Straub, *Combustion* Jan. 1957. In reading papers on hydrazine, the analytical difficulties of estimating oxygen and hydrazine in presence of one another should be borne in mind.

Hydrazine is toxic if inhaled, irritant to the eyes and injurious to the skin; precautions in its use are necessary, but for those who have sufficient knowledge of its properties, it is probably no more dangerous to handle than caustic soda or concentrated sulphuric acid. However, varying opinions are held regarding the seriousness of the health hazards involved in the use of hydrazine, and will be found collected in *Chem. and Ind.* (*Lond.*) 1956, p. 1384.

Estimation of Oxygen in Water. The analytical determination of oxygen in boiler-water cannot be discussed in detail, but some references may be useful. Most of the procedures fall into two groups. The first are based on the principle developed by Winkler, in whose classical method a manganese salt is added to the water followed by alkali to precipitate manganese hydroxide, which then takes up any free oxygen present, producing higher hydroxides; on adding potassium iodide, these liberate their equivalent of iodine, which may be estimated by titration with thiosulphate. In more recent variants of the method, the iodine is estimated photometrically, either by using the absorption of near ultra-violet light by the I_3^- ion, or the blue colour formed when starch is added; the same blue colour is often used for showing the end-point of the titration, but electrometric indication of the end-point is more reliable. Janssen and Smit point out that in methods of the Winkler type, accuracy depends on determining the end-point with a minimum of delay; they describe an electrical circuit which achieves what is needed, and show that by quick working interference can largely be avoided in cases where normally it might be expected.

The methods of the second group are essentially electrometric and depend on measuring the quantity of electricity produced when oxygen is cathodically reduced to OH^-; the principles governing this method were largely worked out in the laboratory by Tödt, who has also applied them to practical problems; in the United Kingdom interest is attached to the Hirsch cell for the same purpose.*

Papers dealing particularly with the iodine method include those of A Bairstow, J. Francis and G. H. Wyatt, *Analyst* 1947, **72,** 340 (photometric method using starch); H. Barnes, *ibid.* 1953, **78,** 501 (double pipette for addition of reagents); A. H. White, C. H. Leland and D. W. Button, *Proc. Amer. Soc. Test. Mater.* 1936, **36,** II, 697 (electrochemical end-point); J. Arnott, J. McPheat and P. B. Ling, *Engineering* 1950, **169,** 553; 1953, **176,** 103 (increased sensitivity by using carbon tetrachloride as collector); G. W. Hewson and R. Ll. Rees, *J. Soc. chem. Ind.* 1935, **54,** 254T; C. Janssen and G. B. Smit, *Analyt. Chim. Acta* 1957, **16,** 276 (" dead-stop-end-point " method).

Papers discussing the principles and practice of cathodic-reduction measurement include F. Tödt, *Z. Elektrochem.* 1950, **54,** 485; A. M. Peers, *Chem. and Ind.* (*Lond.*) 1952, p. 969; P. Hirsch, *Nature* 1952, **169,** 792; A. G. Downson and I. J. Buckland, *ibid.* 1956, **177,** 712; also P. Hirsch, " Galvanic Oxygen Recorder " 1955 (Mond Nickel Co.).

The colorimetric method based on the use of indigo-carmine falls into neither group. When indigo-carmine is reduced with glucose or amalgamated zinc, the blue colouring matter, possessing the formula shown below, is altered to the yellowish-green leuco-base

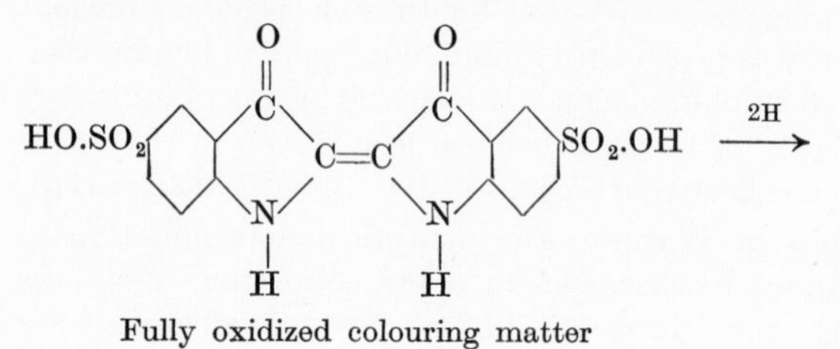

Fully oxidized colouring matter

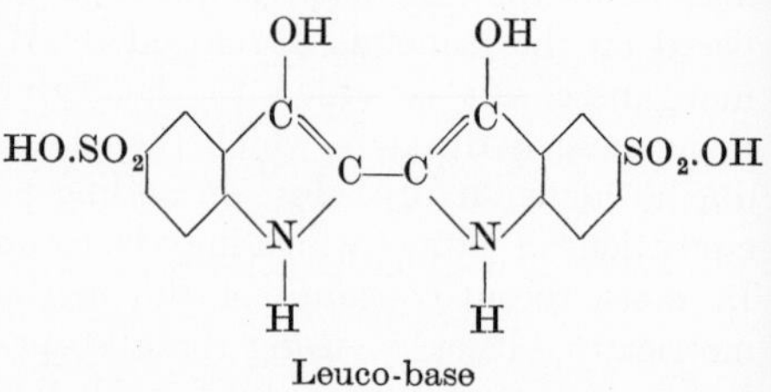

Leuco-base

Up-take of oxygen produces the reverse colour-change, and if the indicator in the reduced state is added to boiler-water, its oxygen-content can roughly be estimated by eye; for more accurate purposes, the colour can be compared with standards. In Russia, where the method is widely used at power stations, these are made by mixing in suitable preparations a picric acid solution (representing a yellow colour similar to that of the leuco-base) with unreduced indigo-carmine. In Great Britain inorganic standards are preferred based on cobalt salt solution (for red), ferric chloride solution (for

* E. C. Potter observes that the electrometric methods rely in the end on some chemical calibration.

yellow) and copper salt solution (for blue). The sequence of colours—produced in presence of glycerine—is (with increasing oxygen-content) yellow-green → red → purple → blue-green. Details are provided by L. S. Buchoff, N. M. Ingber and J. H. Brady, *Analyt. Chem.* 1955, **27,** 1401; G. P. Alcock and K. B. Coates, *Chem. and Ind.* (*Lond.*) 1958, p. 554. Russian procedure is described by R. L. Babkin, *Eleckt. Stantsii* 1954, **25,** No. 1, p. 16.*

In general, the difficulties of oxygen-estimation are not merely due to the very low concentrations of oxygen which require measurement, but also to the possible presence of oxidizing substances like hypochlorites (which with some methods would be included in the oxygen-estimation and make the result too high) and reducing substances like sulphites (which with some methods would make the result too low). These and other practical difficulties are authoritatively discussed by Potter, whose development of a method suited to the requirements of the (British) Central Electricity Generating Board deserves close study (E. C. Potter, *J. appl. Chem.* 1957, **7,** 285, 297, 309, 317; also (with J. F. White) p. 459 and (with G. Whitehead) p. 629).

Other papers deserving study include J. Vesterbel, A. Berger and V. Royer, *Bull. Centre Belge d'Étude et de Documentation des Eaux* 1950, Nos. 8 and 9. K. Wickert and H. Wiehr, *Z. analyt. Chem.* 1953, **139,** 181; 1953, **140,** 350.

Apart from chemical or electrochemical measurements of oxygen-content, a simple method for ascertaining the state of the water consists in mounting a piece of abraded soft iron in the feed-water circuit close to an inspection window; if oxygen is present even in small amounts, the appearance of the iron becomes noticeably altered. This simple device is in wide favour in Germany; the same principle was employed in a boiler fitting introduced many years ago by a Scotch firm, but which did not seem to gain popularity in Great Britain.

Recommendations regarding Alkali-content. However low the oxygen-content is kept, it is generally thought advisable to make the water alkaline, if it is not already alkaline as a result of softening or conditioning treatment. Where there is no fear of caustic cracking, sodium hydroxide will serve to confer the desired alkaline reaction, but where that risk exists, it is well to add sodium phosphate (with sodium hydroxide, if necessary) in such a manner as to ensure that the ratio of Na_2O to P_2O_5 stands between those represented by Na_3PO_4 and Na_2HPO_4; in this way, both an upper and a lower limit is set for the alkali content, as explained on p. 457.

In cases where sodium hydroxide is to be the source of alkali, it is sometimes recommended that the water should be in such a state that, after cooling, a sample gives a definite pink colour when a drop of phenolphthalein is added; this possibly means a pH value of about 8·4, but many authorities (e.g. Johnson, p. 447) would wish the value kept higher. (See also p. 455.)

Potter makes the pertinent remark that the term pH lacks significance

* Application to see the English translation (No. 734) should be made to Central Electricity Generating Board, Technical Information Series, Sumner Street, London, S.E.1.

in boiler operation, where temperatures distort the pH scale in an unfamiliar way, but adds that if the boiler-water has a pH of 10·5 at ordinary temperature, he would generally feel happy. At most power stations the boiler-water, after addition of phenolphthalein and methyl orange, is titrated with nitric acid. At about pH 8·5 the colour of the phenolphthalein is discharged (titration figure " P "), and at about pH 4·5 the methyl orange changes colour (combined titration value " M "). Of these P is considered to represent " free alkali " and M " total alkali ". Whatever the real meaning of these terms—their practical value is that the boiler-water chemist soon accumulates sufficient experience to tell from them whether the water is in its normal state or whether there is some exceptional and perhaps dangerous feature (E. C. Potter, Priv. Comm., July 22, 1958).

Hamer recommends that the alkali-content, obtained by titrating with acid using methyl orange as indicator, shall be 15% of the total solids, although 10% may sometimes be enough if the water is fully de-aerated (P. Hamer Priv. Comm., July 22, 1954).

Organic Additions. For many years various organic substances have been used in boilers either as constituents of proprietary boiler compounds and boiler fluids, or deliberately added to the make-up water. Boiler compounds often contain sodium carbonate, phosphate and sometimes hydroxide, along with tannin, glucose or dextrin; some of the older ones are said to have contained silicate, but the advisability of its introduction may be doubted.

Tannins have been used widely for the prevention of pitting and general corrosion in boilers, especially in locomotive boilers (p. 462). The tannins, however, are a large class of compounds—some more efficacious than others. Moreover different tannins act in different ways; some serve to absorb the last trace of oxygen and probably these act most efficiently in alkaline water;* others serve to produce protective films—probably containing iron tannate; others again affect the nucleation of dissolved salts and help to ensure that dissolved matter is thrown down as a sludge rather than as a scale. Thus a mixture of tannins may be advisable. The " tannin blender " may sometimes be a person whose decisions are not always based solely on scientific principles, but his experience should not be despised.

Amines, which are used both for steam-heating circuits and power boilers, should here be mentioned. They act in two different ways. The water-soluble class, like cyclohexylamine and morpholine, can render carbon dioxide harmless, although they do not remove it, since it is regenerated when the condensate is heated again in the boiler. They are better than ammonia for neutralizing acid, possibly because they have organic groups to attach the molecules to the surface by absorption; also they produce less attack on copper. Enough should be added to raise the pH value of the condensate to the range 8·4 to 9·0, according to T. J. Finnegan, *Corrosion* 1957, **13,** 405t.

* Pyrogallol, to which some of the tannins are related, has long been used for removing traces of oxygen during the purification of gases; this requires the presence of alkali.

The less soluble amines, like octadecylamine, hexadecylamine and dioctadecyclamine, do not remove carbon dioxide, but build up a film which, under some conditions, possesses protective properties. These amines are being extensively used in America (J. J. Maguire, *Industr. engng. Chem.* 1954, **46,** 994). Their possible application in Great Britain is discussed by P. Hamer, *J. Inst. heating and ventilating Engrs.* 1956, **23,** 476. See also W. J. M. Cook, *Ann. Rep. appl. Chem.* 1952, **37,** 42, esp. p. 45; E. W. F. Gillham, *Trans. Inst. mar. Engrs.* 1957, **69,** 140.

Effect of Excessive Alkali Concentrations. The fact that iron can be attacked by alkali seems to be regarded in some quarters as a mysterious happening. Actually the reaction of iron with alkali to form sodium ferroate is no more mysterious than the reaction of zinc to form sodium zincate, although fortunately in the case of iron the reaction only becomes serious at elevated temperatures and rather high alkali concentration. However, in boilers even transient conditions of high alkali concentration may destroy a protective scale by dissolving magnetite or other oxide.

Zinc exists in acid solutions as Zn^{++} and in strongly alkaline solutions as $[ZnO_2]^{--}$, the anion of sodium zincate Na_2ZnO_2. Doubtless in the intermediate pH range, simple cations and complex anions co-exist, but only at small concentrations, owing to the limited solubility of zinc hydroxide. Omitting complications due to existence of $[HZnO_2]^-$, we can picture the triple equilibrium

$$Zn^{++} + 4OH^- \rightleftharpoons [ZnO_2]^{--} + 2H_2O$$

$$\searrow\!\!\nwarrow \qquad\qquad \nearrow\!\!\swarrow$$

$$Zn(OH)_2 + 2OH^-$$
$$\text{(solid)}$$

Dilution will shift the equilibrium from right to left, and addition of extra alkali will shift it from left to right.

The anodic reaction during the corrosion of zinc can be written in acid and nearly neutral solutions as

$$Zn = Zn^{++} + 2e$$

but in strongly alkaline solution as

$$Zn + 4OH^- = [ZnO_2]^{--} + 2H_2O + 2e$$

The corresponding cathodic reaction will generally be the liberation of hydrogen in acid and alkaline liquids, but mainly the reduction of oxygen in nearly neutral solution.

The case of iron is complicated by its two valencies. We have sodium *ferroate,** Na_2FeO_2 or $Na_2O.FeO$, containing divalent iron, and sodium *ferrite*, $NaFeO_2$ or $Na_2O.Fe_2O_3$ containing trivalent iron. The complex anions are $[FeO_2]^{--}$ and $[FeO_2]^-$ respectively.

* " Hypoferrite " would perhaps be a more logical name, but " ferroate " appears to have gained acceptance. It must not be confused with " ferrate ", in which iron has a valency of 6.

The anodic reaction of iron in acid or nearly neutral solution can be written

$$Fe = Fe^{++} + 2e$$

followed under certain conditions by

$$Fe^{++} = Fe^{+++} + e$$

whereas in strongly alkaline conditions, it is

$$Fe + 4OH^- = [FeO_2]^{--} + 2H_2O + 2e$$

followed to a small extent by

$$[FeO_2]^{--} = [FeO_2]^- + e$$

The corresponding cathodic reaction will generally be the liberation of hydrogen in strong acid and strong alkali, but mainly the reduction of oxygen in nearly neutral liquid.* The fact that anodic attack on iron in strong alkali produces mainly ferroate, but some ferrite, was shown in early work by G. Grube and H. Gmelin, *Z. Elektrochem.* 1920, **26,** 459.

Now even in strong alkali, there must be some simple cations present, the equilibria being conveniently written

$$Fe^{++} + 4OH^- \rightleftharpoons [FeO_2]^{--} + 2H_2O$$

and

$$Fe^{+++} + 4OH^- \rightleftharpoons [FeO_2]^- + 2H_2O$$

If anodic attack continues, sufficient $[Fe^{++}]$ and $[Fe^{+++}]$ will be formed to exceed the solubility-product of magnetite, which will then be thrown down. We might possibly expect to obtain a protective film of magnetite. However, the nucleation rate of magnetite is low;† if the solubility product is only slightly exceeded, the crystal nuclei produced will be limited in number, the magnetite being deposited on existing crystals in preference to the formation of fresh nuclei. Thus we obtain a small number of crystals of appreciable size covering up only part of the surface, instead of a host of tiny crystals constituting a protective film. Moreover the crystals need not necessarily be thrown down in contact with the metal at the point of attack. On the contrary, if alkali has become concentrated in a crevice, being more dilute outside it, the ferroate and ferrite formed in the crevice may diffuse outwards and be precipitated as magnetite outside, since dilution shifts the equilibria in the right-to-left direction and thus generates simple cations. The idea that the non-protective magnetite formed during the caustic cracking of boilers is due to the interaction of ferroate and ferrite is largely due to Weir.

The behaviour of iron in liquids of different pH values at high tempera-

* The evolution of hydrogen from nearly neutral solutions is possible on grounds of energy, but the E.M.F. generated by the cell $Fe \mid H_2$ is small, and the hydrogen-evolution type of attack is more likely to produce a protective film, since the anodic and cathodic reactions proceed at contiguous points; in the oxygen-reduction type they often occur on well-separated areas, owing to a non-uniform oxygen-supply, so that the secondary product, formed out of contact with the metal, is non-protective.

† The fact that most wet reactions producing magnetite lead to the precipitation of relatively large crystal-grains, instead of a host of sub-microscopic ones, is a sign of the low nucleation-rate.

tures does not greatly differ from that of zinc at room temperature. In both cases, there is attack by definitely acid and alkaline liquids, with a minimum corrosion-rate at a pH value which in the case of zinc lies not far from pH 11; there is some uncertainty regarding its position in the case of iron. It is fortunate that the rapid attack shown to occur in alkali at high temperature does not occur at ordinary temperatures, where iron is being safely used in industry for handling alkaline liquids.

The concentration of alkali at which attack on iron becomes serious depends on temperature, but Pourbaix, working at the boiling-points of the various solutions, and 1 atmosphere pressure, states that corrosion of iron wire is very slow below 100 g./l., but increases linearly between 200 and 650 g./l. The corrosion is of the hydrogen-evolution type; in Pourbaix's experiments, hydrogen was passed through the liquid and corrosion only occurred when his measurements showed the potential of the iron to be negative to that of hydrogen in the same solution, so that the cell Fe | NaOH | H_2 exerted an E.M.F. in a direction favourable to corrosion. The solution obtained was green and, although the main constituent was ferroate, it is likely that ferrite was also present, since the liquid produced a black deposit of Fe_3O_4 on all surfaces with which it came in contact, and rendered the iron specimen itself dull black or bluish. This shows that the soluble compounds can accumulate in the liquid until it is super-saturated with respect of magnetite (M. Pourbaix, *Bulletin Technique Association des Ingenieurs sortes de l'Université libre de Bruxelles* 1946, p. 67; 1947, p. 109).

Now most boiler water is kept alkaline. The concentration normally aimed at in the body of the boiler is such as to cause little corrosion. The alkali can, however, become dangerous if it becomes locally concentrated, as (1) in leaky seams, especially those of riveted boilers, (2) below porous crusts or deposits, especially in modern boilers working at high out-put, (3) at points especially favourable to the formation of bubbles, where the concentration of alkali will constantly increase.

Concentration in leaky seams is not confined to old-fashioned, riveted boilers; it has been met with in modern plant at flanged joints of pipes outside the boiler proper. If a leaky seam has an exit to the outer air wide enough to permit escape of steam but not of liquid water, the liquid which enters the broader parts of the seam must become enriched in alkali, and the dangerous range of concentration may ultimately be reached. Leakage is particularly likely if the rivet holes have been badly aligned or workmanship otherwise faulty. This same misalignment may produce stresses in the metal, so that the attack takes the form of stress-corrosion cracking. Stresses can be produced in other ways—notably by injudicious caulking, perhaps carried out in an attempt to rectify the effects of bad workmanship. It should be emphasized that it is *non-uniform* stresses which provide the situation most favourable to *caustic cracking*; reasons for this are suggested in Chapter XVI. Packing pieces in boilers often suffer extensive cracking, although the stresses introduced during caulking are not high; they are, however, far from uniform. Burrs at the edge of rivet holes sometimes contribute to cracking.

Weir points out that caustic cracking requires high temperature, high alkali concentration and tensional stress. In his tests, 10% sodium hydroxide produced it but 5% did not. The attack did not always take the form of cracks; at 250°C. he noticed pitting with fissures crossing grains, but intergranular cracking became more common at lower temperatures (C. D. Weir, *Chem. and Ind.* (*Lond.*) 1953, p. 1077).

Concentration under deposits of salts, porous scale or loose scale-particles from another part of the system can occur in any type of boiler, but is most likely to occur when evaporation is rapid. Every bubble of steam forming at a point on a tube must cause the water under the flanks of the growing bubble to become more concentrated than the main mass of liquid, but if the surface is smooth and compact, the concentration will again become uniform when the bubble detaches itself. If, however, the surface is covered with porous matter, there will be only partial mixing, and the water in the pores will remain slightly more alkaline than the water outside. If then further bubbles are formed successively at the same spot—as generally happens—the alkali concentration may ultimately reach the dangerous range.

In the case of a badly constructed, or old, riveted boiler, with leaky seams and stresses present, the alkali, after it has reached the danger level, will eat its way along the grain-boundaries producing a number of branching cracks. The reason why corrosion under stress often takes the form of cracking is discussed on p. 666. In the present case, the attack follows the grain-boundaries probably because the iron is here subject to local stresses, and thus more anodic than in the grain-interiors; there is, however, special attack on the cementite constituent of pearlite islands—an effect noticed independently by Weir and by Cottell.* The main soluble anodic product is sodium ferroate, but probably there is a trace of sodium ferrite, and in due course these will throw down magnetite, regenerating the alkali almost in full, so that the attack can proceed further. If the first magnetite crystal were to be precipitated exactly at the apex of the crack, the attack would presumably be stifled; but this will rarely occur and, when once a magnetite crystal has been formed elsewhere, deposition of fresh magnetite upon it is likely to continue in preference to the formation of fresh nuclei. Cross-sections of the metal from cracked boilers show much magnetite in some of the cracks, but not in all of them. The magnetite only occurs at the tips of cracks which appear to have become dormant—presumably because the chance formation of a magnetite nucleus at the tip has stifled further attack, and the damage has at that point ceased.

The atomic hydrogen formed by the first stage of the cathodic reaction, may partly turn to molecular hydrogen on the spot and be evolved as gas, but some of it may dissolve in the metal; when the concentration becomes sufficient, this atomic hydrogen may diffuse inwards, and be converted to molecular hydrogen at small cavities in the interior, thus building up high pressures. How far internal hydrogen-pressure aids in the disruption

* See important studies of causes of grain-boundary sensitivity by N. Collari and P. Virdis, *Il Calore* 1959, N1–2.

caused by strong alkali is a matter on which opinions still differ; possibly future work will show it to be playing a larger part in boiler damage than the direct eating away of the metal along the grain-boundaries.

A number of instructive case histories of cracking, with discussion of probable causes, is provided by the British Engine Boiler and Electrical Insurance Co. Tech. Reports Nos. **1**, **2** and **3** (1952–56). Various points of view, presented by R. J. Glinn, C. D. Weir, G. A. Cottell, M. Werner, C. Edeleanu, G. M. Sellar and A. A. Berk, will be found in *Chem. and Ind.* (*Lond.*) 1953, p. 1075.

In his papers, Berk describes an ingenious apparatus designed to detect embrittling properties in a water by allowing it to become concentrated through evaporative leakage; a steel specimen is kept bent so as to produce tensile stress at the place where concentration will occur in the water; this is designed to show cracking if the concentration produces free alkali. The A.S.T.M. designation is D807–52. It is described by W. C. Schroeder and A. A. Berk, *U.S. Bur. Mines Bull.* **443** (1941); see also *Trans. Amer. Soc. mech. Engrs.* 1938, **60,** 35 (with R. A. O'Brien); 1943, **65,** 701.

Corrosion due to alkali accumulations below deposits in tubes will not, in general, take the form of active intergranular cracking, but the metal may be eaten away, with precipitation of magnetite, and the hydrogen penetrating inwards will weaken and embrittle the metal—as described on p. 442.

Prevention of Caustic Cracking by Stress Relief. Although caustic cracking has been met with in welded construction, the danger can usually be greatly reduced by stress relief, generally carried out at 600–650°C. for about one hour; such treatment is desirable for other reasons (see also Chapter XVI). For riveted boilers, water-treatment is needed.

The Co-ordinated Phosphate Method. The most natural way of guarding against trouble due to alkali is to prevent the concentration ever reaching the dangerous level. Such a result may be achieved by adding phosphate in such proportions that the ratio of $Na_2O : P_2O_5$ is slightly below 3 : 1 (the value corresponding to Na_3PO_4). If this is arranged, then when evaporation has proceeded sufficiently, solid $Na_3\ PO_4$ will be deposited, and the pH will, if anything, slightly fall; it cannot rise to the value corresponding to a saturated solution of trisodium phosphate (Na_3PO_4)—thus avoiding entry into the dangerous range. Conversely, if, owing to a miscalculation, the ratio exceeds 3 : 1, progressive evaporation will increase the pH value, and ultimately a dangerous situation may arise. In practice—to avoid deposition of solid in feed-pipes—it will be best to introduce the phosphate as Calgon, and, since this produces the acid phosphate by hydrolysis in the boiler, addition of some sodium hydroxide may also be needed; the quantity required depends, of course, on the amount of alkali already in the water.

This ingenious method, referred to by different writers as the Co-ordinated-Phosphate Method or the Zero Caustic Method, has been described by T. E. Purcell and S. F. Whirl, *Trans. electrochem. Soc.* 1943, **83,** 343; an able discussion by A. A. Berk, *Chem. and Ind.* (*Lond.*) 1953, p. 360 should receive study. Another method based on the use of sodium phosphate is described by R. Rath, *ibid.* 1953, p. 600.

The opinion has been heard that in certain circumstances the co-ordinated phosphate method will prove too expensive; that is a point which each boiler-owner must settle for himself. It has also been argued that the method needs better chemical control than will normally be available in a small boiler-house. Undoubtedly the first calculation regarding the quantity of chemicals which will have to be added should be made by someone with chemical knowledge and boiler experience. But the routine tests, required to ensure that the ratio is being kept in the safe region, can be carried out by anyone who can titrate alkali with acid.* The ratio should not fall too far below 3 : 1; otherwise pitting, instead of cracking, may arise.

Potter states that the real difficulty in applying the method to a large plant is that there are large differences in composition in different parts of the system and that, however skilled the chemical staff may be, the method may become unworkable under fairly common conditions (E. C. Potter, Priv. Comm., July 18, 1958).

The Nitrate Method. A cheaper way of avoiding caustic cracking, which has been used with success on certain American railways, depends on the addition of sodium nitrate. It is being used at some power-stations in Great Britain, although doubts have been expressed regarding its suitability for the highest pressures. The ratio of the sodium nitrate added to the total alkalinity obtained by titration with methyl orange as indicator, but expressed as NaOH, should exceed 0·4. Since the adoption of this inhibitor on the Chesapeake and Ohio Railroad the cracking of boilers, which had previously caused much anxiety, has become a very rare occurrence; the evidence is not quite straightforward, since between the period over which there was no water-treatment (ending in 1938) and the period of nitrate treatment (starting in 1941) came about two years of treatment with an organic inhibitor; moreover, it is possible that other precautions besides water-treatment may have been introduced as part of the campaign against cracked boilers. However, the figures do show clearly that nitrate treatment combined with reasonable mechanical precautions has practically eliminated caustic cracking under conditions prevailing on that American railway. Authoritative particulars will be found in papers by A. A. Berk, *Industr. engng. Chem.* 1948, **40,** 1371; *Trans. Amer. Soc. mech. Engrs.* 1951,

* The trained power-station chemist is of course well acquainted with the principles involved, but for others it may be helpful to show how relatively simple is the chemistry involved. A solution slightly *more acid* than NaH_2PO_4 turns methyl orange (M.O.) red, whereas one slightly *more alkaline* than Na_2HPO_4 turns phenolphthalein (P.P.) red. Thus if the water contains Na_3PO_4 without NaOH or Na_2HPO_4 (which means that it is exactly on the limit of the safe region), the amount of acid needed to *dispel* the red colour of P.P. is exactly half that needed to *produce* the red colour of M.O. It is, therefore, only necessary to perform a double titration, using phenolphthalein for the first stage and methyl orange for the second. If the volume of acid needed to dispel the red of P.P. is slightly less than the *additional* volume needed to produce the red of M.O., the water is " safe ". For colorimetric estimation of total phosphate, see F. J. Matthews, " Boiler Feed Water Treatment " 1935, p. 219 (Hutchinson).

73, 859; *Chem. and Ind.* (*Lond.*) 1953, p. 360; cf. R. Ll. Rees, *ibid.* 1952, p. 1213; 1953, p. 1086. An excellent assessment of this and other treatments is provided by C. D. Weir and P. Hamer, *ibid.* 1952, p. 1040.

There is no general agreement as to the manner in which nitrates prevent caustic cracking, but to those who attribute the cracking to hydrogen atoms diffusing inwards through the steel and producing huge pressures at internal cavities, an explanation will at once suggest itself; in presence of nitrate the cathodic reaction, instead of developing atomic hydrogen, will reduce the nitrates to ammonia. It is almost certain that hydrogen is the cause of trouble arising when alkali accumulates below deposits of solid substances in tubes, and in that case Kaufman and his colleagues (p. 444) actually showed in laboratory experiments that nitrate, if present, did prevent the formation of hydrogen.

Some authorities hesitate to recommend nitrate as a cure for caustic cracking, having in mind the fact that similar intergranular cracking can be produced by hot nitrate solutions. However, nitrate cracking is most pronounced in acid solutions (the nitrates which cause it most easily are nitrates of weak bases, like calcium and ammonium nitrates, which will show an acid reaction by hydrolysis), whereas in boilers the proposal is to add nitrate to strongly alkaline solution. It is not unusual to find chemicals behaving differently according as the liquid is acid or alkaline; a chromate added to a strong, non-oxidizing acid increases corrosion, whereas the same chromate added to very slightly alkaline solution diminishes it.

The Sulphate-Ratio Method. A scheme for avoiding cracking, once much advocated, depends on maintaining a certain ratio of SO_4^{--} to the total alkalinity expressed as Na_2CO_3; in the form of the treatment recommended by the American Society of Mechanical Engineers in 1925, a ratio of 1, 2 or 3 is stipulated according as the pressure is 150, 150–250 or over 250 lb./in.^{2}; variants of the prescription have subsequently been published by different authorities. The method was tried on American railways, but in 1950, a Committee of the American Railway Engineers Association reported that " the sulfate-hydroxide ratio maintained in steam boilers was found worthless as an inhibitor in railroad service " and suggest that the railroads " should disregard the sodium sulfate-hydroxide ratio recommendations ". About the time when the method was losing its popularity in the U.S.A., it was being widely recommended in Great Britain; but cracking is known to have occurred in various boilers where the recommendations have apparently been observed; Weir and Hamer (above) cite five such cases from Great Britain, as well as a set of boilers in Germany, which rapidly cracked —despite the maintenance of the prescribed ratios.

It is still widely believed that the sulphate ratio, even if it does not always prevent cracking, at least reduces the risk. The evidence available is insufficient either to establish or to disprove this belief. In order to apply a statistical test we should need to know the number of boilers

(*a*) which have *maintained* the ratio and have *cracked*;
(*b*) which have *maintained* the ratio and have *not cracked*;

(*c*) which have *not maintained* the ratio and have *cracked*;
(*d*) which have *not maintained* the ratio and have *not cracked.*

If the four sets of figures were available for a large total of cases, it would be possible, by applying statistical analysis (p. 949), to decide whether maintenance of the ratio diminished danger or not—*provided* that in all other respects the cases were comparable. Actually the figures are not available, and the cases would not be comparable, since a boiler-owner who will take the trouble to control the sulphate-ratio is the sort of man who is willing to take other—and more rational—precautions.

One cannot feel any confidence that maintenance of the sulphate-ratio does diminish the risk of cracking, but, in present state of knowledge, it would be rash to advise anyone who is observing the ratio and who is in fact avoiding trouble, to discontinue his additions of sulphate; better advice would be to train someone on the spot in tests for cracking.

Some authorities, who hold that sulphate does not render a water non-cracking when concentrated by evaporation, think that it tends to seal leaky seams with sodium or calcium sulphate, and thus diminish the chance of the dangerous concentration-process taking place; if so, it would be well to know whether sodium or calcium sulphate is the sealing agent, since, if it is the latter, some calcium must be left in the water. According to this view, the protection by sulphate rests on a different principle from that by nitrate, and there would seem to be justification for those power-stations which (it is understood) are adding both nitrate and sulphate—hoping to obtain benefit from both; if the alleged benefit of each depends on an entirely different mechanism, this plan is statistically sound.*

It is interesting to seek the origin of the belief in the efficacy of sulphate as a preventive for cracking. Probably the clue lies in the existence of naturally alkaline waters containing sodium carbonate in a large region of the Middle West of the United States, and elsewhere the extensive use of zeolite for softening hard water, producing a feed-water rich in sodium bicarbonate, which generated sodium hydroxide in the boiler. Such conditions often led to cracking, but it was found that when the alkaline water was treated with an appropriate quantity of sulphuric acid, no trouble was experienced. This might be attributed to a reduction of $[OH^-]$ or to the introduction of $[SO_4^{--}]$. Opinion seems to have seized on the latter explanation, and favoured the plan of adding, say, sodium sulphate, which would leave $[OH^-]$ almost unaltered, but would increase $[SO_4^{--}]$. Some laboratory work was carried out at the University of Illinois, which is widely held to support the view that a high value of the ratio SO_4/OH rather than a low absolute value of $[OH^-]$ is to be aimed at. Tables showing cracking and non-cracking waters were published, but although the latter

* If the addition of sulphate multiplies the probability of cracking by α, and the addition of nitrate by β, where α and β are both small compared to unity, then the probability when *both* chemicals are used is $\alpha\beta$ (which should be negligibly small), *provided that there is no correlation* between the boilers likely to be protected by sulphate and those likely to be protected by nitrate; this will only be true if the two chemicals operate in fundamentally distinct manners.

did often show higher SO_4/OH ratios, they also seem to have been used in boilers working at lower pressures; these tables are repeated in two reports and make the evidence look more impressive than on closer inspection it is seen to be. Moreover, Straub's own laboratory experiments published in 1942, indicate (to quote his words) " that with low sodium chloride content, the sulphates do not appear to be effective in preventing failure of the test specimens ". For waters containing chloride exceeding 0·6 of the alkalinity, benefits are claimed to be obtained from the addition of sulphate; the reader may care to study Straub's diagram and decide whether there is evidence for this claim. In any case, chlorides are not desirable constituents for boiler-waters.

With every desire to be fair, the Author remains unconvinced of the benefits of sulphate in water. He would, however, recommend every reader to read the reports and form his own conclusions; they are by S. W. Parr and F. G. Straub, *Illinois Univ., Eng. Exp. Station*, Bull. **155** (1926); **177** (1928); **216** (1930). See also later paper by F. G. Straub, R. C. Bardwell and H. M. Laudeman, *Trans. Amer. Soc. mech. Engrs.* 1942, **64,** 393, showing some modification of views. Cf. *Proc. Amer. Railway engng. Assoc.* 1951, **52,** 223.

Problems of New Boilers. When a new boiler arrives for erection, the tubes and other parts are frequently left exposed to the weather and become rusted both externally and internally. In view of the expense of boiler plant, it is difficult to understand the indifference sometimes shown to ordinary protective measures. However, it is sometimes argued that, in any case, the scale should be removed from heating surfaces before the boiler is put into use, and that a little atmospheric rusting will do no harm and may even assist de-scaling.

Removal of scale or rust is generally performed by running inhibited acid through the tubes or vessels requiring treatment, but it requires experienced personnel, and every precaution should be taken to avoid against leaving acid residues and introducing hydrogen into the metal; carelessly performed pickling will destroy metal and may even lead to hydrogen explosions. Nor should acid cleaning be embarked on lightly, especially in the case of old boilers. Cases are known where boilers which had given no trouble over long periods, started to leak after such cleaning; the rust or scale which had long been sealing the leaks, was dissolved by the acid, and trouble followed.

Problems of Shut-down Periods. If operation of a boiler is to be discontinued for a short period, the policy of " wet storage " may be adopted, the whole interior being filled with water containing sodium hydroxide and sodium sulphite, which is introduced at a low point in the system and made to overflow—often through the super-heater; the exact procedure will depend in the geometry of the boiler, and should be carefully thought out; whatever is decided, the superheater must not be neglected. For some types of boilers, the problem is not simple. Laboratory experience teaches that in filling a complicated glass apparatus with water by upward displacement, it is difficult to avoid leaving air-pockets; in the case of a steel apparatus, air spaces will also be left, but will remain unobserved; if too large they may exhaust the sulphite locally and lead to attack at or just below the water-line.

Sometimes chromate has been used instead of sulphite. Hydrazine phosphate is favoured in some places, although it usually leaves ammonia in the water, which may not always be desirable (see also E. R. Woodward, *Power* 1953, **97,** No. 11, pp. 91, 212).

For longer periods of disuse, the boiler is generally emptied and thoroughly drained; trays of silica gel or quick-lime are then placed inside the drums, and at other suitably chosen places, to absorb the moisture which inevitably remains behind. Another way is to have a small fire in the grate and warm up the boiler from the outside; alternatively air externally dried and heated may be circulated through the interior. In some cases a combination of methods may be needed. Vapour-phase inhibitors (p. 531) have sometimes been employed, but it is doubtful whether these would always reach parts of the surface covered with porous material (see " Chemistry Research " 1957, p. 17 (H.M. Stationery Office)). General advice will be found in Amer. Soc. mech. Engrs. Boiler Construction Code, Sect. VII (1951).

Problems of Locomotive Boilers. In locomotive boilers, copper is in contact with steel, but the area of the steel exceeds that of the copper and, if oxygen is absent, there is generally little trouble. Nevertheless the replacement of copper fire-boxes by steel fire-boxes has improved the life of tubes on one railway (O. V. S. Bulleid, *J. Instn. Loco. Engrs.* 1945, **35,** 204).

Grooving is sometimes experienced on tubes close to the copper tube plates, but this may be due to distortion of tubes where they have been altered in section for entry into the tube plates; or it may be caused by mill-scale hammered into the tube, which provides paths for corrosion to bore down into the steel (E. W. Colbeck, *Proc. Instn. mech. Engrs.* 1943, **150,** 90; see also J. W. Jenkin, p. 91).

Tannin has been used with success, not only on railways but in works locomotives, and seems definitely beneficial to the copper which in some waters suffers attack (in presence of pollution copper might sometimes be anodic to steel). The exact part played by the tannin is doubtful, but it probably serves to remove oxygen, as well as favouring the formation of sludge instead of scale; it can probably act as an adsorption inhibitor, and indeed sometimes leads to a visible film, but it is not certain how far this is protective. Blending of tannins may be advisable (p. 452). Experience with quebraco tannin on South American railways where corrosion had given trouble, and the subsequent adoption of tannin treatment on British railways, is described by J. S. Hancock, *J. Instn. Loco. Engrs.* 1947, **37,** 336; W. L. Topham, *ibid.* 1939, **29,** 805.

The South American procedure was described by Barham, who wrote: " Many waters are being softened by the base-exchange system followed by treatment with sodium hydroxide and an anti-foam compound containing tannin. This has proved very successful, although the base-exchange system alone caused corrosion" (R. J. Barham, Priv. Comm., Nov. 8, 1935).

As already stated (p. 458) nitrate has proved much more satisfactory than sulphate for the prevention of caustic cracking on railways in the U.S.A.; treatment with sulphite, waste lignin and tannin (usually quebraco) have also been recommended for this purpose (A. A. Berk, *Industr. engng. Chem.* 1948, **40,** 1371).

Turner has shown that pitting is very liable to occur at breaks in scale on locomotive tubes produced in handling; particularly where tubes have been straightened and bent, pitting may appear at the intersection of "Lüders" lines, and soft water may penetrate new tubes within a few weeks—providing yet another example of the combination of large cathode and small anode. He recommends that new tubes should be pickled or shot-blasted; the latter treatment leaves the surface in compression—a condition unfavourable to corrosion (T. H. Turner, *Proc. Instn. Mech. Engrs.* 1943, **149,** 74, esp. p. 77; see also discussion at Instn. Loco. Engrs. May 17, 1945, p. 204).

Other information about locomotive boilers is provided by J. Leick, *Werkst. u. Korrosion* 1950, **1,** 64; see also *Amer. Railway Eng. Assoc. Bull.* 1950, **490,** 223.

Problems of Marine Boilers. In the British Navy, a boiler compound containing sodium carbonate (39%), disodium hydrogen phosphate (48%) and corn starch (13%) has been developed, and is stated officially to "give very satisfactory service when properly controlled". It is understood, however, that this mixture has not given universal satisfaction when used in other situations, and there is no doubt that the "proper control" is a condition of success. Particularly the presence of chlorides in the water is dangerous, since the combination of chloride and inhibitor tends to intense localized attack (p. 146); leakage of sea-water into naval boiler supplies produces what is known as "scab pitting" during boiler operation, or "soft scab pitting" during shut-down periods. The former occurs at breaks in the scale, and the combination of large cathode and small anode causes serious attack; the latter occurs mainly in super-heaters and super-heater headers which have become wetted by water "primed" from the boiler. Estimation of chloride contents of boiler water should be carried out at reasonable intervals.

Another trouble is air-bubble pitting, which occurs where bubbles collect below the roofs of drums; the oxygen in the air-bubbles is soon used up, but the nitrogen remains and forms a screen for the steel below the flanks of the bubble; if now the water contains dissolved oxygen, differential aeration currents continue to flow, and the combination of small anode and large cathode may cause intense attack. Readers should consult an excellent paper by I. G. Slater and N. L. Parr, *Proc. Instn. mech. Engrs.* 1949, **160,** 341 and in the Admiralty Brochure, "Boiler Corrosion and Water Treatment" BR 1334 (H.M. Stationery Office). Guidance for water treatment is also provided by B.S. 1170 (British Standards Institution).

In trawlers fitted with Scotch marine boilers (fire-tube boilers with wide tubes), it is not unusual to add sea-water intentionally, sometimes along with lime and soda, to the boiler water—apparently with a view to building up a calcium carbonate scale and preventing corrosion. The practice cannot be recommended.

Interesting work on model boilers of the Scotch marine type at Teddington has brought out many important points. The tubes were exposed to brackish water, made by mixing fresh water with different amounts of synthetic sea-water (containing most of the salts present in real sea-water, but no saponin). When the mixture used was slightly acid, pitting occurred readily, but it was greatly reduced if the steel contained copper. When the water was made slightly alkaline, pitting was slight, and in that case copper had no beneficial effect.

Evidently the state of the water when a service boiler is first put into operation may well be all-important, since in the Teddington model boiler, water which started acid became more acid as the test proceeded, whilst water which started alkaline became more alkaline. An explanation may be suggested; in corrosion by sodium chloride, the alkali formed by the cathodic reaction is likely to swamp the slight acidity formed anodically in the pits, but where calcium bicarbonate is present, as in the brackish mixture, the cathodic alkali will be used up in precipitating calcium carbonate, and the anodic acidity will prevail; in seawater previously rendered alkaline, the calcium bicarbonate has presumably been removed by conversion to carbonate, and alkali will accumulate during the corrosion reaction.

The Teddington research should be helpful in dealing with a situation which has arisen through the non-availability of wrought iron tubes—formerly used in this type of marine boiler. These enjoyed lives of 15–20 years—sometimes 40 years—whereas the steel tubes now used often survive only 10 years. The Teddington results suggest methods by which improved performance can be looked for—even with steel tubes—either by increase in the pH value of the water or by the use of steel containing copper. The original papers deserve close study (F. Wormwell, G. Butler and J. G. Beynon, *Trans. Inst. mar. Engrs.* 1957, **69**, 109; also G. Butler and H. C. K. Ison, *ibid.* 1957, **69**, 121).

Hide-out due to deposition of sodium compounds can be a serious matter in marine boilers. Additions of lignin waste and quebraco tannin have been used with advantage up to pressures of 850 lb./in.^{2} at heat-transfer rates not exceeding 85,000 B.t.u./ft.^{2} per hour (J. Leicester, *Trans. Inst. mar. Engrs.* 1957, **69**, 129, esp. p. 133).

Problems of Benson Boilers. The increasing demand for power and threats of fuel shortages make it more necessary than ever to seek the efficiencies associated with the highest temperatures. This is directing increased attention to the corrosion problems of boilers of the Benson type. Here there are obvious difficulties in the use of water-treatment based on solid chemicals, which will sooner or later form encrustations on the tubes; moreover a volatile alkali seems necessary to reach all the places where its presence is needed. In the large German plants of this type, ammonia is used to confer an alkaline reaction and hydrazine to remove the last trace of oxygen. Complete and continuous oxygen-removal is absolutely necessary; otherwise the ammonia will take up copper in the condenser tubes or other parts of the circuit where copper alloys are employed. At one time, it was feared that copper would be attacked by ammonia with liberation of hydrogen—even in absence of oxygen; German experience has shown that, fortunately, there is no up-take of copper provided that the ammonia in the feed water does not exceed 5 ppm. and the oxygen is kept below 0·02 ppm.* In practice, a still lower ammonia concentration suffices. In the Leverkusen plant which gives steam at 610°C., 0·2 ppm. ammonia is usual; hydrazine tends to decompose giving ammonia, and this must be taken into account in calculating the additions. The comparative slowness of combination of hydrazine and oxygen can be largely overcome by passing the water at 100°C. through a catalyst bed of carbon (H. Tietz, *Z. Ver. dtsch. Ing.* 1953, **95**, 802, 825).

A serious risk with Benson boilers is the fear of cracking through sudden temperature changes. This is the more serious because at these very high temperatures austenitic steels must be used, which have lower thermal conductivity and higher coefficient of expansion than ferritic steels (the action of steam on these materials is more rapid than that of air, but austenitic steels of the 18/8 type stabilized with niobium suffer less scaling at 630° than the ferritic type, with 13% chromium, suffer at 550°). At Leverkusen a steam temperature rise of 15°C. per min. on starting, and a fall of 2·6°C. per min. on stopping is considered safe. Quick temperature changes—even in absence of salt—produces stress-corrosion cracking of the transgranular type (p. 680).

Apart from cracking of the metal we have to consider cracking of the scale—which will lead to corrosion. Ferritic steel, if cooled through a temperature of about 570°C., will suffer conversion of ferrous oxide to a mixture of iron and magnetite, and such a change, taking place backwards and forwards, is likely to make the scale non-protective. Hauffe suggests the avoidance of such a conversion by alloy additions (K. Hauffe, *Werkst. u. Korrosion* 1955, **6**, 117, esp. p. 129).

Another trouble occurs at the bends where the tubes become oval and the sharp curvature combined with the alternating stress may cause the protective scale to crack and corrosion to start. However, the main menace is due to salt deposits which have occasionally been produced, despite the

* Copper is attacked by an ammonia solution containing oxygen, giving a violet solution containing $[Cu(NH_3)_4](OH)_2$; fortunately the equilibrium

$$[Cu(NH_3)_4]^{++} \rightleftharpoons Cu^{++} + 4NH_3$$

leaves an appreciable concentration of Cu^{++} ions, and the potential is such that copper cannot liberate hydrogen, unless the ammonia is concentrated, which will shift the equilibrium to the left and reduce the Cu^{++} ion concentration. In this respect, ammonia differs from cyanide, which forms complex anions so stable that copper can liberate hydrogen in absence of oxygen. For kinetics of the attack, see R. W. Lane and H. J. McDonald, *J. Amer. chem. Soc.* 1946, **68**, 1699.

double distillation of the water; a flame-photometer has been designed to detect traces of salt in the water. If once molten salt is deposited on the hot tube, the scale can be fluxed; if the deposit is alkaline, sodium ferrite is formed directly, but Werner considers that, even if it consists of sulphates or chlorides, the volatile hydrogen chloride or sulphur trioxide can be expelled with formation of sodium ferrite. Hence scrupulous care in water purification is needed (M. Werner, *Werkst. u. Korrosion* 1952, **3,** 333; also Priv. Comm., Jan. 17, 1955).

Even with the greatest care to prevent chemical fluxing and mechanical cracking, the scale in Benson boilers becomes an embarrassment when the thickness is excessive, and the advisability of periodical pickling has been discussed.

Whilst at Leverkusen ammonia and hydrazine have been in successful use, there is a Benson boiler plant at Hamburg which has been in operation for 8 years without addition of any chemical at all. Similar success with the " pure water " system has been reported from America. Some surprise has been expressed regarding the excellent results obtained without chemical additions, and it has even been suggested that the raw feed-water used may have given rise on distillation to some volatile organic inhibitor. However that may be, the achievements recorded are a matter for congratulations to the authorities responsible. Nevertheless, if such ambitious " pure-water " policies are to be adopted widely, it will be necessary to place in positions of authority chemists of high standing. Information about the Pure Water System from America and Australia are provided by G. C. Daniels, *Trans. Amer. Soc. mech. Engrs.* 1944, **66,** 475; W. J. Johnson, Corrosion Symposium (Melbourne University) 1955–56, p. 539, esp. p. 547.

High-temperature Water Problems in Atomic Energy Plants. In certain types of atomic energy plant, pure (deionized) water (p. 433) is brought into contact with metal, the water being heated (under pressure) to temperatures far above 100°C. In some cases, the choice of metals is restricted by physical considerations independent of corrosion, and the behaviour of certain materials, such as zirconium and its alloys—also aluminium—is of special interest to the atomic physicist. In other situations, a wider selection is possible and the behaviour of the stainless group of alloys becomes important. Nearly all the materials considered owe their resistance to a protective film; consequently Simnad's laboratory observation that the dissolution rate of ferric oxide in acids becomes higher after severe irradiation—under conditions probably more severe than those existing at an atomic power plant—should perhaps be kept in mind in choosing material, especially if new types of plant have to be considered (M. Simnad and R. Smoluchowski, *J. chem. Phys.* 1955, **23,** 1961; see also p. 339 of this book).

For most materials the corrosion is greatly reduced if the water is free from oxygen, but attack of the hydrogen-evolution type is in many cases possible, and although this may be negligibly slow at room temperature, it is likely to be much faster as the temperature rises. Pourbaix—arguing from electrochemical theory—has pointed out that although the driving

E.M.F. of the cell Metal | Hydrogen *decreases* slightly with rise of temperature, the corrosion-current is likely to *increase* greatly unless considerations of protective films supervene (M. Pourbaix, lecture to European Corrosion Congress, Paris, 1957).

Where freedom exists regarding the choice of materials, Roebuck's observations suggest that there are several which possess the necessary resistance; thus titanium, zirconium, hafnium, platinum, austenitic stainless steel and certain cobalt alloys, all remain unchanged (apart from a clean tarnish film in some cases) up to 680°F. in water; the same materials also have satisfactory resistance to superheated steam at 750°F. Nickel resists water only up to 400°F., copper and aluminium to 300°F. The dissolved oxygen-concentration in the water used was about 1 ml./litre—which would be considered a very high content at a conventional power-station. These results may appear satisfactory, but DePaul has shown that even on stainless steel and cobalt alloys considerable corrosion occurs at 500°F. in crevices (e.g. between rivet heads and plates) if oxygen is present at about 5 to 10 ml./litre; the bad effect of oxygen is much less if a clearance exceeding 5 mils (0·005 inch) can be maintained. In general, the use of dissimilar metals at contacts does not appear to aggravate the trouble very much (A. H. Roebuck, C. R. Breden and S. Greenberg, *Corrosion* 1957, **13,** 71t; D. J. DePaul, *ibid.* 1957, **13,** 75t. S. C. Datsko, U.S. Atomic Energy Report ANL **5354** (1954); abstract in *Corros. Tech.* 1956, **3,** 303).

Special interest attaches to the behaviour of zirconium, aluminium and iron. Of these zirconium has been mentioned on p. 46. In practice zircaloy —an alloy containing 1·5% tin, which serves to counteract the bad effect of nitrogen on corrosion behaviour, along with small amounts of iron, nickel and carbon—is used (J. G. Ball, *J. Inst. Met.* 1955–56, **84,** 239, esp. p. 247).

Commercially pure aluminium is often used for cladding water-cooled reactors up to 100°C. and is reasonably resistant to 200°C. Above this, blisters appear on some, but not all, specimens—apparently due to internal hydrogen evolution—and the corrosion-rate of the blistered specimens is greater than that of the others. There may also be intergranular attack above 200°C., but Lavigne finds that heavy cold-working of the material causes the attack to be of a general, rather than an intergranular, character, although some grains may still be attacked in preference to others. Draley and Ruther find that the trouble is largely prevented if the aluminium contains nickel, presumably because the low hydrogen-overpotential of nickel favours the liberation of molecular hydrogen at the surface, and the movement of atomic hydrogen inwards is avoided. Iron assists the action of nickel. It is recommended that the sum of iron and nickel should be at least 0·8%, whilst the nickel itself should be at least 0·5%. If much silicon is present, this has an adverse effect and the nickel-content must be increased (J. E. Draley and W. E. Ruther, *Corrosion* 1956, **12,** 441t, 480t; M. J. Lavigne, *ibid.* 1958, **14,** 226t. Cf. K. M. Carlsen, *J. electrochem. Soc.* 1957, **104,** 147; F. H. Krenz, *Corrosion* 1957, **13,** 575t).

As regards the attack on iron, it appears that the purity of the metallic phase is not very important, since "armco" behaves similarly to high-purity iron; however, the corrosion-rate of an electropolished or chemically polished surface is slower than that of a roughly pickled surface (D. L. Douglas and F. C. Zyzes, *Corrosion* 1957, **13,** 361t, 433t).

Wanklyn has examined the effect of pH value, flow-rate, heat transfer, bimetallic couples and also various pre-treatments; of these only one treatment—500 hours in lithium hydroxide of pH 10·5–11·5—was reasonably beneficial (J. N. Wanklyn, A.E.R.E. 1956, M/M 116; abstract, *J. appl. Chem.* 1957, **i,** 22).

Further information about behaviour of metals to high-temperature water is provided by M. C. Bloom, M. Krulfeld, W. A. Fraser and P. N. Vlannes, *Corrosion* 1957, **13,** 297t (iron); W. K. Boyd and H. A. Pray, *ibid.* 1957, **13**, 375t (stainless steel and inconel about the critical temperature); J. Hérenguel and P. Lelong, *C.R.* 1956, **242,** 2941; also paper to be presented at the autumn session of the Soc. franc. de Metallurgie; *Rev. Métallurg.* 1956, **53,** 784, esp. fig. 10, p. 789 (aluminium); K. M. Carlsen, *Corrosion* 1957, **13,** 53a; *J. electrochem. Soc.* 1957, **104,** 147 (aluminium at 100°C. and 230°C.); H. Coriou and others, *Rev. Metallurg.* 1959, **53,** 775 (aluminium); A. B. McIntosh, *Chem. and Ind.* (*Lond.*) 1957, p. 687, esp. p. 690 (various materials); also E. C. Potter, *Rep. Progr. appl. Chem.* 1956, **41,** 170; and *Met. Abs.* 1957, **25,** 27 to 31 for convenient summaries.

Problems of Economizers and Air-heaters. The relatively cool parts of a steam boiler plant introduce problems very different from those met with in the hot regions. The economizer where water receives a preliminary heating before entering the boiler system proper usually develops little corrosion on the water-side (provided that the water has been properly treated), but may suffer on the reverse side, if the hot gases, cooled by the cold tube deposit acid moisture on the surface. Rather similar trouble can occur at the air-heater. It clearly will not occur if the water or the air is sufficiently warm on entering the economizer or air-heater to avoid condensation, and at first sight this seems an easy solution, since the dew-point, as commonly understood, occurs at a conveniently low temperature. However, products of combustion of coal and most fuel oil contain sulphur dioxide, and if this becomes converted even partially into the trioxide (SO_3), which combines with water to form sulphuric acid (H_2SO_4), itself a hygroscopic substance, condensation, accompanied by serious corrosion, may occur at much higher temperatures. The so-called *acid dew-point* of combustion products containing as little as 0·0025% of SO_3 by volume is recorded by Kear as 132°C.; 0·0085% raises it to 171°C. The rate of corrosion is generally greatest at temperatures lying between 20 and 45°C. below the acid dew-point.

If the acid dew-point can be brought low enough, it might be found possible for the water or air to enter the economizer or heater sufficiently hot to avoid condensation, and corrosion should then be absent. Even if condensation cannot be entirely avoided, the acid dew-point should be brought as low as possible, since Flint and Kear find that the corrosion-rate increases steadily as the acid dew-point of the gaseous mixture increases. This means that the concentration of sulphur trioxide must be minimized.

Kear has shown that the presence of carbon smoke in the flue gases, after an initial increase in the corrosion-rate, causes a reduction in the dew-point and in the corrosive properties of the gases; apparently the carbon particles remove the sulphur trioxide by a process of physical adsorption. However, the situation is not simple, since it was found that at or immediately below the dew-point, the presence of carbon smoke accelerates corrosion, whilst above the dew-point, where in clean gas there should be no corrosion at all, there is considerable corrosion in presence of carbon. Apparently, the carbon particles, carrying adsorbed acid, strike the surface and remain in contact long enough to cause sensible attack. Whether it will be possible to develop a method of avoiding trouble based on incomplete combustion in that part of the gas stream which is to heat economizers or air-heaters cannot at present be stated; such a solution need not, of course, involve the emission of black smoke into the outer air.

Our knowledge of a condensation in the neighbourhood of the acid dew-point is largely due to work at the British Coal Utilization Research Association by G. Whittingham, *J. appl. Chem.* 1951, **1,** 382; D. Flint and R. W. Kear, *ibid.* 1951, **1,** 388; R. W. Kear, *ibid.* 1951, **1,** 393; P. F. Corbett, D. Flint and R. F. Littlejohn, *J. Inst. Fuel* 1952–53, **25,** 246; H. D. Taylor, *Trans. Faraday Soc.* 1951, **47,** 1114. Cf. J. R. Rylands and J. R. Jenkinson, *Proc. Instn. mech. Engrs.* 1949, **158,** 405; A. Marsden, *Chem. and Ind.* (*Lond.*) 1958, p. 85.

Coit considers that dew-point corrosion by exhaust gases can be controlled (1) by the use of corrosion-resistant materials, such as certain types of stainless steel or Hastelloy C, (2) by using low-sulphur fuels, or (3) by keeping the surface above the acid dew-point (R. L. Coit, *Trans. Amer. Soc. mech. Engrs.* 1956, **78,** 89).

Of these three methods, the first would increase capital cost, the second would increase running costs, whilst the third would reduce overall efficiency. In this country, the use of fuel containing sulphur appears unavoidable, but considerable interest arises from the Rendle and Wilsdon's projects for injecting controlled amounts of ammonia into the gases, so that the trouble can be overcome without use of expensive fuel or loss of efficiency. Developments on these lines will be closely watched (L. K. Rendle and R. D. Wilsdon, *J. Inst. Fuel* 1956, **29,** 372; also Brochure published by Shell-Mex and B. P. Ltd., "Control of Low Temperature Flue Gas Corrosion").

Kear has experimented on the use of pyridine and other coal-tar bases for reducing corrosion by flue gases, but it is rather doubtful whether the results obtained are directly applicable to power stations (R. W. Kear, *J. appl. Chem.* 1954, **4,** 674).

The behaviour of various cast alloys containing iron, chromium and nickel to flue-gas atmospheres—both oxidizing and reducing—is described by J. H. Jackson, C. J. Slunder, O. E. Harder and J. T. Gow, *Trans. Amer. Soc. mech. Engrs.* 1953, **75,** 1021.

There is also the possibility of preventing, to some extent, the formation of sulphur trioxide. The combustion process produces mainly sulphur dioxide and its conversion to trioxide is considered by Harlow to occur in the hotter parts of the boiler, for instance on the super-heater surfaces; " high-temperature deposits " (ferric sulphate with much coal ash) formed on the super-heater should be distinguished from the " low temperature deposits " produced, as a result of condensation, at the economizer and air-heater. The actual catalyst which brings about conversion of SO_2 to SO_3 is believed to be ferric oxide, which under some circumstances is converted to ferric sulphate; the oxide may be derived from rust. The sulphate deposits on the super-heater tubes are responsible for the adherence of coal ash in a sintered or slagged condition, and Harlow thinks that if the inner layers of sulphate could be avoided the ash formations would not occur. He discusses various methods of avoiding the formation of sulphur trioxide and believes that the fine ash introduced when pulverized fuel is used serves a useful purpose in inhibiting the catalytic action of rusted mild steel. He concludes " pulverized fuel appears to be the best established answer to the problem ", as shown by the very limited extent of trouble in plants where pulverized fuel is used both in this country and in America (W. H. Harlow, *Proc. Instn. mech. Engrs.* 1949, **160,** 359; discussion pp. 369–379).

Other paths have been explored. Some years ago, Tolley compared the catalytic oxidation of sulphur dioxide to trioxide on bare steel, aluminium-sprayed steel and " aluminized " steel (obtained by spraying with an aluminium-cadmium alloy, and then heating to evaporate the cadmium and allow the aluminium to alloy with the iron); the last-named process gave a constant catalysis rate much lower than that which developed on bare steel—due doubtless to the low catalytic power of the alumina present on the surface. These experiments, although carried out without special reference to oil-fired boilers, are not without practical interest in that connection; but they do not suggest that catalysis can be abolished entirely by the use of coatings of aluminium or its alloys (G. Tolley, *J. Soc. chem. Ind.* 1948, **67,** 369; W. E. Ballard, *Proc. Instn. mech. Engrs.* 1949, **160,** 374).

Perhaps the ammonia system already mentioned will prove the best solution for oil-fired plant, but another proposal—the addition of zinc naphthenate to the oil is awakening interest; zinc waste products can also be added to solid fuel, and in both cases the zinc smoke obtained reduces the amount of sulphur trioxide produced (R. W. Kear, *Corros. Tech.* 1956, **3,** 125).

Corrosion due to Sodium Salts in Fuels. Owing to the frequent presence of salt in British coal, hydrogen chloride occurs to a considerable extent in flue gases, often reaching 0·02%, whereas sulphur trioxide rarely exceeds 0·01%. The presence of hydrogen chloride increases the corrosion at temperatures around 115°C. (i.e. well below the acid dew-point); immediately below the acid dew-point there is little effect; it produces a great increase of attack below the water dew-point. The chlorine content of fuel may reach 1%, and as it is generally present as sodium or potassium chloride, its presence may favour a type of deposit on the super-heaters which is bonded by an alkali matrix (R. W. Kear, *Fuel* 1954, **33,** 119; G. Whittingham, *Corros. Tech.* 1954, **1,** 182).

The presence of sodium in oils is also a cause of trouble—particularly in gas-turbines. Such oils produce sodium sulphate in the ash, and although pure sodium sulphate appears to be harmless, a small amount of sodium sulphide can form a eutectic of metal and metal sulphide at the metallic surface; when this has once started, reduction of sulphate to sulphide proceeds by an autocatalytic reaction, so that deterioration then proceeds apace. Thus sodium sulphate ash containing carbon is dangerous. The autocatalytic cycle, however, appears sometimes to start spontaneously. As in most autocatalytic reactions, there is a lack of reproducibility even between carefully controlled experiments. The cycle has been written

$$Na_2SO_4 + 3R = Na_2O + 3RO + S$$
$$M + S = MS$$
$$Na_2SO_4 + 3MS = 4S + 3MO + Na_2O$$
$$4M + 4S = 4MS$$

where M represents a divalent metal and R a reducing agent.

If these equations are correct, each " cycle " increases the amount of sulphide four times. For details, the reader should consult the paper of E. L. Simons, G. V. Browning and H. A. Liebhafsky, *Corrosion* 1955, **11,** 505t.

Condensers

Introductory Remarks. About forty years ago, condenser tubes probably gave more corrosion trouble than any other item of engineering equipment. Today, except in polluted water, they cause little anxiety, although, with the employment of increased water-velocities, problems may well recur in the near future. During certain periods of the 1914–18 war, the retention of important ships in port for the replacement of condenser tubes is said to have caused the British Admiralty as much concern as the enemy fleet; the overcoming of the difficulties was due partly to the careful researches of Bengough, May and their colleagues, partly to improved design of condensers, and partly to the introduction of new materials, such as copper-nickel alloys and aluminium-brass.

The situation does not justify a lengthy discussion of condenser corrosion, but to help those who may meet problems, references to the literature may be given. The earlier reports describing the researches mentioned above are those of G. D. Bengough, with R. M. Jones, W. E. Gibbs, R. H. Smith, O. F. Hudson, R. Pirret and R. May, *J. Inst. Met.* 1911, **5,** 28; 1913, **10,** 13; 1916, **15,** 37; 1919, **21,** 167; 1920, **23,** 65; 1924, **32,** 81; 1928, **40,** 141. Subsequent surveys have been provided by R. May, *Trans. Inst. mar. Engrs.* 1937–38, **49,** 171; 1938, **50,** 194, and still later by P. T. Gilbert and R. May, *ibid.* 1950–51, **62,** 291. An able discussion of both marine condenser tubes and heat-exchanger tubes is due to P. T. Gilbert, *ibid.* 1954, **66,** 1. See also I. G. Slater, L. Kenworthy and R. May, *J. Inst. Met.* 1950, **77,** 309. S. F. Dorey, *ibid.* 1953–54, **82,** 497, esp. p. 499. M. Cook, *ibid.* 1950, **77,** 646; 1954–55, **83,** 433, esp. pp. 436, 438. L. B. Pfeil, *Trans. Inst. mar. Engrs.* 1954, **66,** 169. C. Breckon and J. R. T. Baines, *ibid.* 1955, **67,** 363. H. S. Campbell, *J. Heating Ventilating Engrs.* 1956, **23,** 469, esp. p. 473. P. T. Gilbert, *Engng. and Boiler-house Rev.* 1959, **74,** 12.

Early in the century, the main materials used for condenser tubes were $\alpha\beta$-brasses (particularly 60/40 Cu/Zn) which were popular on the Continent, and α-brasses such as 70/30, more usual in the United Kingdom. The former were stronger, but the corrosion cells between the β-phase as anode and the α-phase as cathode sometimes raised doubts regarding their suitability. In general the structure of $\alpha\beta$-brass in condenser tubes consists of α-phase grains surrounded by β-phase envelopes, and the attack, concentrated on the β-phase, and digging deeply into the metal, may cause damage far greater than the amount of chemical change would seem to indicate.

This type of attack on $\alpha\beta$-brass is not confined to condenser tubes. Muntz metal ($\alpha\beta$-brass) has been used for sheathing wooden piles in certain estuaries in New Zealand, where at one state of the tide, fresh river water flows over heavier sea-water, producing the cell

Muntz Metal | Salt Water | Fresh Water | Muntz Metal

The brass in the salt water is anode, since the copper cation concentration is there kept low by the formation of complex anions such as $[CuCl_2]^-$ or $[CuCl_4]^{--}$, giving opportunity for the corrosion of the β-phase of the brass in the lower parts of the piling. As to whether the effect of the corrosion was serious or not depended on the structure, which in turn depended on the heat-treatment originally given to the alloy. Where this has been such as to produce a continuous net-work of β-phase completely isolating the α-grains from one another, there was disintegration of the alloy within two years. Where the proportion of β-phase was insufficient for this effect, the alloy was resistant (W. Donovan and T. E. Perks, *J. Soc. chem. Ind.* 1924, **43,** 72T, 75T).

α-brasses tubes are free from this particular trouble, provided that the heat-treatment has been correct; incorrect treatment may leave small amounts of β-phase between the grains even in a 70/30 Cu/Zn brass, which, if brought to equilibrium, should consist exclusively of α-phase. An α-brass, containing tin, which appears to give a more resistant film than simple copper-zinc alloys, is called *Admiralty Brass* and has the composition 70/29/1 Cu/Zn/Sn. At one time tubes made from these alloys occasionally underwent longitudunal splitting—obviously a greater danger than gradually developing leakage. Such splitting (now rarely experienced) is due to internal stresses, which can be avoided if the mandrel and die are correctly proportioned; or they can be eliminated by suitable annealing (R. H. N. Vaudrey and W. E. Ballard, *Trans. Faraday Soc.* 1921–22, **17,** 52; H. Moore and S. Beckinsale, *J. Inst. Met.* 1922, **27,** 149; 1923, **29,** 285; R. J. Anderson and E. G. Fahlmann, *ibid.* 1925, **34,** 271).

Deposit Attack. During the period when brass tubes were commonly used, the water-velocity was generally low in land condensers, and the tubes often lasted well, probably owing to the formation of a protective skin by a mechanism similar to that described on p. 124. Sometimes they lasted well on ships also; occasionally, however, the tubes of marine condensers developed intense localized corrosion at places where foreign particles had lodged. When the foreign particles were pieces of coke or other carbonaceous matter, they were considered to act as cathodes of a corrosion-cell, and the attack caused no surprise. Often, however, they were non-conducting-bodies, such as small stones or shell-fragments, wood or seaweed, cotton-waste or even small fishes or jelly-fishes. In such cases, the corrosion was not at first understood. It is now fairly certain that the particles acted as oxygen-screens, greatly increasing the probability that any discontinuities that might arise spontaneously in the protective skin would fail to heal themselves, owing to the poor oxygen-supply (pp. 128, 208). If corrosion once developed, the differential aeration currents would be expected to produce intense attack on the area screened—thus explaining the observations; for the " aerated " cathode was much larger than the anode, and the water-movement provided ready replenishment of oxygen on the main

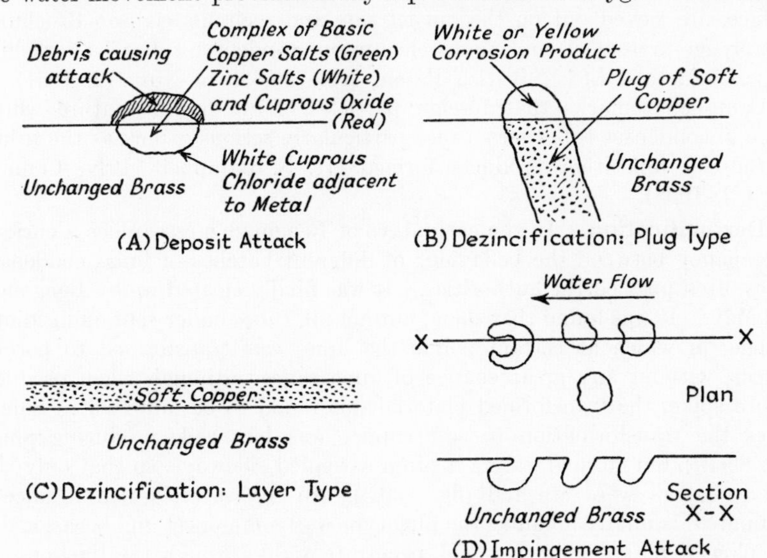

Fig. 89.—Types of Corrosion on Brass condenser tubes (H. S. Campbell).

part of the surface. The fact that corrosion often continued after the foreign body which caused it had been swept away was doubtless due to the fact that the corrosion-products, once formed, could themselves act as oxygen-screens; the stagnant conditions existing in the channels which threaded the products would prevent oxygen from reaching the metal except by a slow diffusion process; where the corrosion-product contained cuprous compounds, this would absorb oxygen chemically and provide still more efficient screening. Much of the localized attack, which often resulted in definite pitting (fig. 89(A)) probably took place during shut-down periods when the water was motionless.

Whatever the exact mechanism of the corrosion set up by foreign bodies and deposits, the obvious remedy, enjoined by Bengough, was to keep the tubes clear of such bodies. Tubes should be cleaned with relatively soft brushes; a hard brush might damage the protective skin and start attack. G. D. Bengough's earlier reports contain many sensible recommendations about keeping tubes free from *débris*. His "Notes on the Corrosion and Protection of Condenser Tubes" (Institute of Metals), published in 1925, still possess value today.

Recently a most ingenious method has been introduced for cleaning tubes without interruption of the operation of the condenser. Although the main object may be to improve the heat transfer, it is fairly certain that it also helps to avoid the on-set of corrosion trouble. The method, due to Taprogge, consisting of introducing into the cooling-water stream a number of balls made of sponge rubber of a density similar to that of the water, and having a diameter, when uncompressed, slightly *larger* than the bore of the tube; they pass through the tubes, automatically wiping the surface, are sieved out on the out-let side and recirculated (see Brochure, "Taprogge system of continuous cleaning of condenser tubes" (Yorkshire Imperial Metals, Ltd.); British Patent No. 700,833 of Nov. 12, 1951).

Campbell remarks that foreign particles of an organic nature which lodge in condenser tubes may cause particularly serious attack as the result of the decomposition products formed (H. S. Campbell, Priv. Comm., April 2, 1958).

Dezincification. In the early days of Bengough's researches a curious discrepancy between the behaviour of different batches of brass condenser tubes often puzzled the authorities. It was finally cleared up by Bengough and May. It was found that some, but not all, tubes underwent an insidious change, in which at certain points the brass was transformed to porous copper, without any great change of appearance, although when prodded with a spike, the transformed material was found to be quite soft. Sometimes the transformation to soft copper was localized, producing plugs (fig. 89(B)), but in acid waters it often extended sideways, so that only the surface layers were affected (fig. 89(C)). In sea-water, which if uncontaminated, is mildly alkaline, the plug type was commonest and occasionally the plug of porous copper would penetrate right through the thickness of the tube, ultimately causing leakage; sometimes, indeed, the plug might be forced out by the water pressure, leaving a sizeable hole. The corrosion-product accompanying dezincification would consist mainly of basic zinc chloride—virtually free from copper compounds. In contrast, those tubes which had not suffered this "dezincification" process—as it came to be called—developed a green corrosion-product containing basic copper chloride, $CuCl_2.3Cu(OH)_2$.

Bengough and May discovered that the tubes which remained immune from dezincification contained traces of arsenic, and that it was possible to prevent dezincification by adding arsenic to brass made from non-arsenical materials. It is today common practice to add 0·02 to 0·06% arsenic to brass condensers alloys (including aluminium brass, discussed later); since this plan has been adopted, dezincification has ceased to be a serious menace. The original paper deserves study (G. D. Bengough and R. May, *J. Inst. Met.* 1924, **32,** 81, esp. pp. 169, 184. See also L. Kenworthy and W. G. O'Driscoll, *Corros. Tech.* 1955, **2,** 247).

Two other elements standing in the same group of the Periodic Table, namely phosphorus and antimony, resemble arsenic in preventing dezincification. However, the addition of phosphorus for the purpose cannot be recommended, since recent researches by Bem show that, at least in the case of aluminium brass, the presence of phosphorus in amounts exceeding 0·02% renders the material susceptible to intergranular corrosion. Arsenic can produce the same effect if present in excessive amounts (e.g. 0·1%) but is very much less dangerous than phosphorus (R. S. Bem, *The Engineer* 1958, **206,** 756. Cf. W. Lynes, *Amer. Soc. Test. Mater.* 1941, **41,** 859).

There are two possible theories of dezincification. The more obvious idea is that zinc is removed by anodic action leaving vacant sites where zinc atoms had existed—thus explaining the porosity of the copper. The other view is that both elements pass into the liquid, but that, after a sufficient concentration of copper ions has been reached, re-deposition of copper at the cathode occurs as quickly as dissolution of copper at the anode. The effective reactions in either case, are the same, namely:—

$$Zn = Zn^{++} + 2e \quad \text{(anodic reaction)}$$

and

$$O + H_2O + 2e = 2OH^- \quad \text{(cathodic reaction)}$$

Thus the driving E.M.F. should be the same for both mechanisms, but in the first mechanism the resistance must soon become high since zinc ions have to thread their way through very fine pores in the copper; it is not easy to see how the first mechanism can persist after the first few atomic layers of brass have become dezincified.

Probably the first mechanism is important at the out-set. The action usually seems to start at points abnormally rich in zinc, such as the grain-boundaries, where, as already explained, specks of β-phase may be present even in 70/30 brass which should at equilibrium consist wholly of α-phase; brasses contining only 15% of zinc generally escape dezincification. Zinc-rich places will be anodic to the rest, and both copper and zinc will start to pass into the liquid; when copper compounds have sufficiently accumulated, metallic copper may be redeposited at a little distance from the point of dissolution and the second mechanism can then continue indefinitely. Possibly arsenic is adsorbed or deposited by mutual replacement at just the points where the nuclei of redeposited copper would first be formed, but its mode of action is still not known with certainty. Some interesting observations are recorded by F. W. Fink, *Trans. electrochem. Soc.* 1939, **75,** 441; see also U. R. Evans, p. 446; W. H. Bitner, p. 448.

At high water-velocities, arsenic is of limited use, since other modes of attack become possible; nor is arsenic effective in $\alpha\beta$-brasses.

The redeposition mechanism—favoured by Bengough in his early work—has received strong support from a study of electropolished sections carried out in Lacombe's laboratory; the subject of the research was a brass surface which had undergone dezincification localized as a pit. Sections through the pit at right angles to the main surface showed the red metallic copper to be present not as a continuous porous mass (as would be expected if it had been residual), but as discrete layers parallel to one another; on sections cut parallel to the main surface copper layers appear as roughly concentric rings. The French experimenters consider that both copper and zinc are attacked, the zinc being reprecipitated as hydroxide (or, it might perhaps be suggested, as basic chloride) and the copper as metal which is plated out all over the surface of the still unchanged brass. The redeposited copper is, however, still porous and apparently fresh corrosion sets in below it, so that after a time the copper layer is undermined and copper deposition then starts upon the new surface of unchanged brass, at an appreciable distance below the former deposit. In this way a succession of parallel copper layers are formed (M. Pruna, B. Le Boucher and P. Lacombe, *Revue de l'Institut Français du Petrole et Annales des combustibles liquides* 1951, **6**, 145, esp. p. 154).

Impingement Attack and Corrosion-Erosion. The measures indicated above usually served to prevent serious trouble, in absence of pollution, at the water-speeds which were customary in condensers at the time when Bengough's researches started; but water-speeds increased as the years went by and they naturally tend to be higher on ships than on land, since on ships space is limited but water-supply unlimited. Thus shortly after Bengough had successfully prescribed for deposit attack, a new form of corrosion became important.

Local break-down of protective films may occur where large air-bubbles impinge on the tube near the inlet ends of the tubes and break up into small bubbles on impact; indeed with the older tube materials, local turbulence can cause attack even in the absence of bubbles. It is often supposed that the impact damages the film, and certainly this may happen if the water contains suspended sand particles. It is possible that even an air bubble can momentarily indent the metal (perhaps within the elastic range) so that the less elastic oxide (or other film-substance) fails to expand sufficiently to cover the whole of the momentarily increased surface, and cracks are formed in the film. Another possibility is that the breaking up of large bubbles into a swarm of small ones results in a vast increase of interfacial energy, which may be reduced if particles of oxide leave the metal and pass away clinging to the small bubbles. Careful microscopic study under cinematograph conditions might distinguish between these and other possibilities, but today it can only be stated that the impingement of big bubbles does produce damage to the film. The impingement often keeps occurring at the same points, producing pits which usually under-cut the metal on the down-stream side, and often show a curious horse-shoe shape; a pit of this form is characteristic of impingement attack; it is usually clean and free from corrosion-product (fig. 89(D)).

Although impingement attack is usually related to air-bubbles in the water, it can be produced in absence of such bubbles when the water-speed is very high or when the water contains suspended sand. All these cases may be included in the more general terms, *Corrosion-Erosion* and *Conjoint Action*—of which further examples are provided in Chapter XVIII.

One method of overcoming the trouble is to avoid impingement of the big bubbles; small bubbles seem to be relatively harmless. Something has been done by improved design of condenser to accomplish this. However, the main solution of impingement attack has been based on the introduction of alloys upon which the skin repairs itself when damaged. The power of aluminium to build protective films, mentioned on p. 64, is shared by many alloys containing it in relatively small amounts.

The introduction of aluminium into brass has produced a material relatively inexpensive and wonderfully resistant to impingement attack; the usual composition is 76–78% copper, 1·8–2·3% aluminium, the remainder being mainly zinc; it is advantageous to introduce 0·04% arsenic to prevent dezincification. The excellent performance of this alloy is due to its protective film, but for that reason it has sometimes given poor results in situations where there is continuous scouring by sand. Another possibility which should be kept in mind is intergranular cracking in situations where the alloy is to be used in cooling-systems employing a water containing ammonia. In unpolluted water, it shows admirable resistance and is much used in ships and has given good results in an oil refinery (S. van der Baan, *Corrosion* 1950, **6,** 14).

The early history of the introduction of aluminium brass is something of a romance; memories are not all in agreement, but the facts—so far as he could ascertain them—have been set down by the Author (U. R. Evans, *Chem. and Ind.* (*Lond.*) 1951, p. 706, esp. pp. 707, 708).

The fact that resistance to impingement attack depends on film-repair, is shown clearly in the early experiments of May. Potential measurements were used to follow the state of the film, since the potential drops suddenly if a film is damaged, exposing bare metal. May found that aluminium brass, when scratched, suffers a drop of potential, which however quickly rises again to a noble value, indicating self-healing of the film; this recovery does not, in general, occur on the ordinary brasses. Violent impingement of bubbles produces no perceptible lowering of potential on aluminium brass, but a rapid fall of potential on brasses free from aluminium (R. May, *J. Inst. Met.* 1928, **40,** 141, esp. p. 152, and footnote: also *Metal Ind.* 1930, **37,** 378).

Another series of very reliable, but more expensive, condenser alloys, have copper and nickel as essential constituents. The first material of this class which came into wide use was a 70/30 Cu/Ni alloy, but it was gradually discovered that the presence of iron improved resistance; it is customary today to incorporate 0·4–1·0% iron and 0·5–1·5% manganese. If the iron and manganese content are each raised to 2% exceptionally good resistance to abrasion from silt is obtained, whilst the general corrosion-resistance is similar to that of standard material with 0·7% iron and 1% manganese, although the resistance to polluted water is lowered (H. S. Campbell, Priv. Comm., April 2, 1958).

In view of the shortage of nickel, successful attempts have been made to develop materials in which iron plays a major role. Work by the British Non-ferrous Metals Research Association during the second world war led to an alloy with 5% nickel and 1·2% iron, well suited for water-trunking, fire mains and other pipes which carry sea-water within ships. An alloy with 10% nickel and 2% iron was produced which would be suitable for condensers, although, owing to the satisfactory performance of aluminium brass, it has not been extensively used in this country. In the U.S.A., considerable amounts of alloys with 10% nickel and 0·7 to 1·5% iron are in service.

Another material which has given good results in trials of limited extent, is a bronze containing 10 to 12·5% tin. Inglis states that the 12·5% tin alloy is only slightly inferior to the 30% nickel alloy, and actually superior where there is severe abrasion from silt (N. P. Inglis, *Chem. and Ind.* (*Lond.*) 1957, p. 184). Other information is provided by G. L. Bailey, *J. Inst. Met.* 1951, **79,** 243; see comments by J. R. Freeman and A. W. Tracy, *ibid.* 1951, **79,** 478, also P. T. Gilbert's report of 1953 (p. 471 of this book).

Corrosion of Condenser Tubes by Polluted Water. Troubles due to pollution of harbours and estuaries constitute problems not completely solved today, and affect power-station boilers situated in such localities, as well as ships. The question of condenser corrosion should be borne in mind when deciding the position for the intake of cooling water at a power-station; on ships it is undesirable that the condensers should contain water drawn from a polluted harbour during periods when the engines are idle; when once a good protective film has been formed by normal working in the open sea, there is a better chance of resisting the effects of pollution.

Different materials behave differently to pollution and sand-scouring. At one power-station where the intake was unfortunately placed near the outlet of a sewage works, considerable trouble was caused from localized corrosion; here copper-nickel tubes behaved better than aluminium brass. The attack by polluted water is sometimes intergranular in character, and once started, has been known to continue when the polluted water has been replaced by clean water. For various opinions, see W. McClimont, *Trans. Inst. mar. Engrs.* 1954, **66,** 14; H. C. Bones (p. 15), F. L. LaQue (p. 16), F. Latimer (p. 12), U. R. Evans (p. 12).

Tin bronze has been advocated for polluted waters, but here its behaviour has not been consistently good; it withstands sand-scouring fairly well and has proved excellent in resisting certain waters found in mining districts, which are acid owing to sulphuric acid derived from sulphide minerals.

Altogether generalization is difficult, and the relative behaviour of different materials may vary according as the corrosion is due to the presence of bacteria in the tubes, or the presence of dangerous compounds (usually sulphur compounds) which frequently have a bacterial origin. Much useful information will be found in papers by T. H. Rogers, *J. Inst. Met.* 1948–49, **75,** 19; 1949–50, **76,** 597; and L. Baker, *Trans. Inst. mar. Engrs.* 1953–54, **65,** 26.

Two compounds require special mention. *Hydrogen sulphide* may occur in estuarine waters, especially in the late summer, when organic matter largely uses up the dissolved oxygen, so that sulphate-reducing bacteria become active. The hydrogen sulphide causes severe corrosion leading to perforation in ordinary brass tube. Hydrogen sulphide is probably the most potent cause of rapid corrosion in polluted waters; laboratory experiments have shown that it can produce serious attack in the absence of organic sulphur compounds. Nevertheless the importance of organic sulphur should not be overlooked.

Serious results can arise from the presence of *Cystine* in polluted water. This substance is di-α-amino-β-thiopropionic acid and can be written Cys—S—S—Cys, when Cys stands for the group

$$-CH_2-\underset{\displaystyle NH_2}{\underset{|}{CH}}-COOH$$

Cystine is a hydrogen acceptor, passing to *Cysteine*, Cys—SH, by the reaction

$$\text{Cys—S—S—Cys} + 2H = \text{Cys—SH} + \text{Cys—SH}$$

The violent localized attack associated with cystine films is due to the efficiency of cystine as a cathodic depolarizer; it takes up the hydrogen produced by the cathodic reaction, and corrosion can thus continue indefinitely even in absence of oxygen. It is not impossible that the cysteine produced by absorption of hydrogen may be conveyed by water movement to the surface and there be reconverted by oxygen to cystine; if so, the cysteine can be regarded as an oxygen-carrier (U. R. Evans, *J. Inst. Met.* 1948–49, **75**, 998).

Cystine is present in many sea-weeds and can be released by bacterial action; it is present in off-shore and in-shore waters, but in especially large amounts in harbours and estuaries where organic material is being broken down. Possibly it should not be called a corrosive substance but rather a dangerous inhibitor. If present in suitable amounts in sea-water, it produces a protective film of a copper-cystine complex on 70/30 brass, but this film, if it becomes thick, tends to blister and break down spontaneously. At any thickness it can break down under impingement attack, producing the dangerous combination of large cathode and small anode; the depth of attack produced by an air-jet impinging on a brass specimen previously covered with a film by immersion in sea-water containing cystine was found by Rogers to be much more rapid than that on an unfilmed specimen; the air-jet apparatus used in these tests is described on p. 810.

Corrosion by Carbonaceous Films. It was noted on p. 125 that the presence of a carbonaceous film on copper water-pipes can set up pitting owing to the combination of large cathode and small anode. There is reason to think that this may occur also on condenser tubes, and although apparently it is not a common cause of trouble, engineers would do well to look out for its occurrence. A case is reported by C. Breckon and J. R. T. Baines, *Trans. Inst. mar. Engrs.* 1955, **67**, 363, esp. p. 368.

Other References

Much information about boiler problems is contained in two collections of papers published in *Bull. Centre Belge d'Étude et de Documentation des Eaux* 1951, No. 11, and 1953, No. 19; the first includes a paper on pitting in British power-stations by E. W. F. Gillham and R. Ll. Rees; see also R. Ll.

Rees, *Chem. and Ind.* (*Lond.*) 1952, p. 1213; E. W. F. Gillham and R. Ll. Rees, *Combustion* 1952, **23,** April, p. 39.

Specifications for water-treatment are reviewed by T. H. Turner, *The Engineer* 1954, **197**, 918, who has given much time to presiding over the bodies responsible for the documents.

Excellent résumés of condenser problems are provided in the papers of Gilbert mentioned on p. 471.

The belief prevailing in some quarters that the corrosion of boilers may be connected with thermo-galvanic currents which can be generated by a cell consisting of two identical electrodes held at different temperatures may make it advisable to give some references to theoretical and experimental papers on the subject. These include E. D. Eastman, *J. Amer. chem. Soc.* 1928, **50,** 292; N. E. Berry, *Corrosion* 1946, **2,** 261; R. M. Buffington, *ibid.* 1947, **3,** 613; H. H. Uhlig and O. F. Noss, *ibid.* 1950, **5,** 140; V. V. Gerasimov and I. L. Rosenfel'd, *Akad. Nauk CCCP* 1957, Report No. 1, p. 29.

Two papers presented to the Instn. Chem. Engrs. on Jan. 6, 1959, reveal recent experience. The first (by J. Arthurs, J. A. Robins and T. B. Whitefoot) concerns an industrial power station; here sodium sulphate had been added from 1948 to 1955 to prevent caustic cracking at bolted joints; it was discontinued in 1955, and no cracking has been experienced. From 1957 onwards hydrazine was introduced (sulphite had previously been used); its handling presented no problems, and the internal condition of the boiler has been " very satisfactory ". The second paper (by R. Ll. Rees and F. J. R. Taylor) describes standardization by the (British) Central Electricity Board. The concentration of alkali prescribed is equivalent to a pH of 10·7–10·9 at 25°C.—which is suitable for precipitating calcium phosphate; pH values as such, however, are not laid down.

Recent publications include a symposium on high-purity water (*Corrosion* 1958, **14,** 414t–434t); a study of boiler-tube corrosion by E. C. Potter, *Chem. and Ind.* (*Lond.*) 1959, p. 308; a study of morphology and growth-law for films produced on iron exposed to steam by J. Päidassi and D. Fuller, *C.R.* 1958, **246,** 604, 759; and two new researches by M. C. Bloom and M. B. Strauss, *J. electrochem. Soc.* (in the press). Readers should also look out for any fresh work which may be published from the British Non-ferrous Metals Research Association on condenser-tube alloys suitable for polluted waters.

CHAPTER XIII

ATMOSPHERIC CORROSION

Synopsis

Atmospheric corrosion can be divided into three types, " Dry ", " Damp " and " Wet ", according to the extent to which atmospheric water vapour enters into reactions at the metal surface. The first type, in which moisture is generally believed to play no significant part, includes (i) the development of invisible (protective) oxide-films which, indirectly, exert appreciable influence in other atmospheric corrosion phenomena, (ii) the tarnishing of certain metals (especially copper, silver and their alloys) in air polluted with certain sulphur compounds. Both processes are related to high-temperature oxidation and for that reason have already been discussed in Chapters II and III respectively. Only a short section, recalling the conclusions reached in those chapters, is included in the present chapter, which deals essentially with the remaining two types—" Damp " and " Wet " corrosion.

After a discussion of some features common to both types—such as the effect of initial weather conditions on subsequent behaviour —the two types are considered separately. *Damp Atmospheric Corrosion* requires moisture in the atmosphere, and suddenly becomes more serious when the humidity exceeds a *critical value*, often about 70%. It is favoured by the presence of volatile acidic substances in the air and, particularly in the case of iron and steel, by the presence of certain disperse solid particles. Experiments are described to indicate that corrosion is most severe in those cases where a hygroscopic substance is present on the surface of the metal; this is sometimes a corrosion-product. Examples of these principles are provided by the *fogging* of nickel in air containing moisture and sulphur dioxide, and by the atmospheric attack on copper under similar conditions. The *bronze disease* which on occasions causes damage in museums also receives consideration. The part played by settled dust particles in the production of rust-spots on iron is then discussed, and the manner in which rusting—once started—may spread over the surface; this leads to a description of *filiform attack* and a consideration of its possible causes. The corrosion sometimes produced indoors through the burning of coal-gas containing sulphur next receives attention, and the manner in which electrical instruments can be affected by corrosion is briefly noticed.

Passing to *Wet Atmospheric Corrosion*, the type generally met with out of doors, we first consider the behaviour of drops lodged at crevices; the effect of air-pollution on outdoor corrosion is discussed, with special reference to problems introduced by discharge from power station chimneys; attention is also called to the rapid corrosion observed near surf beaches in the tropics. The improvement in the atmospheric resistance of steel by the presence of copper is then discussed, and the causes of the remarkable—if somewhat inconsistent—behaviour of wrought iron. The exposure tests carried out at various places to compare the behaviour of materials next receive attention. The behaviour of copper and bronze out of doors is described.

The choice of various materials for use under different conditions is then discussed, and consideration is given to the effect of contact with building materials, notably alkaline materials like cement and mortar. The behaviour of aluminium in structures is discussed at some length. Emphasis is laid on the importance of avoiding crevices, water pockets and condensation traps.

Methods of affording temporary protection against atmospheric attack during storage or transport receive attention in this chapter, but the permanent protection by means of paint coats or coatings of other metals is reserved for Chapters XIV and XV. The chief methods for obtaining temporary protection depend on (1) reduction of atmospheric humidity, (2) the use of volatile inhibitors, (3) wrapping in suitable material and (4) application of temporary coatings which can be removed with a solvent when the articles are put into service. These are discussed in turn.

Types of Atmospheric Corrosion Changes

General. The term " Atmospheric Corrosion " comprises the attack on metal exposed to the air—as opposed to metal immersed in a liquid. It covers, however, a wide range of conditions. In ordinary life a metallic surface may be exposed to:—

(*a*) low humidities (below the critical humidity range at which attack becomes rapid, as explained later);

(*b*) medium and high humidities (above the critical humidity range and extending up to saturation);

(*c*) rain: unsheltered exposure to the open air.

Conditions (*a*) and (*b*) are experienced indoors and are most usually determined by the degree of heating—(*a*) being characteristic of continually heated rooms, whilst (*b*) is characteristic of unheated or poorly heated rooms, or out-door situations shielded from rain. It should not be assumed that situation (*c*) will always produce more damage than (*b*), or that in situation (*a*) the metal will completely escape attack.

Atmospheric corrosion in its widest sense, comprises three corresponding groups of phenomena:—

(1) *" Dry " Oxidation or " Dry " Atmospheric Corrosion.* Here atmospheric water vapour is either virtually absent or, if present, seems to play

no essential part in the reaction. In the absence of atmospheric pollution, all the common metals develop films of oxide which are invisible at ordinary temperature, although they can be conspicuous at elevated temperatures. In the presence of traces of gaseous pollution, copper, silver and certain other non-ferrous metals undergo a visible film-formation, even at ordinary temperatures; this is generally known as *tarnishing*. In like circumstances, iron and steel surfaces remain bright, provided the critical humidity is not exceeded.

(2) "*Damp*" *Atmospheric Corrosion* needs the presence of water vapour in concentrations above a certain minimum value (the "critical humidity") and also traces of gaseous or solid pollution, in the presence of which the corrosion-rate may be greatly stimulated when the critical humidity is exceeded; the pollution often lowers the critical humidity level.

(3) "*Wet*" *Atmospheric Corrosion,* which needs exposure to rain or other sources of (liquid) water.

Tarnishing. Although tarnishing was discussed in Chapter III, its main characteristics may conveniently be recalled. About 1924, Vernon established the close connection between tarnishing and dry oxidation; he showed that tarnishing can obey the same parabolic law which, as found by Pilling and Bedworth about the same time, is obeyed during oxidation at elevated temperature. In pure air, parabolic growth does not occur at room temperature, and only invisible films are formed, thickening by one of the log laws. If, however, sulphur is present, defects in the oxide-lattice becomes much more frequent, for reasons already discussed (p. 75) and visible films are produced; on copper the parabolic law ($dy/dt = k/y$) is obtained even at room temperatures, the value of k being proportional to the sulphide-sulphur content of the atmosphere. Tarnishing proceeds in practically dry air, and moisture may actually retard the process if hydrogen sulphide is present in only very small amounts; moisture increases attack by hydrogen sulphide (perhaps introducing a different form of attack more nearly analogous to damp corrosion), when hydrogen sulphide is in excess. Further details on this subject will be found in early papers by W. H. J. Vernon *Trans. Faraday Soc.* 1923–24, **19,** 839; 1927, **23,** 113; U. R. Evans, *Trans. electrochem. Soc.* 1924, **46,** 247, esp. p. 264; L. R. Luce, *Ann. Phys.* (*Paris*) 1929, **11,** 167, esp. p. 198; K. Fischbeck, *Z. Elektrochem.* 1933, **39,** 316.

Damp Atmospheric Corrosion. When the atmosphere contains large amounts of water vapour, we obtain corrosion having a closer resemblance to that studied in Chapter IV. There is, however, this important difference —that under immersed conditions water is in excess, and the corrosion-rate is often controlled by the rate of oxygen-replenishment, whereas under damp conditions oxygen is in excess, and the corrosion-rate is often decided by the humidity of the air.

On some metals, there may be appreciable destruction of material, notably below rust-spots on steel above the critical humidity. In other cases, the change is merely superficial, but the alteration of colour and loss of reflectivity is usually disturbing to the eye and sometimes (e.g. on reflectors) interferes with usefulness. Measurements of loss of brightness (both specular reflexion and scattered light) due to atmospheric corrosion (in some cases out of doors) are provided by W. H. J. Vernon, *Trans. Faraday Soc.* 1923–24, **19,** 859 (copper, zinc and brass); *J. Inst. Met.* 1932, **48,** 121 (nickel); L. Kenworthy and J. M. Waldram, *ibid.* 1934, **55,** 247 (tin and Britannia metal); N. D. Pullen and B. A. Scott, *Trans. Inst. Met. Finishing* 1956, **33,** 163, esp. p. 172 (aluminium and plated surfaces).

Wet Atmospheric Corrosion. When metal is exposed to rain, the mechanism of the attack becomes close to that developed under immersed conditions, although when a thin stream of rain-water flows continuously over a metal surface, the replenishment of oxygen will be better than when a metallic specimen is immersed in water or when water flows through a pipe. A drop of rain-water held by capillarity, at a place where two plates come close together, provides a situation analogous to that of a drop placed on a horizontal metal place and surrounded with moist air so as to prevent evaporation; even if, in the open air, the original drop evaporates (or falls off) the next one will probably lodge, for reasons of geometry, at exactly the same position.

Effect of Initial Weather Conditions on Atmospheric Corrosion. It is a strange fact that the weather conditions prevailing on the day at which a metallic surface is first exposed may affect its performance months later. This was first noticed by Vernon in his work on copper and zinc exposed indoors in London. Copper exposed in winter when the atmosphere is polluted develops a film which thickens parabolically with a relatively high rate-constant and development of interference colours, whereas if exposed in summer, when the atmosphere is relatively pure, the thickening-rate falls off much more quickly, and interference colours do not appear. The fact that the winter film is non-protective and the summer film protective is not in itself surprising, in view of the lattice-defects known to be introduced by sulphur. What might not have been expected is that the specimen first exposed in summer continues to exhibit film-growth without any break in the relationship even when winter arrives and that it continues free from interference colours.

Rather similar principles were established for zinc which, after initial irregularities, gains weight at a constant rate. The rate is not the same for specimens started on different dates, but once the exposure has fairly started, the rate remains unaffected by alterations in the weather (W. H. J. Vernon, *Trans. Faraday Soc.* 1927, **23,** 113; esp. pp. 128, 136).

In a general way, this has been confirmed in America by Ellis, who exposed specimens of zinc out of doors at Middletown (Ohio) for 28 days, and correlated the *loss* of weight with the weather conditions during the period. It was found that the number of hours of rain-fall during the first five days, and also the number of hours at which the relative humidity was close to 100% during the first five days, greatly influenced the total weight-loss over the 28-day period; rain-fall and periods of high humidity *after* the first five days had little effect (O. B. Ellis, *Proc. Amer. Soc. Test. Mater.* 1949, **49,** 152).

Rather similar results were found on steel exposed in America by Schramm and Taylerson; all the specimens were exposed for 12 months each, but exposure of different specimens was commenced in different months of the year; those started in summer suffered much less at the end of their 12-month period than those started in winter (G. N. Schramm and E. S. Taylerson, *Amer. Soc. Test. Mater.*, Symposium on Outdoor Weathering of Metals and Metal Coatings, 1934).

Sanyal's tests in India show a similar effect of initial conditions for steel, but not for zinc (p. 505).

Mayne, examining steel specimens, painted on a rusty surface, and exposed on a roof affected by smoke from domestic fires in winter but not in summer, found that specimens painted in June attained a longer life for the paint-coat than those painted in December. A considerable amount of soluble ferrous sulphate is present in the rust in December, and very little in June (when, presumably, the ferrous sulphate formed in winter will have been washed away). An important factor—emphasized by Mayne—is the greater conductivity of the moisture present in the rust if sulphate is present. The facts are recorded by J. E. O. Mayne, *J. Iron St. Inst.* 1954, **176,** 143; *J. Oil Colour Chem. Assoc.* 1957, **40,** 183, esp. p. 185.

Recently the Author, using a rubber stamp of the pattern familiar in addressing or dating documents, stamped rows of letters on abraded steel sheet with a ferrous sulphate solution; the letters (formed by ferrous sulphate crystals) were visible after drying. The specimens were then exposed out-of-doors at Cambridge, and it was found that rusting commenced at the rows of letters, spreading outwards over the surface. In other cases, polystyrene lacquer (unpigmented, or carrying a pigment which left the film translucent) was applied over the surface, and rusting then quickly started at the letters and developed so quickly that the lacquer or paint was pushed away after a few weeks, although the area situated well away from the letters remained unrusted; this supports Mayne's views. The rust-development consisted of periods when the " rust " was orange-brown alternated with periods when it was black; the black colours usually appeared after rain. The first rust is doubtless formed by oxidation and hydrolysis of ferrous sulphate, and during wet spells, the cell

iron | ferrous sulphate solution | ferric hydroxide

will reduce the ferric rust to black magnetite, which during the following dry period is oxidized again to ferric rust. Thus more and more iron is converted into bulky rust at places where the ferrous sulphate exists, explaining the pushing away of the lacquer. A salt conferring conductivity on the moisture is needed for the electrochemical reaction. No doubt, metallic iron could react chemically with the lowest layers of ferric rust to give magnetite, but further conversion depends on the facts that magnetite is an electronic conductor and ferrous sulphate solution an ionic conductor. This permits the ferric hydroxide to take up iron by the cathodic reaction, being converted to magnetite, until finally the whole of the orange rust becomes black magnetite. Other salts (chlorides, etc.) can doubtless act in the same way, explaining the fact (p. 505) that rust containing traces of chloride promotes further rusting whilst pure rust does not. No such mechanism is possible for zinc, explaining why zinc, which is more rapidly corroded than iron under conditions of partial or complete immersion in salt solution, is less attacked under atmospheric conditions, except in very acid environment—such as tunnels.

Damp Atmospheric Corrosion

The Principle of Critical Humidity. An important principle, established by Vernon, states that damp corrosion only becomes important when the relative humidity of the atmosphere exceeds a certain critical value; presumably this represents the humidity above which some hygroscopic body present on the surface (or formed during the attack) is able to collect water from the atmosphere, so that the corrosion can continue by a mechanism more like that met with under immersed conditions. The principle, of which examples will be presented later, is discussed in papers by W. H. J. Vernon, *Trans. Faraday Soc.* 1927, **23,** 113, esp p. 162 (foot-note); 1931, **27,** 255, esp. p. 264; 1935, **31,** 1678, esp. pp. 1681, 1692; *Trans. electrochem. Soc.* 1933, **64,** 35; J. C. Hudson, *Trans Faraday Soc.* 1929, **25,** 176, esp. p. 207; W. S. Patterson and L. Hebbs, *ibid.* 1931, **27,** 277; W. S. Patterson and J. H. Wilkinson, *J. Soc. chem. Ind.* 1938, **57,** 445; 1941, **60,** 42; G. D. Bengough and L. Whitby, *Trans. Inst. chem. Eng.* 1935, **11,** 176.

In contrast with tarnishing, which is most marked when the sulphur of the polluted air is present as hydrogen sulphide, damp corrosion largely depends on the presence of sulphur dioxide or sulphur trioxide. Much of the sulphur in the freshly formed combustion-products of coal exists as *dioxide* but it must be recollected that it is derived from the combustion of pyrites—which combustion also furnishes ferric oxide, a catalyst for the oxidation of sulphur dioxide to *trioxide.* During " smog " periods, it is likely that the particles of grit thrown out from locomotives or from low furnace-chimneys carry much sulphuric acid, and are capable of inflicting serious damage on Metals as well as on Men. It is believed that the Beaver Committee reports of 1953 and 1954 paid too little attention to sulphuric acid. Doubtless in the gas phase of the air most of the sulphur is present as dioxide—as shown by the numbers quoted by A. R. Meetham, " Atmospheric Pollution: its origin and prevention " 1956, p. 134 (Pergamon Press); also *Quart. J. roy. meteorolog. Soc.* 1950, **76,** 359, esp. p. 367. J. H. Coste and G. B. Courtier, *Trans. Faraday Soc.* 1936, **32,** 1198. See also work of Harlow, Whittingham, Keir and others quoted on pp. 468–470.

The term *relative humidity* may require definition. We may regard the atmosphere as being " saturated " at a given temperature when it contains just so much moisture as to stand in equilibrium with a *level* surface of pure water at that temperature (evaporation and condensation balancing one another); then if the partial pressure of moisture in an " unsaturated " atmosphere is $x\%$ of that in a saturated atmosphere at the same temperature, the relative humidity of the former is said to be $x\%$.* The critical value

* The *mass* of water vapour expressed as a percentage of the *mass* of water vapour in a saturated atmosphere is called the *percentage saturation.*

of the relative humidity above which corrosion becomes rapid is often found to lie about 70%, but in some circumstances may be much lower (p. 497).

Effect of volatile acids and alkalis on Damp Corrosion. The effect of hygroscopic corrosion-products on subsequent corrosion was brought out in the Author's early work on metal exposed to air containing acid or alkaline vapours at concentrations far higher than would occur even in the most polluted industrial atmosphere. These were exaggerated conditions, but exaggeration often serves a useful purpose in providing clear evidence for fundamental principles, and a summary of these early results may perhaps be helpful; details will be found in the original paper (U. R. Evans, *Trans. Faraday Soc.* 1922–23, **19,** 201).

The metals were placed in the upper portion of "two-storied" vessels of the type commonly used as "desiccators", whilst the liquid was placed in the lower part; the liquids included concentrated sulphuric acid (for the "blank experiments"), pure water, water saturated with carbon dioxide, hydrogen sulphide or sulphur dioxide, concentrated hydrochloric acid and concentrated ammonia. The temperature was not kept constant, and some condensation occurred at nights.

The specimens (copper, zinc, lead, tin, nickel, aluminium, iron and steel) kept over concentrated sulphuric acid remained bright and unaffected after a month; a steel razor-blade remained unblunted. Those placed over water became slightly dulled in some cases (lead, zinc and copper), but unchanged in others (tin, aluminium, and α-brass), whilst iron showed tiny dots or brown rust.* Saturated carbon dioxide produced the same effect as water —but sometimes rather more quickly. (The idea that carbon dioxide is the main cause of every-day atmospheric corrosion is wrong; the effects once ascribed to carbon dioxide are really due to other acidic constituents of the air, such as sulphur dioxide or hydrogen chloride. Vernon concluded from his experiments that carbon dioxide in normal atmospheric concentration actually diminishes atmospheric attack on iron.)

The specimens placed over hydrogen sulphide underwent quicker changes. Iron became covered with a very thin coating of rather loose dark-brown rust within two weeks. Copper began to darken within a few minutes, and developed interference tints within half an hour; after a day it had turned greyish or bluish black, but even after two weeks, the change was only superficial. Brass showed iridescence locally, and lead soon became dull, but tin and aluminium remained changed.

Over sulphur dioxide, changes were still more rapid; iron and steel became dark within a few hours, and nearly black after a day. After two weeks there was a brown or black deposit about 0·1 mm. thick, with local excrescences of yellow-brown ferric hydroxide, which was definitely *wet*, despite the fact that the specimens were not immersed in the liquid; the

* Crennell has recently reported that steel kept four months over water in a desiccator-type vessel shows only "a few isolated specks of rust mainly at the edges" (J. T. Crennell, *J. appl. Chem.* 1958, **8,** 270. See comments by W. H. J. Vernon, ibid. 1958, **8,** 469, and by U. R. Evans, *Chem. and Ind.* (*Lond.*) 1958, p. 681.

dark deposit showed the reaction of a sulphate (indicating oxidation of the sulphur from the 4-valent to the 6-valent state); it contained ferric iron as well as ferrous, although the latter preponderated. The razor-blade lost its edge altogether. Zinc slowly developed a moist white paste—about 0.2 mm. thick after 2 weeks—which contained much sulphate and some sulphite. Nickel quickly became black, and after 2 days began to shed copiously a pale green liquid; after two weeks there was a deposit 0·5 mm. thick, consisting mainly of matter soluble in hot water, with a small dark insoluble residue; the liquid contained sulphite and sulphate. The other metals suffered only superficial change, copper becoming dull reddish-brown, whilst brass was dull yellow, grey and purplish in different places; lead darkened slightly; tin became yellow-grey, whilst aluminium remained unchanged.

Over concentrated hydrochloric acid, zinc became very wet and was rapidly attacked, sheet specimens being eaten through in two weeks; a thick colourless syrup ran off the specimens. This phenomenon was undoubtedly due to the hygroscopic character of zinc chloride; probably for the same reason, brass became wet, whereas copper remained dry, being often only superficially attacked; a dull red, brown or black coat overlay a white layer (probably cuprous chloride); one specimen developed a thick greenish deposit locally. Iron specimens developed a whitish-grey frosted appearance but remained rust-free until they were taken out and exposed to ordinary damp air (free from hydrogen chloride), when the frosty substance (presumably ferrous chloride) quickly absorbed water and oxygen, yielding copious rust. Lead quickly developed interference colours and finally became dark grey; tin became grey, but remained dry. The behaviour of aluminium was variable, but most samples developed a feathery growth, which grew outwards rapidly; one specimen of aluminium wire of less than 1 mm. diameter developed in 8 days a growth extending outwards to a distance of 1 cm.; the feathers probably start from weak points in the otherwise protective film.

Ammonia produced results which contrasted strongly with those observed over the acids. Iron suffered no change (a razor-blade remained unblunted), whilst copper rapidly darkened, soon shedding a violet-blue liquid which showed the reactions of nitrite—and probably contained the ammine $[Cu(NH_3)_4](NO_2)_2$. Brass shed the same violet liquid, and became brittle (cf. p. 967), crumbling in the fingers after 2 weeks, whereas copper underwent no embrittlement. Nickel was variable; some samples remained almost unchanged, others darkened, and sometimes shed the violet-blue liquid, which was found to contain copper—a not uncommon impurity in the nickel available at that time.

One feature of all these tests is that in cases where the specimens became damp or wet, there was rapid attack. It is necessary to ask what caused this absorption of water from an atmosphere which, in most cases, was definitely " unsaturated ".* In the case of zinc over hydrochloric acid, the

* The tension of aqueous vapour above concentrated hydrochloric acid is *less* than that above pure water.

hygroscopic body was clearly zinc chloride, but where metals became moist or (as in the case of nickel) definitely wet over sulphur dioxide solution, it is necessary to postulate catalytic oxidation to sulphur trioxide—which oxidation is confirmed by the presence of sulphates in the corrosion-product. Copper placed over ammonia caused a similar catalytic oxidation of ammonia to the state of nitrite.

The results also emphasize the protective character of the films on tin and aluminium, although these may break down locally on some specimens in hydrogen chloride.

The manner in which the oxide-film breaks down in hydrogen chloride vapour has been studied in Feitknecht's laboratory at Berne by hanging a specimen over hydrochloric acid in a cylinder, so that the vapours gradually diffuse upwards over it; in that way, the successive steps in the break-down can be separated, the products being identified by means of X-rays, and the morphological changes followed by means of the electron microscope. On iron and nickel the oxide breaks down at isolated points, whereas on zinc and cadmium, the whole film undergoes change, producing a smooth layer of basic salt. On copper the original cuprous oxide film thickens, and becomes converted in part to cuprous chloride; still later a vigorous attack sets in at isolated points, leading to $CuCl_2.3Cu(OH)_2$. Feitknecht regards the mechanism as electrochemical, with the oxide-film as cathode and the small areas of metal exposed at breaks as anodes; the interaction between the OH^- ions, formed by the cathodic reduction of oxygen, and the metal ions, formed by the anodic reaction, leads to hydroxide or basic chloride. The rusting of iron, which depends on the amounts of hydrogen chloride and water vapour present, was studied in some detail. For details the reader should consult the original papers (W. Feitknecht, *Helv. chim. Acta*, 1946, **29,** 1801; *Chimia* 1952, **6,** 3; also International Conference on Surface Reactions, Pittsburgh, 1948, p. 212; W. Feitknecht and K. Maget, *Helv. chim. Acta* 1949, **32,** 1639, 1653, 1667; A. Kleiner, Dissertation (Berne) 1952; H. Ziegler, Dissertation (Berne) 1946).

Fogging of Nickel. The experiments just described—carried out under " exaggerated " conditions—may assist understanding of the behaviour of metals towards the far smaller amounts of impurities normally present in the atmosphere. In the days before chromium-plating became common, nickel was much used as a protective covering for steel, and was often effective in preventing attack on the steel basis; however, the nickel itself developed a dull, creamy film, which was unattractive and for some purposes interfered with utility. This change, known as *fogging*, was the subject of an instructive research by W. H. J. Vernon, *J. Inst. Met.* 1932, **48,** 121; *Chem. and Ind.* (*Lond.*) 1943, p. 314; *Nature*, 1951, **167,** 1037.

Vernon exposed specimens of nickel in air containing sulphur dioxide and moisture. When the relative humidity was below 70%, the metal remained bright, but above that level changes were observed, proceeding in two steps. Soon after exposure started, a haze appeared on the surface, which could easily be removed by wiping with a cloth—the original brightness of the surface being thereby restored. Later, however, more vigorous treatment was found necessary, and finally a film was formed which could only be removed by abrading the metal. In the early stages, the haze consisted of free sulphuric acid with nickel sulphate, whereas in the later stages, a solid basic nickel sulphate was present. Thus the catalytic oxidation of sulphur from a valency of 4 to 6 is again met with. Vernon found that if the surface was exposed to light, the change proceeded at about double the speed of that attained in the dark—a significant difference!

Although atmospheric water vapour is needed for fogging, it does not occur on specimens exposed in the open air, where doubtless the rain washes away the sulphuric acid and nickel sulphate as they are formed. Fogging is essentially an indoor trouble, and is likely to occur most rapidly in towns where sulphur dioxide is present in the air.

Alloying with copper was found by Vernon to confer no benefit, but alloying with chromium gave considerable improvement. In modern practice, however, a thin overlay of chromium serves to overcome the trouble. Indeed, what we describe in every-day life as chromium-plated steel is usually steel plated with nickel to protect it from corrosion, with a thinner external layer of chromium to protect the nickel from fogging.

Some analogy to the fogging of nickel is found in the loss of reflectivity of tin or Britannia metal exposed to an indoor town atmosphere, since here also a water-wash suffices to restore the reflectivity in the early stages, whereas in later stages, soap must be used; details are provided by L. Kenworthy and J. M. Waldram, *J. Inst. Met.* 1934, **55,** 247.

Attack on Copper by Air containing Sulphur Dioxide and Moisture. Another example of a critical humidity level placed at about 70% is provided by Vernon's researches on copper (W. H. J. Vernon, *Trans. Faraday Soc.* 1931, **27,** 255, 582. Cf. the results on iron described by W. H. J. Vernon, *ibid.* 1935, **31,** 1678, esp. figs. 9 and 18 which bring out the striking effects produced above the critical humidity by remarkably small amounts of sulphur dioxide).

Specimens exposed to air containing 10% sulphur dioxide underwent only very small weight-increments after 30 days when the relative humidity was 50% or 63%, but a very considerable increase when it was 75%—still more when it was 99% (fig. 90). The cause of this sudden rise in the rate of attack at about 70% was established. The hygroscopic powers of the corrosion-products, after prolonged desiccation, were studied at the various humidities; they were found to increase sharply between 50% and 75%. It was also found that the products from high-conductivity copper were more hygroscopic than those from arsenical copper, explaining why Vernon's arsenical copper was more resistant than his arsenic-free copper, as had been found both in laboratory and in field tents (see, however, American results reported on p. 518). The surface condition was also found to be important. Experiments in air containing 1% sulphur dioxide and 99% relative humidity showed a far greater rate of attack if the surface had been sand-blasted than if it had been prepared with fine emery; Vernon attributed this to an enhanced number of catalytically active centres on the sand-blasted surface—a suggestion which appears reasonable.

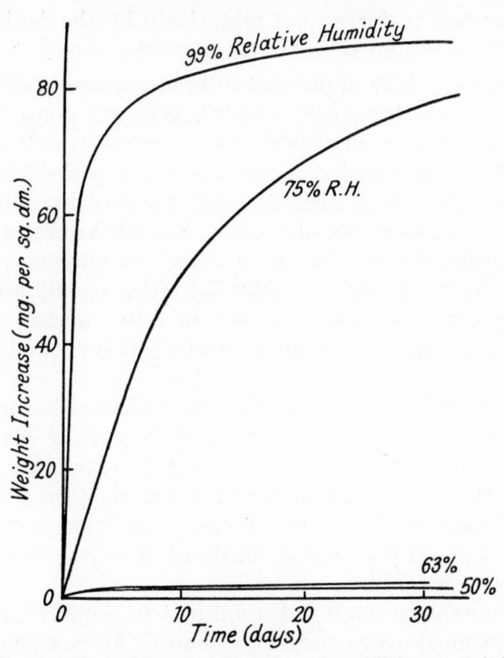

FIG. 90.—Corrosion of Copper in air of various humidities containing 10% sulphur dioxide (W. H. J. Vernon).

An interesting feature of Vernon's work is the effect of varying the sulphur dioxide concentration, whilst keeping the relative humidity at 99% (fig. 91); his analyses of the corrosion-product showed that it is almost exactly $CuSO_4$ at the point of slowest attack, marked N, whereas to the left of it it contains excess $Cu(OH)_2$ and to the right of it excess H_2SO_4. This may arouse surprise, but if the catalytic formation of sulphate from sulphur dioxide consists in attachment of an SO_2 molecule to two oxygen atoms adsorbed on adjacent sites on the copper surface, producing a $CuSO_4$ molecule, a mechanism appears for the start of a copper sulphate film (doubtless including water of crystallization) which would be epitaxial (crystallographically continuous) with the metal. Such continuity is likely to favour protective properties in a

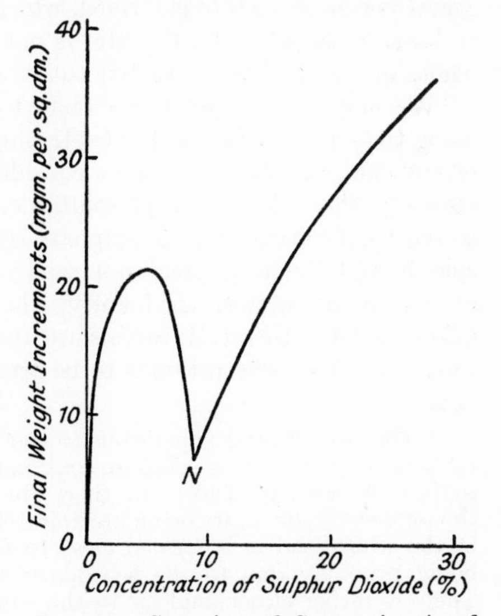

FIG. 91.—Corrosion of Copper in air of 99% relative humidity and various concentrations of sulphur dioxide.

film. The suggestion is thrown out tentatively by the Author, and is not free from difficulties; but no alternative explanation comes to mind.

Bronze Disease. Most of the ancient bronzes preserved in our museums carry a green patina of basic salt, which is generally considered to add to their beauty. This patina may fairly be described as extremely " protective ", having prevented corrosion damage for a period of perhaps 2000 years (during part of which time the articles may have been buried in desert sand or exposed to air out of doors). Yet it is a sad fact that occasionally there are outbreaks of " bronze disease " in museums, during which local corrosion burrows down rapidly into the metal, often producing perforation in a few years and sometimes in a few weeks. This occurs if the articles are exposed to fumes of an acid capable of producing a soluble copper salt.

A serious outbreak occurred at the Fitzwilliam Museum, Cambridge, shortly after the 1939–45 war, among certain objects of beauty and value which had been sent in 1939 to a locality selected to minimize risk of damage from " enemy action ". At the end of the war they came back and were replaced in the museum cases; the disease soon broke out. The trouble was apparently due to acetic acid absorbed from the wood shavings in which they had been packed.

In seeking an explanation it is helpful first to consider the manufacture of white lead pigment (basic lead carbonate). It has long been known that exposure of metallic lead to carbon dioxide (in moist air) produces only a superficial conversion to basic carbonate, since the film, formed *in situ*, protects the metal below it. If, however, it is exposed to moist air containing acetic acid vapour as well as carbon dioxide, relatively soluble acetate or basic acetate is formed, which by a secondary change is converted to basic carbonate, and the film is not in that case protective; in time the whole of the lead is converted to the white compound.*

We may now apply these facts to explain bronze disease. It would seem that the acetic acid, penetrating through occasional cracks in the original patina (in the case under consideration, mainly basic copper chloride), attacked the copper, giving soluble copper acetate, which was converted to basic carbonate or basic sulphate by the acid constituents of the atmosphere, and the acetic acid liberated afresh proceeded to produce further attack on the bronze. Evidently, when such action has once bored deeply into the metal, the acetic anions are at the bottom of the rapidly developing pits, and if the penetration is to be arrested they must somehow be fetched out.

* The case is partly analogous to the production of lead chromate by anodic treatment of lead in a solution containing potassium chromate and excess of sodium chlorate (p. 240); but there the desirable non-protective character of the pigment is due to its being precipitated at a distance from the metal, whereas in the white lead it is formed close to the metal, its non-protective character being probably due to the secondary and porous nature of the compound. There is rather closer analogy to the work of Vernon, who showed that whilst air containing carbon dioxide in contact with copper produced no basic carbonate, this was quickly produced when acetic acid vapour was also present (W. H. J. Vernon, *J. chem. Soc.* 1934, p. 1857).

In the case under review, they were fetched out by an electrical potential gradient; cathodic treatment was applied locally at the site of each pit, the electrolyte being held on a zinc anode cut to form a " nib ", which could be applied to every corrosion-site in turn. Each pit-site received cathodic treatment, first with hydrochloric acid, then with phosphoric acid and finally with sodium carbonate as electrolyte. The advantage of the local treatment is that the main part of the surface remains unwetted, and the original patina, as a whole, undergoes no change; thus the appearance of the specimen remained the same as before the disease had broken out—except in cases where the holes had advanced deeply into the metal before the treatment was undertaken. (Most alternative methods of treating bronze disease cause changes in the colour of the entire specimen.)

The method was worked out by the author to deal with this particular outbreak. It might not be applicable to all future outbreaks of the disease, or to bronzes of different composition. About 500 objects, mostly bronzes from Cyprus, Greece and Egypt, were successfully treated by Mr. Rayner, of the Fitzwilliam Museum. The method failed only in one case—a bronze of Roman origin (U. R. Evans, *Chem. and Ind.* (*Lond.*) 1951, p. 710; supplemented by information from N. Rayner, who stated in 1957 that there had been no recurrence of the disease during the six years which had elapsed since the treatment was applied).

Effect of Vapours from Wood and Organic Materials. The case just cited is merely a single example of damage which can be produced on metals through corrosive substances originating from wood. It is sometimes stated that only certain woods, or woods improperly seasoned, can cause trouble, but this is not the case; even birch—a wood not normally acid, and probably safer than most others—gives off acetic acid at high humidity and elevated temperature. The menace of acids from wood (also from certain glues and plastics) is clearly greatest in confined spaces, the metals most liable to attack being steel, lead, cadmium, zinc and possibly magnesium. A useful table showing the attack by acetic and formic acids, phenol and formaldehyde upon different metals when under humid conditions is published in " Chemistry Research " 1956, p. 18 (H.M. Stationery office). See also " Forest Products Research " 1955, esp. p. 46 (H.M. Stationery Office); V. E. Rance and H. G. Cole, " Corrosion of Metals by Vapours from Organic Materials ", 1958 (H.M. Stationery Office).

Influence of Dust on Rusting. An important feature of Vernon's classical work concerned the part played by dust in the indoor rusting of iron. In a general way, it had been known that dust, settling on steel, promotes rusting; the upper surfaces of steel articles tend to develop rust spots more readily than lower surfaces upon which settlement of dust particles is unlikely.* Vernon showed the intimate relationship that subsists

* The author's first contact with metallic corrosion occurred when, as a boy, he was taken to the laboratory of Prof. C. V. Boys, who showed him his well-fitting tool cabinets and explained " it isn't the damp that rusts the tools; it's the dust ". It is sometimes stated that the heat-treatment of tools increases

between the action of certain kinds of particles and both the state of gaseous pollution and the state of humidity of the atmosphere. He classified suspended solids as:—

(1) *harmless* (producing no rust)—silica particles were used in his experiments;
(2) *intrinsically corrosive* (producing rust where they settle)—characteristically, ammonium sulphate particles;
(3) *indirectly corrosive*—carbonaceous particles which, by adsorbing acidic sulphur gases, can profoundly stimulate corrosion in an unsaturated atmosphere.

Vernon found that clean specimens of iron shielded from suspended particles by means of a muslin screen remained free from rust in an atmosphere in which unscreened specimens (otherwise similar) rusted freely. A specimen removed from the screen after eleven months and then exposed in the ordinary way continued bright for several weeks, while freshly cleaned specimens rusted. Isolated and widely separated rust-points then revealed the breakdown of the film and the beginning of attack; but the marked difference in the total amount of rust on pre-exposed and " new " specimens persisted during continued exposure.

Evidently during the eleven months behind muslin the invisible protective film was becoming more perfect. Vernon showed in later experiments that the gain in weight due to the growth of an invisible film persisted, at a diminishing rate, over quite long periods. The film probably keeps breaking down at infrequent intervals and at isolated points; if no dust or liquid moisture is present, such points repair themselves with invisible oxide, and no rusting occurs. If, however, a particle of ammonium sulphate or sodium chloride happens to settle at a point of break-down, rusting may develop before healing can occur; once started, rusting will continue, since the first rust formed can absorb water even from unsaturated air, and the rusting will then spread laterally—as described later.

This capacity of rust to absorb moisture and set up something akin to wet corrosion, even though the rust may still present a dry appearance, is of great importance. Vernon performed experiments in which iron was exposed to air containing 0·01% sulphur dioxide, with or without the presence of contaminating solid particles on the metal surface, the humidity being gradually increased from zero upwards. As long as the humidity was low, the metal remained bright, but there was a sudden outbreak of corrosion at about 60% relative humidity; a second very marked increase occurred about 80%, which may represent the point at which the water in the capillaries of the rust-gel becomes sufficiently " free " to produce further rusting—a view largely due to Patterson and Hebbs; the lower critical humidity may represent the point at which sufficient water has been absorbed to produce attack in presence of sulphur dioxide. These matters

their liability to corrode, but according to Binaghi, the contrary is the case (R. Binaghi, *Métaux et Corros.* 1949, **24**, 216, esp. p. 222). See also p. 712 of this book.

deserve study in the original papers (W. H. J. Vernon, *Trans. Faraday Soc.* 1923–24, **19,** 886; 1927, **23,** 159; 1935, **31,** 1668, 1678; *Trans. Inst. chem. Eng.* 1939, **17,** 171; *Korros. Metallsch.* 1938, **14,** 213; *Chem. and Ind.* (*Lond.*) 1943, p. 314; W. S. Patterson and L. Hebbs, *Trans. Faraday Soc.* 1931, **27,** 277, esp. p. 282. See also recent views by E. Ll. Evans and E. G. Stroud, *Chem. and Ind.* (*Lond.*) 1957, p. 242, who quote Russian opinions).

The function of the corrosive constituents of dust may be partly to provide an electrolyte (largely regenerated when rust is formed*) and also to produce hygroscopic bodies which can absorb water from unsaturated air. They can also act in other ways. On stainless steel, dust particles may produce local screening from oxygen so that the protective film fails to be kept in repair—a view advanced by H. T. Shirley during the discussion of a paper by W. H. J. Vernon, Chem. Eng. Group. Soc. chem. Ind., Feb. 5, 1937.

Spreading of Rust. Patterson studied the corrosion of iron exposed over water, and found at first an acceleration with time, then a retardation and finally a fairly constant rate of attack. If rusting starts at isolated points, and spreads out—let us suppose—in expanding circles, the rate will tend to increase as the circles become bigger, since the length of their perimeters, along which growth is occurring, becomes greater; when the circles begin to meet one another, this length must decline—thus explaining why acceleration is followed by retardation. When once the whole surface is covered with rust, the only possible change is a thickening of the rust layer; in due course the corrosion-rate becomes roughly constant, since rust is not very protective (under out-door conditions the rate of rusting sometimes tends to fall off as time passes—as shown in Hudson's tests, although here also a fairly constant rate is ultimately reached).

In Patterson's experiments, the extension of attack on mild steel produced rust patches of irregular form—as previously noticed by Vernon—whereas on pure iron, the expanding patches were nearly circular. This concentric expansion has been studied optically by Canac, and a mathematical discussion of the relationship between area covered and time has been provided by the Author (p. 939 of this book). The references are W. S. Patterson, *J. Soc. chem. Ind.* 1930, **49,** 203T; F. Canac, *C.R.* 1933, **196,** 51; U. R. Evans, *Trans. Faraday Soc.* 1945, **41,** 365.

Moisture Films on a Metallic Surface. Before the spreading of rust can be understood, the surface condition of metal exposed to an unsaturated atmosphere requires study. Bowden and Throssell have examined this, using (1) a microgravimetric method, (2) an optical method based on

* When sodium chloride acts on iron, the anodic and cathodic products (ferrous chloride and sodium hydroxide) interact in presence of oxygen to give rust, and the sodium chloride is theoretically regenerated in full. When a carbon particle with sulphur dioxide adsorbed settles on iron, the sulphuric acid, formed by catalytic oxidation, acts on the iron, giving ferrous sulphate, which is oxidized by air to the ferric state; the subsequent mechanism suggested on p. 485 involves no consumption of SO_4^{--}.

polarized light—somewhat similar to that used by Tronstad and Winterbottom in determining oxide-film thicknesses (pp. 790–792). Both methods led to the same conclusions—namely that even at 90% relative humidity the thickness of the water-film was not greater than about two molecules, and that it was only about one molecule at 60% relative humidity. This conclusion was reached in work both on gold, which was probably free from an oxide-film, and on aluminium, which undoubtedly carried one; it was also confirmed on platinum, silver and zinc blende (ZnS); contrary to common belief, the amount of water vapour adsorbed on a clean surface is no greater than the amount of an organic vapour. However, a small contamination with a hygroscopic substance can alter the situation completely; 10^{-7} g./cm.^{2} of potassium hydroxide on the surface would take up 5 molecular layers of water from an atmosphere of 50% relative humidity before the establishment of equilibrium, whilst 25 molecular layers would be built up if the relative humidity were 90%. This probably explains the much higher values for water up-take published by earlier authors, and at once shows why minute particles of hygroscopic bodies, settling on iron, can start rusting. The paper deserves study (F. P. Bowden and W. R. Throssell, *Proc. roy. Soc.* (*A*) 1951, **209,** 297; *Nature* 1951, **167,** 601. Cf. W. H. J. Vernon, *ibid.* 1951, **167,** 1037).

Filiform Attack. It has been remarked that corrosion on steel does not always spread as an expanding circle. Sometimes it advances outwards along thread-like tracks of the form shown schematically in fig. 92. Filiform attack—as it is called—first attracted attention on lacquered steel, when it was found to spread under the lacquer; this occurred in tobacco tins, and acid fumes from the tobacco were held responsible. Later it was found on bare steel and on many other metals; the tracks produced on bare steel are usually shorter than those formed under lacquer, but are similar in character. The tracks often run fairly straight for a certain distance and then sharply turn off through an angle. If the advancing head of a track happens to approach an earlier-formed part of its own length, or if it approaches a track which has started from another point of origin, it changes direction, either turning away sharply or running alongside the old track, without touching it; hardly ever does it intersect its earlier path.*

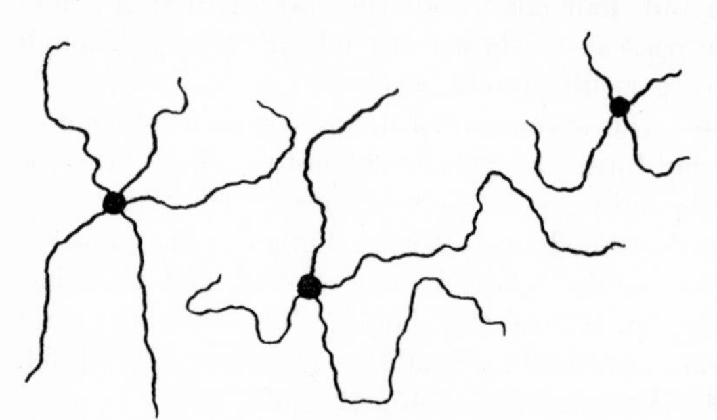

FIG. 92.—Filiform (thread-like) attack (schematic).

* Tracks on safety razor-blades are apparently exceptional in this respect, as they do cross one another—a feature emphasized by J. F. Kayser, *Chem. and Ind.* (*Lond.*) 1952, p. 1126. T. Geiger (*Schweiz. Arch. angew. Wiss.* 1956, **22,** 16) describes cases where there is apparent crossing of two filaments, but careful inspection shows that really they approach very closely and then diverge.

Filiform attack has been studied by several experimenters, notably by G. F. Sharman, *Nature* 1944, **153**, 621; F. Hargreaves, *J. Iron St. Inst.* 1952, **171**, 47; M. van Loo, D. D. Laiderman and R. R. Bruhn, *Corrosion* 1953, **9**, 277; W. H. Slabaugh and M. Grotheer, *Industr. engng. Chem.* 1954, **46**, 1014; R. St. J. Preston and B. Sanyal, *J. appl. Chem.* 1956, **6**, 26; A. Bukowiecki, *Schweiz. Arch. angew. Wiss.* 1955, **21**, 165; J. F. Kayser, *Metal Treatm.* 1952, **19**, 111. See also general discussion in *Chem. and Ind.* (*Lond.*) 1952, p. 1126; further information is provided in "Chemistry Research" 1954, p. 18; 1955, p. 18; 1956, p. 17 (H.M. Stationery Office).

The study of filiform corrosion on steel requires that the surface shall be inoculated at scattered points with some salt, and a method due to Vernon has been used with success by Preston and Sanyal. The salt chosen is powdered, dried, sieved through muslin and suspended in carbon tetrachloride; a specimen is dipped in the suspension for a few seconds and emerges with particles sporadically distributed over the surface. It is then exposed to air of the required relative humidity. At low humidities, no change is produced; the attack occurs only above a critical value—about 58% in the case of sodium chloride. The change is assessed by the gain in weight. With sodium chloride, the rate of weight-increase increases as the humidity rises, being roughly twice as great at 97% as at 94% (in both cases, however, it falls off with time). In contrast, ammonium chloride produces a maximal effect at about 70%, the effect being definitely less at 97%. Certain very hygroscopic salts, such as the chlorides of lithium, calcium and magnesium, produce effects at humidities well below 50%—a point to be borne in mind where iron articles have to be stored in atmospheres likely to contain traces of these salts; Preston and Sanyal suggest that when protective measures for stored articles are under discussion, the atmospheric pollution in the store should be analysed.

In the case of sodium chloride on steel at 99% relative humidity, general corrosion soon occurred around each nucleus particle, but filiform growth started on the 5th day; in contrast, calcium sulphate produced the filiform type after 12 hours. In these experiments the width of the tracks varied appreciably even on the same specimen, but the width of any particular track was fairly uniform over its whole length.*

The corrosion occurring along the track produces a perceptible trench in the metal (0·0002 to 0·0003 inch has been mentioned as the depth) with deeper pits at intervals. The tracks on iron consist of a blue head within which corrosion is actively proceeding, producing ferrous compounds, and a red body representing parts formed earlier where corrosion has come almost to a stand-still, so that the ferrous compounds have been oxidized to ferric. Some tracks show bright interference colours. The rate of advance varies; Hargreaves mentions 0·03 to 0·08 mm. per day as typical, although sometimes in cold weather the advance may temporarily cease.

The mechanism suggested by Preston and Sanyal is that the original salt nucleus absorbs water, and on the tiny area covered by the saturated

* Another authority has described a case where the widest track had 400 times the breadth of the narrowest.

solution, anodic and cathodic spots are formed close together—as happens in the opening stages of drop corrosion (p. 118); the products then interact to give ferrous hydroxide, regenerating the salt; the ferrous hydroxide absorbs oxygen and the surface of the solution covers itself with a membranous envelope of hydrated ferric oxide. Now the corrosion process continuing inside the envelope involves an expansion (which, it may perhaps be suggested, is due in part to the cathodic evolution of hydrogen), and sooner or later the envelope will burst at its weakest point, and the exuding liquid will form a " head " outside the original site of the attack; this will soon set up corrosion, cover itself with a new membranous envelope and in due course, when fresh internal pressure has developed, the new envelope will burst. The sequence of events (outburst, corrosion and membrane formation) is repeated an indefinite number of times, and the effects combine to form a continuous thread.*

The tracks usually continue in a fairly straight line, until diverted suddenly, probably by some obstacle; they run more uniformly if the surface has been polished. Various explanations for the straightness of the tracks and the sudden diversions have been offered, but the following seems the most probable. Part of the cathodic reaction proceeds outside the membrane on the area around the " head ", since in presence of salt the invisible layer of liquid is sufficiently thick to carry the corrosion current, and the good supply of oxygen creates the necessary conditions for a cathodic change; the corresponding anodic reaction will occur on the area within the membrane. This will mean that alkali is present on the metal all round the head, but there will clearly be more on the sides than in front, since at the sides it will have been accumulating for a longer time. Now if a small defect were to arise in the membrane at the sides, the interaction between alkali and iron salts would probably repair it; this is less likely to occur at the front where there is least alkali. Thus the track, having started to move in a given direction, will often continue in that direction. We can similarly understand why, if the head chances to approach its own body (or a track from a different point of origin), the alkali left by the cathodic reaction at a former stage is sufficient to repair any defects in the envelope at the point of approach, explaining the sudden change of direction already mentioned. If the moisture-film as a whole is acid, the cause of the diversion is not strictly speaking the formation of an alkaline zone along the flanks of the tracks, but a zone of diminished acidity, which will be less favourable for the advance, may be expected.

Although in laboratory experiments, it is convenient to inoculate specimens with salt nuclei, it seems fairly certain that in practice filiform attack can start from inclusions in the metal. Hargreaves mentions a certain

* E. Ll. Evans makes the pertinent comment that most of the ideas of filiform growth have been too specific, quoting the early belief that lacquer plays an essential part in the mechanism; it is now known that filiform growth occurs on surfaces carrying no lacquer or other organic coating. He adds " a general theory for filiform spread of corrosion, should be valid not only for Fe but Al, Mg and Zn as well ".

specimen of polished steel which showed many tracks starting at slag inclusions; it was re-polished, soaked in alcohol to wash out damaged inclusions, dried and exposed under conditions which formerly had produced tracks; very few new tracks appeared.

In service, filiform corrosion is most serious where it occurs under lacquer or enamel. It has been observed inside motor-cars, beneath the black finish, at places where the finish was thin, also on nickel-plated, chromium-plated or enamelled steel, but apparently not on tin-plate. Experiments by Preston and Sanyal showed that filiform corrosion of the long thread-like type met with in service could not be produced in the laboratory at 100% relative humidity; only ordinary heavy rusting occurred at the nuclei. It was necessary to reduce the humidity and facilitate the access of water vapour to the steel. In one case, a nitrocellulose lacquer on steel was broken by a pin-point; on exposure to air of 80% relative humidity, filament growth started within seven days. In another experiment, where sodium chloride was present below the lacquer, they placed drops of water outside it for 20 seconds and then removed them; subsequent exposure to 88% relative humidity produced filiform growth, but it was absent in cases where the liquid drops had been left in position.

Indoor Corrosion related to Gas Heating. It has sometimes been suggested that the ravages of atmospheric corrosion could largely be averted if all the coal which today is burnt directly in domestic grates was used to produce coal-gas, which would then be burnt in heating dwelling houses. This book is not the place to discuss how far a type of coal suited for the domestic grate is suited for the gas-works, but it must be pointed out that, whereas the increased use of coal-gas for heating might well reduce outdoor corrosion, it would in some situations increase indoor corrosion owing to the sulphur still present in the gas as carbon disulphide, thiophene or other organic sulphur compound. The law only concerns itself with sulphur present as hydrogen sulphide; this must be " absent " from town gas—which in practice probably means that it is reduced to about 1·5 ppm.* However, the sulphur present as carbon disulphide and thiophene remains. When coal-gas is burnt in a room not provided with a flue, the oxides of sulphur derived from the sulphur present as carbon disulphide, thiophene and other organic sulphur compounds, in conjunction with the large amount of water vapour produced by the combustion, can cause serious damage to metal objects in the room and particularly to steel windows.† The use of gas-heating in churches has produced much damage, especially to organs. Some organ builders now refuse to instal or even overhaul organs unless gas-heating is replaced by some other system. Experiments have shown

* Methods of estimation are described by C. G. T. Prince, *J. appl. Chem.* 1955, **5**, 364. See also J. J. Priestley, *Chem. and Ind.* (*Lond.*) 1955, p. 1372. For mechanism of removal of hydrogen sulphide by bog iron ore, see L. A. Moignard, *ibid.* 1957, p. 1005.

† Most of sulphur is burnt to sulphur dioxide, but some trioxide is formed in gas flames. See A. Dooley and G. Whittingham, *Trans. Faraday Soc.* 1946, **42**. 354.

that the pollution generated during a service with doors and windows closed reached " London smog " levels (E. Ll. Evans, Priv. Comm., July 11, 1958).

Even if the gas were completely free from sulphur, the moisture, condensing on unprotected metal-work, would cause damage, but where the condensate becomes acid, the damage, except to lead, is far worse; if there is a good flue, the trouble is largely overcome, except in rooms situated near the top of a building, where the products of combustion enter the room in some wind conditions.

Much work has been carried out in connection with the corrosion of gas-burning appliances by their own combustion-products; the damage to other metal-work, and particularly that of steel windows and cupboards—seems to have received insufficient attention of late either from the Gas Authorities or from Architects, who could do something to reduce the trouble by suitable design. Some useful research work and other information is, however, provided by J. W. Cobb, J. W. Wood and E. Parrish, *Trans. Instn. Gas Engrs.* 1934–35, **84,** 246; F. Taylor and J. W. Wood, *ibid.* 1939–40, **89,** 339; R. F. Littlejohn and G. Whittingham, *J. chem. Soc.* 1952, p. 3304; A. H. Raine, *Rep. Progress Appl. Chem.* 1946, **31,** 45, esp. p. 53; J. B. Reed, *ibid.* 1950, **35,** 24; esp. p. 33; J. Welsh, *Chem. and Ind.* (*Lond.*) 1951, p. 1114; H. G. Pyke, *Gas World* (*London*) 1948, **128,** 309; G. G. T. Prince, *J. appl. Chem.* 1955, **5,** 364; R. J. Bennie, Corrosion Symposium (Melbourne University) 1955–56, p. 430.

Corrosion of Electrical Equipment and Scientific Instruments. The efficiency of electrical instruments—if not sufficiently protected—may be affected by corrosion in several ways. Any attack on the coils of meters will reduce the cross-section of the wires, thereby increasing resistance and rendering the readings inaccurate. Even more disastrous may be the effect of corrosion-products at contacts, since, when a low potential difference is operating, they may prevent current from flowing altogether. The tarnish-films containing sulphur possess some conductivity of their own, so that, for some purposes and on some types of contact, they may perhaps cause relatively little trouble. The less conspicuous film of basic sulphate caused by damp corrosion may cause more disturbance, if it remains in position, since a basic sulphate may be expected to be an insulator; however, where the switch contacts are designed to slide over one another, there is some prospect that such layers, presumably soft, may be removed.

It has to be remembered that sometimes the formation of a thin oxide-film at a switch is actually welcome in preventing metallic bridging and welding between the two surfaces. For "light duty applications", involving currents measurable in milliamperes but potential differences possibly reaching 250 volts, the oxide must be thick enough to prevent these occurrences but not so thick as to cause a serious increase in contact resistance. Hunt has here supplied some useful information. On palladium and its alloys, an oxide-film is formed on heating above 400°C., but is decomposed at about 800°C.; on ruthenium, oxide is produced about 600°C. and decomposes about 1000°C.; platinum and iridium remain free from oxidation up to their melting-points. Certain palladium alloys have special value in avoiding metallic transfer. Often hardness is desired, since if dust or dirt particles become embedded, the contact is rendered imperfect; platinum alloys with 10 to 20% iridium, 4% ruthenium or 8% nickel are all in use, largely on account of their hardness (L. B. Hunt, *Platinum Metals Review*, 1957, **1**, 74).

Black deposits are sometimes found on platinum contacts. They arise if benzene or similar hydrocarbons have been adsorbed; the heat of the arc removes the hydrogen, leaving carbon. This can cause a resistance which for light-duty contacts is highly objectionable; it does not occur on metals which carry oxide-films, as benzene is not sufficiently firmly adsorbed. For heavy currents, however, such carbonaceous deposits may actually be useful, since the arc, which on clean platinum would always be struck at the same place, becomes better spread out (*Platinum Metals Review* 1957, **1**, 134).

The subject of corrosion at electric contacts is too large to be discussed here; those interested should consult M. Cocks, *Proc. phys. Soc.* [*B*] 1954, **67**, 238; E. B. Moss, *Trans. Soc. Instrum. Tech.* 1951, **3**, 143; P. Quinn, *Metallurgia* 1955, **52**, 115; A. Keil and C. L. Meyer, *Elektrotech. Zeitsch.* 1952, **73**, 31; A. Keil, *Werkst. u. Korrosion* 1952, **3**, 263; U. R. Evans, *Metal. Ind.* (*Lond.*) 1948, **73**, 10. See also the two books entitled " Electric Contacts ", by Ragnar Holm (Gebers, Stockholm), and L. B. Hunt (Johnson Matthey) respectively, and the book on " Physics of Electric Contacts " G. F. Llewellyn Jones (Clarendon Press).

There are various other sources of corrosion trouble in instruments. Phenolic resin mouldings may evolve ammonia which corrodes brass, nickel or silver; flux residues and unsuitable metallic contacts (p. 213) are possible causes of intensified attack (P. Mabb, *Corros. Tech.* 1956, **3**, 217, 232). See also discussion of " Effect of Vapours from Wood and Organic Materials " on p. 493.

A meeting dealing with corrosion in instruments was organized in 1955; papers by U. R. Evans, J. G. Walford, B. W. Balls and C. Lamond, will be found in *Trans. Soc. Instrum. Tech.* 1955, **7**, 123–129; discussion, p. 133.

The growth of metallic " whiskers ", which is only remotely related to corrosion phenomena, causes trouble in telephone and other electrical equipment; see K. G. Compton, A. Mendizza and S. M. Arnold, *Corrosion* 1951, **7**, 327. See also p. 652.

Wet Atmospheric Corrosion

Drop Corrosion under service conditions. A drop of sodium chloride solution placed on a horizontal iron or zinc plate, and prevented from evaporating, produces anodic attack in the centre, the cathodic periphery remaining immune (p. 119). In the open air, the continued existence of an isolated drop on metal is exceptional; either it evaporates, or merges with other drops and runs off. However, at places where two curved metal surfaces come to within a few millimetres of one another without touching, drops are apt to lodge, and even if the first one runs off, another will probably be formed at the same point; thus if the water is saline, anodic attack will develop, concentrated on a limited area. This probably explains why sheets of corrugated galvanized iron piled upon one another on the ground near the sea sometimes suffer serious attack at points where they approach each other, although the same galvanized iron, used as roofing in the same locality, resists the sea-air reasonably well; in the absence of crevices, any zinc chloride produced at a point at one moment would probably be precipitated by the sodium hydroxide formed later, and conditions are relatively favourable to the formation of a protective film.

Outdoor Corrosion. In general, corrosion of metal exposed out of doors is promoted by the presence of acid fumes—such as are present in industrial and urban districts where coal is burnt—or by the presence of salt, for reasons already explained in connection with drop corrosion. In absence of salt or acid, rain-fall—even if heavy—produces relatively little attack, but in Great Britain, an industrial island, there is probably no locality where acid and chlorides are entirely absent. Rain will frequently contain acid, but in urban districts the moisture condensing on metal in the evening will often develop much higher acid concentrations than rain falling from a height. Thus where conditions are favourable to condensation of moisture, some of the most intense corrosion may be met with at relatively sheltered places; on the fully exposed places, the rain quickly washes away any acid produced by condensation. Schikorr states that a dew sample collected in 1935, apparently near Berlin, was essentially N/50 sulphuric acid; a sample collected on the lower side of a glass roof contained Cl^- as well as SO_4^{--}. Nevertheless, the steel protected from rain was less corroded than steel exposed to rain. American experience—cited by Copson —is the opposite. It is clear that the damage produced at a sheltered situation will vary very much according as it is favourable to the condensation of moisture and to the absorption by the moisture-film of sulphur dioxide, which will probably be catalytically oxidized to sulphuric acid. Further information will be found in papers by G. Schikorr, *Z. Elektrochem.* 1937, **43,** 697; *Arch. Metallk.* 1948, **2,** 223; R. L. Copson, *Proc. Amer. Soc. Test. Mater.* 1945, **45,** 555.

Effect of Air Pollution on Outdoor Corrosion. Most of the atmospheric corrosion in Great Britain and in Germany is due to pollution with sulphur oxides derived from the burning of coal. The relationship between the corrosion-rate for steel and zinc and the sulphur dioxide content of the air is very close. At places where the pollution is high, the corrosion-rate is high; at a given place where contamination varies with the season, the corrosion is most rapid when the sulphur dioxide content is highest. Schikorr's curves showing the variations with time of (1) sulphur content, (2) humidity and (3) corrosion-rate at three places in or near Berlin are impressive; the correlation between sulphur-content and corrosion-rate is marked. Hudson's data also deserves close study (J. C. Hudson, *Times Scientific Review*, Spring 1953, p. 14; J. C. Hudson and J. F. Stanners, *J. appl. Chem.* 1953, **3,** 86; G. Schikorr, *Arch. Metallk.* 1948, **2,** 223; 1949, **3,** 76; *Arch. Eisenhuttenw.* 1943–44, **17,** 147; *Schweiz. Arch. angew. Wiss.* 1958, **24,** 33. See also C. J. Regan, *Chem. and Ind.* (*Lond.*) 1953, p. 1238).

There has been much spoken and written about the prevention of air pollution—mostly in connection with its detrimental effect upon human health. If the air of towns could be cleaned, not only the inhabitants but also the metal-work, building stone and textiles would benefit. Most of the discussion has been concerned with the elimination of soot, and since carbonaceous particles often carry adsorbed sulphur compounds, avoidance of soot might lessen corrosion, although it has been argued that, if the number of soot particles were diminished, the sulphur dioxide per particle would be increased. It seems, however, that sulphur oxides other than those adsorbed on soot can represent a major menace to metal, and probably to human health. When the pyrites which accompanies coal is burnt, most of the sulphur is oxidized to sulphur dioxide, but some to sulphur trioxide, which gives sulphuric acid by combination with moisture; a mist of fine sulphuric acid particles is very difficult to condense.

In London a considerable proportion of the coal is burnt at a few large power-stations, and the products of combustion are discharged from high chimneys and to some extent dispersed. From time to time, processes for absorbing the sulphur oxides have been put into operation at certain stations, and the purification figures obtained are impressive; unfortunately, the purifying process generally cools the gases, and instead of rising, they flow horizontally after emerging from the chimney, or even, if the washed mixture contains very fine water droplets in suspension, actually fall towards the ground, owing to the cooling associated with evaporation. Thus the benefits of purification are not so great as the figures suggest. Nevertheless it seems important to persevere with schemes for the treatment of exit-gases, not only in the interest of the health of men and materials, but for the value of the sulphur which today is being wasted. Some gas-cleaning processes enable the valuable sulphur to be recovered in marketable form such as solid ammonium sulphate or liquid sulphur dioxide. The promising Simon-Carves process is discussed by S. R. Craxford, A. Poll and N. J. S. Walker, *J. Inst. Fuel*, 1952–53, **25,** 13. See also Simon-Carves and Borough of Fulham, British Patent 525,883, Sept. 6, 1940. Other processes are mentioned by A. Parker, *J. roy. Soc. Arts* 1950–51, **99,** 85, esp. p. 100. Evidence based on aerial reconnaissance and photography showing how unwashed gases rise from chimneys as plumes whilst washed gases fail to do so is published in *The Engineer* 1959, **187,** 30.

Another process suited for plants too small for recovery schemes, in which the acid is neutralized with ammonia, so that only the relatively harmless ammonium sulphate is discharged into the air, was mentioned on p. 469.

Methods of preventing air pollution lie outside the scope of this book, but those interested in this all-important subject should consult the two reports of the "Committee on Air Pollution", 1953 and 1954 (H.M. Stationery Office); Proceedings of the Edinburgh Conference of National Smoke Abatement Society (1947), esp. the contributions of G. Nonhebel and G. Lessing. C. J. Regan, *Chem. and Ind.* (*Lond.*) 1953, p. 1238; W. A. Damon, *ibid.* 1955, p. 1268; Sir H. Beaver, *ibid.* 1955, p. 220; G. E. Foxwell, *ibid.* 1955, p. 128; W. Mitchell, *ibid.* 1954, p. 102; R. Lessing, *ibid.* 1951, p. 421; N. M. Potter, *ibid.* 1946, p. 218; R. Ll. Rees, *J. Inst. Fuel*, 1952–53,

25, 350; P. J. McLaughlin, *Weather* 1954, **9,** 88; R. S. Scorer, *ibid.* 1955, **10,** 106; W. C. Turner, *ibid.* 1955, **10,** 110; A. R. Meetham, " Atmospheric Pollution: its origin and prevention " (Pergamon Press).

In Great Britain, chlorides probably play a smaller part in atmospheric corrosion than sulphur oxides. This applies to most metals; except on aluminium and lead corrosion is generally more serious in industrial districts than at the seaside. However, a combination of chlorides and sulphur compounds is the worst of all; the presence of chlorides can stimulate the attack by sulphur acids, and since coal often contains salt, the products of combustion are rendered more dangerous through the presence of hydrogen chloride. The reactions* which lead to the formation of hydrogen chloride have been studied by K. H. Brinsmead and R. W. Kear, *Fuel* (*London*) 1956, **35,** 84.

Corrosion in Hot Countries. In the neighbourhood of surf beaches in the tropics, high rates of corrosion have been recorded—sometimes five times those experienced in British industrial districts. Tests carried out, especially in Nigeria, at different distances from the sea show how rapidly the corrosion-rate falls off as the amount of salt reaching the specimens diminishes; even screening from the sea wind can greatly cut down corrosion. In one case, recorded by Ambler, the corrosion-rate was 5·6 g./dm.^{2}/month at 50 yards from the open sea, but only 0·25 at a spot 2 miles inland. At some places the corrosion of ferrous metals and zinc may be proportional to the rate of deposition of salt on a damp textile surface.

Good accounts of the situation in Nigeria are provided by H. R. Ambler and A. A. J. Bain, *J. appl. Chem.* 1955, **5,** 437; H. R. Ambler, *Nature* 1955, **176,** 1082. Information about corrosion in the tropics generally (particularly the Pacific Islands) is furnished by K. G. Compton, *Trans. electrochem. Soc.* 1947, **91,** 705. Compare experience in S. Africa described by W. J. Copenhagen, " Corrosion of metals and modern anti-corrosive practice " (Report published in connection with Public Works of S. Africa) May 1954.

Very interesting results are shown by exposure tests carried out near Kanpur in India. Although corrosion-rates on steel were considerable in the rainy months around July and also in winter, practically no corrosion occurred in March and April; moreover, specimens in tests started in March or April and continued for 12 months suffered less than those started in one of the other months and continued for 12 months—an example of the importance of initial conditions in determining the corrosion-rate. (Curiously enough—in contrast with results in Great Britain and America—the initial conditions of exposure had no influence on the corrosion-rate of zinc, which in India was corroded much more slowly than steel.) The *average* relative humidity of a month was little guide to its corrosive character;† actually

* Probably atomic hydrogen is important, starting the reaction

$$H + NaCl = Na + HCl$$

† The fact that humidity in itself is no guide to corrosion-rate is shown by comparisons between measurements on the beach at Lagos, Nigeria, and a site in the jungle 30 miles inland, where the humidity is similar, the chloride content less, and the corrosion-rate far smaller (H. R. Ambler, *Nature* 1955, **176,** 1082).

at Kanpur there are only two months when this exceeds the critical value of 70% (July and August), but corrosion of steel occurs in most of the other months, being largely due to dew and rain; of these dew produces most corrosion, as the concentration of electrolytes in the condensed moisture becomes higher when there is no draining away or flushing by rain. Another instructive feature of the Indian tests was the close correlation between the *weight-increases* of specimens *before* cleaning and the *weight-losses after* removal of corrosion-products; this removal was carried out either with 5% sulphuric acid under cathodic protection or in hydrochloric containing tin chloride and arsenious oxide (Clarke's solution). The report deserves study: see B. Sanyal and G. K. Singhania, *J. sci. ind. Res.* (*New Delhi*) 1956, **15B,** 448.

In air that is relatively free from salt and sulphur acids, the corrosion-rate of steel tends to fall off with time, even though there is high humidity; this indicates that rust can be protective. In presence of high salinity or pollution with acid fumes the slackening of corrosion-rate with time is much less marked, doubtless because the large amount of soluble matter formed keeps the structure open and porous. American writers stress the different protective qualities of different rust samples—emphasizing the fact that dark rust is the most protective (F. L. LaQue, *Proc. Amer. Soc. Test. Mater.* 1951, p. 495; C. P. Larrabee, *Steel* 1946, **119,** No. 11, p. 134; *Corrosion* 1953, **9,** 259).*

Does rust promote rusting ? The idea that rust can be protective seems at first inconsistent with the wide-spread belief that " rust promotes rusting ". The fact seems to be that only if the rust contains ferrous salts does it stimulate fresh attack; ferrous chloride may be present in rust formed near the sea and ferrous sulphates in rust formed in towns. The difference is brought out in a recent experiment described by Buckowiecki. He placed a patch of rust containing about 1% of chloride on a steel specimen, exposed it to air of 95% relative humidity at 20°C. for a week, removed the rust by light brushing and found fresh corrosion set up where the rust patch had been; corresponding experiments conducted with rust almost free from electrolyte showed practically no attack where the rust patches had been. Earlier experiments on rust under paint had pointed to the same effect. Britton placed drops of distilled water on an iron surface producing patches of rust; the specimen was then painted and exposed out of doors; no serious break-down of the paint-coat occurred at the site of the rust, but corresponding experiments in which drops of salt solution had been used showed premature failure at the places where these had been applied. Similarly, Mayne found that winter rust, which contained ferrous sulphate, caused premature break-down of paint applied over it, whereas summer rust, which was relatively free from ferrous salt, allowed

* Vernon has replotted the means of Hudson and Stanner's data, and his curves bring out a falling off of the initial high corrosion-rate during the first year, after which there is a tendency to straighten out between 1 and 5 years. Larrabee's curves show a similar tendency to become straight, but rather later —generally after about 3 years (W. H. J. Vernon, Priv. Comm., July 18, 1958).

much longer life for the coating. The experiments cited are described by A. Bukowiecki, *Schweiz. Arch. angew. Wiss.* 1957, **23,** 97, esp. p. 103; U. R. Evans and S. C. Britton, *J. Soc. chem. Ind.* 1930, **49,** 173T, esp. p. 178T; J. E. O. Mayne, *J. Iron St. Inst.* 1954, **176,** 140, 143. The reasons are indicated on p. 485.

Effect of Copper in Steel. The early exposure tests carried out by the American Society for Testing Materials pointed to the good effect of copper in steel in increasing its resistance to atmospheric corrosion. In judging this early work, it should be remembered that present-day steel—even where there has been no intentional addition of copper—usually contains appreciable amounts of copper introduced with scrap or pyrites residues. In Buck's early work, carried out when nearly copper-free material was doubtless more common than today, he found this to corrode about twice as fast as that to which copper (0·15–0·34%) had been added (D. M. Buck, *J. Industr. engng. Chem.* 1913, **5,** 447).

Although considerable improvement can be obtained by increasing the copper-content to 0·15%, there seems to be no advantage in going beyond 0·3% copper. Steels containing copper and phosphorus together resist better than those containing copper alone—as emphasized by American and German authorities. Even better results can be obtained by copper in conjunction with chromium and other elements in relatively small quantities. Although not to be compared with 18/8 chromium/nickel stainless steels, these " low-alloy steels " provide an enhanced resistance to atmospheric corrosion at comparatively little added cost. They require painting, but are likely to suffer far less than unalloyed steels if the original paint-coat should become damaged, so that the steel is left unprotected for a period before fresh paint is applied. Some of the results obtained by Hudson and Stanners after 5 years' exposure at Sheffield on 60 varieties show that several varieties of low-alloy steel resisted three times as well as unalloyed mild steel during atmospheric exposure (also twice as well when immersed in sea-water). Of the twelve steels which behaved best, all except three contained chromium. The composition and performances of five of them are stated in Table XIII. The two best (A) and (B) contained nickel and molybdenum, but they only slightly surpassed (C) and (D), which were free from those expensive constituents. (E) is a widely used commercial product, whilst (F) is unalloyed steel inserted for comparison.

TABLE XIII

RATE OF CORROSION OF LOW-ALLOY STEELS (J. C. Hudson and J. F. Stanners)

	Composition	Rate
(A)	0·8 Cr, 3·0 Ni, 0·5 Mo	1·4 mils per year*
(B)	0·1 Cu, 3·1 Cr, 0·1 Ni, 0·4 Mo	1·4 ,, ,, ,,
(C)	2·8 Cr, 1·5 Al, 0·8 Si	1·5 ,, ,, ,,
(D)	0·5 Cu, 2·4 Cr,	1·6 ,, ,, ,,
(E)	0·5 Cu, 1·0 Cr, 0·16 P, 0·8 Si	2·0 ,, ,, ,,
(F)	(Unalloyed)	5·9 ,, ,, ,,

* 1 mil = $\frac{1}{1000}$ inch.

The cause of the good effect of the alloying constituents is still a matter for discussion. Chromium and aluminium are generally regarded as contributing the protective character of their own oxides (cf. p. 64). Copper is considered by Copson to act by producing a more protective form of rust, containing its sulphur as a sparingly soluble basic copper sulphate, perhaps similar to that constituting the protective patina on copper; he considers that the rust in unalloyed steel contains a freely soluble sulphate, which is quickly washed away by the rain. However, the facts as known today do not seem to rule out the possibility that the function of copper is to precipitate any hydrogen sulphide formed by cathodic reduction of sulphur dioxide present in the condensed moisture, which hydrogen sulphide, if not removed, would stimulate the anodic reaction (p. 239).

Several authorities have emphasized the fact that the rust on copper steel is darker than the less protective rust formed on steel nearly free from copper, and there is a tendency to regard darkness in rust as evidence of protective character. Undoubtedly some dark rusts confer a modicum of protection, but if the darkness arises from an intermediate layer of magnetite, the symptom is an unfavourable one; a situation giving rise to ferrous compounds at the metallic surface which only become fully oxidized to the ferric state at an appreciable distance from it, is less favourable to building up a protective film than where ferric oxide or hydroxide can be formed in close contact with the steel over the whole surface.

The subject is a complicated one and deserves study in the original papers, particularly those of J. C. Hudson and J. F. Stanners, *J. Iron St. Inst.* 1955, **180,** 271; F. L. LaQue, *Corrosion* 1950, **6,** 72, esp. p. 76; F. L. LaQue and J. A. Boylan, *ibid.* 1953, **9,** 237; C. P. Larrabee, *ibid.* 1953, **9,** 259; H. R. Copson, *Proc. Amer. Soc. Test. Mater.* 1945, **45,** 554; A.S.T.M. 1952, Preprint **70**; C. Carius and E. H. Schulz, Mitt. Forschungs. Institut Vereinigte Stahlwerke " Uber der Einfluss des Kupfers auf den Rostvorgang gekupferten Stahles an der Atmosphäre und in verschiedenen Wassern " (1929); K. Daeves, K. F. Mewes and E. H. Schulz, *Korros. Metallsch.* 1943, **19,** 233; K. Werny and R. Eschelbach, *Werkst. u. Korrosion* 1950, **1,** 16; W. H. J. Vernon, *J. Iron St. Inst.* 1950, **165,** 290; R. Franks, *ibid.* 1955, **181,** 341. See also discussion, *ibid.* 1957, **186,** 94. C. F. Pogacar and E. A. Tice, *Corrosion* 1951, **7,** 76, discuss the use of low-alloy steels in coke plants. J. C. Hudson, *J. Iron St. Inst.* 1951, **169,** 13, 250 describes uses as sleepers and in coal wagons; B. J. Kelly, *Corrosion* 1951, **7,** 196, discusses uses in railway hopper-car body sheets, and H. F. Brown, *ibid.* 1950, **6,** 268, general application on the railways.

Alloys of the stainless steel class resist very much better than low-alloy steels. Molybdenum and tungsten contents further improve resistance; those stainless steels which do not contain molybdenum or tungsten often darken in urban air and occasionally suffer spotting or pitting. Some interesting five-year tests are described by H. T. Shirley and J. E. Truman, *J. Iron St. Inst.* 1948, **160,** 367; see also discussion by W. H. J. Vernon, *ibid.* 1950, **165,** 290; T. H. Turner, p. 292; S. C. Britton, p. 293.

Tests on steels containing different amounts of chromium at Kure Beach

Niagara Falls and New York show that the corrosion as judged by weight-loss decreases steadily as the chromium increases, and becomes extremely small at 12·5% chromium; however, it requires about 18% to make the steels " virtually stainless ". Inspections of stainless steel which has been in service on buildings in New York and elsewhere for 6 or 7 years have shown that it is often in good condition, although sometimes there is a dark film and on rare occasions pitting. In a marine atmosphere the alloys containing molybdenum develop less rust than the others. (See Symposium on atmospheric weathering of corrosion-resisting steel 1946 (Amer. Soc. Test Mater.), especially papers by W. O. Binder and C. M. Brown, p. 1, H. A. Grove, p. 18 and W. Mutchler, p. 29.)

Corrosion of Ancient Iron. The complaint that modern iron does not resist corrosion as well as the " good old stuff " is no new one. Pliny, 1900 years ago, told how an iron bridge constructed across the Euphrates by Alexander the Great received some replacements in Pliny's own time; the new material was soon noticed to be rusting, whereas the original Macedonian iron remained in good condition.

Ancient iron erections in Eastern countries survive uncorroded today, but portions which have been cut off and placed in rather more corrosive atmospheres (e.g. in Europe) have rapidly been attacked. The famous Delhi pillar erected in A.D. 310 and moved to its present site in A.D. 1052 is still rust-free, but the atmosphere is free from pollution and the humidity is generally low; the sulphur content of the iron is only 0·006%, but Hudson thinks that its good condition is due primarily to the non-corrosive atmosphere rather than to the composition of the iron. Details regarding the pillar, with discussion of the causes of its resistance, are provided by J. C. Chaudhari, *Joswa* 1957, Vol. **5**, No. 1.

In general two factors may contribute to the longevity of ancient iron. First the material itself, having been produced by reduction of the ore with charcoal, should be more free from sulphur than modern material—whether steel or wrought iron—produced by methods involving coal. Secondly, the atmosphere will be relatively free from sulphur acids in many cases, and may actually be alkaline. In most cities situated in different parts of the world, the air contains both sulphur oxides and ammonia; the white deposit which disfigures window glass in Great Britain is largely ammonium sulphate. However, whereas in industrial regions the acid constituents are in excess, the ammonia may often be in excess in the towns of hot countries, and the air, although polluted, is non-corrosive. Causes for the good preservation of ancient or mediaeval iron are discussed by K. Daeves, *Stahl u. Eisen* 1940, **60**, 245; J. C. Hudson, *Nature* 1953, **172**, 499.

Wrought Iron. The corrosion behaviour of the wrought iron as made in Great Britain is definitely *different* from that of steel; for some purposes, iron may be preferable to steel, for others inferior, but statements that the two materials behave alike are untrue; possibly such statements refer to materials sold under the name of wrought iron, but not possessing the essential characteristics of British wrought iron. The remarks in this chapter refer to British iron made by puddling, which contains much slag, and has usually been " piled ". They do not refer to Swedish iron, which contains little slag and has usually not been piled, nor to American Aston-Byers iron, which is essentially low-carbon steel into which slag has been intentionally introduced. Undoubtedly the latter two materials possess value for certain purposes, but their corrosion-resisting properties are not outstanding; in fact, Hudson's tests suggest that they are inferior to ordinary mild steel. Even very pure electrolytic iron shows no advantage over steel in long-period atmospheric exposures; in the Cambridge outdoor tests pure iron took longer to become rust-covered, but, when once rusty, corroded rather more quickly than the particular steel used for comparison.

In the British process of wrought-iron manufacture, " pigs " of cast iron are placed on the hearth of a reverberatory furnace lined with iron oxide, and when the flames play upon the hearth, the carbon of the cast iron combines with the oxygen of the oxide; the " puddler " stirs the whole mass with a long iron bar, and when finally he moves the balls of pasty material from the furnace, they consist of a mixture of metallic iron and

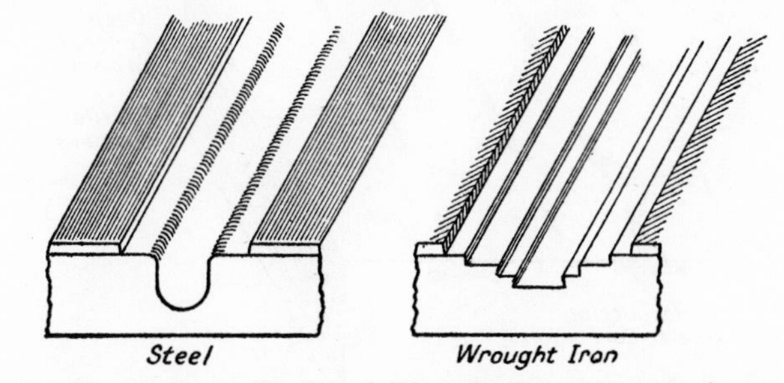

93.—Anodic attack on Steel and Wrought Iron at a gap in protective coat.

slag—much of which is squeezed out under the " shingling " hammer; then the balls may be rolled or hammered into shape. Frequently, however, the material undergoes " piling "; the rolled plate is cut into pieces of convenient size, and about six are placed on the top of one another to form a pile which is strongly heated, and passed through a rolling mill. The various components weld together, the intervening oxide on the interior faces being largely squeezed out. This gives " once-piled iron "; it is possible to cut it up again, and repeat the process, so as to obtain " twice-piled iron ".

Wrought iron, thus produced, consists of definite zones which behave very differently from one another when subjected to corrosive conditions. Sometimes highly laminated steel shows a similar zonal effect, but in general, well made steel corrodes at a similar rate in all directions. A rather exaggerated example of the difference between the two materials, based on early work at Cambridge, is represented in fig. 93. Specimens of steel and wrought iron plate were coated with tar paint and a scratch-line was traced on the face of each, exposing the metal. They were then subjected to anodic attack with the same current for the same time. The total corrosion was determined by Faraday's law and was therefore nearly the same in both cases; but in the case of the steel it penetrated uniformly from the scratch producing a groove of roughly semi-circular section, whereas on the wrought iron the groove was stepped, being broader but less deep than that on the steel. Not all varieties of wrought iron show this stepped outline equally well, but there is always some tendency for the zones to behave differently.

When wrought iron is placed in dilute acid, this eats its way inwards along certain zones, producing a characteristic wooden appearance; attack along the face is confined to certain points, but it proceeds readily along susceptible zones parallel to the surface (fig. 94). In the case of atmospheric exposure, the same thing occurs, but the bulky solid corrosion-product, being now formed internally along the susceptible planes, may cause a swelling, since the volume occupied by the corrosion-product is greater than that of the metal destroyed in producing it. Since the oxygen-supply at the interior is scanty, black magnetite is likely to be formed instead of red

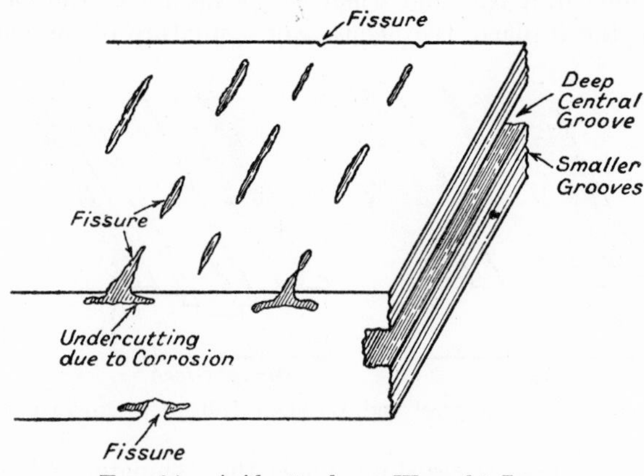

FIG. 94.—Acid attack on Wrought Iron.

rust, and the deterioration may even escape notice until finally the swelling produces disintegration. In the case of iron bars, which have originally been rolled in two directions at right angles, the effect may be to cause the iron to split into rectangular blocks—a phenomenon shown clearly by certain old iron railings on the sea-coast. In industrial situations, old iron-work (ladders, etc.) often shows swelling and disintegration, but at exceptional places where (perhaps through bolting or attachment to some other member) swelling has been rendered impossible for geometrical reasons, the corrosion has ceased to penetrate and the material has locally remained sound. Wrought iron in the roof of the Tropical Palm-house at Kew Gardens, erected about 1840, has suffered corrosion and distortion at places where the structure has provided traps for condensed moisture; many panes of glass have been broken owing to the dimensional changes. Similarly damage has been caused by the expansion of wrought iron in the dome of St. Paul's Cathedral. The zonal corrosion of wrought iron has much in common with the layer corrosion of light alloys (p. 678).

An interesting example of wrought iron corroding with different velocities in different directions is provided by the finger-post at the " Cross Hands " crossing, near Chipping Campden; this was erected in 1669, and the four thin iron hands, with outstretched fingers, point along the four roads. At present (1958) the hands are painted white, and it is possible that the iron has carried a protective covering from its early years. Despite this, some of the fingers have corroded and disappeared completely along the whole or part of their length; others are intact, whilst the flat surfaces, representing the palms and backs of the hands, are also in good condition. Probably on those fingers which have disappeared, microscopic discontinuities had been left in the covering at the tips, so that corrosive influences reached the Q-zones (see below), and proceeded laterally along them at a rate which must have been of the order of an inch per century. Any corrosion starting at discontinuities in the covering on the flat surfaces must have penetrated far more slowly—otherwise the hands would be showing perforation.

It is evident that, if the fact that wrought iron contains susceptible and non-susceptible zones is borne in mind whilst designing, erecting and maintaining metallic structures, there may often be advantage in using wrought iron. It is necessary to take great care to protect—with paint

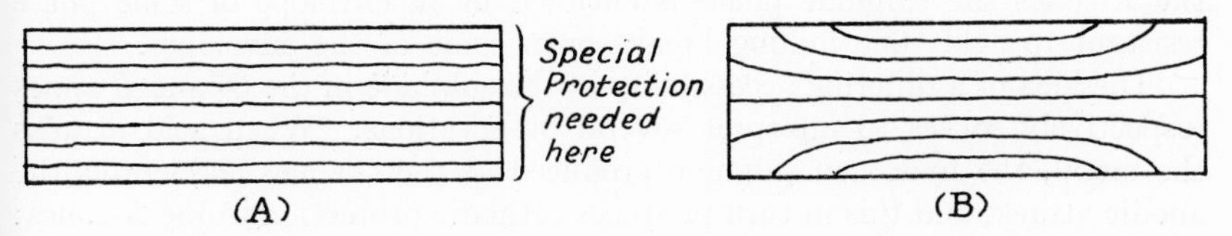

FIG. 95.—Zonal arrangement on Wrought Iron, (A) Zones parallel to face, (B) Zones deflected near sides of bar (J. P. Chilton and U. R. Evans).

or otherwisc—the sheared edges where the susceptible zones would otherwise meet the atmosphere (fig. 95(A)). If this is done, the faces may need less frequent repainting than if steel had been used, provided that in the material employed the zonal system lies accurately parallel to the external surface. Where through careless manufacture, the zones cut the faces near the corners, the results will be less good (fig. 95(B)). The advantage of wrought iron will only be achieved if these principles are intelligently applied. In the past it has been a matter of chance as to whether wrought iron has been erected in a suitable or unsuitable manner, and thus—whilst some structures have lasted for long periods in a most remarkable manner—others have behaved no better, possibly worse, than steel would have done. In Hudson's tests (p. 515), wrought iron exposed without any covering was corroded more slowly than steel low in copper, but behaved no better than steel to which copper had been added.

Laboratory work has been carried out at Cambridge to decide the causes of differences of behaviour of the various zones. Preliminary work, first by the Author and then by Lewis, showed clearly that two sorts of zones are present, but Chilton, who carried out a more extensive investigation, discovered a third zone. A summary of the work is given below, but those interested should consult the original papers by U. R. Evans, *J. Soc. chem.*

Ind. 1928, **47,** 62T; K. G. Lewis and U. R. Evans, *Rep. Iron and Steel Inst. Corr. Comm.* 1934, **2,** 209; J. P. Chilton and U. R. Evans, *J. Iron St. Inst.* 1955, **181,** 113; 1957, **185,** 497; 1957, **186,** 98. They should also study the extensive Italian work described by N. Collari, *Il Calore* 1942, No. 11; 1946, Nos. 2, 7; 1948, No. 11; 1951, No. 7; 1957, No. 4, who reaches similar conclusions, and compare the views of F. N. Speller "Corrosion: causes and prevention" (McGraw-Hill); see also *Corrosion* Nov. 1958, p. 98. Apparently much of his discussion refers to Aston-Byers iron—a fact which doubtless explains any apparent discrepancies. The 1935 edition of Speller's book contains more discussion of this matter than the 1951 edition.

Chilton calls the three types of zones, *Q*- (quickly corroding), *R*- (resistant) and *V*- (very resistant) zones. Of these, the *Q*- and *R*-zones were found to differ in the accessibility of the sulphur; the aggregate sulphur content of both was found to be about the same, but the amount of hydrogen sulphide evolved by the action of 6N hydrochloric acid on coarse millings taken from the *Q*-zones was much higher than in that evolved by coarse millings from the *R*-zones; with fine millings the difference was less. Now the slag present can be seen to consist of two phases, and it would appear that on the *R*-zones the sulphide phase is enclosed in an envelope of some phase resistant to acid; fine milling breaks open some of the envelopes.

The idea of a differing accessibility of the sulphide in the *Q*- and *R*-zones respectively serves to interpret several observations. When acid attacks the metal, the hydrogen sulphide produced at the *Q*-zones accelerates the anodic attack, and this in turn produces cathodic protection to the *R*-zones; the characteristic wooden appearance, due to selective attack, is thus explained. When however an acid liquid is used which already contains hydrogen sulphide, the attack on the *R*-zones is naturally stimulated, and accordingly the contrast between the R- and *Q*-zones is found to diminish, although zones of a third type (the *V*-zones) now remain unattacked.

Similar results are obtained on anodic attack in N/100 sodium chloride; the *Q*-zones are attacked more rapidly than the *R*-zones, unless hydrogen sulphide is present, in which case the contrast is diminished.

The assumption that the sulphide in some parts of the iron is surrounded by slag envelopes receives direct support from the excellent photographs of Collari's latest paper, which show granules of iron sulphide "incapsulated" in the slag grains; the phenomenon is not peculiar to wrought iron, since Whiteley has published photographs of silicate globules in steel with manganese sulphide embedded in them. Collari's photographs also bring out clearly the existence of resisting and non-resisting zones alternating with one another (N. Collari, *Il Calore* 1957, No. 4, especially figs. 1, 2, 3 and 4; J. H. Whiteley, *J. Iron St. Inst.* 1948, **160,** 365, esp. fig. 10).

The fact that silicate may under some circumstances surround sulphide is doubtless connected with relationships existing between the values of the interfacial energies of the three phases; as to why the sulphide is embedded in certain zones and not in others could doubtless be discovered if a detailed research were conducted into the various metallurgical operations. It is open to question whether just such a research would be justified today, since most authorities consider that the manufacture of wrought iron by the puddling process is likely to cease in the not distant future (see below).

As already stated, the *V*-zones can continue to resist acid containing hydrogen sulphide after a period long enough for the *R*-zones to be eaten away. The cause of the enhanced resistance was revealed through collaboration with A. Berger of Brussels, whose skilful micro-analysis showed that the *V*-zones really consist of narrow bands of alloy steel—the nickel and copper contents being extremely high. British wrought iron has no intentionally added alloying constituents, but nearly all scrap now contains alloying metals. Consequently puddled iron before piling generally contains small amounts of copper and nickel—sometimes 0·2% of each; in the *V*-zones, the contents were found in one case to reach 0·7% of copper and 2·3% of nickel; probably if it had been possible to analyse the exact centre of a *V*-zone, still higher figures would have been recorded. Thus in wrought iron, although no alloying constituents have been paid for, effective barriers of highly resistant alloy are actually present.

V-zones do not occur on unpiled iron, and must be formed during piling —probably in the following way. When the pieces are heated before piling, the iron becomes oxidized on each face, whilst the nickel and copper remain in the metallic state, so that the part of the iron just below the oxide layer becomes definitely enriched in the valuable constituents. When the pile passes between the rolls, the oxide is squeezed out (the presence of silicates renders it more fusible) and the two metallic walls weld together to give a single *V*-zone rich in nickel and copper; if the welding is incomplete, a "double-walled" *V*-zone emerges. There is no enrichment of nickel or copper in an *R*-zone.

The wrought-iron industry is threatened by the scarcity of skilled "puddlers", and it seems possible that some years hence puddled iron—as made today—may become unobtainable. Mechanical puddling processes have been tried, with varying success. If puddled iron ceases to be available, the possibility of piling steel would remain. Some authorities foresee difficulty in ensuring that the components of the pile would weld up properly, owing to the absence of slag; but it has been found possible to pile Swedish iron, which contains practically no slag. Piled steel would not be identical with wrought iron as made today, for it would contain no *R*-zones. On the other hand, the valuable *V*-zones should be present—assuming that the technical difficulties could be overcome—and by careful selection of scrap which chanced to be rich in alloying constituents, a useful material might be obtained.

For further discussion of these possibilities, Chilton's paper should be consulted.

Comparison between Different Materials

Exposure Tests in Great Britain. Several series of exposure tests have been carried out by J. C. Hudson, first for the British Non-ferrous Metals Research Association and afterwards for the Iron and Steel Institute Corrosion Committee which later became a committee of the British Iron and

Steel Research Association. His untiring energy has provided a vast amount of reliable data regarding the behaviour of different materials towards different types of atmosphere. The earlier work on non-ferrous materials was carried out on small specimens at five exposure stations representing respectively (1) *rural*, (2) *suburban*, (3) *urban*, (4) *industrial* and (5) *marine* atmospheres; they included specimens exposed in the *open air* where the damage was assessed by (*a*) *loss of weight* on plate specimens (after removal of corrosion-products) or (*b*) the *loss of electrical conductivity* on wire specimens; also specimens *screened from rain* in Stevenson Screens (boxes designed for meteorological instruments, constructed with double roofs and doubly louvred sides) where the assessment was based on (*c*) *gain in weight* of wire specimens and also (*d*) *loss of tensile strength*. His later work on ferrous materials was based on large, heavy plates; the majority were exposed in the painted condition, and the conditions of testing are described on p. 578; but much information has also been obtained regarding the behaviour of unpainted specimens. The tests on low-alloy steels have already been reported (p. 506).

The non-ferrous exposures were carried out on 16 materials; the results obtained at Birmingham, where the corrosion was more rapid than elsewhere, are shown in Table XIV. There was less difference between the various materials than might be expected; the most rapidly corroded materials (zinc and brass) suffered less than five times the damage of the most resistant tested (nickel-chromium alloys). Tables XV and XVI show that industrial and urban atmospheres produce much more damage than rural ones—the marine and suburban atmospheres being intermediate.

The tensile tests on non-ferrous materials are not included in Tables XIV and XV. They showed general correlation with the electrical resistance tests,

TABLE XIV

AVERAGE THICKNESS CORRODED AT BIRMINGHAM EXPRESSED IN MILLIONTHS OF AN INCH PER YEAR (J. C. Hudson)

	Weight-increase of Wire in Stevenson Screens	Electrical Resistance Tests on Wire completely exposed	Weight-loss of Plates completely exposed
80/20 nickel-chromium alloy	111	311	85
Lead	—	—	145
Arsenical copper	126	353	154
High-conductivity copper	116	302	158
70/30 nickel-copper alloy	376	656	243
Cadmium-copper (0·8% Cd)	120	350	179
Nickel	245	565	230
Aluminium bronze (3·5% Al)	170	410	229
Silicon bronze	—	342	228
80/20 Copper-nickel	204	361	213
Tin bronze (6·3% Sn)	151	305	243
70/30 brass	172	594	316
High-purity zinc	280	—	376
Ordinary zinc	—	—	388
60/40 brass	—	—	408
Compo wire	719	721	—

except in the case of the brasses, which often underwent dezincification (p. 474); the redeposited copper seems to contribute to the electrical conductivity, but not to the strength—as is easily understood.

TABLE XV

AVERAGE THICKNESS CORRODED ON THE 16 NON-FERROUS MATERIALS SHOWN IN TABLE XIV EXPRESSED IN MILLIONTHS OF AN INCH PER YEAR (J. C. Hudson)

	Weight-increase of Wire in Stevenson Screens	Electrical Resistance Tests on Wire completely exposed	Weight-loss of Plates completely exposed
Cardington (rural)	32	156	74
Bourneville (suburban)	67	248	119
Wakefield (industrial)	90	368	170
Birmingham (urban)	190	395	221
Southport (marine)	71	345	144

TABLE XVI

CORROSION OF IRON AND ZINC IN ONE YEAR EXPRESSED IN MILLIONTHS OF AN INCH (J. C. Hudson)

	Ingot Iron	Zinc
Llanwrtyd Wells (rural)	1888	118
Calshot (marine)	2104	121
Motherwell (industrial)	3137	180
Woolwich (industrial)	3472	146
Sheffield (industrial)	3888	576
Dove Holes Railway Tunnel (average of two sides)	2953	3529

Table XVI, which shows the relationship between iron and zinc, may provide a link between Hudson's studies of non-ferrous and ferrous materials respectively; except in the railway tunnel, zinc is attacked far less rapidly than ingot iron, which, however, is somewhat less resistant than steel, as shown in Table XVII; the latter table indicates the improvement obtained by introducing copper into steel.

TABLE XVII

RELATIVE CORROSION-RATES OVER FIVE YEARS ON DESCALED SPECIMENS (J. C. Hudson, "Corrosion of Iron and Steel" 1940, p. 82 (Chapman & Hall))

	Calshot (marine)	Sheffield (industrial)
Steel with 0·02% Cu	100	100
,, ,, 0·2% Cu	92	83
,, ,, 0·5% Cu	87	82
Ingot iron	120	106
Staffordshire wrought iron	95	78
Scottish wrought iron with 0·12% Cu added	93	66

Details of Hudson's work will be found in his papers and reports. References dealing with non-ferrous materials are J. C. Hudson, *Trans. Faraday Soc.* 1929, **25,** 177; *J. Inst. Met.* 1930, **44,** 409. Hudson's work on ferrous metals is published in the Five Reports of the Iron and Steel

Corrosion Committee 1931, 1934, 1935, 1936, 1939,* and in his book "The Corrosion of Iron and Steel" (Chapman & Hall). See also J. C. Hudson, *J. Iron St. Inst.* 1948, **160,** 276; *Nature* 1955, **176,** 749. J. C. Hudson and J. F. Stanners, *J. appl. Chem.* 1953, **3,** 86; *J. Iron St. Inst.* 1955, **180,** 271.

Exposure Tests in U.S.A. Another collection of valuable data is based on exposure tests carried out for 20 years at various American stations. The earlier results are conveniently collected in a report of the American Society for Testing Materials dated February 27, 1946, and entitled "Symposium on Atmospheric Corrosion Tests on non-ferrous metals". It contains papers by E. A. Anderson on zinc, W. A. Wesley on nickel, A. W. Tracy on copper, G. O. Hiers on lead and tin, also by E. H. Dix and R. B. Mears on aluminium alloys. A more recent report brings the information up to the end of a 20-year testing period, and includes papers by H. R. Copson on 24 metals and alloys, C. J. Walton and W. King on aluminium alloys, E. A. Anderson on zinc, A. W. Tracy on copper and H. R. Copson on nickel (*Special Tech. Pub.* **175** (1956)).

These papers should be studied in the original reports, but certain points may be brought out here. Copson, reporting on 20-year tests at seven stations, finds copper to be better than brass and much better than manganese bronze. Aluminium lost less weight than copper but its loss of strength was similar; certain aluminium alloys containing copper suffered greater loss of weight and strength than unalloyed aluminium. Nickel suffered greater loss of weight and strength than copper at certain stations, but actually gained strength at others, where the weight-loss (both on nickel and on copper) was low. Lead also gained strength (as compared with the control specimen) in New York and would seem to be a desirable material in urban districts. Zinc lost strength in towns and in localities where the mists or dews were acids; the composition, in Anderson's tests, made little difference to the rate of deterioration.

The aluminium-base alloys, also exposed for 20 years, were found to lose strength rapidly at first, but then the loss became slow and constant (see p. 523). The deterioration-rate as measured by depths of attack also fell off with time. Excellent resistance was shown by some of the clad products. Under marine conditions, the clad varieties of heat-treatable alloys also resisted well, but unprotected alloys containing copper, if improperly quenched or aged, became very susceptible to intergranular attack. Anodic coatings were found more protective than dipping treatments; indeed anodized alloys, when further coated with paints pigmented with zinc chromate or aluminium, afforded admirable protection for 20 years in marine situations and 22 years in urban situations.

The results obtained at the famous testing station at Kure Beach also deserve close study. Some of the time-corrosion curves are not straight—an important point, since it shows that hasty deductions from short-time exposures may well lead to serious error. Sheet zinc corroded quicker at Kure Beach in the opening year than at Middletown, Ohio, but the corrosion-

* A sixth report is in the press.

rate fell off with time, and had become slow after 8 years, whereas that at Middletown remained nearly constant. In contrast, the corrosion of low-copper steel tended to become faster with time at Kure Beach, and to fall off at Middletown. The various curves are reproduced by J. L. Kimberley, *Corrosion* 1957, **13,** 385t.

Behaviour of Copper and its Alloys Outdoors. The changes observed on copper and bronze exposed out of doors differ from those on most other metals in that a type of corrosion which produces a protective green patina is often desired on aesthetic grounds. Indeed, complaints are sometimes heard regarding the long period which must elapse before new copper roofs develop the pleasing green associated with the roofs of old buildings; on a new roof the deposit may for some years be a dark aggregate of sulphide, oxide and soot. This has caused effort to be devoted to the development of processes for stimulating corrosion of the type which leads to the green patina (W. H. J. Vernon, *J. Inst. Met.* 1932, **49,** 153; J. R. Freeman and P. H. Kirby, *Metals and Alloys* 1934, **5,** 67; G. L. Craig and C. E. Irion, *ibid.* 1935, **6,** 35).

Vernon and Whitby have studied the composition of deposits formed naturally on copper exposed out of doors. On roofs 70 to 300 years old the deposit is mainly $CuSO_4.3Cu(OH)_2$ (brochantite) with a little $CuCO_3.Cu(OH)_2$ (malachite). On newer roofs the basicity is less and occasionally normal copper sulphate may be present. Near the sea $CuCl_2.3Cu(OH)_2$ is found, but in seaside towns, the content of basic sulphate greatly exceeds that of basic chloride. On ancient bronze, buried over many centuries in deserts containing salt and soda, basic chloride and carbonate occurs in the deposit. Basic and normal nitrate have been found —it is understood—near power-stations. Information on these matters is supplied by W. H. J. Vernon and L. Whitby, *J. Inst. Met.* 1929, **42,** 181; 1930, **44,** 389; W. H. J. Vernon, *ibid.* 1932, **49,** 153; 1933, **52,** 93; *J. chem. Soc.* 1934, p. 1853; J. C. Hudson, *J. Inst. Met.* 1929, **42,** 198; C. G. Fink and E. P. Poluskkin, *Trans. Amer. Inst. min. met. Engrs.* 1936, **122,** 90; H. Stäger, *Korros. Metallsch.* 1935, **11,** 73, esp. p. 87.

The main trouble arising from the use of copper in architecture is that water descending from a copper roof can greatly stimulate corrosion if it flows over galvanized iron or aluminium (cf. p. 205). Less dangerous but highly disfiguring are the green stains produced on stone, cement or porcelain exposed to water which has previously been in contact with copper or bronze. The stains can be removed with acid, and it is stated that the trouble virtually ceases when a satisfactory patina has been formed on the copper or bronze surface (S. Baker and E. Carr, *Chem. and Ind.* (*Lond.*) 1957, p. 1332, esp. p. 1334).

The care and cleaning of bronze statues, the removal of stains from pedestals, and the cleaning of museum exhibits is discussed by J. F. S. Jack (*Museums J.* 1951, **50,** 231); J. W. Martin (Ministry of Works, Tech. Paper **12,** 1949); A. Lucas, " Antiques " (Arnold) 1924, p. 80; H. J. Plenderleith, " The Conservation of Antiquities and Works of Art " (Oxford University Press). See also the pamphlets on the " Care and Cleaning of Sculptures "

(Royal Society of British Sculptors) and on " Cleaning and Restoration of Museum Exhibits " (British Museum). American experience has been placed on record by C. G. Fink and C. H. Eldridge in " Restoration of Ancient Bronze (Museum) Objects " (Metropolitan Museum, New York).

The actual destruction of metallic copper during atmospheric exposure is generally slow. Tests carried out at four localities in Connecticut lasting 20 years suggest that there is little to choose between the various grades of copper, and that arsenic—in contrast with British results—has no retarding influence (cf. p. 490) (D. H. Thompson, A. W. Tracy and J. R. Freeman, " Corrosion of copper: results of 20 years tests ", A.S.T.M. 1955; A. W. Tracy, *Corrosion* 1951, **7**, 373).

Comparison of Materials. Some excellent surveys of the behaviour of various metals have been provided by G. Schikorr in his section of " Passivierende Filme und Deckschichten " 1950, p. 231 (Editors: H. Fischer, K. Hauffe and W. Wiederholt; publisher, Springer) and in a report on " Blech in Konstruktion u. Festigung " Jan. 17, 1956; also *Z. Electrochem.* 1936, **42,** 107; 1937, **43,** 697; *Korros. Metallsch.* 1941, **17,** 305; *Metalloberfläche* 1947, **1,** A115; 1958, **12,** B33 (the latter paper dealing with the rusting of cast iron); *Schweiz. Arch. angew. Wiss.* 1958, **24,** 33.

His observations on the behaviour of different materials in German atmospheres deserve summarizing:—

Iron in presence of rain, develops ochre-coloured rust which in a few days becomes dull brown; this is sometimes powdery and easily broken off. In certain localities rust develops during the winter and falls off in spring in pieces measuring perhaps 1 $dm.^2$. In relatively unpolluted atmospheres, especially if chlorides are present, the rust may adhere and since the thickness of the rust-layer may be seven times as great as that of the metal destroyed, the corrosion may appear far more advanced than is really the case. On copper steel, the rust is smoother and darker.*

Aluminium develops a protective film on the parts freely exposed to rain even in towns, but is sensitive to lime and cement-mortar; in parts protected from rain, a white floury product accumulates, which is found to contain sulphate, whilst the condensed moisture contains sulphuric acid.

Lead is sensitive to alkalis and should not be laid in contact with lime mortar or Portland cement. It is attacked by the lower fatty acids given off by wood and by many varnishes and plastics.

Cadmium is stable to alkalis and condensed moisture, but is sensitive to the sulphurous constituents of the air, and to the lower fatty acids.

Zinc is also sensitive to the sulphurous constituents of the air; it is sensitive to alkaline materials and should not be laid in contact with cement or lime-mortar.

Copper is sensitive to ammonia, which, even in small amounts, causes season-cracking in (unannealed) brass (p. 691).

* J. C. Hudson adds the comment, " Pitting or intensified local attack usually occurs beneath adherent rust, whereas atmospheric corrosion generally leads to a fairly smooth uniformly corroded surface when the rust is of the granular type."

Nickel is less resistant to atmospheric attack than is sometimes supposed, developing first acid and then basic sulphate (cf. p. 489).

Magnesium alloys have greatly improved in corrosion-resistance during recent years; manganese additions counteract the bad effect of iron and nickel. As in the case of aluminium, the corrosion is most serious in the parts sheltered from rain.

Some interesting calculations of the cost of corrosion-damage to various metals used for architectural purposes have been provided by Godard, who has collected numbers for the outdoor corrosion in many localities of aluminium, copper, lead and zinc. The figures in general assume uniform corrosion which—as is frankly pointed out—is not a valid assumption for aluminium, except in some rural atmospheres. The conclusion is that for marine atmospheres, lead and copper cost less than galvanized iron. In industrial atmospheres, lead is the most economical—its resistance being doubtless connected with the low solubility of lead sulphate. In rural atmospheres, aluminium appears to be the cheapest (H. P. Godard, *Engineering Journal* (*Montreal*) 1953, **36,** 844).

Lead fails to resist in special situations where abrasive dusts keep removing the lead sulphate layer (W. J. Cotton, *Corrosion* 1955, **11,** 469t).

Effect of Contact Materials on Corrosion. It has already been stated that the behaviour of materials depends not only on the atmosphere to which they are exposed, but also on the non-metallic materials with which they come into contact. Lead—despite its resistance to sulphur acids—is sensitive to organic acids given off by wood, and also to alkaline materials; it should not be placed in contact with mortar or Portland cement.

Aluminium is also somewhat sensitive to alkaline materials, although probably some of the cases attributed to attack by cement or mortar may really have been caused by the crevices between the aluminium and the cement or mortar rather than to the alkalinity. Oxygen-deficiency and retention of moisture in crevices both contribute to corrosion in crevices. Nevertheless contact with Portland cement, lime and plaster can set up corrosion, which a recent German research shows to be of a mild character; aluminium which is to be in contact with mortar should be coated—preferably with bitumen (E. Fischer and H. Vosskühler, *Aluminium* 1957, **33,** 606; F. C. Porter, *Corros. Prev. and Control,* May 1959, **6,** pp. 36, 41).

Cases of rapid failure of aluminium have been caused by contact with copper, or to wetting with water which has passed over copper or nickel; the accidental use of an alloy containing copper instead of a copper-free aluminium alloy has also proved disastrous. Contacts with steel plates and bolts require special consideration (p. 192), especially where insulation is impossible. In one procedure the steel bolts are cadmium-plated, the plates zinc-sprayed and the contact surface coated with wet paint, the assemblage being finally finished with a three-coat painting scheme.

Zinc is another alkali-sensitive metal, and should not be laid in contact with cement or lime-mortar; bitumen-impregnated felt is employed to separate zinc roofing from its support; in general, zinc roofs are more likely to suffer perforation from attack starting on the lower side than from that starting above.

Corrosion of Aluminium in Structures. The remarkable resistance of aluminium in certain situations has led to a wide advocacy of its general use. In some circles the belief exists that under no circumstances does aluminium require painting; this belief has probably been fostered by the advertising policy adopted by certain concerns, although the wiser aluminium firms do not claim that painting is never needed, but state (what is undoubtedly true) that if a paint-coat is applied to aluminium, repainting will be required far less frequently than on iron. Aluminium is often superior to steel as a basis for paint, and it is particularly unfortunate that its owners should be discouraged from painting it.* Aluminium protected by paint or other suitable means (with special attention to sheared edges) is in many situations a remarkably resistant material. The painting of aluminium is discussed on pp. 587–588, the cladding on p. 601, metal-spraying on p. 599 and anodizing on p. 249. In the present section, we are mainly concerned with the behaviour of unprotected material.

The situation is complicated by the fact that the ordinary man (and even some engineers) apply the name " aluminium " indiscrimately to the unalloyed metal and to all its alloys. These behave very differently among themselves. Those containing copper (although stronger than the others) are the least resistant to corrosion; if wrongly heat-treated they become liable to intergranular corrosion, or in other cases to layer corrosion with swelling and foliation. Under tensile stress there may be stress-corrosion cracking; cladding often prevents this, but since the cladding produces no striking alteration in appearance (so that its very existence may remain unknown to the owner of the metal), this excellent result has served to confirm the belief that " aluminium " can always be left unprotected. Other alloys, which are stronger than unalloyed aluminium, although weaker than copper-bearing alloys, are relatively resistant to corrosion. Clark, discussing behaviour at a large chemical works, states that the Al/Mg, Al/Mg/Si, Al/Mn alloys and also 99·5% aluminium all possess similar corrosion resistances; he estimates a life of 7 years for bare sheet under unfavourable conditions, as compared with only 2 years for bare galvanized iron (W. D. Clark, *J. Inst. Met.* 1955–56, **84,** 33).

In comparing aluminium and steel in an ordinary situation, a distinction must be made between the exposed portions and recesses. Unpainted steel will probably show red rust in a few days on the exposed surface, whereas aluminium may remain almost unchanged in appearance and will probably suffer very little attack on the exposed surface over long periods. However, at the crevices formed where aluminium comes close to another material—or even to a second aluminium surface—it is sometimes found to be severely attacked; it matters little whether this attack is due to crevice corrosion of the differential aeration type, or whether it is due to contact with cement, lime or other alkaline material; the point is that the aluminium is attacked

* In some cases, however, *unsuitable* painting may make matters worse; see p. 588.

most at just those places where the attack is least likely to be observed. If the "aluminium" is really an aluminium-copper alloy—particularly if it has been wrongly heat-treated—the attack may take the form of layer corrosion, and severe foliation may slowly develop; here again the change may escape observation until the damage has become serious. To summarize, steel produces a conspicuous red corrosion-product, generally on those parts most visible to view; aluminium may produce an inconspicuous white product at the points least accessible to inspection.*

Aluminium roofs on certain famous buildings have lasted for long periods without detriment; Panseri suggests that this is largely due to good design, which has avoided the accumulation of water in recesses; environment must also play a part (C. Panseri, *J. Inst. Met.* 1938, **63,** 15; *Korros. Metallsch.* 1939, **15,** 24; C. Panseri and A. Gragnani, *Aluminio* 1954, **23,** 627).

The lower sides of roofs sometimes corrode appreciably when the upper sides remain in good condition. An example, due apparently to condensation combined with industrial fumes, is described by P. Brenner, F. E. Faller and E. Höffler, *Aluminium* 1956, **32,** 6, 64.

In ordinary structures, the crevice corrosion of aluminium should be capable of being avoided by simple measures such as filling the crevices with jointing compound containing chromate, and then coating the external surface with chromate paint. The jointing compound DTD 369A is commonly recommended; it contains barium chromate, kaolin and a long-oil resin varnish, and can be applied—if desired—with a brush.† There are numerous proprietary joint-sealing compounds available, often containing chromate—some of which appear to be rather specially suitable for aluminium. One of these is understood to be based on castor oil, and possesses only slow drying properties—which should reduce the risk of cracks forming in the jointing compound itself, as a result of thermal changes or settlement. Another is based on synthetic rubber (S. G. Clarke, Priv. Comm., June 5, 1958).

Generalization is difficult and, in any particular case before adopting any particular treatment, all its possible effects should be considered. For instance, if zinc, cadmium or magnesium fittings are to be used, tests should be carried out to ascertain whether any vapours given off by the sealing compound can attack these metals. It is conceivable that a jointing compound containing chromate might in some circumstance produce unsightly staining, whilst one rich in hydrocarbons might cause deterioration of rubber, if that material is included in the design. No such consequences are, however, known to the Author, and it would seem that in general the

* This does not, of course, mean that crevice corrosion is unimportant on iron and steel; Hudson and Wormwell (p. 528) quote many examples of swelling and breaking due to voluminous rust in crevices.

† Recent work shows that compounds based on slightly more soluble chromates, such as potassium zinc chrome, zinc monoxychromate and calcium chromate, give better protection at joints than that based on barium chromate. Media which remain soft (e.g. cyclopentadiene + drying oil) give better results than those which harden (linseed oil + alkyd resin) (Ministry of Supply: unpublished work).

advantages of sealing the joints greatly exceed possible risks of leaving them unsealed. In doubtful cases, advice should be sought, and advantage taken of the information in the possession of organizations representing the interests of users, such as the Building Research Station, Garston, Watford, and also the National Chemical Laboratory (formerly the Chemical Research Laboratory), where an extensive Corrosion Bibliography is maintained.

If crevices are filled in, and protection applied where appropriate, trouble should be generally avoided. But clearly such measures will only be taken if the advisability of taking them is publicized. As long as the fiction that aluminium is so resistant as to need no precautionary measures is maintained by salesmen, it will be impossible to obtain the best results from this extremely valuable material.

A survey of aluminium bungalows (of which over 65,000 have been built in Great Britain since 1945), conducted by the Building Research Station, is described by Jones; his report deserves the closest attention. He recognizes that " the resistance of aluminium and certain of its alloys to simple atmospheric corrosion is high " and observes that (among the non-ferrous metals under discussion) " aluminium is in a class of its own . . . in that it has been applied on a large scale for structural purposes ". He describes, however, a good deal of corrosion trouble including " severe distortion through the growth of corrosion products and the creation of unsightly surfaces ". The major trouble has arisen with extruded high-strength alloy made from scrap, which has " developed, sporadically, severe laminar attack ". The immediate conclusion is that " at present the high-strength copper-containing alloys cannot safely be recommended for building purposes ". His paper with its illustrations deserves careful study; it deals also with examples of corrosion in other metals used in building (F. E. Jones, *Chem. and Ind.* (*Lond.*) 1957, p. 1050, 1435; see also E. Ll. Evans, p. 1436; R. W. Nuttall, p. 1436).

The increasing use of aluminium for windows introduce several questions. It has often been stated that aluminium windows require no painting, and general observation justifies this view so far as the portions washed by rain are concerned; there is sometimes appreciable attack on the portion sheltered by some projection, although it is seldom serious. The crevices which cannot be seen at all represent places more liable to corrosion. Some years ago a brochure sent out by a manufacturer advised that all aluminium surfaces which are to be in contact with masonry, mortar or steel-work should be coated; a zinc chromate jointing compound was one of the coatings suggested. This advice—which has not always been repeated in more recent trade publications—appears to be sound, and although examples could probably be quoted of aluminium surfaces which have remained unchanged after prolonged contact with masonry, it seems best to avoid risks; crevice corrosion is a chancy affair, since its initiation depends on coincidence between a susceptible spot in the metal and the exact position where oxygen-replenishment is, for geometric reasons, minimal. In his authoritative work, Brimelow advises that all aluminium-alloy lugs, screws and those surfaces of a frame or sub-frame that will be anchored in direct contact with masonry, timber or steel should be protected with a thick coating of bituminous or zinc chromate paint or a proprietary water-resistant jointing compound, to prevent corrosion. Steel anchors, screws and other fittings should be similarly painted or given a protective coating of zinc. The question of filling in crevices becomes particularly important in large buildings where a considerable fraction of the wall may consist of glass, and the possibility of dimensional changes due to the different thermal expansion of the various materials has to be borne in mind. In addition to the chromate jointing compound mentioned on p. 200, there are numerous proprietary compositions, some of them rather specially adapted for aluminium, although also suitable for crevices between other materials. These include putty-like mixtures with an oil-content higher than that of ordinary putty. For certain purposes, compositions which set slowly or incompletely are more valuable than those which harden rapidly (the latter may themselves sometimes develop cracks when ground-movements or dimensional changes occur); one of these are understood to be based on castor oil. There are also sealing compounds based on synthetic rubber. The choice between these may often depend on considerations other than corrosion (E. I. Brimelow, " Aluminium in Building (Macdonald); S. G. Clarke, Priv. Comm., June 17, 1958).

Outdoor Tests on Aluminium. The parts of an aluminium surface which are fully exposed to the weather undergo appreciable attack at first but this soon becomes slow and settles down to a roughly constant rate of wastage, which for many purposes can be neglected. Curves showing loss of strength or depth of attack obtained in the careful American tests organized from New Kensington become asymptotic to a line which is set at an angle to the time-axis; earlier predictions (from another source) that it would become asymptotic to a horizontal line (which would mean that if the aluminium is sufficiently thick its life is infinite) are seen to be incorrect.* For many purposes, however, provided that there is no crevice attack, the aluminium used on a structure may be expected to be in good condition when the time comes to dismantle that structure for some reason unconnected with corrosion. Figs. 96 and 97 show the type of curve obtained in the American tests; the results on copper are included for comparison. The papers contain much additional data and should be consulted in the original (C. J. Walton, D. O. Sprowls and J. A. Nock, *Corrosion* 1953, **9**, 345; W. W. Binger, R. H. Wagner and R. H. Brown, *ibid.* 1953, **9**, 440; C. J. Walton, F. L. McGeary and E. T. Englehart, *ibid.* 1957, **13**, 807t; C. J. Walton and W. King, *Amer. Soc. Test. Mater. Spec. Tech. Pub.*

* The values for the depth of attack, h, obtained after short exposures are often consistent with an equation $dh/dt = K_1(h_\infty - h)$, which would indicate a maximal depth, h_∞, incapable of being exceeded however long the experiments last; but they are also consistent with $dh/dt = K_1(h_\infty - h) + U$ indicating a slow attack of velocity, U, continuing indefinitely. The first term will swamp the second in the early stages, but will become negligible in comparison with the second in the later stages. Extrapolation from short-time results cannot distinguish between the two possibilities.

175 (1955). Cf. U. R. Evans, *J. Inst. Met.* 1952–53, **81,** 738; F. A. Champion, p. 739).

Some well-planned tests on experimental huts erected in industrial and marine situations by an aluminium firm to decide certain practical points involved in the use of aluminium in building showed after some years results

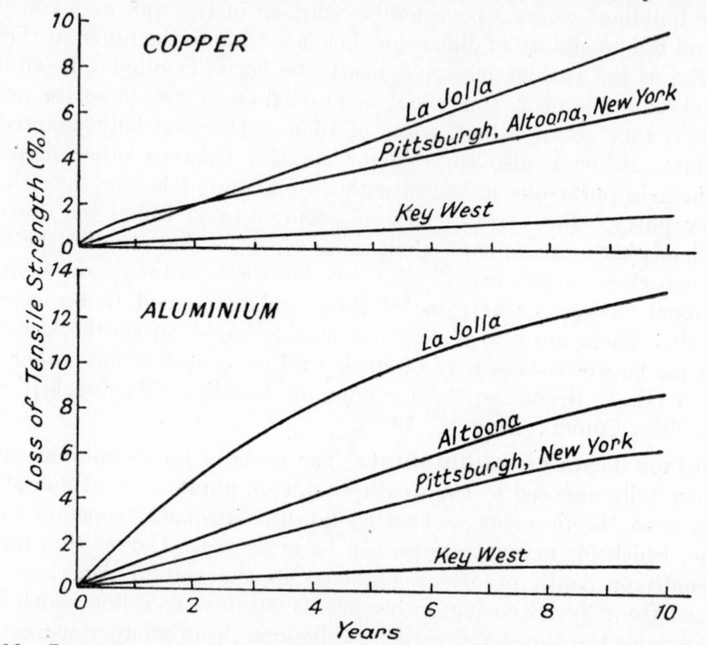

FIG. 96.—Loss of tensile strength during exposure of annealed Copper and 99% Aluminium (a stronger alloy containing 1·2% Manganese gives essentially the same curve) (C. J. Walton, D. O. Sprowls and J. A. Nock).

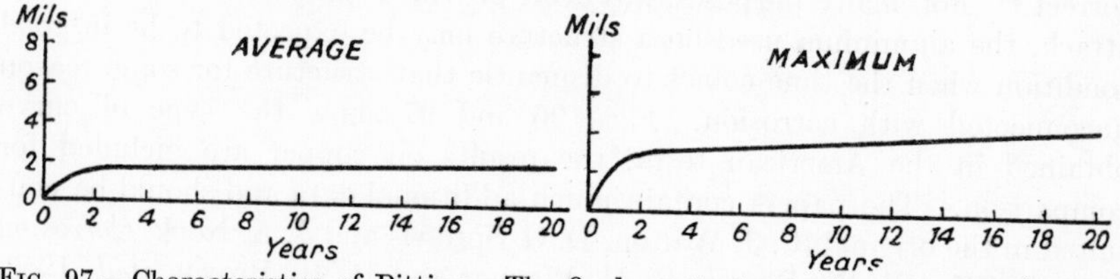

FIG. 97.—Characteristics of Pitting. The final constant rate for 99% Aluminium at New Kensington is of the order of 1 mil. per century, as given by the deepest pits on 9 × 12 inch panels. Larger areas might show deeper pits, and pitting is more rapid on the coast.

which were on the whole encouraging, although there was corrosion at places where the aluminium alloys had been left (intentionally) in contact with steel purlins; in the marine situation a window hinge seized completely after three months at contacts between brass and steel with aluminium. The results are described objectively, and suitable precautions suggested, by E. H. Laithwaite and E. W. Skerrey, *J. appl. Chem.* 1957, 7, 216.

A valuable series of tests on different aluminium alloys exposed for about

3 years in marine, industrial and country atmospheres in Holland, including riveted, bolted and welded specimens, is published, with good photographs, by E. M. J. Mulders, W. G. R. de Jager and J. W. Boon, Metaalinstituut T.N.O., 4e Aflevering, Publicatie 25d (1957).

Aluminium in Statuary. The absence of narrow crevices may explain why aluminium statuary has often resisted the weather so well, in contrast with stranded cables (see below). The statue " Eros " (cast from unalloyed aluminium) was erected in London in 1893 and has undergone very little attack from the highly polluted atmosphere—as shown by inspections carried out about 1937 and 1947; although the form of this work of art is by no means devoid of re-entrant angles, there are probably no very narrow capillaries running far into the material. About 1953, it was decided to replace the lead-lined bowl of the fountain below the statue by a bronze bowl and to make the water-circulation continuous; presumably the water will become acid and may be expected to dissolve copper; it is to be hoped that none of it will come into contact with the aluminium statue, owing, for instance, to spray carried upwards by wind eddies. Friese-Greene mentions spray from the fountain as the reason why, in one part, the lower sides had (in 1947) changed slightly more than the upper sides.* The state of the statue at different times has been described by R. S. Hutton and R. Seligman, *J. Inst. Met.* 1937, **60**, 67; G. H. Friese-Greene, *Light Metals* 1947, **10**, 508. See also *Research* 1953, **6**, 124.

Aluminium in Cables. The performance of aluminium in stranded conductors used in power transmission provides a contrast with that in statuary. In making the comparison it must be remembered that we are no longer dealing with castings (which may well have better protective skins) and that in some cables there will be dissimilar metals in contact, since steel, usually galvanized, is introduced to give strength. Champion and Skerrey state that in a pure marine atmosphere, the attack mainly occurs in the interior where the steel and aluminium wires are in contact, whereas in a mixed atmosphere the pitting of the aluminium occurs mainly on the outside, where industrial contamination predominates. The attack has sometimes been attributed to the cell aluminium | iron; but the steel is generally galvanized, and at first there is anodic attack on the zinc coating with some attack on the iron below; it is not impossible that the dissolved iron deposits magnetite on the aluminium, providing the micro-cells aluminium | magnetite. How far bimetallic couples are the cause of the deterioration in service is still a little doubtful, although in laboratory experiments aluminium coupled to steel was attacked three times as fast as aluminium coupled to aluminium. In service, the rate of deterioration is greatest in the industrial regions, and may be unconnected with the presence of the steel strands. It is, however, greater on line conductors than on earth conductors, since the former are more apt to collect the products of atmospheric pollution. A great deal of the trouble must be

* E. Ll. Evans suggests that the greater attack on the lower sides may have been due to the fact that they were not washed by the rain (cf. p. 502).

due to the retention of water in the capillary crevices between the strands or within adhering foreign matter. The best remedy is grease, which may increase the life by about 10 years. The subject is discussed in detail by J. S. Forrest and J. M. Ward, *J. Instn. elect. Engrs.* 1954, **101,** 271; see discussion by P. J. Ryle, E. W. Skerrey, A. M. Evans, J. Hérenguel, U. R. Evans and others. See also F. A. Champion and E. W. Skerrey, *Light Metals* 1952, **15,** 286; J. Hérenguel, *Métaux et Corros.* 1948, **23,** 242. Instructive pictures of external and internal corrosion in overhead conductors are provided by J. A. Airey, *Corr. Prev. Control* April, 1958, **5,** p. 44.

Aluminium in Transport. Situations can be envisaged where performance would be less good than the curves of figs. 96 and 97 would predict. Trouble might arise on a surface carrying a tight adherent layer of particles having a reducing character and capable of acting as an oxygen-screen. Whether for this reason or another, the use of unpainted aluminium alloys on railways in Great Britain has been mainly restricted to the London Underground sections, where good results have been obtained with an alloy specified to be low in copper; another precaution observed is the liberal employment of chromate jointing compound in the construction. For the main railway lines where coal fuel is general and carriages are apt to be bombarded continuously with grit particles—some of which will adhere—aluminium coaches have generally been painted, and there has been a tendency to revert to steel (T. H. Turner, letter of Dec. 3, 1956; Chief Mechanical Engineer, London Transport Executive, letter of Jan. 27, 1958; see also W. J. Hair, *Corros. Tech.* 1956, **3,** 8, esp. p. 11).

Copper-content for Aluminium Alloys. Regarding the amount of copper which should be tolerated in aluminium-magnesium and aluminium-magnesium-silicon alloys, the published facts show some inconsistencies, but there is reason to think that the upper limit accepted today is too high, at least for aluminium-magnesium-silicon alloys in the fully heat-treated conditions; in the solution-treated condition copper is less harmful.* The persons responsible for specifications may in some cases have been misled by the results of unsuitable tests. At the start of the Second World War, the upper limit fixed in certain German specifications covering marine and aircraft uses was 0·05%. Tests were then carried out based on salt solution containing hydrogen peroxide, which seemed to show that copper was harmless up to 0·2%, and the German authorities raised the limit to 0·1%. Such a test, however, is clearly unreliable, since redeposited copper would catalyse the decomposition of hydrogen peroxide, so that the high-copper alloys were probably subjected to a less aggressive medium than the low-copper alloys; later tests based on salt containing hydrochloric acid showed that, whereas the nearly copper-free alloy suffered only slight intergranular attack, that containing 0·11% copper showed extensive intergranular attack, whilst a copper-content of 0·22% caused almost complete disintegration of the crys-

* There is reason to believe that British manufacturers are today supplying alloys with a copper-content well below the maximum permitted by specification. In cases where specifications allow 0·10%, contents of 0·02 to 0·04% are common and 0·09% relatively rare.

tals. German salt-spray tests gave similar results; an alloy with 0·046% copper behaved much worse than one with 0·010%.

Certain apparent anomalies need not discourage those who advocate a low copper-limit in both classes of alloys. For instance Brenner and Roth's figures seem to show that the alloy with 7% silicon, tested after 3 days' ageing at 100°C., behaves better if 0·9% copper is present; but it is doubtful whether this alloy should ever be subjected to such a heat-treatment. Again, Porter's tests on panels attached for 3 years below the boarding-platforms of London buses should not be construed into the idea that copper is a desirable impurity. Here—as so often—the attack was more serious on the shielded (upper) side than on the lower side, which was exposed to the air and to splashings from the road; what was perhaps unexpected was that alloys with about 4·3% copper as an essential strength-giving constituent showed no intergranular attack, whereas aluminium-magnesium-silicon alloy with about 0·1% copper as an unwanted impurity suffered marked intergranular penetration; unalloyed aluminium with 0·05% copper showed pitting, but practically no penetration between grains. It is not unlikely that an aluminium-magnesium-silicon alloy of really low copper-content would have shown no intergranular attack, but the matter requires further experimentation.

The fact that the well-spread attack on the alloys containing 4% copper produced a penetration less deep than the intergranular attack on the alloys with 0·1% copper and also caused a rather smaller total destruction of metal deserves to be noted; it would, however, seem unwise to adopt a 4% copper alloy for use in vehicles—at least if there is to be welding, since as pointed out on p. 214, almost every combination of time and temperature exists near a weld, and somewhere the combination most dangerous for corrosion may be expected to occur. The facts—so far as they are known—demand close study from engineers. German and other experience has been collected and well presented by M. Whitaker, *Metal Ind.* (*Lond.*) 1952, **80,** 183, esp. pp. 248, 263. See also P. Brenner and W. Roth, *J. Inst. Met.* 1948, **74,** 159; F. C. Porter, " Symposium on the Protection of Motor Vehicles from Corrosion ", Soc. chem. Ind. 1958, p. 64, esp. fig. 4 and Tables II, IV and V.

Further data regarding the atmospheric behaviour of aluminium. Much useful information will be found in the papers of H. P. Godard, *Corrosion* 1955, **11,** 542t; P. Brenner and G. J. Metcalfe, *J. Inst. Met.* 1952–53, **81,** 261, discussion p. 738; M. Reinhart and G. A. Ellinger, *Corrosion*, April 1956, General News Section, p. 84; G. H. Friese-Greene, *Light Metals* 1947, **10,** 508; E. E. Pool, *Chem. and Ind.* (*Lond.*) 1954, p. 957; J. Hérenguel, *Métaux et Corros.* 1948, **23,** 242. See also the long illustrated article on aluminium in buildings and bridges in *Aluminium* 1957, **33,** 441.

Magnesium alloys. In general, alloys based on magnesium should not be exposed without chromating treatment and painting (p. 588), although the alloy with aluminium and zinc shows some remarkable properties, and cases are known where it has been used without paint and has yet withstood corrosion well. Much depends on design—which should avoid traps for water, especially salt water (see p. 200). Contacts of any magnesium alloy with steel bolts should, where possible, be avoided by means of insulating washers; where actual insulation is impossible, a special type of nut, known as a " galvanut ", designed to increase the length of the electrolyte path, can improve the situation by reducing the intensity of attack (Magnesium Electron, Ltd., U.K. Patent Application 25,725/55 dated Sept. 8, 1955).

Avoidance of Crevices, Water Pockets and Condensation Traps. An admirable paper on the corrosion of metals in buildings by Hudson and Wormwell gives numerous examples where the formation of bulky corrosion-product in crevices has led to bulging and sometimes to breakage. Examples include railings, door-hinges, gutter-brackets, roofing, joists and cornices. Most cases could have been avoided by better design or by the use of putty or red-lead paste to fill up crevices on steel (jointing-compound on aluminium). If architects apply the principles emphasized in this paper, and that of Jones (p. 522 of this book), much trouble to house-owners will be avoided (J. C. Hudson and F. Wormwell, *Chem. and Ind.* (*Lond.*) 1957, p. 1078; see also P. E. Halstead, p. 1132, dealing with metals in contact with concrete).

Evidently the avoidance of corrosion in structures depends quite as much on suitable geometrical design as on choice of materials. In buildings, all crevices and pockets where water can collect must be avoided; the water may result not only from leakage but also from condensed moisture trickling down the walls (sometimes the original condensation occurs on the underside of a roof). Designers should also avoid those small horizontal ledges where dirt and moisture can accumulate (W. E. Ballard and others, *Chem. and Ind.* (*Lond.*) 1954, p. 957; J. C. Hudson, *ibid.* 1954, p. 640; also *Times Science Rev.* Spring 1953, p. 14; H. T. Rudolf, *Corrosion* 1955, **11,** 347t).

Everything must be done to prevent condensation of moisture—especially if the air contains products of combustion. Corrosion often occurs behind heat-insulating materials, and it has sometimes been recommended that these should be made water-repellent; such a plan may prove disastrous, as pointed out by Dolbey, since water collects at the metal surface, or within the colder zone of the insulating layer which lies below the dew-point, and then runs down until it meets a horizontal ledge, causing rapid corrosion. Dolbey recommends that the insulating material—instead of being made water-repellent—should deliberately be given good wicking properties, so that water is withdrawn from the metal, parallel to fibres of the insulating material, until it reaches the warm inner surface of the insulation and there evaporates. His article deserves study (N. L. Dolbey, *The Engineer* 1956, **201,** 636).

Precautions against corrosion in the wall cavities of steel houses are laid down in Nat. Building Studies Spec. Rep. **16** (1953) (H.M. Stationery Office). The corrosion of the frame-work and roof members is described by S. G. Clarke and R. St. J. Preston, *The Builder*, 1949, **177,** 332.

The corrosion of steel due to condensed water in steel houses does not merely cause destruction of material. The voluminous character of the corrosion-product may lead to swelling or buckling. Steel windows may become unopenable or unshuttable, long before perforation or breakage is approached. However, a thick zinc coat obtained by hot dipping, followed by painting, should prevent serious trouble to windows except under very acid conditions.

Severe crevice corrosion in the roof of petroleum storage tanks may be caused by accumulation of moisture condensed on the roof. Methods of avoiding it by improved design are discussed by H. N. Boas, Corrosion Symposium (Melbourne University) 1955–56, p. 511.

Many other examples could be given of trouble which could have been avoided if the menace of corrosion had been borne in mind when some structure or vehicle was being planned. Wormwell and Butler urge that the Consultant should be called in at the design stage. LaQue considers that half the rusting on automobiles could be avoided by suitable holes to drain off trapped water (F. Wormwell and G. Butler, *J. Inst. Heating and Ventilating Engrs.* 1955–56, **23,** 461; F. L. LaQue, *Corrosion* 1957, **13,** April, N.A.C.E. News Section, p. 84).

Protection against Atmospheric Corrosion

Permanent Protection. Steel surfaces which are to be permanently exposed usually receive coats of paint or of non-ferrous metal. These matters are discussed in Chapters XIV and XV.

Temporary Protection. Articles which require protection merely during a period of storage or transport call for different treatment. These include machine parts, where a paint coat would interfere with their use in service. They can, however, be seriously damaged by slight rusting in store, so that temporary protection is needed. Later when the time arrives for assemblage, other hazards are introduced, notably from perspiration. If gloves are impossible, goods which have been touched by the hands should be washed with a 5% solution of water in methanol soon after the handling (E. Ll. Evans and E. G. Stroud, *Chem. and Ind.* (*Lond.*) 1957, p. 242, esp. p. 245; the question of corrosion by perspiration is discussed by S. G. Clarke and E. E. Longhurst, Selected Government Reports 1951, no. 29, p. 127; S. J. Eisler and H. L. Faigen, *Corrosion* 1954, **10,** 237; K. J. Collins, *Brit. J. industrial Med.* 1957, **14,** 191. For other problems of packaging, storage and transport see J. J. Ferrigi, *Chem. and Ind.* (*Lond.*) 1957, p. 280; G. Schikorr, "Korrosion VIII", 1954–55, p. 29).

Methods of temporary protection include:—

(1) Reduction of the atmospheric humidity in the store-room.
(2) Use of volatile inhibitors.
(3) Wrapping in suitable paper.
(4) Temporary protective coatings which can be removed by a solvent when the articles are put into service.

These will be considered in turn.

Reduction of Atmospheric Humidity. If the atmosphere is kept below the critical humidity, serious change of the metal surface is avoided. The difficulty is to know what limit to aim at, since the effective dew-point is reduced by sulphur acids in the air and particularly by dust-particles of hygroscopic salt. It is usual to prescribe 50%, but Preston and Sanyal (p. 497) point out that saturated solutions of certain chlorides stand in equilibrium with air at much lower humidity values (12% for LiCl; 31% for $CaCl_2$; 33% for $MgCl_2$); special steps would be needed at places where such hygroscopic salts as these are likely to be present in the dust. In general 30% would seem a safe lower limit, and doubtless this could be raised in cases where the articles stored carry an oil- or grease-film—perhaps to 45%, the figure recommended by F. L. LaQue, United Nations Scientific Conference on the Conservation and Utilization of Resources 1949, Vol. II, p. 227. See also A. Bukowiecki, *Schweiz. Arch. angew. Wiss.* 1957, **33**, 97.

Duly's measurements of the behaviour of steel in contact with (solid) sea salt at different values of the relative humidity are of interest in connection with this matter. No visible corrosion and no increase of weight was noticed if the relative humidity was kept at 30% or lower, but they were detectable at 35%; both became increasingly marked at higher values (S. J. Duly, *J. Soc. chem. Ind.* 1950, **69**, 304).

Special problems arise in articles where a metal piece is in contact with organic material; metal fittings on leather and zip fasteners on clothing provide examples; in early days, brass fasteners on white woollen sweaters were attacked by sulphite bleach and caused green stains to appear on the articles in the shop windows. It is understood that the use of more resistant alloys—doubtless with other precautions—has resolved this difficulty.

Stainless steel parts stored in contact with asbestos or leather packings may suffer corrosion if the conditions are not quite dry, owing to absence of the oxygen needed for film-repair at the point of contact (*Corros. Tech.* 1954, **1**, 91).

Nails or fasteners in wood preserved in zinc chloride have also given trouble. Corrosion by vapours evolved from wood, glue and resin is mentioned on p. 493.

Zinc or galvanized wire products suffer *white rusting* under condensation conditions; it is accelerated by sulphur dioxide, hydrogen chloride or organic acids given off by some types of adhesive tape. This trouble is discussed on p. 654. Proper storage conditions provide the best remedy, but pretreatment in certain chromate baths increases the resistance (P. T. Gilbert and S. E. Hadden, *J. Inst. Met.* 1950–51, **78**, 47; R. A. Neish, *Corrosion*, 1954, **10**, 440; see, however, less hopeful reports from K. Daeves, *Draht* 1957, **8**, 334).

Silica gel is convenient for the de-humidification of air, as it can be rejuvenated by heat. Where articles are to be packed, silica gel can be included in the box—a transparent window being left for inspection. The type of silica gel containing blue cobalt chloride (which turns pink on hydration) is often advocated; the colour-change will indicate exhaustion or the gel. However, it seems that cobalt chloride may be a stronger dehydrating agent than silica, and that the colour-change occurs before the silica is entirely exhausted (F. G. Jaubert, *C.R.* 1949, **228**, 826).

The protection of machines for shipment and storage has been described by Foley. First Scotch tape is wound round the machine; then a solution of some organic film-forming substance is applied from a spray gun; this contains a webbing agent to give long cobweb-like filaments designed to bridge over the gaps in the grid of tape; next a solution of the film-former free from webbing agent is applied. The solvent vapours are blown out and silica introduced through holes left for the purpose; the holes are then patched, a transparent window being left for inspection at one place (J. E. Foley, *Corros. Mat. Prot.* Sept.–Oct. 1947, p. 6; D. W. Harbour, *Corr. Prev. Control.* 1954, **1**, 288; J. Feasey, *ibid.* 1954, **1**, 418).

A spectacular example of preservation by controlled humidity is provided by the ships' interiors in reserve fleets. The U.S. Navy is relying on 30% humidity in the " sealed zones " of " inactive " ships, and this has proved effective in preventing corrosion, as shown by inspections of ships opened after 3 years; the cost is stated to be low (G. C. Wells, *Corrors. Mat. Prot.* Sept.–Oct. 1948, p. 4).

Similar principles are being applied in merchant ships for the air conditioning of the holds.

Calculations of the amount of warm air needed for the prevention of rust and mildew on walls and ceilings of rooms are provided by H. Netz, *Werkst. u. Korrosion* 1953, **4**, 2.

Volatile Inhibitors. In cases where removal of moisture is inconvenient, the introduction of a volatile inhibitor into a space can prevent corrosion to metallic surfaces reached by the inhibitor molecules. If the solid inhibitor is sprinkled as a powder on the floor of some iron or steel article, it will prevent rusting on parts not touched by the powder. There is some difference of opinion as to whether in such a case the molecules reach the remote parts of the surface by evaporating and recondensing, or whether they reach them by surface migration; in some cases, either route would be possible, since substances which are volatile may well be surface-mobile. The distinction is not without interest, and it would seem possible to decide the mechanism by experiments in which the specimens are suspended by very fine threads, the solid inhibitor being sprinkled on the floor of the containing vessel; it should be ascertained whether the protection is as good as in other experiments where similar specimens are suspended at the same distance from the solid inhibitor powder, but connected to the floor by broad strips which would facilitate surface diffusion.*

The general view is that the inhibitor volatilizes, condenses on the remote surface, being then either adsorbed, inhibiting the corrosion, or dissolved in the moisture-film (if present), which then becomes non-corrosive.

One widely used inhibitor of this nature is based on dicyclohexyl-ammonium nitrite; presumably the base acts in a manner similar to the amines used as pickling restrainers (p. 411), whilst the nitrite group acts like sodium nitrite (p. 151); it has the same toxicity for small animals as sodium nitrite. Baker states that di-isopropyl-ammonium nitrite is another promising inhibitor, but for impregnating paper, dicyclohexyl-ammonium

* E. Ll. Evans adds the comment, " The emphasis on the possibility of surface migration . . . is perhaps excessive. For a material of appreciable vapour pressure it seems barely necessary to invoke surface migration."

nitrite, which is less volatile and less soluble in water, may be more suitable. Clearly the volatility must be chosen to suit circumstances. If there is risk of the substance escaping from the container, a less volatile substance should be chosen than where escape is unlikely but the path leading to recesses long. The vapour pressures of the more important inhibitors are recorded by R. B. Turnbull, *Chem. and Ind.* (*Lond.*) 1957, p. 446, esp. p. 448.

Work at the Chemical Research Laboratory, Teddington, has shown that carbonates of these bases are in many cases admirable inhibitors; cyclohexylamine dicarbonate has prevented rusting on cast iron or steel at 90–95% relative humidity. It may be placed on trays or open cartons, or in some situations distributed as powder by means of an air-blast. The carbonate has been used for the preservation of idle boilers (p. 462), including steam-heating systems, but doubts have been expressed as to whether it can prevent corrosion which has fairly started during the winter from continuing during the summer period of idleness. It has, however, been recommended for arresting attack which has started on stainless steel turbine blades under humid, stand-by conditions, and for addition as a powder to containers (including fire-extinguishers) likely to be kept in an unheated store. Another plan is to use a volatile benzoate, where the inhibitive properties characteristic of sodium benzoate (p. 172) are obtained at the remote regions; Vernon has stated that *n*-butyl benzoate gives good results under conditions of fluctuating humidity.

It should be noted that the protection does not extend to all metals. Thus cyclohexyl-ammonium carbonate protects mild steel, cast iron, aluminium, tin-plate and lead, but increases the corrosion of copper and many of its alloys, as well as magnesium. Dicyclohexyl-ammonium nitrite protects mild steel, aluminium, tinplate and copper, but attacks lead and magnesium. The behaviour of zinc, cadmium and solder towards both substances depends somewhat on conditions.*

The subject should be studied in the papers of W. H. J. Vernon, *J. roy. Soc. Arts*, 1948–49, **97**, 578, esp. p. 603; E. G. Stroud and W. H. J. Vernon, *J. appl. Chem.* 1952, **2**, 178; H. L. Bennister, *Research* 1952, **5**, 424; H. R. Baker, *Industr. engng. Chem.* 1954, **46**, 2592; C. A. Rhodes, *Corros. Prev. and Control* 1957, **4**, No. 4, pp. 37, 42; A. Wachter, T. Skei and N. Stillman, *Corrosion* 1951, **7**, 284; W. Paul, *Werkst. u. Korrosion* 1956, **7**, 189. See also " Chemistry Research " 1951, p. 28; 1952, p. 23; 1953, p. 17; 1954, p. 19; 1955, p. 18; 1956, p. 19 (H.M. Stationery Office).

The related matter of the use of cyclohexylamine or morpholine in steam containing carbon dioxide, to prevent corrosion of steam pipes, is mentioned on pp. 182, 998.

Paper Wrapping. The protection of articles by wrapping is no new

* Recent work shows that dicyclohexylamine nitrite fails to prevent corrosion at places where a " dry " mixture of NaCl, $MgCl_2$, $CaCl_2$ and Na_2SO_4 has been dusted on a horizontal surface of metal in a tube merely plugged with cotton wool, whereas cyclohexylamine carbonate gave considerable protection to steel and magnesium whilst increasing attack on copper (Ministry of Supply: unpublished work).

device, but the provision of suitable paper is a problem still not completely solved. Clearly the paper should be low in chloride—a requirement which is not easily fulfilled. The paper should be dried *before* being used for wrapping, but complete elimination of moisture is not easy; as Schikorr has found, metal pieces wrapped in paper containing even a trace of water, and *afterwards* placed in an oven—in the belief that the heat will complete the drying—actually become wet, since the bulky metal, having a heat-capacity exceeding that of the thin paper, may remain cool after the paper has become hot, so that the moisture from the paper condenses upon the metal (G. Schikorr, see p. 534).

The situation can be greatly improved by introducing an inhibitor into the paper; a volatile inhibitor may be expected to give general protection, even though the paper only touches the metal at a few points. Paper impregnated with dicyclohexyl-ammonium nitrite has given good protection to steel, but is less effective for most non-ferrous metals; it is said to be useless for cadmium. Non-volatile inhibitors should protect at points where the paper touches the metal, and may perhaps be expected to diffuse a certain distance through a film of condensed moisture. Work at Teddington has shown that paper (or other wrapping) impregnated with sodium benzoate gives good results in warm, humid atmospheres where specimens wrapped in untreated paper rust severely. Thus at Lagos (Nigeria), where steel gudgeon pins with brass ends became heavily rusted when wrapped in ordinary cellophane, they were completely protected by cellophane impregnated with benzoate (W. H. J. Vernon, *J. Soc. chem. Ind.* 1947, **66,** 137, esp. p. 139; “Chemistry Research” 1938–46, p. 17 (D.S.I.R.)).

For aluminium, paper carrying sodium benzoate is stated to be too alkaline (G. W. Walkiden, *Chem. and Ind.* (*Lond.*) 1956, p. 656).

Useful general information about corrosion in packaging is provided by E. G. Stroud and W. H. J. Vernon, *J. appl. Chem.* 1952, **2,** 166, 173. See also discussion at Birmingham on packaging and storage, *Chem. and Ind.* (*Lond.*) 1956, p. 656; especially papers by E. Ll. Evans and E. G. Stroud (*ibid.* 1957, p. 242), J. J. Ferriggi (p. 280), F. A. Paine (p. 288) and R. B. Turnbull (p. 466). The general treatment of “Corrosion in Packaging” is provided in a Supplement to “Corrosion Prevention and Control” 1956, **3,** April, pages Siii to Sxiv.

For wrapping metals like silver which are liable to tarnishing, requirements are rather different, and attention has been given to paper impregnated with salts of such metals as copper, lead or zinc, which would absorb hydrogen sulphide; copper phosphate and copper chromate have been found to catalyse the oxidation of hydrogen sulphide by air—a matter studied at Teddington. Chlorophyll has been advocated; it has some power of absorbing hydrogen sulphide, but some investigators consider that it merely imposes a physical barrier (“Chemistry Research” 1955, pp. 18–20; 1957, p. 17 (H.M. Stationery Office). L. L. F. Deadman, *Chemical Products* 1955, **18,** 223).

Schikorr has developed a simple test for ascertaining whether a sample of wrapping paper is likely to cause corrosion. A piece of the paper is rolled round a glass tube and a clean weighed wire is wound round the paper; the whole is exposed to air at 92% relative humidity at 37°C. for 4 days; the rust is removed and the wire reweighed. Some types of paper cause almost no rusting, but serious corrosion occurs if it contains chloride in excess of 0·05% sodium chloride or acidity detectable by Congo Red. Schikorr discovered one paper which contained no chloride or acidity but which was corrosive owing to its high sulphate content. Waxed paper is considered by Schikorr to owe its value to inhibitive substances present in the wax and not to exclusion of water vapour. Filter-paper impregnated with wool-fat (from a 10% solution) caused far less corrosion than the same paper unimpregnated. Further information is provided by G. Schikorr, *Werkst. u. Korrosion* 1953, **4**, 81; "*das Papier*" 1953, **7**, 18; 1954, **8**, **431**; 1956, **10**, 142.

Temporary Protective Coatings. The plan of keeping a metal surface greasy to prevent corrosion has been known since ancient times, but the various oily and greasy substances are not all equally suitable. Oils and greases containing hydrocarbons alone seem to be less effective than those containing polar substances; crude petroleum jelly gives better protection than the refined variety—doubtless owing to some minor constituent. Lanoline, derived from wool-grease, is more effective still; it is conveniently applied as a solution in white spirit or naphtha; the protective character of lanoline has been attributed to the polar and unsaturated character of its constituents (p. 997). If dust is present, the lanoline film must not be too thin, or it will be pierced by the dust particles. A 12% solution applied by dipping at 22–24°C. produces a film which protects steel against mild weather conditions, but to withstand the conditions of an English winter needs a 28% solution in white spirit. Specifications DEF 2331 and DTD 121D demand 30 to 35% ("Chemistry Research" 1956, p. 19 (H.M. Stationery Office)).

By introducing polar groups into petroleum products, improved protective properties are obtained. The so-called petroleum sulphonates are apparently derived from an alkyd-benzene-sulphonic acid, with the $-SO_3H$ and alkyd groups in para-positions to one another. The improvement achieved by adding sodium petroleum sulphonate or calcium petroleum sulphonate to ordinary mineral jelly was made evident in tests by Hoar and Smith; they also show that mineral jelly containing 25% lanolin is a better protective than simple mineral jelly and that further improvement is obtained by using wool-grease instead of lanoline. High-melting mineral jellies are slightly more protective than the ordinary jelly, but in that case the addition of the sulphonates are unfavourable. In discussing their results, Hoar and Smith point out that for iron surfaces carrying invisible oxide-films, the groups effective in providing protection are $-SO_3H$, $-COOH$ and $-OH$, whereas for oxide-free surface, as in pickling, the best groups are $>S$, $\genfrac{}{}{0pt}{}{NH_2}{>N}\!\!>CS$ and $\genfrac{}{}{0pt}{}{-S}{-O}\!\!>CS$ (T. P. Hoar and G. C. Smith, *J. Inst. Petroleum*, 1950, **36**, 448; cf. E. J. Schwoegler and L. U. Berman, Abstract, *Met. Abs.* 1955, **22**, 1180).

The adsorption of an organic sulphonate (actually calcium dinonylnaphthalene-sulphonate) has been studied by means of compounds containing a radioactive isotope of calcium by van Hong, S. L. Eisler, D. Bootzin and A. Harrison, *Corrosion* 1954, **10**, 343.

Many different types of temporary protective compositions are on the market, suitable for many different purposes. Where the greased articles are likely to be roughly treated, the film should possess hardness and thickness; some of the best mixtures for such purposes contain resins. For thin metal pressings the coatings should be flexible. The various types are discussed by G. T. Dunkley, *Sheet Met. Ind.* 1950, **27**, 599.

In many cases, inhibitive pigments are included in temporary rust preventives, and these often enhance their protective character; zinc chromate is usual, but even sodium dichromate or potassium chromate have been recommended. A 3-years outdoor trial at Farnborough showed that lanoline containing zinc chromate—with naphtha and white spirit as solvents—gave much better protection than unpigmented lanoline, which permitted rusting.

The various compositions are used in different ways; some are applied hot, so that the film solidifies by cooling, whereas others are applied cold and become relatively hard as the solvent evaporates. Brochures issued by the manufacturers often contain valuable information based on practical experience; on occasions the advice of manufacturers as to the best composition for some particular purpose may be sought with advantage.

Reference may be made to the Symposium on Temporary Protectives, recorded in *J. Inst. Petrol.* 1950, **36,** No. 319. p. 422; No. 320, p. 475. See also " Chemistry Research " 1956, p. 19 (H.M. Stationery Office).

Certain oil-products play an important part in preventing atmospheric corrosion. These include the water-repellent oils, which serve to remove the adhering water drops from wet metal surfaces (e.g. surfaces which have been washed to remove mud). Another class includes the anti-corrosive emulsifying oils mentioned as inhibitors of corrosion on p. 169. Dipping in an oil-emulsion is a simple operation, suited for articles which are already wet, and it is possible to add inhibitive chemicals to the aqueous phase. There is another advantage; the article as it emerges from the liquid, carries a grease-film affording considerable protection against corrosion, which might otherwise start after handling, owing to the corrosive constituents of perspiration (see also p. 529).

A useful survey of temporary protective measures for goods in storage and transport, with special reference to American specifications, is provided by G. Schikorr, *Werkst. u. Korrosion* 1955, **6**, 9; *Metalloberfläche* 1955, **9**, A84, A104.

Other References

Apart from Hudson's book and numerous reports (pp. 515, 516), two short but excellent surveys of Atmospheric Corrosion have been provided by Vernon in his Jubilee and Cantor Lectures (W. H. J. Vernon, *Chem. and Ind.* (*Lond.*) 1943, p. 314; *J. roy. Soc. Arts* 1948–49, **97**, 578, 593, esp. pp. 584, 602).

CHAPTER XIV

PROTECTION BY PAINTS AND NON-METALLIC COATINGS

Synopsis

Chapters XIV and XV are devoted to methods of avoiding or mitigating corrosion troubles by means of protective coatings. In the present chapter ordinary paints and non-metallic coatings receive consideration; Chapter XV is devoted to metallic coatings, and will, for convenience, include metal-pigmented paints.

After an explanation of the meaning of terms, the present chapter opens with a discussion of the economic aspects of protective painting. This is followed by a brief review of the four main methods of applying paint (brushing, spraying, dipping and pouring), the four mechanisms of drying (solidification, evaporation, oxidation and polymerization) and the four principal constituents of oil paints (drying oil, drier, pigment and thinner).

The theory of protective painting next receives consideration; the popular notion that paint-coats protects metal by shutting out corrosive substances is shown to be inadequate as an explanation of the facts. Protection is really due, in different cases, to (1) sluggish ionic movement through the coat, (2) the presence of inhibitive pigments or (3) the formation of inhibitors by the degradation of metallic soaps; a fourth protective mechanism, met with in metal-pigmented paints which provide a modicum of cathodic protection at any gaps in the coating, is deferred to Chapter XV.

Next, the causes of break-down in paint-coats are considered, including decomposition of the vehicle by light, peeling and softening due to cathodically formed alkali, blistering (which can arise from several causes), the pushing up of paint by voluminous rust formed under the coat and finally the effect of mill-scale. This leads on to the subject of the pre-treatment of surfaces before painting, including the removal of scale and rust either mechanically, thermally or chemically, and the production of phosphate-coatings on the surface which form a favourable basis for the application of paint; wash-primers, which greatly improve adhesion, are then considered.

After that the choice of painting systems comes up for discussion; inhibitive pigments for the primer and flaky pigments for the outer coatings receive consideration in turn; there is also discussion of tar and bituminous paints, synthetic resins and chlorinated rubber paints. Exposure tests at home and abroad receive attention, and there is brief treatment of the influence of total paint-thickness and the number of coats.

We then pass to marine painting, where the situation is different from that on land; drying oils, with their soap-forming properties —so valuable for protection inland—are less suitable for ships' painting owing to attack by cathodically formed alkali; there is a brief reference to anti-fouling problems. The problems involved in the painting of galvanized surfaces, aluminium alloys and magnesium alloys then receive attention, including pre-treatment processes.

Finally, there is a brief consideration of protective coatings other than paint, including thick layers of plastic or rubbery material, glass linings, vitreous enamels, ceramic coatings and cermets designed to withstand high temperatures.

Introductory Remarks

Use of Terms. Compositions applied in a liquid condition to form a film on a metallic surface which subsequently hardens to a solid coating are commonly called *Paints.* The term is often restricted to compositions providing opaque films which are usually coloured; the opacity (and/or colour) is usually obtained by the presence of a very fine powder or *pigment*, although in some of the black paints based on bitumen or tar, there is no deliberate addition of that kind. Compositions carrying resinous substances but no pigments are often referred to as *varnishes* and *lacquers.* These terms are used loosely, but some authorities would like to see the word varnish applied to coats which harden by absorption of oxygen, and lacquer to those which dry by evaporation.

Economics of Painting. Much of the cost caused to the world by corrosion is connected with the cost of painting, and particularly with the fact that painting has to be repeated at regular intervals. The period which can elapse between successive paintings and also the relative costs of paint and its application will vary greatly in different parts of the world. Let us consider a situation where repainting is needed every three years. If it were possible to increase the interval to, say, four years, there would be a large saving on the average annual outlay; since the labour costs involved in painting commonly exceed the cost of the paint, it would be economical to purchase a much more expensive paint, if by doing so the user was *quite* sure of obtaining his extra year of service. So long as the user had only the salesman's assurance about the relative performances of different paints, he might hesitate to base exact calculations on them; there now exists, however, reliable information on this subject, provided by disinterested and highly qualified investigators, and it should be possible for users, especially in countries where the economic situation has become difficult, to consider seriously the adoption of superior materials which (despite their greater cost) will lead to a considerable reduction in the annual expense.

The mean annual charge (C) for upkeep depends on the interval (t) which can be allowed to elapse between paintings, the cost of paint and other materials (M) used on each occasion and the cost of labour (L), including therein the expense of erecting scaffolding, preparing the surface, applying the paint, and all incidental costs connected with the interruption of normal work.

It is necessary to calculate ΔM, the maximum increase in M, which could justifiably be accepted if thereby the period between painting was increased from t to $(t + 1)$ years; anything less than ΔM then represents a financial gain, anything more than ΔM a loss; at a value of ΔM which leaves the costs unchanged

$$\frac{M + L}{t} = \frac{(M + \Delta M) + L}{t + 1}$$

or

$$\Delta M = \frac{M + L}{t} = C$$

In other words, it would be profitable to pay for paint and materials any extra sum less than the present total annual outlay, provided that the extra year's life were really obtained. This calculation neglects interest charges; allowance for interest could easily be made, but would not, in normal times, greatly affect the situation.

In 1934, Blom stated that in the Swiss electrical Industry the period between paintings was 10 years, and the cost of painting, etc. 5 or 6 times that of the paint. Let us therefore write

$$t = 10; \quad L = 5M; \quad \Delta M = C = 6M/10$$

In Switzerland, therefore, it would have been advantageous to increase the price paid for paint and materials by anything up to 60%, if it was certain that t would thereby be raised from 10 to 11 years (A. V. Blom, *Bull. Assoc. Suisse Elect.* 1934, **25,** 365).

In Great Britain, t will be much lower, say three years. If, for the sake of argument, we keep L as $5M$, we obtain $\Delta M = C = 6M/3$. Any increase up to 200% would then appear justifiable, if thereby the interval between painting would really be raised from 3 to 4 years. If $L = 3M$, which Fancutt and Hudson regard as a minimal value, increases up to 133% would save money in the long run. (See also J. C. Hudson and F. Fancutt, " Protective Painting of Structural Steel ", p. 37.)

The policy of increasing immediate costs in order to obtain ultimate saving is perhaps more acceptable today than when Blom made his calculations, but there are still to be found people who will argue, " Upkeep costs are very high; let us try and find a cheaper paint."

An engineering approach to the estimation of costs of maintenance painting is provided by R. F. Williams and J. H. Cogshall, *Corrosion* 1957, **13,** 3t.

In view of the prevailing high labour costs, economy depends largely on the prospects of reducing the number of coats. Heierman emphasizes the need for paints designed to give heavier coatings free from defects. He indicates the importance of higher structure viscosity, and suggests a solution of the problem by the adoption of gelled vehicles, possibly containing aluminium alcoholates (F. J. M. A. Heierman, *Verfkroniek* 1957, **30,** 285).

This is probably a sound suggestion; some of the synthetic paints on the market give thinner coats than old-fashioned oil paints. Whether, however, the present emphasis on thickness in coatings as a *sine qua non* of protective value is entirely justified appears to the Author open to question (see also p. 582).

Methods of Applying Paint. A metallic surface can be coated by (1) brushing, (2) spraying, (3) dipping or (4) pouring. Spraying the paint on to the surface as a cloud of tiny droplets from an atomizer saves labour and reaches recesses not easily reached by a brush; the procedure, however, is more likely to shut in moisture if a metallic surface is being coated early in the morning before the condensed film (" dew ") has evaporated; a brush-painter will usually emulsify much of that moisture-film, and the water, entering the paint as tiny particles, ultimately reaches the outer surface and evaporates. Some paint-mixtures which are suitable for application by brush are not suitable for spraying, and it may be necessary to add more thinner than is really advisable. In spraying a considerable fraction of the paint which leaves the nozzle fails to reach the " work " (i.e. the article to be painted). Electrostatic spraying, in which the spraying particles are directed upon the work by means of a strong electric field, has obvious attractions in some situations. Various papers on spraying techniques, by A. L. Newcomb, J. Muirhead, T. Cowland, J. Webb and E. E. V. Sharpe will be found in *Trans. Inst. Met. Finishing* 1937, **34,** 387–442.

Coating by immersion of the whole article in a paint-bath introduces risks that recesses will escape covering, and clearly not all compositions are suited for that method of application. The same is true of the method based upon pouring the paint over the surface to be coated.

Mechanisms of Drying

General. The paint, varnish or lacquer is applied as a liquid and dries to form a solid. The drying may, however, depend on one of four different types of physical or chemical change: (1) solidification on cooling, (2) evaporation of a solvent, (3) absorption of oxygen from the air, combined with polymerization or condensation processes, carried out either (*a*) slowly at room temperatures as in the " air-drying paints " or (*b*) quickly at elevated temperatures as in " baking-varnishes " (the mechanisms are suggested on pp. 995–996), (4) polymerization or condensation without absorption of oxygen (as stated on p. 541, this occurs with only a few baking-varnishes, such as those containing butylated amino-resins).

Solidification Method. In the coating of pipes (p. 280), they are frequently lowered into a hot black mixture based on coal tar or bitumen, which is adjusted so as to be sufficiently fluid at the temperature of application, but solid at room temperature; the pipe is kept immersed until it has attained the temperature of the bath and is then withdrawn; the film solidifies on cooling. Similarly, hot compositions may be applied by brush —a method used in coating the hulls of canal craft.

Evaporation Method. It is also possible to dissolve the film material in a volatile liquid, applying it to the metallic surface by spray or brush; a solid film is left when the solvent has evaporated. Such films are thinner and less protective than those obtained from a hot mixture but they often adhere better; if it is decided to use a hot mixture, it may be good policy first to obtain a thin film by evaporative drying which will serve as a basis for the hot coating.

Many paints and lacquers based on plastics and resins dry by evaporation. One of the earliest series to attain prominence were those based on solutions of nitrocellulose in amyl acetate or other solvent, which provided the amateur painter with an easy means of obtaining smooth coatings, either coloured or transparent. " Cellulose finishes " were at one time widely adopted in the automobile industry, but the term soon began to be used somewhat loosely. Of far greater significance to corrosionists are the paints based on vinyl compounds. Polystyrene, in particular, can be dissolved in xylene or other solvent to produce either a clear lacquer or a pigment-carrying paint; numerous pigments can be used, but one of the valuable features of polystyrene is its power to carry a large pigmentation of metallic zinc without becoming too stiff for application—a matter discussed in the next chapter.

In nearly all these cases, a " plasticizer " is added, an organic compound having a higher boiling-point than the solvent, and which will remain in the film after the solvent has evaporated. Some examples are mentioned on p. 993. Without a plasticizer, paints or lacquers based on plastics or synthetic resins would in many cases be too brittle, but the exact function of the plasticizer is a matter upon which different opinions have been expressed. Some hold that the molecules of plasticizer occupy key positions at internal cavities which would otherwise act as stress-raisers, either filling the cavities as a whole or at least rounding off their sharp ends which would provide danger-points. Others consider that plasticizers act as internal lubricants, which ensure that the film will deform under stress instead of fracturing. The utility of dioctyl phthalate as a plasticizer for vinyl compounds has been attributed to the non-linear shape of the molecule. It is possible that one of the functions of a plasticizer is to reduce the number of linkages between the chains which are not connected with primary valency bonds; this view is advanced by P. B. Stickney and L. E. Cheyney, *J. Polym. Sci.* 1948, **3**, 231.

Oxidation Method. All oil paints dry by absorbing oxygen from the air. The drying oils contain glycerides of unsaturated fatty acids, which are less stable than the corresponding saturated compounds, although the idea that oxygen is taken up at the double bonds and then serves to join the molecules together has been abandoned. However, perhaps in the manner suggested on p. 995, the molecules do join together to form a three-dimensional net-work in which translational motion of the original molecules is lost, so that the film—originally liquid—becomes solid.

A film of linseed oil without admixture, if spread out on a surface and exposed to air, will ultimately solidify, but this is a very slow process. If compounds of certain metals which exist in more than one state of oxidation are present in the film, the drying is greatly accelerated; such metals are lead, manganese and cobalt. The mechanism is discussed briefly on p. 996.

The process can be further accelerated by raising the temperature. At one time the article on which the coating had been applied, generally by spraying, was placed in an oven; today much time can be saved by the use of radiant heat; instead of $\frac{1}{2}$ to 2 hours in the oven, radiation-heating for 4 to 6 minutes may suffice. There are certain other advantages; the radiation is absorbed at just the place where it is required, and it is possible for the film of baking-varnish to reach the appropriate temperature quickly without much heating of the interior of the metal; this is a great advantage in coating those materials (e.g. certain aluminium alloys) which undergo undesired mechanical changes on heating in the range needed for polymerization or condensation; for coating such materials, it is, of course, desirable to select a baking-varnish which hardens at a relatively low temperature. The radiation procedure is described by H. Favart, *Trans. Inst. Met. Finishing* 1954, **31,** 332; discussion on p. 342. L. Walter, *Metal finish. J.* 1956, **2,** 91. See also " Notes on Drying Paints by Radiant Heat " (Brochure issued by Imperial Chemical Industries).

Air-drying paints and baking-varnishes often contain synthetic resins modified to make them compatible with oil (see p. 1004).

Oil paints only give good protection if properly selected and applied.

Duckworth, discussing engineering tenders, has expressed himself as " appalled at the low standard of initial protection offered." Too often do the documents rely on such phrases as " to be given one coat of red oxide before dispatch "—which means, in practice (he points out), application over mill-scale and rust. " There seems to be no desire to deliver equipment in a condition which would reflect credit on the supplier." He appeals to manufacturers to adopt a system of surface preparation and priming which would " produce repeat orders from satisfied customers " (R. A. Duckworth, *The Engineer* 1959, **207,** 141). See also p. 559.

Condensation without oxygen-up-take. Certain polymers based on butylated amino-resins dry by a mechanism different from that met with in linseed oil. They are used on refrigerators and washing-machines, but will not greatly concern the corrosionist. Details are provided by H. R. Touchin, *J. Oil Col. Chem. Ass.* 1956, **59,** 353.

Components of Oil Paints. An oil paint may be said to have four essential constituents, although others, such as resins and metallic soaps, are often added by paint manufacturers to confer desirable physical properties; these are the drying oil, drier, pigment and thinner.

(1) Of *Drying Oils,* the most important are linseed and tung oils. The raw oils are usually treated before being introduced into paints. The heating of linseed oil with driers yields the so-called *Boiled Oil,* whilst protracted heating without driers in pots covered to exclude air produces polymerization and yields *Stand-Oil* or *Litho-Oil.* By blowing air through heated oil,

without the addition of drier, we obtain *Blown Oil*, which is rich in polar groups, such as OH.

Tung Oil differs from linseed oil in possessing two double bonds as close together as possible, $—CH_2—CH{=}CH—CH{=}CH—CH_2—$. The " conjugation " of the double bonds accounts for its exceptional drying properties. Tung oil, however, is less important today than some years ago, as manufacturers prefer the material known as *dehydrated castor oil* which also contains conjugated bonds. Castor oil consists largely of the glyceride of hydroxy-oleic acid, and by the removal of OH and H (giving water) from adjacent carbon atoms, the " dehydrated " oil is obtained. It should be noted that even linseed oil, if heated first at 290°C. in nitrogen and then at 180–200°C. in presence of sulphur dioxide, acquires quick-drying properties similar to those of tung oil, the double bonds moving to the conjugated positions (H. I. Waterman, C. van Vlodrop and M. J. Pfauth, *Research*, 1948, **1**, 186). An early view of the drying process was provided by A. V. Blom, *Kolloid-Zschr.* 1936, **75**, 223; for more detailed mechanism, see L. A. O'Neill, *Chem. and Ind.* (*Lond.*) 1954, p. 384; see also p. 995 of this book.

(2) The *Drier* is usually a naphthenate or linoleate of manganese, lead and/or cobalt.* Compounds of all three metals may sometimes be present in a single paint, since they affect drying in rather different ways, as explained below. Doubtless the metal in the drier is oxidized to the higher state by the oxygen of the air and then passes on its oxygen to the glycerides. Fifty years ago oxides or inorganic salts of the metal were used in this way as " driers ", but it was later found that, if the mixed salts of the various fatty acids present (as glycerides) in linseed oil were used, they became better incorporated in the oil medium, improving the drying. In recent times, other organic salts of the three metals have come to be used—notably the naphthenates.† The amount of drier needed is very small; for instance, in the paints used in the tests of the Iron and Steel Corrosion Committee, the drier contents of the white and red lead paints corresponded to 0·05% lead and 0·003% manganese. " Octoates " (salts of 2-ethyl-hexoic acid) are also used (J. Bakker, *Verfkroniek* 1957, **30**, 118).

The different driers seem to accelerate different parts of the process, cobalt accelerating the oxidation stage, but not the subsequent polymerization, whereas lead acts most powerfully in the later stages. This doubtless explains why mixtures of driers are needed for the best results. It is stated that too much cobalt or manganese drier will cause the paint to harden on its surface, and the skin thus formed may prevent the escape of the solvent or thinner; with lead this does not occur.

The drying time of a paint depends on the pigment as well as the oil-

* Cerium salts have been used, but possess certain disadvantages; see P. R. Deeleman, *Verfkroniek* 1958, **31**, 150.

† The by-product of oil refining commonly known as *naphthenic acid*, is really a mixture of acids largely derived from compounds containing rings of five carbon atoms with —COOH groups attached; the " naphthenates " of the three metals mentioned (each a mixture of salts) have proved useful as driers. See p. 991.

vehicles. With some pigments, it increases greatly on storing the paint. Hermann has found that the drying times of stand-oil paints containing 0·08% cobalt as dryer made 4 days previously lay between 6 and 9 hours for all pigments tested; however, after 6 months' storage, the drying times which, in white lead paints, remained unchanged at 6 to 6½ hours, had risen to 30 hours for yellow iron oxide and to over 100 hours for antimony white; this deterioration is generally attributed to disappearance of the drier either by adsorption on pigment particles or by precipitation (F. J. Hermann, *Verfkroniek* 1948 **21**, 283).

(3) The *Pigment* may be chosen to confer specific anti-corrosive properties, as in the case of red lead or calcium plumbate, but there are other pigments such as iron oxide or titania which have no power of inhibiting the corrosion reaction as such, although they may affect the situation in other ways, either by decreasing permeability for water or, as in the case of some varieties of titania, increasing the risk of chalking (p. 551). The choice of pigment for different purposes is considered later, but reference may here be made to pictures of the pigment particles, both loose and dispersed in paint-films, taken with the electron microscope (S. H. Bell, *Verfkroniek* 1958, **31**, 52; F. H. J. v. d. Beek, *ibid.* 1958, **31**, 62).

(4) The *Thinner* is generally needed because a simple mixture of oil and pigment would be too thick for convenient application. The addition of a volatile liquid, which evaporates after the spreading out of the film, provides the consistency desired. Today white spirit, a petroleum distillate boiling at a temperaturé-range somewhat higher than petrol, is generally used. Turpentine was at one time prized as a thinner; although more expensive than white spirit, it is stated to be a better solvent for the lead soaps formed in white lead paints. Solvent naphtha can act as thinner for some paints. The advantage of using a mixture of thinners is urged by H. W. Talen, *Verfkroniek* 1951, **24**, 241.

Theory of Protective Painting

Causes of the Protective Action. Many people still believe that the function of a paint-coat is to exclude from the metal the water and oxygen needed for rusting. However, paint- and varnish-films are by no means so impervious to water and oxygen as is commonly supposed. Figures from various sources, conveniently collected by Mayne, suggest that the rates at which water and oxygen could pass through a paint- or varnish-film (provided that they were both consumed by the metal as quickly as they arrived) is sufficient to support rusting at a rate similar to the rate of rusting which occurs on *unpainted* steel exposed to an industrial atmosphere. Some authorities feel doubtful as to whether the experiments quoted by Mayne really indicate the permeability of paint-coatings as they exist on metals, but he is probably right in saying that " paint films are so permeable to water and oxygen that they cannot inhibit corrosion by preventing water and oxygen from reaching the surface of the metal ". Readers should study Mayne's paper and the experimental papers cited therein (J. E. O. Mayne, *Research* 1952, **6**, 278).

Other papers concerned with the passage of water or other substances through paint, the state of water within paint and the effect of pigmentation on permeability, include those of C. F. Drake and H. W. Keenan, *J. Oil Col. Chem. Ass.* 1946, **29**, 273; W. W. Kittelberger and A. C. Elm, *Industr. engng. Chem.* 1946, **38**, 694; R. Houwink, *Verfkroniek* 1947, **20**, 172; C. Kalauch, *Werkst. u. Korrosion* 1950, **1**, 400; " Chemistry Research " 1950, p. 11 (O.S.I.R.); A. M. Thomas, *J. appl. Chem.* 1951, **1**, 141; J. van Loon, *Verfkroniek* 1952, **25**, 68; A. V. Blom, *Werkst. u. Korrosion* 1954, **5**, 425; " Korrosion " VIII, 1954–55, p. 1 (Verlag Chemie); F. A. Long and L. J. Thompson, *J. Polymer. Sci.* 1955, **15**, 413. The mechanism of gas diffusion (through polymers) is discussed by C. M. Crowe, *Trans. Faraday Soc.* 1957, **53**, 692, and gas-solubility by P. Mears, *ibid.* 1958, **54**, 40.

The object of painting is to *prevent the starting of attack* rather than to slow down corrosion at a point where it has already started; probably we are less concerned with corrosion-velocity than corrosion-probability. If this is agreed, it may be actually advantageous to use a paint-film which is pervious to oxygen, since oxygen depresses corrosion-probability, whilst increasing the conditional velocity (p. 138). When once rusting has fairly started on a painted surface, it is almost hopeless to stop it. Whilst, however, permeability to oxygen may be a welcome characteristic, the reverse is likely to be true of water permeability. A paint which can take up much water is more likely to allow the initiation of attack than one which takes up little water, since the water may interfere with oxygen-replenishment. Whether or not that is the reason, it does seem that the type of pigmentation which reduces water-uptake greatly prolongs the life of a paint. This is clearly shown by work by Miss Brasher on paints made by adding different pigments to a vehicle consisting of tung oil and ester gum. When the water-uptakes were plotted against the time needed for failure in a laboratory test, the results fell on a single curve. It is not certain that the same results would be obtained with all vehicles, but the results do suggest that a high water-uptake may be inconsistent with long paint-life (D. M. Brasher, *Electroplating and Metal Finishing* 1956, **9**, 280, esp. fig. 8; D. M. Brasher, Priv. Comm., March 25, 1958).

In some cases, special causes can be suggested for the anti-corrosive action of paints. For instance, where the paint contains a chromate pigment of appreciable solubility, the water passing through the paint-film should become non-corrosive, since chromates are powerful inhibitors (Chapter V). In other cases, an inhibitor may be formed by reactions proceeding within the paint-films; oil paints pigmented with the lead or zinc pigments (oxide or basic salts) produce lead or zinc soaps, and, although these are not immediately inhibitive, they produce inhibitive substances by their degradation. Both these cases are discussed later in this chapter (pp. 546, 547), whilst another special cause—cathodic protection by paints containing metallic pigments—is discussed in the chapter which follows (p. 615). However, it is a fact that a paint containing only inert pigments, and which is incapable of acting in any of the ways suggested, can nevertheless greatly postpone the starting of corrosion of the metal studied, and probably retard it after it has started. The reason for this will now be discussed.

Protection due to sluggish Ionic Movement. Mayne attributes the anti-corrosive action of " inert " paints (i.e. those which neither contain nor produce inhibitive substances) to low ionic conductivity. Table XVIII quotes some numbers collected by him, which make clear that the diffusion of sodium chloride is very slow in cases where the water-diffusion is quite rapid (J. E. O. Mayne, *Research* 1952, **5,** 278, esp. p. 282).

TABLE XVIII

DIFFUSION OF SODIUM CHLORIDE AND OF WATER EXPRESSED IN GRAMS/CM.^{2}/YEAR (J. E. O. Mayne)

Film Substance	Sodium Chloride	Water
Alkyd resin varnish	0·000040	0·825
Phenolic resin varnish	0·000004	0·717
Polyvinyl-butyral	0·000002	0·897
Polystyrene	0·000192	0·485

Experiments carried out by Mayne on films of polystyrene, linseed oil and coumarone/tung-oil varnish indicate that it is the *anions* which have difficulty in passing through the films; it is believed that if the film-substance contains ionizable groups, such as —COOH (which may be present even in polystyrene), the film will acquire a charge which is negative in relation to water and thus the passage of anions is opposed, making it difficult for an electrolyte to reach the metal. For details, the original papers should be studied (J. E. O. Mayne, *J. Oil Col. Chem. Ass.* 1949, **32,** 481; 1951, **34,** 473).

If iron coated with polystyrene is placed in sea-water, rusty nodules gradually appear at certain points (evidently the anodes) whilst large blisters, filled with strongly alkali liquid (with a NaOH concentration exceeding normal) appear at others (evidently the cathodes). Presumably the current is at these points carried inwards by Na^+ ions. If, however, the film is made positive in relation to the water, by adding a long chain amine to the polystyrene, immersion in sea-water gives blisters full of iron salts; the current is here being carried by the Cl^- ions, whilst the film has become impervious to Fe^{++} ions (J. E. O. Mayne, *J. Oil Col. Chem. Ass.* 1957, **40,** 183, esp. p. 186).

Perhaps the most interesting evidence is provided from a case where the electrolyte is not corrosive but inhibitive. Mayne found that unpainted iron placed in dilute sodium hydroxide remained bright and unchanged; if the iron was coated with polystyrene, and then placed in the alkaline liquid, rust appeared on the iron, suggesting that the water had penetrated the coat but not the inhibitive OH^- ions.

The diffusion of water vapour through polymers, which obeys Fick's law in polyvinyl acetate but not in polyvinyl alcohol, is discussed by F. A. Long and L. J. Thompson, *J. Polymer Sci.* 1955, **15,** 413. The passage of ions across a membrane is discussed by N. W. Rosenberg, J. H. B. George and W. D. Potter, *J. electrochem. Soc.* 1957, **104,** 111.

In a general way, paint-films which show high electrical resistance when immersed in water provide better resistance to corrosion than those showing low resistance, and numerous methods for the electrical evaluation of paints have been based on this fact. These are not all of equal merit, but reference may be made to papers by E. K. Rideal, *Trans. Soc. Engrs.* 1913, **4,** 249; E. Ritter, *Korrosion u. Metallsch.* 1929, **5,** 64; W. P. Digby and J. W. Patterson, *The Engineer* 1934, **157,** 586, 610. L. H. L. Kooijmans, *Verfkroniek* 1936, **9,** 107; R. C. Bacon, J. T. Smith and F. M. Rugg, *Industr. engng. Chem.* 1948, **40,** 161. Unpublished work by J. E. O. Mayne and C. C. Maitland should receive study when available.

Protection due to Inhibitive Pigments. It is clear that an inhibitive pigment, to be effective, must be sufficiently, but not excessively, soluble. Potassium chromate, introduced into a paint, would doubtless be effective in stopping outdoor corrosion so long as it was present, but would soon be washed away by the rain. Lead chromate, owing to its very low solubility, might fail to prevent corrosion even at the out-set. However, chromates of intermediate solubility might succeed in preventing corrosion both at the start and for long periods afterwards. Mayne found that water which had stood in contact with zinc tetroxychromate for 14 days became practically non-corrosive to iron; the weight-loss of iron partly immersed in it for 102 days was less than 1/100 of that produced by water which had not been in contact with a chromate pigment. Contact with lead chromate (normal or basic) failed to render the water non-corrosive; possibly any anti-corrosive action produced by lead chromate pigments in oil paints should be attributed to the degradation-products of lead soaps (see p. 547), and not to the chromate present (J. E. O. Mayne, *J. Oil Col. Chem. Ass.* 1951, **34,** 473).

The chromates form a large family of compounds, since several metals form a number of chromates differing in basicity, whilst double chromates (containing an alkali metal as well as a heavy metal) are also known. Cole and Le Brocq have made a detailed study of the solubility relationships of chromates containing zinc and cadmium, with and without sodium, potassium or ammonium, whilst Cole has correlated the solubility data with the power to inhibit corrosion when used in priming paints on magnesium alloys, aluminium alloys and steel. He concludes that " the best all-round protection was given by strontium chromate. Fair to good protection was given by potassium zinc chrome, sodium zinc chrome, zinc monoxychromate, zinc tetroxychromate and calcium chromate. Three complex cadmium chromes, three lead chromes and barium chromate all gave poor results." The results were considered to afford some support for the idea that an intermediate solubility range is desirable; on steel and one magnesium alloy the best results were obtained in the solubility range 0·06–0·46%, but the other magnesium alloy seemed to require a higher range (0·46–6·6%). In all cases, the same vehicle (based on litho-oil and courarone) was used, and the test employed was based on intermittent spraying with sea-water. It is not certain that the same order of merit would be obtained with another vehicle or under different conditions of testing, but the two papers are of high quality and deserve careful study (H. G. Cole and L. F. Le Brocq, *J. appl. Chem.* 1955, **5**, 149, 197; see comments by J. E. O. Mayne and others, *Chem. and Ind.* (*Lond.*) 1956, p. 152). For encouraging laboratory results with zinc molybdate, see A. K. Chaudhury and S. C. Shome, *J. Sci. ind. Res.* (*New Delhi*) 1958, **17A**, 30.

Protection due to Degradation Products of Soaps. It has long been known that metallic soaps, or soap-like products, are formed in oil paints containing certain lead or zinc pigments (oxides or basic salts). Mayne found that distilled water kept for 22 days in a beaker coated with a lead soap (prepared by heating litharge and linseed-oil fatty acids in an inert solvent) lost its corrosive property; steel specimens could be kept partly immersed in such water for 118 days without attack on the immersed portions. Soaps made from zinc, calcium, strontium and barium also appeared to inhibit corrosion, but not those from tin, aluminium, iron, copper or chromium. These results seemed at first sight to suggest that the linoleates of lead, zinc, calcium, strontium and barium are inhibitors, but later van Rooyen, working with Mayne, found that this is not the case; lead linoleate, prepared in the absence of oxygen and treated with air-free water, was found to be nearly insoluble, but the presence of small amounts of oxygen increased the apparent solubility from 0·002 to 0·070%—the figure obtained after 8 days of contact between lead linoleate and water in the presence of air. A saturated solution of lead linoleate prepared in the absence of air was found to be corrosive in presence of air. It became evident that it was not the almost insoluble linoleate, but the more soluble degradation-products which were inhibiting the corrosion, and van Rooyen proceeded to analyse the extracts of lead and calcium soaps, using chromatography and other techniques to separate and identify the numerous organic acids present. These were found to be mainly formic, azelaic ($COOH.(CH_2)_7.COOH$) and an unsaturated hydroxy-acid derived from pelargonic ($CH_3.(CH_2)_7.COOH$) with smaller quantities of acetic, propionic, butyric and suberic ($COOH.(CH_2)_6.COOH$). Immersion tests showed that lead and calcium formates were corrosive, but that the lead and calcium salts of azelaic, suberic and pelargonic acids were inhibitive at pH 4·6. A synthetic mixture was made up, based on the results of the analysis, and this was found to possess inhibitive properties (J. E. O. Mayne, *J. Oil. Col. Chem. Ass.* 1951, **34**, 473; J. E. O. Mayne and D. van Rooyen, *J. appl. Chem.* 1954, **4**, 384).

The exact reason why the lead and calcium salts of certain acids should inhibit rusting cannot be stated with certainty, but presumably at weak spots in the invisible oxide-film the relatively soluble ferrous salts which (in absence of inhibitor) would diffuse out into the liquid and yield rust by oxidation and hydrolysis at a perceptible distance, where it cannot afford protection, are retained on the surface until the transformation to the ferric state is complete—thus healing the weak spots. Possibly the dibasic acids are adsorbed, and then hold the ferrous ions either as ferrous salts of the acid in question or as loose adsorption compounds, so that the oxidation to the ferric state occurs very close to the metal. Mayne and van Rooyen suggest that the lead or calcium cation may provide catalytic acceleration of the oxidation of the primary ferrous compounds, thus facilitating the repair of the air-formed film. Against this explanation stands the fact that lead soaps, provided chlorides are absent, inhibit the corrosion of aluminium, as established by M. J. Pryor, R. J. Hogan and F. B. Patten, *J. electrochem. Soc.* 1958, **105**, 9.

Such an objection may not be fatal, since iron in commercial aluminium increases the tendency to corrosion, and it might be argued that if the passage of iron cations outwards through the alumina film could be arrested, commercial aluminium would be as resistant as the purer varieties. Such an argument, however, would require to be backed by experiment, and other explanations suggest themselves. Research in Poland has shown that the fatty acids inhibit stress-corrosion cracking of steel in hot concentrated ammonium nitrate.* This can hardly be due to catalysis, since in a hot oxidizing solution possessing an acid reaction, the conversion of ferrous to ferric salts would proceed readily without a catalyst. Rather it seems to be connected with the fact—also demonstrated in the Polish work—that the polarization curves become steeper when fatty acids are present, by an amount which increases steadily with the length of the carbon chain; the potential shift needed to produce a given current density, either at a cathode or at an anode, increases in rectilinear fashion with the number of carbon atoms in the acid; the current density needed to produce the jump of potential to a passive value is smaller if fatty acids are present than if they are absent. If we assume that the true polarization curves on the bare iron is the same in all cases, and that the function of the attached chains is to reduce the uncovered area so that the real current density exceeds the apparent value, the results would seem to explain themselves—at least qualitatively.† For, unless the chain extends, immobile, at right angles to the metallic surface, the coverage must increase with the chain-length. Whether or not this explanation (thrown out by the Author) appeals to the reader, the Polish papers deserve study (M. Smialowski and T. Ostrowska, *Bull. de L'Acad. Pol. des Sciences*, 1954, **2**, 345; 1955, **3**, 29).

The work of Mayne and van Rooyen suggest a reason for inhibition brought about by such pigments as red lead, white lead and basic lead sulphate in the acid environment which is believed to exist in an oil plant. However, some of the lead pigments possess inhibitive properties in the absence of drying oil. Lewis found that specimens of iron shaken with water and air corroded far less if litharge was suspended in the water; red lead or white lead showed a definite but smaller effect. Mayne obtained complete prevention of corrosion with litharge. Other experiments, in which dilute salt solutions were used, showed very clearly that litharge is an inhibi-

* To prevent misunderstanding it should be pointed out that it is the fatty acid and not the lead salt which prevents cracking in nitrate solution, whereas it is the lead salt and not the fatty acid which prevents corrosion by water.

† If the measured polarization represents the *IR* drop connected with the approach resistance (p. 1026), the rectilinear relation between the potential-rise and the number of carbon atoms explains itself quantitatively also.

tor; the work of Pryor confirms this. The inhibition by the solution of litharge may be due to the comparatively high pH value (9·2 in Lewis' work, but only 7·4 in Mayne's experiments) or to a reserve of alkalinity, as emphasized by Pryor. Whether the pH value would remain so high in a paint-film exposed to the atmosphere in view of acid formed by decomposition of the vehicle or arising from coal smoke in the air may be doubted, but a basic pigment would certainly prevent conditions from becoming as acid as in its absence. Perhaps the litharge serves to keep in repair the air-formed film, being reduced to metallic lead in invisible quantity. There is still doubt as to how far inhibition by lead paints should be ascribed to the power of the pigment to neutralize acid, how far to the catalytic activity of the pigment favouring the conversion of ferrous products to less soluble ferric products at the metallic surface, and how far to adsorption of the degradation-products identified by Mayne and van Rooyen. Another possibility is that they serve to destroy sulphur dioxide—the most corrosive constituent of urban and industrial air; a specimen carrying a coat of red lead paint (without covering coat) becomes pale after a few years in a soot-free atmosphere, the red lead turning into lead sulphate. The reader should study the literature and form his own opinion (J. E. O. Mayne, *J. Soc. chem. Ind.* 1946, **65,** 196; *J. Oil Col. Chem. Ass.* 1951, **34,** 473. K. G. Lewis and U. R. Evans, *J. Soc. chem. Ind.* 1934, **53,** 25T; S. C. Britton and U. R. Evans, *ibid.* 1939, **58,** 90, esp. p. 92; M. J. Pryor, *J. electrochem. Soc.* 1954, **101,** 141).

Whether the true inhibitor in a lead oil paint is the pigment or a degradation-product of the soap, it becomes clear why old-fashioned red lead which contained free litharge produced a better anti-corrosive paint than the modern " non-setting " red lead, which is relatively free from litharge; the new type, however, is more convenient, since the rapid interaction between litharge and oil caused the older type to set hard in the pot. Lewis' experiments showed clearly that litharge, as such, is a better inhibitor than red lead; this would, however, be understandable if soap-formation is responsible, since soaps are formed more easily with litharge than with red lead. The opinion that red lead containing free litharge gives better protection than non-setting red lead has long been held by engineers; Britton's tests (p. 578) proved them right.

If the inhibitive properties of lead oil paints are due to degradation products of the soaps—and not directly to the pigment—we may expect that, for the protection of steel exposed to the acid air of industrial districts, oil paints will prove superior to paints based on synthetic resin of a type which can yield no soaps. On the whole, the extensive tests of Hudson, Fancutt and others suggest that on land oil paints *are* superior to the more modern synthetic paints, although, since the former usually give thicker coats, this superiority can be explained in more than one way. However, when we pause to compare oil paints and synthetic paints for marine purposes, the former are found to be inferior, because the vehicle is readily softened or loosened by the alkali, which is the cathodic-product during corrosion by salt water.

If the true cause of inhibition by oil paints is to be found in the degradation-products of a soap, it might be possible to make a cheap inhibitive paint by adding such a degradation-product (ready-made) to a paint pigmented with a pigment having no inhibitive properties of its own; for instance, the addition of lead azelate (or perhaps calcium azelate, which would be even cheaper) to a common iron oxide paint, might confer inhibitive properties upon it, so that rusting would be prevented even at places where the steel was laid bare by a scratch-line engraved through the coat. Bird, assisted by C. E. Evans, has designed experiments to explore this possibility. Numerous paint-mixtures were applied to steel, engraved with a scratch-line so as to expose the metal, after which a drop of distilled water (or in some series, N/1000 sodium chloride) was applied; with a non-inhibitive paint rusting quickly occurred in such a test. It was found possible, however, by the addition of lead azelate, to produce paints which prevented rusting at the scratch-line. Although most of the compositions which achieved this result would not be favoured by paint technologists for reasons unconnected with corrosion, the results suggest that the principle of adding a ready-made inhibitor is a sound one; there is no apparent reason why it should not be applied by paint technologists to provide cheap paints which would be acceptable in general properties and yet thoroughly inhibitive.

One of the requirements, however, is that the paint should be sufficiently porous to allow the water to soak in and dissolve the inhibitive compound. If the coating is made too porous, there is a risk of rusting at specially susceptible points situated on the paint-coated area; if it is made insufficiently porous, rusting will occur at the scratch-line. Clearly a compromise is needed. In Bird's tests, the best results were obtained with cellulose nitrate as vehicle, regulated amounts of titania being added to confer a modicum of porosity; this prevented rusting, although in the absence of titania there was invariably rusting at the scratch-line. It was advisable to plasticize the vehicle with dibutyl phthalate.

The research, however, was purely exploratory and designed to test a principle rather than obtain a technologically perfect composition. (Unpublished information from the Ministry of Supply.)

Break-down of Paint-coats

General. It is commonly considered that failures of paint to prevent rusting are due to the paint first deteriorating under chemical or physical influences, and thus leaving the metal unprotected. More often, however, the slow corrosion which can occur even when the paint is apparently perfect yields products which damage the paint-coat chemically or physically, so that corrosion is self-accelerating. Thus, as explained below, a drop of salt water lodged in a crevice between two painted surfaces may develop alkali around its edge which will destroy the vehicle of an oil paint. Again, under free atmospheric exposure, rust slowly formed at an exceptional point (where a paint-coat is slightly defective or the steel specially susceptible) will occupy a volume far larger than the metal destroyed, and will push up the coating locally; since an increase in area occurs if, say, a circular patch of flat paint is to be raised into a hemisphere, the paint-coat usually cracks, leaving the metal unprotected. This only occurs if the rust is formed inside the coating, as in atmospheric attack where there is abundance of oxygen. Painted steel fully immersed in stagnant water, where the replenishment-rate of oxygen is slow, usually develops its rust outside the coat; it is surprising how large a volume of rust sludge can be formed outside a coat without causing it to leave the metal, although doubtless a detailed examination would reveal the fact that the coat only remains attached at certain microscopic areas which have escaped attack.

Evidently there are conditions when it is truer to say that corrosion causes paint break-down rather than that paint break-down causes corrosion. Nevertheless, there are other situations—notably in warm and/or damp climates—where the paint perishes first and corrosion follows. This may be connected with destruction of paints by moulds, mildews, bacterial attack or even animals, but such troubles are well known to the paint technologist who can introduce ingredients calculated to minimize them; provided that the ingredients introduced are not substances which are themselves likely to favour corrosion, the subject need not concern us further.

The action of sunlight in decomposing the vehicle of paints requires brief notice. If the vehicle and pigment are transparent, the former may perish, and the latter, left unsupported, is liable to be rubbed off; with ordinary paints (especially those carrying white pigments) the phenomenon is known as *chalking*.

White paints are rather especially liable to chalking, but this should not be attributed entirely to the transparency of the pigment, since light-absorption by the pigment-material seems to play an essential part. Paints containing titania suffer chalking because both forms of TiO_2 (anatase and rutile) absorb ultra-violet light, apparently becoming reduced to Ti_2O_3 with release of atomic oxygen which attacks the vehicle (C. Renz, *Helv. chim. Acta* 1921, **4**, 961; A. E. Jacobsen, *Industr. engng. Chem.* 1949, **41**, 523; E. Lund, *Verfkroniek* 1958, **31**, 14. For chalking of paints containing zinc oxide, see three papers in *J. Oil Col. Chem. Ass.* 1950, **33**, 471–501 by J. R. Rischbieth; G. Winter and R. N. Whittem; C. T. Morley-Smith.

Black paints prepared from bitumen or tar are also sensitive to sunlight (p. 571). In both cases, the introduction of flaky pigments of a substance possessing high reflectivity, such as aluminium, greatly reduces the trouble.

If no precaution is taken against the destruction of vehicle by sunlight or heat, the ultimate exposure of the steel may sometimes lead to serious corrosion. That this does not occur more frequently is due to the fact that, at inland situations in the tropics, where the relative humidity is low and industrial pollution is absent, little corrosion occurs even on a surface which has never been painted. Near the sea, however, the state of affairs is different, and if for any reason the paint has perished, corrosion may be very rapid.

Having briefly noted the cases where paint-destruction leads to corrosion, we shall now pass to cases where corrosion develops paint-destruction.

Alkaline Peeling and Softening of Paint-coats. In an early research, the Author placed drops on N/2 sodium chloride on horizontal steel specimens carrying single coats of paint. Except in rare cases where the paint was entirely inhibitive, the phenomena observed were similar to those observed on unpainted steel, but considerably slower; alkali was gradually formed at the periphery of the drop and ferrous chloride in the centre (p. 118). With nitrocellulose and linseed-oil paints, where the vehicle was capable of being attacked by alkali, the paint-coat became so loose that it could be removed by gentle rubbing with the finger; often the detachment occurred over an area exceeding the original area covered by the drop, since alkali had crept between paint-coat and metal, loosening the former.

It was clearly the cathodically produced alkali which was the cause of deterioration—not the sodium chloride or the anodically produced ferrous chloride. In some of the paints carrying flaky pigment, where the relatively large area of contact between each individual flake and the steel basis made it possible for flakes to remain attached even at a relatively advanced stage of attack, it was found that the paint in the centre of the drop, although rust-stained, remained clinging to the metal; in contrast, around the periphery of the drop the paint was easily detached, although the steel was here quite uncorroded. Moreover, zinc and calcium salt solutions caused practically no peeling, since here the cathodic product was less strongly alkaline. Where the paint was sufficiently inhibitive as to prevent corrosion altogether, there was no peeling or softening—showing that it was not the sodium chloride which was causing the deterioration of the paint. Corrosion was necessary for the softening; if a scratch-line was made through the coat with a scribe before the placing of the drop, corrosion and therefore softening proceeded more readily. It was found possible to reduce the softening and loosening by the addition of copal, which rendered the vehicle less susceptible to alkali (U. R. Evans, *Trans. electrochem. Soc.* 1929, **55**, 243).

The softening and loosening of the films by the cathodically formed alkali probably involve two distinct, but closely related, phenomena. Any vehicle containing saponifiable groups will be dissolved or softened by alkali; thus the glycerides present in linseed oil would in its original (wet) form react with sodium hydroxide giving sodium linoleate; even after oxidation and polymerization the film contains sufficient saponifiable matter to soften appreciably. However, alkali has the power of creeping over metal; a steel sheet specimen partly immersed in sodium chloride solution becomes wetted above the water-line by the cathodically formed alkali creeping upwards; a drop of salt water placed on a horizontal steel surface becomes surrounded with a damp area for the same reason. The power of alkali to creep over a metal surface is not in doubt.* If there is a film of paint

* Probably OH^- ions have a greater attraction than H^+ ions for metal, which perhaps accounts for the fact that metals are often less attacked by acid and more attacked by alkali than would be expected, since at the metal–liquid interface OH^- ions are more abundant and H^+ ions less abundant than in the body of the liquid.

or varnish clinging to the surface, the total interfacial energy of the situation where metal and paint are separated by a film of alkaline liquid may be lower than that existing when the film is attached to the metal. The diminution in energy may be sufficient to make detachment a phenomenon needing no application of work, so that creepage proceeds spontaneously and the film becomes separated from the basis. Clearly this is most likely to occur if OH^- possesses an affinity not only for the metal but for the film-substance—in other words, if the film-substance contains saponifiable groups. Thus, on the whole, the paints which tend to " soften " are those which tend to " loosen ".

It is, however, possible to have loosening without perceptible softening. One black paint, much used by marine engineers 30 years ago, was found normally to afford satisfactory protection to steel against sea-water; but occasionally, for no apparent reason, it was found to peel off rapidly, leaving the steel bare over patches many inches across; that was probably due to the presence of imperceptible scratches, just sufficient to allow a little corrosion with the production of a trace of alkali, which then crept between metal and coating, loosening the latter. The correctness of this explanation was indicated by laboratory experiments; steel specimens were covered with the paint in question, and scratch-lines were made through the coating, a drop of salt water being then placed on each scratch-line. After 24 hours, it was found that vigorous rubbing dislodged the paint over an elliptical area around the scratch-line, although it remained adherent elsewhere; unscratched specimens resisted stripping, and the paint appeared unaffected, apart from a microscopic ridge following the drop-margin—apparently the opening stage of detachment.

Clearly to avoid loosening and softening, we must use media free from saponifiable groups; paints compounded from such media should be preferred for marine use. The vinyl compounds (including polystyrene) provide examples of alkali-resistant media. On ships which are to receive cathodic protection, involving the formation of alkali in large amounts on certain parts of the hull, alkali-resistant coatings of some sort are absolutely essential, since oil paints or others susceptible to alkali will rapidly be removed. This is now generally recognized (p. 301), but it is still not always appreciated that for most purposes where sea-water will reach painted metal, paints made from vehicles susceptible to alkali should be avoided —even if no electric current is to be applied from an external source.

It is undoubtedly true that oil paints frequently give good service at sea, since in some situations there may be no accumulation of alkali at any one point. Imagine a point, P, on a painted surface over which drops of sea-water are running down; whilst the edge of a drop happens to be over P, alkali will be formed there, but the next moment the centre of the drop will be over P, and ferrous chloride will be the product; on the average, the cathodic and anodic products will be formed in equivalent amounts, and no alkali will accumulate. If the drops merge to form a continuous water layer of uniform depth, the chance of alkali accumulation will be still smaller.

Consider, however, a place where two plates come close together, providing an opportunity for a drop to lodge in a definite position determined by the geometry, so that, when the first drop has been removed by evaporation, jolting or other cause, another will be caught at exactly the same place; in these circumstances, ferrous chloride will always be produced at the centre and alkali at the edge of the drop-position—so that alkaline peeling may be expected. This probably explains the service cases mentioned above; if vehicles resistant to alkali had been chosen, such unexpected failures would not have occurred.

The necessity for alkali-resistant paints on ships seems to be recognized today in the U.S. Navy (see p. 584), and the experimental work performed in Great Britain points in the same direction. Mayne has summarized the position thus:—

" When the paint is to be exposed to sea-water . . . sodium hydroxide is the cathodic product; the early work of Evans has indicated that under these conditions saponification of the vehicle may soften the coat. Fancutt and Hudson examined the relative protective value of 127 paints when immersed in sea-water and it can be seen from the ' Merit Figures ' of the different media that protection was associated with the alkali-resistance of the vehicle. Thus for protection against sea-water linseed oil is unsuitable " (J. E. O. Mayne, *J. Oil Col. Chem. Ass.* 1951, **34,** 473, esp. p. 477, discussing F. Fancutt and J. C. Hudson, *J. Iron St. Inst.* 1946, **154,** 273P).

Blistering. Some paints can develop blisters even in the absence of corrosion. Mayne applied films of plasticized polystyrene to glass and immersed the specimen in water; blisters appeared and increased in size with time; they contained no liquid. The blistering was more severe in distilled water than in sea-water, but the distribution was affected by the choice of plasticizer. He attributes such blistering to the absorption of water by the film, which, in the absence of constraint, would cause an increase of volume; if the film is firmly attached to the basis, internal stresses will be produced, but at any point of low adhesion, these stresses can be relieved if the film detaches itself and rises as a blister; salt in the water is likely to oppose the entry of liquid into the film-substance, since the separation of water from a solution (which thereby becomes more concentrated) involves an increase in free energy; thus salt renders blistering less severe—as Mayne found. Conversely, any water-soluble substances within the film should help to draw the water in, and favour blistering. Kittelberger and Elm postulate the presence of water-soluble compounds in the film in their explanation of blistering, but the presence of such compounds in some of the films which suffer blistering appears improbable, and the known facts can be otherwise explained. In contrast, soluble substances can play an important role in producing blisters of another type. When steel coated with plasticized polystyrene is placed in salt water, blisters are formed which are found to contain strongly alkaline liquid; in Mayne's measurements the alkali content varied from 0·59N to 1·31N. Such blisters probably appear at cathodic points where alkali is formed, and the alkali, creeping between metal and film substance, loosens the latter. In addition to the relatively large transparent blisters marking the cathodic points, Mayne observed small rust-nodules at the anodic points. The loosening effect of the alkali, which allows the film to leave the metallic surface as water is imbibed, probably depends on the principles already discussed in connection with alkaline peeling. Gay suggests that water-soluble inclusions present in the paint-film suck in water by osmosis, but Mayne's experiments point to electric endosmose, rather than ordinary osmosis, as the cause of the movement of water through the film. The distinction is important, since if electric endosmose is the operating factor, there must be corrosion before any blistering can occur, a current being needed for the passage of water* (J. E. O. Mayne, *J. Oil Col. Chem. Ass.* 1949, **32**, 504, 505; 1950, **33**, 312, 538; P. J. Gay, *ibid.* 1949, **32**, 488; W. W. Kittelberger and A. C. Elm, *Industr. engng. Chem.* 1947, **38**, 695; D. M. Brasher and T. J. Nurse, *J. appl. Chem.* (in the press).

Recent observations by the Author suggest that hydrogen-evolution below a paint-coat is a more important cause of blistering than is generally recognized.

Effect of Rust below Paint. To obtain protection, paint should be applied directly to a clean metallic surface. This advice may seem too obvious to deserve printing, but probably a large fraction of the paint applied to metal today is used to give coats which at some points at least remain separated from the metal by rust or scale. Strictly speaking, it is doubtful whether paint is ever applied directly to the metal, since there is almost always an invisible oxide-film; invisible films, however, do not seem to detract from the protection afforded by the paint, but rather improve it. In contrast, the thick mill-scale left after rolling and the scale left after welding are both generally unfavourable; the same is often true of rust-patches produced before painting. Of these two disturbing influences, rust will be considered first.

The rust produced by pure water on iron should consist essentially of hydrated ferric oxide without soluble iron salt. This appears to be comparatively harmless. Britton applied drops of Cambridge rain-water to steels and allowed them to dry, leaving circles of rust; paint was then applied over the rust, after which the specimens were exposed out of doors at Cambridge. Little additional failure was observed at the sites of drops of distilled water, although the rust left by drops of salt solution produced rapid break-down. Mayne has established the fact that paint applied over rust at Cambridge in the summer months behaves far better than that applied in the winter months, and the superiority of summer-painted over winter-painted specimens can still be seen 5 to 7 years after the application of the paint. He attributes the difference to ferrous sulphate shut into the

* The corrosion need not occur on the basis steel. If steel coated with zinc-pigmented paint is immersed in salt water, anodic corrosion of the zinc may produce cathodic alkali on the steel below the coat; in this case, the Author is inclined to think that osmosis does play a part, water being sucked in by alkali and the loosened coat rises up in blisters; hydrogen, evolved by the cathodic reaction on the steel, may help.

rust below the paint. It is well known that the moisture condensing in winter in towns is strongly acid with the products of combustion of fuel, and ferrous sulphate will be formed on the metal; sulphur compounds adsorbed on soot particles setting on the surface will doubtless help. However, rain falling in spring or early summer from fairly high levels will be relatively pure and will wash away the soluble iron salts; if shortly afterwards paint is applied, it will shut in little SO_4^{--}. A reason for the bad effects of soluble iron salts shut in under paint is suggested on p. 485. Mayne attributes the effect of soluble salts shut in below paint to a short-circuiting of the resistance of the paint-film; a path of low resistance through the liquid threading the rust is provided between anodic and cathodic areas. Thus, despite the fact that the ions do not readily move through a paint-vehicle, formation of bulky rust below paint continues apace, so that the paint-coating is forced outwards and soon fails (J. E. O. Mayne, *Official Digest of Federation of Paint and Varnish Production Clubs* 1957, **29**, 706; *J. Iron St. Inst.* 1954, **176**, 140).*

It is found, in practice, that where steel must be painted without complate removal of rust, red lead paint gives fairly good results; it is generally considered that the new calcium plumbate paints are even better. However, whilst good protection can occasionally be obtained by painting over rust, it cannot be taken for granted; the results obtained after painting a rusty surface depend on the atmospheric conditions prevailing in the period preceding painting, as well as on the paint chosen.

Effect of Scale below Paint. The coating of mill-scale present on freshly rolled steel often appears continuous, except near sheared edges, where it generally flakes off, and at places where the plate or sheet has been bent, or bored with holes intended to take bolts or rivets. Even where the scale appears continuous, invisible gaps may be present. Probably a truly continuous layer of scale would serve as an additional paint-coat and increase protection, but in practice such continuity is rarely attained. Within a given small area, the inception probability of rusting is probably less on a plate covered with mill-scale than on a scale-free plate. However, the velocity of attack at the exceptional places where gaps in the scale-coat exist is increased by the presence of the mill-scale. Engineers have frequently observed that places where identification numbers have been painted outside the scale-layer on a steel plate shortly after rolling withstand corrosion better than the main part of the plate where the scale had been pushed off by under-rusting during the early period of exposure to the weather (i.e. before the protective paint-coat was applied over the whole surface). This observation has led them to suggest that if the whole area of the plate had been painted immediately after rolling before any rusting had occurred (the paint being applied to the outer surface of the mill-scale), the protection would have been excellent everywhere.

* Dust particles in paint-films lower electrical resistance and anti-corrosive properties. The need for thick coats—often urged today—may be explained on the basis of dust particles (C. Graff-Baker, *Chem. and Ind.* (*Lond.*) 1958, p. 590).

There is, however, a flaw in the argument; the chance of a gap in the scale occurring within the relatively small areas represented by identification numbers is itself small, and the excellent performance in most cases is not surprising; if the whole plate had received paint applied outside the scale, similar excellence would have been achieved over large regions, but at certain points, representing gaps in the scale, there would have been intense rusting, pushing away the paint locally and permitting the development of pits far more objectionable than uniform attack.

The result of applying paint on the outer surface of mill-scale is particular disastrous if small areas (" holidays ") are left unpainted. In the earlier Cambridge exposure tests (p. 577), it was always the rule to leave a strip unpainted at the upper end of the inclined specimen. On this strip, rusting always set in soon after exposure—whether the steel carried scale or had been descaled before paint was applied. On specimens where the steel carried scale the corrosion was found to penetrate along the interface between metal and scale, and bulky rust was formed at this level, so that the scale, along with the paint-coat or coats outside it, soon began to flake off; clearly inhibitive pigments in the paint were of no avail in preventing this, since the inhibitor was separated from the point of attack by the scale; nor was there any advantage, in such a situation, in applying several coats of paint instead of only one. In contrast, those specimens where the paint had been applied to steel previously descaled (usually by grinding), the tendency for the rusting from the unpainted strip to extend below the paint was much less than on the scale-bearing steel, and this was particularly the case when the paint contained an inhibitive pigment. After about 7 years exposure at four different sites, representing rural, marine, urban and mixed atmospheres, and comprising thirty different one-coat, two-coat or three-coat systems, a comparison between each specimen painted over the intact scale with the corresponding specimen painted after removal of the scale showed that, in nearly every case, the latter was definitely in better condition; in no case did the scale-covered specimen show any superiority (S. C. Britton and U. R. Evans, *J. Soc. chem. Ind.* 1939, **58,** 90, esp. p. 92).

Similar results were obtained by J. C. Hudson (p. 578) using thicker and larger specimens, although here the plan of leaving an unpainted area was not adopted. A plate of copper-bearing steel, pickled to remove mill-scale before being painted and exposed at Sheffield, was still in excellent condition after 6 years, whereas a specimen carrying the same paint applied over the scale had suffered break-down at numerous small points. Specimens of Swedish iron exposed 9 years at Sheffield told the same story, the pickled specimen behaving far better than the unpickled one.

If steel carrying apparently intact mill-scale is exposed out of doors, rusting soon starts at invisible gaps, and under some weather conditions, extends sideways below the scale; the voluminous character of the rust forces the scale away from the metal, so that under favourable circumstances, the whole surface may become free from scale after some months; it is, of course, very rusty but most of the rust can be removed by vigorous brushing, which also removes any mill-scale that still exists outside the layer rust; the mill-scale on an unweathered specimen is for the most part tightly attached, and cannot be removed with an ordinary wire-brush.

However, the weather conditions on the day when exposure starts greatly affect the chance of removing scale by under-rusting. In 1935 Thornhill exposed at Cambridge two series of specimens of the same steel in the same surface condition, starting the exposure, however, on two different dates; they behaved differently from one another, and very differently from those exposed some years earlier by Lewis. The apparent removal of scale by under-rusting can be very deceptive. In some of the Cambridge specimens rusty warts were observed covering up the scale over appreciable areas, but when the rust was removed by wire-brushing it was found that, except at the centre of each wart, the scale was still intact, and sufficiently adherent to resist removal on further brushing. Apparently, however, even in this condition some attack below the scale is occurring, and it may happen that—during weather-periods of violent temperature-fluctuation—the scale on such specimens will suddenly peel off over almost the whole area. Thermal fluctuation is especially effective in causing sudden scale peeling; in the case under review the peeling occurred after a series of sunny days (R. S. Thornhill and U. R. Evans, Rep. Corr. Comm. Iron Steel Inst. 1936, **4,** 175; K. G. Lewis and U. R. Evans, *ibid.* 1936, **3,** 173, 177).*

The unpredictable behaviour of mill-scale, and its dependence on the weather conditions prevailing at the start of exposure, is of practical importance, since in the past, engineers have often relied on weathering followed by wire-brushing to remove the scale. The type of scale on the steel also affects the ease of removal by weathering, just as it affects its removal by pickling (p. 406), so that two steel plates exposed under identical conditions may shred their scale at different rates. Even if the weather conditions are ideal for quick undermining and removal of mill-scale, it remains a fact that the time needed for the under-parts of a structure, which are sheltered from the weather, may be inconveniently long, and if the owner of the structure waits for these parts to shred their scale before he starts the painting, the more exposed parts will have reached an advanced stage of corrosion. If the period allowed is too short, some parts will carry numerous small gaps in the scale, presenting a particularly unfavourable condition for painting—as shown by the tests of Lewis, described below. For these and other reasons, reliance on weathering as a method of scale-removal is open to grave objection—as has been forcibly pointed out by J. C. Hudson in his book, " The Corrosion of Iron and Steel " (Chapman & Hall).

Comparison of Specimens carrying different amounts of Scale. The deterioration of the surface through partial removal of mill-scale by weathering was brought out by Lewis' tests at Cambridge. Different specimens of originally scale-covered steel were weathered for 0, 1, 2, 3, 4, 6, 8 and 12 months respectfully; they were then wire-brushed to remove

* *Frosty* nights have been known to bring off *paint*-coats which were already loosened and almost ready to peel; see S. C. Britton and U. R. Evans, *J. Soc. chem. Ind.* 1930, **49,** 173T, esp. p. 180T.

loosened scale and loose rust, painted and re-exposed. For comparison, specimens were included which carried their scale intact (not wire-brushed before painting), along with other specimens completely descaled by pickling.

It was found that after 33 months' total exposure, specimens weathered for 4 months or longer retained their paint-coats firm and unchanged, although some of those weathered for 4 months still carried small areas of scale on the day of painting; pickled specimens, and those weathered sufficiently long to be completely scale-free, behaved equally well. Specimens weathered for 1 or 2 months, which carried appreciable areas of scale adherent at the time of painting (10–25% of the whole), behaved badly —worse in fact than unweathered specimens which carried an almost intact scale layer; the undermining of the paint-coat observed on specimens weathered for 1 month before painting mainly started from the edge of the unpainted areas; but specimens weathered for 2 or 3 months before being painted showed rusting or slight rising of the paint-coat on the face at points remote from the unpainted area. The bad effect of extensive areas of scale-residue may be due to corrosion-couples set up between scale and metal, but there are other possibilities; it is more difficult to remove rust-residues (which probably contain ferrous sulphate) from small holes in the scale than from scale-free plates; also there is a chance that parts of the scale, although sufficiently firmly attached to resist removal by wire-brushing, may in fact be partly undermined, and may start to lift later, perhaps at a time of temperature fluctuation, pushing away the paint also.

Pre-treatment of Surfaces before Painting

Removal of Scale and Rust. The observations just recorded show that, for successful protection by paint, removal of scale and rust is desirable, but that methods based on weathering followed by brushing are unreliable, and may easily make matters worse if the weathering is not continued sufficiently long. It should be emphasized that mill-scale, unless previously loosened by under-rusting, cannot be detached by ordinary wire-brushing; the idea, once prevalent, that such removal is possible is due to the fact that wire-brushing often *burnishes* the scale, producing a metallic sheen which may mislead the operator into the belief that the underlying metal has been exposed; comparison between the appearance of a truly descaled specimen and one carrying burnished scale will at once show up the difference between the two. Other methods must be sought for genuine descaling.

Mechanical Methods. The cleaning of steel (i.e. removal of scale, rust, old paint and miscellaneous dirt) may be carried out by dry (mechanical) methods, thermal methods, or wet (chemical) methods. Among the mechanical methods, sand-blasting and grit-blasting are most important, and represent the best method of dealing with firmly adherent mill-scale. Various degrees of abrasive blasting are recognized in industry, the most thorough of which cleans the surface completely and leaves a white matt surface. However, a lesser degree of abrasive blasting can produce considerable benefit, since one of the greatest dangers in painting is the presence of flakes of mill-scale partly undermined by rusting, but still adhering strongly at certain points, so that they are not removed by wire-brushing. If left in position, these become loose after a short atmospheric exposure, and rise up, pushing off the paint outside them. The application of sufficient abrasive blasting to remove that danger may appreciably prolong the life of subsequently applied paint-coatings, although clearly complete cleaning is preferable.

Of the types of abrasive blasting, sand-blasting is out of favour owing to risks of silicosis; the hazard is greatly reduced in the so-called wet sand-blasting, where the sand is wetted with water, and this method is in extensive use in some countries. It has occasionally been proposed to use a solution of sodium dichromate and/or tri-sodium phosphate instead of water, so as to provide the new surface with a protective film at the moment of its formation, but this does not seem to be common practice.

Grit-blasting is to be preferred to sand-blasting from the health standpoint; moreover, whilst sand-blasting often reduces the fatigue strength of the metal, a suitably chosen system of grit-blasting may even improve it (p. 599). The material is frequently referred to as "steel grit" or "steel shot", but most of it is really cast iron, having a composition such as would produce a grey iron if used for a casting, but actually showing a structure similar to white iron owing to the rapid cooling of the particles. True steel grit is obtainable, but is used only for special purposes. Information regarding grit-blasting is provided by J. H. Frye and G. L. Kehl, *Trans. Amer. Soc. Met.* 1938, **26,** 192, and regarding wet sand-blasting by J. E. Harris and W. A. D. Forbes, *Trans. Instn. nav. Archit.* 1946, **88,** 240, esp. p. 249. The types of abrasive used—especially before vitreous enamelling—are discussed by D. W. S. Hurst and J. Bradshaw, *Metal finish. J.* 1958, **4,** 276.

The self-contained *Vacublast* system, which recovers the abrasive material, and largely avoids flying particles, is coming into favour; the cleaning gun is a blast cabinet in miniature, the abrasive being directed on the work through an inner cone and sucked back by vacuum in the outer cone. The method is somewhat slow and there has been difficulty in adapting it for sharply curved surfaces, but this is being overcome. Although more expensive than some other methods, the use of *Vacublast* is extending and likely to extend further.

It is believed by some that sand-blasting (possibly grit-blasting also) renders the surface very susceptible to corrosion, and it has even been suggested that the surface should deliberately be allowed to rust, followed by wire-brushing; the idea is that the most susceptible (disarrayed) portions of the metal will be converted to rust and thus eliminated. However, the Cambridge experiments showed that specimens cleaned by sand-blasting before painting behaved as well or better than those cleaned by grinding or pickling, and that the rusting and brushing of a sand-blasted surface before painting made performance worse. Hudson's experience has been similar as regards steel to be exposed to the atmosphere. (For the question of descaling of ships' plates, see p. 583.)

Thermal Methods. Another process, which does not remove the whole of the contaminating matter but provides a surface more suitable for painting than a weathered surface carrying patches of rust and scale, is known as *Flame-Cleaning*. A row of closely spaced oxy-acetylene burners is moved steadily over the surface, desiccating the rust and leaving much of the adherent matter brittle, so that, on subsequent brushing, a great deal is removed; the surface should be left with a dry porous film to which the paint will generally adhere well. Experience is needed in deciding the correct rate at which the flames should travel, so as to leave the contamination loose and brittle without welding it to the surface. Flame-cleaning is probably inferior to abrasive blasting, and may introduce dangers to health if applied to surfaces carrying old coats of lead paint; it has, however, one advantage—that the paint can be applied to a warm surface, thus eliminating any risk of condensed moisture. For details, see F. Fancutt and J. C. Hudson, " Protective Painting of Structural Steel " 1957, p. 24 (Chapman & Hall).

Chemical Methods. Wet processes include acid pickling and its variants, already discussed on pp. 404–420, where the dangers of hydrogen-cracking received attention. Pickling is easier to carry out than might be imagined; large plates and long poles have been treated in suitably shaped tanks. The fear has been expressed that traces of acid left on pickled metal will cause trouble under the paint; actually such traces will soon cease to consist of acid, becoming ferrous sulphate or chloride, according to the acid employed, but ferrous salts below paint can be dangerous—as already indicated. In the past, it was sometimes recommended that pickling should be followed with a sodium hydroxide bath, so as to neutralize any acid that may be lurking in crevices. This advice is open to objection. The residue in the crevices will contain iron salts, and an alkaline bath is likely to produce a membranous precipitate at the mouth of the crevices, thus shutting in the contamination. Moreover, alkali is more prone to creep than acid, and if shut in below paint, it is likely to produce loosening and softening. If an alkali bath is felt to be necessary, lime is preferable to soda, since in such a case there is hope that any residues may be converted to harmless calcium carbonate before the paint is applied.

The ascertain the dangers of faultily conducted pickling, comparative experiments were conducted at Cambridge. The complete sequence tested consisted of (1) pickling, (2) washing, (3) treatment in an alkali bath, (4) washing, (5) drying, (6) painting and (7) exposure out of doors. On certain specimens, all the operations were conducted as carefully as possible, but on others certain " intentional " faults were introduced. Either (*a*) the washing was incomplete, (*b*) the drying was incomplete or (*c*) the alkali bath was omitted. In the earlier stages of exposure no bad results could be observed from the " faulty " procedures, but after 4 years it was observed that the paint on specimens where the drying had been incomplete and those where the soda had not been properly rinsed off, were behaving worse than the others. On practical structures, the consequences of incomplete washing and drying are, of course, most serious at places where there are concave corners or crevices between plates.

If it is possible to follow pickling in sulphuric acid by a short dip in hot phosphoric acid—as in the Footner process (p. 405)—the dangers inherent in pickling largely vanish, and indeed the iron phosphate film left on the metal adds appreciably to the protection afforded by the paint. This process, followed by painting whilst the steel is still hot, probably represents the best method for dealing with steel in cases where large numbers of plates or poles of uniform size have to be prepared and painted; methodical planning should eliminate difficulties and make the costs reasonably low. For small and miscellaneous jobs, its advantages are less marked. Information about the Footner process as applied at the rolling mill will be found in *Metallurgia* 1952, **45,** 253. The nature of the film, which undergoes alteration on air-exposure, becoming more basic and less soluble, is discussed by M. Donovan, J. W. Scott and L. L. Shreir, *J. appl. Chem.* 1958, **8,** 87.

Authoritative views on the relative merits of various cleaning processes —based on careful observation—are provided by J. C. Hudson and his colleagues. Descaling by weathering may take 2 years, and during all this time the steel is corroding—perhaps at a rate exceeding 5 mils per year in an industrial district; if the scale is not fully removed, the pattern of subsequent paint break-down corresponds exactly to that of the breaks in the scale. Flame-cleaning is satisfactory if suitably applied, but it involves the human element and (according to the 1951 figures) is rather more expensive than grit-blasting, as well as less effective; mechanical methods of descaling appear to be better than hand methods. Pickling, if used, should follow the Footner method.

The effect of surface preparation is shown in Table XIX.

TABLE XIX

LIVES OF PAINT-COATS AT SHEFFIELD IN YEARS (J. C. Hudson)

	2 Coats Red Lead and 2 Coats Red Iron Oxide	2 Coats Red Iron Oxide
Weathered and hand-cleaned	2·3	1·2
As rolled. Scale apparently "intact"	8·2	3·0
Pickled	$>9{\cdot}6$	4·6
Sand-blasted	$>10{\cdot}3$	6·3

For details, see J. C. Hudson, *J. Oil Col. Chem. Ass.* 1945, **28,** 27; J. C. Hudson, *Trans. Inst. Met. Finishing* 1952–53, **29,** 124; J. C. Hudson and W. A. Johnson, *J. Iron St. Inst.* 1951, **168,** 165; *Schweiz. Arch. angew. Chem.* 1958, **24,** 33; J. F. Stanners, *Corros. Tech.* 1954, **1,** 188. The effects of different pre-treatment methods are presented by F. Fancutt, Iron St. Inst. spec. Rep. No. **31** (1946). The costs and merits of pickling and grit-blasting (in an American situation) were compared by F. J. Farnell, *Metal Finish.* (*N.Y.*), 1949, **47,** 69.

Phosphate-coatings. The bluish-grey coating on steel emerging from the hot phosphoric acid bath in the Footner process consists of phosphates of iron—largely ferrous phosphate, although ferric phosphate appears on air-exposure; there is also a gradual conversion of primary ferrous phosphate to tertiary phosphate which is less soluble, just as calcium carbonate is less soluble than bicarbonate. The film affords some protection and a basis for paint. However, phosphates of iron are not ideal for this purpose, since the oxidation of ferrous to ferric compounds by the air with subsequent reduction by the metallic iron to the ferrous condition will involve volume changes. Better results might be expected from a phosphate-coat containing a metal in which only one state of oxidation is stable; whether for this reason or another, coatings consisting mainly of zinc or manganese phosphate do provide superior results.

Zinc forms three phosphates $Zn_3(PO_4)_2$, $ZnHPO_4$ and $Zn(H_2PO_4)_2$. The solid bodies contain water of crystallization. Machu tabulates the following phases:—

$$Zn_3(PO_4)_2.4H_2O$$
$$ZnHPO_4.3H_2O$$
$$Zn(H_2PO_4)_2.2H_2O$$

and describes the first as insoluble and the last as easily soluble in water. The tertiary phosphate $Zn_3(PO_4)_2.4H_2O$ dissolves in phosphoric acid (much as $CaCO_3$ dissolves in carbonic acid giving a bicarbonate), but if the pH of the liquid is subsequently raised, a solid is likely to be precipitated. The case of manganese phosphates is slightly more complicated; the acid phosphate $Mn(H_2PO_4)_2.2H_2O$ is partly decomposed by water to give $MnHPO_4.3H_2O$ and free phosphoric acid. Here again the tertiary phosphate $Mn_3(PO_4)_2.7H_2O$ is sparingly soluble in water, but easily soluble in acids (W. Machu, "Die Phosphatierung" 1950, pp. 14–17 (Verlag Chemie)).

If now we take a clean surface of iron and immerse it in phosphoric acid previously saturated with either tertiary zinc or tertiary manganese phosphate, the attack upon the iron, besides giving an iron phosphate as the anodic product, is likely also to throw down a solid zinc or manganese phosphate owing to the rise in pH produced by the cathodic reaction; this should occur whether the cathodic reaction is the liberation of hydrogen (consuming H^+ ions) or the reduction of oxygen or oxidizing agent (producing OH^- ions), since both changes raise the pH value. It would not be surprising for the iron phosphate to assume an epitaxial relationship with the metallic iron from which it is directly formed, and for the zinc or manganese phosphate to continue the crystal structure of the iron phosphate. If suitable crystallographic relationships exist between the three main phases (iron, iron phosphate, zinc or manganese phosphate), the admirable adhesion and protection are easily understood. It cannot be claimed that the crystallography of phosphating is sufficiently understood, but it is not unlikely that the unique properties of the phosphate layers produced under favourable conditions will ultimately be correlated with crystal habit. Photographs published by Holden show that pickling of the steel just before phosphate treatment can completely alter the size of the phosphate crystals composing the film; this is highly suggestive. Holden describes the layer of crystalline, water-insoluble metal phosphates, as "integral with and tightly adherent to, the basis metal". Wusterfeld's observation that the crystal structure of the phosphate coat is no finer on fine-grained than on coarse-grained metal is probably not really inconsistent with this idea. The reader should examine the evidence and form his conclusions (H. A. Holden, *J. Electrodep. tech. Soc.* 1948–49, **24,** 111, esp. fig. 1, opposite p. 120; A. Wusterfeld, *Arch. Metallk.* 1949, **3,** 253).

Evidently a careful adjustment of the acidity of the bath is needed for successful production of suitable films; if the acid content is too high, the bath will act simply as a pickling bath and no protective layer will be produced. Machu, quoting Rathje, states that the film produced by the zinc bath contains tertiary zinc phosphate along with varying amounts of secondary and tertiary iron phosphate, whilst those produced in the manganese bath contain secondary and tertiary manganese phosphate along with various ferrous phosphates (secondary predominating). A wealth of information, particularly regarding the porosity of films formed under different conditions, will be found in the excellent book of W. Machu, " Die Phosphatierung " (Verlag Chemie); also " Korrosion VIII " (1954–55), p. 68 (Verlag Chemie); see also W. Rathje, *Ber. dtsch. chem. Ges.* 1941, **74,** 357.

Accelerators for Phosphating. As long as the cathodic reaction is the evolution of hydrogen or the reduction of oxygen (always a sluggish process), the formation of the phosphate layer may be expected to proceed slowly. The process may, however, be speeded up by the introduction of an accelerator into the solution. This is generally some substance that enters into the cathodic reaction, so that it becomes possible to apply the solution by spraying, or to obtain an adequate coating from a short dip, instead of the prolonged immersion and elevated temperature needed in an " unaccelerated " bath. Thus Machu quotes 30–45 min. at 95–98°C. for the unaccelerated zinc bath, and 60 min. for the manganese bath; accelerators can reduce these times to 3–10 min., and he estimates that in Germany 90% of the work was (in 1950) carried out in the accelerated baths, which in his opinion last longer, preserve their composition better and provide superior protection. There is a strong feeling against accelerated baths in some quarters, and probably some of the accelerators used in the early days produced highly undesirable effects; copper salts undoubtedly can produce acceleration, but their presence appears undesirable in baths intended to provide protection. Many of the baths quoted by Machu contain oxidizing agents, such as nitrates, nitrites and chlorates; nitrate and chlorate are often used in combination, and gives less trouble than nitrite, which requires replenishment at intervals. Interest has been shown in organic nitro-compounds—which have been the subject of many patent applications; nitroguanidine is described as one of the strongest accelerators and is effective at a low concentration (0·3%)—which rather suggests that it may be acting as a catalyst, stimulating the cathodic reaction without being destroyed in the process. It is a little surprising to find that some reducing agents also produce acceleration, notably sodium bisulphite; its action is not fully understood* but it is unlikely to become widely adopted as an accelerator, since it promotes the accumulation of iron in the bath, maintaining the

* Machu suggests that these bodies convert anodic regions into cathodic regions.

(soluble) ferrous condition and preventing the elimination of the iron as (insoluble) ferric phosphate. There has been controversy about the relative merits of baths based on zinc and manganese; an accelerated process involving manganese phosphate is advocated in *Metal finish. J.* 1956, **2,** 106, 112.

Tests carried out at Woolwich have shown that an accelerated manganese phosphate process gives as good results as the corresponding non-accelerated bath; in baths based on zinc phosphate, acceleration is disadvantageous. This authoritative and comprehensive research deserves study (J. F. Andrew, S. G. Clarke and E. E. Longhurst, *J. appl. Chem.* 1954, **4,** 581, esp. p. 583).

Commercial Processes. Numerous proprietary processes, available today, bear names such as *Parkerizing*, *Bonderizing* and *Walterizing*. Some of the *Jenolizing* processes are apparently of this character; others are intended for application to rusty steel structures on the site. *Granodine 33* is a process for securing adhesion of paint to zinc-coated steel.

In commercial phosphating, a different solution is used according as the film is intended to afford a basis for paint, or to be covered merely with oil. In both cases, the advantage may be a double one; the protection is improved, and the paint or oil is held in place. Naturally the composition of a proprietary solution is not in general disclosed, but the patents on which the different procedures are based (up to about 1950) will be found in Machu's book. It is a not uncommon practice to follow phosphating with the application of dilute chromic acid or a chromate.

Phosphate processes are largely used today for motor-car, cycle and refrigerator parts, as well as smaller articles such as nuts, bolts and washers. Oxide-scale must first be removed by pickling or grit-blasting, and then the surface must be thoroughly cleaned in an organic solvent, by vapour degreasing (p. 624) or in a hot alkaline wash. The articles suspended by hoods from a conveyer usually pass through successive vessels where cleaning, phosphating and drying takes place; small articles are treated in barrels or tipping baskets. Another method suitable for treating bright, clean surfaces, is to apply the solution from a spray gun in the same manner as paint is applied, under about 40 lb./in.^{2} pressure and then, after 2 to 5 minutes, wash it off by hosing and dry by air-blast or in the oven. Details are provided by H. A. Holden, *J. Electrodep. tech. Soc.* 1948–49, **24,** 111.

Some years ago, the introduction of a procedure in which complete motor-car bodies were dipped into a phosphate bath aroused some public interest, but a mechanical spray treatment probably represents the more efficient form of process, providing the greatest amount of output from a given floor-space and minimizing the bulk of solution which must be kept in play. The body passes in turn through alternate spraying-zones and draining zones, the length of each spray-zone deciding the length of time allotted to the treatment. The jets are arranged to wet all parts of the surface—including the recesses. A chromate dip after phosphating is usual. A sequence of operations is (1) hand cleaning, (2) high-pressure water wash, (3) alkali cleaning, (4) rinses, (5) phosphate-treatment (nitrate-chlorate accelerator), (6) rinses, (7) inhibitive rinse in chromic and phosphoric acid,

(8) draining and drying in oven (R. A. Barltrop and G. Murrey, Symposium on Protection of Motor Vehicles from Corrosion (Soc. Chem. Ind.) 1958, p. 90). The general subject of the finishing of motor-cars (phosphate-treatment and subsequent enamelling) is discussed by J. T. O'Reilly, *Trans. Inst. Met. Finishing* 1954, **31,** 314; H. Silman, *J. Electrodep. tech. Soc.* 1949–50, **25,** 175, esp. p. 178).

The use of phosphating on bearing-surfaces to help in maintenance of a film of lubricant is mentioned on p. 749.

Advantages of phosphating. Experiments have been performed on specimens which, after being phosphated and painted, are inscribed with a scratch-line by means of a sharp point cutting through the paint and exposing the steel; they are subsequently exposed to salt-spray or other corrosive conditions, side by side with similar specimens carrying no phosphate coat; comparison after a suitably chosen period affords evidence that the tendency for corrosion to spread under the coat is greatly reduced by the chemical pre-treatment.

It has, of course, to be remembered that phosphating represents an extra operation, and, before deciding to adopt a phosphating process, the question must be asked whether it would not be better to provide an additional paint-coat. No general answer can be given, but in situations where the type of failure commonly experienced consists of under-rusting from gaps or scratches in the paint-coat (the paint being pushed away by the voluminous rust), phosphate-treatment deserves serious consideration.

Some of the information on this subject comes from quarters which are not disinterested and may have to be accepted with reserve. Fortunately objective tests carried out by competent persons representing the users' point of view are available, and in general these provide satisfactory evidence of improved paint performance where preliminary phosphate treatment has been applied. They also provide some information regarding the relative merits of zinc and manganese baths, and of accelerated and unaccelerated baths (J. F. Andrew, S. G. Clarke and E. E. Longhurst, *J. appl. Chem.* 1954, **4,** 581; S. G. Clarke and E. E. Longhurst, *J. Iron St. Inst.* 1952, **170,** 15; G. Schikorr, *Arch. Metallk.* 1949, **3,** 82. From the proprietors' side may be mentioned papers by van M. Darsey and W. R. Cavenagh, *J. electrochem. Soc.* 1947, **91,** 351; H. A. Holden, *J. Electrodep. tech. Soc.* 1948–49, **24,** 111; R. F. Drysdale, *Trans. Inst. Met. Finishing* 1953–54, **30,** 124. The photographs of F. C. B. Marshall, *J. Oil Col. Chem. Ass.* 1939, **22,** 242, are impressive).

An exceptional situation where phosphating is believed to make performance worse is one where painted steel will in any case rarely be used; if the painted surface is to come into contact with acid, this is likely to dissolve the phosphate and loosen the coating (F. Kolke, *Farben-Zeitung* 1931, **36,** 2235; 1934, **39,** 331).

Points important for successful phosphating include the need for careful degreasing before the phosphate dip—particularly if a cold bath is to be used; also the washing off of the remains of the phosphate solution before painting, especially if the work contains crevices and the solution contains accelerators. Useful information about the weights of coatings obtained in different ways is provided by A. R. King *Paint Manufacture* 1952, **22,** 325, 463, esp. Table III, p. 326; see also description of " Rotodip " procedure on p. 463.

The improvement in paint-adhesion due to phosphate treatment is brought out by Britton and Angles, who have developed a method for measuring adhesion (p. 814). In some cases, the gripping power is improved ten times (S. C. Britton, *Sheet Met. Ind.* 1948, **25,** 1185; S. C. Britton and R. M. Angles, *J. Iron St. Inst.* 1951, **168,** 358).

An alternative method of obtaining good adhesion for paint is the provision on the steel of a very thin tin coat (0·000008–0·00003 inches). A comparison between the two methods has been provided by Silman and Wernick, who state that the phosphate method gives the better keying, but is more liable to be damaged by impact. The tin coating is certainly porous, and probably cathodic to the steel, but the pores become sealed with rust plugs under atmospheric conditions; the method is inadvisable for continuous immersion in salt water (H. Silman and D. Wernick, *J. Electrodep. tech. Soc.* 1949–50, **25,** 175, partly quoting S. C. Britton). See also p. 637 of this book.

Phosphating treatment of rusty surfaces. The phosphating processes hitherto described are applicable to surfaces from which scale, rust and other contamination have previously been removed; they are carried out under cover—generally in a factory; the value of such processes for many purposes is generally acknowledged. However, there are also processes intended for treatment of surfaces carrying rust, the object being to transform objectionable rust to harmless iron phosphate. These can be applied to rusty structures on the site, or can be used for treatment of small articles without any elaborate series of tanks. It is understood that the preparations sold for this purpose contain phosphoric acid, sometimes a pickling restrainer and sometimes a wetting agent. There are several such mixtures on the market, but the instructions for use vary. In some cases the makers recommend a thorough wash with water after the treatment, so as to remove the excess of acid; in other cases, no such washing is demanded, and if that plan be adopted care must be taken not to apply too much of the liquid, since accumulation of acid residues in crevices may have undesirable effects.

Experience with liquids intended for the phosphating of rusty surfaces varies somewhat. It is understood that good results have been obtained for the removal or conversion of the small amounts of rust frequently present on steel parts after storage, but for the treatment of steel structures which have developed thick rust through long exposure to the open air, there is less general agreement. Cases are known where the parts treated before painting in the manner indicated have behaved less well than parts which received no pretreatment. It may be unfair to generalize, since there are many different preparations and doubtless the manner of application affects the results; but, in the present state of knowledge, it would seem better to remove rust and scale mechanically, or by pickling, and then—if further treatment before painting appears necessary—to apply one of the phosphate processes intended for rust-free surfaces.

Fancutt and Hudson print the compositions of two surface-washes intended for application on the site (as opposed to application in a dipping-tank indoors). One contains, in addition to phosphoric acid and water, butyl cellosolve and a little chromic acid; the other has phosphoric acid, water, butyl cellosolve and a wetting agent. The same investigators mention that several proprietary washes are available. Their opinion, which is shared by several other authorities, is that these washes possess value in overcoming the adverse effect of the slight rusting which sometimes appears on bright articles (free from mill-scale) after storage under unfavourable conditions; but that for badly rusted surfaces, the results obtained will always prove greatly inferior to those yielded by pickling or grit-blasting —as was shown by exposure tests at Sheffield (F. Fancutt and J. C. Hudson " Protective Painting of Structural Steel " 1957, pp. 29–31 (Chapman & Hall)).

Wash Primers (Pre-treatment Primers, Etch Primers). The introduction of a method of improving the adhesion of paints based on the so-called Wash Primers is a development of some importance. The credit for the invention is essentially due to American investigators, and there are several types of wash primers, varying considerably in composition from one another; within each type, however, the composition must be kept within a rather narrow range. Most of these are two-package systems, the contents of the two packages being mixed at the last moment, or at least within a few hours of use; one-package compositions have been put forward, but it is generally felt that the two-package systems give better results.

A typical wash primer consists of a basic zinc chromate pigment dispersed in a solution of polyvinylbutyral resin. Just before application, it is mixed with a thinner containing phosphoric acid. One type can be either stoved at 250°F. (121°C.) for 10 minutes or dried in cold air in about an hour. Finishing coats can then be applied, which may with advantage be based on urea-modified alkyd resin. Tests by Halls show the adhesion to be excellent. He considers that the wash primer can take the place of a chromate treatment on aluminium, or alternatively the phosphate treatment on steel. This would be a welcome development on both materials, since the articles could first be degreased with an organic solvent (perhaps trichlorethylene), then receive the wash primer and finally be painted—without entering any aqueous bath at all (E. E. Halls, *Ind. Finishing* 1951, **3,** 251, 300. L. J. Coleman, *J. Oil Colour Chem. Ass.* 1959, **42,** 10).

Wash primers vary among themselves in performance, and seem to be more satisfactory on aluminium, and perhaps steel, than on some other metals; even on aluminium they are considered by some authorities to be inferior to anodizing and on steel to phosphating. Nevertheless, their value is not in doubt.

The name " wash primer " has been criticized; it is argued that the film left behind is too thin to deserve the name of primer and can contribute little to the total protection. Its main function is to provide good adhesion, and if this is achieved, it may be more valuable than a thicker film which readily parts company either with the metal below or the layer above. Some prefer to speak of an " etch primer " and others of a " pre-treatment primer "; the choice of name is surely a matter of secondary importance.

The reactions involved in the setting process are not simple. Rosenbloom considers that the chromate is dissolved by the phosphoric acid, and oxidizes the metal, producing an inorganic film, probably oxide. The changes can be followed by movements of potential, which is at first high owing to hexavalent chromium, then drops as the chromate is reduced, and finally rises again as the film becomes perfected, ennobling the metal. The inorganic film may itself contribute to adhesion, but the improvement is attributed largely to an organic film of complicated character formed when the polyvinyl butyral is oxidized by the chromic and phosphoric acids (H. Rosenbloom, *Industr. engng. Chem.* 1953, **45**, 2561).

An important research carried out at Washington has greatly extended our knowledge of the mechanism. The zinc tetroxychromate used in wash primers has a structure which is capable of housing considerable amounts of zinc oxide; zinc phosphate tetrahydrate is precipitated at active areas on the metallic surface and stifles the corrosion which would otherwise occur (I. J. Kruger and M. C. Bloom, report summarized in *Corrosion Prevention and Control.* Nov. 1955, **2**, p. 25).

Choice of Painting Systems

Paints for Inner and Outer Coats. Having decided on the preliminary preparation of the surface, it is necessary to select paints for the various coatings. The plan generally adopted (except in cases where baking varnishes or other special types are demanded) is to choose a paint containing an inhibitive pigment for the lowest coat (the *primer*), and to cover this with two or more outer coats designed to protect the primer from chemical change and mechanical damage. For instance, the primer may be red lead or calcium plumbate, and the remaining coats some form of iron oxide.

The need for external coats is particularly marked in the case of red lead. If a red lead coat was left without further covering, it might easily become damaged by abrasion, owing to its poor physical properties, or coverted into lead sulphate by the sulphur oxides present in urban air. Only if it is covered with a paint of desirable physical properties, will full advantage of the red lead inhibitor be obtained. Iron oxide in linseed oil is often used for the outer coats and has the advantage of cheapness; the flaky type of haematite (p. 571) provides a mechanically robust finish; oils toughened by heat-treatment or admixture with resinous substances are useful for the external coat.

Inhibitive Primers. As regards pigments for the priming coat, red lead still enjoys prestige as an inhibitor, but is probably much less used today than early in the century. It is heavy to apply, dries rather slowly, yields a coat easily damaged, whilst some types possess limited keeping powers after mixing (p. 549).* However, Hudson and Stanners (p. 579) find a mixture of red lead, white lead and asbestine (2 : 2 : 1) to be practicable and effective and to possess better keeping qualities than neat red lead.

Some interesting American observations on red lead as a primer may here be quoted. The best behaviour was obtained on surfaces cleaned by pickling or sand-blasting; the results were reasonably good on surfaces carrying the scale nearly intact, but poor where scale had been partly removed by rusting and wire-brushing. On this last type of surface—which too often confronts the painter in practice—red lead in linseed oil (9 parts raw to 1 part " bodied ") gave better protection than red lead in a linseed/glyceryl-phthalate medium.†

The yellow calcium plumbate, recently introduced as a pigment, is related to red lead, which can be regarded as lead plumbate. It is already in wide use and has won praise in many quarters; it seems to possess inhibitive properties fully equal to those of red lead and is less heavy; moreover, paints made from it possess better keeping properties. It is usual to include aluminium stearate in the paint to prevent settling. Read believes that calcium plumbate polarizes both the anodic and cathodic areas—depositing calcium carbonate on the latter; on extraction with water, the pigment gives up 0·5% of soluble matter, mainly calcium hydroxide (N. J. Read, *J. Oil Col. Chem. Ass.* 1950, **33**, 295; *Corros. Tech.* 1956, **3**, 119; *Verfkroniek* 1958, **31**, 277).

The Author is inclined to suggest that interaction between two anodic products may occur, effectively stifling the anodic attack. Calcium plumbate is usually written Ca_2PbO_4, and the anodic reaction might discharge PbO_4^{----} or alternately produce Fe^{++}. This suggests the formation of Fe_2PbO_4, which might be sufficiently insoluble to seal the anodic points. It is more likely that $2Fe^{++}$ and PbO_4^{----} would interact to give the still more sparingly soluble Fe_2O_3 and also PbO—the latter being itself an excellent inhibitor.

Paints pigmented with metallic lead have from time to time been put on the market, and the best of these seem to possess inhibitive powers similar to those of red lead; after spreading out on the metallic surface, they probably yield lead soaps, by action of the vehicle on the lead in presence of air, and these in turn produce inhibitive degradation-products; if air is needed for the reaction, their keeping powers should be better than those of red lead.

Lead cyanamide, another relatively new introduction, probably also provides soaps and inhibitive degradation-products. Opinion is not unani-

* Even non-setting red lead paints show gradual increase of consistency on storage if made with raw or boiled linseed oil; if stand-oil is used, the consistency first increases, then returns almost to the original value, followed by a slight fresh increase (H. W. Talen, *Verfkroniek* 1951, **24**, 210).

† This refers to an industrial atmosphere. For sea-water, paints based on vinyl resins or balanced coumarone/coal-tar, pigmented with red lead, seem to be best. In general, paint should not be applied below 40°F. (4°C.), but vinyl paints and some other synthetics can be applied down to 5°F. (− 15°C.) (G. Diehlman and E. L. Beenfeldt, *Corrosion* 1951, **7**, 88).

mous about its merits, but the paints made from it seem to give less trouble through settling than some other lead paints (*Engineering* 1953, **175**, 314).

Zinc chromate, usually diluted with a considerable amount of iron oxide, is used as a primer on steel, and the combination seems to possess inhibitive properties similar to those of red leads. Several of the mixtures found by Hudson and Stanners to give good results as primers in their tests contain chromates. A mixture of zinc tetrahydroxychromate with zinc oxide was the best; but zinc chromate with lead basic sulphate was also good (p. 580). The use of a chromate on steel carrying rust-patches is open to objection. Moreover zinc chromate is not well suited for highly acid conditions (F. Fancutt, *Chem. and Ind.* (*Lond.*) 1955, p. 1492, esp. p. 1499).

Flaky Pigments. Although the ordinary powdery form of iron oxide pigment is satisfactory for the outer coat, the flaky form known as *micaceous iron ore* (or *flaky haematite*) produces a more robust coating; the flaky particles lie parallel to one another, providing a sort of armour; moreover, the paths through the oil vehicle, along which sulphur compounds or other undesirable substances must thread their way before reaching the priming coat or the metal, will clearly be lengthened if the pigmentation provides flakes lying parallel to one another. Hudson obtained excellent protection from two coats of micaceous iron ore, without primer, in tests lasting 15 years at Birmingham (J. C. Hudson, *J. Iron St. Inst.* 1951, **169**, 153). The function of flaky pigments is discussed by H. Rabaté, *Peintures, Pig. Ver.* 1954, **30**, 572.

Other paints with flaky pigments display similar valuable properties. Aluminium paints of this type are much used for outer coatings, and are particularly valuable on oil tanks where their power to reflect the sun's rays keeps the oil relatively cool. This calls for a product of high leafing power, usually obtained by ball-milling with stearine (the mixture used often contains stearic, palmitic and some oleic acid), so that the individual flakes are covered with a film of stearic acid and aluminium stearate, not easily removable with solvents (this film is objectionable, if it is desired to obtain metallic contact between the particles, as in the production of primers; see p. 615). Wray states that protection deteriorates if the leafing characteristics are poor, or the pigment concentration too low. Adulteration with mica also gives bad results. Particulars regarding the manufacture and use of the material are provided by J. D. Edwards and R. I. Wray, "Aluminium Paint and Powder" (Reinhold); R. I. Wray, *Corrosion* 1954, **10**, 50; G. W. Wendon, *Paint Manufacture* 1949, **19**, 265.

Tar and Bitumen Paints. Black paints, whether based on coal-tar products, residual products from the oil refinery or natural bitumen, are largely used, especially in industrial districts where oil paints fare badly in the contaminated atmosphere; conversely, in pure country air they sometimes fare worse than oil paints, being liable to decomposition when exposed to light; the light-sensitivity, however, is reduced by incorporating flaky aluminium. These paints are apt to soften and run down a vertical surface if they become hot, and their black colour increases heat-adsorption in sunshine; on the other hand, in cold weather they are liable to crack.

Thus one of the problems in compounding black protective compositions is to secure a maximal range of temperature over which a coat maintains suitable physical properties. If this range covers the fluctuations of temperature which occur in service, they give excellent protection, and it seems likely that in some of the paints available unidentified inhibitive substances are present which can prevent attack even where the steel is laid bare at a scratch-line. The paints can be made water-repellent by introduction of suitable ingredients, and a very thick coating can be obtained with a single application.

One disadvantage, however, met with in some of these black compositions—especially those based on coal-tar products—is that they contain substances which inhibit the drying of oil paints; thus an oil paint applied over a tar primer may remain " wet " for a long period. Conversely, it is unwise to apply a tar paint over a red-lead primer unless a long period for the complete hardening of the red lead has been allowed; if the time allowed is too short and if the black outer coat develops a net-work of cracks—as often occurs—the cracks may extend down into the red lead coat, so that the whole combination becomes ineffective. Thus the manufacture and use of these products requires judgment, but considerable progress has been made in recent years and there are situations where they give admirable protection. Their use on buried pipes (p. 280), and on ships (pp. 553, 585) is discussed elsewhere. Other information regarding bituminous and tarry coats is provided by R. St. J. Preston, Chemistry Research Spec. Rep. No. **5** (H.M. Stationery Office); W. F. Fair, *Corrosion* 1956, **12**, 605t. The constitution and structure of coal-tar pitch are discussed by L. J. Wood and G. Phillips, *J. appl. Chem.* 1955, **5**, 326. The brittle fracture of tars and bitumens, with special reference to road-making materials, is discussed by P. J. Rigden and A. R. Lee, *J. appl. Chem.* 1953, **3**, 62.

Interesting results on tar and bitumen paints emerge from the tests of Hudson and Stanners. The unpigmented mixtures, on the whole, gave rather disappointing results, but great improvement was achieved by pigmentation. Unpigmented blends of gilsonite with a mixture of drying oils failed more quickly in a marine atmosphere (Brixham) than in a city (Derby); similar blends carrying pigments were in excellent condition after 6 years. Again coatings obtained from solutions of bitumen in naphtha or white spirit (presumably dried by evaporation) gave poor results without pigment, but pigmentation improved the lives from 0·8 to 5·2 years at Derby and from 0·7 to 4·6 years at Brixham. These numbers refer to two-coat painting schemes. Some other tests where the black mixtures were applied over red lead gave poor results, since the red lead was not allowed to dry long enough and the outer coat often showed poor adhesion (J. C. Hudson and J. F. Stanners, *J. appl. Chem.* 1955, **5**, 173, esp. p. 184).

Rather similar conclusions about the need for an adequate interval between the application of the two paints had been reached in the Cambridge tests (S. C. Britton and U. R. Evans, *J. Soc. chem. Ind.* 1932, **51**, 211T, esp. p. 215T).

Some of the coal-tar coatings used in the U.S.A. for underground structures contain, besides coal-tar pitch and heavy oils from the tar, bituminous coal itself, along with mineral filler (N. T. Shideler, *Corrosion* 1957, **13,** 392t).

An interesting type of protection is the black stoving enamel obtained by mixing gilsonite and linseed oil along with a small amount of drier. It is stoved at 300–450°F. (150–230°C.) for ½ to 1½ hours, and has been largely used in the cycle industry. The surface may be phosphated before application.

Synthetic Resin Paints. Most paints manufactured today contain synthetic resins, and some are stated to be "100% synthetic"; the distinction between the synthetic and natural product is not so sharp as might perhaps be thought, since it is possible to "modify" a synthetic resin by attachment of groups from a drying oil, so that a single molecule can be "synthetic" and "natural" in different parts. Probably the real advantage of the synthetic basis is reproducibility. There is no reason why different batches of a synthetic resin, produced in a factory under controlled conditions, should vary among themselves, but it is common knowledge that there is a difference between linseed oil obtained from different countries and doubtless between different crops from a single country. A useful account of modern organic finishes is provided by H. Hollis, *Trans. Inst. Met. Finishing* 1953–54, **30**, 31.

The modern paints generally dry faster than the old-fashioned oil paints—some of which required days for proper drying in certain situations. They tend, however, to give thinner films. Hudson and Stanners state that "as a rule paints containing synthetic resins must include higher percentages of solvent than linseed-oil paints if equally good brushability is to be achieved"; a thinner coat is the result. But the difference is not great. Of the paints used in their tests at Derby, mentioned above, those based on linseed oil (stand-oil blended with refined oil) had an average thickness of about 3·3 mils, whilst for those based on "other media" the thickness was about 2·7 mils; these "other media" included coumarone, as well as an alkyd resin prepared from phthalic anhydride, glycerine, linseed stand-oil and linseed-oil fatty acids. Hudson and Stanners are inclined to attribute the longer lives of the true oil paints (about twice that of the others—on the average) to the greater thickness. If that is the only reason, it should surely be possible to thicken the modern paints slightly, so as to provide better service; Mayne, however, has suggested an alternative explanation based on soap-formation (see p. 547).

Synthetic resins have largely displaced natural products for certain types of stoving enamels. Other important classes of synthetic paints which dry by evaporation of the solvent—such as the polystyrene paints—have already received notice (p. 540).

Often the coatings contain more than one type of synthetic resin. Larson, for instance, gives formulae for air-drying paints containing both polyvinyl and alkyd resins, and "roller coatings" designed for drying in the oven (5–15 min. at 175–190°C.) containing both polyvinyl and urea-formaldehyde resins; in both cases the solvent contains xylene and methyl-isobutyl-ketone, whilst cyclohexanone is added as high boiling solvent (also dioctyl phalate as plasticizer in the "roller coatings"); the air-drying paint contains some naphthenate drier and both paints are pigmented with titania (rutile) with a small amount of antimony oxide in the air-drying paint (V. L. Larson, *Verfkroniek* 1958, **31**, 108, esp. p. 111).

Two classes of material likely to achieve importance in protective painting are the epoxy- and the polyester-resins. To obtain drying of the coat at ordinary temperature, it is often necessary to mix two, or perhaps three, components at the last moment—which may be inconvenient; but if that disadvantage is accepted, drying can be made as fast as anyone could desire; occasionally the difficulty may be that the mixture is liable to set in the pot or (worse still) in the spraying pistol, before application is complete.

The earlier epoxy-resins were divided into two main classes which set by the addition of amides or amines respectively, but recently other "curing agents" have been introduced. Some of the earlier amide curers supplied proved to have limited keeping properties, apparently owing to hydrolysis by absorbed water vapour; the earlier amine curers possessed better keeping properties, but were toxic and produced dermatitis in sensitive persons (sometimes also inducing sensitivity towards other substances which are not normally regarded as liable to produce dermatitis). Lately, however, new curing agents, some of them amines, have been introduced which largely overcome the objections. (For precautions and treatment methods for dermatitis, see L. B. Bourne, *Chem. and Ind.* (*Lond.*) 1957, p. 578. For new non-amine hardening agents, see F. J. Allen and W. M. Hunter, *J. appl. Chem.* 1957, **7**, 86.)

Paints and lacquers based on unsaturated polyesters generally require two additions to produce the curing reaction, namely a cobalt compound (e.g. cobalt naphthenate) and an organic peroxide (e.g. benzyl peroxide or, for low-temperature curing, methyl-ethyl-ketone peroxide or dicyclohexyl peroxide); the cobalt compound can be added to the resin as supplied by the manufacturer without serious loss of keeping power, but the peroxide must be added at the last moment.

Coatings based on epoxy-resins and unsaturated polyesters possess in most cases a fundamental advantage over (say) the phenol-formaldehyde resins in that no water or other volatile constituent is evolved during the curing stage (the reaction by which the resin is manufactured *does* involve expulsion of water). There would seem better prospects of obtaining a truly water-tight paint in cases where no water has to escape during curing.

These vehicles are particularly suited to metal-pigmented paints, and are therefore discussed further in the next chapter (p. 616). However, epoxy-resin paints carrying ordinary pigments are already in use, but are generally considered to behave less well in pure water than in sea-water. Some show a welcome resistance to alkalis and acids. McFarland states that baked epoxy-resins possess a much better resistance to alkalis, organic solvents and most acids than the air-dried products, although the resistance of the latter often improves a few days after application. Epoxy-resin coats have been considered for situations where glassy enamels are today used, but their resistance to most chemicals is no better and sometimes worse;

they are more resistant to chipping but less resistant to abrasion; the claims of the two classes are compared by N. S. C. Miller, *J. Oil Col. Chem. Ass.* 1957, **40**, 478.

Some references to epoxy-resins may be welcome to readers. These are:—

G. H. Ott, *Métaux et Corros.* 1948, **23**, 41; H. W. Talen and T. Hoog, *Verfkroniek* 1951, **24**, 313; R. N. Wheeler, *Trans. Inst. Met. Finishing* 1953–54, **30**, 158; F. P. Hiron, W. H. Rudd and J. J. Zonsveld, *Verfkroniek* 1954, **27**, 122; G. C. Bates, *Metal finish. J.* 1955, **1**, 97; D. H. Nicholson, *Corros. Tech.* 1956, **3**, 4; R. McFarland, *Corrosion* 1956, **12**, 187t; E. S. Paice, *Chem. and Ind.* (*Lond.*) 1957, p. 674. See also general discussion, *Chem. and Ind.* (*Lond.*) 1957, p. 1253. Combinations of epoxy-resins with coal tar for coating the bottom of crude oil tanks are discussed by R. M. Carter, *Corrosion* 1957, **13**, 270t. Combinations with polyamide resins (an important group, of which " nylon " is a familiar example) are discussed by D. E. Floyd, D. E. Peerman and H. Wittcoff, *J. appl. Chem.* 1957, **7**, 250. The cause of the good adhesion of epoxy-resins is authoritatively discussed by N. A. de Bruyne, *J. appl. Chem.* 1956, **6**, 303.

Polyester resins are discussed by L. H. Vaughan, *Chem. and Ind.* (*Lond.*) 1956, p. 996; L. W. J. Damen, *Verfkroniek* 1956, **29**, 213; J. C. Bevington, *Proc. roy. Soc.* (*A*) 1957, **239**, 420; P. Maltha, *Verfkroniek* 1957, **30**, 79, esp. p. 83.

Chlorinated rubber. Probably the most important use of chlorinated rubber is in connection with zinc-pigmented paints (p. 615), but other paints based on it possess valuable properties. Careful manufacture is required and the user requires to exercise judgment in obtaining his product from the right source. In the early development of this class of paint, pains were take to overcome certain techical difficulties, and the chlorinated rubber paints achieved a deserved reputation; this possibly encouraged other concerns to start making the paints, and some of the products which have come on the market more recently appear to have given disappointing results. Probably the presence of hydrogen chloride has been the cause of trouble in some cases. The work of Dyche-Teague deserves mention; he improved the stability by vacuum-distillation after chlorination, which removed the last traces of hydrogen chloride; he also succeeded in obtaining chlorinated rubber solutions of lower viscosity than those previously available (the late F. C. Dyche-Teague, quoted by V. D. Johnson, Priv. Comm., Feb. 4, 1958; see also British Patent 305,968 of 1927, and U.S. Patents 1,819,316 and 1,826,275 of 1931. Cf. D. L. Davies, *Verfkroniek* 1959, **32**, 82).

Bonded Plastic Layers. Relatively thick coatings of polyvinyl chloride sheet can now be bonded to sheet steel by a method developed by the British Iron and Steel Research Association. The coating is adherent and durable, and the process should enable steel to be used in situations where it has hitherto been regarded as unsuitable (S.S. Carlisle, *Sheet Met. Ind.* 1957, **34**, 431).

High-temperature Paints. At ranges of temperature where most organic compounds would decompose, two classes of paints are attracting attention. One contains *butyl titanate* and may be pigmented with aluminium where the greatest heat-resistance is required, or with zinc where corrosion-resistance is more important (A. B. Cox and G. Winter, *Paint Manufacture* 1955, **25**, 146; G. Sachs, *Verfkroniek* 1956, **29**, 140; R. Sidlow, *J. Oil Col. Chem. Ass.* 1956, **39**, 415). The other is based on *silicone resins* (*Paint Manufacture* 1946, **16**, 247; D. Cannegieter, *Verfkroniek* 1953, **26**, 112; P. A. J. Gate, *Trans. Inst. Met. Finishing* 1957, **34**, 18; S. L. Chisholm and N. N. Rudd, *Corrosion* 1957, **13**, 473t, esp. p. 475t; *Verfkroniek* 1957, **30**, 327; also J. Fischer's chapter of F. Tödt's book, " Korrosion und Korrosionschutz " (de Gruyfer, Berlin)).

These two vehicles, when heated, instead of decomposing to form solely gaseous products, leave non-volatile titania or silica. However, some of the more conventional paint systems can withstand temperatures up to about 300°C.—as shown in Hudson's tests. One is a stoving paint system based on a 100% phenolic resin medium having the lower coats pigmented with zinc oxide and barytes, and the finishing coat pigmented with micaceous iron ore. Another system relies on Portland cement in stand oil (F. Fancutt and J. C. Hudson, " Protective Painting of Structural Steel " 1957, p. 78 (Chapman & Hall)).

Other methods for protection at high temperatures depend on ceramic coatings and cermets (p. 593).

Outdoor Exposure Tests on Painted Specimens

Historical Note. In order to ascertain the best manner of using paint to prevent the corrosion of steel, extensive programmes of exposure tests have been carried out in all parts of the world. The amount of information thus amassed has been so great that, if it were properly utilized, the cost of corrosion and preventive measures could today be greatly reduced. Although the different tests have been carried out on specimens differing in thickness, shape and size and often exposed in a different manner, there is general agreement between the conclusions emerging from the various series.

The pioneer work in regard to atmospheric conditions was carried out by the American Society for Testing Materials whose reports on the subject date back to 1913; the marine tests of the Institution of Civil Engineers, largely carried out by the veteran corrosion investigator, J. A. N. Friend, were started in 1920. The tests organized from the Author's laboratory at Cambridge, mainly carried out by S. C. Britton, only started in 1929; they have involved over 3000 small specimens. The Iron and Steel Institute Corrosion Committee started a series of tests on much larger specimens a few years later. These were carried out under the energetic leadership of J. C. Hudson. Later, the Committee became the Corrosion Committee of the British Iron and Steel Research Association, and some of the recent series of tests were carried out in conjunction with F. Fancutt, who had already accomplished distinguished research in the field of protective painting, and in co-operation with the paint industry. Meanwhile in many other countries, series of exposure tests had been started, and one of Hudson's many services to the World has been to establish friendly relations with workers abroad, bringing their results on appropriate occasions to the notice of readers in English-speaking countries.

In general, the behaviour of painted iron or steel will depend on

(1) the *metallic basis* (especially the content of copper, sulphur and phosphorus);
(2) the *surface condition* (especially the presence of scale, rust, salts, moisture, dirt, dust and old paint);
(3) the choice of the various *paints coats*, the selection and thickness of primer, along with the number of covering coats, being specially important;
(4) the *atmosphere* or *environment* to which the painted surface will be exposed.

Some of the programmes have introduced all these four variable, but the third has received most attention. Reference to the recent papers is provided in the proper place, but the earlier reports and papers may here be mentioned. The early American tests will be found in *Proc. Amer. Soc. Test. Mater.* 1913, **13**, 332; 1914, **14**, 1, 259; 1915, **15**, 1, 190, 215; 1917, **17**, 1, 451, with subsequent reports issued annually. The British marine tests appear in fifteen reports on the " Deterioration of Structures by Sea-Water ", issued by the Institution of Civil Engineers annually between 1920 and 1935.

Tests organized from Cambridge. The author's tests were designed to answer certain questions which he believed, rightly or wrongly, to be of crucial importance. A short description of the results may be permissible, since, having been viewed by him directly, the specimens do constitute the main basis of the opinions offered in this chapter. They are described by S. C. Britton and U. R. Evans, *J. Soc. chem. Ind.* 1930, **49**, 173T; 1932, **51**, 211T; 1936, **55**, 337T; 1939, **58**, 90; *Trans. electrochem. Soc.* 1933, **64**, 43.

These tests were carried out on steel 1·1 mm. (0·044 inch) thick; the specimens measured 7·5 × 5·7 cm. (3 × $2\frac{1}{4}$ inches), an unpainted strip being left at the top; they were exposed at an angle of 12° to the horizontal on the frame shown in fig. 98. The four main exposure stations comprised

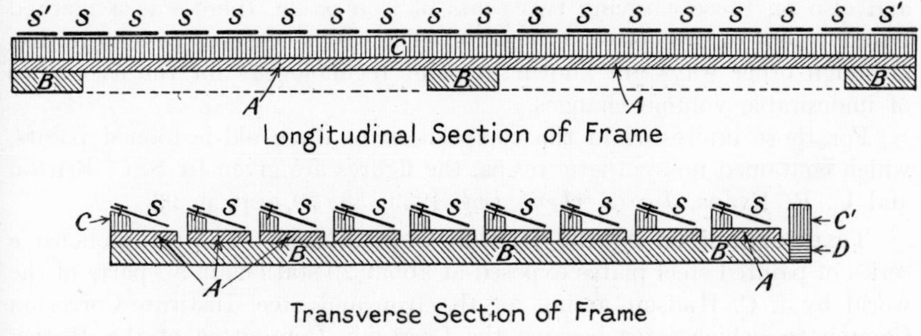

FIG. 98.—Cambridge frame for painted specimens (S) attached to wood (A, B, C, D).

an urban atmosphere (London), a marine atmosphere (Selsey Bill), a rural atmosphere (Grantchester Meadows) and a mixed atmosphere, with much coal smoke in winter but little in summer (Cambridge). The London and Cambridge specimens were exposed for 7 years, but the Selsey Bill collection had to be withdrawn after 5 years; at Grantchester Meadows, where corrosion was slow, some of the exposures continued for 9 years.

London air was much more prejudicial to good protection than the sea air of Selsey Bill, where the exposure rack occupied a position exceptionally free from coal-smoke. Corrosion at Cambridge was 2·9 to 3·5 times as fast as at Grantchester Meadows—about a mile away—doubtless owing to sulphur from coal smoke in Cambridge and its essential absence at Grantchester Meadows.

One two-coat combination gave almost perfect protection in London for 7 years; the primer was made from red lead in linseed oil and the outer coat from micaceous iron ore in a proprietary vehicle. The corresponding two-coat system with ordinary iron oxide in linseed oil as the outer coat was slightly less good; a similar three-coat system (red lead followed by two coats of iron oxide) gave admirable protection. However, three coats of iron oxide showed signs of failing in London—showing that iron oxide is less good than red lead as a priming coat; in Cambridge three coats of iron oxide gave almost perfect protection for the same period of 7 years, provided that the steel had been descaled; as already stated, specimens carrying scale below the paint almost always behaved worse than those without scale.

Among priming coats, red lead, which, as just stated, was superior to iron oxide, was further improved by the deliberate addition of litharge. A mixture of zinc chromate and iron oxide gave rather better results than red lead in London and was equally good in Cambridge; a priming coat pigmented with metallic lead gave good results when used below iron oxide, but when tested as a single coat gave worse protection than most others.

Some of the paints were designed to avoid the volume changes which many vehicles undergo when they dry; on the whole linseed oil tends to contract, and tung oil to expand; a mixture of tung oil and boiled linseed oil was therefore made up, adjusted so that the expansion and contraction would neutralize one another. When this mixture was substituted for refined linseed oil on specimens carrying the red lead/iron oxide combination, and also on those carrying two coats of iron oxide, there was a marked improvement of protective properties. The result deserves passing notice, although other ways are known to paint technologists for the avoidance of undesirable volume changes.

For those interested in the composition of these old-fashioned paints, which contained no synthetic resins, the figures are given by S. C. Britton and U. R. Evans, *J. Soc. chem. Ind.* 1939, **58**, 90, esp. p. 93.

Tests organized by the British Committees. The comprehensive series of painted steel plates exposed at about 20 stations in all parts of the world by J. C. Hudson, acting for the Iron and Steel Institute Corrosion Committee (which later became the Corrosion Committee of the British Iron and Steel Research Association) are described in detail in the Five

Reports of the Iron and Steel Institute published in 1931, 1934, 1935, 1936 and 1938,* and in papers by J. C. Hudson, *J. Iron St. Inst.* 1942, **145**, 87P; J. C. Hudson and T. A. Banfield, *ibid.* 1948, **158**, 99; J. C. Hudson and J. F. Stanners, *J. appl. Chem.* 1955, **5**, 173; J. C. Hudson, *Schweiz. Arch. angew. Wiss.* 1958, **24**, 46. The practical conclusion drawn from these tests and from general service experience are summarized, in a form well suited to the needs of owners of steel-work, in a small book by F. Fancutt and J. C. Hudson, " Protective Painting of Structural Steel " 1957 (Chapman & Hall).

The book mentioned should be studied, and if possible Hudson's reports and papers also. Only a few of the salient points can be set down here. Fancutt and Hudson have summarized the situation in four words " Prepare adequately; prime inhibitively ". They emphasize the fact—too often ignored—that correct surface preparation of the steel may be more important than the correct choice of paint; " even the best paint schemes will not perform well on a badly prepared surface "; in contrast, an inferior painting applied to well-prepared steel sometimes gives surprisingly good results. Nevertheless they deprecate the use of cheap paint as false economy, in view of the fact that in any case the cost of paint is much less than the cost of painting (see p. 538 of this book). Likewise they lay stress on the need for adequate thickness of paint (5 mils or 0·005 inch), favouring in most of their practical examples a number of relatively thin coats to obtain the necessary thickness. With the paints available today, this is possibly sound advice (see also p. 582); but at least three, more often four, coatings will be needed to give the recommended thickness.

The composition of the steel (unless stainless steel is used) influences results far less than the factors just mentioned, but the presence of small amounts of chromium, copper and/or nickel in steel confers the advantage that if gaps in the paint-coat are accidentally left (or are produced by rough treatment), or if a painted surface is not repainted promptly after the paint has broken down, low-alloy steels will suffer much less at the unprotected spots than ordinary mild steels.

As regards methods of preparing the surface, the paint-lives shown in Table XIX (p. 562) make clear the advantage of removing the scale by pickling, or better still sand-blasting, before painting.

In regard to pickling procedure, Footner's duplex method (p. 405) showed only slight superiority to ordinary pickling in sulphuric or hydrochloric acid as carried out by Hudson; had pickling in these two acids been carelessly preformed, the advantage of the phosphoric-acid after treatment might well have been accentuated. Where abrasive blasting is used to clean the steel, the authors again stress the importance of a minimal thickness of 5 mils for the paint-coat; otherwise, they consider, there is a danger of the peaks of the rough surface protruding from the paint.

Flame-cleaning is described as " relatively expensive " but has the advantage that the equipment is mobile; it is stated to give results inferior to those obtained by grit-blasting or pickling.

* A sixth report is in the press.

The Authors recommend that painting should follow descaling with as short an intervening period as possible; 7 days' unpainted exposure out of doors, which caused appreciable rusting, reduced the paint-life, although a month's storage indoors, which caused no gross rusting, did little harm.

As regards choice of paint for the priming coat, Hudson's early tests emphasize the good performance of red lead in linseed oil—a paint-type covered by British Standard Specification 2523 (1954). Two coats of such paint covered with two coats of red iron oxide have protected steel for 13 years in the industrial atmosphere of Sheffield and 17 years in the marine atmosphere of Calshot.

Later results showed that several pigmentations with linseed oil as vehicle gave even better results, as brought out by the figures obtained at Derby (industrial) and Brixham (marine); these are shown in Table XX.

TABLE XX

LIVES OF PAINT-COATS AT DERBY AND BRIXHAM (J. C. Hudson)

	No. of years needed to produce rust over 0·5% of surface	
	at Derby	at Brixham
Red lead (complying with B.S. 2523)	4·5	$>$7·2
Basic lead sulphate, 40; zinc tetrahydroxychromate, 40; asbestine, 20	$>$7·0	6·9
Zinc oxide, 40; zinc tetrahydroxychromate, 40; asbestine, 20	$>$6·9	$>$7·9
Basic lead sulphate, 60; zinc chromate, 20; asbestine, 20	6·0	$>$8·4
Red lead, 40; white lead, 40; asbestine, 20	5·2	$>$8·4

Fancutt and Hudson mention also the finishing coats favoured by authorities in different parts of the world; these include red iron oxide, white lead (often tinted, in practice) and various mixtures of aluminium and micaceous iron ore. They consider that tar and bitumen paints are effective as finishing rather than priming paints, and indicate that their performance can be materially improved by incorporating aluminium.

As regards seasons for painting, they prefer the period between April to September and consider it desirable to avoid December and January. Their comparison between the results of application by brush and spray shows slight superiority for sprayed coats, but they suggest that this may be due to the fact that they were slightly thicker. There is a chapter on the re-painting of old steel-work.

Tests and Experience in other Countries. Much useful information was brought together in 1955 at a Symposium organized in London by the Corrosion Group of the Society of Chemical Industry, and attended by distinguished guests from many countries. Only a few features can here be summarized, and readers should consult the references quoted below—which include for convenience certain foreign papers presented elsewhere. The discussion at the Symposium deserves special study (*Chem. and Ind.* (*Lond.*) 1955, p. 1648).

The Swedish tests have brought out the importance of paint-thickness

—a factor also emphasized by Hudson (p. 579); however, the Swedish results indicate that it is essentially the thickness of the priming coat which matters. Keeping the total thickness constant at 40 μ, but increasing the part due to the priming coat, the life increases steadily. Nylén and Trägårdh recommend two priming coats and two finishing coats for strongly corrosive atmosphere, two primers and one finishing coat for moderately corrosive atmosphere, and one of each for mildly corrosive atmospheres (P. Nylén and K. F. Trägårdh, *Chem. and Ind.* (*Lond.*) 1955, p. 1574; K. F. Trägårdh, *Corros. Tech.* 1954, **1**, 164).

The Belgian tests have brought out a number of important points. One series showed that the substitution of red lead for iron oxide in an oil paint increased the durability about three times after 10 years; it is false economy to use a poor primer, and good top coats cannot compensate for a bad selection of the lowest coat. A second series seemed to indicate that modern compositions (based on synthetics) were less good than classical primers such as red lead in linseed oil, or a mixture of red lead and iron oxide in stand-oil and linseed oil. However, a third series showed that certain alkyd primers pigmented with suitable mixtures of pigments could yield excellent results, especially in aggressive atmospheres; a recommended mixture was zinc chromate and basic lead sulphate in alkyd, with small amounts of zinc oxide and asbestos. van Rysselberghe and Bermane point out that open-air exposure tests will never be completely satisfactory, and raise the question as to whether the expense of such tests is justified, or whether more fundamental work, designed to elucidate the mechanism of protection, would not provide better value. Readers should study these points in the original papers (M. van Rysselberghe and D. Bermane, *Chem. and Ind.* (*Lond.*) 1955, p. 1587; P. Erculise, *Nature* 1952, **169**, 603).

At the same symposium, the fundamentals of paint protection was discussed by the French authority, Dechaux. He considers that at pores in paints, anodic and cathodic action produce acid and alkali respectively; the alkali precludes the use of saponifiable materials. It is possible, in suitable circumstances, to plug the pores with chalk, magnesia, or with rust; to obtain " rust-plugging " the pigment must either be an oxidizing agent and/or an alkali. With red lead present, the anodic reaction produces the higher iron oxides close to the metal. Under conditions where four coats pigmented with iron oxide last only 4–6 years, two coats of red lead followed by two coats of iron oxide last 11 years; once again the superiority of red lead as a primer is brought out (G. Dechaux, *Chem. and Ind.* (*Lond.*) 1955, p. 1535).

A paper by Bigos gave a valuable review on protective practice in the U.S.A. and brought out some points not commonly appreciated. He emphasized that, whilst broken scale is very detrimental in highly corrosive environments, it may be relatively harmless in mildly corrosive atmospheres (one assumes that firmly attached scale, not scale loosened by under-rusting, was in his mind). As regards different atmospheres, he states that American industrial atmospheres are very corrosive in the opening stages of exposure, but the initial rate (5 mils per year) soon drops to $\frac{1}{2}$ mils per year; in a marine atmosphere the initial rate is lower, but there is less decline with time, so that after 10 years' exposure the corrosion in industrial and marine atmospheres may be much the same (J. Bigos, *Chem. and Ind.* (*Lond.*) 1955, p. 1503).

The results of the Dutch Tests are reported periodically in a series of special reports published by the Metaalinstituut T.N.O. Afdeeling Corrosie (Delft); a brief discussion may be convenient at this point. The main facts quoted come from Report No. **32**, published in 1955. The tests are carried out on specimens of sand-blasted steel bent in the middle, so as to give two limbs, one horizontal and the other inclined at 135°; each limb is further divided into four parts, the two right-hand parts carrying one priming coat, and the left-hand parts two priming coats; the parts nearest to the bend also carry two layers of a covering coat pigmented with zinc white and titania. With red lead as a primer, the application of a covering coat caused rust to appear *more quickly* than where it was absent; two red lead coats gave better protection than one. Iron oxide gave poor results when used alone as primer, but it was possible to dilute the red lead with 50% of iron oxide without deterioration,* provided that raw linseed oil was used; stand-oil was slightly less satisfactory and a " water-resistant varnish " furnished definitely inferior results; these remarks refer to red lead tested without covering coats.

Most of the Dutch results are in good agreement with those arrived at elsewhere, but in one respect there is apparent divergence. At the meeting organized at Brussels in June 1954 by the A.B.E.M., Lobry de Bruyn stated that sand-blasted surfaces gave better results if they were allowed to rust before painting; most of the British speakers disagreed, and the discrepancy may be due to differences in the locality or the season at which the rusting took place; Mayne's results (p. 556) suggest that these factors would have a marked effect.

Thickness of Coats. It is today customary to emphasize the need for maintaining a minimal thickness of paints. Possibly with the paints available to day, it is necessary to build up to a certain thickness, but hope may be expressed that if for the moment we must insist on a minimal thickness, this will not discourage efforts to develop new paints which will provide protection even when thin. It seems to be thought that if the amount of paint applied is too small, peaks of bare metal will be left uncovered, the paint only filling the intervening valleys. This picture seems unrealistic; it is rather unlikely that the peaks will remain completely uncovered; possibly with many paints, only a transparent vehicle will extend over the peaks, the pigment particles collecting in the valleys; however, it should not be impossible to adjust the interfacial energies of the four phases concerned (metal or metal oxide, vehicle, pigment and air), so that at equilibrium (i.e. when the total energy is minimal) even the peaks are covered with pigment. Doubtless this would produce a finish less pleasant

* It has been found elsewhere that by judicious use of extenders, the red lead content can be reduced to about 30% without loss of effectiveness (P. J Gay, *Chem. and Ind.* (*Lond.*) 1955, p. 1656).

to the eye, and for some purposes otherwise unsatisfactory, since an uneven surface tends to harbour dirt and (on ships) impedes movement, but, from the point of view of avoidance of corrosion, it should be possible to evolve a paint giving a thin coat possessing the necessary properties. Even with existing paints, the necessary thickness might perhaps be reduced if dust could be excluded—a view expressed by C. Graff-Baker, *Chem. Ind.* 1958, p. 590.

Number of Coats. The question of the number of coats is a different one from that of total thickness. If several coats are applied on the top of one another, it is unlikely that accidental gaps in one will coincide with gaps in the next; the same is true of defects caused by loose bristles from brushes, grit particles and the like. Thus, for a fixed total thickness, a large number of thin coats is less likely to involve defects penetrating all the coats than a smaller number of thick ones; however, the multiplication of coats demands more labour and therefore greater expense, besides requiring a longer time for the completion of the job. Where the paints chosen are such that thickness is desirable for its own sake, it will be well to make the paints sufficiently stiff that the required total thickness can be reached without demanding an excessive number of coats.

Marine Painting

Treatment of Ships. There are three operations in the external painting of ships: (1) the preparation of the surface, (2) the application of a paint intended to prevent corrosion, (3) the application of a coat intended to prevent fouling (i.e. the attachment of marine plants and animals which, if allowed to occur, will reduce the ship's speed and increase the fuel bill). Of these, the first is too often left to chance, whilst the third only concerns the corrosionist indirectly, because the copper and mercury compounds used in the anti-fouling preparations may themselves interfere with protection against corrosion. It should be noted in passing that the periodical return to dry dock for renewal of the anti-fouling coats is convenient in allowing the corrosion situation to be kept under observation.

Preliminary Treatment before painting. The preparation of the ships' plates before painting is of the highest importance. The British Admiralty have long been pickling plates intended for important work. However, the practice, long prevalent in many commercial ship-yards, of allowing the scale to "weather off" by under-rusting, through exposure in the yards or on the stocks during construction, has often worked well, and could be made to work still better if the stacking of plates awaiting their turn were carried out systematically so that every plate obtains its due share of weathering influences. The scale on two sides of a rolled plate is often somewhat different, and weathers off at different rates, and this fact should be taken into account.

Hudson has found that—in contrast with the painting of a land-structure —better results are obtained if a freshly descaled surface is allowed to rust before it is painted. This suggests that descaling by weathering—although unsatisfactory for land structures—might be made tolerably good for ships. However, various points have to be borne in mind. An acceleration in the rate of ship-building may bring it about that the plates, which, at a slow building-rate, would have been almost free from scale at the time of the first painting, will under the new conditions still carry scale over most of the surface; the combination of large cathode (the scale-covered area) and small anode (the breaks in the scale) may then lead to pitting. An unusually prolonged period of fine weather which militates against rapid under-rusting of the scale might have the same effect. Another change for the worse might be brought about by changes in the finishing temperature at the rolling mills, leading to a different type of scale (cf. p. 406). Thus the plan of descaling by weathering cannot be regarded as reliable. If it is adopted, care should be taken to see that loose scale, rust and other matter are removed before paint is applied.

The advisability of removing the scale from ships' plates before painting is emphasized by the American Authority, Ffield, who points out that the serious trouble is not general corrosion, but pitting, set up electrochemically, either by (1) stray currents from a shore source (see p. 299), (2) currents set up by connection with a second material (e.g. the bronze propeller) or (3) those flowing between bare steel and mill-scale. The pictures accompanying his article deserve study (P. Ffield, *Corrosion* 1952, **8**, 29, 69).

Choice of Paints. The first requisite of a marine paint—a vehicle which resists softening or loosening by the alkali formed by the cathodic reaction—has already been mentioned (p. 554). In cases where cathodic protection schemes are involved, this is already recognized as essential; it is gradually coming to be accepted in other cases. The American vinyl scheme mentioned below is an indication of the trend of opinion.

As regards pigmentation, red lead was at one time much used on ships, but at least for the submerged portion it has lost favour. Hudson and his colleagues have tested numerous combination of pigments in marine paints, mostly on steel plates attached to rafts. Some of the best combinations contain basic lead sulphate; this probably interacts with the vehicle to give soaps which suffer degradation to inhibitive compounds similar to those isolated by van Rooyen (p. 547). One excellent paint—understood to have been used with success for the bottom of the liner *Queen Elizabeth*—contains basic lead sulphate, aluminium powder, Burnt Island red* and barytes in a vehicle compounded of phenol-formaldehyde resin, stand-oil and tung oil (J. C. Hudson, *Times Science Rev.* 1953, p. 15; *Chem. and Ind.* (*Lond.*) 1954, p. 640; *Trans. Inst. Met. Finishing* 1952–53, **29**, 124. See also F. Fancutt and J. C. Hudson, *J. Iron St. Inst.* 1944, **150**, 269P; 1946, **154**, 273P).

In surveying the situation, Kingcome states that for the upper portion of ships, paints containing red lead and graphite have been widely used in the past, but in recent years there is a movement towards alkyd resin paints containing zinc chromate, which are less costly, dry more quickly and are harder. For use under water, Hudson's paints receive mention, and also

* Mainly consisting of iron oxide.

an American plan based on a vinyl resin pigmented with red lead, to give a dry-film thickness of 4 to 6 mils, covered with an anti-fouling coat containing cuprous oxide and rosin, the total thickness being 10 mils; this scheme provides excellent protection against corrosion, a smooth surface and admirable anti-foul properties (J. C. Kingcome, *Corr. Prev. Control* 1954, **1**, 207, 411).

A system based on the co-polymer of vinylidene chloride and acrylonitrile, used on American submarine tanks and in other situations, is described by W. W. Cranmer, *Corrosion* 1956, **12**, 245t.

For the wind-and-water zone of the hull, the paints needed are different from those used for covering the permanently immersed and the permanently unimmersed areas. Black paints based on tar and bitumen have long been used at this level, and give satisfactory results. For vessels of the British reserve fleet, the zone in question has been treated with one of the Cambridge cementiferous paints, E412 (p. 619), covered with two coats of chlorinated rubber paint; E412 was chosen " because it was a cementiferous paint the Admiralty had experience with; it had good anti-corrosive properties and it had no organic medium which could be damaged by alkali " (L. T. Carter and J. T. Crennell, *Trans. Inst. nav. Archit.* 1955, **97**, 413, esp. pp. 424, 450; cf. U. R. Evans, p. 435).

Anti-Fouling Measures. This book is not the place to discuss the composition of anti-fouling mixtures, the relative merits of different toxic substances and the composition of a vehicle which will release them neither too fast nor too slowly. But since the magnificent work of Harris, Pyefinch and their colleagues, who applied expert biological and chemical knowledge to a problem previously tackled empirically, was conducted for the Iron and Steel Institute Corrosion Committee, it is fitting to mention the main papers. Two reports from the Marine Sub-Committee were printed in *J. Iron Inst.* 1943, **147**, 339P; 1944, **150**, 143P. See also K. A. Pyefinch, *ibid.* 1945, **152**, 229P; H. Barnes, M. W. H. Bishop and K. A. Pyefinch, *ibid.* 1947, **157**, 429; J. E. Harris and W. A. D. Forbes, *Trans. Inst. nav. Archit.* 1946, **88**, 240.

Although the prevention of fouling is no simple problem, and may not depend solely on the solubility of the toxic substances present, it is their solubility and solution-rate which is likely to affect corrosion; it is, therefore, worth noting that solubility in sea-water may be different from that in distilled water. Ragg states that mercuric oxide is four times as soluble in Baltic Sea Water as in distilled water, the yellow form being 15% more soluble than the red. Of the copper compounds tested by him (Cu_2O, CuO, CuCN, CuCNS), the cyanide, CuCN, was the most soluble in sea-water (M. Ragg, *Farbe u. Lack* 1950, **56**, 435).

Metallic zinc is found to possess a modicum of anti-fouling power, but it is doubtful whether this is of practical importance; probably a layer of zinc sprayed from a pistol on a steel hull would prove no substitute for a well-formulated anti-fouling composition containing copper and possibly mercury. Various opinions are quoted by E. W. Skerrey, *Chem. and Ind.* (*Lond.*) 1957, p. 1275; J. C. Hudson, p. 1390; L. Kenworthy, p. 1390.

Painting of Non-ferrous Metals

Pre-treatment of Galvanized Iron. One of the problems of painting galvanized iron is the difficulty of obtaining adhesion. If a corrugated galvanized roof is painted immediately after erection, the paint may soon peel off. At one time owners were advised to allow the bright zinc to become dull before paint was applied; this certainly improved adhesion, possibly because the porous zinc corrosion-product provided keying, but it was often inconvenient to apply paint at exactly the right moment, especially as one part of the roof would probably reach a suitable condition more quickly than another. If painting is deferred too long, rust patches appeared where the zinc coating has been eaten through, and it may then be difficult to arrest the rusting by any ordinary paint, although paints pigmented with metallic zinc have given fair results in such situations (p. 615). Proposals have been made to accelerate corrosion of the zinc in various ways, so as to produce a surface which can receive paint soon after erection; one wash advocated some years ago contained a copper salt, whilst another contained hydrochloric acid in ethyl and butyl alcohols; to the author, it appears unwise to run the risk of shutting in traces of copper salt or chlorides under a paint-coat.

Phosphate-treating of the zinc has here proved welcome (p. 565), but an alternative treatment is provided by the Cronak Process—a dip in an acidified solution of sodium dichromate; the short time needed (perhaps 5 to 10 seconds at ordinary temperature) is a great attraction. Clarke and Andrew recommend 150–200 g./l. sodium dichromate with 5–9 g./l. concentrated sulphuric acid; if the acid-content is too low, the films produced are too thin; if it is too high, adhesion is poor, perhaps owing to hydrogen evolution. The iridescent films produced under optimal conditions protect against humid conditions and saline spray; brown films are less satisfactory. The film normally contains 20% water and dehydration reduces the protective value; storage under normal conditions before painting does not render the film less protective, although it reduces the solubility of the film-compound. The film may be about 0·00002 inch thick and contains both trivalent and hexavalent chromium. Some authorities regard it as a reservoir which steadily releases inhibitive CrO_4^{--} ions; Clarke suggests that it acts also as an inert barrier. The matter is discussed from different angles by S. G. Clarke and J. F. Andrew, *J. Electrodep. tech. Soc.* 1944–45, **20**, 119; H. A. Holden, 3rd Internat. Conf. Electrodep. 1947, p. 57; see also *Industrial Finishing* 1951, **3**, 574; F. Taylor, *J. Electrodep. tech. Soc.* 1944–45, **20**, 206; E. E. Halls, *Metallurgia* 1948, **37**, 299.

When the question of pre-treatment has been settled, it is necessary to select a paint for the lowest coat. The value of metallic zinc at this level is recognized but there is some disagreement as to the relative merits of (1) zinc-rich paint (p. 615), and (2) a paint pigmented with a mixture of metallic zinc and zinc oxide. Old galvanized roofs which have started to rust have been successfully treated with a mixture of zinc oxide and flake aluminium followed by an outer coat containing two flaky pigments—

aluminium and micaceous iron ore; as already stated, some success in arresting rust has also been obtained by application of zinc-rich paint.

It is, of course, desirable to find a paint which can be applied directly to a new galvanized roof, without the need of phosphate- or chromate-treatment. There is increasing reason to hope that calcium plumbate fulfils these requirements; another plan—favoured in the U.S.A.—is the application of a vinyl acetate emulsion paint, which contain emulsifiers and wetting agents so that any grease present is emulsified, and good adhesion obtained. It is too early, however, to say whether a complete solution of the difficulty can be found on these lines. Fuller information is provided by P. Costelloe and E. Page, *Trans. Inst. Met. Finishing* 1958, **35**, 107; J. F. H. van Eijnsbergen, *Verfkroniek* 1958, **31**, 243.

Paints for Aluminium. In the painting of aluminium and its alloys, special attention should be given to crevices and recesses. In structures, jointing compounds (stiff pastes usually containing chromate pigments) are usually included between the components, and paint applied around the junctions; the painting of the freely accessible surface is less important—sometimes perhaps unnecessary. Statements by salesmen that aluminium never requires painting are to be deplored—especially since the type of man for whom such statements are intended fails to distinguish aluminium from its alloys, or between material which is clad, sprayed, anodized or untreated. However, if appropriate protection is given, renewal of the paint on aluminium can be carried out at much less frequent intervals than would be necessary on steel.

Aluminium and its alloys often receive pre-treatment before painting, either by anodization or one of the dipping processes described on pp. 250–252; the main object should be to ensure that the recesses receive full treatment. At first sight this requirement seems easiest to fulfil with a simple dip, since during anodizing the current density is lowest in the recesses; however, as the more prominent parts become coated with insulating oxide, the current will be diverted to the recesses, so that ultimately all parts should receive their coating, provided that the surface is free from grease and that there is no entrapped air.

The paints used on aluminium and alloys are generally pigmented with zinc chromate or tetrahydroxy-chromate. Rigg and Skerrey in their outdoor tests at industrial, marine and rural stations found that these two pigments are outstanding and definitely better than iron oxide, which, however, serves for mild conditions (J. G. Rigg and E. W. Skerrey, *J. Inst. Met.* 1952–53, **81**, 481).

Different opinions have been expressed regarding the use of lead paints for aluminium. Some writers praise calcium plumbate, whilst others hold that all lead compounds should be avoided. Rigg and Skerrey found a red lead primer definitely harmful on light alloys (worse on magnesium than on aluminium). The experiments of Pryor suggest possible reasons for the conflict of experience; he found that watery extracts from metallic lead or litharge deposited metallic lead on aluminium and accelerated its corrosion, whereas extracts from red lead or white lead neither deposited metallic lead nor accelerated corrosion; the reason for the absence of deposition of metallic lead was the relatively low pH; lead linoleate was found to be an excellent inhibitor for aluminium and to inhibit the galvanic corrosion normally set up when that metal is coupled to lead (M. J. Pryor, R. J. Hogan and F. B. Patten, *J. electrochem. Soc.* 1956, **103**, 206C).

Under service conditions, however, it may often be difficult to foresee whether or not the situation will favour deposition of metallic lead, and at the moment it seems inadvisable to use lead paints without careful consideration of the situation.

Clark regards aluminium as superior to steel as a basis for paint; given correct choice of paint, this is probably true. He recommends zinc chromate in linseed oil for the priming coat, the covering coat being selected on consideration of cost and aesthetic appeal (W. D. Clark, *J. Inst. Met.* 1955–56, **84**, 33).

Despite what has been said above, there may be situations where aluminium is best left unpainted—at least on the exposed faces; (the filling in of crevices with paint or jointing compound is generally a wise precaution). Champion points out that if the paint becomes an absorbent poultice, attack may be encouraged; certain alloys should be left unprotected in certain situations. His findings are summarized on p. 594.

Complaints have been heard about poor adhesion of paint to an ordinary aluminium surface even where there has been careful degreasing. Chemical treatment or anodizing often improve matters. However, paints vary in this respect and it is claimed that epoxy-resins adhere well to aluminium alloys which have been vapour-degreased (p. 624), but not otherwise treated (P. A. Dunn, *Light Metals* 1952, **15**, Jan., p. 38; April, p. 131).

Probably the best method of obtaining adhesion is by the employment of a wash primer (p. 568).

Paints for Magnesium. The protection of magnesium alloys is a different matter from that of aluminium. The unprotected metal readily evolves hydrogen from salt solutions, and simple painting does not prevent this. The metal first needs chemical treatment—generally in a chromate bath—and then painting—generally with a chromate pigment in the innermost coating.

Of the two important classes of magnesium alloys, containing aluminium and zirconium respectively, the first receives most immediate benefit from the chromate treatment, but if a suitable paint is applied after the treatment, both receive effective protection.

When magnesium is placed in a chromate solution, a small amount passes into solution and a protective film (probably chromium chromate, of composition roughly $Cr_2O_3.CrO_3$) is produced. This only occurs satisfactorily if the pH value is correct, and an efficient bath should incorporate a buffer system. In the case of baths containing dichromates and ammonium salts, the formation of dichromate from chromate prevents the pH from falling too low, whilst the escape of ammonia into the air prevents it from rising too high—as explained by L. F. Le Brocq and H. G. Cole, *Métaux et Corros.* 1949, **24**, 177.

The following classification of processes used in Great Britain is partly based on the booklet, "Pretreatment of Elektron Alloys" (Magnesium Elektron Co.), Jan. 1952:—

(1) An **Acid Chromate bath,** containing sodium dichromate and nitric acid; this is used for rough castings and sheet (W. O. Kroenig and G. A. Kostylev, *Korros. u. Metallsch.* 1932, **8**, 147; J. Hérenguel, *Métaux et Corros.* 1938, **13**, 82).

(2) The so-called **R.A.E. half-hour bath,** used boiling, and containing potassium dichromate, ammonium dichromate, ammonium sulphate and ammonia in amounts chosen to give the particular pH appropriate to the alloy requiring treatment; this is used for machine sections in cases where dimensional tolerances are strict (H. Sutton and L. F. Le Brocq, *J. Inst. Met.* 1935, **57**, 199; E. G. Savage, *J. Electrodep. tech. Soc.* 1938–39, **15**, 79).

(3) A **Chrome-manganese bath** containing sodium or potassium dichromate with manganese sulphate and magnesium sulphate, used cold or hot (satisfactory films appear in 3 to 10 minutes at the boiling-point); this is used for various alloys, especially those containing zirconium.

(4) The **Selenious Acid bath**; this is used for repairing damage which has occurred on films originally produced by chromate baths. The selenium process, due to Bengough and Whitby, consists of immersion in a bath containing selenious acid and sodium chloride, which produces a coating of selenium by simple replacement; the bath can be used cold, and 5 to 10 minutes usually suffices; for certain alloys a bath containing sodium selenite and phosphoric acid is preferable. The coating is to some extent self-healing; if a scratch is made in it, the hydrogen selenide produced by decomposition of magnesium selenide at the pores, deposits fresh selenium on the exposed surface. Precautions against poisoning by hydrogen selenide are necessary, especially if the treatment is to be carried out in a confined space (G. D. Bengough and L. Whitby, *J. Inst. Met.* 1932, **48**, 147; *Trans. Inst. chem. Eng.* 1933, **11**, 176; L. Whitby, *J. Inst. Met.* 1935, **57**, 250; J. Frasch, *Métaux et Corros.* 1938, **13**, 91, esp. p. 94; H. Fournier, *Métaux* 1934, **9**, 506; J. Cournot and L. Halm, *C.R.* 1937, **204**, 1941).

Electron diffraction studies of the various coatings are provided by H. G. Hopkins, *J. Inst. Met.* 1935, **57**, 227, and S. Yamagudi, *J. app. Phys.* 1954, **25**, 1937, who includes coatings from fluoride baths, discussed below.

Many attempts have been made to anodize magnesium. Magnesium oxide is relatively soluble, but the fluoride has a low solubility and can form the basis of protective coatings. The German *Flussal* process uses a bath containing fluorides of potassium and ammonium along with secondary phosphates of either potassium or ammonium. The process—along with other German processes—is described by W. Machu, "Nicht-metallische anorganische Uberzüge" (Springer) 1952, pp. 113–140, esp. p. 120. Several of the French processes involve the use of alternating current, and are discussed by J. Frasch, *Métaux et Corros.* 1938, **13**, 115, 132, 209; 1939, **14**, 9, 83; J. Pomey, E. Fourquin and X. Hardy, *Journées Lutte Corros.* 1938, p. 298C.

Of the electrochemical processes now in favour in this country, the best consists of A.C. treatment in 10% ammonium bifluoride under an E.M.F. gradually increased until 90–120 volts is reached; the current diminishes as the coating is formed. This process (sometimes described as anodizing) is essentially a cleaning operation for the removal of pre-existing films and miscellaneous contamination which might set up corrosion. The fluoride film left behind has limited protective character, but can serve as a " key " to increase the adhesion of the paint; alternatively, the process can be followed by immersion in an ordinary chromating bath which dissolves the fluoride film and leaves another, more protective, film in its place.

A practical objection to many of the electrochemical treatments for magnesium is the high E.M.F. demanded. A relatively new American low-voltage process for chromate coating will be watched with interest; it is described by M. Lorant, *Metal finish. J.* 1955, **1**, 250.

Wash primers (p. 568) have also been used as a preliminary treatment for magnesium alloys.

Of the paints to be applied to magnesium after the pre-treatment, the lowest should be pigmented with chromate, the next with aluminium, whilst the outer-most should be chosen to give the required colour and other properties; water-repellence and high gloss are valuable features, as they prevent the attachment of soot or other contamination liable to start corrosion. All paints intended for magnesium should be free from lead and iron compounds. The vehicle should be resistant to alkali. Sometimes an alkyd resin is used, but epoxy-resins are now favoured—especially the type using an amide curer. One system, stated to produce a coating very flexible, resistant to abrasive wear and possessing admirable adhesion, is obtained by applying an epoxy-paint pigmented with zinc chromate, stoving for 15 min. at 180°C., then applying a coating pigmented with aluminium and stoved under the same conditions; a black finishing coat stoved at 200°C. for 30 min. completes the scheme. All screws, bolts and joints should be coated before being brought into position (M. Hardouin, *Metal Ind.* (*Lond.*) 1956, **87**, 408; W. F. Higgins, *Chem. and Ind.* (*Lond.*) 1958, pp. 218, 1604).

Coatings other than paint

Thick Organic Coatings. Relatively thick layers of plastic or rubbery material—whether natural or synthetic—are often used for lining tanks of other material, with a view to preventing access of corrosive liquid to the metallic phase. The two requirements are that it shall adhere well to the steel, and withstand the liquid. Sometimes adhesion presents no difficulty, but often special treatment is needed. Rubber, for instance, will not readily adhere to a steel basis, but does so if the steel is first brass-plated (p. 620). Stuart has shown that adhesion of rubber vulcanized in contact with 70/30 brass is due to bonding by a sulphide-film which adheres only at grains possessing a certain crystal-orientation (N. Stuart, *Plastics* (in the press). See also S. Buchan, " Rubber to Metal Bonding "(Crosby Lockwood)).

Information about the reaction of plastics and other lining materials with various liquid is obtainable from Tables (see below). These mostly refer to stagnant liquids; moving liquids with grit particles in suspension are likely to damage protective films in a manner that the tables do not predict; even bubbles or vacuum cavities may alter the situation, since cavitation-damage is not unknown on plastics.

Methods of applying plastic coats—including flame spraying—are discussed by G. A. Curson, *Corrosion Prevention and Control* 1956, **3**, Jan. p. 30.

The remarkable resistance of fluorine compounds towards many reagents (p. 353) naturally suggests their use for coating metals; the coatings obtained are often both water-repellent and oil-repellent. The difficulty is to ensure that the coating is non-porous. In the case of polychlortrifluorethylene, the effective porosity towards nitric acid (and doubtless other reagents) depends on the previous heat-treatment; the presence of crystalline matter causes macro- or visible cracks; amorphous material contains only micro-cracks, too small to give perceptible leakage of nitric acid. These matters are discussed by F. H. Garner, S. R. M. Ellis and J. C. Gill, *J. appl. Chem.* 1956, **6**, 407; the use of polytetrafluorethylene dispersions for metal finishing is propounded by E. M. Elliott, *Trans. Inst. Met. Finishing* 1956, **33**, 355.

Instead of using steel tanks or pipes lined with plastic or other organic coating, it is possible to make the whole vessel of plastic; this requires a greater thickness, and may prove expensive; the fact that plastics conduct heat much less well than metals and often soften below their own melting-points (as in the case of some types of polythene, which otherwise possess many attractive properties) introduces limitations. Their employment lies outside the scope of this book, but tables showing the resistance of different plastics and the like to various liquids—organic and inorganic—may be quoted. These include F. Ritter, " Korrosionstabelle nicht-metallischer Werkstoffe " (Springer); Dechema Werkstofftabelle (Verlag Chemie); E. Rabald, " Corrosion Guide " (Elsevier).

Cement Coats. The use of cement mixtures in protecting steel depends, at least in part, on the production of alkali during setting; they are relatively useless in presence of sodium chloride. The subject has been discussed on pp. 302–308.

Glass Linings and Vitreous Enamels. Glass vessels have long been used at chemical works for large-scale operations involving acids, and in other situations glass-lined steel vessels, despite obvious disadvantages, have found application. Developments in glass-lined pipes are discussed by S. C. Orr, *Corrosion*, April 1950, Topic Section, p. 1; couplings of hastelloy, inconel or stainless steel are recommended. Thin linings for steel vessels can be obtained through the use of vitreous enamels; these are comparable to the organic enamel coatings provided by baking-varnishes, but probably more impervious and more resistant to many reagents. The material is essentially a boro-silicate glass containing fluorine (generally made by fusing together borax, felspar, quartz and cryolite), which is finely ground, suspended in water or organic solvent, and applied to the surface of the article by dipping or spraying. The articles are then dried by warming, and finally heated in a furnace sufficiently to melt the enamel, so that the particles flow together, yielding a continuous coat. The operation may then be repeated to give a second coat. The technology of enamelling is well discussed by W. Machu, " Nicht-metallische anorganische Überzuge " 1952, pp. 272–374 (Springer).

The ordinary vitreous enamel, familiar on domestic jugs, basins and cooking-ware, is not specially resistant to reagents. Resistance to acids is improved by increasing the silica and adding titania or zirconia (oxides possessing acidic rather than basic properties). Cobalt oxide and/or nickel oxide, is added to the ground-coat to improve adhesion. Some authorities believe that iron displaces metallic cobalt or nickel as a " tree ", thus providing an interlocking bond; measurements of electrical resistance provide evidence of the real existence of the tree, but there is a tendency to attribute the improved adherence to surface roughening of the steel or the presence of ferrous oxide crystals at the metal–enamel interface. A variety of views are expressed by R. M. King, *J. Amer. ceram. Soc.* 1943, **26**, 41, 358; A. G. Eubanks and D. G. Moore, *ibid.* 1955, **38**, 226; F. D. Lynch and A. L. Friedberg, *ibid.* 1955, **38**, 257.

The need for good adhesion is evident. Clearly if pieces of enamel chip off through temperature changes, or through application of small bending stresses or trivial impacts, corrosion becomes inevitable. Trouble arises from differences in coefficients of expansion. The ordinary enamel has a coefficient almost exactly equal to that of cast iron, but the acid-resisting types have lower coefficients and steel higher ones; thus there may be difficulty in coating steel directly with acid-resisting enamel; however, by applying a series of layers, each having a slightly lower expansion and a higher acid-resistance than the last, the situation can be improved (L. Stuckert, *Chem. Fabrik* 1933, **6**, 245; E. E. Giessinger, *Chem. and Ind.* (*Lond.*) 1935, p. 23, esp. p. 25).

Acid-resisting enamels are usually richer in silica and poorer in bases than ordinary enamels; boron, a normal component of ordinary enamels, should be avoided. In contrast to the more easily extractable metals, aluminium is sometimes a useful component of acid-resisting enamels. The state of alumina in silicate glasses is a matter of some interest and has been discussed by A. A. Appen, *J. appl. Chem. U.S.S.R.* 1953, **26**, 9 (English translation, p. 7).

Probably the most serious trouble in enamelling iron is gas evolution from the steel basis. At one time, the gas responsible was considered to be carbon monoxide, but now hydrogen is generally regarded as the cause of the typical defects. It may be evolved during the firing process, causing " reboiling ", or afterwards, causing " fish-scaling ". Clearly great care should be exercised with the pickling of the metal before enamelling, but something may be done by adding silica (15%) and alumina (10%) to the enamel to keep the structure open, thus facilitating the escape of gas (A. G. Rion, *Bull. Amer. Ceram. Soc.* 1954, **33**, 16; C. A. Zapffe and C. E. Sims, *Metals and Alloys* 1941, **13**, 444; *Foundry Trade J.* 1940, **63**, 225, 305, 368; 1941, **64**, 11, 92, 164; Y. Lindblom and S. E. Dahlgren, *Metal finish. J.* 1955, **1**, 517).

For vitreous enamelling in chemical plant construction, see H. W. Cremer and G. Brearley, *Chem. and Ind.* (*Lond.*) 1957, p. 374, esp. p. 376.

Coloured panels of enamelled aluminium are being used in building; the aluminium is pre-treated in caustic soda and nitric acid, then sprayed with powdered frit of the desired colour, which is fused about 540°C. (*Metal finish. J.* 1957, **3**, 195, 418; 1959, **5**, 146).

Ceramic and other Coatings for High Temperatures. The use of new types of ceramic coats for avoidance of oxidation is attracting much attention. These are of particular interest in connection with the frictional heat of high-speed aircraft which may raise the temperature into a range where aluminium and magnesium alloys deteriorate. One of the most successful American formulae is based on a glass high in barium and free from alkali, containing chromic oxide and enameller's clay. This is ground, suspended in water and sprayed on to a surface, previously roughened by sand-blasting. After drying it is fired at 1850°F. (1010°C.) for 3 to 10 minutes. If successfully applied the coating withstands 1650°F. (900°C.) for 500 hours and is resistant to acid and alkali. (T. A. Dickinson, *Ceramic Age*, July 1952, **60**, 15; *Metal finish. J.* 1956, **2**, 89; B. Zick, *ibid.* 1956, **2**, 51).

The numerous attempts made to coat or replace the alloys exposed to severe conditions—in a gas-engine exhaust or aero-engine exhaust (where the lead bromide from leaded fuel adds to the dangers), jet-engine coverings and the like—by means of cermets and ceramic materials generally, have enjoyed varying success. *Cermets* are combinations of oxides, carbides, silicides, borides or similar hard compounds, bonded with metals; they can be used at temperatures where metals would creep but often display poor resistance to thermal and mechanical shock; those based on carbides have poor resistance to oxidation. Success varies with the service conditions and the composition chosen. Mixtures of alumina with metallic chromium, or titanium carbide with cobalt, chromium or silicon-iron have given satisfactory results; other refractories tried are zirconia, titania, silica or even iron oxide; titanium silicide is a material with possibilities. One process for the coating of ingot iron or low-alloy steel uses a chromium-boron-nickel powder bonded with an enamel frit and clay. Higher operating temperatures can be withstood by pure oxide coatings; coatings of alumina, zirconia or titania can be applied by hot-spraying; the oxide in rod form is heated by a premixed oxy-acetylene flame and projected as droplets in a high-velocity air blast. Surveys of the subject, with references to articles describing the various processes, have been provided by S. W. Ratcliffe, *Rep. Progr. appl. Chem.* 1954, **39**, 50; D. B. Binns, *ibid.* 1955, **40**, 42. For use on jet flame tubes and pistol engine exhausts, see R. Tait, *Chem. and Ind.* (*Lond.*) 1957, p. 506, esp. p. 508.

At one time it was considered necessary to apply cermets by flame-spraying. It has been shown, however, that the chromium-boron-nickel coating mentioned above can be applied by normal ceramic methods as a slip or watery suspension; dipping or spraying may be used, followed by firing; a continuous metallic coat is formed, which brazes on to the surface of the steel (D. G. Moore and J. R. Cuthill, *Bull. Amer. ceramic Soc.* 1955, **34**, 375).

Other References

Fancutt and Hudson's book has been mentioned (p. 579). Two earlier books on anti-corrosive painting—also short—still deserve attention. One is M. Ragg's " vom Rost und vom Eisenschutz " (Union deutsche Verlagsgesellschaft), and the other L. A. Jordan and L. Whitby's " Preservation of Iron and Steel by means of Paint " (Research Association of British Paint, Colour and Varnish Manufacturers).

Much practical information will be gained from " Finishing Handbook and Directory " (publishers of " Product Finishing "), and the report on "Metal Finishing" produced by the Anglo-American Council on Productivity.

Numerous specifications exist for pigments, vehicles, paints and testing methods, and new ones keep appearing. The catalogue of the British Standard Institution should be kept under study.

For many purposes, the reader will wish to consult general books on paint which make no special reference to its use in preventing corrosion. Among these may be mentioned H. W. Chapfield, " Paint and Varnish Manufacture " (Newnes), and N. Heaton, " Outlines of Paint Technology " (Griffin); also for the technical reader with limited scientific back-ground " An introduction to Paint Technology ", published by the Oil and Colour Chemists' Association, and based on refresher lectures delivered in 1946 for men demobilized after war service.

A paper by Champion should receive attention before deciding to leave aluminium alloys unpainted. He considers that high-copper alloys (H14 and H15) should always receive paint (preferably preceded by metal-spraying) in urban, industrial and marine atmospheres; if possible, they should be painted in rural atmospheres also. In contrast, he considers that alloys containing less than 0·1% copper (H10 and H30) can be left bare in rural and urban atmospheres, whilst the parts washed by rain can be left bare in industrial atmospheres also; in marine atmospheres, painting is necessary (F. A. Champion, *J. Oil Colour Chem. Ass.* 1958, **14,** 730; esp. Table V, p. 740. For composition of alloys, see B.S. 1476 (1955 revision)).

This authoritative pronouncement may be compared with the more sweeping claims sometimes met with, such as the statement that the ability of aluminium " to resist corrosion even in the severest of marine conditions . . . eliminates the otherwise necessary task of painting and other costly maintenance work " (*The Engineer* 1959, **207**, 214).

CHAPTER XV

METALLIC COATINGS AND METAL-PIGMENTED PAINTS

Synopsis

The present chapter is devoted to the protection of one metallic material—generally iron or steel—by covering it with a thin layer of a second metal or alloy.

It opens with a review of the various methods of application; these provide coatings differing physically and chemically from one another in a degree which far exceeds the difference existing between paint-coats obtained (say) by brushing and spraying respectively. The procedures, however, are only described in so far as is necessary to bring out factors favourable or unfavourable to protection against corrosion; those who desire to study the technology of coating processes should consult the books mentioned at the end of the chapter. The methods available include *dry processes* (dipping in molten metal, heating in powder, deposition from vapour, spraying with molten globules and cladding), *wet processes* (electroplating with external E.M.F. and chemical deposition without external E.M.F.) and *metal-pigmented paints* (including paints made from zinc and aluminium powders in organic vehicles, as well as cementiferous paints carrying zinc particles set in a cement-like matrix); iron paints, which may perhaps come to provide a method of surface preparation before ordinary painting, are mentioned at this point. The electrodeposition of alloys is also discussed, with special reference to some relatively recent developments.

The origin and effects of imperfections in coatings then receive attention. Since some of these arise from deposition on dirty surfaces, it is necessary to consider methods of preliminary cleaning before plating, as also fluxing before hot-dipping. Methods for determining the frequency of pores and the effective porosity are discussed, along with the question of damage in service. The relation between thickness of coating and service life is considered. Stresses in coatings and the manner in which they cause cracking or detachment now come up for notice. The electrochemical situation arising at gaps in coatings which are either cathodic or anodic to the basis metal is then discussed. The chapter ends by indicating the uses to which the various metals are put in providing protective coatings, with further emphasis on the need for adequate thickness; an attempt is made to compare the merits of different metals and alloys as coating materials.

Dry Methods of Application

Deposition by Dipping in Molten Metal. If an article consisting of a material of high melting-point (such as steel) is dipped, or made to pass, through another metal in the fused state (tin, zinc or aluminium), it may, under favourable conditions, emerge carrying a layer of that metal, which soon solidifies. If the article was in a correct surface condition when it entered the molten bath, a uniform coating can thus be produced. The preliminary cleaning method varies according to the metal to be applied, and is indicated later in the chapter; but success mainly depends on the removal of an oxide-film either in a reducing atmosphere or by means of a liquid flux.

Between the basis metal and the main deposit, an alloy layer is generally found; sometimes there may be several alloy layers. On tinned steel, the tin-iron alloy layer separating tin from steel is very thin; at one time indeed an alloy layer was believed to be absent in this case, but its presence was revealed by the examination of sections cut at an angle—a method which is widely used for increasing the apparent thickness.

On zinc, the layer or layers of zinc-iron alloy are liable to be thick, and their brittleness introduces a risk of fatigue cracks starting in the alloy layer and later extending into the basis metal if the coated article is subjected in service to vibration or alternating stress. The brittleness depends, however, on structure as well as composition. The alloy layer on zinc can be made thin by reducing the time of contact between steel and zinc. Moreover, by adding 0·10 to 0·24% aluminium to the molten zinc, the formation of the ordinary zinc-iron alloy layer can be diminished or even suppressed altogether—probably because a much thinner film of a constituent containing iron, aluminium and zinc is produced, which effectively isolates the iron from the zinc. However, the effect of aluminium depends not only on the aluminium-content, but also on time; with short periods of immersion, aluminium in the bath retards the rate of attack on the iron, whereas with long periods it increases it. For details the readers should study the original papers (W. L. Hall and L. Kenworthy, *Sheet Metal Ind.* 1947, **24**, 741; H. Bablik, "Galvanizing" 1950, pp. 204–223 (Spon); H. Bablik, F. Götzl and R. Kukaczka, *Werkst. u. Korrosion* 1951, **2**, 163; H. Bablik and F. Götzl, *Metal finish. J.* 1956, **2**, 365; M. L. Hughes and D. P. Moses, *Metallurgia* 1953, **48**, 105).

In contrast, suitable annealing of a zinc-coated steel article can convert the whole coating to an alloy layer, which, although inclined to develop numerous small cracks when bent, is as a rule surprisingly adherent. In some corrosive atmospheres, this *galvannealed* steel appears to be more resistant than normal zinc-coated steel (S. E. Hadden, *J. Iron St. Inst.* 1952, **171,** 121; cf. W. Rädeker, quoted by G. Schikorr, *Schweiz. Arch. angew. Chem.* 1958, **24,** 40. Under immersed conditions, the situation is different).

Deposition by Heating in Metallic Powder. By packing small steel articles in a closed drum filled with metallic zinc powder and heating, preferably in a furnace with a reducing atmosphere, each article becomes covered with a layer that is essentially alloy, the composition becoming less rich in zinc as we pass from the exterior inwards. Generally it is well to mix oxide powder with the metallic powder to prevent the particles from welding together. In the analogous process for aluminium, ammonium chloride is added to deal with the protective oxide-film which otherwise surrounds the particles and prevents the metal from reacting. Sherardizing (a method of obtaining a zinc-iron alloy layer) is discussed on p. 654, whilst methods of obtaining aluminium-iron alloy layers receive consideration on p. 64.

In view of the brittleness of the acid-resisting irons rich in silicon, the prospect of diffusing silicon into the surface layers (only) of ordinary iron or steel articles (the interiors of which will retain their normal mechanical properties) has long seemed tempting. Interest attaches to recent work in Russia, where it has been found possible to carry out the diffusion of silicon at room temperature using an atmosphere of chlorine. The coatings obtained are stated to resist 10% sulphuric, hydrochloric and phosphoric acid (N. S. Gorbunov and A. S. Akopdzhanyan, *J. appl. Chem. U.S.S.R.* 1956, **29**, 713, 717 (English translation, pp. 635, 659)).

Deposition from Vapour. The possibility of introducing a metal from a vapour phase is illustrated by the *Chromizing* of steel in chromous chloride vapour, which yields an iron-chromium alloy coating. In an early process, introduced by Becker and others, the chromous chloride vapour was made by passing a mixture of dry hydrogen chloride and hydrogen over ferrochrome or chromium at about 950°C., and was then brought into contact with the heated steel. Many variants are possible. In one of them the iron and steel articles are packed in refractory material previously impregnated with chromous chloride; on heating, the vapour reacts from iron to give ferrous chloride and chromium; the latter diffuses inwards, producing an alloy layer integral with the body of the article and not liable to detachment. In some forms of the process the chromium content on the exterior may exceed 13% and sometimes reaches 30%, so that the layer, which is reasonably ductile, can provide protection against nitric acid of concentrations which would rapidly attack uncoated steel. The process has been successfully applied in condensers and air heaters and also serves for the coating of small articles such as screws, nuts and bolts. The kinetics of the reactions have been studied by T. P. Hoar and E. A. G. Croom, *J. Iron St. Inst.* 1951, **169**, 101. Information is provided by P. Galmiche, *C.R.* 1950, **230**, 89; G. Becker, E. Hertel and C. Kaster, *Z. phys. Chem.* (*A*) 1936, **177**, 213; G. Becker, K. Daeves and F. Steinberg, *Stahl u. Eisen* 1941, **61**, 289; D. W. Rudorff, *Metal Ind.* (*Lond.*) 1941, **59**, 194; R. L. Samuel and N. A. Lockington, *Metal Treatm.* 1951, **18**, 354, 407, 440, 495, 543; 1952, **19**, 27, 81; H. Kalpers, *Brennstoffe-Wärme-Kraft* 1951, **3**, 416. See also *Chem. and Ind.* (*Lond.*) 1950, p. 448; 1951, p. 716.

Further developments make use of mixtures containing aluminium and/or silicon, and yield alloy coats possessing resistance to high-temperature oxidation and many chemicals.

Other methods of deposition from vapour depend on different principles.

Cobalt, tungsten or chromium can be deposited by heating in the vapour of the appropriate carbonyl which usually decomposes in contact with a surface held at 450–600°C.; the matter is discussed by J. J. Lander and L. H. Germer, *Metal Ind.* (*Lond.*) 1947, **71**, 487.

An interesting method of depositing tungsten on nickel depends on an oxidation-reduction cycle. A tungsten source is heated above 2000°C. and the nickel article at about 700°C., in an atmosphere containing hydrogen and water vapour. The hot tungsten is oxidized to a gaseous oxide—probably WO_2—which becomes reduced to the metallic form on the nickel surface, possibly by atomic hydrogen formed at the high temperature. The theory is not completely understood, but the facts are described by M. G. Charlton and G. L. Davis, *Trans. Inst. Metal Finishing* 1957, **34**, 28.

During early physical researches on the effect of passing an electric discharge through low-pressure gas, the experimenter often noticed a metallic deposit on the glass near the cathode. Later, methods were developed for the intentional coating of surfaces placed near the cathode of a discharge tube; the process is generally known as *sputtering*. 2000 volts D.C. is a convenient E.M.F. The particles projected from the cathode consist largely of neutral atoms, travelling at velocities similar to that of the thermal velocity of atoms at the boiling-point of the cathode material; Tolansky considers that " there is virtually evaporation at local points on the cathode ". The vacuum requirements for sputtering are not stringent; 0·1 mm. of mercury may suffice.

A related process, known as *thermal evaporation*, requires a pressure below 10^{-4} mm., perhaps 10^{-5} or even 10^{-6}. It has been used to produce aluminium films on large telescope mirrors. The metal to be evaporated may be a bead on a hot wire filament, or a disc on a hot strip filament, and the high vacuum is needed to ensure that the mean free path of particles exceeds the distance between the molten metal and the surface to be coated. The particles emitted are of atomic size.

Details of both processes, which have already found commercial application in producing the initial deposits on wax records, the preparation of optical mirrors, and the coating of jewellery, plastics and optical parts, are provided by S. Tolansky, *J. Electrodep. tech. Soc.* 1951–52, **28**, 155. The electrical resistance of the deposits, which exceeds that of bulk metal, is discussed by J. Riseman, *Trans. N.Y. Acad. Sci.* 1957, **19**, 503.

If either of these processes is to be used for producing layers designed to give protection against corrosion, the question of porosity requires serious consideration. The " sputtered " deposits are often discontinuous, and sometimes consist of aggregates of microscopic but well-shaped crystals. The porosity of the thermally evaporated films depends on the conditions of formation, as shown by work on the deposition of iron, carried out in Uhlig's laboratory. Iron is unlikely to be used as a protective coating, but presumably the same variation in porosity would be met with in other metals. Evaporated films produced by heating pure iron wires *in vacuo* were found to be highly surface-rich, having roughness factors of 5 to 11; those produced from iron wire wrapped on electrically heated tungsten were much more compact, with roughness factors as low as 1·7 (H. C. Gatos and H. H. Uhlig, *J. electrochem. Soc.* 1952, **99**, 250; H. H. Uhlig and T. L. O'Connor, *ibid.* 1955, **102**, 562; *J. phys. Chem.* 1957, **61**, 402).

Spraying. Various patterns of pistols have been designed to provide a spray of tiny globules of molten metal, capable of being directed on to a surface to be coated. The globules—whether or not they are still molten at the moment of impact—flatten out to form flakes; under ordinary conditions of commercial working, they are probably mostly liquid when they impinge, unless the pistol is being held unusually far from the surface to be coated.* The result is a layer of overlapping scaly particles, with their planes parallel to the metal surface. The outer part of the layer is so porous that a drop of an aqueous solution placed on the surface may spread out as on blotting paper, but there is a possibility that the inner part next to the metal may be less porous (p. 636). For protection against atmospheric corrosion, porosity is not necessarily a disadvantage, since corrosion-products will interlock with the pores, largely filling them up, and remain attached to the surface instead of being easily detachable, as would be the case if they were formed on a smooth surface; thus there is a tendency for the corrosion-rate to diminish with time. Again, a porous sprayed surface forms an admirable basis for a suitable paint, the vehicle of which will penetrate the pores, obtaining firm anchorage. Obviously a vehicle must be chosen which does not attack the metal forming the sprayed coating. The power to anchor paint and retain corrosion-products is one of the advantages of sprayed coatings over other types of coating; another advantage is the applicability to large structures which can be sprayed *in situ*.

The surface to be sprayed must be cleaned by grit-blasting or sand-blasting (the latter being less desirable owing to danger of silicosis); this abrasive treatment is necessary, not only to remove scale and rust, but also to provide anchorage for the sprayed metal. There is no alloying, and the attachment is purely geometrical—due to a dovetailing of the coating metal into the basis metal. The grit must be angular; when once it has become rounded, it should be replaced.

The spraying pistols are of three main types, according as the metal is reeled in as *wire*, sucked in as *powder* or poured in *molten* from a crucible. There has been some rivalry—especially between the wire and powder procedures; probably, however, each has its part to play in our economy. In every case, successful protection depends on attention to small details, in the preparation of the surface as well as during the actual spraying, and the owner of metal-work will do well to employ firms known to possess experience of providing protection under the particular corrosive conditions

* Ballard states that the particles going down the central zone of the sprayed jet are molten when they strike the surface. Matting and Raabe have endeavoured to measure the temperature of the particles, both by optical and calorimetric methods, and find it to be close to the melting-point of the material applied (or the *solidus* temperature in the case of an alloy). Ballard feels doubts about the accuracy of the optical method (W. E. Ballard, Priv. Comm., April 29, 1958; also *Proc. phys. Soc.* 1945, **157**, 67. A. Matting and W. Raabe, *Schweissen u. Schneiden* 1956, **8**, 369).

in which he is interested. As mentioned on p. 7, mere compliance with a specification will not in itself ensure success.

For details regarding the processes the reader may be referred to the following papers by Authorities well qualified to speak for the different designs:—

W. E. Ballard, *Chem. and Ind.* (*Lond.*) 1955, p. 1606; *Trans. Inst. Metal Finishing* 1953, **29**, 174. F. A. Rivett, *Chem. and Ind.* (*Lond.*) 1955, p. 1612; H. F. Tremlett and W. A. Johnson, *ibid.* 1955, p. 1616; F. R. Himsworth, *ibid.* 1955, p. 1618; A. P. Shepard and R. J. McWaters, *Corrosion* 1955, **11**, 115t; W. B. Meyer, *ibid.* 1949, **5**, 282; W. E. Stanton, *Corros. Tech.* 1956, **3**, 311. An excellent survey is given in Zinc Development Association, Tech. Memo, No. **12**. See also British Standard Specification No. **2569** (1955), and discussion by F. A. Champion, *Electroplating* 1955, **8**, 180, 189.

Most patterns of pistol contain what is in effect a blow-pipe flame, capable of melting the metal fed into it—whether as powder or as wire; in the latter case, the wire is propelled by cog wheels, usually driven by a compressed-air motor. The flame is surrounded by a sheath of compressed air, so that the molten metal, instead of falling under gravity as large drops, is carried forward at high speed as a swarm of tiny particles which strike the surface to be coated and flatten out on impact; as far as possible, they should be made to impinge vertically; the particles at the margin of the shower of particles, which inevitably strike the surface at an angle, tend to bounce off.* In the simplest form of operation, the pistol is held in the hand and moved over the surface until the whole is covered. There are, however, mechanical installations for the large-scale coating of structural steel, in which the parts to be treated are moved forward automatically, so as to come under the influence of banks of sprays; such methods have considerably cheapened the procedure of coating large surfaces; some calculations made in 1949 suggested that the cost of the fully mechanical job was then less than half that of the hand process.

The two metals usually applied by spraying are zinc and aluminium. The aluminium used—whether it be introduced as wire or powder—must be free from copper, and precautions must be taken against the risk of using miscellaneous scrap containing small amounts of copper-bearing alloy. Lead and cadmium are also easy to spray—but their poisonous character demands very special precautions; lead coats are useful for protecting steel against acid-laden air. The spraying of tin provides coats thicker than those easily obtained in other ways, but oxidizing conditions must be avoided. Since these coatings are mainly used on containers for products in which sealing of pores cannot be expected, the thickness must be adequate to give complete cover.

* Ballard emphasizes the distinction between grit-blasting, where the particles should be made to impinge at an angle of 30° to the normal, and spraying, where normal impingement should be aimed at. Where abrasive removal of the scale is desired, a certain gliding action is needed to scarify the scale; normal impact has little descaling effect. Where, as in metal spraying, adhesion of the particles is desired, normal impact is the only way to achieve this (W. E. Ballard, Priv. Comm., April 23, 1958).

Although the ordinary sprayed coat carries no alloy layer between metal and coating, it is possible to obtain alloying by subsequent annealing. In the *aluminizing* process, steel is sprayed with aluminium, which is then covered with bitumen, and heated in a furnace. The bitumen serves to protect the aluminium from oxidation until iron has diffused outwards; when once formed, the iron-aluminium alloy is very resistant to oxidation and can protect the body of the steel against hot gases in service. If the aluminium used contains $\frac{1}{2}$ to $\frac{3}{4}$% of cadmium, the bitumen can be dispensed with; no cadmium remains in the final coating.

An invention due to Clayton, known as *Peen-Plating*, may conveniently be mentioned at this point. Articles to be coated (tacks, nails, hose-clamps, washers, fasteners, hinges, coat-hooks and the like) are placed in a modified tumbling barrel, containing the metal to be applied in powder form, along with some " impacting material " such as steel grit, shot, wire cuttings, fused aluminium-oxide chips or glass balls (the impacting material is not needed in the coating of steel nails); water and certain chemical " promoters " must also be present. Steel articles which are to be coated with zinc should first be degreased and acid-pickled; in some cases they then receive a very thin copper coating by immersion in hot copper sulphate containing an inhibitor, but frequently the copper coating can be omitted. During the tumbling, coatings are built up on the articles by the successive welding of small pieces of the second metal and generally show a laminated structure. Peen-plating has thus a certain relation to spraying, but the necessary pressure is obtained by impact between the article and impacting material (or between one article and another), whereas in spraying the metallic globules themselves possess sufficient momentum to develop the appropriate pressure. To obtain the welding needed for peen-plating, the oxide must be removed without appreciable attack on the metal, and this is the function of the " promoter ", which may contain an acid (or in some cases an alkali) along with an inhibitor; frequently a non-ionic inhibitor is suitable or a cationic inhibitor containing both polar (hydrophilic) and non-polar (hydrophobic) groups. An interesting account of peen-plating, with an estimate of costs, is provided by G. H. Jenner and T. P. Hoar, *Trans. Inst. Metal Finishing* 1956–57, **34**, 253. See also R. Pottberg, *Metal finish. J.* 1957, **3**, 358. Further information is provided in the patent specifications of E. T. Clayton, U. S. Patent 2,640,002 (May 26, 1953); E. T. Clayton and R. Pottberg, British Patents 740,075, 740,104 (1955).

Cladding. If a " sandwich " consisting of a thick central plate of strong but corrodible material, interposed between two thinner plates of resistant (but usually weaker) metal, is rolled under suitable conditions, the three components weld together, and are reduced in cross-section to the same extent; the result is a thinner plate or sheet consisting of the strong material " clad " with relatively thin layers of the resistant material. In general, there will be no alloying, but, if the manufacturing process has been conducted at too high a temperature, or if the " clad " article becomes heated in service, a constituent of the central zone may diffuse into the external layers; thus in a plate of aluminium alloy clad with copper-free aluminium undue heating in service may cause the copper from the central zone to diffuse into the outer zones, considerably reducing the corrosion-resistance of the latter (p. 670). Cladding is mainly used for light alloys, but steel also can be clad with nickel or stainless steel. Nickel-clad steel is much used in chemical industry in situations where solid nickel would be too expensive (E. Warde, *Chem. and Ind.* (*Lond.*) 1956, p. 762); the welding of nickel-clad vessels is described by P. J. Gegner, *Corrosion* 1956, **12,** 261t.

Aluminium kettles are being clad with super-purity aluminium in order to avoid the attack which occurs on unclad kettles when the supply water contains copper; this is not a practical proposition for saucepans which are often cleaned with steel-wool abrasive—an operation which would scrape off a soft layer of pure aluminium (F. A. Champion and E. E. Spillett, *Sheet Metal Ind.* 1956, **33**, 25).

Wet Methods of Application

General Principles of Electroplating. In contrast to hot-dipping, electroplating usually gives a deposit of a single metal without intermediate alloy layer, although in some instances a slow inter-diffusion between basis and coating can occur. Sometimes the metal deposited is purer than the anode. The article to be coated is made the cathode in a solution containing a compound of the metal to be deposited; the current is provided from an external source of E.M.F. If the anode consists of the metal which is being deposited on the cathode, the bath is automatically replenished, provided that the current efficiency at the anode is equal to that at the cathode; this is not often exactly the case, and in practice some adjustment of composition by additions of salt, acid and/or water may occasionally be needed; the current efficiencies at anode and cathode are usually high, but not necessarily 100%.

In certain cases it may be necessary to use an insoluble anode, such as graphite, in which case the anodic reaction consists of the evolution of oxygen and the production of acid, which will periodically require to be neutralized; a soluble compound containing the metal to be deposited will also have to be added from time to time. Such a procedure is not adopted if it can be avoided, owing to the increased power costs; with an insoluble anode a relatively high E.M.F. is required to do " chemical work ", whereas with a soluble anode, the essential change is merely a transfer of metal from anode to cathode, so that a small E.M.F. suffices.

Sometimes, however, where it is intended that the anode shall dissolve quantitatively—or at least with sufficient efficiency to replenish the material deposited on the cathode—it does in fact become passive; energy is then used up in the production of oxygen, and an increased E.M.F. is needed to maintain the current; moreover the bath becomes depleted of metal. Passivity sets in easily when nickel-plating is conducted from a sulphate bath; largely to prevent passivity, most of the nickel baths used today contain chlorides. Special types of anodes, made of nickel containing small amounts of sulphur and copper, can be used to avoid passivity.* These dissolve, in favourable circumstances, in a uniform and economical fashion. Anodes are often rolled to an oval section; such a shape is found to minimize the amount of metal left over for remelting as scrap, when the time arrives to replace an old anode by a new one. It is common practice to put the anodes in bags in order to avoid particles consisting of undissolved inclusions reaching the cathode surface, where they would cause imperfections in the coating. However, small particles easily pass through the fabric of any bag, and continuous filtration is advisable for the best results.

The crystallographic structure of electrodeposits is a fascinating subject, but cannot be discussed here in detail. Fischer has made a careful examination of the way in which different crystal forms are produced in presence of different ions; although the deposits which he has studied include types of structure which would be unacceptable to the electroplater, the principles emerging from his work—and that of several other pure-science investigators —should be helpful to all (H. Fischer (with H. Matschke and P. Pawlek), *Z. Elektrochem.* 1950, **54**, 459, 477; 1951, **55**, 92. T. Erdey-Grúz, *Z. phys. Chem.* (*A*) 1935, **172**, 157; N. Thon, *C.R.* 1933, **197**, 1312, 1606. A. Portevin and M. Cymboliste, *Trans. Faraday Soc.* 1935, **31**, 1211; *Rev. Métallurg.* 1933, **30**, 323. G. Bianchi, *Annali di Chimica* 1955, **45**, 99. A. Glazunov, *Z. phys. Chem.* (*A*) 1933–34, **167**, 399. D. J. Macnaughtan and A. W. Hothersall (with G. E. Gardam and R. A. F. Hammond), *Trans. Faraday Soc.* 1928, **24**, 387; 1933, **29**, 729, 755; 1935, **31**, 1168; *J. Electrodep. tech. Soc.* 1928–29, **4**, 31, 81; 1949–50, **25**, 203. D. N. Layton, *ibid.* 1951–52, **28**, 239. W. Blum and H. S. Rawdon, *Trans. Amer. electrochem. Soc.* 1923, **44**, 397. V. Kohlschütter, *Z. Elektrochem.* 1924, **30**, 74; 1927, **33**, 272, 290; 1932, **38**, 213 (with F. Jakober, F. Uebersax, A. Torricelli and A. Good). H. Fischer and H. F. Heiling, *Trans. Inst. Metal Finishing* 1954, **31**, 90; R. Weil and H. J. Read, *Metal Finish.* (*N.Y.*) 1956, **54**, No. 1, p. 56).

Recent work by Gerischer on the deposition of silver will interest the pure scientist. He has studied the two steps (1) the penetration of ions through the phase boundary to give adsorbed atoms (*Adatoms*), and (2) the fitting of those atoms in stable positions relative to the lattice. Of these the second stage is the sluggish one and determines the rate at which the deposition can proceed (it should be noted that silver is a normal metal, and that the state of affairs might prove different if an abnormal metal

* "The addition of traces of sulphur to nickel, or in special types of commercial anodes, the addition of sulphur in conjunction with nickel oxide and copper, results in profound and little understood changes in the mechanism of electro-dissolution of the nickel. Such anodes dissolve as a rule most uniformly, giving a minimum of insoluble sludge. Presumably the main factor in overcoming passivity is a very thin film of nickel sulphide at the grain boundaries. Excessive preferential attack at the grain boundaries must, however, be avoided; otherwise the grains are undermined and may detach themselves from the main body of the anode. Besides being wasted the grains may find their way, even through a diaphragm bag, to the cathode, causing the deposition of most ghastly nodules" (P. Hersch, Priv. Comm., April 30, 1957. See also A. J. Brookes, 10th B.I.S.R.A. Chemists' Conference, *J. Iron St. Inst.* 1957, **185**, 497; W. A. Wesley, *Bull. Inst. Metal Finishing* 1956, **6**, 87).

like nickel was being studied). The paper deserves study (H. Gerischer, *Z. Elektrochem.* 1958, **62**, 256).

Polarization and Throwing Power. Given a smoothly dissolving anode, no chemical work is performed and the E.M.F. needed for plating is usually small—as already explained. It is smaller for metals of the " normal group ", where the potentials needed for anodic attack or cathodic deposition lie close to (but on different sides of) the equilibrium potential; it is higher for those of the " abnormal group ", including nickel, where the polarization is considerable (p. 907). However, the anion has an important influence, the polarization being often smaller in chloride than in sulphate solution; sulphamate baths usually give very low polarization. The adoption of a bath with low polarization will reduce the E.M.F. needed for the plating process, and thus minimize power costs, but for coating articles of complicated shape, a bath of very low polarization will cause most of the deposition to occur on the prominent parts of the cathode, and the surface in the recesses will receive little covering, on account of the high resistance of the liquid path leading to them. Thus it may be better to choose a solution with appreciable polarization; in such a case, the high current density at the prominences shifts the potential in a direction unfavourable to deposition, and thus diverts current to the recesses. The solution is then said to have *throwing power*, and clearly if the object of the deposit is to protect against corrosion, throwing power is desirable, since the recesses must receive adequate covering. The fact that zinc deposited from a cyanide bath covers the recesses better than zinc from a sulphate bath is probably due to the higher polarization in the first case. However, other factors besides polarization affect throwing power; the covering of the recessed portion is naturally favoured if the liquid has high conductivity. Excellent discussions of throwing power are provided by G. Bianchi, *La Chimica e l'industria* 1953, **35**, 414 and by G. E. Gardam, *J. Electrodep. tech. Soc.* 1949–50, **25**, 77.* A new method of expressing results is suggested by R. V. Jelinek and H. F. David, *J. electrochem. Soc.* 1957, **104**, 279.

Plating from Solutions containing the Metal as Anion. If a silver nitrate solution is used for the deposition of silver, the deposit produced consists of a limited number of disconnected silver crystals—not a continuous coating; so soon as a few nuclei have been produced, it is easier for the metal deposited to continue building these crystals than to start fresh ones; thus we obtain a crystalline deposit, probably loosely adherent and certainly discontinuous—which can provide no sort of protection to the underlying metal. The deposition of silver from a nitrate solution is in regular operation at silver refineries, where the only object is to transfer silver from anodes of crude silver to the cathodes, leaving the impurities behind, and to do this with the minimum consumption of energy—for which purpose the simple salt solution with its low polarization is admirably suited. But in

* Gardam points out that a cell widely used for the assessing of throwing power is open to criticism on dimensional grounds, giving results which must depend on the scale of the experiment.

the plating shop the coarse crystalline deposit would be utterly unsuitable, and a complex bath must be used—despite the greater expenditure of power resulting from the higher polarization.

If in place of a nitrate bath, we use a solution containing a complex cyanide, $K[Ag(CN)_2]$ or $Na[Ag(CN)_2]$, usually with excess of potassium or sodium cyanide and some carbonate, the coating is continuous and the structure extremely fine. Many other metals (gold, copper, zinc and cadmium) are plated from complex cyanide baths, which yield finer deposits than can easily be obtained from a solution of an ordinary salt (e.g. sulphate).

Other useful plating baths contain the metal as anion. Complex nitrites are used in the deposition of palladium, whilst tin can be deposited from a stannate bath. Again bright, fine deposits of chromium are obtained from baths containing chromic acid along with sulphuric acid, where the greater part of the chromium is present as CrO_4^{--} or $Cr_2O_7^{--}$ anions, and comparatively little as Cr^{+++} cations. Attempts to deposit chromium from baths based solely on trivalent chromium have yielded coarse crystalline deposits, unsuited for protective purposes; the nearest approaches to success have been the development of baths containing complex oxalates, and here once more we find chromium in the anion (H. T. S. Britton and O. B. Westcott, *J. Electrodep. tech. Soc.* 1933, **8**, Part 5, p. 1. W. H. Wade and L. F. Yntema, *Trans. electrochem. Soc.* 1938, **74**, 461).

A little thought will show that deposition from anions is likely to give more fine-grained deposits and better throwing power, than deposition from cations. In a silver or chromium plating bath, before the E.M.F. is applied, a given $[Ag(CN)_2]^{--}$ or CrO_4^{--} anion is just as likely to be moving towards the article to be plated as away from it. When the plating E.M.F. is turned on, there is a general tendency for the anions to move *away from* the article; however, since most of the current is carried by other ions (H^+, SO_4^{--} and perhaps Cr^{+++} in a chromium-plating bath; Na^+, CN^-, CO_3^{--} and OH^- in a silvering bath) a $(CrO_4)^{--}$ or $[Ag(CN_2)]^{--}$ ion will still occasionally reach the cathode surface, where it will pass probably only a brief existence. Either it is pulled away as a whole, or the outer components (O^{--} or CN^-) are pulled away, the central Cr or Ag atom being left attached to the metal. Now this is an exceptional occurrence, requiring considerable activation energy—which explains why the polarization is high, the throwing power good and (what is less welcome) the current efficiency lower than that commonly obtained in baths containing simple salts. Furthermore, since the exceptional occurrence will rarely take place at the point which is crystallographically most favourable to deposition, we shall tend, instead of depositing metal on existing crystals, to start new crystals, thus obtaining the fine structure which the plater desires. The subject is complicated, and the picture just presented may be an over-simplification, but it does suggest a reason for the fact that, whilst deposition from cations gives the refiner what he wants, deposition from anions is often more welcome to the plater.

In depositing from anions, we are deliberately aiming at occurrences which are unusual, but not impossible. Clearly in such a case, careful control of the conditions is necessary; particularly we must adjust current density and temperature. A very important factor is the concentration of any ion which plays no direct part in the deposition process but which, by carrying a large part of the current, moderates the general movement of the complex anions away from the cathode, and thus favours their occasional impingement on the cathode surface. On the whole, a higher temperature is used for depositing a metal from a complex bath than for depositing the same metal from a bath in which it exists as cations; and this would certainly be expected since the anions have to reach the cathode as a result of thermal movement, moving *against* the potential gradient.

The explanation suggested is supported by the fact that polarization in cyanide solution is usually much higher than in simple salt solutions. Early work by Foerster showed that the deposition of copper from copper sulphate solutions, could be carried out rapidly at potentials quite close to the reversible potential Cu | $CuSO_4$, provided that the solution was in rapid motion (in stagnant solution, there was concentration polarization); in a complex cyanide solution, however, deposition was only possible if the potential was shifted well below the equilibrium value, and the polarization (which was here activation polarization) increased rapidly with current density. However, the argument is not quite conclusive, since the copper is in the cupric state in $CuSO_4$ and the cuprous state in $K[Cu(CN)_2]$; moreover Glasstone states that in the deposition of silver from $K[Ag(CN)_2]$ baths, there is little polarization—other than concentration polarization (S. Glasstone, *J. chem. Soc.* 1929, p. 690, esp. p. 694).

It would seem, therefore, that other factors—besides polarization—play a part in determining the good deposits obtained from complex salt solutions. One may be the attachment of cyanide ions, acting like the atoms of a catalyst poison, at just those points where the silver atoms would normally attach themselves, thus diverting deposition to other places, so that a new crystal is frequently started; hence we obtain a swarm of little crystals instead of a few big ones, and even when the deposit is still thin it covers the whole cathode surface.

Other substances which keep low the concentration of simple cations by complex formation are useful in plating. E.D.T.A., the well-known sequestering agent (p. 1000), and its derivatives, promise to provide alternatives to cyanide in plating, to regulate crystal-growth and produce brightening, besides rendering certain impurities harmless and improving the corrosion of the anode. The health hazard is much smaller than in the case of a cyanide bath. The E.D.T.A. can be precipitated from expended baths and used again (R. L. Smith, *Chem. and Ind.* (*Lond.*) 1956, p. 1284).*

* S. C. Britton adds the comment that, although chelating agents are certainly promising, some difficulties have been experienced, probably associated with their decomposition at electrodes. A tin-zinc plating bath based on the tri-sodium salt of N-hydroxyethyl-ethylenediamine triacetic acid has been described but has not yet reached industrial application (A. E. Davies and R. M. Angles, *Trans. Inst. Metal Finishing*, 1955–56, **33**, 277).

In the case of chromium plating, the potential gradient will tend to make the various anions (CrO_4^{--}, $Cr_2O_7^{--}$) move away from the cathode surface, but especially at higher temperatures (chromium-plating proceeds best from a warm bath), thermal movement will occasionally allow one to come into contact. It was pointed out in Chapter VII that in anodization the anions move *with the field* and are likely to attach themselves to the anode through the (negative) oxygen. In chromium-plating, the opposite is the case; an anion must move *against the field*, if it is to reach the cathode, and will then attach itself, perhaps only momentarily, through the (positive) chromium. It may then be pulled off again as a whole by the field, but sometimes the oxygen may be pulled off alone, leaving chromium or a lower oxide on the metal; or hydrogen ions may be driven on to the surface removing the oxygen as water. Since many alternative reactions are possible, a low current efficiency must be expected; much of the current is expended on the evolution of hydrogen. However we do obtain a deposit of chromium metal, containing oxide and also a considerable amount of hydrogen; some authorities (p. 415) regard the deposit originally produced as a hydride. Since the formation of chromium metal is not connected with the orderly movement of cations but the random movement of anions, a bright deposit (almost amorphous) may be expected, and is indeed obtained.

As already mentioned, a careful adjustment of conditions is needed in such cases, particularly the concentration of other mobile ions which can carry part of the current. Whether for this or some other reason, platers attach great importance to preserving the correct ratio of CrO_3 to H_2SO_4, and various opinions have been advanced regarding the optimal value. The ration of CrO_3 to H_2SO_4 *by weight* now favoured is about 100 : 1, although sometimes a higher ratio, up to 130 : 1, is adopted.* Much information regarding the composition of plating baths is provided by W. L. Pinner and E. M. Baker, *Trans. electrochem. Soc.* 1929, **55**, 315; J. W. Cuthbertson, *ibid.* 1931, **59**, 401; W. Blum, P. W. C. Strausser and A. Brenner, *Bur. Stand. J. Res.* 1934, **13**, 331; M. Cook and B. J. R. Evans, *J. Electrodep. tech. Soc.* 1934, **9**, 125, esp. p. 132; N. A. Isgarischew, *Korros. Metallsch.* 1930, **6**, 157; G. E. Gardam, *J. Electrodep. tech. Soc.* 1929, **4**, 113; H. Silman, *Trans. Inst. Metal Finishing* 1954–55, **32**, 43; H. Silman, Priv. Comm., April 25, 1957.

Recently other substances have been used instead of (or additional to) sulphuric acid; these include fluorides, fluosilicates and borofluorides. The so-called " self-regulating " baths have attracted much attention. Details of these developments will be found in Silman's paper, quoted above.

The picture of chromium-deposition from anions, presented above, is by no means generally held. Some authorities believe that hexavalent chromium is reduced first to trivalent (possibly then to divalent) and subsequently to metallic chromium. The potentials at which the various reactions should take place under reversible conditions are conveniently collected by E. Deltombe, N. de Zoubov and M. Pourbaix, *Cebelcor Rapport*

* In early papers, the ratio was expressed as CrO_3 molarity to H_2SO_4 normality.

Technique No. 41 (1956). See also summary of views by E. S. Spencer-Timms and C. Williams, *Rep. Progress appl. Chem.* 1955, **40,** 162.

Plating from Solutions containing the Metal as Cations. On some metals reasonably good deposits can be obtained from simple salt solutions; copper, for instance, gives satisfactory coatings from copper sulphate solutions containing sulphuric acid, but certain precautions are needed. If a steel article is to be copper-plated, the deposition should be started from a cyanide bath, otherwise copper will be thrown down in objectionable form by simple replacement.

$$Fe + Cu^{++} = Cu + Fe^{++}$$

—a happening impossible in a cyanide bath where the Cu^{++} and Cu^{+} concentrations are kept extremely low, owing to formation of complex anions.

The abnormal metals (iron, cobalt, nickel) can be deposited from simple sulphate or chloride baths without difficulty; probably the high polarization characteristic of these metals is a helpful factor.

It is only possible to obtain smooth deposits on the cathode and smooth dissolution of the anode if care is taken to ensure uniformity of bath composition over the surfaces of both electrodes. Tait expresses the matter by saying that the problem of obtaining smooth, sound deposits at high current densities (i.e. high rates of deposition) depend on the rate at which the denuded layer formed at the cathode can be " hurried " to the anode for replenishment and the concentrated layer at the anode " hurried " to the cathode for deposition. The interchange must be rapid, systematic and uniform. Ordinary methods depending on bubbling of gas or agitation with an ordinary stirrer are too chancy and uncertain. The interchange requires to be regarded as an engineering problem, and in cases where it has been solved very rapid deposition rates have been found possible. The principle has so far been applied mainly to electroforming, but the matter deserves study in connection with electroplating (W. H. Tait, Priv. Comm., April 17, 1957; also British patent application 25,552/54 of Jan. 1955).

Addition Agents. Deposition from a simple salt solution (e.g. a sulphate bath for zinc or cadmium, a fluoborate bath for tin or lead) generally requires an " addition agent " in the bath, to prevent the formation of rough or tree-like deposits. Suitable substances have been discovered—often empirically—and the list presents a diversity of compounds. The addition agent for zinc may be aluminium sulphate or dextrin; that for cadmium a mixture of gelatine, diphenylamine and α-naphthol. For lead, ordinary glue has proved a useful addition. When tin is deposited from a sulphate bath, a mixture of gelatine and β-naphthol is a suitable addition agent; gelatin alone has little smoothing action but acts as a protective colloid to keep the β-naphthol in solution. Cresol-sulphonic acid, also added to stabilize the electrolyte against atmospheric oxidation, may itself contribute to the improvement of the deposit (A. W. Hothersall and W. N. Bradshaw, *J. Electrodep. tech. Soc.*, 1937, **12**, 113; 1939, **15**, 31; Instructions for Electrodepositing Tin, Tin Research Institute, 1956; R. M. Angles, K. W. Caulfield and R. Kerr, *J. Soc. chem. Ind.* 1946, **65**, 430).

Many addition agents are proprietary mixtures of undisclosed composition. Even with complex salt baths, addition agents are often advantageous; caffeine, for instance, occurs in some recipes for cyanide cadmium baths.

More than one view is taken regarding the action of addition agents, but it is probably connected with adsorption of the agent on active points of growing crystals where deposition would otherwise occur, so that growth is blocked and deposition diverted elsewhere. In many cases, the agents come to be built into the deposit. Certain bath constituents which are not generally referred to as addition agents seem to act in an analogous way. The classical work of Macnaughtan, Hothersall, Gardam and Hammond (p. 657) on electrodeposited nickel showed that small amounts of oxide or basic salt included in the deposit can reduce the grain-size and influence the hardness. If the pH is regulated so as to introduce a suitable amount of basic matter, it is possible to obtain a fine-grained deposit and avoid the risk of grain-coarsening on annealing.

Many addition agents have a marked *levelling action*, so that the surface after plating is actually smoother than the original surface of the unplated article. This valuable property has been studied quantitatively by Watson and Edwards who adopted as a standard surface a " micro-groove master " —as used in the gramophone record industry. If the solution contained a levelling agent, the depth of the grooves diminished as plating continued. The difference between the average groove depths before and after plating divided by the average thickness of deposit was taken as a measure of levelling power. In a nickel-plating bath, thiourea showed the highest levelling figures, but quinoline ethiodide and coumarin, which showed lower figures, gave brighter deposits. In all cases, the amount added must be controlled between certain limits. If the amount added is too small, the levelling power is low (sometimes negative) and the deposit dull; if it is too high, the levelling power is again low, and the deposit dull, sometimes dark and brittle. This research made the cause of levelling fairly clear. The organic body was found to be incorporated in the deposit formed at the projection, where there is ready replenishment, but to a much smaller extent in metal deposited in the grooves; since the organic body discourages deposition (doubtless by adsorption on the most favourable sites), the effect will be to cause more deposition in the grooves than elsewhere. This outstanding paper deserves study (S. A. Watson and J. Edwards, *Trans. Inst. Metal Finishing* 1956–57, **34**, 167).

Addition agents have been studied by Shreir who has made several useful suggestions to explain his observations. He thinks that some diminution of crystal-size may result if an agent is absorbed on the edges of a growing crystal, so that deposition which would occur thereon is diverted elsewhere. Where, however, there is rapid adsorption and desorption, resulting in a condition where all possible sites become uncovered for part of the times, deposition will occur in a random manner, resulting in a smooth, finely crystalline deposit; the activation polarization is naturally increased (E. G. Neal and L. L. Shreir, *Trans. Faraday Soc.* 1956, **52**, 703, esp. p. 708; see also L. L. Shreir and J. W. Smith, *ibid.* 1954, **50**, 393).

Brighteners. Some of the substances added to plating baths play a notable part in increasing the brightness. Small amounts of carbon disulphide have long been used in cyanide silver baths for that purpose, but it is probable that the true brightener is not carbon disulphide but a decomposition product—possibly thiourea; certainly the addition of thiourea, or some other compound containing sulphur, to a silver-plating bath, improves the brightness and is easier to control than the addition of carbon disulphide. Thiourea also serves as a brightener for copper deposited from a sulphate bath, being used in conjunction with an organic wetting agent.

Copper deposited from a cyanide bath containing gelatine consists of alternate bands rich respectively in copper and gelatine. A similar banded structure is found in nickel deposited from " bright " baths. The brightener for nickel may be *anionic* (an aromatic sulphamate or sulphonamide) or *cationic* (an aldehyde, ketone or aromatic amine, or a salt of zinc or cadmium which probably deposits colloidal hydroxide or basic salt). A combination of the two types may give a deposit which is brighter and more ductile than that obtained with a single brightener. Information regarding the structure of bright deposits is provided by G. Grube and V. Reuss, *Z. Elektrochem.* 1921, **27**, 45, esp. p. 49; J. A. Henricks, *Trans. electrochem. Soc.* 1942, **82**, 113; W. R. Meyer and A. Phillips, *ibid.* 1938, **73**, 377. The character of brightness is discussed by W. L. Pinner, G. Soderberg and E. M. Baker, *ibid.* 1941, **80**, 539, esp p. 546, and the general subject by S. Wernick, " Electrolytic Polishing and Bright Plating of Metals " (Redman).

In an excellent discussion of bright plating, Hoar brings out the theoretical relationhip between bright plating and bright dipping. In general, lack of smoothness in cathodic deposits is due to the fact that deposition tends to occur preferentially at sites provided by incomplete rows of atoms. It is from such positions of incompleteness that atoms are preferentially removable in an etch-bath, leaving the etch figures; bright dip baths are liquids which remove atoms selected at random, usually through the intermediary of a solid film, so that etching is avoided.

Some brightness can be obtained by introducing a second metal into a plating bath, probably because the foreign atom disarrays the lattice and reduces the probability of deposition on crystallographically selected sites; nickel-cobalt alloys have long been known to produce brighter deposits than plain nickel, although today cobalt brighteners are not in favour, being more expensive than organic agents; in other cases (as in the deposition of tin-nickel alloys) an intermetallic compound of complex structure, not built up of simple layers, and therefore not so prone to selective site deposition, is plated on the cathode. Co-deposition of oxide confers only limited brightness; colloidal particles of oxide or hydroxide are too big.

In Hoar's opinion, the best addition agents are:—

(1) *Small mobile molecules* which readily attach and detach themselves, covering perhaps 90% of the surface at any moment, but which are constantly

deserting it and re-attaching themselves in haphazard fashion, thus leaving sites vacant for random, non-selective deposition of metals.* (Formaldehyde, ammonia, cyanides, fluoride ions, iodide ions are considered to act in that way; thiourea, probably formed from the carbon disulphide added to silver-plating baths as brightener, may also be included in this class.)

(2) *Large molecules*, having the power of being adsorbed preferentially at just those points where metal deposition would occur (this class includes gelatine, β-naphthol and various sulphonic acids).

For details see T. P. Hoar, *Trans. Inst. Metal Finishing* 1952–53, **29**, 302.

It is known that the brightening agent is to some extent incorporated in the coating. Lustrous nickel coatings from baths containing napthalene trisulphonate are found to contain carbon and sulphur. The situation is demonstrated most clearly when the brightener is a coloured substance; if fuchsin has been used as a brightener in nickel-plating, it is recovered when the metallic deposit is dissolved in hydrochloric acid, the amount being a linear function of the fuschin concentration in the plating bath (J. L. Dye and O. J. Klingenmaier, *J. electrochem. Soc.* 1957, **104**, 275).

Many plating technologists classify the substances added into (1) *levellers* (or *carriers*) which can be tolerated in large amounts without ill effects to the deposit; they fill up depressions but never confer full brightness, (2) *brighteners proper*, which give full brightness when added in very small amounts in presence of a leveller, but the quantity is critical; they produce cracking or peeling when present in excess. Fishlock regards cobalt salts, aryl-sulphonic acids, sulphonamides and sulphonimides as levellers, whilst cadmium and zinc salts, formates, aldehydes and ketones are brighteners. This classification appears to have a scientific basis—as shown by studies of the effect of two classes of compounds on potential, carried out by C. C. Roth and H. Leidheiser, *J. electrochem. Soc.* 1953, **100**, 553; see also D. J. Fishlock, *Chem. and Ind.* (*Lond.*) 1956, p. 977. Further information on bright and semi-bright nickel-plating baths is furnished by T. E. Such, *Trans. Inst. Metal Finishing* 1954–55, **32**, 26. Bright tin deposition from sulphate baths with wood tar dispersed in octyl-sulphuric acid is discussed by C. A. Discher and F. C. Mathers, *J. electrochem. Soc.* 1955, **102**, 387; A. M. Harper, A. Mohan and S. C. Britton, *Trans. Inst. Metal Finishing* 1956–57, **34,** 273, also International Tin Research Council, Ann. Rep., 1957, p. 19.

Brightness in nickel deposits does not seem to be associated with a small grain-size, but is produced where the free surface of the grains is flat and parallel to the general surface; certain orientations are associated with internal stress and hardness (D. J. Evans, *Trans. Faraday Soc.* 1958, **54**, 1086).

The combination of electropolishing followed by bright plating holds out attractions to manufacturers, and should result in great reduction of expensive hand-polishing. With proper control the results should become more reproducible than is possible with hand processes, where the " touch " must vary from one worker to another.

* i.e. bodies with small values of A and E in the expression $Ae^{-E/RT}$ for the energy needed to cross a potential barrier.

Some bright baths in use today are thought to produce excessive internal stress in the deposit; cases of cracking and peeling are too common. However, research and control should overcome this. There is also a feeling that some bright baths produce deposits less protective than the matt or semi-matt deposits obtained from older baths and finished by hand or mechanically. Even if that is the case, it may still be possible, by modifying the bright baths, to obtain good results.

Periodic Reversal. Another device for improving smoothness and soundness of deposits is today attracting attention. This is a periodic reversal of the current so that the article, for short periods, becomes the anode; the anodic treatment preferentially removes excrescences and any portions of the metal where the structure is unsound. The principle appears reasonable, but some authorities are slightly critical of the results. Those interested should study the paper of the originator, G. W. Jernstedt, *Metal Finish.* (*N.Y.*) 1947, **45**, Feb., p. 68, along with the authoritative appraisals of W. Blum, *Trans. Inst. Metal Finishing* 1953–54, **30**, 182, esp. p. 192, and W. Pfanhauser, *La chimica e l'Industria* 1953, **35**, 639. See also A. Hickling and H. P. Rothbaum, *Trans. Inst. Metal Finishing* 1956–57, **34**, 53.

Deposition from Non-aqueous Solution. Certain metals, which cannot be deposited from aqueous solution, owing to the fact that the potential needed for deposition is so negative that all the current would be expended in hydrogen-evolution, can sometimes be deposited from non-aqueous baths. The deposition of aluminium from a solution of aluminium chloride in ether containing also lithium aluminium hydride is discussed by J. H. Connor and A. Brenner, *J. electrochem. Soc.* 1956, **103**, 657; R. J. Heritage, *Trans. Inst. Metal Finishing* 1954–55, **32**, 61. Recent work at the Bureau of Standards has led to the deposition, from organic baths, of beryllium, also beryllium-boron, magnesium-boron, aluminium-zirconium-boron and aluminium-titanium alloys (W. E. Reid, J. M. Bish and A. Brenner, *J. electrochem. Soc.* 1957, **104**, 21; G. B. Wood and A. Brenner, p. 29; H. Connor, W. E. Reid and G. B. Wood, p. 38).

Deposition from solution without external E.M.F. A simple but usually unreliable method of obtaining a deposit on an article is to immerse it in a solution of a salt of a more noble metal. A clean steel article immersed in a suitable solution of copper salt quickly coats itself with copper, by the reaction,

$$Cu^{++} + Fe = Cu + Fe^{++}$$

Although an attractive appearance can be obtained (if certain addition agents are present), the coat has virtually no power to prevent rusting. In some waters it may even stimulate corrosion; not merely is the attack intensified, being concentrated upon breaks in the coat, but the total amount of rust formed appears to increase, probably because copper is a more efficient cathode than iron oxide (p. 190). A little thought will show that the presence of invisible gaps in a coat formed by simple replacement is almost inevitable, since the cathodic deposition of copper depends on the anodic dissolution of iron, and must therefore become slow before the gaps are filled up. For a very soft metal, however, this process of " simple replacement ", followed by surface treatment designed to make the coating flow over the gaps, may possess practical significance; it has, in fact, been used for the tinning of small steel articles. In other cases, deposits obtained by simple replacement are useful as a basis for true electrodeposits applied later with an external E.M.F.; " immersion deposits " of zinc are used in preparing aluminium articles before plating with nickel and chromium (p. 657).

The main objection to simple replacement is avoided if the current is generated by contact with another metal. For instance, the deposition of tin on steel can be brought about from an acid or an alkaline bath by the use of an aluminium contact piece as anode; good distribution of the deposit depends on satisfactory positioning of the contact piece—with care to avoid an uncoated gap at the point of contact itself. The method may be usefully employed during the normal electro-deposition of highly recessed articles where the recesses would otherwise remain almost uncovered. Aluminium alloy pistons are usually coated with tin (0·0001–0·0002 inch thick) by immersion in alkaline stannate; the tin layer aids " running in " and prevents " scuffing " (D. E. Weimer and J. W. Price, *Trans. Inst. Metal Finishing* 1953–54, **30**, 95).

The deposits obtained by simple replacement are greatly improved by movement of the liquid; Piontelli has obtained adherent and compact deposits of silver, copper, lead, thallium or cadmium on a zinc rod rapidly rotated in the appropriate solution (R. Piontelli, *J. electrochem. Soc.* 1940, **77**, 267).

The scientific mechanism of mutual replacement has been the subject of much study. Important papers are those of O. Erbacher, G. Jensen-Hellmann and A. Mellin, *Z. Metallk.* 1949, **40**, 249; R. Piontelli, *Métaux et Corros.* 1948, **23**, 131. Radioactive tracer methods have been used by M. Simnad, A. Spilners and L. Yang, *Trans. Inst. Metal Finishing* 1954, **31**, 82.

Greater practical importance is attached to methods which deposit a metallic layer without causing any anodic attack on the basis metal or contact material. This is possible if the liquid contains a powerful reducing agent. A familiar example is the silvering of glass by placing it in contact with a solution containing ammoniacal silver nitrate and a reducing agent, such as sugar or Rochelle salt. Either of these is a sufficiently powerful reducing agent to make the production of metallic silver a " possible reaction " (i.e. one accompanied by a drop in free energy). Nevertheless, the activation energy involved in the production of a silver nucleus at a point situated in the body of the liquid is high, so that deposition occurs preferentially at an interface, e.g. a glass surface. Hence the deposition of a silver mirror becomes possible.

Similar mixtures have been developed to provide a metallic deposit on the surface of another metal. Such a method has attractions for articles of intricate shape—e.g. for coating the insides of narrow tubes, where electroplating with an external current presents difficulties. In applying this procedure to the deposition of nickel the reducing agent is sodium hypophosphite; it is possible to obtain a hard layer consisting of a nickel-phosphorus alloy, which is said to possess a corrosion-resistance superior to that of unalloyed nickel, but is somewhat brittle. For plating on noble metals, deposition may have to be initiated by contact with a base metal, but, once started, is self-stimulating. Some metals like lead and tin cannot thus be coated, so that soldered joints will remain uncovered. It is too early to say how far such methods will come to replace conventional plating methods for ordinary purposes. Information is provided by A. Brenner (with C. E. Riddell, D. E. Couch and E. K. Williams), *Bur. Stand. J. Res.* 1946, **37**, 31; 1947, **39**, 385; 1950, **44**, 109; A. McL. Aitkin, *Metal Finish. J.* 1956, **2**, 269; G. Gutzeit and E. T. Mapp, *Corros. Tech.* 1956, **3**, 331; C. H. de Minjer and A. Brenner, *Plating* 1957, **44**, 1297. For " Electroless " deposition of chromium, see H. J. West, *Metal Finish.* (*N.Y.*) 1955, **53**, July, p. 62, and of palladium, R. N. Rhodes, *Trans. Inst. Met. Finish.* 1959, **36**, 82.

The structure of the nickel-phosphorus deposit is found to be unrelated to the basis and independent of thickness; it appears to be lamellar and remarkably free from pores. It has been described as amorphous and " liquid-like " (A. W. Goldenstein, W. Rostoker and F. Schossberger, *J. electrochem. Soc.* 1957, **104**, 104).

An important example of deposition without current is that of titanium. Coatings of this metal would be most welcome owing to its remarkable chemical resistance. Promising results have been obtained by Straumanis. If a piece of titanium is placed in a molten bath consisting of sodium and/or potassium chloride containing the correct quantity of oxygen, it breaks up, giving a black dispersion. This is because the absorption of oxygen swells the lattice and alters the coefficient of expansion, so that the whole disintegrates into small particles of titanium containing oxygen. If an article of iron or steel is introduced into the dispersion, the liquid has sufficient reducing properties to clean the surface and the titanium particles now adhere to it, providing a deposit. However, there are some advantages in starting with a titanium powder already containing the right amount of oxygen, the whole process being then conducted below a helium atmosphere. It is found that the powder made by sieving commercial titanium sponge usually contains 3–5% of oxygen in solid solution, and is suitable for the process. The best coatings are obtained from a titanium-oxygen alloy containing 95% of titanium atoms. A special research by Straumanis has shown that the deposits are formed by the direct impact of the titanium particles; a small amount of exchange occurs between iron and titanium (Fe and $TiCl_3$ giving $FeCl_2$ and Ti), but at the most it accounts for only 4% of the titanium deposited. Moreover, deposits have been produced on aluminium crucibles and thimbles where an exchange reaction is out of the question. Titanium has also been deposited on copper. In general, adhesion is good and coated specimens can be bent without detachment; they are resistant to nitric acid and also to copper sulphate, although if the cleaning has been insufficient, red specks appear (M. E. Straumanis, with A. W. Schlechten and Y. P. Huang, *Metall* 1956, **10**, 901; 1958, **12**, 501; also (with S. T. Shih), *J. electrochem. Soc.* 1956, **103**, 395; 1957, **104**, 17).

Application by Means of Metal-pigmented Paint

Zinc-rich Paints with Organic Vehicles. Although paints pigmented with a mixture of metallic zinc and zinc oxide have long been known, the content of metallic zinc in the older paints was too small to allow the coating to act as an electrically continuous conductor. If particle-to-particle contact is to be obtained, a high pigmentation is required; for zinc—according to Mayne's measurements—there must be about 95% of metal in the coat after the thinner or other volatile constituent has evaporated. With ordinary vehicles such as linseed oil, it is impossible to reach this level of pigmentation without making a mixture too thick for convenient application, but many years ago Dyche-Teague achieved success using as vehicle chlorinated rubber made by a special process (U.S. Patent 1,826,275, Oct. 6, 1931).

Later, Mayne carried out electrical measurements on various zinc-rich paints. He obtained excellent results by using, as vehicle, polystyrene dissolved in xylene or other volatile hydro-carbon; a plasticizer should be added. When a richly pigmented paint is spread on a steel surface and exposed to air, the xylene evaporates, leaving a mass consisting of zinc particles in electrical contact with one another and with the steel basis; for practical purposes, such a coating can be regarded as a continuous conductor, although the specific conductivity is low compared to that of a zinc coating obtained by hot-dipping or electrodeposition.

Mayne's experiments in sea-water lasting 20 months showed effective cathodic protection provided to the steel basis, even at places where it had been laid bare by a scratch-line penetrating the coat. There is, however, some tendency to develop blisters in salt water, especially at places where alkali will be developed on the steel (cathodic) surface below the zinc (anodic) coat; Pass states that the blisters always contain alkali, and that there is little or no blistering in distilled water* (J. E. O. Mayne and U. R. Evans, *Chem. and Ind.* (*Lond.*) 1944, p. 109; J. E. O. Mayne, *J. Soc. chem. Ind.* 1947, **66**, 93; A. Pass, *J. Oil Col. Chem. Ass.* 1952, **35**, 241; 1954, **37**, 483 (with J. Blowes and M. J. F. Meason)).

Paints pigmented with Aluminium. All types of aluminium powder commercially obtainable consist of particles carrying a highly protective skin, which may consist of oxide or, in the case of the flaky type of aluminium, of aluminium stearate. Metal-to-metal contact is not easy to obtain, but Bird has made some experiments on grinding the particles in xylene by the use of a ball-mill, producing a paste which is then mixed with the vehicle. His object was to rupture the oxide-film surrounding each particle; the flattening of a sphere to a disc involves an increased surface area, and the original oxide-film must develop gaps.

* The alkali will tend to loosen the film; there is some evidence that the cathodic evolution of hydrogen at the interface between the zinc and the steel can push up the paint as a blister.

If a paint made in this way is spread out as a coat on steel and becomes hard before the discontinuities in the film on the individual particles have been healed by the formation of fresh oxide, we should obtain metal-to-metal contact of the particles with the steel and with one another. Such a result was not attained when the vehicle consisted of polystyrene; it is possible that, in a polystyrene vehicle, the metallic particles, becoming freshly oxidized on the surface, are pushed away from one another owing to the expansion, so that conductivity is lost. With a vehicle based on epoxy-resin, which gains strength more rapidly, better results were obtained, especially when the pigment consisted of an aluminium-zinc alloy; 1%, 5% and 10% zinc have been tried and 5% appears a suitable composition. Probably the improvement is due to the better electronic conductivity of the film surrounding the grains, which becomes higher when the alumina contains foreign atoms.

Some laboratory experiments have given encouraging results. Scratch-lines traced through a paint-film consisting of alloy powder in epoxy-resin, before the steel specimen was placed in sea-water, caused very little rusting —much less than with a conventional paint. Presumably the steel exposed at the scratch-line received sacrificial protection, but this is not absolutely certain. Moreover, it is too early to say whether protection at gaps in a coating could be obtained in service.

Somewhat better results have been obtained from a mixed pigment consisting of aluminium powder with a small amount of zinc powder. In such a case, the zinc probably provides cathodic protection during the opening periods, until time has elapsed for break-down to develop at points on the oxide-film separating the aluminium from the water. After this, anodic attack on the aluminium probably provides the cathodic protection to the steel and the attack on the coating is less rapid than on paints pigmented only with zinc; there is also less risk of blistering.

Interesting results have been obtained by using polyester resin as the vehicle containing aluminium-zinc alloy pigments. Here there is protection to steel at a scratch-line, and since the coatings are very hard, the risk of scratch-lines appearing in service is reduced. The drying is rapid and several coats can be applied in a day. The main practical objection to epoxy-resin and poly-ester paints is that the addition of the curing agent (or its equivalent) must be made at the last moment; otherwise the mixture may harden in the paint pot or (what may be more disastrous) in the spraying pistol. In the case of epoxy-resins, the curer to be added at the last moment is an amine or an amide; some of the amines have toxic vapours or cause dermatitis, whilst some of the amides have bad keeping properties. In the case of polyester resins, two additions are needed, a cobalt compound (often naphthenate) can be added at an early stage, but an organic peroxide (benzoyl peroxide or methyl-ethyl-ketone peroxide) must be added at the last moment (see pp. 993, 1005).

The results obtained with all these aluminium-pigmented paints have been somewhat variable; different laboratories—or even different workers in the same laboratory—have reported different results, probably owing to small differences in the techniques or materials employed. This type of paint has not been tried in service. A patent application has been made by C. E. Bird (Application No. 1181/55).

Different uses of aluminium in paint have received notice in Chapter XIV; these include the valuable properties of flaky aluminium in outer coats—especially for protecting light alloys—and the incorporation of aluminium in heat-resisting coatings of the silicone and titanate classes.

Iron Paints. There are other possibilities in metal-pigmented paints. If iron powder is mixed with polystyrene lacquer, it is possible to apply the paint to weathered steel which carries (on different parts) mill-scale, rust and perhaps old paint, covering up all these contaminating substances; the result is a surface which, although rough and unsightly, is free from rust and on which some other metal-pigmented paint (carrying, for instance, a mixture of aluminium and zinc in epoxy-resin) can be applied. The stiff mixture of iron and polystyrene (perhaps described more accurately as a plaster than as a paint) can also be used to fill up holes which have been eaten by corrosion through steel sheet. Relatively little work has been done on this method of preparing surface for painting, but it appears to hold out some promise. It possesses this advantage over pickling and grit-blasting, that, instead of removal of material during the cleaning process, there is actual addition of metal and appreciable increase of thickness. It could be applied cheaply under conditions where grit-blasting or pickling would be expensive or difficult. Whether the iron-polystyrene mixture would adhere well to surfaces carrying certain kinds of contamination is uncertain, but in the limited tests carried out from the author's laboratory, no difficulty has been experienced, even when the coating was applied to a surface with areas still carrying old paint of a different type. Here again a patent application has been made by C. E. Bird (Application No. 1182/55).

Another paint containing aluminium has been developed in Russia for the prevention of corrosion in heat-exchange systems. Baked coatings of the phenol-formaldehyde type pigmented with flake aluminium are said to allow a heat-transfer which is 94–97% of that provided by clean bare metal and far greater than those obtained on rusty steel. A convenient summary is provided in *Corros. Tech.* 1956, **3**, 285.

Metal-pigmented Paints with Cement-like Vehicles. The cementiferous paints were produced in the early 40's in the author's laboratory. It is well known that a pasty mixture of zinc oxide with zinc chloride solution, or a paste of magnesium oxide with magnesium chloride solution, possesses cement-like properties; either mixture, after being moulded into any desired shape, sets to a hard mass consisting of basic chloride. The zinc cement was at one time used in dentistry, and the magnesium cement was in favour for the floors of houses until it was discovered that steel pipes laid in it suffered serious corrosion.

If, instead of using zinc oxide, metallic zinc powder is made into a paste with magnesium chloride solution, the attack on the zinc should lead to magnesium hydroxide as cathodic product, which could then interact with the magnesium chloride to give the cement-like basic magnesium chloride;

alternatively, it could interact with the zinc chloride formed by the anodic reaction, to give the cement-like basic zinc chloride. In either case, assuming the metallic zinc to be present in great excess at the outset, we should be left with a mass of metallic zinc particles in metal-to-metal contact with one another and held together in a cement-like matrix. Instead of magnesium chloride, a solution of barium chloride can be used; indeed several chlorides produce analogous effects. A variety of cement-like compounds may be formed in different cases—a matter studied by Mayne and Thornhill.

Masses containing metallic zinc, a suitable salt (chloride, or in some cases chlorate, which is rapidly reduced) and excess of iron powder were explored by the author with a view to obtaining quick-setting metallic compositions which when hard possessed metallic properties (some were magnetic). It was soon discovered that the main practical value of the reactions under study was for the production of a paint which, when dry, would contain metallic zinc particles in metal-to-metal contact. Several of such paints were made, varying in composition and designed for use in different situations. Table XXI shows the composition of three of the best cementiferous paints. The first would be used in situations where the largest possible zinc-content is desired; the second was formulated for an industrial situation where it was desired to keep the zinc-content to a minimum; the last one has been used by the British Admiralty, as stated on p. 585, especially at the wind-and-water zone.

The cementiferous paints are essentially priming coats, which set quickly, giving a layer on which other coats can be applied. The cementiferous layer becomes hard and adherent, but is extremely porous, and the protection is electrochemical rather than mechanical; however, it provides a good basis for coatings of organic paint. Cementiferous paints can be applied directly to wet surfaces—as on a freshly docked ship; the ease with which they are spread out, along with the fact that brushes and pots can be cleaned " under the tap ", tends to increase their popularity. Against this must be set the fact that the cement-matrix is slowly dissolved by sea-water; it is essential to protect it by a relatively water-tight middle coating. Since the paints were worked out in the author's laboratory, it may be best to attempt no personal appraisal, but to quote the objective conclusions reached by Pyefinch—as a result of tests conducted in the sea at Millport for the Marine Corrosion Sub-Committee of the British Iron and Steel Research Association. After a year, a three-coat scheme, consisting of Cementiferous Paint (E412) covered with a coating of commercial anti-corrosive paint and one of commercial anti-fouling composition, showed no rusting, whereas a three-coat scheme consisting of two coatings of commercial anti-corrosive paint and one of commercial anti-fouling composition showed definite rusting, and more blistering than occurred when the lowest coat was cementiferous (K. A. Pyefinch, *J. Iron St. Inst.* 1948, **158**, 229. For laboratory studies, see J. E. O. Mayne and R. S. Thornhill, *ibid.* 1948, **158**, 219. The general uses of the paints are conveniently summarized in a brochure issued by the Zinc Development Association, " Zinc Dust in Protective Coatings " (1956), p. 16).

TABLE XXI

COMPOSITIONS OF THREE CEMENTIFEROUS PAINTS

E378	112 grams zinc dust in 20 c.c. of 44% $MgCl_2.6H_2O$
E390	56 grams zinc dust with 6 grams ferric oxide (Indian Red) and 6 grams china clay in 20 c.c. of 31% $MgCl_2.6H_2O$
E412	336 grams zinc dust with 6 grams of zinc oxide, 6 grams of china clay and 7 grams of barium phosphate in 70 c.c. of 22% $MgCl_2.6H_2O$

An improved class of cementiferous paints, due to Mayne, is made by mixing zinc dust with a solution of potassium phosphate (K_2HPO_4 or K_3PO_4). These paints perform well even when applied on surfaces carrying rust as well as mill-scale. A rusty specimen, covered with such a paint-coat and placed in sea water, developed no fresh rust in six months—even when the steel had been exposed by a scratch-line traced through the paint; a similar specimen covered with cementiferous paint of the oxy-chloride type gave inferior results—possibly due to the appreciable solubility of the oxy-chloride matrix in sea-water (J. E. O. Mayne, *J. Iron St. Inst.* 1950, **164**, 289).

Both types of cementiferous paints have proved useful in repairing and protecting sheet metal on motor-cars which had become perforated by corrosion.

A type of paint consisting of metallic zinc in a zinc-lead-silicate matrix is recommended for protecting cargo tanks by C. G. Munger, *Industr. engng. Chem.* 1957, **49**, July, p. 59A.

Deposition of Alloys

General. The simultaneous deposition of two metals on a cathode to give a solid solution of uniform composition (or perhaps a true intermetallic compound, as is obtained in tin-nickel alloy plating) involves some difficult electrochemical problems. At very low current densities (i.e. near the equilibrium potential) it might be expected that only the more noble metal would be deposited; but as the current density is raised, conditions become less unfavourable for the deposition of the less noble metal, and indeed if the noble metal is the more liable to polarization, the proportions of the two in the deposit may become similar at manageable current densities. In general, however, it is necessary to introduce constituents into the bath which will bring the effective deposition-potentials close together, thus improving the prospects of obtaining simultaneous deposition in acceptable proportions.

Since the ratio of the two metals in the deposit is generally different from that in the liquid, the composition of the latter will alter with time, and it is not always possible to maintain the proper conditions by using soluble anodes consisting of the alloy itself. Sometimes by using separate anodes of the two constituents of sizes adjusted to deliver the required amounts of metal, reasonable constancy is obtained. Sometimes there may be recourse to insoluble anodes with periodical replenishment with salts of the two metals; this involves the disadvantages mentioned on p. 602.

General discussions of the alloy deposition are provided by C. L. Faust, *J. electrochem. Soc.* 1940, **78,** 383; and J. R. I. Hepburn, *J. Electrodep. tech. Soc.* 1943–44, **19,** 1. The theoretical principles involved are ably propounded by R. Piontelli, *Metallurg. ital.* 1948, **40,** 3; 1955, **47,** 204.

Copper-Zinc Alloys. The deposition of brass on steel has attracted practical interest, in connection with rubber coatings; rubber adheres badly to steel, but much better to brass, apparently owing to formation of a sulphide film by interaction between the copper and the sulphur present in the rubber; a copper-rich brass is generally preferred (for discussion, see N. Stuart, *Plastics* 1956, **21,** 308).

Clearly if steel can be plated with brass, the problem of protection with rubber is greatly simplified. This cannot easily be accomplished from a solution of simple salts; a bath containing zinc and copper sulphates would probably yield a deposit of unalloyed copper.* If the plating-bath contains potassium cyanide, the deposition potentials of both metals are shifted in a negative direction; but the potential of copper is shifted furthest, owing to the stability of complex ions such as $[Cu(CN)_2]^-$, so that the deposition of an alloy becomes a practical proposition. It may even be possible to use brass anodes in maintaining the brass composition, provided that the current density is not too high. Ferrocyanides must be excluded. Many of the formulae for brass-plating baths include ammonia, but this tends to escape when the temperature rises, and it is convenient to use a less volatile base such as monoethanolamine. Information about relations between the compositions of bath and deposit are provided by S. G. Clarke, W. N. Bradshaw and E. E. Longhurst, *J. Electrodep. tech. Soc.* 1943–44, **19,** 63; *Metal Ind.* (*Lond.*) 1944, **64,** 393.

The introduction of brass-plating baths free from poisonous cyanide would be welcome, and some Indian researches deserve attention, although they do not seem to have reached the stage of industrial application. These show that the ethanolamine complexes of copper and zinc can give smooth, adherent deposits of widely varying copper-zinc ratios, and that alkaline baths containing copper, zinc (as zincate) and glycerine allow both metals to be deposited at the same potential, so that alloy-plating becomes possible (S. K. Ray, *Bull. cent. electrochem. Res. Inst. Karaikudi* 1951, **1,** No. 1, p. 11).

Other Copper Alloys. The plating of copper-tin alloys from alkaline baths containing sodium cuprocyanide and stannate is discussed by S. Baier and D. J. Macnaughtan, *J. Electrodep. tech. Soc.* 1935–36, **11,** 1; and from pyrophosphate/cyanide baths by W. H. Safranek and C. L. Faust, *Plating* 1954, **41,** 1159. The alloys with 10% tin have a pleasant golden colour and, protected by lacquer, are used as an ornamental finish. When a shortage of nickel was stimulating a search for substitutes, the (British) Committee for the Co-operative Development of Alternative Finishes

* In electro-analysis, the quantitative deposition of copper, free from zinc, from a mixed solution is ensured by the addition of nitric acid, which will be reduced at a potential on the positive side of that needed for zinc deposition.

(C.C.D.A.F.) reported that a 10–12% tin bronze inner coating (0·001 inch thick) followed by a chromium coating (0·0004 inch thick) gave under conditions of atmospheric exposure the best results of the finishes tested (*Metal finish. J.* 1955, **1,** 16). (See also Tin Alloys, p. 622.) Accounts have also been given of the deposition of copper–lead alloys from cyanide-tartrate baths by A. L. Ferguson and N. W. Hovey, *J. electrochem. Soc.* 1951, **98,** 146; copper-nickel-zinc alloys from cyanide baths by C. L. Faust and G. H. Montillon, *ibid.* 1934, **65,** 361; 1935, **67,** 281; 1938, **73,** 417; and copper-nickel alloys from thiosulphate baths by D. C. Gernes and G. H. Montillon, *ibid.* 1942, **81,** 231.

Iron-Zinc Alloys. The fact that steel coated with zinc by hot-dipping becomes more resistant to corrosion in certain atmospheres after an annealing process designed to produce an alloy-coat (p. 596) lends special importance to recent work on the electrodeposition of iron-zinc alloy coatings. Recent work at the B.I.S.R.A. Laboratories (Swansea) show that the alloys thus produced possess varied and attractive properties. The coating containing only 6% zinc possesses high reflectivity, good adhesion to steel, and, being deposited preferentially in hollows, tends to render the surface smoother; it is hard, does not tarnish indoors, but is unsuited for use in the open air. The alloy with 63% zinc is thought to be capable of resisting outdoor corrosion; it can be buffed and chromium-plated; the alloys with 33–55% also have considerable corrosion-resistance (S. Jepson, S. Meecham and F. W. Salt, *Trans. Inst. Metal Finishing* 1954–55, **32,** 160).

Tin Alloys. Coatings containing *tin and nickel* can be deposited from a bath containing chlorides of tin and nickel along with sodium and ammonium fluorides. The deposit consists essentially of an intermetallic compound, and the composition is almost constant (about 65% tin); it is hard and bright with a faint rose-pink lustre; the cause of its brightness has been suggested on p. 610. The alloy is resistant to atmospheric tarnishing and confers considerable protection even in atmospheres polluted with sulphur dioxide or hydrogen sulphide. It is unaffected by alkalis, most neutral salt solutions (apart from slight tarnishing in certain cases), nitric acid and sulphurous acid; sulphuric acid produces slight darkening; other liquids resisted include many which attack unalloyed tin and nickel. Plated articles remain in perfect condition after three months in a domestic kitchen and withstand salt, mustard, vinegar and lemon juice. On the other hand, the alloy is cathodic to iron, and thin coatings which are threaded by pores fail to protect steel; if an under-coating of copper (0·0005 inch thick) is present, the combination becomes protective against atmospheric corrosion, despite the pores in the alloy layer, which seem then to become plugged with corrosion-product, darkening but remaining free from rust stains. Details of application and uses are provided by N. Parkinson, *J. Electrodep. tech. Soc.* 1950–51, **27,** 129; S. C. Britton and R. M. Angles, *ibid.* 1950–51, **27,** 293; 1952–53, **29,** 26; *Tin and Uses,* Oct. 1955, No. 33, p. 10; A. E. Davies, *Trans. Inst. Metal Finishing* 1954, **31,** 401; E. S. Hedges and J. W. Cuthbertson, *Chem. and Ind.* (*Lond.*) 1952, p. 1250; F. A. Lowenheim, W. W. Sellers and F. X. Carlin, *J. electrochem. Soc.* 1958, **105,** 338.

The *tin-zinc* alloy coatings (about 75% tin) are deposited from a hot bath containing tin as stannate and zinc as cyanide, along with free alkali and free cyanide. Anodes of the same composition are employed. Articles thus coated find application in radio and television sets, being generally regarded as alternatives to cadmium-plated articles; they are used in certain parts of aircraft, automobiles and cycles. The coating can be passivated in 2% hot chromic acid and provides a suitable basis for paint.

Another use of the coatings is concerned with mitigating bimetallic corrosion. Exposure tests in marine and industrial atmospheres showed that aluminium structures joined with steel bolts suffered less corrosion at the bimetallic junction if the steel had been coated with the tin-zinc alloy; after 6 months the bolts were still easily unscrewed; the corresponding results with zinc- or cadmium-plating on the bolts were less good.

Surfaces covered with tin-zinc layers are easy to solder and permit the use of non-corrosive flux—an obvious advantage. The coatings cannot challenge zinc coats solely as a protection against weather, but offer advantages in situations of prolonged humidity where they would probably provide service equal to that of cadmium and at lower cost. Particulars of production, and tests on these coatings, are furnished by J. W. Cuthbertson, *J. Electrodep. tech. Soc.* 1950–51, **27,** 13; *Tin and Uses* 1949, p. 5; 1951, p. 3; S. C. Britton and R. M. Angles, *Metallurgia* 1951, **44,** 185; S. C. Britton and R. W. de Vere Stacpoole, *ibid.* 1955, **52,** 64.

The *tin-cadmium* alloys, deposited from a fluoride-fluosilicate bath or alternatively from a stannate-cyanide bath, are discussed by A. E. Davies, *Trans. Inst. Metal Finishing* 1955–56, **33,** 74, 85.

The *tin-copper* alloy (Speculum-plating*) is today being used as a coating on both sides of the Atlantic, and although discoloured in outdoor exposure, retains its silver-like colour well in most indoor atmospheres; it resists the action of many foodstuffs. It is deposited from a bath containing tin as stannate, copper as complex cyanide, with free alkali and sodium cyanide, but separate tin and copper anodes must be used. Particulars are provided by R. M. Angles, F. V. Jones, J. W. Price and J. W. Cuthbertson, *J. Electrodep. tech. Soc.* 1946, **21,** 19.

In depositing speculum metal on intricate objects, it may be difficult to keep the composition everywhere within the correct limits; there is such a thing as " composition throwing power " as well as " thickness throwing power "—a point brought out by J. W. Cuthbertson, *J. Electrodep. tech. Soc.* 1950, **25,** 81.

Tungsten Alloys. The high resistance of tungsten to acids would make a process of depositing metallic tungsten on other metals extremely welcome. Unfortunately this cannot be accomplished from aqueous solution. Processes exist, however, for the deposition of tungsten alloys with either cobalt, iron or nickel, due mainly to the work of Holt and his colleagues. They obtained a cobalt-tungsten alloy with 50% tungsten from a bath containing cobalt sulphate, sodium tungstate and citric acid in the mole

* The bright alloy of composition 40/60 Sn/Cu has long been used for reflectors and is called *Speculum Metal.*

ratio 2 : 2 : 3, adjusted to pH 7 with ammonia; bright deposits were obtained, which showed no tendency to tarnish. The iron alloys contained 53% tungsten, but the nickel alloys only 10 to 35% (L. E. Vaaler and M. L. Holt, *Trans. electrochem. Soc.* 1946, **90,** 43; W. E. Clark and M. L. Holt, *ibid.* 1948, **94,** 244; M. H. Lietzke and M. L. Holt, *ibid.* 1948, **94,** 252).

Explanations for the ability to deposit tungsten if (and only if) cobalt, iron or nickel are co-deposited, along with experimental information about the deposition of tungsten-cobalt alloys, are provided by T. P. Hoar and I. A. Bucklow, *Chem. and Ind.* (*Lond.*) 1955, p. 1061; *Trans. Inst. Metal Finishing* 1954–55, **32,** 186.

Imperfections in Coatings: their Causes and Effects

General. If a metallic coating is continuous and uniform in thickness, its behaviour in a corrosive environment should be similar to that of a solid plate of the same material, although, since the coating may differ physically and chemically from the solid plate, it must not be assumed that the rate of attack will be exactly the same. However, if gaps exist in the coating, or are produced by rough usage or corrosion in service, we have to consider the possibility of the cell Coating Metal | Basic Metal. Sometimes corrosion of one metal may be accelerated and that of the other prevented or diminished by electrochemical action. Before considering the electrochemistry of the subject, it is necessary to know why the gaps exist.

Preliminary Cleaning of Surfaces. Unless the surface to be coated is perfectly clean, either gaps in the coating or poor adhesion must be expected. For the removal of scale and rust, either pickling (p 404.) or mechanical methods such as shot-blasting (p. 599) are available. Where the metal is to be applied from a spraying pistol, the surface must be grit- or sand-blasted in the final stage to provide anchorage for the coating. In other cases, pickling is customary, but the dangers of hydrogen absorption should be remembered; before the plating of springs, for example, anodic pickling—at least in the final stages—is often preferred (p. 408). Certain platings tend to shut in the atomic hydrogen, which would otherwise escape, whilst in the application of hot-dipped coats, hydrogen produced in pickling may be evolved during or just after the dipping process, causing the coat to rise in blisters (this is rather specially liable to occur with Terne-plate coats (p. 651)).

When a coating is to be applied by electrodeposition the surface, after the removal of scale and rust, is usually polished by mechanical or electrical methods, but careful degreasing is needed before the plating operation. This can be carried out in several ways, and frequently various methods are combined. Thus we may have a three-stage procedure, consisting of (*a*) removal of excess grease by dipping in benzine or naphtha, (*b*) vapour-degreasing, usually in trichlorethylene and (*c*) an alkaline cleaner, usually containing sodium silicate, phosphate, carbonate, aluminate or mixtures of them all, for removing the final traces (S. Wernick, *J. Electrodep. tech. Soc.* 1934–35, **10,** 164; P. D. Liddiard, *Chem. and Ind.* (*Lond.*) 1941, pp. 684, 686, 713).

Vapour degreasing has obvious attractions, the article being placed in a tank above boiling trichlorethylene or other solvent which condenses on the surface and drops down, carrying the dissolved grease with it; condensation ceases when the article reaches the boiling-point, so that the system is less suited for light articles than heavy ones, unless provision is made to keep the former cool; in any case, water-cooled coils are advisable, to minimize the escape of vapour. Corrosion by hydrochloric acid, formed from the trichlorethylene by action of moisture, especially in presence of light, was once a serious menace. Today trichlorethylene is usually sold slightly alkaline and, as a further precaution, vapour phase inhibitors are sometimes added. However, certain alloys can cause catalytic decomposition, liberating hydrogen chloride, and it is best to test the liquid from time to time to ensure that it is alkaline; a test is described by R. B. Turnbull, *Chem. and Ind.* (*Lond.*) 1957, p. 446.

Another trouble may be the production of sulphuric acid by reaction of the solvent with sulphur compounds derived from cutting oils or lubricants.

Details of cleaning processes are provided by E. E. Halls, *Metal Treatm.* 1940, **6,** 131; C. B. F. Young, *Iron Age* Oct. 21, 1937, p. 40; W. Burden, *Met. Cleaning Finishing* 1935, **7,** 504, 509; M. Cook and B. J. R. Evans, *J. Electrodep. tech. Soc.* 1933–34, **9,** 125; H. Silman, *ibid.* 1942–43, **18,** 21; 1943–44, **19,** 131, esp. p. 136. Useful information in practical experience is provided in the report of a " Discussion on Metal Degreasing ", *Metal Ind.* (*Lond.*) 1957, **90,** 189, 212, 229, 251.

Although the removal of thick oxide is essential before coating, an invisible oxide is usually present on metal before electroplating or, in certain procedures, actually produced after introduction; for some purposes, platers like to make iron articles passive by anodic action, under conditions where oxygen is freely evolved, before starting cathodic deposition. The question as to what happens to the invisible oxide has aroused some discussion. Doubtless in an acid bath, the oxide on iron will be removed by reductive dissolution before deposition occurs, and in some other cases, it may be reduced to metal. Russian work with radioactive tracer technique indicates that the partial reduction of a film may leave a surface condition better suited to the production of a fine-grained, continuous, adherent and protective coating than many methods of surface preparation (L. I. Kadaner and A. Kh. Masch, *Zhur. fiz. Khim.* 1956, **30,** 1983).

For metals to be coated by hot-dipping, the oxide is removed by fluxing. In the " galvanizing " (zinc-coating) of steel sheet, a layer of molten flux may be kept on the surface of the bath of molten zinc at the place where the articles enter it. In another procedure, flux is applied to the articles as a preliminary operation and dried by heating before the articles enter the zinc bath. In the manufacture of tinplate, the sheets often enter the bath through a flux layer, and leave it through palm oil. Both for zinc and tin, the flux mixtures usually contain zinc chloride and frequently ammonium chloride (which is believed to form the complex $ZnCl_2.NH_3$); sometimes foaming agents and other additions are present.

For coating steel with aluminium, special methods of cleaning the steel before dipping are needed. The small amount of oxidation which would occur during air-exposure on the way from a pickling vessel to the molten bath is enough to prevent success. In one American process now operated on a large scale, the articles to be coated receive alkaline cleaning, acid pickling, washing and drying; they then pass through a molten salt mixture understood to contain cryolite, sodium chloride and aluminium fluoride, thence direct into the aluminium bath—itself covered with a layer of the molten mixture—and then back into the salt-mixture; they are cleansed with water, rinsed with acid, and finally with hot water.

In another method, which has now been in successful operation for about ten years, the articles are heated in an oxidizing furnace atmosphere at about 450–650°C. to give a blue oxide-film, which is then reduced at about 850–900°C. in hydrogen (cracked ammonia is suitable), and thence directly into the molten aluminium; the aluminium often contains 6% silicon—an addition which produces thinner but more uniform coatings with improved bending properties. The somewhat porous iron produced by the reduction of the oxide probably unites readily with the molten metal.

In a British process being developed at the B.I.S.R.A. Laboratory, Swansea, for the continuous coating of strip, the strip, after passing over rollers through a vapour-degreaser, scrubber and pickling vessels, is covered with glycerine, and thence passes direct into the molten metal; the glycerine takes fire on meeting the liquid aluminium, and this maintains non-oxidizing conditions. The bath contains silicon to restrict the growth of an alloy-layer. The process is also being used for wire (D. P. Moses, G. G. Popham and M. L. Hughes, *Metallurgia* 1952, **45,** 70; 1953, **48,** 105; M. L. Hughes and R. Humphreys, *Sheet Metal Ind.* 1953, **30,** 955; see also *Metal finish. J.* 1957, **3,** 273).

Gaps in Coatings. Even when cleaning has been carefully carried out by normal procedure, there may be a tendency for metal to run away from certain spots which after cooling are found to be uncovered. Thus in the tinning of copper containing oxide-inclusions, these generally remain uncoated. The situation is improved if the copper, before tinning, receives cathodic treatment in sodium hydroxide, so as to reduce the oxide-inclusions; however, the best plan is to employ oxide-free copper. The presence of cuprous oxide in a sample of copper is revealed by amalgamation with acid mercuric chloride; like molten tin, the mercury refuses to adhere to the inclusions, which are thus shown up as black spots (L. W. Haase, *Metallwirtschaft* 1935, **14,** 32; E. J. Daniels, *J. Inst. Met.* 1936, **58,** 199; W. D. Jones, *ibid.* 1936, **58,** 193).

The graphite present in cast iron also makes a special preparation of the surface necessary for the production of continuous tin coatings; mechanical cleaning, e.g. by shot-blasting, followed by treatment in special fused salt baths provides the required uniform surface. Some steels also are not satisfactorily wetted by tin after normal cleaning procedures; such " difficult steels " are those which have been cold-worked with a lubricant and close-annealed with the resulting formation of a thin non-reactive skin. Removal of this surface layer may be effected by burning off, pickling in oxidizing acids or by allowing the steel to rust, followed by pickling. Details of the methods used are given by W. E. Hoare, "Hot Tinning" (Tin Research Institute).

Very thin coatings of tin on tinned steel (tinplate) are full of pores; the number decreases as the thickness increases. In electrodeposited coatings also, porosity decreases as thickness increases but the factors involved are not quite the same. Electrodeposition will start at certain points, and spread laterally as well as increasing in thickness. The larger the amount of metal deposited, the smaller is the chance of a given point remaining uncovered; thus the fraction of the surface uncovered, and also the number of pores, are found to decrease as the mean thickness of the coating increases. In many quarters, it has become customary to specify a minimal coating thickness, with a view to avoiding porosity. Now there are other good reasons for prescribing minimal thickness (p. 630), but to demand a thick deposit simply to avoid porosity would seem to be wasteful and likely to discourage useful research. It should be possible, in time, to devise electrochemical conditions which would avoid porosity even in very thin coats—namely conditions which cause deposition to start at many points close together instead of a few points far apart. At one time platers used to employ a "striking bath" in which a very high current density was applied in the opening stages of plating and then reduced to normal level. The function of the initial high current density (which would have given unsound deposits if continued) was probably to provide a larger number of starting points for the deposit. There may be other ways of doing this, better suited to present-day conditions. "Striking" seems to have been largely abandoned, and there might be difficulty in applying it in its original form to modern baths—as pointed out by S. Wernick, *J. Electrodep. tech. Soc.* 1950–51, **27,** 77.

A distinction must be drawn between the number of pores per unit area and the total porosity (the fraction of unit area which is bare). Pores in a coating which is cathodic to the iron basis can be counted in four different ways:

(1) *The Ferricyanide Test* in which the plated specimen is exposed to a solution of sodium chloride (6%) and potassium ferricyanide (0·05%), so that a blue spot appears at each pore in tinplate or nickel coatings on steel; these can easily be counted. The mixture can be applied on paper and the pattern of blue spots can be filed as a permanent record (P. W. C. Strausser, A. Brenner and W. Blum, *Bur. Stand. J. Res.* 1934, **13,** 519).

(2) *The Hot-water Test* in which the specimen is immersed for some hours in water of high purity at 90–95°C. which produces rusty spots at the pores in tinplate (D. J. Macnaughtan, S. G. Clarke and J. C. Prytherch, *J. Iron St. Inst.* 1932, **125,** 160; A. W. Hothersall and R. A. F. Hammond, *Trans. electrochem. Soc.* 1938, **73,** 449).

(3) *The Salt-spray Test* in which the specimen, exposed to a spray of sodium chloride for 50–100 hours, develops rust spots at pores.

(4) *The Hydrogen-Peroxide/Salt Test* in which the specimen is placed in a solution of sodium chloride (3%) containing hydrogen peroxide (1·5% by

volume of the "20-volume" reagent); in ten minutes rust-spots appear at pores in nickel-plated steel and can easily be counted (S. C. Shome and U. R. Evans, *J. Electrodep. tech. Soc.* 1950–51, **27,** 65).

The four tests do not always agree; the ferricyanide solution—especially if the ferricyanide concentration is unduly high—may eat its way through nickel or other coating at points where the coating is thin but not absent, and thus seems to indicate pores which do not, in fact, exist. The hot-water test may give too few pores, because the conductivity is very low, and rust at very small pores, taking time to develop, may be overlooked. The hydrogen peroxide solution produces practically no attack on the nickel at the concentration recommended, and one cause of error is thus avoided. The numbers of pores, as counted by Shome, are definitely less than those obtained with ferricyanide, but of the same order of magnitude; the two sets of counts are reproduced in Table XXII, which also illustrates the fact that the pores become more infrequent with increase of the coating thickness. The nickel was deposited on cold-rolled steel, previously degreased.

TABLE XXII

FREQUENCY OF PORES IN NICKEL-PLATING OF DIFFERENT THICKNESSES DEPOSITED FROM A BATH CONTAINING $NiSO_4.7H_2O$ (220 g./l.), $NiCl_2.6H_2O$ (20 g./l.) AND H_3BO_3 (30 g./l.) (pH 4·0) (S. C. Shome and U. R. Evans, *J. Electrodep. tech. Soc.* 1950–51, **27,** 69)

Average Thickness	No. of Pores per sq. cm. Hydrogen Peroxide Test		Ferricyanide Test	
0·6μ	141, 181, 153, 142, 173	Av. 158	152, 144, 201, 173, 161	Av. 166
1·3μ	41, 71, 50	Av. 64	55, 76, 81	Av. 71
2·5μ	31, 25	Av. 28	41, 53	Av. 47
5·0μ	17, 28	Av. 22	33, 39	Av. 36

Recent work at Swansea attributes the different results obtained by the three methods to failure of liquid to enter small pores, especially when the surface tension is high. The pore-count is increased (1) by adding a wetting agent, (2) by evacuating the vessel *before* introducing liquid, (3) by applying pressure *after* introducing liquid (D. Eurof Davies and D. D. Jones, unpublished work).

The total porosity (the proportion of the area still left bare) has been measured in several ways. One is to remove the coating from the basis, and estimate the combined area of the channels threading it by measuring the rate of leakage of gas through the holes under a standard pressure-difference; alternatively a photographic method can be used. The gas-leakage method is largely due to Thon, and has obvious attractions—provided that no gaps are produced during removal. Ogburn and Benderly who have—on behalf of the American Platers' Society—compared the gas-leakage and photographic methods regard them as complementary, but state that both have serious limitations for measuring the protective values of deposits (N. Thon, L. Yang and S. Yang, *Plating* 1953, **40,** 1011, 1135; N. Thon and D. Dean, *ibid.* 1954, **41,** 503; F. Ogburn and A. Benderly, *ibid.* 1954, **41,** 61; D. T. Ewing and J. M. Tobin, *J. electrochem. Soc.* 1956, **103,** 545).

The importance of the total porosity is that (assuming no further damage to the coating in service), it represents a measure of the maximum corrosion rate attainable under conditions where there is no limitation placed by the cathodic reaction (e.g. by the rate of supply of oxygen). An attempt was made some years ago to determine this maximum corrosion-rate electrically. Shome measured the current set up when a specimen of nickel-plated steel was joined to a coil of copper gauze presenting an area large compared to the iron area laid bare at the pores; the liquid was 3% sodium chloride containing sufficient Rochelle salt to prevent precipitation of iron. The total amount of iron passing into solution in 20 minutes agreed well with that calculated from the coulombs passing between the specimen and the copper gauze—as shown in Table XXIII. Both fell off in marked degree as the thickness increased.

TABLE XXIII

Total Leakage of Iron through Nickel Coatings of different thickness (S. C. Shome and U. R. Evans, *J. Electrodep. tech. Soc.* 1950–51, **27,** 59)

Average Thickness	Micrograms of Iron passing into Solution: Determined electrically	Determined by colorimetric estimation of iron
0·6 μ	373	367
1·25 μ	293	295
2·5 μ	193	197
5·0 μ	140	143
10·0 μ	100	100
15·0 μ	64	65
20·0 μ	38	37
30·0 μ	6·5	6·2

At the time when these measurements were made it was believed that numbers were proportional to the total porosity, but it is probably safer to regard them simply as measures of the "leakiness" of the system. Recent work by Britton and M. Clarke suggests that the resistance of the electrolyte in the pore-channels exercises a degree of control on current-generation greater than was originally supposed. With resistance control, the leakage current through a coating of fixed thickness should certainly increase with an increase of total porosity, but it will also decrease with the greater length of pore-channel if the thickness of coating is increased. There is also some doubt as to whether in some of Shome's experiments complete freedom from oxygen control was obtained (S. C. Britton, Priv. Comm., July 8, 1958; S. C. Britton and M. Clarke, *Trans. Inst. Metal Finishing* 1958–59, **36,** 58).

The work of Britton and Clarke should receive careful study. However, nearly all the conclusions drawn from Shome's work are unaffected by the suggested change in the interpretation of the numerical results. He studied numerous factors which earlier investigators have regarded as causes of porosity and found that some of them seem to play little part. For instance, during nickel-plating the presence or absence of hydrogen bubbles on the cathode surface hardly affected the number of pores. Platers are, of course, right in their desire to avoid hydrogen clinging to the surface; bubbles are bound to interfere with uniform deposition and cause depressions, but they do not seem to be the cause of pores penetrating the whole thickness. On the other hand, if the pores are the sites of inclusions, which do not " take " the nickel, then a short reversal of the current, after deposition has taken place for some minutes, might concentrate attack on the iron exposed at pores, and undermine the inclusion. Shome tried the effect of such a reversal and found it to reduce the number of pores appreciably; but it did not eliminate them altogether.

Shome also found that a plating of cobalt contained only one-tenth the average number of pores per unit area present in nickel of equal thickness deposited from a similar bath. However, attempts to reduce the porosity and thus increase protection by depositing a thin layer of cobalt followed by a thicker deposit of nickel were unsuccessful.

It appears to be important to keep all plating-baths free from suspended impurities, by continuous filtration; removal of objectionable matter by means of activated carbon and similar precautions are to be warmly supported. But Shome's work suggested that at least with the older plating baths the main factor affecting porosity was roughness of the surface on which deposition was to take place. Accidental scratch-lines present on steel before the deposition of nickel or cobalt produced blue lines in the ferricyanide test, showing that the film was here leaky. Certain specimens were engraved with scratch-lines made under a loaded needle and it was found that pores appeared along the scratch-lines, the number per unit length increasing with the load on the needle; annealing after scratching, but before plating, made little difference to the number of pores—which shows that it was surface irregularity, and not internal stress, which caused the pores along the scratch-lines. Roughening by anodic etching in hydrochloric acid also increased porosity. The importance of the surface condition of the surface on which deposition occurs is emphasized in the researches of Thon and of Steer. The pertinent papers are S. C. Shome and U. R. Evans, *J. Electrodep. tech. Soc.* 1950, **26,** 137; 1950–51, **27,** 45, 65; A. T. Steer, *ibid.* 1949–50, **25,** 125 (also paper to 3rd Int. Electrodeposition Conf. 1947). N. Thon, L. Yang and D. Keleman, *Plating* 1950, **37,** 280, 631; 1951, **38,** 1055. The scratching machine used by Shome had been described by R. B. Mears and E. D. Ward, *J. Soc. chem. Ind.* 1934, **53,** 382T, but was subsequently improved by A. U. Huddle; see also U. R. Evans, *Trans. Inst. Metal Finishing* 1956–57, **34,** 139, esp. figs. 1, 2 and 3.

It should be added that cases have been brought forward where porosity occurs not at the depressions but at the prominences of rough surfaces, and also cases where plating on smooth surfaces has fared worse than that on rough surfaces; these latter may be cases of bad adhesion followed by cracking; for types of deposit showing poor adhesion, anchorages afforded by roughness may be useful. The situation is evidently complicated, and the recent introduction of levellers which produce preferential deposition in depressions requires to be taken into consideration.

Damage produced during Service. It has been remarked that performance in service has often little relation to measurements of porosity, so that tests by the ferroxyl or hot-water methods are almost useless in gauging quality; the use of modified salt-spray tests, the application of corrosive slurries and exposure to condensation in atmospheres containing sulphur dioxide have all received some recommendation as alternatives (*Plating* 1957, **44,** 763; S. C. Britton and D. G. Michael, *Trans. Inst. Metal Finishing*, 1952–53, **29,** 40; *Sheet Metal Ind.* 1955, **32,** 576; J. Edwards, *Trans. Inst. Metal Finishing*, 1958, **35,** 55; J. H. Hooper, *ibid.* 1958, **35,** 79).

The poor correlation between porosity and service life is not surprising, since clearly the bare area will increase during exposure. Apart from cracking due to internal stress followed by detachment—which will be considered later—corrosion itself may produce bare spots not present on the freshly plated surface. Attack generally starts locally and spreads out, as indicated on p. 115; it may happen that the corrosion may penetrate through a thin coating, reaching the basis metal before the lateral spreading has caused the different corroding areas to meet one another. If the plating material is uniform, so that the corroded volumes are represented by microscopic expanding hemispheres, the time needed to penetrate and set up a bimetallic couple should be roughly proportional to the square of the thickness for a thin coating; but for a thick coating, provided that the starting points are sufficiently close together to allow the corroding areas to merge quickly, the time needed may for rough purposes be regarded as directly proportional to the thickness.

It is today generally assumed that the life of a coating is proportional to thickness, and this is borne out by observations on carefully prepared material carrying coatings of uniform thickness and exposed in tests carried out under geometrically simple conditions. Particularly impressive are the graphs published by Hudson for the lives of coatings plotted against thickness. The points for aluminium, zinc and cadmium fall very close to three (different) straight lines (J. C. Hudson, *Corros. Tech.* 1947, **4,** 349, esp. fig. 21 on p. 353).

If the coatings are not uniform, the life will often be determined by the thinnest places. On hot-galvanized iron, it has been stated to be impossible to produce uniform thickness below a certain thickness—perhaps 0·7 oz./ft.^{2}; if this statement is accepted, it follows that iron with only 0·5 oz./ft.^{2} will enjoy a life far less than five-sevenths of the life enjoyed by iron carrying 0·7 oz./ft.^{2}, uniformly distributed.

There are clearly practical limits to the rule that life is proportional to coating thickness. Some old galvanized material carrying a thick coat was found to be in splendid condition after 15 years' service in the Scilly Islands, whilst new galvanized iron erected in the same situation was rusty within a year of erection; it is fairly certain that the old iron had not carried a coating with an average thickness fifteen times that of the new, but it is likely that the new had been delivered with certain places very thinly coated.

However, the longevity of old roofing sheet and fencing wire at various places in Australia (33 to 100 years)—which exceeds the life of modern products in the same localities—is attributed by Swaine to the greater coating thickness of the older material. It is likely that the galvanized iron produced in the old days really did carry heavier coats than is supplied today, partly because the iron surface was rougher than is usual today, and partly because there was no squeezing or draining away of excess zinc —two factors emphasized by Bartlett.* It is not easy to attribute the superiority of the old product to the composition of the coating. Gilbert, for instance, states that impurities—other than copper, which is slightly beneficial—have little effect on life, assuming that the thickness of the coating is not affected (U. R. Evans, *J. Inst. Met.* 1928, **40,** 99, esp. p. 118; D. J. Swaine, *Chem. and Ind.* (*Lond.*) 1951, p. 764; 1953, p. 799; W. V. Bartlett, *ibid.* 1957, p. 1404; P. T. Gilbert, *J. appl. Chem.* 1953, **3,** 174).

It is impossible to emphasize too strongly the importance of a uniform coating-thickness if full benefit is to be obtained from the material deposited. This is particularly true of zinc-coated wire. Very uniform thickness is possible on wire electroplated with zinc by the Tainton process (p. 652), and reasonable uniformity for wire coated by passage through molten zinc if it leaves the bath vertically. However, hot-galvanized wire has been met with having a very variable thickness, combining the disadvantages of excessive thickness and excessive thinness; where a coating is too thick, it will break easily on being bent or wound round supports, especially if an alloy coating is present in large amounts; where it is thin, it will be eaten through quickly by corrosion. Material with variable coating-thickness does not give the consumer the benefit of the zinc for which he has paid. The importance of adequate thickness in galvanized cisterns is brought out on p. 639.

Much useful information regarding zinc-coated iron and steel products (including wire, wire-rope fencing, sheets and hardware) will be found in a compilation issued by the American Society for Testing Materials (1956); it contains 21 " specifications ", 3 " recommended practices " and 5 " methods of tests ".

It is generally agreed that the porosity of nickel-plating increases with exposure owing to the progress of corrosion. This is shown by Wesley and Knapp's observations on thin nickel foil, which, when exposed to the atmosphere, developed pitting which led to perforation; presumably something of the same sort occurs in a nickel layer attached to steel. We are not so much concerned with the original pores in the deposit as with what Pierce and Pinner (see below) call *activity centres*; these can be shown up by exposing specimens to anodic attack at a low E.M.F., which does not remove nickel from the main portion, but only from the activity centres; the test is said to give results in agreement with service tests on automobiles driven long distances in winter.

The effect of chromium over nickel is brought out in striking fashion by

* S. C. Britton points out that no additions were then made to the zinc to prevent alloy formation.

Wesley's work. For the first few months the chromium was found to be protective, but then suddenly there was a tremendous acceleration of the penetration, which was attributed to a combination of the large chromium cathode and small nickel anodes (cf. p. 636). These and other American papers deserve study (W. A. Wesley and B. B. Knapp, *Trans. Inst. Metal Finishing* 1954, **31,** 267; W. J. Pierce and W. L. Pinner, *Plating* 1954, **41,** 1034; N. Thon, L. Yang and D. Keleman, *ibid.* 1950, **37,** 631; H. Brown, M. Wemberg and R. J. Clauss, *ibid.* 1958, **45,** 144).

Stresses, Cracking and Detachment. Apart from the gaps produced by corrosion, the basis metal can become exposed by cracks developing in the coatings, followed by flaking and detachment. This can, of course, be produced by rough usage—particularly by sharp bending of wire or sheet.

The detachment need not occur at the interface between basis and coating, if there is a plane of greater weakness within one or the other. Hothersall found that the apparently poor adhesion often displayed by deposits of nickel on abraded brass was due to the weakness of the abraded layers of the brass, perhaps increased by absorption of hydrogen; the line of failure passed through this weakened region of the brass, instead of following the interface itself. If the abraded material had been removed by etching before the brass was plated, the adhesion was improved, and probably became equal to the strength of the basis metal or coating metal—whichever was the weaker (A. W. Hothersall, *J. Electrodep. tech. Soc.* 1931–32, **7,** 115; *Trans. electrochem. Soc.* 1933, **64,** 69).

Detachment can only occur if the strain energy is thereby reduced. Presumably the work needed to detach unit area of coating will be almost independent of thickness, whilst the strain energy per unit area (due to internal stresses) will be roughly proportional to thickness; thus for a specimen subject to constant external stress, a certain thickness must be reached before detachment can occur. This probably explains why the external shearing force needed to detach a chromium deposit from a steel rod, *decreases* as the thickness of the deposit increases, as shown by the measurements of E. Zmihorski, *J. Electrodep. tech. Soc.* 1947–48, **23,** 203, esp. fig. 3, p. 207. The bad effect of internal stress on fatigue strength is discussed on p. 724 of this book.

The peeling of nickel deposits is often due to internal stresses introduced by organic contamination, which may be lignin from the plating tanks, or dextrin and gelatine from the anode bags; such substances are best destroyed by potassium permanganate (E. T. Richards, *Metal Ind.* (*Lond.*) 1953, **83,** 166).

The existence of stress in a deposit is most simply demonstrated by plating on one side (only) of a thin piece of foil suspended vertically; the foil will curl in one direction or the other, according as the stress is tensional or compressional; by reflection from a light source on to a scale, a rough measurement of the stress is possible. Other more refined methods have been devised, the subject being conveniently surveyed by Hammond, who points out that the volume changes occur in the deposit *shortly after deposition*; on a rigid basis, shrinkage would give a contractile or tensile stress, whilst expansion would give a compressional stress; on a thin flexible basis, the whole bends in such a direction as to relieve the stress, whilst if the deposit is non-adherent, it peels off (R. A. F. Hammond, *Trans. Inst. Metal Finishing* 1953–54, **30,** 140).

There are clearly advantages in a *null method.* An apparatus in which metal is deposited on a strip cathode held rigidly at the lower end, the tendency to curl being exactly balanced by an electromagnetic restoring force, the magnitude of which provides a measure of the stress, is described by T. P. Hoar and D. J. Arrowsmith, *Trans. Inst. Metal Finishing* 1956–57, **34,** 354; *Metal finish. J.* 1958, **4,** 166.

The sign of the stress varies. Hothersall stated that generally iron, cobalt, nickel, copper, silver and cadmium deposits show tensional stresses, whereas zinc and lead show compressional stresses. Even in the same metal, however, the results vary with the conditions of deposition. Nickel shows most tensile stress when deposited at low temperatures and low current densities. Saccharin reduces the tensile stress, and may even eliminate it; Marchese reports that large amounts replace it by compressive stresses, but

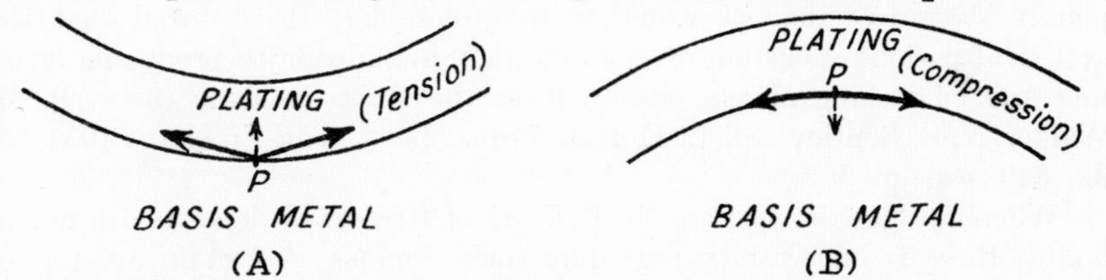

FIG. 99.—Internal stresses in an electrodeposit which (A) tend to detach the deposit if tensional, but (B) press it against the basis if compressional (with tension in the basis metal).

that the results are erratic. The superposition of A.C. on D.C. diminishes the stress in nickel deposited from the Watts bath (p. 657), but does not eliminate it. Certain brightening agents increase the stress, as also does hydrogen peroxide, which is sometimes added to prevent the formation of hydrogen bubbles on the cathode during deposition; it is claimed that certain proprietary wetting agents prevent pitting without causing tensile stress.

Great interest is being taken in the possibility of eliminating—or even reversing—the stresses in nickel deposits by the use of sulphamate baths —originally developed in Italy. Certain addition agents are required for the best results. Diggin states that a sulphamate bath containing naphthalene-1 : 3 : 6-trisulphonic acid yields deposits with stresses which are tensional at 31°C. and compressional at 49°C. (A. W. Hothersall, Symposium on Internal Stresses in Metals and Alloys (Institute of Metals) 1947–48, p. 107; R. Piontelli and G. F. Patuzzi, *Metallurg. ital.* 1942, **34,** 245; V. J. Marchese, *J. electrochem. Soc.* 1952, **99,** 39; M. B. Diggin, *Trans. Inst. Metal Finishing* 1954, **31,** 243; R. C. Barrett, *Plating* 1954, **41,** 1027).

A picture of the effect of stresses on adhesion, due to Kushner, is reproduced in fig. 99. A tensional stress in the deposit tends to bend the surface in the manner suggested in exaggerated form in fig. 99(A); the two tensional forces acting on point P are not quite in the same line, and there will be a small resultant force tending to detach the plating. A compressional stress in the deposit (balanced by tensional stress in the metallic basis) will bend the surface as suggested in fig. 99(B) and tends to keep the deposit pressed against the basis. Nevertheless if there is an area of bad adhesion, compressional stress can relieve itself by raising a blister (J. B. Kushner, *Metal Finishing N.Y.* 1956, **54,** No. 4, pp. 48, 57).

Thus, whilst tensional stresses are undoubtedly dangerous, it should not be assumed that compressional stresses are always beneficial. Probably for fatigue resistance they should be welcomed, since they then serve to annul the tensional half-cycle of the (applied) alternating stress. For ordinary purposes, it has been argued that if nickel is laid down in compressional stress, a crack can neither start nor be propagated; experiments seem to confirm this. On the other hand, the amount of energy needed to detach a deposit is theoretically lower if the detachment relieves internal stress —whether that be tensional or compressional. Moreover, compressional stresses in the coating have usually to be balanced by tensional stresses in the adjacent layers of the basis metal. In many circumstances a completely stress-free deposit would seem preferable. It is stated that the system of periodical current reversal (p. 612), in addition to producing levelling and smoothing effects, also reduces the stress in some cases (P. M. Walker, N. E. Bentley and L. E. Hall, *Trans. Inst. Metal Finishing* 1954–55, **32,** 349, esp. p. 362).

Whereas the existence and bad effects of stresses in deposits are not in doubt, there is uncertainty regarding their origins. A certain amount of stress in electro-deposits would be expected in cases where epitaxial relationships exist between deposit and basis. Hothersall in early papers discussed the conditions under which the deposit continues the crystal-structure of the basis metal, so that grain-boundaries in the deposit are the prolongations of grain-boundaries in the basis; this is usually possible when the two metals crystallize in the same system, and sometimes when they do not (as in tin deposited on copper). He considered that adhesion will only be good if the crystalline structure of the basis is continued; thus included matter (e.g. colloidal hydroxide in nickel) which prevents such continuity might prevent satisfactory adhesion. Electron-diffraction work in Finch's laboratory showed that the orientation of crystals of the basis usually determines that in the deposit, and that where this is not the case (as in nickel deposited on tin) adhesion is poor. Certain apparent disagreements between the Hothersall and Finch groups of research workers were satisfactorily cleared up when it had been agreed that the question of continuity must be judged on a cross-section and not on the front view of the plated metal (A. W. Hothersall, *Trans. Faraday Soc.* 1935, **31,** 1242; *J. Electrodep. tech. Soc.* 1934–35, **10,** 143; 1936–37, **12,** 181; G. I. Finch, C. H. Sim and A. L. Williams, *Trans. Faraday Soc.* 1936, **32,** 852, esp. p. 855; 1937, **33,** 564; G. I. Finch, *J. Electrodep. tech. Soc.* 1936–37, **12,** 184; D. J. Macnaughtan, p. 182; C. H. Desch, p. 184. Cf. A. Goswami, *Trans. Faraday Soc.* 1958, **54,** 821).

The belief has existed, based rather on theoretical argument than experiment, that this epitaxial continuation will only occur if the discrepancies between the inter-atomic distances are such as to cause only a limited degree of misfit between the atoms in the two metals. Apparently the facts do not always bear out the predictions of theory. James found that iron was deposited on copper with its (110) plane on the (100) plane of the basis, the misfit being 12%; nevertheless oriented overgrowth was obtained. On gold, however, where misfit should be small or absent, the orientation of the iron was usually found to be random. Experimental evidence suggests that crystallographic misfit due to epitaxy is not the main cause of the stress present in electrodeposits. The stresses often seem to develop after plating had ceased—in which case epitaxial causes cannot be involved. Indeed, Sadek has found stresses in copper deposited on a copper basis—where the misfit explanation is excluded (J. A. James, *Trans. Faraday Soc.* 1955, **51,** 833; H. Sadek, M. Halfawy and S. G. Abdu, *J. electrochem. Soc.* 1955, **102,** 226).

A more probable explanation ascribes the stresses to changes in the state of hydrogen deposited with the metal. This may exist as interstitial hydrogen atoms (or partly as protons + electrons, as explained on p. 398); such hydrogen has only a small effect on the lattice dimensions, but even if the alteration of lattice dimensions were large, it could hardly produce strains in the deposit, since at the moment of deposition the atoms are free to take up whatever distances would represent minimal strain energy. However, the atomic hydrogen may produce high-pressure molecular hydrogen at cavities existing within the coating or between the coating and basis metal. If this production of molecular hydrogen occurs *after* the deposition has been completed, and if the metallic coating were not firmly glued to the basis, it would alter in dimension; in practice it *is* glued down, so that the alteration in dimensions is restricted, and internal stresses remain in the metal, which can only be relieved if the layer becomes detached; hence energy considerations predict that detachment will occur with less performance of work than would otherwise be the case. If the production of molecular hydrogen in cavities occurs *during* deposition, but the high pressure gas escapes by leakage into the air *after* deposition is complete, then internal stresses will be set up in the opposite direction. This may explain why the sign of the stress varies.

It should also be noted that the shape of the cavities may affect the sign; if these run at right angles to the plane of the metal, an escape of high-pressure hydrogen will tend to bring the opposite walls closer together, bringing about the effect of an internal tensile stress; if they exist parallel to the plane of the metal, the internal tension will be produced at right angles to the metal, leading to compressional stresses in the plane of the metal. When the hydrogen pressure raises a blister or escapes by producing a crack, stress will be relieved.

The situation is, however, complicated and deserves further experimental and theoretical study.

Snaveley's work on hydrogen in a chromium deposit, which is best regarded as present as an unstable hydride, is discussed on p. 415.

Gaps in Cathodic Coatings. Let us consider the electrochemical situation at a gap in a coating which is cathodic to the basis metal. It is sometimes thought that a cathodic coating containing gaps is worse than useless, since (it is argued) intensified corrosion will occur at gaps, owing to the combination of large cathode and small anode. Electrochemical principles, however, suggest that such intensification should occur in certain circumstances but not in others. Common observation has made us familiar with cases where no intensification occurs at gaps in a cathodic coating. For instance, a poorly nickel-plated cycle handle-bar soon shows rust-spots, but the penetration of attack into the steel is slow, and the loss of thickness is almost certainly less than the general loss of thickness which would occur on an unplated steel handle-bar—even if aesthetic considerations sanctioned the use of the latter.

Intensification arises at gaps only if the resistance of the liquid is so small that remote parts of the covering can effectively support the cathodic reaction. It is most likely to occur when the plated surface is fully immersed in a liquid of high conductivity, and when the plating is a metal which under cathodic conditions will be kept free from oxide; this, in effect, means a noble metal like copper, as explained on p. 191.

Some early experiments at Cambridge on steel strips carrying sprayed coatings of copper and nickel provide examples. The coating was broken by sharp bending of the strip, so as to lay bare the steel, and they were exposed to acid fumes. Copper-coated steel exposed to vapours from concentrated hydrochloric acid suffered local corrosion which was more intense than that which the vapour produced on uncoated steel over the whole surface. Voluminous rust produced between steel and copper at the bends pushed away the coating, so that gradually the damage became more extensive (presumably the intensity of attack diminished). A similar detachment of coatings by under-rusting was noted in air containing sulphur dioxide and moisture, both on copper-coated and nickel-coated specimens, but definite intensification was not recorded in these cases; the conductivity of the liquid film was probably lower.

Marked intensification of attack, amounting to perforation of the steel near the edge, was observed on copper-coated steel carrying a broken coating, after 91 days immersion in N/2 sodium chloride solution. However, under fully immersed conditions, the rust was formed outside the coating at gaps, and detachment of the coating by voluminous products formed below it did not occur.

Some other results obtained in the same research are less easy to explain. Steel carrying a nickel coating—not broken by bending—was sprayed daily with N/100 sulphuric for 37 days, with intermediate exposure to the laboratory air; it remained practically unaltered. The same was true of steel carrying a zinc coating, which would presumably be anodic, whilst steel carrying a copper coat suffered a little rusting, although the main change recorded was " superficial discoloration of the copper ". There is no doubt that here the outer parts of the sprayed coatings of nickel or copper were porous, but van Rooyen has suggested that possibly the innermost layers of thick sprayed coatings are non-porous. Another possibility is that at the out-set anodic attack upon the steel at the bottom of the pores proceeds at high current density owing to the proximity of the nickel or copper; the steel then becomes passive. Various other observations on the behaviour of coatings of different metals, varying in thickness, are described in the same paper (U. R. Evans, *J. Inst. Met.* 1928, **40,** 99).

There are several situations where definitely porous coatings of a cathodic metal confer substantial protection to steel. Porous lead coatings are effective in industrial regions (they are less effective in rural or marine regions). Rust generally appears at such pores soon after exposure, but later ceases to develop; it is generally believed that the pores become plugged with lead sulphate (J. C. Hudson and T. A. Banfield, *J. Iron St. Inst.* 1946, **154,** 229P, esp. p. 244P).

If we accept the idea of plugging with lead sulphate, it would seem that both metals are attacked at the outset. Thus, whatever the polarity of the iron | lead couple, the lead is not at that stage sufficiently anodic to protect the iron, nor the iron sufficiently anodic to protect the lead. It is indeed possible that lead may be slightly anodic to iron when moisture containing sulphuric acid is condensing on it. If a continuous covering of lead sulphate were formed, attack on the lead would cease, but if discontinuous crystals are formed (and experience from chemical industry suggests that that can sometimes happen), deposition of lead sulphate upon them will keep the Pb^{++} concentration lower than would be the case in moisture free from SO_4^{--} ions, and the potential will be shifted in the negative direction; it is doubtful whether it could move sufficiently far for the lead to confer cathodic protection to the iron.

A very thin porous coating of tin applied to steel before painting prolongs the period which elapses before rust appears (p. 567), despite the fact that tin is cathodic to steel under ordinary atmospheric conditions. Britton has offered a reasonable explanation for the absence of intensified attack; he points out that the paint should prevent the external part of the tin coating from acting as cathode since movement of ions through a paint vehicle does not easily occur—as Mayne has shown (p. 545). If the only parts of the tin surface available for the cathodic reaction are the walls of the pores threading the exceedingly thin coating, the cathodic area will probably be little larger than the anodic area, and there is no reason to expect intensification (S. C. Britton, *J. Oil Col. Chem. Ass.* 1950, **33,** 125).

There is another possible reason. Corrosion of steel, which starts at sensitive spots, may be delayed or prevented if these are covered up; unless the pores in a tin coating tend to coincide with the sensitive spots on the steel, it is reasonable to expect that a discontinuous tin coating will depress the probability of the inception of rusting.

Gaps in Anodic Coatings. It is often stated that a zinc coating protects steel "sacrificially"; where the steel is exposed at a gap, zinc (it is argued) is anodically attacked and steel cathodically protected. If this is the true basis of protection by zinc, we are protecting a relatively cheap material, steel, by the more expensive one, zinc. That is not necessarily unsound economy; it is often financially prudent to protect a buried steel pipe by sacrificing a magnesium anode, because steel as a buried pipe-line is something far more valuable than steel as an ingot, owing to the expense of pipe-manufacture, transport and burial; the cost and inconvenience of replacing a section of a pipe-line which has developed perforation at some inaccessible spot may be very high, and, if the magnesium anode serves to prevent this, the sacrifice is not made in vain.

Nevertheless it is probably untrue to say that the main mode of protection of steel by zinc coatings is sacrificial, except in the opening stages and at the edges of galvanized steel sheets, or at places where the coating has been damaged in manufacture or transport.

The main reason for covering steel with coatings of zinc or aluminium is that a protective film is easily formed on these metals in environments which would produce non-protective rust on steel. Zinc and aluminium are metals of fixed valencies, and the solid corrosion product is formed in physical continuity with the metal; iron has two valencies, and often the relatively soluble ferrous compounds first formed are converted to the less soluble ferric compounds (e.g. rust) at an appreciable distance, so that the rust, even if adherent in part, will be discontinuous and non-protective. In alkaline environments (near cement and mortar) both zinc and aluminium are attacked, and cannot be used to protect steel; in acid environments (e.g. railway tunnels), zinc may be attacked more quickly than steel and zinc coatings have short lives. In industrial or urban districts where condensed moisture is acid, galvanized coatings require painting after erection (p. 586), but in such situations the zinc has probably served a useful purpose in protecting steel sheet during transport and erection.

When galvanized iron is used for cisterns, tanks or pipes, another principle is involved. If the zinc coating chances to be discontinuous, as near sheared edges or at drilled holes, the protection is sacrificial at first; zinc suffers anodic attack, whilst steel receives cathodic protection. If the water contains calcium bicarbonate, a chalky deposit is soon thrown down, since the rise of pH produced by the cathodic reaction will produce CO_3^{--} ions,

$$HCO_3^- + OH^- = CO_3^{--} + H_2O$$

and the solubility product of $CaCO_3$ is soon exceeded. The chalky layer cannot transmit electrons and the cathodic reaction will slow down—causing the anodic attack on the zinc to slow down also. Probably the slow anodic attack on the zinc which continues will be sufficient to prevent anodic attack on those points of iron still exposed at pores in the chalky layer; even if an iron ion should enter the liquid in one of these pores, it is likely to interact with the chalk and with oxygen to yield rust, and will not escape into the body of the liquid; in this way, we obtain the protective form of rust discussed on p. 161; the layer of chalk may gradually become chalky rust, or in certain situations it may be converted to zinc oxide or hydroxide. In all cases, the layer remains clinging to the metal, affording a reasonable degree of protection.

This only occurs in waters containing calcium bicarbonate. Galvanized iron is not well suited for very soft waters. For waters of medium hardness, a thicker coating of zinc is needed than for waters of great hardness. Fig. 100 suggests the reason. If, at the outset, there is a small area of bare steel (fig. 100(A)) it is necessary to cover this with calcium carbonate, and if the water is very hard, the anodic destruction of a small amount of zinc will suffice to cover this original bare area and any small marginal extension of the bare area owing to the attack on the zinc (fig. 100(B)). However, if the water is less hard (or if it contains much Na^+ as well as Ca^{++}), the amount of zinc which has to be destroyed before the chalky film is effective over the original area may involve the exposure of a large fresh area of steel, and the rate of fresh exposure may outrun the rate of covering with chalk (fig. 100(C)). The only way to avoid this is to increase the thickness of the zinc coat (fig. 100(D)), so that the same destruction of zinc involves the laying bare of a smaller area of steel (fig. 100(E)). The matter has been studied and discussed by S. C. Britton, *J. Soc. chem. Ind.* 1936, **55,** 19T.

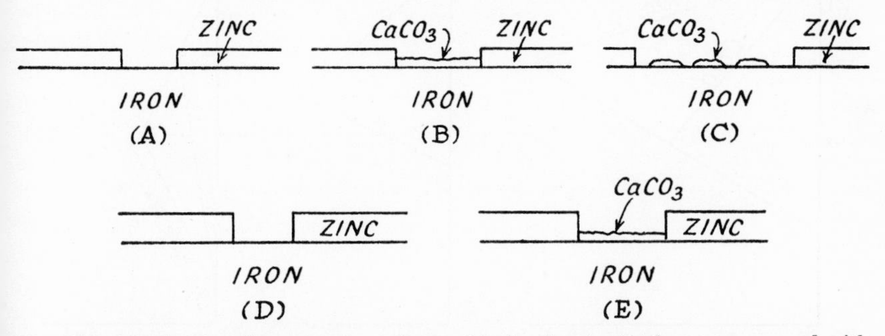

FIG. 100.—Steel exposed at a gap in a zinc coat in hard water (A) becomes covered with $CaCO_3$ without serious extension of the gap (B). In softer water a great destruction of zinc (i.e. a great extension of the gap) would fail to cover the steel (C), and a greater thickness of zinc (D) is needed to produce the desired covering (E) (diagrammatic representation of S. C. Britton's results).

In most waters, the polarity of the cell zinc | iron is reversed at high temperatures, and the zinc becomes the cathode. Thus protection of exposed steel from rusting is not achieved in hot-water systems, and care must be taken not to leave sheared edges or other bare spots exposed to the water. The probable cause of change of polarity is suggested in fig. 101(A and B). The potential of zinc in water is that at which the anodic reaction (the passage of zinc ions into the liquid at gaps in the surface oxide-film) occurs at exactly the same rate as the cathodic reaction, which involves the passage of electrons through the film; it is given by the intersection point of the two polarization curves. At ordinary temperatures the potential of zinc falls *below* the potential of iron, so that zinc is anodic to iron—which can receive cathodic protection, provided that the area exposed is small compared to that of the zinc. However, at high temperatures zinc oxide becomes a better conductor, and the cathodic curve becomes less steep, so that the potential, indicated by the intersection point, now lies *above* that of iron; the zinc is cathodic and incapable of affording protection. This potential reversal has been studied experimentally by P. T. Gilbert, *J. electrochem. Soc.* 1952, **99**, 16.

The anodic attack on aluminium is highly localized. Thus although the anodic polarization curve starts from a lower level than that of zinc, it is much steeper and the intersection point may lie higher. Thus zinc has been used to confer cathodic protection to aluminium (p. 189). The effective potential of aluminium depends on the liquid, being, as usual, more active (more negative) in chloride solutions, which stimulate the anodic reaction (p. 238). Thus in salt water, aluminium is effectively anodic to steel, and will confer cathodic protection on it, provided the area of steel exposed is not too great, whereas in most supply waters, aluminium is either cathodic to steel or insufficiently anodic to provide a protective current (fig. 101(C)).

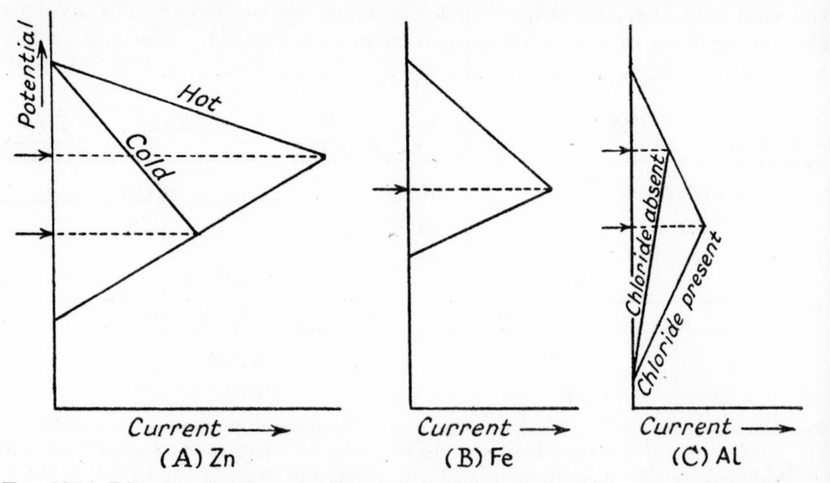

FIG. 101.—Diagram showing why Zinc is usually anodic to Iron at low temperatures, but cathodic at high temperatures where zinc oxide becomes a better electronic conductor. Zinc gives protection at gaps against cold water but not generally against hot water. Aluminium gives protection against salt water but not generally against fresh water.

This has been shown in early experiments on steel strips carrying sprayed aluminium coatings. They were bent to break the coat, exposing the steel, and immersed in water. In Cambridge supply water (containing calcium bicarbonate, but practically no chloride) rusting of the steel along the exposed area commenced in 3 hours, whereas in N/2 sodium chloride solution, the steel showed no rust at all even after 31 days. Zinc-sprayed specimens were protected in both waters, but the attack on the zinc in the chloride solution was far more rapid than that on the aluminium; when the zinc was nearly exhausted, the specimens started to rust, and this occurred in 20 to 27 days, according to the thickness of the coat. It was concluded that, where there is risk of a gap in the coating, zinc *must* be used for the fresh water; but for the salt solution, although either metal gave protection at the out-set, the aluminium coating was preferable, as the protection would last much longer.

Where there was no gap in the coat, the aluminium-sprayed specimen suffered no attack in supply water. This may have been due to the pores becoming blocked with corrosion-product, or to the fact that the pores did not reach the steel. The immunity of the steel exposed to chloride solution at the bends must have been due to cathodic protection; the gaps produced by bending the steel were far too broad for any question of blockage with corrosion-product; similar gaps allowed rusting by the supply water to develop readily (U. R. Evans, *J. Inst. Met.* 1928, **40,** 99, esp. p. 113).

Sometimes the cathodic protection produced by sprayed aluminium coatings takes time to develop. When any metal carrying an oxide-film is brought into contact with a liquid, time is needed for microscopic break-down points to grow into definite corroded area. With aluminium, break-down only occurs under conditions where replenishment of oxygen is slow (p. 211); specimens of aluminium sheet partly immersed in N/10 potassium chloride show potentials rising with time, indicating film-repair, whilst zinc, iron or steel specimens under the same circumstances show falling potentials indicating film break-down (U. R. Evans, *J. chem. Soc.* 1929, p. 111, esp. pp. 115, 121).

In 1931 Britton exposed plates of steel sprayed with aluminium of different grades and thicknesses at Grantchester Meadows, south-west of Cambridge, where the atmosphere was free from smoke and salt, and condensed moisture was likely to possess little conductivity; this represented conditions unfavourable for cathodic protection. After three months, scratch-lines were traced through the coating, so as to expose the steel, which developed typical rust after two days. Brown rust remained in the scratches for three months; thereafter, it gradually darkened and finally disappeared, its place being taken by a dark stain (probably indicating reduction to magnetite). Seven years later, when the specimens were taken in, the steel exposed at the scratch-lines was practically uncorroded and quite free from rust. This was a clear case of cathodic protection. It is believed that the break-down of the alumina film occurred after ferrous hydroxide had been formed as primary corrosion-product of the exposed iron; ferrous hydroxide would absorb oxygen, being converted to rust, and the channels between the flaky particles would become anaerobic; the alumina film would not be kept in repair and gradually the aluminium would become anodic and generate a current sufficiently strong to protect the steel cathode, and also reduce the rust to magnetite. The part played by iron corrosion-products in causing aluminium to become an effective anode has become apparent since Bird's work (C. E. Bird and U. R. Evans, *Corros. Tech.* 1956, **3,** 279). Britton's exposure tests are described by S. C. Britton and U. R. Evans, *J. Soc. chem. Ind.* 1932, **51,** 211T, esp. p. 217T; 1936, **55,** 337T, esp. p. 340T; 1939, **58,** 90; U. R. Evans, *Chem. and Ind.* (*Lond.*) 1956, p. 195).

The question has been raised as to whether a sprayed aluminium coating is in sufficiently good electronic contact with the steel basis to act as an anode and confer cathodic protection at gaps. The contention is that, since every aluminium globule arriving from the spraying pistol is surrounded by an oxide-film, there will be no contact. Such an argument is not conclusive. The area of the film which surrounds a spherical particle will not suffice to cover the same particle after it has been flattened into a disc by impact. Thus electrical contact between sprayed aluminium and steel is likely to be established, and also between one aluminium flake and another. In the Author's experience the contact is sufficiently good, and this is confirmed by others. Walkiden estimates the resistance between sprayed aluminium and steel basis as only about 0·1 ohm for a 1 cm.^{2} area, whilst Tolley states that the specific conductivity of sprayed aluminium is about one-fifth of that of the cast metal. In view of these numbers from authoritative sources, and confident statements from Mayne and van Rooyen who have carefully studied the matter, the author has no misgivings regarding the electrical contact between coating and basis. In case, however, readers wish to follow the discussion, the following references are provided: W. F. Higgins, *Chem. and Ind.* (*Lond.*) 1956, p. 77, 382; U. R. Evans, p. 195; G. Tolley, p. 464; G. W. Walkiden, p. 528; J. E. O. Mayne and D. van Rooyen, 1957, p. 718; G. Tolley, p. 1045.

A matter to be considered more seriously is the question as to whether cathodic protection plays a part in the normal functioning of the sprayed coatings. Some authorities seem to attribute protection to the stifling of gaps in the coating with corrosion-product. Where the coating is not broken by bending, or interrupted by tracing of a scratch-line, the plugging of pores probably plays a large part in the protection—which, as already stated, appears to occur in some cases where the coating is cathodic to the basis metal; (the corrosion-products themselves often arise from electrochemical action between the two metals). However, in some of the cases of cathodic protection described above, the breadth of the gaps in the coating must have been many hundred times the thickness of the coat and the mechanical plugging of such gaps is out of the question.

In view of conflicting statements—some of them thrown out without any real attempt to establish the truth by experiment—the author wishes to restate certain facts based on personal observation. (1) There exists sufficient contact between sprayed aluminium and steel to provide the current needed for cathodic protection, provided that aluminium is definitely anodic to steel in the liquid wetting the surface. (2) Cases where steel has remained unrusted at broad gaps in the coat can only be explained by cathodic protection; this does not, of course, exclude the possibility that mechanical blocking plays a useful part in cases where the gaps are narrow. (3) There are many cases where no cathodic protection is conferred, either (*a*) because aluminium is cathodic or insufficiently anodic to the steel or (*b*) because the conductivity of the liquid path is too small, or (*c*) because the gaps are too wide. Conditions of continuous wetting with an appropriate liquid of high conductivity are more favourable to cathodic protection than atmospheric conditions where the surface periodically dries up.

Protection at Gaps in Cadmium Coatings. Cadmium coatings are much used on steel, not only for protecting it from attack but also for preventing serious bimetallic action where steel will be in contact with an aluminium alloy. There has been some hesitation to use it on high-tensile alloy steels owing to dangers of hydrogen embrittlement; this matter is discussed on pp. 414, 415. In many other situations, however, cadmium-plating has proved satisfactory.

The steel exposed at a gap in a cadmium coating is usually found to escape rusting. In a salt water, this has sometimes been attributed to plugging with basic chloride or other product. However, the protection at gaps has been observed in demineralized (salt-free) water, where explanations based on basic chlorides cannot apply.

The normal electrode potential of cadmium is less negative than that of iron, and if a cadmium-iron couple were immersed in a solution containing both ions in equivalent amounts, the cadmium would be the cathode, so that electrochemical protection of the iron would not be expected. If, however, the couple is placed in water containing neither iron ions nor cadmium ions, the case is different, and activation energy becomes important. The two metals, at least near the junction, will establish the same potential towards the water, and at this common potential it is likely that the cadmium will pass as ions into the liquid rapidly, since the activation energy is relatively low, whereas iron, with its high activation energy, will pass in more slowly than would be the case if it were not joined to cadmium; thus the junction will afford a considerable degree of cathodic protection to the iron. The matter is discussed further on p. 651.

Uses of Different Metals as Coatings

Noble Metals. The high cost of metals like gold and platinum limits their use to thin layers and special cases. The risk of intensified corrosion at discontinuities is a real one; many years ago steel nibs plated with gold were offered for sale; the gold soon became worn away at the tips, providing the fatal combination of large cathode and small anode, and the venture was a failure. Increase of hardness would reduce risks for some purposes, but the improvement is not always very great; recent measurements in Russia show that the number of revolutions of a brass roller needed to wear away a deposit was only 1·6 times as great for gold containing nickel as for unalloyed gold (N. P. Fedot'ev, N. M. Ostroumova and P. M. Vyacheslavov; abstract, *Met. Abs.* 1954–55, **22,** 455).

Rhodium, a harder material, has proved its utility for reflectors and electrical contacts. Although somewhat less resistant than gold (it is appreciably attacked in air containing high concentrations of sulphur dioxide), it withstands ordinary atmospheric pollution. It is usually deposited from baths containing sulphates, phosphates or both—sulphate being better for thick deposits, for which a sulphamate bath also merits consideration (E. A. Parker, *Plating* 1955, **42,** 882. A. H. Stuart, *Electroplating* 1948, **1,** 88. A. M. Weisberg, *Mater. and Meth.* 1953, **37,** No. 3, p. 85. F. H. Reid, *Trans. Inst. Metal Finishing* 1956, **33,** 105. J. Fischer and F. Leonhard, *Metall* 1956, **10,** 608. Also *Platinum Metals Review* 1957, **1,** No. 1, p. 14).

Rhodium-plated reflectors remain unchanged when stored under conditions where silver tarnishes. Rhodium is valuable for electric contacts where very low voltages are involved and where any interference with the passage of electrons would be disastrous; in radio-frequency or audio-frequency work, alteration of resistance may introduce spurious signals, whilst oxide or tarnish films can cause partial rectification (L. B. Hunt, *Metal Ind.* (*Lond.*) 1947, **71,** 339, esp. p. 341).

Platinum is usually deposited from a sodium hexahydroxyplatinate bath, but a relatively thick, compact layer produced by mechanical cladding may be safer; platinum-clad anodes have been used in the cathodic protection of ships (p. 300). The electrodeposition processes are described by E. C. Davies and A. R. Powell, *J. Electrodep. tech. Soc.* 1937, **13,** Paper 26; R. H. Atkinson, Paper 25; also *Trans. Inst. Met. Finishing* 1958, **36,** 7.

Other metals of the platinum group are used for special purposes. Palladium can be deposited from a bath containing the complex nitrite $Na_2[Pd(NO_2)_4]$ which allows soluble anodes to be used; thick deposits can be obtained from a chloride bath, but this needs insoluble anodes and a divided cell (R. H. Atkinson and A. R. Raper, *J. Electrodep. tech. Soc.* 1932–33, **8,** Part 10; 1934, **9,** 77). Ruthenium can be deposited from a solution of $Ru(NO)Cl_3$ containing hydrochloric acid; conditions are described in *Platinum Metals Review* 1957, **1,** No. 1.

The use of silver for the plating of table-ware is well known. Like gold, it is deposited from a cyanide bath, where it is present as $K[Ag(CN)_2]$ (or $Na[Ag(CN)_2]$); this is usually prepared by dissolving solid silver cyanide in potassium (or sodium) cyanide, which must be present in excess, to guard against formation of anodic films; the amount needed is highest at high current densities. The baths always contain carbonate (produced by action of carbon dioxide on the cyanide), and this increases conductivity and throwing power, but the power consumption is not necessarily diminished, as it introduces an increased risk of anode films. One formula contains potassium nitrate. The composition of silvering baths is discussed by E. J. Dobbs, *J. Electrodep. tech. Soc.* 1935–36, **11,** 104; G. B. Hogaboom, *ibid,* 1927–28, **3,** 73; S. Glasstone and E. B. Saniger, *Trans. Faraday Soc.* 1929. **25,** 590; N. E. Promisel and D. Wood, *Trans. electrochem. Soc.* 1941, **80,** 459.

Silver-plating is usually completed in a bath containing carbon disulphide, thiourea, sodium thiosulphate or other brightener. The advantage of avoiding poisonous cyanides has stimulated research on cyanide-free baths, based, for instance, on thiourea or guanidine (H. Gochel, *Z. Elektrochem.* 1934, **40,** 302; L. I. Gilbertson and F. C. Mathers, *Trans. electrochem. Soc.* 1941, **79,** 439; D. K. Alpern and S. Toporek, *ibid.* 1938, **74,** 321; C. W. Fleetwood and L. F. Yntema, *Industr. engng. Chem.* 1935, **27,** 340).

Copper can also be applied electrically, and where the basis is steel, it may be started from a cyanide bath (p. 605) and continued from a sulphate bath; gelatine or alum is sometimes added to improve the deposit. Deposition from a pyrophosphate bath gives a dense deposit, which Russian investigators attribute to a layer of $P_2O_7^{----}$ ions anchored on each crystal nucleus and blocking its growth, so that the crystal-structure remains fine (A. E. Ukshe and A. I. Levin, *J. appl. Chem. U.S.S.R.* 1955, **28,** 388; English translation, p. 363).

In general, the danger of intensified attack at pores in thin electro-deposited coatings is a real one (p. 636). If steel is to be protected by copper, it may be safer to apply a thick compact layer by cladding; here the nature of the bond is important—a matter discussed by H. Bröking, *Metalloberfläche*, 1947, **1**, 101.

Tin Coatings. Tin is applied to steel, and also to copper and its alloys, by dipping in molten tin, by electrodeposition or by spraying. The largest use of tin coatings is represented by *tinplate* (sheet steel coated with tin by hot-dipping or electrodeposition), but there are other important applications.

For hot-dipped coatings, the metal is cleaned, coated with flux and dipped into molten tin. Coatings on copper or steel have a layer of alloy of tin with the base metal surmounted by a layer of tin. The coatings are used for industrial and domestic food and beverage handling equipment, for dairy plant and for some electrical equipment, including wire.

In a variant of the process, known as " wiping ", the tin is applied in molten, stick, or powder form to the prepared and heated surface over which it is wiped by suitable means. This process is useful for local application or for the coating of one side of a sheet or vessel; it has the disadvantage that much of the unalloyed tin layer of the coating may be removed, leaving on copper a thin coating which may have a high proportion of alloy.

Practical guidance on the control of the hot-dipping process is given in the booklet " Hot Tinning " by W. E. Hoare, published by the Tin Research Institute. " Roller-coating " processes for the hot tinning of steel strip are briefly discussed by S. S. Carlisle, *Sheet Metal Ind.* 1937, **34**, 405, esp. p. 415.

For the production of electrodeposited tin coatings, several electrolytes are available. The special conditions of high-speed plating on rapidly moving strip for the production of tinplate have called for the development of special plating baths which are covered by patents. Examples are the " Ferrostan Bath ", based on stannous sulphate and phenolsulphonic acid, the " Halogen Bath ", based on stannous chloride and with alkaline fluoride. The two acid baths require the action of organic " addition agents " (see J. W. Cuthbertson, *J. Electrodep. tech. Soc.* 1950–51, **27**, 13).

For still plating three types of plating solution are mainly used: (1) an *acid sulphate* bath containing stannous sulphate, free sulphuric acid and crude cresol-sulphonic acid with gelatin and β-naphthol as addition agents, (2) an *alkaline* bath containing tin as a stannate and (3) an *acid fluoborate* bath containing organic addition agents. In plating from the alkaline stannate bath, twice as many ampere-hours are required to deposit the same amount of tin as are required in a stannous salt bath. The alkaline bath possesses, however, advantages in requiring no addition agent and in demanding less thorough prior degreasing of the metal to be coated. Potassium stannate and potassium hydroxide have some advantage over the sodium compounds since the greater solubility of the potassium stannate permits deposition at higher current densities. The lower cost of the sodium compounds has, however, kept them in use where the greater rate of plating is not needed. Stannite must be excluded, as it causes spongy deposits and the solution of the anodes is controlled to avoid its formation. For tin anodes, the required condition is obtained by either subjecting them initially for about one minute to a current density considerably above that used in normal working or by lowering them slowly into the bath with the current flowing through them. Too high a current density may result in complete passivity and special alloy anodes are available to extend the upper operational limit; these are commonly used with the potassium stannate bath to take advantage of its higher rate of plating.

Electrodeposited coatings are used in electrical equipment and for several of the purposes for which hot-dipped coatings are also used. They have the advantage over hot-dipped coatings in permitting a considerably increased range of thickness. In electrical applications tin coatings have the advantage of easy solderability, thus avoiding the use of corrosive fluxes; they show good resistance to the vapours from wood, insulating materials and plastics which can be destructive to zinc and cadmium (see p. 493).

Ranges of coating thickness recommended for various purposes are prescribed in B.S. 1872. The various electrodeposition processes are discussed by S. Baier and R. M. Angles, *J. Electrodep. tech. Soc.* 1938–39, **15,** 4; A. W. Hothersall and W. N. Bradshaw, *ibid.* 1936–37, **12,** 113; 1938–39, **15,** 31; A. W. Hothersall, S. G. Clarke and D. J. Macnaughtan, *ibid.* 1933–34, **9,** 101; N. Parkinson, *ibid.* 1950, **26,** 169; R. Kerr, R. M. Angles and K. W. Caulfield, *J. Soc. chem. Ind.* 1947, **66,** 5, 7; M. M. Sternfels and F. A. Lowenheim, *Trans. electrochem. Soc.* 1942, **82,** 77; C. E. Glock, *ibid.* 1943, **84,** 249; F. A. Lowenheim, *Tin and Its Uses,* April 1955; G. Hänsel, *Z. Elektrochem.* 1935, **41,** 314; P. Fugassi, *J. electrochem. Soc.* 1953, **100,** 121C. A pyrophosphate bath is recommended by J. Vaid and T. L. Rama Char, *J. electrochem. Soc.* 1957, **104,** 282. A booklet, "Instructions for Electrodepositing Tin" issued by the Tin Research Institute provides general guidance to the practical aspects of the methods.

The methods of production of *tinplate* (mild steel carrying a very thin coating of tin) have undergone may changes in the present century. Most plants today use a continuous process for the production of the steel base involving a steady reduction of thickness so that the steel, starting as a slab, finally becomes a thin band or coil which can be carried forward to the coating operation.

At one time the tin layer was always applied by hot-dipping and this process is still used for part of the production; the sheets (previously cut to appropriate sizes) are passed through a flux into molten tin from which they emerge through palm oil, passing between rollers which control the thickness of the coating.

The shortage of tin which arose during the Second World War stimulated interest in the use of electrodeposition processes; these had previously attracted attention mainly as a convenient means of coating the continuous steel strip which was already available. Electrodeposition could produce coatings much thinner than those obtainable by hot-dipping and its adoption promised to enable supplies of tinplate to be maintained during the period of tin shortage at some sacrifice of existing standards of quality. The problems of large-scale production were so rapidly and successfully overcome by American industry that electrolytic tinplate obtained and has retained a large and growing share of the tinplate output.

The process now used is to pass the steel strip continuously through cleaning media and then through electroplating baths in which tin is plated at high current density; the baths used have been mentioned earlier (p. 645). The plated strip, carrying a matt white deposit, is carried on through rinsing tanks to a furnace where heating by radiation, conduction or high-frequency methods momentarily raises the temperature of the tin coating just above its melting-point; this is known as *flow-brightening*. The strip is then passed through a chromic acid or a chromate bath, with or without applied current (the passivation process); it finally receives an oil film either by passage through a mist of oil droplets or by spraying an oil-water emulsion. The flow-brightening process produces the lustrous appearance which the hot-dipping process had caused to be associated with tinplate and also facilitates rapid soldering, a matter of some importance in the production of cans. An alloy layer is produced during the process; details are discussed by C. J. Thwaites, *J. Iron St. Inst.* 1956, **183,** 244. (The application of flow-brightening to small electro-tinned parts by immersing them in a suitable hot oil is described by C. J. Thwaites and W. E. Hoare, *Tin and Its Uses*, Summer 1957, No. 39.)

The passivation process improves tarnish-resistance and retention of easy solderability; it also prevents discoloration of the tinplate during the stoving of lacquers. The oiling process assists handling of the sheet in fabrication processes and also gives some added protection against atmospheric influences.

The thickness of coating produced by the electrolytic process has been limited to the range 0·000015–0·00006 inch. whereas for hot-dipped tinplate the thickness of coatings ranges from 0·00006 inch to 0·0008 inch. For canning foodstuffs, the range of thickness is commonly 0·00003–0·0001 inch; the more heavily coated grades of hot-dipped tinplate are used for permanent hardware such as dairy and kitchen equipment, gas meters, returnable containers and petrol tanks. Because of its lower range of coating thickness electrolytic tinplate is generally less corrosion-resistant than hot-dipped tinplate. It is, however, less expensive and possesses sufficient corrosion-resistance for many purposes, especially when lacquering or decoration, often applied for other reasons, give additional protection.

Details of the development of the electrolytic process are provided by P. R. Pine, *Trans. electrochem. Soc.* 1941, **80,** 631; K. W. Brighton, *ibid.* 1943, **84,** 227; C. E. Glock, *ibid.* 1943, **84,** 249; G. E. Stolz and W. G. Cook, *ibid.* 1943, **84,** 259. Detailed accounts of plants may be found in *J. Iron St. Inst.* 1948, **159,** 297; C. Frenkel, *J. Electrodep. tech. Soc.* 1945–46, **21,** 129; W. E. Hoare, *J. Inst. Prod. Eng.* 1951, **30,** 104. For general accounts of tinplate production and qualities, reference may be made to W. E. Hoare and E. S. Hedges, " Tinplate " (Arnold); W. E. Hoare, " Tinplate Handbook " (Tin Research Institute) and W. E. Hoare, *Trans. Inst. Metal Finishing*, 1954, **31,** 172. " Tin and its Alloys", edited by E. S. Hedges (Arnold).

Tin coatings are not normally used to protect steel out of doors; they find their largest use in industrial and domestic containers and equipment for food and beverages, for which their general good resistance to corrosion and the harmlessness to health of the corrosion-products fits them. The protection which they can give to the underlying metal has, however, considerable practical interest.

Tin is usually anodic to copper and will protect it at breaks in a coating. The copper-tin alloy layer is, however, strongly cathodic to tin and may be anodic to copper. Exposure of the alloy which may be local at the outset in thin wiped coatings leads to corrosion of the tin with the production of dark markings; the protection of the copper becomes doubtful, although attack upon the copper is rarely intensified (S. C. Britton and D. G. Michael, *J. appl. Chem.* 1957, **7**, 349).

The polarity of the iron/tin couple has special practical importance because of the great use made of tinplate to make cans for foodstuffs. In solutions of mineral salts, iron is anodic to tin, and steel carrying a porous tin layer will rust on exposure to the weather and may become perforated if left immersed in a salt solution; the perforation is the result of a large tin cathode and small iron anode. However, in the interior of cans containing certain natural products, the polarity is reversed; most fruit acids (e.g. citric) form complex anions containing tin and, in fruit juice, the potential of the tin is strongly displaced in the negative direction—like the potential of copper in a cyanide solution. Thus steel exposed at pinholes in the tin does not suffer intensified attack but receives a certain (incomplete) protection through sacrificial corrosion of the tin. Some additional protection is given to the steel by the dissolved tin, an action attributed by Hoar and Havenhand to the precipitation of hydrogen sulphide which would otherwise stimulate attack on the local anodes in the steel (cf. p. 878).

The tin coating thus confers protection to the steel against localized attack and perforation. The most serious result of corrosion inside a can is, however, not perforation, but the production of hydrogen swells; the hydrogen evolved causes the pressure in the can to increase and, although there is no danger to health, the can becomes unsaleable because the customer cannot distinguish the result from that caused by dangerous decomposition of the contents. So long as the tin coating is substantially complete, the rate of corrosion and of evolution of hydrogen is generally low; the cathode area is very restricted. In consequence, for a large range of products packed in unlacquered cans, and for some in lacquered cans, the time required to produce hydrogen in quantity sufficient to produce a swell depends substantially on the thickness of the tin coating. Exposure of the steel base over a somewhat greater area than that of the original pores may, however, result in a sharp increase in the rate of corrosion. The nature of the steel base dictates how great this increase will be.

The early work of Morris and Bryan pointed to pronounced differences in the behaviour of steel bases; this was followed by valuable investigations by Hoar and others. Statistical analysis of the results of tests of a large number of uncoated steels (of the now-outmoded pack-rolled type) in citrate buffers showed that low corrosion-rates were associated with a high copper-content and a low sulphur- or phosphorus-content; the correlation of sulphur- and phosphorus-contents did not permit a clear distinction to be made, but consideration of the possible mechanisms producing the effect seemed to point to sulphur as the responsible element. Further experiments on the same general lines but with several types of fruit packed in cans showed that increase of copper-content of the steel and decrease of the phosphorus-content improved its resistance to corrosion by some fruits; for other fruits there was no apparent effect and generally the sulphur-content of the steel appeared to be immaterial. Other work, carried out by Hartwell, indicated that for some other fruits, a high copper-content was harmful, but confirmed that reduction in phosphorus-content was generally helpful. Much experience from planned tests and general practice has largely confirmed that increasing copper-content of the steel can be beneficial or harmful according to the nature of the packed product, but that increasing content of non-metals is nearly always harmful and that increasing contents of "tramp elements" (nickel, chromium, molybdenum, arsenic) are also to be viewed with suspicion. Since it is impracticable to make tinplate steels specially for each product packed, American practice has become to prescribe special steels for the products most liable to hydrogen swelling. A typical steel has upper limits of 0·015% of phosphorus, 0·05% of sulphur, 0·01% silicon, 0·06% of copper and close control on "tramp elements". A second grade selected for moderately corrosive products may have up to 0·2% of copper; the phosphorus is limited to 0·02% and sulphur to 0·05% but the content of "tramp elements" is not limited. A third type, for the least corrosive products, is similar to the second grade but a higher phosphorus content (0·07–0·11%) is permitted (R. R. Hartwell, see p. 650).

Bulk composition of the steel is not the only influence; several metallurgical features have come under suspicion. In particular, it has been shown that adverse changes can take place in the surface layer of steel during the close-annealing process which intervenes between rolling and coating. For electrolytic tinplate, the light pickling given before plating and the slight degree of alloying during flow-brightening are insufficient to remove this layer which is left to promote earlier hydrogen swelling. The presence of such a layer can readily be detected since it exhibits an abnormally long induction period when the steel is exposed to corrosion by strong hydrochloric acid; the remedy of attention to annealing furnace atmospheres is being applied.

Many cans are lacquered internally. Originally lacquers were applied to prevent the discoloration of some fruits through alteration of the colouring matter through tin or iron. Now lacquers are frequently used to supplement the thinner tin coatings which have come into use and to prevent changes of flavour caused by traces of iron; there remain, however, some products for which the use of an unlacquered can is needed to preserve colour and flavour.

It might be expected at first sight that lacquering would reduce corrosion and the tin is in fact well protected. However, many years ago it was found that lacquering could increase the liability to hydrogen swells or perforation. Probably the reduction of corrosion of the tin and the consequent relative absence of tin salts allowed attack on the steel to proceed unchecked, especially at the seam area where the forming operation was liable to break the lacquer coating and the tin coating at the same places.

As early as 1935, Morris reported that by spraying cans of double-lacquered tinplate with a quick-drying lacquer after fabrication he was able to secure improved performance. Since then improvements in lacquers and lacquering process, including the application of a special strip of lacquer on the seam after fabrication, have been widely adopted with extremely beneficial results.

References to earlier work that deserves study are T. N. Morris and J. M. Bryan, *Trans. Faraday Soc.* 1933, **29**, 395; T. P. Hoar and D. Havenhand, *J. Iron St. Inst.* 1936, **133**, 239P; T. P. Hoar, T. N. Morris and W. B. Adam, *ibid.* 1939, **140**, 55P; 1941, **144**, 133P; W. B. Adam, *Chem. and Ind.* (*Lond.*), 1938, p. 682; C. A. Edwards, D. L. Phillips and D. F. G. Thomas, *J. Iron. St. Inst.* 1938, **137**, 223P; R. R. Hartwell, " Surface Treatment of Metals " (Amer. Soc. Metals) 1941, p. 69.

For descriptions of research and practice since the advent of electrolytic tinplate see R. R. Hartwell, " Advances in Food Research " (Academic Press, N.Y.) 1951, Vol. III, p. 327; A. R. Willey, J. L. Krickl and R. R. Hartwell, *Corrosion* 1956, **12**, 433t; R. R. Hartwell, Report of 3rd International Congress on Canned Foods, Rome, 1956, p. 133; H. Liebmann, p. 111; E. L. Koehler, *Trans. electrochem. Soc.* 1956, **103**, 486; E. L. Koehler and C. M. Canonico, *Corrosion*, 1957, **13**, 227t.

A general review including a suggested explanation of the corrosion machanism is provided by H. Cheftel and J. Monvoisin, La Corrosion des Boites de Fer-blanc dans l'Industrie des Conserves, Bulletin 12 of Etabl. J. J. Carnaud, 1954. The selection and application of lacquers is discussed by T. G. Green and M. Thomas, *Trans. Inst. Metal Finishing* 1953–54, **30**, 112; W. E. Allsebrook, *Paint Manufacture* 1954, p. 384; Saunders, *Tiddskr. Hermetikindustr.* 1956, **42**, 251.

Lead Coatings. Where lead is to be used to protect steel tanks against sulphuric acid, no risks of porosity must be accepted and autogenous attachment of sheet lead to the steel may be advisable. Steel sections coated with lead, suitable for the construction of pickling tanks, are manufactured. For other purposes, thin coatings may be acceptable; as pointed out (p. 637) pores in lead coats on steel exposed to an industrial atmosphere seem to become plugged with lead sulphate, although lead coats on buried pipes which give useful protection for a few years seem to accelerate corrosion when once gaps have appeared (p. 279). Lead may be applied by spraying (with precautions demanded by its poisonous character) or deposited from a fluosilicate bath, with glue as addition agent; a pyrophosphate bath is being examined in India (J. Vaid and T. L. Rama Char, *J. electrochem. Soc.* 1957, **104**, 460).

Application of lead to steel by hot-dipping is unsatisfactory, unless a bonding agent is present; the iron may be immersed in antimony chloride solution to precipitate antimony, and then passed into molten lead. However, it is better, if hot-dipping is to be adopted, to apply an alloy containing tin, which adheres better than unalloyed lead. *Terne plate* is made by dipping steel sheet into lead containing 15–30% tin. It is used for roofing, and in industrial districts where sulphur acids are present in the air, it probably resists better than galvanized iron. It serves also for packing dry goods for shipment, petrol and sometimes paints.

Lead-tin coatings are applied by hot-dipping to the steel bodies of fire-extinguishers and to copper wash-boilers; they are also used on the flue parts of gas appliances where the contribution of the lead to resistance to sulphur dioxide is valuable.

Cadmium Coatings. The advantage of cadmium as a coating is connected with the fact that the metal itself is in some circumstances resistant, owing to the formation of a protective film,* but it is not intrinsically noble, so that there is no acceleration of the attack upon the basis metal if gaps occur in the coating. Indeed there is usually cathodic protection upon a steel basis (p. 643).

An important use of cadmium coatings is in the prevention of unfavourable electrochemical corrosion, when dissimilar metals have to be used in contact, especially for high-tensile steel parts on aircraft which unavoidably must be in contact with light alloys; it is considered by some authorities to prevent frettage (p. 748) and the corrosion fatigue which follows frettage. Considered empirically, the evidence appears to justify its use, but different opinions are held and the whole subject is complicated by fears about hydrogen embrittlement (pp. 413–415). Cadmium can be deposited by spraying, or by dipping in the molten metal, but is generally electrodeposited from a cyanide bath with a suitable addition agent. Details are provided by H. Marston, *J. Electrodep. tech. Soc.* 1934–35, **10**, 57; S. Wernick, *ibid.* 1928–29, **4**, 101; 1930–31, **6**, 129; *Trans. Faraday Soc.* 1935, **31**, 1237.

Cadmium-plating is largely used in electrical instruments where its easier solderability gives it some advantage over zinc. Differing views have been expressed on the relative corrosion-resistance of zinc and cadmium for this application. Both metals are corroded by acid fumes given off by certain types of insulating tape and wood, and by the fumes (not always acidic) given off by certain synthetic resins. Possibly the varying experience with the two metals is due to the differing effects on them of particular vapours. Chromate treatment, which improves resistance generally, does not prevent attack by the vapours from insulating tapes at elevated temperatures. Further information is provided by P. T. Gilbert and S. E. Hadden, *J. Electrodep. tech. Soc.* 1949–50, **25**, 41; G. Schikorr, *Metalloberfläche* 1951, **5**, A177. The advantage of chromate treatment is described by R. H. Wolff, *Metal Finishing* (*N.Y.*) 1955, **53**, No. 4, p. 48.

* In liquid containing small amounts of chloride, the sparing solubility of the basic cadmium chlorides may be helpful in stifling corrosion. If the anodic product, $CdCl_2$, and the equivalent amount of cathodic product (2NaOH) were to interact to give $Cd(OH)_2$, the concentration of chloride in the liquid would remain unaltered. If they give a basic chloride, there is a gradual replacement of corrosive Cl^- by inhibitive OH^- in the liquid film close to the metal.

The use of cadmium-plating on condensers is attended by the danger of short-circuiting through the growth of " whiskers ".*

Zinc Coatings. Zinc can be applied in five ways:—

(1) *Electrodeposition* can be conducted from a sulphate bath containing an addition agent, but it may be better to use a cyanide bath containing excess of alkali, since this possesses a better throwing power; frequently a brightener is added—sometimes methyl ethyl ketone or a terpene derivative —and often aluminium sulphate. The deposits thus obtained are as bright as cadmium-plating; a thicker coat is needed for protection, but the plating-time need not be increased since a high current density is permissible. Details of the baths are provided by R. M. Burns, *Trans. electrochem. Soc.* 1939, **75**, 127; J. L. Bray and F. R. Morral, *ibid.* 1940, **78**, 309. R. O. Hall and C. J. Wernland, *ibid.* 1941, **80**, 407; R. Spears, *J. Electrodep. tech. Soc.* 1937–38, **14**, 127; J. S. Jones and P. M. Walker, *ibid.* 1941–2, **17**, 155; H. Fischer and H. Baermann, *Korros. Metallsch.* 1938, **14**, 356.

Whilst for articles with concave surfaces, electrodeposition is sometimes criticized on the groups of insufficient protection in recesses (the parts where hot-dipping often gives the thickest coats), electrodeposition appears well suited for the coating of wire, which has no concavities; it provides uniform coatings of thickness adjustable within wide limits. In the Tainton process, the electrolyte is an acid solution of zinc sulphate produced directly from the ore and suitably purified; it is run into long narrow troughs along which the wire passes continually; the current passes between the wire as cathode, and anodes of lead containing 1% silver. The contacts delivering current to the wire are made of copper and are protected with rubber (stuck on with rubber solution) except at the tips. Zinc naturally soon starts to be deposited on the copper tip, but grows out as a tree, and from time to time, the operatives knock it off. After the wire emerges, duly plated, it is washed, dried and polished with rotating blades tipped with tungsten. This and other processes are described by E. H. Lyons, *Trans. electrochem. Soc.* 1940, **78**, 317; H. Roebuck and A. Brierley, *J. Electrodep. tech. Soc.* 1945–46, **21**, 91; R. S. Brown, Priv. Comm., Jan. 14, 1958.

(2) *Hot Dipping* is used largely for sheet and also for manufactured hollow ware. The molten zinc is often controlled thermostatically at a temperature between 430 and 460°C., according to requirements, with fluxing as described on p. 624. For purposes of avoiding the brittle alloy layer, attempts were made some years ago to keep the time of immersion short by passing the steel through a shallow layer of zinc floating on the top of a deep layer of lead; this method yielded only a thin coating of zinc, and the products were unsatisfactory. The converse arrangement (a shallow layer of lead at the bottom of a deep zinc layer) is regarded, however, as

* " Whiskers " are not really a type of corrosion, but some references to their formation on various metals and probable causes may be helpful. See G. W. Sears (with S. S. Brenner and R. V. Coleman), *Acta Met.* 1953, **1**, 457; 1955, **3**, 361, 367; 1956, **4**, 268; 1957, **5**, 131. A general survey of whiskers is given in S. M. Arnold, *Proc. Amer. Electroplaters Soc.* 1956, p. 26 (Bell Monograph 2635). See also p. 501 of this book.

sound practice, as it aids the removal of dross. For particulars of technical procedures, see Zinc Development Association, Memo No. 1 (1953); W. L. Hall and L. Kenworthy, *Sheet Metal Ind.* 1947, **24,** 741.

A good method of avoiding brittleness is to add a regulated amount of aluminium to the zinc which, given an appropriate time and temperature of dipping, prevents the formation of the alloy layer. It is impossible to have a layer of flux floating on the zinc if aluminium is to be present, and in a process carried out in Monmouthshire for the continuous production of thin galvanized sheet of great flexibility, the steel is cleaned in a reducing furnace atmosphere and passes directly into the molten zinc containing aluminium. The most flexible quality produced is said to withstand bending double without injury to the coat, whilst a cost calculation made from the user's standpoint suggests that the expense of the zinc coating is less than that of applying a single coat of paint (J. Lewis, Priv. Comm., April 1, 1958).

In the Sendzimir process used for coating strip in America, the steel is oxidized by heating (which also removes residues of lubricating oil), next reduced in a controlled atmosphere (which also anneals it) and finally plunged into the molten zinc which readily unites with the spongy layer left by the reduction; brittle alloy layers are avoided by adding aluminium to the zinc bath (K. Oganowski, *Iron Age* 1950, **165,** June 8, p. 71).

Hot-dipping is much used for the production of zinc layers on corrugated roofing sheet. Roofs made of galvanized iron should be painted after erection (p. 586); calcium plumbate paints as primers have received commendation in some quarters. Hot-galvanized iron is also used for cisterns and tanks for hot and cold water supplies, being suitable for hard waters but not, in general, for soft ones (p. 639). A useful discussion of the corrosion-behaviour of hot-galvanized storage tanks is provided by I. L. Newell, *Corrosion* 1953, **9,** 46. The procedure as used for steel window frames is described by E. F. Pellowe, " Hot Galvanizing of Steel Windows " (brochure); see also his paper to the Hot Dip Galvanizing Conference, Copenhagen, July 1950.

The importance of adequate thickness in all these uses has already been emphasized.

Although electrodeposition processes which give coatings of uniform thickness are often preferred for wire, hot-dipping is still used, and a brittle layer can be avoided if the wire, as it leaves the lead annealing bath, passes through an alkaline flux containing cyanide; the surface thus becomes very slightly carburized, and the carburized layer on taking up zinc gives a ductile alloy layer; such wire can be wrapped around a mandrel having a diameter equal to that of the wire without cracking or flaking.

(3) *Spraying* processes have been described on p. 599. They have given admirable results on ships, pylons and all sorts of structures. Painting should usually follow spraying. A primer must be chosen which is compatible with zinc; this is often an oil-paint pigmented with zinc chromate and iron oxide; the vehicle may contain linseed and tung oils along with synthetic resins. It is usually followed by one or more outer coats.

(4) *Painting with Zinc-rich Paint* has been described on p. 615. This procedure is definitely inferior to metal-spraying; the coats are not mechanically robust and the adhesion in certain environments has been criticized. Nevertheless, regarded as a priming coat, and covered with other paints, it gives most satisfactory results, and is also useful for touching up damage in coats of zinc applied by other processes. In view of claims that it can be applied over rust, it should be recalled that the behaviour of all paint applied over rust varies with the weather conditions prevailing at the time of application—as indicated by Mayne's work (p. 485); probably traces of rust shut in under zinc-rich paint do no harm, provided that ferrous salts are absent, but the idea that zinc-rich paint can in all circumstances be spread over a thickly rusted surface is wrong. The introduction of zinc-rich paint is a notable addition to the anti-corrosion materials, and it is a pity if excessive claims by a few over-enthusiastic salesmen have interfered with the appreciation of its real merits. The relative merits of zinc-rich paints and red lead, based on experience on German railways, are discussed by J. C. Fritz, *Werkst. u. Korrosion* 1955, **6**, 521.

(5) *Sherardizing* is much used for small articles such as hooks, bolts and window latches. They may be packed in a powder mixture of zinc and zinc oxide and heated for some hours at 370–450°C. For springs the temperature should be lower, the treatment being correspondingly prolonged. During the heating, the articles acquire an alloy layer; it is considered that the iron diffuses into the zinc, not vice versa. The presence of the oxide prevents coalescence of the zinc particles—even above the melting-point of zinc. On exposure of such articles to corrosive atmospheres, the iron of the alloy layer becomes yellow or brown, which may mislead observers into thinking that rusting of the steel basis has started. Apart from this change, which is harmless but disfiguring, the process gives good protection against many types of atmosphere. (Details will be found in Zinc Development Association, Memo No. 10 (1948).)

The author's comparative tests on steel coated by spraying, sherardizing, hot-dipping and electrodeposition, and exposed to intermittent spraying either with N/2 sodium chloride or N/100 sulphuric acid, showed that the alteration of weight was smaller on the sherardized specimens than on any others. There was good evidence that a protective layer was being built up on a sherardized surface, since the change of weight during a second series of sprayings was much less than during the first. There was some evidence of the building up of protective films of corrosion-products on zinc coatings produced by the other processes, at least when intermittent spraying alternated with dry periods. Much more work would be needed to establish the situation, but the reader may care to study these early experiments, and draw his own conclusions (U. R. Evans, *J. Inst. Met.* 1928, **40**, 99; esp. pp. 126, 127.)

White Rusting. One of the troubles experienced by exporters of galvanized articles, particularly barbed fencing wire, is the "white rusting" which occurs during storage or shipment under damp conditions, especially if chlorides or acid fumes are present. It may be doubtful whether this problem has been entirely solved for wire products, although Gilbert and Hadden report fair results from chromate treatments. In the case of galvanized sheet, a large American concern has almost eliminated trouble, at least for goods transported in closed cars or trucks, by adopting the Neish process, in which sodium dichromate and water-glass are applied to the surface; the water-glass acts mainly as a vehicle for holding the inhibitor in position, but also plays its part in reacting to produce a film integrally united to the surface, although there is no visible coating (R. B. Mears, Priv. Comm., Nov. 4, 1955; R. A. Neish, U.S. Patent 2,665,232, Jan. 5, 1954; P. T. Gilbert and S. E. Hadden, *J. Inst. Met.* 1950–51, **78,** 47. See, however, K. Daeves, *Draht* 1957, **8,** 334, who finds some of the chromating processes of limited value).

Chromate passivation is generally stipulated for zinc coatings intended to withstand tropical conditions; it is discussed on p. 586. See also P. Skeggs, *J. Electrodep. tech. Soc.* 1951–52, **28,** 15.

Aluminium Coatings. Various methods of applying aluminium have already been mentioned. The commonest procedure is metal-spraying (p. 599), but steel hot-dipped in aluminium is now being produced on a large scale (p. 625), whilst the cladding of aluminium-copper alloy plates with copper-free aluminium layers has proved a valuable method of protection. Alloy coats are produced by heating steel articles in a powder mixture (p. 597). Electrodeposition from aqueous solution is impossible, but some interesting advances have been made in plating from organic baths (p. 612).

In cases where an aluminium coating will give protection at the widest gap likely to occur, an aluminium coating is likely to last longer than a similar zinc coat (p. 640), but there are conditions where aluminium will not protect at gaps at all. The long life of sprayed aluminium coatings on steel even in bad atmospheres is impressive—as shown by Britton's outdoor tests carried out with several grades and thicknesses of aluminium in different atmospheres. These showed that it was advantageous to apply a coating of ordinary oil-plant outside the sprayed aluminium layer. In London, the coatings carrying no paint were beginning to disintegrate after 7 years' outdoor exposure, although there was still no rust; however, those carrying a single paint-coat showed no disintegration, and would probably have survived a considerably longer period without deterioration. Conversely, paint keeps its colour better if applied on an aluminium-sprayed surface; even in soot-free air, paint applied direct to iron becomes dark owing to iron compounds moving outwards and becoming deposited on the surface; this does not occur if the paint has been applied to aluminium-sprayed steel—or to stainless steel (see S. C. Britton and U. R. Evans, *J. Soc. chem. Ind.* 1939, **58,** 90, esp. p. 92; also H. S. Ingham, *Corrosion* 1957, **13,** 252t).

The complaint has been heard that some modern paints give bad results on an aluminium-sprayed surface, and that such a surface is best left unpainted unless a full three-coat schedule is possible. It is conceivable that the vehicle of some quick-drying paints may attack aluminium; it is also likely that some paints produce too thin a coating to cover completely the prominences. The Author believes, however, that anyone who wishes to use a single coat of paint with a view to increasing the life of a sprayed coating of aluminium or zinc should have no difficulty in purchasing something suitable.*

The behaviour of aluminium-sprayed coats on hydrogen cylinders used during the Second World War for balloon barrages in all parts of the world —a severe test, well withstood—is summarized by U. R. Evans, *Chem. and Ind.* (*Lond.*) 1951, p. 706, esp. p. 709. For the use of aluminium spraying on trawlers, see W. E. Ballard, *Metallurgia* 1939, **19**, 139.

Gilbert has compared zinc and aluminium coatings for protection against atmospheric corrosion. He estimates that a zinc layer of 2 oz./ft.^{2} (i.e. 3·3 mils) will last 7 to 10 years in industrial districts, whereas a similar *thickness* of aluminium, weighing only 0·7 oz./ft.^{2}, should last more than 10 years; he considers coats produced by the powder process as good as those produced by the wire process if the composition is correct. In rural atmospheres, a 3-mil coating of either metal would probably last 30 to 40 years. For marine conditions he thinks that zinc would last longer than aluminium, which appears to fail in 15 to 20 years. Lead coats give good results in heavily polluted conditions but 3 oz./ft.^{2} (3 mils) is needed for long life (P. T. Gilbert, Lecture to Assoc. Belge pour l'Etude, l'Essai et l'Emploi des Matériaux, June 1954).

Nickel and Chromium Coatings. The method of producing a bright finish on motor and cycle parts, furniture and the like, based on a relatively thick coating of nickel followed by a thinner coating of chromium to prevent fogging of the nickel, has been mentioned on p. 450; the composition of the chromium bath was discussed on p. 607. Recent improvements are discussed by Silman, who remarks that the chromium coats are usually fissured and often add little to the protection of the basis metal. If produced at high temperatures and low current densities, the coatings become highly protective but are then no longer bright. A compromise is obtained by working at 60°C. and 400 amp./ft.^{2} which gives brightness and good protection, at some sacrifice of throwing power (H. Silman, *Metal finish. J.* 1955, **1**, 11).

The nickel-plating bath has undergone many changes. In early days, a solution of nickel ammonium sulphate was often used and gave excellent deposits, but plating was very slow; any attempt to use higher current densities introduced the risk that the anodes would become passive. The addition of chlorides to prevent passivity and the control of pH by addition of boric acid led to the famous "rapid" plating bath of Watts; this now classical formula was published in 1916. Later other ingredients, such as sodium fluoride and sodium sulphate, were introduced, but even in 1934, Cook and B. J. R. Evans, discussing methods for the automobile and cycle

* In case the reader wishes to know what was used by Britton, the composition may be given. Britton's red paint contained 14 grams of red iron oxide in 12 c.c. of boiled linseed oil, 3 c.c. of turpentine and 1·5 c.c. of liquid drier. This is, of course, an old-fashioned type and is not quick-drying, but if only one coat has to be applied, the speed of drying becomes unimportant.

industry, still recommended a Watts-type bath.* Modern baths contain addition agents, brighteners and the like. The exclusion of impurities is important; nitrates, arsenic compounds and many organic colloids are detrimental; the latter can be destroyed by means of permanganate, excess of which is in turn destroyed by means of hydrogen peroxide. Papers discussing the effect of bath composition on the quality of deposit include those of M. Cook and B. J. R. Evans, *J. Electrodep. tech. Soc.* 1933–34, **9,** 125, esp. p. 131; D. J. Macnaughtan and R. A. F. Hammond, *ibid.* 1928–29, **4,** 95. D. J. Macnaughtan and A. W. Hothersall, *ibid.* 1929–30, **5,** 63; A. W. Hothersall and G. E. Gardam, *ibid.* 1936–37, **12,** 81; 1950–51, **27,** 181; W. A. Wesley (with J. W. Carey and E. J. Roehl), *Trans. electrochem. Soc.* 1939, **75,** 209; 1942, **82,** 37. For chemicals suitable for plating, see Brit. Stand. Spec. **564** (1934); also Brochure issued by Mond Nickel Co., "Nickel Plating for Engineers" (1956).

The deposition of nickel, chromium or other metals on aluminium and its alloys involves special problems, owing to the oxide-film present. Various methods have been evolved to overcome this difficulty, the most popular being based on dipping the aluminium article in a sodium zincate solution, which gives a deposit of zinc by simple replacement, on which the electro-deposition of other metals can take place. This method in its simplest form proved unreliable, and a process was developed in Norway by Vogt, based on electrodeposition (in turn) of zinc, brass, nickel and usually chromium; in some forms of the process a heat-treatment was necessary, but in a modified form worked out by Edwards and Swanson for the British Non-ferrous Metals Research Association, no heat-treatment is necessary—at least for certain alloys of aluminium. The article receives preliminary cleaning in trichlorethylene, then cathodic treatment in alkali; a dip in a mixture of sulphuric and nitric acid follows, and then further cathodic treatment; next comes the dip in sodium zincate solution, giving a deposit of zinc by simple replacement, which is at once dissolved away in 50% nitric acid; the aluminium is now free from its original oxide-film and two very short platings in a zinc and brass bath (for 20 sec. and 10 sec. respectively) produce a surface suitable for the ordinary nickel-plating, which is generally followed by chromium-plating. The process sounds complicated but most of the operations are short; Wernick states that it has proved very successful in industry (J. Edwards and C. J. Swanson, *Trans. Inst. Metal Finishing* 1952–53, **29,** 190. See also W. Bullough and G. E. Gardam, *J. Electrodep. tech. Soc.* 1946–47, **22,** 169; G. L. J. Bailey, *ibid.* 1950–51, **27,** 233; S. Wernick and R. Pinner, *Sheet Metal Ind.* 1955, **32,** 35, 113, 189, 273, 345, esp. p. 120. The industrial aspects of the matter are ably discussed by A. W. Wallbank, *J. Electrodep. tech. Soc.* 1951–52, **28,** 209).

Interest attaches to a recent paper by Atkinson, who describes baths which allow direct deposition of copper on aluminium without any zincate

* The bath contained nickel sulphate, nickel chloride and boric acid, with pH kept between 5·5 and 5·7 by daily additions of sulphuric acid or basic nickel carbonate. The nickel content was 50 g./l., the chloride content 7·5 g./l. and the boric acid 20 g./l.

treatment. The bath first used consisted of copper oxalate dissolved in ammonium oxalate, but it was then found that the addition of boric acid, with replacement of part or all of the ammonium oxalate by ammonium pyro-phosphate, improved the results; the addition of triethylamine is recommended (J. T. N. Atkinson, *J. electrochem. Soc.* 1958, **105,** 24).

Comparative behaviour of different metallic coatings on outdoor exposure. Twelve-year exposure tests have been carried out by Hudson at seven stations at home and abroad, on steel specimens carrying seven different metals and alloys, applied in six different ways to give three different thicknesses. These provide valuable and objective information regarding the relative merits of different systems. In addition to purely metallic schemes, composite schemes consisting of metallic coatings followed by paint were included. Only a summary of the results can here be provided; the original papers deserve study (J. C. Hudson and T. A. Banfield, *J. Iron St. Inst.* 1946, **154,** 229P; J. C. Hudson and J. F. Stanners, *ibid.* 1953, **175,** 381).

Comparison between the different metals shows that in non-industrial atmospheres, coatings of metals cathodic to steel tended to fail earlier than those of anodic metals. However, in the highly polluted atmosphere of Sheffield, coatings of lead and lead-tin alloy, which are cathodic to steel, seemed to be as good as any; a lead sulphate film was quickly developed and rust-spots appeared but remained unimportant after long periods (it is remarkable that a lead layer which permitted rusting of over 1% of the area after 4 years at Sheffield, allowed 50% rusting after only 2·2 years in the marine climate of Calshot). The anodic coatings showed results which depended on the thickness; thus the thinner coatings of zinc and aluminium had failed after 5 years, but the thicker coatings of the same metal were in some cases still in good condition after 12 years. Cadmium was inferior to zinc or aluminium at Sheffield—a result in agreement with earlier tests in American cities (salt-spray tests in the laboratory had predicted that cadmium should be better than zinc, but such a forecast turned out to be misleading).

Among methods of application, cementation was found to be suitable for producing thin zinc coats but less suitable for aluminium. Electrodeposition was recommended on the grounds that the thickness could be accurately controlled; equal weights of zinc deposited from sulphate and cyanide baths gave much the same protection. Hot-dipping gave best results with zinc, but was not equally good with the other metals—although assessment was difficult. The results showed metal-spraying to be a valuable method for all the metals and alloys tested; in the case of the cathodic metals there was more rust-staining than on a hot-dipped or electroplated coating; for anodic metals, the protection, on a basis of equal weights, was comparable to that given by the other methods. At Sheffield, it was found possible to protect flat steel surfaces against rusting for ten years, either by sprayed aluminium ($\frac{3}{4}$ oz./ft.^{2}), zinc (3 oz./ft.^{2}) or lead (3 oz./ft.^{2}); these weights correspond to 3, 5 and 3 mils respectively.

The general question as to whether a metal cathodic or anodic to the basis (usually steel) should be chosen will depend on circumstances. An anodic coating must itself be corroded to some extent and is therefore generally unsuitable when a lustrous appearance is required or when corrosion-product might interfere with some mechanical function or electrical contact, or might contribute undesirable impurities to any product coming into contact with the metal. Where a cathodic coating is used, it must have the thickness and receive the care in application necessary to reduce porosity to a minimum; sometimes considerations of weight, as in aircraft, or of space, as in screw threads, may limit the thickness of coating permissible, so that an anodic coating will be chosen. Frequently, the size and physical structure of the metal to be protected will dictate the choice of coating metal and process of application. Maintenance and liability to mechanical damage have also to be considered; if a structure is to be maintained by painting after a period of service or if the coating is liable to suffer physical damage, an anodic coating will generally be preferred. Care should always be taken to base the choice on the conditions of service expected; the results of tests made with full exposure to the weather may not always provide good guidance to behaviour in sheltered positions where the hygroscopic nature of corrosion-products play an important part and the emission of vapours from non-metallic components may contribute to the nature of the environment. Anodic coatings are generally more prone to suffer from these special conditions, but it must be remembered that corrosion at pores in cathodic coatings can also be severe in presence of condensed moisture and that the need for complete covering of the basis metal is no less than it is in conditions of full exposure. It is evident that in selecting the type of coating many factors have to be taken into consideration and no rigid rules are possible; knowledge, experience and intelligent testing are necessary to obtain the best results.

Other References

Those desiring to study any particular type of coating in greater detail should consult such works as R. M. Burns and W. W. Bradley, " Protective Coatings for Metals " (Reinhold). W. Machu " Moderne Galvanotechnik " (Verlag Chemie). H. Silman, " Chemical and Electroplated Finishes " (Chapman & Hall). E. S. Hedges, " Protective Films on Metals " (Chapman & Hall). H. Bablik, " Grundlagen des Verzinkens " (Springer): English translation by C. A. Bentley (Spon). W. E. Hoare and E. S. Hedges, " Tin Plate " (Arnold). W. E. Ballard, " Metal Spraying and Sprayed Metal " (Griffin). See also " Recent Developments in Plating " (Int. Rev. of Metal Finishing) edited by I. S. Hallows (1954), and the chapter by J. Fischer on metallic coatings in F. Tödt's " Korrosion und Korrosionsschutz " (de Gruyter, Berlin). For scientific principles of electrodeposition, see H. Fischer, " Elektrolytische Abscheidung und Elektrokristallisation von Metallen " (Springer).

Recent papers discuss levelling as revealed by radio-isotopes (S. E. Beacom and B. J. Riley, *J. electrochem. Soc.* 1959, **106,** 309) and platinum-plating on zirconium at Oak Ridge (*Platinum Met. Rev.* 1959, **3,** No. 1, p. 8).

CHAPTER XVI

INTERGRANULAR CORROSION AND STRESS CORROSION

Synopsis

The tendency for foreign atoms to congregate at grain-boundaries, even in cases where no second phase is present, has been brought to notice in Chapter X. In the present chapter, it is shown how this can lead to *intergranular corrosion*, which produces more damage than general corrosion. Causes of special susceptibility may be due to the impoverishment of some protective constituent in the grain-boundary net-work or to a new anodic phase. Examples are provided by stainless steels and zinc die-casting alloys.

Next *stress-corrosion cracking*, produced by tensional stress and corrosion acting together, receives attention. The name should strictly only be used where the damage exceeds that produced by stress and corrosion acting separately. Stress corrosion frequently produces cracking along grain-boundaries, but sometimes the cracks cross the grains. Atomic rearrangement (which at high temperatures might occur throughout the grains) may at an intermediate temperature be confined to grain-boundaries where the structure is a loose one; the change may render the boundary zones different—mechanically and electrochemically—from the bodies of the grains. In a material where, in the absence of corrosion, a tensile stress would produce gliding within the grains (so that the material is ductile), the grains may be pulled apart when corrosive agencies are present, owing to the destruction of some of the material which would normally hold them together. Often the amount of matter destroyed is very small, so that we are dealing with mechanical breakage made possible by corrosion; in other cases, the anodic attack eats its way along planes crossing the grains, and the stress serves largely to decide the direction in which the fissure shall advance, so that we are dealing with transgranular corrosion directed by stress.

Examples are given of stress-corrosion cracking from aluminium alloys (largely intergranular), stainless steel (usually transgranular) and magnesium alloys (either intergranular or transgranular, according to circumstances). We next come to *layer corrosion*—a term which covers a number of phenomena; a common type of layer corrosion is really a variety of intergranular attack acting in material with flattened grains, so that the damage largely extends parallel to the surface. Nitrate cracking of steel and the cracking met with near welds in gas equipment then receive consideration.

Evidence of currents flowing between grain-boundaries and grain-interiors is next presented—based on aluminium and copper alloys. The possible role of hydrogen in stress-corrosion cracking is discussed along with other views on corrosion-cracking mechanisms. The chapter ends with a discussion of testing methods.

Intergranular Corrosion

Influence of Traces of Impurity on Intergranular Behaviour. In Chapter IV emphasis was laid on the fact that corrosion usually starts locally at points of disarray; sometimes it spreads out and becomes general; sometimes it remains localized and produces pitting; sometimes it heals up. In Chapter X, it was pointed out that, although the places where corrosion starts could in some cases be identified with imperfections in the structure, yet the immediate cause was the foreign atoms which, differing in size from the other atoms, tend to settle at dislocations. The fact that in certain materials special attack occurs at the grain-boundaries cannot be attributed simply to the discontinuity in the crystal structure, since this occurs at the grain-boundaries in *all* materials. It is evidently due to the influence of impurities. The counter-argument that this cannot be the case since intergranular attack is met with on exceptionally pure materials is seen on examination to be unsound. The amount of impurity needed to fill up, or settle around, all the holes existing at grain-boundaries is very small, and if a material contains only just enough impurity to do this, the attack may become concentrated more exclusively at the grain-boundaries than if a larger amount of impurity were present, since in the latter case attack falls *also* at places of disarrayed structure within the grains. If, however, the material is of extreme purity, even the attack along the grain-boundary should be avoided, and this is found to be the case—as shown, for instance, by the work of E. C. W. Perryman, *J. Metals* 1953, **5**, 911.

Perryman studied aluminium of different iron contents exposed to alkali. The specimens had been heat-treated at 690°C. for 3 days and either (1) water-quenched, or (2) allowed to cool in the furnace, reaching room temperature after 70 hours. The results, reproduced in Table XXIV, show clearly that some iron is needed for grain-boundary attack; in the furnace-cooled

TABLE XXIV

BEHAVIOUR OF ALUMINIUM EXPOSED TO 0·3N SODIUM HYDROXIDE FOR 3 DAYS (E. C. W. Perryman)

Iron Content	Water-quenched	Furnace-cooled
0·001%	No grain-boundary attack " Differential attack " (i.e. attack varying from grain to grain)	No grain-boundary attack Differential attack
0·009%	Some grain-boundary attack Differential attack	Heavy grain-boundary attack No differential attack
0·021%	Distinct grain-boundary attack No differential attack	Some grain-boundary attack Slight differential attack
0·037%	Heavy grain-boundary attack No differential attack	Little grain-boundary attack Distinct differential attack

specimens, the susceptibility declines again when the iron-content exceeds 0·02%.

In 0·3N sodium hydroxide, the grooves produced at the grain-boundaries were shallow ones—the breadth being greater than the depth; in 10% hydrochloric acid the breadth was roughly equal to the depth. This may be an example of the fact that alkalis often dissolve oxide-films on aluminium more quickly than do acids—for which there is other evidence. Perryman considers that the iron acts as a cathode, and, since we are concerned with the hydrogen-evolution type of attack, that seems quite possible. On the other hand, the presence of iron in solid solution might, by distorting the lattice, reduce the activation energy needed for aluminium atoms to pass from the metal into the liquid as ions. If the argument is correct, it might be expected that on quenched specimens, where the iron will be retained in solid solution, the grain-boundaries would be anodic, whilst in furnace-cooled specimens, where the iron will be present as a separate phase, they should be cathodic; that is essentially what was found in earlier work by R. B. Mears and R. H. Brown, *Industr. engng. Chem.* 1941, **33,** 1001, esp. p. 1004.

The segregation of impurities at grain-boundaries in extremely pure metal is no *ad hoc* assumption, having been proved by the autoradiograph researches of F. Montariol, P. Albert and G. Chaudron (*C.R.* 1952, **235,** 477). Other evidence is provided on p. 380 of this book.

Die-casting Alloys. Intergranular corrosion is a serious problem in zinc-aluminium die-casting alloys exposed to steam—or to a salt-laden atmosphere. The effect is generally attributed to the distribution of impurities, which produces an electrochemical contrast between grain-boundaries and grain-bodies. It is interesting to know that similar alloys, when rolled into sheet, suffer exfoliation or layer corrosion; this can here be regarded as a special form of intergranular attack, acting along the edges of flakes representing grains flattened by the rolling (see pp. 678–680). In some cases, suitable annealing followed by slow cooling can prevent the trouble. Roberts states that zinc free from aluminium does not suffer intergranular attack, even in the presence of lead. Aluminium renders it susceptible, even if the amount is insufficient to produce a second phase; if a second phase is present, it is the zinc-rich phase which is susceptible. Magnesium improves the situation; lead makes it worse (C. W. Roberts *J. Inst. Met.* 1952–53, **81,** 301; K. Löhberg, *ibid.* 1952–53, **81,** 680. Cf. R. Piontelli and F. Cremascoli, *Z. Metallk.* 1941, **33,** 245; E. Andres and K. Löhberg, *ibid.* 1941, **33,** 208; 1942, **34,** 73).

Wolf states that German die-casting alloys exported to the Tropics before the war contained copper as well as zinc and aluminium; now copper is omitted or severely limited, and this prevents intergranular corrosion, provided that zinc of purity exceeding 99·99% is used. Ageing at 100°C. for 6–8 hours is found to give dimensional stability under tropical conditions. Laboratory tests described by Pelzel, however, appear to show that an alloy containing 0·3% Al and 1·6% Cu is absolutely resistant to intergranular corrosion after 3 days in a moist atmosphere at 95°C. (W. Wolf, *Metall* 1955, **9,** 655; E. Pelzel, *ibid.* 1954, **8,** 169).

The intergranular attack on lead cable sheaths by stray currents, which causes penetration to be so much more rapid than simple application of Faraday's Law would predict, is discussed on pp. 265–268.

Effect of Impoverishment. Owing to the structural disarray, atomic re-arrangement can take place along grain-boundaries at temperatures too low to produce re-arrangement in the grain-interiors. Consider an alloy which at high temperatures consists, under equilibrium conditions, of a single phase, but at room temperatures of two phases. If heated for some time at the high temperature, to produce the single phase, and then quenched, the single-phase structure may survive, although it is now metastable. If the quenched material is afterwards heated at a temperature somewhat below that at which a single phase would become the stable condition, there is a re-arrangement of atoms which, if continued sufficiently long, will lead to the stable system;* thus particles of the second phase may appear within the grains of the major phase. If the heating is conducted at a lower temperature, where the atomic mobility is insufficient for re-arrangement within the grains, some re-arrangement (which may or may not lead to definite particles of the second phase) can occur at the grain-boundaries. Now this will often leave a zone of material close to the grain-boundaries impoverished in some constituent of the alloy, and as a result either less susceptible or more susceptible to attack.

A case where it is made *less* susceptible is provided by work in Lacombe's laboratory on very pure aluminium containing small amounts of iron (along with traces of copper and silicon which, however, do not concern us). Here the attack by the acid mixtures used in that laboratory produces pits, the frequency of which increases as the iron content increases. There are numerous pits along sub-grain boundaries, and considerable attack on grain-boundaries, since the iron atoms will have settled into the holes at these points. Such segregation of iron necessarily leaves an impoverished zone on either side of the boundaries; this zone remains entirely free from pits, as shown in the beautiful photographs of G. Wyon and P. Lacombe (Phys. Soc. Conference on "Defects in Crystalline Solids", 1954, p. 187).

The opposite case—where the impoverished zone is *more* susceptible to attack—is exemplified by the behaviour of stainless steel near a weld-line; this has been mentioned on p. 214. When 18/8 stainless steel is heated in the range 500–800°C. and then immersed in a corrosive liquid (many ordinary river waters are sufficiently corrosive for the purpose), preferential attack on the grain-boundaries occurs and crystals are ultimately dislodged; given time, the material may disintegrate to a powder. This is generally attributed to the fact that chromium, having a high affinity for carbon, separates as particles of chromium carbide along the grain-boundaries, but not (at that temperature-range) elsewhere. Associated with the grain-boundaries is a zone impoverished in chromium, the element responsible for the protective type of film, so that preferential attack is not surprising.

* Some alloys, heated for a shorter period, assume an intermediate state in which new atomic groupings tends to anchor dislocations and thus prevent gliding. This is the basis of *temper hardening*.

The main methods of combating the trouble were mentioned on p. 214. These are (1) the addition of stabilizers, like niobium or titanium, which have a still higher affinity for carbon, (2) the use of a very low-carbon alloy or (3) heat-treatment after welding, designed to bring all chromium carbide back into solution, followed by quenching. The first plan is the one most usually adopted; it is necessary to add titanium or niobium in amounts equal respectively to 4 and 10 times the carbon-content. After such additions, it is commonly assumed that the absolute value of the carbon-content does not matter—which appears to be true of most liquids; however, in 10% sulphuric acid, the attack is doubled as the carbon-content is raised from 0·04% to 0·10%. Thus, even in presence of titanium or niobium, the carbon-content may remain a matter of some importance (H. T. Shirley and J. E. Truman, *J. Iron St. Inst.* 1952, **172,** 377).

The vexed question as to how far it is practical to overcome the trouble by producing alloys of very low carbon-contents—avoiding the additions of titanium or niobium—is discussed by G. E. Speight and D. J. D. Unwin, *Rep. Progr. appl. Chem.* 1949, **34,** 117; J. J. Heger and J. L. Hamilton, *Corrosion* 1955, **11,** 22; see also *Chem. and Ind.* (*Lond.*) 1950, p. 427.

The statement of the carbide-precipitation theory presented above may be slightly over-simplified. In stainless steel containing molybdenum, the separation of the hard, brittle, non-magnetic *sigma phase* becomes important. (This can exist in plain iron-chromium alloys when the chromium is very high, but it can be formed at much lower chromium-contents when the alloy contains 3 to 4% molybdenum.) Since the sigma phase contains more chromium and molybdenum than the matrix from which it separates, the regions around the sigma particles are presumably left impoverished and therefore locally susceptible to attack by certain reagents. Thus the corrosion-resistance sometimes declines when the sigma phase appears. It seems, however, that a sigma-network causes sensitivity only to nitric acid, whereas a carbide-network allows severe intergranular attack by other acids (D. Warren, *Corrosion* 1959, **15,** 221t, Cf. L. Smith and K. W. J. Bowen, *J. Iron St. Inst.* 1948, **158,** 295; H. T. Shirley, *ibid.* 1953, **174,** 242).

Many attempts have been made to study the precipitation of chromium carbide along the grain-boundaries microscopically, but it is possible that some of the earlier photo-micrographs claiming to be pictures of " carbide particles " were really pictures of holes produced by etching reagents—possibly representing, however, the former sites of such particles. There is no doubt, however, that the particles can be shown up by photography. Lacombe has used an electropolishing bath developed in the author's laboratory, to render visible the chains of carbide particles along the grain-boundaries without any etching (M. Pruna, B. le Boucher and P. Lacombe, *Revue de l'Institut français du Petrole et Annales des combustibles Liquides* 1951, **6,** 145, esp. fig. 18. Also P. Lacombe, *Métaux, Corrosion et Industries* 1956, **31,** 337. The electropolishing method is described by U. R. Evans and D. Whitwham, *J. Electrodep. tech. Soc.* 1946–47, **22,** 24).

The carbides can be shown up most clearly by means of the reflection type of electron microscope, which avoids the uncertainties attaching to the replica technique needed for the transmission type; Sykes considers that the appearance supports the view that intercrystalline corrosion is associated with chromium impoverishment at grain-boundaries (C. Sykes, *J. Inst. Met.* 1955–56, **84,** 287, esp. fig. 3, Plate XLVII).

The action of nitric acid on 18/8 stainless steel containing titanium, under conditions existing at a chemical works, has been described by Heeley and Little. The main surface resists well, but near welds there may be shallow grooving along the zone which has reached 600–750°C. and sharper fissuring along the zone which has reached 1300°C. Treatment for 2 hours at 880°C. will generally prevent the susceptibility caused by heating at 600–750°C., but may actually increase the susceptibility after heating at 1300°C., unless the titanium/carbon ratio is raised to 7 : 1, above which there is a slight improvement. Even for dealing with the susceptibility produced at 600–750°C. the ratio should exceed 4 : 1, which is the theoretical figure calculated on the assumption that the whole of the titanium is available to precipitate carbon as TiC; in practice, allowance must be made for the titanium present in combination with nitrogen (E. J. Heeley and A. T. Little, *J. Iron St. Inst.* 1956, **182,** 241).

Stress-corrosion Cracking

Relative Strength of Grain-boundaries or Grain-bodies. As explained in Chapter X, one of two things may occur when tension is applied to a bar of crystalline metal. *Either* the grains may be pulled apart leading to intergranular fracture, *or* gliding may occur within the grains, so that deformation occurs without fracture, and the material behaves in a ductile fashion. The former occurrence is only usual at high temperatures where the intergranular material assumes properties close to those of a liquid; at low temperatures, gliding occurs within the grains—thanks to the dislocations present in all ordinary material. It is largely on this account that metallic material can be used without disaster even under service conditions where, owing to slight error of dimensioning or positioning, parts of a structure receive, at the out-set, more stress than was intended; for instance, if a number of parallel members are intended to share a load, but one, being a trifle too short, receives more than its fair proportion at the out-set, it will pull out slightly until the stress re-distributes itself in healthy fashion. If the overshort piece were to crack, the extra load would then fall on the others, and they would all fail in turn. The same is true of a single member which, owing to slight departure from the intended shape or position, receives at the out-set excess stress at one part. Only those materials which deform instead of cracking under temporary excess load can be seriously considered for engineering purposes, but some materials, which deform suitably in the absence of corrosive influences, may crack if these influences are present.

In general, the decision between deformation and intergranular rupture will be made in the opening stages, even though the cracking (if it occurs at all) may develop very slowly. Consider (fig. 102(A)) a boundary VWX separating two grains P and Q situated at the free surface of a bar which is subjected to tension. Clearly if rupture is to occur at all, it will commence at a place, such as WV, where there is a straight pull on the boundary; if a crack starts at VW (fig. 102(B)), the stress concentration at W may allow the crack to continue towards X, although at the out-set conditions in this part would have been less favourable to cracking, owing to the angle between the stress-direction and boundary-direction (apart from the fact that cracks start less easily at points in the interior than on the surface). If, on the other hand, gliding occurs in the plane KL, in such a way as to make the grain Q slightly longer and narrower (fig. 102(C)), it will relieve the stress along VW; slight additional stress will fall on adjacent grains, but in the general case, that is not likely to initiate cracking.

In absence of corrosive influences, gliding may be expected on any useful material, instead of intergranular rupture, but if, along the boundary VW, there is a material anodic to the main phase, the removal of this material by corrosion may allow stress to start cracking along VW, and in that case

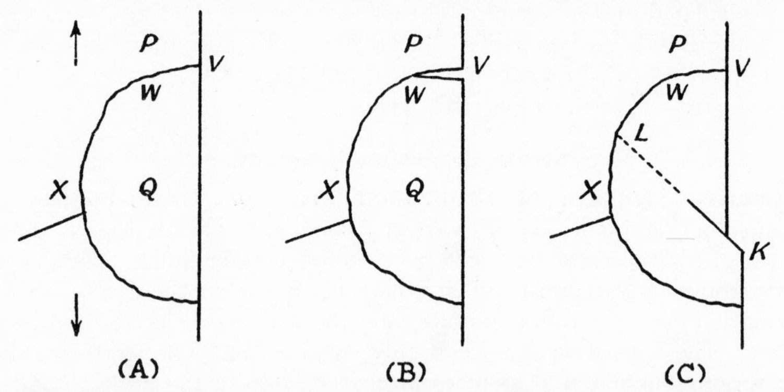

FIG. 102.—Diagram illustrating decision between intergranular cracking along VW (where there is a straight pull) and gliding along LK (which would relieve stress at VW).

it many continue in due course along WX. Clearly a continuous layer of anodic material along VWX will constitute the most dangerous situation, but even if, owing to the metallographic changes being restricted to certain parts of the grain-boundaries, we have a discontinuous path of susceptible material, there is still a risk of rupture; for, when the chemical influence has eaten its way along a section of susceptible material and has been held up by chemically sound material, the stress-intensification may be sufficient to cause the resistant bridge holding the grains together to break, so that the damage can continue. Conversely, if the mechanical cracking is held up, the electrochemical attack may demolish the strong bridge, especially as the tensional stress will usually shift the potential in a direction favourable to anodic attack. Thus stress and corrosion, acting together, can produce cracking, where either acting separately would produce none; but there can be a gradual passage from essentially mechanical cracking aided by corrosion, to corrosion aided (and directed) by stress.*

* If the precipitation of a new phase during heat-treatment has occurred along potential slip-planes within the grains, instead of along the grain-boundaries, the stress-corrosion cracking may be transgranular, not intergranular.

If the re-arrangement along the grain-boundaries is such as would cause a volume change, a new dangerous factor is introduced. Consider that the changes proceed at certain points along a grain-boundary, producing a new phase which (if unstressed) would occupy a larger volume. It might be considered that this would force the grains apart, producing holes at intervening points (fig. 103(A)); such holes would weaken the boundary like the perforations between postage stamps. This, however, will only occur in extreme cases, since small holes are unstable (nucleation energy is needed for the starting of a cavity just as for the starting of a crystal). In general, the effect will be that the atoms at the points in question will be unnaturally far apart; equilibrium will be set up between the compression at the points in the new phase and tension at the intervening parts of the boundary

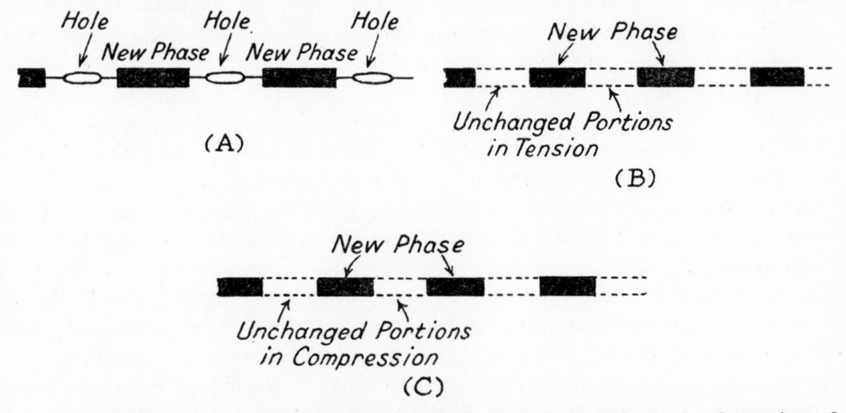

FIG. 103.—Diagram showing stress produced along grain-boundaries by formation of new phase at certain points (only). If the new phase occupies an increased natural volume, it will, in the extreme case (A) produce holes at intervening points, but in the general case (B) merely tension at right angles to the boundary at the unchanged parts. If the new phase occupies a diminished volume, the converse effect (C) is produced.

(fig. 103(B)). When tension is applied from an external source, the compressional stresses in the new phase assist the applied force to elongate the material, whilst the fact that at the intervening points we start with the atoms unnaturally far apart reduces the work needed to produce rupture at those places. The net effect is that the intergranular strength is diminished. Curiously enough, the same argument applies in systems where the change has produced a diminution in volume, although now there is tensional stress at the points of change and compression at the intervening parts (fig. 103(C)).

It would seem, therefore, that, when the heat-treatment of a material has been such as to produce some sort of atomic rearrangement at certain points (only) along the grain-boundaries but not within the grain-bodies, stress-corrosion cracking may be aided, not only by the electrochemical contrast between the boundaries and interiors, but also by the mechanical contrast. There is little doubt that the volume-changes postulated above do really occur. In cases where the structural changes occur, not at the grain-boundaries, but along certain planes within the grains, similar arguments apply.*

Probably the most important cases of stress set up by volume changes are those where the cathodic reaction produces hydrogen, some of which may combine at once to give hydrogen molecules (so that bubbles are seen issuing from the corrosion-cracks) whilst others diffuse through the material and generate high-pressure hydrogen at internal cavities (pp. 403, 417). How far this is an important factor in stress-corrosion is a matter to be discussed later. It seems to provide an explanation of cases where corrosion cracking occurs almost without visible corrosion-product.

Relative Importance of Intergranular Corrosion and Stress Corrosion. Intergranular attack, working inwards between the grains, causes considerably more loss of strength than the same total destruction of metal uniformly distributed over the whole surface; thus there is a risk that, after a long period of corrosion, a member made of a material subject to preferential intergranular attack, may fail when suddenly subjected to a load which originally it would have been well able to withstand. This is important in connection with metal-work exposed to the atmosphere, where corrosion, in absence of suitable protective measures, may be proceeding all the time, but where dangerous stresses are applied only on infrequent occasions.

Nevertheless, intergranular corrosion, where the attack is distributed over practically all the grain-boundaries cutting the surface, is generally less dangerous than stress corrosion, which occurs where stress acts continuously in a corrosive environment, producing cracks following a few paths only (frequently, but not invariably, intergranular paths). This limitation of damage to a few paths, running roughly at right angles to the direction of the load, causes far more rapid propagation of fissures through the material, particularly since, in some cases, part of the propagation consists of purely mechanical tearing.

There are many materials which are susceptible to intergranular attack but not to stress-corrosion cracking—a point rightly emphasized by Champion. As he points out, stress corrosion is only a serious problem on a limited number of materials—and then, in most cases, only after an incorrect heat-treatment. It must, however, be pointed out that the materials in which it causes anxiety include some which, on account of their strength, the engineer particularly desires to use. The reader should study

* In the aluminium-copper alloys, the reorganization within the grains is believed to proceed in four steps sometimes written

$$\text{Solid solution} \longrightarrow GP_1 \longrightarrow GP_2 \longrightarrow \theta' \longrightarrow \theta \text{ (or } Al_2Cu)$$

It is believed that some steps involve an expansion and others a shrinkage. Probably the re-organization proceeding along grain-boundaries also involves volume changes, but it is unlikely that it proceeds by the same steps. Indeed Langer's electron-microscope observations show the absence of θ' at grain-boundaries, and he suggests good reasons for this absence (E. Langer, *J. Inst. Met.* 1955–56, **84**, 471).

the paper of F. A. Champion, *J. Inst. Met.* 1954–55, **83,** 385, and the discussion (same volume p. 554).

It may now be convenient to discuss stress corrosion as met with in various materials.

Aluminium Alloys. If the stress corrosion of light alloys is to be understood, a few words are necessary regarding the manner in which these alloys respond to heat-treatment—intentional or accidental.

Many of the stronger alloys owe their strength to age-hardening or temper-hardening properties. Most of the aluminium-copper alloys, including H15 (B.S. specifications 1470–1477), which also contains magnesium, silicon and manganese, form homogeneous solid solutions when heated at a high (*solution-treatment*) temperature. If cooled so slowly as to maintain equilibrium, another phase (or phases) should appear. If the alloy is quenched from the solution-treatment temperature, the single solid solution is preserved, but on prolonged storage at room temperature atomic re-arrangement occurs which may be regarded as the *first step* towards the attainment of the equilibrium conditions; the re-arrangement occurs first at dislocations where the structure is already loose, and the groups of atoms produced probably serve to anchor the dislocations, thus hardening and strengthening the material.* If it is desired to produce hardening more quickly, it is possible to heat at an intermediate (*temper-hardening*) temperature, when atomic re-arrangements (not necessarily the same as those involved in age-hardening) occur quickly, and give rise to a rapid increase of strength—often a greater increase than is obtained by age-hardening at room temperature. Temper-hardened alloys are today generally described as " artificially aged "—in contrast to " naturally aged " alloys which have attained strength by storage at ordinary temperature.

Provided that the temper-hardening is carried out for the recommended time at the recommended temperature, there is in most materials no need to fear special corrosion trouble. If the temperature is slightly too low, or the time much too short, the changes may proceed further at the grain-boundaries than within the grain-bodies, since the structure is loosest along the boundaries, and an electrochemical contrast between boundaries and bodies is to be expected. If the grain-boundaries develop material anodic relative to the grain-bodies, intergranular attack is likely, and as the ratio of cathodic to anodic area will be large, attack is likely to be intense. Furthermore the mechanical properties of the boundary may be altered, and in some cases this alteration, combined with the electrochemical contrast, may lead to stress-corrosion cracking.

Some of the aluminium-magnesium alloys, which are not regarded as useful temper-hardening materials, do nevertheless undergo atomic re-arrangements along the grain-boundaries if they become heated in service, and then become susceptible to intergranular corrosion and stress corrosion. An extreme example is provided by an alloy containing 9% magnesium

* Any attempt to cause deformation in the metal would involve moving of the dislocation away from these atomic groupings—and require enhanced force; see p. 376.

which, although resistant in the homogeneous condition, becomes susceptible to stress corrosion after being warmed—even at 60°C., a temperature which may be reached through exposure to the sun (M. Hansen, *J. Inst. Met.* 1939, **64,** 77).

Obviously such an alloy is unsuitable for general use, but the alloys with 7% and 5% magnesium have been used in rivets for aircraft, and, although far less susceptible than the 9% alloy, have been found occasionally to suffer intergranular attack. The 5% alloy is the less susceptible of the two, and is generally preferred today; normally it is quite safe, but if subjected to tropical temperatures for long periods, rivets become liable to intergranular attack. Cold-working of the rivets before they are heated at tropical temperature is found to facilitate the precipitation of the β-phase (or more properly the β'-phase), which is regarded as the cause of preferential attack along grain-boundaries (G. J. Metcalfe, *J. Inst. Met.* 1946, **72,** 487).

In the case of the aluminium-magnesium alloys, it has long been customary to attribute the intergranular sensitivity to a net-work of the β-phase (essentially the compound Mg_2Al_3, although there is some departure from that composition). At one time, it was thought that the alloy only became susceptible to stress-corrosion cracking if there was a continuous net-work of this phase. Careful studies in the laboratory of the British Non-ferrous Metals Research Association suggest that this belief represents an over-simplification. Alloys heated below 212°C. require a continuous net-work for susceptibility, but when the alloys have been heated above that temperature, corrosion-cracking may occur even when the phase is present as discrete " lakes "—which shows that some other factor is at work, as already suggested (E. C. W. Perryman and S. E. Hadden, *J. Inst. Met.* 1950, **77,** 207, esp. p. 216).

In the case of aluminium-copper alloys—including H15—the cause of susceptibility appears to be different. Several authorities emphasize the fact that if a phase rich in copper ($CuAl_2$) is deposited as discrete particles along the grain-boundaries, a continuous zone impoverished in copper will exist adjacent to the grain-boundaries and is likely to be anodic towards the main part of the grain-bodies, which will still contain plenty of copper; one authority attaches importance to the $CuAl_2$ particles acting as cathodes. The theories as originally put forward may require a little modification. There is no doubt that the denuded layer postulated in these theories can really exist,* but the conditions which favour its development do not seem to be those which produce most susceptibility to stress-corrosion. Moreover, the production of $CuAl_2$ does not seem necessary for susceptibility, nor is it

* Evidence for the real existence of a denuded zone along the grain-boundaries of the alloy containing 4% copper is provided in electron-micrograms by G. W. Thomas, reproduced by H. K. Farmery and U. R. Evans, *J. Inst. Met.* 1955–56, **84,** 413; Plate LXVII. Evidence for its existence in aluminium-manganese alloys is provided by A. Robillard and P. Lacombe, *C.R.* 1954, **238,** 1814; on some alloys, the reagents used in etching micro-sections cause less attack on the denuded zone than elsewhere, leaving the region relatively bright and producing what was described in Gayler's classical work as the " light phenomenon " (M. L. V. Gayler, *J. Inst. Met.* 1946, **72,** 243, esp. p. 247).

certain that this compound possesses the chemical and electrochemical properties postulated by some authors; these objections may not be serious, since the precipitation of any other copper-rich phase would leave an impoverished layer. On the whole, it seems likely that the denuded layer may, in some circumstances, play a real part in causing intergranular susceptibility, but once more the facts suggest that some other factor is at work.

The denuded layer theory was developed independently in America and in Russia, and the papers of both groups of authors deserve careful study. The Russian work was concerned mainly with intergranular attack, and attaches much importance to the electrochemistry of a three-phase system. When, for instance, an aluminium-copper alloy is heated so as to give, say, $CuAl_2$ particles along the grain-boundaries, we have to consider three phases: (1) the denuded layer, (2) the $CuAl_2$, (3) the unchanged solid solution of the grain-bodies. It is thought that the most serious situation will arise when (1) is anodic both to (2) and to (3). Detailed discussion will be found in papers by G. W. Akimow and N. D. Tomashow, *Korrosion u. Metallsch.* 1932, **8,** 197; 1937, **13,** 114; 1939, **15,** 157. The American work, concerned more directly with stress corrosion, is described by E. H. Dix, *Trans. Amer. Inst. min. met. Engrs.* 1940, **137,** 11 (see also E. H. Dix and H. H. Richardson, *ibid.* 1926, **73,** 560); R. B. Mears, R. H. Brown and E. H. Dix, " Symposium on Stress-corrosion Cracking of Metals " 1944–45, p. 323 (A.I.M.E.—A.S.T.M.).

There may be, however, a more general reason why a heat treatment which initiates atomic re-arrangement along a grain-boundary should leave groups of atoms anodic to the rest, if that treatment is discontinued before the re-arrangement is complete along the whole length of the boundary. As already explained, such a change should leave the material close to the boundaries in a state of stress and—quite apart from any change of composition—much of the material along the grain-boundaries is likely to become anodic relative to the grain-interiors.

There is yet another possible reason, which although somewhat speculative, may deserve consideration. Any re-arrangement which proceeds spontaneously must involve a passage from a somewhat unstable initial state to a more stable final state, but before reaching the final stable arrangement, the atoms must pass through an intermediate, highly energy-rich condition (if this were not the case, no activation energy would be needed, and the re-arrangement would take place rapidly, even at low temperatures; the fact that some heating is needed shows that the intermediate state is rich in energy). Now if before the transformation is complete we suddenly cool down the alloy, some energy-rich material will be retained, and clearly if the alloy later comes into contact with a corrosive liquid, this energy-rich material at the grain-boundaries will pass into the liquid more readily than the matter forming the grain-interiors, since it already possesses its activation energy. We are thus led to expect that, in such circumstances, the attack will tend to follow paths along grain-boundaries.

The amount of chemical or electrochemical destruction of material needed to substitute the pulling apart of grains for gliding within the grains need not be large. Farmery, studying the stress-corrosion cracking of aluminium alloys containing 7% magnesium or 4% copper, heated at a temperature chosen to produce intergranular sensitivity, found that at stresses well below the proof stress (i.e. stresses which, in a non-corrosive environment, could be withstood indefinitely), the grains were pulled apart rapidly if the tension was applied in a solution of sodium chloride containing a small amount of sodium bicarbonate.* Now the peculiar feature of the fracture was its brightness and the relative absence of corrosion-product; some of the facets produced recalled an electropolished finish. Inspection of the fractures almost suggested that the failure had been a purely mechanical one, and yet there was good evidence to prove that electrochemical factors were involved. Farmery found that the life could be shortened by junction with copper (a cathodic metal) or lengthened by junction with zinc (which acts as an anodic metal). Indeed, by applying a strong cathodic current, fracture could often be prevented altogether (H. K. Farmery and U. R. Evans, *J. Inst. Met.* 1955–56, **84,** 413; see also discussion, p. 513).

This had indeed been shown by earlier workers. Edeleanu arrested the progress of a crack which had already advanced half-way across a specimen of aluminium-magnesium alloy by making the specimen strongly cathodic. Again Gilbert and Hadden stopped cracking by shutting off oxygen—which clearly shows that the action is not purely mechanical; this method, however, is not effective in all conditions; Farmery found that specimens pre-treated in aluminium chloride solution (which removed the protective film normally present on such material) could not be prevented from cracking by exclusion of oxygen (C. E. Edeleanu, *J. Inst. Met.* 1951–52, **80,** 187, esp. p. 189; P. T. Gilbert and S. E. Hadden, *ibid.* 1950, **77,** 237, esp. p. 242).

Farmery made a study of a "Differential Stress Cell" comprising two pieces of the same material immersed in the same liquid—one piece being stressed and the other unstressed; this provided a current, the stressed electrode being the anode, and the strength of the current increased as the stress is increased. The matter possesses possible importance in connection with the situation near welds; where small areas carrying internal stress are in close proximity with larger areas relatively free from stress, there is a threat of the dangerous combination of large cathode and small anode.†

* The time required for fracture increased as the stress decreased. Over the certain range, there seemed to be a rectilinear relation between the logarithm of the failure time and the stress applied. Some authorities claim that there is a critical stress below which fracture does not occur. In Farmery's work on the Al–Cu alloy, failure could be produced at very low stresses (e.g. one quarter of the yield stress) if time was allowed. The apparent discrepancy is probably due to the fact that Farmery used a constant-load method, whereas those who have obtained a critical value used a constant-strain method; for instance, their specimens were bent into loops, so that the cracking of the tensile side relieved the tension; under such conditions, failure may be avoided indefinitely if the strain is below a certain critical value. Some discussion of this matter is provided by J. P. Fraser, G. G. Eldredge and R. S. Treseder, *Corrosion* 1958, **14,** 517t.

† The importance of the Differential Stress Cell may extend to ferrous materials, and may help to explain the fact that caustic cracking of steel is

(The possibility of stress-relieving treatment here suggests itself, but the treatment must be chosen to avoid conditions likely to enhance the intergranular susceptibility of the material.)

The use of aluminium alloys hardened with magnesium silicide is widely advocated today in order to avoid corrosion trouble. These are generally weaker than the copper-bearing alloys, but if silicon in excess of the amount needed to produce Mg_2Si is present, reasonable strength is obtained; manganese and chromium are often added, or chromium and copper; artificially aged alloys are stronger than naturally aged alloys.* The alloys containing copper deteriorate in mechanical properties after corrosion, the elongation suffering much more than the ultimate tensile strength; the deterioration is most marked on artificially aged alloys (M. Whittaker, *Metal Ind.* (*Lond.*) 1952, **80,** 263).

Although stress corrosion in aluminium alloys is normally intergranular, the incidence varies considerably, and the worst results are produced when there is susceptibility on a few boundaries only. Conversely if it can be distributed over all the boundaries and also over gliding planes, the damage is much less serious. Brenner has shown how the addition of 0·2% of chromium to an alloy containing 7% magnesium distributes the damage over boundaries and gliding planes, thus greatly reducing the risk of cracking (P. Brenner, *Z. Metallk.* 1953, **44,** 85, esp. figs. 7, 8 and 10).

An important paper describing stress corrosion on aircraft is provided by R. N. Hooker and J. L. Waisman, *Corrosion* 1954, **10,** 325.

Effect of Heat-treatment on Susceptibility of Aluminium Alloys. It is well established that susceptibility to stress corrosion varies with heat-treatment—as indeed do the mechanical properties. Unfortunately, in some alloys, at least, the treatment which gives the best mechanical properties may not be the one which provides the best chemical resistance.

In general, naturally aged alloys are more resistant to attack than artificially aged alloys, which carry zones of localized precipitation, and corrosion tends to be concentrated on these zones. This is well brought out in Italian work on an alloy containing 5% copper. If it has been underaged, the zones occur along the grain-boundaries; if it has been aged so as to produce maximal strength, they occur along slip-planes; if it has been over-aged, the precipitation is distributed over the entire grain matrix. It is recommended that ageing should be continued slightly beyond the point at which maximal mechanical strength is obtained—so as to reduce the risk of stress corrosion or intergranular corrosion. The stress-corrosion tests were carried out in sodium chloride containing hydrogen peroxide—which is open to objection for reasons explained later; but corrosion tests

difficult to bring about in the laboratory in experiments on uniformly stressed specimens, but occurs easily with a complex stress system comprising stresses of different magnitude and character on different parts, as shown by E. W. Colbeck, S. H. Smith and L. Powell, *Proc. Instn. mech. Engrs.* 1943, **149,** 63; U. R. Evans, p. 89.

* H. K. Farmery remarks that if too much silicon is added, the material develops grain-boundary brittleness on artificial ageing; manganese or chromium additions prevent this.

were also carried out in sodium chloride acidified with hydrochloric acid, the hydrogen-evolution being measured; the general conclusions may probably be accepted (M. Paganelli, *Alluminio* 1955, **24,** 335).

The most extensive study of the effect of heat-treatment is published in a classical paper by W. D. Robertson. He also used the sodium chloride/hydrogen peroxide mixture as reagent, and the measurements refer to an alloy which is apparently not widely used in Europe for engineering purposes (24 S, containing 4·55% Cu, 0·24% Fe, 1·50% Mg, 0·66% Mn and 0·15% Si). It is likely, however, that many other alloys would show a similar behaviour (W. D. Robertson, *Trans. Amer. Inst. met. Engrs.* 1946, **166,** 216).

Robertson's tests were carried out in two stages; the specimens were subjected to corrosion when bent into an arc giving a fibre stress equal to

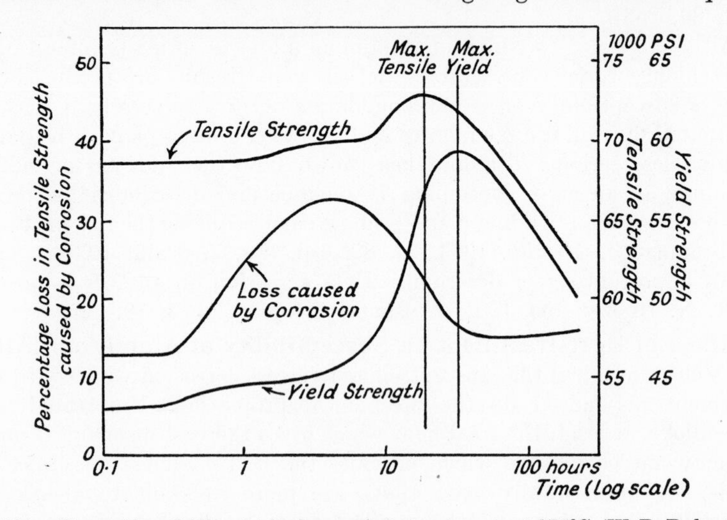

FIG. 104.—Properties of Alloy 24 S after various treatments at 175°C. (W. D. Robertson's results, replotted).

80% of the tensile yield-stress; they were then broken in an ordinary tensile testing-machine, the loss of strength due to corrosion being adopted as a criterion of damage.

A few of the results, re-plotted to bring out certain features, are presented in fig. 104. It is seen that, if 175°C. is adopted as the ageing temperature, a time of ageing which gives the maximum yield stress produces no greater susceptibility to stress-corrosion cracking than would be obtained by longer treatment; if, however, the treatment is stopped at the point which will give the maximum tensile strength, the susceptibility to corrosion weakening is greater. If the furnace-man (perhaps fearing the drastic falling off of tensile strength which would occur with an over-long treatment), makes the treatment too short (which will still give him a tensile strength not far short of the best obtainable), the weakening produced by corrosion may become serious (fig. 104).

A given change of properties can be obtained equally well either by a short treatment at high temperature or by a long treatment at low temperature. If the logarithm of the time needed to produce an effect is plotted against the reciprocal of the absolute temperature, straight-lines are obtained, as shown in fig. 105.* A study of this diagram shows that at all temperatures, the time which gives the maximum strength always lies well within the limits which produce corrosion weakening; indeed, the line showing maximum strength before corrosion lies uncomfortably near the line showing minimum strength after corrosion (in forming his judgment, the reader should not forget that time is plotted on a logarithmic scale).

It would be highly desirable that other alloys should receive similar study—preferably without the use of hydrogen peroxide. If, as is feared, the results establish the principle that susceptibility becomes serious when the treatment has been carried out at too low a temperature or for too short a time, this should be made generally known. It is not sufficiently appre-

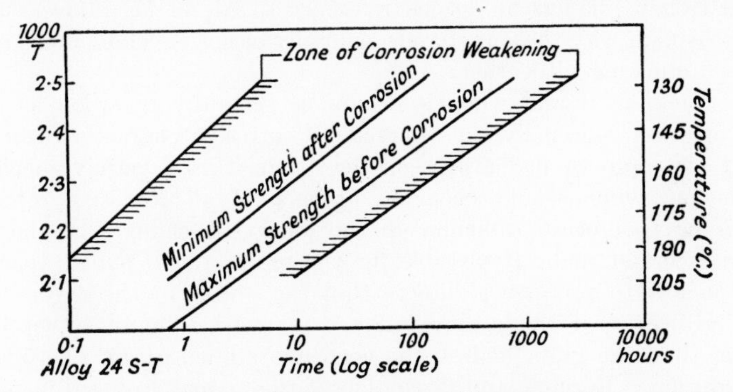

FIG. 105.—Properties of Alloy 24 S after various combinations of time and temperature (W. D. Robertson's results, replotted).

ciated today that, for some materials at least, the attainment of a correct heat-treatment is analogous to jumping over a hidden ditch and landing on safe ground on the far side; anyone who jumps short will meet with disaster (U. R. Evans, " Corrosion: a Symposium " 1955–56, p. 143, esp. p. 164 (University of Melbourne)).

Stress-corrosion Cracking of Aluminium Alloys containing Zinc and Magnesium. For a quarter of a century it has been known that very high strength can be obtained by adding zinc and magnesium to

* The straightness of the graphs arises from the fact that the chance of an atom attaining the activation energy, W, needed for some change is proportional to $e^{-W/RT}$, so that the time needed for a definite proportion of the atoms to undergo change becomes proportional to $e^{+W/RT}$ and on plotting its logarithm against $1/T$, a straight line is to be expected. The fact that the lines in fig. 104 are not quite parallel shows that the value of W corresponding to the different types of atomic re-arrangement is not the same. The stress-corrosion process is itself probably a process requiring activation energy, but on plotting the logarithm of the reciprocal life against $1/T$, the points are not found to fall on a single straight line, rather suggesting that there is more than one mechanism operative (O. Lissner, *Z. Metallk.* 1952, **43,** 147).

aluminium; the alloys developed usually contain copper and manganese, whilst most of the recent ones have contained chromium—for reasons explained below. Possibly because of the great resistance to gliding in these materials, there is often a marked tendency for the grains to be pulled apart by stresses left after fabrication or introduced during assembly. Sometimes aircraft components made from the earlier forms of the alloys developed visible cracks during storage or assembly, and this naturally caused alarm regarding the use of such materials generally, although it was generally found that, if no cracks developed in the early life of a component, it proved reliable. It has already been explained that the decision between intergranular rupture and harmless gliding is likely to be reached as soon as stress (internal or applied) is put upon the material, and that, provided that intergranular rupture is avoided at the out-set, a very small amount of gliding should be sufficient to relieve the situation. The argument presented on p. 665 has no special reference to Al–Zn–Mg alloys, but may serve to explain why these materials generally either develop cracks quickly or escape cracking altogether.

The cracking, if and when it occurs, is generally regarded as due to stress corrosion caused by the appreciably corrosive character of ordinary air, but the opinion has also been held that it is a purely mechanical phenomenon which would occur in absence of all corrosive influences. There is no theoretical difficulty in the idea that changes at the grain-boundaries might make it possible for the grains to be pulled apart at a stress too low to produce gliding within the grains in these very strong alloys. However, if the grain-boundary material is more susceptible to corrosion than the grain bodies, the corrosive influences are likely to help the disruption. Enquiry and discussion of this matter are still in progress and the reader is advised to study new papers which may be published regarding the part played by corrosion.

Meanwhile it seems that today the majority of authorities favour the idea that the cracking is due to stress corrosion; Dix in his remarkable Campbell Memorial Lecture of 1950, definitely described the cracking observed on experimental structures built from earlier alloys of this type as stress-corrosion cracking. He attributed failure to high residual tensile stresses, which may have been set up in three ways—all easily avoidable by reasonable precaution. These were (1) Unsuitable assembly methods. (2) Unsuitable working procedure adopted for fabrication. (3) Quenching; although rapid quenching induces compressive stresses in the surface layers, which are favourable, a subsequent removal of material may expose regions carrying high tensile stresses (E. H. Dix, *Trans. Amer. Soc. Metals* 1945, **35,** 130; 1950, **42,** 1057, esp. pp. 1105, 1112, 1121, 1124; J. A. Nock, *Metals and Alloys*, 1944, **20,** 922; W. L. Fink, J. A. Nock and M. A. Hobbs, *Iron Age*, 1945, **156,** Nov. 1, p. 64).

In the same lecture, Dix described the successful efforts to overcome the trouble by the addition of chromium—an achievement largely due to J. A. Nock. The important effects of chromium are (1) an alteration of the shape of the grains, which are equiaxed and regular in absence of chromium, but become elongated and irregular when chromium is present, (2) inhibition of precipitation along the grain-boundaries and the production of precipitation elsewhere. Such changes would be expected to reduce the risk of stress-corrosion cracking, at least under tension applied longitudinally, but it is still not quite certain why chromium brings about these metallurgical changes. Dix states that, although chromium does not completely eliminate the susceptibility to stress-corrosion cracking, it makes the susceptibility " comparable with that of other high strength aluminium alloy products which have given long and satisfactory service in important engineering structures ", and adds, " Over seven years of commercial use of the products of this alloy, no important incident of stress-corrosion cracking has occurred in service." The material in question (75 S) contains 0·25% chromium, the main constituents being 5·6% zinc, 2·5% magnesium, 1·6% copper, with a maximum of 0·3% manganese.* Dix describes numerous laboratory experiments—mostly carried out by C. J. Fiscus at stresses higher than those likely to be encountered in service—showing the smaller liability to corrosion cracking in comparison with a similar alloy free from chromium.

An Australian report, written from the stand-point of engineers in the aircraft industry, should receive consideration. It states that intercrystalline corrosion can occur in 75 S which has been solution-treated but not artificially aged; if there has been correct ageing, it is less susceptible, but it will still crack in an intercrystalline manner if it is in an undesirable state of internal stress, or is subjected to external stress in a corrosive environment. Discussing methods of composition adjustments designed to render the alloy free from stress corrosion, they say that chromium " is favoured by the Americans. Uniformly distributed chromium in amounts 0·18 to 0·4% can be very effective ", adding that " British alloys employ manganese for the same purpose in amount 0·3 to 1·0% " (H. E. Arblaster and P. F. F. Thompson " Corrosion: a Symposium " (University of Melbourne), 1955–56, p. 379, esp. p. 392).

The Australian statement quoted above should not be taken to mean that no interest is taken in chromium by British manufacturers; several British products contain chromium, and the effects of chromium and other elements have received careful study in British laboratories. It has been found that 0·25% chromium is beneficial, but larger amounts confer no additional advantage (and indeed are undesirable on mechanical grounds). The chromium-containing alloys suffer only mild surface corrosion under conditions where the chromium-free alloys suffer pitting and intergranular attack. Chadwick and his colleagues confirm Dix's observation that chromium diminishes stress corrosion by producing elongated grains, and it is now found to be most effective in the presence of manganese. When both metals are present the length of the grains may be ten times the breadth. Copper is beneficial, apparently because it favours attack at the centre of the grains as well as along the boundaries, an intermediate zone being left

* The British specification DTD 683 covers a rather wider range of alloys than the American 75 S but is generally similar.

unattacked, so that in effect intergranular failure is diminished. It should be noted that some of these results depend on laboratory tests with sodium chloride/hydrogen peroxide mixtures, and there is difference of opinion how far such tests represents service behaviour. Nevertheless the papers deserve close study (M. Cook, R. Chadwick and N. B. Muir, *J. Inst. Met.* 1951, **79,** 293; R. Chadwick, N. B. Muir and H. B. Grainger, *ibid.* 1956–57, **85,** 161; see discussion by H. K. Farmery (**85,** p. 542), P. R. Sperry (p. 543), W. M. Doyle (p. 542), C. Smith (p. 539) and authors' reply (p. 544).

Valuable information is provided by Hérenguel regarding an alloy with 6% zinc and 2% magnesium. After noting that the alloy made from very pure materials fails at stresses close to the elastic limit much more quickly than commercial material—he studied the effect of various minor constituents, and found that beryllium, boron, molybdenum and niobium accelerate the deterioration, whilst improvement is obtained by adding copper or silver (which enter into solid solution) or chromium and manganese (which are barely soluble); combinations of copper and chromium hold promise (J. Hérenguel, *Rev. Métallurg.* 1947, **44,** 77).

Directional Effects in Stress Corrosion. If the improved performance obtained by adding chromium and/or manganese to aluminium alloys depends on alteration of the shape of grains, it will clearly vary with the direction of stress-application. Recent investigators have distinguished between the results obtained on applying stress to a rolled slab in the three possible directions, but in some of the earlier researches the importance of directionality was not appreciated, and it is possible that the claims for the beneficial influence of certain elements in certain alloys may have been too general and too sweeping.*

Most of our information regarding the effect of direction concerns the aluminium-copper-magnesium alloys. Liddiard and Bell cut specimens from extruded material representing different directions relative to the extrusion axis, and found that the results varied with the direction—especially in the case of material containing appreciable amounts of manganese; the susceptibility to stress corrosion was marked when the specimen was stressed in direct tension normal to the direction of extrusion (E. A. G. Liddiard and W. A. Bell, *J. Inst. Met.* 1953–54, **82,** 426, esp. Table II, p. 428).

Layer Corrosion. Material which is in a state to suffer ordinary intergranular attack when the structure is nearly equiaxed, will show a rather different phenomenon, if it has been rolled out, flattening the grains. Although annealing will alter the structure, there often remains a tendency for the new grains to be longer in the longitudinal than in the transverse direction. Moreover, all inclusions and segregates present in the original casting will be transformed into streaks or flakes by the rolling, and these will not necessarily disappear on annealing. Thus there is a tendency for

* Sutton states that the advantages claimed for chromium-additions are not obtained if the alloy is stressed in the " short transverse " direction (H. Sutton, Priv. Comm., April 29, 1958).

corrosion starting downwards at right angles to the face, to change direction and proceed along planes parallel to the rolled surface.

If such material is exposed to the atmosphere, the corrosion-product formed along the planes parallel to the surface will in general fill a larger volume than the material destroyed; it will therefore swell the material and push apart the unaltered intervening layers; this disintegration into lamellae (*exfoliation*) will open paths for the continued advance of the attack. Tensile stress in a longitudinal direction will not help the advance, and although transverse stresses may do so, they may not be needed, since the swelling caused by the voluminous character of the products is sufficient to cause the stresses needed for propagation of the damage. The author has been inclined to regard certain types of layer corrosion as stress corrosion where the corrosive action produces its own stress; but this view is not universally accepted.

In aluminium alloys, especially those with copper as the main alloying constituent, layer corrosion is sometimes met with on rolled sheet or extrusions after heat-treatment carried out at too low a temperature or for too short a time. The danger is greatly reduced by suitable heat-treatment. The presence of suitable cladding or sprayed metallic coating on the face may afford cathodic protection under damp conditions, especially where the moisture has a high conductivity owing to salt or acid; such types of protection are of doubtful efficacy if the metal-work becomes nearly dry at intervals, and metal-spraying of the face is unlikely to arrest layer corrosion which has already started. In any case the cladding or metal-spraying must consist of a material which behaves anodically to the plate or sheet; for alloys containing copper, a covering of unalloyed aluminium will serve, but for those containing magnesium it is usually necessary to use an aluminium-zinc alloy with 1 to 5% zinc.

There are many different ways in which corrosion can penetrate along planes parallel to the face in aluminium alloys. Some of these are essentially cases of grain-boundary attack, modified by the fact that the grains are, or have been, elongated in the plane of rolling. There is, however, another type of attack along planes parallel to the face which deserves notice. It may happen that, at an early stage of rolling, a slab may stick to both rolls, so that its two halves open out like the jaws of a crocodile; if the two halves are subsequently brought together they may seem to weld up, but owing to the formation of oxide during the period when the jaws were open, the welding is often imperfect; if the material is one that is susceptible to stress corrosion, the cracks starting from points on the face which would otherwise advance across the specimen, will be deflected sideways parallel to the surface. Specimens of such material, subjected to ordinary stress-corrosion tests under longitudinally applied force, will often give exceptionally long lives, but the numbers obtained are not really a sign of high quality, and the corroded specimens, although able to withstand steady longitudinal stresses, tend to break when twisted.

Again, if the material has been produced from an ingot in which there has been segregation of copper in certain parts, then, after rolling or extrusion, there will be layers alternatively rich and poor in copper. Corrosion cracking will tend to follow these layers; it will proceed along the copper-rich layers in cases where the copper has been precipitated as a separate phase, but it will follow the copper-poor layers in cases where the copper remains in solid solution, so that the copper-poor zones are anodic with the copper-rich zones as cathodes (E. A. G. Liddiard, Priv. Comm., Nov. 24, 1955).

Tate has discussed the treatment of aircraft skin which has suffered exfoliation owing to layer corrosion. If the exfoliation is serious the skin may have to be replaced, but if it is only slight it may be possible to remove the corroded portion entirely with sharp scrapers, after which chromic acid solution should be applied. Steel wool should never be used, as detached steel particles may set up galvanic cells; adherent carbonaceous matter can act in the same way (B. Tate, *Chem. and Ind.* (*Lond.*) 1957, p. 506).

The analogy between layer corrosion of light alloys and the zonal corrosion of wrought iron (p. 511) will be evident, but there is one difference. The zonal corrosion of wrought iron frequently starts at sheared edges, whereas the layer corrosion of light alloys rarely starts here, but more often at points on the face, penetrating downwards for a short way (probably along grain-boundaries) and then turning into a plane parallel to the surface.*

Some of the published photographs of layer corrosion, showing the material splitting up into thin flakes and completely disintegrated, are impressive and indeed alarming, but such cases are not common; they generally represent material which has the wrong composition or has received the wrong heat-treatment. Nevertheless, extruded material, even when correctly heat-treaed and exposed to the atmosphere for some years, sometimes starts to show foliation, connected with corrosion along grain-boundaries and bands parallel to the extrusion direction. Cases were brought out in the tests of G. J. Metcalfe, *J. Inst. Met.* 1952–53, **81,** 269. See also E. A. G. Liddiard and W. A. Bell, *ibid.* 1953–54, **82,** 426. Description of layer corrosion (foliation) of light alloys in laboratory tests is provided by H. Vosskühler, *Werkst. u. Korrosion* 1950, **1,** 143, 179, 310, 357. See also F. Bollenrath and W. Bungardt, *Z. Metallk.* 1942, **34,** 160.

Cracking of Stainless Steel. The intergranular corrosion of stainless steel often observed near weld-lines and apparently connected with layers denuded of chromium through to the precipitation of chromium carbide, was described on p. 214. A different type of damage, essentially transgranular, is met with, when stainless steel is stressed in concentrated chloride solution; this does not depend on a heat-treatment calculated to throw the chromium out of solution. Most investigators have used concentrated magnesium chloride in studying it. In contrast with the stress-corrosion cracking of aluminium alloys which is essentially a mechanical failure helped

* H. K. Farmery suggests that a peripheral coarse-grained skin normally prevents layer corrosion from starting from the edges of extruded sections; by a heat-treatment designed to produce an elongated grain-structure in susceptible sheet material (Al/4% Cu), he has obtained exfoliation from the edges as well as the faces.

by chemical action, stress-corrosion failure of austenitic stainless steels seems to be essentially an electrochemical eating away of the metal along a narrow front, the function of the stress being probably to increase the distance between the atoms near the tip of the advancing crack—which will reduce the activation energy needed for them to pass into the liquid. This will have the effect of directing the attack along a path roughly at right angles to the applied load.

The matter has been studied by Hoar and Hines, who divide the life of specimens into two periods. The first is similar to that obtained in absence of stress on tin (p. 124), and consists of a thickening of the film by anodic formation of oxide, until finally the accumulated acid causes a break-down; Hoar and Hines followed the thickening of the film by watching the movement of the potential. The film-thickening and break-down does not need the application of stress, but stress is needed for the events of the second period, which consists in the development of a crack running across the grains and finally ending in fracture.

The potential movement during this second period allows a rough indication of the current flowing between the tip of the crack as anode and the face outside the crack as cathode. By means of subsidiary experiments in which different cathodic current densities were applied from an external source through a separate anode, Hoar and Hines determined the current density needed to bring the potential to the value observed during the cracking; in that way, it was calculated that the current flowing during the advance of the crack would be sufficient to produce the advance at the rate actually observed. The results are consistent, therefore, with the view that the anodic action is cutting a pathway through the metal without any mechanical tearing.

Even in this case it is clear that both electrochemical and mechanical factors are involved; the advance of the cracks can be halted—either by cathodic protection, or by discontinuation of the tensile load. Nevertheless, the part played by corrosion appears to be more important, and that played by stress less important, than in Farmery's work on aluminium alloys.

Some difference of opinion exists, however, regarding the reason for the course taken by the fissure. Edeleanu thinks that the corrosion tends to follow the quasi-martensite which is found to exist within the mainly austenitic material, but Hoar and Hines do not accept that view in its entirety. It may be that the deformed material at the tip of an advancing crack is sufficiently anodic to the rest to ensure that the attack is concentrated on the tip, rather than on the walls, which doubtless cover themselves with a protective film; the removal of atoms from distended material probably requires considerably less activation energy than from material with accurately arrayed atoms. Thus narrow fissures are produced instead of rounded pits. The reader should consult the original papers, and form his opinion (T. P. Hoar and J. G. Hines, *J. Iron St. Inst.* 1956, **182,** 124; 1956, **184,** 166; C. Edeleanu, *ibid.* 1953, **173,** 140; M. F. Baldy and R. C. Bowden, *Corrosion* 1955, **11,** 417t; H. H. Uhlig, A. White and J. Lincoln, *Acta Met.*

1957, **5,** 473; *J electrochem. Soc.* 1958, **105,** 325; K. W. Leu and J. N. Helle, *Corrosion* 1958, **14,** 249t.

Although stress corrosion in austenitic stainless steel is usually transgranular and branching, other types of attack are met with. Where carbides have been precipitated along grain-boundaries, intergranular attack may occur locally; it seems likely that the intergranular cracks are the first to be formed, and then act as stress-raisers for the transgranular cracking. In other cases, where carbides are absent, pitting has occurred and provided stress-raisers for transgranular cracking. Illustrations are provided by W. G. Renshaw, *Corrosion* 1956, **12,** 477t.

Service cases of stress-corrosion cracking on stainless steel associated with chloride are becoming all-too numerous. Examples, including gas coolers, water de-aerators, sterilizing baskets, food cookers and apparatus for handling wet salt, are described by H. R. Copson and C. F. Cheng, *Corrosion* 1957, **13,** 397t; cases involving hot water and steam installations are discussed by W. L. Williams, *ibid.* 1957, **13,** 539t.

Particular attention is now being directed to the risks of stress-corrosion cracking in certain situations in atomic-energy plants or in high-pressure boiler plants. Cases of failure seem generally to be due to chlorides but they may not always be introduced from the water; it is believed that traces of chloride present in crevices in the metal surface, perhaps introduced from trichlorethylene or similar degreaser, may have been the cause of certain serious break-downs.

The risks of corrosion cracking of stainless steel by steam has been examined by Edeleanu and Snowden, who regard the danger as a real one when high stresses are involved. In highly superheated steam, such as 600°C. and 1500 lb./in.^{2}, no stress-corrosion cracking occurs if the steel surface is clean; in presence of chloride contamination and oxygen, oxidation will occur and life will be shorter than that due to ordinary creep. At lower temperatures, particularly below 400°C., stress corrosion can occur with 1500 lb./in.^{2} steam, but probably only if the steel is contaminated with such substances as chlorides or caustic alkali. The cracking with chloride contamination is rapid near the dew-point, but becomes slow when there is 20°C. super-heat; it needs the presence of oxygen, but caustic cracking occurs in absence of oxygen. The contamination with chlorides and caustic may occur on evaporating surfaces, and particularly in restricted places where impurities can become concentrated; a carry-over of impure water from the boiler is a possible source of contamination. The cracks formed both with chloride or caustic contamination are of the form previously found in boiling chloride solutions (C. Edeleanu and P. P. Snowden, *J. Iron St. Inst.* 1957, **186,** 406).

Cracking in Magnesium Alloys. Sometimes a stress-corrosion crack may be either intergranular or transgranular according to circumstances, and instructive examples are provided by the alloys of magnesium. One of these, which contains aluminium, zinc and manganese as essential constituents, with iron as an impurity, suffers transgranular cracking after one heat treatment and intergranular cracking after another; the transgranular cracking is probably connected with the phase FeAl deposited on the basal plane of the hexagonal crystals, whereas the intergranular cracking is connected with $Mg_{17}Al_{12}$ deposited at the grain-boundaries. For details, the reader should consult an interesting paper by D. K. Priest, F. H. Beck and M. G. Fontana, *Trans. Amer. Soc. Met.* 1955, **47,** 473; discussion, p. 487.

Ideas about the propagation of a stress-corrosion crack may be helped by studying some photographs taken in W. D. Robertson's laboratory during the fracture of magnesium single crystals under tension in absence of corrosion. One of these shows a narrow band of disarrayed material which extends far ahead of an advancing crack. The two edges of the band consist of twinned material, presumably not greatly disarrayed, but the central part of the band is severely distorted, as shown by the displacement of rows of etch-pits which outside the band are straight and parallel to one another. It would seem likely that such a disarrayed region should be readily attacked, and the fact that hydrochloric acid attacks the band preferentially during the metallurgical etching used to show up the structure (causing the band to appear as a white streak under an illumination which leaves the rest of the surface dark) indicates that this is the case. It is easy to see that in general, when once an incipient crack has appeared on a stressed specimen in a corrosive environment, the deformation of the material along the region which constitutes the continuation of the incipient crack could produce a band of susceptible material capable of rapid removal. Although no corrosive influences were applied during the cracking of the specimen, the paper will be helpful to all interested in stress-corrosion cracking (R. E. Reed-Hill and W. D. Robertson, *Acta Met.* 1957, **5,** 728, esp. fig. 3, p. 730).

Nitrate Cracking. It has long been known that steels sometimes develop intergranular cracking when subjected to tensile stress in a hot nitrate solution. Nitrates of weak bases, giving a solution which is acid owing to hydrolysis, are more liable to cause trouble than sodium nitrate, and most of the experimental work on the subject has been based on a solution containing both calcium and ammonium nitrates. Although some steels are much more susceptible than others, it does not seem possible to associate cracking—or immunity from cracking—with any particular constituent. Parkins (see below) has shown that a steel can become susceptible or non-susceptible to nitrate cracking according to its heat-treatment. By subjecting a steel which was originally non-susceptible to a long anneal below the critical point—say, 72 hours at 700°C.—he rendered it susceptible. Such a treatment produces carbide particles along the grain-boundaries, and after cooling the metallic iron between the carbide particles was found by means of X-rays to be in a state of stress; the stress probably arose from the differences in the coefficients of expansion of carbide and metal. Whatever the cause of the stresses, Parkins has shown that the time required for fracture, as a result of tensile stressing in hot nitrate, decreases steadily as the lattice-distortion increases. There is little doubt that the strain present (doubtless by reducing the amount of activation energy needed for the iron atoms to pass as ions into the liquid) encourages attack along grain-boundaries rather than into grain-bodies.

Polish investigations also suggest that carbide formed along the grain-boundaries renders steel susceptible to stress-corrosion cracking (in boiling 40% ammonium nitrate), but show that the martensitic structure present in specimens quenched from 910° in water and tempered at 200° is associated with great susceptibility—probably owing to internal stress; specimens quenched in oil are resistant (M. Smialowski, E. Gasior and C. Bieniosek, *Bull. Acad. Polonaise des Sciences et des Lettres* 1950, Vol. I, No. 2, supplement).

Certainly external stress is not needed for the development of cracks. Keating described how steel boxes used for storage or transport of relatively dry crystalline nitrate develop intergranular attack without the application of appreciable stress in service; but this attack, distributed over most of the grain-boundaries, is less serious in its results. However, when a steel surface is exposed to tensile stress in contact with hot nitrate solution, the internal stresses along grain-boundaries running at right angles to the load will assist the applied stresses in tearing grain from grain, and the electrochemical action will demolish any bridges which could hold up the advance of the cracks. When once one particular crack has fairly started, the stress-concentration at its tip will cause it to advance more quickly than its competitors, and the fracture will spread through the thickness of the metal. The cathodic reactions are doubtless the reduction of nitrate on the faces outside the crack, and the anodic reaction will be the passage of iron into the liquid as ions, giving (in effect) iron nitrate at the tip of the crack. At low temperatures, a slimy mass of hydrated ferroso-ferric products collects on the surface, and apparently interferes with the replenishment of nitrate at the cathodic areas, thus preventing the destruction; at high temperatures, the product is anhydrous magnetite, itself a good electronic conductor, and cracking proceeds apace. These observations—due to Wooster and Nockolds—help to explain why a hot nitrate solution readily produce cracks, but not a cold one. Readers should study the excellent paper by R. N. Parkins, *J. Iron St. Inst.* 1952, **172,** 149, and supplement this with information supplied by F. H. Keating, Symposium on Internal Stress in Metals and Alloys, 1948, p. 321 (Inst. Met.). See also U. R. Evans, quoting W. A. Wooster and S. R. Nockolds, *J. Iron St. Inst.* 1941, **143,** 159; E. Herzog, *Métaux et Corros.* 1949, **24,** 29; *J. Iron St. Inst.* 1953, **174,** 140, discussing Parkins' results and quoting earlier work with Portevin; J. T. Waber, H. J. McDonald and B. Longtin, *Welding J.* 1945, **24,** 268S. For caustic cracking at bauxite extraction works, see F. A. Champion, *Chem. and Ind.* (*Lond.*) 1957, pp. 71, 967).

Nitrate cracking has much in common with caustic cracking (p. 456); in both cases, the damage is intergranular, occurs mainly at elevated temperatures, with considerable production of magnetite. There are, however, differences. In nitrate solution destruction seems to be directed on to the metallic phase, not the carbide, whereas on steel from boilers which have suffered caustic cracking the carbide phase of pearlite areas is found to be preferentially attacked. This feature has been observed by three different experimenters (Weir, Cottell and Quarrell) using different methods.

Cracking of welded steel at Gas-Works. Numerous cases have arisen in gas-purifying plant where cracking has occurred near welded joints; the cracks run either parallel or perpendicular to the weld-lines, their direction being clearly dictated by residual stresses left after welding; nevertheless the cracking only occurs in a certain chemical environment, namely places where there is an alkaline condensate containing ammonia, ammonium sulphide and cyanide (or other cyanogen compound). It seems that those steels which are susceptible to nitrate cracking are particularly liable to gas-works cracking (*Trans. Instn. Gas Engrs.* 1951–52, **101,** 305; *Instn. Gas Engrs.,* Copyright Pub. **517** (1957). C. E. Pearson and R. N. Parkins, *Welding Research* 1949, **3,** 95r. Also *Gas Journal* 1957, Nov. 27, p. 464; discussion, p. 468).

Extensive work by Parkins in the field and in the laboratory has shown that it is possible to overcome the trouble by thermal stress-relief; alternatively shot-peening (bombardment with shot) or hammer-peening (beating with a round-headed hammer) may be used—a treatment capable of replacing tensile by compressional stresses, and of closing up any fissures which would otherwise act as stress-raisers.

Parkins considers peening less reliable than the annealing, although he states that its benefits may persist after the compressed layer has been removed by general corrosion; if localized, peening may even start cracking, doubtless by providing stress-raisers (it may be suggested here that a systematic study of peening processes, varying shot-size and shot-velocity, on the lines adopted by Gould in connection with corrosion fatigue, might establish a reliable procedure; see p. 727).

The modern system of low-temperature stress-relief around a weld-line, about which the experience of practical men greatly differs, is also discussed by Parkins. This method, which depends on moving a series of gas-burners mounted on a trolley above the weld-line, closed followed by water-sprays, so as to obtain heating and rapid cooling along certain zones on either side of the weld, is convenient where introduction of welded work into a furnace would be impossible or costly. It involves an entirely different principle from furnace-annealing, since instead of attempting to loosen the atoms so that they can rearrange themselves freely in unconstrained arrays, the process depends on producing dimensional changes in certain zones of the material; such changes may reduce the longitudinal stress remaining after cooling, but can sometimes (it is thought) increase the transverse stresses. It cannot be regarded as so reliable as furnace stress-relief (1 to 3 hours at 600–650°C.). For details, see R. N. Parkins, *British Welding J.* 1955, **2,** 495; K. Wellinger, F. Eichhorn and F. Löffler, *Schweissen u. Schneiden* 1955, **7,** 7; H. G. Kurz, *ibid.* 1955, **7,** 291; W. Liebig, *ibid.* 1955, **7,** 355.

Possibility of Stress-corrosion Cracking in Pre-stressed Concrete. A situation demanding vigilance concerns the risk of corrosion cracking of wires in pre-stressed concrete, especially in cases where for mechanical reasons chlorides have been incorporated in the mixture. The matter is discussed in Chapter VIII, but it may be well to recall that calcium chloride is sometimes employed to accelerate the early stages of hardening. It cannot be assumed that chloride will be harmless when the steel is under tensile stress. It is satisfactory to know that researches at the Building Research Station are in progress, and the Engineer is advised to study the results immediately they become available.

Currents passing between Grain-boundaries and Grain-bodies. Stress corrosion of the intergranular type is most likely to start if a potential difference exists between the boundary zone and the grain interiors. Later, as a fissure develops other factors may play a part in generating the electric currents which pass between the anodic area at the crack-tip and the cathodic area on the face outside—such as the accumulation of acidity in the crack, the shift of potential caused by the stress concentration at the tip and the local breakage of films (e.g. oxide) which remain protective on the crack walls, but keep cracking at the tips. However, at the out-set a potential difference—due to some chemical or physical peculiarity of the boundary zone—is probably necessary, and its existence has been skillfully demonstrated by Dix, Mears and Brown (E. H. Dix, *Trans. Amer. Inst. Met. Engrs.* 1940, **137,** 11; R. B. Mears, R. H. Brown and E. H. Dix, Symposium on Stress Corrosion Cracking of Metals, 1944–45, p. 323, esp. p. 325 (A.S.T.M. —A.I.M.E.).

In this method two identical specimens are cut from a coarse-grained sheet; on one all the grain-boundaries are protected with an insulating, impervious coating, so that, on immersion in a liquid, only the grain-bodies are exposed to it, whilst on the other the bodies are coated but a zone about 1 mm. wide left along the grain-boundaries. Clearly the genuine boundary zone showing peculiar properties is much narrower than 1 mm., but the measurements do suffice to show that, on materials susceptible to grain-boundary attack, the grain-boundary net-work is definitely anodic to the grain-interiors. The absolute measurements of the potential have little significance, and this would be the case even if the width of the gaps left on the specimen representing the boundary material could be reduced to atomic dimensions.* It may be noted, however, that on an aluminium alloy con-

* It seems to have been thought in some quarters that the working E.M.F. of the corrosion-cell can be obtained by first measuring the potential of a specimen with only the boundaries exposed and then that on a specimen with bodies exposed, and subtracting one from the other. Now where two electrodes represent reversible systems (such as $Cu \mid Cu(NO_3)_2$ and $Ag \mid AgNO_3$) it is possible to obtain the E.M.F. of the combination ($Cu \mid Cu(NO_3)_2 \mid AgNO_3 \mid Ag$) by subtraction, although even here the E.M.F. declines, through polarization, when current flows. However, for irreversible cells the subtraction method becomes invalid. For instance, on a specimen with grain-boundaries exposed, the potential measured is a compromise potential (or " mixed potential ") due to two (different) anodic and cathodic reactions proceeding at different places on the grain-boundaries, whilst that measured on the specimen with boundaries protected is a compromise potential due to local anodes and cathodes on the grain-interiors. However, in the actual corrosion process current mainly flows from the boundary as anode and the body as cathode; local anodes on the cathodic area and local cathodes on the anodic area hardly operate.

taining 4% copper which had been quenched in cold water from 500°C. and then aged 16 hours at 150°C., the grain-boundary potential towards 0·1N potassium chloride was 0·20 volt on the anodic side of the grain-body potential.

More interest attaches to the current obtained when two specimens, one with the boundaries exposed, and the other with the bodies exposed, were placed in a highly conducting solution and joined through a low-resistance meter; in one experiment 2·1 m.a. was obtained from specimens having an area of 50 cm.^{2} exposed to sodium chloride containing hydrogen peroxide; it would be interesting to repeat the experiment with sodium chloride kept saturated with oxygen.

Bakish and Robertson have carried out rather similar work on copper alloys. Very rightly they emphasize the importance of distinguishing between the purely crystallographic effect of grain-boundary conditions (such as would occur in pure copper) from the combined effect of structure and composition (such as is present in alloys). The grain-boundaries were isolated by preparing a section and placing a drop of resin on the centre of a grain and working it towards the boundaries, by means of a nylon tip attached to a micro-manipulator, so as to leave the boundaries uncovered. It was fully recognized that the potential values have no absolute significance, the gap being wide compared to the true boundary zone, but the interesting point was brought out that in ferric chloride solution the potential at the boundaries is at first anodic towards that of the grain-bodies, but after 8 hours the two become the same, the value being that of polycrystalline copper. The reason is possibly that given on p. 391. The observations agree with the fact that, starting with a flat surface, grooves are first developed by the attack along the boundary, but after a time they cease to deepen, and the copper is uniformly dissolved away, maintaining an equilibrium angle at the boundary.

70/30 brass in ferric chloride behaves similarly; the potentials of boundary and body are different at first but later become the same; accordingly, the etching produces grooves along the boundary, but there is no continuous penetration. In ammonia the grain-boundaries are anodic to the grain-bodies in presence of oxygen, but cathodic in its absence—which may help to explain the different behaviour of certain brasses according as they are immersed in ammonia solution or exposed to ammonia vapour (p. 692).

Much of the work was devoted to the behaviour of an alloy of the composition Cu_3Au in ferric chloride. When different crystal faces were exposed, they were attacked at different rates, and the results support the idea that only octahedral faces are really stable—as in the case of copper (see p. 391). In absence of stress, ferric chloride produced circular areas of corrosion at points, some of which may have been imperfections formed during the original growth of the crystal. When the crystal was strained these point sites became the nuclei of cracks. In absence of growth-imperfections cracks were nucleated preferentially at pores situated within the deformed material present at clusters of slip-planes. Macroscopic fracture proceeded by the growth of cracks, accompanied by local plastic deformation; finally isolated cracks joined together and the specimen broke into two parts. In the absence of corrosion, the crystals were usually ductile, and suffered gliding along octahedral or dodecahedral planes when the resolved shear stress became sufficient; cracking occurred only in presence of ferric chloride; if, after a crack had been initiated, ferric chloride was removed, plastic deformation and typical ductile failure occurred. Brittle failure could be prevented by cathodic protection through contact with copper.

It is believed that when ferric chloride acts on the Cu_3Au the copper is removed, leaving a gold sponge, which acts at cathode, being about 0·2 volt nobler than the alloy, so that attack penetrates into the alloy; in CuAu this does not occur as there is too much gold to provide effective continuous paths of copper, which is removed only from near the free surface of the specimen—as explained on p. 350 in connection with parting limits (R. Bakish and W. D. Robertson, *Acta Met.* 1955, **3,** 513; 1956, **4,** 342; *J. electrochem. Soc.* 1956, **103,** 320).

The work of Graf on the behaviour of gold-copper alloys of different composition in aqua regia supports the idea that the attack is stimulated by electrochemical action so that the susceptibility to stress-corrosion cracking requires the presence of gold to act as cathode. He thinks that both elements pass into the liquid and gold is then reprecipitated; where reprecipitation is prevented by complex formation (as occurs in gold-silver alloys exposed to cyanide solutions), there is no susceptibility to cracking. In the gold-copper system there is a maximum susceptibility when 20 to 30% of the atoms consist of gold (which corresponds to the Cu_3Au alloy used in the Bakish–Robertson research); alloys containing little gold have longer stress-corrosion lives, doubtless because the cathodic stimulation is weak, whereas those where over 50% atoms are gold also have long lives—doubtless because the parting limit has been passed. Graft's work, which covers both one-phase and two-phase alloys, deserves study (L. Graf, *Metallforschung* 1947, **2,** 193, 207; *Z. Metallk.* 1949, **40,** 275. L. Graf and J. Budke, *ibid.* 1955, **46,** 378; L. Graf (with K. Matthaes), *Werkst. u. Korrosion* 1957, **8,** 261, 329).

Corrosion-rate during Stress-corrosion. The researches of Gerischer deserve study because they show how the rate of corrosion increases during the stress-corrosion cracking; the results should be compared with those obtained in corrosion fatigue (p. 717). Copper-silver alloy wires were kept in a solution containing sodium sulphate, sulphuric acid and a small amount of copper sulphate. The increase of length was observed during the process—as well as the corrosion-rate, as measured by the current passing between the wire as anode and a copper cylinder surrounding the wire as cathode. In the final stages before fracture, the lengthening proceeded jerkily, and each sudden elongation was accompanied by a sudden spurt of current. Rather similar phenomena were observed on unalloyed copper wires, but they are better marked on material forming a protective film, which is broken at each sudden elongation (H. Gerischer, *Z. Elektrochem.* 1957, **61,** 276; *Werkst. u. Korrosion* 1957, **8,** 394; also Priv. Comm., Dec. 16, 1957).

Role of Hydrogen in Stress-corrosion Cracking. In many of the cases of stress corrosion hitherto discussed, hydrogen is certainly produced. During the stress corrosion of aluminium-magnesium alloys, bubbles may be seen emerging from the cracks. In such cases, there are two possible cathodic reactions, the reduction of oxygen and the evolution of hydrogen. The former can only take place on the face outside the cracks, thus involving a considerable resistance for the liquid path joining anode and cathode. Hydrogen can be produced close to the anodic tip, and the cell resistance must be much lower; on the other hand, the E.M.F. is also lower, and since the cathodic alkali is produced close to the point of anodic attack, obstructive solid matter is likely to be produced. It has been shown (p. 672) that if oxygen is shut off from the solution, cracking is prevented—at least on some types of material; if the cathodic reaction consists of oxygen-reduction, the two products, formed at points well separated, will not cause any obstruction. It appears that, speaking generally, most of the attack is of the oxygen-reduction type.

Nevertheless, the production of some hydrogen is not in doubt, and it also seems to occur in caustic cracking; presumably the hydrogen formed by the cathodic reaction first consists of single atoms, which then proceed to form molecules. Some of the atomic hydrogen might, however, diffuse into the metal and become converted to molecular hydrogen in internal cavities. If this happens at a cavity which lies ahead of the direction of an advancing crack, the pressure developed in that cavity will co-operate with the applied tensile force—already intensified by geometrical considerations at the tip of the crack—to accelerate the advance of the crack. The idea that internal hydrogen pressure developing in cavities provides much of the mechanical breaking force would explain the clean character of the fissures, which are often free from solid corrosion-products, despite the fact that corrosion is clearly involved, since cracking can be hastened by an anodic current and slowed down or stopped by a cathodic current.

The hydrogen mechanism would also explain why the addition of nitrate stops cracking in hot sodium hydroxide solution, since the cathodic reaction becomes the reduction of nitrate instead of the formation of hydrogen. The fact that gas-works cracking occurs in presence of alkali, sulphide and cyanide provides some support for a mechanism based on hydrogen pressure developing in the harmless micro-cracks which are believed by some authorities to exist near welds, but which will only start to extend if hydrogen pressure develops within them; in the latter event the direction of extension will be dictated by the stresses left by welding, since the hydrogen exerts stress equally in all directions. At first sight, it may seem surprising that iron should evolve hydrogen from an alkaline liquid, but the presence of cyanide, which keeps the Fe^{++} ion concentration low and thus maintains a highly negative value for the potential $Fe \mid Fe^{++}$, establishes conditions as favourable to the discharge of hydrogen ions as those existing in acid solutions. Normally, the hydrogen would probably be evolved as gas-bubbles but sulphide would largely prevent this happening on the external surface and a build-up of pressure in internal cracks is easy to understand.

The addition of sulphide or selenide to the liquid in which stainless steel is suffering stress-corrosion cracking greatly stimulates the advance of the cracks. This seems to provide evidence in favour of the hydrogen mechanism. On the other hand, sulphides and selenide additions do not seem to help the stress-corrosion cracking of aluminium alloys.

One objection has been raised to the hydrogen mechanism in that several of the cases of stress-corrosion cracking mentioned are definitely intergranular, whereas typical hydrogen cracking (it is asserted) is transgranular. This objection does not seem to carry weight. Some authorities consider hydrogen cracking to be commonly intergranular; it seems that it can be either transgranular or intergranular—according to the character of pre-existing cavities in the metal, which serve for the collection of high-pressure hydrogen, as shown by the observations and photographs of J. Duflot, *Rev. Metallurg.* 1952, **49,** 35.

The evidence in favour of the hydrogen mechanism is somewhat indirect, and opinion varies greatly as to the extent to which internal gas-pressures contribute to cracking; Zappfe (p. 420), assigns it a major role. The authors considers that it is important in some cases, although it can hardly be the sole factor, since stress-corrosion cracking is met with on noble materials in liquids which would seem incapable of liberating hydrogen; for instance, gold-silver alloys (33% gold) suffer stress-corrosion failure in about one minute in nitric acid (L. Graf, *Metallforschung* 1947, **2,** 193).

Other views of Stress-corrosion Cracking. It is widely believed that the cause of the propagation of stress-corrosion cracking is that the stress keeps breaking the film at the bottom of a crack. Such a view has long been held by Champion, and a careful study of potential measurements by H. L. Logan provides data which can be explained in such a way. Logan found that when the air-formed films were removed from surfaces of various materials by abrasion in argon, the potentials subsequently measured were more negative than those of surfaces prepared in air; in the case of an aluminium alloy, the potentials were −1·43 volts after argon preparation (relative to the saturated calomel electrode) but −0·67 volt when tested in the ordinary way. Appreciable changes in potentials of notched specimens, stressed in tension, occurred at just above the stresses at which the corrected stress-strain curves deviated from the straight elastic modulus line, and these changes were of the same order of magnitude as those obtained by the argon-preparation. Logan concluded that stress corrosion depends on electrochemical action between the filmed portion as cathode and the film-free portion as anode (H. L. Logan, *Bur. Stand. J. Res* 1952, **48,** 99; cf. J. C. Chaston, *Nature* 1948, **161,** 891. A description of the generation of current between two specimens of the same material, one kept film-free by abrasion and the other left free to develop a protective film will be found in an early paper by U. R. Evans, *J. Inst. Met.* 1925, **33,** 27, esp. p. 31).

The work of Hoar and Hines clearly shows that the break-down of a film (p. 681) either by electrochemical or mechanical action, is often necessary before cracking can develop. However, it is a little difficult to accept the mechanism as a general theory of crack propagation; if stress alone could keep the film permanently disruptured at the tip of the crack, an aluminium alloy should suffer stress-corrosion cracking in the absence of oxygen, the cathodic reaction being then of the hydrogen evolution type. This is not generally the case. Moreover, some of Farmery's experiments are difficult to explain on the film-cracking theory. A specimen of aluminium magnesium alloy in a susceptible condition was subjected to stress-corrosion cracking until the crack had advanced half-way across the specimen, and cracking was then arrested by the application of a cathodic current; after 30 minutes the current was stopped, but the crack did not resume its advance, although 15 hours later fresh cracks appeared elsewhere. In another experiment, the crack was allowed to advance about one-third of the way across the specimen, and was again arrested by a cathodic current for 30 minutes; the cathodic current was then discontinued and the specimen was subjected to an increased load; yet it remained unbroken after 48 hours. It seems certain that if mechanical film-rupture at the tip of the crack was the determining factor for propagation, the advance of the cracks would have been resumed after the discontinuation of the protective current—at least when the load was increased. If, however, the anodic production of acidity is what maintains the cracking, when once it has started, it is easy to account for the results; cracking—once started—will continue for the same reason that pitting—once started—continues (p. 123).

Season Cracking of Brass. Many years ago, much trouble was caused by the cracking of brass cartridge cases stored in eastern cities where the air was hot and contained ammonia. This was traced down to internal stresses left after fabrication, which were largely tensional in the surface layers of the parts affected, although balanced by compressional stresses in the central zone. The chemical action on the intergranular material allowed the grains, initially in tension, to contract, leaving definite fissures between one and another. The trouble was met with on other highly worked brass articles besides cartridge cases and processes were worked out for minimizing susceptibility to cracking by annealing; by careful choice of time and temperature, the stresses could often be dispelled without loss of the highly desirable work-hardness. The recommended procedure varies between different types of brass, and is laid down in the classical papers of H. Moore and S. Beckinsale, with C. E. Mallinson, *J. Inst. Met.* 1920, **23,** 225; 1921, **25,** 35; 1922, **27,** 149; 1923, **29,** 285.

Some doubt has been expressed as to whether ammonia alone can produce this effect. It seems fairly certain that oxygen and water-vapour are needed, and that small amounts of carbon dioxide accelerate it; probably larger amounts prevent it (G. Edmunds, E. A. Anderson and R. K. Waring, S mposium on Stress-corrosion Cracking of Metals 1944–45, p. 7 (A.S.T.M.—A.I.M.E.)).

Cracking has occurred in pipes of unalloyed copper laid in cement floors containing a foaming agent which gives off ammonia (P. E. Halstead, *Chem. and Ind.* (*Lond.*) 1957, p. 1133; A. H. Goodger, Priv. Comm., June 1, 1959).

In brass the attack may possibly start at zinc-rich places (perhaps even

β-phase) which may exist in nominally α-brass in the unannealed state —a suggestion already made in connection with another type of attack (p. 472). It has been asked why ammonia in the air is more apt to cause damage than acid fumes (stress-corrosion cracking is not unknown on brass exposed to products of combustion of coal-gas containing oxides of sulphur, but ammonia is the most usual cause). Probably in all cases attack starts on the grain-boundaries, but in an acid environment, copper ions accumulate in the intergranular trenches produced and this shifts the potential in a direction unfavourable to anodic attack along the grain-boundaries. Thus in the acid environment, attack soon becomes general, and the severance of grains does not normally occur. On the other hand, if the moisture deposited on the metal is ammoniacal, the simple copper cations will be almost completely converted to complex ions such as $[Cu(NH_3)_4]^{++}$, and anodic attack along the boundaries can continue.

The fact that ammonia cracking can occur on single crystals has been used as evidence that other factors are involved; it would be interesting to know whether the cracks in single crystals follow sub-grain boundaries. Russian investigators favour explanations based on reduction of strength by electric double layers (V. V. Skorchelletti and V. A. Titova, *J. appl. Chem. U.S.S.R.* 1953, **26,** 41 (English translation, p. 37)).

Stress-corrosion cracking of brass has been noticed in refrigerators, owing to traces of ammonia given off from albumins in the food; German work has indicated that the presence of 0·9% silicon greatly reduces susceptibility (H. Steinle, *Metall* 1955, **9,** 492).

It should be noted that exposure to air containing ammonia may produce different results from immersion in an ammonia solution, and that different types of brass behave differently. Work at the British Non-ferrous Metals Research Association has largely established the facts, but not—it seems —fully explained them. In a moist ammoniacal atmosphere, α-brass can suffer intergranular penetration even in the absence of stress, but the penetration is accelerated by stress; in $\alpha\beta$- and β-brass the penetration in the absence of stress is slight, but stress causes severe transgranular cracking across the β-phase grains. When tested under tension with part immersion in concentrated ammonia solutions, extruded β-brass specimens fail by intergranular cracking on the immersed part of the specimens, but α- and $\alpha\beta$-brass are hardly attacked below the water-line; above the waterline the cracking is transgranular on most alloys, but 70/30 brass shows a mixture of transgranular and intergranular cracking. A possible explanation of this contrasting behaviour of the two portions (respectively immersed and exposed to air) is suggested on p. 687. The facts are provided by A. R. Bailey, S. Morris and A. J. K. Wiesiolek, *Metal Ind.* (*Lond.*) 1953, **83,** 497; A. R. Bailey and W. H. Lowther, *ibid.* 1954, **85,** 126; A. R. Bailey and C. Robins, *Rev. Métallurg.* 1956, **53,** 105; E. C. W. Perryman, *J. Inst. Met.* 1954–55, **83,** 369. See also convenient review by M. E. Whitaker, *Metallurgia* 1948, **39,** 21, 66.

A mercury test is often used to predict whether a brass will suffer season-cracking or not. Sometimes an article (perhaps selected at random from a pile of mass-produced articles) is dipped into acid mercurous nitrate solution which precipitates a film of mercury on it; if desired, the film can be thickened by a subsequent dip in a pot of liquid mercury. An article carrying internal stress will develop cracks, and since internal stresses favour season-cracking, the pile from which the sample was taken then becomes suspect for use in ammoniacal climates. However, not too much importance must be attached to the results of the mercury test, since the pattern of cracks developed by mercury is utterly different from that developed by ammonia vapour—as shown clearly by T. A. Read, J. B. Reed and H. Rosenthal, Symposium on Stress-Corrosion Cracking of Metals 1944–45, p. 90, esp. fig. 1 (A.S.T.M.—A.I.M.E.). See also F. Aebi, *Z. Metallk.* 1956, **47,** 421.

The stress-corrosion cracking of brass can occur also under very different conditions. Breckon and Gilbert emphasize the need for the stress-relief of brass pipes after bending. Even if this is carefully carried out, dangerous assembly stresses can be introduced by misalignment. Stress corrosion in copper-base alloys may be either inter- or transgranular; corrosion fatigue is almost always transgranular (C. Breckon and P. T. Gilbert, *Metal Ind.* (*Lond.*) 1958, **93,** 114).

Work on the brittle intergranular fracture of β-brass containing aluminium is described by S. Harper, *J. Inst. Met.* 1956–57, **85,** 415.

Testing Methods for Intergranular Corrosion and Stress-corrosion Cracking

Tests for Intergranular Susceptibility. It is often desired to discover whether a material (owing to faulty composition or unwise heat-treatment) has become susceptible to intergranular corrosion. The tests depend on the use of some reagent which exerts a specific action on the material at the grain-boundaries which if present causes susceptibility. In the early days of austenitic stainless steel, Hatfield used copper sulphate acidified with sulphuric acid, which attacks the impoverished net-work—if present—so that the grains of a susceptible steel became dislodged; given time, the whole may fall to powder, each grain representing a grain of the alloy. Despite criticisms, Hatfield's is still regarded by many authorities as the most useful test available, but it may be well, after the dip in acidified copper sulphate solution, to bend the specimen and thus reveal any penetration which has occurred.

The test can be rendered more sensitive if, after the specimen has been placed in copper sulphate solution containing sulphuric acid, the loss of electrical conductivity is determined. For the reagent digs down along the grain-boundaries if the material is susceptible to intergranular attack, and causes a notable loss of conductivity even if the total destruction of material is small; evidently the change in electrical properties is a better criterion than the loss of weight (J. R. Mangin, Dissertation, Zürich; abstract *Stahl u. Eisen* 1954, **74,** 1615).

A promising new test consists in immersion in acid ferric sulphate. This reagent digs rapidly down along the edges of grains in alloys in the susceptible conditions and dislodges them, so that a small amount of corrosion, in the strict sense, produces a large amount of attack. The test has one advantage over the nitric acid test described below, namely that the products of attack do not accelerate the reaction—so that four specimens can be tested in a single flask. It shows up only the type of susceptibility due to the precipitation of chromium carbide, whereas nitric acid reveals that due to the sigma phase. Thus the two tests are likely to be useful for different purposes. For details see M. A. Streicher, *Amer. Soc. test. Mater. Bull*, April 1958, No. **229**, p. 77.

Tests based on hot nitric acid are today very commonly employed. The type of intergranular attack produced by nitric acid differs in some respects from that produced by other reagents, and the use of tests based on hot nitric acid to gauge the suitability of alloys for general use in the chemical industry has sometimes led to the choice of the wrong material. A material which behaves satisfactorily under a variety of service conditions, and which withstands acid copper sulphate, may nevertheless suffer severe attack in hot nitric acid. Even where the material is intended to withstand nitric acid in service, the frequently-prescribed test based on boiling 65% or 70% nitric acid requires to be used with caution. It is found that the chromic acid, produced by the action of nitric acid on the steel, accelerates the attack—so that small differences in the manner of carrying out the test may greatly influence results. Those desiring to use this test should study. the researches and recommended procedures of H. T. Shirley and J. E. Truman, *J. Iron St. Inst.* 1952, **171**, 354; *J. appl. Chem.* 1954, **4**, 273.

The nitric acid test, as usually conducted, requires a long period of exposure. Streicher's method of preliminary evaluation by means of anodic etching in N/10 oxalic acid, has, therefore, aroused considerable interest, since it furnishes results in 10 to 20 min. as opposed to perhaps 10 days in nitric acid. The test is probably more valuable for detecting susceptibility arising from carbide separation than that from sigma-phase separation (M. A. Streicher, *Werkst. u. Korrosion* 1954, **5**, 363).

Stainless steel in service sometimes comes into contact with insulating material from which chlorides may be leached out. For such cases, interest attaches to a test described by A. W. Dana and W. B. Delong, *Corrosion* 1956, **12**, 309t.

Tests for Stress-corrosion Cracking. Testing methods for susceptibility to stress corrosion are conveniently reviewed by G. F. Sager, R. H. Brown and R. B. Mears, Symposium on Stress-corrosion Cracking of Metals, 1944–45, p. 255 (A.S.T.M.—A.I.M.E.) and by H. P. Godard and J. F. Harwood, *Corrosion* 1955, **11**, 93t.

The methods of applying the stress fall into two groups, *Constant Strain Methods* and *Constant Load Methods*. As an example of a Constant Strain Method, the apparatus preferred by Farmery is shown in side view, plan, section and perspective in fig. 106(A), (B), (C) and (D); two other types in common use are shown in fig. 106(E) and (F). As an example of a Constant Load Method, Farmery's apparatus is shown in fig. 107 (H. K. Farmery, Priv. Comm., July 2, 1957).

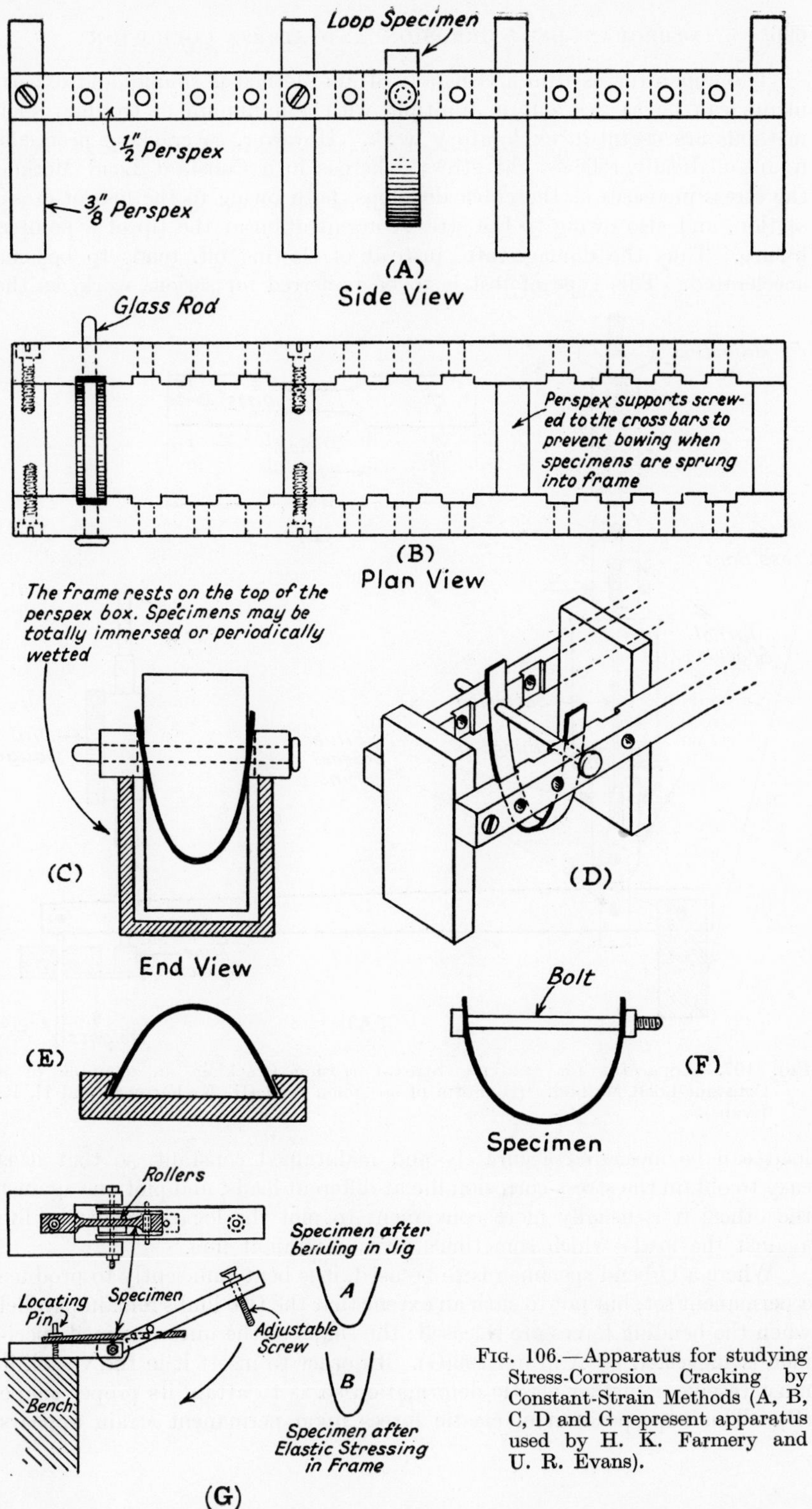

Fig. 106.—Apparatus for studying Stress-Corrosion Cracking by Constant-Strain Methods (A, B, C, D and G represent apparatus used by H. K. Farmery and U. R. Evans).

The apparatus used in a Constant Strain Method is simple, and experiments can be carried out in multiple with a minimum of trouble; such methods are useful in exploratory work. However, as cracking proceeds, it automatically relieves the stress, whereas in a Constant Load Method the stress increases as the crack develops, both owing to the loss of cross-section, and also owing to the stress-concentration at the tip of a pointed fissure. Thus the damage-rate, instead of slowing off, tends to become accelerated. This type of test is to be preferred for serious work, as the

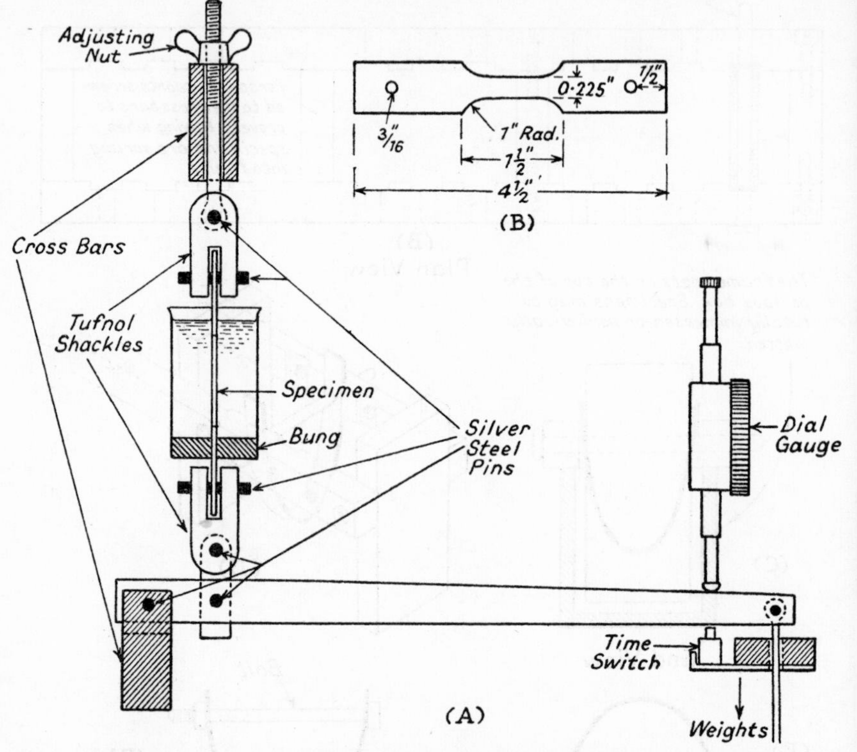

FIG. 107.—Apparatus for studying Stress-Corrosion Cracking—an example of a Constant-Load Method. (B) Form of specimen used (H. K. Farmery and U. R. Evans).

load can be measured accurately and maintained constant, so that it is easy to obtain the stress-corrosion life at different loads, and plot one against the other; it is usually more convenient to plot the logarithm of the life against the load—which sometimes gives a straight line.

Where a U-bend specimen is to be used, it is bent sufficiently to produce a permanent set, but not to such an extent that the two limbs remain parallel when the bending forces are released; the shape of the unconstrained specimen is suggested by A in fig. 106(G). In order to insert it in the vessel, it must receive a further elastic deformation so as to attain its proper shape (B). This superposition of elastic forces upon permanent strain appears important for stress-corrosion cracking. Farmery's apparatus for the reproducible bending of a specimen through a constant angle is also shown in fig. 106(G).

The need for elastic stress superimposed on plastic strain in a loop test is probably the laboratory counterpart of the fact that in service—notably in caustic cracking—damage occurs at the places where stressing is very non-uniform, some parts having suffered permanent deformation with consequent stress relief and other parts being under elastic constraint. The importance of the combination of elastic and plastic stress is emphasized by A. E. Schuetz and W. D. Robertson, *Corrosion* 1957, **13,** 437t, esp. p. 442t. Other forms of loop testing are discussed by O. Schaeber, *Z. Metallk.* 1940, **32,** 210; see also F. Bollenrath and W. Bungarth, *ibid.* 1942, **34,** 160. A carefully thought-out constant-strain test is described by J. P. Fraser, G. G. Eldredge and R. S. Treseder, *Corrosion* 1958, **14,** 517t, 524t.

In all stress-corrosion tests, a scatter of results between " identical experiments " is to be expected. It is usual to record the average life, but this is hardly satisfactory—especially if some specimens remain unbroken at the end of the experimental period. Instead of using the " mean ", we can determine the " median ", or, as it is now often called, the " half-life " —the period needed for the breakage of half the specimens; this criterion, which reduces the time needed for the completion of the experimental series, has been adopted by G. Schikorr and G. Wassermann, *Z. Metallk.* 1949, **40,** 201.

It remains to discuss the liquid which is to surround the stressed specimen under test. Immersion in simple salt solution, such as sodium chloride, gives very slow results, but spraying with 3% sodium chloride is found to produce cracking much more quickly; the stress-corrosion life of aluminium-magnesium alloys decreases with rise in the salt concentration up to 10%, after which no further decrease occurs (E. C. W. Perryman and S. E. Hadden, *J. Inst. Met.* 1950, **77,** 207, esp. p. 218).

For studying the mechanism of cracking in aluminium alloys containing magnesium or copper, Farmery found the addition of sodium bicarbonate to sodium chloride to produce a convenient acceleration of cracking. However, since this does not serve to produce rapid cracking in all alloys, it is not suitable for assessing the relative susceptibility of materials of different classes. The function of the bicarbonate is probably to destroy the cathodic alkali, which would otherwise swamp the weak acidity produced at the anodic points; since the accumulation of acidity at these points is believed to be necessary for the continuation of attack, factors which prevent its destruction will stimulate cracking. The function is the same as that of the calcium bicarbonate used by Porter and Hadden (p. 123) in producing the pitting of aluminium, but in the case of a deep, narrow crack as opposed to a shallow pit, it is not necessary to produce a mound of porous solid over the anodic point in order to prevent mixing of the anodic and cathodic products; thus calcium ions are not needed and sodium bicarbonates serves. The testing liquid adopted contained 0·5N chloride and 0·005N bicarbonate (H. K. Farmery and U. R. Evans, *J. Inst. Met.* 1955–56, **84,** 413, esp. p. 417).

Many investigators obtain acceleration by adding either hydrochloric acid or hydrogen peroxide to the sodium chloride solution. These stimulate the cathodic reaction, and since the control is largely cathodic, the advance of the anodic crack becomes far more rapid. For studying the mechanism of cracking, such stimulation can be defended, but where the object is to compare the susceptibility of different alloys—or of the same alloy after different heat-treatments—neither addition can be regarded as suitable; the results of the two additions do not always agree, and it is unlikely that either reproduces the effect of an unaccelerated field test. The hydrochloric acid addition probably changes the cathodic reaction from oxygen-reduction to hydrogen-liberation, and it might be expected that the acceleration obtained would vary with the hydrogen-overpotential of the material. Nevertheless, the method may serve to distinguish a material which has become susceptible through unsuitable heat-treatment from the same material in the non-susceptible state; when the hydrogen evolved is plotted against time, the susceptible material gives a curve showing more pronounced acceleration of the gas-evolution than the non-susceptible (T. Marshall and G. J. Schafer (with H. J. Todd), *J. appl. Chem.* 1958, **8**, 303; 1959, **9**, 38).

Hydrogen peroxide is unsatisfactory, since it suffers catalytic decomposition in presence of heavy metals—notably copper; a comparative test between an aluminium-copper and an aluminium-magnesium alloy carried out in salt solution containing peroxide is likely to be invalid, since, although the composition of the liquid may be the same in the body of the solution, it will be different at the surface of the two alloys; even for a comparison between two alloys containing copper, the reagent seems unsuitable, since the catalytic decomposition will vary with the form in which the copper exists in the alloy.*

A solution containing sodium chloride and potassium chromate is free from this objection; such a test was developed in early work at New Kensington and appeared to place specimens which had received different thermal treatments in the same order of merit as atmospheric exposure tests (G. F. Sager, R. H. Brown and R. B. Mears, Symposium on Steam-corrosion Cracking of Metals 1944–45, p. 255, esp. p. 266 (A.S.T.M.–A.I.M.E.)).

Since the essential function of hydrochloric acid, hydrogen peroxide or potassium chromate, is to stimulate the cathodic reaction, and thus (since control is cathodic) to accelerate the anodic reaction and the progress of cracking, the same object may be achieved (for some purposes more conveniently) by applying a known current from an external source, the specimen being made the anode. If there is no passivity, a given current will produce the same total destruction of metal for all specimens, but if the material is in a susceptible state, the damage will be concentrated at the fissures, whereas on non-susceptible material the destruction will be spread out. Attachment to a copper plate, without applied E.M.F., also produces

* Recent work at Chicago shows that on alloys containing copper the hydrogen peroxide concentration falls off rapidly with time, whereas pure aluminium causes no appreciable loss (W. H. Colner and H. T. Francis, *J. electrochem. Soc.* 1958, **105**, 377, esp. p. 378).

acceleration. Both methods have been used at Cambridge (D. Whitwham, Dissertation, 1948; H. K. Farmery, Dissertation, 1955).

A discrepancy between the results obtained in salt solution containing hydrogen peroxide and hydrochloric acid respectively is brought out in Chadwick's studies of Al–Mg–Si alloys, whilst the discrepancy between the results of accelerated tests with service behaviour is emphasized by Parker, who is probably right in saying that such tests have led to a " distorted view " of the situation (R. Chadwick, N. B. Muir and H. B. Grainger, *J. Inst. Met.* 1953–54, **82,** 75; R. T. Parker, *ibid.* 1951, **79,** 482; I. G. Slater, *ibid.* 1951, **79,** 483).

Criticism directed against the use of unnatural testing fluids appears more reasonable than that directed against high testing stresses, since under exceptional conditions the working stress may greatly exceed that for which allowance has been made. Stresses are additive, and a member may carry (1) internal stresses due to cold work of the original material, (2) structural stresses introduced in assembly if there have been slight dimensional errors or misalignment (say, of rivet-holes), (3) the service stresses characteristic of normal working conditions, and (4) exceptional stresses of short duration due to sudden acceleration. If these all happen to be tensional, the total may be close to the proof-stress or yield-point. It is well, therefore, to carry out tests, not only at low stresses, but also at a range extending up to the 0·1% proof stress—since " most engineering materials must be expected to be stressed locally well above their yield-points by riveting, caulking, bending etc." (E. A. G. Liddiard, *J. Inst. Met.* 1954–55, **83,** 555).

Laboratory and field methods for studying sulphide corrosion cracking of the type prevalent in the sour oil-fields are described by J. P. Fraser, G. G. Eldredge and R. S. Treseder, *Corrosion* 1958, **14,** 517t.

Great interest attaches to a test carried out at an aircraft construction works on finished aircraft components, which are dipped periodically into 3% sodium chloride by an automatic mechanism—remaining 10 minutes in the liquid and then 50 minutes in air; the cycle is continued for 33 to 45 weeks. It is stated that such a test comes closest to representing service failures (R. N. Hooker and J. L. Waisman, *Corrosion* 1954, **10,** 325).

Other References

Three Symposia provide much information. These are:—

Symposium on Stress Corrosion Cracking of Metals (1944) (Report published jointly by American Society for Testing Materials and American Institute of Mining and Metallurgical Engineers, 1945).

Symposium on Internal Stresses in Metals and Alloys (Inst. Met., 1948).

Symposium on Stress-corrosion Cracking and Embrittlement, organized by the Electrochemical Society, 1954 (Report, edited by W. D. Robertson, and published by Wiley, New York, and Chapman & Hall, London, 1956).

Attention is directed to an admirable survey of mechanical and metallurgical factors in wet corrosion by P. Lacombe, *Métaux, Corrosion, Industries* 1956, **31,** 337. A method showing up stress distribution on a specimen by covering it with a brittle varnish, applying stress and comparing the distance between cracks with those on a calibration piece subjected to known stress, is described by F. Hiltbold, *Schweiz. Arch. angew. Wiss.* 1958, **24,** 56.

A subject important in stress-corrosion cracking is described by D. McClean in his book, " Grain Boundaries in Metals " (Clarendon Press).

Readers should study, when they become available, the papers to be presented at Pittsburgh to the National Conference on the Physical Metallurgy of Stress-Corrosion Cracking organized for April 2 and 3, 1959, by the Metallurgical Society, A.I.M.E.

Recent papers include a study of the electrochemistry of stress corrosion (D. K. Priest, *J. electrochem. Soc.* 1959, **106,** 358) and several discussions of the stress corrosion of stainless steel, including behaviour in alkaline liquids (J. N. Wanklyn and D. Jones, *Chem. and Ind.* (*Lond.*) 1958, p. 888), the effect of applying anodic and cathodic currents (J. G. Hines and T. P. Hoar, *J. appl. Chem.* 1958, **8,** 764), methods for studying sulphide cracking (J. P. Fraser, G. G. Eldridge and R. S. Treseder, *Corrosion* 1958, **14,** 517t), and the relative importance of carbide- and sigma-phases in promoting intergranular susceptibility (D. Warren, *ibid.* 1959, **15,** 213t, 221t).

CHAPTER XVII

CORROSION FATIGUE

Synopsis

A distinction is first drawn between *Stress-Corrosion Cracking* (caused by steady tensile stress in a corrosive environment) and *Corrosion-Fatigue Cracking* (caused by alternating or cyclic stress in a corrosive environment); in service alternating stress and tensile stress are frequently superimposed, giving a fluctuating or pulsating stress. The behaviour of metal towards alternating or cyclic stress in the absence of corrosion (*Dry Fatigue*) then receives discussion, leading to the idea of a *fatigue limit*—a stress-range below which cracking does not occur; no such limit can be expected when corrosion is present. Two-stage experiments in which a period of dry fatigue is followed by corrosion fatigue are described, which show that, if material along gliding planes is to become specially susceptible to chemical or electrochemical attack, it must be "caught on the move".

There is then a brief account of the early history of corrosion fatigue, including Haigh's discovery of the phenomenon and McAdam's classical experiments, especially those showing corrosion-fatigue cracks starting from corrosion pits. Next, the main types of machines for studying corrosion fatigue receive attention. Further two-stage experiments are described, in which the corrosion-fatigue stage is followed by a dry-fatigue stage. These show that if the corrosion is discontinued at a time when only a single crack has developed, and dry fatigue continued, the final life may be shorter than if corrosion had been continued to fracture (since many cracks then develop—producing a situation less weakening than one crack). It is concluded that protection should be applied from the start of the service life of a stressed member.

Methods of prevention or mitigation are then discussed, including numerous types of protective coating (with due consideration of the possible dangers of internal stresses present within the coat), water-treatment, peening, nitriding and cathodic protection.

Readers without previous knowledge of the subject may find it useful to study this chapter in conjunction with the simple treatment provided by A. J. Gould, Internat. Conf. on Fatigue of Metals (Instn. mech. Engrs. and A.S.M.E.) 1956.

Relations between Fatigue, Corrosion Fatigue and Stress Corrosion

Stress Corrosion and Corrosion Fatigue. The phenomenon of stress-corrosion cracking—associated with steady tensile stress acting in a corrosive environment—is met with only on certain materials, and, in general, only after incorrect heat-treatment. In contrast, corrosion fatigue—associated with alternating or fluctuating stress* in a corrosive environment—is met with on almost any material liable to ordinary corrosion. It is sometimes thought that the two types of conjoint action can be distinguished by the character of the fracture—that stress-corrosion cracking is intergranular, whilst corrosion fatigue is transgranular. This is not always the case; the stress-corrosion cracking of magnesium alloys, and that of stainless steel in concentrated magnesium chloride, is mainly transgranular, although in the former case it may be intergranular after certain heat-treatments, whilst in the latter case cracks may follow the grain-boundaries for short distances. Conversely, corrosive fatigue is apparently intergranular on lead; even in steel, although mainly transgranular, cracking can follow grain-boundaries for short distances, where those boundaries chance to run in an appropriate direction (D. Whitwham and U. R. Evans, *J. Iron St. Inst.* 1950, **165**, 72, esp. figs. 3, 4 and 5).

The distinction between stress corrosion and corrosion fatigue is not so sharp as is sometimes supposed. In service, the stress frequently fluctuates about a mean which is not zero, and for this reason corrosion tests under such conditions are sometimes performed in industrial laboratories; the situation is equivalent to a symmetrical alternating stress oscillating about zero as mean, with a steady tensile stress superimposed, and thus represents a condition intermediate between corrosion fatigue and stress-corrosion cracking.

Dry Fatigue. Consider a bar of metal subjected to reversed stress in absence of corrosive influences. Below a certain limit, called the *Fatigue Limit*, which is often about one half of the ultimate tensile strength as commonly determined, no serious change occurs in the material; during the opening stage some gliding may be observed in certain grains, but later the stressing causes only elastic deformation, without serious evolution of heat. At stresses above the fatigue limit, however, the effect is to cause some of the gliding planes gradually to develop cavities, which in turn join up and produce cracks. The life to fracture at different stresses is generally shown by a curve, such as fig. 108, where the horizontal scale represents the

* The following definitions are suggested by A. J. Gould:—

(1) *Alternating Stress* is stress which alternates in sign between positive (tensional) and negative (compressional) values.

(2) *Reversed Stress* is a special case of Alternating Stress where the maximal and minimal values are equal, but of opposite sign.

(3) *Fluctuating or Pulsating Stress* is stress which varies cyclically but does not change its sign.

(4) *Repeating* (*or Repeated*) *Stress* is a special case of Fluctuating Stress, where one extreme value of the stress is zero.

In general, types (1), (3) and (4) are asymmetric and (2) symmetrical.

logarithm of the number of cycles endured before fracture occurs. It is not certain whether there is a sharp elbow at the point X or whether the curve is here rounded off. Moreover, it is not at all certain whether the life should be regarded as infinite below the fatigue limit, or whether merely the probability of cracking within a reasonable space of time becomes

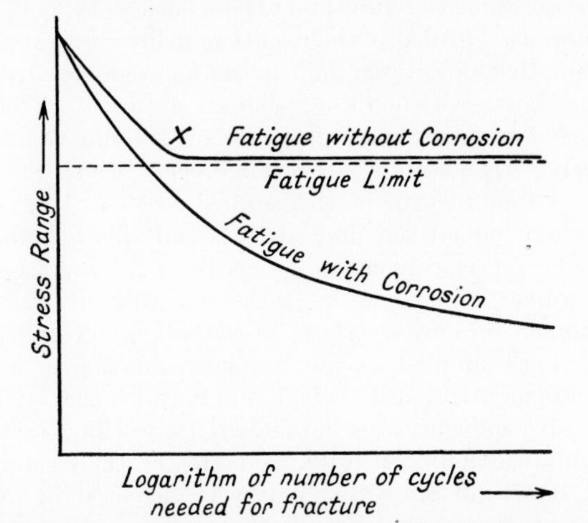

FIG. 108.—Fracture under alternating stress with or without corrosion.

extremely small—which would mean, in fact, that there is no absolute limit. The reason for the doubt prevailing today arises from the considerable scatter between the lives obtained in duplicate experiments carried out under identical conditions on specimens intended to be identical. It would require an extremely large number of long experiments to decide between the contrary opinions held on quite fundamental issues.

Effect of an Oxidizing or Corrosive Environment. Whatever may be the truth about the existence of an absolute fatigue limit under non-corrosive conditions, it is certain that none exists when corrosive action is present; the situation in a corrosive environment is suggested by the lower curve of fig. 108; failure may be obtained even at a low stress-range if the alternating stress continues for a long period. Moreover, on many non-ferrous materials, fatigue tests carried out in air under conditions which are *not intended to be corrosive* yield similar drooping curves, failure being obtained at low stresses if sufficiently long time be allowed. Now the ordinary air fatigue experiment is really a corrosion fatigue experiment carried out under rather mildly corrosive conditions, and there has been speculation as to whether experiments carried out with absolute exclusion of chemical influences, such as sulphur dioxide, water vapour and oxygen, would not reveal the existence of a genuine fatigue limit on non-ferrous materials. Undoubtedly very prolonged experiments would be needed to place the matter beyond doubt, and on most materials no study extended over the necessary period has yet been undertaken. In the case of aluminium, however, and its alloy containing 7% magnesium a study of the curves in very long-continued experiments does seem to suggest the real existence of a limit (T. T. Oberg, *Metal Progr.* 1951, **60,** No. 1, p. 74; cf. C. A. Stubbington, *J. Inst. Met.* 1955–56, **84,** 524, esp. fig. F, p. 525; T. Broom, J. H. Mollineux and V. N. Whittaker, *ibid.* 1955–56, **84,** 357, esp. fig. 7, p. 361; cf. C. L. Bore, *J. roy. Aero. Soc.* 1956, **60,** 331).

It would be desirable to repeat and extend such work, both on aluminium and other materials. Probably the results in many cases would be negative and would show that no fatigue limit exists for some non-ferrous materials even under non-corrosive conditions; indeed, future work may show that even on ferrous materials, where a fatigue limit is today commonly assumed to exist, it merely represents a stress-range below which the life is so long that the risk of fracture can be disregarded.

In cases where no fatigue limit exists—and this includes all cases of corrosion fatigue—it is convenient to speak of an *Endurance Limit*, the highest stress-range which can be withstood without fracture for some particular number of cycles—*which should always be specified*; 5×10^7 cycles is sometimes adopted as the standard duration of a test.

It has been demonstrated by Gough and Sopwith that with experiments in which corrosive influences are minimized by working *in vacuo* or in an inert gas, endurance limits are obtained greater than those obtained in the open air; on 70/30 brass the limit was increased by 26% when the material was tested in partial *vacuo*. The original papers deserve study. Thompson has increased the fatigue life of copper by performing his experiments in nitrogen; Wadsworth has obtained a large increase by working in high *vacuo* (H. J. Gough and D. G. Sopwith, *J. Inst. Met.* 1932, **49**, 93; 1935, **56**, 55; 1946, **72**, 415. N. Thompson, N. Wadsworth and N. Louat, *Phil. Mag.* 1957, **1**, 113, esp. p. 123; N. Wadsworth, *J. appl. Mech.* 1957, **24**, 161).

It should be noted, however, that with copper and some of its alloys, specimens kept wet with water may sometimes enjoy *longer* lives than those tested in air—despite the possibility of corrosion on the wet specimen; this has been observed by several experimenters. The explanation is simple—the water serves to keep the specimen cool; the beneficial effect of water is mainly observed at high stresses, where the life would, in any case, be too short for corrosion to do much harm. Titanium, which is highly resistant to corrosion fatigue, shows a similar effect; the curve in distilled water is found to fall above that in air, probably as a result of cooling (N. P. Inglis, *Chem. and Ind.* (*Lond.*) 1957, p. 188; J. B. Cotton and B. P. Downing, *Trans. Inst. mar. Engrs.* 1957, **69**, 311, esp. fig. 10, p. 318).

Mechanism of Dry Fatigue. If a specimen having a plane surface is subjected to a low tensile stress, the deformation takes the form of slip (p. 372), and gliding within a grain may produce what appears, under a low-power microscope, to be a single ledge, but is seen, under the electron microscope, to be a series of very small steps close together (fig. 109) (W. A. Wood, Int. Conference on Fatigue, *Instn. mech. Engrs.* and A.S.M.E., 1956, Session 6, Paper 4).

If a similar specimen is subjected to alternating stress, it might be expected that gliding should occur to-and-fro on the same planes without serious change. In ideal material, containing no defects other than the dislocations needed for the gliding, this might indeed happen; in the interior of the mass, instead of the dislocations becoming piled up at the grain-boundaries (as happens under a steady tensile stress), they would move to-and-fro across the grain, producing no new effect.

However, with material as we know it, this is surely too much to expect. Any roughness developing on a gliding-plane will be cumulative, since it will interfere with smooth gliding, and thus produce further roughness. It is not surprising that the slip-lines which appear on the grains (following those planes of dense atomic packing which happen to lie nearly at an angle of about 45° to the stress axis) tend to thicken as time goes on; the

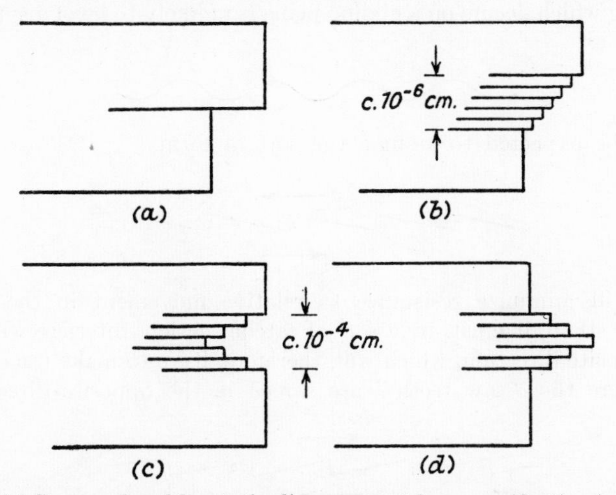

FIG. 109.—(*a*) Step produced by static slip as seen at low magnification in the optical microscope. (*b*) The same as seen in the electron microscope. (*c*) Notch contour and (*d*) Peak Contour, as produced by fatigue (W. A. Wood).

thickening is due to the development of gliding on fresh planes close to, and parallel with, the original ones. Presumably the planes originally adopted have become roughened or otherwise rendered unsuitable as paths of gliding, so that gliding must either cease, or occur on new planes.

Sooner or later, when the planes most suitable for smooth gliding without rupture have been utilized, the decision has to be made, whether gliding (to-and-fro) shall be replaced by elastic deformation (to-and-fro) or whether cracking shall start. Since clearly a form of relative movement which causes the two portions to separate—however locally—must demand more work than the smooth movement of dislocations characteristic of ideal gliding (which, as pointed out on p. 374, involves very little work), the initiation of cracking will only occur at relatively high stresses. Below a certain stress-range (the fatigue limit) only non-destructive elastic deformation will continue; at higher stresses, when once a crack, even on an atomic scale, has started, it is likely to extend, owing to stress-intensification. Thus the rate of damaging, starting slowly, will be self-accelerating; according to one school of thought, the accumulation of damage should obey the Compound Interest Law.*

If the stress is insufficient to start cracking (and the stress required may vary somewhat from specimen to specimen), gliding will cease and the deformation will become elastic; thereafter, there is no further (plastic) damage, and the heat evolved in any cycle is very small so that the energy demanded is, under ideal conditions, almost negligible, as indicated by the very narrow hysteresis loop; below the fatigue limit, therefore, life should be long—some would say, infinite.

The manner in which gliding produces a change likely to obstruct further gliding is well illustrated in the extrusion effect observed by Forsyth. Any roughening which occurs on a gliding plane is unlikely to be of a symmetrical type such as

but may be expected to assume the sort of form

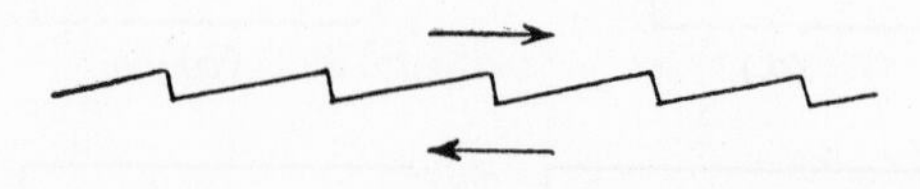

such as will minimize resistance to relative movement in the direction indicated. If so, it must, by a sort of ratchet action, interfere with gliding in the opposite direction, which will, therefore, have to make use of another plane, where the " saw teeth " are sloped in the opposite direction.

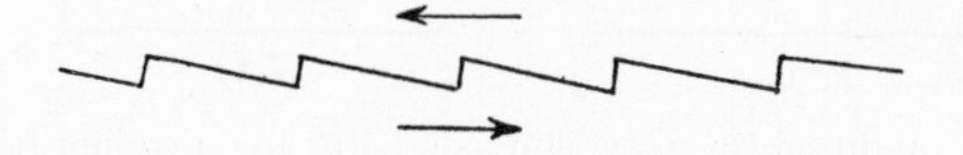

If we develop planes best suited for movement in the two directions as two parallel bands it follows that the slice of material situated between the bands must, on the application of alternating stresses, move as a whole in a single direction, and should be extruded as a ribbon at the free surface (fig. 110). Forsyth has observed such ribbons, less than 0·1 μ thick, issuing from the surface. He has obtained this effect on age-hardening alloys which are capable of local softening through over-ageing and also on unalloyed metal in the work-hardened condition. Thompson, however, has reported it on copper annealed *in vacuo*. Wood has also obtained it on annealed metal and has studied the phenomenon by means of taper photographs. His results have led him to the view that extrusion phenomena are really brought about by a mechanism rather similar to the ratchet action crudely suggested above.

The ratchet action should not only produce extrusion at one end of the

* Cases, however, are known, where cracks formed early in a fatigue test, fail to advance further. For criterion for propagation, see W. J. Harris, *Metallurgia* 1958, **57,** 193.

ribbon, but also the formation of fine crevices ("intrusion") at the other. Such intrusion effects have been studied on copper by Hull, who was able to observe them in the very early stages (less than 1/100 of the estimated life). Hull thinks that these crevices may well constitute the start of the fatigue cracks. His work also confirms the view that local softening is not a necessary factor in extrusion (D. Hull, *J. Inst. Met.* 1957–58, **86**, 425).

The extrusion is, however, accentuated by local softening of an age-hardened or work-hardened alloy, as Forsyth's experiments show; the reason probably is that, if the effect is restricted to a few regions where softening

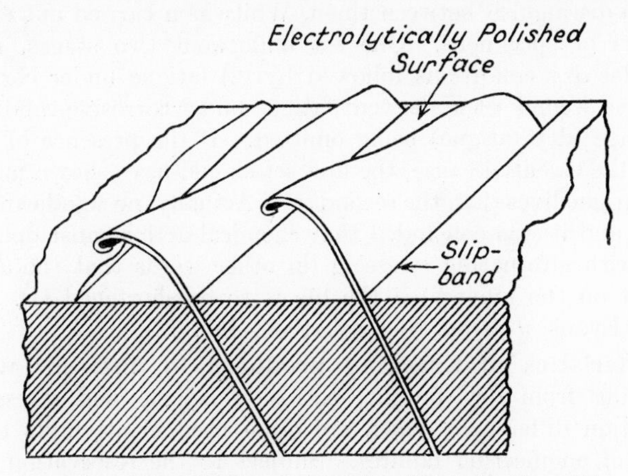

FIG. 110.—Slip-band extrusion (P. J. E. Forsyth).

has taken place, the rate of extrusion—other things being equal—is likely to be much greater.

The experimental facts are stated by P. J. E. Forsyth, *Nature* 1953–54, **171**, 172; *J. Inst. Met.* 1954, **82**, 449; P. J. E. Forsyth and C. A. Stubbington, *ibid.* 1954–55, **83**, 395; 1956–57, **85**, 339; E. A. Calnan and B. E. Williams, *ibid.* 1955–56, **84**, 318; P. J. E. Forsyth, Internat. Conf. Fatigue of Metals, *Instn. mech. Engrs.* and A.S.M.E., 1956, Session 6, Paper 5; N. Thompson, N. Wadsworth and N. Louat, *Phil. Mag.* 1956, **1**, 113, esp. p. 125; W. A. Wood, *Phil. Mag.* 1958, **3**, 692, also Priv. Comm., Nov. 7 and Nov. 28, 1957. Another helpful approach to fatigue is provided by E. Orowan, *Rep. Progr. Phys.* 1948–49, **12**, 185.

Mechanism of Corrosion Fatigue. It has been emphasized that the corrosion of disarrayed material takes place more readily than that of perfectly organized material; corrosion also occurs more quickly at high temperatures than low ones. Under fatigue conditions both disarrayed material and high local temperatures must almost certainly exist along the gliding planes. It is not surprising that, at a stress-range where, for practical purposes, only elastic changes would occur in the absence of corrosion, relative movement continues—ultimately leading to crack-formation—if corrosive influences are present to destroy the interlocking material which would otherwise hold up gliding and substitute elastic deformation.

Thus corrosion fatigue can be attributed *either* to

(1) the disarray of material produced by cyclic stress; *or* to
(2) the rapid relative movement of the atoms (which can be regarded merely as a high local temperature if the movement is haphazard, but may even be more liable to facilitate chemical action if movement becomes enhanced preferentially in certain directions).

These two theories are fundamentally different. The first attributes the preferential chemical attack to *potential energy*, and the second to *kinetic energy*. To distinguish between them, Whitwham carried out experiments on two sets of specimens. One set underwent two stages, namely (1) fatigue under dry conditions followed by (2) fatigue under corrosive conditions. The second set underwent fatigue under corrosive conditions only, the first stage (dry fatigue) being omitted. If the presence of disarrayed material is the essential cause, the first set should have shown much shorter corrosion-fatigue lives than the second set. Actually, no significant difference was found, and it was concluded that chemical action must operate simultaneously with alternating stressing (in other words that the atoms must be " caught on the move ") if the life is to be shortened (D. Whitwham and U. R. Evans, *J. Iron. St. Inst.* 1950, **165**, 72).

Characteristics of Cracking produced in different ways. The cracks arising from dry fatigue, corrosion fatigue and stress corrosion display certain differences which are often used in attempts to establish the causes of engineering failures. Subject to the reservation that there are exceptions to most generalizations, the following statement of the situation, provided by Gould, may be helpful to the reader:—

Dry Fatigue gives *lone* cracks, mostly transgranular, very fine if unetched, but broad if etched, since etching exposes the collateral distressed (slip) material; very sharp-ended.

Corrosion Fatigue gives cracks which are mostly transgranular, and usually occur in *families*; these cracks grow in width as the process extends, often ramifying during their continued extension.

Stress-corrosion Cracking generally gives branching, intergranular cracks, sharp-ended; but on some materials (notably stainlesss steel) it gives transgranular cracks, which may also show fairly sharp ends in sections.

In general, stress-corrosion cracks—especially those connected with caustic soda—are distinguished by their branching character and sharp ends which stand in sharp contrast with the blunt ends of the corrosion-fatigue cracks as seen in micro-sections. This distinction is emphasized both by Cottell in relation to the caustic cracking of boilers, and by Champion when discussing the processing of bauxite in caustic soda. Gould thinks that corrosion-fatigue cracks would really be found to be sharp-ended, if they could be traced to their very ends, which doubtless often lie outside the plane of the section. This may well be true, but the " apparent bluntness " of cracks in micro-sections as commonly prepared is generally regarded as valuable evidence of corrosion fatigue. Certainly Cottell's photographs of cracking in boilers emphasize the contrast between (*a*) the blunt-ended cracks produced by corrosion fatigue and (*b*) the sharp branching intergranular cracks due to caustic cracking (stress-corrosion cracking). He writes, " One would expect that the examination of a considerable number of specimens would be bound to reveal transition types but I have never come across one. The only transition characteristic seems to be a fairly common one, this being the presence along the flanks of (*a*) type cracks of traces of the (*b*) type " (British Engine Boiler and Electrical Insurance Co., Vol. III (1957); also G. A. Cottell, Priv. Comm., June 29, 1957. A. J. Gould, Priv. Comm., July 10, 1957. F. A. Champion, *Chem. and Ind.* (*Lond.*) 1957, p. 967; T. P. Hoar and J. G. Hines, *J. Iron St. Inst.* 1956, **182**, 124, esp. p. 138; 1956, **184**, 166, esp. p. 170).

Some authorities regard a broadening of the cracks near their mouths as an important feature of corrosion fatigue. It should be emphasized, however, that the form of cracks varies greatly with conditions; for instance, the cracks produced by air fatigue in copper tend to be intergranular at high stresses and transgranular at low stresses, both types appearing at the intermediate range of stress (D. S. Kemsley, *J. Inst. Met.* 1956–57, **85**, 420).*

The causes of stress-corrosion cracking have been discussed in Chapter XVI, but reasons may be suggested for the differences between the forms of cracks produced in dry fatigue and corrosion fatigue. Suppose that a single crack has developed to the stage shown in fig. 111(A). If dry fatigue is operating, this one crack will certainly continue in preference to the starting of a fresh crack, since there will be much greater stress-intensification

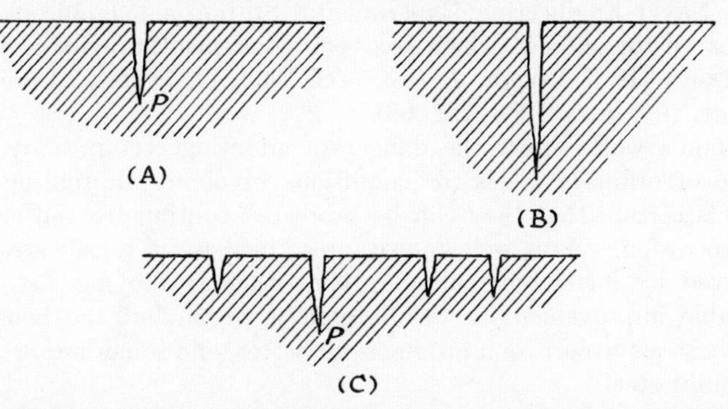

FIG. 111.—(A) Early stages of cracking in dry fatigue or corrosion fatigue. (B) Later stage in dry fatigue. (C) Later stage in corrosion fatigue (schematic).

at *P* than elsewhere; thus we arrive at the long lone crack shown in fig. 111(B). If, however, corrosion fatigue is at work, electric currents must be flowing from the external face as cathode, where oxygen can be replenished, and the tip of the crack, where the metal is distorted, facilitating the escape of cations into the liquid, as anode; the combination of large cathode and small anode produces, as usual, great intensity of attack. It is, however,

* H. K. Farmery remarks that on some Al–Zn–Mg alloys also, the air fatigue cracks are transgranular at low stresses and intergranular at high stresses.

clear that the resistance of the liquid path increases steadily as the crack lengthens; sooner or later, therefore, the anodic reaction at the tip of the original crack will be too slow to balance the cathodic reaction on the face outside, and inevitably fresh corrosion-fatigue cracks will develop. Thus families of cracks arise as indicated schematically in fig. 111(C). In the situation shown in fig. 111(C), the stress-intensification at the tips of the numerous cracks is much less than at the ends of a single crack (fig. 111(A)), explaining certain results established by Simnad (p. 719). The stress-intensification produced at the end of a crack and its magnitude in different cases can be studied by photo-elastic methods; plastic is cut to the required shape, to represent the cracks, and held under stress. Examination under polarized light indicates the stress-distribution, shown by bands of different colours; such observations show that families of cracks produce less stress-intensification than lone cracks.

The Facts of Corrosion Fatigue

Early Work. The discovery of corrosion fatigue during the First World War was made by Haigh, when seeking the explanation of the frequent failure of paravane towing ropes, which were kept in a state of vibration whilst exposed to sea-water (B. P. Haigh, *J. Inst. Met.* 1917, **18,** 55; *Trans. Inst. chem. Eng.* 1929, **7,** 29; B. P. Haigh and D. Jones, *J. Inst. Met.* 1930, **43,** 271).

Between 1926 and 1930 (and at intervals later) McAdam carried out at the U.S. Naval Engineering Experimental Station a magnificent series of researches on the subject, which has been conveniently reviewed by Gough and by Dorey (H. J. Gough, *J. Inst. Met.* 1932, **49**, 78; S. F. Dorey, *Trans. Instn. Nav. Architects*, 1933, **75**, 208).

McAdam's work shows the danger of adopting certain alloy steels in the place of ordinary steels for conditions involving alternating stresses, unless it is certain that they can be protected continually and completely against corrosion. As is well known, great increase in tensile strength can be achieved by introducing alloying constituents; alloying also produces considerable improvement in dry fatigue strength, but the behaviour of most alloy steels to corrosion fatigue is no better, and sometimes worse, than that of mild steel.

For instance, let us compare a steel containing 1·5% Ni, 0·73% Cr and 0·28% C with a plain 0·16% carbon steel (hardened and tempered); the tensile strengths are 62·0 and 29·3 tons/in.^{2} respectively, whilst for an endurance in air of 5×10^7 cycles the fracture stresses are $\pm$ 30·3 and $\pm$ 16·0; thus in both cases the strength has been roughly doubled by introducing the alloy constituents. However, the corresponding endurance stresses in a fresh water are $\pm$ 7·2 and $\pm$ 8·9 tons/in.^{2} respectively—the alloy steel being slightly the worse of the two; in salt water the numbers are $\pm$ 4·0 and $\pm$ 6·2. These and other measurements show that the type of alloying which confers high mechanical strength may utterly fail to confer resistance to corrosion fatigue.

Better results, however, are obtainable on materials which possess resistance to stressless corrosion. 14% chromium steel (stainless cutlery steel), which has a tensile strength and air fatigue strength lower than those of the alloy steel just mentioned, has much better resistance towards corrosion fatigue, the stresses being 16·0 both in fresh and sea-water; Monel metal gives 11·6 in fresh water, and 12.5 (a higher value, surprisingly) in sea-water.

When Gough provided his masterly summary of McAdam's work in 1932, the corrosion-fatigue nature of failures in marine propeller shafts and rudders, steering arms and axles of motor vehicles, boilers and super-heater tubes, pump shafts, rods and bodies and many other situations, was made apparent. At the present time serious problems of a corrosion-fatigue nature exist for drill-pipes in the oil-fields, rock drills in mining districts, locomotive axles (especially where exposed to brine leakage from refrigeratory

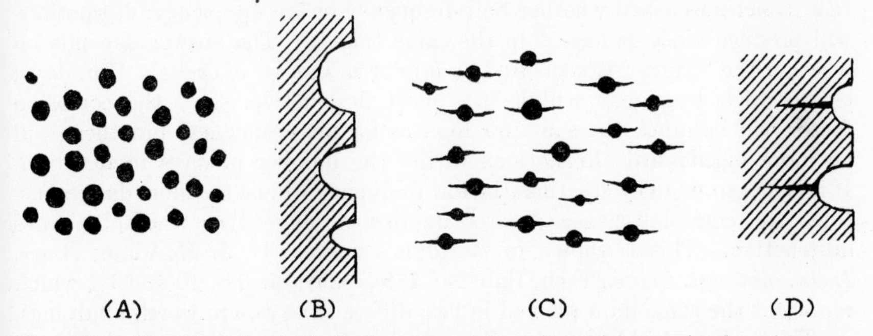

FIG. 112.—Corrosion of rods in fresh water, giving pits which are circular when seen from front (A) and saucer-shaped when seen in section (B). Cracks produced when alternating stress is present (seen from front in C and in section in D) (D. J. Adam and G. W. Geil).

cars), and have caused anxiety in connection with aircraft, winding-cables for mines, rolls at rolling mills and in many other situations.* In some of these cases, energetic work on protective methods has removed the dangers for the time being, but this success has tended to lull those responsible for design and maintenance into thinking that the danger has ceased to exist. Recently, there has been an awakening to the urgent needs for more research.

* Cases of corrosion-fatigue cracking in steam pipe systems at power stations are discussed in a report issued by the British Electrical Authority in 1949, where intermittent thermal stress is considered a more potent cause of trouble than vibration. Numerous cases of cracking in various parts of boiler systems, probably due to fatigue or corrosion fatigue, are described by G. A. Cottell and others in British Engine, Boiler and Electrical Insurance Co. Report 1954, Vol. II, pp. 80, 130, 138, 162, 165, 221, 245 and 252. The case on p. 165 (cracking of chromium-plated crackshafts) was probably due to residual tensile stress in the plating and hydrogen in the steel, and advice is given to bake chromium deposits at 300–425°C., and not at 200°C., which makes matters worse. This advice is largely based on work by H. L. Logan and C. E. Heusner; see also p. 724 of this book. The question of corrosion fatigue on ships' hulls is discussed, in connection with a paper by A. J. Gould, by W. A. D. Forbes, A. J. T. Gibbons and R. J. Brown (*J. Iron St. Inst.* 1950, **165,** 294–5).

One of the interesting features arising from McAdam's work is the connection between pitting and cracking. For many of his tests, McAdam used a fresh water which, in the absence of stress, produced on steel rods numerous pits, circular when viewed on the surface of the rod (fig. 112(A)) and hemispherical or saucer-shaped when viewed in section (fig. 112(B)); in cases where alternating stress was present, cracks developed laterally from opposite sides of the semi-circles, as seen on the surface (fig. 112(C)) and passed down as roots, as seen in the section (fig. 112(D)). The relation between pitting and the initiation of cracks is brought out in Gough's work on aluminium; it appears that a few large pits have a greater effect than a host of small ones, as would be expected, and that fracture is essentially due to preferential corrosion along the previously formed slip-bands (H. J. Gough, *J. Inst. Met.* 1932, **49**, 17, esp. p. 78).

Another feature brought out in McAdam's work is the effect of frequency. It is sometimes asked whether high-frequency or low-frequency alternations will produce most damage "in the same time"? The answer depends on whether the "time" is measured in *days* or in *number of cycles*. If in days, clearly high frequency will be the most destructive, since the corrosion factor will be much the same for high as for low frequency, but there will be more mechanical alternations within the number of days in question. If it is measured in cycles, then the low frequency will be the more dangerous, since the corrosion has a greater number of days over which to cause destruction. This is shown in McAdam's curves (D. J. McAdam, *Amer. Instn. min. met. Engrs.*, Tech. Pub. No. **175** (1929); his figs. 10 and 11, which represent the same data plotted in two different ways, are worth studying).

The reader would probably do well first to study the lecture by H. J. Gough (*J. Inst. Met.* 1932, **49**, 17) before going on to the individual papers of D. J. McAdam, *Proc. Amer. Soc. Test. Mater.* 1926, **26**, II, 224; 1927, **27**, II, 102; 1928, **28**, II, 117; 1929, **29**, II, 250; 1930, **30**, II, 411; 1931, **31**, II, 259. *Trans. Amer. min. met. Engrs.* 1929, **83**, 56; 1932, **99**, 282; also his papers with R. W. Clyne and G. W. Geil, *Bur. Stand. J. Res.* 1934, **13**, 527; 1940, **24**, 685; 1941, **26**, 135; *Proc. Amer. Soc. Test. Mater.* 1941, **41**, 696.

The resistance of many materials to corrosion fatigue varies greatly with the state of heat-treatment. This is rather particularly true of certain steels much used in mining drill rods—whether plain carbon steels or steels containing chromium and molybdenum. Failure usually occurs at the part of the rods near the junction of the heat-treated and untreated ends, at the place where spheroidization starts (D. S. Kemsley, *Nature* 1955, **175**, 80; also "Mining and Quarrying Engineering", Sept. 1956; R. W. Manuel, *Corrosion* 1947, **3**, 415).

In view of the behaviour of copper mentioned on p. 704, the use of copper alloys to withstand alternating stress under corrosive conditions naturally suggests itself. It appears that aluminium bronzes containing iron and nickel, although weaker than stainless iron or steel under dry fatigue, are stronger under corrosion fatigue conditions (J. McKeown, D. N. Mends, E. S. Bale and A. D. Michael, *J. Inst. Met.* 1954, **83**, 69).

Machines used for Corrosion-fatigue Studies. Even at the time when McAdam's work was being published, there were practical men in his own country arguing that, if the results which he was publishing were correct, the stressed members in certain appliances, vehicles or structures should have failed which, in point of fact, had not failed. Reasons for apparent discrepancies between laboratory experiments and service experience are always worth discussion, and in this case a probable explanation can be tentatively suggested.

There are two main types of fatigue-testing machines; the push-pull type where a rod is subjected alternately to tension and compression, and the rotating-bar type where a rod or wire, suitably bent into a very slightly bow-shaped form, is made to rotate, so that any particular point on the surface is in tension when it is on the convex side and in compression when it comes round to the concave side. McAdam used such a machine, applying a water-stream so as to sweep the whole surface. Now with this arrangement, there is a current due to the cell

Metal on tensional side (Anode)	Water	Metal on compressional side (Cathode)

Presumably this current helps the corrosion. No such current is generated when a push-pull machine is used, since at any one moment the whole area is in tension or the whole in compression; there is never an occasion when one part is in tension and the other part in compression. Simnad has shown (p. 390) that the potential of a single point on a bowed wire alters, in marked degree, when the wire is rotated, being more negative when the point is in tension than when it is in compression. Thus the currents mentioned cannot be neglected. There are other reasons why, even in dry fatigue, rotating-bar machines produce different results from push-pull machines, but probably the main cause operating in corrosion fatigue is the presence of the differential stress currents just mentioned (cf. Farmery, p. 672 of this book). Whatever the causes, Gould's experiments have established the fact that the endurance of steel under push-pull conditions in sea-water is consistently higher than under rotating-bar conditions (A. J. Gould, *J. Iron St. Inst.* 1949, **161,** 11, esp fig. 3; for the effect of stress on potential see U. R. Evans and M. T. Simnad, *Proc. roy. Soc.* (*A*) 1947, **188**, 372, esp. p. 383; *J. Iron St. Inst.* 1947, **156**, 531; H. K. Farmery and U. R. Evans, *J. Inst. Met.* 1955–56, **84**, 413).

Clearly, there are cases where the rotating-bar test is the appropriate one, for instance on locomotive axles; but in other cases, the use of rotating-bar machines will give results which are unduly alarming if applied to cases when there are no rotating bars in service. Testing should be conducted as far as possible in a machine representing service conditions.

There are various forms of rotating-bar machines suitable for corrosion-fatigue research, based closely on the forms used for dry fatigue research. In the cantilever type (fig. 113), the rod is supported at one end, the other being pulled slightly downwards by means of a weight. If the diameter is uniform, the stress will be maximal at the supported end—an inconvenient situation. McAdam overcame this by using a tapered specimen calculated to make the stress uniform (within 1·5%) over a length of 1·5 inches—the maximal value being reached at the central point of that length.

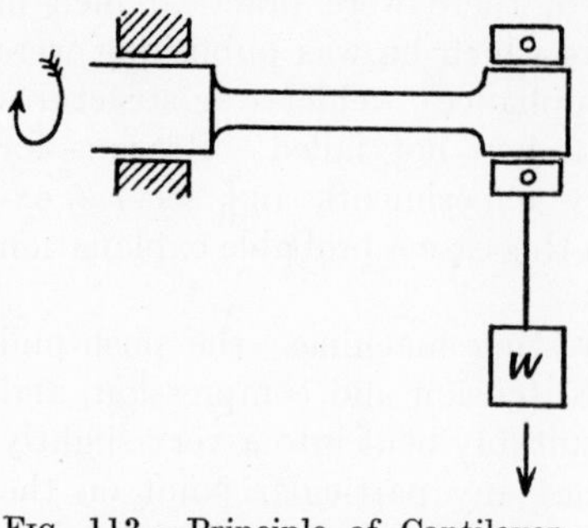

Fig. 113.—Principle of Cantilever Fatigue Testing-Machine (schematic).

If (fig. 114) the rod is supported at two places, with two equal weights suspended from collets at two other places (or a single weight distributed equally between these two points of suspension), the well-known four-point loading is obtained, which gives approximately uniform stress over the length between the two weights—a very real advantage. For wire also machines are available which give uniform stress over a considerable length, as in the Kenyon machine (fig. 115) where a wire is bent into a bow in a

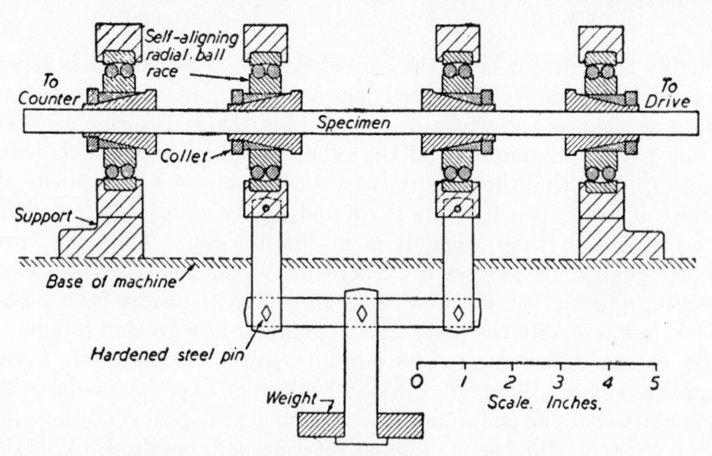

Fig. 114.—Fatigue Testing-Machine giving uniform stress over considerable length (schematic).

vertical plane by means of four-point loading, most of the uniformly stressed area being immersed in the corrosive liquid. In the Haigh-Robertson machine, the wire is again bent into a bow, but in a horizontal plane and by forces applied only at its two ends, so that the stress is not uniform, except over a short length near the centre.

In most of the Cambridge researches, the corrosive liquid has been applied at a chosen point only, in such a way as to give a reproducible water-line. Gould, using horizontal rod specimens, applied the liquid, by means of a drip feed, to a sling of cotton tape, which passed under the specimen and just out of contact with it; Huddle (fig. 116) used a wick held taut in a plastic bow, with a screw adjustment, capable of bringing the string to just such a distance from the rod that a narrow ring of liquid was carried around the periphery when the rod was rotating; Simnad used a similar arrangement, but substituted for the wick a glass rod waxed over its entire surface except for an unwaxed streak on the side nearest to the wire—which formed a pathway for the liquid. The liquid adopted was generally a solution of potassium or sodium chloride—but in some of Simnad's experiments N/10 hydrochloric acid was used.

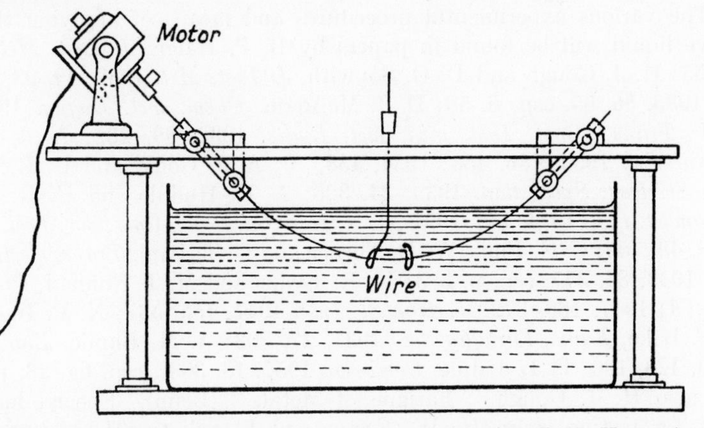

FIG. 115.—Corrosion-fatigue Testing-Machine for wire (J. N. Kenyon).

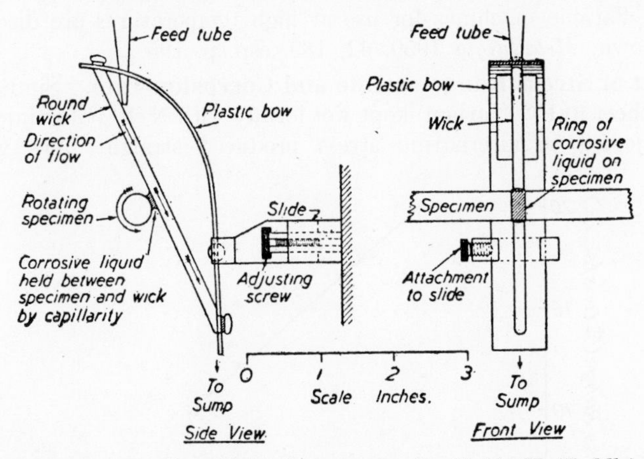

FIG. 116.—Feeding device for corrosive liquid (A. U. Huddle).

These arrangements proved convenient and provided a sharp and reproducible boundary between the wet and dry area. Most other experimenters have preferred to keep wet a large area of the specimen. In Gerard and Sutton's work, 3% sodium chloride was applied as a spray of fine droplets, whilst in much of McAdam's work, carried out on a cantilever machine, fitted with tapered specimens, the liquid was applied diagonally so as to sweep the surface from the outer to the inner fillet. Inglis and Lake—and also Binnie—have used drip feeds.

There are certain advantages in keeping a large uniformly stressed area in contact with the corrosive liquid, as in the Kenyon machine, since if only a small area is wetted, there is a chance that sensitive spots—distributed over the surface as a whole—may nevertheless be absent on the small area in question. There is little advantage in having a big area kept wet if the highest stress is only reached on a small part of it.

The various experimental procedures and means of applying the corrosive liquid will be found in papers by B. P. Haigh, *J. Inst. Met.* 1917, **18**, 55; H. J. Gough and D. G. Sopwith, *J. Inst. Met.* 1932, **49**, 93, esp. p. 97; 1935, **56**, 55, esp. p. 58; D. J. McAdam, *Chem. met. Engng.* 1921, **25**, 1081; *Trans. Amer. Inst. min. met. Engrs.* 1932, **99**, 282; A. J. Gould, *Engineering* 1933, **136**, 453; 1934, **138**, 79; A. J. Gould and U. R. Evans, *Iron St. Inst. Spec. Rep.* 1939, **24**, 325; A. U. Huddle and U. R. Evans, *J. Iron St. Inst.* 1944, **149**, 109P; J. N. Kenyon, *Proc. Amer. Soc. Test. Mater.* 1940, **40**, 705; C. G. Fink, W. D. Turner and G. T. Paul, *Trans. electrochem. Soc.* 1943, **83**, 377, esp. p. 383; U. R. Evans and M. T. Simnad, *Proc. roy. Soc.* (*A*) 1947, **188**, 372; *J. Iron St. Inst.* 1947, **156**, 531; N. P. Inglis and G. F. Lake, *Trans. Faraday Soc.* 1931, **27**, 803; A. M. Binnie, *Engineering* 1929, **128**, 190; F. L. Laque, *Corrosion* 1957, **13**, 303t, esp. fig. 23, p. 311t. See also H. J. Gough, "Fatigue of Metals" (Benn). Special machines exist for testing strip (H. W. Foster and V. Seliger, *Mech. Engineering* (*U.S.A.*) 1944, **66**, 719); these could probably be adapted for corrosion fatigue. Fatigue machines for use at high temperatures are discussed by J. McKeown, *Metallurgia* 1950, **42**, 189, esp. p. 195.

Effect of Stress Range on Life and Corrosion-rate. Simnad's tests on steel sheet (0·19% carbon) kept wet locally with N/10 potassium chloride and subjected to alternating stress produced straight lines when the

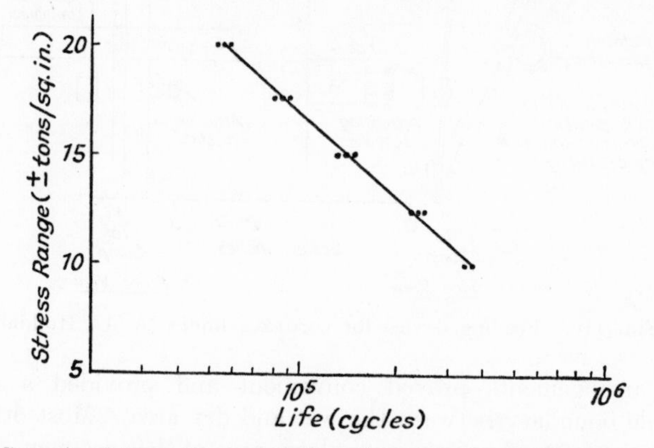

FIG. 117.—Corrosive-fatigue lives plotted on logarithmic scale against Stress Range (M. T. Simnad and U. R. Evans).

logarithm of the life was plotted against the stress-range (fig. 117). All these measurements were made below the fatigue limit (which was 21 tons/in.^{2} —as obtained on wire coated with petroleum jelly containing zinc chromate, to prevent accidental corrosion); thus breakage only occurred when corrosion —as well as stress—was present. The rate of corrosion of the iron (which remained constant over a considerable time period) was found to rise with rising stress (fig. 118). The fact that stress increases corrosion-rate cannot

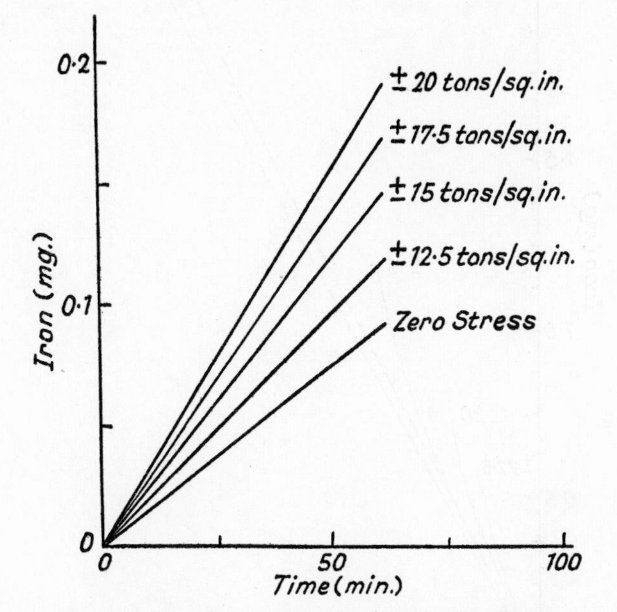

FIG. 118.—Initial rates of Corrosion in M/10 Potassium Chloride showing accelerating effect of stress (M. T. Simnad and U. R. Evans).

be explained by an increased surface due to cracks advancing into the metal, since it was noticed in the opening stages of the experiment before sensible cracking would have occurred. Nor can it be attributed entirely to the removal of obstructive corrosion-products by the alternating stress; special experiments in which rust was intentionally removed from time to time showed that this did indeed increase the rate of attack, but the quantity of rust remaining attached to the specimen itself increased with the stress-range. Nor can it be attributed to cracking of the original air-formed film by the stressing, since the effect of stress was met with in tests in N/10 hydrochloric acid; a marked rise occurs after a time—earlier at high stresses than at low ones (fig. 119). The results are, however, explicable on the view that metal in motion on the gliding-planes requires less activation-energy to pass into the solution than motion-less metal, increasing the corrosion and facilitating the propagation of cracks (U. R. Evans and M. T. Simnad, *Proc. roy. Soc.* (*A*) 1947, **188**, 372; *J. Iron St. Inst.* 1947, **156**, 531).

Effect of Short Periods of Corrosion Fatigue followed by Dry Fatigue. Perhaps the most important feature of Simnad's work was his two-stage experiments in which he exposed specimens to corrosion fatigue in N/10 potassium chloride, washed and dried them, and then exposed

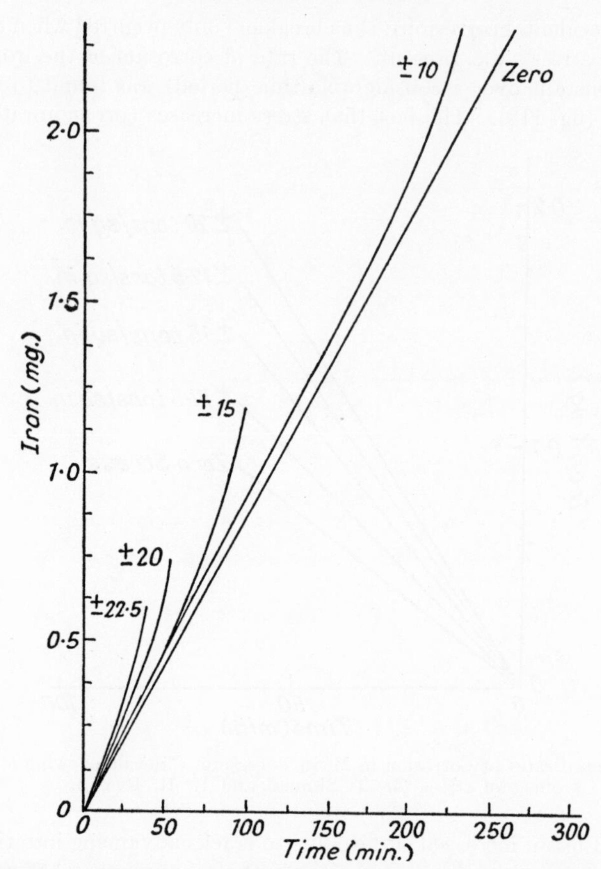

FIG. 119.—Rates of Corrosion in N/10 Hydrochloric Acid showing effect of Stress. The figures on curves are the stress-ranges in tons/in.[2] (M. T. Simnad and U. R. Evans).

them to fatigue in air, in absence of corrosive substances (in one special set of experiments, the specimens were, in the second stage, wetted with potassium chromate—to inhibit accidental corrosion). Simnad found that when the first (corrosion) stage was shorter than a certain " critical period ", the life in the second stage was long, probably because the corrosion-fatigue cracks had not had time to start before corrosion was discontinued. (If the corrosion-stage had been omitted altogether, the life should, according to one view of fatigue, have been " infinite ", since the experiments were carried out below the fatigue limit.) If, however, the corrosion and fatigue period appreciably exceeded this critical period, the air-fatigue life became extremely short—probably owing to the formation of a single corrosion-fatigue crack which acted as a stress raiser, and extended into the metal during the air-fatigue period.

If the corrosion-fatigue period was further increased, the total life became longer again. This may seem surprising, but an explanation soon suggests itself. As stated on p. 710, the advance of a single corrosion-crack into the metal must sooner or later slow down, owing to the increasing resistance of the liquid path connecting the cathodic area (the outer surface) with the anodic area (the tip of the crack); at that stage new cracks will start at other places on the surface. If corrosion is discontinued and dry fatigue substituted when there is only one crack, the life will be short owing to the stress-intensification at its tip. If corrosion continues until many cracks have appeared, the life will be longer since, as already stated, a family of cracks is less weakening than an isolated crack.

TABLE XXV

LIVES OF STEEL CORROSION-FATIGUE SPECIMENS (M. T. Simnad and U. R. Evans)

[All figures should be multiplied by 10^5 to obtain lives expressed in number of cycles or by 0·28 to obtain lives expressed in hours]

Liquid	Stress (tons/in.^{2})	Critical Period	Life if Corrosion is continued up to Fracture	Worst moment for stopping Corrosion	Life if Corrosion is stopped at worst moment	Remaining Life after stoppage of Corrosion	Remaining Life if Corrosion had been continued
N/10 KCl	± 20	0·9	5·2	2·3	3·8	1·5	2·9
N/10 KCl	± 17·5	2·0	9·0	4·0	6·1	2·1	5·0
N/10 KCl	± 15	4·1	13	6·4	10	3·6	6·6
N/10 KCl	± 12·5	9·1	21	10	14	4·0	11·0
N/10 HCl	± 20	1·5	3·5	2·4	3·0	0·6	0·9
N/10 HCl	± 15	3·3	5·7	no appreciable shortening of life by stopping corrosion			
N/10 HCl	± 10	9·1	13·2				

The figures shown in Table XXV bring out the fact that, if corrosion is once allowed to start, and subsequently protective measures are applied at an unfortunate moment, the remaining period of survival may be shorter than if corrosion had been allowed to continue up to fracture, being sometimes reduced by a half. Such figures should not be applied too confidently to practical cases; the liquids used by Simnad were ones which would not be met with in service, the conditions of wetting (a sharply-defined ring of liquid around the wire specimen) were unnatural, whilst the frequency—chosen to give results in a reasonably short space of time—was exceptionally high (6,000 R.P.M.). Nevertheless they do serve to indicate the necessity, in practical engineering service, of applying protective measures *at the outset* (they should not be deferred until after a trial run). If, at a later stage, it is discovered that some place has been left unprotected (whether by original inadvertence, or by subsequent damage to a protective coating), earnest consideration must be given to the possibility that a belated application of " protective measures " may shorten whatever life would otherwise remain. Before any decision, the part must be most carefully examined for incipient cracks, and in cases where failure would be disastrous, renewal of the part may be the only safe course.

The matter would seem to deserve more attention from Engineers than it receives. In recent years, several large congresses have been organized to study Fatigue, and imposing discussions of Fatigue Life presented; perhaps one short paper on corrosion fatigue may be included in the programme, as a concession to those who urge that corrosion is important.* Yet if, in service, corrosive influences as well as alternating stresses are present, they can make complete nonsense of all the theoretical predictions about Life which have ignored corrosion, unless suitable protective measures are applied from the first. If the time allotted to corrosion fatigue at fatigue congresses is a measure of interest in a subject affecting public safety, the situation is a serious one. Possibly it is felt that until more definite knowledge of corrosion fatigue is available—capable of providing a basis for the calculation of Life—the subject is not ripe for discussion; if that is so, then more research work is urgently needed.

Prevention of Corrosion Fatigue

General. Our knowledge regarding measures to be taken against corrosion fatigue is unsatisfactory. Three kinds of information may be used in selecting a protective scheme:—

(1) *Past Experience.* If it is known that, in the past, a certain protective method has protected material A against environment X, this gives some confidence in adopting it for protecting A against X on the next occasion, although trivial alterations of conditions may sometimes have unexpected effects. But to decide therefrom whether A, thus protected, will withstand environment Y, or whether material B will withstand environment X, involves an element of guess-work, even if B and Y are not very different from A and X.

(2) *Field Corrosion Tests without Stress.* Much information is available about the manner in which various protective coatings prevent corrosion in the absence of stress (see Chapters XIV and XV). Since stress-less testing involves no machines, and requires only the simplest set-up, it is not difficult to accumulate information. Now corrosion fatigue cannot proceed without corrosion, and it may be argued that, if a thoroughly reliable protective scheme can be found, the menace is eliminated. In applying stress-less tests to fatigue conditions, however, it is necessary to ensure:—

(*a*) that the alternating stress does not crack the coat or open out pores which otherwise would be unimportant.

(*b*) that the protective coat (which may be a metal) does not itself carry

* Some of these summary papers—despite the ungenerous ration of time often allowed to their authors—do provide useful introduction to students starting the subject. Attention is directed to the excellent paper by A. J. Gould, quoted on p. 701.

tensile stresses, which might start fatigue cracking in the basis metal even in the absence of corrosion.

(3) *Laboratory tests under Corrosion-fatigue Conditions.* Laboratory tests may be useful in ascertaining whether conditions (*a*) and (*b*) are fulfilled; being carried out at high stress and high frequency, they occupy little time in providing results. But as a means of establishing the number of cycles which the material, protected in the same manner as the test specimen, would endure in service before fracture, laboratory tests are useless. The engineer, in dealing with dry fatigue, is used to saving time by adopting high frequency, and often tends to assume that the number of cycles withstood in service (at lower frequency) will be much the same as that withstood in the laboratory test. Whether or not such assumptions are justified for dry fatigue, it is dangerous in corrosion fatigue to use unnaturally high stress or high frequency, since the corrosive agency has not the same time time to act. Doubtless the degree to which different protective schemes prolong the life in accelerated laboratory tests may provide tentative information regarding their relative merits in service, but to assume that the multiplication of life will be the same in service as in the test is wrong. A protective scheme which increases the life from an hour to a day in the laboratory test, will not necessarily increase it from a month to two years in service.

It is clear that laboratory tests on the effects of protective coatings, carried out at stress-ranges and frequencies more like those met with in service, are needed. Each test will occupy a long time, and a research programme will demand whole batteries of machines, but until such information has been acquired, the situation must be regarded as hazardous.

Nevertheless, by the common-sense use of the three kinds of information suggested above, decisions are being taken today which have not often led to disaster. The author possesses no figures, but he is inclined to guess that fewer break-downs are due to the choice of a wrong protective scheme than to the omission of protection altogether, or to a gap accidentally left in the protective coating at a danger-point.

Protection by Coatings. The important work on light alloys carried out at the Royal Aeronautical Establishment under Dr. Sutton about 1935 still deserves quotation in connection with the prevention of corrosion fatigue, although the material then used on aircraft was probably in a different condition from that favoured today. Specimens were tested on cantilever machines both in air and also in salt-spray. Against salt-spray, lanolin coatings and even cadmium-plating availed little, but zinc-plating gave substantial protection. Coatings of organic resins and enamels afforded a high degree of protection where the alloy had previously been anodized. The best results were obtained by first anodizing the material in a chromic acid bath and then applying a coating of synthetic resin stoved for 2 hours at 100°C. (the low temperature being doubtless chosen to avoid increased risk of stress-corrosion cracking through changes in the alloy). After such treatment, the material endured 10^7 cycles in salt-spray at $\pm$ 12·2 tons/in.^{2} whilst at $\pm$ 9·5 tons/in.^{2} the endurance was 5×10^7 cycles; both these stress-ranges exceeded those withstood by uncoated material tested in air (9·1 and 9·0 tons/in.^{2} for 10^7 and 5×10^7 cycles respectively), and lay far above the numbers for uncoated material in salt-spray (3·3 and 3·0). The original paper deserves study (I. J. Gerard and H. Sutton, *J. Inst. Met.* 1935, **56**, 29).

In view of the use of anodizing as a basis for synthetic resin in Gerard and Sutton's work, it should be pointed out that anodizing by itself often gives bad results; Gould—using a different type of alloy—found it to be unsatisfactory; Gerard and Sutton themselves found very low values (4·8 and 4·5) for anodized specimens carrying only a lanolin coat.

A deterioration of air-fatigue behaviour by anodizing was noted by Inglis and Larke, who attributed it to local stress-intensification caused by small surface defects in the anodized layer; these were revealed by microscopic examination. Hoar, however, suggests that it is due to grain-boundary attack, producing nicks at the grain-boundaries; such nicks, which have been clearly shown up by Lacombe's work, are likely to be effective stress-raisers. Whichever view is correct, it would seem that the conditions of anodizing may be important in deciding whether anodizing will be beneficial or detrimental. Further work is called for, but the existing data and the various interpretations deserve study before anodizing is adopted, even though additional coatings (besides anodizing) are to be applied (N. P. Inglis and E. C. Larke, *J. Inst. Met.* 1954–55, **83**, 117; T. P. Hoar, *ibid.* 1954–55, **83**, 556. P. Lacombe, *Trans. Inst. Met. Finishing* 1954, **31**, 1, esp. fig. 5).

There is some reason to believe that anodizing in the chromic acid bath gives better fatigue resistance than anodizing in the sulphuric bath. Russian tests showed that the (dry) fatigue strength declined after anodizing in sulphuric acid but rose after anodizing in chromic acid; however, the alloys used in the two cases were apparently not the same and the film-thicknesses very different. British work has shown that, although anodizing in sulphuric acid causes a deterioration in fatigue resistance, this can be largely restored by subsequent sealing in hot water; probably the improvement is connected with the fact that sealing converts a slight tensile stress into a slight compressional stress—presumably through the expansion accompanying hydration. This change in the stress-direction was brought out in the work of Bradshaw and Clarke, and the results fit well with later work in the same laboratory by Hammond and Williams (p. 725), who showed that the conversion of tensile to compressional stress greatly improves fatigue behaviour (A. V. Shreyder, A. V. Byalobzhesky, Z. T. Zagetsenko and B. V. Serebrennikov, *Metallovedeniye i Obrabotka Metallov* 1956, **4**, 14; W. N. Bradshaw and S. G. Clarke, *J. Electrodep. tech. Soc.* 1945, **24**, 147, esp. p. 161; S. G. Clarke, Priv. Comm., March 14, 1958).

Sutton's experiments were mainly carried out on aluminium alloys of the H15 type (containing copper, with magnesium, silicon and manganese) —apparently hardened by ageing at room temperatures. Similar alloys are still used today, but are commonly artificially aged, which provides a little additional strength. There is also a preference for alloys containing zinc as well as copper, when very high strength is required; in contrast, alloys containing magnesium and silicon but no copper (H10 or H30 type*) will be chosen in situations where the highest strength is not necessary but where corrosion is feared. It is highly desirable to have on record test results, analogous to those of Gerard and Sutton, but carried out on modern alloys after the heat-treatment usual today. There is some reason to think that much of the required information already exists locked away in files, and probably its non-appearance in journals may be due less to a policy of secrecy than to lack of time for writing up. It is, however, unfortunate that, when publication occurs, the information disclosed is sometimes incomplete. For instance, the excellent paper by Inglis and Larke on H10 alloys refers to specimens painted in accordance with a "carefully prepared schedule"; perhaps this only means the type of protection advocated in Chapter XIV but—probably owing to no fault of the authors—the vagueness deprives the paper of much of its value.

The H10 alloy (with 0·58% Mn, 0·61% Mg. and 0·94% Si) was studied in the fully heat-treated condition. It was found that for unpainted specimens the endurance strength over 10^8 cycles in 3% sodium chloride was little more than one quarter that in air. However, the endurance of the painted material in the salt solution was somewhat better than that of the unprotected material tested in air (N. P. Inglis and E. C. Larke, *J. Inst. Met.* 1954–55, **83**, 117; W. H. L. Hooper and N. P. Inglis, *Chem. and Ind.* (*Lond.*) 1954, p. 1334, esp. p. 1343).

It would seem that the painting schemes available for light alloys today, carefully applied, are capable of meeting the menace, although it must not be forgotten that paint-coats undergo slow deterioration in service—a factor which laboratory tests do not generally introduce. The main danger, however, is that the application of paint may be made *too late*, or *with insufficient thoroughness*, or that *maintenance may be inadequate*. This may not always b the fault of the inspectorate and maintenance staff, who, with present designs, sometimes experience the greatest difficulty in even seeing the places most requiring to be watched—a matter which designers should bear in mind.

Ceramic coats improve fatigue strength considerably, and some useful information is provided by W. J. Plankenhorn, *J. Amer. ceram. Soc.* 1954, **37**, 281.

Metallic coats have been discussed in connection with corrosion set up at gaps (p. 636); since in service the stress usually fluctuates about a mean that is not zero, the coats must be chosen to deal with such fluctuations. It used to be considered that coatings which were anodic to the basis metal were safe, and those which were cathodic were dangerous, since the intense corrosion started at a gap in a cathodic coat might develop under stress into a crack. Thus Kenyon found that the plating of steel with a noble material like bronze actually shortened the life by reason of fatigue cracking set up by galvanic action at imperfections in the plating; no such deterioration

* Manganese is optional in H10 and compulsory in H30; H. K. Farmery remarks that H30 is now usually preferred.

occurred when the wire had first been hot-galvanized before being plated with copper (J. N. Kenyon, *Proc. Amer. Soc. Test. Mater.* 1940, **40**, 705).

However, the presence of stresses and cracks in the coating metal and its power to affect the menace of hydrogen in the basis metal may be quite as important as electrochemical polarity. Thus whilst zinc applied in a proper condition should provide a modicum of cathodic protection at gaps in a coat, a zinc coat applied to steel by hot-dipping may have a thick alloy layer which easily develops cracks; these may extend into a steel during alternating stressing, and, even in the absence of corrosion, the fatigue life will be short. It is possible to apply zinc by spraying, if the preliminary roughening does not cause too much deterioration of fatigue strength,* or by electroplating if hydrogen embrittlement can be avoided. Cadmium is in favour in some quarters, but may shut in hydrogen left by previous pickling, and the pickling requires careful specification as well as the plating. It might be thought that the use of anodic pickling instead of an acid dip would avoid trouble, but cases are known where anodic treatment itself leaves poor fatigue properties. In the case of chromium, the coating may itself contain much hydrogen; in one (French) method, this is afterwards removed by what is in effect mild anodic treatment (Bertin, "Chrome dur" 1948, p. 33).

It is important to avoid discontinuities in nickel-plating for electrochemical reasons, but the main menace of nickel-plating is the fact that it often carries tensile stresses. If these can be eliminated by the use of special baths or otherwise, its value would be greatly increased. Many years ago Barklie showed that the presence of a lead layer between steel and the nickel coat serves to insulate the steel from any cracks which may appear in the nickel and which would otherwise extend into the basis, but this may not be a practical solution under modern conditions. The pertinent information is provided by R. H. D. Barklie and H. J. Davies, *Proc. Instn. mech. Engrs.* 1930, Vol. I, p. 731.

Today a different method is available, namely to use nickel-plating baths of special composition (not always fully disclosed, but probably in many cases containing nickel sulphamate) which produce coatings containing compressional, not tensional, stresses.

The tensional stresses in chromium-plating on steel are particularly detrimental to fatigue behaviour—even in the absence of corrosion, and here heat-treatment after plating is largely used to remove the dangers. It is important to use the right conditions. Treatment at temperatures below about 440°C. actually increases internal stresses and depresses the fatigue limit. Above 440°C. the stresses are reversed, becoming compressional, and the fatigue behaviour greatly improved; Hammond's measurements show that it exceeds that of unplated steel. Another method of overcoming the difficulty is to shot-peen after plating, which also, according to Almen's

* See R. C. Miller and A. W. Brunot, *Welding J.* 1954, **33**, 275S, who compare the effects of rough threading with those of grooving and roughening. Grit-blasting is recommended by W. L. Williams, *Proc. Amer. Soc. Test. Mater.* 1949, **49**, 683.

data, provides a fatigue strength superior to unplated steel. Engineers should study the original papers of C. Williams and R. A. F. Hammond, *Trans. Inst. Metal Finishing* 1955, **32,** 85; J. O. Almen, *Product Engineering* (*New York*) June 1951, p. 109; E. R. Gadd, Internat. Conference on Fatigue of Metals (Inst. mech. Engrs. and A.S.M.E.) 1956; L. Mehr, T. T. Oberg and J. Teres, *Monthly Rev. Amer. Electroplaters Soc.* 1947, **34,** 1345.

The latest work of Williams and Hammond brings out a rectilinear relationship between the fatigue strength of the unplated steel and the change of fatigue strength produced by chromium-plating. A weak steel has its fatigue strength increased by plating, and a strong steel has its fatigue strength diminished. Their curves make it possible to predict from the fatigue strength of the unplated steel the value of the fatigue strength after plating. A similar rectilinear relationship has been obtained for nickel; it is possible to reduce the internal tensile stress to zero, or even to replace it by compressional stress, by adding sodium naphthalene trisulphonate to a plating bath of the Watts type.* This may bring the fatigue strength of a very weak steel above that of unplated metal; in the case of a strong steel, the fatigue is raised above that material plated in a bath free from sulphonate, although still lower than that of unplated metal (C. Williams and R. A. F. Hammond, *Trans. Inst. Metal Finishing* 1957, **34,** 317; R. A. F. Hammond, Priv. Comm., March 14, 1958).

In these ways, it may be hoped to overcome the trouble due to internal stresses, but the need to avoid porous coats, or coats which develop discontinuities, through corrosion or otherwise, in service, will remain—especially if the metal surface is likely to become wet in service, or even moist where there is salt in the atmosphere.

It is possible that peen plating (p. 601) may prove helpful in overcoming the troubles mentioned above; here the coating is carried out under conditions unlikely to introduce hydrogen (H. K. Farmery, Priv. Comm., June 10, 1958).

Apart from plating, electropolishing, if employed, may reduce the fatigue strength, which is rather unexpected since the process should remove roughness and stress raisers generally. The fact that a finely ground surface may have a considerably higher fatigue strength than an electropolished surface has been explained by arguing that the former produces a host of stress raisers close together, whereas the latter may leave only a few; it has similarly been argued that the good fatigue behaviour of a peened surface is not due to the absence of stress raisers but to their abundance in mutual proximity. However this may be, electropolishing has been found to have certain advantages over manual polishing, since it always removes the same amount of metal, whereas no two polishers produce identical results. One method of improving the poor fatigue qualities of an electropolished surface depends on the process known as " vapour honing " (p. 408).

* The use of sulphamate baths diminishes the internal stress and seems slightly to improve resistance to fatigue, but it is doubtful whether any superiority of performance over that obtained with a Watts-type bath is sufficient to compensate for the greater cost and increased liability to pitting (D. A. Fanner and R. A. F. Hammond, *Trans. Inst. Metal Finishing*, Winter 1958–59, **36,** 32).

Thermal treatment has also been studied, but the limits of good results appear to be rather narrow; no benefits are obtained below 450°C., whilst deterioration sets in about 550°C. (A. T. Steer, J. K. Wilson and O. Wright, *Aircraft Production* 1953, **15**, 242).

Water Treatment to prevent Corrosion Fatigue. If a cooling water which is causing corrosion fatigue can be made non-corrosive, there is a good prospect of avoiding the trouble, although, in border-line cases, a protective film capable of withstanding stress-less corrosion might break down under stress. Two classes of inhibitors have aroused special interest, the chromates and the emulsifying oils. The former are useless where there are reducing agents present—as in some oil-field situations. They have, however, their fields of application; the use of zinc yellow ($4ZnO \cdot K_2O \cdot 4Cr_2O_3 \cdot 3H_2O$) for dealing with corrosion fatigue caused by salt water dripping from railway refrigerator cars may serve as example (C. G. Fink, W. D. Turner and G. T. Paul, *Trans. electrochem. Soc.* 1943, **83**, 377).

The action of chromates in preventing or postponing the corrosion-fatigue cracking of wire, has been studied in the laboratory by Gould, who followed the potential of steel wires immersed in water containing chromate and chloride. The potential remained high as long as the film survived; in absence of stress, if the chromate concentration was sufficient, it remained permanently high and the wire remained unattacked. With stress, if the chromate was too low, the potential rose at first, in accordance with Hoar's principle (p. 124), then suddenly collapsed, indicating film break-down; after this cracks developed in the wire, leading to failure—sometimes quickly. It was evident that each test consisted of two periods—the " film life " (up to the potential tumble) and the " wire life " (up to fracture); these could be separately measured. Both these lives increased, at constant chloride concentration, with rising chromate, up to a certain value; they decreased, at constant chromate, with rising chloride and rising stress (A. J. Gould and U. R. Evans, *Iron St. Inst., Special Reports* 1939, **24**, 325).

Emulsifying oils have been used extensively in Germany, but are useless in strong saline waters which prevent emulsification. Certain wetting agents are efficacious in reducing corrosion fatigue because they ensure uniform attack over the whole surface, even though the damage, as measured by loss of weight, is increased (R. B. Waterhouse, " Fatigue of Metals ", refresher course 1955, p. 105 (Institution of Metallurgists)).

Prevention of Corrosion Fatigue by Peening. It has long been known that liability to dry fatigue depends on surface conditions. Small irregularities (burrs, tool-marks and the like) act as stress raisers, whilst residual stress remaining in the external layers, if tensile, will shorten the life. The same features will also accelerate corrosion fatigue, and in addition small irregularities, whether asperities or cavities, may serve as susceptible points where corrosion can start; such points sometimes become the sites of pitting in stress-less corrosion, but of cracking under stress.

Surface treatments which tend to level out roughness and leave compressional stress should have the opposite effect. It has long been known that bombardment with small particles at high velocity, known as "peening", improves the dry fatigue life; the particles used are frequently chilled iron, of a composition which, if used for castings, would produce a grey iron casting, but which, cooled down quickly (as occurs if the material consists of small shot particles) may show a structure similar to white iron; the material was at one time being sold under the name of "Steel Shot".

It has for some time been believed that peening conferred resistance to corrosion fatigue as well as to dry fatigue. To ascertain whether this was true, Gould carried out corrosion fatigue tests on high-carbon steel bars which had been peened in seven different ways, using shot of different sizes driven on to the surface at various air pressures; one series of corrosion fatigue tests was carried out with very dilute sulphuric acid (intended to represent the acidity of the moisture which condenses on steel in industrial districts) and another with sea-water. Specimens with a finely ground finish of high quality were tested for comparison. All the peened bars gave higher endurances than the finely ground finish, but differed considerably among themselves; at fairly high stress-range, the "best" procedure gave about nearly ten times the life of the "worst". Favourable results were obtained with large shot at low pressure, or small shot at high pressure; it would seem necessary to obtain a fairly thick layer of compressed material. One interesting point is that the surface obtained with fairly heavy peening could subsequently be roughened by a short abrasion with angular grit without loss of corrosion-fatigue resistance; this might prove useful if it was desired to apply protective coats on a peened surface, which would normally provide poor anchorage in the absence of roughening (A. J. Gould and U. R. Evans, *J. Iron St. Inst.* 1948, **160**, 164).

The poor corrosion fatigue life shown by a finely ground finish of high quality in Gould's work raises the question as to whether this is necessarily a dangerous finish. No definite information regarding corrosion fatigue seems to be available. For dry fatigue, it would seem that fine grinding may not reduce strength if carried out with extreme care, but, in the form carried out commercially today, this finish is best avoided; the same is probably true of corrosion fatigue, particularly as grinding is likely to rub in foreign bodies, e.g. iron particles into stainless steel or an aluminium alloy. Some information is provided by D. N. G. Cledwyn-Davies, *Proc. Instn. mech. Engrs.* 1955, **169**, 83.

The good effect of peening has been explained in several ways: (1) the provision of a zone in compressional stress, (2) the closing up of cracks that could act as stress raisers, (3) the provision of a large number of small stress raisers close together—a situation which, as explained on p. 710, produces much less stress-intensification than a single fissure. If the latter explanation is the correct one, it becomes easy to see why grinding, unless carried out with extreme care, may produce a few scours deeper than others and thus lead to poor fatigue resistance. Further evidence is provided by Brown, who subjected steel to rotary bending both in air and in water; peening by itself caused considerable increase in fatigue life on steel, whilst peening followed by stress-relief caused a further increase, although presumably this would remove the supposedly beneficial compressional stresses (N. B. Brown, abstract, *J. Iron. St. Inst.* 1951, **169**, 199).

This work deserves repetition and extension. If the benefits of peening are due to a layer in compressional stress, they may not last long in service, since the layer in question will soon be corroded away. Such a possibility will not be brought out in laboratory tests carried out at high frequency in order to obtain quick results; the damage to the compressed layer caused by corrosion in the course of a test to destruction may be much less than under service conditions where the frequency is lower and the life longer. Such considerations require to be borne in mind in applying laboratory tests to a practical case, but if the benefits of peening depend on some other factor than the layer in compression, there is less reason to mistrust the results of laboratory tests.

Nitriding. Special steels containing elements possessing an affinity for nitrogen, such as chromium and molybdenum (sometimes aluminium or vanadium) acquire, when heated in anhydrous ammonia at about 500°C., a rim of hard material in a state of internal compression; the process, generally known as *nitriding*, has sometimes been described as *chemical peening*. The compression arises from the fact that the nitrides of the metals mentioned are very voluminous, and yet hard, so that the metallic matrix by which they are surrounded is left in a state of compression. Even before nitriding these special " nitriding " steels have a considerably higher air fatigue limit than ordinary mild steel; the ammonia treatment improves it further. For corrosion fatigue, nitriding is necessary to bring out the resistance of the material. Thus the material containing 1·58% Cr, 0·87% Al, 0·33% Mo and 0·26% C, tested for $1 \cdot 7 \times 10^8$ cycles in Tees water withstood $\pm$ 25 tons/in.^{2} when nitrided, but only $\pm$ 5 tons/in.^{2} if not nitrided; ordinary mild steel sustained $\pm$ 2 tons/in.^{2}, whilst 18/8 stainless steel containing 1% tungsten sustained 11·1% tons/in.^{2}, when tested in the fully hardened condition (N. P. Inglis and G. F. Lake, *Trans. Faraday Soc.* 1931, **27**, 803; 1932, **28**, 715).

Cathodic Protection. Relatively early observations by Haigh and others show that corrosion fatigue could be prevented on galvanized iron in service, even where gaps in the zinc left parts of the surface uncovered. This certainly suggests cathodic protection. Laboratory experiments by Gould (p. 716) and Simnad (p. 717) establish the fact that such protection can occur. In N/10 potassium chloride, Simnad found that very low cathodic currents slightly *shortened* the life, probably by reducing the number of cracks and thus increasing stress-intensification. However high current densities produced a lengthening of the life, and at a certain current density, which was higher for a high stress-range than a low one, the specimens remained unbroken at the end of 20,000,000 cycles—a remarkable result, although not necessarily easy to utilize in service. In acid liquids, cathodic protection, although causing some slight increase of life, did not prevent failure; it is worth noticing, however, that the amount of iron passing into solution was reduced to zero by a cathodic current (the current needed at 20 tons/in.^{2} was less than twice that needed at zero stress);

perhaps the iron dissolved out in the crack was redeposited as metal on the face outside. Evidently there are limitations to the use of cathodic protection.

Stuart's work at Cambridge suggested that contact with zinc during corrosion fatigue appreciably increases the corrosion-fatigue life in nearly neutral solution, but that under acid conditions there is little improvement. This is not surprising; indeed some shortening of the life, due to hydrogen absorption, had been expected when the research started, but was not obtained under the conditions of the laboratory test. American work has brought out some interesting results; pickling in warm sulphuric acid does appreciably reduce the fatigue strength, but an inhibitor such as diortho-tolyl-thiourea minimizes the deterioration, which is attributable more to notch formation than to hydrogen absorption. Such results refer to mild steel, and some of them to a small stress-range—limitations which should be borne in mind in applying the results to service conditions. Certainly for strong alloy steels, the presence of hydrogen is today greatly feared as a cause of cracking, whilst Jackson reports that in plain-carbon spring steels (hardened and tempered) the effect of a hydrogen charge may be catastrophic; the material cracks as soon as a fatigue test (without corrosion) is started (N. Stuart and U. R. Evans, *J. Iron St. Inst.* 1943, **147**, 131P; G. L. Kehl and C. M. Offenhauer, *Trans. Amer. Soc. Metals* 1940, **28**, 238; J. S. Jackson, Internat. Conf. on Fatigue of Metals (Instn. mech. Engrs. and A.S.M.E.) 1956).

Other References

A sound understanding of Dry Fatigue is important for those who would understand Corrosion Fatigue. An important paper by W. A. Wood and R. B. Davies (*Proc. roy. Soc.* (*A*) 1953, **220**, 255) deserves special study. Other useful papers are those of W. A. Wood, *Bull. Inst. Met.* 1955, **3**, 5; F. P. Bullen, A. K. Head and W. A. Wood, *Proc. roy. Soc.* (*A*) 1953, **216**, 332; R. F. Hanstock, *J. Inst. Met.* 1954–55, **83**, 11; N. F. Mott, *Proc. roy. Instn. Great Britain* 1954, **35**, 666; *J. Iron St. Inst.* 1956, **183**, 233, and A. K. Head, *J. Mech. Phys. Solids* 1953, Jan. p. 134. See also A. J. Kennedy, *Nature* 1956, **178**, 810, for views on creep and fatigue. Books to be studied include R. Cazaud, " Fatigue of Metals " translated by A. J. Fenner (Chapman & Hall). The Published Reports of Conferences on Fatigue deserve attention. That organized at Melbourne in 1946 included papers on Corrosion Fatigue; one was held at Cambridge (Mass.) in 1952, another at Oxford (see account in *Nature* 1955, **175,** 980), whilst an International Conference held in 1956 includes three papers by A. J. Gould (on Corrosion Fatigue), W. A. Wood and P. J. E. Forsyth, which deserve close study. The report on the Refresher Course on Fatigue of Metals organized by the Institution of Metallurgists in 1955 includes a paper by G. Forrest which describes the effect of surface finish, notches and other stress raisers, and a useful general account of Corrosion Fatigue by R. B. Waterhouse. A symposium on fatigue and work-hardening was organized by the Royal Society in 1957; the papers are published in *Proc. roy. Soc.* (*A*) 1957, **242,**

145–227. A valuable survey of Corrosion Fatigue comes from P. T. Gilbert (*Metallurgical Reviews* 1956, **1**, 379). The avoidance of fatigue as influenced by design is discussed by R. Cazaud, " Atomes " 1948, **3** (Jan.), p. 9; R. A. MacGregor, W. S. Burn and F. Bacon, *Trans. N.E. Cst. Instn. Engrs. Shipb.* 1934–35, **51**, 161.

Russian work on corrosion fatigue is reviewed in *Corr. Prevention and Control* 1955, **2**, Dec., p. 37. Information regarding the influence of temperature is provided by I. Cornet and S. Golan, *Corrosion* 1959, **15**, 262t. The hardening accompanying (dry) fatigue is studied by T. Broom and R. K. Ham; *Proc. roy. Soc.* 1959, **251**, 186. French tests (running to 10^8 cycles) on methods for preventing corrosion-fatigue on low-alloy steel suggest that nitriding, or zinc-coating followed by acid chromate treatment, is effective; a pre-constraint treatment which leaves biaxial compression, followed by zinc-coating with acid chromate treatment, seems to give the best results (A. Royez and J. Pomey, *Rev. Métallurg.* 1959, **56,** 122).

CHAPTER XVIII

OTHER TYPES OF CONJOINT ACTION

Synopsis

It is first explained that, under conditions where oxidation or wet corrosion normally stifles itself owing to the formation of a film, such stifling may fail to occur if the film is continuously being removed or damaged. Thus, without special assumptions, it is to be expected that chemical and mechanical influences, acting conjointly, will produce far more damage than if they act separately; this is found to be the case.

Nevertheless, the two main examples of conjoint action discussed in the present chapter are probably less simple than the case just indicated; few authorities regard them merely as the chemical or electrochemical corrosion of metal kept constantly clean and unprotected by mechanical action. The two examples are

(1) the *fretting corrosion* of two surfaces moving to and fro over one another in presence of oxygen, and

(2) the pitting produced by the *collapse of vacuum cavities* in water, or the *impingement of air-bubbles* on a metallic surface which normally protects itself with a film; the damage due to vacuum cavities is often known as *cavitation erosion.*

In all these cases, there is disagreement about the extent and manner in which chemical and mechanical factors are involved.

In approaching the subject of fretting corrosion, it is necessary first to consider the *wear* produced by *unidirectional* movement of one surface relative to another, before proceeding to the *frettage* connected with *oscillatory* movement; the manner in which damage is altered in character and increased in quantity by the presence of oxygen is then discussed; there is a short section on the wear of colliery ropes. Passing to immersed conditions, one must consider the formation of *vacuum cavities* or *vapour cavities* (often really low-pressure air-cavities), and then the manner in which they inflict damage when they collapse (*cavitation*); we then pass to the *impingement attack* caused by the impact of bubbles containing air at roughly atmospherical pressure, and to the pitting caused by water jets. After that, some special cases of conjoint action require attention, such as the localized damage caused where sand-particles in a water-stream keep impinging, with consequent damage to what would otherwise be a protective film. The cracking or disintegration due to the internal production of voluminous corrosion-products is also considered in the present chapter, although certain examples—such as the layer corrosion of light alloys and the swelling of wrought iron—have already been described in Chapters XIII and XVI. Finally, attention is directed to the change of potential produced when an oxide-film is locally scraped away, with the possibility thus provided for severe electrochemical attack.

General Considerations

Self-stifling of Dry Oxidation and its failure to occur when the film is removed mechanically. In early chapters, we met with cases where a metal starts to oxidize rapidly, but soon develops a protective film, so that the rate of destruction becomes negligibly slow; it is this self-protection which enables a " base " metal to be employed in service without prior application of a protective coating. A base metal exposed to reasonably dry air (R.H. below about 40%) generally undergoes no visible change. If at the out-set it is free from oxide, it will quickly develop an invisible film, but oxidation will cease before there is change visible to the eye. For some metals, such as iron, this is not true of humid air.

Supposing that oxidation by nearly dry air, gives a curve of the form suggested in fig. 120(A), it is clear that if, at certain moments (P_1, P_2, P_3 . . . etc.) the film is removed or completely broken up, rapid attack will be resumed (fig. 120(B)). If the damage is inflicted at sufficiently frequent

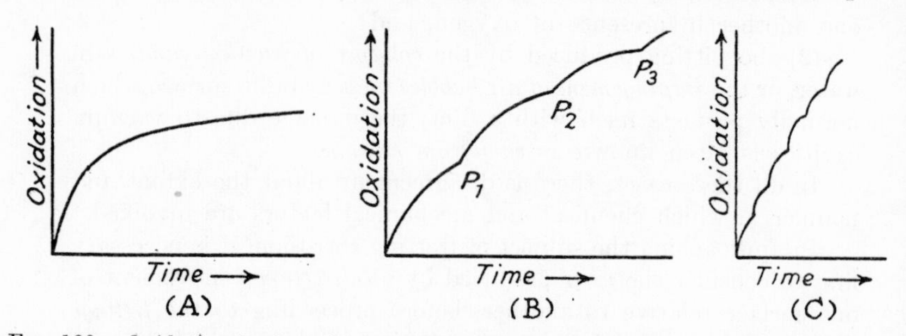

Fig. 120.—Oxidation by dry air with (A) film producing a stifling; (B) film broken at P_1, P_2, P_3; (C) film broken frequently, producing a nearly straight line.

intervals, we get a curve (fig. 120(C)) barely distinguishable from a straight line—of which the steepness may approach the steepness of the initial part of curve A, provided that the damage on each occasion is sufficiently complete.

This argument is independent of the nature of the film. As shown in Chapter II, there is little doubt that the film produced by dry air is an oxide. However, whatever the exact chemical composition of the oxidized layer may be—provided it is capable of being rubbed off or rendered non-protective more easily than a similar layer of unchanged ductile metal can be removed—mechanical treatment combined with chemical change must cause greater damage than either influence acting alone.

This would be true even if there were no rise of temperature or disarraying of the structure. In fact, however, the large local rise of temperature which often accompanies mechanical damage will increase the rate at which chemical action can take place, and will increase the limiting value of the film-thickness which could (in absence of further damage) be reached before thickening becomes slow. Thus a temperature-rise, although not essential to the argument, may sometimes play a part in accelerating damage.

Moreover, the fact that certain mechanical treatments leave the surface atoms disarrayed and capable of leaving the surface with only slight increment of energy must increase chemical reactivity. It might be argued that the decreasing oxidizing velocity shown in fig. 120(A) is due not merely to the formation of a film over the whole surface, but to the exhaustion of the supply of surface atoms in a specially reactive state; but the similarity between the curves representing the progress of oxidation on annealed and hard-rolled surfaces of the same metal suggests that in practice film-formation is the main reason for the falling off of the initially rapid oxidation-rate.* Even if the exhaustion of disarrayed material is accepted as a contributory cause, the conclusion is the same; for any mechanical action, applied continuously or intermittently, besides removing oxide, is likely to leave the metal in the disarrayed, chemically reactive condition, and once more we reach the opinion that conjoint (chemical-mechanical) action should produce more damage than would be expected from either chemical or mechanical action working alone.

Self-stifling of Wet Corrosion and its failure to occur when the film is removed mechanically. Under immersed conditions, several metals continue to corrode at a rate which cannot be neglected (e.g. iron or zinc in chloride solution). Other metals, like aluminium, behave differently; in a chloride solution, aluminium suffers corrosion at a rate which first increases with time, then fluctuates and finally becomes slow, as shown by Champion's curves (fig. 121); these were obtained by a gasometric method based on that of Bengough (F. A. Champion, *Trans. Faraday Soc.* 1945, **41,** 593).

A detailed interpretation of these curves has been offered by Champion, but they are here introduced to indicate that the air-formed film originally present on the aluminium takes some time to break down, and that there is a period of rapid corrosion before a new—more reliable—film is developed. Clearly, if owing to mechanical action, a protective film is continuously removed or interrupted, persistent rapid attack may be expected.

Varieties of mechanical damage on a film-covered surface. A surface may suffer damage, either by something striking it, or by something pressed against the surface and then moved parallel to it. In the first case, the impacting body may be (1) *solid,* as when particles of shot or grit are hurled against a surface during peening, or (2) *liquid,* as in the impact of droplets on turbine blades, (3) *gaseous,* as when air-bubbles entrained in a

* Vernon found that the oxidation-curves for hard-rolled and annealed aluminium of the same purity were practically the same; aluminium samples of different purities in the same state gave different curves (W. H. J. Vernon, *Trans. Faraday Soc.* 1927, **23,** 113, esp. p. 152).

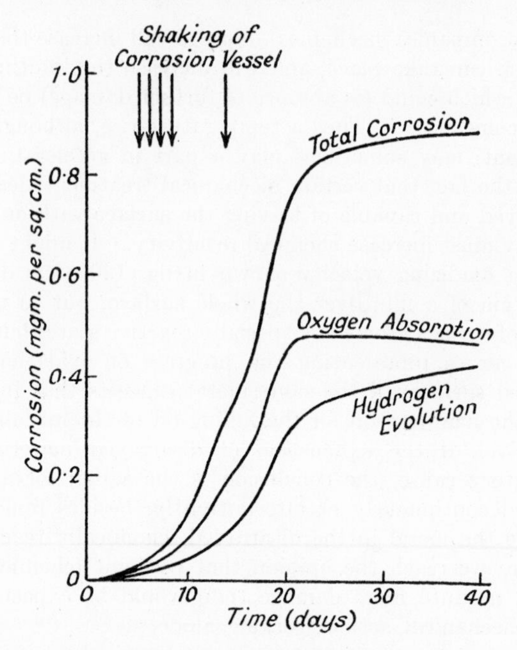

FIG. 121.—Corrosion of pure Aluminium in N Potassium Chloride solution showing separately the attack connected with hydrogen liberation and oxygen liberation (F. A. Champion).

water-stream, strike the surface, and rebound (often breaking up in the process) or (4) *vacuous*, as when an unstable cavity containing water-vapour and perhaps low-pressure gas, collapses at or close to the surface, often sending out a compression wave. It is possible that the turbulence in a water-stream plays a great part in the damage connected with the collapse of cavities—a view held by Callis. The special damage caused by the impact of air-bubbles under conditions where they break up into small ones, may be due to the complicated system of stresses produced, some of them very well suited to break up the film and others to pull off fragments.

Now a local impact on film-free metal surface is likely to dent the surface; if the stress exceeds the elastic limit, a permanent depression in the metal will be produced; if not, the metal will spring back to its original shape. It is not likely that metallic fragments will be dislodged unless the metal is a very brittle one, or has become very brittle through cold-work produced by numerous former impacts. If, however, the metal carries a film of oxide, a substance lacking the ductility of metal, the denting of the metal is likely to cause cracking of that film; in cases where the stress applied exceeds the elastic limit of the metal, the cracks in the oxide will remain; in cases where the stress in the metal is purely elastic, cracks in the oxide may close up to some extent, but the oxide is unlikely to weld together completely and will probably remain leaky. Continued impacts at the same point are likely to permit chemical action at that point to continue, and thus we see that an impact which would produce limited mechanical damage to the metal may lead to serious chemical damage.

The case of movement parallel to the surface is a little different. If two plates which appear "smooth" to the eye are pressed against one another, they only come into true contact at a limited number of points where microscopic projections exist on one or other surface; if one plate is made to slide over the other, the conditions at each projection are not unlike those produced when a weighted needle is pulled over a plate, leaving a groove as it moves, owing to the deformation of the metal. The formation of a visible scratch-line on a ductile metal does not necessarily involve the dislodgement of much material, if no film is present. In cases where a film is present on the metal, it has been sometimes imagined that the scratching removes that film, but often this does not seem to occur. Where the oxide (or other film-substance) is harder than the metal, it appears to be broken up and pushed into the soft metal, rather than chipped off. However, since the area of the curved groove must be greater than the area of the flat surface which existed before scratching, the oxide cannot cover the whole, and gaps must be left—especially at the two margins of the scratch (see fig. 28(*b*), p. 109). The idea that the film can be removed bodily by abrasion with emery-paper may have arisen from the fact that the abrasion of heat-tinted metal destroys the interference colours and renders the metal silvery, but this proves nothing; the breaking up of the film into disordered fragments will destroy the optical conditions needed for the production of colours, and experiments based on the transfer method described on p. 111 suggest that most of the oxide is broken up and pushed into the metal. However, although abrasion may fail to remove the oxide, it will certainly interrupt the film and render it non-protective. Given time, new films would re-form at the interruptions, but if mechanical action persists along certain lines, chemical change is likely to continue unchecked.

It would thus appear that the relative movement of two surfaces—one or both metallic—might allow chemical action to continue where otherwise it would stifle itself, and that in an extreme case chemical attack will persist without appreciable mechanical removal of material. In general, however, relative movement will displace a certain amount of material, and usually the material removed will contain both oxide and metal; on a surface prepared by abrasion followed by mechanical polishing, there will be a mixed zone of metal and oxide, and in such a case it would be impossible to remove the oxide without removing metal. The removal of both will keep the metal clean and capable of reaction, but it is difficult to distinguish sharply between the parts of the damage due to mechanical and chemical action respectively, since the first makes possible the continuation of the second; the damage due to conjoint action will not be capable of being expressed as the sum of two terms representing the mechanical and chemical contributions respectively, but will generally contain terms involving both mechanical and chemical factors, so that it becomes meaningless to say that the damage is $x\%$ mechanical and $(100 - x)\%$ chemical (cf. p. 754).

Wear and Frettage

Wear. Our knowledge of the wear caused by relative motion of two smooth, but not frictionless, surfaces, sliding one over the other, is mainly due to the extensive work of Bowden and his colleagues, with later work from Hirst's laboratory. Neither set of researches has any direct connection with corrosion, but no one considering conjoint action can disregard the knowledge gained from them. The principle papers of the Bowden group are by F. P. Bowden with S. H. Bastow, D. Tabor, J. W. Menter, K. E. W. Ridler, T. P. Hughes, M. A. Stone, L. Leben, G. K. Tudor, J. S. McFarlane, J. E. Young, G. Rowe, P. H. Thomas, R. F. King, K. R. Eldredge and W. H. Pascoe, *Proc. roy. Soc.* (*A*) 1931–32, **134,** 404; 1935, **151,** 220; 1936, **154,** 640; 1938–39, **169,** 371, 391; 1939, **172,** 263; 1946–47, **188,** 329; 1947–48, **192,** 247; 1950, **202,** 224, 244; 1950–51, **204,** 514; 1951, **208,** 311, 444; 1952, **212,** 439, 477, 485; 1954, **223,** 29, 225; 1955, **229,** 181, 198; 1956, **235,** 210. Other views were expressed at the Discussion Meeting on Friction, *Proc. roy. Soc.* (*A*) 1951, **212,** 439–520. See also F. P. Bowden, *Research* 1950, **3,** 147; *Endeavour* 1957, **16,** 5, and the book by F. P. Bowden and D. Tabor, " Friction and Lubrication of Solids " (Clarendon Press). The papers of the Hirst group are quoted on p. 738.

Bowden's researches have shown that sliding is not a continuous process, but consists of alternations between " sticking " and " slipping ". The sticking is probably due to welding at projections where the two surfaces are in true contact; this makes relative movement temporarily impossible until the accumulated stress becomes sufficient to break the bridge and produce a sudden rapid slip, often accompanied by a very high temperature. Under some conditions, relatively fusible metals may be heated to their melting-points, whilst on other metals a rise to 500° or even 1000°C. occurs locally and momentarily. Since the breaking of the bridge need not necessarily follow the exact plane along which the two parts originally united, transfer of material from one surface to the other becomes possible—and is, in fact, observed. After a copper surface has been made to slide over steel, fragments of copper can be detected adhering to the steel surface. McFarlane and Tabor have shown that there is a close connection between the coefficient of friction and the coefficient of adhesion.

Now the presence of an oxide-film will greatly hinder welding, and experiments have shown that if the two metals have been carefully cleaned *in vacuo*, the friction becomes unusually high; it falls off as soon as a trace of oxygen is admitted. Sulphide, chloride and iodide films have a similar effect in reducing friction. Indeed some of the substances intentionally added to lubricating oils to reduce friction contain sulphur and chlorine compounds and may act by producing films which prevent welding. Sulphides and chlorides usually have lower melting-points than the metals, and this may be an advantage; in the case of oxide-films, there is a danger of fusion of the metal below a solid film, with undesirable results—a possibility discussed by Finch (see below).

Different substances added to improve lubricating oils may act in different ways; Beeck thinks that the good effect of tricresyl phosphate, present in the right amounts, is that it corrodes away prominences of the metal, thus increasing smoothness; excessive amounts increase the wear—as would indeed be expected.

Again, the presence of a regulated quantity of an organic acid greatly improves the lubricating qualities of a hydrocarbon oil by acting on the metal to form a soap, which constitutes the true lubricant. Usually good lubrication is only obtained where soaps can be formed; hydrocarbons without fatty acid additions are generally poor lubricants. Bowden reports that zinc, cadmium, copper and magnesium slide smoothly when lubricated with paraffin oil containing 1% of lauric acid, whereas platinum, nickel, chromium and aluminium, which resist soap-formation, show " stick-slip ". Lauric acid, which melts at 44°C., will lubricate steel up to 110°C., which probably represents the melting-point or softening-point of an iron laurate.

Clearly all these additions require to be made with discretion. An excess of acid—such as may be formed accidentally in oil by oxidation—produces undesirable results and inhibitors are often added to prevent such oxidation. Again, lubricating oils which have been treated with chlorine compounds to reduce the friction during service, may leave the surface in a state liable to rusting, if the surfaces are subsequently exposed in an oil-free condition to damp air—a matter studied by Bukowiecki. Further information on these subjects is provided by F. P. Bowden, *J. Inst. Petroleum* 1948, **34,** 654; E. B. Greenhill, *ibid.* 1948, **34,** 659; J. N. Gregory, *ibid.* 1948, **34,** 670; B. S. Wilson and F. H. Garner, *ibid.* 1951, **37,** 225; O. Beeck, " The Chemical Background for Engines Research " 1943, pp. 260, 273 (Interscience Publishers); G. I. Finch, *Proc. phys. Soc.* (*B*) 1950, **63,** 785; A. Bukowiecki, *Schweiz. Arch. angew. Wiss.* 1951, **17,** 182; J. Pomey and F. Loury, *Métaux et Corros.* 1949, **24,** 135; G. H. Denison and P. C. Condit, *Industr. engng. Chem.* 1945, **37,** 1102; C. F. Prutton, D. Turnbull and D. R. Frey, *ibid.* 1945, **37,** 917; W. Hirst and J. K. Lancaster, *Proc. roy. Soc.* (*A*) 1954, **223,** 324; E. C. Greenhill, *Trans. Faraday Soc.* 1949, **45,** 625, 631. See also Symposium on the Protection of Motor Vehicles from Corrosion 1958 (Soc. Chem. Ind.), esp. papers by W. G. Stevenson and N. E. F. Hitchcock.

Considerable information about the transfer of material from one surface to another during mutual rubbing is derived from work in Hirst's laboratory. In the research of Kerridge and Lancaster, a pin of brass was pressed against a rotating ring of the hard alloy, stellite. The brass had been irradiated in an atomic-energy pile, and then left for a sufficient period to allow the short-life isotopes to disappear, so that the main radioactivity left was due to a zinc isotope of half-life 250 days. In this way it was possible to measure the transfer of brass to the stellite. It was found that there are two stages; first the brass is transferred in discrete amounts to the stellite surface, and, as a result of successive transfers at neighbouring points, builds up patches of film—which patches spread in area and increase in thickness. When a patch has grown to about 50 times the amount of a single discrete transfer, the bond becomes loosened, owing to alternate compression and tension, and the patch is detached, as a roughly rectangular flake; these flakes accumulate to form *wear débris*. Papers deserving study are those of M. Kerridge and J. K. Lancaster, *Proc. roy. Soc.* (*A*) 1956, **236,** 250; J. F. Archard and W. Hirst, *ibid.* 1956, **236,** 397; 1956–57, **238,** 515; J. K. Lancaster, *Proc. phys. Soc.* (*B*) 1957, **70,** 112; W. Hirst and J. K. Lancaster, *J. appl. Phys.* 1956, **27,** 1057.

Frettage. Frettage differs from wear in that it occurs at places where no relative motion between two surfaces has been intended by the designer, but where in fact a slight oscillatory motion of microscopic amplitude does occur. For instance, two surfaces forming parts of a machine or vehicle, bolted together tightly so that in theory they constitute a single member, may in practice, owing to vibration or other cause, undergo this small relative motion. Again flat metal sheets packed on one another may during transport undergo slight relative movement to and fro, and this causes damage which is generally considered to be akin to the fretting corrosion of machines. The chief geometrical difference between the conditions causing wear and frettage is that wear occurs when there is relative movement in a single direction over long distances so that the detritus produced does not accumulate at a single point, whereas in frettage the relative position of the two surfaces remains stationary—apart from the barely visible oscillation —so that accumulation becomes possible. The rate of relative movement is generally much slower in frettage than in wear. The chief chemical difference is that the presence of oxygen, traces of which *militate against* damage in *wear*, *increases* damage due to *frettage*, which in presence of oxygen is called *fretting oxidation* or *fretting corrosion.*

It should be noted that frettage often occurs in vehicles or machines when they are being transported. Automobiles—after delivery by rail or in a larger road-vehicle—sometimes show frettage between the leaves of laminated springs and in ball-races of bearings; the damage is much more serious than the damage due to wear which occurs when the cars are driven under their own power. In aircraft, frettage has been found between plates of bolted or riveted assemblages or at the edges of lugs carrying locating pins, and in such cases small cracks often originate from the fretted region —a matter considered later. Stacks of aluminium sheet or food containers transported by road suffer fretting corrosion where they touch. Damage can occur to contacts in electrical switching gear subject to vibration. In rotary fatigue-testing machines frettage sometimes sets in on the specimen where it is held in the grips, so that fracture ensues—not at the point where the investigator designed it to occur—but at a point where no damage had been intended at all.

An excellent review of Fretting Corrosion has been prepared by R. B. Waterhouse, *Proc. Instn. mech. Engrs.* 1955, **169,** 1157; much useful information is provided by F. W. Fink, Soc. Automotive Engrs. Jan. 1954. An authoritative account of the wear of metals—including frettage—is due to F. T. Barwell, *J. Inst. Met.* 1957–58, **86,** 257, esp. p. 263. Experimental papers deserving study include:—J. S. Halliday and W. Hirst, *Proc. roy. Soc.* (*A*) 1956, **236,** 411; J. S. Halliday, Conference on Lubrication and

Wear (Inst. Mech. Engrs. Oct. 1957); K. H. R. Wright, *Proc. Instn. mech. Engrs.* 1952–53, **IB**, 556; *Corros. Prev. Control* 1954, **1**, 405; L. Grunberg and K. H. R. Wright, *Proc. roy. Soc.* (*A*) 1955, **232**, 403; H. H. Uhlig, W. D. Tierney and A. McClellan, Amer. Soc. Test. Mater. Spec. Tech. Pub. **144** (1953); J. R. McDowell, *ibid.* **144** (1953); I.-Ming Feng and H. H. Uhlig, *J. appl. Mech.* 1954, **76**, 395, 401; D. Godfrey, E. E. Bisson and J. M. Bayley, N.A.C.A. tech. notes **2180** (1950) and **3011** (1953); H. C. Gray and R. W. Jenny, *Trans. Soc. Automotive Engrs.* 1944, **52**, 511. See also the classical work of G. A. Tomlinson, *Proc. roy. Soc.* (*A*) 1927, **115**, 472; G. A. Tomlinson, P. L. Thorpe and H. J. Gough, *Proc. Instn. mech. Engrs.* 1939, **141**, 223; A. Thum and F. Wunderlich, *Z. Metallk.* 1935, **27**, 277.

Laboratory Studies of Frettage. The machine used by Tomlinson, Thorpe and Gough in their classical experiments comprised a coaxially

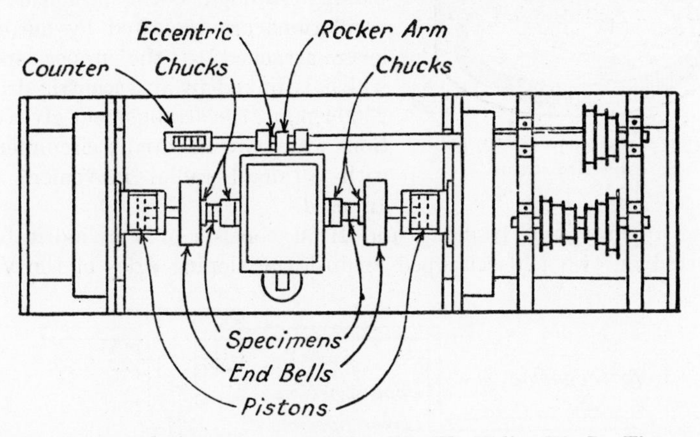

Fig. 122.—Fretting Apparatus (schematic) (H. W. Uhlig, W. D. Tierney and A. McClennan).

arranged assembly of three rings, with the centre one oscillating relatively to the two outer ones. This work demonstrated that frettage needs relative motion; where the vibration was so small as to be taken up by an elastic straining of the surface, no damage was observed. The slip during the first few cycles of an experiment was generally more extensive than the constant distance reached later on, but during this initial period of easy slip there was no damage, although later, when adhesion between the surfaces had started to restrain the amount of slip, damage developed.

The machine developed in Uhlig's laboratory (fig. 122) contained two cylindrical specimens of 1 inch diameter (fig. 123) one being subjected to oscillating motion. Provision was made for a variation of frequency between 56 and 3,000 cycles per minute, whilst the amount of relative slip could be established by means of an adjustable cam at any distance up to 0·008 inch. On steel (0·15% carbon), the rate of corrosion decreased during a brief running-in period, later becoming constant; the weight-loss produced after 67,800 cycles was found to be proportional to the length of the slip.

The effect of increasing the pressure was first to increase the weight-loss suffered during that period and then to reduce it, evidently because the length of the slip became small when the surfaces were pressed tightly together.

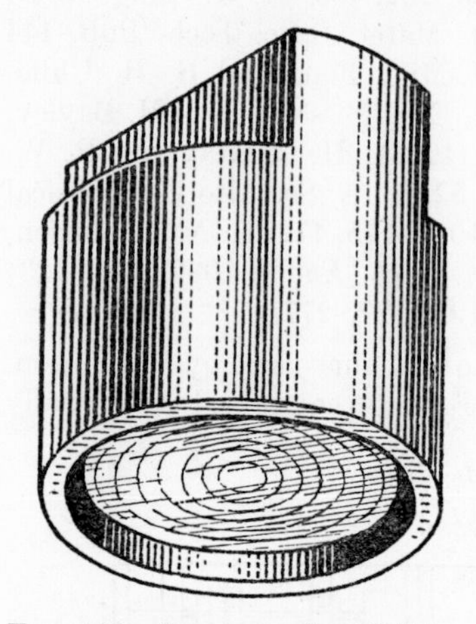

FIG. 123.—Specimen for Uhlig's Fretting Apparatus.

An apparatus designed by Tomlinson for his work with Thorpe and Gough, and used later by Wright and Rolfe, is shown diagrammatically in fig. 124. The upper specimen takes the form of an annulus of width 1 mm. and a mean diameter of 1·5 cm., which is made to oscillate relatively to the lower specimen. The load is applied by means of a rod passing through both specimens, whilst oscillations are imparted by means of a lever attached to the upper specimen which is linked to an eccentric driven at 2300 r.p.m.; the arrangement gives oscillations which are essentially circumferential, with a small radial movement superimposed.

The apparatus of Halliday and Hirst consists of a cylindrical sleeve supported on two pads clamped against the sloping sides of the V block

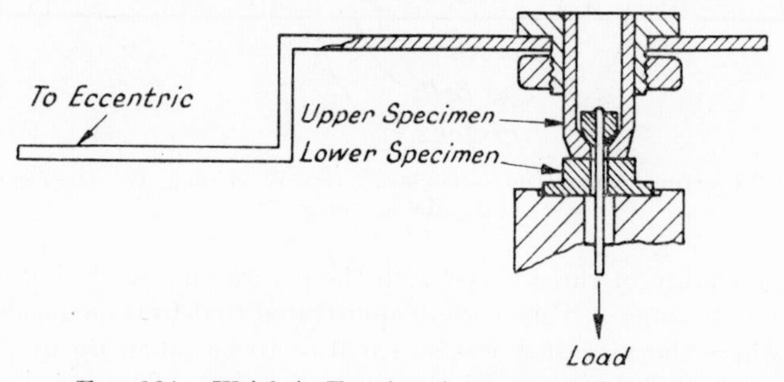

FIG. 124.—Wright's Fretting Apparatus (schematic).

(fig. 125), and loaded by weights hung on a shaft passing through the sleeve; the oscillations are provided by a vibration generator, and the measure of the frictional forces involved is obtained by the relation between the voltage and current used to drive it. They also measure the electrical resistance of the contacts, either whilst the specimens are vibrating or during vibrationless periods (J. S. Halliday and W. Hirst, *Proc. roy. Soc.* (*A*) 1956, **236**, 411).

In Uhlig's work, weight-loss served as a measure of damage. Wright assessed damage by careful lapping of the lower surfaces and noted the weight-loss associated with the disappearance of the scars. This greatly increased the sensitivity of the method, since 10^{-5} cm. scars correspond to a weight-change of 0·2 mg., but it should be noted that the measured weight change after the lapping is not proportional to the amount of metal destroyed by the frettage; if the frettage has produced trenches of semi-circular cross-section, the weight-loss after lapping should be proportional to the square root of the destruction during frettage. A similar assessment was used by Halliday and Hirst, who made certain assumptions about the shape of the scars.

Halliday and Hirst obtained different results according as the amplitude was large (exceeding 300 μ) or small. At the smaller amplitudes studied, the thin oxide-films initially present (stated to be Fe_3O_4), afforded protection against welding between the two metallic surfaces for a few cycles, so that a mild plastic flow occurred, producing scars smoother than the original surface. Later the film broke down and welds appeared; on continued rubbing loose oxide-particles became visible (first black, later red) and finally the surfaces became extremely smooth.

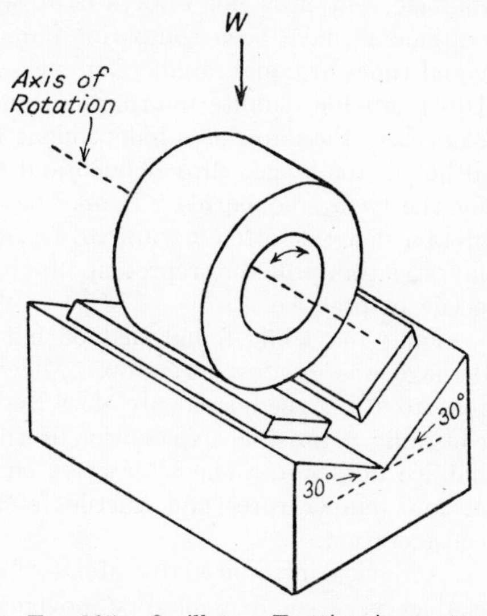

Fig. 125.—Oscillatory Fretting Apparatus (schematic) (J. S. Halliday and W. Hirst).

With large amplitudes, the entire regions where the surfaces were in true contact became torn, with the formation of large welds between the surfaces. Score marks appeared, having lengths corresponding to the amplitude of vibration, and corresponding in position to metallic fragments found adhering to the opposite surfacc—apparently transferred from one surface to the other by a " ploughing " action. A small amount of loose shiny *débris* appeared, and subsequently loose particles, first large and black (with diameters 1 to 0·1 μ), and later red (with particle size down to 100 Å). Both were identified by electron diffraction as α-Fe_2O_3, but it could perhaps be suggested, in passing, that the larger, black ones may have contained magnetite at their centres.

Their measurements of friction gave some surprising results. Large increases in the friction only occurred at amplitudes above about 300 μ, and the higher the load, the longer was the period of high friction; later the coefficient of friction fell to very low values of about 0·05—values not normally met with in unlubricated conditions except on such plastics as *teflon*. For low values of the amplitude and load, the initial period of high friction was very short, and the coefficient of friction soon fell to about 0·05. Halliday and Hirst attribute the low friction to loose *débris* rolling about and acting " somewhat in the manner of ball bearings "; they showed that by stopping the experiments and clearing out the *débris*, a high value for the friction was again obtained—if only temporarily.

Halliday and Hirst explain the " apparently pernicious nature of fretting corrosion " by supposing that it is " usually observed as a phenomenon concentrated upon a very small part of the surface whereas in unidirectional motion the surface regions supporting the load change continually; the wear is therefore distributed over the whole surface and appears less severe ".

The work deserves most careful study, but the engineer must use his judgment as to whether the particular type of fretting damage met with in his own experience can be regarded as being frictional wear concentrated on a small area.

Several authors have stated that frequency has no effect on the fretting damage, but it is not always clear whether those who have made these statements have been comparing damage produced in experiments lasting equal times or equal numbers of cycles. On this point, however, Feng and Uhlig provide definite information, and confirm that—for a fixed number of cycles—the damage is independent of frequency for slips of 0·0004 inch, although for longer slips (0·0036 and 0·0091 inch), the damage is greatest for the *lower* frequencies. It may be doubted whether slipping over these greater lengths often occurs in service; the experiments of some other investigators probably represent slipping over distances greater than would occur in practice.

Feng and Uhlig found that on mild-steel specimens run in dry air the damage was greatest at —150°C., decreased gradually as the temperature rose to 0°C., then suddenly decreased again and above 50°C. remained constant; a possible explanation is that liquid water acts as a lubricant but ice does not. The alternative suggestion that brittle fracture occurs at low temperatures and ductile fracture at high temperatures deserves consideration.

A device introduced by McDowell deserves special attention since it enables very small oscillatory movements to be applied—thus avoiding a defect common in some testing-machines, which produce slipping over distances greater than would occur in service. The specimen is a bar placed in an ordinary rotary fatigue machine under 4-point loading (p. 714), so that it undergoes very slight alternating bending as the bar rotates. A pad is screwed to a key fixed securely to the bar at a slight distance and is pressed on to it by means of an adjustable spring. If there were no bending, the pad would always press against the same point, but, since the bar is first slightly convex and then slightly concave at the side where the pad is attached, there is a very small movement to and fro, which can lead to frettage (J. R. McDowell, Amer. Soc. Test Mater., Fretting Symposium, Spec. Tech. Report **144,** (1953), p. 24).

Effect of the Atmosphere on Fretting Damage. Most recent investigators agree that the damage is greater in presence of oxygen than in its absence, despite the failure of some early attempts to prevent damage by excluding oxygen. Sakmann and Rightmire found less damage *in vacuo* or in protective atmospheres such as nitrogen and hydrogen, than in air.

Feng and Uhlig found the damage to steel in nitrogen to be only about one-sixth that in oxygen; in the former case the *débris* was metallic iron, whereas in presence of air oxide was formed.

The *débris* on steel is generally reddish-brown, and has been called *cocoa*. When oxygen-access is not intentionally restricted, α-Fe_2O_3 seems to be the main constituent—as has been demonstrated in the X-ray work of Cornelius and Bollenrath and that of Dies—also by the electron-diffraction studies of Wright. However, when the oxygen-supply is cut down, a black *débris* is formed, which Wright found to be magnetite or possibly γ-Fe_2O_3 (the difficulty of distinguishing the patterns has been mentioned on p. 25). It has been assumed in some quarters that any black product observed must be metallic iron, and some experiments which have been widely exhibited on a cinematograph film, are often interpreted in that sense; a chromium-steel ball was fretted against the underside of a glass slide, so that the changes could be observed through a microscope; the colour of the *débris* at the point of formation was certainly black, and the brown product was secondary, formed from the black one. However, it does not follow that the black product is metallic iron; Wright finds that the black *débris* is not metal but oxide. Readers should compare the evidence provided by B. W. Sakmann and B. G. Rightmire, N.A.C.A., tech. Note **1492** (1948); D. Godfrey, N.A.C.A. Report **1009** (1951); I-Ming Feng and H. H. Uhlig, *J. appl. Mech.* 1954, **21,** 395; I-Ming Feng and B. G. Rightmire, *Chartered mech. Engr.* 1956, **3,** 157; I-Ming Feng and B. G. Rightmire, *Proc. Instn. mech. Engrs.* 1956, **170,** 1055; K. H. R. Wright, *ibid.*, 1952–53, **IB,** 556; 1956, **170,** 1062; H. Cornelius and F. Bollenrath, *Arch. Eisenhüttenw.* 1940–41, **14,** 335; K. Dies, *ibid.*, 1942–43, **16,** 399; J. J. Broeze, *Engineering* 1952, **173,** 693; N. E. Promisel and G. S. Mustin, *Corrosion* 1951, **7,** 377, esp. p. 380.

Several investigators report nitrogen in the *débris*, but it is not always quite clear whether this comes from the atmosphere or the metal (K. Dies, see above. Also H. Schottky and H. Hiltenkamp, *Stahl u. Eisen* 1936, **56,** 444).

Mechanism of Fretting Damage. Presumably if air is restricted, the damage follows a course similar to that responsible for wear, and indeed the comminution of metal to metallic powder may play a part even when oxygen is present; Sakmann and Rightmire (see above), applying a piezo-electric device, noted that the slip was not continuous, but showed stick-slip characteristics. They also noted that, although the damage was generally greater in presence of oxygen than *in vacuo*, the frictional force was generally less—thus bringing out an analogy to wear. Clearly any metallic powder thus produced may be oxidized further from the point of formation, or later in time, explaining the occurrence of oxide in the *débris*. However, it would not explain why oxygen increases the damage to the metal.

One possible explanation is that the oxide formed is in many cases harder than the metal; alumina is harder than aluminium, and ferric oxide harder than low-carbon iron. If this is the explanation, the rate of damage should increase as the hard oxide *débris* accumulates and acts increasingly as abrasive. Curves correlating damage and time do sometimes show portions where the damaging-rate increases with time, and these could be explained in the manner suggested. However, in some experiments the damaging-rate seems to become constant, or even to decline with time. The periods of constant damaging-rate may be connected with the fact that, when once sufficient abrasive particles exist to cover the whole surface, further increase in their number cannot worsen the situation further; or again they may represent a state in which the abrasive particles are being expelled from a crevice as quickly as they are formed. If the explanation suggested is correct, fretting damage, even in presence of oxygen, should disappear in those combinations where the metal or alloy is harder than its oxide or oxides. An attempt to realize such a situation might be worth making, since, if successful, it would suggest a way of avoiding frettage trouble, whilst, if entirely unsuccessful, it would enable us to discard an erroneous and unhelpful theory.

Hirst emphasizes the manner in which hard particles can locally increase the stress. This occurs in unidirectional, as well as in oscillatory, motion. The detritus formed in a bearing in the space between journal and sleeve occupies a greater volume than the voids produced in the metal, so that the force per unit area must increase, and the two surfaces may ultimately lock together. Clearly any force which breaks the lock must produce damage to the metal.

Much depends, however, on the amplitude. Halliday pictures the sequence of events thus. First, frettage breaks down the natural oxide-films and leads to welding between the two metallic surfaces. Next, in absence of oil, loose oxidized *débris* is produced. At small amplitudes, this rolls about, acting rather like balls in a bearing, and the friction becomes surprisingly low ($\mu < 0{\cdot}1$). If the amplitude is increased, sliding tends to replace rolling, and this produces damage by abrading the surface; when the surfaces are mild steel and an aluminium alloy, it seems that the damage is largely due to abrasion by hard alumina particles. Paraffin oil inhibits corrosion, but produces little true lubrication and prolongs the opening period during which welding can occur; in presence of air, this period is short (J. S. Halliday, Conference on Lubrication and Wear, Oct. 1957, paper 39 (Instn. mech. Engrs.)).

One fact which seems to favour an explanation based on abrasion by hard particles is the effect of water vapour on frettage. Generally water favours oxidation processes, but Wright (p. 739) has found that frettage damage is frequently less at 50% relative humidity than at 0% or 100%. If a soft hydroxide is formed instead of hard oxide—even temporarily—it is easy to understand why there is so little damage at 50%. At 100%, one must postulate a different mechanism, based on rather rapid rusting, the rust being constantly rubbed off, exposing fresh metal.

Against the theory of abrasion by oxide must be set the fact that McDowell (p. 742) obtained bad frettage between two surfaces of hardened tool steel. Moreover cases have been described where the very hard oxide produced by an aluminium surface causes no more damage than the soft oxide originated from a zinc surface, the mating surface being hard steel in both cases (A. J. Fenner, K. H. R. Wright and J. Y. Mann, Internat. Conf. on Fatigue of Metals, 1956, Session **4**, Paper 8 (Instn. mech. Engrs.))

In contrast, Wright's study of the fretting corrosion of a series of cast-irons differing in hardness shows that the severity of attack falls off in marked degree as the hardness is increased; in the absence of lubrication, the softest iron tested suffered seven times as much damage as the hardest; in lubricated tests, the performance of the softer irons was greatly improved (K. H. R. Wright, Instn. mech. Engrs. Conf. on Lubrication and Wear, Oct. 1957, Paper 26).

Evidently much depends on working conditions, and particularly on the possibility of escape for the hard *débris.* Halliday and Hirst point out that in a shaft fitting closely into a sleeve, where *débris* (occupying a greater volume than the original metal) will accumulate, abrasive wear will be accentuated, ultimately leading to jamming or fatigue failure; in contrast, a gear coupling, where the components are free to move about, will suffer less serious damage, although there will be a loss of fit (J. S. Halliday and W. Hirst, *Proc. Instn. mech. Engrs.* 1956, **170,** 1062).

The practical engineer must carefully consider the service situation and, before deciding to apply the results of any particular set of laboratory results, make sure that they really represent the problem with which he personally is faced.

A second mechanism can be suggested for frettage oxidation under dry conditions; one can picture the oxide-film being rubbed off easily, exposing the metal to fresh oxidation, so that destruction continues unchecked. This would explain periods of constant damaging-rate by the principle suggested in fig. 120 (p. 732), whilst cases in which the damaging-rate falls off with time would be explained on the supposition that, when once the crevice has become filled with *débris*, the replenishment of oxygen at the central portions becomes a slow process. At such a stage, it is difficult to picture a uniform supply of oxygen to the whole surface, although doubtless channels would exist, leading to certain points; thus oxidation is likely to penetrate inwards into the metal more quickly at some places than others, producing, not a uniform film, but a mixed zone of metal and oxide, similar to that formed during Vernon's work on abraded surfaces (p. 57). Now this mixture of metal and oxide may well be more friable than the unchanged metal, and consequently it is soon broken up, giving access to the still ductile metallic zone below it. Such an explanation of the accelerating action of oxygen is favoured by K. H. R. Wright (Priv. Comm., Dec. 16, 1955); cf. U. R. Evans, *Proc. Instn. mech. Engrs.* 1956, **170,** 1061.

Despite some uncertainties, the following classification of cases, which is partly based on that of Fink, appears to represent the situation, and introduces some features from both theories.

(1) Where the atmosphere is *inert* (free from oxygen) and the motion *unidirectional* (or at least with a high amplitude of reciprocation, as would occur in some reciprocating machinery), there is little damage; metallic powder may be produced, but it can easily escape.

(2) Where the atmosphere is *inert*, but the motion consists of *oscillations of small amplitude*, the *débris* is trapped, but since it is free from hard oxide, the damage is still limited.

(3) Where the atmosphere is *air*, the *débris* produced by the oscillatory movement contains much *hard oxide*, and wear is accelerated by its abrasive action; at the same time, the oxygen, penetrating into the abraded surface, produces a friable mass containing oxide as well as metal, which mass is more easily disintegrated than the original ductile material. Thus the damage is considerable.

Frettage as a Cause of Fatigue. It has been noticed by engineers —particularly on aircraft—that fatigue cracks develop in areas which also show severe frettage. The conclusion is often drawn that frettage is the cause of fatigue. This is not the only possible interpretation of the undeniable association of the two forms of damage; it is conceivable that conditions which favour frettage also favour fatigue-cracking, so that the two have a common cause. Even if we could accept the view that frettage and fatigue are not cause and effect, but two phenomena arising from a common cause, that would not diminish the importance of frettage to the engineer, since any area which is observed to be undergoing frettage must be looked upon as one likely to develop fatigue cracks unless some suitable adjustments are made. Frettage is more conspicuous in its early stages than fatigue, although the final result of a fatigue crack is more disastrous. Thus, observation of a frettage area should be regarded as providing a warning signal that fatigue may be expected to occur—or may even have started already.

Nevertheless, the question as to whether frettage is a cause of fatigue, or merely an accompaniment, has something more than academic importance and deserves investigation. Until recently, the only systematic research on the subject had been that of Warlow-Davies, who described his paper as " Brief results of preliminary experiments "; he put forward his conclusions tentatively, making it clear that further work was needed; unfortunately he never carried out this further work, and the praise-worthy caution which Warlow-Davies exercised in reporting his results has not been observed by some who have quoted them.

Warlow-Davies' method consisted of a two-stage test. First the specimen was submitted to fretting by being gripped in a split clamp which was caused to oscillate, by connection to an eccentric device driven from an electric motor, the maximum slip-length being $1{\cdot}37 \times 10^{-3}$ inch between test-piece and clamping surface (assuming rigidity of parts); subsequently it was subjected to an ordinary alternating bending stress in a rotary fatigue machine, and the lives compared with those of similar specimens which had undergone no preliminary frettage. The lives appeared to be reduced by the frettage to an extent which was probably significant, but was not large. For instance, one million cycles in the frettage stage reduced the fatigue strength by about 13%—a reduction similar to that produced by two days' immersion in natural water in the fatigue tests of McAdam and Clyne (p. 712). Warlow-Davies' conclusion that the losses are sufficiently serious to " warrant a more thorough investigation " appears reasonable, and it is a pity that about 15 years elapsed before this was carried out (E. J. Warlow-Davies, *Proc. Instn. mech. Engrs.* 1941, **146,** 32; 1942, **147,** 86; see discussion on pp. 83–85).

Some years ago, Waterhouse examined typical examples of frettage produced in engineering service to ascertain whether the contour of the cavities formed by fretting was such as to provide stress raisers for the initiation of fatigue cracks. Visual examination showed that the cavities could generally be assigned to two types; the first was characterized by shallow dish-like depressions and the second by small deep holes, respectively. It was assumed that the first type arose when the abrasive oxide *débris* had been able to escape from the initial point of attack and had started abrasion in the vicinity—spreading the damage but rendering it less intense. The second type was due to entrapment of *débris*, so that the abrasive particle bored downwards, producing holes; as already stated, the volume of oxide exceeds that of the steel destroyed in producing it, and the interiors of such holes must be under considerable pressure.

Talysurf records were made of the surfaces, and it was shown that, at the edge of holes of the second type, the metal had been forced up into a rim—a sign of the pressure existing in the hole, which was calculated to reach about four times the normal tensile strength. These results certainly do seem to indicate that some of the cavities produced by fretting should be capable of initiating fatigue cracks. Some specimens were examined with a magnetic crack-detector, which, however, did not reveal any obvious cracks in the vicinity of the cavities; supersonic methods of crack detection were found to be unsuitable for articles of such irregular shape (R. B. Waterhouse, *Proc. Instn. mech. Engrs.* 1955, **169,** 1157, 1171).

Recently Field and Fenner have obtained definite evidence of the enhanced effect of combined fatigue and frettage. Their apparatus is a push-pull machine in which a flat specimen with parallel sides undergoes stress fluctuating between two levels of tensile stress. Frettage can be produced by two bridge-shaped clamps which bear on the specimens. When the clamps are present, the life of the specimen is greatly reduced, showing that frettage can enhance the damage due to fatigue. Removal of the clamps after they have been in position for a time exceeding $\frac{1}{5}$ of the life obtained in the absence of a clamp, fails to increase the longevity—which shows that the injurious effect of frettage occurs during the initial stage when there is still metal-to-metal contact, and before a layer of oxidized *débris* has been built up. Subsequent tests with non-metallic clasps (resin-bonded alumina) showed severe fretting damage but little reduction in fatigue life (A. J. Fenner and J. E. Field, *Rev. Métallurg.* 1958, **55,** 475).

Avoidance of Frettage Damage. Since no frettage occurs if there is no relative movement of the surfaces, or if the movement which occurs causes no deformation beyond the elastic limit, two contrasting methods of prevention would seem to be available; we may clamp the pieces so tightly as to *prevent* any slip; or we may apply lubricant, to *encourage* the slip but avoid deformation.

The first method is achieved by increasing the load. Uhlig's work, already quoted, showed that as the load was increased, damage increased to a maximum, and then decreased to zero, because movement was prevented. Such a method has been applied in practical cases. Rolfe tells how the tapered pinion of a certain coupling had suffered fretting. The loading was increased by recessing the middle portion of the taper so that the area of contact was reduced and the stress per unit area increased—eliminating the trouble. Had the increase in pressure been insufficient to prevent movement altogether, the damage would probably have been aggravated (R. T. Rolfe, *Allen Engineering Review* 1953, **30,** 13; 1955, **34,** 24).

Nevertheless it seems doubtful whether the problem can, in all cases, be solved in this way. Johnson has shown theoretically, and confirmed practically, that, although the area over which slip can occur is decreased by increasing the normal load, it is difficult to eliminate slip completely at points where the absence of slip would involve great stress concentration. The designer should try to avoid such stress concentration by the introduction of generous fillets (K. L. Johnson, *Proc. Instn. mech. Engrs.* 1955, **169,** 1167, esp. p. 1169).

Slipping can sometimes be prevented by coating the surfaces with another metal. The plan usually adopted is electroplating with copper, tin, silver or gold; Wright states that fretting between a cast aluminium alloy housing and the bearing was successfully eliminated by tin plating. If the coefficient of friction is insufficient to prevent slip, the plated surfaces themselves suffer fretting and the plating is soon worn away, unless, of course, the metal serves as a lubricant (lead and indium are said to act in that way).

Cadmium, another easily fusible metal, is greatly used to prevent frettage and the accompanying fatigue cracking, especially on aircraft. Great reliance is placed on its presence—based rather on experience than on understanding of its manner of action.

Platings may, however, function in another way; if the elastic modulus is low, the whole of the movement may be kept within the interposed layer, and frettage corrosion is avoided. Rubber gaskets have been used, with some success, and depend on the same principle. A bonded gum-rubber film is stated to prevent slip even at an amplitude of 0·002 inches (W. E. Campbell, *Corros. Tech.* 1954, **1,** 204).

If slipping cannot be entirely prevented, it may be best to lubricate the surface. Perhaps the most remarkable anti-friction material is polytetrafluoethylene (*teflon, fluon*) which has a coefficient of friction against steel of only 0·06; Wright finds that there is no fretting at an interface between these two materials. Molybdenum sulphide has been much advocated to prevent trouble; it works well in the laboratory, but not always equally well in practice. There is some conflict of opinion regarding its use—undoubtedly due to the fact that it behaves differently under different service conditions. Some have obtained no benefit by adding it to oil, whilst others recommend such a procedure. One method of application in molasses followed by burning on is advocated by some and condemned by others. Some recommend bonding in resin. The information is largely empirical, and experiments representing service conditions seem necessary in each case; it is worth while, however, to note the analogy between the crystal structures of molybdenum sulphide and graphite—both lubricants having the atoms arranged in layers (*Engineering* 1954, **178,** 171, with comments by F. G. Kay, p. 229; E. W. H. and R. F. Strohecker, *Corros. Tech.* 1954, **1,** 206).

Waterhouse draws attention to the possibilities of two-component films; with Schulman he found that, on phosphated steel, treatment with an oil-in-water emulsion containing both a strongly polar water-soluble molecule and a weakly polar oil-soluble molecule which have a strong affinity for one another, resisted the on-set of fretting corrosion in a machine of the Tomlinson type for a period longer than similar phosphate-coated specimens lubricated with conventional machine-oil; such two-component films reach equilibrium much more quickly than single-component films. It is, however, still undecided whether these and other methods, which are encouraging in the laboratory, will succeed in service (J. H. Schulman, R. B. Waterhouse and J. A. Spink, *Kolloid-zsch.* 1956, **146,** 77).

Obviously by choosing a material which is resistant to abrasion, the danger of damage by hard particles may be reduced, but the character of the oxide formed must be taken into account. Thus chromium-plating, which is very hard, and which can be obtained in a form containing a network of cracks designed to hold lubricant, so that the effective friction is low, nevertheless fails to prevent fretting damage—possibly owing to the hardness of the oxide.

On the whole, frettage tends to fall off on steel as hardness increases. Surface treatments, including case-hardening and nitriding, as well as shot-blasting (to introduce compressional stresses), have been advocated by different authorities for reducing damage in certain circumstances; it is doubtful, however, whether a general solution of the problem can be found on these lines.

Anodizing has been recommended as a means of preventing frettage on aluminium alloys, but if the main danger of frettage is that it increases liability to fatigue, it should not be forgotten that some types of anodizing themselves render a surface susceptible to fatigue. Phosphating has been considerably used—apparently with success—for lubricating surfaces; the lubricant is held in place and chemical action of an undesired character is reduced. The picture sometimes presented is that of the lubricant being held in between the phosphate crystals, or of the surfaces being kept apart by the composite layer; neither of these ideas seem consistent with the properties of phosphate crystals. Midgley's work suggests a more acceptable explanation. When a surface is phosphated, it is etched to produce a series of interconnecting surface channels, with intervening plateaux which become covered with phosphate crystals. On subjecting the surface to wear, the phosphate crystals are soon removed, leaving the plateaux to form the bearing surface, whilst the channels serve to entrap the lubricant and prevent it from being squeezed out by the pressure of the bearing surfaces. If at the out-set the surface carries scratch-lines, the chemical action is first directed on strained portions, the crystals being deposited close to the scratches; it may be suggested that this preferential conversion of strained material to phosphate is the cause of the improved chemical resistance. The paper deserves study from mechanical engineers (J. W. Midgley, *J. Iron St. Inst.* 1957, **185,** 215).

Since fretting oxidation depends on the presence of oxygen, a rational method of prevention would seem to be oxygen-exclusion. This, however, is not easy to bring about. Some authorities attribute the good effect of oil or rubber to oxygen-exclusion, but the solubility of oxygen in oil is not negligible, and the improvement due to rubber can be explained in other ways.

There is no universal panacea for fretting corrosion, but a number of partial remedies are available—some suitable for one situation and some for others. A careful study of each case of frettage is desirable before deciding on the appropriate treatment.

Wear of Colliery Ropes. The deterioration of the wire-ropes used for winding in coal-mines is apparently related to frettage, since Mayne showed in laboratory tests that periodical cleansing, designed to remove the abrasive products, diminished the rate of wear. The presence of lubricating oil, preferably as thick as can be conveniently applied, improved the performance, perhaps by diminishing chemical action; tests in which the specimens were sprayed with sodium chloride solution (with or without ammonium sulphate) suggested that asphaltic oil was best, although ordinary cylinder oil was improved by the addition of oleic acid. McClelland has advocated galvanized wire, and has recommended replacement of a rope if it has lost 10% of its strength through corrosion fatigue or 20% through wear or corrosion; replacement is also needed if the outer wires have lost 40% of their thickness or have become loose. Trent has studied the chains of pits which appear along cracks in the brittle martensitic layer formed by the rapid heating and cooling associated with friction; he ascribes them to martensite-sorbite cells, or, probably to a smaller extent, to differential aeration (J. E. O. Mayne, *Engineering* 1939, **148,** 157; A. E. McClelland, *Iron Coal Trades Rev.* 1943, **146,** 151; E. M. Trent, *J. Iron St. Inst.* 1941, **143,** 401P, esp. p. 410P. J. Firminger, *Corros. Prev. Control* 1956, **3,** April, p. 25).

A conference held in 1950 brought out the fact that over half the breakages of mine-ropes are due to corrosion or corrosion fatigue, although the formation of hard martensite through friction of the rope against fixed objects, leading to fatigue, and also work-hardening due to excessive bearing pressure, are other important causes. It was stated that corrosion could be controlled or prevented by the use of good zinc coatings in conjunction with careful lubrication during manufacture and in service; corrosion fatigue required the same precautions, but the suppression of kinetic shocks and any form of repeated stressing, is also important. Frettage between the strands, producing material having the appearance of rust (analogous to the " cocoa " mentioned on p. 743) has also been met with; the outside of a rope suffering internally from frettage often remains clean and bright ("Proceedings of Conference on Wire Ropes in Mines", Inst. of Mining and Metallurgy (1950–51); especially contributions from A. E. McClelland, R. S. Brown, I. A. Usher and L. W. Sproule. The corrosion of wire rope caused by mere traces of chloride (0·01%) in hemp has been discussed by A. Bukowiecki, *Schweiz. Bauzeitung* 1958, **76**, Heft 30.

Cavitation Damage and Impingement Attack

Cavitation Phenomena. If water flows through a glass tube of varying diameter, the pressure must vary at different points on its path, and often the water is seen to be milky at points where the pressure is decreasing, becoming clear again later; the milkiness is due to small bubbles, which were at one time described as "vacuum cavities" and were believed to contain nothing except water vapour; it is probably more correct to call them bubbles of low-pressure air. Pure water free from dissolved gas and from micro-bubbles can be subjected to a considerable tension before any sort of cavity is formed. In general, however, nuclei of gas are present in micro-cavities in the walls of the channel through which the water is flowing, or on surface of any solid in the path of flow; frequently there may be invisible micro-bubbles in the water itself. The fact that water seems often to contain micro-bubbles of gas even when unsaturated is in itself a matter demanding explanation; it may be that the gas is absorbed on microscopic solid particles, or perhaps the existence of traces of surface-active impurities lends stability to these tiny bubbles since the ratio surface/volume will necessarily be large.* However, the matter need not be discussed here, since in practice sufficient nuclei for cavitation phenomena are probably provided by small amounts of gas contained in pores opening into the surface of the solid swept by the water-stream.

Where the pressure is high, the gas will be confined to these pores, but if it becomes low, then it will expand, protruding as tiny bubbles from the mouths; if the water contains considerable dissolved gas, and becomes super-saturated at a point of low pressure, part of this will pass into the protruding bubble, which will grow further; finally its size reaches the stage at which part becomes detached and is swept away by the water-stream, leaving just enough gas in the pore to act as nucleus for the next bubble.

Now whether the bubbles arise thus or in some slightly different manner, their growth is a fairly smooth process, whereas their collapse is in the nature of an explosion, being accompanied by a characteristic sound, and capable of causing appreciable damage if a succession of bubbles collapse on, or close to, a solid surface. It is likely that a bubble continues to expand after it has reached a size appropriate for equilibrium with the concentration of dissolved gas present, since the atoms, which had been moving away from the bubble centre during growth process, continue to move outwards owing to their inertia; when this movement ceases, the

* The air-nuclei in freshly poured in water are numerous; but they become less numerous on standing (K. S. Iyengar and E. G. Richardson, *Nature* 1958, **181**, 1329).

bubble, abnormally large and unstable, starts to contract with increasing speed. The contraction will usually be arrested before the volume has become zero and a wave of compression is then sent out through the water in all directions; such a wave is capable of causing damage to any solid surface close at hand.

Many attempts have been made to calculate the pressures which can be generated by the collapse of a cavity. The earlier calculations often assumed that the cavity, containing nothing but vapour, was completely annihilated —which is probably not the case; some of them neglected the need for the dissipation of latent heat, which would set a limit to the rate of collapse —an error pointed out by Silver; however, even when allowance had been made for heat-dissipation, Silver calculated that bubbles of ordinary size generate 9 tons/in.^{2}, whilst small ones may momentarily reach higher pressures. There is no doubt that very high pressures are, in fact, reached, since injury is inflicted on glass and bakelite—where chemical corrosion is unlikely; presumably they exceed the compressional strength of glass and bakelite (R. S. Silver, *Engineering* 1942, **154,** 501; F. von Schwarz, *Z. Metallk.* 1941, **33,** 236; H. Schröter, *Mitt. Ver. dtsch. Ing.* 1933, **77,** 865).

On ships, cavities are liable to be formed on the suction side of the propeller, particularly if the surface is rough. The cavities produced in this low-pressure region may travel and collapse on reaching a region of higher pressure, damaging any metal surface on which the collapse takes place (or near to which it takes place, if we consider the damage to be caused by a wave of compression). Similar trouble may arise in other branches of engineering service. The damage may be amazingly rapid. Ride quotes a case where a $\frac{1}{2}$-inch cast-iron liner of a Diesel engine was perforated after 300 hours' operation. Bondy has described how, " after a destroyer had rushed for several hours at maximum speed, the armour plates above the propeller were pierced by a hole of the dimensions of about one square foot " (R. N. Ride, Symposium on Corrosion (Melbourne University) 1955–56, p. 267, esp. p. 269; C. Bondy and K. Söllner, *Trans. Faraday Soc.* 1935, **31,** 835, esp. p. 837).

The damage to marine propellers has been described by Dorey, who states that cast iron and cast steel are " prone to have local regions of porosity " and develop " a deeply pitted form of attack ", whilst bronzes " exhibit eroded regions having the appearance of areas that have been subjected to intense local sand-blasting " (S. F. Dorey, *J. Inst. Met.* 1953–54, **82,** 497, esp. p. 500; cf. P. Ffield, L. M. Olosher and A. J. O'Neill, Paper to (American) Society of Naval Architects and Marine Engineers, May 1956, p. 20).

Callis, who has special experience on these matters, states that true cavitation (due directly to the collapse of vacuum cavities) is rarely seen on undamaged marine propellers today, but they frequently show large clearly defined areas of roughness, particularly towards the tips, which does not appear to be associated with cavitation-cavity collapse. The cause is probably to be ascribed to an action differing only in degree from that known as impingement attack (p. 476).

He relates the trouble to the existence of *irregular flow*, and this seems very probable, since a true steam-line flow of water over the surface is incapable of setting up a concentration cell, which will only arise if the direction of flow has a component at right angles to the surface.

The term " irregular flow " is used, since more specific conditions are involved than are generally defined by non-laminar or " turbulent " flow. The conditions envisaged involve fluctuations in velocity and/or incident angle of the water flowing relative to a small element of the surface immersed in the liquid, where the limits of the range of one or both of these factors embrace a critical condition, peculiar to each material, below which an oxide-film can form, and above which a film cannot form, or is removed if present. It is considered possible that these conditions may exist in the region of propeller blade tips covered by sheet cavitation.

The possibility of the creation of momentary local ion concentration cells must also be envisaged, and an overall effect arising from differences between the electrochemical characteristics of the phases (in multiphase alloys) exposed at the attacked area, must not be overlooked.

Callis is examining an alternative explanation of the damage set up when sea-water carrying air-bubbles impinges upon a metal. Suppose that the velocity is above that at which the maintenance of a protective film is possible, then the arrival of an air-bubble at the surface results momentarily in a state of stagnation of the liquid behind the bubble, and permits of the formation of a corrosion film which is promptly removed when the bubble is swept away. The presence of air-bubbles, in other words, provides the condition of irregular flow as defined above. The significance of bubble size and the proportion of free air, to which reference is frequently made, is believed to lie in the fact that these factors together constitute a factor of time, in that at any small element of surface exposed to the flow, the two factors more or less define the time interval between the arrival at the point of successive air-bubbles, that is, the time for a cycle of film-formation and film-removal. The maximal damage-rate will occur when the time interval between the arrival of successive bubbles is not longer than that necessary for a film to form, and this appears to be of the order of 0·03 seconds.

The mechanism described is not unlike the idealized type of conjoint action described at the opening of the chapter (p. 732). Callis considers that the conception of the influence of irregular flow may rationalize and provide a general explanation (1) of the moto-electric effect of Müller and Konopicky (p. 130), (2) of impingement attack, (3) of the wastage of marine propeller tips (possibly as a result of sheet cavitation) and (4) of true cavitation erosion. The mechanism suggested is at present under experimental test (G. T. Callis, Priv. Comm., April 29, and May 2, 1958).

Mechanical and Chemical Effects of Cavity Collapse. Some of the cases of cavitation damage must be ascribed to purely mechanical causes —notably the damage caused to glass or bakelite in water; another possible example—the damage caused to metal in toluene—is mentioned later (p. 755). It is easily understood that the water-hammer effect connected with the sudden collapse of low-pressure cavities would produce dents, and, on brittle substances, fracture; the fact that cavitation seems to produce small fragments is perhaps a little surprising, but mechanisms have been suggested by engineers and need not here be discussed. It is, however, fairly certain that, if the metal is receiving mechanical damage, whether by denting or fragmentation, the surface will locally be left free from any protective film which otherwise would have restrained chemical action, and rapid chemical destruction becomes a possibility. If a succession of bubbles collapses at the same point on a surface,* we may expect the chemical action proceeding at that point to be prevented from stifling itself, and pitting is likely. The relative magnitudes of the damage produced by mechanical action alone (e.g. in toluene) and that produced by mechanical and chemical action conjointly (e.g. in water) cannot be foreseen, but sometimes the possibility of chemical influences intervening may seriously increase the damage. As in the case of frettage, it is not possible to divide the damage into two parts and say that $x\%$ is mechanical and $(100 - x)\%$ chemical, since the mechanical damage facilitates chemical action by destroying protective films, whilst chemical action facilitates mechanical destruction—e.g. when a pit undermines the surface layers of the metal.

Experiment shows that the intervention of chemical action can, in some circumstances, greatly enhance cavitation damage. This is true of ordinary corrosion, without any applied E.M.F.; when an E.M.F. is applied, the damage is increased if the specimen is made the anode, but diminished or abolished if it is made the cathode—an important point demonstrated by Petracchi and confirmed by Foltyn. Petracchi used a venturi or nozzle-type apparatus with a channel 20×5 mm. in section, and studied the behaviour of specimens exposed to the water just " down-stream " of the zone of cavitation. His results on brass, conveniently tabulated by Callis are reproduced in Table XXVI.

Petracchi's curves also deserve study (G. Petracchi, *Metallurg. ital.* 1949, **41,** 1; V. Foltyn, *Strojirenstvi* 1952, **2,** 402).

Callis, comparing the resistance of materials to cavitation damage and impingement attack, reaches the conclusion that a purely mechanical explanation is inadequate to explain the results obtained in service. He writes, " There is an encouraging consistency about the relative resistance of similar materials in different tests, and it is usually found that materials with a high resistance to corrosion under stagnant conditions, e.g. in sea-water with a normal dissolved oxygen content, behave well in tests for erosion resistance. Confining attention to cast copper base alloys, the aluminium bronzes and high tensile brasses with high aluminium contents

* It is generally assumed that in a water-stream the bubbles will follow roughly the same course and impinge at the same point. Such an assumption appears not unreasonable. However, only a small proportion of bubbles impinging produce a pit—as shown by Knapp's results. He considers that " only a small fraction of the travelling cavities are of the right diameter to have the possibility of producing a pit and only a few of these will collapse in the narrow band around the stagnation zone where they can acquire enough energy to cause damage " (R. T. Knapp, N.P.L. " Symposium on Cavitation in Hydrodynamics " 1956, Paper **19,** p. 10 (H.M. Stationery Office)).

TABLE XXVI

EFFECT OF CATHODIC AND ANODIC CURRENTS ON DAMAGE BY WATER STREAM (G. Petracchi)

Current Density (mA/cm.²)	Weight Less (mg.)		
	2 hours	10 hours	20 hours
Nil	0·2	1·5	4·0
0·02 (Anodic)	1	5	9·5
0·1 (Anodic)	1·5	7	—
1·0 (Anodic)	>12	—	—
0·1 (Cathodic)	Nil	Nil	Nil

show the highest resistance, followed by high tensile brasses with aluminium contents of 2% or less, and the lowest resistance is shown by gun metals, silicon bronzes and pure brasses."

He then considers a typical case of damage associated with cavitation near the tip of the suction side of a marine propeller, the cavitation being induced by local edge distortion, and speaks of "the deep narrow cavities, roughly circular but non-uniform in cross-section", which tend to turn in a direction parallel to the water-flow; he adds the comment: "it is difficult to imagine that a cavitation void could penetrate to the bottom of one of the holes, to implode finally on the very bottom of so tortuous a channel". He refers to other cases in which there is a complete absence of distortion, and a tendency for the soft alpha brass to stand out from the beta matrix; these characteristics are "to be seen on the most slender projections in the most eroded area, and all these features are suggestive of a process of corrosion and not a mechanical battering by high-velocity slugs of water" (G. T. Callis, N.P.L. "Symposium on Cavitation in Hydrodynamics" 1955, Paper **18** (H.M. Stationery Office)).

Higgins' success in preventing or reducing the pitting of cast-iron propellers by means of magnesium protectors (p. 759) provides fairly direct evidence that the trouble is chemical or electrochemical rather than mechanical.

Some of the contributors to the N.P.L. Symposium, however, express divergent opinions regarding the relative importance of mechanical and chemical influences. Eisenberg quotes experiments at the California Institute of Technology in which the damage obtained in toluene below a helium atmosphere was not appreciably different from that observed in water. However, Wheeler, whilst agreeing that there is little difference where precautions are taken to avoid corrosion below the aqueous medium, reports that under other conditions severe pitting is produced by water, the rate-loss measured being rapid; "in toluene", he adds, "nothing approaching this has ever been observed here".

Wheeler has endeavoured to separate the results of mechanical and chemical action by filtering the liquid in which a steel specimen has been undergoing cavitation damage. He estimated the soluble iron salts found in the filtrate and took them to represent the result of corrosion, whilst the dark material left on the filter-paper was provisionally assumed to be composed of fragments of metallic iron broken off by erosion; these assumptions led to the conclusion that, in a 10-minutes' test in oxygen-free water, 14% of the damage was caused by corrosion, whilst in air-saturated N/10 potassium chloride as much as 68% was chemical. It seems likely that corrosion may have had even more effect than these figures suggest, since the black material did not readily dissolve in acid, and required for dissolution 10 minutes heating in 50% hydrochloric acid containing nitric acid; it cannot, therefore, have been entirely metallic iron. If the slowly soluble matter was magnetite—formed by interaction of the soluble anodic and cathodic products—this part ought to be included in the percentage attributed to chemical corrosion.

Although there is no reason to doubt the general conclusion reached in Wheeler's important work—namely that chemical influences play a very important part in cavitation damage—it is fair to future investigators to point out that filtration does not provide a satisfactory method of separating the products of corrosion and erosion. If this simple method had been used in the early researches on the corrosion of iron partly immersed in stagnant chloride solution without any application of mechanical force, the product left after filtering the liquid (rust or magnetite, according to the oxygen-supply to the vessel) would have been ascribed to mechanical erosion. Nevertheless, used by Wheeler as an exploratory method, the filtration procedure did give some valuable information.

Wheeler has made some calculations suggesting that high temperatures may be produced locally and temporarily at points where the cavities collapse; the fact that interference colours appear on specimens in cavitation experiments is consistent with this view, although it does not furnish proof of high temperatures, since bright colours can be produced during the wet corrosion of specimens at room temperatures (p. 91). However, it is fairly certain that some rise of temperature will occur as a result of the energy released, and if the temperature should rise from the range where the log laws prevail into that where the parabolic thickening becomes possible, the effective oxidation-rate may be greatly increased; thus—when there is mechanical removal of the oxide at short intervals—the rate of destruction of the metal will be greatly increased also. On iron the change of growth-law has been found to occur at 200°C. or 300°C., according to the surface condition (p. 835). If the cavitation energy fails to raise the temperature above the change-point, the effect on the rate of destruction will be less serious. It is understood that work on the local temperature rise is in progress.

At the N.P.L. Symposium, Rasmussen provided a comparison between the behaviour of metals and non-metals (*perspex* and ebonite). Since non-metals as well as metals suffered damage, Rasmussen—although expressing himself with admirable caution—inclines to the view that the mechanical "hammer effect" is the main cause of the observed erosion. It is possible, however, to interpret Rasmussen's results differently. The photographs suggest that the damage inflicted on perspex and ebonite, which is certainly mechanical, is of a different type to that produced on metal, and the measurements show that the loss of weight is considerably smaller. Rasmussen emphasizes the fact that if the liquid contains air-bubbles the damage is lessened, and explains this by the fact that a water–air mixture is more compressible than air-free water. In the case of aluminium, the damage falls off steadily as the air-content is increased, becoming negligible when 4% of air is present; this might be due to film-repair made possible when excess of air is present everywhere. On cast iron, the presence of air does not reduce the damage to zero; here Rasmussen recognizes that rust is being formed and removed, so that the loss of weight is due to a chemical effect, the mechanical action having been stopped by the effect of air-bubbles (N.P.L. "Symposium on Cavitation in Hydrodynamics" 1956 (H.M. Stationery Office); especially contributions by P. Eisenberg, G. T. Callis, W. H. Wheeler, R. E. H. Rasmussen).

Shalnev obtained deeper cavitation pitting on cast iron in synthetic sea-water than in fresh water, and drew the conclusion that the chemical factor was important, although on brass and bronze he considered mechanical properties more serious (K. K. Shalnev, *C.R. Acad. Sci.* (*U.R.S.S.*) 1951, **78**, 33).

Testing Methods for Cavitation. The earlier demonstrations of cavitation damage were largely carried out in water flowing through channels of suitably varying cross-section, so that cavities were formed and collapsed, inflicting damage; this is generally referred to as the "Venturi-type test". The advantage of such a form of test is that in service, the formation and collapse do not necessarily occur at the same point, and it is possible in the Venturi-type test to inspect the damage caused by collapse separately from any damage inflicted at the point of formation; often, however, the latter is negligible.*

A convenient testing method described by Schumb depends on vibration produced magnetically; a cylindrical specimen with a concave, saucer-shaped depression at its lower end is attached to a vertical nickel tube which is set into vibration by means of an alternating magnetic field (perhaps 9,000 cycles per second); the method depends on the magneto-strictive effect in nickel. The concave face is immersed during the period of vibration in corrosive liquid—e.g. sea-water. Cavities are formed when the specimen moves upwards, producing a negative pressure, and collapse when it descends, so that the material suffers damage. The method has been criticized by Speller and LaQue for failing to bring out in the laboratory the benefits obtained by adding chromate, which is found to be effective in service as an inhibitor; also for failing to predict the favourable results obtained in service from austenitic cast iron. It has, however, been used in two important researches by Beeching, and, in a different form, by Wheeler

* Poulter obtained serious destruction on glass or quartz immersed in liquid by applying a high pressure and suddenly releasing it; it is doubtful whether this has much bearing on cavitation damage as experienced in service, but his paper deserves study (T. C. Poulter, *J. appl. Mech.* 1942, **9**, A31).

(p. 757). One possible criticism is that formation and collapse of cavities occur at the same point, but Beeching obtained correlation between results obtained by the vibration test and the Venturi-type test respectively. Another objection sometimes urged is that the vibration test is essentially a test of hardness; however, in Wheeler's hands, it has certainly served to reveal the importance of chemical action. Readers should compare the views of W. C. Schumb, H. Peters and L. H. Milligan, *Metals and Alloys* 1937, **8,** 126; R. Beeching, *Trans. Instn. Engrs. Shipb. Scot.* 1941–42, **85,** 210; 1946, **90,** 203, 239; F. N. Speller and F. L. LaQue, *Corrosion* 1950, **6,** 209; Y. Bonnard and E. Josso, *Métaux et Corros.* 1948, **23,** 116.

Rasmussen (p. 757) used two apparatus, one involving the flow of liquid through a channel of shape arranged to suck in air at a certain point through a series of small holes, giving bubbles which later impinged on a metallic specimen. In the other apparatus, a flat round disc immersed in water rotated on a vertical shaft; it was smooth except for two circular holes close to the margin at diametrically opposite points; specimens of the material under test were inserted flush with the surface at points " downstream " of the marginal holes, at positions such that cavities formed at the holes collapsed on the surface and inflicted damage.

Prevention of Cavitation Damage. There has long been a difference of opinion as to whether immunity from damage—particularly on propellers—should be attained by choice of resistant materials, or by geometrical design; it is possible to choose shapes which minimize the production of air-cavities or cause them to collapse at places where damage would not have a serious consequence. On one occasion it was found possible, by re-designing a large ship's propeller, to reduce the damage to the equivalent of 0·08 inch per 100,000 miles, whilst other ships carrying propellers of the older design were suffering 12·5 inch per 100,000 miles (F. McAlister, *Trans. Instn. Engrs. Shipb. Scot.* 1941–42, **85,** 244).

By rounding off the edge of propellers, it is possible to avoid many of the effects of cavitation, notably the erosion and holing of the metal and the noise (" singing ") of the propeller. There is, however, also a decrease of resistance to the movement of the propeller, which resistance is necessary for the propulsion of the ship; thus the change of shape might reduce the mechanical efficiency (E. G. Richardson, *Endeavour* 1950, **9,** 149).

These difficulties have largely been solved by a wise combination of the favourable factors. Callis, who has emphasized the need for the choice of chemically resistant materials, also recognizes the part played by shape; cavitation-erosion, he states, is now rarely seen on propellers, as it is avoided " by informed design " (G. T. Callis, *Metal Ind.* (*Lond.*) 1951, **79,** 167, esp. p. 168).

Beeching, using the vibration method, has found a general reduction of damage on nickel brasses as the nickel content is increased, and advocates the use of brasses containing 15% nickel for large marine propellers, and a content of 20–30% for small high-speed propellers; this conclusion has been criticized by A. J. Murphy and G. T. Callis, *Trans. Instn. Engrs. Shipb. Scot.* 1946, **90,** 236; see reply by R. Beeching on p. 244.

Meanwhile, the demonstration that cathodic protection can almost eliminate damage under some laboratory conditions holds out distinct hope. Much will be required to be done before the method can be systematically applied in practice. Some trials on magnesium protectors attached to motor-tugs fitted with cast-iron propellers have given promising results. These are described by Higgins, who considers that they support the views that the pitting of cast-iron propellers is primarily a chemical process and that it can be prevented by cathodic protection, although the limited life of the anodes is a factor to be taken into account. Duff has reported similar success on vessels fitted with bronze propellers (R. I. Higgins, R.C.I.R.A., *J. Res. Dev.* Dec. 1957, **7,** 129; H. G. Duff, *Corros. Tech.* 1958, **5,** 250).

It is possible that cathodic protection has been used already, without the fact being fully recognized. It has long been customary, on ships fitted with bronze or brass propellers, to fix a block of zinc around the propeller shaft; the principal object in mind has been to counteract the cathodic action of the copper alloy which would otherwise accelerate attack on the ships' plates, but the zinc probably serves to protect the propeller itself. Again, it has been found that the cavitation of the cast-iron propellers usually fitted to trawlers can be greatly diminished by spraying with aluminium, the effectiveness of which is probably due in part to cathodic protection (see p. 642).

Impingement Attack. The damage inflicted by bubbles near the inlet end of condenser tubes (p. 476) was at one time considered to be something quite different from that inflicted by the vacuum cavities developed by propellers. However, there is probably no sharp distinction. It is now believed that the vacuum cavities are really low-pressure air-bubbles, and that the collapse does not involve annihilation; there is, indeed, considerable evidence that the so-called collapse consists of oscillations, the bubble becoming alternatively larger and smaller. This idea of cyclic expansion and contraction—for which Rayleigh's work provided evidence—has in recent years come into prominence (E. Crewdson, *The Engineer* 1953, **195,** 122; W. D. Chesterman, *Proc. phys. Soc.* (*B*) 1952, **65,** 846).

However, there is a real difference between (1) the bubbles containing air at pressures not far below atmospheric which damage condenser tubes and (2) the low-pressure cavities which damage propellers. In the former case, the bubbles bounce off instead of collapsing, and, if they are small, no great damage may result. If they are large, they break into many parts and considerable damage results; the break-up must cause a complicated stress situation, and if the metal is deformed thereby—even within the elastic range—there may be breakage of the protective skin, which will be the more serious since, in a condenser tube, a train of bubbles will generally impinge at the same point. Where the bubble fragments leave the metal, portions of the film-substance may stick to them and be sucked away from the metal; this is most likely to occur if the film adhesion is weak, as may occur if the film has been formed by cations moving outwards, and leaving cavities between film and metal (cf. p. 48). Thus the impingement of bubbles of relatively high-pressure air may introduce a new factor very liable to damage the film—a factor absent in the collapse of low-pressure cavities. Against this the presence of oxygen in the bubbles may help repair, and the presence of two opposing factors may explain the apparent disaccord between the results reported by those who have followed the movement of potential during impingement; sometimes it moves sharply in the negative direction, indicating mechanical destruction of the film, but where this does not occur, there may be a movement in the positive direction, indicating that repair is prevailing.

Despite these differences, the two forms of damage should be regarded as two extreme cases of a single series of phenomena—of which the middle members are less frequently met with in service.

Impingement attack in condenser tubes can generally be avoided by using a brass containing aluminium (p. 477); such material yields a film more resistant to damage and better capable of self-repair than ordinary brass, although it is not very satisfactory in contaminated harbour waters containing hydrogen sulphide, organic sulphur compounds or sand in suspension. Alternative materials are discussed on p. 478.

The chief characteristics of the pits made by impingement of bubbles is their undercutting of the surface layers of the metal at the forward end (fig. 89, p. 473)—probably connected with the scouring action of the bubbles which removes corrosion products. The horse-shoe ground-plan shown by some but not all the troughs is usually explained by assuming a deflection of the bubbles by adherent particles such as rust; Bengough and May actually watched the formation of horse-shoe grooves on sections of brass tubes held in slightly larger glass tubes (G. D. Bengough and R. May, *J. Inst. Met.* 1924, **32,** 81, esp. pp. 204, 209, 220).

Miscellaneous Cases of Conjoint Action

Effect of Sand. In many situations, water running through a plant (e.g. through cooling jackets or pipes) carries sand in suspension which may cause chemical corrosion to continue at places kept bare of film by the abrading action of the sand particles, although stifling occurs elsewhere. The aspect is such that the damage may be ascribed erroneously to simple mechanical abrasion; whereas in fact it is due to conjoint action. Some of the operations in beet sugar factories involve water carrying suspended sand, and corrosion troubles have often been met with (H. Claassen, *Korrosion u. Metallsch.* 1936, **12,** 305).

Troubles at Hydro-electric Stations. The high water velocities at hydro-electric stations are liable to produce wastage of various kinds. Trouble has been reported both in the pipe-lines and in the water turbines. Some of this is doubtless simple mechanical abrasion by suspended particles, but if the water is at all corrosive there will usually be conjoint action. Ample opportunities exist for cavitation damage, whilst some of the wastage is attributed by Passerini to differential aeration; oxygen is constantly replenished where the metal is washed by rapidly moving water, but may become exhausted in recesses. There are also risks of stray electric currents, or electric currents set up by differences of temperature (L. Passerini, *L'energia elettrica* 1929, **6,** 168; 1932, **9,** 894; G. Scarpa, *ibid.* 1933, **10,** 210; N. Faletti, *ibid.* 1934, **11,** 277; R. Auerbach, *Kraftwerk,* 1931, p. 15).

Here again the remedy seems to rest mainly in improved design, although choice of materials may help. In certain designs of water turbine, cavitation set up in the clearance space causes wastage at the back of the blade, and this can sometimes be remedied by guard rims on the low-pressure side, which do not indeed prevent cavitation, but deflect it to a region where it is less harmful (H. Mueller, *Z. Ver. dtsch. Ing.* 1935, **79,** 1165).

Damage in Steam Turbines. At the high-pressure end of a steam turbine where the steam is dry, there may be little damage, unless much salt is entrained. In the middle regions, where drops exist in the steam, damage is often serious, and will be increased by salt particles sticking to the blades; even here mechanical hardness may be of more importance than chemical resistance as normally understood, since if the metal is sustaining mechanical damage no protective skin can survive. An example comes from Honegger's early work; in the hard condition, 14% chromium steel gave good performance—better than 5% nickel steel—but in the soft condition it was no better than brass. One plan is to make the blades of stainless steel, with only the edges in the hard state (E. Honegger, *Elektrotech. Zeitsch.* 1929, **50,** 465; *Brown-Boveri Rev.* 1927, **14,** 95; see also *ibid.* 1934, **21,** 25; H. Zschokke, *Korros. Metallsch.* 1937, **13,** 386).

Ferritic steels containing 6% or 12% chromium (with titanium, vanadium and niobium for creep-resistance) have some advantages over austenitic materials (L. S. Robson, *The Engineer* 1957, **203,** 978).

Success depends on sensible design coupled with good choice of materials, but generalization is difficult; sometimes a corrosion-resistant material has been welded on to the blade at the place liable to attack; if that plan is adopted, the risk of bimetallic corrosion must receive due consideration. For marine turbines, highly alloyed steels have given satisfaction (T. H. Burnham, *Trans. Inst. mar. Engrs.* 1934, **46,** 1, esp. p. 10; L. M. Davis and J. M. Mousson, *Metal Progr.* 1939, **35,** 349).

In all parts of the system, there is danger of rusting during periods of rest, through condensation of moisture, and it may be well to blow warm air through the system when shutting down.

Cavitation in feed pumps at Central Stations are discussed in *Corrosion* Feb. 1953, Conference and Exhibition Section, p. 34, where it is made clear that 5% chromium in steel greatly improves resistance. Cavitation in oil is discussed by R. A. Schaefer, J. F. Cerness and. H. A. Thomas, *Trans. Inst. Metal Finishing* 1954, **31,** 454.

Attack by Moving Drops. In various situations (aircraft, sea-plane floats and steamer hulls) water-drops will strike a metal surface at high velocity, and if the water is salt, the attack may be far more rapid than static corrosion; destruction is evidently due to conjoint action, and laboratory studies have shown discs 0·5 mm. thick to be perforated in 36 days; pore-free material shows the greatest resistance (G. Welter, *Engineering* 1938, **145,** 521; M. Vater, *Z. Ver. dtsch. Ing.* 1937, **81,** 1305). The relationship between drop-impact attack and cavitation is brought out by A. Keller, *Schweiz. Arch. angew. Wiss.* 1957, **23,** 346.

Armouring based on fibre-glass and epoxy-resin has been advocated for protection of the splash zone on off-shore structures, and for various situations in the oil industry (W. F. Oxford, *Corrosion* 1957, **13,** 615t).

Disintegration by the internal production of voluminous corrosion-product. It has been noticed (p. 510) that wrought iron exposed to the atmosphere with the emerging points of the *Q*-zones insufficiently protected sometimes undergoes serious swelling, especially near the sea coast. This is characteristic of atmospheric attack, where the bulky product will be formed internally. It is, however, by no means confined to marine atmospheres; in the neighbourhood of an acid works, the author has found strips of ferrous material so swollen and rotten that it could be split apart like shale by means of the fingers.

Such swelling is less common under immersed conditions where the solid corrosion-product is often thrown down outside the metal, but where for special reasons a bulky material is formed within a microscopic crevice running into the metal, the result may be very serious. If the metal is strong enough to withstand the pressure exerted by the volume increase, the crevice will become plugged and the action gradually cease; but if once the expansive forces start to lever open the original crevice into a definite crack, this process will continue indefinitely. Cases have arisen at sulphuric acid works where pipes have burst through bulky iron salts formed within the pores (R. Hay, Priv. Comm., Oct. 31, 1927).

It is not certain whether such cases should be ascribed to the conversion of iron into bulky iron sulphate, or to interconversion between the hydrates. The equilibrium diagram of the system $FeSO_4$–H_2O is not simple;* but, neglecting the existence of two hydrates with $2H_2O$ and $4H_2O$ which are sometimes formed within a narrow range of temperature, it may be said that a super-saturated solution ought (under reversible conditions) to deposit $FeSO_4.7H_2O$ below about 57°C. and $FeSO_4.H_2O$ above 64°C. In presence of sulphuric acid the equilibrium temperature of the change

$$FeSO_4.7H_2O \rightleftharpoons FeSO_4.H_2O + 6H_2O$$

occurs at a lower temperature, and it is not unlikely that in a pipe carrying liquid at a temperature which varies with time, there may be transformation backwards and forwards from one hydrate to the other. The numbers given for the specific gravities in the literature suggest that a considerable decrease of the volume of the solid phase should accompany the left-to-right change. Thus if the liquid liberated is free to enter or escape from a crevice where solid is entrapped, there will be a sudden expansion on cooling into the range where the heptahydrate is stable. If a monohydrate crystal is wedged in a crevice, it is not unlikely that the expansion on cooling may crack the metal.

In quite different situations, the pressure developed by the expansion

* For details see J. W. Mellor, " Comprehensive Treatise of Inorganic and Theoretical Chemistry ", 1935 edition, Vol. **19,** pp. 246–251.

accompanying corrosion changes can plug crevices or alternatively expel the products. When iron or steel articles have lain in the sea for some time and are brought out, washed, dried and exposed to air in a room, they develop little globules of moisture at points on the surface; over each globule, membranous rust develops, and when the globule has evaporated, the rust-membrane remains, resembling, when viewed under the microscope, a tiny globe of amber glass. The phenomenon is probably due to the anodic corrosion-product, ferrous chloride, remaining in invisible crevices after the surface has apparently become dry; this is hygroscopic, and the absorbed water expels the globule. The formation of such globules in cooling-water jackets is generally regarded as proof that salt-water has been used for cooling purposes.

Mixtures of iron filings and ammonium chloride solution are used for sealing joints, and, although the practice is not for all purposes to be recommended, it is understood that tight joints can thus be obtained, owing to the expansion accompanying rusting.

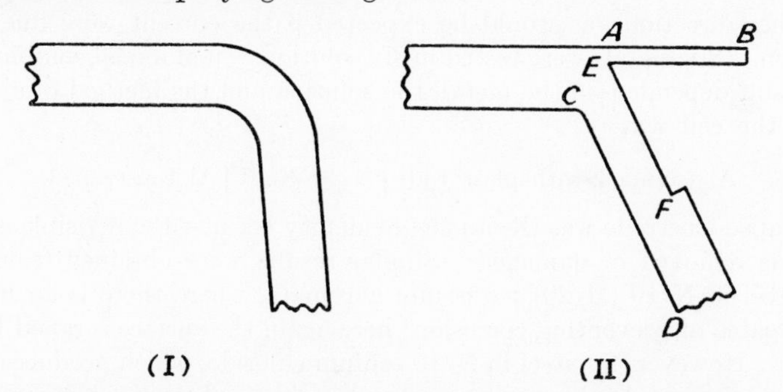

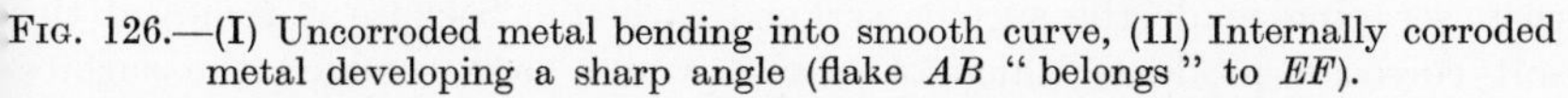

FIG. 126.—(I) Uncorroded metal bending into smooth curve, (II) Internally corroded metal developing a sharp angle (flake *AB* " belongs " to *EF*).

In early laboratory work at Cambridge on the corrosion of zinc sheet in chloride solution, it was hoped to follow the progress of attack by measuring the loss of thickness. Surprisingly it was found that often even after conspicuous corrosion had developed, the thickness had not diminished and sometimes had increased. This was not due to corrosion-products adhering to the external surface but to internal corrosion-products. The corrosion in such material sometimes proceeds along planes parallel to the surface—as is proved when, at a certain stage of attack, the corroded sheet is taken out and subjected to a bending force. In contrast to an uncorroded specimen, which gives a smooth bend (fig. 126(I)), the corroded specimen may break at a sharp angle (fig. 126(II)) and certain layers which properly belong to one limb remain as the prolongation of the other; thus the flake *AB* really represents the surface layers of the limb *CD*, which have broken off owing to corrosion following the plane *EF*.

Further information about rust globules on iron and flake attack (exfoliation) on zinc will be found in the original papers (U. R. Evans, *J. Soc. chem. Ind.* 1926, **45,** 37T, esp. p. 42T; 1928, **47,** 55T, esp. p. 61T).

Electrochemical Effects of Local Abrasion. It has been stated in connection with the pitting produced by cysteine in brass condenser tubes that the intensity was due to the combination of small anode (the break in the film) and the large cathode (the copper-cysteine compound forming the film, which, if not interrupted as a result of impingement, would be protective). Many other examples could be cited of electrochemical action set up by mechanical damage. This has long been realized, and the electric currents flowing between abraded and unabraded surfaces of the same metal were studied in early work (U. R. Evans, *J. Inst. Met.* 1925, **33**, 27).

In this research, a differential abrasion cell was used, consisting of two sloping electrodes of the same material clamped in the same liquid, so that the surface of one of them could be scraped; they were joined to a milliammeter, and it was found that when the left-hand electrode (only) was scraped, the current passed through the metal in one direction and when the right-hand electrode (only) was scraped, a similar current passed through it in the other direction—as would be expected if the current were due to the abrasion. Six metals were tested in six solutions—making 36 combinations. The result depended on the metal, the solution and the method of abrasion. When the cell was

Al (scraped with glass rod) | N/10 NaCl | Al (unscraped)

the scraped electrode was the anode, evidently because the invisible alumina film was removed or damaged. Similar results were obtained from a cell with steel in N/10 (M/20) potassium chromate, where there is an invisible film capable of preventing corrosion; here again the surface scraped became anodic. However, for steel in N/10 sodium chloride, which produces visible rust, scraping made the surface scraped cathodic, because it removed the soft corrosion-product, which otherwise would have shielded the surface from oxygen. The results of other experiments could be interpreted in a similar manner; removal of an invisible protective film rendered the electrode anodic, whilst removal of a soft visible corrosion-product rendered it cathodic.

To test the correctness of this interpretation, other experiments were carried out in which a quill pen was used in the place of a glass rod; it had been found that the quill, although capable of shifting the soft corrosion-product, failed to produce visible scratches on uncorroded steel, left only slight marks on aluminium, but was able to produce conspicuous scratches on lead. In those cases where the effect of scratching with glass had been to render the surface cathodic (by removing a soft product), the quill produced a current not much lower than that produced by glass. However, on steel in sodium chromate, where glass abrasion rendered the surface anodic, the quill, being incapable of effectually damaging the oxide-film, produced no current; in the case of lead, which was sufficiently soft to be scratched by a quill, so that the film was certainly damaged, the anodic current was more than half that produced by glass.

It is easy to understand that persistent, local abrasion of a metallic surface in a liquid which is normally resistant owing to film-formation, will cause a small anode surrounded by a larger cathodic area and thus lead to intense corrosion; where the surface as a whole would be attacked, local abrasion will produce little effect.

Instructive results have been obtained by Akimov and G. B. Clark, who measured the electrode potential of various metals in 3% sodium chloride (1) without special precautions, (2) whilst the surface was kept " clean " by continuous abrasion with a carborundum rod. It was found that this " cleaning " shifted the potential of the certain metals of the " A " groups (Be, Al, Nb, Cr, Mo) by about 500 mv.; these metals carry highly protective films. In the case of the transition elements (Fe, Co, Ni), the shift was somewhat smaller, whilst for some of the B-Group Metals (Zn, Cd), it was smaller still. Measurements were recorded for other metals and o her reagents, and deserve study; they are conveniently summarized by G. W Akimov, *Corrosion* 1955, **11,** 515t, esp. pp. 525t, 527t. Cf. study of effect of strain on potential by J. C. Giddings, A. G. Funk, C. J. Christensen and H. Eyring, *J. electrochem. Soc.* 1959, **106,** 91.

Hatwell has designed an apparatus in which materials can be abraded with a rotating stone in argon in a closed vessel, and the potential measured against a saturated calomel electrode. Unusually negative numbers are obtained. Aluminium after argon abrasion shows −1·58 volt in 3% sodium chloride, but −0·80 volts if the abrasion is followed by 15 minutes exposure to air; for uranium in 3% sodium chloride, the numbers are −1·05 and −0·66 respectively; for zirconium in 3% sodium chloride, −0·99 and −0·45; whilst for 18/8 Cr/Ni stainless steel in 3% sodium nitrate, they are −0·15 and + 0·18 volt (H. Hatwell, *C.R.* 1953, **236,** 1881).

Such numbers serve to indicate the danger of local abrasion in a corrosive environment which would be likely to produce a small anode surrounded by a large cathode, with possibilities of a considerable E.M.F. and intense corrosion.

In interpreting experiments involving abrasion, Courtel's work on grinding *in vacuo* may prove useful. Steel was ground in the electron diffraction camera and at once examined at grazing incidence. When there was a good vacuum, there were no sparks, but they appeared if the pressure exceeded 1·5 mm. mercury; rings representing γ-Fe_2O_3 or Fe_3O_4 then soon appeared, the metallic iron rings becoming gradually more diffuse (R. Courtel, *Métaux et Corros.* 1950, **25,** 188).

Alternate Corrosion and Abrasion of Ferrous Material. Some interesting work at Derby deserves notice. Specimens were exposed out of doors, and abraded at suitable intervals; since corrosion-rate varies with weather the unit of time adopted was the " standard week "—the period needed to produce a certain standard weight-gain on the specimen. Each specimen was subjected to corrosion for a given number of standard weeks between each abrasion. Except for grey cast iron, corrosion-resistance was reduced by abrading away the rust, and the corrosion-rate was increased as the frequency of abrasion was increased; evidently the rust is to some extent protective, the corrosion-rate being slowest when it is left undisturbed. Cast iron behaved differently from steel. Grey cast iron showed a good resistance to abrasion alone; its resistance to ordinary corrosion was similar to that of steel, but its position in the order of merit of the materials tested rose when there was a combination of abrasion and corrosion. Three white cast irons showed good resistance to corrosion, to abrasion and to a combination of the two (J. Dearden and J. D. Swindale, *J. Iron St. Inst.* 1957, **185,** 227).

Other References

The bibliographies provided in the paper on frettage by R. B. Waterhouse (*Proc. Instn. mech. Engrs.* 1955, **169,** 1157) and in the " Symposium on Cavitation in Hydrodynamics " 1955 (H.M. Stationery Office) include many papers deserving careful study. In the papers delivered to that symposium, the mechanism of the formation of cavities is authoritatively discussed although the chemical consequences of their collapse are overlooked by some authors; an admirable review of the symposium is provided by E. I. Brimelow, *Engineering* 1956, **181,** 14. A short section on chemical wear and fretting is included in F. T. Barwell's " Lubrication of Bearings " (Butterworth).

Recent Canadian work on cavitation shows how damage depends on temperature, pressure and wettability, being maximal at temperatures where the number of oscillating bubbles is maximal; this points to a method for control of the damage, which is regarded as a conjoint (mechanical-chemical) effect (W. C. Leith, *Engng. J.* (*Montreal*) March 1959). The mechanical aspects of cavitation have been summarized by D. J. Godfrey, *Chem. and Ind.* (*Lond.*) 1959, p. 686.

QUANTITATIVE SECTION

CHAPTER XIX

VELOCITY MEASUREMENT AND TESTING

Synopsis

The first eighteen chapters have presented information which has been mainly qualitative—despite the occasional introduction of very simple equations, such as that embodying the parabolic law of film-growth. The remaining chapters deal with the quantitative aspects of corrosion. The present chapter is devoted to the experimental methods used in scientific measurement and empirical testing. Chapters XX and XXI will then present the theory of the kinetics of film-growth and electrochemical corrosion respectively, expressed in terms of elementary mathematics—such as is usually learnt at school but sometimes forgotten on leaving The last chapter deals with irreproducible phenomena, where the concepts of probability and statistics must be introduced; no prior knowledge of these subjects is assumed.

The present chapter starts by considering why quantitative research is of interest both to the pure scientist and to the engineer —although for different reasons; it then describes methods of measurement used for pure-science and engineering purposes, including the testing of protective coverings.

The pure-science section starts by considering the methods of obtaining growth-curves for films; thickness can be assessed by weight-increase or electrometric measurements; examples are given of the kind of apparatus which has been employed, with the suggestion that it should not be too slavishly copied, since every new research demands a new design. The methods for stripping and transfer of films also receive attention. Optical methods for studying film-thickness, based either on interference colours or the changes produced in polarized light on reflexion at a surface, are then briefly discussed. Next, procedures suitable for obtaining corrosion–time curves under conditions of partial or complete immersion are indicated—based either on weight-loss, gas-absorption or chemical estimation of metal in the corrosion-product. Methods are described for studying potential; these include (1) the variation of potential at different points on a corroding surface, (2) the movement of potential with time, and (3) change of potential with applied current (" polarization curves ").

Passing to tests designed to aid the practical engineer or plant-designer, some general remarks on field tests lead to the natural desire for acceleration, with a few examples of the almost ludicrous attempts made in early days to satisfy that desire. The rational basis of acceleration is indicated—the intensification of natural factors without the introduction of new ones; the omission of " idle periods " can also shorten the time needed to obtain results. Spray tests and condensation tests receive attention, and some general remarks are made about tests designed to decide the choice of material for some particular industrial purpose. The testing of coatings then comes up for consideration, including thickness, porosity, adhesion and hardness—some of which have already been discussed; this is followed by a brief discussion of the industrial testing of inhibitors under service conditions. The chapter ends with thediscussion of fundamental reasons for the failure, in some circumstances, of small-scale laboratory experiments to predict behaviour in a large-scale plant.

The Need for Quantitative Measurements

Importance of Measurements in Pure Science. The pure scientist wishes to understand the mechanism of phenomena. For him, measurement is needed if he is to decide which of a number of rival theories is the right one. Probably several of them will account for the facts qualitatively; a quantitative study, however, may enable most of them to be discarded.

There are generally two stages in the process of discovering the correct mechanism. First, we ask whether the measurements obey an equation of the form which would be expected if a certain theory is right; if they do not, that theory may be dismissed. Obedience, however, does not settle the question, since a given set of measurements may be consistent, within experimental error, with two (or more) equations which look quite different when written in algebraic form. Moreover, the same equation can often be reached by different arguments starting from two (or more) assumed mechanisms; thus the parabolic, inverse logarithmic and direct logarithmic equations of film-growth can each be derived by simple mathematical arguments starting from essentially different premises. In the second stage of the process, therefore, we enquire whether the numerical values of the constants involved in the equations, calculated from a given mechanism, agree with those experimentally established, and if this is found to be true of only one mechanism, that mechanism may be accepted, at least provisionally, and the others discarded.

For instance, the rate-constant which occurs in the parabolic law of film-growth, can, on the assumption of Wagner's mechanism, be calculated from the electrical properties of the film-substance, and the surprisingly good agreement with the values determined by experiment provides evidence that Wagner's mechanism at least follows the right lines; if it were completely wrong, there would be no reason to expect that the calculated and observed values would even be of the same order of magnitude.

Again, if the electrochemical mechanism of corrosion by salt solutions is correct, the electrical currents which are found to be flowing between different regions of the metallic specimen should be equivalent, in the sense of Faraday's Law, to the measured corrosion-rate. The close agreement between the calculated and observed values affords direct evidence of the essential correctness of the mechanism. Had the mechanism not been electrochemical, the calculated and observed values would not, in general, have been even of the same order of magnitude.

It should be emphasized that this agreement between calculated and observed values has been obtained in corrosion research without any arbitrary assumptions regarding the values of the quantities introduced into the expression. In this respect, the kinetics of corrosion processes have reached a higher scientific level than the kinetics of many other processes in physical chemistry, where agreement between a series of pairs of values (one series representing the observed and the other the calculated values) is only obtained by *assuming* agreement in one or more pairs, so as to obtain the numerical value for the constants appearing in the equation representing the mechanism.*

Evidently, where agreement between certain Calc and Obs numbers has to be *assumed* for the purpose of the calculation, the second stage of the test has not been applied at all, and even the first stage has been applied in an unsatisfactory manner, unless a real study of the degree of agreement with alternative equations has been carried out by someone acquainted with the Theory of Errors and possessing practical experience of Curve Fitting.

There is no need to press the criticism too far, but, in view of the justifiable uneasiness felt today about the use of "Calc and Obs" methods in some branches of kinetics, it must be emphasized that the agreement between Calc and Obs values in corrosion work involves none of these arbitrary assumptions, and may fairly be claimed to afford convincing evidence of the essential correctness of the mechanisms. In other words, the scientific basis of the kinetics of corrosion processes is more satisfactory than that of the kinetics of certain other phenomena—commonly regarded as "purer" science.

Importance of Measurements in Practical Problems. To the man whose interest in corrosion arises from the trouble caused by it in industry

* For instance, a paper on some kinetic process may contain a table showing "Obs" and "Calc" values in parallel columns, with an agreement which may at first sight strike the reader as impressive. Closer scrutiny will probably reveal one, two or perhaps three sets of numbers, placed in square brackets, where agreement is *exact*; these are the numbers where agreement has been *assumed* in obtaining the values of the one, two or perhaps three constants (k_1, k_2, k_3) occurring in the equations—which values are needed for obtaining the other "Calc" numbers. When this is realized, the fact that the other Calc numbers are of the right order of magnitude ceases to be so impressive, and the question suggests itself; even though the remaining Calc numbers agree with the equation under test within experimental error, might they not also agree, within experimental error, with the numbers obtained from some completely different equation corresponding to a different mechanism?

or engineering, but who cares nothing about the scientific mechanism, the need for quantitative data arises in a different way. Such a man wishes to know how long a sheet or plate of a given thickness will withstand some atmosphere or liquid before it becomes perforated by corrosion, or how long a stressed member will withstand a corrosive environment before it fractures. At first sight, it seems easy to gratify his wishes, by empirical measurement of corrosion-rates on representative materials exposed to a variety of conditions, and it would seem that, for answering his questions, no scientific understanding of mechanism is needed. However, a little consideration will show that reliance on empirical data provides no satisfactory answer to such questions. Corrosion intensity depends on a combination of many factors (for instance, time, temperature, flow-rate and the concentration of each constituent of metallic, liquid and gaseous phases). If ten values (each) of ten variables are to be studied empirically, we need ten thousand million experiments to cover all possible combinations; even if certain unusual combinations were omitted from the programme, an empirical survey of the field would be out of the question. There are not enough qualified experimenters to carry out the enormous task, nor would it be right to use the competent men available for such soul-destroying work.

Even if we assume that the quantitative measurements had been made and recorded, they would probably be incapable of answering the questions raised by industrial conditions. The geometrical simplicity rightly adopted in the laboratory and on the exposure station rarely exists in service, and one must apply empirical results obtained in the simple situation to a new situation of a much more complicated character. Given an understanding of the mechanism, this should not be impossible; but if only empirical information is available, the answer to a practical question is guesswork. Thus, even for severely utilitarian purposes, some scientific understanding is absolutely necessary.

The widely prevailing idea that corrosion problems can be solved in advance by constructing tables filled with empirically determined values of corrosion-rates involves other difficulties. If a material is uniformly corroded by a certain liquid, and if the attack obeys a simple law (e.g. if the amount of metal destroyed is proportional to time), then it is an easy sum in arithmetic to calculate the loss of thickness to be expected in a certain time. If the law relating damage (Q) to time (t) is more complicated, the table will require to record more than one value for each combination of metal and liquid; if, for instance, the law was

$$Q = k_1 t + k_2 t^2 + k_3 t^3$$

we should need three columns giving values of k_1, k_2 and k_3 for each combination of metallic substance and corrosive liquid. Even this would not be impossible, but when we consider cases where corrosion is localized as pits or trenches, the tabulation of data in useful form becomes still more difficult and complicated.

Many of the tables published to guide the industrial chemist provide numbers of loss of thickness expressed as inches per year (" *ipy* values "), apparently calculated from the weight-loss on the assumption that the corrosion will be uniformly spread out; if in fact it is localized, the loss of thickness at the worst affected points will greatly exceed the *ipy* value.

Attempts to meet the difficulty by introducing " pitting factors " are unsatisfactory. It is arguable that, for some purposes, it is best to provide tables of a different type—which, whilst containing few numerical data, furnish qualitative guidance, based on practical experience, as to whether under industrial conditions a given material does, or does not, stand up sufficiently well to a given chemical or a given environment as to justify its adoption in some particular plant. Such information is provided in the tables of E. Rabald's " Corrosion Guide " (Elsevier).

There are, however, cases where the provision of numbers is essential. The well-known " Corrosion Hand-book " (Electrochemical Society), edited by H. H. Uhlig with sections contributed by individuals—each an authority on his subject—contains much valuable quantitative data as tables and diagrams, with expert advice in the text. Further information—quantitative and qualitative—is furnished in R. J. McKay and R. Worthington's " Corrosion Resistance of Metals and Alloys " (Reinhold). The German Dechema Werkstoffe-tabellen (Verlag Chemie) provide a mine of information.

Quantitative Methods in the Pure Science of Corrosion

Methods for obtaining curves showing growth of films. The amount of an oxide-, sulphide- or similar film on metal can be estimated *gravimetrically* by determining the increase in weight, *electrometrically* by measuring the coulombs needed to reduce the film-substance, or *optically* by following certain changes which accompany film-growth. The optical method allows the film-thickening on a single specimen to be followed, so that a single specimen provides a curve showing thickness plotted against time. The electrometric method requires a number of specimens, which have been exposed to oxygen (or other non-metal) for different times; each specimen provides one point only on the curve relating the growth of the film with time. This is also true of the gravimetric method as generally used, but it is possible, by the use of special balances to study the weight-increase as a specimen undergoes oxidation at constant temperature, and in that case a complete curve is obtained from a single specimen. Methods in which each specimen gives a single point are time-consuming, and since the points often do not fall accurately on a curve, it may be necessary to carry out several experiments for each value of the time, plotting the average values of the weight-increment. However, when once a smooth curve has been obtained, it does represent *the behaviour of the material*, and *not that of a single specimen.* The electrometric method has special value where several layers, representing (say) different oxides, are present, since they can often be identified and estimated separately.

Complications due to pre-existing oxide. Metallic specimens which have received their preliminary preparation in air always carry invisible oxide-films, and with most methods of preparation there is a shattered or disarrayed layer of metal. On mechanically polished surfaces, there exists the so-called *Beilby film*, consisting of metal, oxide and polishing material intimately mixed; its structure is regarded by some investigators as vitreous, but probably it should be regarded as composed of distorted crystal fragments of dimensions exceeding, but comparable to, the interatomic distance. An abraded surface probably carries a *débris* of coarser blocks; often there is a mixed layer of oxide and metal on the surface, due either to penetration of oxygen downwards between the blocks at the temperatures attained during the abrasion, or to the fact that the invisible oxide present on the surface before abrasion has been pushed in instead of being rubbed off (there is fairly definite evidence of the pushing of hard oxide into softer metal when a weighted needle is drawn over an oxidized surface, and it is reasonable to think that this pushing-in process would occur during abrasion with emery or carborundum). However, the confused metal-oxide complex, which is so often present on a metallic surface, possesses practical importance, and oxidation studies of specimens prepared by abrasion have provided valuable information which could not have been obtained in any other way (W. H. J. Vernon, E. A. Calnan, C. J. B. Clews and T. J. Nurse, *Proc. roy. Soc.* (*A*) 1953, **216,** 375).

For the study of oxidation-kinetics, it has seemed to several investigators advantageous to start with an oxide-free surface obtained by heating in hydrogen, the specimen being then brought to the temperature chosen for the oxidation process without intervening exposure to air. Some authorities have suggested that such hydrogen-reduction may leave a layer of spongy metal, but there is no sign of this in the subsequent oxidation behaviour, provided the temperature has been reasonably high.* It seems likely that on metals where oxidation consists of cations moving outwards through a film, reduction in hydrogen will consist of cations moving inwards, so that the structure of the metal produced by reduction should be reasonably compact. The hydrogen-reduction method was used for preparing specimens before Winterbottom's optical studies of oxidation (p. 791), where a spongy layer would presumably have been detected. The same method was used in the Cambridge researches on iron by Eurof Davies (p. 836) and on copper by Mills (p. 859).

Gravimetric measurement of Film-thickness. If oxidation for the appropriate time produces a weight-increase of W grams on a specimen of A cm.², the thickness of the oxide-film expressed in cm. is clearly equal to $\frac{W}{AD}\cdot\frac{M}{(M-m)}$, whilst the thickness of the layer of metal destroyed is $\frac{W}{Ad}\cdot\frac{m}{(M-m)}$, where D and d are the densities of oxide and metal respec-

* In one early research where the invisible oxide film on copper was removed cathodically at ordinary temperature, an invisible film of spongy metal was certainly produced and exercised a marked influence on subsequent behaviour, conferring the power of developing, at room temperature, oxide-films thick enough to display several orders of interference colours. See U. R. Evans, *J. chem. Soc.* 1925, p. 2484.

tively, M the molecular weight of the oxide and m the weight of metal in M of oxide (m is the atomic weight of copper in CuO and twice the atomic weight in Cu_2O).

Cathodic measurement of Film-thickness. In the usual electrometric method a specimen carrying an easily reducible oxide receives cathodic treatment at a constant current (of the order of milliamperes), obtained from a rather high E.M.F. (perhaps 100 volts) placed in series with a high resistance (of the order of 10^5 ohms); clearly this resistance is high compared to the internal resistance of the reduction cell, and the E.M.F. high compared to changes in the potential during reduction, so that the current, I amperes, remains constant—irrespective of happenings in the cell; the number of Coulombs is It, where t is the time in seconds up to the appearance of the potential change which indicates the completion of reduction. The film-thickness is then given by JIt/FAD where J is the

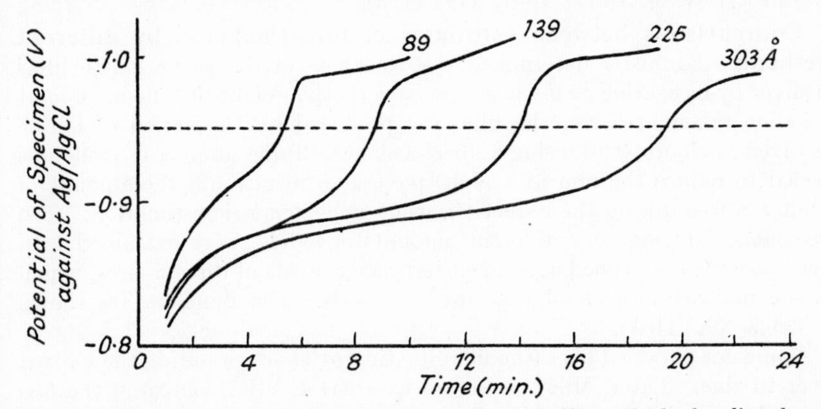

FIG. 127.—Electrometric reduction curves for ferric oxide films; the broken line shows the level taken as end-point (D. Eurof Davies, U. R. Evans and J. N. Agar).

equivalent weight of the film-substance, i.e. the molecular weight divided by the number of electrons involved in reducing each molecule, and F is Faraday's number. For the equation

$$Fe_2O_3 + 6H^+ + 2e = 2Fe^{++} + 3H_2O$$

J is $M/2$. It is also $M/2$ for the reduction of cuprous oxide to the metallic state

$$Cu_2O + 2H^+ + 2e = 2Cu + H_2O$$

but for the reduction of silver iodide

$$AgI + e = Ag + I^-$$

J is equal to M. The potential change is usually somewhat gradual, as indicated in fig. 127, and the inflection-point, where the rate of alteration with time is greatest, is taken as the end-point. Where there are several components in the film (different oxides and/or sulphides), there are several plateaux on the curve, and sometimes the amounts of the various components can be separately determined.

Electrometric methods have been used mainly for the study of films on silver, copper and iron; references are given on p. 776. Recently the principle has been applied to estimate the thickness of invisible oxide-films on tin and to follow their development during exposure to air; the high over-potential of hydrogen-evolution on tin makes possible the quantitative reduction of the oxide in the sense of Faraday's Law—a necessary condition for success (F. W. Salt and J. G. N. Thomas, *Nature* 1956, **178,** 434; see also *Metal finish. J.* 1957, **3,** 66.)

The method, although mainly used for films of interference-colour thickness, appears to be applicable to the thinnest films possible. Eurof Davies has detected films about 5 Å thick on silver which has been heated at 800°C. in oxygen; (these can be regarded as chemi-adsorbed oxygen or two-dimensional oxide-films). The contrast between the potential-time curves with those obtained on silver cleaned by heating in hydrogen is impressive D. Eurof Davies, *Nature* 1957, **179,** 1293).

Comparisons between estimates of film-thickness by different methods. Bannister determined the thickness of the silver iodide films on silver by measuring *on the same specimen* (1) the weight-increment (weight of iodine taken up when the film was produced by the action of iodine dissolved in chloroform), using a micro-balance, (2) the number of coulombs needed to reduce the film in a N/10 potassium nitrate, (3) the amount of iodide set free during the reduction, using nephelometric estimation. Ten specimens, carrying very different amounts of iodide, were examined. In every case there was good agreement between the sets of three figures, which gives confidence in each of the three methods. The numbers are shown in Table XXVII(A).

Bannister obtained his cathodic reduction by short-circuiting the iodized silver to zinc. Later, Miley, using an external E.M.F., compared the *loss* of weight of oxidized iron with the coulombs produced during the partial cathodic reduction of the ferric oxide film in ammonium chloride—obtaining good agreement. Price and G. J. Thomas used a similar method to estimate sulphide films on silver, obtaining three estimates from each specimen by observing (1) the gain of weight when the sulphide films were produced by exposure to ammonium sulphide vapour, (2) the number of coulombs needed for cathodic reduction to silver in ammonium chloride solution, (3) the loss of weight which accompanied the reduction. Here again three values were obtained from *each specimen*, and excellent agreement was observed on four separate specimens (Table XXVII(B)).

The fact that in all these cases the electrometric measurements tallied well with measurements obtained by other methods shows that the principle is sound. However, the procedure used by the earlier workers has been considerably modified by later ones. Miley worked in cells open to the air and used an ammonium chloride solution as electrolyte; he adopted a rather high current-density, so that the time of reduction was short and the errors involved were kept relatively small. Later experimenters have preferred a lower current density which allows a more precise determination of the end-point. At low current densities, Miley's procedure would introduce serious

TABLE XXVII

EXAMINATION OF SINGLE SPECIMENS BY THREE DIFFERENT METHODS

(A) *Silver specimens carrying iodide films* (U. R. Evans and L. C. Bannister)

Specimen No.	Weight-increase mg./cm.^{2}	Electro-metric	Nephelo-metric
1	0·162	0·161	0·167
2	0·128	0·130	0·113
3	0·106	0·096	0·113
4	0·059	0·059	0·058
5	0·040	0·042	0·040
6	0·040	0·035	0·037
7	0·018	0·020	0·017
8	0·014	0·014	0·014
9	0·004	0·006	0·004
10	0·002	0·003	0·004

(B) *Silver specimens carrying sulphide films* (L. E. Price and G. J. Thomas)

Specimen No.	Weight-increase (mg., absolute)	Electro-metric	Weight-decrease
1	3·60	3·61	3·46
2	3·93	3·81	3·83
3	4·80	4·90	4·60
4	3·93	4·02	3·86

inaccuracies due to (1) cathodic reduction of dissolved oxygen; (2) destruction of the film through reductive dissolution connected with local currents at weak points. Closed cells, with careful exclusion of oxygen, were first adopted by Campbell and U. B. Thomas.

Various solutions have been used by different experimenters; since the potential of reduction depends on pH, there is an advantage in using buffer systems—a point emphasized by Mills (p. 859). Ammonium chloride solution, which was used successfully by Miley and Eurof Davies, was found by Mayne and Hancock to destroy invisible films without any applied cathodic current. Doubtless this destruction was due to reductive dissolution at places of spontaneous cracking; Eurof Davies had annealed his iron in hydrogen which probably removed internal stresses liable to produce spontaneous cracking. Miley (who used abraded specimens) employed a high current density; above a certain value of the current density, errors due to local cells set up at cracks will disappear for reasons analogous to those brought out by Bianchi in connection with cathodic protection (p. 896). Below this value, part of the cathodic current needed to destroy the film will be provided by local anodes, making the electrometric measurements too low.

Oswin and Cohen have adopted a sodium-borate/hydrochloric-acid buffer solution of pH 7·5 for the electrometric estimation of films on iron, but consider that the accuracy of such a method cannot exceed $\pm$ 15%.

A phosphate buffer was used for oxide on copper by Mills and for oxide on silver by Eurof Davies. For films on copper Lambert and Trevoy favour potassium chloride purified by preliminary electrolysis, whilst Bouillon,

Piron and de Lil prefer sodium hydroxide for films containing cuprous oxide, cupric oxide and cupric hydroxide.

Eurof Davies in his study of films on iron in the apparatus of fig. 129 (p. 778) obtained satisfactory agreement between micro-gravimetric and electrometric measurements when only a ferric oxide film was present; on specimens where there was a magnetite layer below the ferric oxide, no agreement was obtained, since the magnetite was not reduced under the conditions observed. Mills obtained good agreement between weight-increment and electrometric values for films on copper. The references (with numbers of the pages carrying the tables showing the agreement between different methods) are:—U. R. Evans and L. C. Bannister, *Proc. roy. Soc.* (*A*) 1929, **125,** 370, esp. p. 379; H. Miley and U. R. Evans, *J. chem. Soc.* 1937, p. 1295, esp. p. 1298; L. E. Price and G. J. Thomas, *Trans. electrochem. Soc.* 1939, **76,** 329, esp. p. 332; W. E. Campbell and U. B. Thomas, *ibid.*, 1939, **76,** 303, esp. p. 312; D. Eurof Davies, U. R. Evans and J. N. Agar, *Proc. roy. Soc.* (*A*) 1955, **225,** 443, esp. p. 450; T. Mills and U. R. Evans, *J. chem. Soc.* 1956, p. 2182, esp. p. 2186; H. G. Oswin and M. Cohen, *J. electrochem. Soc.* 1957, **104,** 9, esp. p. 15; R. H. Lambert and D. J. Trevoy, *ibid.*, 1958, **105,** 18; F. Bouillon, J. Piron and M. de Lil, *Nature* 1956, **178,** 1406.

Hancock and Mayne have used de-aerated potassium chloride solution for cathodic measurement of the invisible film formed on iron exposed to air over calcium chloride for 48 hours, finding it to be 40 Å, 36·5 Å and 26 Å thick on pickled iron, pickled steel and hydrogen-reduced iron respectively; the film is reinforced and thickened by 20–120% when the specimen is subsequently placed in a solution of an inhibitor such as sodium benzoate, acetate, borate, carbonate or hydroxide (P. Hancock and J. E. O. Mayne, *J. chem. Soc.* 1958, pp. 4167, 4172 (2 papers)).

Anodic Measurement of Film-thickness. For noble metals like copper and silver, which can easily be reduced to the metallic state, or for iron where the reduction from ferric oxide to ferrous ions proceeds smoothly in ammonium chloride solution, electrometric estimation depends on making the film-covered specimen the cathode. For aluminium, cathodic reduction is impossible, but Hart has developed an anodic procedure. It is known (p. 243) that the film produced on aluminium anodized in, say, ammonium tartrate grows rapidly up to a certain thickness and then almost ceases to grow; the thickness attained is proportional to the potential drop across the film, being about 14·5 Å per volt. If, now, an aluminium specimen already carrying a film (of thickness y, expressed in Ångstrom units) is made an anode at a low potential, some current will pass; but if the potential is raised in steps, the current being measured at each potential value, there will be a sharp break in the curve when the potential reaches the value $y/14{\cdot}5$ at which further thickening over the whole surface becomes possible. Hart used an electrically driven rotary switch to assist the step-wise regulation of potential, which was increased by 0·2 volt every 2 minutes. The principle involved is simple, but various complicating factors have to be taken into account, and those who contemplate adopting the method should study the original paper (R. K. Hart, *Proc. roy. Soc.* (*A*) 1956, **236,** 68; cf. a rather similar method developed independently by M. S. Hunter and P. Fowle, *J. electrochem. Soc.* 1956, **103,** 482).

Apparatus used for studying film-growth in oxygen. Anyone planning a research should design apparatus appropriate to that research, and not copy apparatus designed for a different purpose. It may, however, be convenient to describe some apparatus used in recent work at Cambridge, for general guidance rather than exact imitation.

In studying the oxidation of iron, Eurof Davies (p. 776) used the set-up shown in fig. 128. The specimen was contained in a tube of clear quartz, which was surrounded by a movable tubular electric furnace capable of being placed around any part of the tube at will. Arrangements were made

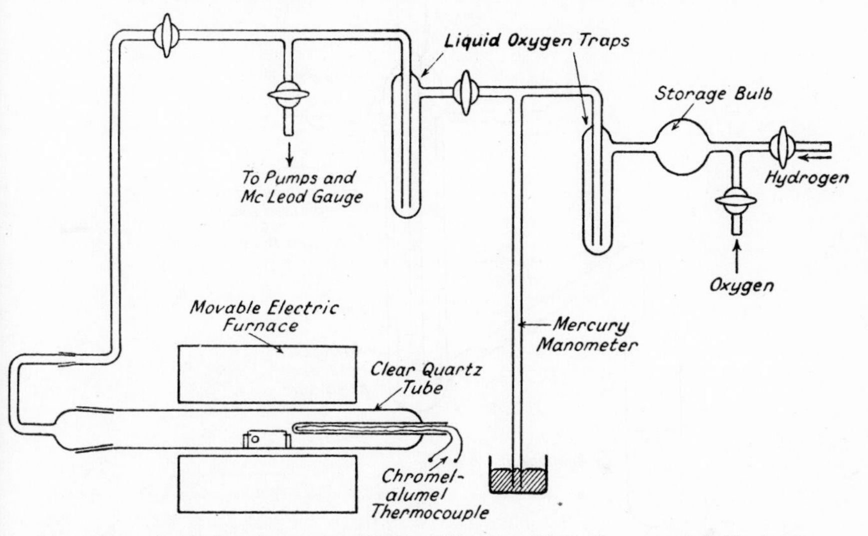

FIG. 128.—Oxidation Apparatus (D. Eurof Davies, U. R. Evans and J. N. Agar).

to fill the tube with carefully purified hydrogen, and to evacuate it at any time before admission of gas.

The specimen, degreased in acetone, and then exposed for 2 hours in a Soxhlet extractor containing pure benzene, was dried, and exposed to carefully purified hydrogen at 10 cm. pressure at 470° for two periods of 10 minutes, with intervening evacuation, cooled *in vacuo*, kept in a desiccator for 24 hours and weighed on a micro-balance. It was then taken up to 360°C. *in vacuo* and the air-formed film removed by reduction in hydrogen for two periods of 10 minutes. After evacuation and cooling to the chosen experimental temperature, oxygen at 1/5 atm. was admitted. When the time chosen for oxidation was complete, the apparatus was again evacuated, the specimen being rapidly brought to the cool end of the tube by means of a magnet. When it had reached room temperature, it was taken out and re-weighed; the gravimetric measurement was in error by the small amount of the oxygen in the invisible film present at the time of the first weighing.

The ferric oxide layer of the film was then measured electrometrically. The surface was covered with two layers of polystyrene varnish except for a 1 cm.^{2} circle and an area at the top for electrical connection. The oxide present on the uncoated circle was estimated by cathodic reduction at 17 μA/cm.^{2} in N/10 ammonium chloride, using the apparatus of fig. 129; a silver/silver-chloride electrode was used as anode, the anodic reaction being the deposition of chlorine ions. The specimen was suspended from a platinum hook, with the exposed circle opposite the tubulus of the silver/silver chloride reference electrode.

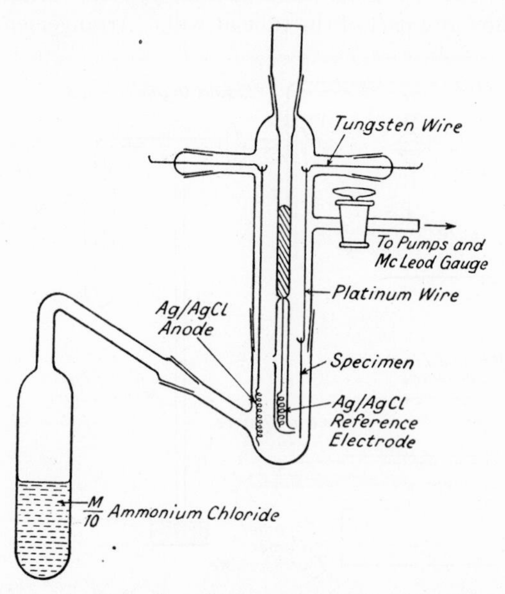

FIG. 129.—Electrometric Reduction Apparatus (D. Eurof Davies, U. R. Evans and J. N. Agar).

The ammonium chloride solution was first de-aerated in the side arm by freezing it in liquid oxygen, evacuating to 5×10^{-6} mm. of mercury and then thawing the solution; this process was twice repeated, the cell being isolated from the oil-diffusion pump during the thawing. Hydrogen was then admitted until the pressure reached 1 atm. and the solution was transferred into the reduction cell by rotating the side arm, so that the liquid, flowing down under gravity, covered the exposed surface.

For specimens carrying very thin films, the procedure was modified. The specimen was first suspended from a glass hook and then at the chosen moment was released by rotating the hook. In dropping, it engaged a copper support so that electrical circuit was completed at the instant of immersion.

The potential was at first measured every minute, but every 15 seconds when the end-point approached; typical potential-time curves are shown in fig. 127 (p. 773), the inflection-point being taken as end-point. Magnetite, when present, could be seen as a dark stain on the metal after the ferric oxide colour-film had disappeared. If no magnetite was present, the specimen was bright after the reduction, and there was generally good agreement between the gravimetric and electrometric estimates of the ferric oxide-film —indicating that the error mentioned above was not important.

The films were also transferred to glass by the method described on p. 785, and studied chemically; they were also examined by X-rays and electron diffraction. All methods of examination gave good agreement.

For details see D. Eurof Davies, U. R. Evans and J. N. Agar, *Proc. roy. Soc.* (*A*) 1954, **225,** 443.

Apparatus for studying oxidation in a gaseous mixture. In cases where a mixture of gases is to be used, a mixing chamber is needed; the

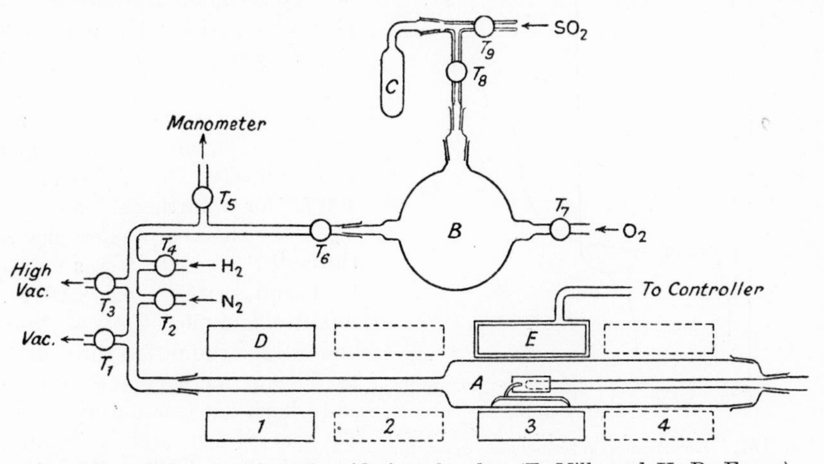

FIG. 130.—Mixing vessels and oxidation chamber (T. Mills and U. R. Evans).

apparatus used by Mills (p. 781) for studying the oxidation of copper in oxygen containing small amounts of sulphur dioxide is shown in fig. 130, which, however, does not include the three purification trains for the oxygen, hydrogen and nitrogen. It was convenient to have two separate furnaces, each carried on a railway, so that either of them could be brought to position 3 (i.e. the portion of the tube containing the specimen). Furnace *D* was wound so as to provide a temperature of about 400°C. and was used for the preliminary reduction of the invisible film in hydrogen, whilst Furnace *E* was used for the oxidation proper, and provided temperatures over the experimental range (88–172°C.).

The procedure for preparing gas atmospheres containing known concentrations of sulphur dioxide was as follows. The reservoir *B* and the smaller measuring vessel *C* were evacuated to below 10^{-3} mm. mercury, taps T_6 and T_8 were closed, and sulphur dioxide (from a siphon) admitted to *C* through T_9, a mercury escape-trap in the sulphur dioxide line indicating when C was filled to atmospheric pressure. T_9 was then closed, T_8 opened and again closed, thus proportioning the sulphur dioxide between B and C (C was much smaller than B, and the amount left in it could be neglected), oxygen was then admitted through T_7 until the pressure in the reservoir reached atmospheric value. If a double or treble concentration of sulphur dioxide was desired, the " proportioning operation " could be repeated, but other concentrations were easily obtained through replacement of C by a measuring vessel of larger or smaller volume; the *B*7 joint facilitated quick replacement, and the volumes of the various vessels were obtained by weighing the water needed to fill them.

After the specimens had been placed in the oxidation chamber, air was pumped out as quickly as possible, a value below 5×10^{-5} mm. mercury being regarded as a satisfactory vacuum. Initially furnace D was at position 1 and E at 2. Tap T_3 was closed and hydrogen admitted up to 50 cm. mercury. Furnace E was then moved to 4 and D to 3, and the specimen was subjected to hydrogen at 400°C. for a further 15 minutes. Then the chamber was evacuated through T_3, furnace D was moved to 1 and switched off. After 5 minutes pumping, T_3 was closed and nitrogen admitted through T_2 to 50 cm. mercury pressure—so as to flush out residual hydrogen. T_2 was closed and a vacuum applied through T_3. When the temperature had dropped to within 20°C. of the desired temperature for oxidation, furnace E was moved to 3, and in about 15 minutes the temperature was found to be steady at the oxidation temperature; a further 30 minutes was allowed for conditions to settle down, then the prepared mixture was quickly admitted from the reservoir so as to bring the pressure to the desired value as indicated by the mercury manometer. The period of oxidation was measured by a stop-watch, started when T_6 was opened. When the chosen period was complete, E was moved to 2 and the oxidation chamber evacuated through T_3. The specimen was cooled *in vacuo* to 50°C., then the chamber was filled with nitrogen to hasten the cooling to room temperature, after which the film-thickness was determined electrometrically in a buffer solution containing equi-molar amounts of NaH_2PO_4 and Na_2HPO_4 (pH 6·9) with a trace (10^{-4} mole/litre) of potassium chloride to allow the functioning of a silver/silver-chloride electrode. In some cases,

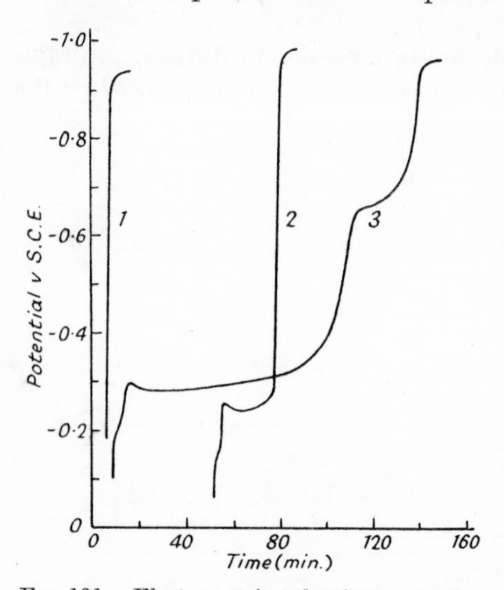

FIG. 131.—Electrometric reduction curves for Copper. Curve 1 was obtained from film-free Copper (freshly reduced in hydrogen); curve 2 from a specimen carrying mainly Cu_2O; surve 3 from a specimen carrying Cu_2O and CuO (T. Mills and U. R. Evans).

other reference electrodes were used, but their potentials were always measured against a saturated calomel electrode. Some curves (obtained in the absence of sulphur dioxide) are shown in fig. 131. For details, see T. Mills and U. R. Evans, *J. chem. Soc.* 1956, p. 2182.

Methods for drying or humidifying gases. In figs 128 and 130, the purification and drying trains (being of conventional pattern) have been omitted. Hart, in his research on aluminium carried out at Cambridge contemporaneously with those of Eurof Davies on iron and Mills on copper, dried his oxygen by passage over silica gel, soda lime and two tubes of magnesium perchlorate and a liquid-air trap; after passing through the oxidation vessel containing the specimen (which was thermostated at 20 ± 0·05°C.) the oxygen escaped through a trap partly filled with silicone oil, a flow-rate of 1 litre per hour being maintained throughout the experiment. For his work on humid oxidation, the apparatus of fig. 132 was

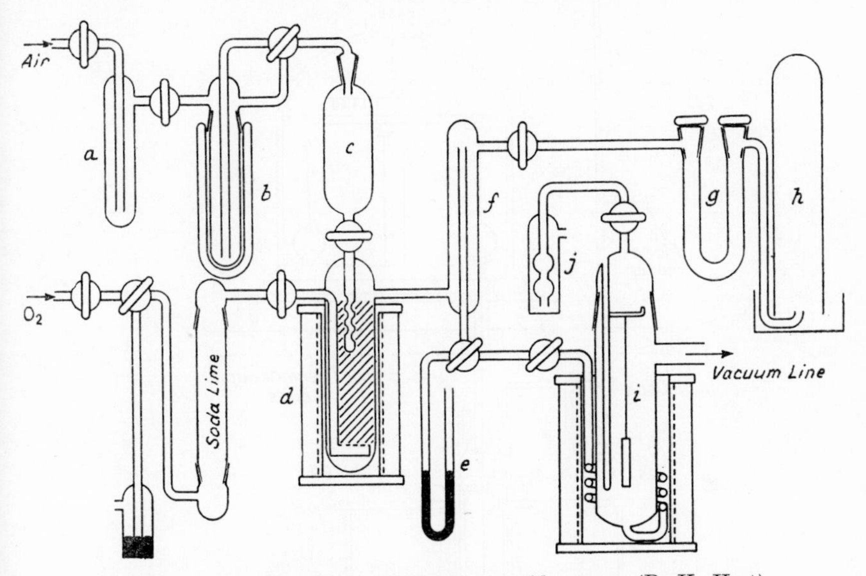

FIG. 132.—Apparatus for oxidation in humid oxygen (R. K. Hart).

used. Oxygen from a gas-train passed upwards through the humidifier, *d*, which contained spun glass wool, meeting a downward flow of twice-distilled water (previously degassed by alternate boiling and freezing *in vacuo* in *b*); this was admitted from *c* at the prescribed rate, being replaced by air freed from carbon dioxide by passage through 40% sodium hydroxide in *a*. The desired humidity of the gas which passed from *d*, through trap *f*, into the thermostatically controlled oxidation-chamber, *i*, was regulated by choosing the temperature and flow-rate, and the value was checked by measuring the mass of water vapour contained in a measured volume of gas (R. K. Hart, *Proc. roy. Soc.* (*A*) 1956, **236**, 68).

Other methods of controlling humidity are known, and that developed by Vernon and Whitby would probably be found convenient for most purposes. It consists of mixing desiccated and saturated air in any desired proportions, the saturated air being obtained by admitting steam into an air stream and condensing out the excess moisture under thermostatic conditions; they point out that, as a means of obtaining saturation, passage of air through vertical water columns is inefficient (W. H. J. Vernon and L. Whitby, *Trans. Faraday Soc.* 1931, **27**, 248).

Apparatus for continuous weight-increment curves. Where it is desired to obtain from a single specimen a continuous curve showing the increase of oxidation with time, very different apparatus is needed, and the

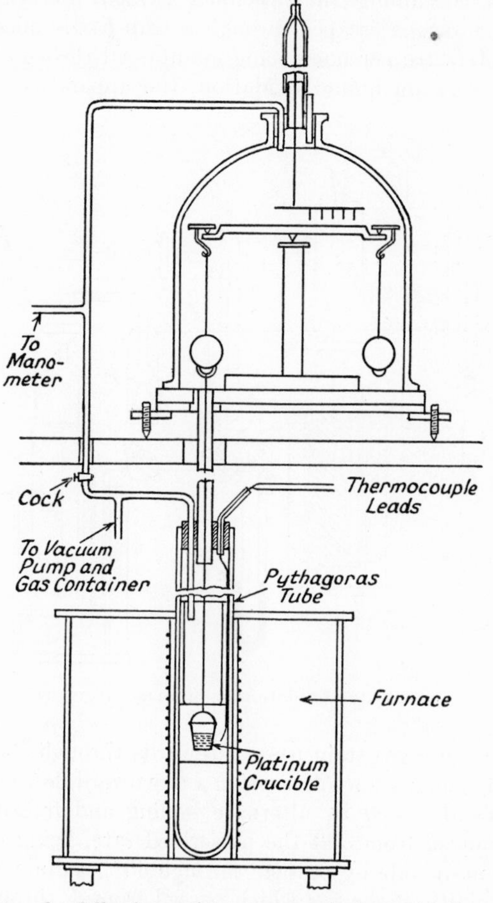

Fig. 133.—Apparatus for following weight changes during oxidation (J. White; used later by R. F. Tylecote).

forms have varied greatly with the purpose desired. The work of Gulbransen, which amongst other things showed the minute increase in weight which occurs when oxide-free iron is exposed to oxygen at ordinary temperatures, developing an invisible film, has become classical. His balance has also been used in high-temperature work, notably in the researches of Smeltzer on aluminium. The beautiful apparatus of Chévenard, used by Bénard for his studies of the early stages of oxidation on heated iron, also deserves special mention. For more general purposes, the apparatus of White, shown in fig. 133, may prove useful. This was originally designed for studying, not oxidation, but the dissociation of ferric oxide into magnetite and oxygen; it has, however, been adopted for work on the oxidation of copper by Tylecote. The balance used by White was a sensitive assay balance suitably modified, and placed in a bell-jar containing the same gas-mixture as the furnace.

The papers in which these various forms of apparatus deserve careful study from anyone wishing to weigh specimens in the oxidation apparatus; he may decide to design an apparatus to suit his special purpose. The references are:—E. A. Gulbransen, *Trans. electrochem. Soc.* 1942, **81,** 327; 1942, **82,** 375; *Rev. sci. Instrum.* 1944, **15,** 201; W. W. Smeltzer, *J. electrochem. Soc.* 1956, **103,** 209; P. Chévenard, X. Waché and R. de la Tullayne, *Métaux* 1943, **18,** 121; J. Bénard, *Bull. Soc. chim. Fr.* 1949, p. D89; J. White, *Iron St. Inst. Carnegie Schol. Mem.* 1938, **27,** 1; R. F. Tylecote, *J. Inst. Met.* 1950–51, **78,** 327. A ring furnace for studying the oxidation of 39 specimens simultaneously is described by W. J. Day and G. V. Smith, *Industr. engng. Chem.* 1943, **35,** 1098; see also H. W. Maynor and R. E. Swift, *Corrosion* 1956, **12,** 293t.

Gas-absorption methods of obtaining continuous curves. Another method of following the progress of oxidation depends on the drop of pressure which occurs when a specimen is heated in a closed vessel containing a

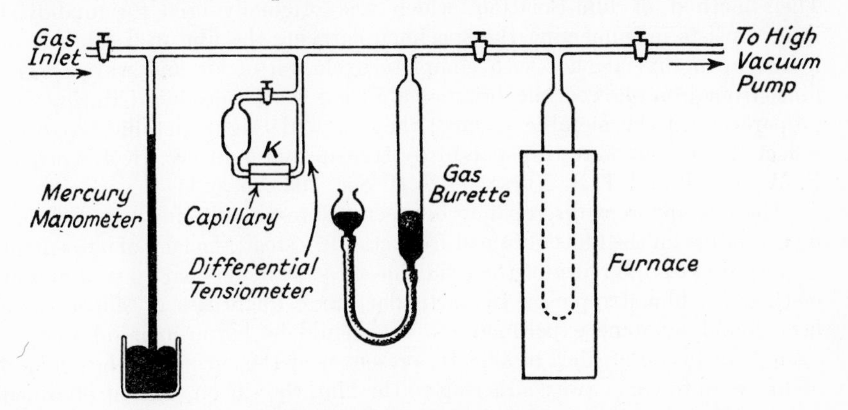

FIG. 134.—Manometric apparatus for measuring oxidation at constant pressure (H. J. Engell, K. Hauffe and B. Ilschner).

limited amount of oxygen; clearly the method is most sensitive at low pressure, where a small amount of oxidation will cause a big movement on the pressure gauge. Methods depending on gas absorption have been used in several distinguished researches, including those of F. J. Wilkins and E. K. Rideal, *Proc. roy. Soc.* (*A*) 1930, **128,** 394, 407; and of J. A. Allen and J. W. Mitchell, *Disc. Faraday Soc.* 1950, **8,** 309.

In the older forms of the method, the results did not show oxidation at constant pressure. In the later form, illustrated in fig. 134, a micro-burette enables the volume of gas absorbed to be measured at approximately constant pressure; the constancy of pressure is obtained by means of a differential tensiometer with a capillary containing diethyl phthalate; if the liquid is at the null-point, the pressure is correct; small movements from the null-point serve to indicate the gas-volume taken up. This apparatus is due to H. J. Engell, K. Hauffe and B. Ilschner, *Z. Elektrochem.* 1954, **58,** 478. Another technique is described by J. V. Cathcart, L. L. Hall and G. P. Smith, *Acta Met.* 1957, **5,** 245.

Stripping of Oxide-films. Yet another method of studying oxidation depends on the stripping of the film and estimation of the metal in it. In the case of films formed on alloys, this can tell us whether one of the constituents tends to become concentrated in the film, or to be excluded therefrom. (Electron diffraction gives some information on the same point, although in thick films it may tell us little about the inner layers.) Vernon, Wormwell and Nurse showed that in the films obtained from abraded or polished 18/8 stainless steel the Cr/Fe ratio was higher than that existing in the alloy—suggesting that the polishing operation tends to concentrate chromium in the outer part of the surface. If the polishing powder used had contained chromic oxide, the surface-enrichment in chromium was naturally greater, but a significant degree of chromium-enrichment was obtained even on surfaces polished with alumina (the films isolated then contained aluminium); only a trace of nickel was found. Their method of film-isolation, which was originally used for unalloyed iron, consists in immersing the specimen carrying the film in a solution of iodine in methyl alcohol with complete exclusion of air and water. The films from iron showed the original abrasion marks produced during the preparation of the metallic surface; they showed bright metallic lustre by reflected light but were transparent by transmitted light (W. H. J. Vernon, F. Wormwell and T. J. Nurse, *J. chem. Soc.* 1939, p. 621).

There is apparent discrepancy between the results obtained by different investigators on the films obtained from stainless steel; the lack of agreement may be due to difference of the oxidizing treatment, but partly to different methods of film-stripping. In particular, large quantities of silicon have been found by some experimenters. It should be borne in mind that, if a single inclusion of silica or silicate, present near the surface of the original metal, were to come away adhering to the film, the silicon-content obtained on analysing the film-sample might well be high, even if there were no silicon in the film-substance proper; it seems rather unlikely, however, that the variation of composition can thus be explained away. Readers may care to form their opinion after comparing the various papers (W. H. J. Vernon, F. Wormwell and T. J. Nurse, *J. Iron St. Inst.* 1944, **150,** 81; E. M. Mahla and N. A. Nielsen, *J. electrochem. Soc.* 1948, **93,** 1; T. N. Rhodin, *Corrosion* 1956, **12,** 123t, 465t).

Methods of Transferring Films. The Cambridge method for the transfer of films from heat-tinted metal to glass, plastic or other transparent support may here be mentioned, since the chemical analysis of such films was used as an auxiliary method by Eurof Davies. (The transfer has more

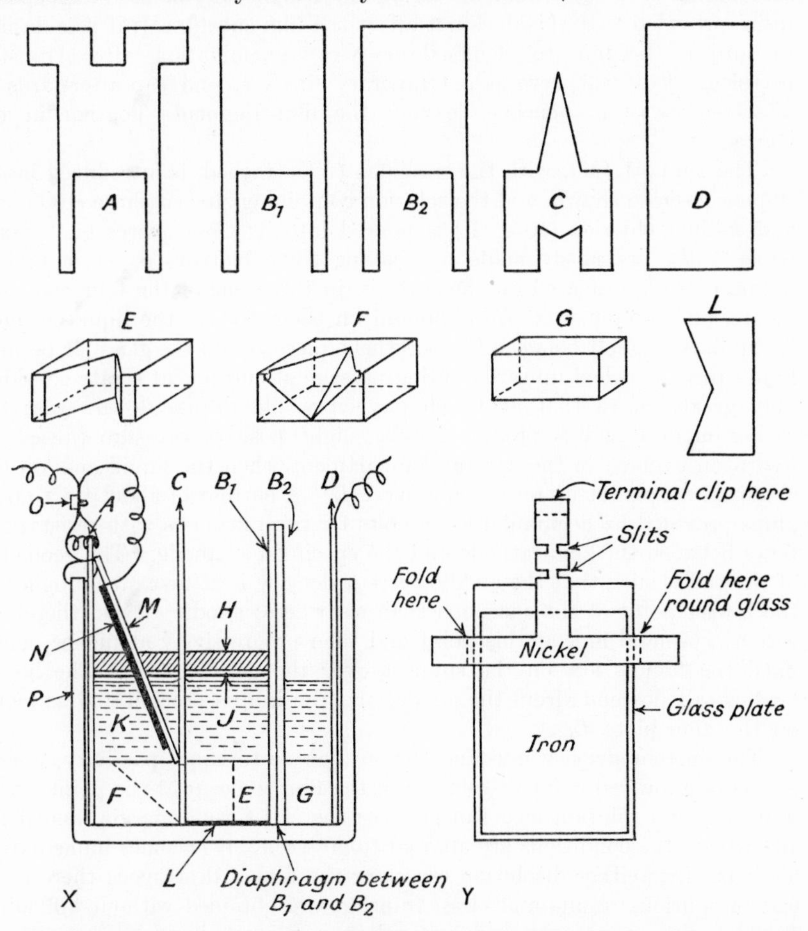

FIG. 135.—Apparatus for transfer of oxide-films from heat-tinted meta to glass or plastic. N, iron specimen, generally 3·8 × 2·5 cm., 0·14 mm. thick. P, glass jar measuring (internally) 5·0 cm. long, 3·3 cm. broad (dimension at right angles to paper) and 5·0 mm. deep. A, B_1, B_2, C, polythene pieces cut from sheet 0·6 mm. thick. J, wax-impregnated paper. H, paraffin wax (poured on molten). D, zinc cathode. The diaphragm (held in position between B_1 and B_2) consists of thick blotting-paper. The zinc pieces L serve to precipitate metallic iron from the anodically formed ferrous salts, and the zinc–iron couple soon starts to evolve hydrogen (U. R. Evans and R. Tomlinson).

often been used for studying the stress-conditions and colours of the films —particularly the tendency of the duplex films to curl up owing to differential stress, and the manner in which the colours vary according to the method of observation (p. 22).) The cell used for the transfer is shown in fig. 135.

The specimen is attached to the glass or plastic support, the side carrying the film to be transferred being next to the support; the side remote from the support should be freed from visible film before attachment. If the attachment is to be permanent, an organic cement may be used (e.g. epoxy-resin, provided that this does not affect the film-substance); if it is desired to subject the film to chemical or X-ray examination after transfer, petroleum jelly will serve as a temporary fixative, and can afterwards be dissolved away in benzene, leaving the film-fragments floating in the benzene.

The support (*M*), with the specimen (*N*) attached, is introduced in the sloping position shown, and the solution containing zinc sulphate (100 g./l.) and sodium chloride (50 g./l.) is poured into it; three pieces of " waxed paper " (*J*), previously made by soaking thick blotting paper in molten paraffin wax are floated on the surfaces in three out of the four compartments, and molten wax (*H*) is poured on them to seal the liquid surface. A capillary gap left between the polythene piece *A* and the glass jar permits liquid to be expelled upwards during passage of current, at a rate considerably greater than that at which oxygen would diffuse downwards; the expulsion of liquid is produced by a slight positive pressure caused by hydrogen evolved in the central compartment when the liquid acts on the zinc-iron couple set up on the zinc piece *L*. A current of about 0·1 to 0·12 amps provided by accumulators (8 volts for iron) in series with resistances, flows between the zinc cathode and the specimen as anode. The geometry of the cell is such that the anodic current density is at first much higher at the lowest point of the specimen than elsewhere; anodic attack, therefore, eats away metal first at this point and then progressively at higher points until the film is left on the support over the desired area. The heavy ferrous salt descends from the anode, and the iron is precipitated as metal on the zinc plate *L*.

The current density must not be so high as to cause passivity, nor so low as to allow reductive dissolution of the film to occur at the point where iron, film and solution meet; in practice there is a wide margin, and if for any reason the conditions are allowed to go wrong, it becomes immediately obvious, since either disolution ceases or the film is destroyed; there is no fear of spurious results and clean transfers are obtained without difficulty. The geometry of the cell and internal fittings must, however, be adhered to; otherwise bad results are likely to be obtained with iron; with nickel (where reductive dissolution does not occur), more latitude is possible. Various small points require attention. For instance, if petroleum jelly is used as temporary adhesive, there is risk that accidental stresses from the spring terminal clip may gradually lever the specimen away from the support during the process (especially if the temperature rises during the flow of current); to prevent the transmission of such stresses, a limp wire connection may be provided, as in fig. 135 (X), or alternatively, slits may be cut in the thin nickel foil which serves to convey current to the specimen, so as to reduce its rigidity, as in (*Y*). For details, see U. R. Evans and R. Tomlinson, *J. appl. Chem.* 1952, **2**, 105.

Methods of Measuring Film-thickness based on Interference Colours. For films producing interference colours, which depend on the mean distance between the inner and outer surface of the film, the colour can itself be used in estimating film-thickness; by exposing metal to oxygen or air for different times and comparing the colour with that of standards, an accurate time-thickness curve can be obtained.

Pioneer work on these lines was carried out by Tammann, who obtained rough measurements of the thickness by matching the colour produced by a film on the metal (viewed by *reflected* light) with the colour produced by an *air-film* between glass (viewed by *transmitted* light). The thicknesses of air-films needed to give different colours had been tabulated by Rollett, and the thickness of the air-film found to match the colour of the metal divided by the refractive index (n) of the oxide (or other film-substance) was taken to represent the thickness of the film on the metal.

The reason why the colour of the air-film as viewed by *transmitted* light was adopted is shown in fig. 136. If an air-film is viewed by reflected light (fig. 136(A)), one beam is reflected at an interface with a denser medium

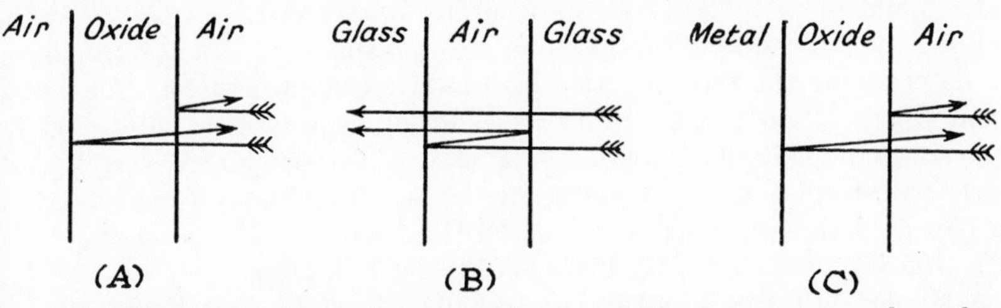

FIG. 136.—Interference colours produced (A) by air-film between two glass plates viewed by reflected light; (B) by air-film between two glass plates viewed by transmitted light; (C) by oxide-film attached to metal viewed by reflected light.

and the other at an interface with a light medium—which difference by itself will introduce a phase-difference of π between the two beams. With transmitted light (fig. 136(B)), no such phase-difference arises. Passing to an oxide-film on metal (fig. 136(C)), it would seem that, provided that no phase-difference is introduced between the two reflections from the inner and outer surfaces of a film adhering to metal, the figures for air-film colours by transmitted light provide the valid comparison. If, for instance, the film is of such a thickness that the green light reflected from one surface is exactly out of phase with that reflected at the other, the light will have a reddish colour, since red is complementary to green. In the case of air-films this will occur when the thickness becomes $\frac{\lambda_G}{4}, \frac{3\lambda_G}{4}, \frac{5\lambda_G}{4}, \frac{7\lambda_G}{4}$, where λ_G is the wave-length of green light in air. The wave-length in the film-substance will be $\frac{1}{n}$ that in air, so that we shall then get the 1st, 2nd, 3rd, 4th other reds at thicknesses of $\frac{1}{4}\frac{\lambda_G}{n}, \frac{3}{4}\frac{\lambda_G}{n}, \frac{5}{4}\frac{\lambda_G}{n}, \frac{7}{4}\frac{\lambda_G}{n}$, explaining why we have to divide by the refractive index, n.

The argument, however, is inaccurate, since a specific phase-change is known to occur at the metallic surface. Further complications are due to the fact that n itself varies with wave-length, and by the fact that the colour is not determined solely by the position of the wave-length where there is maximal interference, but is influenced by the intensity at wave-lengths on either side of the maximum. Thus, whilst colours can undoubtedly form the basis of the measurement of thickness, the simple rule adopted by Tammann should not be used. Bannister (p. 775), studying the thickness of iodide-films on silver, obtained excellent agreement between micro-gravimetric, electrometric and nephelometric results, but poor agreement with the measurements derived from colour by application of Tammann's rule. He did, however, use the other measurements to calibrate the colour-scale, and having once made a series of coloured standards of known film-thickness, was able to use observation of the colours produced on silver immersed in iodine solution for different times to determine the law governing growth, which turned out to be represented by the simple parabolic equation.

It should be possible to obtain accurate measurements of thickness by studying the light reflected from a film-covered specimen in a spectrometer, and determining the wave-length where interference is greatest. This was accomplished in early work by Constable, but he was more interested in catalysis than oxidation kinetics, and used a surface giving very broad interference bands—not well suited for the accurate study of growth-laws; the papers, however, deserve study (F. H. Constable, *Proc. roy. Soc.* (*A*) 1927, **15**, 570; 1927–28, **117**, 376, 385; 1929, **125**, 630).

A similar principle has been applied by Charlesby and Polling in an accurate study of the films formed on anodizing tantalum (p. 249). The tantalum anode develops several orders of brilliant colours, and by the use of a spectrometer the interference bands can be followed through eight orders. The thickness of the film is proportional to the E.M.F. used to produce it, and it was found that values of the E.M.F. needed to produce the various interference bands for any particular wave-length are spaced out at equal intervals—which means that the thickness needed to produce the $(n + 1)$th interference of light of wave-length λ exceeds that needed to produce the nth interference by an amount which is independent of n; this was true for all values of λ studied. They thus arrived at a means of measuring film-thickness without a prior knowledge of the specific phase-change (which itself was found to vary with λ) and established accurately the relationship between thickness and E.M.F. (16·0 Å per volt) quoted in Chapter VII. The paper deserves close study (A. Charlesby and J. J. Polling, *Proc. roy. Soc.* (*A*) 1954–55, **227**, 434, esp. fig. 5, p. 442).

Causes of Sequence of Interference Colours. Fig. 137 serves to explain why the " character " of the colours occurring at different " orders " is not quite the same, and why the sequence of colours does not exactly repeat itself. Interference may be expected whenever the effective paths travelled by light reflected at the two surfaces differ by an odd number of *half* wave-lengths; thus, if we could neglect the specific phase-change, it would occur when the film-thickness differed by an odd number of *quarter* wave-lengths. Consider a film thickening progressively on a metallic surface, viewed by white light. So long as it is very thin, the interference band will be in the ultra-violet, and there will be no colour (Stage A). When a certain thickness is reached, the blue light reflected from one surface will be out of phase with that reflected from the other surface, and the deficiency of blue light will make the appearance yellow (Stage B). When a greater thickness is reached, the longer green waves suffer interference, producing a reddish mauve (Stage C); still greater thicknesses brings the interference

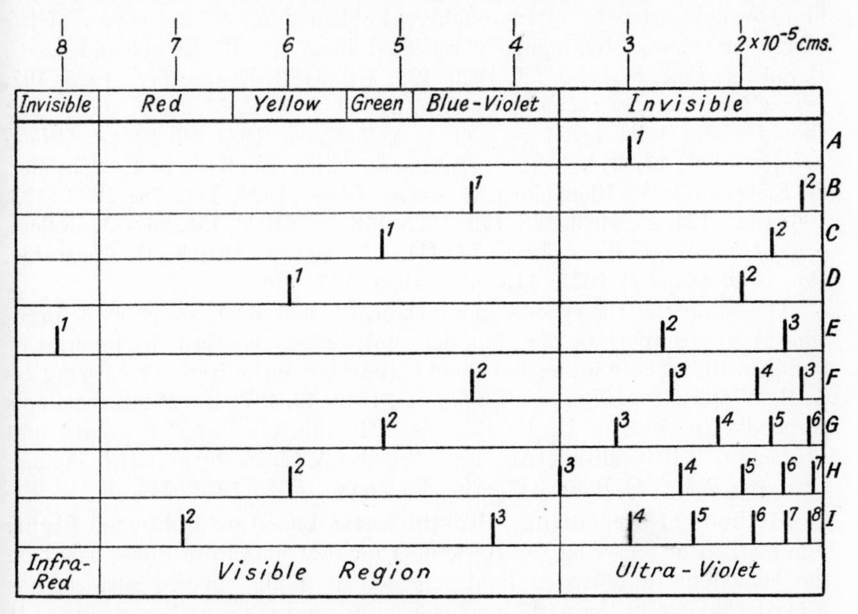

FIG. 137.—Production of Interference Colours. *A* represents a very thin (invisible) film giving interference band in ultra violet—hence no colours. In *B*, *C*, *D*, the first band advances across the visible spectrum, producing first order colours. It leaves the visible region before the second band enters it—explaining the silvery hiatus (*E*). The second band then produces second-order colours (*F*, *G*, *H*) but is more closely followed by the third band, explaining the formation of a second-order green; there is no green at the end of the first order.

into the yellow region, leaving the reflection blue (Stage D). Finally the first interference band passes out of the visible part of the spectrum before the second has entered it and we get a *silvery hiatus* (Stage E). Then the 2nd band enters, giving us second-order yellow (Stage F), second-order red (Stage G) and second-order blue (Stage H). However, the 3rd band follows the 2nd more closely than the 2nd followed the 1st, and will enter the blue-violet region whilst the 2nd band is still in the red, yielding a green at the end of the second-order colours (Stage I); this is a departure from the colour-sequence noted in the first-order colours, from which green should be absent, provided that the film-thickness is uniform.

The reason for the unequal spacing of the bands is simple. If interference occurs when the thickness y is $\frac{\lambda}{4n}, \frac{3\lambda}{4n}, \frac{5\lambda}{4n}, \frac{7\lambda}{4n}$ etc., it follows that a given thickness will produce interference of light having wave-lengths $4ny$, $\frac{4}{3}ny$, $\frac{4}{5}ny$, $\frac{4}{7}ny$ etc.—so that the values converge.

The fascinating subject of interference colours lies outside the scope of this book, but the author's experiments and views will be found in his papers—which also explain the weakness or absence of first-order colours in the case of films composed of a very transparent substance on metal of high reflectivity, and show why the colour of scattered light from a film-covered surface is often roughly complementary to the colour of the surface as viewed by regularly reflected light (U. R. Evans and L. C. Bannister, *Proc. roy. Soc.* (*A*) 1929, **125,** 370; U. R. Evans, *ibid.*, 1925, **107,** 228; *Chem. and Ind.* (*Lond.*) 1926, p. 211 ($\lambda/2$, $3\lambda/2$, $5\lambda/2$ . . . on p. 212 should read $\lambda/4$, $3\lambda/4$, $5\lambda 4$ . . . etc.); *Kolloidzschr.* 1934, **69,** 129; *J. Colloid Science* 1956, **11,** 314. Cf. G. Tammann (with W. Köster, E. Schröder, G. Siebel and W. Rienäcker), *Z. anorg. Chem.* 1920, **111,** 78; 1922, **123,** 196; 1922, **124,** 25; 1923, **128,** 179; 1925, **148,** 297; 1926, **156,** 261; A. Rollett, *S.B. Akad. Wiss. Wien*, 1878, **77,** III, 177, esp. p. 229; F. H. Constable, *Proc. roy. Soc.* (*A*) 1927, **115,** 570; 1926, **117,** 376).

The subject of the specific phase-change (which itself varies with wave-length) is an important one, but does not directly concern the kinetics of film-growth. Those interested should study the authoritative treatment by A. B. Winterbottom *Trans. Faraday Soc.* 1946, **42,** 487; also Winterbottom's Appendix (p. 802) of U. R. Evans's "Metallic Corrosion, Passivity and Protection", 1946 edition (Arnold). See also A. Charlesby and J. J. Polling, *Proc. roy. Soc.* (*A*) 1955, **227,** 434; L. Young, *ibid.*, 1958, **244,** 41.

Methods of measuring Film-thickness based on Polarized Light. The method of following the thickening (or destruction) of films, based on the behaviour of polarized light reflected at a film-covered surface, has proved valuable in the past—particularly in connection with passivity. It possesses fresh possibilities for the study of the kinetics of film-growth, since it is non-destructive and should allow the tracing of continuous time–thickness curves. The disadvantage of the method is its difficult theoretical basis, which requires knowledge of the optics of surfaces not commonly taught today; moreover, a lengthy mathematical procedure is needed to translate laboratory measurements into film-thicknesses.

When polarized light is reflected at a metallic mirror, the two components (parallel and perpendicular to the place of incidence) suffer a change of phase and a change of amplitude, but not to the same extent, so that the reflection causes a relative phase-change and a relative amplitude-change. If the surface carries a film, each of these relative changes is altered by an amount which depends both on the film-thickness and on the refractive index of the film-substance, providing us with two equations with two unknowns, so that in favourable circumstances their solution should reveal the values of both thickness and refractive index. Very thin films—only a few Ångstrom units in thickness—produce a large effect, and the special value of the technique is for the measurement of these very thin films.

It is necessary to measure the two relative changes, and this is accomplished in a spectrometer developed in Tronstad's laboratory at Trondheim and described by A. B. Winterbottom's Appendix to U. R. Evans' "Metallic Corrosion, Passivity and Protection", 1946 edition, p. 809 (Arnold).

The principle of the apparatus is shown in fig. 138. Monochromatic light is plane-polarized by the polarizer P and is reflected at the specimen to be studied; this would normally become elliptically polarized owing to

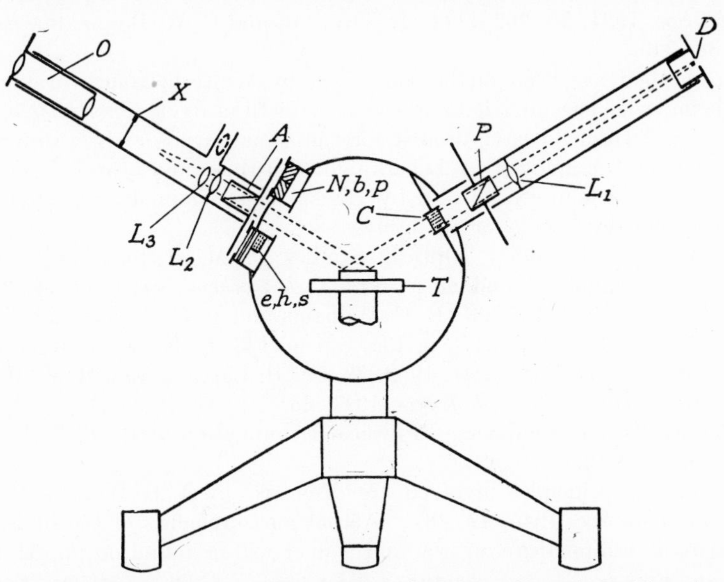

FIG. 138.—Principle of polarized-light method of studying films (A. B. Winterbottom). D, diaphragm; L_1, L_2, L_3, lenses; P, polarizer; C, quarter-wave plate; *Nbp* and *ehs*, half-shade systems; A, analyser; X, cross-hairs; O, Ramsden ocular.

the relative changes mentioned above, but in fact the compensator plate C is adjusted to cause the light which reaches the analyser A to be plane-polarized. If the polarizer and analyser azimuths and the retarding power of the compensator plate are known, the two relative changes are obtained from equations given in Winterbottom's appendix (see above).

The method is based upon the optical work of Drude and was applied by Freundlich to detect the invisible oxide-films formed by the action of dry air on pure iron mirrors. It was then used by Tronstad in a brilliant series of researches on passivity, in which he traced the thickening of films during anodic passivation and their loss of thickness when the electrode was subsequently made the cathode.*

Another feature of Tronstad's classical work at Trondheim was the

* An interesting observation was that on making the mirror an anode and cathode alternately a number of times, the film-thicknesses tended to become greater after each successive anodic treatment. In some cases a thickness was ultimately reached at which interference colours could be seen under suitable illumination. Clearly alternate reduction and oxidation will produce a layer which is both reactive and porous, permitting access to the more compact metal

measurement of the thickness of layers responsible for passivity. The main conclusions are briefly stated on p. 235; references to the principal papers may, for convenience, be given here:—

L. Tronstad, *Z. phys. Chem.* (*A*) 1929, **142**, 241; 1931–32, **158**, 369; *Trans. Faraday Soc.* 1933, **29**, 502; 1935, **31**, 1151; *Nature* 1931, **127**, 127; L. Tronstad and C. G. P. Feachem. *Proc. roy. Soc.* (*A*) 1934, **145**, 115; L. Tronstad and T. Höverstad, *Z. phys. Chem.* (*A*) 1934, **170**, 172; *Trans. Faraday Soc.* 1934, **30**, 362, 1114; L. Tronstad and C. W. Borgmann, *ibid.*, 1934, **30**, 349.

The method has been further improved by Winterbottom, working at Trondheim, who has used it to study the growth of oxide-films with time; Williams and Hayfield have used it for the same purpose. In a different form it has been employed by Leberknight, Lustman and others in Mehl's laboratory at Pittsburg, and also by Haas. Here again it seems best to let the authorities speak for themselves:—

A. B. Winterbottom, "Optical Studies of Metal Surfaces" 1955 (F. Bruns Bokhandel, Trondheim); *Trans. electrochem. Soc.* 1939, **76**, 326; *J. sci. Instrum.* 1937, **14**, 203; E. C. Williams and P. C. S. Hayfield, *Inst. Met. Monograph* No. **23** (1957), p. 131, esp. p. 143; C. E. Leberknight and B. Lustman, *J. Opt. Soc. Amer.* 1939, **29**, 59; B. Lustman and R. F. Mehl, *Trans. Amer. Inst. min. met. Engrs.* 1941, **43**, 248; G. Haas, *Ann. Physik* 1937–38, **31**, 245; B. Chalmers, "Physical Examination of Metals" (1939), Vol. I, pp. 130–138 (Arnold).

The optical principles involved are discussed by A. B. Winterbottom, *Trans. Faraday Soc.* 1946, **42**, 487. Actual measurements of the invisible films at room temperature on iron and copper will be found on pp. 52, 53.

Methods of measuring Film-thickness based on Radioactive Isotopes. A very promising method used in recent years has depended on introducing into the film-forming medium a small proportion of a radioactive isotope of an element which is to constitute either a major or minor constituent of the film. Measurement of the radioactivity of the film by a counter method will serve to show the amount of radioactive substance per unit area, and, if the ratio of the two isotopes is known, will indicate the total amount of the element in question; this will provide a measure of the thickness if the composition is otherwise known, or of the composition if the thickness is otherwise known. The elements thus estimated include chromium, carbon, sulphur and phosphorus, and the principle has been successfully used—mainly for films produced on iron immersed in chromate or other inhibitor, or on copper immersed in polysulphide or thio-urea solution or on aluminium anodized in a sulphuric acid or other bath—at Teddington, Ottawa, Madrid, Pittsburgh and New Kensington (D. M. Brasher and A. H. Kingsbury, *Trans. Faraday Soc.* 1958, **54,** 1214; M. Cohen

below, so that the layer of *active material* becomes thicker after each cycle. This gradual building up of film-thickness by alternate anodic and cathodic treatment has long been familiar, on a much grosser scale, in accumulator-plate manufacture, where, on a plate of compact lead, a thick layer of active material is ultimately built up.

and A. F. Beck, *Z. Elektrochem.* 1958, **62,** 696; J. Llopis, J. M. Gamboa and L. Arizmendi, Instituto de quimica fisica, Ann. Report No. 1 (1958); M. T. Simnad, *Inst. Met. Monograph Series* 1953, **13,** 23; R. C. Plumb, *J. electrochem. Soc.* 1958, **105,** 154C).

Methods of Measurement used in Wet Corrosion. In wet corrosion, the " yard-stick " for the progress of attack is often the loss of weight after exposure of a series of specimens for different times; plotting will yield a corrosion–time curve—which may be a straight line, as in Borgmann's work on zinc in potassium chloride solution (p. 89). In general, it is well to perform all experiments in duplicate or triplicate, involving a great deal of work, but the curve finally obtained is then representative of the material—not of a single specimen. In the case of corrosion localized at a limited number of pits, good reproducibility is not to be expected (p. 913).

The apparatus can be of the simplest character. A set of beakers or square jars containing the specimens immersed, wholly or partially in the liquid, will serve. If the gas chosen is air, the beakers or jars can be placed on a shallow tray containing water, with a large bell-jar over the whole; this prevents evaporation and for rough purposes it may suffice to raise the bell-jar once a day to renew the air. It is better, however, to place them in a sealed box through which humidified oxygen or air is passed, the box being itself surrounded by thermostated water (U. R. Evans, *J. chem. Soc.* 1929, p. 111, esp. p. 119).

Where the specimens are to be studied under conditions of partial immersion—a severe test owing to the rapid replenishment of oxygen at the meniscus, and the possibility of intense water-line attack in certain liquids—it may be convenient to clamp the specimens at the dry part above the water-line; in that case there is no contact between specimen and glass, and consequently no crevice attack; the specimens may be rectangular plates or cylindrical rods. For many purposes, plate specimens placed in tall narrow beakers sloping at an angle of perhaps 70° to the horizontal, and covered with liquid up to the desired level, can be used; but that plan will introduce crevices between glass and metal at the bottom.

For total immersion, a similar method can be used, with liquid completely covering the specimen. Suspension by glass hooks, passing through holes bored in the plate specimen, is adopted in some laboratories, but special corrosion phenomena at the hole are liable to occur, and the specimens may swing about.

If possible, all experiments should be carried out in a water thermostat, an air thermostat or a thermostated room.

Cleaning of specimens before experiments. The preliminary surface preparation requires consideration. Abrasion is often used to remove the original oxide, but to some extent pushes this into the metal instead of rubbing it off. Machining (used in Bengough's work, described below, for disc specimens) is free from that objection. Pickling in acid is often preferred, but may leave a hydrogen charge; if inhibited acid is used, an organic film may remain on the surface. Sand-blasting gives a consistent corrosion-pattern (p. 92) but is liable to leave particles of embedded sand.

Degreasing is always advisable. The preliminary stages can be carried out by immersing in solvent, but the layer of liquid which adheres to the specimen on withdrawal should at once be sucked off by pressing between sheets of clean blotting paper, and not allowed to evaporate—which would redeposit on the metal all the oily matter contained in it. Vapour-degreasing or treatment in a Soxhlet should generally follow. Different solvents are preferred by different workers, and the best choice may depend on the previous history of the metal. Acetone, benzene, xylene, carbon tetrachloride have their advantages and disadvantages; the first two are very inflammable, and the last may contain hydrochloric acid; also its heavy vapours are toxic to an extent which seems to vary from person to person.

The specimens will always carry invisible oxide when the first weighing is carried out. It may be well to leave them 24 hours in a desiccator before the first weighing is carried out, so that the oxide at least reaches a fairly steady state.

Cleaning of Specimens after Experiments. After the specimens have been exposed to the corrosive solution for the desired period, corrosion-products must be removed, and the specimen washed, dried and reweighed; for drying, alcohol (free from acetic acid) and acetone are sometimes helpful. For the removal of the corrosion-product, inhibited acid is generally recommended; preliminary experiments are necessary to find the lowest concentration which will remove all the products without seriously attacking the metal. It is usual to expose a weighed "blank" (uncorroded plate) to the inhibited acid for the period needed to clean the experimental plates, ascertain its weight-loss and subtract that amount from the weight-loss of the others. In the case of iron plates partly immersed in salt solution, the loose rust comes away on mere rubbing, but the interference-colour film of adherent matter near the water-line requires acid for its removal. It is arguable that it might be neglected, as its weight is small and it does not represent corrosion of a destructive character. The fact that it contains oxygen as well as iron means that a small error is introduced if it is not removed; possibly its removal may involve a greater one.

Hudson, in removing rust from steel specimens exposed out-of-doors, uses Clarke's solution, made by dissolving antimonious oxide (2 parts by weight) and stannous chloride (5 parts) in concentrated hydrochloric acid (100 parts); it is used cold, but should be kept stirred during the de-rusting of specimens. For zinc specimens he first removes loose products with a wooden scraper and then immerses them for one minute, with scrubbing, in cold 10% acetic acid.

Opinions differ about methods of removing corrosion products from non-ferrous metals, and each experimenter would be well to try out the various reagents for himself. A German report recommends 5% sulphuric acid for copper, and 5% hydrochloric acid for aluminium. For zinc, ammonium acetate and hot chromic acid have been used by different workers, but Stroud finds that the first removes only thin layers and that the second attacks the metallic zinc when there is Cl^- or SO_4^{--} in the corrosion-products; he has used cold chromic acid containing silver and strontium hromates which precipitate Cl^- and SO_4^{--}, and in this ingenious manner lissolves the products without attacking the basis metal; the liquid should e kept stirred (E. G. Stroud, *J. appl. Chem.* 1951, **1**, 93; cf. *Z. Metallk.* 936, **28**, 20, esp. p. 22; S. G. Clarke, *Trans. electrochem. Soc.* 1936, **69**, 31).

Masking of Specimens. In some cases, it may be desired to expose nly a limited part of a specimen, masking the sheared edges which behave lifferently from the face, and usually also the back. In the case of a vertical pecimen, the object may be to coat everything except a small square area n the front side, so that the whole can then be supported close to the vertical vall of a square cell by means of a clip above the water-line; the crevice etween wall and specimen introduces no complication if the back is coated.

At Cambridge, polystyrene lacquer has been used as masking agent, or stoving varnish of the phenol-formaldehyde type where heating is per-nissible; if varnishes are insufficiently impervious to water, they can be overed up with molten paraffin wax. If the varnish chosen is one which s wetted by wax, it is easy to arrange that the molten wax, spread by a rush nearly up to the edge of the varnished zone, will automatically flow p to the edge—but no further, since wax does not wet metal. Other nasking agents have been recommended, and the mixture of gutta percha nd paraffin wax developed during classical work on plating, is still used oday (D. J. Macnaughtan and A. W. Hothersall, *Trans. Faraday Soc.* 1930, **26**, 163).

For some purposes a masking agent resistant to aggressive reagents will e desired, or one capable of being used at elevated temperature. Recent vork at Teddington indicates that polytrifluorochloroethylene (*P.T.F.C.E.*, *el-F*, *Hostaflon*) possesses useful properties, but a number of coats appear o be necessary, followed by baking at 280°C. For certain materials (e.g. luminium alloys) this would cause structural changes and alterations in orrosion-resistance. The method, however, will have uses and should be tudied (G. Butler and E. C. Seabrook, *Chem. and Ind.* (*Lond.*) 1958, p. 155).

Gas-absorption methods of obtaining Corrosion-Time Curves. The progress of corrosion can be followed by measuring the oxygen absorbed, he hydrogen evolved or both. This enables a complete curve to be obtained rom a single specimen. The apparatus is somewhat elaborate, but when nce it has been made, results can be produced quickly; alternatively, xperiments can be continued over long periods.

Bengough's arrangement is shown diagrammatically in fig. 139(A). The pecimen is a disc, finished usually by machining or abrasion, and supported n three glass points at a chosen distance below the surface of the liquid —generally potassium chloride solution. The space above the liquid is illed with oxygen, and the rate at which oxygen is taken up is indicated n a gas-burette; since a perceptible time is taken for the oxygen to pass lown through the liquid to the metal surface, whether by convection or liffusion, there may be a small time-lag between the progress of corrosion nd the gas-burette measurement, but for experiments lasting over months r years, the error should be imperceptible. There is occasionally crevice corrosion at the points of support, but it is doubtful whether this affects the measurements. The whole apparatus is placed in a water thermostat.

Where the attack is solely of the oxygen-reduction type, the readings of the gas burette should be an exact measure of the corrosion, but if some hydrogen is evolved, it is necessary to measure separately the corrosion of the oxygen-reduction and hydrogen-liberation types. For that reason, a platinum wire is provided, which can be heated electrically, when desired. At some chosen time, after the uncorrected reading has been taken, the wire is heated so as to burn the hydrogen in the excess of oxygen; the new reading, obtained after allowing time for the temperature to become uniform, will allow the observer to estimate the corrosion of the hydrogen-liberation type, and then, by deduction, that of the oxygen-reduction type.

A slight modification of the Bengough apparatus was introduced by Borgmann who used an air-thermostat and buried the vessel in a box filled

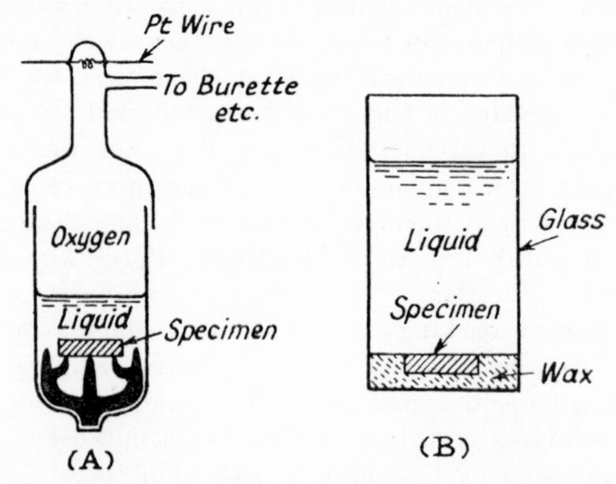

Fig. 139.—(A) Principle of Bengough's apparatus for measurement of corrosion-rate by oxygen-absorption. (B) Mears's modification.

with metal turnings—so as to provide a large heat-capacity; the object was to avoid possible disturbance due to radiation from the thermostat heater, which might affect the convection process in the liquid and thus the replenishment of oxygen at the surface of the specimen. Another modification is due to Mears, who mounted the specimen flush with a wax surface, to prevent convection currents set up by corrosion-products falling over the edges of the disc (fig. 139(B)). For details, the relevant papers should be consulted:—

G. D. Bengough, J. M. Stuart and A. R. Lee, *Proc. roy. Soc.* (*A*) 1927, **116,** 425, esp. p. 438; *Chem. and Ind.* (*Lond.*) 1933, p. 195, esp. p. 197; G. D. Bengough and F. Wormwell, *Rep. Corr. Iron St. Inst.* 1935, **3,** 123; 1936, **4**, 213; C. W. Borgmann and U. R. Evans, *Trans. electrochem. Soc.* 1934, **65**, 249, esp. p. 253; U. R. Evans and R. B. Mears, *Proc. roy. Soc.* (*A*) 1934, **146**, 153, esp. p. 159.

A different type of method, also based on oxygen-absorption, is described by Delahay. Air-saturated solution flows steadily through a cell containing the specimen into a vessel containing a dropping mercury electrode which allows the oxygen to be estimated; the oxygen lost is a measure of the corrosion (P. Delahay, *Corrosion* 1951, **7**, 146).

In corrosion by acids where the hydrogen-evolution type of corrosion proceeds rapidly, simple methods of collecting and measuring the gas in burettes suggest themselves, but if the reaction is violent, temperature-control may become difficult. Moreover the agitation of the liquid by the gas-evolution affects replenishment of acid, and the geometrical design of the apparatus may influence the results.

An ingenious method has been evolved by Bloom for measuring the corrosion produced by water or aqueous solution on steel or stainless steel at elevated temperatures (e.g. 316°C.); the hydrogen is allowed to diffuse outwards through the walls of a capsule made from thin-walled tubing which has been filled with the chosen liquid, sealed by pinching and spot-welding at the two ends and then introduced into the experimental vessel. The corrosion is calculated from the rate of pressure-change caused by the hydrogen effusing outwards through the walls into the gas-space, which may contain helium (M. C. Bloom and M. Krulfeld, *J. electrochem. Soc.* 1957, **104**, 264; also, with W. A. Fraser and P. N. Vlannes, *Corrosion* 1957, **13**, 297t).

Experiments with relative motion of metal and liquid. It is often desired to conduct experiments under conditions where there is relative movement of metal and liquid. Complete stagnancy may mean that the corrosion-rate is controlled by the rate of supply of some essential material, e.g. oxygen; it is generally of greater interest to know what the corrosion-rate would be if the oxygen replenishment-rate was so high as to allow the control to be exercised by the chemical or electrochemical reactions proceeding at the metallic surface. When such reactions are capable of proceeding very rapidly, however, control by the replenishment of material may still prevail even under conditions of fairly rapid movement—as in King's work on corrosion by mixtures of acids and oxidizing agents (p. 332); with still more rapid movement, reactional velocity may exercise partial or complete control in such cases.

The various forms of rotor apparatus developed by Bengough and his successors at Teddington are suitable for the study of corrosion at high relative velocity between metal and liquid (see also p. 131). In King's laboratory at New York, work has been carried out on rapidly-rotating specimens, whilst Twiss and Gutterplan, testing aluminium discs at rotation speeds up to 12,000 r.p.m. in various waters, use Morton flasks with ribbed sides which causes the liquid to move outwards and upwards, finally tumbling down on to the top of the specimen, with great turbulence and introduction of air (F. Wormwell, *J. appl. Chem.* 1953, **3**, 164; M. B. Abramson and C. V. King, *J. Amer. chem. Soc.* 1939, **61**, 2290; C. V. King and M. Shack, *ibid.*, 1935, **57**, 1212; S. B. Twiss and J. D. Guttenplan, *Corrosion* 1956, **12**, 263t. Another apparatus for measuring corrosion velocity in flowing water is described by F. Wormwell and T. J. Nurse, *J. appl. Chem.* 1952, **2**, 685).

Methods involving chemical determination of metal in the corrosion-product. Instead of using weight-loss, oxygen-absorption or hydrogen-evolution for following the progress of corrosion, the metal in th corrosion-product may be estimated; in appropriate cases, the solid product may be filtered off from the soluble metallic salt and estimated separately In general, the estimation will be carried out at the end of an experiment so that each specimen will provide only a single point on a corrosion–tim curve. Where, however, the product is wholly soluble, it may sometime be permissible to remove, at intervals, aliquot parts of the liquid and estimat the metal in them, thus obtaining a complete curve from one specimen Such a procedure raises several questions—such as the replacement of th portion withdrawn by fresh liquid; the answers must generally be though out by the experimenter.

Methods involving estimation of small traces of metal. Fo the opening stage of corrosion (and for later stages where attack is slow) sensitive reagents are required for detection and estimation. The use o sodium diethyl-dithiocarbamate as a sensitive reagent for copper and othe metals has been described by Atkins; this will often detect the corrosio produced in a few seconds, whilst in some cases the test can be made stil more sensitive by extracting the metal derivative in a small volume of chloro form or carbon tetrachloride. The reagent will serve to show the actio of sea-water on most heavy metals, including zinc, copper, iron, tin, lea (W. R. G. Atkins, *Trans. Faraday Soc.* 1937, **33**, 431). The micro-estimatio of copper with triphenyl-methyl-arsonium thiocyanate is described b K. W. Ellis and N. A. Gibson, *Analyt. Chim. Acta* 1953, **9**, 363.

For iron $\alpha\alpha'$-dipyridyl is often used, but isonitroso-dimethyl-dihydro resorcinol, which allows iron to be detected at 0·02 ppm., is recommende by S. C. Shome, *Analytical Chemistry* 1948, **20**, 1205. Orthophenanthroline which is suitable for slightly acid solution, is preferred by M. Stern *J. electrochem. Soc.* 1955, **102**, 609; for details see E. B. Sandell, " Colorimetri Determination of Traces of Metals " 1950, p. 375 (Interscience Publishers)

E.D.T.A. (p. 1000) has proved useful for estimating aluminium, zirconiun or gallium, the method being accurate but not exceptionally sensitiv (G. W. C. Milner, with J. L. Woodhead, P. J. Phennah and G. A. Barnett *Analyst* 1954, **79**, 363, 475; 1955, **80**, 77; 1956, **81**, 367).

Aluminium can be determined colorimetrically by means of ammoniun aurin tricarboxylate (A. C. Rolfe, F. R. Russell and N. T. Wilkinson *J. appl. Chem.* 1951, **1**, 170). A method based on ferron in buffered solutio is described by H. Green, *Metallurgia* 1958, **57**, 157. The sensitive metho described by E. B. Sandell " Colorimetric determination of traces of metals " (Interscience Publishers) 1950, p. 152, can be used in presence of chromates The colorimetric determination of chromium is discussed by L. Meites *Analyt. Chim. Acta* 1958, **18**, 364.

These micro-analytical methods may be useful in measuring the initia corrosion-rate which can be very different from that ultimately attaine (F. Tödt, " Korrosion VIII " (1954–55), p. 6 (Verlag Chemie)).

Other Measurements

Local Loss of Thickness. Where the corrosion is localized, the total destruction of metal is far less important than the *pit-depth*. Simple instruments have been designed for measuring the position of a pit bottom relative to the general surface. One of these is essentially a micrometer provided with a fine needle-point connected to an electric circuit; when the point makes contact with metal at the bottom of the pit, a signal is shown on an ammeter (R. S. Thornhill, *J. Iron St. Inst.* 1942, **146**, 73P; esp. p. 90P; T. J. Summerson, M. J. Pryor, D. S. Keir and R. J. Hogan, Amer. Soc. Test. Mater. spec. tech. Pub. **196**, p. 157).

In other cases, the probability that corrosion will start at all within a certain area may be more important than the velocity attained if it does start. This matter is discussed in the statistical chapter (p. 935).

Potential distribution. The distribution of potential throughout the liquid with a travelling electrode—as carried out by Agar—is discussed on p. 861 in connection with the measurement of corrosion currents. The special case of the potential drop at the cathodic and anodic areas of iron specimens partly immersed in salt solutions was much studied in early researches, by means of tubuli with ends lightly pressing against chosen regions of the metal; the results showed that the difference between the readings at cathodic and anodic areas was far greater in dilute than in concentrated salt solutions. The experiments, perhaps still possessing something more than historical interest, are described by A. L. McAulay and F. P. Bowden, *J. chem. Soc.* 1925, p. 2605; U. R. Evans, L. C. Bannister and S. C. Britton, *Proc. roy. Soc.* (*A*) 1931, **131**, 355.

Such methods probably suffer errors in the fact that the tubuli screen the metal locally from current if they touch the surface, and introduce an *IR* drop if they do not; they may also interfere with diffusion, and in the case of solutions containing an inhibitor may set up attack at places where it would not otherwise occur. The improved types of tubuli designed by Piontelli should be considered by anyone wishing to work on these lines. Two of these are shown in fig. 140. The method of introducing the tubulus through a hole in the otherwise flat surface has obvious attractions, but several points will require consideration by anyone proposing to adopt it —notably the advisability for annealing after making the hole, and the possibility of an altered grain-size in the metal surrounding it. There is some distortion of the current lines (shown broken) and of the equipotentials when the hole is left open (fig. 140(*a*)), but the disturbance is overcome by nearly plugging the opening (fig. 140(*b*)). The flat-ended tubulus shown in fig. 140(*c*)) produces some disturbance when out of contact with the metal, but the lines become straight when it is in contact (fig. 140(*d*)). The first of these two methods have been successfully used in Piontelli's laboratory for tracing anodic and cathodic polarization curves in a solution of a salt of the metal; the apparatus is shown in fig. 141. Some of Piontelli's polarization curves are shown in fig. 178 (p.) 907. For further details see R. Piontelli, *Ricerca sci.* 1955, **25**, 750; *Z. Elektrochem.* 1955, **59**, 778.

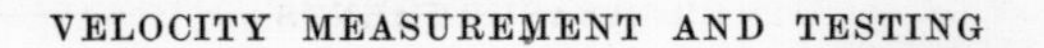

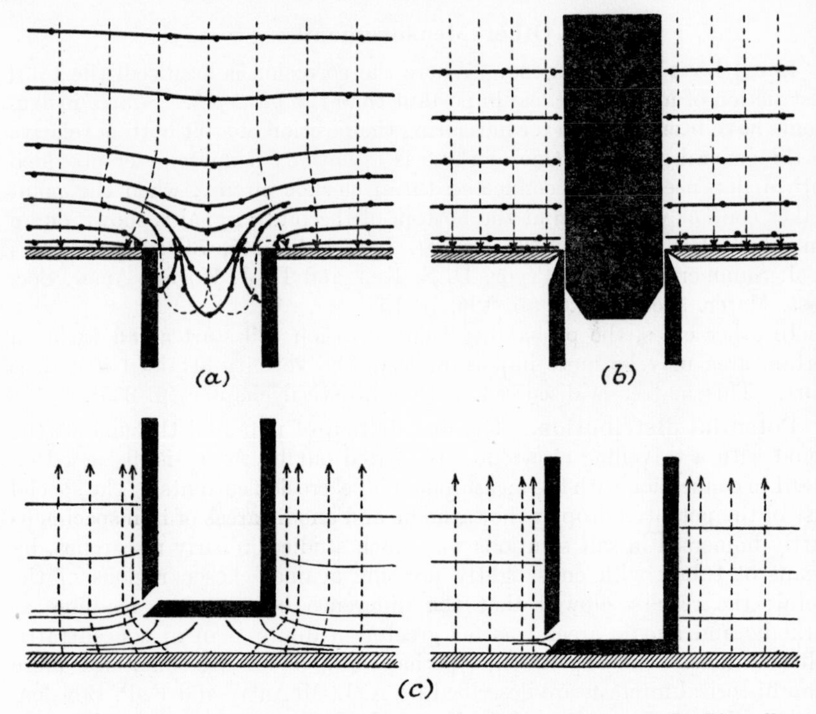

FIG. 140.—Improved forms of tubuli for potential measurement (R. Piontelli).

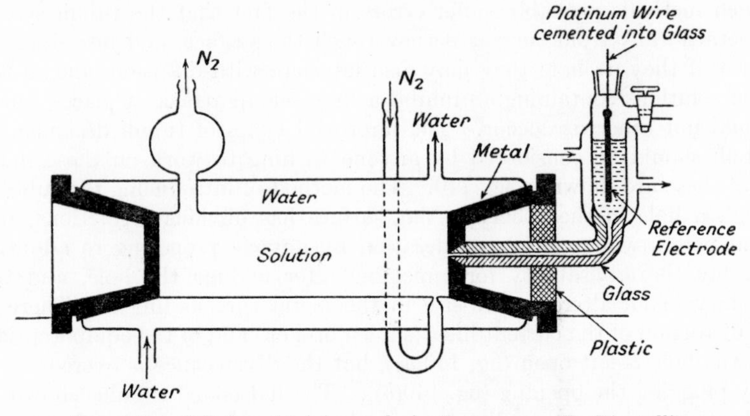

FIG. 141.—Cell for obtaining polarization curves (R. Piontelli).

Potential-Time Curves. The compromise potential at which anodic and cathodic reactions proceed at the same pace, possesses some interest, and its drift with time shows whether corrosion is tending to develop or stifle itself. In the first case, the potential will fall with time, and in the second will rise with time; this arises from the fact that a specimen carrying an oxide-film interrupted by only small gaps will show a higher potential than one carrying an oxide with many gaps or large gaps—as explained in p. 899.*

The compromise potential is often measured during the progress of an experiment in which corrosion-velocity of a specimen is being measured simultaneously (fig. 142). The corrosion vessel may be connected by a liquid path (which may be filter-paper strip, or an inverted U-tube filled with agar jelly having a suitable salt-content) to a second vessel containing a liquid chosen to eliminate or diminish errors due to junction potentials (usually concentrated potassium chloride); it is then possible to connect a calomel or silver/silver-chloride electrode with this second vessel, and read the potential from time to time on a potentiometer; a valve potentiometer is today generally preferred.

If, in this procedure, the liquid-junction potential is not entirely abolished, the error introduced—being constant in time—will not matter, provided that the object is merely to study drift of potential with time. A more serious risk, when the liquid surrounding the specimen is supposed to contain no chloride, is that chloride from a calomel electrode will gradually syphon or diffuse over into it; it may be well to interpose further intermediate vessels, and to minimize junction potentials by using as reference electrode silver (or mercury) in contact with a *dilute* solution of potassium chloride (or sulphate) saturated with the appropriate silver or mercury salt.

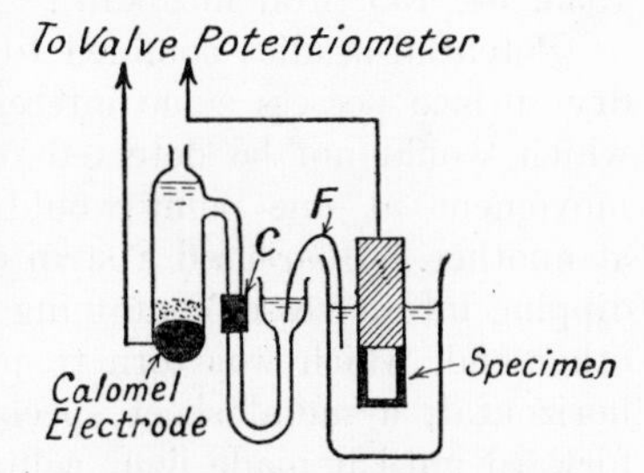

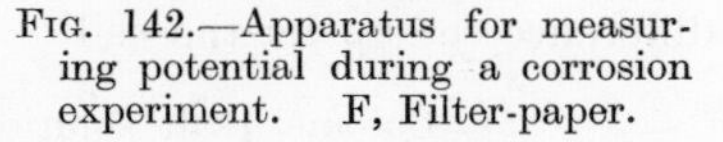
FIG. 142.—Apparatus for measuring potential during a corrosion experiment. F, Filter-paper.

The importance of potential-time curves was brought out in an early paper by Callendar. A year later, May, studying aluminium brass, used the movement of potential to show that mechanical damage to the film repairs itself under conditions where the film on ordinary brass fails to repair

* The potential at which the local anodic and cathodic currents balance themselves (in absence of applied current) gives no indication of the natural corrosion-velocity, but merely a general idea as to whether attack is likely to develop or stifle itself. More information regarding the corrosion-resistance of materials may be provided by the measurements of the potentials taken up during anodic attack. A method based on this principle has been developed in comparing materials for use in veterinary surgery. It has been found that the measured E.M.F. of the cell

Material under test | serum | calomel electrode
(anode) (cathode)

if reduced by the amount of the *IR* drop over the liquid in the cell, takes up a value which is surprisingly independent of the current flowing, and which shows a good correlation with the weight-loss due to corrosion. The electrochemical principles involved are not simple but it is possible that the method could be extended to other uses. The original paper must be consulted (E. G. C. Clarke and J. Hickman, *J. Bone Jt. Surg.* (British Number) 1953, **35B**, 467).

itself. Later Bannister obtained time–potential curves from iron, steel and various alloys of the stainless steel type, and found that they gave some information regarding the protective character of the film (L. H. Callendar, *Proc. roy. Soc.* (*A*) 1927, **115**, 349; R. May, *J. Inst. Met.* 1928, **40**, 141, esp. pp. 147, 152; L. C. Bannister and U. R. Evans, *J. chem. Soc.* 1930, p. 1361).

Since that time numerous potential–time curves have been published. Some materials which show rising potentials in presence of oxygen, show a fall in its absence. The potential of aluminium in a salt solution greatly varies according as oxygen is present or absent; stainless steel behaves in the same way in dilute acid and in salt solutions, and even unalloyed iron in alkaline solutions. References include H. Endo and S. Kanazawa, *Sci. Rep. Tohoku Univ.* 1933, **22**, 537 (aluminium); F. Müller, *Z. Ver. dtsch. Ing.* 1938, **82**, 844; H. H. Uhlig, *Trans. Amer. Inst. min. met. Engrs.* 1940, **140**, 387; I. D. G. Berwick and U. R. Evans, *J. appl. Chem.* 1951, **1**, 576 (stainless steel). H. L. Lochte and R. E. Paul, *Trans. electrochem. Soc.* 1933, **64**, 155 (iron in alkali).

Potential studies confined to very small wetted areas on an otherwise dry surface possess some interest; they bring out fluctuations with time which would not be detected on a large immersed surface, where an up-movement at one point would be cancelled out by a down-movement at another. Hoar used a strip of filter paper 5 mm. broad, with one end dipping into a vessel containing a salt solution, but supported so that the other end, which was torn to provide ragged wisps of wet cellulose, was horizontal; a metal sheet specimen with its surface vertical was moved forward until it made light contact with one of the wisps. The potential difference set up by the cell

Calomel electrode	salt solution	filter-paper wisp	chosen point on the metal sheet

was balanced on a potentiometer and its movement with time studied. The form of the curve obtained depended on material, liquid and surface condition. Pure iron and mild steel showed rising curves in contact with M/10 sodium phosphate but falling curves with M/10 potassium sulphate or chloride, whereas stainless steel (13% chromium) showed a rising curve with N/10 potassium chloride if well finished, but a falling one if tested in a coarsely finished condition. Most of these curves were smooth, and the potential had become nearly constant after 1 to 2 minutes. However, iron previously immersed for 18 hours in M/5 potassium chromate, then washed, dried and tested in contact with paper soaked in M/10 potassium sulphate gave potential–time curves showing violent fluctuations; different curves were obtained at five points on the same specimen; the fact that the rises and falls appeared at different moments at the five points suggests why potential–time curves obtained on a relatively large area of an immersed specimen in the ordinary way may fail to show such violent fluctuations. Hoar's results make clear the dynamic character of film-development, with periods of repair alternating with break-downs. Such a method could probably be extended to provide valuable information regarding corrosion mechanism. The work is described by T. P. Hoar and U. R. Evans, *J. Iron. St. Inst.* 1932, **126**, 379, esp. p. 382.

Potential-Current Curves. The experiments described above concern movements of potential in space and time. Movement with applied current constitute what are generally known as *polarization curves*—some of which have been discussed in connection with anodic passivity in Chapter VII. There are two experimental procedures, the *potentiostatic* and the *galvanostatic*. In the potentiostatic method, the potential is kept fixed by a suitable electrical circuit at some chosen value, and the current density recorded when it has become constant; then a new potential is imposed, and the corresponding value of the current density is found. After sufficient data has been obtained, a curve relating current to potential can be plotted. In the galvanostatic method, a constant current is imposed, and the potential recorded when it has become constant; then a new value of the current is imposed and the new potential recorded, and so on. The i–V curve obtained by the two methods is under favourable conditions identical. A simple form of circuit for the potentiostatic method is indicated in fig. 56 (p. 229), whilst for the galvanostatic method a battery providing a high E.M.F. in series with a high resistance should send a current through a cell which will remain constant—independently of happenings within that cell. In both cases, much more elaborate circuits are frequently used.

On stainless steels, it is very important to know the range of potential over which the steel is passive, since this range determines the reagents which the steel will resist. It is known that if the potential is too high, the protective skin will pass into solution as a chromate; if the potential is too low, reductive dissolution will occur; in either case corrosion will set in. Since different additions to the alloy affect the limits of the safe range in different ways, it is necessary to ascertain whether the particular potential level likely to be fixed by the reagent (regarded as a redox system) will fall within the range favourable to the passivity of any particular sample. In principle, the method used by Olivier (p. 229) would serve, but Edeleanu's application of a potentiostat is convenient and his papers deserve study (C. Edeleanu, *J. Iron. St. Inst.* 1957, **185**, 482; 1958, **188**, 122).

Tests for Practical Corrosion Problems

Field Tests. The only reliable method for predicting the life of a material in service (or the period which can safely be allowed to elapse between renewals of paint) is to expose specimens under the conditions which the material will have to withstand in service. If it will be exposed to the weather, specimens must be fixed to frames on a roof or in a field. If it will be immersed in the sea, attachment to a raft is indicated. If material is required for a new tank at a chemical works, plates sunk in an existing tank in operation, containing the liquid at the working temperature, may be used. Some authorities include all such tests under the term *Field Tests.* Despite their differences, they can all claim the same advantages; they also suffer from the same disadvantages—the most obvious being the long period which must elapse before even tentative conclusions can be drawn.

Many field tests have already been described in this book. J. C. Hudson's extensive out-door tests on non-ferrous materials provide one example (p. 514). He used various criteria of corrosion damage: gain-in-weight with products adhering, loss-in-weight with products removed, loss of strength or loss of electrical conductivity. Other examples are the numerous tests on steel specimens carrying paint or metallic coatings (pp. 576, 658). However, some general remarks on field tests may be introduced at this point.

In atmospheric tests on plate specimens, the angle of slope has to be settled. Vertical specimens tend to screen one another from the weather unless hung in a long row—which may involve differences in atmospheric conditions at the two ends of that row. The rain runs off quickly from a vertical surface, whereas on horizontal specimens, the rain collects as a "puddle" on each specimen, thus having a longer time to act; this constitutes a more severe test but a somewhat less reproducible one, since, unless the specimens are accurately level, the nature of the puddle may vary from one specimen to another. Preliminary tests at Cambridge favoured the use of specimens sloping gently to the horizontal, and such specimens, which will not shelter one another, can be attached to a frame placed on any flat ground or flat roof. The Cambridge rack is shown in fig. 98 (p. 577).

Another question to be decided is the size of specimens. One school of thought favours large specimens and neglects changes near sheared edges where conditions are undoubtedly exceptional; another argues that we are mainly interested in behaviour under the most severe conditions, which exist near the edges, and that if observations at the centre are without significance, it is wasteful to use large specimens. It would seem that both types of test are desirable, and the fact that, where the effect of the same factors have been studied in the two types, the conclusions reached have generally been much the same, reassures us in accepting conclusions where only one type has been used.

Nevertheless, caution is necessary in applying the results of field tests to service problems. In tests, the conditions are geometrically simple, and corrosion damage may be much more rapid in practical cases, owing to retention of moisture in crevices after the main surfaces are dry. Again rain-water collected on a large roof of slate or glass, may pick up sulphur compounds or salt in large amounts and may then fall off on to a small area of metal which obtains far more than its proper "ration" of pollution.

Furthermore, isolated specimens attached to a stand in a field—or to a raft in the sea—escape the stress conditions which may be present when the same material is used as part of a building or a ship. Similarly short pipe-lengths buried in soil are subjected to different stress-conditions from operating pipe-lines, and escape the effect of long-line currents (p. 268). Thus the order of merit of materials obtained in tests on small specimens often differs from that obtained in service—a point brought out by the tests of G. N. Scott, *Proc. Amer. Petroleum Inst.* 1934, **15**, Section IV, p. 30.

There are other factors which cause tests on isolated specimens to give misleading results. If we wish to know which, of a number of materials, is best suited for use in constructing a tank in a chemical plant where reactions involving acids or other corrosive liquids will be carried out, we might think of hanging small specimens in a similar tank where the industrial operation is already in progress; such a method, however, is likely to introduce crevice effects absent in service where the whole plant would be made of the material in question, besides excluding stress effects present in service. Moreover, unless steps are adopted to avoid electrical connection between the various specimens, the conclusions drawn about relative merits may be very misleading; to give a single example, unalloyed aluminium may be attacked less than an alloy by a reagent when the materials are exposed separately, but when connected together may suffer severe anodic attack, conferring cathodic protection on the alloy.

These warnings are not intended to discourage the organization of field tests, which, given time, are probably the most practically useful form of testing, but to emphasize the fact that earnest consideration is necessary if misleading results are to be avoided; testing should not be entrusted to someone who is incapable of clear thinking.

Since the corrosive properties of different atmospheres differ widely, it has been proposed, when subjecting any material to a field test for corrosion-resistance, to expose beside it a similar specimen of steel carrying 0·2% copper, and then express the weight-loss of the new material as a percentage of that of the copper steel. Copper steel is chosen in preference to nominally "copper-free" steel, since today practically all steel contains small but variable amounts of copper, and the effect of variations in copper content is small around 0·2% but large around 0·01%. The suggestion has aroused much discussion and certain features of the original proposals have not escaped criticism, but it appears to be a sensible one (E. E. White, *Chem. and Ind.* (*Lond.*) 1955, p. 952; see comments by F. D. Murphy, p. 1090 and F. L. LaQue, p. 1486).

The desire for an accelerated test. The objection commonly raised against field tests is the long time which must elapse before useful conclusions can be drawn. The desire for a quick answer is natural, but some of the attempts to satisfy the demand have been ridiculous, being apparently based on the assumption that the order of merit of materials is the same in all environments. It has long been known that acids attack metals rapidly, and an "acid test" was at one time advocated for deciding resistance to corrosion; it was quite useless, since the order of merit of a series of materials in acid is different from that in natural waters or in air; the addition of 13% chromium to steel, which confers resistance to many waters and atmospheres, actually increases attack by dilute hydrochloric acid; phosphorus increases the resistance of copper steel to the atmosphere, but the acid test fails to reveal this (K. Daeves, *Arch. Eisenhuttenw.* 1935–6, **9**, 37).

Corrosion can also be accelerated by making the specimen an anode with an external E.M.F. and this has been advocated as a quick method of assessing resistance. However, in cases where Faraday's Law is obeyed, the corrosion produced by a certain number of coulombs, expressed in gram-equivalents, should be the same for all materials. The result gives no indication of the " natural " corrosion to be expected in service. The test does serve to show up materials which tend to go passive, but there is no guarantee that the material which becomes passive during anodic treatment in some standard liquid will be passive under the average service conditions. This sort of " electrochemical test " now only possesses historical interest, but those wishing to study the matter may turn to R. J. Anderson and G. N. Enos, *Proc. Amer. Soc. Test. Mater.* 1924, **24**, II, 725; W. D. Bancroft, *Industr. engng. Chem.* 1925, **17**, 336; W. H. Thornton and J. A. Harle, *Trans. Faraday Soc.* 1925–26, **21**, 23, with criticisms by W. H. J. Vernon, p. 32, A. J. Allmand, p. 32, and U. R. Evans, p. 33.

Not all electrochemical testing methods, however, are open to the criticism just made. In Tödt's corrosion meter, no current is applied from the outside, but the power of the material to generate a current (i.e. to corrode) is measured. A specimen is immersed in the corrosive liquid along with a platinum electrode and the current generated by the cell thus formed is measured; application of Faraday's Law enables the corrosion-rate to be calculated. For the comparison of rapidly corroding materials, the test may not be discriminating, since the corrosion will probably be fixed by the rate at which oxygen is replenished (or hydrogen liberated) at the surface of the platinum cathode, and may be almost unaffected by the anode material. The test is serviceable where control is anodic rather than cathodic. If we are testing the relative suitability of different inhibitors for addition to a corrosive water, the falling off of current observed when a given inhibitor is added to the water is likely to be a pointer to its suitability in service (F. Tödt, *Z. Elektrochem.* 1928, **34**, 586, 591, 853; 1934, **40**, 536; 1949, **53**, 132 (with R. Freier and W. Schwarz); 1950, **54**, 485; 1952, **56**, 165, " Korrosion VIII " (1954–55), p. 6 (Verlag Chemie); E. Schumann, *Ber. Korrosionstagung* 1933, **3**, 47, esp. p. 52).

The real objection, however, to most schemes for acceleration is the introduction of some factor not present in service—an acid, an anodic current or a cathode of dissimilar metal. It would be better to retain the factors present in service, but to intensify them; the replacement of air by oxygen at ordinary or elevated pressure does speed up attack—as shown in Bengough's work. At high pressures, however, the corrosion-rate begins to fall off again, and this passivity, whether partial or complete, will probably occur at different pressures for different materials. In some cases, a corrosion test carried out in high-pressure oxygen would not reflect service behaviour.

Air-conditioner Tests. Interesting results have been obtained in accelerating atmospheric corrosion by the omission of idle periods during which no corrosion is occurring, and by increasing the rate of replenishment of the sulphur compounds which are the cause of rapid attack, without increasing their concentration. Here no unnatural features are introduced; only waste of time is avoided.

Specimens placed in the plant erected for the conditioning of air entering a building, are simultaneously exposed to water-spray and a rapidly driven current of air introduced from outside; in this situation they often suffer very rapid corrosion. Here, the corrosive bodies to which the metal is exposed are much the same as on a wet day out of doors, but their replenishment is far more rapid. Moreover the dry spells, during which there is possibly little corrosion, are omitted. However, it must not be assumed that this omission introduces no error. Outdoor corrosion is usually most rapid at partly sheltered places where the damp surface is able to pick up sulphur oxides from the atmosphere without them being washed away by rain; such a situation is hardly represented in air-conditioning plant tests. Moreover, metal exposed outdoors develops periodically a damp, and probably hydrated, corrosion-product, which during intervening periods dries up; this periodic desiccation will certainly affect its character one way or the other way, although whether it becomes more protective by consolidation or less protective by cracking, cannot always be foreseen. In an air-conditioning plant, working day and night, the periodic drying will be omitted, and the order of merit of a number of materials (or of a given material carrying different coatings) may be different from that obtained in service after a longer period. In an air-conditioner which is closed down over the week-end, the necessary dry period may be introduced, but this should not be accepted as rendering the test valid without careful study of the conditions; corrosion-products may " look dry " and yet contain much water. The use of a working air-conditioner is, therefore, open to some doubt; it is, however, believed that a small plant similar in design but erected for the special purpose of testing specimens might give reliable information far more quickly than is possible with the ordinary form of exposure test.

Spray Testing. Exposure in a spray cabinet has long been used in corrosion-testing—particularly for assessing the quality of protective coatings. The specimens are placed in a chamber in which the air contains spray from an atomizer; the solution sprayed on them is either sodium chloride solution, sea-water, dilute acid, or, as in a recent American variant of the test, salt solution acidified with acetic acid. In general, the specimens are kept wet all the time (even if the spray is periodically turned off, they have little chance to dry up completely); conditions would be more realistic if truly dry periods were introduced, but possibly alternate wet-and-dry periods can be more readily provided by means of a test cycle (p. 811). In any case, it is optimistic to hope that the spray test will predict behaviour except under service conditions closely resembling those existing in the spray chamber.

Attempts to assess the promise of newly introduced materials with the spray test have led to regrettable results. When cadmium-plating was introduced, spray tests appeared to show that cadmium coatings were superior to zinc coatings, which caused cadmium to be used in situations where zinc would have been better.

The failure of salt-spray testing to predict the results of field behaviour it is well brought out by F. L. LaQue, *Proc. Amer. Soc. Mater.* 1951, **51**, 495, esp. fig. 29 on p. 529; see also *Materials and Methods* 1952, **35,** No. 2, p. 77, with discussion in No. 3, p. 77. An excellent account of forms of nozzle, settling-rate and other factors is provided by W. Blum and L. J. Waldron, "Corrosion Handbook" 1948, pp. 970–978 (editor, H. H. Uhlig; publishers, Wiley: Chapman & Hall). See also F. Champion, *Metal Ind.* (*Lond.*) 1952, **80**, 123, who recommends *intermittent* spraying; also G. Bianchi (with A. Mora and U. Gandolfi), *Metallurg. ital.* 1952, **44**, 9, 330; 1955, **47**, 551, who discusses the effect of shapes of specimen and position in the cabinet.

Comparisons of tests with natural and synthetic sea-water are described by T. P. May and A. L. Alexander, *Proc. Amer. Soc. Test. Mater.* 1950, **50**, 1131. Various formulae for synthetic sea-water are given in the "Corrosion Handbook" p. 1121 (editor, H. H. Uhlig; publishers, Wiley: Chapman & Hall); F. L. LaQue *Corrosion* 1957, **13**, 303t; also in DIN 50900 (now under revision, but the formula given in the 1951 edition is to be retained). See also F. E. Cook, H. S. Preiser and J. F. Mills, *Corrosion* 1955, **11**, 161t, esp. p. 163t; *J. Iron St. Inst.* 1948, **158**, 463, esp. p. 477; F. Wormwell, G. Butler and J. G. Beynon, *Trans. Inst. mar. Engrs.* 1957, **69**, 109, esp. p. 112. Few of these formulae take account of organic ingredients such as saponin or cystine, and in some cases the content of calcium bicarbonate is left to chance, although this may be the most important constituent (cf. p. 816).

A recent editorial criticism of the salt-spray test deserves notice. It is pointed out that the test has failed to become reproducible in 40 years. For assessing anodic coatings like zinc or cadmium, a thickness measurement is said to give almost as much information, whilst for cathodic metals like copper or nickel the spray test does nothing more than show up the porosity (N. Hall, *Metal Finish.* (*N.Y.*) 1956, **54,** April, p. 47).

Possibly this is rather pessimistic. Although highly misleading in deciding whether a coating of metal X is better than a coating of metal Y, the spray test may be useful, when once X has been chosen, in deciding whether the coating has been carefully applied; the test has value as a precaution against fraud or careless workmanship.

Moreover, attempts are being made to improve the test by adopting different liquids; one formula favoured in the U.S.A. contains 5% sodium chloride with sufficient acetic acid to bring the pH down to 3·2—3·5. The acidified solution is said to give results on plated automobile parts in 16–48 hours instead of the 450–2300 hours needed with the solution formerly used. McMaster states that the results accord with service behaviour. Hooper is slightly less optimistic, and considers that none of the tests are entirely satisfactory where the basis material is a copper alloy. He does, however, consider the acidified salt test the most useful for general purposes, and states that if the spray cabinet is heated to 95–100°F. by radiation, the test can provide information about quality within 72 hours or less. He greatly prefers it to the ordinary 5% salt-spray test which, he says, appears to have no value for assessing the coating systems investigated by him (Cr–Ni and Cr–Ni–Cu deposits on steel), although useful for showing up gross discontinuities in Ni coatings of about 0·5 mil. thickness. The sulphur dioxide test he considers to have a limited field of application, its only real merit being quickness. The acidified salt test is described in a Report of Committee B3 of the Amer. Soc. Test Mater. 1954, **54**, 147. See discussion by W. D. McMaster, *Metal Finish.* (*N.Y.*) 1956, **54**, Nov. pp. 48, 55; *Plating* 1955, **42**, 904; also J. H. Hooper, *Chem. and Ind.* (*Lond.*) 1957, p. 1640; *Trans. Inst. Metal Finishing* 1958, **35**, 79.

Tests designed to decide choice of material. In determining which of a number of metals is to be adopted in some industrial situation (or if a non-resistant material like steel must be used, in determining which of a number of protective processes is to be adopted), it would seem advisable to design a test which represents the geometry of the industrial situation, suitably simplified, using liquids (and/or gases) such as will occur in service, but applied in a manner designed to give rapid results without the introduction of new factors. Such a test assumes the services of someone possessing understanding of corrosion principles to design the apparatus and interpret the results; comparison of the new materials with standard materials, the service behaviour of which is already known, is advisable. The specimen to be tested must be in the same state of heat-treatment as will be used in service, and, if welded, must be welded by the process to be adopted in service (G. N. Flint, *Chem. and Ind.* (*Lond.*) 1956, p. 762).

The results of such a comparative test, whether qualitative or quantitative, will have no absolute significance, and probably such a specially designed test would have no validity in litigation; its object is merely to enable a Corrosion Specialist to judge whether the adoption of some material may be expected to avoid corrosion trouble. The conditions (temperature, stirring, etc.) must be the same for all materials to be compared, and should be close to service conditions, which require careful study before the test is designed. If intermittent wetting occurs in service, this should be provided in the test, either by attaching the specimens to a frame which is alternately raised and lowered, or by fixing them in a vessel in which the liquid level is raised and lowered—a procedure which may cause less disturbance. If the metal is to be partly immersed in practice, this can easily be arranged in the test. If it is to be totally immersed in practice, it should be totally immersed in the test, but before total immersion is assumed, a careful inspection should be carried out to ascertain whether in service air-pockets may not exist in metal vessels which are intended (and probably believed) to be kept completely full.

An example of a simple test devised to comprise the most dangerous situations likely to occur in service is provided by Raine's method of testing aluminium cable sheathing materials in different waters. The specimen is a bent strip, partly immersed in water, and freshly aerated water is allowed to drip on and trickle over the part exposed to air. Different regions of the material are subjected to (1) conditions of complete immersion, (2) crevices, (3) water-line conditions, (4) drip-contact and (5) dried-up residues (P. A. Raine, *Chem. and Ind.* (*Lond.*) 1956, p. 1103, esp. fig. 1, p. 1104).

Where the metal is to be in contact with running liquid, the conditions of flow must be suitably represented in the test; a careful study must be made of the type of flow (turbulent or stream-line) which occurs in service, and if both types are present at different points, such a combination (which may produce differential aeration or other effects) should probably be introduced into the test. For liquids flowing through pipes, the value of the Reynolds number may be more important than the absolute velocity; Japanese work has shown that the law governing the action of ammonia, containing an ammonium salt and oxygen, upon copper, changes suddenly when the Reynolds number exceeds a certain value (S. Uchida and I. Nakayama, *J. Soc. Chem. Ind.* (*Japan*) 1933, **36,** 416B, 635B).

Impingement and Jet Tests. When in service it is likely that entrained air-bubbles will impinge on the walls of a tube, as in condensers (p. 476), a jet test with water containing such bubbles will give fairly good prediction regarding the relative resistance of different materials; it is important to arrange that the bubbles break into smaller ones at the point of impingement. Such a test was worked out by G. D. Bengough and R. May, *J. Inst. Met.* 1924, **32,** 81; esp. fig. 10, p. 136. See also I. G. Slater, L. Kenworthy and B. May, *ibid.*, 1950, **77,** 309, esp. p. 326; R. May and R. W. de Vere Stacpoole, *ibid.*, 1950, **77,** 331.

For other purposes, instead of a succession of bubbles, a broad continuous jet of air is forced under constant pressure from a nozzle and caused to impinge obliquely on the specimen immersed in water. This produces pits rapidly and serves to discriminate good from bad material; such an apparatus has been described by H. W. Brownsdon and L. C. Bannister, *J. Inst. Met.* 1932, **49,** 123.

There has been some discrepancy between the results obtained from jet tests carried out by American workers at Kure Beach and those obtained in British laboratories; as a result of co-operative work, however, the causes have largely been cleared up, and the results brought into harmony (P. T. Gilbert and F. L. LaQue, *J. electrochem. Soc.* 1954, **101,** 448).

Condensation Tests. Under many conditions of atmospheric corrosion, the damage is essentially due to the condensation of acid moisture. Condensation tests are then appropriate—such as have been developed under Vernon at Teddington. The Beaker Test—described by Preston and assessed by a Sub-Committee of the B.I.S.R.A. Corrosion Committee—fulfils requirements in a delightfully simple manner. The specimens are placed in the upper part of a large beaker containing a specified amount of sulphur dioxide solution at the bottom, further quantities being added at prescribed intervals; the beaker is placed above an electric heater, but a lead cooling coil is wrapped round the upper part, so that moisture keeps condensing on the specimens. This represents the sort of conditions present in outdoor situations sheltered from rain (B.S. 1391 (1952)).

Testing of Protective Coatings

General. The condensation test just mentioned—and also a special form of salt droplet developed at the Armaments Research Establishment which has been assessed by the same Sub-Committee—is largely designed for the testing of coatings. Where the material is sheet which may undergo rough usage in service, it is usual to apply a standard degree of damage to the coating in an indenting apparatus which produces a " dome " of standard form, and also to trace scratch-lines through the coating. Attention is given to the spread of corrosion from the places where the coating has been damaged. The results obtained with this test demonstrate the advantage of phosphating steel before painting (R. St. J. Preston, *J. Iron St. Inst.* 1948, **160**, 286; Methods of Testing (Corrosion) Sub-Committee, *ibid.*, 1948, **158**, 463; 1952, **171**, 255; see also " Chemistry Research " 1957, p. 18 (H.M. Stationery Office)). Another condensation test is described by T. E. Lloyd, *J. Metals.* 1950, **2**, 1092. The form used by S. C. Britton and D. G. Michael, *Trans. Inst. Metal Finishing* 1953, **29**, 40, has received commendation. See also A. Kutzelnigg, *Werkst. u. Korrosion* 1958, **9**, 429. A form suitable for testing chromium-plated articles is described by J. Edwards, *Trans. Inst. Metal Finishing* 1958, **35**, 55.

Tests-cycles. For testing painted specimens under conditions of intermittent wetting, an apparatus called the *Weatherometer* is often used. The specimens are placed against the inner side of the walls of a cylindrical enclosure; at the centre of the enclosure is fixed a rotating stand carrying (1) a spray which plays on each specimen in turn, (2) a heating lamp which then dries it, often (3) an ultra-violet lamp and sometimes (4) a douche. Thus each specimen undergoes a cycle of changes, such as would be experienced in outdoor exposure. The object of the ultra-violet lamp is, of course, to bring about " chalking " and other changes which the shorter waves present in sunlight may produce in the vehicles of many paints.

Whilst for heavy specimens it is easiest to let them remain stationary and rotate the spray and lamps, it is possible, for lighter specimens, to make them pass in turn before stationary sprays and lamps; in some of the earlier researches designed to compare different types of steel in the unpainted condition, small specimens were attached to an endless moving belt passing over a series of pulleys. The cycle of operations could be arranged to simulate any desired conditions of atmospheric exposure. Interesting results were obtained by the use of a test solution containing ingredients which had been proved, in Vernon's work, to play an important part in atmospheric exposure; the liquid recommended contained sulphuric acid (N/500), sulphur dioxide (0·02% by volume), ammonium sulphate (N/50,000) and was nearly saturated with carbon dioxide; it gave good agreement with field-tests and brought out the improvement in corrosion-resistance produced by copper in steel. Earlier experiments with intermittent spraying, applied by hand or mechanically, were described by U. R. Evans and S. C. Britton, *Rep. Corr. Comm. Iron St. Inst.* 1931, **1**, 139; J. C. Hudson, *ibid.*, 1931, **1**, 211; W. H. Hatfield and H. T. Shirley, *ibid.*, 1931, **1**, 156; W. A. W. Schroeder, *ibid.*, 1934, **2**, 185; W. H. Hatfield, H. T. Shirley, T. Swinden, W. W. Stevenson, J. C. Hudson and T. A. Banfield, *ibid.*, 1936, **4**, 159; T. Swinden and W. W. Stevenson, *J. Iron St. Inst.* 1940, **142**, 165P; W. H. J. Vernon, *ibid.*, 1940, **142**, 183; also *J. Soc. Chem. Ind.* 1947, **66**,

140; W. J. Hair, *Corros. Tech.* 1956, **3,** 8; J. C. Hudson, " Corrosion of Iron and Steel " (1940), p. 235 (Chapman & Hall).

The vexed question as to the best type of lamp for tests of the weatherometer type is discussed between T. R. Bullett and R. J. Bran, *Metal finish. J.* 1956, **2,** 198.

A rotor apparatus for testing paints is described by Vernon. A circular ebonite holder rotated, normally at 1500 r.p.m., by an electric motor, carries six specimens mounted flush with the ebonite. Tests in sea-water discriminate rapidly between good and bad ships' paints. This apparatus proved useful during the Second World War for eliminating coatings with bad adhesive power (W. H. J. Vernon, *J. Soc. chem. Ind.* 1947, **66,** 137).

An interesting discussion of accelerated testing of paint-systems took place at the " Fatipec " Congress of 1953 (*Verfkroniek* 1953, **26,** 275).

The detachment of paint in service often depends on developments unconnected with any lack of physical adhesion between coating and metals; these may be the formation of rust below the coat which locally pushes it away from the metal or the formation of alkali by the cathodic reaction which creeps between metal and coating, loosening the latter (p. 552). However, the measurement of physical adhesion, as such, is desirable, since, if it is low, detachment may occur in service independently of any corrosion change. The various quick adhesion tests available are compared by W. V. Moore, *Metal finish. J.* 1957, **3,** 241, whilst an ingenious bullet method has been described by W. D. May, W. D. P. Smith and C. I. Snow, *Trans. Inst. Metal Finishing* 1956–57, **34,** 369.

A bending test for electrodeposited coatings is described by Edwards. A plated strip is bent round a curved former and the curvature at which cracks start is observed (J. Edwards, *Trans. Inst. Metal Finishing* 1958, **35,** 101).

Other tests for protective coverings developed at Teddington by Wormwell and Brasher depend on following the manner in which potential, electric resistance and the capacitance change with time, during the immersion of a painted specimen in sea-water or other liquid. The curves connecting resistance with time (or capacitance with time) show a sudden increase of steepness which serves to indicate the onset of rusting—the method being free from the personal factor associated with the ordinary visual observations of the condition of painted specimens (F. Wormwell and D. M. Brasher, *J. Iron St. Inst.* 1949, **162,** 129; 1950, **164,** 141; discussion *ibid.*, 1950, **165,** 299; D. M. Brasher and A. H. Kingsbury, *J. appl. Chem.* 1954, **4,** 62; D. M. Brasher and T. J. Nurse, *ibid.* 1959, **9,** 96. Cf. R. C. Bacon, J. J. Smith and F. M. Rugg, *Industr. engng. Chem.* 1948, **40,** 161; J. N. Wanklyn and D. R. Silvester, *J. electrochem. Soc.* 1958, **105,** 647).

Thickness of Coatings. Apart from corrosion tests on coated specimens, direct measurements of thickness are advisable, since thickness often governs service life, for reasons discussed on p. 630. The meter described by Plomp for paint-coats is purely mechanical, consisting of a needle which by means of appropriate screws is first brought into light contact with the outer surface of the paint, and then, penetrating the paint reaches the metal below. The positions are signalled by the lighting or extinction of a lamp on an electric circuit; the thickness is read off on a scale. Many thickness meters are magnetic and are applicable only for coatings on steel or magnetic metals. A comparison of two of them has been provided by F. Fancutt and J. C. Hudson, *Paint Technology* 1948, **13,** No. 146. See also *Verfkroniek* 1953, **26,** 221; H. W. van der Hoeven, *ibid.*, 1954, **27,** 88; H. Plomp, *ibid.*, 1953, **26,** 324. A magnetic balance is described by E. S. Spencer-Timms, *J. Electrodep. tech. Soc.* 1944–45, **20,** 139. Cf. C. E. Richards, *J. Soc. chem. Ind.* 1937, **56,** 343T. A magnetic method for sprayed coats is described by R. E. Mansford, *Chem. and Ind.* (*Lond.*) 1956, p. 824.

The thickness of metallic coatings can be measured by non-destructive methods depending on supersonic reflection, thermal conductivity, electrical conductivity, electric induction effects, X-rays and γ-rays, as well as the magnetic methods already mentioned. Some interesting comparisons are made by J. G. Kerley, *Corrosion* 1947, **3,** 467. Another method depends on back-reflection of β-radiation (P. Skeggs, *J. Electrodep. tech. Soc.* 1952, **28,** 15).

In a promising thermo-electric method of estimating thickness, two probes (one kept hot by means of a coil, the other being cold) are pressed down on the coated surface at a definite distance; the current, which depends on the thickness of the coat, is amplified and measured. The apparatus has been designed in the B.N.F.M.R.A. laboratory, and patent protection has been sought. The device is described by A. R. Heath, *Metal finish. J.*, 1955, **1,** 145.

Destructive methods may depend on the loss of weight produced when the coating is dissolved by a liquid chosen to dissolve the coating metal and leave the basis metal unattacked. In the rougher form of the method suitable for routine purposes drops are placed on the coated surface or a jet of corrosive liquid directed on to it; in either case, if the rate of attack of the liquid on the coating metal, under the strictly standardized conditions, is known, the thickness is indicated by the time needed for the basis metal to be laid bare. The B.N.F. Jet test, which depends on this principle, is well known and has wide application. Details are provided by S. G. Clarke, *J. Electrodep. tech. Soc.* 1936–37, **12,** 1, 157; K. W. Caulfield and W. E. Hoare, *Sheet Metal Ind.* 1949, **26,** 753; J. Edwards, *Trans. Inst. Metal Finishing* 1954, **30,** 17. See also brochure " The B.N.F. Jet Test " (British Drug Houses).

When, as in hot-dipped coatings, there is an alloy layer (or layers) below the outer coating, it is best to use anodic attack with a constant current, the potential being measured; changes in the potential value indicate that one or other layer has been completely removed. The thickness of any particular layer can be roughly calculated from Faraday's Law, and that for the complete coating obtained accurately by loss of weight at the end of the operation. The method has been used for measuring the thickness of zinc and alloy layers on hot-galvanized wires by Britton, and more generally by Francis and by Waite; a rather similar principle has been used by Thwaites and Hoare in following the alloying which accompanies flow-brightening on tinplate (p. 647). In Britton's work on galvanized wire, the procedure recommended for ascertaining whether the wire carries the specified thickness of coat is to subject the specimen to a known current for a time calculated to dissolve such a thickness; it is then removed, wiped with cotton wool, and immersed in 10% copper sulphate for 5 seconds. If no bright red copper is deposited, it is concluded that no steel has been laid bare and that the coating conforms to specification (S. C. Britton, *J. Inst. Met.* 1936, **58,** 211; H. T. Francis, *J. electrochem. Soc.* 1948, **93,** 79; C. F. Waite, *Plating,* 1953, **40,** 1245; C. J. Thwaites and W. E. Hoare, *J. appl. Chem.* 1954, **4,** 236; C. J. Thwaites, *J. Iron St. Inst.* 1956, **183,** 244, esp. p. 252; cf. C. T. Kunze and A. R. Willey, *J. electrochem. Soc.* 1952, **99,** 354; W. Katz, *Stahl u. Eisen* 1955, **75,** 1101).

At one time, the " Preece test " was much used for ascertaining whether the coating on galvanized wire was sufficiently thick. This depends on the number of one-minute dips in copper sulphate solution which can be withstood before a firmly adherent copper deposit is formed on the exposed steel base. Unfortunately copper sulphate attacks unalloyed zinc faster than zinc-iron alloy, so that a comparison of electro-deposited, hot-galvanized and galvannealed coatings of equivalent thickness* would, if subjected to the Preece test, show results unfair to the first-named process. The Preece test is perhaps more useful for providing information about uniformity than average thickness. Schikorr suggests that the loss of weight at the moment when adherent red copper first appears would be a better criterion than the number of dips. Criticisms of the test are provided by E. C. Groesbeck and H. H. Walkup, *Bur. Standards J. Res.* 1934, **12,** 785; O. F. Hudson, *J. Inst. Met.* 1936, **58,** 223; H. Bablik, " Galvanizing " (translation by C. A. Bentley) 1950, p. 494 (Spon); H. Bablik, F. Götzl and E. Nell, *Metalloberfläche* 1953, **7,** A66; G. A. Ellinger, W. J. Pauli and T. H. Orem, *Proc. Amer. Soc. Test. Mater.* 1953, **53,** 125; G. Schikorr, *Draht* 1954, **5,** 217.

Other papers on thickness testing include H. H. Egginton, *J. Electrodep. tech. Soc.* 1947–48, **23,** 191 (general); G. Howells, *Corros. Tech.* 1954, **1,** 233 (tin); H. F. Beeghly, *J. electrochem. Soc.* 1950, **97,** 152 (tin); W. E. Hoare and J. B. Gustin, *Sheet Metal Ind.* 1955, **30,** 1042 (tin); J. Hérenguel, *Rev. Métallurg.* 1954, **51,** 36 (zinc); S. G. Clarke and J. F. Andrew, *J. electrodep. tech. Soc.* 1946–47, **22,** 1 (cadmium).

Measurement of Other Properties of Coatings. Porosity testing of nickel coatings has been discussed on p. 626. Methods used for tin or tin alloy coatings are described by S. C. Britton and D. G. Michael, *Trans. Inst. Metal Finishing* 1952–53, **29,** 40, and by J. Pearson and W. Bullough, *J. Iron St. Inst.* 1948, **160,** 376; also for copper, nickel and chromium layers on zinc-base die-castings by H. K. Lutwak, *Trans. Inst. Metal Finishing* 1952–53, **29,** 349.

Adhesion testing was mentioned on p. 567. References include S. C. Britton and R. M. Angles, *J. Iron St. Inst.* 1951, **168,** 358; K. Wolf, *Verfkroniek* 1954, **27,** 57; A. Brenner and V. D. Morgan, *Proc. Amer. Electro-*

* In the *galvannealing* process, the wire on leaving the galvanizing tank passes through a muffle furnace where it is heated to 670°C. for about 15 seconds —which greatly increases the thickness of the alloy layer.

platers Soc. 1950, **37,** 51; J. M. Cowan, *Electroplating* 1954, **7,** 79; C. Williams and R. A. F. Hammond, *J. Electrodep, tech. Soc.* 1954, **31,** 124; H. C. Schlaupitz and W. D. Robertson, *Plating* 1952, **39,** 750, 764, 862, 932; W. D. May, N. D. P. Smith and C. I. Snow, *Trans. Inst. Metal Finishing* 1957, **34,** 369. The adhesion of enamels is assessed by a method described by W. H. F. Tickle, M. K. Knowles and H. Crystal, *Metal finish. J.* 1955, **1,** 377.

Hardness testing for paint films has been discussed by H. A. Gardner, *Verfkroniek* 1947, **20,** 170; J. Hoekstra and J. A. W. van Laar, *ibid.*, 1952, **25,** 13; see also Committee Report, *ibid.*, 1952, **25,** 131.

Tests based on loss of brightness. For surfaces which are normally bright, loss of reflectivity is regarded far more seriously than loss of weight; this may consist of loss of total light reflected, or a scattering of light, reducing the specular reflection. Loss of reflectivity was used as a criterion of corrosion-damage in early work by W. H. J. Vernon, *Trans. Faraday Soc.* 1923–24, **19,** 839, esp. p. 851; and by L. Kenworthy and J. M. Waldram, *J. Inst. Met.* 1934, **55,** 247. A reflection meter was designed and applied to the study of tarnishing by W. P. Digby, *The Engineer* 1935, **159,** 219, 254, whilst the effect of corrosion of the diffusion of light has received detailed study from F. Canac, *C.R.* 1933, **196,** 51; 1934, **199,** 1117; 1935, **201,** 330.

Tests for Inhibitors in Service. For many purposes, especially in the oil industry, an inhibitor may be added to the liquid in a well, a tank, a pipe-line or a process plant. It is naturally desired to ascertain whether it is being effective in stopping corrosion. The procedure of hanging weighed specimens at the chosen position for a period, then removing and re-weighing them has long been used, but there are many objections to it. The type of method coming into favour (especially in the U.S.A.) involves a wire or strip specimen which can be fixed in the pipe, vessel or tower through which the liquid is running, the decline of conductivity, caused by the decrease of cross-section due to corrosion, being recorded continuously or at intervals; in one form of the method, the specimen is made one arm of a Wheatstone bridge. The advantage of such a principle is that the specimen can be left in position for several consecutive periods during which different inhibitors (or no inhibitor) are used, so that their efficacy can be compared. A corrosion meter based on changes in electrical resistance has been designed by Marsh and Schaschl, and should have wide application not only in industry, but also in scientific studies (A. J. Freedman, A. Dravnieks, W. B. Hirschmann and R. S. Cheney, *Corrosion* 1957, **13,** 89t; W. L. Denman, *ibid.*, 1957, **13,** 43t; W. L. Terrell and W. I. Lewis, *ibid.*, 1956, **12,** 491t; G. A. Marsh and E. Schaschl, *ibid.*, 1956, **12,** 534t; 1958, **14,** 155t).

An experimental tower to evaluate inhibitors for cooling water systems is described by W. L. Denman and C. B. Friedman, *Corrosion* 1957, **13,** 179t.

Failure of Laboratory Conditions to reproduce Service Results

General. The practical engineer usually distrusts laboratory results. In general, his distrust is well founded. Some of the causes of discrepancy are connected with the facts that the laboratory set-up is not a faithful representation of the service plant or structure on a reduced scale, and the materials, liquids and gases present in the laboratory model differ appreciably from those of full-scale working. This cause is easily understood, and could, no doubt, in many cases be eliminated. There are two other causes of a more fundamental character which receive little consideration from the laboratory worker, and which are less easy to deal with. The various causes of discrepancy must be considered in turn.

Discrepancies between solid materials used in a test model. It is easy to stipulate in writing that the materials used in a test must represent those to be used in service both as regards composition and structure, but difficult to carry out the stipulation in practice. Suppose that the practical structure has welded joints, it would seem right to use the same welding process for the small-scale model; in that case, however, the proportion of the surface affected by the temperature-changes and stresses which accompany welding will be far higher in the model than in the working structure. It is not easy to see how such discrepancies can be avoided.

Discrepancies between corrosion fluids met with in service and in the laboratory. Nearly all fluids met with in practice have numerous constituents. How many of these should be represented in the solution made up for laboratory tests? Natural sea-water contains a very large number of constituents, inorganic and organic. At one time it was customary, in making up synthetic sea-water for laboratory testing, to introduce a large number of salts and also one organic constituent of natural sea-water, namely saponin. Today it is usual to omit saponin from the formula. It is possible that its effect on corrosion is not sufficient to justify inclusion, but it is difficult to deny that cystine affects corrosion by polluted sea-water; there is a case for including cystine in artificial sea-water, unless the test is concerned solely with corrosion proceeding in some unpolluted part of the sea. Pollution is, however, catered for in another way; hydrogen sulphide is sometimes added to the water used for testing condenser alloys.

Another point requiring consideration is the function of calcium bicarbonate in corrosion by sea-water. The fact that sea-water is generally less corrosive than sodium chloride solution is generally ascribed to the formation of a "chalky layer" by the cathodic reaction; if so, calcium bicarbonate is the constituent which most deserves to be introduced in controlled amount into any synthetic solution designed to represent sea-water in a test.

The solution approved by the British Standard Institution for the purpose contains only four salts, $NaCl$, Na_2SO_4, $MgCl_2$ and $CaCl_2$; doubtless the distilled water used will always contain carbon dioxide, so that the ions of $Ca(HCO_3)_2$ will be present, but not in closely controlled amounts; there is no stipulation that the water used shall be "equilibrium water" (i.e. water in equilibrium with the carbon dioxide of the atmosphere).

This four-salt formula has been derived from the first report of the B.I.S.R.A. "Methods of Testing (Corrosion) Sub-Committee" which also presents a seven-salt formula, including $NaHCO_3$ (0·1 g./l.) and $CaSO_4$ (1·3 g./l.) as well as $NaCl$, $MgCl_2$, $MgSO_4$, KCl and KBr. Here the ions of $Ca(HCO_3)_2$ are present in regulated amounts—as seems desirable.

It is, however, possible that the constituent of the " chalky layer " which slows down the corrosion-rate by hindering the cathodic reaction is really $Mg(OH)_2$—not $CaCO_3$; if this could be established, it would weaken the argument in favour of a rigid standardization of the $Ca(HCO_3)_2$ content of the test liquid. In considering this point, reference may be made to two analyses published by Hudson of the white deposits found on specimens; simple immersion in sea-water (at Gosport) gave a deposit with 57% $CaCO_3$, 3% $CaSO_4$ and 3% " magnesium carbonate ", whereas immersion under cathodic polarization (at Devonport) gave a deposit with 47% $CaCO_3$, 13% " magnesium carbonate " and 33% " magnesium oxide and hydroxide " (J. C. Hudson " The Corrosion of Iron and Steel " 1940 edition, pp. 158, 175 (Chapman & Hall)).

The two formulae referred to above will be found in *J. Iron St. Inst.* 1948, **158**, 463, esp. pp. 477, 482; see also B.S. Specification (provisional) **1391** (1947). Other references to formulae for synthetic sea-water will be found on p. 164 of this book. The German formula (to be retained in the new edition of DIN 50900, expected in 1959) will prescribe 28 g. NaCl, 7 g. $MgSO_4 \cdot 7H_2O$, 5 g. $MgCl_2.6H_2O$, 2·4 g. $CaCl_2.6H_2O$ and 0·20 g. $NaHCO_3$, and thus standardizes the effective calcium bicarbonate content (G. Schikorr, Priv. Comm., July 2, 1958).

Discrepancies due to Dimensional Considerations. Consider a corroding structure, with current passing between the cathodic and anodic areas; the E.M.F. provided by the corrosion-reaction will be used up partly in polarization at the two areas and partly in the IR drop over the path joining the areas. Let us reduce all lengths to N^{-1} times their original value, so as to produce a faithful small-scale model. The areas shrink in the proportion N^{-2}, and if the polarization (which depends on current per unit *area*) is to remain the same, the current must change from I to IN^{-2}. However, the resistance is increased in the proportion N, since the cross-section of the paths joining anode to cathode is multiplied by N^{-2}, whilst their length of the paths is multiplied by N^{-1}. Thus to keep the IR drop the same, the current must change from I to IN^{-1}. If the same liquid is retained for the small-scale model, the polarization and IR drop cannot *both* remain unaltered. Hoar and Agar, however, point out that, if the conductivity of the liquid used for the small-scale test is N^{-1} that of the liquid used in the full-scale plant, the polarization and IR drop can be preserved unchanged; only in that case will the model forming a perfect small-scale reproduction of potential distribution in the plant. This change of conductivity can be obtained by altering the concentration *approximately* N^{-1} times (not exactly N^{-1} times, since the activity coefficient varies with concentration). Provided that the other essential factors are not affected by the alteration of concentration, this ingenious suggestion completely solves the difficulty. Unfortunately an alteration of salt-concentration generally alters the oxygen-solubility, and thus affects the polarization if the cathodic reaction is oxygen-reduction; if the alteration is serious, it is difficult to see how the small-scale model can faithfully reproduce happenings in the large-scale plant. Nevertheless, the plan should greatly reduce the discrepancy and the original paper deserves study (J. N. Agar and T. P. Hoar, *Disc. Faraday Soc.* 1947, **1,** 158).

Discrepancies due to Statistical Considerations. If the liquid contains corrosive and inhibitive constituents nearly balancing one another, so there is a finite chance of break-down in a given area, it is possible that the probability of break-down in a small test-specimen may be very small and that on the full-scale plant very large. In such a case, the laboratory experiment cannot be expected to give a correct prediction of large-scale behaviour. This matter is further considered in Chapter XXII. See also U. R. Evans, *Corrosion* 1957, **13,** 833t.

Other References

A useful book on testing methods is F. A. Champion's " Corrosion and Testing Procedure " (Chapman & Hall). See also F. L. LaQue, " Design and Interpretation of Corrosion Tests," his Edward Marbury Lecture, 1951 and his paper for the Corrosion Symposium, Melbourne University 1955–56, p. 169. The indexes of the reports of the American Society of Testing Materials provide information on specific problems; see also the reports of the B.I.S.R.A. " Methods of Testing (Corrosion) Sub-Committee ", *J. Iron St. Inst.* 1948, **158,** 463; 1952, **171,** 255.

A useful list of specifications of atmospheric testing methods is provided by A. Kutzelnigg, *Werkst. u. Korros.* 1956, **7,** 65.

A machine for producing reproducible scratches on metals with diamond indenter, ball or gramophone needle pressed down under controlled load and moving at determined rate has been designed at Swansea; details, when available, will command attention (D. Eurof Davies and D. D. Jones, unpublished work).

Those interested in the electrodic processes involved in the electrometric estimation of film-thickness may find help in a paper by B. Kabanov, R. Burstein and A. Frumkin, *Disc. Faraday Soc.* 1947, **1,** 259.

CHAPTER XX

VELOCITY OF FILM-GROWTH

Synopsis

In the qualitative discussion of oxidation and film-growth (Chapters II and III), the main types of growth-laws were briefly mentioned, namely the *rectilinear* law, the *parabolic* law (which provides moderate slackening of growth-rate with thickness) and certain others, grouped together as the *log laws* (which can bring growth practically to a stand-still in the later stages); although the different log laws are algebraically distinct, they produce curves which in certain circumstances may be difficult to distinguish.

The present chapter will suggest theoretical foundations for these various laws. If the film grows by uniform movement of material through the film-substance, a general equation is obtained of which one limiting case is the simple parabolic equation; this represents the situation at temperatures high enough to allow thermal movement of the ions, so that the effect of a potential gradient in the film is directive, not causative; the other limiting case is the inverse logarithmic equation, representing the situation at lower temperatures where movement of the ions would be impossible without help from an electric field. The simple parabolic equation must be replaced by the mixed parabolic equation if the boundary reactions cannot easily keep up with the maximum velocity which the rate of movement through the film would permit; in extreme cases this becomes the rectilinear equation. For very thin films where the transfer of electrons, not the movement of ions, is the controlling factor, we get the direct logarithmic equation. At slightly greater thicknesses, and at a temperature too low to permit movement through the oxide lattice but capable of permitting movement through pores, we should expect the asymptotic equation when the pores are self-stifling, and the direct logarithmic equation when they are mutually stifling. If vacancies left by cations moving outwards collect to form cavities at the base of the film, we may expect a new logarithmic equation, of which the direct logarithmic equation is a special case. Finally, if the oxidation starts at discrete nuclei and spreads outwards over the surface, we may expect sigmoid forms.

Of these, the inverse logarithmic, asymptotic and direct logarithmic equations are low-temperature forms; it will be shown that, under certain circumstances, one can be regarded as an approximation to another—explaining the difficulty sometimes experienced in deciding between them experimentally, and justifying the course adopted in Chapter II of grouping them together as the " log laws ".

Nearly all the equations predicted by theory have been realized experimentally, but, since not all the possibilities have been explored theoretically, it is quite to be expected that experimental results will occasionally be obtained which do not fit any one equation. For instance, the three mechanisms leading to the three low-temperature equations might proceed simultaneously—in which case intermediate forms would be expected.

It will be shown how, in the case of the parabolic law, it is possible to predict the value of the growth constant from a knowledge of the electrical properties of the substances involved; the correctness of the predictions provides substantial support for the mechanism suggested.

At this point it becomes necessary to discuss the variation of the rate-constants with temperaturc—a matter possessing obvious practical importance, besides theoretical interest, since it provides a measure of the activation energy needed for the oxidation changes. Then the acceleration produced by such additions as moisture and sulphur compounds will be discussed. Finally, the causes and effects of film break-down and leakage receive attention.

General

The two types of Semi-conductors. In Chapter II, it was explained that film-thickening can occur by the movement of metallic cations *outwards* through the film, or of anions (particularly oxygen ions) *inwards*. The former will occur when there are vacant cation sites, as in cuprous oxide, where the metal content is slightly less than corresponds to the formula Cu_2O, and where there are believed to be a certain number of cupric ions (Cu^{++}) present—which will maintain electrical neutrality despite the vacant cation sites; such oxides are known as *p-semi-conductors*. There are other oxides containing less oxygen than the formula would demand; they are known as *n-semi-conductors*. In some cases these contain vacant anion sites and thickening then occurs by oxygen moving inwards. Some *n*-semi-conductors, however, are believed to owe their departure from the stoichiometric composition to the presence of excess cations occupying interstitial positions and not to vacant anion sites. Zinc oxide, which contains less oxygen than the formula ZnO suggests, is generally regarded in that way, and it is considered that zinc oxide films thicken by zinc ions moving outwards by way of the interstitial positions.

Thus for practical purposes, there are three classes of oxide-films:*

(1) *p*-conductors which thicken through cations moving outwards, taking advantage of the vacant cation sites:

Example, cuprous oxide.

* Logically, a fourth class might be expected, with anions moving inwards, taking advantage of interstitial anion sites. No example can today be suggested, and, in view of the size of the anions, it is doubtful whether one will be discovered.

(2) those n-conductors which thicken through cations moving outwards, taking advantage of interstitial sites:

Example, zinc oxide.

(3) those n-conductors which thicken through anions moving inwards, taking advantage of vacant anion sites:

Example, α-ferric oxide.

It is a curious fact that some of the mathematical treatments apply equally to all three classes.* Thus in cases where the growth of the film is controlled simply by movement of material across it, there is no distinction between the cases where cations move outwards and anions inwards—so long as the film is continuing to thicken uniformly and remains unbroken. As explained later, however, direction of movement does influence the building up of conditions which may ultimately bring about sudden breakdown of the film; until this occurs the direction of movement does not affect the growth-law.

In any film containing vacant sites, the passage of material across it depends on a vacancy becoming filled by a neighbouring ion moving into it, thus leaving a new vacancy, which is filled in its turn. In effect, the vacancies themselves travel across the film in the opposite direction to the general movement of the ions; for some purposes, it may be convenient to assign mobility values to the vacancies.

Random thermal movement and directed movement under a potential gradient. It is clear that at high temperatures thermal movement of vacancies may be occurring in a solid block of oxide. Under uniform conditions, such movement occurs in all directions in a haphazard manner. In a film of oxide clinging to metal a gradient will usually exist, causing motion in one direction to predominate, and it is this fact that allows the oxidation to continue, although at a rate diminishing with increasing film-thickness. Two kinds of gradient are possible. If the oxygen atoms adsorbed on the outside of the film capture electrons from the metal, owing to the electron affinity mentioned in Chapter II, an electrical potential gradient may be set up by the negative charge on the oxygen ions and the positive charge left on the metal; this will be such as to cause movement to predominate in a sense favourable to continued oxidation (i.e. to make cations move outward and cation vacancies inwards, or to make oxygen anions move inwards and anion vacancies outwards). Alternatively, if at the outer surface of a growing oxide-film equilibrium is maintained with the oxygen of the gas phase and if equilibrium is also maintained with metal at the inner surface, the composition will be kept different at the two surfaces, so that then a concentration gradient (or more strictly, a chemical potential gradient) exists, and movement will proceed in such a direction as to reduce this gradient. Each type of movement is due ultimately to the same cause, and, in absence of control by boundary reactions, the driving force should be the same in both cases—a matter

* However, certain differences in the behaviour of the n- and p-oxides are emphasized by T. B. Grimley and B. M. W. Trapnell, *Proc. roy. Soc.* (*A*) 1956, **234**, 405.

considered later. Under some circumstances both types of gradient can operate simultaneously.

Cation or anion mobility. As already stated, the mathematics of film-thickening are often identical whether the particles moving through the film are cations or anions. However, the secondary effects may be different. Consider the case when cations move outwards (and their vacancies inwards). Oxygen atoms become adsorbed on the surface of the oxide-film, and by attracting electrons from the metal, become converted into oxygen ions (the passage of electrons across the film in cases of variable valency probably occurs by exchange of electrons along ionic chains; thus a cupric ion obtaining an electron from its neighbour, converts that neighbour into a cupric ion which can now accept an electron from the ion behind—and so on). The oxygen ions can now attract metallic cations from the outer layer of the oxide into new places where they constitute, with the oxygen ions, an additional oxide layer. This leaves vacancies at places previously occupied by these cations and such vacancies migrate inwards under the electrical or chemical potential gradient, and may *either* (*a*) accumulate at the metal–oxide interface, *or* (*b*) enter the metal and become annihilated at the end of dislocations (which are either prolonged or contracted thereby). If they accumulate at the interface, they may produce cavities—shown by the work of Pfeil (p. 34) and that of Dunnington, Beck and Fontana (p. 49) on iron. Such cavities must reduce the area of contact between metal and oxide, and thus complicate the growth-law, ultimately causing oxidation to become abnormally slow (see p. 836). Alternatively, the film over the cavities may suddenly become leaky, and rapid oxidation may suddenly start again; these sudden breaks in the curves have been noticed on p. 50. Another possibility is that the cavities are closed up by shrinkage of the oxide, as shown by W. J. Moore, *J. chem. Phys.* 1953, **21**, 1117.

When anions move inwards, the formation of cavities is unlikely; in contrast, compressional stresses may often be expected in such cases. Duplex films on iron, consisting of magnetite and ferric oxide layers, when stripped from the metal, curl up into tight rolls (p. 43) indicating compressive stress in the ferric oxide, but not in the magnetite. This confirms the belief, arrived at independently on different evidence, that magnetite grows by cation movement and ferric oxide by anion movement. Compressive stresses in a thin film are likely to render it more protective, keeping the atoms abnormally close together and pressing together the walls of any fissures. When, however, the film reaches a certain thickness, break-down may occur, which will relieve the stress but cause a sudden spurt in the oxidation-rate.

Classification of Cases. In the sections which follow, it will be shown how, starting from the possible alternative assumptions, such as (1) movement of ions through the film-substance as a whole, (2) movement through pores, and (3) movement restricted to certain parts of the surface, simple mathematical equations can be derived, most of which have been established experimentally. To simplify the expressions, constants (K_1, K_2, K_3 . . .

k_1, k_2, k_3 . . .) are frequently introduced—often without explanation, since their meaning should be obvious; all of these represent constants which, at fixed temperature, are independent of time. Of other symbols used, A and a are constants introduced in special cases, whilst θ is a variable, being a function of y, the film-thickness.

1. Uniform Passage through the Film-substance

(*a*) **General Equation representing passage across a uniform film.** For the moment it will be assumed that the potential gradient is electrical in character; the effect of a chemical potential gradient will be considered later. Provided that temperature and effective area also remain constant, the rate of thickening will be proportional to the average rate of movement of a typical vacancy across the film; it will fall off as the thickness y increases, since the potential gradient is proportional to $1/y$. If W_a is the activation energy needed for an ion to move across an energy barrier into an adjacent vacancy (so that the vacancy moves into the position formerly occupied by the ion), the chance of an ion possessing the minimal requisite energy will be $e^{-W_a/kT}$ where k is Boltzmann's constant. In absence of a gradient, movement will occur equally in all directions. If a gradient is present, proportional to $1/y$, the chance of movement in one direction will increase to $e^{(-W_a+K_1/y)/kT}$ whilst the chance of movement in the opposite direction will diminish to $e^{(-W_a-K_1/y)/kT}$. Thus after a sufficient time the effective movement will be proportional to $A(e^{\theta} - e^{-\theta})$ where $A = e^{-W_a/kT}$ and $\theta = K_1/ykT$. The thickening rate is proportional to the effective movement

$$dy/dt = AK\left[1 + \theta + \frac{\theta^2}{2} + \frac{\theta^3}{6} + \frac{\theta^4}{24} + \ldots\right] - \left[1 - \theta + \frac{\theta^2}{2} - \frac{\theta^3}{6} + \frac{\theta^4}{24} - \ldots\right]$$

In two limiting cases, this general equation may be replaced by simpler forms known as the simple parabolic and inverse logarithmic equations respectively. The first corresponds to temperatures high enough to ensure frequent movement of vacancies even in the absence of a gradient, so that the function of the gradient is directive rather than causative; the second may be expected, under suitable circumstances, at temperatures low enough to make movement of the ions over the potential barriers almost impossible in the absence of a gradient.

(*b*) **Simple Parabolic Equation, representing control by passage across the film at temperatures where the ions are mobile.** At a fairly high temperature, y will soon become sufficiently large to justify neglect of terms involving θ^3 and higher powers. The 1st and 3rd terms of the expansion cancel, so that, neglecting the 4th and later terms, we obtain

$$\frac{dy}{dt} = 2AK\theta = \frac{2AKK_1}{kT}.\frac{1}{y} = \frac{K_2}{y}$$

or

$$\frac{1}{2}y^2 = K_2 t + K_3$$

This is the simple parabolic equation. Under the hypothetical conditions where the movement is supposed to be solely due to an electrical gradient, it merely represents the fact that the current carried by the ions obeys Ohm's Law.

The parabolic equation has been verified for practically all metals over an appropriate range of temperature, but it is generally replaced by one of the log laws at low temperatures and sometimes by a rectilinear law at high temperatures; it may fail—being usually replaced by the mixed parabola—at low oxygen pressures. For certain metals, notably aluminium, obedience to the parabolic law occurs over a somewhat restricted range of conditions.*

(*c*) **Inverse Logarithmic Equation, representing control by passage across the film at temperatures where ions are normally immobile.** The other limiting case of the general equation is most likely to be realized at low temperature, and particularly in metals of fixed valency, where vacant sites are relatively infrequent and immobile, so that the chance of their movement in absence of help from a potential gradient is negligible. Movement can be expected when the potential gradient is large, i.e. when the film is thin, but film-thickening will become extremely slow after a certain thickness has been reached—a point emphasized by Cabrera and Mott. In such a case one of the two exponential terms (namely, the term representing movement *against* the potential gradient) can be neglected, and the equation becomes

$$\frac{dy}{dt} = A'e^{\theta} \quad \text{where } \theta \text{ represents } K_1/ykT$$

or $$dt = \frac{1}{A'}e^{-\theta}\,d\theta . \frac{dy}{d\theta}$$

But $$\frac{d\theta}{dy} = \frac{d}{dy}\frac{K_1}{kT}\frac{1}{y} = -\frac{K_1}{kT}y^{-2}$$

so that $$\frac{dy}{d\theta} = -\frac{kT}{K_1}y^2 = -\frac{K_1}{kT}\theta^{-2}$$

giving $$dt = -\frac{K_1}{A'kT}\theta^{-2}e^{-\theta}\,d\theta = K_4\theta^{-2}e^{-\theta}\,d\theta$$

Thus $dt/d\theta$ is the product of two variables θ^{-2} and $e^{-\theta}$ but, of these two, the relative alteration produced in θ^{-2} by a small increment of θ will be small compared to the relative alteration of $e^{-\theta}$ (by a " relative " alteration is meant the absolute alteration expressed as a fraction of the original

* Smeltzer found the oxidation of high-purity aluminium (electropolished) in the range 400–600°C. to be somewhat complicated. After a formative stage governed by the parabolic law the oxidation becomes slow (W. W. Smeltzer, *J. electrochem. Soc.* 1956, **103,** 209). Other investigators report parabolic growth in the range 350–450°C. and rectilinear growth at 500–550°C. (E. A. Gulbransen and W. Wysong, *J. Phys. Chem.* 1947, **51,** 1087; cf. T. B. Grimley and B. M. W. Trapnell, *Proc. roy. Soc.* (*A*) 1956, **234,** 405).

value). For $\frac{(d/d\theta)\theta^{-2}}{\theta^{-2}}$ is $-2\theta^{-1}$ whereas $\frac{(d/d\theta)e^{-\theta}}{e^{-\theta}}$ is 1. Thus, provided that $\theta \gg 1$ which we are assuming to be the case, we can regard θ^{-2} as approximately constant and write $dt \sim -K_5 e^{-\theta}\, d\theta$ where K_5 is a new constant.

It is not always legitimate, in such cases, to integrate from $t = 0$, $y = 0$, even if the experiment were to start with a film-free metal; for most cases of film-growth show divergencies at the outset from the equation which ultimately comes to be obeyed. It is safer to start integrating at a time, t_0, sufficiently late for a steady state to have been established; let θ_0 and y_0 be the values of θ and y at $t = t_0$. Approximately, we then obtain

$$t - t_0 = K_5(e^{-\theta} - e^{-\theta_0})$$

or

$$K_5 e^{-\theta} = t - t_0 + K_5 e^{-\theta_0}$$

which is

$$-\theta + \log_e K_5 = \log_e (t - t_0 + K_5 e^{-\theta_0})$$

$$= \log_e \left(K_5 e^{-\theta_0}\left(\frac{t - t_0}{K_5 e^{-\theta_0}} + 1\right)\right)$$

$$= K_6 + \log_e (a(t - t_0) + 1)$$

where a represents $1/K_5 e^{-\theta_0}$ and K_6 represents $\log_e K_5 e^{-\theta_0}$.

Thus

$$\theta = K_7 - \log (a(t - t_0) + 1)$$

which becomes

$$\frac{1}{y} = K_8 - K_9 \log_e (a(t - t_0) + 1)$$

where K_8 must be $1/y_0$ since when $t = t_0$ the last term is zero. Thus we can write

$$\frac{1}{y} = \frac{1}{y_0} - K_9 \log_e(a(t - t_0) + 1)$$

This is *inverse logarithmic equation.* It has been found to be obeyed during the oxidation of aluminium by dry oxygen at ordinary temperatures (R. K. Hart, *Proc. roy. Soc.* (*A*) 1956, **236,** 68) and during the anodic oxidation of aluminium and zirconium (A. Charlesby, *Proc. phys. Soc.* (*B*) 1953, **66,** 317, 533; *Acta Met.* 1953, **1,** 340, 348; J. J. Polling and A. Charlesby, *Proc. phys. Soc.* (*B*) 1954, **67,** 210).

The proof of the inverse logarithmic equation just offered was put forward by the author (U. R. Evans, *Rev. pure appl. Chem.* (*Melbourne*) 1955, **5,** 1). An earlier proof, based on assumption of control by a boundary reaction, due to N. Cabrera and N. F. Mott, *Rep. Progr. Phys.* 1948–49, **12,** 163, is more often quoted.*

* Cabrera and Mott argue that the fraction of the potential drop available for helping ions over the energy barrier at the metal–oxide interface is $\frac{\alpha V}{y}$, where α is the distance from the outermost site in the metal to the summit of the barrier and V the voltage across the film. If film-growth is controlled by the rate at which ions can cross this interface, an equation of the type $dy/dt = Ae^{B}$ would be expected, leading to the inverse logarithmic equation, either by the proof given in the text or by an alternative proof offered by Cabrera and Mott. The expression $\alpha V/y$ assumes that electric properties in

The oxidation of aluminium exposed to oxygen, where the low E.M.F. is that of the cell aluminium/oxygen, seems to obey the same law as anodization, where a high E.M.F. can be applied from an external source, leading to a thicker film. In both cases, the thickness at which the growth of the continuous film practically ceases is 14·5 Å per volt. (In some anodizing baths, however, an outer, porous film may be produced—a matter considered in Chapter VII.)

(*d*) **Mixed Parabolic Equation, representing partial control by an interface reaction.** Even at the time when the simple parabolic equation was put forward, it was realized that only under special circumstances was oxidation controlled solely by the transport rate, and that in the general case the reaction at one or both interfaces ought to be introduced. In 1924, the author obtained a more general equation by combining the equations representing respectively the law of mass action at the boundary and transport across the film. If, for simplicity we assume that oxygen is moving inwards through the film, its concentration in excess of that value which would represent equilibrium with the metal being C_i at the inner surface of the film and C_o at the outer surface, then the transport-rate should be proportional to $(C_o - C_i)/y$ and the oxidation-rate to C_i.

In a steady state

$$dy/dt = K_p(C_o - C_i)/y$$

and

$$dy/dt = K_c C_i$$

where K_p is a physical constant and K_c a chemical constant.

---

the boundary layer are the same as in the bulk of the oxide, but if we were dealing with an electrostatic field or an electrokinetic potential distribution involving only ohmic resistances, any difference of properties could be allowed for by writing $k\alpha V/y$, where k is a constant independent of V or y; in that case an equation of the type $dy/dt = Ae^B$ would still emerge. However, obedience to Ohm's Law would be associated with parabolic thickening, and in the case under consideration the resistance of the boundary layer cannot be regarded as ohmic; thus the present author finds difficulty in accepting the expression $\alpha V/y$. Some of the other assumptions implicit in the Cabrera–Mott proof have been questioned by Hauffe. In the anodization of tantalum the idea of boundary-reaction control is rejected by Vermilyea, who presents evidence for the view that the energy barriers controlling ionic flow are situated, not at the interface, but within the film; similar conclusions have been reached by Young. Dewald, studying the anodization of indium-antimony alloys considers that at high fields, the control is exercised by passage through the film, but at low fields the rate-controlling process is located at the interface, since different faces are oxidized at different rate (the (332) face ten times as fast as (110) or ($\bar{3}\bar{3}\bar{2}$)). If a difference between the oxidation-rates on different faces is accepted provisionally as evidence for control by an interface reaction, then it would seem that the oxidation of aluminium (which obeys the inverse logarithmic law) is not controlled by the interface reaction, since the rate is the same for different crystal faces both during natural oxidation and anodization. The reader should consult the references quoted and form his own opinion. (D. A. Vermilyea, *J. electrochem. Soc.* 1956, **103,** 690; J. F. Dewald, *ibid.* 1957, **104,** 244; L. Young, *Trans. Faraday Soc.* 1957, **53,** 841; R. K. Hart, *Proc. roy. Soc.* (*A*) 1956, **236,** 68 ; S. J. Basinska, J. J. Polling and A. Charlesby, *Acta Met.* 1954, **2,** 313; K. Hauffe, *J. Chim. physique* 1956, p. 855. The references to Cabrera and Mott, and to U. R. Evans, will be found in the text.)

Elimination of C_i gives

$$\frac{dy}{dt} = \frac{K_p K_c C_o}{K_c y + K_p}$$

or

$$\frac{1}{2}K_c y^2 + K_p y = K_p K_c C_o t + K'$$

more simply written

$$K_1 y^2 + K_2 y = K_3 t + K',$$

which approximates to rectilinear form when t and y are small and to a simple parabola when they become large.

The same equation was obtained by Fischbeck and Jost by picturing two resistances in series, one, representing transport across the film, being proportional to thickness, and the other, representing movement across an interface, being independent of it. Thus

$$\frac{dy}{dt} = \frac{K}{R_1 y + R_2}$$

or

$$\frac{1}{2}R_1 y^2 + R_2 y = Kt$$

(the analogy with electrical resistance has been criticized). Fischbeck found the equation to be obeyed during the action of steam on iron, and later Wagner and Grünewald confirmed it for copper exposed to oxygen at low pressure. It is evident that low oxygen pressures—which involve the risk that oxygen molecules in the appropriate state will not be supplied at a rate sufficient to replenish the oxygen lost by reaction with cations arriving through the film—are likely to cause a sensible departure from the simple parabola; this was found in early work by Wilkins and Rideal. Recent Spanish measurements of the formation of sulphide films on copper exposed to polysulphide or thiourea solution are interpreted in terms of the mixed parabolic equation.

The papers quoted above are those of U. R. Evans, *Trans. electrochem. Soc.* 1924, **46,** 247, esp. p. 269; K. Fischbeck *Z. Elektrochem.* 1933, **39,** 316; W. Jost, " Diffusion und chemische Reaktion in festen Stoffen " 1937, p. 32 (Steinkopf); C. Wagner and K. Grünewald, *Z. phys. Chem.* 1938, **40,** 455; F. J. Wilkins and E. K. Rideal, *Proc. roy. Soc.* (*A*) 1930, **128,** 394, 407. Cf. J. P. Baur, D. W. Bridges and W. M. Fassell *J. electrochem. Soc.* 1956, **103,** 273; J. Llopis, J. M. Gamboa and L. Arizmendi, Report No. 1 (1958), Instituto de Quimica fisica, C.S.I.C.

(*e*) **Rectilinear Equation, sometimes representing complete control by an interface reaction.** One limiting case of the mixed parabolic law is of course the simple parabolic law, whilst the other is the rectilinear law $dy/dt = K'$ or $y = K't + K''$. This is sometimes obeyed for oxidation at high temperatures, especially in furnace atmospheres where the oxidizing constituent (which may be carbon dioxide, water vapour or oxygen) is present in limited amount; here movement in the gas phase, which must clearly affect the rate of replenishment of material needed for the external boundary reaction, may influence the rate of oxidation. Hauffe and

Pfeiffer, discussing the oxidation of iron in carbon monoxide/dioxide mixtures of composition chosen to produce a ferrous oxide film free from overlying magnetite or ferric oxide film, point out that undoubtedly growth would become parabolic if the film was allowed to become sufficiently thick, but that with fairly thin sheet (0·5 mm.) the whole cross-section may become oxidized at 900–1000°C. before appreciable departure from rectilinear form is detected; at 850°C. growth is parabolic (K. Hauffe and H. Pfeiffer, *Z. Metallk.* 1953, **44**, 27).

Rectilinear growth might also be expected on light, voluminous metals, where the oxide, occupying less space than the metal destroyed, should be porous and non-protective—as pointed out by Pilling and Bedworth in their classical work (p. 39); they actually obtained a straight line for the oxidation of calcium, but it is doubtful whether the facts are equally simple for all light metals. In any case, owing to the heat evolved, it is difficult to maintain the temperature constant. Further complications are introduced by the rapid conversion of oxide into hydroxide or carbonate by traces of water or carbon dioxide and also by the volatility of certain oxides.

Even in cases where growth is strictly rectilinear, it is unlikely that it really represents the chemical velocity constant. Dunn and Wilkins suggest that, even in metals which form porous oxides, a thin oxide layer of constant thickness would remain next to the metal in a state of tension but free from rifts (being braced by the metal), and that the constant rate may represent movement of material across that layer of constant thickness (N. B. Pilling and R. E. Bedworth, *J. Inst. Met.* 1923, **29**, 529; J. S. Dunn and F. J. Wilkins, " Review of Oxidation and Scaling of heated solid Metals " 1935, p. 67 (D.S.I.R.)).

It should be noted, however, that departure from the parabolic law (in extreme cases giving rectilinear growth) may occur in heavy metals at high temperatures, if the oxide-scale keeps cracking away from the metal. When studying the oxidation of high-purity nickel, Gulbransen and Andrew observed a marked departure from parabolic growth above 1000°C., which they attributed to faulty adhesion of the scale (E. A. Gulbransen and K. F. Andrew, *J. electrochem. Soc.* 1957, **104**, 69C. See also p. 857 of this book).

If, as suggested in Chapter II, the oxidation of nickel at these high temperatures is due partly to inwards movement of oxygen, the stresses set up—of which there is some evidence (p. 43)—would account for the film cracking when a sufficient thickness is reached; if it is wholly outwards, the cavities left below the film would lead to faulty adhesion.

Tungsten obeys the rectilinear law under high oxygen pressures between 600° and 850°C., but above 850°C. the oxide sublimes—which may account for an observed departure from the linear rate. Near the edges of cold-sheared sheet, there is more rapid oxidation, with considerable exfoliation; the material breaks into thin plates, each consisting of metallic tungsten embedded in oxide, and this sandwich effect results in an irregular increase of oxidation as the surface area involved increases (J. P. Baur, D. W. Bridges and W. M. Fassell, *J. electrochem. Soc.* 1956, **103**, 266).

Clearly the mixed parabolic equation becomes equivalent to the rectilinear equation at small thicknesses ($R_1 y \ll R_2$) and to the simple parabolic law at great thicknesses ($R_1 y \gg R_2$). It is thus easy to understand the results of Bénard who, studying the oxidation of iron at 1000°C., obtained a curve which, after initial irregularities, was straight for two minutes, but finally settled down to parabolic form (J. Bénard, *Bull. Soc. chim. France* 1949, p. D.89).

(*f*) **Direct Logarithmic Equation, representing control by electronic (as opposed to ionic) transport.** Hitherto it has been assumed that *electrons* can move across the oxide film so easily that it is the passage of *material* which is the sluggish, rate-controlling process. At low temperatures, especially on metals of fixed valencies, this may not always be the case. The normal conductivity of an oxide is often surprisingly high at a high temperature but becomes very low at low temperatures, and we have to consider whether some other mechanism is available for electron-transfer. It happens that electricians have long been puzzled by the fact that the current flowing between two plates of metal carrying thin films but pressed against one another, is, at low temperatures, far greater than would be calculated from the " conductivity " of the film-substance. This is generally attributed to the *tunnel effect*, connected with the circumstance that electrons, commonly regarded as small particles, also possess properties associated with wave-motion. This " dual personality " shown by electrons introduces behaviour not commonly exhibited by material particles.* For instance, a material particle of mass m approaching an energy barrier will only surmount it if the kinetic energy ($E = \frac{1}{2}mv^2$, where v is the component of the velocity in a direction normal to the barrier) is greater than the increase of potential energy at the top of the barrier. In the case of an electron, there is a finite probability of passing through the barrier even if E is less than the potential energy increase, but the probability diminishes with the breadth of the barrier (y). For a rectangular barrier of form , the proportion of electrons passing across it is given by $e^{-4\pi y\sqrt{2m(\phi-E)}/h}$, where ϕ is the height of the barrier, and h is Planck's constant. If, therefore, the oxidation-rate of a film is controlled by the rate at which electrons can pass across it, then

$$dy/dt = K_1 e^{-K_2 y}$$

where K_2 represents $4\pi\sqrt{(\phi - E)2m}/h$ or

$$e^{+K_2 y} dy = K_1 dt$$

Integration gives $$e^{K_2 y} - 1 = K_1 K_2 t$$

assuming (as is probably here justifiable) that the law has been obeyed

* Light, which is commonly regarded as wave-motion but possesses some of the properties of small particles, exhibits an analogous irregularity. If two parallel glass plates are placed at a distance of the order of the wave-length of incident light, which arrives at such an angle that, according to classical optics, it should be totally reflected, a fraction of it will be transmitted through the second plate, although no transmission occurs if this second plate is removed.

from the outset and that the metal was initially free from film ($y = o$ when $t = o$). Under these conditions,

$$e^{K_2 y} = K_1 K_2 t + 1$$

or

$$y = \frac{1}{K_2} \log_e (K_1 K_2 t + 1)$$

usually written

$$y = K \log_{10} (at + 1)$$

where a has the dimensions $[\text{Time}]^{-1}$. The thickening-rate should here be almost independent of temperature.

The proof offered above is an over-simplification, since the barrier will not in fact be rectangular, but, except at very weak electron currents, Holm considers that the error involved will not be serious. Further information can be obtained in books of R. Holm, "Electric Contacts" 1946, pp. 115–119 (Gebers, Stockholm), and of J. Frenkel, "Wave Mechanics" 1932, Vol. I, pp. 111–113 (Clarendon Press).

The relation reached is known as the *Direct Logarithmic Equation.* It is met with also on thicker films, where the Tunnel Effect cannot operate, and alternative mechanisms leading to it are suggested on pp. 834, 837.*

A tunnel-effect mechanism of film-growth was suggested by Mott in 1939. It was criticized by Lustman on several grounds; it predicts, for instance, that logarithmic thickening should be almost independent of temperature—which seemed to conflict with the facts then available. However, the tunnel-effect theory has recently been revived in a new form by Hauffe and Ilschner, who point out that, since the equations for electron passage and material passage are different, the curves will, in general, cut. Now the rate of oxidation will be controlled by whichever process is more sluggish, and if the cutting occurs at an experimentally accessible point, we may expect to obtain control by electron passage in the very early stages, and by ion passage at slightly greater thicknesses; that is, we may expect obedience to the direct logarithmic law at very small thickness and to the inverse law at slightly greater thicknesses.

When this prediction from theory was published, Scheuble's experimental work on the oxidation of nickel at 200°C. was quoted as a good example of direct logarithmic growth of films with thicknesses running up to 18 Å, but no direct experimental evidence was available for the transition from the direct to the inverse logarithmic law. Shortly afterwards, however, a research by Hart in the author's laboratory provided a probable example. Hart was studying the oxidation of aluminium in humid oxygen. He found definite obedience to the inverse law in the later stages, and definite disobedience in the early stages, where the measurements—although somewhat limited in number—were consistent with the direct logarithmic law. Hart's results are shown in fig. 143; they seem to provide evidence for growth according to (1) the direct logarithmic law at very low thickness, (2) the

* Another derivation of the Direct Logarithmic Equation, differing from that of Cabrera and Mott in assigning control to the chemi-adsorption step, is proposed by P. T. Landsberg, *J. chem. Phys.* 1955, **23,** 1079.

inverse logarithmic law at greater thicknesses. The direct logarithmic equation has recently been found to govern the oxidation of silicon and germanium exposed to " room air " by Archer, who calls it the Elovich equation. Papers deserving study include N. F. Mott, *Trans. Faraday Soc.* 1939, **35,** 1175; B. Lustman, *Trans. electrochem. Soc.* 1942, **81,** 359, esp. p. 372; K. Hauffe and B. Ilschner, *Z. Elektrochem.* 1954, **58,** 382. The experiments on nickel are described by W. Scheuble, *Z. Phys.* 1953, **135,** 125, those on aluminium by R. K. Hart, *Proc. roy Soc.* (*A*) 1956, **236,** 68, esp. p. 81, and those on silicon and germanium by R. J. Archer, *J. electrochem. Soc.* 1957, **114,** 619.

The detailed researches on zinc by Vernon, Akeroyd and Stroud led to curves which obeyed the direct logarithmic equation. The author (who has had the privilege of examining the original graphs of the Teddington

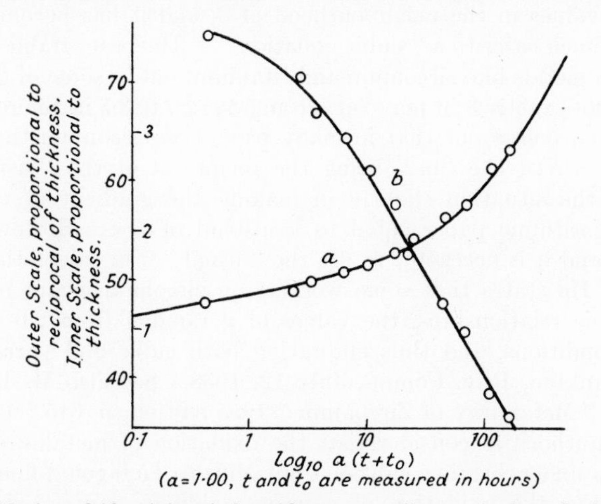

FIG. 143.—Oxidation of Aluminium in humid oxygen. Curve *a* is plotted with thickness as ordinate, and curve *b* with reciprocal thickness (R. K. Hart).

investigators) agrees that, objectively considered, this represents the equation of " best fit " for the experimental data. Other possibilities are not excluded, nor is this surprising since, as shown later, each of the " log-law " equations can, in certain circumstances, serve as an approximation to another; it is, therefore, not inconceivable that future work may show that zinc obeys the direct law for very thin films and the inverse law for slightly thicker films, but there is no experimental evidence for this at present. Moore and Lee find that the logarithmic law gives place to the parabolic law above 370°C. (W. H. J. Vernon, E. I. Akeroyd and E. G. Stroud, *J. Inst. Met.* 1939, **65,** 301; W. J. Moore and J. K. Lee, *Trans. Faraday Soc.* 1951, **47,** 501).

Recent work on hafnium indicates that, over the range 350–710°C., the direct logarithmic law is obeyed in the initial stages of growth, which becomes parabolic in the later stages; over the range 900–1200°C. the rectilinear law prevails (W. W. Smelzer and M. T. Simnad, *Acta Met.* 1957, **5,** 328).

(*g*) **Cubic Equation,** probably representing a *transition state.* If, in a set of experimental values log y is plotted against log t, we often obtain a straight line. This is necessarily the case if the equation is a simple power-law $y^n = kt$; in the case of the parabolic law where $n = 2$, the gradient of the line obtained when y is plotted against t on double-logarithmic paper will be $\frac{1}{2}$, but this will only be the case if we can neglect the integration constant k' of the full equation $y^2 = kt + k'$. Assuming that k' can be neglected, the gradient will be a measure of the protective character of the film; the higher the value of n, the more completely does the oxidation stifle itself. The general equation (p. 823) will provide possibilities of obtaining values of n exceeding 2, although as the inverse logarithmic law is approached the points, plotted on double-logarithmic paper, will no longer fall on a straight line. Under conditions where the protectivity is greater than the parabolic law would predict, we may, over limited ranges of experimental time, get values in the neighbourhood of 3, and it has become customary to ascribe such cases to a " cubic equation ". The best established example come from metals like zirconium and titanium, but in some of the published work n is not exactly 3; it may vary from (say) 2·7 to 3·3 in different instances.

Wanklyn points out that in many papers on zirconium the equation is written $y = Kt^n$, the " n " being the reciprocal of the " usual " n; this simplifies the situation slightly in making the gradient of the curve on double logarithmic paper equal to n instead of $1/n$. However, the K is modified, and it is necessary to use the " usual " n in calculating activation energies. He states that some workers on zirconium claim parabolic and others cubic relationships, the values of n ranging from 2·0 to 3·5 under various conditions, and thus embracing both cubic and parabolic ranges. (J. N. Wanklyn, Priv. Comm., July 12, 1958. See also B. Lustman and F. Kerze, " Metallurgy of Zirconium " 1955 edition, p. 615 (McGraw-Hill)).

Some authorities consider that the oxidation of metallic copper obeys a cubic law under certain conditions, but Mills (p. 858) could find no evidence of this. For the oxidation of cuprous oxide to cupric oxide, however, Hauffe and Kofstad obtained numbers which are most satisfactorily expressed by a cubic equation, whilst the most recent work on the oxidation of zirconium shows close agreement with the cubic law.

Many authorities are inclined to regard the cases where the experimental facts can be conveniently expressed by a cubic equation as representing an intermediate region of the general equation where *n happens* to be about 3. Such views are held by J. N. Wanklyn, and are sympathetically regarded by D. L. Douglas and F. C. Zyzes, *Corrosion* 1957, **13,** 361t, esp. p. 372t.

Certain writers explain the experimental facts by assuming that oxidation is proceeding simultaneously by two different equations, whilst others again consider that the cubic law is sufficiently well established in its own right to demand a special mechanism. At least four such mechanisms have been offered, which the reader may care to study and compare. The relevant papers are those of N. Cabrera and N. F. Mott, *Rep. Progr. Phys.* 1948–49, **12,** 163, esp. p. 177; W. E. Campbell and U. B. Thomas, *Trans. electrochem. Soc.* 1947, **91,** 623; T. N. Rhodin, *J. Amer. chem. Soc.* 1950, **72,** 5102; K. Hauffe and P. Kofstad, *Z. Elektrochem.* 1955, **59,** 399; *Werkst. u. Korrosion*

1955, **6,** 117; 1956, **7,** 642; W. Jaenicke, " Passivierende Filme und Deckschichten " 1956, p. 176 (edited by H. Fischer, K. Hauffe and W. Wiederholt; published by Springer); H. H. Uhlig, *Acta Met.* 1956, **4,** 541, esp. p. 550. Evidence for the cubic law on zirconium is presented by M. W. Mallett, W. M. Albrecht and R. E. Bennett, *J. electrochem. Soc.* 1957, **104,** 349; R. G. Charles, S. Barnartt and E. A. Gulbransen, *Trans. Met. Soc. A.I.M.E.* 1958, **212,** 101.

2. Movement through Pores

General. At relatively low temperatures movement through the oxide lattice can be neglected, except for very thin films where the high electric field helps the movement of ions. There is, however, no reason to rule out the possibility of movement through discrete pores, such as might exist (1) along the lines where three grains meet, (2) along lines of disordered atoms such as might occur where two slip-planes intersect, (3) along the axes of screw dislocations, or (4) along lines representing edge dislocations; the word " pore " may be used to include any line along which movement can occur at temperatures too low for thermal movement within the body of the grains. If the pores are broad enough to admit oxygen molecules, fresh oxidation to the metal at the base of the pores will occur, but in most cases this will probably plug up the pores in course of time. Narrower pores (perhaps really lines of disarray) may still permit movement of ions, and if the oxidation mechanism depends on oxygen ions moving inwards, the compression developed may well plug the pores, provided that the metal is one producing an oxide that occupies a bigger volume than the metal destroyed. Even where the thickening is due to cations moving outwards, there is a chance of such movement bringing ions into positions where a pore, previously active, becomes blocked. Thus there is a possibility that N, the number of pores per unit area, will diminish as W, the amount of oxygen absorbed per unit area, increases, and presumably, in absence of complicating factors, $-dN$ will be proportional to dW. The fact that both zinc and copper in certain surface conditions exhibit a range of temperature over which the corrosion-rate *declines* with rising temperature—as shown respectively by Vernon and by Lustman—gives support to the idea that pore-blocking (or something equivalent) is occurring; a rise in temperature must involve increased movement of atoms (or ions) and we can only conclude that that movement is doing something which obstructs oxidation instead of promoting it. The experimental evidence for oxidation-rate declining with rising temperature in the case of zinc is presented by W. H. J. Vernon, E. I. Akeroyd and E. G. Stroud, *J. Inst. Met.* 1939, **65,** 310, esp. fig, 10, p. 315, and in the case of copper by B. Lustman, *Trans. electrochem. Soc.* 1942, **81,** 359, esp. fig. 8, p. 366.

Two separate cases demand consideration. We may assume that the formation of fresh oxide can only produce blockage in the pore in which it is formed, or we may consider the chance that the compressional stress produced can cause blockage of neighbouring pores. These two cases (*self-blockage* and *mutual-blockage*) lead to two different equations.

(*h*) **Asymptotic Equation,** representing *Self-blocking Pores.* If the rate of oxygen-uptake is determined by the number of pores still remaining open, so that

$$dW/dt = k_1 N$$

and if, as already suggested

$$-dN = k_2\, dW$$

$$-\frac{dN}{dt} = -\frac{dW}{dt}\cdot\frac{dN}{dW} = k_1 k_2 N = k_3 N$$

$$\frac{1}{N} dN = -k_3\, dt$$

$$\log_e (N/N_0) = -k_3 t$$

where N_0 is the value of N at $t = 0$.

or

$$N/N_0 = e^{-k_3}$$

$$\frac{dW}{dt} = k_4 e^{-k_3 t}$$

$$W = k_4(1 - e^{-k_3 t})$$

assuming that $W = 0$ when $t = 0$.

This is a true asymptotic equation, for W can never exceed k_4, however great t becomes.

It was once thought that the outdoor exposure of aluminium alloys obeyed some such law, and that the corrosion-damage came to a stand-still, so that there was some value of the corrosion-damage which would never be exceeded—however long the exposure-period might be. American tests in the open air have, however, shown that the corrosion-rate, which first falls off rapidly, finally becomes constant at a low value, which for some practical purposes could probably be neglected (C. J. Walton, D. O. Sprowls and J. A. Nock, *Corrosion* 1953, **9,** 345). The character of asymptotic forms is discussed by U. R. Evans, *Nature* 1949, **164,** 909; *J. Inst. Met.* 1952–53, **81,** 738; and by F. A. Champion, *Trans. Faraday Soc.* 1945, **41,** 593; *J. Inst. Met.* 1952–53, **81,** 739.

(*i*) **Direct Logarithmic Equation,** representing *Mutually-blocking Pores.* If the compressional stress in one pore exercises a pressure on neighbouring pores, there is a chance that some of these may become blocked; in considering the effect of dW on $-dN$, we now have to take into account the value of N, since this will decide how many pores exist within an effective distance of the first pore. Thus

$$-dN = k_2 N\, dW \quad \text{or} \quad \log_e (N/N_0) = -k_2 W$$

Thus if, as before, the rate of oxygen-uptake is determined by the number of pores

$$\frac{dW}{dt} = -k_1 N = k_3 e^{-k_2 W}$$

we have

$$e^{+k_2 W}\, dW = k_3\, dt$$

which gives, assuming that the metal is oxide-free at the out-set, and that the equation is obeyed from $t = o$, $W = o$,

$$k_3 t = \frac{1}{k_2}(e^{k_2 W} - 1)$$

which can be expressed $Kt + 1 = e^{k_2 W}$

or $$k_2 W = \log_e (Kt + 1)$$

It is conveniently written $W = K' \log_{10} (at + 1)$

This is the *direct logarithmic equation*, already met with on p. 830; as explained later, it sometimes gives curves similar to the asymptotic and inverse logarithmic equation; it is, however, not truly asymptotic, since it is not possible to suggest a value of W which will never be surpassed, however great t is made.

The logarithmic law is found to be obeyed by iron and copper (p. 27) at temperatures lower than the range over which the parabolic law is obeyed. Agreement between the results of different workers is at first sight incomplete and there are also some slight differences regarding interpretation, although probably results and opinions are not really irreconcilable. Vernon, Calnan, Clews and Nurse (p. 57), working with abraded iron, showed that above 200°C., oxidation consisted of the formation of an oxide-film free from metal, since the increase in weight of the specimen agreed with the weight of oxygen in the film obtained on stripping by the iodine method; below 200°C., the two measurements did not agree, pointing to the penetration of oxygen into the interior. The growth-law, which conformed to the parabolic law above 200°C., suddenly switched to the logarithmic law below 200°C.; Vernon and his colleagues concluded that the parabolic law represents the growth of a uniform film, whilst the logarithmic law represents penetration into the interior. Davies, Evans and Agar, using " as-rolled " iron (unabraded) with the surface oxide reduced by hydrogen-treatment, found the transition from parabolic to logarithmic law to occur somewhat above 300°C., and there was evidence that oxidation, starting at weak points, spread sideways under the film instead of downwards into the interior. The minor differences between the Teddington and Cambridge results were doubtless due to the different methods of surface preparation. It may be pertinent to mention Païdassi's work at higher temperatures; he obtained different behaviour according as the specimens had been heated in hydrogen at 700°C. and cooled to the experimental temperature or heated to that temperature in air (J. Païdassi, *Acta Met.* 1956, **4,** 227).

A reason can be suggested for the passage from one law to the other. At relatively high temperatures, where atomic re-arrangement is possible, any pores (weak points in the film due to disarray) will speedily be healed up, but this will not put a stop to oxidation, since passage through the oxide-lattice can take place at these temperatures, leading to thickening by the parabolic law. At low temperatures, passage through the main part of the film is impossible, but healing of weak points by atomic rearrangement will be less rapid. However, there is the possibility that the expansion which occurs when metal is transformed to oxide may tend to close up neighbouring leakage paths—leading to the logarithmic law, as already shown.*

The reader may care to study the experimental papers and consider the slightly different interpretations; the references are D. Eurof Davies, U. R. Evans and J. N. Agar, *Proc. roy. Soc.* (*A*) 1954, **225, 443** (iron); W. H. J. Vernon, E. A. Calnan, C. J. B. Clews and T. J. Nurse, *ibid.*, 1953, **216, 375** (iron); U. R. Evans and J. Stockdale, *J. chem. Soc.* 1929, p. 2651 (iron, copper, nickel); W. H. J. Vernon, E. I. Akeroyd and E. G. Stroud, *J. Inst. Met.* 1939, **65,** 301, esp. p. 319; B. Lustman, *Trans. electrochem. Soc.* 1942, **81,** 354, esp. p. 366.

3. Conditions where only part of the area is concerned in film-thickening

(*j*) **New Logarithmic Law,** representing the *effects of cavities or obstructions.* Even where at the out-set the whole oxide-film is capable of transmitting ions (or vacancies), the area available for such transmission may shrink with time; this may occur where vacancies (or alternatively particles of the oxide of some alloying constituent) collect below the main film at the metal–oxide interface. Since the film-thickening is no longer uniform, it is best to represent the progress of oxidation by the weight-increase per unit area (W).

A cationic vacancy migrating inwards may either (1) remain at the metal–oxide interface, (2) pass into the metal and become attached to a dislocation, or (3) pass through the metal (if sheet) and join a cavity on the far side, as demonstrated in the experiments of Dunnington, Beck and Fontana (p. 48). Vacancies remaining at the interface may collect to form cavities, and reduce the available area to ϕ times its original amount ($\phi < 1$), so that dW/dt is reduced from $k_8\, dy/dt$ to $k_8\phi\, dy/dt$ where dy/dt represents the thickening rate at places where no cavities have been formed, and k_8 is the ratio of weight per unit area to thickness. The chance that a vacancy may be retained at the interface is presumably proportional to the number of sites available, so that

$$-\,d\phi = k_9\phi\, dW$$

or

$$\log_e \phi = -\,k_9 W + k_{10}$$

giving

$$\phi = k_{11}e^{-k_9 W}$$

so that

$$dW/dt = k_8\phi\, dy/dt = k_{12}t^{-1/2}e^{-k_9 W}$$

assuming that y falls off according to the parabolic law.

* Research at Teddington has indicated that a similar growth law holds for the thickening of air-formed films on iron immersed in chromate solution after exposure to air. If we neglect an initial period which increases with the duration of the previous air-exposure, the amount of chromium introduced into the film increases with time according to the equation

$$Q = V \log\,(at + 1),$$

where the value of a is about 1·0 when time is expressed in minutes. The experimental data and curves will be found in "Chemistry Research" 1955, p. 14 (D.S.I.R.).

Thus $$e^{+k_9W}\, dW = k_{12}t^{-1/2}\, dt$$
or $$e^{k_9W} = 2k_9k_{12}t^{1/2} + 1$$
assuming the relation to hold from $t = o$, $W = o$.

We obtain $$k_9W = \log_e (K(t^{1/2} + 1/K)).$$
which could be written $W = k' \log_e (a(t^{1/2} + k''))$.

If $1/K \ll t^{1/2}$ this becomes

$$k_9W = \log_e (K^2t)^{1/2}$$

which is $$W = \frac{1}{2k_9} \log_e (K^2t)$$

and if $K^2t \gg 1$, this is indistinguishable from the direct logarithmic equation of section (*i*). Possibly some cases where the direct logarithmic equation has been established experimentally are explicable by cavity formation. However, there is no reason why $1/K$ should always be small, and Mills, studying the oxidation of copper, has obtained results inconsistent with the direct logarithmic law but consistent with the new relation,

$$W = k' \log_c a(t^{1/2} + k'')$$

Experimental details are described by T. Mills and U. R. Evans, *J. chem. Soc.* 1956, p. 2182. The theoretical aspects of cavity formation below films are discussed by A. Dravnieks and H. J. McDonald, *J. electrochem. Soc.* 1948, **94,** 139; K. Sachs, *J. Iron St. Inst.* 1947, **187,** 93, esp. p. 100; cf. A. S. de Brasunas, *Metal Progr.* 1952, **62,** No. 6, p. 88.

The effect of particles of oxides of a second metal collecting at the main metal–oxide interface should have a similar effect. The well-known work of Portevin, Prétet and Jolivet (p. 63) on the iron-aluminium alloys, where there is increasing departure from the parabolic law as the aluminium content increases, may perhaps receive an explanation on these lines.

(*k*) **Sigmoid Equations, representing lateral growth.** Although in most cases a fairly continuous film of two-dimensional oxide is soon produced over the whole metallic surface, the formation of three-dimensional crystals of appreciable thickness may in some cases require considerable nucleation energy, but when once formed may spread laterally over the surface. The experimental examples mostly refer to low pressures; beautiful micrographs of iron oxidized by Bardolle and Bénard and by Gulbransen, McMillan and Andrew show embryo oxide crystals in forms related to the crystal structure of the metal, so that their orientation varies from grain to grain. The papers, which deserve study, are J. Bardolle and J. Bénard, *Rev. Métallurg.* 1952, **49,** 613; E. Gulbransen, W. R. McMillan and K. F. Andrew, *Trans. Amer. Instn. min. met. Engrs.* 1954, **200,** 1027.

Lateral growth may be recognized in the growth-curves by the form being concave upwards (when w is the ordinate and t the abscissa). The simplest proofs of the laws depend on statistical methods and are deferred to Chapter XXII. It is there shown (p. 939) that for films spreading in circles at constant rate from nuclei which appear simultaneously at time $t = o$, the fraction of the area covered at time t is $(1 - e^{-kt^2})$, whereas if the nuclei keep appearing sporadically in time as well as space on the ever-diminishing uncovered fraction, then the fraction covered is $(1 - e^{-kt^3})$. If the oxidation, once started at a point proceeds there at uniform rate, so that dw/dt is proportional to the area uncovered, we have $dw/dt = k'(1 - e^{-kt^n})$, where n is 2 or 3; thus we get a curve which is concave upwards at the outset, but finally becomes straight. If the thickening rate falls off according to the parabolic law, the curve may be concave upwards at first and finally convex upwards; this is known as a *sigmoid* curve.

Although in practice the oxide does not spread out in circles, the general form of the curves should be that indicated. Curves which are concave upwards near the origin have been produced on copper at low pressures by Bloomer and by Wagner and Grünewald. Tylecote's curves on copper are largely sigmoid, being concave upwards at the outset and becoming parabolic later. The curve for a magnetite layer growing below a ferric oxide layer was found to be concave upwards by Davies, Evans and Agar (p. 836) and was seen to be due to the magnetite spreading laterally, since the ferric oxide layer exhibited different colours according as it was backed by magnetite or by metal.

Bloomer's early demonstration of a concave-upward curve was only published in 1945 in a paper by D. H. Bangham, *J. Sci. Instrum.* 1945, **22,** 230. See also R. F. Tylecote, *J. Inst. Met.* 1952–53, **81,** 681; C. Wagner and K. Grünewald, *Z. phys. Chem.* 1938, **40,** 455.

Relationships between the Growth Equations

Analysis of curves. A set of curves obeying the various equations is printed in fig. 144, but the shapes can vary greatly according to the relative

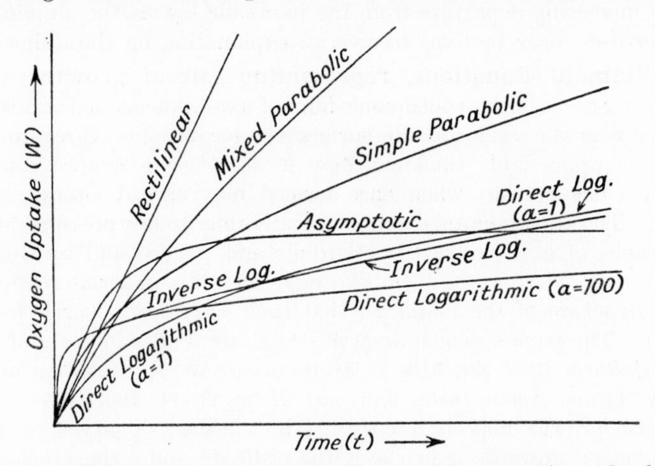

FIG. 144.—General forms of curves corresponding to various equations (the shape may vary according to relation between the constants).

values adopted for the constants. To emphasize this point two curves both representing the direct logarithmic equation have been drawn, one corresponding to $a = 1$, and the other to $a = 100$; the values of K are adjusted in the two cases to bring the curves close enough for comparison. It will be noticed that one of them runs very close to the inverse logarithmic curve in the early stages, whilst the other runs close at a later stage. Neither the inverse logarithmic curve nor the direct logarithmic curve with $a = 1$ show any signs of becoming asymptotic to a horizontal line within the area covered by the diagram.

The rectilinear curve has limited significance in a practical situation; a metal which gives a straight line in a laboratory test carefully carried out under strictly isothermal conditions might in service give an accelerating curve (concave upwards) owing to the rise in temperature, unless the conditions were unusually favourable to heat-removal.

The testing of a curve obtained experimentally to ascertain whether a

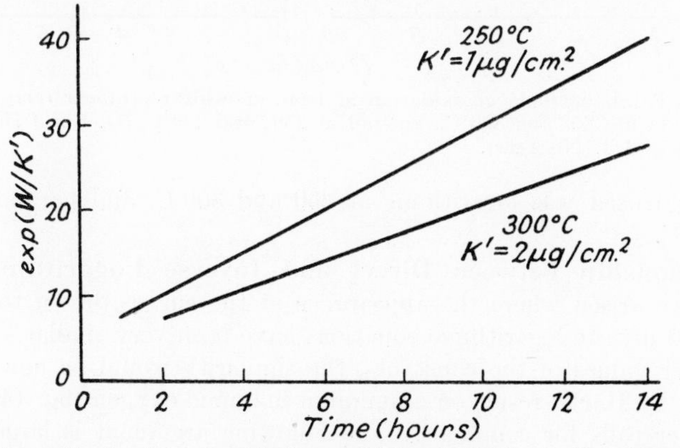

FIG. 145.—Logarithmic plots of oxidation at 250° and 350°C. (D. Eurof Davies, U. R. Evans and J. N. Agar).

(In figs, 145, 146 and 147 W represents the oxygen-content of the α-Fe_2O_3 film in μg./cm.^{2}.)

certain equation is obeyed requires some care. We may consider a case where oxidation is measured by the weight-increase, w. Obedience to the simple parabolic law $w^2 = kt + k'$ may be indicated if a straight line is obtained on plotting w^2 against t; it will not necessarily pass through the origin, since k' may not be negligible. If k' is small, w may be plotted against t/w and this also serves to test for the mixed parabola,

$$K_1 w^2 + K_2 w = K_3 t.$$

Obedience to the direct logarithmic law $w = K \log_e (at + 1)$ is verified by ascertaining whether values of K exist which give straight lines on plotting $e^{w/K}$ against t. Obedience to the inverse logarithmic law requires that there is a value of K giving a straight line when $e^{-1/Kw}$ is plotted against t. Alternative procedures can be suggested; the direct logarithmic law can be tested by ascertaining whether a value of a exists which will give a straight line when $\log (at + 1)$ is plotted against w. Figs. 145 and 146 taken from the paper of Davies, Evans and Agar (p. 836) show clearly that the oxidation

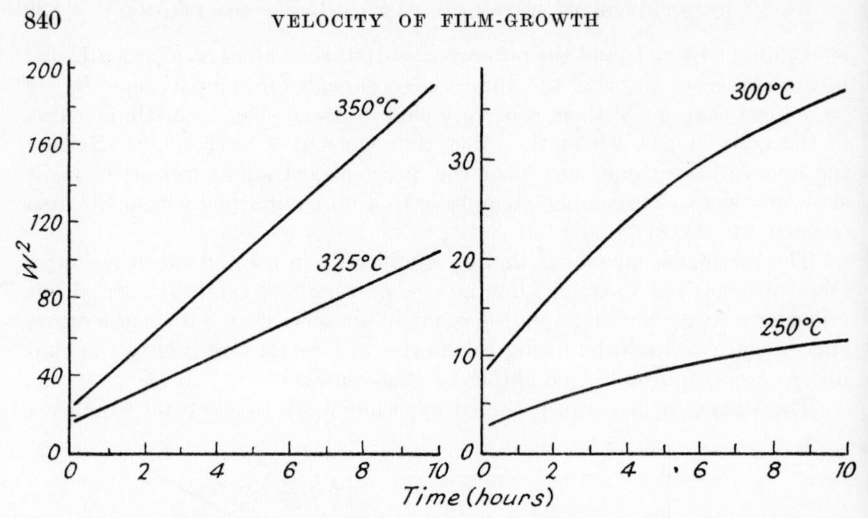

Fig. 146.—Parabolic plots of oxidation of Iron, showing rectilinear relation between W^2 and t at 325° and 350°C., but not at 250° and 300°C. (D. Eurof Davies, U. R. Evans and J. N. Agar).

of the iron used was logarithmic at 250 and 300°C. and parabolic at 325 and 350°C.

Relationship between Direct and Inverse Logarithmic Laws. Cases have arisen where the appearance of the curves drawn to obey the direct and inverse logarithmic equations have been very similar. Although with other values of the constants, the similarity would be much smaller, as shown by Hart's result on aluminium in humid oxygen (fig. 143, p. 143), the matter calls for comment. The following argument is largely due to B. Ilschner (private communication).

If we write the inverse logarithmic equation

$$\frac{1}{y} = K_8 - K_9 \log_e (a(t - t_0) + 1)$$

and consider a time-range where t is large compared to t_0 and also to $1/a$, this becomes

$$\frac{1}{y} \sim K_8 - K_9 \log_e at$$

so that

$$y \sim \frac{1/K_8}{1 - \dfrac{K_9}{K_8} \log_e at}$$

For a range where

$$|\log_e at| \ll \frac{K_9}{K_8}$$

this becomes

$$y \sim \frac{1}{K_8}\left(1 + \frac{K_9}{K_8} \log_e at\right)$$

so that the inverse logarithmic equation has become a good approximation of the direct logarithmic equation.

Relationship between Direct Logarithmic Equation and Asymptotic Equation. Under other circumstances the asymptotic equation becomes a good approximation to the direct logarithmic equation. This is seen if we write the equations $W = k_1(1 - e^{-k_2 t})$ and $W = k_1 \log_e (kt + 1)$ in expanded form

$$W = k_1[k_2 t - \tfrac{1}{2}k_2{}^2t^2 + \tfrac{1}{6}k_2{}^3t^3 - \tfrac{1}{24}k_2{}^4t^4 + \ldots]$$

$$W = k_1[k_2 t - \tfrac{1}{2}k_2{}^2t^2 + \tfrac{1}{3}k_2{}^3t^3 - \tfrac{1}{4}k_2{}^4t^4 + \ldots]$$

Clearly at values of t sufficiently small to justify neglect of the third and later terms, the two equations become experimentally indistinguishable (U. R. Evans, *Nature* 1949, **164,** 909).

This provides another reason why it is difficult to judge from short-time experiments, whether there is, or is not, some limiting value for the oxidation, which will never be exceeded, however long the experiment lasts (cf. p. 523).

Mode of expression of the laws. The direct and inverse logarithmic laws have appeared in the literature in many forms; some of these are open to objection on dimensional grounds. The statement

$$y = K \log (t + 1)$$

is inacceptable, since t and 1 have different dimensions and the equation would give curves of different shapes according to the unit of time chosen. It should be altered to

$$y = K \log (at + 1)$$

where a has the dimensions $[\text{Time}]^{-1}$. It may happen that a assumes a numerical value close to 1·0 when some particular unit of time is adopted, but this does not justify the omission of a.

In cases where $at \gg 1$, the form $y = K \log at$ becomes justifiable; it seems, however, unwise to simplify it further into $y = K' + K \log t$ where K' represents $K \log a$, although this would not cause the shape of the curve to alter with a choice of time unit, since if a larger time unit is adopted $\log a$ increases to the same extent as $\log t$ decreases.

Possibility of several mechanisms proceeding simultaneously. The term " log laws " was used in Chapter II to cover the three laws, the direct logarithmic, the inverse logarithmic and the asymptotic; as shown above, the three equations, may, over limited ranges of time, give curves which are experimentally indistinguishable. They are all low-temperature laws, applicable to cases where the haphazard thermal movement of ions within the oxide-lattice is so sluggish that movement through the film can only occur in other ways. The inverse logarithmic law should be obeyed if this movement is solely due to ions being *pulled through the lattice* by the high electrical potential gradient which exists when the film is very thin. The asymptotic law should be obeyed if the movement occurs *through pores,* which gradually become diminished in number through *self-blockage.* The direct logarithmic law should be obeyed if it is assumed that the congestion arising in a pore tends to block neighbouring pores, i.e. that there is *mutual blockage.* In practice, it may well be that there is still some movement through the lattice when pore-movement becomes important, and that both self-blockage and mutual blockage contribute to the diminution of the number of pores with time. Thus the relations between W and t may fall between those indicated by the three equations, but since it is often experimentally difficult to distinguish between the three pure-bred equations, it will be even more difficult to analyse a hybrid in such a way as to assign responsibilities to the three mechanisms. This would seem to justify the bunching of the three equations in Chapter II under the single heading of "log laws".

Limitations on Obedience to the various Growth-laws. If the simple parabolic law $dy/dt = k/y$ is to be obeyed on an originally film-free specimen from the out-set (when $y = 0$), the initial value of dy/dt must be infinite. Now the boundary reactions could never proceed at infinite velocity, so that in fact any case of parabolic oxidation which starts with a film-free surface must conform to the "mixed" equation, although in some cases the initial, nearly straight, portion may be so short as to be incapable of experimental observation.

There is no *mathematical* reason why the direct logarithmic and exponential equations should not be obeyed from the moment when $t = 0$, $y = 0$, although some of the *physical* interpretations suggested for them would seem only to be applicable to conditions where a continuous uniform film already exists. The question, however, arises as to whether the inverse logarithmic equation expressed in the form

$$1/y = 1/y_0 - K_9 \log_e (a(t - t_0) + 1)$$

is valid at very high and very low values of t. That it does, in fact, fail to be obeyed under moist conditions for very thin films on aluminium was indicated by the measurements of Hart, and a physical explanation, based on the ideas of Hauffe and Ilschner, has been offered (p. 830). However, there is no mathematical absurdity introduced in the assumption that the equation holds from the lowest values of t. If the range of validity includes $t = 1/a$, we can adopt $1/a$ as the value for t_0 (the arbitrary point from which we agree to measure all our times); the equation then reduces to $1/y = 1/y_0 - K_9 \log_e at$ and it is clear that for very small values of t, the logarithmic term becomes a large negative quantity, so that $1/y_0$ can be neglected; $1/y$, therefore, becomes a large positive quantity, and it is evident that y tends to zero as t tends to zero. Obedience to the equation from the outset involves, for an originally film-free surface, no mathematical absurdity, although there may be physical objections to it.

In contrast, there is no escape from the statement that the equation must become invalid at very large values of t, since at a finite value of t the logarithmic term will become equal to $1/y_0$, making y infinite; beyond that point, $1/y$ becomes negative—which has no physical meaning. This failure of the law must arise, whatever algebraic proof is adopted. With the particular proof offered on p. 824, the point where the argument fails is quite clear. Everything depends on θ being large compared to unity, that is y being small compared to K_1/kT. As soon as that assumption ceases to be justifiable, the whole argument breaks down. This point should be borne in mind by those who today take the inverse logarithmic law as being for practical purposes asymptotic—arguing that by waiting long enough dy/dt will become sufficiently small to satisfy reasonable requirements. If these requirements are fulfilled before the assumption just mentioned becomes unjustifiable, it is legitimate to regard the inverse logarithmic relationship as equivalent to an asymptotic law; if not, the assumption may lead to false conclusions.

Protective character of Films obeying different Growth-laws. The parabolic equation is sometimes regarded as a sign of a "protective" film, and undoubtedly a metal which gives a parabolic graph is more resistant than one which gives a rectilinear graph. Indeed a low value for the parabolic rate-constant at the temperature of proposed service may justify the choice of a metal for practical use at that temperature. However if, at a given film thickness, the rate of oxidation is, say, ten times the maximum value regarded as tolerable, it is necessary (under conditions of obedience to the parabolic law) to wait until the film has thickened to ten times its present value before conditions become satisfactory; in practice, since thick films are liable to rupture, satisfaction may never be attained at all. In contrast, when one of the log laws is obeyed, a relatively small increase of thickness (perhaps 10 Å) may cause a sensational decrease in the oxidation-rate. That is one reason why the addition of alloying elements which cause the appearance of a new phase at the base of the film and thus occasion a departure from parabolic growth in favour of approximately logarithmic growth may be more advantageous than the additions which, dissolving in the film-substance, reduce the number of defects and thus depress the value of the parabolic rate constant, without changing the form of the growth-law.*

State of present knowledge. Theoretical arguments starting from different premises, have led us, without *ad hoc* assumptions, to a number of equations. Others could be reached by taking into account other possibilities, such as the building up of space-charges, which have interested many authorities. Nearly all the equations arrived at above have been realized experimentally—sometimes *after* they had been obtained mathematically. However, there remain many reasonable premises which have not yet been used as starting-points for theoretical prediction; if experimenters sometimes obtain results which correspond to no published equations, that does not justify distrust in the theoretical reasoning. On the

* J. N. Wanklyn suggests that this may be one way of regarding Huddle's "squeeze-energy" theory. If we raise the energy barrier for ionic movement, we raise the temperature at which logarithmic growth gives place to parabolic growth. This may be more valuable than a reduction in the number of defects by the Hauffe principle. He also points out that in the equation $dy/dt = Ae^{-Q/kT}$, a reduction in the number of defects influences A, but an increase of squeeze energy influences Q and may be more effective. In connection with Wanklyn's suggestion, it may be pertinent to recall that in the oxidation of iron, Teddington work on abraded specimens placed the transition from logarithmic to parabolic growth at 200°, whilst Cambridge work on hydrogen-cleaned specimens placed it at about 300° (p. 853).

contrary the accord between experiment and theory has been most impressive; the agreement between calculated and observed values assembled by Wagner for the parabolic rate-constant (Table XXVIII, p. 849) provide examples. In other cases, further work—mathematical and/or experimental—will be needed before the situation is cleared up, but there is no reason why theory and practice should not be completely reconciled.

The present situation may be tentatively summarized thus:—

(1) When the oxidation-rate is controlled purely by movement through the film, without boundary-reaction control and without changes in effective area, a *general equation* is obtained, with two limiting cases:—

(*a*) At *high temperatures*, where there would be free movement in the absence of a potential gradient, the effect of which is merely directive, we obtain the *simple parabolic law*.

(*b*) At *low temperatures*, where there would be no movement in the absence of a gradient, we obtain the *inverse logarithmic law*.

(2) Other controlling factors modify these equations:—

(*a*) Partial control by a boundary reaction replaces the simple parabolic law by the *mixed parabolic law*.

(*b*) Complete control by a boundary reaction (or by movement through an unbroken layer of constant thickness below a porous film) gives the *rectilinear law*.

(*c*) Control by electron-transport, instead of material transport, as the "sluggish stage", replaces the inverse logarithmic law by the *direct logarithmic law*.

(3) If movement occurs, not through the lattice of the film-substance but by pores, which tend to become blocked as the film thickens, then

(*a*) When the pores are self-blocking, we may expect the *asymptotic* law.

(*b*) When the pores are mutually blocking, we may return to the *direct logarithmic law*.

(4) If the effective area varies with time, there will be further changes:—

(*a*) Where at the base of the film cavities accumulate (or perhaps particles of a second oxide phase), we get a *new logarithmic law*, of which the direct logarithmic law is a special case.

(*b*) Where three-dimensional oxidation starts at nuclei and spreads outwards, we get curves which are *concave upwards* in the early stages, and often *convex upwards* in the later stages (sigmoid form).

The cases where an alloy forms more than one phase in the resulting film are being worked out by Wagner, whose results deserve study (C. Wagner, *J. electrochem. Soc.* 1956, **103**, 627).

It should be noticed that small factors affect the decision as to which growth-law is obeyed. It has long been difficult to reconcile the statements made by different reliable observers regarding the oxidation-behaviour of copper. The apparent discrepancies may be due to differences in contents of the minor constituents or in the method of surface-preparation. The work of Löhberg and Wolstein suggests that the rolling treatment is important; they state that after small reductions by rolling the oxidation at first obeys an exponential law, later becoming parabolic; after heavy reduction, a parabolic law holds good from the first and the same is true of material which has been recrystallized after rolling (K. Löhberg and F. Wolstein, *Z. Metallk.* 1955, **46**, 734).

The argument has been heard that Pfefferkorn's discovery of thin spine-like out-growths of oxide makes nonsense of existing mathematical theories, which include no reference to such out-growths; a certain tinge of satisfaction may be detected in those advancing such arguments which doubtless provide a convenient excuse for not studying the tiresome mathematics. However, the weight of oxide comprised in these extremely thin spines must be extremely small, and it is doubtful whether its neglect would affect the experimental numbers; moreover, even in the event of the weight of the spines being comparable to that of the film, it does not follow that they would alter the form of the growth-law.

If we are dealing with an oxide which grows by " outward " movement and the temperature-range in which parabolic law is valid, the form of the law would certainly remain the same even if the spines were taken to account. Presumably the spines are formed at places where a line of loose structure (perhaps an edge dislocation) penetrates the film, so that cations move rapidly outwards along this line, and on arrival at the outer end of the film become converted to oxide, pushing outward as a slender pillar the oxide which has previously been formed. The rate of oxidation due to such movement will be $\alpha k_l/y$, where α is the (small) fraction of the total area having the necessary loose structure, and k_l the (abnormally high) value of the parabolic constant appropriate to the loose structure. The rate of oxidation over the main surface will be $(1 - \alpha)k_m ly$ where k_m is the (lower) value of the parabolic constant appropriate to the dense film. Thus,

$$\frac{dy}{dt} = \frac{\alpha k_l}{y} + \frac{(1 - \alpha)k_m}{y}$$

which is of the usual form, $dy/dt = K/y$. The law is therefore unchanged.

This would not be true of all growth-laws or all forms of out-growth, and it certainly behoves the theoretical corrosionist to take cognisance of all the results which his colleague in the laboratory is obtaining. For the moment there seems no need to introduce revolutionary changes into our system of oxidation kinetics—and that view is also held by K. Hauffe, " Oxydation von Metallen und Metallegierungen " 1956 edition, p. 72 (Springer).

Calculation of the Numerical Value of the Rate-constant. It is not possible to calculate from theoretical principles the values of the constants involved in all the various laws, but in the case of the parabolic law, $dy/dt = K_2/y$, the value of K_2 is capable of calculation from the mobility or diffusivity of the moving particles. In an early paper, Wagner accomplished this for the oxidation of copper. He started from the assumption that the particles move partly under an electrical potential gradient and partly under a chemical potential gradient (C. Wagner, *Z. Phys. Chem.* (*B*) 1933, **21,** 25).*

Method based on the concept of a purely electrical gradient. Hoar and Price obtained Wagner's expression for K_2 by assuming the gradient to be purely electrical; this is not strictly true, and the validity of their argument has been challenged. However, their proof is so much simpler than that of Wagner that it is profitable to ask whether conditions exist under which their assumption will introduce no error.

There is an essential difference between movement under the two types of gradient. An electrical potential gradient imposed on a film of constant composition produces a net movement in one direction by decreasing the energy required for an ion to cross a barrier in that direction and increasing the energy required for passage in the contrary direction. A chemical potential gradient imposed on a film in absence of an electric field increases the number of ions available to move in one direction, reducing that in the other, so that there will be a net movement in the first direction. However, in oxidation, the driving force will be the same, and also the energy dissipated as heat will be the same, whether the movement occurs under an electrical gradient, chemical gradient or both.

Validity of the Method. Consider a number of beads threaded on long frictionless vertical strings. A bead released at the top will fall with acceleration, g, and at the bottom its kinetic energy will be converted to heat-energy equal to mgh, where m is the mass of the bead and h the height of the string. Now consider that the strings (still frictionless) are horizontal, so that gravity does not operate, but a magnetic field is imposed which (the beads being magnetic) produces an acceleration equal to g; again the heat produced on impact after distance h will be mgh. If we incline the string and adjust the field so that the acceleration (now due partly to gravity and partly to magnetism) is still g, the heat produced remains mgh.

Now suppose that we surround the string with a viscous liquid, so that, except in the opening stages, there is no acceleration, since the bead velocity settles down at such a value that the heat produced in overcoming the viscous resistance exactly balances the loss in potential energy as the bead moves through the field. *One and only one value of the velocity* will fulfil this condition, since the viscous resistance increases with velocity and (for a given viscous liquid) the limiting velocity attained will be the same, provided that the total energy to be transformed into heat (equal to mgh, if we neglect the opening stages during which there is acceleration) is the same in all cases. It does not matter whether the movement is due to a gravitational field, a magnetic field or both; indeed the statement would

* The equations of Bardeen, Brattain and Shockley suggest that both types of gradient contribute almost equally to the movement; they say " the effect of the electrostatic field in the oxide is to double the current to be expected from diffusion alone ". They are referring to copper in air about 1000°C. Probably this view is not held widely today, but time may yet justify it (J. Bardeen, W. H. Brattain and W. Shockley, *J. chem. Phys.* 1946, **14,** 714).

remain true even if there were no field at all, the beads being propelled by means of internal microscopic motors driving screw-propellors with the power adjusted to involve a total energy output equal to *mgh* during the journey.

The same principle will apply to particles crossing an oxide-film. To produce a given increment of thickness, the total number which must cross will be the same, and the total energy change will be the same, whether the propulsion depends on an electrical potential gradient, a chemical-potential gradient or both. The mean velocity attained should be approximately the same in all cases—subject to certain assumptions stated below—and this conclusion is not invalidated by the fact that a chemical-potential gradient is not a " field " in the true sense of the word. The assumptions are that (1) the initial stages and all changes comprised in the term " boundary reaction " can be neglected, (2) the resistance forces are independent of the nature of the field; this might not be the case if the particles possessed dipole moments and seriously non-spherical shapes, so that, under a very high field, they would present different resistances according to the nature of the directing forces. In practice, that will not occur with the particles and field-strengths operative under conditions of oxidation where the simple parabola may be expected, and the neglect of the initial acceleration is also justifiable, since in cases where boundary events exercise control, the simple parabola must be replaced by the mixed parabola.

Thus, *for the conditions under study, the Hoar-Price proof appears to provide a legitimate basis for calculating the value of the growth constant.* For some other purposes, it would be advisable to go further into details. Particularly the existence of layers in which the anions and cations are not present in equivalent quantities, thus producing a space-charge, requires attention; such layers may occur at the boundaries on thick films, but on thin films might embrace the whole (cf. H. J. Engell, K. Hauffe and B. Ilschner, *Z. Elektrochem.* 1954, **58**, 467, 478).

Calculations of the Expression for the Rate-constant in terms of Electrical Magnitudes. The Hoar-Price method is valuable in making clear the manner in which resistance either to ions or to electrons can restrict oxidation. A feature of film-growth is that it requires *both* (1) the passage of electrons through the film, and *also* (2) the passage of cations *or* anions *or* both (it is not necessary that both types of ions should cross the film, but if they do both participate in the conductivity, they will co-operate, not oppose one another, since, although they move in opposite directions, they carry opposite charges). The state of affairs is closely analogous to a primary electric cell having its terminals joined by a wire. Ions flow through the liquid and electrons flow through the wire; if, by snipping the wire, we stop the flow of electrons, the flow of ions through the cell will cease also; if, by emptying out the liquid, we stop the flow of ions, the flow of electrons through the wire will cease also. Here we have two resistances *in series*, the internal or ionic resistance (R_i) and the external or electronic resistance (R_e), the total resistance being $R_i + R_e$. Similarly, for the film-growth, if electronic movement should cease, ionic movement would cease also, and *vice versa*; here again we must *add* the ionic resistance and electronic resistance to obtain the total resistance.

The specific conductivity κ of oxide is the current which will flow across a cube of edge-length 1 cm. when 1 volt is applied to opposite faces. In the most general case, electrons, cations and anions may all contribute to this current in proportions represented by the three transport numbers n_E, n_C and n_A, where $n_E + n_C + n_A = 1$. The specific electronic conductivity is $n_E\kappa$ and the specific ionic conductivity is $(n_C + n_A)\kappa$. The electronic resistance R_e of a film of area A cm.² and thickness y cm. will be $\dfrac{y}{An_E\kappa}$, whilst the ionic resistance R_i will be $\dfrac{y}{A(n_C + n_A)\kappa}$. Thus

$$R_e + R_i = \frac{y}{A\kappa}\left(\frac{1}{n_E} + \frac{1}{n_C + n_A}\right) = \frac{y}{A\kappa n_E(n_C + n_A)}$$

since $n_E + n_C + n_A = 1$. By Faraday's Law, $dy/dt = \dfrac{JI}{FAD}$ where J is the equivalent weight of the oxide, D its density, F Faraday's Number and I the current flowing.

By Ohm's Law

$$I = \frac{E}{R_e + R_i} = EA\kappa n_E(n_C + n_A).\frac{1}{y}$$

Hence

$$\frac{dy}{dt} = \frac{EJ\kappa n_E(n_C + n_A)}{DF}.\frac{1}{y}$$

Thus in the parabolic equation, $dy/dt = K_2/y$, the rate-constant, K_2, expressed in terms of electrical magnitudes, is $EJ\kappa n_E(n_C + n_A)/DF$ (T. P. Hoar and L. E. Price, *Trans. Faraday Soc.* 1938, **34,** 867).

If the electrons can pass across the film very easily, so that n_E is nearly unity, and if the whole of the ionic movement is due to cations ($n_C \gg n_A$), the expression becomes $J\kappa n_C E/DF$. These conditions are fulfilled for cuprous oxide at 1000°C. and, since the E.M.F. of the cell copper | oxygen was determined in early work by Treadwell and the value of n_C by Dünwald and Wagner, it is possible, after applying some corrections for oxygen-pressure, to calculate the value of K defined by the equation $d\eta/dt = K/y$, where η is the weight of oxide expressed as equivalents per cm.². Wagner thus obtained results which agreed well with the value determined by Feitknecht; the calculated value was 6×10^{-9} and the experimental number 7×10^{-9}. The pertinent references are C. Wagner, *Z. phys. Chem.* (*B*) 1933, **21,** 25; H. Dunwald and C. Wagner, *ibid.* (*B*) 1933, **22,** 212; W. D. Treadwell, *Z. Elektrochem.* 1916, **22,** 414; W. Feitknecht, *ibid.*, 1929 **35,** 142.

In cases where the E.M.F. of the cell Metal | Oxygen has not been directly measured, it can be obtained from the free-energy change, which is ζEF, where ζ is the valency number.

It may often be easier to measure the self-diffusion coefficient than the transport number, and if the activities of metal in the film at the inner and outer surfaces of the film, under conditions of equilibrium with the metallic phase and gaseous phase respectively, are a_i and a_0, the "rational rate constant", K_r (the equivalents passing each second across 1 cm.^{2} of film when the film is 1 cm. thick), can be obtained from a formula due to Wagner. This formula, which is valid only when n_E is close to unity, is

$$K_r = C\int_{a_0}^{a_i} (D_C^* + bD_A^*)d \log_e a$$

where C is the concentration of metallic or non-metallic ions expressed in equivalents per cm.^{3}, b is the ratio of the anionic and cationic valencies (neglecting signs), whilst D_C^* and D_A^* are the self-diffusivities of cation and anion respectively. For a film of thickness y, the number of equivalents passing will be K_r/y, which is of course much higher than K_r, since y is small compared to 1 cm.*

Comparisons between Calculated and Observed Values. Table XXVIII, due to Wagner, shows the remarkable agreement between calculated and observed values of the rational rate constant K_r.

TABLE XXVIII

Calculated and Observed Values of Rational Rate-constants expressed in equiv. cm.$^{-1}$sec.$^{-1}$ (C. Wagner, "Atom Movements" (1951), p. 153 (Amer. Soc. Metals))

Reaction	Temperature, etc.	Calculated	Observed
$2Ag + S = Ag_2S$	220°C.	2 to 4 × 10^{-6}	1·6 × 10^{-6}
$Cu + \frac{1}{2}I_2 = CuI$	195	3·8 × 10^{-10}	3·4 × 10^{-10}
$Ag + \frac{1}{2}Br_2 = AgBr$	200	2·7 × 10^{-11}	3·8 × 10^{-11}
$2Cu + \frac{1}{2}O_2 = Cu_2O$	1000° & 8·3 × 10^{-2} atm.	6·6 × 10^{-9}	6·2 × 10^{-9}
$2Cu + \frac{1}{2}O_2 = Cu_2O$	1000° & 1·5 × 10^{-2} atm.	4·8 × 10^{-0}	4·5 × 10^{-9}
$2Cu + \frac{1}{2}O_2 = Cu_2O$	1000° & 2·3 × 10^{-3} atm.	3·4 × 10^{-9}	3·1 × 10^{-9}
$2Cu + \frac{1}{2}O_2 = Cu_2O$	1000° & 3·0 × 10^{-4} atm.	2·1 × 10^{-9}	2·2 × 10^{-9}

Himmel, Mehl and Birchenall, measuring self-diffusion by radio-active methods, have confirmed the Wagner mechanism, obtaining good agreement between the observed and calculated values of the constants, not only for the growth of oxide on a metallic substrate, but also for the conversion of one oxide to another, e.g. wüstite to magnetite (L. Himmel, R. F. Mehl and C. E. Birchenall, *J. Metals* 1953, **5**, 827).

In considering the agreement between the "Calculated" and "Observed" values, the reader should note that there has been *no arbitrary assumption of the value of any quantity used in the calculations.* If the mechanism under consideration were wrong, there would be no reason to

* Hauffe emphasizes the need for correct application of the self-diffusion coefficients. "Where we have a p-conducting layer, then it is necessary to determine the self-diffusion coefficients of the cations in this layer which is *in equilibrium with the corroding atmosphere at the same pressure as during the oxidation.* In the other case where we have a n-conducting layer, then the determination of the self-diffusion coefficient is right when the other phase is not the corroding gas but the metal or the alloy."

expect that the " Obs." and " Calc." numbers would even be of the same order of magnitude (see p. 768).

Kinetics of oxidation of metal forming two or more oxide-layers. Treatment of cases where two layers appear during oxidation is complicated, unless it happens that one can be regarded as very porous compared to the other; Pilling and Bedworth thought that on copper the effect of the outer cupric oxide layer could be disregarded and that movement through the inner cuprous oxide controlled the oxidation process; this may have been true under the conditions of their work. There has been some divergence of opinion on the kinetics of the more general case; cobalt has been a metal much discussed. Readers should study the individual views (N. B. Pilling and R. E. Bedworth. *J. Inst. Met.* 1923, **29**, 529, esp. p. 556. G. Valensi, *Rev. Métallurg.* 1948, **45**, 205; *Metallurg. ital.* 1950, **42**, 77. E. A. Gulbransen and K. F. Andrew, *J. electrochem. Soc.* 1951, **98**, 241. O. Kubaschewski and B. E. Hopkins, " Oxidation of Metals and Alloys " (1953), pp. 1, 137 (Butterworth). K. Hauffe, chapter in " Passivierende Filme und Deckschichten " 1956, p. 217 (edited by H. Fischer, K. Hauffe and W. Wiederholt; published by Springer)).

Effect of Temperature

Activation Energy. If W_a is the energy needed for a particle to pass from one position of minimum energy over an energy barrier to the next position, the probability of a particle attaining such energy is given by $e^{-W_a/kT}$. This, in a general way, applies to ions passing from one site in the oxide-lattice to the next one. If the parabolic rate-constant is proportional to $e^{-W_a/kT}$, its logarithm plotted against $1/T$ should give a straight line. If movement through the oxide depends on vacant lattice-sites, energy is needed to form a vacancy and energy is needed for a neighbouring ion to move into it, but, since the concentration of particles possessing the requisite energy is in each case given by an expression of the form $e^{-W/RT}$, we may still expect* the rectilinear relationship between $\log K_2$ and $1/T$. From the gradient of the line it is possible to calculate the activation energy of the oxidation process but there is some disagreement as to the physical meaning to be attached to the number thus obtained. Some regard it as the work needed to produce sufficient distortion to the lattice to enable the particle to squeeze its way from one position of low energy to another.

A free-energy change ΔW can be split up into $\Delta H + T\Delta S$ where ΔH is the heat and ΔS the entropy. Such a procedure has been widely adopted

* If the activation energy needed for the formation of a vacancy is W_f and that for movement of a neighbouring ion into a vacancy is W_m, the separate probabilities of the combined events are proportional to $e^{-W_f/kT}$ and $e^{-W_m/kT}$ respectively, so that the combined probability of both events occurring at a point within a given element of time will be proportional to the product $e^{-(W_f+W_m)/kT}$—assuming always that the two probabilities are *uncorrelated*. If the parabolic rate-constant (K_2) is proportional to this expression, then $\log K_2$ plotted against $1/T$ should give a straight line.

by those interested in the kinetics of chemical reactions in solution or in a gas-phase; the old Arrhenius method of writing the reactional velocity $Ae^{-W/RT}$ where A depends on the number of molecular collisions, and $e^{-W/RT}$ represents the chance that the energy will be sufficient for the collision to lead to reaction does not satisfactorily explain all cases, and the writing of $e^{-W/RT}$ in the form $e^{\Delta S/R}.e^{-\Delta H/RT}$ has been found helpful by many physical chemists, but not by all. Moelwyn-Hughes writes, " It is difficult to see what, apart from a certain uniformity of notation, is being gained by its adoption. There is, moreover, no general agreement among workers in the field of kinetics on the meaning of the term 'entropy of activation' " (E. A. Moelwyn-Hughes, " Physical Chemistry " 1957, p. 1238 (Pergamon Press); cf. Sir C. N. Hinshelwood, *Chem. and Ind.* (*Lond.*) 1957, p. 1642).

The concept of " entropy of activation " has been widely used in discussions of oxidation kinetics, and, however much the reader may agree with Moelwyn-Hughes, he will be forced to familiarize himself with the notation —if only because he will find so much of the numerical data tabulated in terms of it. Gulbransen and Andrew express the parabolic rate-constant involved in the oxidation of nickel, as

$$K(p_0^{1/6})e^{\frac{\Delta S^\circ/3+\Delta S^*}{R}}.e^{\frac{-(\Delta H^\circ/3+\Delta H^*)}{RT}}$$

where $\Delta S^\circ/3$ and $\Delta H^\circ/3$ represent entropy and heat for vacancy formation, whilst ΔS^* and ΔH^* represent the entropy of activation and heat of activation for diffusion. K represents $\frac{2}{4^{1/3}}\gamma \nu a^2 \Omega N$, where γ is a constant depending on the geometry of the jumps, ν the vibration frequency in the appropriate direction, a the distance between sites, Ω the volume of oxide formed per ion and N the number of ions per c.c.

Frequently the entropy of activation is negative—a matter of some practical importance in connection with the effect of temperature on oxidation-rate.* The meaning of these negative values has been regarded in different ways by Gulbransen and by Moore. The latter has provided a collection of ΔH^* and ΔS^* values, reproduced in Table XXIX. It will be noticed that zirconium appears in two sections.

Those interested in the heat and entropy contributions should consult E. A. Gulbransen and K. F. Andrew, *J. electrochem. Soc.* 1950, **97,** 383; 1951, **98,** 241; 1954, **101,** 128; *Annals New York Acad. Sci.* 1954, **58,** 830; W. J. Moore, *J. electrochem. Soc.* 1953, **100,** 302; C. Zener, *J. appl. Phys.* 1951, **22,** 372; W. W. Smeltzer, *J. electrochem. Soc.* 1956, **103,** 209.

In recent years there have been endeavours to supply a physical meaning for the term " Entropy of Activation ". The concept has been used in interpreting an important research at Exeter into the oxidation of magnesium about 350–600°C. In dry oxygen, the results are consistent with the idea that the thinnest oxide-film (possibly in tension and continuous)

* A negative value of ΔS^* does not, of course, mean that the oxidation decreases as temperature rises, since it occurs in an exponent.

TABLE XXIX

ACTIVATION HEATS AND ENTROPIES (W. J. Moore)

	ΔH^*	ΔS^*
p-type oxides with cation vacancies		
Cu | Cu_2O at 800–1000°C.	33·8†	+ 0·5†
Ni | NiO at 500–1000°C.	34·7	− 17·0
Co | CoO at 700–1000°C.	36·3	− 10·0
Fe | FeO | $Fe_{34}O_4$ at 700–950°C.	22·4	+ 8·6
V | V_2O_3 at 400–600°C.	30·7	− 7·2
Mo | MoO_2 at 350–450°C.	36·5	+ 0·5
n-type oxides with interstitial metal		
Zn | ZnO at 360–400°C.	28·5	− 22·6
Al | Al_2O_3 at 350–450°C.	21·4	− 28·6
Be | BeO at 750–950°C.	50·3‡	− 7·5‡
Cr | Cr_2O_3 at 700–900°C.	66·3	+ 12·0
Zr | ZrO_2 at 500–900°C.	32·0	− 7·1
W | WO_3 at 400–500°C.	45·7	+ 11·0
Ta | Ta_2O_5 at 250–450°C.	27·4	− 11·0
n-type oxides with anion vacancies		
Zr | ZrO_2 at 200–425°C.	16·8	− 27·3
Ti | TiO_2 at 350–600°C.	24·3	− 22·8

† G. Valensi, however, considers ΔS^* to be negative, namely — 0·3; his number for ΔH^* is 35·5.

‡ These are the figures of D. Cubiccioti. E. A. Gulbransen and K. F. Andrew find ΔH^* 59·5 and ΔS^* — 3·4.

is protective, but that after a certain thickness has been reached cracks appear and extend down to a certain distance from the metal, so that the oxidation-rate rather suddenly increases and then remains constant, since the oxidation is now controlled by passage through an uncracked layer of constant thickness (see p. 40). After a time a second acceleration occurs, attributed to the cracks extending right down to the metal, which is appreciably volatile and starts to burn in the gas phase.

When moisture is present, the induction period which must elapse before the " breakaway " (at which the oxidation-rate increases to the constant value) is shortened, whilst the rate corresponding to this rectilinear period is increased. The water probably converts O^{--} ions to OH^- ions and, in order to preserve electrical neutrality, vacancies must be introduced at half the cation sites, thus distorting the structure and reducing the value of the activation energy, E, in the expression $Ae^{-E/RT}$. Since E occurs in the index, this might be expected to cause a very large increase in the oxidation-velocity, but for the fact that A, which contains the entropy factor $e^{\Delta S/R}$ is greatly diminished, for the distortion will diminish the disorder caused when a magnesium ion moves to the top of a potential barrier. Thus the increase of velocity caused by 3·2% water in the oxygen is only about ten-fold. The paper will be helpful in understanding the oxidation of other metals like titanium and zirconium which form protective films that sometimes break down, and should be studied carefully (S. J. Gregg and W. B. Jepson, *J. Inst. Met.* 1958–59, **87,** 187; cf. A. R. Ubbelohde, *Disc. Faraday Soc.* 1957, **23,** 128).

Plotting of the Temperature Effect. In cases where the parabolic law is obeyed over the range of temperature, without complicating factors, it should be possible to obtain straight lines on plotting $\log_{10} Q_t$ against $1/T$, where Q_t is the quantity of oxygen taken up in some standard time t. The gradient of the line enables us to calculate the activation energy.* Such straight lines were established in early work on copper and brass by Vernon and by Dunn, and similar relationships have been found for many metals by subsequent investigators. On iron, Portevin and his colleagues found two lines intersecting at about 925°C.—apparently owing to the $\alpha \longrightarrow \gamma$ transformation (W. H. J. Vernon, *J. chem. Soc.* 1926, p. 2273, esp. p. 2276; J. S. Dunn, *J. Inst. Met.* 1931, **46,** 25, esp. p. 36. A. Portevin, E. Prétet and H. Jolivet, *Rev. Métallurg.* 1934, **31,** 101, 186, 219; *J. Iron St. Inst.* 1934, **130,** 219, esp. p. 237).

Where there is a change from one growth-law to another, there will clearly be a break in the curve. This was first noted by Vernon, who found that on plotting $\log Q_t$ against $1/T$, two straight lines were obtained, cutting at 200°C., the temperature at which, on his material, logarithmic oxidation gave place to parabolic. It should be noted here that, since the manner in which Q depends on t varies with the growth-law, the angular relationship between the two sections must be expected to vary according to the standard time chosen. This was actually found to be the case in plotting the Davies–Evans–Agar data, where the high-temperature section was steepest when 2, 4, 6 or 8 hours was adopted, but definitely less steep at 15 or 30 min. (fig. 147). The fact that, owing to a different previous treatment of the iron, the passage from one law to another occurs at a higher temperature than in Vernon's work, does not affect the principle involved. The reader may like to compare the diagrams quoted, by referring to W. H. J. Vernon, *Trans. Faraday Soc.* 1935, **31,** 1668, esp. p. 1673; D. Eurof Davies, U. R. Evans and J. N. Agar, *Proc. roy. Soc.* (*A*) 1954, **225,** 443, esp. p. 457.

Dravnieks has studied the parabolic rate-constants of the various metals at $0{\cdot}6T_M$ where T_M is the melting-point on the absolute scale. The crude numbers seem to bear little relation to the other properties of the metals but he proceeds to calculate certain " reduced diffusion rates " with instructive results. He imagines one mole volume of oxide made into a cube, and placed on the surface of the metal, so that the N metal ions provide $N^{2/3}$ paths, each $N^{1/3}$ ions long. He calculates the diffusion-rate r' referred to equal numbers of paths and jumps, and then divides r' by F_0, the free energy of oxide-formation—thus obtaining r, the diffusion-rate for unit driving force. The values of r now begin to show a degree of regularity. For divalent oxides of the same lattice types, r increases as the cation-size decreases, whereas in the trivalent and tetravalent oxides, it increases with increase of cation size. The values of $\log r$ are found to fall nearly on a straight line when plotted against $\Delta F^*/RT$, where $\Delta \mathbf{F}^*$ is the free energy of activation. The values of $\log r$ vary from 7·7 to 11·7, indicating

* A rapid method of assessing the activation energy is put forward by P. Kofstad, *Nature* 1957, **179,** 1362.

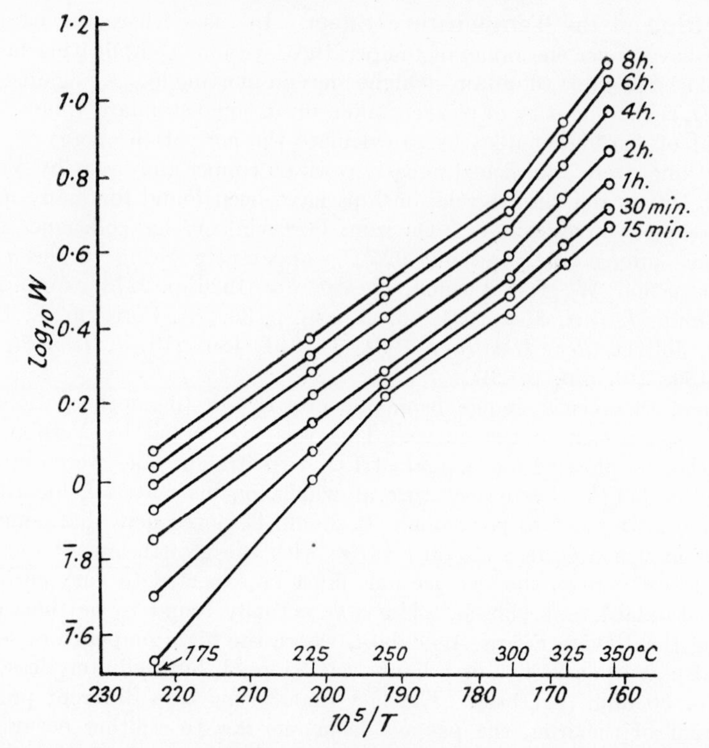

FIG. 147.—Influence of temperature on oxidation-rate of Iron (D. Eurof Davies, U. R Evans and J. N. Agar).

diffusion-rates which extend over four orders of magnitude. Various other apparent regularities are discussed in the paper, which deserves study (A. Dravnieks, *J. electrochem. Soc.* 1953, **100**, 95).

Conditions for Film Break-down and Leakage

General. It has been mentioned that films sometimes break down spontaneously during oxidation, so that the oxidation-rate suddenly increases; also that films which have ceased to grow in air (especially the invisible films formed at ordinary temperatures) break down in contact with water. In seeking the conditions needed for such break-down, it is necessary to distinguish between the two cases where the film grows (1) by inward movement of an anion (e.g. oxygen) and (2) by outward movement of a metallic cation. It should be observed that in some cases both forms of movement occur, and theoretically this should always be possible when the film is very thin and the electric field very strong; for even though the two activation energies, W_a and W_c demanded for the site-to-site jumps of the two ions be very different, the rates of movement (proportional to $e^{(-W_a+K_1/y)/kT}$ and $e^{(-W_c+K_1/y)/kT}$ respectively) may both be appreciable. Suppose that W_a is greater than W_c, then, at a certain thickness, K_1/y will become too small to overcome the effect of W_a and anion movement will practically cease; at another (greater) thickness, cation movement will cease also and thereafter the film will not appreciably thicken further. Thus there should be a range of thickness over which movement occurs almost exclusively in one direction.

Cases where the Anion moves Inwards. Provided that the Pilling–Bedworth ratio exceeds unity, *thin* films formed by inward movement of oxygen are likely to be protective. They will carry considerable compressional stresses, due to the excess of M/D over m/d (see p. 39) and if the surface layers of the metal before oxidation carry internal stresses, these will be inherited by the film and may either increase or decrease the stress set up by the volume discrepancy; possibly around surface blemishes, there will be increase of stress at some points and decrease at others. However, so long as the film is thin, its spontaneous collapse will at most points be impossible; external work would be needed to damage it.

At a certain limiting thickness, this immunity will vanish. Consider a small region of area a, where the adhesion of the film is slightly lower than elsewhere. Let the adhesional work per unit *area* be W_A, and the compressional energy per unit *volume* in the film be W_C. Evidently detachment and break-down can only occur spontaneously when $ayW_C > aW_A$. In other words, when y exceeds W_A/W_C, the area can detach itself (perhaps as a microscopic blister) and break up under the compressional force. (Rather similar considerations determine crack-formation in films of electro-deposited metal, carrying internal stresses—a matter discussed on p. 632.)

The best established case of inward movement of anions during oxidation is provided by titanium. Here the film formed at low temperatures is a thin, dense, grey layer, but at high temperatures it is thick, porous, yellow-brown scale, composed of layers of oxide, which, in the words of Jenkins (quoted at greater length on p. 44) " have been twisted and shattered like natural rock strata ". It becomes clear why the thin invisible film normally present on titanium prevents corrosion by aqueous reagents, why, in contrast with some other metals, there is little or no tendency for break-down to occur at surface defects or places of internal stress, and why the thicker film formed at higher temperatures gives very little protection.

Cases where the Cation moves Outwards. Different factors prevail where the film grows by cations moving through it to take up positions on the outer surface. Apart from some stress arising from epitaxial considerations in the thinnest films, the film-substance should be almost free from constraint, and the causes of break-down mentioned above do not operate. However, the cations moving through the metal leave vacancies at the base of the film, and, although some of these may be absorbed at dislocations, others will join together to form cavities, and sooner or later, as the film thickens, it will locally loose touch with the metal. This lack of support is most likely to cause break-down at places where the metal has suffered some sort of surface treatment which has left a complicated system of internal stresses—some tensional and some compressional—in equilibrium (see p. 109). At a point where the stress gradient is high, the removal of metal into the film will upset this equilibrium and a very small resultant stress will be applied to the locally unsupported film, which being very thin, is likely to break. Suppose, for instance (fig. 27, p. 109) that the metal, before oxidation, was tensional near the surface and compressional below. After the film-formation, the tensional layer has partly vanished, being replaced by a cavity. Evidently now the metal is predominantly in compression, and in its efforts to expand will apply tension to the unsupported film, which will rupture. A new film will form, which, if sufficient internal stresses remain, will rupture in its turn. It is to be expected that this "crack-heal process" will continue until internal stresses have been sufficiently exhausted. Experimental evidence for crack-heal is provided on p. 176.

With the destruction of the material at the exceptional points where the stress gradient is high, the cracking will become gradually less frequent; this probably accounts for Mears' results on inception probability at scratch-lines, which, in a weakly inhibitive liquid, falls off steadily as exposure to air becomes longer (p. 52). However, a specimen of metal which has almost ceased to develop cracks when exposed to dry air may start once more when placed in water. The criterion for spontaneous cracking in such a case requires to be considered. Imagine a circular area of radius r where the film is unsupported, and suppose that the metal carries net compressional stress. Cracking—perhaps across a diameter—is only possible if the reduction of strain energy in the metal, say $\pi r^2 \omega_s$ exceeds the increase of interfacial energy $2r.y.\omega_i$. In a given environment ω_i (the interfacial energy per unit area of the crack) is constant with time, but ω_s (the diminution of strain energy per unit area which results from cracking) decreases on each successive occasion of cracking. Cracking cannot occur unless $\pi r^2 \omega_s > 2r\omega_i.y$, and in practice, owing to the activation energy required, it will probably cease *before* y reaches the value $\pi r \omega_s / 2\omega_i$. However, if now the specimen is plunged into water, ω_i will be smaller, and cracking may be resumed. This possibly explains why corrosion in water starts at places where surface defects or internal stresses are present (p. 108).* If the water contains anions compatible with the passage of metal cations into the liquid (Cl^- ions fulfil this condition for most metals) the film will never be repaired, and corrosion will set in. If it contains inhibitive ions, the film will ultimately be repaired, but the cracking will continue (even at places where the stress gradient is comparatively low) until ω_s has been reduced to correspond to the new value of ω_i.

At low temperatures, it seems probable that oxidation due to crack-

* The fact that the breaking-strength of a material is sometimes decreased when it is wetted with a liquid which has no solvent action has long been known. It is easier to cleave mica when wetted with water. Benedicks has studied this effect in detail; he finds that the bending load needed for the breaking of sugar is reduced to one quarter or less if it is wetted with a saturated sugar solution; reduction of strength is also brought about by several organic liquids. The breaking-strength of gypsum is reduced by various organic liquids without solvent power (C. Benedicks and R. Härdén, *Arkiv för Fysik* 1951–52, **3,** 407; also *Rev. Métallurg.* 1948, **45,** 9).

heal at regions of internal stress will die away asymptotically (not necessarily following the particular asymptotic equation developed in section (*h*) on page 834)—as the internal stresses become used up. However, at higher temperatures, internal stresses in the oxide may be produced continuously as the film is formed—either due to the strain involved in epitaxial growth or, in the case of films formed by oxygen moving inwards, due to the fact that the oxide produced would, if unconstrained, occupy a larger volume than the metal destroyed. In some cases the strain may result from disparity between two oxide-layers, as illustrated by the tendency of duplex films on iron (with a Fe_2O_3 layer over Fe_3O_4) to roll up, or by the manner in which fragments of duplex films on copper (with a CuO layer over Cu_2O) sometimes fly off as saucer-shaped particles. In all these cases, since the work needed to detach the film is proportional to the area detached whilst the relief of strain-energy is proportional to the volume, there must be some value of the thickness (y_F) where flaking (sufficient to relieve most of the strain-energy, but not necessarily amounting to complete detachment of the film over the area in question) can occur spontaneously. Thus film break-down may be expected at intervals sufficient to produce this thickness y_F, and after each break-down the oxidation will suddenly become very rapid—as has been found to occur (pp. 46, 49, 177). If the temperature is sufficiently high, y_F is reached in a time so short that only the second term of the mixed parabola is important, and the effect is to produce roughly rectilinear oxidation at a speed ($y = Kt/R_2$) equal to the starting speed of the mixed parabola (p. 826).

$$\tfrac{1}{2}R_1y^2 + R_2y = Kt$$

If the breaks occur at such long intervals that $R_2y_f \ll R_1y_f^2$, the time needed to reach y_f is $\frac{R_1y_f^2}{2K}$ and the effective thickening-rate of the (approximately) rectilinear growth will be relatively slow, being expressed by

$$\frac{y}{y_f} = \frac{t}{R_1y_f^2/2K} \quad \text{or} \quad y = \frac{2K}{R_1y_f}t$$

Wanklyn suggests another way in which initial parabolic growth would later be replaced by rectilinear growth. Suppose the propagation of a crack, once formed, proceeded more smoothly than the nucleation of fresh cracks—a not unreasonable assumption since the end of a crack is a stress-raiser—then after cracking has once started, oxidation is likely to obey a roughly rectilinear law (J. N. Wanklyn, Priv. Comm., July 12, 1958).

Effect of water and sulphur compounds on the growth-law. It has long been known that moisture and sulphur compounds—separately or in combination—can greatly increase the oxidation rate—especially on materials which normally form protective films. Early work by Hatfield (Table XXX) brings this out.

In some cases the character of the growth-curve changes. Hart (p. 831) found that on aluminium in dry oxygen at 20°C., the inverse logarithmic law is obeyed, leading to practical cessation of attack when a certain thickness

TABLE XXX

WEIGHT-INCREASE IN 42 HOURS AT 900°C. EXPRESSED IN MG./CM.^{2}
(W. H. Hatfield, *J. Iron St. Inst.* 1927, **115**, 486)

	Mild Steel (0·17% C)	Stainless Steel (17·7% Cr, 8·1% Ni)
Pure air	55·2	0·40
Pure air with 5% H_2O	74·2	3·24
Pure air with 5% H_2O, 5% SO_2	152·4	3·58
Pure air with 2% SO_2	65·2	0·86

is reached; small temperature increases do not sensibly alter the oxidation rate. However, in oxygen of 80% relatively humidity at 25°C., the direct logarithmic law is obeyed for about 10 hours, and is then replaced by the inverse law, ending, not with practical cessation of growth, but with a slow continuous increase; after seven days exposure the films were $1\frac{1}{2}$ times as thick as those produced in the same time in dry oxygen.

The effect of traces of sulphur dioxide on the oxidation of copper was studied by Mills. He first made careful measurements in pure oxygen; the initial oxidation-rate slowed down, coming practically to rest, and then suddenly oxidation started again; it was believed that the sudden leakiness, developing in the first (pseudomorphic) film after it had become protective, was due to recrystallization. Both stages obeyed the logarithmic law at relatively low temperatures; no evidence was found for the cubic law reported by some other investigators (p. 832). At 172°C. growth was parabolic at first, and then slowed down, apparently for reasons suggested on p. 836; the facts agreed with those obtained at 180°C. by Dighton and Miley who found parabolic growth at first, slowing down to give a logarithmic relationship (T. Mills and U. R. Evans, *J. chem. Soc.* 1956, p. 2182; A. L. Dighton and H. Miley, *Trans. electrochem. Soc.* 1942, **81**, 321. Cf. J. A. Allen, *Research* 1952, **5**, 487).*

Mills found that small amounts of sulphur dioxide added to the oxygen produced two results. The rate of oxidation was greatly increased, doubtless through the introduction of extra lattice defects, and the time needed for the break-down of the first-formed (pseudomorphic) film was shortened—as can easily be understood, since atomic rearrangements occur more readily in faulty structure. This acceleration increased with the amount of sulphur dioxide added, but only occurred if the sulphur dioxide reached the unoxidized

* Williams and Hayfield studying the oxidation of electropolished copper by means of polarized light find smooth growth, conforming to the logarithmic law, for the first minute, followed by a period of jerky growth, finally resulting in renewed smooth growth conforming to the logarithmic law but with a different gradient when log t is plotted against y. If we assume that recrystallization spreads from nuclei appearing sporadically at different points at different moments and that, at a place where the atoms are re-arranging themselves, the permeability becomes abnormally great, the jerkiness of the growth which, the authors feel, is not due to experimental error, can be understood (E. C. Williams and P. C. S. Hayfield, *Inst. Met. Monograph,* No. 23 (1957) p. 131, esp p. 143 and figs. 12 and 13).

metal surface; if an oxide-film had been formed by previous exposure to air before the sulphur compound arrived, there was no acceleration—explaining earlier results (p. 76). The second result was the formation of cuprous sulphate in the film, in an amount which was a rectilinear function of the cuprous oxide present; no cuprous sulphate was produced until the cuprous oxide had reached a certain amount, but thereafter the amount of sulphate was proportional to the excess of oxide over that limit, the proportionality being independent of the sulphur dioxide content of the gas phase. The cuprous sulphate did not affect the rate of oxidation. Mills interpreted his results in the light of work on adsorption upon cuprous oxide published by W. E. Garner, T. J. Gray and F. S. Stone, *Disc. Faraday Soc.* 1950, **8,** 246; F. S. Stone and P. F. Tiley, *ibid.*, 1950, **8,** 254. Readers may care to study the original paper (T. Mills and U. R. Evans, *J. chem. Soc.* 1956, p. 2182) in connection with early work by W. H. J. Vernon, *Trans. Faraday Soc.* 1927, **23,** 113; F. H. Constable, *Proc. roy. Soc.* (*A*) 1929, **125,** 630.

Other References

Many aspects of oxidation not considered in this chapter—notably effects attributable to space charges—are discussed in a number of authoritative text-books. These include K. Hauffe, " Oxydation von Metallen und Metallegierungen " (Springer); K. Hauffe, " Reaktionen in und an festen Stoffen " (Springer); O. Kubaschewski and B. E. Hopkins, " Oxidation of Metals and Alloys " (Butterworth); " Passivierende Filme und Deckschichten " (edited by H. Fischer, K. Hauffe and W. Wiederholt; published by Springer). Useful summaries of oxidation theory up to 1951 and 1956 respectively are provided by K. Hauffe, *Werkst. u. Korrosion* 1951, **2,** 131, 221, 243, and by M. T. Simnad, *Industr. engng. Chem.* 1956, **48,** 586. The stand-point of the solid-state physicist is well presented by T. B. Grimley in Chapter 14 of W. E. Garner's " Chemistry of the Solid State " (Butterworth).

Film-growth under immersed conditions often follows the same laws as film-growth on metal exposed to dry air or oxygen. Examples will be found on p. 101 (footnote) and p. 825, but others merit attention. A logarithmic growth-law for the formation of an oxide-film on stainless steel immersed in aqueous solution containing oxygen or an oxidizing agent was suggested by the measurements of I. D. G. Berwick and U. R. Evans, *J. appl. Chem.* 1952, **2,** 576, and more definite evidence has been provided by M. Stern, *J. electrochem. Soc.* 1959, **106,** 376.

CHAPTER XXI

VELOCITY OF WET CORROSION

Synopsis

The electrochemical mechanism of wet corrosion has been described qualitatively in the earlier part of the book. Chapter IV dealt mainly with the corrosion connected with currents flowing between different areas on a single metal, Chapter VI with corrosion phenomena produced by contact between dissimilar metals, Chapter IX mainly with corrosion produced when acids or alkalis displace hydrogen, whilst various examples of corrosion acting conjointly with stress, strain or surface abrasion were introduced into Chapters X, XVI, XVII and XVIII. Although in many cases the qualitative evidence furnishes proof of the real existence of the electric currents, the electrochemical mechanism only became completely established after it had been shown that the strength of the current flowing is, in the cases studied, equivalent, in the sense of Faraday's Law, to the measured corrosion-rate. In the first section of the present chapter, evidence is presented for this equivalence; the argument does not involve any *ad hoc* assumption of arbitrary values for the constants appearing in the equations.

Next, the factors which decide the magnitude of the electric currents are considered, with the help of a graphical construction; this shows why, for instance, corrosion-velocity often first rises with salt-concentration and then falls. Then corrosion of the hydrogen-evolution type is considered; here also satisfactory agreement between observed and calculated values of the corrosion-rate has been achieved.

Attention is now turned to stimulation of corrosion by an applied anodic current, and its diminution or prevention by a cathodic current. Anodic attack has already been discussed in Chapter VII, and its quantitative treatment presents no difficulty provided that the applied current is high compared to the "natural" corrosion-current; the predictions from Faraday's Law are then often fulfilled with fair accuracy, assuming that passivity does not set in. The process of passivation is capable of mathematical expression, both as regards the dependence of passivation-time upon initial current density and also the manner in which the current flowing diminishes with time as passivity develops. The protection afforded by a cathodic current, which has already been discussed qualitatively in Chapter VIII, now receives quantitative consideration; special attention is given to the criteria which are generally believed to show the point at which the current applied to a protected structure becomes sufficient to stop corrosion completely.

A section on potential measurements and their significance follows, with special reference to potential-time curves. Finally, recent developments in thermodynamic and kinetic electrochemistry are considered, and their applications to problems of metallic corrosion—including the value of Pourbaix diagrams and the significance of Piontelli's division of the metals into two classes differing in electrochemical behaviour.

Comparison between Observed and Calculated Corrosion-rates (Oxygen-absorption Type of Attack)

Partly Immersed Zinc. The case of zinc will first be discussed since, owing to the fixed valency, it is simpler than that of iron.

That an electric current was really passing through the liquid between the anodic and cathodic regions of zinc partly immersed in salt solution

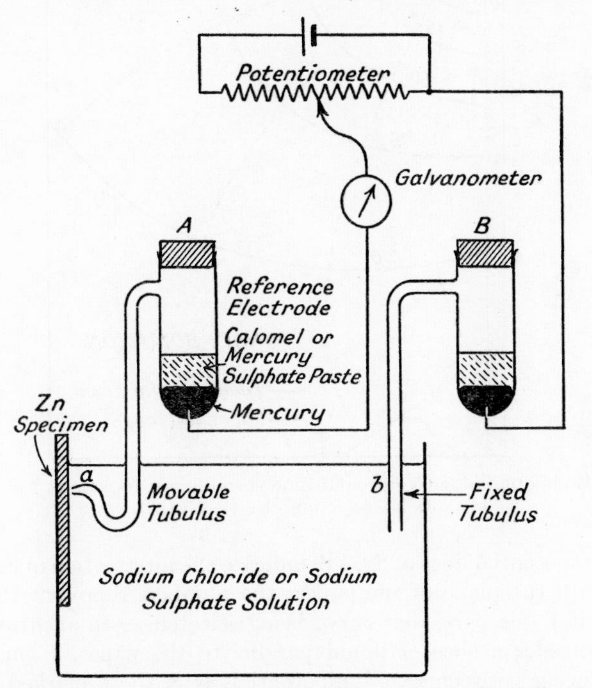

FIG. 148.—Apparatus for measuring corrosion currents on Zinc (J. N. Agar and U. R. Evans).

had been shown qualitatively in early work at Cambridge (p. 90). In 1939, just before the war interrupted this pure-science research, Agar succeeded in measuring that current, and compared it with the corrosion-rate as measured by weight-loss and chemical estimation in the corrosion-product. He used (fig. 148) rectangular specimens of zinc sheet in sodium chloride (or sulphate) with two calomel (or mercurous sulphate) electrodes connected by tubuli, one (B) being fixed at a point a few inches distant from the zinc, while the other (A) could be moved about at will by racking in any of three directions; the three co-ordinates defining the position of the tubulus-tip were read off on three scales. In that way, equipotential surfaces in the liquid could be traced (fig. 149), and, the conductivity of the liquid being known, the current flowing could easily be calculated on the basis of Ohm's Law.

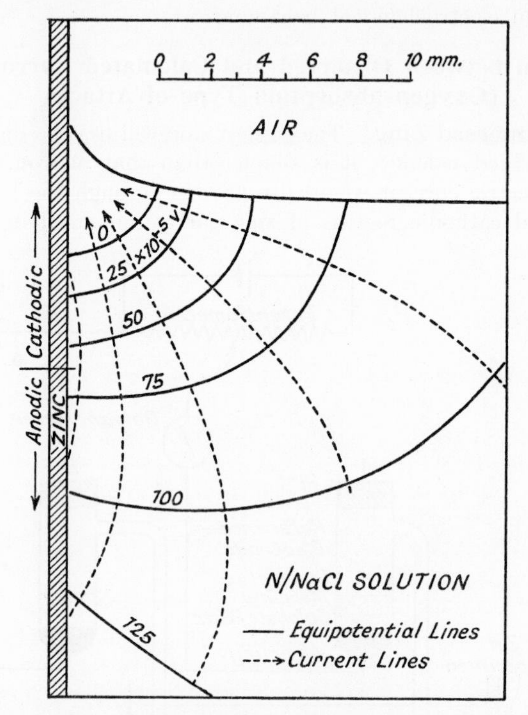

FIG. 149.—Equipotential curves around Zinc corroding in N Sodium Chloride solution (J. N. Agar and U. R. Evans).

The equipotential lines of fig. 149 indicate the intersection of equipotential surfaces with the plane of the paper; the numbers represent the potential (as millivolts), one particular curve being selected as an arbitrary zero. If now we consider a slice of liquid parallel to the paper, 1 cm. thick, the current flowing between two equipotentials (e.g. those marked 50 and 75) is approximately

$$\frac{L\kappa}{h}(V_1 - V_2)$$

where L is the average length of the two equipotentials, h the average distance between them, κ the specific conductivity of the salt solution, whilst V_1 and V_2 are the voltage numbers on the two curves. The current, divided by Faraday's number, gives the corrosion velocity in gram-equivalents per second. The currents measured were strong enough to account for the greater part of the corrosion actually observed, and probably corresponded to the whole of the attack leading to loose corrosion-product. A small amount of adherent matter produced above the water-line was probably due to the upward creepage of sodium hydroxide formed by the cathodic reaction.

Agar's numbers show discrepancies of 4 to 9% between Obs and Calc values, but he considers that in general agreement better than 10% cannot be expected. He points out that " successive measurements tend to disturb the corrosion distribution " and would today recommend a concentration of effort on getting " a good set of equipotentials at about half the total period of immersion "—or perhaps rather before half-time since the corrosion-rate falls off slightly with time. He suggests that electrical observations at 6–7 days should give a corrosion-rate somewhere near the average for 14–16 days (J. N. Agar, Priv. Comm., March 3, 1959. See also J. N. Agar's research in an article on " Cambridge work interrupted by the war " by U. R. Evans, *J. Iron St. Inst.* 1940, **141**, 219P, esp. 221P).

Partly Immersed Iron. The case of iron in sodium chloride solution is slightly more complicated, because the valency of iron is variable. Ferrous hydroxide is not precipitated as such, and the solid product obtained is, according to the conditions, red-brown rust ($FeO(OH)$), black magnetite (Fe_3O_4) or sometimes a green ferrous-ferric compound. The accurate measurement of the corrosion-currents was carried out at an earlier date on iron than on zinc, being accomplished by Hoar in 1932. Success was made possible by the use of a mild steel of special quality, provided by the late Dr. W. H. Hatfield; this was so uniform in composition that in a fairly concentrated salt solution the distribution of corrosion was always the same. Thus, a specimen could be cut along the line separating what would later become the corroded and uncorroded areas and the two parts mounted almost in their natural relative positions, but just out of contact; attachment to a glass plate served to hold them correctly in place in a vessel containing the appropriate salt solution.

At some salt-concentrations it was possible to measure the current directly by joining the two segments through a low-resistance milliammeter; but this procedure, although simple in principle, was not very accurate; nor could it be employed except within a limited range of concentration. A more satisfactory method was to join the segments to an external source of current (fig. 150) and to study, by means of a tubulus opening close to the cathodic area and joined to a calomel electrode and potentiometer, the relationship between cathodic potential and the strength of current flowing; this is expressed by curve C of fig. 151. Having once obtained such a curve, it became easy, by measuring the cathode potential on an *uncut* specimen, to read off on the graph the current which must flow to give that potential; (the anodic curve (A) was also measured, but was less helpful, being insufficiently steep). Thus the corrosion-rate obtained by application of Faraday's Law could be compared with that measured directly from loss

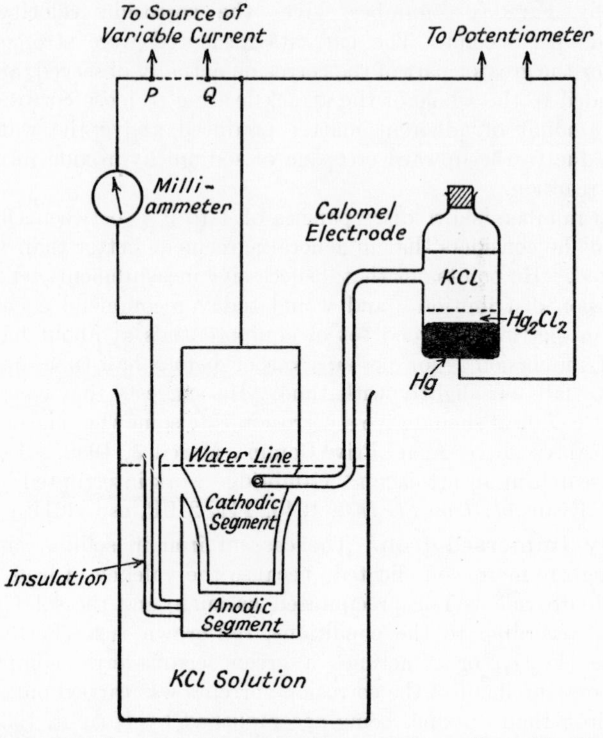

FIG. 150.—Apparatus for measuring corrosion-currents on Iron (T. P. Hoar and U. R. Evans).

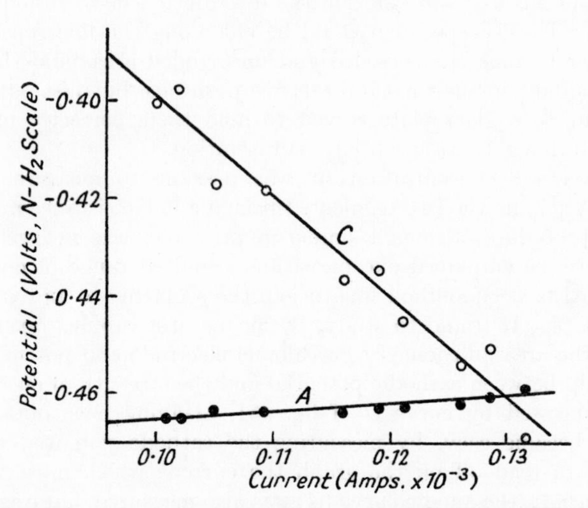

FIG. 151.—Polarization curves obtained on Steel in N/20 Potassium Chloride solution. Note that the current-range studied does not extend to zero (T. P. Hoar and U. R. Evans).

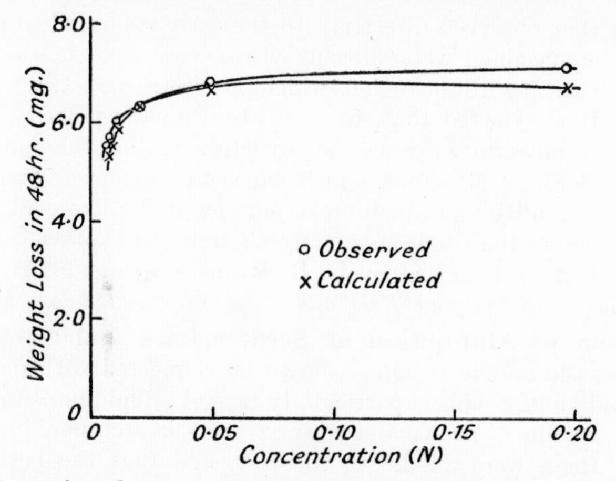

FIG. 152.—Comparison between observed and calculated values (T. P. Hoar and U. R. Evans).

of weight; the agreement (fig. 152) was surprisingly good (U. R. Evans and T. P. Hoar, *Proc. roy. Soc.* (*A*) 1932, **137**, 343).*

Fully Immersed Iron. On vertical specimens partly immersed in potassium or sodium chloride, the cathodic area is the zone lying just below the water-line where the oxygen needed for the cathodic reaction can readily be replenished. Hoar's results suggest that most of the cathodic reaction occurs within the meniscus itself; the anodic area, where attack occurs, is further down. On a horizontal, fully immersed surface, no part is specially favoured as regards oxygen-supply (although replenishment is slightly better at the edges than elsewhere); the anodic and cathodic areas move about, so that in the end the whole area may become corroded, each point having been anodic at some time or other. The current here is much smaller than on partly immersed specimens, but Noordhof, by means of a tubulus consisting of two concentric tubes leading, respectively, to two silver/silver chloride electrodes, succeeded in measuring it, after suitable amplification, and found that the electrically calculated corrosion-rate was in reasonable

* It seems possible that Hoar's measurements of potential include a considerable amount of " Approach Resistance Polarization " (p. 1026), since it is otherwise difficult to explain the approximate straightness of the graphs obtained at both cathodic and anodic areas. Activation polarization should not give a straight line except very close to the equilibrium conditions, where the net current is the difference between two exponentials. If, however, at any moment, the cathodic and anodic reactions are proceeding at certain points only, straight-line curves may be expected (p. 1026). However that may be, the accuracy of the method for estimating the current is not appreciably affected, since Hoar was matching the potential obtained with a known current applied from an external source against the potential obtained with an unknown current connected with the corrosion process. If the potential is the same in the two experiments, it may safely be concluded that the unknown current is equal to the known (external) current; this conclusion need not be vitiated by the inclusion of an *IR* term in both cases.

accord with that observed directly. In the early stages when only certain regions of the specimen were showing visible corrosion, it was found that these were identical with the regions which the electrical instrument indicated as anodic. It is believed that, in such cases, anodic polarity is generally dependent on physical looseness of structure rather than any chemical peculiarity. Groups of atoms, which can detach themselves most easily, pass into the liquid by an anodic reaction; when the loosened material on any small area is exhausted, anodic attack shifts to another region. G. S. Noordhof's work is described by U. R. Evans, *Corrosion* 1951, **7**, 238; see also *Proc. Internat. Congress Pure and Appl. Chem.* (*London*) 1947, **5**, 743.

Corrosion of Aluminium at Scratch-lines and Blemishes. In certain cases the anodic regions seem to be connected with damage to an invisible oxide-film. This is particularly true of aluminium, which builds a very protective film. Brown and Mears, tracing scratch-lines on aluminium, found that these were anodic to the rest, and that the current flowing corresponded to the corrosion-rate. Unintentional blemishes, such as can often be seen on commercial aluminium sheet, were also noticed to form the seats of anodic attack. Brown and Mears cut a piece of aluminium into two pieces, blocking out all the blemishes on one piece with a wax–resin mixture and applying the same mixture to the main surface of the other so as to leave only the blemishes bare; on joining the two pieces, immersed in liquid, through a milliammeter, the current flowing was found to be equivalent to the corrosion-rate (R. H. Brown and R. B. Mears, *Trans. electrochem. Soc.* 1938, **74**, 495; 1942, **81**, 455).

Corrosion of Iron at Scratch-lines. On iron, corrosion only occurs selectively at a scratch-line if the liquid is made just sufficiently " inhibitive " to keep the main part of the film in repair. If we trace a scratch-line on iron which has been exposed for some days to dry air after being cleaned by abrasion, and then place upon it filter-paper soaked in sodium bicarbonate solution of a selected concentration standing between the higher (passivating) range and the lower (corroding) range, rust will generally appear along the scratch-line, but not (on satisfactory specimens) elsewhere. In 1935 the author proved that current was really passing through the wet filter-paper between the scratch-line (as anode) and the uncorroding region on both sides (as cathodes) by means of the " dielectrode " shown in fig. 153(A); this was brought down into contact with the filter-paper at various chosen positions, the edge of the dielectrode being always kept parallel to the scratch-line. The dielectrode consisted of two vertical rectangles of oxidized copper sheet surrounded by filter-paper soaked in the bicarbonate solution and joined by flexible wires to a central-zero micro-ammeter. The two $Cu \mid Cu_2O$ electrodes were virtually non-polarizing and a very small fraction of the current flowing horizontally along the main filter-paper was diverted through the central-zero instrument. The current flowed from the scratch in both directions, but died away at a distance from the scratch-line—as shown in fig. 153(*b*).

Although in such an apparatus the micro-ammeter readings represent only a fraction of the corrosion current flowing through the filter-paper, they

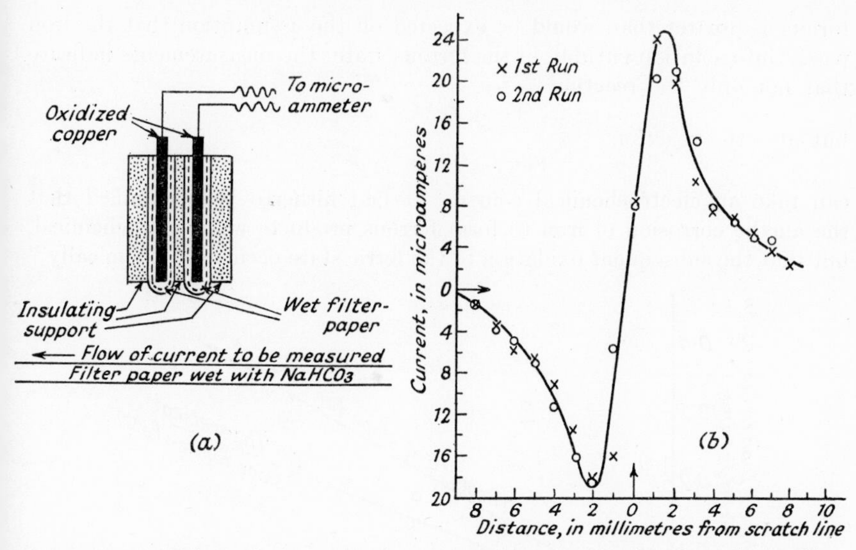

FIG. 153.—(*a*) Dielectrode for measuring corrosion-currents flowing through wet filter paper covering Iron. (*b*) Current flowing through dielectrode (U. R. Evans).

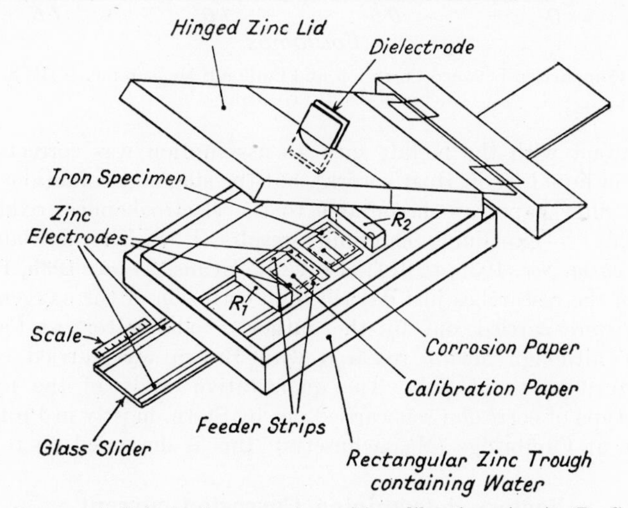

FIG. 154.—Improved dielectrode apparatus with calibration circuit (R. S. Thornhill and U. R. Evans).

should be proportional to the latter if the Cu | Cu_2O electrodes are not polarized. Thornhill, who evolved a greatly improved apparatus (fig. 154), calibrated the dielectrode on a strip of wet filter-paper through which known currents from an outside source were made to flow. In that way the measurements recorded in fig. 155 were obtained. The correlation between the total coulombs generated and the weight of iron corroded show that the former is greater than would be expected on the assumption that the iron passes into solution entirely in the ferrous state; the measurements indicate that not only the reaction

$$Fe = Fe^{++} + 2e$$

but also the reaction

$$Fe^{++} = Fe^{+++} + e$$

can take an electrochemical course. It had hitherto been assumed that the anodic corrosion of iron to form ferrous products was electrochemical, but that the subsequent oxidation to the ferric state occurred " chemically "

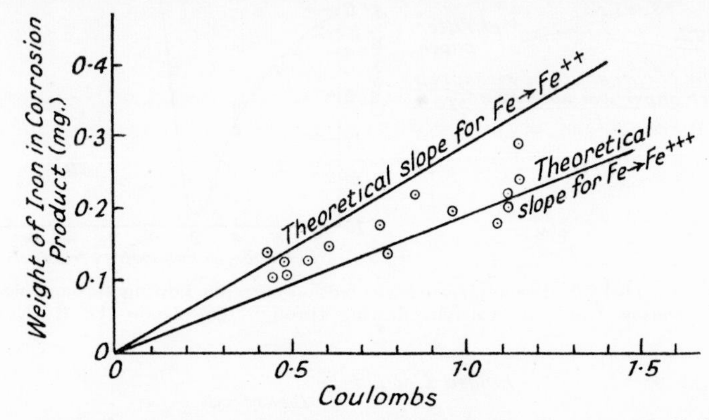

FIG. 155.—Comparison between Corrosion and Coulomb Measurements (R. S. Thornhill and U. R. Evans).

out of contact with the metal; such an assumption was correct for iron immersed in liquid (as in Hoar's work), but the stagnant conditions existing in sodden filter paper are favourable to the electrochemical oxidation of Fe^{++} to Fe^{+++}—explaining Thornhill's results (R. S. Thornhill and U. R. Evans, *J. chem. Soc.* 1938, p. 614; also U. R. Evans, *Nature* 1935, **136,** 792).

Most of the researches just described on corrosion of the oxygen-absorption type were carried out by the author's collaborators at Cambridge (England), although Mears' research with Brown was carried out after Mears' return to the U.S.A. The quantitative study of the hydrogen-evolution type of corrosion was carried out by Stern, mostly in Prof. Uhlig's laboratory at Cambridge (Massachusetts); this is discussed on p. 879.

Factors determining Corrosion-current

Graphical Method. If the strength of the corrosion-current was that obtained by dividing the E.M.F. by the cell resistance, metals would corrode much more rapidly than is found to be the case; fortunately the current is greatly reduced by polarization. Whenever a current passes between metal and liquid, the potential shifts in such a direction as to make conditions less favourable to the flow of current. Thus, if current flows between anodic and cathodic areas of a specimen, the potentials approach one another and the E.M.F. diminishes, so that there is a certain value of the current—that corresponding to the intersection of the two polarization curves—which can never be exceeded, however low the resistance may be. With corrosion in concentrated salt solution (and also in acid) the conductivity is generally high, and it may often be permissible to take the intersection of the curves as a measurement of the corrosion-current (fig. 156(A)). Division by Faraday's number then leads to the corrosion-rate in gram-equivalents per second.

It is the current density, not the absolute current, which decides polarization, and thus, to predict the corrosion-rate from first principles, we need to know, besides the equations of the two polarization curves, the respective areas of the anodic and cathodic regions. These are not always available, but even when the calculation of the corrosion-rate from first principles is impossible, the graphical method is useful. The arguments which follow are independent of the form of the polarization curves, which for simplicity are shown as straight lines on the diagrams.

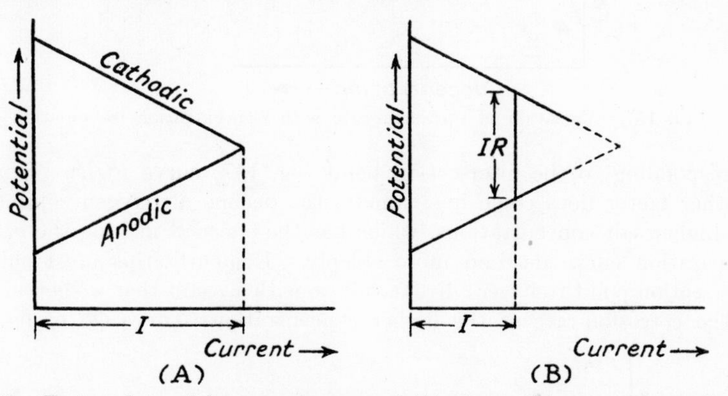

FIG. 156.—Factors determining corrosion-current (A) when resistance can be neglected, (B) when resistance is appreciable.

In dilute solutions it is necessary to retain an appreciable E.M.F. for pushing the corrosion-current over the (considerable) resistance of the liquid path. The corrosion-current will no longer be the abscissa of the intersection-point, but will assume that value I which will make an intercept between the two polarization curves equal to IR, where R is the resistance of the circuit (essentially that of the liquid path, since the resistance of the metal can generally be neglected). Clearly this intercept (fig. 156(B)) represents an E.M.F. exactly sufficient to force current I over resistance R.*

Consider now Hoar's work on steel specimens partly immersed in solutions of potassium chloride of different strength. At a very low concentration a large intercept is necessary, and the current is small. If we double

* In the case of steel specimens partly immersed in potassium chloride solution, the difference of potential between the cathodic zone at the water-line and the anodic zone lower down, was found in early work to be about 140 mv. at M/1000, about 30 mv. at M/100, about 12 mv. at M/10 and about 5 mv. at M concentration (U. R. Evans, L. C. Bannister and S. C. Britton, *Proc. roy. Soc* (*A*) 1931, **131,** 355, esp. p. 366).

the concentration, R declines, so that a smaller intercept suffices (the necessary intercept is not halved, however, since I increases). As the concentration steadily increases, the intercept needed steadily decreases, and the corrosion-current increases. If no other factor were to be introduced, it would be expected that the corrosion-current, and therefore the corrosion-rate, would increase with concentration, becoming asymptotic to a value

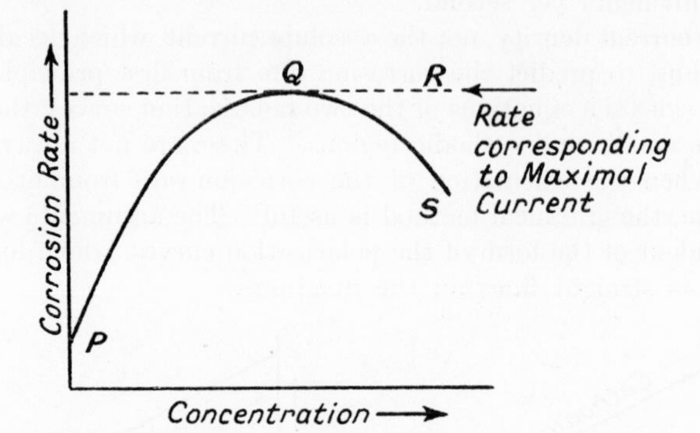

Fig. 157.—Variation of Corrosion-rate with Concentration (schematic).

corresponding to the intersection point (fig. 157; curve PQR). However, another factor does come in—namely, the decline of oxygen solubility at the higher salt concentrations, which has the effect of making the cathodic polarization curve descend more steeply. Evidently this must shift the intersection point to lower current values, with a result that we get a decline of the corrosion-rate at the higher concentrations (curve QS of fig. 157).

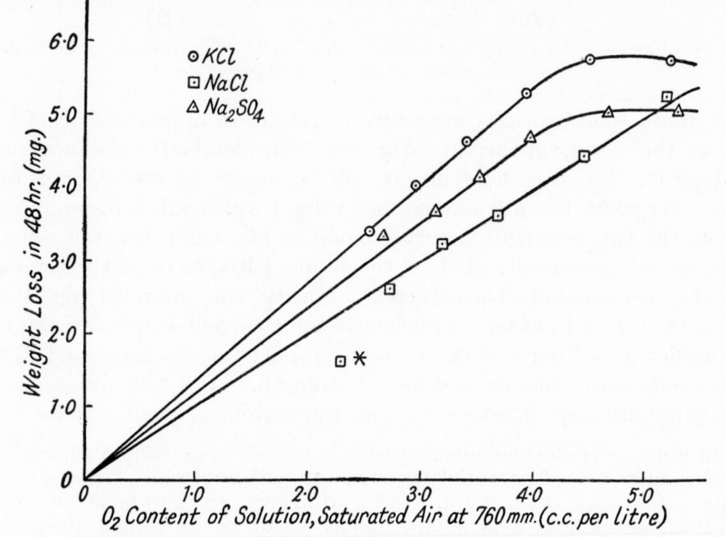

Fig. 158.—Relation between Oxygen-solubility and Corrosion-rate on Iron (T. P. Hoar and U. R. Evans).

Slight further complications are caused by the formation of complex anions at the highest salt concentration which tends to shift downwards the anodic curve.

The effect is that the corrosion-rate first rises with increase in concentration, reaching a maximum about 0·3 to 0·6N (the position of the maximum varies with the different forms of iron employed) and then declines again. At concentrations *below* that corresponding to the maximum, it is mainly the conductivity of the solution which controls the rate of attack; at concentrations *above* the maximum, it is the oxygen-solubility. Hoar plotted the corrosion velocities obtained in solutions of potassium chloride, sodium chloride and sodium sulphate of various concentrations against oxygen-solubility, and obtained (fig. 158) straight lines at the lower oxygen-solubilities (corresponding to the *higher* salt-concentrations); at high oxygen-solubility (*low* salt-concentrations), he obtained marked departure from straightness, because here control was no longer exerted by the concentration of oxygen (U. R. Evans and T. P. Hoar, *Proc. roy. Soc.* (*A*) 1932, **137,** 343).

Calculation of Corrosion-rate from Oxygen-supply. In cases where corrosion-rate is controlled solely by oxygen-replenishment, it should often be possible to calculate the rate from purely physical considerations. It is well known that the rate of diffusion of dissolved matter across a layer of liquid of cross-section A and thickness h is

$$\frac{(C_1 - C_2)AD}{h}$$

where C_1 and C_2 are the concentrations at the two surfaces and D the diffusion coefficient. If the diffusing substance is considered to be used up almost as soon as it arrives so that $C_2 \ll C_1$, this becomes nearly equal to C_1AD/h. Now Bengough and Wormwell (p. 96) found that the corrosion-rate of horizontal steel discs in N/10 potassium chloride solution under oxygen depended on the depth of immersion below the water-level, provided that the depth was not too great. It is possible that in such cases the corrosion-rate might be calculated from the rate of replenishment of oxygen, although in some situations diffusion across the clear liquid may not be the only factor involved; special effects at both boundaries would require to be considered in a full theoretical treatment. It is believed that of the gas molecules striking the surface of a stagnant liquid, only a few actually enter the liquid (the proportion increases if the liquid is in violent motion) —so that it cannot be assumed that the liquid at its upper surface is necessarily kept saturated with oxygen. Further complications may be introduced by the accumulation of corrosion-products which may affect the rate of passage of oxygen in some cases. None of these factors, however, would make the theoretical treatment really difficult, and calculated values might well be found to agree well with observed values. Such calculations would, of course, involve no electrical quantities, but agreement, if obtained, would not mean that the corrosion mechanism of fully immersed specimens is not electrochemical (the work of Noordhof, described on p. 865, strongly supports its electrochemical character). It would merely mean that the corrosion-rate is limited by the rate of oxygen-arrival, and that the potentials of the anodic and cathodic areas adjust themselves so as to produce a current exactly sufficient to use up the oxygen as it arrives—obedience to Ohm's Law being of course maintained.

Under conditions of complete immersion in stagnant liquid, the corrosion-rate falls off with increase of salt-concentration, owing to the diminished oxygen solubility. The maximum of the corrosion-rate of iron in potassium chloride (which Hoar found to be at 0·3 to 0·6N for partly immersed specimens, where a meniscus zone existed on which oxygen replenishment was easy) was found by Bengough (p. 96) to occur between 0·001N and 0·01N for fully immersed specimens, the variation depending on the cross-section of the vessel and other factors. The corrosion-rate is greatly affected by vibration and thermal disturbances in the liquid when the specimens are fully immersed, but much less affected when they are partly immersed.

The calculation of corrosion-rate of partly immersed specimens from purely physical considerations presents greater mathematical difficulty than that of a fully immersed surface parallel to the liquid, but the difficulties have been elegantly overcome by Bianchi, who took advantage of the close analogy existing between diffusion under a concentration gradient (fig. 159(1)) and migration under an electrical potential gradient (fig. 159(2)). If two electrodes forming opposite sides of a cell of irregular shape containing a solution of a salt of the electrode metal are kept at constant potentials V_1 and V_2, it is possible, by exploring the interior of the liquid with a tubulus by a method analogous to that of Agar (p. 862), to trace the course of the equipotential lines; the current lines will cut the equipotentials at right angles. Alternatively the *conjugate* form may be used (fig. 159(3))—with the electrodes at the sides—when the pattern is preserved, but the current lines now become equipotentials and *vice versa*. By substituting concentration for potential and conductivity for diffusivity, the desired information about diffusion-rate and distribution of concentration is obtained. Absence of polarization is necessary for success and Bianchi found lead electrodes in a solution of lead sulphamate to fulfil this requirement; they provided what was practically an Ohmic System.

Now in adopting that method to ascertain the oxygen-supply to a vertical partly immersed metal specimen, the procedure would seem, at first sight, to be the construction of a cell (fig. 159(4)) bounded at the top with a horizontal electrode carefully curved at the left-hand end so as to represent the shape of a meniscus, and with a vertical electrode on the left-hand side; the electric current passing between them should represent the oxygen diffusion from air to metal in the corrosion experiment. However, the difficulties in dealing with the two electrodes so very close together, with risks of short-circuiting and possibly errors due to polarization, appear formidable. By adopting the conjugate arrangement, in the form indicated on fig. 159(5), these troubles are neatly avoided, and Bianchi obtained impressive agreement between the oxygen-replenishment rate thus calculated, and that corresponding to the corrosion-rate in an actual corrosion experiment on zinc. The procedure could not be used in all cases; it virtually ignores the complications due to corrosion-products producing physical obstruction to the diffusion, or chemical removal of oxygen, and it assumes that oxygen diffusing to the anodic area takes part in local cathodic reactions

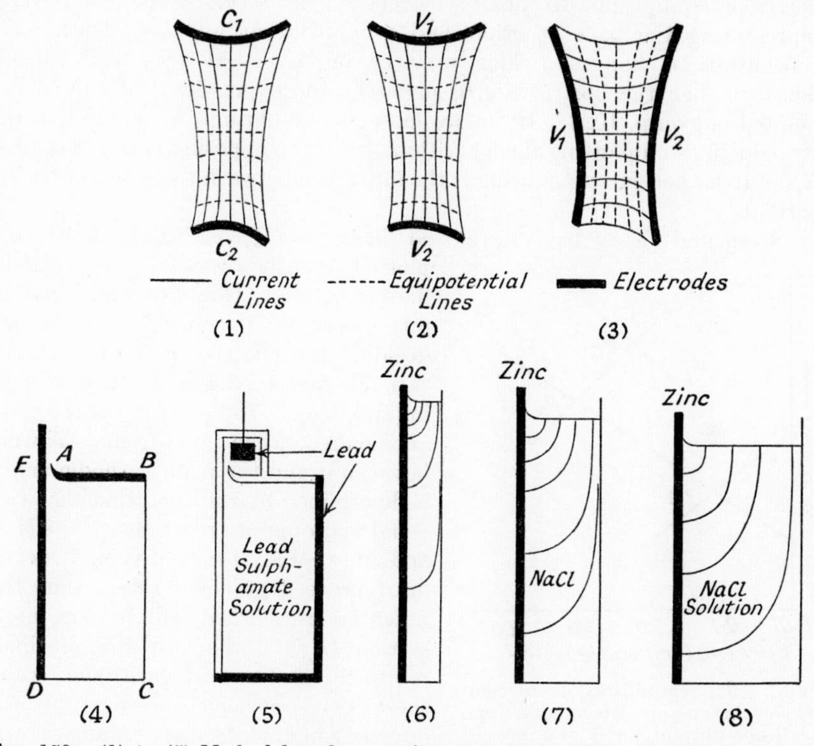

FIG. 159.—(1) to (5) Method based on conjugate curves. (6) to (8) Effect of shape of vessel (G. Bianchi).

(which may or may not be true). However, as shown by a study of equipotentials in the model, most of the passage of oxygen in actual corrosion experiments must occur close to the meniscus, and the effect of these disturbing factors is not serious. Bianchi's work on this and other corrosion problems deserves careful study (G. Bianchi, *Metallurg. ital.* 1953, **45,** 123, 323).

Further development of the Graphical Method. The corrosion-velocity of a metal in a solution possessing sufficiently good electrolytic conductivity to justify neglect of the *IR* intercept is given by the intersection of the cathodic and anodic polarization curves as shown in fig. 156, where for convenience they are drawn as straight lines. Now Hoar's experimental measurements (fig. 151, p. 864) show that when *I* is plotted against potential the points do in fact fall close to a straight line. As already suggested, this may be due to the fact that much of the measured polarization is really due to the resistances of the bottle-neck approaches to the small points where, at any given moment, the cathodic and anodic reactions are taking place. If the polarization belonged to the activation-polarization type, straight lines would be expected on plotting, not I, but log I against potential (p. 1024). Provided that we are considering a region sufficiently far from equilibrium to justify neglect of the second exponential term representing the back reaction, and provided that we also neglect local concentration changes, (which may be important at very high current density), then the polarization is given by the expression $b \log (i/i_0)$—as pointed out on p. 1023. If we are close to equilibrium, we can obtain the straight lines on plotting the logarithm of the two opposing currents against E, and the net current flowing is the difference between these two opposing currents.

Stern and Geary have developed diagrams for conditions where log I plotted against E would give straight lines (fig. 160). They use semi-logarithmic paper, and consider the factors deciding the effective potentials at the cathodic and anodic areas on a corroding specimen. The potential at the main anodic area, when no current is flowing between it and the main cathodic areas, is determined by the condition that two equal and opposite reactions ($M \rightarrow M^{++}$ and $M^{++} \rightarrow M$, for a divalent metal) must proceed at equal rates; thus the potential established will be the intersection (P_A) of the two straight lines representing (on a logarithmic scale) these two opposing currents; at the intersection point, the value of the current flowing will be the exchange current. Similarly the potential at the main cathodic area, in the absence of current flow between anodic and cathodic areas, would be the intersection (P_C) of two other straight lines (representing perhaps $O_2 \rightarrow OH^-$ and $OH^- \rightarrow O_2$ respectively); here the current flowing will be an exchange current having a different value from that at P_A. If now we imagine the main anodic and cathodic areas joined so that the corrosion process becomes possible (with the anodic reaction $M \rightarrow M^{++}$ and the cathodic reaction $O_2 \rightarrow OH^-$), we obtain the corrosion-current by producing the line representing $M \rightarrow M^{++}$ at P_A and that representing $O_2 \rightarrow OH^-$ at P_C until they intersect at P_X; the ordinate of P_X represents the potential of the corroding specimen, whilst the abscissa represents (on the logarithmic scale) the corrosion-current, from which the corrosion velocity can easily be obtained. Stern and Geary acknowledge the possible effect of concentration changes at high currents, and the reader should also bear in mind the influence of the ohmic resistance of bottle-neck approaches. The paper

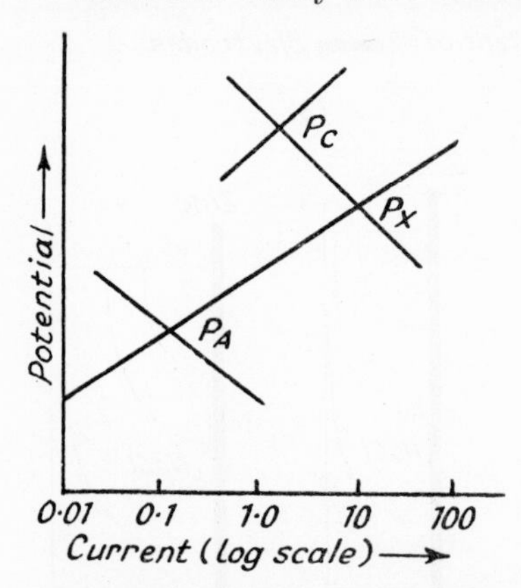

FIG. 160.—Graphic Construction showing factors determining Corrosion-Current and Corrosion-Potential, assuming obedience to Tafel's Law (M. Stern and A. L. Geary).

deserves study (M. Stern and A. L. Geary, *J. electrochem. Soc.* 1957, **104,** 56; see also M. Stern, N.A.C.E. Symposium, Buffalo, May 9–10, 1956; *J. electrochem. Soc.* 1957, **104,** 600, 645).

Current flowing between bimetallic couples. Waber has considered mathematically the current flowing between two metals in contact which act as cathode and anode respectively, under various geometrical conditions. The current flowing (and hence the corrosion-rate) will in general be limited both by polarization and resistance, but (other things being equal) polarization will predominate at small sizes and resistance at large sizes—a feature also brought out by Hoar and Agar. This provides another example of the danger—mentioned in Chapter XIX—of using small-scale laboratory experiments to make predictions regarding behaviour in large-scale plant unless the situation has been thoroughly pondered in the light of dimensional analysis. The papers discussing different geometrical cases are J. T. Waber (with M. Rosenbluth and B. Fagan), *J. electrochem Soc.* 1954, **101,** 271 (co-planar electrodes with negligible polarization); *ibid.*, 1955, **102,** 344 (co-planar electrodes, one infinite, having equal polarization parameters); *ibid.*, 1955, **102,** 420 (semi-infinite co-planar electrodes); 1956, **103,** 64 (influence of electrolyte film thickness); *ibid.*, 1956, **103,** 567 (very thin films, limiting case); other parts are in the press. Cf. T. P. Hoar and J. N. Agar, *Disc. Faraday Soc.* 1947, **1,** 158.

Cathodic and Anodic Control. In Hoar's work on steel in N/20 potassium chloride solution (fig. 151, p. 864), the cathodic curve was steep and the anodic one nearly horizontal. This is clearly a case of cathodic control, i.e. control by oxygen-replenishment. Anodic control is also possible, the two cases being shown schematically in figs. 161(*I*) and 161(*II*).

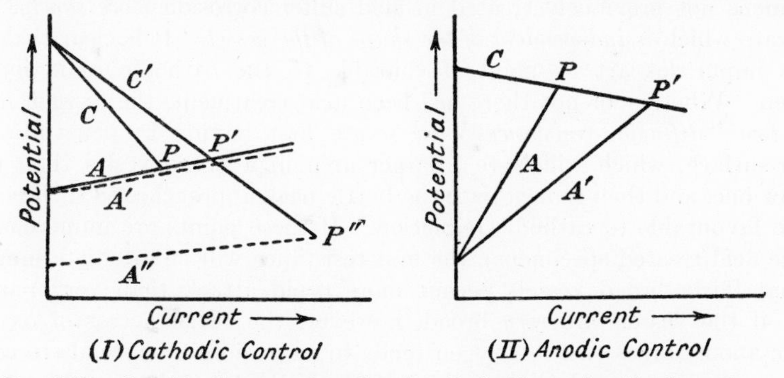

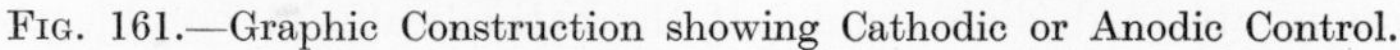

FIG. 161.—Graphic Construction showing Cathodic or Anodic Control.

Clearly in the first case, any factor which facilitates the anodic reaction, even to the extent of halving the slope of the anodic curve, will nevertheless hardly increase the corrosion-current (shown by the intersection); in contrast, a factor which diminishes the steepness of the cathodic reaction will greatly increase the corrosion-current (from the abscissa of P to that of P') and thus increase the corrosion-rate. For anodic control (fig. 161(*II*)), the converse relationship holds good.

Cathodic control by limitation of oxygen-supply is commonly met with, e.g. in the corrosion of iron in chloride solution. Anodic control—where the polarization curve is swung upwards—is a less common phenomenon: examples are provided by steel in certain chloride–carbonate mixtures, which give partial but not complete inhibition; also by magnesium in certain chloride–fluoride mixtures (L. J. Benson, R. H. Brown and R. B. Mears, *Trans. electrochem. Soc.* 1939, **76,** 259).

Even when the anodic curve is almost horizontal, any addition to the liquid which shifts it bodily downwards (as from AP' to $A''P''$ on fig. 161(I) will increase the velocity (from P' to P''). An example may be provided by the effect of sulphide on the attack upon iron in a chloride solution, where the S^{--} ions not only catalyze the anodic reaction, but keep the Fe^{++} concentration low (S. C. Britton, T. P. Hoar and U. R. Evans, *J. Iron St. Inst.* 1932, **126,** 365, esp. p. 370).

Influence of impurities in corrosion of the oxygen-reduction type. It was pointed out on pp. 89, 310 that, whereas impurities greatly affect the corrosion of zinc by acids where the cathodic reaction is hydrogen-evolution, they have only a small effect on zinc partly immersed in a chloride solution where the cathodic reaction is oxygen-reduction. Bianchi has discovered that the effect of impurities on the corrosion-rate in such a case may depend on the shape of the experimental vessel. He has studied the behaviour of zinc specimens previously treated in acid, which causes the noble impurities present in the zinc to accumulate on the surface, and has then exposed them partly immersed in a neutral sodium chloride solution, either in narrow vessels or in broad vessels (fig. 159(6) (7) and (8), p. 873); in general, corrosion proceeds more quickly in a broad vessel than in a narrow vessel. Zinc specimens not previously treated in acid suffer corrosion *more slowly,* and at a rate which is *independent of the shape of the vessel.* It is clear that the noble impurities act as points favourable to the cathodic reduction of oxygen. Whether or not there has been acid treatment, the oxygen must pass *two "diffusion-resistances" in series,* first having to penetrate the water-surface, which will have a larger area in a broad vessel than in a narrow one, and then to negotiate the bottle-neck approaches to such points as are favourable to cathodic reduction. If these points are numerous (as on the acid-treated specimens), the first resistance will be rate-determining, so that fairly broad vessels permit more rapid attack than very narrow ones; if the vessels are *very* broad, however, the easier access of oxygen to the anodic areas (where oxygen tends to be an inhibitor) tends to cause the corrosion-rate to decline again. If the cathodically favourable points are rare (as on specimens not treated in acid), the bottle-neck approaches will be rate-determining, and the corrosion will be slower and independent of the shape of the vessel—explaining Bianchi's results (G. Bianchi, *Metallurg. ital.* 1953, **45,** 123, 323).

Bianchi has also critically surveyed earlier work by Tödt, Delahay and also his own colleagues, Piontelli and Poli. It is clear that the nature of the cathodic surface may be important or unimportant, according to the conditions. He also emphasizes the fact that of the two steps in the reduction [(1) $O_2 \rightarrow H_2O_2$, (2) $H_2O_2 \rightarrow OH^-$], the first shows little overvoltage (on platinum) and the second much overvoltage, so that in practice they both occur at nearly the same potential; this may explain the fact that some experimenters have failed to detect the intermediate product, hydrogen peroxide, except on a few metals—such as zinc. The paper, along with other papers quoted in it, deserves careful study (G. Bianchi, *Atti. Collegio Ingegneri Milano*, 1956, 5–6; G. Bianchi and B. Rivolta, *Chimica e Industria* 1954, **36,** 358; G. Bianchi, *Corrosion et anticorrosion* 1957, **5,** 146; F. Tödt, *Z. Elektrochem.* 1928, **34,** 586; P. Delahay, *J. electrochem. Soc.* 1950, **97,** 198, 205; 1952, **99,** 414, 546 (with L. J. Stagg); R. Piontelli and G. Poli, *Atti Accademia d'Italia* 1942, **13,** 903).

Comparison between Observed and Calculated Corrosion-rates (Hydrogen-evolution Type of Attack)

General. In applying the graphical method to attack by acids, it is usually possible to take the intersection point of the two polarization curves as a measurement of corrosion-rate; an IR intercept is not needed, as the anodic and cathodic points are close together, making R low.

If attack by non-oxidizing acid was controlled solely by the cathodic reaction (formation of hydrogen), and if the cathodic polarization curve was the same on all surfaces, then the order of attack on different metals would be the order of their electrode potentials, as suggested schematically in fig. 162; metals standing on the positive side of hydrogen in the Table of Normal Potentials (p. 312) should suffer no attack by these acids, after a short period needed to produce a certain very low concentration of ions of the metal in question.

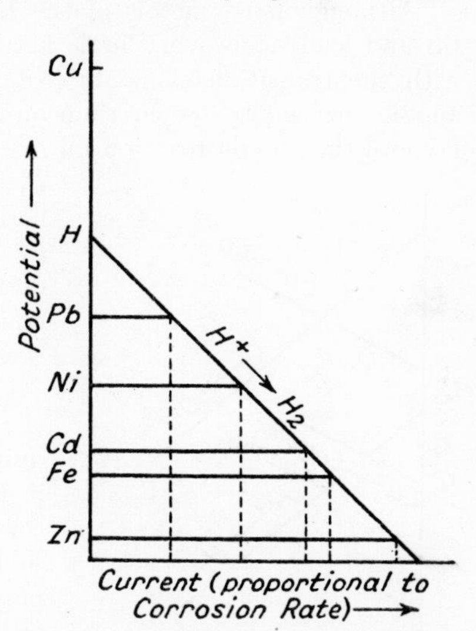

FIG. 162.—Hypothetical rates of acid attack which would occur if cathodic polarization was the sole controlling factor and was the same for all metals.

However, this is far from being the case. Pure zinc has a much higher overpotential than iron, and the cathodic curve is therefore steeper. Thus zinc dissolves in dilute sulphuric acid more slowly than iron, despite its more negative potential (fig. 163(a)). The corrosion-currents on the two metals are given by the intersection points, P_1 and P_2 respectively. Impurities in the zinc, or the addition of a trace of platinum salt to the acid, will greatly increase the attack; contact with platinum, a metal of extremely low hydrogen overpotential (giving therefore a much less steep cathodic curve), greatly increases the corrosion-rate of zinc, moving out the intersection point to P_3.

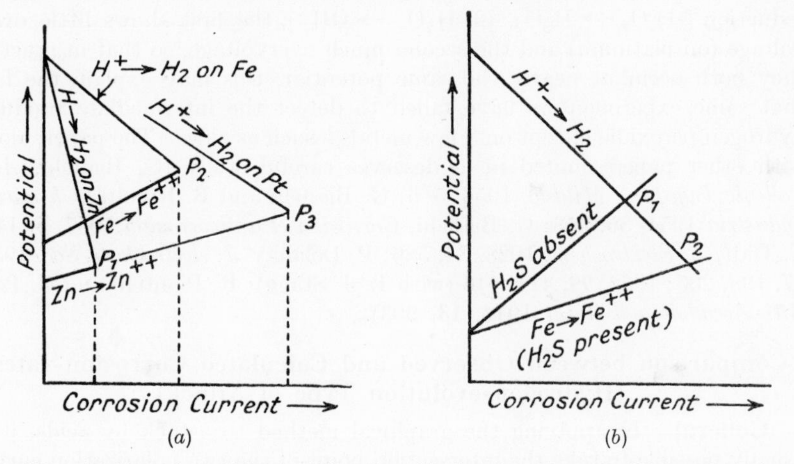

FIG. 163.—(*a*) Corrosion of Zinc and Iron in acid, showing effect of addition of platinum salt (schematic). (*b*) Effect of hydrogen sulphide on the acid corrosion of Iron (schematic).

Although most metals of the B groups (copper, silver, zinc, cadmium, tin and lead) show very little activation polarization, this is not the case with the transition elements (e.g. iron, cobalt and nickel), and here the anodic curves are steeper than on zinc; the presence of hydrogen sulphide favours the anodic reaction on iron, and shifts the intersection point from P_1 to P_2 (fig. 163(*b*)), thus increasing the corrosion-rate; but traces of tin or copper in the liquid (the copper possibly being derived from copper in the steel) precipitate the sulphide and the corrosion-rate declines (pp. 316, 878).

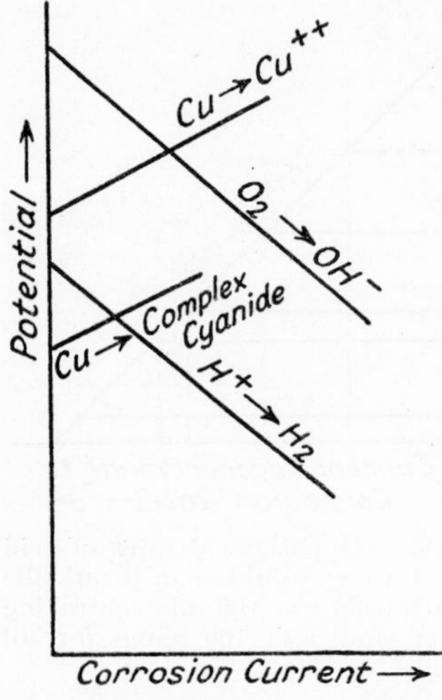

FIG. 164.—Effect of Cyanides on corrosion of Copper (schematic).

If the acid is diluted, the hydrogen curve is shifted downwards (58 mv. for each pH unit), and the corrosion-rate becomes slower; it might be expected that in most metals, corrosion would be impossible in alkaline solutions. However, the formation of complex ions (such as $[ZnO_2]^{--}$) shifts downwards the anodic curve also; on zinc the E.M.F. available for the corrosion process in alkali is comparable to that in acid, both curves being shifted downwards.

Cyanides, despite their alkaline reaction, can often attack more vigorously than acids, by keeping low the concentration of simple cations. Thus copper which is not attacked by dilute sulphuric acid in the absence of oxygen, is attacked by potassium cyanide solution (fig. 164).

Iron in Acidified Salt Solution. Corrosion in absence of oxygen has been studied in two important papers by M. Stern, *J. electrochem. Soc.* 1955, **102,** 609, 663.

He first studied the corrosion of pure iron in sodium chloride, acidified with hydrochloric acid to give various pH values, in complete absence of oxygen. The corrosion was uniform, producing no pitting or intergranular attack, but after prolonged action some crystal faces were found to be more attacked than others. The corrosion-rate was determined by withdrawing, at intervals, known quantities of solution and estimating the iron colorimetrically with orthophenanthroline (p. 987); from this the equivalent corrosion current I_{corr} was calculated on the assumption that the corrosion was entirely electrochemical. Careful measurements were made of the manner in which the corrosion-rate fell off as an external cathodic current was applied. The corrosion-potential was also measured, both in the absence and the presence of applied cathodic current. At low values of applied current, their relations varied with pH, but at higher values, the various curves relating E to log I at different pH values tended to merge into a common straight line—thus showing obedience to the Tafel equation. Now electro-chemical principles (p. 1023) indicate that, in a region of constant Tafel slope b, the over-potential η should be given by $\eta = b \log (I/I_0)$ where I_0 is the exchange current, obtained by extrapolation of the Tafel curve to the reversible hydrogen potential. In general, I is the sum of the externally applied current and the current due to local corrosion cells, but in the special case where no current is applied, the corrosion-potential, measured from the reversible value of the hydrogen electrode at the pH used in the experiment, is

$$E_{corr} = - b \log \frac{I_{corr}}{I_0}$$

Since b is known from the slope of the straight portion of the curves obtained at different pH values, the measurement of E_{corr} provides a method of calculating I_{corr}. Excellent agreement was obtained, as shown in Table XXXI.

TABLE XXXI

Observed and Calculated Values of the Corrosion (of the Hydrogen-evolution type) in acidified 4% NaCl (M. Stern)

pH value	0·96	2·0
Exchange current density i_0	0·10 microamp/cm²	0·11 microamp/cm²
Tafel slope, b	0·100 volt	0·100 volt
Corrosion-potential, E_{corr}	0·203 volt	0·201 volt
Current calculated from electrical measurements	10·5 microamps	11·0 microamps
Current equivalent to the corrosion-rate obtained by chemical analysis	11·1 microamps	11·3 microamps

The manner in which the electrode potential changed with an applied current I_{ext} also agreed well with that predicted by

$$\eta = - b \log \frac{I_{ext} + I_{corr}}{I_0}$$

Stern points out that this electrochemical method of calculating the corrosion-rate is not universally applicable, being subject to many limitations, including the necessity for (1) the complete absence of " depolarizer " (e.g. oxygen),* (2) the elimination of concentration polarization and resistance polarization. The particular procedure adopted by him also requires that the surface should exhibit the Tafel relationship between potential and log I.

In his second paper, Stern studied the effect of small quantities of alloying elements on the attack upon iron by acidified sodium chloride (pH 1 and 2), or by 0·1M citric or malic acid. Sulphur, phosphorus and carbon were found to stimulate corrosion, whereas copper and manganese were effective in counteracting the detrimental influence of sulphur. These results may profitably be considered in connection with those of Hoar and Havenhand (p. 316).

A method based on principles somewhat similar to those of Stern was presented by G. Okamoto, M. Nagayama and N. Sato at the 8th meeting of C.I.T.C.E. (1956).

Bonhoeffer and Jena had previously studied the behaviour of eleven types of iron in sulphuric acid, measuring the potential when different currents were applied; the gradient of the straight lines relating potential to current density can be regarded as a " polarization resistance ", which, in Stern's opinion, is not an Ohmic resistance but arises from the fact that the difference between two exponentials may provide a straight lines (see p. 1023 of this book). The polarization is found to be a rectilinear function of the exchange current and also of the reciprocal of the corrosion current. This is important as suggesting a way of calculating corrosion-rate from simple electrical measurements (K. F. Bonhoeffer and W. Jena, *Z. Elektrochem.* 1951, **55,** 151; M. Stern, *Corrosion* 1958, **14,** 440t).

Recently, with Weisert, Stern has developed a promising method which is fairly general. It depends on the measurement of the current needed to shift the potential through a certain small distance (not exceeding 10mv., otherwise the curve connecting E and I will cease to be straight). For a system controlled by activation polarization, the corrosion current is obtained by applying

$$\frac{\Delta E}{\Delta I} = \frac{\beta_A \beta_c}{2{\cdot}3(I_{\text{corr}})(\beta_A + \beta_c)}$$

where β_A and β_c are the Tafel slopes of the logarithmic local anodic and cathodic curves; for systems where the corrosion is controlled by concentration polarization (e.g. oxygen-diffusion), $\beta_c \gg \beta_A$, giving

$$\frac{\Delta E}{\Delta I} = \frac{\beta_A}{2{\cdot}3\ I_{\text{corr}}}$$

If β_A and β_c are known, the corrosion current can be calculated with reasonable accuracy. Even if they are not known, a fair assessment can be made. β_c is theoretically infinity where corrosion is controlled solely by diffusion of (say) oxygen to the cathodic area, but with activation

* A method designed for neutral solutions containing oxygen has been put forward by H. J. Engell, *Arch. Eisenhuttw.* 1958, **29,** 553.

polarization it may be as low as 0·06, whilst β_A varies in practice from about 0·06 to 0·12. Despite the apparently wide variation of β values, the magnitude of $\beta_A\beta_c/(\beta_A + \beta_c)$ varies comparatively little, because β_A and β_c appear both on the numerator and denominator. Thus, even without knowledge of the β values, the corrosion-rate can usually be estimated to within a factor of 2, although experimental calibration may be required in some cases. Data from a number of sources support theoretical predictions. Since the corrosion-rates cover six orders of magnitude, double logarithmic paper is used. The logarithm of $\Delta E/\Delta I$ plotted against the logarithm of the corrosion current (or of the exchange current where equilibrium systems are being studied) shows that the points fall close to a straight line of the slope predicted by theory. The argument should be studied when the paper becomes available (M. Stern and E. D. Weisert, A.S.T.M., in the press).

Question of Discrete Anodic and Cathodic Areas in the Hydrogen-evolution Type of Corrosion. There is some difference of opinion as to the geometrical distribution of the electrodic reactions in corrosion of the hydrogen-evolution type. In the case of commercial zinc placed in acid, the hydrogen gas can be seen coming off from places where the black residue of impurities is collecting, and there is evidence that the cathodic reaction really occurs on that residue (p. 318). On pure metals, and particularly on amalgamated zinc—used in the experiments of Wagner and Traud—it is likely that the cathodic and anodic changes proceed at contiguous points, or perhaps at the same point at consecutive moments. Even if the cathodic and anodic parts of the reaction occur so close together in time and space that a single equation such as

$$Zn + 2HCl = ZnCl_2 + H_2$$

completely expresses the change, it still remains legitimate to divide it into component parts such as

$$Zn = Zn^{++} + 2e \text{ (anodic)}$$

and

$$2H^+ + 2e = H_2 \text{ (cathodic)}$$

The measured potential will then be that value at which the cathodic and anodic components occur at the same rate.

This idea of a mixed potential has been successfully used in Hackerman's laboratory during a study of the corrosion of iron in acid in presence or absence of oxidizing agents. The solution-rate in air-free 2N hydrochloric acid was found to be independent of stirring speed, but in acid saturated with air, it increased 1·6 times when the specimens were rotated. The results suggested that the rate of solution in non-oxidizing acids is controlled by both the anodic and cathodic corrosion reactions, and is independent of ferrous ion concentration. Oxygen, if present, is cathodically reduced and increases the solution-rate, which then becomes dependent on the replenishment-rate—as had been found in the early work of van Name, Whitman and others (see below).

When the corrosion process is controlled by the replenishment of some essential reagent (such as oxygen or hydrogen ions), no electrochemical factors will appear in the expression defining the corrosion velocity; this does not necessarily mean that the mechanism has ceased to be electrochemical, but merely that the rate of the electrochemical change is itself adjusting itself to conform to the replenishment rate of the essential reagent, as explained on p. 872. In general, the corrosion-rate on area A expressed in gram-equivalents can be written as CDA/δ where D is the diffusivity coefficient of the essential constituents, C is the concentration of the essential reagent in the body of the liquid (also expressed in gram-equivalents) and δ the thickness of the stationary diffusion-layer (δ decreases as the stirring becomes more rapid). Unless some provision is made to keep C constant, the corrosion-rate will fall off with time, as the reagent is used up, causing the solution to become increasingly impoverished. The progress of the impoverishment is sometimes expressed $dx/dt = kA(a - x)/V$ or

$$k = \frac{-2{\cdot}3V}{A} \cdot \frac{1}{t} \log_{10}(a - x)$$

where A is the area of the test specimen, $(a - x)$ is the amount of the essential reagent remaining in volume V of the solution at time t, and k is a constant.

The papers mentioned above, with others discussing attack by acid and oxidizing agents, are C. Wagner and W. Traud, *Z. Elektrochem.* 1938, **44,** 391; A. C. Makrides, N. M. Komodromos and N. Hackerman, *J. electrochem. Soc.* 1955, **102,** 363; R. G. van Name and D. U. Hill, *Amer. J. Sci.* 1916, **42,** 301; W. G. Whitman, R. P. Russell, C. M. Welling and J. D. Cochrane, *Industr. engng. Chem.* 1923, **15,** 672. C. V. King, " Pittsburgh International Conference on Surface Reactions " 1948, p. 5; also C. V. King (with M. Hochberg, H. Salzberg and F. S. Lang), *J. electrochem. Soc.* 1950, **97,** 191, 290; 1952, **99,** 295.

The idea of a uniform surface without areas specially reserved for anodic or cathodic reaction has been used in interpreting further work in Hackerman's laboratory on the action of amine restrainers in retarding the attack of hydrochloric acid upon fairly pure *iron*; the authors recognize that the treatment might not be appropriate for *steel* " where secondary phases may have considerable influences ". The results can be explained " on the assumption of a uniform metal surface and uniform adsorption ". All the amines tested show both cathodic and anodic inhibition—the latter usually predominating (H. Kaesche and N. Hackerman, *J. electrochem. Soc.* 1958, **105,** 191).

Anodic Stimulation

Anodic Behaviour of Metals. When a metallic specimen in a corrosive liquid is made an anode by application of an external E.M.F., the corrosion-rate is generally increased, and so soon as the current applied greatly exceeds that corresponding to the rate of " natural " corrosion (which would occur in absence of the applied E.M.F.), the corrosion-velocity can usually be calculated with reasonable accuracy by applying Faraday's Law, using the numbers given in Table XXXII.

TABLE XXXII

Electrochemical Equivalents in grams per coulomb

$Al \rightarrow Al^{+++}$	$0 \cdot 96 \times 10^{-4}$
$Fe \rightarrow Fe^{++}$	$2 \cdot 89 \times 10^{-4}$
$Fe \rightarrow Fe^{+++}$	$1 \cdot 92 \times 10^{-4}$
$Ni \rightarrow Ni^{++}$	$3 \cdot 04 \times 10^{-4}$
$Cu \rightarrow Cu^{+}$	$6 \cdot 58 \times 10^{-4}$
$Cu \rightarrow Cu^{++}$	$3 \cdot 29 \times 10^{-4}$
$Ag \rightarrow Ag^{+}$	$11 \cdot 18 \times 10^{-4}$
$Zn \rightarrow Zn^{++}$	$3 \cdot 39 \times 10^{-4}$
$Cd \rightarrow Cd^{++}$	$5 \cdot 82 \times 10^{-4}$
$Sn \rightarrow Sn^{++}$	$6 \cdot 17 \times 10^{-4}$
$Pb \rightarrow Pb^{++}$	$10 \cdot 73 \times 10^{-4}$

Thus 1 amp. hr. should dissolve 1·04 grams of pure iron (as ferrous ions), 1·22 grams of zinc or 3·86 grams of lead. Nevertheless many cases are known where the corrosion-rate is appreciably less or appreciably greater than the equivalent of the current—as shown in early researches on the anodic attack of many metals reported by H. S. Rawdon and E. C. Groesbeck, U.S. Bureau of Standards, Tech. Paper 1928, **367,** 243.

Departure from the calculated corrosion-rate is met with if a metal passes into the combined state in more than one valency condition; aluminium and magnesium are sometimes attacked as though their valencies were less than 3 and 2 respectively. An important cause of a corrosion-rate greater than that calculated from the current strength arises from the fact that natural corrosion cannot always be neglected. An even greater discrepancy can arise if anodic corrosion eats its way along grain-boundaries, so that the grains themselves drop out unchanged, causing a weight-loss enormously greater than the calculated value. This can occur on relatively pure metal—as shown by extensive studies of lead, copper and silver anodes in nitrite solutions, and of tin and aluminium anodes in oxalate solutions, carried out by F. H. Jeffery (Priv. Comm., 1934).

Cases where the corrosion is below the calculated value are referred to passivity—complete or partial. If a metal becomes completely passive, the current is mainly used in producing oxygen; the corrosion-rate then becomes very small, and frequently assumes a different character; thus an active iron anode usually dissolves quantitatively in the divalent state, whilst a passive iron anode shows a barely detectable corrosion in the trivalent state (or sometimes, in an alkaline solution, in the hexavalent state). Much research work has been devoted to the conditions under which a metallic anode becomes passive and particularly to the relation between the initial current density and the time of passivation. This will now be discussed.

Time of Passivation in Non-stagnant Conditions. At a vertical anode, where the heavy corrosion products can fall under gravity, the conditions needed for film-formation may never be reached at low current densities; even at relatively high current densities, passivity may be avoided at any anode (irrespective of geometry) if the liquid around the anode is kept in a state of violent agitation. In general there will be some current density, ω_0, below which passivity does not occur at all; this may be regarded as equivalent to the rate of removal of the corrosion products, whether by convection or stirring. Above ω_0 passivation sooner or later sets in; the time to establish passivity is often inversely proportional to the current density ω applied, after deducting the "wasted" current density ω_0, as would be expected if a film of some definite thickness were needed for passivity. Thus the time of passivation, t_p, is given by $(\omega - \omega_0)/Q$, where Q is constant.

Such a relationship would be expected without special assumptions, but the first case in which it was established—a gold anode in chloride solution—appears to involve other factors. The facts are not in doubt, having been established by Shutt and Walton and confirmed by Armstrong and Butler, who explained them differently. In normal hydrochloric acid, the straight-line graph relating ω and $1/t_p$ was at the higher current densities almost the same whether there was vigorous stirring or little movement in the liquid; at low current densities, however, the more stagnant conditions gave much shorter values of t_p. Replacement of chloride by sulphate solution shortened the value of ω_0 and also Q, which now represented approximately the amount of electricity needed to form a unimolecular layer of Au_2O_3.

There seem to be two factors at work. At low current densities, passivation is delayed—or even prevented—if the stirring is such as to remove products so quickly that film-formation becomes impossible. However, the influence of the anion (which should not greatly affect the removal of soluble products) is best interpreted if we consider the possibility of alternative anodic reactions, borrowing from Piontelli (p. 907) the idea of a phalanx of adsorbed (oriented) water molecules. If we assume that gold ions cannot easily pass this phalanx, then, in absence of chloride, the anodic potential drop, instead of drawing gold ions into the liquid, will detach hydrogen ions from the adsorbed water molecules (moving them in the direction of the cathode), so that oxygen is left on the metal, which develops an oxide-film and becomes passive. If, however, chloride ions are present, these are adsorbed, interrupting the water phalanx and allowing the gold to pass into solution.*

This somewhat resembles the view taken 25 years ago by Shutt and Walton, who considered that adsorption of Cl^- ions must occur before gold can enter the liquid. If adsorption velocity is proportional to the Cl^- concentration in the liquid, there will be a limiting rate of gold-dissolution proportional to $[Cl^-]$, and if the current density corresponding to this rate is exceeded, OH^- ions begin to play a part, leading to film-formation and passivity; Shutt and Walton found that ω_0 was proportional to $[Cl^-]$—a relation which is thus easily explained. They also found that in N potassium chloride, t_p drops suddenly when the pH exceeds 10·8, and thought that above this point the spontaneous adsorption of OH^- ions can compete with that of Cl^- ions.

* R. S. Thornhill suggests that the gold may form complexes with ions containing chlorine, but considers that in absence of chlorides gold ions, which are easily hydrated, should be able to pass easily from one water cluster to another.

Armstrong and Butler have obtained the same linear relation between ω and $1/t_p$ in unstirred chloride solutions, and explain it by considering diffusion of Cl^- ions across a denuded layer next the anode. The diffusion rate should be proportional to $[Cl^-]$, and if it is sufficient to supply Cl^- ions for anodic dissolutions, passivity should never set in—thus explaining why ω_0 is found to be approximately proportional to $[Cl^-]$. It is not unlikely that this is really the mechanism in the absence of stirring. The various views may all possess value in different situations and deserve study (W. J. Shutt and A. Walton, *Trans. Faraday Soc.* 1932, **28,** 740; 1934, **30,** 914; 1935, **31,** 636; J. A. V. Butler (with G. Armstrong and J. D. Pearson), *ibid.*, 1934, **30,** 1173; 1938, **34,** 806; W. J. Müller and E. Löw, *ibid.*, 1935, **31,** 1291. See also R. H. Roberts and W. J. Shutt, *ibid.*, 1938, **34,** **1455,** who found some of the same relationships when studying a chromium anode).

Time of Passivation in highly Stagnant Conditions. The monumental researches of W. J. Müller and his collaborators were largely devoted to the passivation of horizontal anodes in dilute sulphuric acid under conditions favouring complete stagnancy and accumulation of anodic products, so that even at very low current densities, passivity arrives after a long period. It is probably incorrect to state that there is no limiting current density, since presumably diffusion alone would allow dispersal of the anodic products formed at a very small current density; within the range studied by Müller, however, a finite value of t_p can be assigned to each value of i. In general, the solid which crystallizes out is a sulphate, and this covers up a fraction of the anode surface, causing the absolute current to decline; the true current density on the parts left uncovered, however, increases, so that ultimately the potential associated with this high current density reaches a value at which the formation of an oxide-film and then the evolution of oxygen, become possible; the anode is now passive and the stirring associated with the gas evolution causes the crystalline sulphate to dissolve.

Müller has measured the variation of the current with time and his results show that there are two stages in the covering process—

(1) the deposit starting at certain nuclei extends *laterally* over the surface, forming a layer of nearly uniform thickness;

(2) when covering is nearly complete, lateral growth ceases, and *growth in depth* starts.

The relation between current and time is different in these two stages. Müller worked out equations for both stages, and showed them to be obeyed for several metals when used as anodes in dilute sulphuric acid. It is interesting to note that, at each stage, the same equation is obeyed by iron in sulphuric acid as that obeyed by lead in sulphuric acid, although on iron the passivation process occupies many minutes (the time depending on the initial current density), whereas on lead, owing to the low solubility of lead sulphate, it is complete in a fraction of a second (a special oscillograph was needed for the research on lead).

Unfortunately, Müller's experiments were conducted under conditions which were neither potentiostatic nor galvanostatic; this complicates the analysis of his results. Nevertheless, his mathematical treatment deserves study. Readers of German may like to consult W. J. Müller's own book, " Die Bedeckungstheorie der Passivität der Metalle und ihre experimentelle Begründung " 1933 (Verlag Chemie). A slightly modified presentation of the main features of the mathematical argument has been provided by the author (U. R. Evans, " Metallic Corrosion, Passivity and Protection ", 1946, pp. 51–54).

Anodic Attack of Iron in Sulphuric Acid. Experimental studies under potentiostatic and galvanostatic conditions, carried out by U. F. Franck, have greatly increased our quantitative understanding of anodic passivation. For the case where the layer deposited is a non-conductor, he has worked out simple equations for the movement of potential with time at constant current, and for the movement of current with time at constant potential. Ideally, the final state arrived at should be the same whichever method is used, and indeed the curves relating the values of current and potential after attainment of a steady state obtained by the two methods do lie close together. However, the movements of each variable with time is different according as the layer produced is (1) a non-conductor, (2) an ionic conductor or (3) an electronic conductor. Since each of these can be studied potentiostatically or galvanostatically, there are six possible combinations, but Franck does not consider each of them in detail.

It should be pointed out that often one type of change is followed by a second. Thus on iron in sulphuric acid, non-conducting material (almost certainly ferrous sulphate) is deposited as nuclei and spreads laterally as a layer of nearly constant thickness—in the manner envisaged by Müller; the thickness of the layer may be different for different salts. After a sufficiently high potential has been reached at the small gaps remaining in the layer to permit formation of oxide, the sulphate layer disappears (quickly if stirring is vigorous, the temperature high and the acid dilute, but more slowly if the liquid is relatively concentrated acid so that the solubility of ferrous sulphate is low). The oxide is an electronic conductor, so that the second part of the process (formation of the invisible film responsible for true passivity) obeys different laws.

The lateral spreading of non-conducting compound which is appreciably soluble over an anode surface of total area F_0 will obey the equation $dF_X/dt = K(I - I_k)$ where F_X is the portion covered with a film, I_k the dissolution-rate of the compound expressed as the equivalent current, and I the current flowing. Under galvanostatic conditions, I is constant, so that

$$F_X = K(I - I_k)t$$

Clearly

$$F_0 = K(I - I_k)t_p$$

where t_p is the time of complete coverage, i.e. the time of passivation.

We can write

$$\frac{1}{t_p} = \frac{I - I_k}{F_0 Q}$$

where Q is the quantity of electricity needed for the building of 1 cm.² of deposit, which will vary with the salt deposited, but should be constant for a given salt. Franck measured t_p under several conditions, and obtained straight lines on plotting $1/t_p$ against I. These were displaced laterally when either the stirring-rate or temperature was raised, since either change would increase I_k; however, the displaced curves were all parallel to one another, showing that Q remained constant (i.e. that the material deposited remained the same). The results are shown in fig. 165.

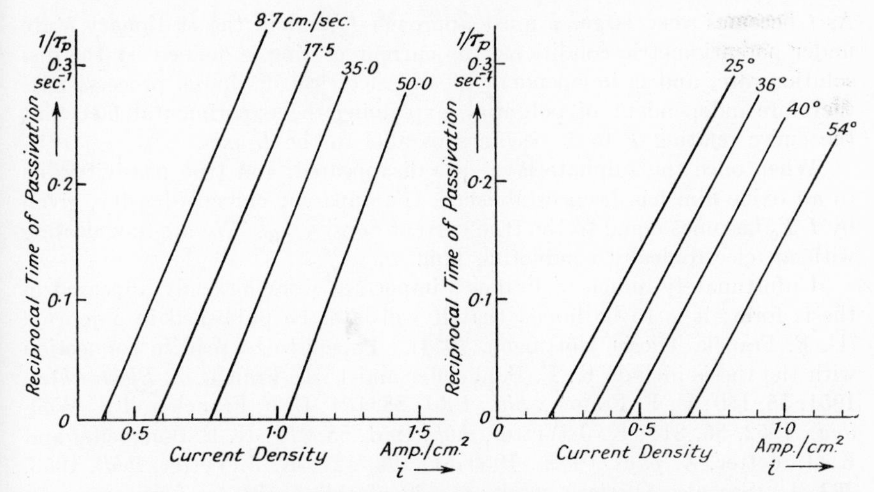

FIG. 165.—Effect of Water-flow and Temperature on relationship between Current Density and Time of Passivation (U. F. Franck).

It is clear that F_X/F_0 increases rectilinearly with t, reaching unity at t_p. The true current density I_W, is $IF_0/(F_0 - F_X)$, and should give a hyperbola when plotted against t. If the potential drop E connected with the current flowing through the gaps in the layer is proportional to the current density, this should also give a hyperbola when plotted against t, and Franck's experimental curve was roughly of hyperbolic form. Theoretically E would rise to infinity at t_p, if the reaction leading the coverage were the only one possible. In practice, however, when E has reached a certain level, other reactions (oxide-formation and oxygen-evolution) become possible, and the rise in E eases off.

Under potentiostatic conditions, I is not constant, but the potential chosen decides the true current density I_W, whilst I is equal to $I_W(F_0 - F_X)/F_0$. If we call the fraction of the surface covered α, this is equal to F_X/F_0 and $I = I_W(1 - \alpha)$ so that $\alpha = 1 - I/I_W$. The corresponding expression $1 - I_k/I_W$ may be written α_*, so that $I_k = I_W(1 - \alpha_*)$. Thus

$$F_0 \frac{d\alpha}{dt} = \frac{dF_X}{dt} = K(I - I_k) = KI_W[(1 - \alpha) - (1 - \alpha_*)]$$

so that $d\alpha/dt = KI_W(\alpha_* - \alpha)/F_0$

Integrating on the assumption that $\alpha = o$ when $t = o$, we obtain

$$\alpha = \alpha_*(1 - e^{-KI_W t/F_0})$$

or

$$t = \frac{F_0}{KI_W} \log_e [\alpha_*/(\alpha_* - \alpha)]$$

so that

$$I = I_k + (I_W - I_k)e^{-KI_W t/F_0}$$

or

$$t = \frac{F_0}{KI_W} \log_e \frac{I_W - I_k}{I - I_k}$$

As t becomes very large, I must approach I_k and in the stationary state under potentiometric conditions, the current passing is defined by the dissolution-rate, and is independent of any electrical discharge process; it is therefore independent of potential, explaining the experimental fact that the curve relating T to E becomes parallel to the E-axis.

When once the sulphate layer has disappeared, and true passivity due to an oxide-film has been established, the apparent current density, given by I/F_0 becomes equal to the true current density I_W. We are now dealing with an electronically conducting film.

Unfortunately, much of Franck's important work has only appeared in thesis form; it is to be hoped that it will later be published in a journal (U. F. Franck, Thesis, Göttingen, 1954). Papers to be read in connection with the thesis include K. F. Bonhoeffer and U. F. Franck, *Z. Elektrochem.* 1951, **55,** 180; U. F. Franck, *ibid.*, 1951, **55,** 154; U. F. Franck and K. Weil, *ibid.*, 1952, **56,** 814; K. J. Vetter, *ibid.*, 1951, **55,** 274; K. F. Bonhoeffer and K. J. Vetter, *Z. phys. Chem.* 1950–51, **196,** 127; K. J. Vetter, *ibid.*, 1953, **202,** 1. See also Olivier's work (p. 229 of this book).

In Franck's studies of iron in sulphuric acid based on the potentiostatic method, the potential was raised by steps, the corresponding values of current being measured on each occasion after attainment of the steady state; at a certain value of the potential, a jump occurs from the "active" state (in which the metal is covered with a layer of crystalline sulphate, but is still slowly dissolving as ferrous ions) to the "passive" state (in which the metal is believed to be covered with an invisible oxide-film and is still more slowly dissolved as ferric ions); the "active" dissolution as ferrous ions may be equivalent to about 0·2 amps/cm.^{2} whereas the "passive" dissolution as ferric ions is far slower, perhaps about 7×10^{-6} amp/cm.^{2}. Franck found that the jump occurred when the potential reached about 0·45 volts (on the hydrogen scale), but on subsequent lowering the potential in steps, activation occurs whilst the potential is still well above 0·45 volts (perhaps about 0·55 volts). This is probably due to the fact that during the lowering of the potential there is no ferrous sulphate layer present so that the composition of the liquid next the metal is much the same as that in the body of the solution; during the ascent, this is not the case, the liquid in the crevices threading the sulphate layer being different in composition from that outside.

The fact that the passivation and activation potentials are not the same, has long been known. As early as 1911, Flade noticed that the potential corresponding to activation varies with the pH of the solution, usually by about 59 mv. for each pH unit. Bonhoeffer and Beinert have pointed out that the potentials measured do not correspond to the values which would be expected if passivity was due to a film of known oxide (FeO, Fe_3O_4, or Fe_2O_3). In the author's view, this would not be expected, since ferric oxide cannot give protection unless excess of oxygen, or an oxidizing agent, is present (see p. 226); thus a much higher potential than that corresponding to ferric oxide, as commonly met with, would be expected, and is indeed obtained. If the avoidance of film destruction by reductive dissolution is to be attributed to an oxygen charge, the dependence on pH is to be expected and also the high range of values actually met with. However, the facts are complicated, and the dependency on pH is different in different cases. Several explanations of the Flade potential and its variation with conditions were offered at the International Colloquium on the Passivity of Metals, held at Heiligenberg, near Darmstadt in 1957. The papers appear in a single number of the *Zeitschrift für Elektrochemie* 1958, **62,** 619–827; see especially those of U. F. Franck, K. G. Weil, K. J. Vetter, Y. M. Kolotyrkin, B. Le Boucher, W. Feitknecht, J. H. Bartlett and others. See also F. Flade, *Z. phys. Chem.* 1911, **76,** 513; H. Beinert and K. F. Bonhoeffer, *Z. Elektrochem.* 1941, **47,** 441. The matter is being discussed, in the light of the established facts of reductive dissolution—a factor too often overlooked—by Pryor, whose paper, when it appears, should receive study (M. J. Pryor, *J. electrochem. Soc.* (in the press)).

Behaviour of Iron in Nitric Acid. The striking contrast between the violent reaction of iron in moderately dilute nitric acid and its indifference to the concentrated acid has received an elegant graphical interpretation from the brothers Pražák. As explained on p. 1042, a metal anode becomes passive when subjected to a current density greater than that corresponding to the greatest rate at which metal can enter solution, so that an alternative reaction (e.g. formation of oxide) becomes inevitable. Where the anodic product is sparingly soluble, the obstruction may be due to a solid salt layer, and in that case the limiting current density may be greatly increased by stirring. In nitric acid, however, the limit set to the current flow is not likely to be caused by a nitrate film; it is only likely to occur when the current imposed is so great that the supply of atoms in the correct energy state to enter the liquid is insufficient. The limiting current density will be increased by rise in temperature but not greatly affected by stirring; if the current exceeds the limiting value, the potential must rise to the level needed for the alternative reaction. In fig. 166, the anodic curve (1) is actually based on the behaviour of iron in sulphuric acid (direct measurements in nitric acid were considered inadvisable owing to possible complications arising from "chemical oxidation"); provided that the limiting current density is caused by the necessity of providing iron atoms with the necessary activation energy to enter the solution as cations, the nature of the anion should not greatly affect the course of the curve.

The cathodic reaction in nitric acid occurs at a higher potential than the evolution of hydrogen, but is likely to descend somewhat as the current increases. For concentrated acid, this descent may be slight (curve 2), especially if stirring is avoided so that autocatalytic conditions prevail (p. 336). In dilute nitric acid exhaustion becomes possible and the descent may be considerable (curve 3). If, for simplicity, we assume that both anodic and cathodic reactions can occur at any point of the metallic surface, so that the anodic area and cathodic area are each equal to the area of the specimen, then, for a specimen measuring 1 cm.^{2}, the numbers printed on the scale can be held to represent either current density or current. It can be seen that the cathodic curve for concentrated acid may cut the anodic curve at P, above the rise caused by the limiting current density and, although there may be attack at first, as is observed (p. 328), much of the current

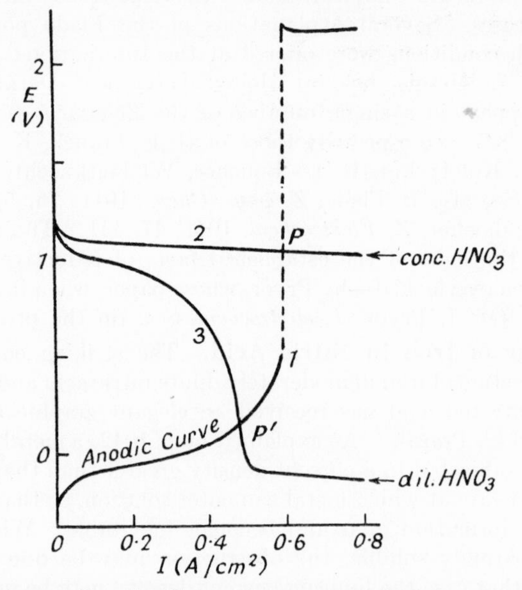

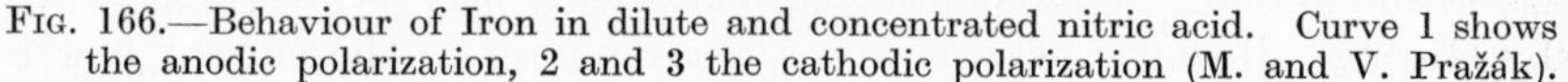
FIG. 166.—Behaviour of Iron in dilute and concentrated nitric acid. Curve 1 shows the anodic polarization, 2 and 3 the cathodic polarization (M. and V. Pražák).

must be devoted to film-formation—so that the reaction soon falls off. For the dilute acid, the curves intersect at P', below the critical current density, and the rapid attack continues unchecked (M. Pražák and V. Pražák, *Coll. Czech. Chem. Comm.* 1956, **21**, 564).

Cathodic Protection

General Principles. Consider a piece of metal immersed in a corrosive liquid of sufficient conductivity to allow the intersection point B to be used as an indication of the corrosion-velocity (fig. 167). Now apply a current from an external anode. Clearly the sum of the current from the external anode and the " corrosion-current " due to local anodes still operating on the surface must exactly balance the cathodic current. Thus if the external current is represented by the length DE, the corrosion-current will be CD, and the potential will be represented by C. If we depress the potential to F, the corrosion-current becomes zero, and the specimen is completely protected. Thus to attain cathodic protection, the *potential of the whole must be brought down to the open-circuit potential of the most active anodic point*; this principle, expressed in two different ways, was published in the same year by T. P. Hoar, *J. electrodep. tech. Soc.* 1938, **14,** 33, and by R. B. Mears and R. H. Brown, *Trans. electrochem. Soc.* 1938, **74,** 519, who arrived at it independently.

Hoar has pointed out that the amount of current needed to bring down the potential to the required level depends on the conditions, and particularly

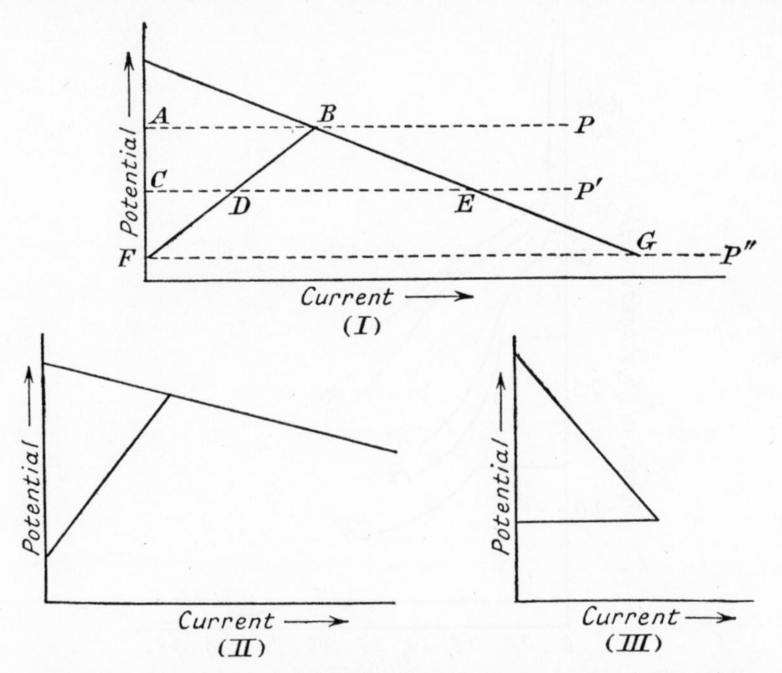

FIG. 167.—Graphical representation of Cathodic protection (schematic).

on the oxygen-supply; if the rate of oxygen-replenishment at the cathodic area is high, the current needed for protection becomes considerable.*

The argument just developed involves several assumptions, namely that the specific conductivity in the corroding liquid is so high that the potential is defined by the intersection of the two polarization curves, and that the protecting current is applied in such a way that all parts of the specimen

* R. S. Thornhill remarks that the current needed for protection depends on the temperature, which has the effect of swinging BG upwards. He mentions experiments on a cell composed of steel and copper electrodes buried 10 yards apart in soil, which showed that the corrosion-current becomes zero when the copper is polarized to the iron potential by means of an external E.M.F.

have equal chances of benefiting from it. These conditions are often fulfilled in laboratory experiments—but rarely in practical cases of cathodic protection, such as a long pipe-line buried in fairly dry soil.

Criterion for Protection based on a break in curve. Under the somewhat artificial conditions just stated, the curve relating applied cathodic current density i and measured potential V may be expected to show a sharp break at the value of i requisite for entire prevention of corrosion. In early work at Cambridge, Britton traced the relation between i and V using various metals as cathode in N/10 potassium chloride kept saturated with oxygen and stirred with an eccentrically mounted glass rod rotating 15 revolutions per minute, the whole being thermostated at 20°C. The

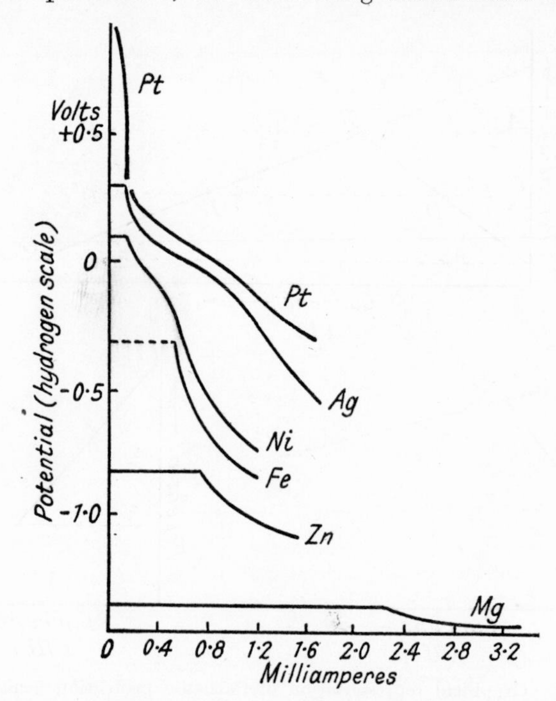

Fig. 168.—Relation between cathodic potential and current in N/10 Potassium Chlordie solution containing oxygen (U. R. Evans, L. C. Bannister and S. C. Britton). The portion of the iron curve shown as a broken line varies with time; see original paper.

curves obtained are shown in fig. 168. When the applied current density was high, corrosion was prevented, but, below a certain " protective " value, corrosion was seen to be occurring; over this region, the curve was horizontal —except in the case of iron, where the situation was more complicated, since the potential reading varied with time.

The horizontality can easily be explained, if we assume that the function of the current is to keep a layer of inhibitive alkali at the surface; for any given stirring-rate, a certain minimal current density would be needed to maintain the alkali swept away by the liquid motion.* If any value of the applied current is too small to protect the whole, corrosion starts and spreads out until the combined cathodic current density on the part still unattacked (the sum of that due to the cathodic reaction of the local cell, namely the reduction of dissolved oxygen, and the current from the external source) reaches the protective value; then the extension ceases. Thus although the proportion of the surface protected from attack may vary with the applied current, the current density on this area, and therefore the potential, will be roughly constant—explaining the horizontal character of this part of the curve. The argument suggested assumes that the corrosion proceeding does not interfere with the supply of oxygen; with iron, such interference is likely to occur, since the ferrous hydroxide, formed by interaction of the anodic and cathodic products, is a greedy absorbent of oxygen—explaining the failure to obtain a satisfactory horizontal limb on iron (U. R. Evans, L. C. Bannister and S. C. Britton, *Proc. roy. Soc.* (*A*) 1931, **131**, 355, esp. p. 366).

The conditions maintained in this laboratory research (carried out to throw light on the scientific mechanism of corrosion), were extremely different from those which would exist in the field, where cathodic protection is being used by engineers to solve a practical corrosion problem. In most practical cases, there is no uniform distribution of the cathodic current supplied from the external source nor are the local cathodic areas evenly distributed; often the protection provided is not due to maintenance of an alkali film, and around buried structures the conductivity of the soil-water is generally too low for *IR* drops to be neglected. Nevertheless, shortly after the publication of the paper quoted (and, it is understood, as a result of it), attempts were made to ascertain the value of the current needed for protection of a working pipe-line by looking for a sharp break on the curve relating current and potential; the sort of method employed seems to have been the application of gradually increasing amounts of cathodic current to the pipe-line from an external ground-bed or sacrificial anode, measuring at each current value the local potential (obtained perhaps with a copper/copper sulphate electrode placed on the ground above the pipe). If the curve of V plotted against I showed a sharp break, this was assumed to represent the value of the current needed for protection. It is not surprising that the results obtained, although sometimes, by good fortune, approximately right, were often seriously in error. This was at first attributed to neglect of the *IR* drop, and attempts were made to alter the method so as to allow for that omission, but the main causes of the discrepancy—connected with the geometrical complexity of the practical situation—received, until recently, insufficient attention. The researches

* In another research it was shown that a higher current density was needed for protection under stirring conditions than under stagnant conditions, and that the provision of grooves on the cathode, which tended to prevent dissipation of the alkali, diminished the absolute value of the current needed for protection, despite the fact that the surface area was increased. See U. R. Evans, *Metals and Alloys* 1931, **2**, 62.

now proceeding at Emeryville, mentioned on p. 292, may go far to rectify the situation.

The reader may care to study papers where various points of view are presented. These include J. M. Pearson (article in " Corrosion Handbook " 1948, p. 923 (Editor, H. H. Uhlig; Publishers, Wiley: Chapman & Hall)); G. R. Kehn and E. J. Wilhelm, *Corrosion* 1951, **7,** 156; S. C. Britton, *ibid.*, 1951, **7,** 403; E. Schaschl and G. A. Marsh, *ibid.*, 1957, **13,** 243t. W. J. Schwerdtfeger and A. C. McDorman, *J. electrochem. Soc.* 1952, **99,** 407; W. J. Schwerdtfeger, *Bur. Stand. J. Res.* 1957, **58,** 145. The Emeryville work is described by E. W. Haycock; *Corrosion* 1957, **13,** 767t.

Criterion of Protection based on Potential between Metal and Liquid. It is commonly stated that a pipe-line is completely protected if the potential is on the negative side of $- 0{\cdot}85$ volt to the Cu | saturated $CuSO_4$ electrode, that is $- 0{\cdot}53$ volt on the normal hydrogen scale; in presence of sulphate-reducing bacteria, more negative values ($- 0{\cdot}95$ volt to Cu | $CuSO_4$, or $- 0{\cdot}63$ volt to hydrogen) appear necessary (p. 285). These numbers are perhaps best regarded as empirical statements of potentials which hitherto have proved satisfactory. If the function of the cathodic current is to plug gaps in a protective coating with calcium carbonate or perhaps magnesium hydroxide (later converted to rust), the potential needed for precipitation will depend on the calcium-content (or magnesium content) of the soil and also on its redox value. Possibly practical experience may be a better guide than theoretical calculations; it should, however, be borne in mind that the amount of experience based on careful inspection of pipe-lines which have received cathodic protection over a number of years is still limited.

In soils free from calcium, magnesium and sodium salts, virtual protection of steel should be set up at a potential where $Fe \rightarrow Fe^{++}$ is exactly balanced by $Fe^{++} \rightarrow Fe$, and this will depend on the concentration of Fe^{++} ions which, in a steady state, is present around the pipe-lines. If it were N/100, we would expect (since the normal electrode potential of iron is $- 0{\cdot}44$ volts) that corrosion would cease at a potential of $- 0{\cdot}50$ volt to hydrogen—which would accord fairly well with general experience (see p. 285). However, if such a concentration is maintained just outside the pipe, there will be a slow diffusion outwards into the soil, and corrosion must take place at such a rate as to replenish the Fe^{++} ions which are thus lost. Probably the loss by diffusion in most soils would be small, and the corresponding corrosion-rate could be neglected. In presence of S^{--} ions, the concentration of Fe^{++} needed for protection at $- 0{\cdot}53$ volt may be incapable of existing, owing to the low solubility product of FeS; this may, perhaps, be the reason for the fact that a more negative potential is needed in presence of sulphate-reducing bacteria. Such theoretical considerations are thrown out as a basis for thought, but they should not be too hastily applied to practical situations.

Criterion of Protection based on Equipotential Surfaces. The only criterion for protection which is satisfactory from the scientific point of view is one which would be difficult to apply in practice. It has been studied in the laboratory by Bianchi, whose papers deserve study as examples of clear thinking. To simplify the situation, Bianchi has mainly studied cases where polarization is excluded. Thus he used lead in a solution of M/2 lead sulphamate which renders polarization negligible; the exploring electrode was a glass tube drawn out to a capillary point containing the same liquid with a lead wire immersed in it—so that liquid-junction potentials were avoided.

Bianchi takes as a model of corroding metal a cathodic surface containing

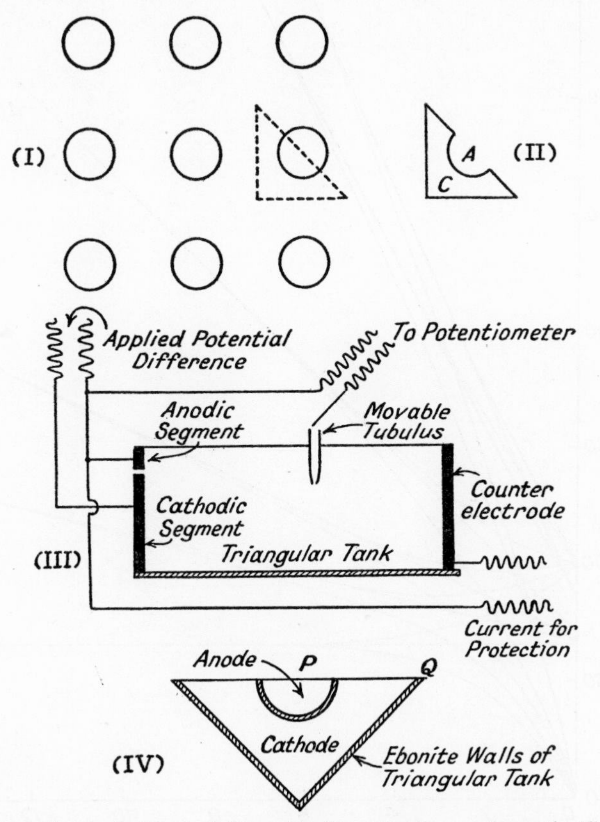

FIG. 169.—Triangular tank for studying cathodic protection—schematic (G. Bianchi)

anodic circles arranged regularly as shown in fig. 169(I); this pattern is merely an infinite repetition of the triangle of fig. 169(II), and to examine the situation he used a triangular trough (fig. 169(III)) with the anodic semi-circle, separated by a small insulating gap from the surrounding cathodic triangle (fig. 169(IV)). Whilst in practice the current flowing between anodic and cathodic areas is due to chemical or physical differences between different areas in the metal or the adjacent liquid, Bianchi (instead of making the anodic and cathodic areas of different metals, e.g. iron and copper) made both portions of lead, but joined them to a source of variable potential, so that a definite potential difference (usually about 100 mv.) could be maintained between them. The counter-electrode which forms the anode of the polarizing current is placed at the other end of the trough, which is filled with lead sulphamate solution, and the movable tubulus of the exploring electrode is used to trace equipotential surfaces in the liquid. If the protecting current is sufficient, the potential followed on a line normal

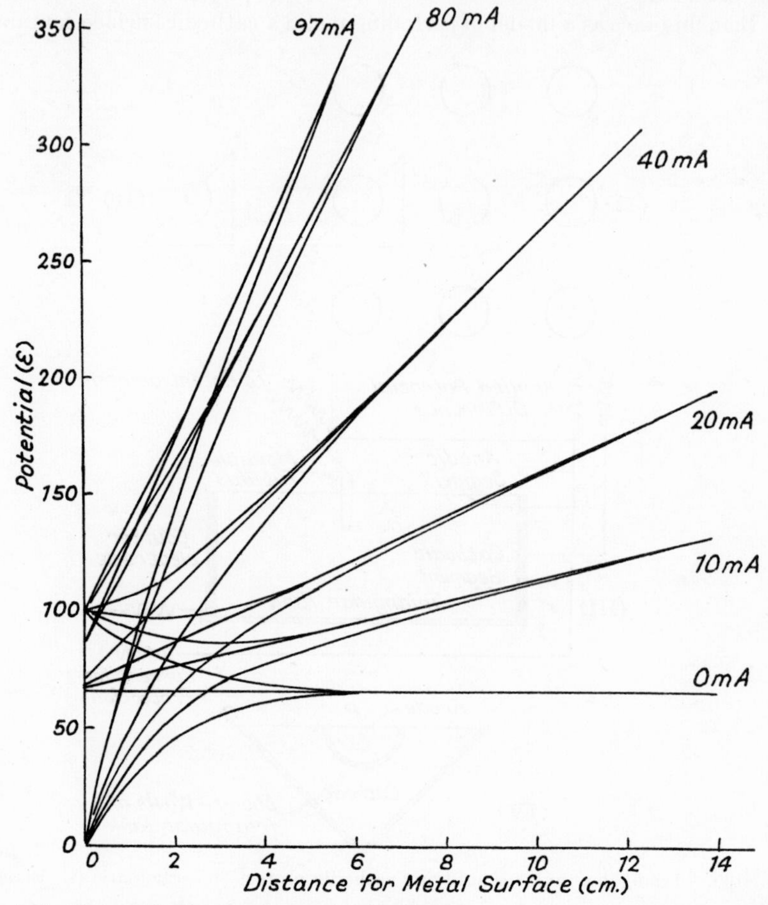

FIG. 170.—Potential distribution on application of current from external source—schematic (G. Bianchi).

to a point on the anode will steadily rise; if it is insufficient, it will first sink and then rise. Fig. 170 shows curves from the central points, P, of the anodic areas (also curves from corners, Q, of the cathodic areas, which start at 0 and blend with the curves from P at a distance from the surface). It is clear that for protection at P, 20 ma. is insufficient and 40 ma. more than sufficient; however to protect the *edges* of the anodic areas, larger currents might be needed. For details the original papers should be studied

(G. Bianchi, *Annali de chimica* 1953, **43,** 173; 1956, **46,** 742. *Ricerca Sci.* 1956, **26,** 427, 2359).

Relation between Protective Current and Corrosion-current. It is thought by some that the current needed to be applied to obtain complete cathodic protection is equal to the corrosion-current which would be flowing if there were no protection. A little consideration will show that this is only true if the corrosion reaction is under cathodic control. Consider (fig. 171), a piece of metal in the ground which is being corroded at a rate decided solely by the rate of oxygen-replenishment at the cathodic area (C), situated (probably) in the upper portion. If now we join this to an external anode (either a sacrificial anode of magnesium, or a ground-bed of iron connected through a source of E.M.F.) and apply a current exactly equal to the corrosion-current, then the cathodic reduction of the oxygen reaching C, instead of being balanced by the anodic attack of iron at A_i, is balanced by the corrosion of magnesium or iron at the external anode, A_e. Since the corrosion had previously been determined solely by the rate

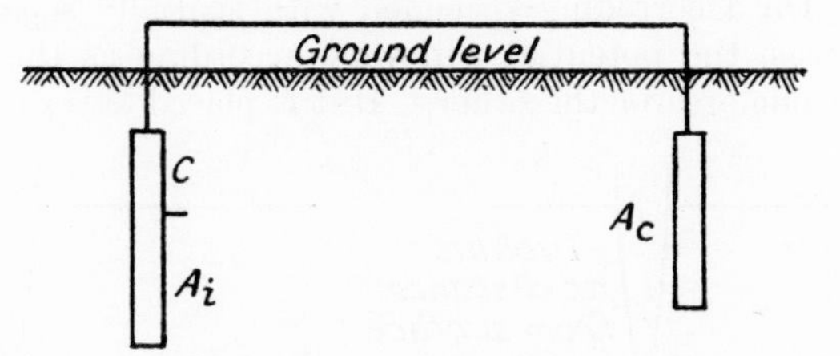

FIG. 171.—Explanation of approximate equality of protective current and corrosion current (schematic).

of oxygen-supply, there is now no opportunity of any corrosion at A_i; protection is complete.

If, however, the control had even been partly anodic, then, when most of the current is diverted to A_e, the total current no longer remains the same, and there is a small current available for corrosion at A_i; for complete protection, the current applied through A_e must *exceed* the former value of the corrosion-current.

Schaschl and Marsh have measured the ratio of protective current to corrosion-current in an extensive series of researches and have collected similar results from the work of others. In neutral or slightly alkaline liquid or soils the ratio slightly exceeds unity (1·1 to 1·3 is usual). For acid liquids, it is usually well below 1·0. Schwerdtfeger, using 0·2% sodium chloride, reports that the reciprocal ratio for 16 sets of experiments ranged from 0·88 to 0·92, giving about 1·1 for the ratio as expressed by Schaschl and Marsh (E. Schaschl and G. A. Marsh, *Corrosion* 1957, **13,** 243t, esp. table 1, p. 247t; W. J. Schwerdtfeger, *Bur. Stand. J. Res.* 1957, **58,** 145, esp. p. 150).*

* R. S. Thornhill remarks that, although the figure is near unity when the anodic curve is flat, it may become very much higher when the cathodic curve becomes flat—as at elevated temperatures.

Some of these principles were brought out in an early paper by Hoar on the tin-iron couple in citric acid or citrate buffer solution, although here the protection produced by joining a tin specimen to the iron was not complete. Hoar wrote, " The cathodic parts of the iron surface can evolve or depolarize hydrogen at a limited rate, and in an uncoupled specimen this is equivalent to the rate of anodic dissolution. But if a tin anode is coupled to the iron, part of the anodic dissolution is transferred to the tin. The cathodic reaction still proceeds at the *same* limited rate (unless the cathodic potential is considerably altered, which in the present case it is not). Hence the anodic dissolution of the iron is diminished by an amount equivalent to the current flowing in the couple " (T. P. Hoar, *Trans. Faraday Soc.* 1934, **30,** 472, esp. p. 481).

A mathematical theory of cathodic protection is provided by C. Wagner, *J. electrochem. Soc.* 1952, **99,** 1; 1957, **104,** 631.

Potential Measurements and their Significance

General. On a corroding specimen with spatially separated cathodic and anodic areas, the potential will vary according as the tubulus tip is placed against one area or the other. If it is placed at a distance from the

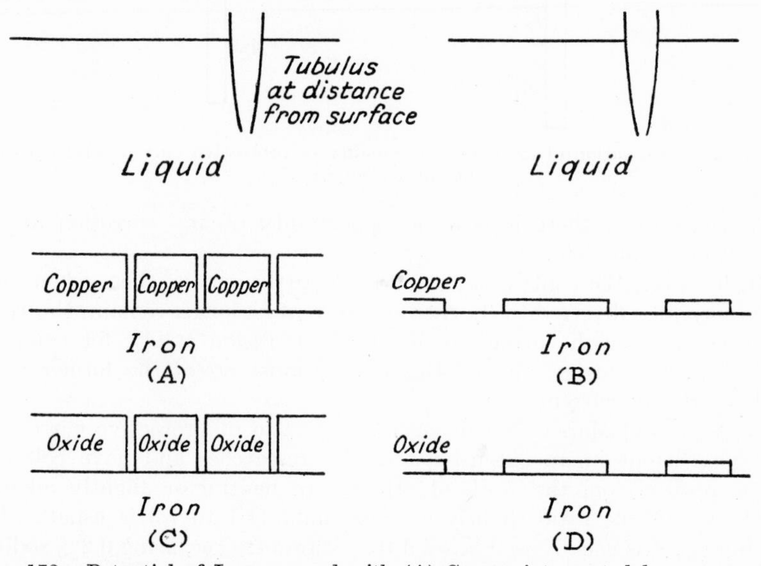

Fig. 172.—Potential of Iron covered with (A) Copper interrupted by narrow pores, (B) Copper interrupted by large gaps, (C) Oxide interrupted by narrow pores, (D) Oxide interrupted by large gaps (schematic).

metal, then the " compromise potential " measured will usually be that represented by the intersection point of the two polarization curves.

If the metal (e.g. iron) is covered with a porous coating of a more noble metal (e.g. copper), the potential as measured will depend on the number, cross-section and length of the pores (the length being equal to the thickness of the coating). If the pores are few and narrow, and the copper coating thick (fig. 172(A)), the potential will be close to that of massive copper in the same liquid. It will differ from that of the iron for two reasons: (1) the *IR* drop in the pores will be considerable, especially if the copper coating is thick (*I* is here the current flowing between copper and iron), (2) the anodic polarization may be high, since the area of iron is small, and even a small value of *I* may represent a large value of current density. However, as the combined cross-sections of the pores increases or as the thickness of the copper coating decreases (fig. 172(B)), both these factors become less important, and the potential as measured will approach that of the iron, especially if, owing to limitations in the rate of oxygen-replenishment, cathodic polarization becomes important. Thus the measurement of compromise potential in such cases may be a useful method of obtaining information about the effective porosity of a conducting coat.

Potential–Time Curves. The principle established applies equally to oxide-films possessing electronic conductivity, although in such cases the *IR* drop over the film must also be taken into account. When an oxide-film is threaded by few and narrow pores, the potential will be similar to that of an electrode of massive oxide in the same liquid (fig. 172(C)). When the pores are broad and numerous, it will be similar to that of film-free metal (fig. 172(D)). Thus supposing that a piece of metal carrying a porous oxide-film is placed in a liquid, the potential will tend to *rise* with time if the film is *self-healing* (i.e. if the corrosion process progressively blocks up the pores with corrosion-product), and to *fall* with time if the bare area is *spreading*. The absolute value of the potential may possess no significance, but its movement with time does indicate whether the material is becoming passive or is developing corrosion (rather as the movement of a barometer with time has greater significance in weather prediction than the absolute value of the pressure recorded).

Potential–time curves have been much used for such purposes. One early application was that of May (p. 477), who used them to show whether mechanical damage to a protective film on a brass was setting up corrosion, or whether it was being automatically repaired. Numerous potential–time curves on stainless steel fully immersed in N/10 potassium chloride were recorded by Bannister, who studied the effect of composition and previous history (including abrasive treatment, thermal treatment and chemical treatment). Stainless steels containing molybdenum gave rising curves under circumstances where others gave falling ones, whilst a surface with a fine finish showed a higher potential (and less tendency to fall) than a similar material coarsely ground. Other experiments showed the benefit of pre-treatment with nitric acid, as previously recommended by Hatfield, and the adverse effect of heating at 500–900°C. Bannister also studied ordinary steel in solutions of chloride, chromate and mixtures of the two; chromate gave rising curves, and chloride falling curves; a high final potential here represented immunity, a middle value slight rusting and a low value profuse rusting. For details, see L. C. Bannister and U. R. Evans, *J. chem. Soc.* 1930, p. 1361.

Since that time, much use has been made of potential–time curves, mainly for the study of materials carrying protective films. Abrasion below sodium chloride or hydroxide was found sometimes to depress the potentials of A-group metals (Be, Al, Nb, Cr, Mo) by about 0·5 volt, doubtless owing to the removal of the protective film—as shown by G. W. Akimov (p. 765). Unpublished work on aluminium by R. C. Plumb should be studied when it appears.

The fact that a change in potential and a change in corrosion-rate do not necessarily move together is evident when it is considered that the introduction of a second element into a metal may accelerate corrosion in two ways, by stimulating either the cathodic or the anodic reaction. In the first case, it will raise the compromise potential, represented by the ordinate of the intersection point, whilst in the second case, it will depress the compromise potential; the corrosion-rate, however, may be increased in both cases. A factor which affects both cathodic and anodic reactions might greatly increase the corrosion-rate and yet leave the potential almost unaffected; an example seems to be provided by the effect of internal stress, which greatly promotes corrosion, but sometimes has only a small affect on the corrosion potential (p. 909). The idea that a shift of potential in the positive (noble) direction will produce a smaller corrosion-rate is shown to be wrong by the effect of adding copper to aluminium. This greatly *increases* susceptibility to corrosion, and yet shifts the potential in the *noble* direction. Unalloyed aluminium corrodes far more *slowly* than an aluminium-copper alloy in most waters when they are unconnected; the latter, however, has a more noble potential and can be protected by electrical connection to unalloyed aluminium—which in that event is corroded more *quickly* than the alloy.

Thermodynamic and Kinetic Considerations

Pourbaix Diagrams. One of the most remarkable developments of recent years has been the construction of diagrams, with pH value and potential as co-ordinates, representing the various chemical and electrochemical equilibria which affect Metallic Corrosion. The curves divide the diagram into regions within which *immunity*, *corrosion* or *passivation* may be expected to prevail. We owe these diagrams to the inventiveness of M. Pourbaix of Brussels. Thanks to his energy and enthusiasm, they have now been constructed for all the important metals.* Each diagram contains a vast amount of information concisely and clearly expressed.

Every reader should study the principles enumerated in M. Pourbaix's book, " Thermodynamiques des Solutions aqueuses diluées " (Meinema, Delft), or J. N. Agar's excellent translation, " Thermodynamics of Dilute Aqueous Solutions " (Arnold). The diagrams in the book embody information available in 1945 regarding copper, iron and chromium, but more recent data regarding these and numerous other metals will be found in reports issued by " CEBELCOR " (Centre Belge d'étude de la Corrosion)

* An Atlas which will bring together the Pourbaix diagrams for all important metals is under preparation by Cebelcor.

and by " CITCE " (Comité de Thermodynamique et de Cinetique Electro-chimiques). The latter committee published two reports in 1950 on lead and silver by M. Pourbaix with P. Delahay and P. van Rysselborghe, whilst among the " Rapports Techniques " issued by Cebelcor the following deserve special notice: **No. 21** (an admirable *vue d'ensemble*, introducing diagrams for Te, Cu, Ag, Zn, Pb, Al, As, Au, Be, Cd, Co, Hg, Sb, Sn, Ti, Tl), **No. 23** (dealing with Ni), **No. 28** (a useful list of *Enthalpies libres de Formation Standards à 25°C.*), **No. 31** (dealing with U); **No. 32** (W), **No. 33** (Te), **No. 34** (Mg and Zn), **No. 35** (Mo), **No. 39** (Mg.), **No. 41** (Cr), **No. 42** (Al), **No. 43** (passivation by chromates, molybdates, etc.), **No. 44** (Cl), **No. 45**

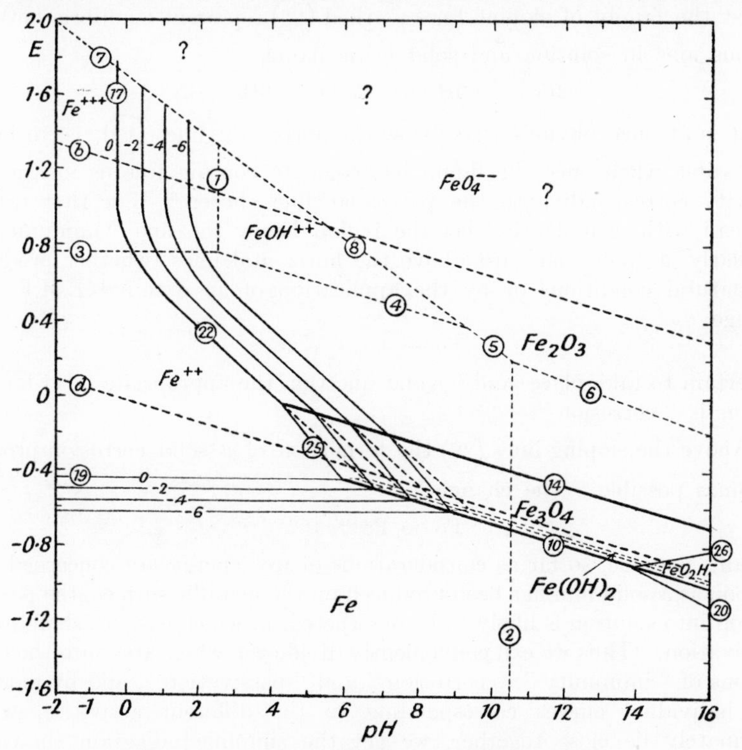

Fig. 173.—Pourbaix Diagram for Iron at 25°C.

(Zr), **No. 46** (As), **No. 47** (B), **No. 48** (Bi), **No. 50** (Tc), **No. 51** (Re), **No. 52** (Ta), **No. 53** (Nb), **No. 55** (Sb), **No. 58** (Ru), **No. 59** (Rh), **No. 60** (Pd), **No. 61** (Os), **No. 62** (Ir) and **No. 63** (Pt).

Each line in a Pourbaix diagram represents the conditions of some equilibrium. A horizontal line represents equilibrium for a reaction involving electrons but not involving H^+ or OH^- ions; a vertical line represents an equilibrium involving H^+ or OH^- ions but not involving electrons; a sloping line represents an equilibrium involving H^+ or OH^- ions and *also* involving electrons.

For instance, in the Pourbaix diagram for iron (fig. 173) the family of horizontal lines marked (19) show the potentials of the electrode equilibrium

$$Fe \rightleftharpoons Fe^{++} + 2e$$

at ferrous ion " activities " of 10^0, 10^{-2}, 10^{-4}, 10^{-6} etc. times Normal, according as the number attached to the line is 0, — 2, — 4, — 6, etc.; they are spaced out at intervals of 59 mv., owing to the operation of the law mentioned on p. 1016. The verticals marked (1) represent the hydrolysis of ferric ions

$$Fe^{+++} + H_2O \rightleftharpoons Fe(OH)_2^{++} + H^+$$

whilst the family of sloping lines marked (22) represent equilibria between ferrous ions in solution and solid ferric oxide

$$2Fe^{++} + 3H_2O \rightleftharpoons Fe_2O_3 + 6H^+ + 2e$$

It is at once obvious that below the horizontal lines (19), corrosion is impossible when once the liquid has come to contain ferrous ions at the activity corresponding to the particular line chosen.* For that reason, we can with confidence label the region below the line " immunity ".† Similarly, at potentials just above the horizontal lines (whether produced by natural conditions or by the application of an external E.M.F.) the change

$$Fe \longrightarrow Fe^{++} + 2e$$

is certain to take place, and beyond question the appropriate label for this region is " corrosion ".

Above the sloping lines (22) the formation of a solid corrosion-product becomes possible. The change

$$Fe \longrightarrow Fe^{++} + 2e$$

remains possible, so far as considerations of free energy are concerned, but so soon as a solid film has been produced on the metallic surface, the passage of iron into solution is likely to be obstructed, in which case we shall obtain passivation. Thus we can conveniently divide the whole area into the three regions of " immunity ", " corrosion " and " passivation ", and by merging the individual curves corresponding to the different activities, which fortunately lie close together, we get the simplified diagram shown on fig. 174 for iron. Similar simplified diagrams for chromium, zinc and copper are represented in figs. 175, 176 and 177. It will be noticed that in all cases there is more than one area marked " corrosion ". The small corrosion area on the extreme right of the diagram for iron represents the formation of ferroates (p. 454) and the larger area on the right of the diagram for zinc to that of zincates (p. 453). The corrosion area at the top of the chromium

* R. S. Thornhill points out that at the equilibrium potential as much iron is dissolved as is redeposited, so that the weight-loss and corrosion-rate are zero. At more positive potentials, corrosion prevails over redeposition, whereas at more negative potentials deposition prevails and there is no corrosion.

† In some of the earliest diagrams, the region now called " Immunité " was labelled " Passivité ".

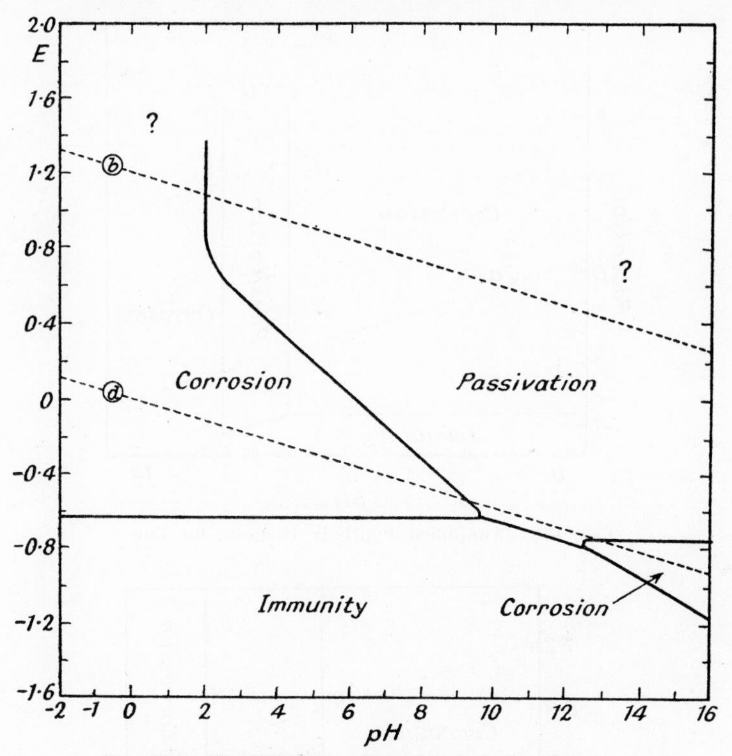

FIG. 174.—Simplified Pourbaix Diagram for Iron.

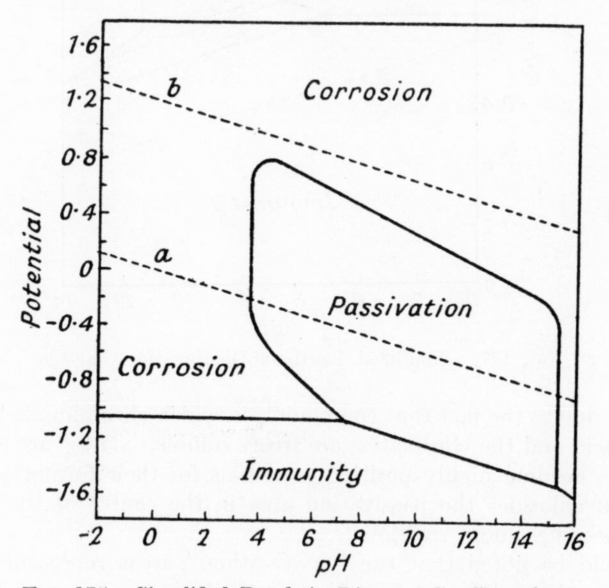

FIG. 175.—Simplified Pourbaix Diagram for Chromium.

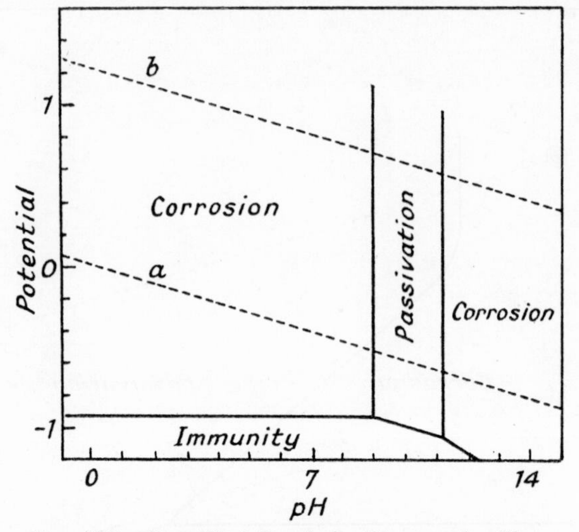

Fig. 176.—Simplified Pourbaix Diagram for Zinc.

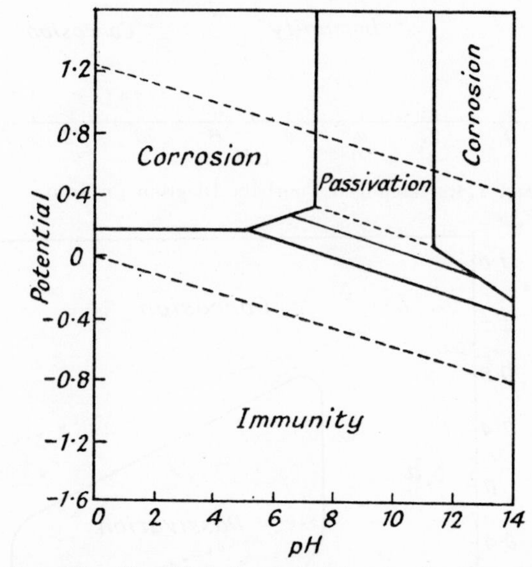

Fig. 177.—Simplified Pourbaix Diagram for Copper.

diagram is due to the fact that compounds in which chromium is hexavalent (chromic acid and the chromates) are freely soluble. These are energy-rich compounds needing highly positive potentials for their formation. In the presence of chlorides the passivation area in the centre of the chromium diagram becomes more restricted.

It should be noted that the " passivation " areas represent a type of protection entirely different from that met with in the " immunity " areas.

Within the latter (provided that metal cations are present in the appropriate concentration) corrosion will be *impossible* for reasons of *energy*, whereas in the " passivation " area corrosion is obstructed for reasons of *geometry*; provided that the film has once been formed, and provided also that the film-substance has appropriate physical characteristics, the corrosion-rate may become neglibibly slow. Thus whereas in the " immunity " area it is true to say that corrosion *cannot* occur, this would be an inaccurate description of the " passivation " area, which merely represents a region where, in general, corrosion does not occur to any extent if the film is protective. In a popular exposition, the matter can be brought home, crudely but clearly, by replacing the words " immunity ", " corrosion " and " passivity " by " can't ", " does " and " doesn't ", respectively.

The extent to which the solid corrosion-product obstructs corrosion reactions may depend on its physical character, and the extensive work on the morphology of corrosion-products by Feitknecht's laboratory (p. 910) will be helpful on this point. Another important factor is the exact position of formation of a corrosion-product. If it is formed by a secondary reaction, so that (at least at some places) it is out of physical contact with the metal, corrosion will not be arrested. On the contrary, the existence at such places of the corrosion-product in the form of loose blisters will tend to maintain, next to the metal, local conditions which, in strict conformity with the Pourbaix diagram, are favourable to corrosion; for instance, within a blister, the conditions may be locally acid, even though, in the body of the solution, the pH would be sufficiently high to predict passivation.

Moreover, in deciding whether a small applied anodic current will or will not raise the potential sufficiently for it to enter the passivation region, various factors must be considered. The nature of the anion is all-important, and the conditions of movement in the liquid. If the conditions are stagnant, and the layer of liquid next to the metal becomes super-saturated with some salt, which then spreads out as isolated crystals (not directly protective), the current density may become far higher than simple calculations would predict—as in the case of W. J. Muller's experiments, where ferrous sulphate crystals raised the true anodic current density (p. 885); as to whether this will bring the potential into the passivation region varies greatly with the character of the metal and with the nature of the anion present—as brought out in Piontelli's work (p. 906).

It would seem that, for a complete understanding of corrosion phenomena, thermodynamic data must be combined with kinetic and crystallographic data. It is believed that a combination of recent researches in Belgium, Italy and Switzerland—along with those in other countries—would go far to provide a quantitative prediction of corrosion behaviour.

Meanwhile, it is necessary, whilst admiring the ingenuity and compactness of the Pourbaix diagrams, to bear in mind that other information is needed before any prediction can be made regarding the life of a metallic structure. Some engineers, noting with joy the thermodynamic foundation of the diagrams, have felt that they have at last means of predicting corrosion-behaviour based on principles with which they are already familiar in connection with heat-engines, and on which they have learnt to place confidence; they have even jumped to the conclusion that it is only necessary to measure the pH value of a soil and the potential-drop existing between it and the metal of a buried structure, by means of one of the instruments which enable a semi-skilled person to make both measurements, in order to reach a decision as to whether the soil will corrode the structure or not. Unfortunately, the situation is not so simple. No such claims have, of course, been made by Pourbaix or his colleagues, and it is not their fault that this brilliant work has aroused in some quarters expectations which may not be entirely justified. Pourbaix has himself dealt with some of the misapprehensions under which the authors of certain documents appear to lie.

An excellent discussion of passivity in the light of Pourbaix diagrams —particularly the diverse behaviour of chromates, pertechnetates, perrhenates and other oxygen-rich salts—has been provided by Pourbaix himself (M. Pourbaix, *Z. Elektrochem.* 1958, **62**, 670).

Polarization Curves. Although the abnormal electrochemical behaviour of the transition elements had been noted by earlier investigators, Piontelli was probably the first to emphasize the difference between the polarization curves furnished by two groups of metals. Only certain examples of his findings can be given here, but the original papers deserve careful study. They include R. Piontelli and G. Poli, *Gaz. chim. ital.* 1948, **78**, 717; 1949, **79**, 210, 214, 538, 642, 863; R. Piontelli and G. Bianchi, *ibid.*, 1950, **80**, 581; R. Piontelli, *ibid.*, 1955, **85**, 665; G. Poli, *ibid.*, 1956, **86**, 526; R. Piontelli, *Ricerca sci.* 1948, **18**, 824; R. Piontelli, G. Bianchi, U. Bertocci and B. Rivolta, *ibid.*, 1956, **26**, 838; R. Piontelli, G. Serravalle and R. Ambrosetti, *Metallurg. ital.* 1955, **47**, 200; R. Piontelli, *J. Chim. physique* 1948, **45**, 115; 1949, **46**, 288; 1952, **49**, C53; R. Piontelli, *Z. Elektrochem.* 1951, **55**, 128; R. Piontelli, G. Bianchi and R. Aletti, *ibid.*, 1952, **56**, 86; R. Piontelli and G. Serravalle, *Trans. Inst. Metal Finishing* 1957, **34**, 293; R. Piontelli, *J. Inst. Met.* 1951–52, **80**, 99 (Autumn Lecture to Institute of Metals); R. Piontelli, *Corrosion* 1953, **9**, 115. References up to 1955 are conveniently collected by R. Piontelli in *Ricerca sci.* 1955, **25**, 750.

The normal metals like zinc and lead can be regarded, without undue over-simplification, as being built up of "ready-made" cations and electrons, so that the anodic dissolution process is a relatively simple affair—the passage of cations from the metallic phase into the liquid, where they become hydrated, the electrons being left behind. Little activation energy is needed for such a change and a considerable anodic current density is possible at potential values only slightly above the equilibrium potential, just as a considerable cathodic current density is possible at values only slightly below it. Piontelli's polarization curves for zinc, in solutions of sulphate and chloride are shown in fig. 178. They are not quite symmetrical about the equilibrium point but the form of each curve depends on the anion; those anions which produce most polarization on the anodic side produce important polarization on the cathodic side. The fact that the cathodic and anodic limbs of the curve are similarly affected by the anion suggests that the activation energy is due to something present on the metal on both sides of the equilibrium point, and not to an obstructive anodic product. Possibly the activation energy represents a squeeze energy—the work needed for the metallic ion to thrust its way through the phalanx of adsorbed particles (e.g. water molecules); this work is smaller when the particles are simple than when they are complicated. Piontelli attributes the different behaviour of the various ions to their varying deformability—the power of the distribution of the electric charges within the ion or atom to become modified with the conditions. The deformability is high for chlorine ions, and Piontelli thus explains the catalytic effect of chlorides for anodic and cathodic processes. The matter is further discussed on p. 239.

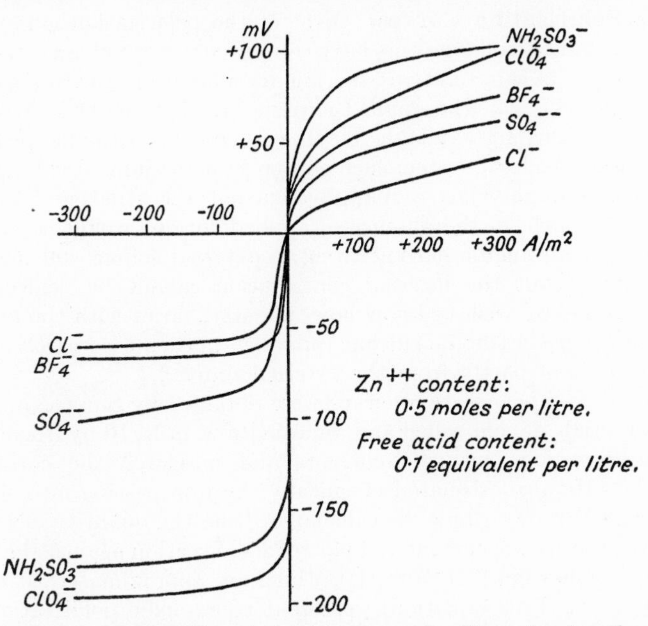

FIG. 178.—Anodic and cathodic polarization curves for Zinc, showing the effect of different anions (R. Piontelli and G. Poli).

Whatever be the cause of the anion effect, it is certain that on a normal metal a very high anodic density will be required to raise the potential to a point where, according to the predictions of the Pourbaix diagram, passivity can set in. Thus the normal metals are not easily rendered passive by anodic treatment.

When we pass to abnormal metals like iron, the polarization is much greater. The smaller values of the exchange currents (the current which is passing equally in each direction at the equilibrium point where the *net* current is zero) suggests an essential difference from the normal metals. It is pointed out on p. 1035 that the separation of the atoms of such metals from one another is less easy than those of the normal metal, and the same is true of the extraction of ions during anodic dissolution. In other words, the activation energy is much higher, and the polarization curves steeper. There is now a real possibility that even with moderate current densities the potential will reach the level where the Pourbaix diagram predicts passivation. This explains why the abnormal type becomes passive so much more easily than the normal type. The table of exchange currents given on p. 1032 suggests that under electrochemical conditions where zinc or copper would pass smoothly into the liquid as ions, the supply of iron or nickel ions would fail, and alternative reactions resulting in the production first of oxide and then of oxygen, will of necessity set in. The presence of chlorides, however, facilitates the electrodic reactions and decreases the chance of reaching the potential where passivity sets in.

True Polarization Curves. Most of the polarization curves studied by Piontelli and his collaborators have concerned a reversible system formed by a metal in its salt solution; the anodic reaction is the opposite of the cathodic reaction, and the applied current has represented practically the whole of the current available for the anodic or cathodic process. If, however, we consider a system such as iron in potassium or sodium chloride solution containing oxygen, and apply a current, the situation is less simple; at the areas which, in the absence of applied current would be anodic, the application of an anodic current from an external source will increase the anodic current but the internal contribution cannot be neglected. For many purposes, we wish to know how potential varies with the *true* anodic current or the *true* cathodic current (provided in each case partly by a local corrosion-cell and partly from the external source).

Such " true " polarization curves were obtained by Simnad in his study of the potentials of cold-rolled and annealed iron in N/10 hydrochloric acid. He applied various cathodic currents and measured the corresponding potentials. He also estimated chemically the iron passing into the liquid; by applying Faraday's Law, he calculated, from the quantity of iron found in solution, the anodic current and plotted its logarithm against the potential (fig. 179). These lines (A_1 and A_2), which are approximately straight (i.e. obedient to the Tafel equation), represent true anodic polarization curves, the highest points being obtained in the absence of current from an external source; the lowest points (at the left-hand end of the diagram) represent nearly complete cathodic protection. Now the corrosion-rate provides a measure not only of the (internally generated) anodic current, but also of an (equal) cathodic current; by adding this to the known value of the cathodic current applied from the external source, Simnad obtained the true cathodic polarization curves (C_1 and C_2). These also are apparently straight, suggesting obedience to the Tafel equation.* The change of direction at the point of protection is well marked in oxygen-free acid (it was not so well marked in oxygen-saturated acid).

* The small number of points on the particular curves shown in fig. 179 hardly justify the assertion that the lines are straight, but other curves obtained under slightly different conditions (in presence of oxygen, or with the use of steel instead of iron) all point to the cathodic curves becoming straight when protection is complete.

Fig. 179, which refers to pure iron in oxygen-free acid, shows clearly that the corrosion-rate of cold-worked metal is greater than that of annealed, and that at all points on A_1 and A_2 the potential corresponding to a given current density is about 22 mv. lower for cold-worked than for annealed iron. The same general features (more rapid corrosion and lower potential for cold-worked than annealed metal) were also noted on steel in oxygen-

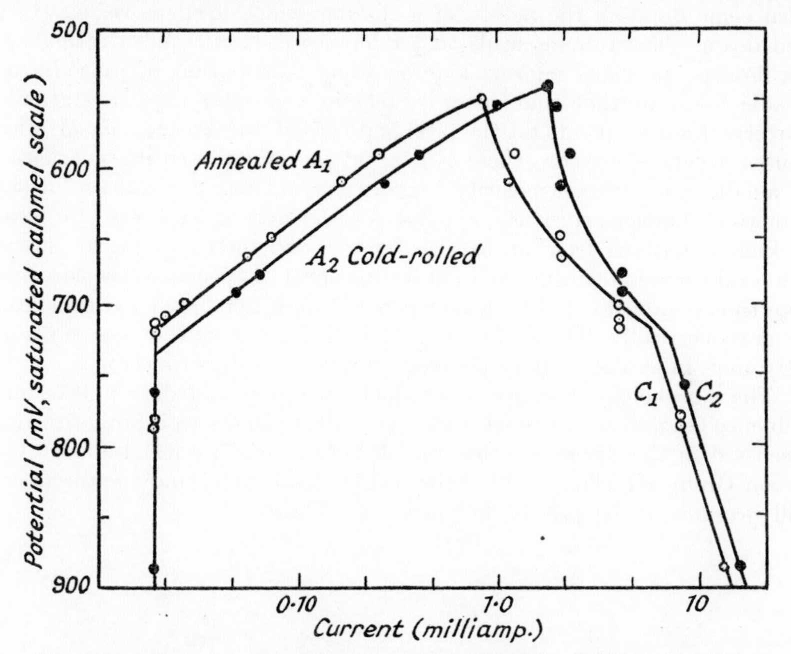

Fig. 179.—Pure Iron in oxygen-free N/10 Hydrochloric Acid. Relation between potential and current (M. T. Simnad and U. R. Evans).

free acid and for both iron and steel in oxygen-saturated acid (M. T. Simnad and U. R. Evans, *Trans. Faraday Soc.* 1950, **46,** 175).

In applying this method to cases where the applied current is anodic, it must be *subtracted,* not added, in obtaining a true cathodic polarization. Similar methods of obtaining " true " polarization curves have been used by J. Elze and H. Fischer, *J. electrochem. Soc.* 1952, **99,** 259, and particularly by T. P. Hoar and R. D. Holliday, *J. appl. Chem.* 1953, **3,** 502.

Other References

There appear to be no text-books devoted essentially to the scientific principles governing the velocity of wet corrosion. The reader will do well, however, to study the writings of Pourbaix (p. 900) and Piontelli (p. 906), already mentioned, and also a treatment of general corrosion theory by R. Piontelli, *Corros. et. Anticorros.* 1957, **5,** 291. Some of the papers included in " Passivierende Filme und Deckschichten " (edited by H. Fischer, K. Hauffe and W. Wiederholt: published by Springer)—notably those of K. J.

Vetter (p. 72), and W. Jaenicke (p. 160)—will repay study; also some of the papers delivered to the Corrosion Symposium (Melbourne University) 1955–56—notably that of G. M. Willis (p. 52).

The ingenious " Corroscope " devised by Francis should receive careful study when full accounts have been published. A cylindrical specimen immersed in corrosive liquid is rotated rapidly, so that different points in turn come opposite the mouth of a tubulus which itself moves slowly up and down. The tubulus leads to a reference electrode and a cathode-ray oscilloscope provides information regarding distribution of cathodic and anodic areas in time and space. Steel in tap-water gives a stationary pattern of anodic points (at the pits) and cathodic areas; but in acid, which causes general corrosion, there is a rapid flicker representing fluctuations of anodic and cathodic polarity. There are obvious possibilities for such a method, particularly if the apparatus could first be calibrated by means of known currents from an external source and then be used to measure the total corrosion-current on a naturally corroding specimen; the corrosion-rate thus obtained could then be compared with that measured chemically or gravimetrically (H. T. Francis, preliminary account in the press for *Corrosion*; fuller descriptions planned, probably for *J. electrochem. Soc.*).

The morphology of corrosion-products, which as stated on p. 905, must influence the decision as to whether passivation can occur, is authoritatively discussed in the Spring Lecture to the Society of Chemical Industry Corrosion Group for 1959 by W. Feitknecht; this lecture, and the discussion, will presumably be printed in *Chem. Ind.* (*Lond.*).

CHAPTER XXII

STATISTICAL CONSIDERATIONS

Synopsis

It is first pointed out that, whereas in the majority of reactions which receive quantitative study from chemists, identical results are obtained on each repetition of an experiment, this is not always the case with corrosion reactions, which often give varying results when performed several times. It is shown that this irreproducibility is to be expected, and accords with a universal principle established centuries ago by the Swiss mathematician, Jacques Bernoulli.*

Different methods of assessing variation are then presented—including the variance, the standard deviation and the coefficient of variation. Certain important types of distribution—including the *Binomial*, the *Normal* and (later in the chapter) the *Poisson*—receive consideration. A warning is sounded against the all-too-common assumption that when the normal law is found to represent with some accuracy the situation near the centre of a distribution, it will serve to predict the situation in the outlying regions. The use of a distribution extending from $-\infty$ to $+\infty$ is unrealistic for corrosion problems, where negative values are in some situations meaningless and very high positive ones impossible.

The question of pit-depths is then considered; here electrochemical principles as commonly accepted seem to preclude the formation of any pit deeper than a certain value within a certain time. The measurement of the deepest pit on a small specimen will sometimes serve as an approximate estimate of the depth of the deepest pit on a large area (a matter of great practical importance in estimating whether a pipe or tank will become perforated within some fixed time), but this is only true if the distribution is of a certain type. In other cases, however, methods based on *ranking* should enable us to make an approximate assessment of the deepest pit on the large area, if we can measure on the small specimen not only the deepest but the 2nd deepest pit, the 3rd deepest and so on.

The concept of *Expectation* is then introduced, and it is explained how this differs from *Probability*. Poisson's distribution is considered, and it is shown how the drop method of studying corrosion enables *Inception Probability* to be measured. It is shown that in

* Jacques Bernoulli (1654–1705) was the eldest of the four members of the Bernoulli family who have left their marks on mathematics; his nephew, Daniel Bernoulli (1700–1782) also contributed to our understanding of Probability—especially towards the end of his long life.

some circumstances (notably in testing the efficacy of inhibitors), experiments on small specimens may be completely misleading as a prediction of the power of the inhibitors to prevent corrosion on large areas; it is indicated how such mistakes might be avoided. As already pointed out in Chapter XIX, there are two quite independent causes for the failure of small-scale experiments to predict large-scale behaviour correctly; in one case the reasons become clear on applying *Dimensional Analysis* and in the other on applying *Statistical Analysis*.

A treatment of the manner in which the film-covered area varies with time when films grow outwards over a surface, starting at randomly disposed nuclei is presented in this chapter (instead of in Chapter XX), since a method based on Expectation enables the result to be obtained far more quickly than by ordinary mathematical procedure.

We then pass to consider cases where comparison between two sets of measurements obtained in presence or absence of a factor (e.g. contact with a second metal) appears to show that this factor is influencing corrosion—favourably or otherwise; the question often arises as to whether the evidence is significant or whether the difference between the two sets of measurements arises merely from the play of chance. It is shown that statistical tests based on the *t-function* are helpful in answering this question. In cases where the factor can be introduced in varying amount (e.g. the content of some impurity or alloying element), the *correlation coefficient* or *r-function* often provides the appropriate answer.

Those who have to apply statistical methods to corrosion problems will do well first to study a recognized text-book of statistics. It is hoped, however, that the present chapter may help the ordinary reader to understand why lack of reproducibility is bound to occur in certain circumstances, and why laboratory experiments may be misleading as a guide to service behaviour unless carefully planned and rightly interpreted. It should also enable the reader to judge critically the conclusions of other people who have adopted statistical arguments. Such a critical judgment is very necessary today, when the atmosphere is not altogether healthy, owing to the ready availability of such aids as work-sheets, tables and distorted-scale graph paper accompanied by ready-made instructions for using them. These are admirable things when used by competent persons merely as labour-saving devices; they may be dangerous when acquired by persons possessing limited statistical sense, who suddenly find themselves able, without any real understanding of the subject, to produce sheets of statistical results, impressive to the more credulous of their friends, but nevertheless seriously misleading. It is essential that those before whom these results are laid should be in a position to judge of their reliability and relevance to the question under consideration.

Distribution of Experimental Results and Application to Corrosion Problems

Reproducibility and Scatter. In ordinary chemical reactions, the result of an experiment, accurately carried out, is reproducible; if the experiment is repeated many times, the same quantitative result should be obtained on each occasion. Reproducibility is achieved because the experimental volume studied contains vast numbers of molecules or ions, and this makes the " scatter " of the results very small—in accordance with Bernoulli's Principle, which is discussed later.

In phenomena where only a limited number of molecules or ions are present in the experimental volume, different volumes may behave differently. Perhaps the simplest example comes not from chemistry but from physics. If a gas is bubbled through a liquid, sub-microscopic particles are entrained in it; these are formed because each rising bubble of air as it breaks through the liquid surface, blows a small " soap-film ", and the extremely tenuous film then bursts and throws off much smaller particles of liquid into the emerging air. The number of ions in the volume of each tiny particle will be low, and although, on the average, each particle contains as many positive ions as negative ones, some individual particles will contain an excess of positive ions and some an excess of negative ones. This is one reason why many of the entrained particles are found to be electrically charged; in some circumstances other factors operate, and have been examined by Harper, but he emphasizes the fact that, in circumstances where the cause of the charges is the limited number of ions in each particle, there is excellent agreement between experimental results and statistical calculations (W. E. Harper, *Phil. Mag. Supplement*, 1957, **6**, Part 24, p. 365).

In corrosion experiments, a scatter of results is sometimes obtained, not because the particles are limited in number, but because the number of points on the area of the specimen where the anodic reaction (or the cathodic reaction) can develop may be quite small, and as a result two specimens, intended to be identical, will often behave differently. Where corrosion is well spread out (with anodic and cathodic reaction proceeding at a large number of points), reproducibility is to be expected, and if the experiment is carefully repeated many times, the result is quantitatively the same on each occasion. But if the attack occurs at a limited number of points, reproducibility is not to be looked for—however accurate the procedure may be. If an experimenter obtains different results on different occasions, this does not prove that he is unskilful; there are, indeed, circumstances where anyone claiming complete reproducibility for his experiments would throw himself open to grave suspicion.

Consider, for instance, experiments on specimens of size 1×1 cm. exposed to a corrosive liquid which is capable of producing (on the average) one corrosion-centre per sq. mm.; then the " expected number " of centres on each specimen will be 100, and—provided that the one active centre does not interfere with the activity of the others by " mutual protection " (p. 113)—we should get approximately (but not exactly) 100 centres on each specimen. By using a liquid containing a controlled quantity of inhibitor, we can reduce the frequency of centres to any desired extent. Suppose that sufficient inhibitor has been added to reduce it to one per sq. dm.; in that case reproducibility is clearly impossible, since the "expected number" is now only 0·01 per specimen, so that about 99% specimens will show no corrosion at all, whilst about 1% of them will have one centre apiece (the chance of two centres per specimen is here too small to deserve consideration).

Even if the amount of inhibitor added had been only sufficient to reduce the number of centres to one per sq. cm. (i.e. one per specimen), it is unlikely that reproducibility would have been achieved. It would then be physically possible for every specimen to develop exactly one centre per specimen, but Poisson's distribution (p. 934) predicts that about 36·8% of the specimens will have no centre, whilst a certain number should have two or even three.

The practical importance of this is seen if we consider a metal standing in contact with a liquid which gives on the average one centre per $dm.^2$. Laboratory experiments on specimens measuring a few square centimetres will almost certainly show no appreciable corrosion, whereas if the same combination of metal and liquid is used in industry for vessels or plant measuring some square metres in area, the chance of corrosion arising *somewhere* is almost a certainty. In such cases, *small-scale laboratory experiments can be completely misleading*, and it is necessary to take statistical considerations into account if we are to avoid disaster.

Bernoulli's Principle. The manner in which reproducibility improves as the number of events increases is well illustrated by considering the tossing of coins or the throwing of dice. Let us define a "success" in the two cases as the obtaining of a "head" or a "six"; if p represents the probability of success in a single throw and q that of failure, then for unbiassed coins, $p = 1/2$, $q = 1/2$ and for true dice $p = 1/6$, $q = 5/6$. Supposing the coin is tossed four times, the chance of four successes is $(1/2)^4$ or 1/16, whereas if the die is thrown four times, it is $(1/6)^4 = 1/1296$. The chance of three successes *followed* by one failure with dice is clearly $\frac{1}{6} \times \frac{1}{6} \times \frac{1}{6} \times \frac{5}{6} = \frac{5}{1296}$, but the total chance of three successes *combined with* one failure is four times that amount, namely $\frac{5}{324}$, since the failure can occur either on the first, second, third or fourth throw. Many readers will know that the chance of n successes in m throws is conveniently obtained by writing out the binomial expansion of $(p + q)^m$:—

$$C_0p^m + C_1p^{m-1}q + C_2p^{m-2}q^2 + C_3p^{m-3}q^3 + \ . \ . \ . + C_mq^m$$

where the coefficient, C_n, represents $\frac{m!}{n!(m - n)!}$. Then the 1st, 2nd, 3rd, etc. term of the series represents the chance of 0, 1, 2, etc. failures; the nth term always represents the chance of $(n - 1)$ failures.

This is the *Binomial Distribution*, and by plotting the probabilities of obtaining different number of successes, we obtain a distribution diagram, with its highest point (or *mode*) at about pm, as would be expected. If, as in the coin-tossing experiment, $p = 1/2 = q$, the curve will be symmetrical; in the general case, it will be asymmetric. Fig. 180(A) represents the distribution of results on repeatedly drawing balls from a bag containing one red and two black balls ($p = 1/3$, $q = 2/3$). When m is small, the diagram consists of large steps but as m is increased the steps become smaller, giving a nearly smooth curve; as m approaches infinity, the curve approaches

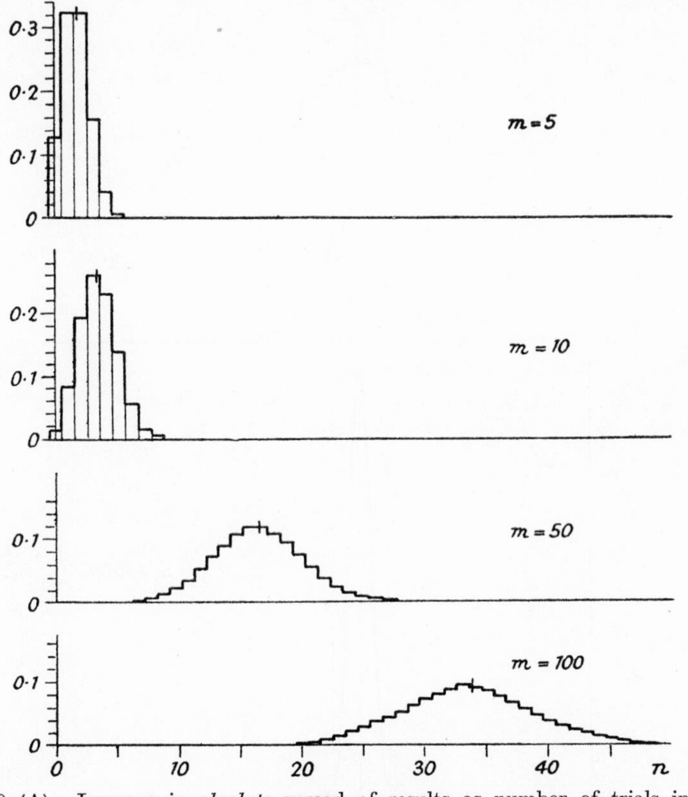

FIG. 180 (A).—Increase in *absolute* spread of results as number of trials increases (T. C. Fry).

complete smoothness. With increase of m, the horizontal spread of the distribution becomes greater, and this would at first sight seem to indicate that reproducibility becomes *worse* as the number of events increases. However, what concerns us in practice is not the *absolute* deviation from the average value, but the deviation expressed as a fraction of the average value. If, instead of using n as the abscissa, we use n/m we obtain a " fractional distribution curve " (fig. 180(B)). A comparison of the two diagrams shows that, whereas the *absolute* spread *increases* as m increases, the *fractional* spread *decreases*. Indeed when $m = 1000$, the results obtained in repeated experiments almost all fall close to 0·33; in other words, we draw a red ball on about one-third of the occasions. Those interested in this subject may care to consult T. C. Fry, " Probability and its Engineering Uses " (Macmillan), Chapter IV.

Certain common fallacies must here be mentioned. It is commonly stated that even if a coin has been tossed 100 times and has fallen head-upwards on each occasion, the chance of obtaining a tail on the next toss remains 1/2. This is perfectly true, provided that it is quite certain that

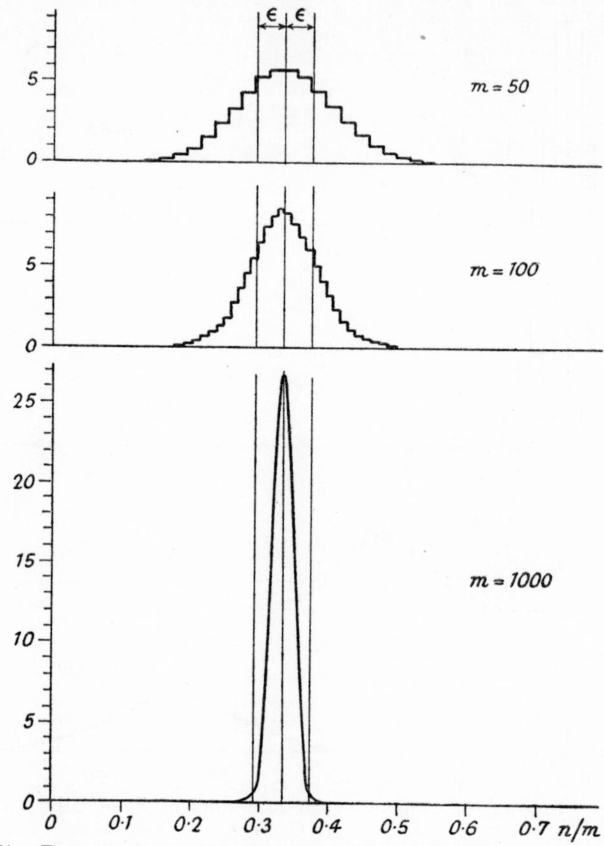

FIG. 180 (B).—Decrease in *fractional* spread of results as number of trials increases (T. C. Fry).

the coin and the tossing are unbiassed. The fact of 100 consecutive heads, however, raises doubts about this *proviso*, suggesting that the coin may have two heads, or is unsymmetrically weighted or is being tossed by a person who has acquired a technique for obtaining heads at will. Unless these and similar possibilities can be excluded, the chance of a tail on the next toss must be assessed at less than 1/2.

There are cases, however, where a monotonous sequence of results may alter the assessment in the opposite direction. If from a bag stated to contain red and black balls in equal numbers, balls are withdrawn, one by one, and not replaced, and if the first drawings have provided a sequence of reds, the chance of a black at the next drawing now exceeds 1/2, since the proportion of blacks in the bag has increased. This assumes confidence in the assurance that the bag originally contained equal numbers of red and black balls. If a very long sequence of reds occurs, the possibility that a mistake has been made, and that the bag contains only reds, deserves consideration.

Application to Corrosion Problems. The significance of the foregoing reasoning to measurements of corrosion is this. Where, as in the attack upon zinc or iron by salt solutions with access of oxygen, corrosion spreads out over considerable areas, and the rate is mainly governed by the replenishment-rate of oxygen to points where the cathodic reaction can take place, then good reproducibility may be expected, provided that the points in question are sufficiently numerous on each specimen. Since the results show good reproducibility where the experimental conditions have been carefully controlled, it would seem that the points in question *are* numerous. Even on the small areas involved in drop experiments (of the order of 1 cm.^{2}), there is good agreement at the outset. The figures for some experiments on drop corrosion carried out in 1924 have already been provided in Table IX (p. 134). In that table each figure represented the *average* of four experiments, but it may be interesting now to show the *individual* measurements (Table XXXIII). It will be noted that, at least in the short experiments, the agreement was remarkably good.*

TABLE XXXIII

INDIVIDUAL MEASUREMENTS OF CORROSION BY DROPS OF SALT SOLUTION ON STEEL IN UNITS OF 10^{-5} GRAMS (U. R. Evans, *J. Soc. chem. Ind.* 1924, **43,** 315T)

	5½ hr.	18½ hr.	48¾ hr.
N/10 KCl . .	65, 65, 66, 66	191, 210, 200, 194	350, 257, 325, 512
N/10 K_2SO_4 . .	71, 73, 77, 77	222, 207, 215, 197	325, 327, 435, 305
N/10 KNO_3 . .	80, 75, 82, 77	208, 264, 232, 235	300, 302, 357, 307

When, however, we pass to experiments on liquid containing an inhibitor, the situation is very different. Here, especially if the experimental area is small (as in drop experiments), some of the specimens may show marked production of rust and others no visible change at all; among those where visible corrosion occurs, there is marked difference between the amount and distribution of attack. This is not surprising because the number of centres from which corrosion is able to develop is small, and Bernoulli's principle would predict poor reproducibility. Even in absence of inhibitor, reproducibility may be poor on such materials as stainless steel

* The greater scatter in the longer experiments was due to the outward creepage of the cathodically formed alkali, thus causing the cathodic area to increase by an amount which differed from one drop to another—probably due to slight differences in the surface condition of different specimens.

or aluminium where the attack is usually localized at a limited number of points.

If the poor reproducibility were due to random errors of measurement, it would be possible to attain accuracy by carrying out experiments in multiple, and taking the average result; in general, the error of the mean is proportional to $1/\sqrt{N}$ where N is the number of measurements carried out, and thus by sufficient repetition it would be possible to get as near to the " accurate " value as anyone could desire.* When, however, the scatter is due, not to errors of measurement, but to the inherently irreproducible character of the phenomenon under study, there is no reason to regard the *mean* (average) values as more " accurate " than any other; properly, regarded as measurements, they are all of about the same accuracy. In such a situation, the mean value loses its special interest for us. Possibly the *mode* (the value obtained more frequently than any other) has greater significance; this is the summit point of the distribution diagram, and will only coincide with the mean if the distribution is symmetrical. In other cases, the *median*, the value which is exceeded in exactly half of the experiments, is useful.† In general, however, the information which we require for practical purposes in cases where reproducibility is poor, is *not a single number* (whether the mean or the mode), but a *distribution diagram*, or estimate of the probabilities, which will show at a glance the chance of getting values within certain limits, or the chance of exceeding some value which is regarded as constituting a danger-point.

Unfortunately, even if we repeat each experiment 20 or 30 times—and in corrosion research few experimenters will contemplate more repetition than this—the prospects of obtaining an accurate distribution diagram are poor. However, by using the data to the best advantage, we can obtain at least some idea of the distribution which would be indicated more accurately by a study of some hundreds of specimens; clearly it is important to use the data in the best possible way, but often today this is not done.

Distribution Diagrams. In many branches of science and technology, the *Histogram* is the generally accepted method of providing information regarding distribution; it is crude and wasteful, but it makes a surprisingly wide appeal today, and has come to be understood by the man in the street. If we are interested in the distribution of heights in a group of men, we can show as a set of vertical lines the number whose heights fall within each of a series of equal intervals of height (each interval being, perhaps, 1 inch broad); the tops of these lines then form a rude distribution curve; instead of lines, rectangles based on the intervals in question seem to be generally preferred, and the assemblage of rectangular blocks constitutes a *histogram* (fig. 181). The distributions of fig. 180 were really histograms.

* Even in such a case the plan of obtaining accuracy by repetition fails if the measuring instrument, instead of giving " random " errors, shows a " bias " towards errors in one particular direction.

† In studying sulphide cracking the *critical strain* (the value at which the probability of failure of a specimen is 1/2) has been used as a criterion by J. P. Fraser, G. G. Eldredge and R. S. Treseder, *Corrosion* 1958, **14,** 517t.

The method is too familiar to need discussion, and when there are a large number of individual measurements to be considered, there is no valid objection to its use. But for indicating the distribution of results in a number of identical corrosion experiments, which at best can represent only 20 or 30 results, it is clearly useless. Suppose, for instance, that we have 24 results, representing the loss of weights of specimens in 24 corrosion experiments intended to be identical. If we divide the range of weight-losses into 8 equal sections, so as to obtain 8 blocks on our histogram, we have on the average only 3 values falling in each block, so that the heights of the blocks vary in big jumps, and the results cannot give any accurate idea of the vertical dimensions. Alternatively, if we choose 6 blocks with 4 specimens (on the average) in each block, or even 4 blocks with 6 specimens each, we obtain greater precision for the block-height, but the number of blocks becomes too small for horizontal accuracy.

The unsatisfactory character of the histogram method arises from the fact that it introduces inaccuracies and wastes data. In constructing a histogram (fig. 181) we treat all points within a certain range as though

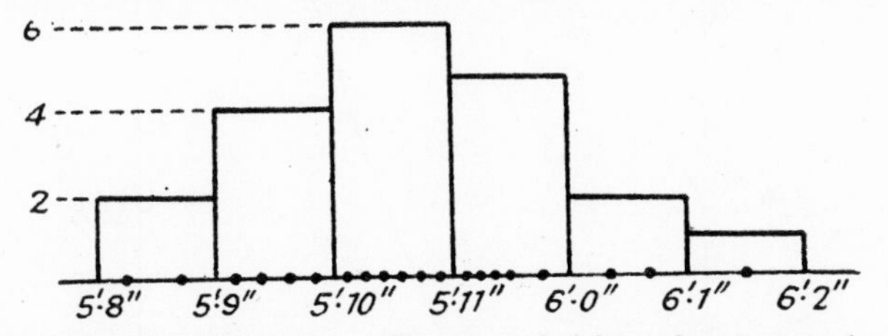

FIG. 181.—Example of a Histogram. The exact heights of each man in the sample is shown by a dot on the horizontal base-line; the number falling within each 1-inch range is represented by the height of the appropriate block.

they had the same value—which is not true. We also neglect the information provided by the exact positions within the range. If the points situated in a given block mostly lie close to the left-hand boundary of the range (as in the range between 5′ 11″ and 6′ 0″ on fig. 181), the histogram obtained is exactly the same as if they had lain close to the right-hand boundary. When we deliberately waste and vitiate data, we cannot hope to convey the greatest possible amount of accurate information in our diagram.

Clearly information regarding a sample of 24 is insufficient to give any precise picture of the population from which it was drawn, or of the probabilities which that hypothetical population represents. Nevertheless we can make better use of it than by plotting histograms. If we plot against the weight-loss, W, the number of specimens $n_{<W}$ with losses less than each value of W, we obtain 24 points, indicating the general course of a *Cumulative Curve* or *Ogive*; such a curve will not pass exactly through all the points, but it will pass between them and fairly close to most. The curve thus obtained gives some idea of the distribution in the population from which our sample of 24 was drawn. It will differ appreciably from that which would be obtained from 240 or 2400 specimens, but at least a hazy picture of the distribution will be presented. To obtain anything approaching

accuracy, more than 24 specimens are certainly needed. Thus fig. 182 constructed from a sample of 94 specimens gives a fair degree of assurance* regarding the shape of the distribution of the "population" from which the sample was taken. By differentiation we may obtain a *Frequency Curve* (similar to fig. 183, p. 922) which reaches its highest point where the cumulative curve is steepest. The meaning of the frequency curve is best explained by saying that the number of specimens falling between x and $(x + dx)$ is represented by $y\,dx$, where y is the ordinate corresponding to the value of x under consideration.

The cumulative or ogive is, however, for many purposes the best way of expressing the data. Fig. 182 records the whole of the available data

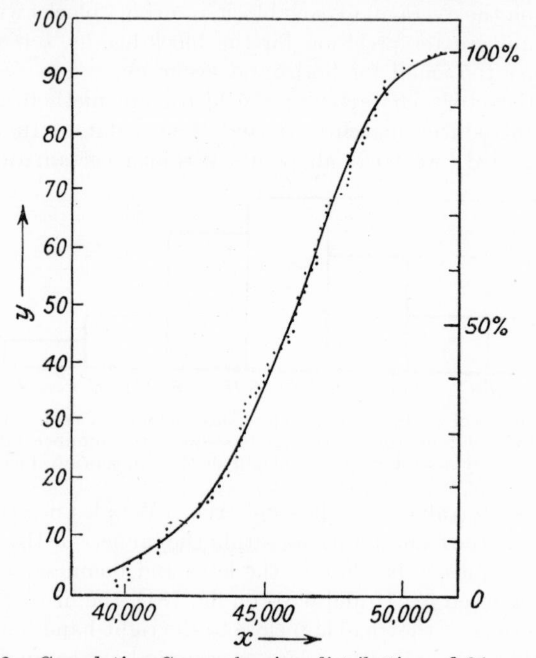

FIG. 182.—Cumulative Curve showing distribution of 94 specimens.

about our 94 specimens quite concisely and shows at a glance, either the absolute value of the number exceeding any chosen value of x (left-hand scale), or, if desired, the proportion exceeding that value (right-hand scale). These are concepts which the man in the street can grasp. The use of the frequency curve requires familiarity with calculus. Moreover, the construction of the frequency curve introduces inaccuracy due to various causes; one is the use of graphical differentiation—a procedure which is often carried out in an unsatisfactory manner, although the accuracy could be greatly improved.

* G. F. Peaker adds the comment, "The degree of assurance can be roughly estimated by dividing the observations, in the order in which they were taken, into odd and even, and comparing the two ogives thus obtained."

Measurements of Variation. Perhaps the most useful method of expressing scatter of results in a group of N experiments, is the *variance*, $\Sigma x^2/N$, where x represents the deviation of each experimental measurement from the mean. If the mean is not already known from the results of other work, it is necessary to estimate it by averaging the results of the N experiments, and in that case a better estimate of the variance is $\Sigma x^2/(N-1)$; however, if N is not too small, the difference between the two expressions is not great, and for the moment we will adopt the simpler form, $\Sigma x^2/\mathrm{N}$. The use of the *squares* of the deviations means that deviations on *either* side of the mean serve equally to increase the variance—as is logical. The variance is an *additive* quantity. If two completely separate and independent causes, e.g. variation of material and inaccuracies in measurement, produce two variances, the variance due to the simultaneous operation of both causes is the sum of the separate variances.

The *standard deviation*, σ, the square root of the variance, $\sqrt{\frac{\Sigma x^2}{N}}$ is also a useful measure of the scatter, but is not additive. The *Coefficient of Variation* is the standard deviation expressed as a percentage of the mean value, $100\sigma/M$. In some ways it is a better indication of the tendency to fluctuate than the standard deviation itself; for the absolute value of σ will " look big " when M is big, and " look small " when M is small.

The " *standardized deviation* ", x/σ or δ, is the deviation of any one measurement expressed as a fraction of the standard deviation, σ. It has no dimensions; if we are measuring a length, it will be the same whether x and σ are expressed in centimetre or inches. The coefficient of variation is also dimension-less.

Normal Distribution. In discussing the binomial distribution on p. 915, it was remarked that the stepped curve gradually becomes smooth as m is increased, and that when $p = 1/2 = q$ the curve is symmetrical. In such a case, the curve representing the binomial distribution becomes a good approximation, in the portions near to the mode, to the well-known normal distribution; it should be mentioned here that the Poisson distribution (discussed on p. 934) may also become almost indistinguishable from the normal distribution near the mode, although it must diverge in the outlying portions, since the Poisson distribution has no negative values, whereas the normal distribution extends to $-\infty$. The relations between these three distributions—and the circumstances when one can be used as a suitable approximation to another—are admirably discussed by T. C. Fry, " Probability and its Engineering Uses " 1928, pp. 213, 238, esp. figs. 24, 25, 26, 27 (Macmillan).

Although the normal distribution has less claim to universality than is sometimes supposed, it is widely employed today and deserves serious study. It is most simply expressed by considering the standardized deviation $\delta = x/\sigma$. The frequency of different values of δ is

$$F = \frac{1}{\sqrt{2\pi}} e^{-\delta^2/2}\, d\delta$$

The normal distribution, expressed as a frequency curve* is shown in fig. 183 and as a cumulative or ogive in fig. 184, where the ordinate represents

$$\int_{-\infty}^{\delta} \frac{1}{\sqrt{2\pi}} e^{-\delta^2/2} \, d\delta$$

Graph paper is obtainable with one scale distorted in proportion to this integral, and if points are plotted on such paper (fig. 185), then, instead of the ogive, we get a straight line—provided that normal distribution is

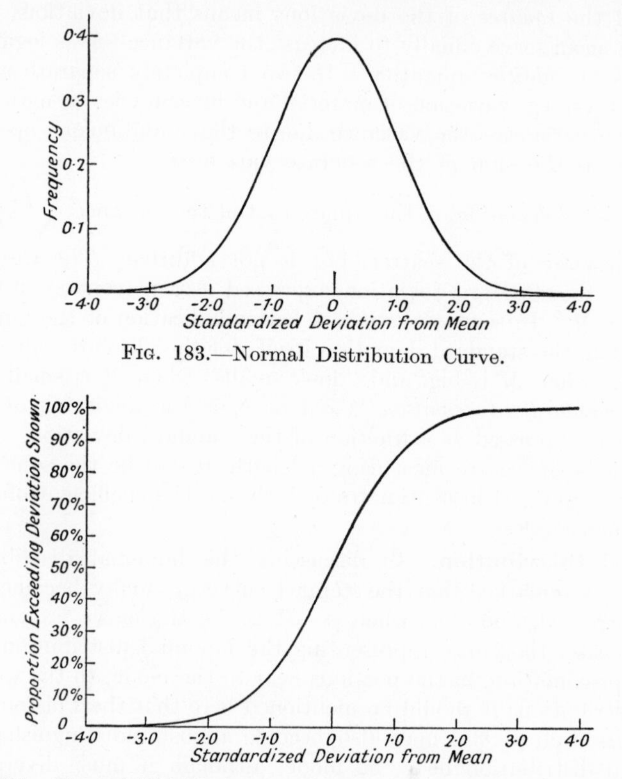

FIG. 183.—Normal Distribution Curve.

FIG. 184.—Normal Distribution expressed as a Cumulative or Ogive.

being obeyed. This is a quick way of ascertaining whether or not a series of values is normally distributed.

The frequency of a quantity which is normally distributed only becomes zero at $-\infty$ and $+\infty$. Since negative values of corrosion velocity are in many cases impossible or even meaningless (unless the corrosive liquid contains the ion of the metal under consideration, in which " negative corrosion " might be taken to mean " deposition "), it is not to be expected

* Note in passing that the total area below the frequency curve is equal to unity (as is always the case) since $\int_{-\infty}^{\infty} e^{-\delta^2/2} \, d\delta$ is $\sqrt{2\pi}$.

that corrosion processes will obey the normal distribution strictly. However, as already remarked, certain other distributions—some of which do *not* extend from $-\infty$ to $+\infty$—become experimentally indistinguishable from the normal distribution in the region of the mode. Supposing that the corrosion data are *really* distributed on one of these other distributions (one which excludes negative values), then a normal equation may still provide a convenient short-hand expression of the facts, suitable to summarize the frequency with which different values will be obtained in the region of the mode—i.e. in the region where the values generally fall. If that is legitimate, it is extremely convenient, since the properties of normal distributions are available in tables for ready reference. It is easy to calculate the mean and the standard deviation, and the proportion of the specimens which will show deviations greater than any particular fractions or multiples of σ can be looked up in the tables; the chance of obtaining

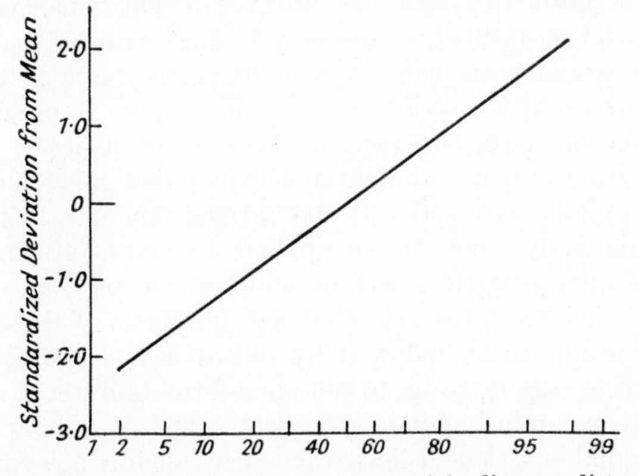

Fig. 185.—Normal Distribution expressed as a straight line on distorted-scale paper.

some abnormally high deviation is a matter of great practical importance, and constitutes one of the chief uses of statistical methods. For deviations not greatly exceeding σ, this method is probably legitimate. It may be sometimes legitimate well beyond σ, but our confidence in this must depend upon the extent to which the more extreme observations support it. In a truly normal distribution, only one in 100 specimens will show a deviation exceeding $2{\cdot}3\sigma$, only one in 1000 a deviation exceeding $3{\cdot}1\sigma$, and so on. If we are certain that the distribution continues to be approximately normal in these regions, it is safe to adopt such estimates, but in general we cannot feel confident on that point.

We may apply the term *quasi-normal* to any distribution which is almost coincident with the normal distribution in the region close to the mode, where most of the experimental values lie, but diverges in the outlying regions, where there are possibly no experimental points at all; these are, however, the regions which interest us, since they represent exceptional, and possibly disastrous, occurrences. Suppose now that one of these quasi-normal distributions represents the *true* distribution, so that the values obtained in a very large number of experiments would fall on it. It is fairly certain that the values obtained in a smaller number of experiments, mostly placed fairly near the mode, will *seem to indicate obedience to the normal law.* If now it is assumed that the distribution is normal in its outlying regions, and if the risk of the exceptional (disastrous) event is calculated on such an assumption, the calculation may be seriously in error.*

For instance, it is known that the fatigue lives of specimens intended to be identical do, in fact, vary among themselves. It is agreed by everyone that the distribution of the lives is not normal, but some authorities incline to the view that the *logarithms* of the lives are normally distributed. In certain cases, such a view is justified for the region near the mode, although in others it has been accepted on insufficient evidence.† Let us consider a case where there is ample evidence that the logarithms of the lives are normally distributed near the mode; are we then justified in assuming a continuation of normality into the outlying region, when calculating the chance that a dangerously short life may be met with on some exceptional specimen? It would seem that such assumptions have in the past been made; the reason why no serious disaster has arisen from such a practice is probably the fortunate circumstance that, in most cases, the departure from normality occurs in a direction which makes for safety.

This book is not concerned with dry fatigue, but it is to be feared that if statistical methods come to be applied to corrosion fatigue or stress corrosion, a similar procedure will be adopted—namely an attempt to fit values representing corrosion behaviour (or functions of these values) on a normal distribution curve, followed by extrapolation (possibly facilitated by distorted-scale paper), so as to obtain information about the chance of an exceptional event in a region regarding which no direct experimental evidence is available. Conclusions reached by such a procedure should be accepted only after careful consideration of the extent to which the most extreme observations available have been used and found concordant with the estimate of the probabilities; clearly caution is particularly necessary in situations where the exceptional event would represent a disaster.

* The point at issue is thus stated on the opening page of Gumbel's new book: " One of the principal notions to be used is the ' unlimited variate '. Here ' common sense ' revolts at once and practical people will say ' statistical variates should conform to physical realities and infinity transcends reality. Therefore this assumption does not make sense '. The author has met this objection when he advocated his theory of floods, at which time this issue was raised by people who applied other unlimited distributions without realizing that their methods rely on exactly the same notion ". Evidently the pot had been calling the kettle black! Gumbel then argues that " this objection is not so serious as it looks, since the denial of the existence of an upper or lower limit is linked to the affirmation that the probability for extreme values differs from unity (or from zero) by an amount which becomes as small as we wish. Distributions currently used have this property " (E. J. Gumbel " Statistics of Extremes " 1958, p. 1 (Columbia University Press)).

† One curve published in a journal of repute carries only four points, but is stated to show that the distribution is essentially normal; 23 specimens were tested, 19 of them to destruction; the results were arranged as a histogram, yielding 4 points, so that the data were not fully utilized.

It should be remembered that many of the circumstances which have brought prestige to the normal distribution in other fields of knowledge hardly apply to corrosion. In many biological subjects a normal curve does seem to represent the observational data with reasonable accuracy. In the early development of the kinetic theory of gases, normal (Gaussian) distribution of molecular movement played a great part; Fry says that gas-kinetics is practically the only problem (in physics) " in which the so-called Normal Law appears as a consequence of an argument which is even approximately sound physically ". None of these cases of normality present any close analogy to corrosion problems. Another legitimate reason for the importance of the normal distribution in branches of knowledge where data are plentiful is that, even where the individual values are distributed in a manner which is far from normal, the distribution of the *means* of groups of such numbers shows approximate conformity to normal distribution and the approximation becomes increasingly close as the numbers in the groups increase. In some branches of knowledge, this is a matter of great practical importance, but it has little or no application in corrosion, where the data are generally scanty.

Against this, an example may be given where normal distribution provides a concise empirical statement of measurements possessing interest to the corrosionist. Scott, studying " an impressive range of soils " has shown that the logarithms of the conductivities are normally distributed, and, in putting this forward, emphasizes the fact that there has been " no resort to theory " and " no employment of advanced statistical methods " (G. N. Scott, *Corrosion* 1958, **14,** 396t).

Distribution within Finite Limits. There is really no need to accept a distribution stretching from $-\infty$ to $+\infty$, since distributions are known which are confined by finite boundaries at one or both ends. Some of these are mathematically related to the normal distribution. As examples may be cited the families of curves developed by Karl Pearson, who noticed that many of the distributions met with in statistical problems (including the normal) comply with the relationship

$$\frac{1}{F}\frac{dF}{d\chi} = \frac{a + \chi}{b + c\chi + d\chi^2} \quad \text{where } F \text{ is the frequency of the variable } \chi$$

and that in some of these the frequency becomes zero at a finite value whilst in others (including the normal) one or both boundaries are placed at infinity. Since, by adjustment of the relationship between the constants a, b, c and d, a vast selection of shapes and sizes of curves becomes available, it should be possible to find an empirical curve which fits the given set of measurements of (say) corrosion velocity. If all that is wanted is to find an equation which embodies a short-hand statement of the facts, the equation to that curve provides exactly what is required. The distributions were developed by K. Pearson, *Phil. Trans. roy. Soc.* 1895, **186,** 343; they are discussed in many statistical text-books, e.g. H. Jeffrey's " Theory of Probability " 1948, pp. 64–68 (Clarendon Press).

If, however, we desire that our equation shall have some sort of scientific basis, it is better to approach the problem first by stating a picturesque derivation of the normal law sometimes used in elementary text-books, and considering whether this could be modified to fit the corrosion situation.

If we imagine a crowd of pedestrians dumped down at a given point on an east–west road and instructed to take N walks each of length l, tossing a coin before the start of each walk to decide whether it is taken to the east or to the west, they will soon spread themselves out in both directions, since only in the unlikely event of a walker taking $N/2$ eastward and $N/2$ westward walks will he find himself back at his starting-point. As N becomes large the distribution approximates to the normal, with the mode at the starting-point, and when N is infinite, the pedestrians will be distributed along the road from $-\infty$ to $+\infty$.

To reconcile this picture with the facts of corrosion, we must exclude negative corrosion-velocities by forbidding walks in (say) the westerly direction, the variation being now introduced by agreeing that the coin tossing shall decide between an eastward walk and a period of rest. The extension to $+\infty$ can be avoided by making l small; N can still be large enough to smooth out the steps, and yet the maximum distance which can be covered even if all N periods have been devoted to walking (and none to resting) remains finite (Nl). Thus we have set a finite boundary and excluded negative readings. There are now N possible positions and the distribution is defined by the statement that the probability of being at r at the end of the period is the $(r + 1)$th term of the expansion of $(p + q)^n$, where p and q are respectively the chances of a rest and a walk.

We can elaborate the procedure by basing the decision between a walk and a rest on the drawing of a single ball from a bag containing reds and blacks in any desired proportion. It may conform to the principles of corrosion to alter the proportion of reds to blacks in the bag with the distance from the origin, since (for instance) the rate of pitting will sometimes vary with the depth of pit already reached. It may also increase realism to have two bags, one for use after a rest period and one after a walk, since the potential needed to start (or re-start) attack is different from that needed to maintain it.

Doubtless further elaboration would be called for, but this sort of argument might lead to a distribution conforming to experimental measurements (since the variable factors which are introduced, not arbitrarily but with a genuine desire to provide for physical realities, are sufficiently numerous to provide all sorts of shapes and sizes) and at the same time possessing a basis founded on the accepted mechanism of corrosion. There would seem to be some promise for investigation on these lines, but they would have to be conducted by someone combining a bent for statistics with a sound knowledge of the electrochemical principles operating in corrosion. Anyone wishing to undertake such a task will find useful information regarding Random Walks in W. Feller's " Introduction to Probability Theory and its Applications " Vol. I, pp. 279–362 (Wiley: Chapman & Hall).

Until some investigation of this sort has been carried out, normal distribution (or some other exponential type) is likely to be assumed in corrosion problems—either owing to respect for tradition or owing to the practical convenience of using the tables and distorted-scale paper available for such distributions. It is necessary to consider how far such procedure will lead us into trouble.

In questioning the use of distributions which stretch from $-\infty$ to $+\infty$ as expressions of corrosion data, some distinction must be drawn between the two outlying regions. The negative region is quite impossible, at least for metal placed in liquid initially free from ions of the metal in question. The high positive region, representing sensationally rapid velocity, is excluded for a different reason. In any corrosion experiment, whether localized or spread out, there is a chance that each individual atom in the metal may come to possess energy far higher than the average; and if it came about that *all* the atoms within a small region possessed such exceptional energies, that region would pass into solution with sensational rapidity; the chance of this happening is, however, represented by a high-order infinitesimal. Theoretically, there is no limit to the corrosion-rate which might be attained in that way, but, in view of the vast number of atoms present even in a microscopically small volume, Bernoulli's Principle will ensure that their energies occur in practically the same proportions in all places, and any *measurable* departure from the proper velocity is not to be expected. We neglect it in practice for exactly the same reason as we neglect the infinitesimal chance that a cup of water, placed in a refrigerator, will start to boil.

The *marked* departure from reproducibility which occurs in cases where corrosion is localized at one or two points and which is usually absent when the anodic and cathodic points are numerous is not due to the limited number of atoms involved, but to the limited number of points involved, and here there is likely to be a maximum velocity which can be approached in an ideal situation, although usually the rates attained will fall short of it by amounts varying from one experiment to another (p. 928).

Distribution of Pit Depths

The use of the Deepest Pit as Criterion. One case in which a limiting velocity is sometimes assumed to exist (often without explicit statement of the assumption) is provided by the measurement of pit depths. It is common practice in dealing with corrosion of the pitting type to state not only the loss of weight of a specimen but also the depth of the deepest pit on it; there is some variation between the value for the deepest pit on duplicate specimens, but no absurd contradiction is generally introduced by the practice, and most authorities believe that it is worth-while to record such values.

In deciding whether the depth of the deepest pit on a small specimen is a sound criterion of the sensitivity of the material to corrosion-damage, it is well to consider two cases. Supposing the distribution of pit depth on a large specimen is of the form shown in fig. 186(A)—roughly asymptotic to the *horizontal* axis—the depth of the deepest pit on a small specimen will vary from one specimen to another; for on a small specimen, the depths must be represented by a stepped curve, such as fig. 186(B), and since the effect of blending a large number of these stepped curves must be to produce the smooth curve of fig. 186(*A*), it follows that the last step on fig. 186(*B*), representing the deepest pit, must fall at different points on the different small specimens. In such a case, a knowledge of the deepest pit on a single small specimen gives us very little guidance to the deepest pit to be expected

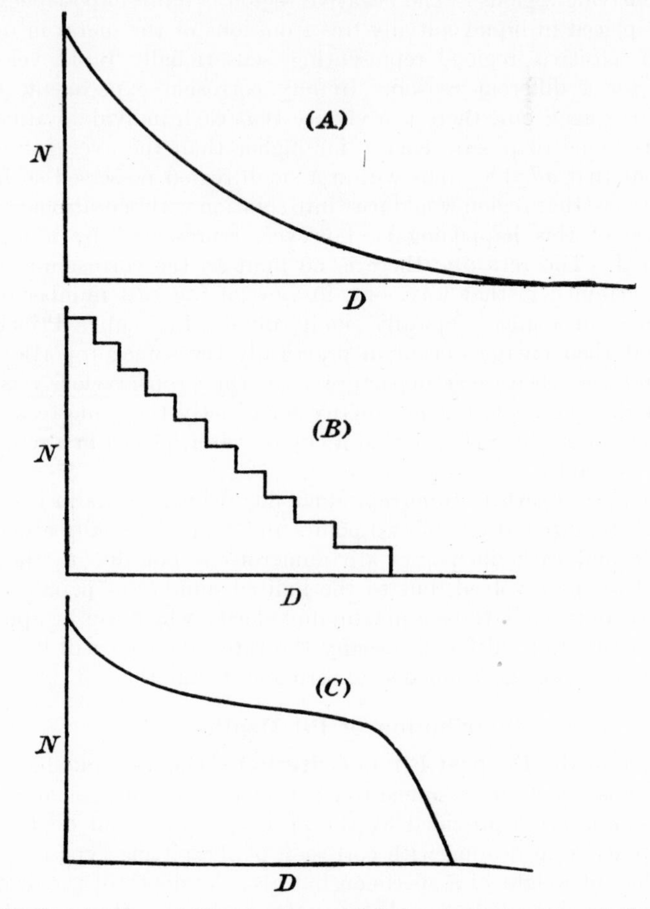

Fig. 186.—Distribution of Pit Depths on infinite specimen (*A* and *C*) and finite specimen (*B*).

on a large area, although if the depth of the second, third, fourth . . . deepest pits are also known, and the area of the small specimen is stated, the information may serve as a basis for prediction—as shown later.

If, however, the distribution curve (fig. 186(*C*)) approaches the *D* axis at a sharp angle, the depth of the deepest pit is likely to be much more useful; in the limiting case, it might become asymptotic to a *vertical* line. If the distribution curve is drawn to *cut* the horizontal axis, this would seem to indicate that there is some depth which (in the time allowed) is never exceeded. In such cases, provided that the "small" specimens are not *too* small, the depths of the deepest pit measured on the various specimens will not be very different from one another, and any of them will serve as a rough indication of the greatest depth to be expected on a much larger area. There is nothing inherently improbable in such a form of curve. If the oxygen-supply were sufficiently rapid to allow the cathodic reaction to proceed at any required pace, the conductivity sufficient to allow oxygen arriving at places distant from any pit to be utilized in the cathodic reaction, and the pits sufficiently far apart to avoid competition for the oxygen arriving, then the rate of pitting would be determined solely by the anodic reaction, and for a given material there will be some rate

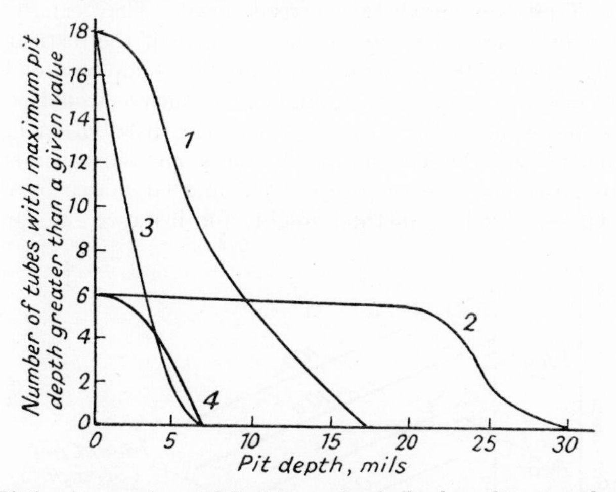

FIG. 187.—Pitting in experimental Scotch marine boiler by salt water (density 2·5 oz. per gallon) containing 4 ml./l. oxygen (curve 1) or 15 ml./l. oxygen (curve 2); curves 3 and 4 refer to 4 and 15 ml./l. oxygen respectively but with pH control (F. Wormwell, G. Butler and J. G. Beynon).

which cannot be exceeded—a rate fixed by the proportion of atoms possessing the necessary activation energy to pass through the energy barrier into the liquid. In these (idealized) conditions, we would expect a curve asymptotic to a vertical line. In making this prediction of a limiting pitting velocity, we are assuming that—owing to the operation of Bernoulli's principle—the distribution of energies under stated electrical and thermal conditions are definitely determined, and we are neglecting the infinitesimal chance already mentioned of finding groups of atoms possessing energies of magnitude greatly exceeding that characteristic of the temperature.

Although figs. 186(*A*) and (*C*) represent idealized forms, the curves representing the distribution of pit depths measured at Teddington on the tubes of a model boiler (reproduced in fig. 187) show that a fairly sharp angle of approach may in practice be obtained, encouraging us to believe that the measurement of the deepest pit on a specimen (the area of which should be stated) will often be of practical value in predicting results—although the information will be more valuable if the depths of the other pits are known. A study of the Teddington data (only a small part of which is included in fig. 187) shows that the type of curve which approaches the horizontal axis at a sharp angle is more common in weakly saline water than in strongly saline water. The Teddington results are published by F. Wormwell, G. Butler, J. G. Beynon and H. C. K. Ison, *Trans. Inst. Marine Engrs.* 1957, **69,** 109, 121 (2 papers); they may be compared with the distribution curves for pit-depths on aluminium in sea-water published by T. J. Summerson, M. J. Pryor, D. S. Keir and R. J. Hogan, Amer. Soc. Test. Mater. Spec. tech. Pub. **196,** p. 157, esp. figs. 1, 2 and 3.

Prediction of Depth of Deepest Pit. It has been pointed out that the depths of pits on metal vary considerably. They can be measured in the laboratory on a relatively small specimen; if the distribution curve resembles fig. 186(*C*), the measurements provide some sort of idea of the depth of the deepest pit on a much larger area, such as would be concerned in a service problem. Even in cases where the distribution more closely resembles fig. 186(*A*), the data obtained from a small area can still be used to give a rough estimate of the deepest pit on a large area by the method suggested below. Such a method might, for instance, be developed to

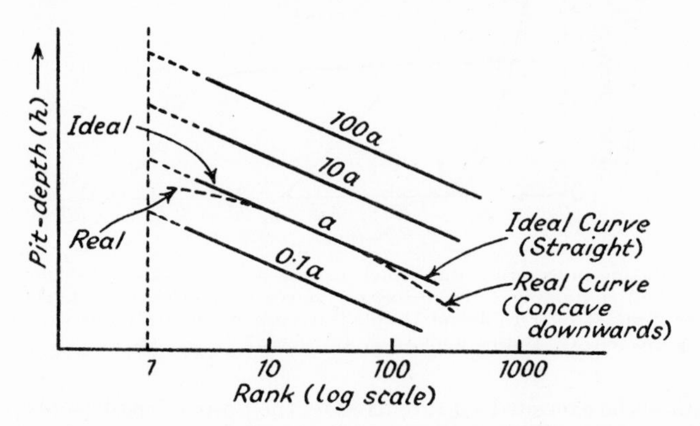

Fig. 188.—Ideal curves representing rectilinear relation between pit-depth and log (rank); real curves are concave downwards (based on ideas by G. G. Eldredge).

predict—from the measurements of the depths of pits carried out on a mile length of pipe-line—the depth of the deepest pit on a ten-mile length of the same pipe in the same soil.

The method suited for dealing with such problems depends on " ranking " the pits, assigning the rank number 1 to the deepest, 2 to the 2nd deepest and n to the nth-deepest. Eldredge, studying pits on oil wells, has found that within certain limits, log J (where J is the rank number) plotted against the pit depth (h) gives a straight line. If, for the moment, we may assume that this is a general relationship, an easy method of calculation suggests itself. Let us compare the curves for areas 0·1a, a and 10a (fig. 188); it is clear that the depth corresponding to $J = 100$ on the curve for a will correspond to $J = 10$ on the curve for $0{\cdot}1a$, where there are only one-tenth the total number of pits, and to $J = 1000$ on the curve for $10a$. In other words, the curves for $0{\cdot}1a$, a, $10a$, $100a$, etc. will be spaced at equal distances apart, and if (within the range of values under consideration) the sloping lines continue to be straight, any vertical line will make equal intercepts between them. If we are justified in drawing this line at $J = 1$, so that each intersection represents the depth of the deepest pit on the area represented by the sloping line it follows that

$$h_A = h_1 + K \log A$$

where h_1 is the average depth of the deepest pit on unit area, h_A that on area A and K a constant. It is true that the equation corresponds to a limitless distribution and implies the possibility that values of h of any desired value could be obtained by making A sufficiently large, which on kinetic grounds is not to be expected. The error, however, is on the right side, and the actual value of h_A is likely to be less than that indicated. The intersection at $J = 1$ is probably the safest estimate of the deepest pit available to us.

Now considerable evidence in support of this relationship has in fact been afforded by the earlier measurements of Ewing on pipe-lines, carried out for the U.S. Bureau of Standards many years before the publication of Eldredge's paper. Ewing's procedure differed from that of Eldredge in the following respect. Instead of measuring the depth of the deepest, 2nd deepest, 3rd deepest . . . nth deepest pit on a fixed area, Ewing measured the average of the deepest pit on a large number of specimens of the same area, and then repeated the same for specimens of other areas; he found that h_A is in fact a rectilinear function of $\log A$, as would be expected if Eldredge's relationship was valid for pipe-lines. Scott had also amassed a large amount of data and had expressed his results by a different equation, but Eldredge shows that some of Scott's results fit Ewing's equation better than they fit Scott's own equation.

In applying the study of pits on a small area to predict the deepest pit on a large one, we are, of course, extrapolating—with all the well-known risks involved in extrapolation. However, as stated above, the error should be on the right side. Eldredge's curves, although reasonably straight in the important range, do tend to be concave downwards, and if this tendency is present on the lines corresponding to large areas, the actual depth corresponding to $J = 1$ will be less than that calculated. The tendency to be concave downwards is a sign that there is some limiting pit depth which (in a given time) cannot be exceeded for electrochemical reasons. If the curves relating h and $\log J$ were straight for the largest areas, it would mean that there is no value of h, however great, which could not be met with somewhere, provided that a sufficiently large area was studied. This is unrealistic, and the tendency to downward concavity may be accepted as something which would be expected and which we have every right to take into account.

Eldredge regards his method as an approximation to the extreme-values method of Gumbel, which has been much employed in recent years in statistical problems concerning floods, droughts, snowfalls, wind-gusts (in aeronautics), vital statistics and strength of materials. Since, however, as suggested below, a physical interpretation can be suggested for Eldredge's logarithmic relationship, it is perhaps best to present it in its own right. This need not deter readers from studying Gumbel's important reports in the original, and applying his methods to corrosion problems in suitable cases. Gumbel's new book, quoted on p. 924 (footnote), may here be helpful.

The reader should study the paper of G. G. Eldredge, *Corrosion* 1957, **13,** 51t; cf. P. M. Aziz, *ibid.*, 1956, **12,** 495. The statistical data of Scott and Ewing regarding pit depths on pipes are quoted in Eldredge's paper. See also G. N. Scott, *Proc. Amer. Petroleum Inst.* 1934, **15,** IV, 18 (A.P.I. Pipe Coating Tests, progress report No. IV); S. Ewing, " Soil Corrosion " (Amer. Gas. Assoc.). The contributions to the Washington Conferences on Underground Corrosion (1933 and 1937) by G. N. Scott, S. P. Ewing, L. M. Martin and R. F. Passano deserve study. Classical papers on " extreme values " methods include R. A. Fisher, *Phil. Trans. roy. Soc.* (*A*) 1921–22, **222,** 309; L. H. C. Tippett, *Biometrika* 1925, **17,** 364; R. A. Fisher and L. H. C. Tippett, *Proc. Cambridge Phil. Soc.* 1927–28, **24,** 180. Gumbel's main report will be found in the U.S. Bureau of Standards Applied Mathematics Series No. 33, 1954.* For application to fatigue problems see A. M. Freudenthal and E. J. Gumbel, *Proc. roy. Soc.* (*A*) 1952, **216,** 309; E. J. Gumbel, *Trans. New York Acad. Sci.* 1956, **18,** 334.

Despite the substantial evidence in favour of the logarithmic relationship presented above, the situation must be regarded as unsatisfactory so long as the relationship is purely empirical, since—in absence of scientific understanding—it is impossible to predict whether or not an empirical relationship can safely be applied in circumstances different from those where it has been established. A possible interpretation of the relationship—admittedly speculative—will therefore be offered for the reader's consideration.

If N is the number (per unit area) of pits which have penetrated to a depth h, and $(N - dN)$ is the number which still exist at depth $(h + dh)$, then if the chance of an individual pit being arrested within the distance dh is independent of N and h, we can write this chance as $p dh$, where p is a constant, giving

$$-\frac{dN}{N} = p\,dh \quad \text{or} \quad h = -\frac{1}{p}\log_e N$$

* The reader should form his own opinion, but two warnings may be permissible. (1) Although observations in the early part of Gumbel's main report are largely general, many sections in the later part refer to situations where the distribution is one of the exponential types and extends to infinity. (2) The experimental evidence about fatigue lives today is not entirely favourable to their log-normal distribution; where, however, that distribution does represent the facts, it is best regarded as a convenient empirical relationship, since suggestions regarding a physical basis have been unsatisfactory.

This immediately explains why a straight line is obtained on plotting the logarithm of the number of pits reaching different depths against those depths.

The assumption that the probability of progress of a pit being arrested within the distance dh is independent of N and h may at first sight appear contrary to the electrochemical mechanism. Certainly the *rate* of pit-deepening will depend on the position of the pit relative to its neighbours, since under cathodic control the various pits will compete for oxygen, and a pit surrounded closely by others will progress more slowly than one which stands apart from the crowd; the mutual protection effect will produce a similar result. It might also be expected that as N becomes smaller (with time), p would decline. However, after the long periods involved in the observations quoted, the variation in pit depths is probably connected with the fact that for some of them progress has been arrested by reaching regions where the escape of metallic cations requires greater activation energy than the production of a film. As explained in Chapter VII, there is always competition between these two alternative types of anodic change; where the metallic phase is disarrayed the cations can escape more easily than where the structure is perfect, and, if a region of perfect structure should be reached, film-formation may take preference. Thus in considering the situation after a long period, it may be justifiable to assume that p is independent of N and h and is decided by the local character of the metallic phase.

This suggests a possible interpretation for the relationships given above; time will show whether it is the right one.

Expectation

The question of Additivity in Probability. A little consideration will show that the simple addition of probability values is not, in general, permissible. If we draw one ball from a bag containing one red and two black balls, the chance of obtaining a red is $1/3$; if we carry out the drawing twice (the ball being replaced before the second draw), the chance of a red is clearly not $2/3$; if it were admitted that the procedure of adding the probabilities was permissible, three draws would give $3/3$ (i.e. a certainty of obtaining a red) and four draws would give $4/3$ (which is meaningless).

The rational procedure for obtaining the probability of obtaining at least one red in a number of draws is this. On the first draw, there is a $1/3$ chance of a red and a $2/3$ chance of no reds. This $2/3$ chance can then be sub-divided into a $1/3 \times 2/3$ chance of a red on the second draw, and a $2/3 \times 2/3$ chance of no reds; we are now entitled to add the $1/3 \times 2/3$ (or $2/9$) chance that, having failed to get a red on the first occasion, we succeed on the second, to the $1/3$ chance of getting a red on the first draw, making the total chance of at least one red $1/3 + 2/9 = 5/9$. The chance of obtaining no reds in two draws is clearly $4/9$, and if a third draw takes place this can be subdivided into $1/3 \times 5/9$, the chance of obtaining a red on the third draw after two failures, and $2/3 \times 4/9$, the chance of a third failure; the total chance of obtaining at least one red in three draws is

$5/9 + 4/27 = 19/27$. Clearly the chance of obtaining no reds in n draws is $(2/3)^n$, and the chance of obtaining at least one red is given by the expression $1 - (2/3)^n$ and is thus represented by the series 1/3, 5/9, 19/27, 65/81, 211/243 . . ., approaching unity asymptotically as n increases; it can never exceed unity.

The Additivity of the Expected Number. It is convenient to find some concept which is susceptible to simple addition, and the *expected number* (commonly called the *expectation*) fulfils this requirement. If we perform a large number of " experiments ", each consisting of a single draw from the bag, the number of reds per experiment will be about 1/3 (it will approximate to 1/3 more and more closely as the number of experiments is made larger). If each experiment consists of two draws, the number of reds per experiment will approximate to 2/3; if it consists of four draws, the number will be 4/3. We call this the expected number or expectation, and it will be noted that the expectation is strictly additive, can exceed unity and is not, in general, an integer. The concept of expectation is useful in a number of highly diverse problems in corrosion and film-growth, of which a few examples will be given.

The Poisson Distribution. Imagine a large sheet of metal containing particles of some undesirable inclusion distributed at random, and imagine that we have ruled out the sheet into centimetre squares. Now suppose for simplicity that *on the average* there is one inclusion per square centimetre. The " expected number " of inclusions in any one square (the average of a very large number of counts) is 1·00. We shall, however, not always find one in each square; sometimes there will be a square without any inclusion, but that will be compensated for by other squares which contain two inclusions or perhaps three—rarely four, and very rarely five or more. We require to know the chance of obtaining 0, 1, 2, 3, etc. inclusions per square, when the expectation is 1·00—or indeed any other value.

The Poisson Distribution, which represents the required information, is not empirical like some of the other distributions mentioned above, but can be derived by theoretical reasoning. Indeed more than one proof is available; the most usual approach is by regarding it as a special case of the Binomial Distribution, where q is nearly unity, p very small but m very large, so that pm (the expectation of " successes ") is still finite. Another proof is provided by T. C. Fry, " Probability and its Engineering Uses " 1928, pp. 220–227 (Macmillan).

Both proofs lead to the conclusion that the probability of obtaining exactly n " successes " in trials where the expected number of success is E, is expressed by the formula

$$p_n = \frac{e^{-E}E^n}{n!}$$

This is only true if all the events are really independent, so that the occurrence of one success does not influence the attainment of another; in the case of inclusions there must be no tendency for them to occur in clusters; in the case of corrosion-pits there must be no " mutual protection ".

Unfortunately, in the case of the distribution of pits, mutual protection generally does arise and complicates the situation. It was pointed out on p. 114 that on a metal plate partly immersed in salt solution the intense corrosion at cut edges sometimes produces an adjacent zone completely free from corrosion. In just the same way, on a plate undergoing pitting, the existence of a vigorous pit at one point makes conditions unfavourable to the development of pits at points around. There are several reasons for this. On aluminium, corrosion is believed to be controlled by the cathodic reaction on the area outside the pits, where there are only limited number of points suited for the cathodic reduction of oxygen. If so, two pits situated close together will compete for the corrosion-current available, and therefore have a smaller chance of developing the autocatalytic mechanism needed for continued corrosion than a single isolated pit. However, there is a more general reason. The high anodic density at a vigorous pit shifts the potential in such a direction as will reduce the available E.M.F., and the whole potential distribution in the liquid will be altered in such a way as to make conditions less favourable to attack at points around. If corrosion at our first pit has fairly started, the initiation of fresh attack in the area around may never occur—even at points where otherwise attack would have set in; it needs a more positive potential to initiate attack than to maintain it.* It is not surprising that the observed distribution of pits on aluminium is different from that predicted by the direct application of the Poisson formula—as has been shown by the careful work of R. B. Mears and R. H. Brown, *Industr. engng. Chem.* 1937, **29,** 1087.

Inception Probability. For some purposes we desire to know the inception probability, p_I, defined by the statement that the probability in an infinitesimal area-element dA is $p_I dA$; in an infinitesimal area-element the chance of the occurrence of two pits becomes a second-order infinitesimal and can be neglected. If there were no mutual protection, the expected number of pits on a finite area A would be $p_I A$, and the chance of obtaining other numbers, whether larger or smaller than $p_I A$, could be predicted—assuming the material to be uniform and exposed to a uniform environment —by applying the Poisson expression. Generally mutual protection makes prediction inaccurate, but $p_I A$ denotes the number of points of attack which *would* develop under idealized conditions where mutual protection was excluded e.g. if the layer of corroding liquid were made very thin.

Such conditions have been studied by Chilton, adapting a method due to Whitton. Chilton covered a polished wrought iron surface with a thin layer of 0·001N sulphuric acid diluted with five times its volume of alcohol; the alcohol soon evaporated leaving an acid film, thin enough to produce

* This is best shown by Brennert's work on stainless steel and tin. On gradually raising the E.M.F. of a cell fitted with a stainless steel anode, the anode potential gradually becomes more positive, until suddenly the film collapses, the current greatly increases and the potential drops to a low value. Similarly, with a tin anode, an excess potential must be applied before the film breaks down and corrosion commences (S. Brennert, *Jernkontorets Annaler* 1935, p. 281; *Korrosion u. Metallsch.* 1936, **12,** 46; Int. Tin Res. Dev. Council Tech. Pub. D2 (1935)).

interference tints, and almost certainly thin enough to exclude mutual protection. The results showed that on wrought iron the inception-probability of the Q-zones was much greater than that of the R-zones,* whilst that of a mild steel used for comparison was intermediate. No numbers were recorded, but it is likely that in some circumstances a method of that sort could be used for obtaining inception probability values by a direct count (J. P. Chilton and U. R. Evans, *J. Iron St. Inst.* 1955, **181**, 113, esp. p. 119; W. I. Whitton, *Trans. Faraday Soc.* 1950, **46**, 927).

The Drop Method. Corrosion by thin films of moisture is a special case, and for ordinary conditions of immersion in a liquid the measurement of inception probability must be carried out, not by direct counting of the points where corrosion is starting, but in an indirect manner. Although mutual protection may invalidate the Poisson formula in the cases where $n > 0$, it cannot affect it in the case where $n = 0$. Since the factorial of zero is unity, the expression $e^{-E}E^n/n!$ then becomes e^{-E}, that is $e^{-p_I A}$, so that by measuring the proportion of small, unconnected areas which develop no corrosion at all, an accurate estimate of p_I can be arrived at, and used to predict the risk of corrosion starting on areas of larger size.

It is convenient to obtain the small areas by producing square drops on a surface, which, being unconnected with one another by any *liquid* path, can be treated as separate experiments.† Mears has described a method of obtaining square drops. Two sets of equi-distant wax lines at right angles are traced on an iron surface, using a 5% solution of paraffin wax in carbon tetrachloride; after the solvent has evaporated, it is possible to flood the surface with any desired liquid and then tilt the specimen momentarily so that the liquid runs away from the waxed lines but remains as square drops covering the unwaxed areas. The flooding and subsequent observation of the drops can be carried out in a closed glass vessel, so that the effect of different gas mixtures around the drops on corrosion probability can be studied. Later Chilton, in the research quoted above, improved the method by using a rubber stamp to print the mesh of wax lines (R. B. Mears and U. R. Evans, *Trans. Faraday Soc.* 1935, **31**, 527).

The method was used by Mears to measure the probability of corrosion by drops of standard size—i.e. the proportion of drops which cause rusting on iron (or visible corrosion on another metal) and to observe how it varies:—

(1) with constituents in the *gas-phase*, when the metal and liquid are kept constant;

(2) with the nature or concentration of the *liquid*, when the metal and gas-phase are kept constant;

(3) with impurities in the *metal*, when the gas-phase and liquid are kept constant.

* For the meaning of the terms Q- and R-zones, see p. 512.

† The fact that there is a *metallic* connection between the drop areas does not invalidate the method, since provided that there is no liquid path between the different drops, the anodic and cathodic currents within each drop must cancel one another and there should be no drainage of electrons from one drop area to the next.

Mears' work with oxygen–nitrogen mixtures has been mentioned on p. 138. As the oxygen-content of the gas-mixtures increased, the inception probability diminished but the conditional velocity (attained in those drops where corrosion did start) increased. He also studied the effect of sulphur dioxide and carbon dioxide; sulphur dioxide increased both inception probability and conditional velocity of steel in pure water; carbon dioxide increased velocity without appreciably affecting probability. The numbers are shown in Table XXXIV (A).

TABLE XXXIV (A)

INCEPTION PROBABILITY AND CONDITIONAL VELOCITY BY DROPS (3 × 3 MM.) OF PURE WATER ON STEEL AT 25°C (R. B. Mears and U. R. Evans)

Atmosphere	Inception Probability (percentage of drops producing Attack)	Conditional Velocity (mg. per drop corroding in 22 hrs.)
80% N_2, 20% O_2	5	0·06
79% N_2, 20% O_2, 1% CO_2	6	0·12
79% N_2, 20% O_2, 1% SO_2	100	1·56

Other experiments were carried out on scratch-lines. Here the specimens were divided by waxed lines into areas measuring 5 × 16 mm., and at the centre of each area a scratch 12 mm. long was engraved with a gramophone needle under a load of 400 grams. The unwaxed areas were then covered with a solution of a salt which is inhibitive when concentrated and corrosive when dilute, the concentration chosen being at the border-line between the corrosive and passivating ranges. The time of exposure to air between the moment of the engraving of the scratch and the placing of the liquid on the surface greatly affected the probability; in one set of experiments carried out in pure oxygen with 0·07M sodium bicarbonate as the liquid, 84% of the scratch-lines developed rust when the time was 0·25 minutes; but the proportion gradually diminished, being only 27% when the time of air-exposure was 1024 minutes—a sign that the invisible film was becoming increasingly protective.

Other experiments gave information about the mutual protective effect. Each area here carried two scratches, one made 2 hours, and the other about ½ minute, before the liquid (0·14M sodium bicarbonate) was applied; the " new " scratch almost invariably produced corrosion, and depressed the probability of corrosion at the " old " scratch, especially when they were close together—as shown in Table XXXIV (B).

The surface condition of the metal was found by Mears greatly to affect probability, and also minor constituents in the metal; sulphur increased

TABLE XXXIV (B)

MUTUAL PROTECTION BY NEIGHBOURING SCRATCH-LINES (R. B. Mears and U. R. Evans)

Distance between the Scratches	Probability of Old Scratch Corroding	Probability of New Scratch Corroding
(New Scratch omitted)	69%	—
4 mm.	44	94%
2 mm.	33	96
1 mm.	19	96

probability, whilst copper, under some conditions, diminished it, probably by precipitating any hydrogen sulphide which would otherwise pass into solution (cf. p. 878). This work on the effect of minor constituents in the metallic phase will be found in a separate paper (R. B. Mears, *Iron St. Inst. Carnegie Schol. Mem.* 1935, **24**, 81).

Mears' principal use of the square-drop method, with drops of fixed size, would appear legitimate—provided that the alteration in the liquid does not perceptibly affect the shape of the drop. The method has, however, been used in more than one laboratory for comparing the effect of drops of different size (keeping the metal, liquid and gas phase constant); anyone tempted to use it for such purposes should bear in mind that small drops differ in shape from large drops, so that other factors besides drop area are altered when a small drop is substituted for a large one—notably the rate of oxygen-replenishment; every drop is thicker in the centre than at the margin, and in small drops the thin marginal portions form a larger fraction of the whole than in large ones.

If it were desired to study the effect of variation of area, other methods, more appropriate than the square-drop method, could probably be developed. It would be possible to determine p_I by using fully immersed specimens completely masked except for a small area A left exposed. By counting the proportion which develop corrosion (P_A), the value of p_I could be calculated from the equation $P_A = 1 - e^{-p_I A}$ and then used to calculate the probability of P_A for other values of A. Such a method, however, has not yet been put into practice.*

It may be asked under what conditions inception probability—as opposed to conditional velocity—is important. In those practical cases where corrosion is certain to start somewhere, corrosion velocity is the quantity which interests us most. In those cases where there is some prospect of avoiding corrosion altogether through the use of inhibitors, inception probability claims attention; Mears' papers include experiments on solutions containing inhibitors as well as corrosive salts. It should not be forgotten, however, that the addition of inhibitor in a quantity insufficient to prevent corrosion may often intensify the attack (p. 145).

For metal exposed to the atmosphere which becomes wetted during intermittent rainy periods, being then covered with a thin layer of water possessing no great conductivity, both probability and velocity may affect useful life. If the metal has been painted, but a few very small gaps have been left in the paint coating, the chance of corrosion starting at these gaps may be more important than the velocity with which, in the event of initiation, it develops.

* The size effect has lately been studied by Hines in relation to the Hancock–Mayne method for ascertaining the concentration of inhibitor needed to prevent corrosion; this depends on finding the concentration at which the potential of a steel specimen subjected to a small anodic current density rises with time to a noble value instead of falling. Hines finds that "for areas up to 6 × 6 inches more inhibitor is required the greater the exposed area. The effect is small, however, and would be covered by using slightly more inhibitor than the text indicated" (J. G. Hines, *Chem. and Ind.* (*Lond.*) 1959, p. 354).

Lateral Growth of Films. The concept of Expectation has proved useful in providing a rapid solution of certain mathematical problems which would otherwise be difficult. For instance, the laws governing the spread of films starting at points on a surface and extending until they meet were mentioned in Chapter XX, but the proof has been deferred to the present chapter, since a statistical method based on Expectation reaches the result more quickly than ordinary mathematical methods. The analogous three-dimensional problem (the growth of grains in a mass of metal) can also be solved rapidly by the statistical procedure; in that case the same final expression has been reached by Johnson and Mehl using an ordinary mathematical method, but the proof was so lengthy that only a summary could be included in their paper (W. A. Johnson and R. F. Mehl, *Trans. Amer. Inst. min. met. Engrs.* 1939, **135,** 416, 451; the full proof is found in an Appendix obtainable from the American Documentation Institute, 2101, Constitution Avenue, Washington, D.C. The statistical method is given by U. R. Evans, *Trans. Faraday Soc.* 1945, **41,** 365).

We will start with the two-dimensional problem; the object is to ascertain the manner in which the uncovered fraction of the area α diminishes with time t, when the film extends from the starting-points as circles* expanding with radial velocity v. The relation between α and t depends on whether fresh nuclei keep appearing on the still uncovered parts of the surface, at a constant *Nucleation Rate*, Ω (best defined by stating that the probability of a nucleus appearing in time element dt on area element δA is $\Omega\ \delta A\ dt$) or whether all the nuclei appear at the moment when the exposure starts without any subsequent appearance of further nuclei (here we must speak of a *Nucleation Density*, ω, best defined by stating that the probability of a nucleus being included in area element δA is $\omega\ \delta A$).

The first case is less simple than the second. On every occasion when a nucleus appears, an expanding circle starts out (as from a point where a rain-drop has fallen on the surface of a lake). Now consider a representative point Q. The expected number E of circles which would pass over Q in time t (the average of the number obtained after an infinite number of counts) can be calculated by computing the elementary contribution dE of an annulus of breadth dr situated at radial distance r from Q, and then integrating for all values of r from zero to vt; circles starting from nuclei developing at distances exceeding vt cannot reach Q within time t. Integration is permissible, because expectation is additive—as explained above.

The annulus in question has an area $2\pi r\ dr$. During a time-period equal to $t - r/v$, any point within the annulus can send out circles capable of reaching Q before time t, so that

$$d\mathrm{E} = \Omega\left(t - \frac{r}{v}\right)2\pi r . dr$$

or

$$E = 2\pi\Omega \int_0^{vt} (tr - r^2/v)\ dr = \pi\Omega v^2 t^3/3$$

* If the growth-rate varies with direction, so that the areas are not circular, the case is more complicated and cannot here be discussed.

The chance that Q can *escape* being crossed by any of the expanding circles which started after $t = 0$ will clearly be

$$p = e^{-E} = e^{-\pi\Omega v^2 t^3/3}$$

Now in actual fact, when once a part of the surface has become covered with a film, no further nucleation can occur within that area, and only one expanding circle can pass over Q. In the case of rain-drops falling on a lake, new (small) circles can start inside the old (large) ones, so that the two cases are not quite the same. However, in calculating the chance that Q will *escape* being reached by any expanding circle within time t, the circumstance that the new (small) circles never start at all on the film-covered part of the surface cannot cause an error; a small (late) circle could not in any case reach Q until after the large (early) one within which it was formed had passed over Q. Thus the fraction of the surface remaining film-free at time t will be

$$\alpha = e^{-E} = e^{-\pi\Omega v^2 t^3/3}$$

which at very small values of t, approximates to

$$\alpha = 1 - \pi\Omega v^2 t^3/3$$

showing that the curves are S-shaped, becoming steeper with time at small values of t, and less steep again as α asymptotically approaches zero.

The second case (nuclei appearing at $t = 0$, but not subsequently) is simple. The contribution dE of an annulus of radius r and breadth dr is clearly $2\pi r . dr . \omega$ so that

$$E = 2\pi\omega \int_0^{vt} r\, dr = \pi\omega v^2 t^2$$

or

$$\alpha = e^{-E} = e^{-\pi\omega v^2 t^2}$$

Once again the curves are S-shaped.

The formulae can be applied not only to oxide-films growing on metal, but also to the crystallization of thin layers of molten metal. In such cases it may be of interest to consider the mean size of the " grains " formed where the growth from different nuclei meet. Where the nuclei appear only at $t = 0$, and not subsequently, this is clearly $1/\omega$. Where they continue to appear, the total number of grains formed during the solidification process is

$$\int_0^\infty \alpha\Omega\, dt = \Omega \int_0^\infty e^{-\pi\Omega v^2 t^3/3}\, dt$$

Writing x for t^3 and A for $\pi v^2\Omega/3$, this becomes

$$\frac{\Omega}{3}\int_0^\infty x^{-2/3} e^{-Ax}\, dx = \frac{\Omega}{3}\frac{\Gamma(1/3)}{A^{1/3}}$$

The mean grain-area is the reciprocal

$$\frac{3A^{1/3}}{\Omega\Gamma(1/3)} = \frac{A^{1/3}}{0{\cdot}893\Omega} = \frac{(\pi v^2\Omega/3)^{1/3}}{0{\cdot}893\Omega}$$

$$= 1{\cdot}137\left(\frac{v}{\Omega}\right)^{2/3}$$

This expression has some interest in connection with the size of spangles formed in a metallic coat obtained by hot dipping, but is only valid if the velocity of outward growth from the nuclei is constant in time and not affected by other factors, such as the rate of removal of heat.

The same gamma-function method can be used for the analogous three-dimensional problem, i.e. the determination of the grain-size in cast metal, but is only valid where v is constant—which will rarely be the case under ordinary casting conditions. The mean grain-volume is found to be $1{\cdot}117\left(\frac{v}{\Omega'}\right)^{3/4}$, where Ω' is defined by the statement that the chance of a nucleus appearing within time dt in volume δV is $\Omega' \, \delta V \, dt$.

Further details of these and related problems are given in the original paper (U. R. Evans, *Trans. Faraday Soc.* 1945, **41**, 365).

Significance Tests

General. One of the important uses of statistics in corrosion research is to decide whether experimental data which at first sight appear to show that some factor stimulates or discourages corrosion can be regarded as significant evidence or whether the apparent effect arises from the play of chance.

The matter can be illustrated by a simple case which is easy of solution although it is rarely met with in service. Suppose that ample experimental data is available regarding corrosion-damage in a certain environment, and that the " lives " of specimens (the period survived before perforation, or in the case of stressed specimens survived before fracture) are known to be distributed, within the range covered by experience, according to the normal law, and suppose also that mean life and standard deviation are known with reasonable accuracy. Then a single experiment is performed with some substance added to the corrosive liquid, and it is found that the life is greater than any recorded before. Does this prove that the substance added is really an inhibitor, or does it arise from the play of chance? The latter supposition is not in itself unreasonable since normal distribution implies that there is a possibility—on rare occasions—of meeting lives much longer than those usually met with; it is therefore not impossible that the substance added was not the cause of the unusually long life. However, provided that the distribution can be taken as normal, it is easy to find from tables (or from curves, such as those of fig. 184) the chance of attaining a standardized deviation equal or greater than that represented by the exceptional life survived. If the probability of this arising simply from the play of chance is found to be small, it seems more reasonable to adopt the alternative explanation and admit that the exceptional life was probably caused by the addition—even though the matter cannot be regarded as definitely proven.

This simple case rarely arises in practice. It is likely that the experimenter will decide to perform several experiments with the substance added (perhaps the same number as those performed without any addition); this will, of course, make a judgment easier. However, even under the best conditions commonly met with, the number of experiments that can be performed with or without the addition is much smaller than is desirable, since each experiment takes a long time to carry out; the experimenter may be in doubt as to the true distribution curve and even of the true value of the standard deviation. It has already been stated that where the number of experiments (N) is limited, the best estimate of the standard deviation is not $\sqrt{\Sigma x^2/N}$, but $\sqrt{\Sigma x^2/(N-1)}$, and this expression* often gives a reasonably accurate idea of σ. A more serious element of the situation is that in such a case, when N is small, it is not accurate to employ the tables of normal distribution in calculating the probability of a given deviation arising simply from the play of chance. The difficulties have, however, been greatly diminished by the use of the statistic known as t, the distribution of which has been studied by W. S. Gosset (who published under the pseudonym " Student "); the various applications of the *t-function* have been discussed by Fisher. The papers and the tables showing the distribution of the *t-function* are: " Student ", *Biometrika* 1908–9, **6,** 1; *Metron (Ferrara)* 1925, **12** (3) 105; R. A. Fisher, *ibid.*, 1925, **12** (3) 90; the tables are reproduced in many statistical text-books.

The form of the *t-function* varies with the nature of the problem facing the statistician, but for the problem generally met with in corrosion research —the question as to whether significance should be attached to any particular value (Δ) obtained for the difference between the means of two sets of experiments (with or without the " addition ", or other factor under study)—the appropriate expression is

$$t = \Delta\sqrt{\frac{N_1 N_2(N_1 + N_2 - 2)}{(\Sigma x_1^2 + \Sigma x_2^2)(N_1 + N_2)}}$$

where N_1 and N_2 are the numbers of experiments in each set.

The curves of fig. 189, drawn by Thornhill from figures published in Student's later (Italian) paper, indicate the chance of exceeding any given value of t after any known number of experiments $N(= N_1 + N_2)$. It is usual to take the view that if the probability of obtaining the observed value is less than 0·05, the hypothesis that the difference between the two sets arises from the play of chance can be rejected and the alternative view that the factor is having a genuine influence accepted. Some authorities, at least for certain purposes, would demand a lower limit, for instance 0·01. Weatherburn states, " If P is less than 0·05 we regard our value of t as significant. If P is less than 0·01 we regard it as highly significant." In this respect, the t-test cannot be regarded as completely objective, but if the limit to be accepted is decided before the calculation has been worked out, that criticism is to some extent removed.

The curves show that if the total number of experiments has been about

* The reason for the appearance of $(N - 1)$ instead of N is that, although N separate experiments have been performed, there are only $(N - 1)$ independent estimates of x; if $(N - 1)$ of the experimental values are disclosed along with the mean of the whole set, the remaining estimate of x is fixed, since there can only be one value consistent with the data already disclosed.

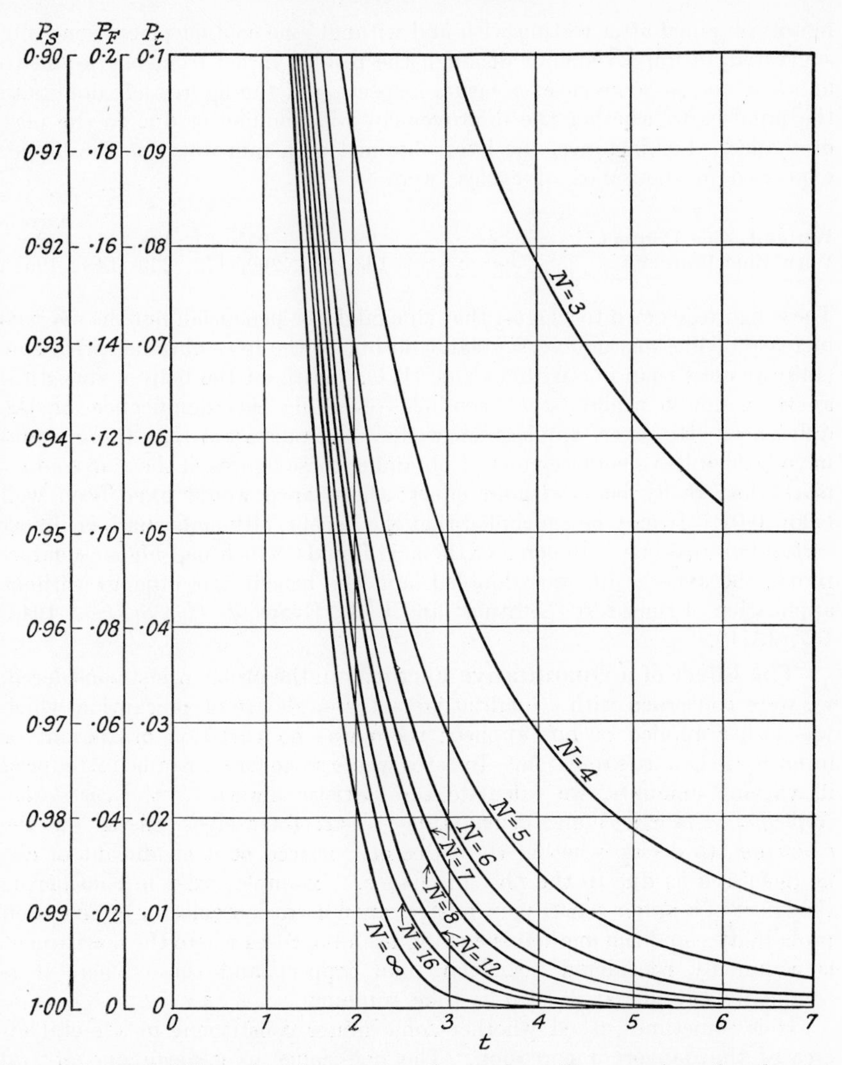

FIG. 189.—Probability of misinterpreting the t-value. For meaning of scales, see p. 949 (Curves due to R. S. Thornhill, based on data by " Student ").

8, the experimental difference between the two sets can only be regarded as significant if t exceeds about 2 and only highly significant if it exceeds about 3. For larger numbers of experiments, slightly lower values of t can be regarded as evidence of significance.

The Effect of a Qualitative Factor. An example of the use of the t-test is provided by Stuart's investigation into the affect of zinc protectors on the corrosion-fatigue life of steel strips. Before the research started, it had been rather expected that zinc protectors would sometimes increase the danger of fatigue failure, by introducing hydrogen. Actually the figures obtained after testing with and without zinc contact pieces generally suggested an improvement, although the results varied from one liquid to another, and in some cases a casual inspection of the figures left doubts in the mind as to whether the improvement was genuine or due to the play of chance. For instance, the lives obtained with rain-water as the liquid, expressed in thousands of cycles, were:—

		Average
Without Zinc Contact	98, 118, 97, 91, 148, 171	120
With Zinc Contact	193, 223, 200, 177, 230, 148	195

These figures seemed to suggest that zinc might be beneficial, but the average increase of life was not very great, and one of the lives obtained with zinc (148) was less than one without zinc (171). Without the help of statistical analysis, doubt would have been felt regarding the significance of the difference. However, application of the *t*-test suggested that there would have been only a slender chance of obtaining these figures, if the zinc contact pieces had really been without effect; the chance would have been well below 0·02. It may be concluded that the benefit, although small, is almost certainly a real one. In some of the acid liquids, which had higher conductivity, the average life was doubled, and the benefit was obvious without application of the *t*-test (N. Stuart and U. R. Evans, *J. Iron St. Inst.* 1943, **147**, 131P).

The Effect of a Quantitative Factor. In the problem just considered, we were concerned with a qualitative factor—a device or precaution which was either applied or not applied; there was no variation of amount or intensity in Stuart's research. In cases where some factor can be introduced in varying amounts, we calculate the statistic known as the *correlation coefficient* (*r-function*) and then apply tests, themselves based on the *t-function*, to decide whether the value of r arrived at is significant or can be dismissed as due to the play of chance. Examples arise in considering the effect of a minor constituent in steel upon corrosion velocity or corrosion probability—and the method is equally useful in cases where the constituent is commonly considered beneficial (like copper) and those where it is commonly considered dangerous (like sulphur).

It is sometimes asked whether some minor constituent in a metal increases the danger of corrosion. The questioner may mean one of two things. He may be asking whether the constituent causes an increase of probability or whether it causes an increase of corrosion-velocity; to simplify discussion, we will assume that he is thinking of corrosion-velocity.

At first sight, this would seem to be a question capable of being answered by direct experiment. However, to carry out a special set of experiments under uniform conditions keeping the content of all the other constituents precisely the same and varying only the element in question, may not be easy. Nor will it be entirely satisfactory to carry out the experiments under conditions which are kept strictly constant for all the materials to be compared; variation of the one element may alter the conditions of heat-treatment needed to obtain the best mechanical properties, so that a test under strictly comparable conditions would then mean that some of the materials were being tested in a state which does not bring out their best qualities. In practice, it may be necessary to make use of a series of materials in which the content of the element in question varies within wide limits, whilst some variation occurs simultaneously in the contents of the other elements and probably in the thermal and mechanical pre-treatments. If then the corrosion-velocities of the series are measured and tabulated against the content of the element, it is possible that a certain tendency for the corrosion-velocity to rise with increasing content may be detected; but exceptions will probably be noted and when the corrosion-velocity is plotted against the content of the element, the points will not fall exactly on a curve; even the order of the materials, arranged according to the contents of the element, may be not quite the same as the order of corrosion-velocities.

In such a case, the suspicion must arise: is the apparent tendency for corrosion-velocity to be high in the samples rich in the element simply due to the play of chance? Even a number of samples containing exactly the same amount of that element would show considerable scatter in the velocities. Non-statistical friends—appealed to for a judgment—will probably express different views: one may say that " I think the velocity is really tending to rise with the content of the element ", whilst another may say " The numbers seem to run just anyhow."

A calculation of the Correlation Coefficient gives an answer to this question which, if not absolutely definite, is objective—when once a decision has been made regarding the appropriate limit to be adopted for the probability needed for " significance ". Once more we adopt provisionally the hypothesis that, in point of fact, the fluctuation in corrosion-velocity has nothing to do with the content of the element, but arises from the play of chance. The probability that the values for the velocity obtained by experiment could arise fortuitously is then calculated, and if it turns out to be very small, we reject the hypothesis, and adopt the alternative view that the increase of corrosion-velocity has at least *something to do* with the content of the element; as shown below, that is not the same thing as saying that the element is the direct *cause* of the high corrosion-rate.

The simplest way of defining the correlation coefficient between two variables X and Y is

$$r_{XY} = \frac{\overline{XY} - \overline{X}.\overline{Y}}{\sigma_X.\sigma_Y}$$

where the $\overline{X}$ represents the average value of X, and $\overline{XY}$ the average of the product XY. The " correlation coefficient " r is thus the difference between the product of the two means and the mean of the products, the value being " standardized " by being expressed as a fraction of the standard deviations. If X and Y are uncorrelated, there is no reason why the product of the means should be greater or less than the mean of the products, but if large values of Y tend to go with large numbers of X, then in obtaining the individual values of XY before calculating the average value $\frac{\Sigma XY}{n} = \overline{XY}$, certain items are introduced which are of a higher order of magnitude, and despite the fact that low values of Y go with low numbers of X, the mean value of the product ($\overline{XY}$) will certainly exceed the product of the means $\overline{X}.\overline{Y}$. Thus the expression is a reasonable measure of correlation; its value extends from -1 to $+1$. Completely uncorrelated factors would give a value of about zero, whilst perfect correlation (where the plotting of Y against X provides a series of points falling exactly on a straight line) will give -1 if Y falls as X rises, or $+1$ if Y rises as X rises.

If we agree to measure X and Y from arbitrary zero-points situated at their respective means, so that $\overline{X}$ and $\overline{Y}$ both become zero, r_{XY} becomes $\frac{\overline{XY}}{\sigma_X \sigma_Y}$ or $\frac{\Sigma(XY)}{\sqrt{\Sigma x^2 \Sigma y^2}}$. This form of expression is found in many text-books.

The calculation of r does not in itself answer the question under discussion since if the number of materials used is small, the experimental value may differ from the " true " value (the value which would be obtained by studying an infinite number of materials). Thus, where there is really no correlation, the true value of r being zero, a calculation may show a value for r appreciably differing from zero. The curves of fig. 190, plotted by Thornhill, may help in deciding whether the experimental value of r is " significant ". If the probability that (even though r was really zero) the limited range of experiments might permit a value of r equal or greater than that actually obtained, is less than 0·05, it is generally agreed that there is genuine correlation. Evidently even where there have been, say, 25 materials tested, values of r less than about 0·33 must be disregarded.

Mears has studied the effect of the content of minor constituents in 25 steels on the Inception Probability. His results (Table XXXV) suggest that steel rich in sulphur tend to have high inception probabilities.

TABLE XXXV

Correlation Coefficients for Inception Probability (R. B. Mears)

	Emery Finish	Lathe-turned Finish
Carbon	+ 0·45	+ 0·26
Sulphur	+ 0·52	+ 0·54
Manganese	+ 0·32	+ 0·47
Silicon	+ 0·19	− 0·13
Phosphorus	− 0·02	− 0·20
Copper	− 0·12	− 0·38

Let us now consider the high value of r connecting sulphur-content and inception probability (Table XXXV). Does this high value prove that a high sulphur-content is the *cause* of high inception-probability? Not necessarily. It is well known that sulphur tends to be segregated in the same part of the ingot as phosphorus. High-sulphur specimens will tend to be

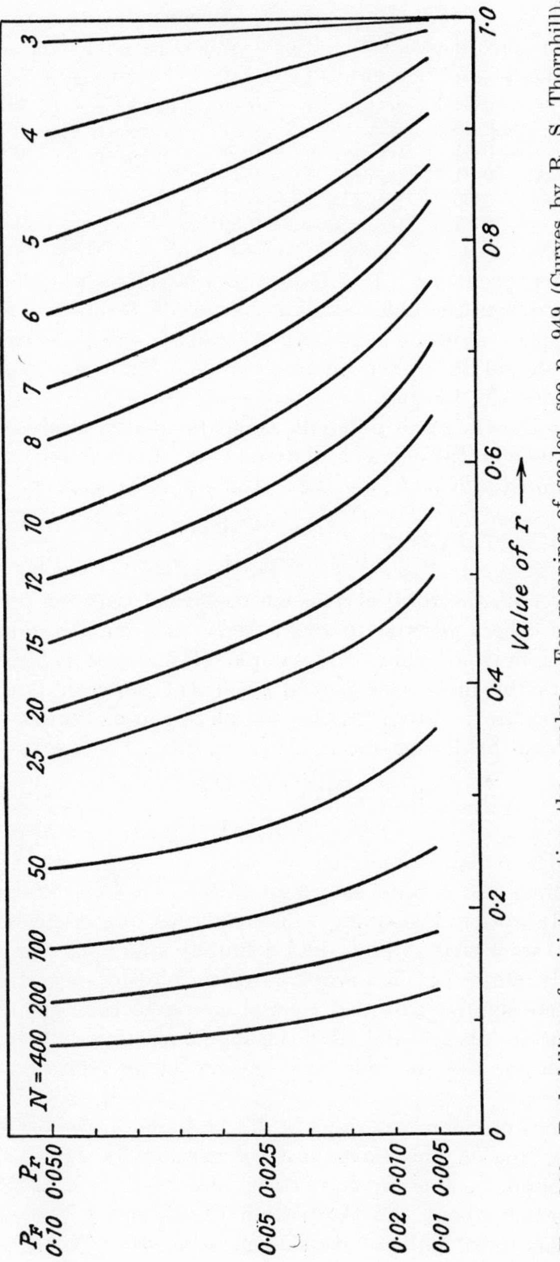

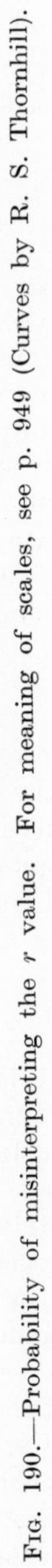

FIG. 190.—Probability of misinterpreting the r value. For meaning of scales, see p. 949 (Curves by R. S. Thornhill).

TABLE XXXVI

TOTAL AND PARTIAL CORRELATION COEFFICIENTS FOR CORROSION-VELOCITY
(T. P. Hoar and D. Havenhand)

[V represents corrosion-velocity; other symbols represent the elements]

Total Coefficient	First-order Partial	Second-order Partial
$r_{SV} = +0{\cdot}20$	$(r_{SV})_{Cu} = +0{\cdot}85$	$(r_{SV})_{Cu.P} = +0{\cdot}79$
$r_{PV} = +0{\cdot}28$	$(r_{SV})_{P} = -0{\cdot}03$	$(r_{PV})_{Cu.S} = -0{\cdot}34$
$r_{Cu.V} = -0{\cdot}51$	$(r_{PV})_{Cu} = +0{\cdot}59$	$(r_{Cu.V})_{SP} = -0{\cdot}90$
$r_{SP} = +0{\cdot}81$	$(r_{PV})_{S} = +0{\cdot}19$	
$r_{Cu.S} = +0{\cdot}66$	$(r_{Cu.V})_{S} = -0{\cdot}89$	
$r_{Cu.P} = -0{\cdot}37$	$(r_{Cu.V})_{P} = -0{\cdot}69$	

high-phosphorus specimens. If it happened that phosphorus favoured the inception of corrosion, then high-sulphur specimens, being also, for the most part, high-phosphorus specimens, would generally be readily corroding steels, and thus we should draw the conclusion, on insufficient evidence, that sulphur promotes the inception of corrosion.

To separate the effects of different elements, after a study of groups of specimens in which all the element contents vary, it is necessary to calculate *partial correlation coeffiicients*, by using the formulae such as

$$(r_{12})_3 = \frac{r_{12} - r_{13}r_{23}}{\sqrt{(1 - r_{13}^2)(1 - r_{23}^2)}}$$

where r_{12} represents the total correlation coefficient between factors 1 and 2, with all the others allowed to roam freely in a manner representative of the population from which the sample of material is drawn, whilst $(r_{12})_3$ represents the first-order partial coefficient between 1 and 2, with factor 3 held constant. From this we can obtain second-order coefficients, with factor 4 also held constant

$$(r_{12})_{34} = \frac{(r_{12})_4 - (r_{13})_4(r_{23})_4}{\sqrt{(1 - (r_{13})_4^2)(1 - (r_{23})_4^2)}}$$

Hoar and Havenhand, studying 36 steels in citric acid, applied this method, obtaining the results shown in Table XXXVI, which refers to Velocity, not Inception Probability. Calculations of significance, applied to these values, show that sulphur does definitely stimulate corrosion whilst copper definitely retards it. The probable reason, already suggested on p. 878, is that hydrogen sulphide, formed by the action of the acid on sulphide inclusions in the steel, is a stimulant of the anodic reaction, but that if copper enters the solution, the sulphide will quickly be precipitated as copper sulphide.

Details of the corrosion experiments will be found in the original papers by R. B. Mears, *Iron St. Inst. Carnegie Schol. Mem.* 1935, **24**, 81; T. P. Hoar and D. Havenhand, *J. Iron St. Inst.* 1936, **133**, 283P, esp. p. 289P, whilst general information about the Correlation Coefficient is provided in the books of Shewhart and of Yule and Kendall, mentioned on p. 950. F. N. David's "Tables of the Correlation Coefficient" (Biometrika Office) may prove useful.

The reader may become puzzled when he compares the values recorded in the tables printed in certain text-books, since he will notice that the numbers for the probabilities indicated by some tables represent half the numbers given in others. This arises from the fact that in certain problems interest is attached to the *combined* chance of the value falling above $+x$ and below $-x$, which in a symmetrical distribution doubles the probability. The smaller numbers are likely to be appropriate in most corrosion problems, but, before they are applied in any particular case, the possibility that they should be doubled ought to receive consideration. Figs. 189 and 190 show both sets of values as P_t (or P_r) and P_F; P_s is $(1 - P_t)$.

Use of the χ^2-Test. In discussing the reliability of the sulphate-ratio treatment for avoiding caustic cracking in boilers, it was stated (p. 459) that a definite decision would require knowledge of the respective numbers of

(1) untreated waters which produced cracking
(2) untreated waters which did not produce cracking
(3) treated waters which produce cracking
(4) treated waters which did not produce cracking.

Unfortunately these numbers do not appear to be known. If, however, statistics become available in this or any analogous case, it will be possible to use the χ^2-test, which is described in most books on Statistics, such as that of Yule and Kendall (see below).*

Other References

No one book covers the whole ground needed for the application of statistical theory to corrosion problems, although the standard works contain much that the Corrosionist will never employ. For certain parts of the subject T. C. Fry's " Probability and its Engineering Uses " (Macmillan) is admirable, whilst R. A. Fisher's two books, " Statistical Methods for Research Workers " (Oliver & Boyd) and " The Design of Experiments " (Oliver & Boyd) are too well known to need commendation. Among elementary books, W. N. Bond's " Probability and Random Errors " (Arnold) is a useful introduction to the subject, but may need to be supplemented, since certain parts of the subject are not included; the chapter on curve-fitting deserves mention in view of the not infrequent publication of graphs which do not appear to represent the " curves of best fit " (a particularly serious matter if logarithmic scales have been employed where a trivial departure from a line may be equivalent to a large error).† Perhaps C. E. Weatherburn's " First Course in Mathematical Statistics " (Cambridge University Press) is the most worthy of recommendation to the beginner; the author states that it is intended for " the student with an average mathematical equipment,

* E. C. Potter has applied the χ^2 method to the question of the effect of oxygen in boiler-water, in a lecture presented to a symposium held at Moscow in 1958. See also *Chem. and Ind.* (*Lond.*) 1959, p. 308.

† See also F. Daniels, " Mathematical Preparation for Physical Chemistry " (McGraw-Hill), Chapter XXI of the 1928 edition.

including an ordinary knowledge of the Integral Calculus ", the aim being " to explain the underlying principles, and to prove the formulae and the validity of the methods which are the common tools of statisticians ". J. F. Kenney's " Mathematics of Statistics " (van Nostrand, Chapman & Hall) has two volumes, the first involving only elementary mathematics and the second being more advanced. G. U. Yule and M. G. Kendall's " Introduction to the Theory of Statistics " (Griffin) is an exceptionally useful book, whilst H. Jeffrey's " Theory of Probability " (Clarendon Press) is held in reverence by those who can spare time and thought for the intellectual aspects of the subject. Other books commanding respect are those of W. A. Shewhart, " Economic Control of Quality of Manufactured Product " (Macmillan), O. L. Davies, " Statistical Methods in Research and Production " (Oliver & Boyd), F. N. David, " Probability Theory for Statistic Methods " (Cambridge University Press), and C. C. Peters and W. R. van Voorhis, " Statistical Procedures and their Mathematical Bases " (McGraw-Hill).

APPENDIX I

CHEMISTRY

General and Inorganic Chemistry

Atoms and Molecules. Scientists have long held the view that matter is not infinitely divisible but consists of small particles or " atoms ". There are many sorts of atoms, varying in size; the atoms of the heavier metals are mostly smaller than those of the lighter metals. For many purposes it is permissible to regard metals as being built up of little spheres packed in contact, their centres being usually 2 to 3 Å apart.* More recent research has shown that sub-division of atoms is possible, using special techniques, but this sub-atomic division (" fission ") is of a special character; thus, so long as a piece of copper contains many atoms, the smaller pieces obtained by cutting it up still possess the properties of copper; but the fragments obtained by fission of single copper atoms are no longer copper, and possess new properties.

In solid copper, the atoms are closely packed; for many purposes, each atom can be regarded as having a fixed position and as being in contact with its neighbours. When copper is heated strongly in the absence of air, it first melts to a liquid; in the molten state, each atom no longer has a fixed position, but can move about, although it is still essentially in contact with other atoms. At a still high temperature, copper boils, producing a gas, in which the atoms are separated by distances large compared to their diameters.

Many non-metallic substances are gaseous at room temperature. Examples are oxygen and nitrogen, the two main constituents of air; in each case, the atoms move about in pairs, and these pairs, known as " molecules ", are separated by distances large compared to the distance between the two atoms forming a pair. The molecules possess rapid translatory motion, and when a small quantity of gas is introduced into a vessel, the molecules soon distribute themselves throughout its interior. The gas pressure p is produced by the moving molecules beating against the walls of the vessel and being reflected. If we halve the volume V of this vessel, and thus double the concentration of the molecules, we double the pressure, provided that we are concerned with an " ideal " gas (in which complications due to attraction between the molecules can be neglected); oxygen and nitrogen come near to being ideal gases. At high temperatures the movement of the molecules is more rapid than at low temperatures, and the beating-rate or pressure will increase; or, if the containing vessel is a cylinder provided with a piston, and it is arranged for the pressure to remain constant, the volume will increase. For an ideal gas, $pV = RT$ where T is the absolute temperature (the temperature on the centigrade scale $+ 273°$) and R the " gas constant ".†

In oxygen and nitrogen, only one type of atom is present, although they are commonly joined in pairs. There is a special kind of oxygen differing from ordinary oxygen in possessing a characteristic odour, in which each molecule consists of three atoms instead of two. This is called *Ozone*, and is written O_3, ordinary oxygen being O_2; ozone is produced by passing a silent electric discharge through ordinary oxygen.

* An Angstrom Unit (1 Å) is 1×10^{-8} cm. or 4×10^{-9} inch. The " atomic radii " of most heavy metals lie between 1·2 and 1·5 Å, so that the distance between centres of neighbouring atoms is between 2·4 and 3·0 Å. See R. W. G. Wyckoff, " Crystal Structures ", Vol. I (Interscience Publishers). The Atomic Radii will be found in Table II.10 and the Ionic Radii in Tables III.2 and III.3.

† If p is measured in atmospheres and V in millilitres (ml.), $R = 82·07$ per gram molecule.

The essential differences between the behaviour of gases, liquids and solids can now be stated. In the *gaseous* state, a small amount of material will expand and assume the volume and the shape of any vessel into which it is introduced. At lower temperatures, where molecular movement becomes less violent, the attraction between molecules is able to cause " condensation " to form *liquid* in which the molecules remain close together; a liquid will not necessarily fill the whole volume of a large vessel into which it is introduced, although it will quickly assume the shape of the *lowest* part of the vessel, since the molecules still possess translatory motion, so that a liquid lacks rigidity. At still lower temperatures, translatory motion becomes unimportant, although the molecules still oscillate about mean positions; thus rigidity appears, and the body becomes *solid* and incapable of taking on the shape of the containing vessel. In some cases, the passage from the liquid to the solid state is gradual, giving a glass; the glassy condition is discussed on p. 971. More often, the passage from liquid to solid occurs sharply at a definite temperature, the freezing-point. Under " reversible " conditions, the freezing-point should be identical with the melting-point, at which the solid becomes liquid on heating; but more often there is a lag in freezing known as " super-cooling ", so that in practice the freezing occurs at a temperature below the melting-point. Except where a glass is formed, the atoms array themselves on freezing in a definite arrangement, forming a " crystal ".

Most metals are crystalline in character, and in copper the arrangement of the atoms can best be described by imagining space divided into little cubes, the centres of the atoms being placed at every cube-corner and also at the centre of each cube face. Iron assumes a similar crystal structure at temperatures above 906° C., but at ordinary temperatures the arrangement can best be described by imagining the centres of the atoms placed at every cube corner and every cube centre. These two arrangements are respectively known as face-centred and body-centred cubic lattices.

Compound and Mixtures. Air is a " mixture ", consisting mainly of nitrogen and oxygen molecules, with smaller amounts of other gases. The physical properties lie between those of nitrogen and oxygen, but since there is about four times as much nitrogen as oxygen in air, it resembles nitrogen in physical properties more closely than it resembles oxygen. Although the molecules O_2 and N_2 are present, air normally contains no molecules in which both oxygen and nitrogen atoms are present together.* However, by passing air through an electric arc, and then suddenly cooling it, molecules such as NO and NO_2 are obtained; these " oxides of nitrogen " possess properties quite different to those of either nitrogen or oxygen; nitrogen peroxide (NO_2), for instance, is a red gas with unpleasant odour. Thus, whereas air is merely a *mixture* of nitrogen and oxygen, nitrogen peroxide is a *compound*, formed by chemical combination of nitrogen and oxygen; oxygen and nitrogen, when uncombined, are known as *elements*.

The essential feature of chemical combination—as opposed to physical mixture—is a change of properties. When sodium (a substance with metallic appearance) is introduced into chlorine (an odorous, poisonous gas), a violent reaction occurs with great evolution of heat; the product is sodium chloride or common salt, which, as is well known, has properties very different from those of sodium or chlorine; elements contain a single kind of atom, whereas compounds contain two or more kinds.

Atomic Weights and Atomic Numbers. Over a century ago, chemists had developed methods of determining the relative weights of the atoms, which were first expressed on a scale where the weight of the hydrogen atom (H) was adopted (arbitrarily) as unity; on this scale, the oxygen atom (O) was 15·87.

* During thundery weather or near power stations small amounts of oxides of nitrogen may be detected in air.

Later, for reasons of practical convenience, the atomic weight of oxygen was taken to be 16·000, but the change made little difference; when O is 16·000, H becomes 1·008. The absolute weight of a single hydrogen atom is now known ($1 \cdot 66 \times 10^{-24}$ gram); the absolute weight of the single oxygen atom almost 16 times as much, and the weight of a molecule (O_2) almost 32 times as much.

The Periodic Classification. About 1864, it was noticed that when the elements were arranged in the order of their atomic weights, there was a tendency for elements possessing certain properties to recur at regular intervals; the length of the " periods " (the number of elements intervening between the recurrence of properties), was observed to be 8 on the first two occasions, but later became 18.

Various ways of arranging the elements so as to bring out the periodic character of their properties have been suggested; a convenient form of " Periodic Classification " is printed in Table XXXVII. In order to display to best advantage the relations between the properties, it is necessary in four places to depart slightly from the exact order of Atomic Weights as obtained from purely chemical measurements. The actual order in which the elements are written down is indicated by the " Atomic Numbers " placed at the top of each square; the experimentally determined " Atomic Weights " are shown in Table XXXVIII, which also serves as a key to the symbols used to represent different atoms.

The four cases where the accepted order of Atomic Numbers differs from the experimentally determined order of Atomic Weights concern the elements of Atomic Numbers 18, 27, 52 and 90 (A, Co, Te and Th), each of which has an atomic weight slightly higher than that of the element which is assigned the next number (19, 28, 53 and 91). These departures from the strict order of atomic weights place argon, potassium, tellurium, iodine, thorium and protoactinium in the same vertical columns as other elements which they closely resemble, whereas otherwise they would all be placed with unrelated elements. Moreover, the properties of cobalt are in many ways such as to place it between iron and nickel. Doubtless at the time when the Periodic Table was drawn up, such tinkering with the experimental results may have been regarded by some chemists as arbitrary, but today it is seen to be justified. An independent method, based on X-rays, has been discovered for measuring the atomic number of an element, which—as explained below—possesses a definite meaning in connection with the structure of the atom. These X-ray studies show that the atomic number for cobalt (27) is really lower than that of nickel, despite the higher atomic weight.

Sub-atomic Structure. The periodic recurrence of properties suggests that the atoms of different elements are not the ultimate constituents of matter, but are themselves built up in a consistent manner from something simpler—that " something " being common to all the elements. It is now believed that the atom, in its electrically neutral state, consists of a positively charged nucleus surrounded by negative " electrons "—the electron being in fact an " atom of electricity ". The number of electrons is equal to the atomic number. The positive charge is attributed to the presence within the nucleus of positive particles called " protons "—a proton being about 1845 times as heavy as an electron. The number of protons must be equal to the number of electrons outside the nucleus, if the atom is to be electrically neutral; thus if the nucleus contained nothing but protons, the atomic weight referred to $H = 1$, should be equal to the atomic number. The fact that the atomic weight is usually higher than the atomic number suggests the presence of a neutral " makeweight "; this is called the " neutron ", a particle of the same weight as a proton but possessing no electrical charge. Hydrogen—alone among the elements—contains no neutrons, and here the atom consists of one proton (representing the nucleus) and one electron. All the other elements contain neutrons as well

TABLE XXXVII

The Periodic Arrangement of the Elements

Group No. . .	O	IA	IIA	IIIA	IVA	VA	VIA	VIIA	VIII	IX	X	IB	IIB	IIIB	IVB	VB	VIB	VIIB	O
Valencies found in the Group . .	0	1	2	3	(3) 4	(2, 3, 4) 5	(2) 3, 4, 6	2 (3) 6, 7	2, 3, 4 6, 7, 8	2, 3, 4	2 (3) 4	1, 2, 3	(1) 2	(1) 3	2, 4	5, −3	6, −2	7, −1	0
	2 He	3 Li	4 Be											5 B	6 C	7 N	8 O	9 F	10 Ne
	10 Ne	11 Na	12 Mg	13 Al											14 Si	15 P	16 S	17 Cl	18 A
	18 A	19 K	20 Ca	21 Sc	22 Ti	23 V	24 Cr	25 Mn	26 Fe	27 Co	28 Ni	29 Cu	30 Zn	31 Ga	32 Ge	33 As	34 Se	35 Br	36 Kr
	36 Kr	37 Rb	38 Sr	39 Y	40 Zr	41 Nb	42 Mo	43 Tc	44 Ru	45 Rh	46 Pd	47 Ag	48 Cd	49 In	50 Sn	51 Sb	52 Te	53 I	54 Xe
	54 Xe	55 Cs	56 Ba	57–71 RARE EARTHS	58 ↵	73 Ta	74 W	75 Re	76 Os	77 Ir	78 Pt	79 Au	80 Hg	81 Tl	82 Pb	83 Bi	84 Po	85 At	86 Rn
	86 Rn	87 Fr	88 Ra	89 Ac	90 Th	91 Pa	92 U	93 Np	94 Pu										

Notes: (1) Element No. 1 (hydrogen) which should have a line to itself, is omitted.
(2) To save space, Elements 57 to 71 (Rare Earth Metals) are placed in a single square.
(3) Elements 95 to 101 (recent discoveries) are omitted; there are doubts about the appropriate positions, and they have no interest to the corrosionist.*
(4) The Transition Elements are boxed in with a thick line in the centre of the table.
(5) The Non-metallic Elements are boxed in with a broken line on the right-hand part of the table.
(6) The Inert Gases (Group O) are placed both on the left and the right of the table; each member which appears on the right is repeated at the left-hand end of the line below.

* Some authorities would group together all the elements between 89 and 101 (also the undiscovered 102 and 103) below the rare earths. See G. T. Seaborg, *Endeavour* 1959, **18**, 5.

TABLE XXXVIII

ATOMIC WEIGHTS AND SYMBOLS OF THE ELEMENTS
(from *J. Chem. Soc.* 1956, p. 4989)

Name	Symbols	Atomic Weight	Name	Symbols	Atomic Weight
Actinium . . .	Ac	227	Neptunium . .	Np	approx. 237
Aluminium . . .	Al	26·98			
Antimony . . .	Sb	121·76	Nickel	Ni	58·71
Argon	A	39·944	Niobium . . . (Columbium)	Nb(Cb)	92·91
Arsenic	As	74·91			
Astatine . . .	At	approx. 210	Nitrogen . . .	N	14·008
			Osmium . . .	Os	190·2
Barium	Ba	137·36	Oxygen . . .	O	16·000
Beryllium (Glycinum) . .	Be	9·013	Palladium . . .	Pd	106·4
			Phosphorus . .	P	30·975
Bismuth . . .	Bi	209·00	Platinum . . .	Pt	195·0
Boron	B	10·82	Plutonium . . .	Pu	approx. 242
Bromine . . .	Br	79·916			
Cadmium . . .	Cd	112·41	Polonium . . .	Po	210
Caesium . . .	Cs	132·91	Potassium . .	K	39·100
Calcium	Ca	40·08	Praseodymium .	Pr	140·92
Carbon	C	12·011	Promethium . .	Pm	approx. 145
Cerium	Ce	140·13			
Chlorine . . .	Cl	35·457	Protoactiniun . .	Pa	231
Chromium . . .	Cr	52·01	Radium . . .	Ra	226·05
Cobalt	Co	58·94	Radon (Radium emanation) . .	Rn	222
Copper	Cu	63·54			
Dysprosium . .	Dy	162·51	Rhenium . . .	Re	186·22
Erbium	Er	167·27	Rhodium . . .	Rh	102·91
Europium . . .	Eu	152·0	Rubidium . . .	Ru	85·48
Fluorine . . .	F	19·00	Samarium . . .	Sa	150·35
Francium . . .	Fr	approx. 223	Scandium . . .	Sc	44·96
			Selenium . . .	Se	78·96
Gadolinium . .	Gd	157·26	Silicon	Si	28·09
Gallium	Ga	69·72	Silver	Ag	107·880
Germanium . .	Ge	72·60	Sodium . . .	Na	22·991
Gold	Au	197·0	Strontium . . .	Sr	87·63
Hafnium . . .	Hf	178·50	Sulphur . . .	S	32·066
Helium	He	4·003	Tantalum . . .	Ta	180·95
Holmium . . .	Ho	164·91	Technetium . .	Tc	approx. 99
Hydrogen . . .	H	1·008			
Indium	In	114·82	Tellurium . . .	Te	127·61
Iodine	I	126·91	Terbium . . .	Tb	158·93
Iridium	Ir	192·2	Thallium . . .	Tl	204·05
Iron	Fe	55·85	Thorium . . .	Th	232·09
Krypton . . .	Kr	83·80	Thulium . . .	Tm	168·94
Lanthanum . .	La	138·92	Tin	Sn	118·70
Lead	Pb	207·21	Titanium . . .	Ti	47·90
Lithium . . .	Li	6·940	Tungsten (Wolfram)	W	183·86
Lutecium . . .	Lu	174·99	Uranium . . .	U	238·07
Magnesium . .	Mg	24·32	Vanadium . . .	V	50·95
Manganese . . .	Mn	54·94	Xenon	Xe	131·30
Mercury . . .	Hg	200·61	Ytterbium . .	Yb	173·04
Molybdenum . .	Mo	95·95	Yttrium . . .	Y or Yt	88·92
Neodymium . .	Nd	144·27	Zinc	Zn	65·38
Neon	Ne	20·183	Zirconium . . .	Zr	91·22

Approximate Atomic Weight of Elements, Numbers 95–101 (*not included in Table XXXVII*). *The figures for* 99 *and* 100 *are still uncertain*

Element	No.	Symbol	Weight
Americium	(95)	Am	243
Curium	(96)	Cm	243
Berkelium	(97)	Bk	249
Californium	(98)	Cf	249
Mendelevium	(101)	Mv	256

as protons in the nucleus. In some of the lighter elements (carbon, nitrogen and oxygen), the numbers of neutrons and protons are equal, so that the atomic weight is double the atomic number; in the heavier elements, the neutrons outnumber the protons, and the atomic weight exceeds twice the atomic number.

The chemical properties of an element depend on the number of electrons surrounding the nucleus—that is, on the atomic number. It is possible for two atoms to exist which contain the same number of electrons and protons, but different numbers of neutrons; both of these atoms will show almost identical chemical properties and nearly the same physical properties; the atomic numbers will be the same, but the atomic weights different. Varieties of an element differing in atomic weights, but possessing almost the same chemical properties, are known as *Isotopes*. Many of the elements as we meet them in our world, are really mixtures of isotopes. That is the reason why the Atomic Weights, as determined chemically, are not—except in a few cases—whole numbers. Chlorine, for instance, as met with in the laboratory, contains a mixture of atoms having atomic weights 35 and 37 respectively; the lighter isotope is present in more than three times the amount of the heavier one, and the atomic weight as measured is 35·46. In chlorine, as in oxygen, the atoms occur in pairs, so that the formula is written Cl_2. In the case of sulphur, the isotopes occur in slightly different proportions in samples drawn from different sources, so that the atomic weight varies by about 0·003, the average value being 32·066.

Many elements exist in nature as mixtures of two or more isotopes, but there are other isotopes which are not commonly met with, but which are produced during the fission of radioactive elements. These occur as bye-products in the operation of atomic piles, and since some of them possess radioactivity in dangerous degree, their disposal adds to the difficulties and hazards of nuclear power-production. Some radioactive isotopes, however, possess genuine utility as markers or tabs for the study of chemical and physical processes. A radioactive isotope will possess practically the same properties as a non-radioactive isotope of the same element. If, therefore, a trace of a compound containing a radioactive isotope is added to a solution of the analogous compound containing the ordinary (non-radioactive) isotope, the two types of atoms will undergo the same chemical and physical changes; since the radioactivity of the one isotope enables it to be detected easily at any particular situation, the presence of the ordinary isotope, there may be inferred from the presence of the radioactive isotope. Also radioactive substances have been used as markers in studying oxidation, because their presence at a given level in the film can easily be ascertained and will serve to distinguish between different mechanisms of oxidation (p. 34). Again, in ascertaining whether the film formed by a solution of a chromate on iron contains chromium or consists of pure iron oxide, the addition of a small amount of chromate containing a radioactive chromium isotope allows the presence and distribution of chromium in the film to be studied (p. 152).

Inert gases. A feature of the Periodic classification is the Inert Gases of Group O (printed in the right-hand column of the Table, but repeated for convenience in the left-hand column, where each inert gas appears one line lower). These gases possess no power of chemical combination, and must consist of atoms so stable and complete that they cannot easily furnish or capture electrons.

They occur at atomic numbers 2, 10, 18, 36, 54 and 86, the interval between their positions consisting therefore of 2, 8, 8, 18, 18 and 32 elements; these four numbers form a series and can be written 2×1^2; 2×2^2; 2×3^2 and 2×4^2.

If we accept the view that the inert gases of Group O have their outer " shells " of electrons complete and stable, it is natural to suppose that the elements of Group IA (which follow the inert gases) must have one electron outside the stable shell, and will acquire structural stability when they part with that one electron; likewise that those of Group VIIB, which are placed just before the inert gases, should acquire structural stability by capturing an extra electron and thus completing their outside shell. This is found to be the case, and explains why sodium (of Group IA) reacts violently with the chlorine (of Group VIIB) to form sodium chloride (NaCl)—a substance far more stable and less active than either of its components. X-ray studies teach us that the structure of sodium chloride is like an assemblage of tiny cubes with sodium and chlorine at the alternate corners, so that each sodium has as its nearest partners 6 chlorines whilst each chlorine has as its nearest partners 6 sodiums; the chlorines and the sodiums are not " atoms " but ions, the chlorine being an " anion "—an atom with an extra electron and carrying therefore a negative charge (Cl^-)—whilst the sodium is a " cation "—an atom which has lost an electron and carries a net positive charge (Na^+). It becomes easy to understand why sodium chloride usually crystallizes in cubes, and why a crystal tends to split into smaller cubes (along " cleavage planes ") when broken up. Moreover, an explanation suggests itself for the stability of the structure, held together by electrostatic forces, each ion being in close contact with six ions of opposite charge. The fact that common salt (sodium chloride) is built up from ions rather than atoms is demonstrated by the fact that if it is dissolved in water, and if an electric E.M.F. is applied, the chlorine moves towards the positive pole and the sodium towards the negative pole.

Valency. Sodium is said to have a *positive valency* of one, since it can donate one electron to another atom, becoming stabilized in the process; conversely chlorine has a *negative valency* of one, becoming stabilized on capturing one electron. Other elements may donate or capture more than one electron and are said to have a valency exceeding one. The main valencies exerted by the elements in each group are printed at the top of Table XXXVII; for the groups following and preceding the Inert Gases, the valency number is precisely what would be expected if we accept the view that stability is achieved when the ion formed has the same number of electrons as an atom of an inert gas. Provided that we confine ourselves to the groups fairly near Group O, we can write down with fair confidence the formula defining the composition of any compound formed by the combination of a metal (an electron donor possessing positive valency) with a non-metal (an electron-acceptor possessing negative valency). Thus the formula of magnesium oxide is MgO, that of calcium chloride $CaCl_2$, aluminium oxide Al_2O_3 and so on. But when we move far away from Group O, the maximum valency indicated by the rule hitherto adopted is not always attained, and lower oxides or chlorides appear, which are often more stable than those formed according to the rule. Thus, although in Groups VIA and VIIA, the oxides CrO_3 and Mn_2O_7—which would be predicted by the rule just suggested—do exist, they are unstable and behave as strong *oxidizing agents*, readily passing on part of their oxygen to any other compound which will accept it (such an acceptor of oxygen or analogous element being called a *reducing agent*); CrO_3 is less stable than Cr_2O_3, and Mn_2O_7 less stable than Mn_2O_3.

The fact that the parting with a large number of electrons may lead to instability is not difficult to understand. Even supposing that the loss of seven electrons from a manganese atom were to lead to a structure comparable to that of the inert gas argon, it has to be remembered that the removal of successive electrons from an atom which is developing a high positive charge will involve work against electrical forces. In groups IA, IIA, IIIA the valencies are usually fixed at 1, 2 and 3, respectively, but when we pass further to the right, the valency becomes variable, the " lower " oxides or chlorides being often more stable than the higher ones; in Groups VIII, IX and X valencies of 8, 9 and 10 are rarely or never attained, and valencies met with are in most cases 2 and 3 (sometimes 2 and 4). The situation displays some irregularities, as shown in Table XXXIX, which presents the formulae of the known oxides of the elements of the first " long family " (19 to 35). The composition of the chlorides is generally obtained by halving the number of metallic atoms present in the oxide or doubling these of the non-metal (thus the chlorides of potassium and calcium are KCl and $CaCl_2$, whereas the oxides are K_2O and CaO); in some cases, the highest chloride (e.g. $CrCl_6$ and $MnCl_7$) does not seem to exist.

Where there are two fairly stable sets of compounds, the lowest set is distinguished by the termination " ous " and the higher set by the termination " -ic "; thus we have ferrous chloride $FeCl_2$, ferric chloride $FeCl_3$; platinous chloride $PtCl_2$, platinic chloride $PtCl_4$.

Whilst in the higher " A " groups, a definite rule of valency seems absent, it re-appears, to some extent, in the " B " groups. The rule obeyed seems to suggest that the atoms of Group X represent a stable structure somewhat analogous to the Inert Gases and that the elements of Groups IB, IIB, IIIB, etc. tend to lose 1, 2, 3 electrons, etc. so as to reach that stable structure. The situation is somewhat less simple, but the rule just stated serves as a rough guide in predicting the formulae of compounds, and generally, but not invariably, proves reliable. The main exceptions are provided by copper and gold, which, in addition to forming the oxides Cu_2O and Au_2O (as the rule would predict), also form higher oxides CuO and Au_2O_3. Moreover mercury, in addition to forming HgO (as the rule would predict), also forms Hg_2O; thallium, in addition to Tl_2O_3, forms Tl_2O, whilst tin and lead, in addition to SnO_2 and PbO_2, form SnO and PbO.*

The variable valency of the metals met with in the centre of the Table, is accompanied by the appearance of colour among the compounds; the compounds of the metals of Groups IA, IIA, IIIA (and some of those of Group IVA) are colourless, but colour becomes common in the succeeding groups. It varies with the valency in any given metal, but confining ourselves to compounds of valency 2, the aqueous solutions of the salts display colours usually attributed to the ions in the hydrated condition; thus Mn^{++} is pink; Co^{++} red (blue if anhydrous), Ni^{++} green, Cu^{++} blue (some of the cupric salts are green, probably owing to the presence of complex ions); in contrast, Zn^{++} is colourless, and so is Fe^{++}—despite the fact that ferrous salt solutions, as ordinarily prepared, are pale green (probably owing to slight oxidation and hydrolysis).

The oxides, most of which are nearly insoluble, display some remarkable variation of tint; chromic oxide, Cr_2O_3 is green, but the higher oxide, CrO_3 (generally known as " chromic acid ") is orange-red, and, unlike most of the oxides of heavy metals, freely soluble in water. The various oxides of manganese include MnO (pale green), Mn_3O_4 (reddish-brown), Mn_2O_3 (brown), MnO_2 (black) and Mn_2O_7 (green). Iron oxides include ferrous oxide, FeO (black), magnetite, Fe_3O_4 (black, magnetic), ferric oxide, Fe_2O_3 (reddish brown); all three iron oxides are capable of considerable variation of composition from that indicated by the formulae (p. 24). We may also note cobaltous oxide, CoO (grey), cobaltic oxide, Co_2O_3 (dark brown); nickelous oxide, NiO (green) and a black higher oxide of somewhat variable composition, usually written Ni_2O_3; cuprous oxide, Cu_2O (red), cupric oxide, CuO (black); zinc oxide, ZnO (white, becoming yellow when heated); mercurous oxide, Hg_2O (brown-black) and mercuric oxide, HgO (red); lead monoxide, known, in its usual form as litharge, PbO (pink-yellow), red lead, Pb_3O_4 (scarlet), and lead dioxide, PbO_2

* There are also intermediate " salt-like " oxides like Pb_3O_4, which probably contain both divalent and tetravalent lead—as suggested later.

TABLE XXXIX

OXIDES OF ELEMENTS OF THE FIRST LONG FAMILY

K_2O										Cu_2O					
	CaO			VO	CrO	MnO	FeO	CoO	NiO	CuO	ZnO		(GeO)		
						Mn_3O_4	Fe_3O_4								
		Sc_2O_3	Ti_2O_3	V_2O_3	Cr_2O_3	Mn_2O_3	Fe_2O_3	Co_2O_3	Ni_2O_3			Ga_2O_3		As_2O_3	
			TiO_2	VO_2	CrO_2	MnO_2							GeO_2		SeO_2
				V_2O_5										As_2O_5	
					CrO_3	MnO_3									
						Mn_2O									

often called " lead peroxide " * (chocolate-brown); titania, TiO_2, zirconia, ZrO_2 and thoria, ThO_2 (all white).

The sulphides show many analogies to oxides. Those of iron and copper show variation from the text-book compositions, owing to vacancies in the crystal structure at sites where cations ought to exist. A mineral known as pyrrhotite occurs in nature and is assigned the formula $Fe_{11}S_{12}$ in many books on mineralogy; it can be regarded as ferrous sulphide, with about one-twelfth of the sites where ferrous ions (Fe^{++}) ought to exist really vacant, whilst about one-sixth of the other sites are filled by Fe^{+++} ions, thus preserving electrical neutrality. Lead sulphide (galena, PbS) is important as an ore of lead; it crystallizes in cubes, the Pb and S particles (probably Pb^{++} and S^{--} ions) being arranged like the Na^+ and Cl^- ions in NaCl. The corresponding sulphide of zinc (ZnS)—also an important ore—crystallizes in two forms; both of them differ in structure from NaCl.

Non-metallic Elements. Most of the elements possess " metallic " properties, being excellent conductors of heat and electricity and good reflectors of light. The elements which lack these properties are mainly found in the top right-hand corner of the table. The distinction, however, is not a sharp one; in Group VB, for instance, we pass from typical non-metals (nitrogen and phosphorus) through arsenic (a non-metal possessing some metallic characters) to antimony and bismuth, which are essentially metals. The " non-metals " are mostly electron-acceptors. Hydrogen is a border-line case; it lacks typical metallic properties, and occasionally acts as an electron-acceptor, as in the hydrides LiH, NaH and CaH_2. More often it is a donor, and combines with non-metallic elements to give compounds in which the negative valency of the non-metallic element determines the formula; thus we have such compounds as hydrogen fluoride (HF), water (H_2O), ammonia (NH_3) and methane (CH_4); and again (from the elements on the line below), hydrogen chloride (HCl), hydrogen sulphide (H_2S), phosphine (PH_3) and silicon hydride (SiH_4). Other compounds deserving attention are hydrogen bromide (HBr), hydrogen iodide (HI) and hydrogen selenide (H_2Se).

Most of those elements that accept electrons from hydrogen will denote them to oxygen, which possesses an exceptionally strong affinity for electrons, and here the positive valency comes into play; thus the highest oxides of carbon, nitrogen, sulphur and phosphorus are CO_2, N_2O_5, SO_3 and P_2O_5. As often occurs at high valency numbers, lower oxides are also known; thus sulphur forms SO_2 as well as SO_3 whilst carbon forms CO as well as CO_2; the oxides of nitrogen and phosphorus include N_2O, NO_2 NO and N_2O; P_2O_3, P_2O_5. The elements of Group VIIB (known as halogens, or salt-formers) have themselves high affinity for electrons, and possess little power of combining with oxygen, although chlorine does form unstable compounds such as Cl_2O, ClO_2 and Cl_2O_7, whilst iodine, in which the non-metallic properties are less strongly developed, forms the relatively stable I_2O_5. Among the oxides of non-metallic elements many, like SO_2, N_2O, NO, NO_2, CO and CO_2, are gaseous at ordinary temperature, but others like P_2O_5, SiO_2 and B_2O_3 are solids even at elevated temperatures. P_2O_5 is used for drying gases, combining with water to form phosphoric acid, H_3PO_4; SiO_2 can be produced in a porous or surface-rich form known as " silica gel ", which also absorbs water, but here the water is held in the " adsorbed state " tightly attached to the surface, and not—it is believed—combined to form a definite compound.

Acids, Salts and Alkalis. Many of the non-metallic oxides react violently with water to yield acids; thus SO_3 yields sulphuric acid, H_2SO_4; SO_2 yields sulphurous acid, H_2SO_3; N_2O_5 yields nitric acid, HNO_3; whilst P_2O_5 yields

* Strictly, the name " peroxide " should be restricted to compounds which yield hydrogen peroxide when treated with acid; according to this criterion, PbO_2 is not a peroxide. Another term requiring explanation is " sesqui-oxide ", which is applied to oxides having the general formula M_2O_3.

metaphosphoric acid, HPO_3, which slowly combines with additional water to give orthophosphoric acid, H_3PO_4.* These acids are miscible with water, and in dilute solution dissociate into hydrogen ions (H^+, sometimes written H_3O^+, since the ion is probably really a proton associated with a water molecule) along with an anion such as SO_3^{--}, SO_4^{--} and NO_3^-. A phosphoric acid solution will in general contain PO_4^{---}, HPO_4^{--}, $H_2PO_4^-$ and unionized H_3PO_4. In " strong " acids, the dissociation (ionization) is sufficient to supply a large amount of H^+, whereas in a " weak " acid, only a low concentration of H^+ is present.

Sulphamic acid, $NH_2.SO_2.OH$, should be noticed in passing since, although possessing little interest in itself, it must be regarded as the parent body of the sulphamates which possess great value to the electro-plater.

By neutralization of acids with oxides of metals, or with hydroxides (see below), salts are produced, such as sodium sulphate, Na_2SO_4, or sodium nitrate, $NaNO_3$. The solid salts can often be obtained by evaporating the solution, and cooling; being generally less soluble in cold than in hot water, the salt " crystallizes out " as the temperature drops, but the crystals sometimes contain " water of crystallization ". Thus we obtain $Na_2SO_4.10H_2O$ from a cold solution of sodium sulphate, but the " anhydrous " (water-free) Na_2SO_4 is usually deposited if crystallization is conducted above 32·4°C.; its solubility, unlike that of the hydrate, decreases with rising temperature. Crystals of the hydrate " effloresce " when exposed to fairly dry air, losing water and yielding a powdery mass of the anhydrous salt.

The sulphate of many " divalent " metals (metals having a valency of 2) crystallize with 7 molecules of water; thus we have $ZnSO_4.7H_2O$; $FeSO_4.7H_2O$; $MgSO_4.7H_2O$ and $NiSO_4.7H_2O$; many of these " heptahydrates " are closely related in crystal structure. Copper sulphate differs from these in forming a pentahydrate $CuSO_4.5H_2O$. Calcium sulphate is met with as the hydrates, gypsum, $CaSO_4.2H_2O$ and hemihydrate $2CaSO_4.H_2O$, but also as anhydrite $CaSO_4$. Plaster of Paris is made by heating gypsum to drive off $\frac{3}{4}$ of its water; the particles of the powder obtained consist mainly of hemihydrate, which can be mixed with water and moulded into any required shape. This liquid between the grains (saturated with respect to the hemihydrate) is super-saturated with respect to the dihydrate, which consequently settles out, so that the whole becomes rigid.

Acids appear when the compounds HF, HCl, HBr and HI (formed by combination between hydrogen and the four " halogen " elements) dissolve in water. Hydrofluoric acid (HF) attacks glass; hydrochloric acid (HCl) is highly corrosive to most metals; hydriodic acid (HI) is unstable, and tends to liberate iodine.

If an attempt is made to concentrate hydrochloric acid by driving off the water through boiling at atmospheric pressure, it is impossible to go above 20·24% HCl. Acid of this composition distils unchanged; more concentrated acid first loses HCl, and finally the same constant-boiling mixture comes over; more dilute acid at first loses water and the concentration rises to the constant-boiling level. The composition of the constant-boiling acid varies with pressure, being 23·2% at 50 mm. and 18·0% at 2500 mm. mercury pressure.

The case of nitric acid is similar. Here the constant-boiling mixture contains 68% HNO_3. However, 100% HNO_3 can be made in various ways, e.g. by adding the appropriate amount of N_2O_5 to relatively dilute acid. True HNO_3 is unstable, and gives off fumes; it can be obtained colourless, but is liable to form NO_2 by decomposition, which confers a yellow or red colour. The so-called " red fuming nitric acid ", however, is produced by bubbling NO_2 through white acid; it is an unstable and highly corrosive substance, and is generally used in a state where a little water is present, which renders it less dangerous.

* Pyrophosphoric acid ($H_4P_2O_7$) also deserves notice; the pyrophosphates are mentioned in the section devoted to condensed phosphates on p. 969.

A mixture of hydrochloric and nitric acids is known as *Aqua Regia*. It has the power of dissolving noble metals like gold which resist either acid when present by itself. This is generally attributed to the formation of chlorine *in situ* by the equation $HNO_3 + 3HCl = NOCl + 2H_2O + Cl_2$. In practice, it is well to add more than three equivalents of HCl to one of nitric acid; A. J. Berry, on the basis of practical experience combined with a study of the literature, recommends a mixture of four volumes HCl (Sp. G. 1·18), one volume HNO_3 (Sp. G. 1·42) and one volume of water.

Alkalis are formed when the oxides of metals of Groups IA and IIA combine with water. Thus the oxides, Na_2O and K_2O, have so great an affinity for water that they are rarely met with in the anhydrous state; combination with water gives the " hydroxides ", NaOH and KOH, generally known as caustic soda and caustic potash; the hydroxides are soluble in water, giving solutions which dissociate or " ionize " into Na^+ (or K^+) and OH^-. The anion OH^- is the essential constituent of alkali, just as H^+ (strictly speaking H_3O^+) is the essential constituent of acid. Likewise, calcium oxide (CaO), (" quick lime "), has a great affinity for water, developing much heat when it gives $Ca(OH)_2$ (" slaked lime "), which, although much less soluble in water than NaOH, yields a strongly alkaline solution. The compound MgO has less affinity for water, but the product, $Mg(OH)_2$, although only slightly soluble, does confer an alkaline reaction upon the solution.

Salts are formed when alkalis are " neutralized " by acid; sodium chloride or sulphate can be produced thus

$$NaOH + HCl = NaCl + H_2O$$

$$2NaOH + H_2SO_4 = Na_2SO_4 + 2H_2O$$

but since both acids and alkalis are largely ionized before their interaction, we can write the changes

$$Na^+ + OH^- + H^+ + Cl^- = Na^+ + Cl^- + H_2O$$

$$2Na^+ + 2OH^- + 2H^+ + SO_4^{--} = 2Na^+ + SO_4^{--} + 2H_2O$$

Elimination of the terms present on both sides converts *either* of these equations to the simple form

$$OH^- + H^+ = H_2O$$

—a general equation which will serve to express the neutralization of *any* " strong " (i.e. highly ionized) acid by *any* " strong " (highly ionized) alkali.

The groups SO_4 and NO_3, which (when charged) form the anions of sulphates and nitrates respectively, cannot exist as uncharged molecules, since sulphur has only six electrons to bestow in assuming a stable state comparable to that of the inert gases, and nitrogen only five. Thus these groups represent ions carrying charges, SO_4^{--} and NO_3^-, and accordingly migrate to the positive pole when a sulphate or nitrate solution is " electrolysed " between platinum electrodes. The fact that the SO_4 and NO_3 groups exist as such in the solid state is brought out by the X-ray study of crystalline sulphates and nitrates; in a solid crystalline sulphate, each sulphur atom has four oxygen atoms situated immediately around it.

Salts are usually regarded as being formed from acids by replacement of hydrogen atoms by those of metal; thus the two sodium sulphates, H_2SO_4 and $NaHSO_4$, represent H_2SO_4 with both or one hydrogen atom replaced by Na. This conception of a salt receives justification from the fact that when metallic zinc is placed in dilute sulphuric acid, hydrogen gas is evolved and the solution on evaporation and cooling deposits crystals of colourless hydrated zinc sulphate, $ZnSO_4.7H_2O$.

The acidic oxides formed by elements standing on the left of the inert gas, and the alkaline oxides formed by elements on their right—are readily soluble in water. Most other oxides, however, are sparingly soluble. In contrast metallic salts (chlorides, sulphates and nitrates) are in most cases freely soluble. Some exceptions exist; silver chloride (AgCl), thallous chloride (TiCl), cuprous chloride (CuCl) and mercurous chloride (HgCl, more correctly written Hg_2Cl_2) are sparingly soluble white precipitates, formed when sodium chloride solution is added to a solution of a silver, monovalent thallium or monovalent mercury salt. Addition of a chloride to a solution of lead nitrate, $Pb(NO_3)_2$ produces a white precipitate of lead chloride, $PbCl_2$, which is not very soluble in cold water although it dissolves on heating the liquid. It should be noticed that whilst cuprous and mercurous chlorides are sparingly soluble, cupric and mercuric chlorides ($CuCl_2$ and $HgCl_2$) are freely soluble. Many other metals have two chlorides, both freely soluble. Sometimes one is much less stable than the other; thus chromic chloride ($CrCl_3$) is stable but chromous chloride ($CrCl_2$) is unstable, readily taking up oxygen or evolving hydrogen, so as to reach the chromic state.

Most sulphates are freely soluble, except those of barium and lead ($BaSO_4$ and $PbSO_4$). There are no sparingly soluble nitrates.

Other salts of interest to the corrosionist include sodium thiosulphate, $Na_2S_2O_3.5H_2O$ (which the photographer calls " hypo ") and sodium hypophosphite, $NaH_2PO_2.H_2O$, a strong reducing agent used in the electroless deposition of metals (p. 614).

A different (older) conception of salts as the combination of two oxides—one basic, the other acidic deserves mention; zinc sulphate was regarded as $ZnO.SO_3$, a combination of zinc oxide (" basic ") and sulphur trioxide (" acidic "). This " dualistic " system is only applicable to the salts of acids containing oxygen, and even in such cases does not represent the true structure of the sulphates, either in solution or in the crystalline form. Nevertheless it is sometimes a convenient method of representation, particularly in mineralogical chemistry; the rock-constituent orthoclase felspar, for instance, is sometimes written $K_2O.Al_2O_3.6SiO_2$. Again salt-like bodies are formed by combination of two metallic oxides—one less basic than the other, and functioning as though it were an acid oxide; yellow potassium chromate, K_2CrO_4, and the orange dichromate, $K_2Cr_2O_7$, can be written $K_2O.CrO_3$ and $K_2O.2CrO_3$ respectively; the idea that K_2CrO_4 is derived from an acid H_2CrO_4 is unrealistic, since it is doubtful whether H_2CrO_4 has been isolated. Nevertheless, the chromates are closely connected with the sulphates both in crystal structure and solubility, the chromates of barium and lead being sparingly soluble.

Hydrides. The hydrides formed by non-metallic elements, include neutral, acid and alkaline compounds. Water is neutral, consisting largely of unionized molecules (H_2O), although at any moment about two in 10,000,000 molecules become ionized, a proton passing momentarily from one molecule to a neighbour giving $(H_3O)^+$ and leaving OH^-; since the hydrogen and hydroxyl ions are formed in equal quantities, this does not affect neutrality. Hydrogen sulphide (H_2S) is slightly soluble in water, yielding a weakly acid solution; the salts, such as Na_2S, being formed from a weak acid and strong base, yield alkaline solutions. In contrast, ammonia (NH_3) is very soluble in water, the solution being distinctly alkaline, owing to combination with water to give NH_4OH which ionizes into NH_4^+ and OH^-. The salts, such as ammonium chloride, NH_4Cl and ammonium sulphate $(NH_4)_2SO_4$, are white crystalline bodies; the former is volatile, splitting up into NH_3 and HCl in the vaporous state, although the two molecules recombine on cooling to give solid NH_4Cl. The cation NH_4^+ shows a general analogy to Na^+ and K^+, and the salts of NH_4 are known as " ammonium salts "; the phosphonium salts, built up from the ion PH_4^+, are somewhat similar.

Cyanides and Complex Salts. The group —CN behaves in a manner recalling the behaviour of the halogens, forming hydrocyanic acid, HCN, which, however, is much weaker than HCl, HBr or HI; consequently solutions of its salts, such as potassium cyanide, KCN, react alkaline. The cyanides of the alkali metals of Group IA are soluble, but those of many heavy metals are sparingly soluble, and are generally thrown down as a precipitate when potassium cyanide is added to the appropriate salt solution. Thus cuprous cyanide and silver cyanide are white precipitates (CuCN and AgCN), recalling cuprous chloride and silver chloride; they redissolve, however, in excess of potassium cyanide, forming soluble complex cyanides with compositions KCN.CuCN and KCN.AgCN, best written $K[Cu(CN)_2]$ and $K[Ag(CN)_2]$, since they ionize into K^+ and a complex anion $[Cu(CN)_2]^-$ or $[Ag(CN)_2]^-$. These " complex cyanides ", with the metal present in the anion,* are valuable in electroplating; others used by electroplaters are $K[Au(CN)_2]$, $K_2[Zn(CN)_4]$ and $K_2[Cd(CN)_4]$. Here the complex ion containing the heavy metal (gold, zinc or cadmium) moves, on electrolysis of the solution, towards the *anode*, but nevertheless the heavy metal is deposited on the *cathode*—usually in a smoother form than would be obtained from a simple salt solution such as gold chloride (p. 605).

The complex cyanides (and indeed complex salts generally) fail to show the " reactions " of the simple salts of the metal; thus a solution of any " simple " silver salt ($AgNO_3$, Ag_2SO_4, etc.) yields a white curdling precipitate of AgCl when sodium chloride solution is added; a solution of $K[Ag(CN)_2]$ gives no such precipitate.

The complex cyanides include two remarkable compounds, once known as " yellow prussiate " and " red prussiate " of potash, but now called potassium ferrocyanide ($K_4FeC_6N_6$) and potassium ferricyanide ($K_3FeC_6N_6$) respectively. These could be regarded as double cyanides containing ferrous and ferric iron, written $4KCN.Fe(CN)_2$ and $3KCN.Fe(CN)_3$ respectively, but in solution the iron is present in the complex anions $[FeC_6N_6]^{----}$ and $[FeC_6N_6]^{---}$ and they are better written $K_4[FeC_6N_6]$ and $K_3[FeC_6N_6]$. Their solutions do not show the reactions of iron salts. Thus whereas simple ferrous and ferric salts give precipitates when treated with sodium hydroxide, the two complex cyanides give no such precipitates. They have, however, reactions of their own; thus potassium ferrocyanide yields a white precipitate of zinc ferrocyanide when added to a zinc sulphate. Copper ferrocyanide, a reddish-brown precipitate, and lead ferrocyanide, a white one, are thrown down in analogous manners. Most interesting of all are the ferrocyanides and ferricyanides of iron; blue precipitates are obtained by interaction of a ferric salt with potassium ferrocyanide or of a ferrous salt with potassium ferricyanide; ferric and ferricyanide yield no precipitate, but a deep greenish-brown coloration; a ferrous salt and a ferrocyanide yield a white precipitate, readily becoming blue through absorption of oxygen from the air.

Although the complex cyanides provide, perhaps, the most remarkable examples of anions containing a metal, many other cases are known. Cuprous chloride, which is almost insoluble in water, dissolves in concentrated hydrochloric acid to give a solution containing a compound best written $H[CuCl_2]$. The double iodide of potassium and cadmium appears to ionize as indicated by $K[CdI_3]$. Even a concentrated solution of the plain cadmium iodide, generally written CdI_2, appears to behave as $Cd[CdI_3]_2$, for the cadmium tends to move towards the anode rather than the cathode; when the solution is diluted, the complex seems to break up and movement towards the cathode becomes predominant. Detailed information regarding complex-ion formation and the extent to which complexes are dissociated is provided in " Stability Constants, Part II, inorganic ligands ", 1958 (Chemical Society, London).

The group CN is known as cyanogen, and free cyanogen, $(CN)_2$ is known as a poisonous gas. Hydrogen cyanide (HCN) is a liquid giving off an exceptionally poisonous vapour. Other cyanogen compounds include the fertilizer, calcium cyanamide, CN.NCa, formed by passing nitrogen over strongly heated

* More than one complex anion may be present in a given solution. Thus a solution of potassium copper cyanide contains $[Cu(CN)_4]^{---}$, $[Cu(CN)_3]^{--}$ as well as $[Cu(CN)_2]^-$ according to H. P. Rothbaum, *J. electrochem. Soc.* 1957, **104,** 682.

calcium carbide, CaC_2, and lead cyanamide, which has been used as a pigment in paints designed to prevent the corrosion of iron.

Salts formed by Combination of two Metallic Oxides. Examples of salts containing a heavy metal in the anion are provided by sodium stannite, sometimes written $Na[HSnO_2]$ and the much stabler sodium stannate ($Na_2SnO_3.3H_2O$, or perhaps really $Na_2Sn(OH)_6$); these contain divalent and tetravalent tin respectively. Again we have calcium plumbate, Ca_2PbO_4, an important corrosion-inhibitive pigment. The chromates and dichromates, which can be regarded as $K_2O.CrO_3$ and $K_2O.2CrO_3$, but behave as $K_2[CrO_4]$ and $K_2[Cr_2O_7]$ respectively, have already been mentioned; these are well-defined bodies, deposited as shapely crystals of fixed composition when a solution, saturated hot, is allowed to cool.

If sodium hydroxide is added to zinc sulphate solutions, a white precipitate, $Zn(OH)_2$, is formed which dissolves in excess of the alkali, forming a clear solution believed to contain a sodium zincate, which may be $Na[HZnO_2]$ or $Na_2[ZnO_2]$ —according to the amount of alkali added. Similarly alkali added to an aluminium salt solution throws down a white flocculent precipitate, dissolving in excess of alkali to give a clear liquid believed to contain an " aluminate ". Other examples could be cited, but in several cases, the complex salt has not been isolated in the solid state—nor is the composition of the complex ion always known with certainty.

Two oxides of the same metal, one exerting basic and the other acidic properties, can combine to form a " salt-like " intermediate oxide; thus red lead (Pb_3O_4) is sometimes regarded as $2PbO.PbO_2$ and magnetite (Fe_3O_4) as $FeO.Fe_2O_3$; both are almost insoluble, and their compositions often depart slightly from those suggested by the formulae.

Magnetite is an example of a group of compounds known as *spinels*, formed by combination of the oxides of a divalent and trivalent metal; they crystallize in the cubic system. The name spinel originally belonged to a mineral having the composition $MgO.Al_2O_3$ and prized as a gem-stone; but it has now been applied to the whole group compounds having the common formula $M''O.M_2'''O_3$, where M'' and M''' are divalent and trivalent metals respectively.* In practice M'' can represent a mixture of divalent metals and M''' a mixture of trivalent metals. Thus between chromite (chrome iron-stone) with composition $FeCr_2O_4$ and magnetite (which can be written $FeFe_2O_4$), we have a range of mixed spinels of intermediate compositions $Fe(Fe_{2-x})Cr_xO_4$.

Magnetite can be regarded as derived from ferrous oxide (FeO), which crystallizes in the cubic system. If one quarter of the sites which should be occupied by Fe^{++} ions are left vacant, electrical neutrality being preserved by the fact that half of the other Fe^{++} sites are really occupied by Fe^{+++} (so that 4FeO becomes $FeO.Fe_2O_3$) we get magnetite. We can remove further Fe^{++} (increasing the number of vacant sites) and finally obtain the cubic γ-Fe_2O_3; the latter, however, is unstable over a certain range of temperature and readily undergoes atomic rearrangement, producing the rhombohedral α-Fe_2O_3. Cubic solid solution intermediate in composition between Fe_3O_4 and γ-Fe_2O_3 are known. Some authorities state that true γ-Fe_2O_3 only occurs in presence of a small amount of water which cannot be removed without transformation to α-Fe_2O_3. Details of different views should be studied in the papers of G. Hägg, *Z. phys. Chem.* (*B*) 1935, **29,** 95; E. J. W. Verwey, *Z. Kristallogr.* 1935, **91,** 65; I. David and A. J. E. Welch, *Trans. Faraday Soc.* 1956, **52,** 1642. See also p. 25 of this book.

Oxides, anhydrous and hydrated. It is sometimes stated that metallic oxides " dissolve " in acids to form salts. In many cases, however, the dissolution of anhydrous oxides proceeds with extreme slowness. The " sesqui-oxides "

* Spinels are of two types, differing in crystalline arrangement of the atoms, the " normal " and the "inverse "; the inverse spinels of which magnetite is an example, have higher electrical conductivity and diffusion rates. See J. O. Edström, *J. Iron St. Inst.* 1957, **185,** 450, esp. p. 453.

(possessing the general formula M_2O_3, where M represents a trivalent metal) are extremely sluggish in their reaction with acids, unless the conditions are such as to allow reduction to the divalent condition (see p. 225). However, the " hydroxides ", formed by precipitation of salts with alkali, will quickly redissolve when acid is added. In many cases they also dissolve in excess of alkali, forming bodies analogous to the zincates and aluminates mentioned above. Only a few metals form hydroxides which are reasonably insoluble in excess of dilute alkali; these include nickel, silver and iron, although even here ferrous hydroxide is appreciably soluble in concentrated alkali.

The hydroxides can be regarded as " hydrates " of the oxides, and include cases where the O^{--} ions are either partly or completely replaced by OH^-ions. Thus aluminium forms (1) the anhydrous oxide *alumina* (Al_2O_3), which is the chief constituent of corundum and emery—also of the gems ruby and sapphire, (2) the partial hydrate, *Boehmite*, ($Al_2O_3.H_2O$, better written $AlO(OH)$), and (3) the complete hydrate, *Bayerite*, ($Al_2O_3.3H_2O$, probably better written $Al(OH)_3$). Another form of $Al(OH)_3$ is known as *Hydrargillite*.

In the case of iron, the oxides and hydroxides can exist in different states of oxidation. Thus a piece of metallic iron, heated strongly in air, develops three layers of oxide, corresponding roughly to the compositions FeO, Fe_3O_4 and Fe_2O_3, with some variation of composition within each layer (p. 24). A ferrous sulphate solution—if quite free from air—gives a white precipitate when alkali is added. This is ferrous hydroxide, $FeO.H_2O$ or $Fe(OH)_2$; it turns green if exposed to air, forming a body containing ferric as well as ferrous iron, and if sufficient oxygen is present, ultimately becomes brown. If a solution containing both ferrous and ferric salt is precipitated with alkali, the compound may be greenish or black, according to conditions; in some cases, black magnetite (Fe_3O_4) is formed. If a solution containing ferric salts alone is treated with alkali, the precipitate is a reddish-brown; ferric hydroxide occurs in different forms, according to its mode of formation. Ordinary rust often consists mainly of $FeO(OH)$, with much water held by capillarity in the porous mass (p. 494).

Details of the oxidation of ferrous hydroxide are furnished by J. E. O. Mayne, *J. chem. Soc.* 1953, p. 129, and by W. Feitknecht and G. Keller, *Z. anorg. Chem.* 1950, **262**, 61; see also the papers quoted by these authors.

Acid and Basic Salts. Whilst the neutralization of NaOH with H_2SO_4 gives a solution from which normal sodium sulphate (Na_2SO_4) can be obtained by evaporation and crystallization (hydrated or anhydrous according to the procedure adopted), the addition of a double quantity of H_2SO_4 provides a solution from which crystals of sodium acid sulphate, or bisulphate, can be obtained; it is usually obtained without water of crystallization ($NaHSO_4$) although $NaHSO_4.H_2O$ is known; gentle ignition gives the anhydro-salt $Na_2S_2O_7$. This particular anhydro-salt is not specially stable, but in the analogous case of the potassium chromates, the orange solution obtained when KOH is treated with twice as much CrO_3 as would yield the yellow K_2CrO_4 deposits on evaporation and cooling, not $KHCrO_4$, but fine crystals of the orange-red dichromate, $K_2Cr_2O_7$; a trichromate $K_2Cr_3O_{10}$ and tetrachromate $K_2Cr_4O_{13}$ are known.

Whilst acid salts have long been recognized as well-established compounds which crystallize well and usually possess definite composition, it was at one time usual to dismiss the so-called " basic salts " as adsorption compounds of variable composition; where colloidal systems are involved (p. 968), that view was probably correct in some cases, but not all. A method for establishing the existence of a true basic salt of definite composition was worked out by W. Lash Miller and F. B. Kenrick, *J. phys. Chem.* 1903, **7**, 259; see also A. Findlay and A. N. Campbell, " The Phase Rule and its Applications ", 1938 edition, p. 267 (Longmans, Green).

Today, large numbers of well-defined basic salts are known, and their crystal habits and conditions of stability have received detailed studies in the laboratory of Feitknecht (see p. 489). Compounds important in connection with corrosion problems include the two zinc hydroxychlorides, $ZnCl_2.4Zn(OH)_2$ and $ZnCl_2.6Zn(OH)_2$, and the well-known pigment zinc tetroxychromate, $ZnCrO_4.4Zn(OH)_2$, which is an important constituent of inhibitive paints.

Pasty mixtures of powdery zinc oxide with zinc chloride solution possess cement-like properties; they can be moulded into shape, and slowly attain rigidity, owing to the separation of a solid basic chloride. Similar setting properties are shown by pasty mixtures of magnesia with magnesium chloride. At one time the magnesia type of cement was largely used for the laying of floors, but it was found that steel pipes embedded in it became rapidly corroded. However, similar basic chlorides formed from mixtures of zinc dust with magnesium (or barium) chloride solution can protect steel against corrosion, provided that the zinc dust is in excess; that is the basis of " cementiferous paints " (p. 617).

In some cases basic chlorides are formed by simple " hydrolysis " (decomposition by water). Stannous chloride ($SnCl_2.2H_2O$) is a colourless solid and anhydrous stannic chloride ($SnCl_4$) a heavy liquid. Both react with water to give solutions which readily become cloudy owing to " hydrolysis " (decomposition by water) producing free hydrochloric acid with either a hydroxide of tin, or perhaps a basic salt; the solution can be clarified by adding excess of free hydrochloric acid. The production of cloudiness is not unusual when a salt of a heavy metal is added to water, but generally a small amount of free acid renders the liquid clear. In some cases, large amounts of acid are needed to dispel the cloudiness—notably in the case of bismuth salts, which interact with water to yield sparingly soluble basic salts, such as BiOCl.

$$BiCl_3 + H_2O \rightleftharpoons BiOCl + 2HCl$$

When excess of water is added to solid bismuth chloride, the left-to-right reaction occurs; sufficient addition of free hydrochloric acid causes the right-to-left reaction. The basic salt BiOCl represents a one-third-way mark on the way to converting the oxide Bi_2O_3 into the chloride ($2BiCl_3$)—just as the acid salt $NaHSO_4$ is the half-way mark in converting H_2SO_4 into Na_2SO_4.

Co-ordination Compounds. The compounds hitherto considered have, with certain exceptions, conformed to the laws of valency. The number of compounds which can be formed by " primary valency " (the transfer of electrons from one atom to another to an extent necessary for the stability of both) is necessarily limited. If we admit the existence of " secondary valency ", based on the sharing of electrons, the number of possible compounds becomes much greater, but in practice considerations of geometry interpose a limitation. It is possible to arrange symmetrically four or six molecular groups around a central atom, and the important class of compounds known as " co-ordination compounds " depends in most cases on such an arrangement. The molecular groups are frequently ammonia or a substituted ammonia (e.g. one of the amines mentioned on p. 997); here the cases where the " co-ordination number " is six are the most important. If the number of ammonia groups in the compound is less than six, it can be made up to six with (H_2O) molecules or even by anions such as Cl^-. In the latter case the Cl^- ions, being part of the closely bound set of six, lose their power to break away (ionize) when the compound is dissolved in water. A typical series of co-ordination compounds, in which only that part of the chlorine printed outside the square bracket is capable of ionizing and showing the reactions typical of Cl^-, is provided by the " ammines " of platinic chloride:—

$$[Pt(NH_3)_6]Cl_4$$
$$[Pt(NH_3)_5Cl]Cl_3$$
$$[Pt(NH_3)_4Cl_2]Cl_2$$
$$[Pt(NH_3)_3Cl_3]Cl$$
$$[Pt(NH_3)_2Cl_4]$$

In all cases the groups inside the square bracket amount to six. The first compound on the list has all its chlorine ionizable; the last one has no ionizable chlorine. Other possibilities arise, however, if water molecules enter the complex such as in

$$[Pt(NH_3)_2(H_2O)_2Cl_2]Cl_2$$

For the corrosionist, the most important cases of co-ordination compounds are met with in copper. If ammonia is added to a blue cupric sulphate solution, green cupric hydroxide is precipitated which redissolves in excess of ammonia, giving a deep violet blue solution,* containing the ammine $[Cu(NH_3)_4]SO_4$. If the precipitate is filtered off and then treated with excess of ammonia, we obtain $[Cu(NH_3)_4](OH)_2$ in solution; the violet-blue liquid thus produced has the power of dissolving cellulose and is used in one process for making rayon. The low concentration of simple Cu^{++} in an ammoniacal solution has the result that copper behaves as a less noble metal towards ammonia than towards acid.

Colloidal Hydroxides and Basic Salts. When a solution of say, potassium chloride, saturated at a high temperature, is cooled down, it becomes super-saturated and usually deposits crystals as soon as a small degree of super-saturation is attained; the concentrations of K^+ and Cl^- ions are so high, that in the course of their random wandering it is likely that sufficient ions will chance to come together in the correct positions to form a stable crystal nucleus. If, however, we consider a hydroxide like $Zn(OH)_2$, considerable super-saturation may be needed before solid appears; this is not surprising, since the equilibrium solubility is so low that even when it is greatly exceeded, the chance of one Zn^{++} and two OH^- coming into the appropriate positions must be small. When the super-saturation is sufficient, solid may be deposited, but the hydroxide thrown down is not always crystalline; in some circumstances, we obtain a gelatinous or flocculent mass consisting of " colloidal particles " clinging together as a loose net-work, but containing much water within the meshes. Such a gelatinous mass is known as a " gel " and in some gels the particles appear to be " amorphous " (non-crystalline); others consist of small particles which X-ray analysis show to be crystalline. In certain circumstances, the colloidal particles may remain separate from one another, suspended in the liquid, so that no solid or gelatinous phase is apparent, but the presence of particles much larger than molecules is detected when a narrow beam of light is directed from one side into the liquid; the light is scattered in all directions, producing the phenomenon known as a " Tyndall Cone ". Such a liquid containing colloidal particles is known as a " sol "—which should be distinguished from a genuine " solution ", a liquid containing dissolved molecules.

The power of the particles to remain separate in a sol instead of becoming flocculated to form a gel is partly due to the fact that they carry a charge and thus repel one another; a positive charge may be caused by adsorbed Zn^{++} ions, and in that case flocculation can be aided if anions carrying more than one negative charge are present, which help to discharge the positive colloidal particles. Thus when a zinc salt is titrated with sodium hydroxide, zinc hydroxide is not thrown down as a visible precipitate until the alkali is definitely in excess of the " theoretical " amount; however, the amount of alkali which can be added before cloudiness appears is greater if the main anion present is Cl^- than if it is SO_4^{--}; probably the double negative charge on the SO_4^{--} ions is specially effective in precipitating the positive colloidal particles. Some

* Many people call this solution " deep blue "; they would only use the word " violet " for a solution which transmitted red as well as blue light. However, the sensation produced by the shortest waves which affect our eyes has long been called " violet ", the still shorter (invisible) rays being called " ultra-violet "; this " violet " sensation requires no admixture of long red rays. Many " violet " dyes (e.g. gentian violet) do transmit red as well as a wide range of blue rays, with absorption in the green region, and, as the visual sensation produced is not dissimilar from that produced by pure spectral violet light, the idea has taken root that violet is necessarily a mixture of blue and red.

colloidal hydroxides can be brought into different sol-forms, carrying either positive or negative charges respectively. Thus ferric hydroxide can be produced either as a positive sol due to adsorbed Fe^{+++} ions or a negative sol due to adsorbed anions (probably OH^-). When the sol is flocculated, a brown-red gel is formed. Rust is essentially a gel of ferric hydroxide, but the particles generally consist of the partially hydrated FeOOH rather than the fully hydrated $Fe(OH)_3$; nevertheless there is much water held by capillary forces in the porous mass. It is believed that even rust which looks dry often contains much more water than the formula FeOOH would suggest. However, our knowledge is incomplete. Considering the importance of rust to the corrosionist, it is remarkable that it has not been more closely studied; further research is badly needed.

The fact that water containing the hydroxide of a heavy metal can become super-saturated many times without depositing solid probably provides the reason why pure water can be corrosive to certain metals, notably zinc. Under conditions where the building up of a protective film of zinc hydroxide might reasonably be expected, this fails, at some points at least, to appear, so that pitting develops (p. 101).

Elements building chains and rings. Certain elements standing in the right-hand top corner of the Periodic Table are peculiar in forming compounds of high molecular weight in which the elements are joined as long chains or sometimes rings. These elements include carbon, silicon and phosphorus, but whereas carbon atoms can link up directly with one another (p. 983), the atoms of silicon and phosphorus are usually joined through oxygen. Thus the structure of the large group of compounds known as the " condensed phosphates " is generally believed to involve the grouping

$$\begin{matrix} & O & & O & & O & \\ —O & P & O & P & O & P & O— \\ & O & & O & & O & \end{matrix}$$

—although in many cases there is uncertainty as to details.

The best developed crystalline phosphates are the orthophosphates of which the sodium salts are $Na_3PO_4.H_2O$, $Na_2HPO_4.12H_2O$ and $NaH_2PO_4.H_2O$; by loss of water, however, we get *pyrophosphates* such as $Na_4P_2O_7$ or *metaphosphates* such as $NaPO_3$. The latter generally exist in polymeric forms, which differ from one another as regards solubility and structural properties according to the mode of preparation; some of them are glassy, and can generally be written $(NaPO_3)_x$, although they in some cases depart slightly from that composition. Glassy sodium metaphosphate was prepared by Proust in 1820 by heating sodium ammonium hydrogen orthophosphate $NaNH_4HPO_4$; the loss of water and ammonia left a mass with a composition roughly $NaPO_3$.* The commercial compound generally known as *Calgon* was at one time known as " hexametaphosphate " and written $(NaPO_3)_6$, but it is now believed to be more complicated. Other condensed phosphates include the triphosphate $Na_5P_3O_{10}$. Many of these various bodies are used in water treatment to prevent scaling (their complex-forming power hinders solid calcium compounds from appearing on the walls of a pipe or container). They are also used as corrosion-inhibitors, for which purpose the presence of a calcium salt is required. Detailed information regarding this large group of compounds is provided by B. Topley, *Chem. Soc. Quarterly Review* 1949, **3,** 345.

* A. J. Berry furnishes these details. Berzelius first obtained sodium pyrophosphate, $Na_4P_2O_7$, by the action of heat on disodium orthophosphate in 1816, and gave its composition correctly in terms of the dualistic formulation. Sodium pyrophosphate was more thorouglhy studied by Clark in 1827. In 1820 Proust obtained a sodium metaphosphate by strongly heating microcosmic salt ($NaNH_4HPO_4$); the loss of water and ammonia left a mass with a composition roughly $NaPO_3$. The various phosphoric acids and their salts were first studied systematically in 1833 by Graham. He prepared the acid pyrophosphate $Na_2H_2P_2O_7$ by heating monosodium orthophosphate, NaH_2PO_4, and on heating more strongly obtained the metaphosphate $NaPO_3$. As regards terminology, the term *pyro* was introduced by Clark, and the term *meta* by Graham; the term *ortho* was not introduced by Graham, but by Odling.

Even more complicated are the silicates, which are important in petrology. Whereas many stratified rocks consist mainly of simple compounds (sandstone is largely SiO_2 and limestone largely $CaCO_3$), the igneous rocks are aggregates of many complicated silicate minerals, each of which may contain numerous metals replacing one another in varying degree, so that the composition of a single mineral may show fluctuation over a wide range. Some may be regarded as derived from such simple acids as H_4SiO_4 or H_2SiO_3, by replacement of H by metals; others seem to be basic salts of $H_2Si_2O_5$.

The silicate minerals which mainly concern the corrosionist are those where the molecular groups are arranged in strings or sheets—a structure which is reproduced in the external form of the particles. Thus asbestos shows a string-like structure, and is much used in the production of fire-resistant cloth and felt; also, impregnated with bitumen, for the protection of buried pipes. *Asbestine*, a fibrous material prepared from a variety of asbestos, is often added to paints to prevent the settlement of heavy pigments in the pot; but the name " asbestine " is also applied to varieties of talc. A sheet structure is met with in mica and talc, which are related to one another; their tendency to form thin flakes is well known. Mica and talc powder are used in paints and bituminous protective compositions. Talc can be assigned the formula $Mg_3(Si_2O_5)_2(OH)_2$, but the magnesium may be partly replaced by other metals; the crystal structure shows sheets of tetrahedra and octahedra, explaining the lamellar tendency. Another mineral showing flaky structure is china clay, which is sometimes written $Al_2Si_2O_9H_4$—probably really $Al_2(Si_2O_5)(OH)_4$; this is also used in certain paint compositions (p. 619). Like many clay minerals it is formed by the degradation of igneous rocks. A very useful mineral of the clay class is *Bentonite*, which is employed, mixed with gypsum, as a back-fill into which magnesium anodes can be embedded for cathodic protection processes. Its value is largely due to its " thixotropic " properties; it will absorb many times its own volume of water forming a mixture which, although rigid enough to withstand small forces becomes liquid in character when violently stirred or shaken; the rigidity assumed when it is left undisturbed is due to the settling of the particles into some arrangement which resists mild deforming forces. Despite its rigidity, the open structure and the large amount of liquid in its meshes allows a high conductivity—hence its utility as a back-fill (see also p. 288).*

The zeolites are a group of porous silicates possessing an open three-dimensional frame-work of $(SiAl)_nO_{2n}$ anions with cations such as Na, K, Ca or Mg in the cavities balancing the negative charge. The structure is sufficiently open to allow water to permeate through the mass, and since the cations present in the water can change places with those in the zeolite, it is possible to soften water containing Ca^{++} and Mg^{++} ions, which are replaced by Na^{+} ions; later the mass is revivified with a solution of sodium chloride. The softness of natural water emerging from a green-sand formation is due to the presence of a mineral of this class (glauconite) in the green-sand. Synthetic zeolites can be made, but in modern water-softening process base exchange depends on resins carrying acidic groupings, which replace Ca^{++} or Mg^{++} ions in percolating water by H^{+} ions (p. 433).

Silicates are also important as components of cements. The setting of cement is a complicated reaction involving the hydration of certain calcium silicates and aluminates, with the liberation of lime, so that the mixture gradually develops alkalinity during the setting process. The production of alkali (a corrosion-inhibitor) during setting is clearly favourable to the protection of steel buried in concrete or other mixtures containing cement; but chlorides are sometimes added to concrete to accelerate the setting reaction, and this may introduce dangers; the matter is considered on pp. 303–306.

Glasses. The compounds of silicon and several elements standing close

* Recent information regarding the structure of clay minerals is provided by C. B. Amphlett, *Endeavour* 1958, **17**, 149.

to it in the Periodic Classification show a tendency to assume the glassy condition. Elements which can form cations of small size and high electrical charge, such as silicon, boron, phosphorus and arsenic, all show this tendency; a few elements like vanadium, capable of forming cations of high charge but standing in a remote part of the table, also form glasses. Most glasses in use today contain more than one element besides oxygen, but silica (SiO_2) and boric oxide (B_2O_3) can assume the glassy condition in the absence of a second oxide.

It is today generally believed that silica, whether crystalline or glassy, contains the silicon and oxygen atoms (perhaps better described as silicon and oxygen ions) joined in an infinite net-work, each silicon being in contact with four oxygen atoms whilst each of the oxygen atoms serve to connect two silicon atoms; this accords with the fact that the valency of silicon is 4 and that of oxygen 2. It is possible to arrange the atoms so that the net-work is regular and continuous, each silicon atom being in a similar position relative to its neighbour as any other silicon atom; in that case the structure is crystalline. Alternatively, the net-work can be made irregular or random, the meshes being of unequal sizes; in that case the silica is vitreous or glassy; when heated it exhibits no sharp melting-point, because the energy required to break down the atomic linkages differs in different parts of the irregular net-work. Commercial glass contains metallic elements—notably sodium and calcium, and sometimes potassium, lead and/or barium. In general, the metals are present as cations in the interstices; since this would upset the electrical neutrality of the system, some of the silicon atoms (really cations) are missing from the net-work, thus providing the necessary excess of oxygen anions to neutralize the metal cations. Some glasses contain boron, and this probably is placed on the net-work, like the silicon, and not in the interstices. A good discussion of glass will be found in the report of Symposium on the Vitreous State held in 1953 in the Department of Glass Technology of Sheffield University.

At one time, a sharp distinction was drawn between (1) " crystalline " substances where the particles (atoms, ions or molecules) were supposed to be arranged in definite orderly array and (2) glassy substances from which all systematic arrangement was considered to be absent and which were regarded as super-cooled liquids. However, the modern view is that even in the liquid state there is a tendency towards the arrangement of particles into orderly array, although above the melting-point the lattice-defects are so numerous that the material can flow under weak forces, such as gravity, and thus assume the shape of the containing vessel. When a liquid is cooled down, two changes may occur: (*a*) the vacant sites may become so few in number that the structure will not subside under gravity, so that the liquid becomes a solid, (*b*) the atomic arrangement may become regular, so that the substance behaves towards light and X-rays in a manner characteristic of a crystal. In many substances, these two changes occur at the same temperature, which constitutes a sharp melting-point, above which the *crystalline solid* becomes liquified. Sometimes the first change occurs more readily than the second, and we get on cooling the liquid a *glassy solid*, which is rigid enough, but possesses optical structures different from that of a true crystal, although there is evidence of tendencies to local crystallinity; the liability of glass to " devitrify " is only too well known to chemists. Sometimes, the second change occurs more readily than the first and we get *liquid crystals*.

Those interested in the properties of glass should consult R. W. Douglas, *Trans. British Ceramic Soc.* 1954, **53**, 748; *J. Soc. Glass Technology* 1947, **31**, 50.

For the corrosionist, glass possesses importance in connection with protective coatings. Occasionally metallic vessels are equipped with a thick glass lining, but the brittleness of the material limits the utility of that method. It is more usual to apply a thin layer of vitreous enamel—essentially a pigmented glass; the pigment must remain solid and undissolved at a temperature at which the glass vehicle becomes liquid.

To withstand highly corrosive liquids, ceramic vessels may be used in the place of metallic vessels. Of the main types of ceramics, porcelain, which has been fired at such a temperature that the mixture of silicate minerals has become sufficiently fused to form a non-porous largely vitreous (glassy) mass is the most useful; the surface usually carries a glaze, or glass layer, which is truly molten at the temperature of firing. Earthenware, where the interior consisting of particles of clay minerals which have been heated sufficiently for them to adhere together but not so strongly as to abolish porosity, is generally unsuitable for dealing with corrosive chemicals; if it is used, a glaze is, of course, essential. Stoneware, mentioned on p. 265 as a common material for the ducts in which telephone cables are laid, is intermediate in character; the body is essentially non-porous but not vitreous; a glaze is normally applied.

Analytical Chemistry

Volumetric Analysis. Given the formula of any compound and the atomic weights of its component element, its exact composition can easily be calculated. Thus sodium chloride with formula NaCl contains 22·99 grams of sodium and 35·46 grams of chlorine in 58·45 grams of sodium chloride; 58·45 is said to be the " molecular weight " of sodium chloride; simple calculation shows that sodium chloride contains 39·3% and 60·7% of sodium and chlorine respectively. Similarly silver nitrate, $AgNO_3$, with molecular weight 169·89 (107·88 + 14·01 + (3 × 16)) contains 63·5% of silver. If a solution containing 58·45 grams of sodium chloride interacts with one containing 169·89 grams of silver nitrate, there should be complete mutual precipitation of the chlorine and the silver; if precipitation is complete, the " filtrate " produced after filtering off the silver chloride should be substantially free from either chlorine or silver.

We thus arrive at a method of estimating the concentration of a sodium chloride solution of unknown strength. It is convenient to use a solution containing the molecular weight, or some simple fraction thereof, expressed in grams, dissolved in exactly one litre of solution. If one gram-molecular weight is present in a litre, the solution is called " molar " (M); if one tenth of the gram-molecular weight, it is " deci-molar " (M/10). Thus 10 c.c. of, say M/10 silver nitrite will require exactly 10 c.c. of M/10 sodium chloride for complete interaction. If, however, we run in M/10 silver nitrate from a burette into 10 c.c. of sodium chloride of unknown concentration and if we find that the end-point, corresponding to complete removal of chlorine as silver chloride, occurs when x c.c. of M/10 silver nitrate has been introduced, the concentration of the unknown solution is clearly $\frac{x}{10} \times \frac{M}{10}$ or $\frac{x}{100}$ times molar.

In order to decide when the end-point corresponding to exact removal of the chlorine has been reached, several methods are available. One depends on the fact that silver chloride is a precipitate which " curdles ", so that, if the sodium chloride is contained in a stoppered bottle and is vigorously shaken between successive small additions of silver nitrate, it is possible to judge whether the next addition has produced a cloud of fresh uncurdled silver chloride or not; when a fresh addition fails to produce a cloud, that is a sign that the end-point has been reached.

This method—which needs some experience—was at one time preferred for the estimation of silver at refineries, using a chloride solution of known concentration. For the corrosionist who wishes to determine the chloride concentration of a liquid, a method depending on an indicator may be easier and perhaps more accurate, since the liquid under examination may contain something which affects the curdling properties of silver chloride. A few drops of potassium chromate solution may be added to the liquid containing chloride and then silver nitrate is run in until the precipitate is no longer white but buff; the buff colour is a sign that white silver chloride is no longer being precipitated, and that the precipitation of silver chromate, a reddish-brown compound, has started; silver chromate, being appreciably more soluble than silver chloride, will only begin to be precipitated after the whole of the chloride has been removed.

The chromate method is regarded by some authorities as inaccurate, but good results can be obtained if the pH value and the chromate concentration are regulated in the manner recommended by A. J. Berry and J. E. Driver, *Analyst* 1939, **64**, 730.

Alternatively, an adsorption indicator may be used. The dyestuff fluorescein renders a silver chloride precipitate pink whilst Ag^+ ions are in excess of the Cl^- ions, since Ag^+ will be adsorbed on the AgCl particles conferring a positive charge on them, so that they attract the coloured anions of the fluorescein. As soon as Cl^- ions come to be in excess, the AgCl particles become negatively charged, the anions are released and the colour turns to yellow.

In estimating the concentration of an unknown solution of an acid, it is usual to " titrate " it with an alkali solution of known strength, that is to run in such a solution from a burette until the acid has been exactly neutralized. In the case of hydrochloric acid and sodium hydroxide, 10 c.c. of M alkali will exactly neutralize 10 c.c. of M acid; in the case of an acid of unknown strength, the concentration is easily calculated from the volume of alkali required to produce neutrality. There is here a choice of indicators which change colour at the end-point, and for titrating a strong acid with a strong alkali, the old-fashioned litmus (which is red in acid and blue in alkali) is still suitable, but is today rarely used. For acid-alkali titrations either phenol-phthalein or methyl orange is suitable in different cases, but many other indicators are available for special purposes. The theory of the colour change is briefly explained on p. 1034.

In neutralizing M sulphuric acid (a dibasic acid), twice the volume of molar alkali is needed to neutralize it, for the equation now is

$$2NaOH + H_2SO_4 = Na_2SO_4 + H_2O$$

This slightly complicates the arithmetic, and since it is convenient to employ solutions of strengths chosen to make equal volumes interact with one another, " normal " solutions are employed instead of " molar " solutions. Normal (N) or deci-normal (N/10) sulphuric acid contain respectively half the amounts of acid present in an M or M/10 solution; thus whereas M H_2SO_4 contains 98·09 ((2 × 1·01) + 32·07 + (4 × 16)) grams per litre, N H_2SO_4 has only 49·04 g./l.; 49·04 is said to be the " equivalent weight " of sulphuric acid. For hydrochloric acid M and N represent equal concentrations, since the molecular weight and equivalent weight are the same.

It might be thought that a normal solution of sodium hydroxide could be prepared by weighing out 40·00 (22·99 + 16 + 1·01) grams of NaOH, dissolving it in water, and making up the total value to 1 litre (1000 c.c.);* also that a normal solution of sulphuric acid could be prepared by weighing out 49·04 grams and making up to a litre. Neither procedure is accurate, since both sodium hydroxide and sulphuric acid absorb water with avidity, and the material in the bottles does not correspond exactly to the compositions NaOH and H_2SO_4. It is better to weigh out sodium carbonate (Na_2CO_3)† a salt which " reacts " alkaline to most indicators, being formed from a strong alkali and a weak acid (carbonic acid, H_2CO_3, the acid present in a solution of carbon dioxide, CO_2)

* For most purposes, 1 litre can be regarded as equal to 1000 cubic centimetres. It has become usual to write ml. instead of c.c., thus indicating a volume which is exactly 0·001 litre.

† The sodium carbonate is best prepared by gently heating the bicarbonate ($NaHCO_3$) until the weight becomes constant. All accurate volumetric analysis requires that the solutions employed shall be either made from " weighable " substances of well-defined compositions, or " standardized " against solutions made from such compounds. For guidance in analytical methods see G. Charlot and D. Bézier, " Quantitative Inorganic Analysis ", translation by R. C. Murray (Methuen), and W. F. Hillebrand and G. E. F. Lundell " Applied Inorganic Analysis " (Wiley).

" reacts " alkaline to many indicators. By titrating with acid of unknown strength, the concentration of the acid solution can be determined, and it can then be used to standardize a solution of sodium hydroxide, if desired. For such a purpose, the choice of indicator becomes important. Litmus can still be used, but in that case the carbonate solution must be kept *boiling* when the acid is added from the burette, so that the carbon dioxide may be expelled as quickly as it is formed by the reaction

$$Na_2CO_3 + H_2SO_4 = Na_2SO_4 + H_2O + CO_2$$

The need for heating is avoided if the indicator chosen is methyl orange—an indicator which is turned red by the strong acid, H_2SO_4, as soon as the latter has been added in such a quantity that it is present in excess, but not by the weak acid, H_2CO_3.

A molar (or deci-molar) solution is a term which has a perfectly definite meaning, since each compound possesses a single molecular weight; however, the " equivalent weight " depends on the reaction under consideration and the statement that a solution is N or N/10 only acquires a definite meaning if it has been made clear what reaction it is expected to undergo. Thus if a solution of ferric chloride ($FeCl_3$) were to be used to estimate an unknown silver nitrate solution, a normal solution would clearly be equal to an M/3 solution, since the molecule contains three chlorine atoms; if it were used to estimate a stannous chloride by a reaction such as

$$2FeCl_3 + SnCl_2 = 2FeCl_2 + SnCl_4$$

a normal solution would be equal to a molar solution, since each molecule furnishes only one chlorine atom available to oxidize the stannous to the stannic condition.

The change just mentioned can be written in ionic form

$$2Fe^{+++} + Sn^{++} = 2Fe^{++} + Sn^{++++}$$

and is an example of an oxidation-reduction reaction. Ferric chloride is the oxidizing agent and stannous chloride the reducing agent; the ferric salt is said to undergo reduction to ferrous chloride, whilst the stannous chloride undergoes oxidation to the stannic chloride ($SnCl_4$). It will be evident that here the words " oxidation " and " reduction " are being used, not to denote the addition or removal of oxygen, but an analogous change involving addition or removal of chlorine or other non-metal.

Oxidation-reduction reactions are used for the volumetric estimation of metals like iron which occur in two states of oxidation. In absence of chlorides, potassium permanganese solution of known strength is run in from a burette to an acidified solution containing ferrous iron. So long as Fe^{++} is in excess, the permanganate is decolorized, being reduced from MnO_4^- to Mn^{++}; when a pink appears, that is a sign that the end-point has been reached.

If chlorides are present, the permanganate method cannot be used, since hydrochloric acid itself reduces permanganate, producing chlorine. Dichromate is in such a case a suitable oxidizing agent and is run in in small quantities; after each addition, a drop of liquid is removed on a glass rod to a porcelain tile on which drops of potassium ferricyanide solution have been placed as a " side indicator "; so long as ferrous iron is in excess, a blue precipitation is produced.

Methods of ascertaining the end-point based on side indicators are time-consuming and unsatisfactory. In titrating a ferrous salt with dichromate, barium diphenylamine sulphonate may be added to the ferrous solution as an internal indicator; so soon as the dichromate is in excess, a purple colour appears.

If the iron contains both ferrous and ferric iron, the total iron can be determined by reducing the whole to the ferrous state and then titrating as above. The reduction can be carried out either by passing sulphur dioxide into the faintly acidified liquid followed by a stream of carbon dioxide to eliminate the sulphur dioxide; or alternatively by adding stannous chloride in slight excess, followed by mercuric chloride to remove the residual stannous chloride. By determining the total iron and the ferrous iron in two separate fractions of the same liquid, the ferric iron is obtained by subtraction. Other methods are described in the analytical text-books.

Colorimetric and Nephelometric Methods. Where a very low concentration of a metallic ion has to be determined (e.g. where some liquid has been in contact with the solid metal for a short time and it is desired to ascertain the amount of corrosion), a reagent is added which gives a colour with that ion, and the intensity of the colour is compared with that of equal volumes of liquid to which different (known) amounts of the ion have been added. With care this should enable the concentration in the unknown liquid to be assessed with fair precision. The method is known as *colorimetric* analysis. Similarly if a small concentration of chloride in a water which has become contaminated with sea-water is to be determined, silver nitrate is added, and the cloudiness compared with that produced in samples containing known amounts of chloride; this is known as *nephelometric* analysis.

In both cases, it is customary to use apparatus based on photo-electric cells, in performing the determination, and doubtless this makes things easier and possibly more accurate. The use of such apparatus, however, in no way dispenses with the need to take certain obvious precautions. The unknown liquid under examination and the known liquids used for comparison must have the concentrations of all their constituents the same except for the metal under estimation, since the presence of a third substance may affect the intensity of the colour. Similarly, the temperature must be the same. The coloration produced by adding thiocyanate to a ferric salt varies greatly with temperature; if the unknown is at a different temperature to the known comparison liquids, an error will be introduced. Finally all reagents must be tested for metal impurities. Some samples of thiocyanate contain traces of ferric hydroxide, and give clear solutions in water, which become pink when acidified; there are clearly possibilities of great error here. For helpful instruction see Chapter XVII of the Charlot-Bézier book mentioned on p. 973 or T. R. P. Gibb's "Optical Methods of Chemical Analysis" (McGraw-Hill).

Metals and Alloys

The Metallic State. It has been stated that solid sodium chloride is built up, not from sodium atoms and chlorine atoms, but from sodium ions and chlorine ions. It is perhaps equally true to say that metallic silver is built up, not from silver atoms, but from silver ions and free electrons. There is one important difference between the two cases; in sodium chloride, the negative chlorine ions occupy fixed positions on the lattice between the sodium ions, whereas in silver, their negative counterparts (the electrons) are mobile—as is demonstrated by the good electrical conductivity of metallic silver.

When sodium chloride is placed in contact with water, both Na^+ and Cl^- ions enter the liquid in the hydrated state; when silver is placed in water, silver ions can, if the electrode potential between water and metal is suitable, enter the liquid, leaving their electrons behind. This is why corrosion of metals is an electrochemical phenomenon; silver has no solubility as atoms—only as ions. If the aqueous phase is a solution of silver nitrate, it is possible to adjust the potential between metal and solution at such a value that the passage of ions from metal to liquid is exactly balanced by passage from liquid to metal; even in this equilibrium condition there is believed to be a rapid passage proceeding in both directions—which accords with the idea that metallic silver contains ions "ready-made".* These matters are considered in Appendix II.

* With certain "abnormal metals", passage in both directions proceeds much more slowly; see p. 1035.

Alloys. Two metals can " alloy " with one another giving either *solid solutions* (mixed crystals) or *intermetallic compounds* (intermediate phases) to which a formula can be assigned; in general, such a formula presents no obvious relationship with the valencies as shown in oxides, sulphides and chlorides—a matter discussed later.

In some alloy-systems, where the two metals have their atoms arranged on the same kind of crystal lattice and where the radii of the two atoms do not differ by more than about 15%, it is possible to obtain a continuous series of solid solutions containing the two metals A and B in any proportion from 0% B to 100% B; in general, the lattice parameter varies gradually as B atoms are introduced in the place of A atoms. In cases where the size relationships are less favourable, the mutual solubility will be limited; solid solutions (mixed crystals) may still exist but only over certain ranges of composition. Thus from O to x% B, there may be a single phase—a solid solution of B in A; from y% to 100% B we get another single phase—a solid solution of A in B; between x% and y%, the alloy will be a mosaic of two phases, namely solid solution A

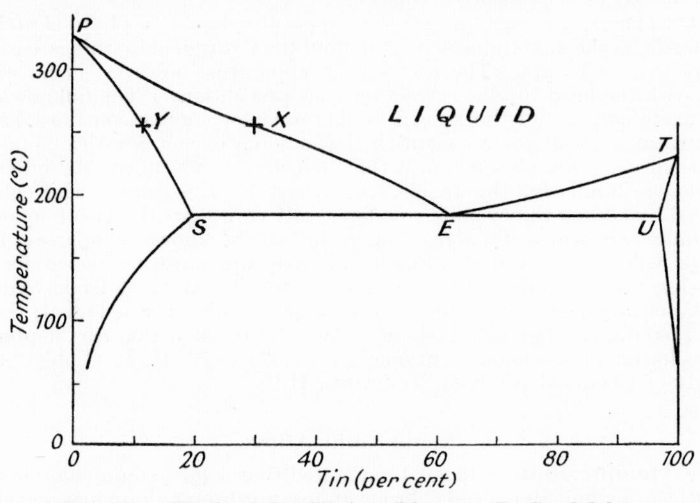

FIG. 191.—Tin-Lead Equilibrium Diagram (D. Stockdale).

containing x% of metal B and solid solution B containing $(100 - y)$% of metal A.

An appreciable amount of mutual solid solubility is often found even when the two metals have their atoms arranged according to different crystal systems. Alloys of tin (tetragonal) and lead (cubic) consist of a single phase when the tin content is sufficiently low or sufficiently high; between certain limits two saturated solid solutions are present. The presence of tin dissolved in molten lead lowers the temperature at which the liquid is in equilibrium with the solid. Pure liquid lead is in equilbrium with pure solid lead at 327°C.; at that temperature the passage from liquid to solid is exactly balanced by the passage from solid to liquid. If the liquid metal contains dissolved tin the rate of passage from liquid to solid is greatly reduced since not all the particles hitting the interface consist of lead, whereas the passage in the opposite direction is altered to a smaller extent, since the solid phase contains relatively little tin.

If a liquid containing 30% of tin is cooled down (fig. 191), it only starts to deposit solid (under equilibrium conditions) at a temperature represented by point X—well below the melting point of pure lead; since the solid deposited (point Y) contains only about 13% of tin, tin accumulates in the liquid, and this accumulation continues as the temperature continues to fall. If equilibrium is preserved between solid and liquid, the composition of the solid is given by the "solidus" line (*YS*), each point on which represents the composition in equilibrium with the liquid of composition represented by the point on the "liquidus" (*XE*) at the same horizontal level. Similarly an alloy of composition 90% tin, 10% lead, deposits crystals of a solid solution rich in tin, of composition represented by the solidus (*TU*), whilst the liquid is represented by the liquidus (*TE*), the temperature sinking as solidification proceeds.

Either alloy on reaching the temperature corresponding to the "eutectic point" (*E*) should start to deposit *both* solid solutions as a eutectic mixture, since liquid metal of the eutectic composition (61·9% tin) is in simultaneous equilibrium with both solid solutions. Thus the structure, after cooling just below the eutectic temperature, is as follows:—

Alloys up to 19·5% tin consist of a single phase of lead containing different amounts of tin in solid solution.

Those with 19·5 to 61·9% have primary crystals of lead (containing 19·5% of tin) embedded in a matrix of eutectic.

That with 61·9% tin is entirely a eutectic mixture.

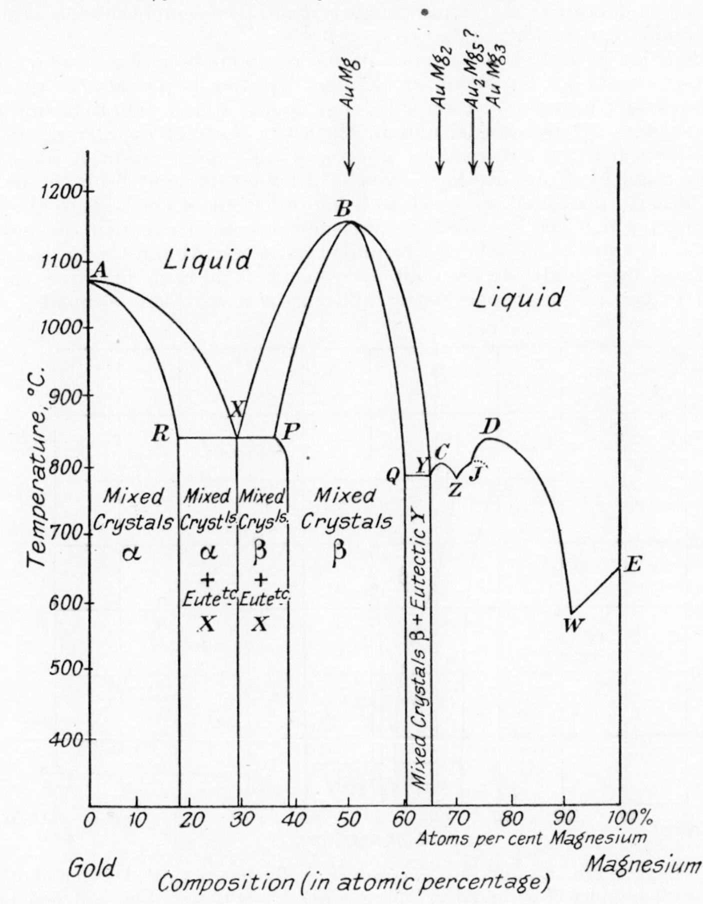

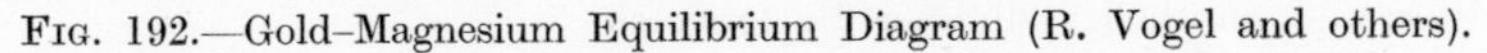

FIG. 192.—Gold–Magnesium Equilibrium Diagram (R. Vogel and others).

Those with 61·9% to 98% have primary crystals of tin (with 2% lead in solid solution) along with eutectic.

Those with 98 to 100% consist of a single phase of tin containing varying small amounts of lead in solid solution.

When the material is cooled to room temperature, the range of solid solubility declines further. If the cooling is sufficiently slow, an alloy containing 19·5% tin will throw out small amounts of a solid solution containing less than 2% lead—as suggested in the fig. 191, which, however, refers to alloys cooled so slowly as to maintain equilibrium; for rapidly cooled alloys the limiting compositions and structure will be modified.

Eutectic mixtures occur in many alloy systems. Their micro-structure varies, but a zebra pattern consisting of alternative streaks of the two phases is common.

Other alloy systems have phases in the centre of the diagram derived from intermetallic compounds. The gold-magnesium system is shown (somewhat incompletely) in fig. 192. The compounds AuMg, $AuMg_2$ and $AuMg_3$ appear at maxima on the curves; their melting-points are lowered if the liquid contains either more gold or more magnesium than corresponds to these compositions, and the solids deposited also depart slightly from these compositions, as shown by the solidus curves for AuMg.

In this alloy system, the formulae of the compounds acting as parents for the various sets of solid solution are simple, but this is not always the case. Even when simple formulae are met with, they appear at first sight to be unrelated to the valencies of the metals, and in the early days of metallography, the formulae seemed to be governed by no law at all. More recently, rules have been discovered by Hume-Rothery, which, although different from the laws of valency hitherto presented, do serve to bring order out of chaos; thus the compounds from which the so-called β-solid solutions are formed in the systems Cu–Zn, Cu–Al and Cu–Sn have the formulae CuZn, Cu_3Al and Cu_5Sn, but these compositions (seemingly diverse) all correspond to a ratio of three valency electrons to two atoms (copper being taken as a monovalent element).

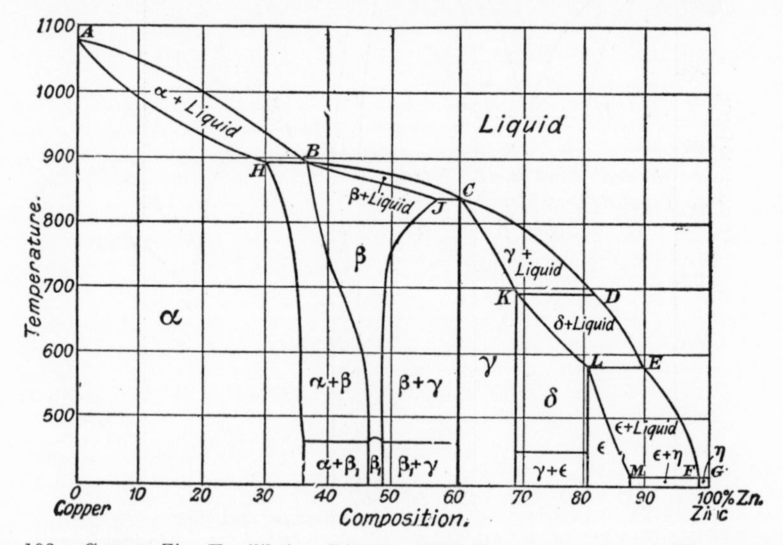

FIG. 193.—Copper–Zinc Equilibrium Diagram (E. S. Shepperd, C. H. Desch and others). A revised diagram will be found in "Metals Reference Book" Vol. I, p. 386, (1955) (Editor, C. J. Smithells; publisher, Butterworth).

In cases where the two alloying metals have very different melting-points, the diagrams may appear to exhibit no maxima. Maxima probably exist on the meta-stable parts of the curves but cannot be realized in a state of equilibrium. An example is provided by the diagram for the copper-zinc alloys (brasses), shown in fig. 193.

The aluminium-copper alloys are important as showing the property of *age-hardening*. Part of the diagram is shown in fig. 194. An alloy with under 4% copper should consist of a single solid solution at 500°C. and, if very slowly cooled, would deposit a second phase, forming a stable two-phase alloy. If " quenched " from 500°C., it remains a single-phase material, but the structure is unstable, and if " aged " (stored at ordinary temperature for some days) or " temper-hardened " (heated for some hours at a temperature well below the range where the single-phase structure would become stable), the hardness and strength increases. This is ascribed to a local re-arrangement of atoms which can be regarded as the initial stages of a movement towards a stable structure; if the temper-hardening is carried on too long, the added strength disappears, since it is the initial stages of the re-arrangement which give the desired improvement of properties. It is generally thought that the groups of re-arranged atoms tend to anchor the dislocations in the structure which would have to move if the material were to be deformed and thus hinder the deformation (p. 372). Most age-hardening or temper-hardening alloys contain other metals—added for different purposes; thus H15 usually contains about 0·85% Mg, 0·9 Si and 1·2% Mn in addition to about 4% Cu.

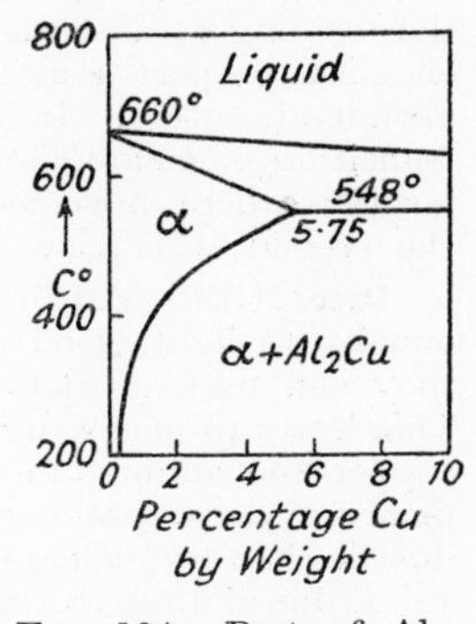

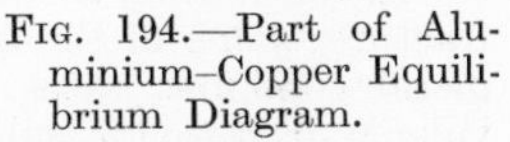

Fig. 194.—Part of Aluminium–Copper Equilibrium Diagram.

Alloys containing Transition Elements. In general, the physical properties of a continuous series of solid solutions vary gradually as one atomic species replaces the other, but in certain alloy-systems there is a sharp break at a definite composition—which can be explained from our knowledge of atomic structure. Considering the elements in order of their atomic numbers (H, He, Li, B, etc.) we see that the atoms are built up by the addition of electrons one by one. These electrons are believed to arrange themselves in shells; in general, a shell is built up around a central nucleus until it is complete (the element being then inert) and after that a new shell is started. However, at certain places in the periodic table a greater stability is obtained if a new outer shell is started before the previous one is complete, so that atoms are formed having unfilled places in an inner shell. The elements 21 to 28 (Table XXXVII, p. 954) all have vacant places in the so-called D-shell or D-band (which only becomes filled at element no. 29, Cu); they are called by physicists the *Transition Elements* (at one time chemists used that term in a more restricted sense to denote the elements of Groups VIII, IX and X). The transition elements are for the most part those which behave abnormally in electrochemical reactions. The atoms of normal elements can easily shed electrons from the outer shell and thus pass smoothly as ions into the liquid, whereas with the transition elements (abnormal metals) the formation of an ion involves a change requiring considerable " activation energy "—as shown by the electro-chemical differences between the two classes (p. 907).

If now we alloy a transition element (having vacancies in the D-band) with a B-Group element (one that easily sheds its valency electrons), electrons from the latter pass into the former, filling up the D-band. When a sufficient quantity of the B-Group element has been added to fill the D-band completely, there is a sudden change in properties—which has been demonstrated in the alloy systems Ni–Cu and Pd–Au (Ni and Pd being transition elements whilst Cu and Au are B-Group elements). At that point, the magnetic susceptibility becomes small, the reflecting power, electrical conductivity and lattice spacing all undergo alterations, whilst the specific heat also shows peculiarities. The matter is important to corrosionists because it has been suggested that not only the physical properties (depending on the interior atoms of the material) but also the corrosion-behaviour (depending on the surface atoms) should change at this composition also. There is nothing unreasonable about such a suggestion, but the existing evidence is rather against the occurrence of an abrupt change in chemical properties at the composition where the physical properties show an abrupt alteration. It has also been suggested that an alloy composed of two transition elements should show a change of chemical behaviour at the composition where movement of electrons from the first exactly fills the D-band of the second; this view is not widely held at present.

Interstitial Additions. Metals have certain characteristic properties: opacity to light, good reflectivity, high conductivity for electricity and heat; they can undergo deformation instead of fracture when subjected to stress. This latter property, however, depends on the crystal-arrangement; metals like copper and aluminium crystallizing in the cubic system, which provides many planes of dense packing along which gliding can occur, are more malleable and ductile than hexagonal metals, like zinc and magnesium, which have few planes suited for gliding, so that a force applied in an unfortunate direction can cause fracture; some of the intermetallic compounds are extremely brittle. When a metal combines with a non-metal to form a compound obeying the simple valency-laws (oxide, sulphide, chloride), it loses most of its " metallic properties " (although certain oxides like magnetite, containing two types of ions which can exchange electrons, possess electrical conductivity).

It is, however, possible for a metal such as iron to take up limited quantities of certain non-metals (carbon, nitrogen, hydrogen) in interstitial positions between the atoms of the lattice, and in that case the metallic characteristics are retained. The mechanical properties are modified in a manner which may or may not be welcome; in general, the resistance to gliding is increased, so that the material may grow stronger but too often becomes appreciably brittle.

Steels (fig. 195). As is well known, the atoms of pure iron are arranged

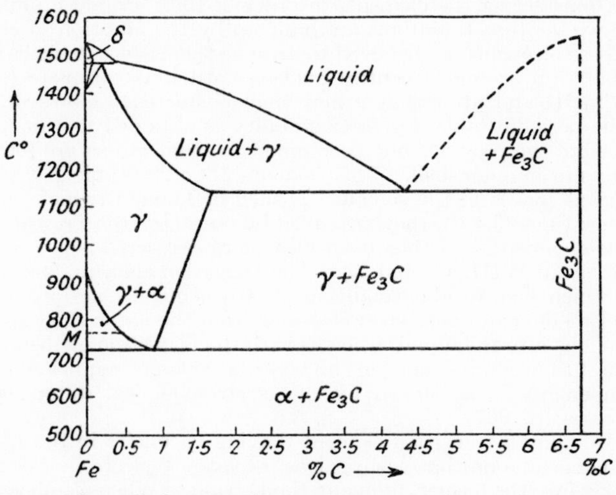

FIG. 195.—Part of Iron–Carbon Equilibrium Diagram. The region where α-iron holds carbon in solid lies close to the vertical axis and is not shown; the maximum solubility is 0.035% at 723°C., decreasing to 0.007% at room temperature.

on a face-centred cubic lattice between 906°C. and 1400°C.; the metal then known as γ-iron. Below 906°C. they are arranged on a body-centred cubic lattice; it is then known as α-iron. The body-centred lattice again becomes stable above about 1400°C.; it is then known as δ-iron. Certain alloying elements cause the γ-range to become narrower and finally to disappear; thus in iron containing over 14% chromium, the α- and δ-ranges merge, and there is no γ-region (fig. 196). Elements which cause the γ-range to become narrower are Cr, W, Mo and Si; others (Ni, Mn, Co) cause it to become broader.

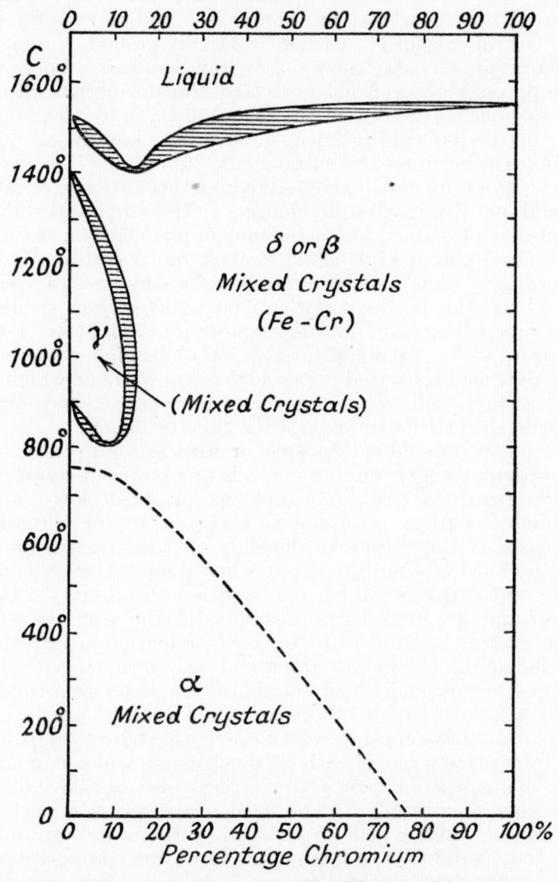

FIG. 196.—Iron–Chromium Equilibrium Diagram.

γ-iron will take up carbon atoms in interstitial positions, but the carbon solubility in α-iron is very small (0·035% at 723°C.). If γ-iron containing less than 0·83% carbon is cooled, it starts to deposit almost carbon-free α-iron at a temperature which decreases with the carbon content, so that as the carbon concentration rises in the remaining γ-phase, the temperature of equilibrium between the *austenite* (γ-iron with carbon) and the *ferrite* (α-iron almost free from carbon) steadily falls. Austenite containing more than 0·83% carbon deposits *cementite*, an iron carbide of formula Fe_3C, and the equilibrium-temperature drops as deposition continues. In either case, when the composition of the remaining γ-phase has reached 0·83% carbon, it changes at 723°C. to a

" eutectoid " of ferrite and cementite. A eutectoid is analogous to a eutectic but is formed in the solid state. Sometimes the eutectoid formed in steel displays the same zebra structure met with in eutectics and is then called lamellar pearlite. The speed of cooling, however, affects the structure obtained; the metallographist distinguishes a number of structures obtained by different cooling-conditions, giving them special names such as *troostite* and *sorbite*; the different structures are associated with different mechanical properties. He also describes the alloys above 0·83% carbon as *hyper-eutectoid* steels and those with less than 0·83% carbon as *hypo-eutectoid* steels; those with low carbon contents—around 0·2% carbon—are commonly known as *mild steels*.

If a steel with about 0·83% carbon is rapidly cooled, the eutectoid transformation occurs at an abnormally low temperature. Very rapid cooling (quenching) suppresses the normal eutectoid transformation altogether, but a different transformation occurs about 350–150°C., and the steel assumes a characteristic needle-like appearance in sections known as " martensite ", associated with great hardness; by subsequent " tempering " (controlled heating) of hardened steel at a temperature chosen to give the desired properties (generally between 230 and 300°C.), martensite changes to troostite, and the steel becomes less hard but also less brittle. The quenching of plain carbon steels may produce internal stresses and perhaps quenching cracks, but if certain alloying elements are present, ordinary slow cooling produces the same result as quenching in carbon steel. Thus the hardening which on plain carbon steels would come about only on quenching, can in steel containing 7% nickel be obtained on cooling at a rate which involves no risk of quenching-stresses. Chromium, which on very slow cooling actually *raises the temperature* at which the eutectoid transformation occurs, *reduces the velocity* of the change, and chromium steels may show martensitic structure even after slow cooling.

Alloying elements are added to steel for several purposes. For the corrosionist the importance of chromium in steel is its power of conferring resistance against high-temperature oxidation and also against low-temperature wet corrosion. Stainless cutlery steel contains 11 to 13% of chromium and 0·25 to 0·30% of carbon; the structure depends on heat-treatment and contains martensite in the hardest conditions, but when tempered may consist of aggregates of ferrite and carbides (which may contain chromium). The steels with 18% chromium and 8% nickel are austenitic in the state in which they are used and have greater resistance to corrosion than the cutlery steel just mentioned. The nickel is commonly described as an austenite stabilizer, and certainly the presence of nickel does enable the single-phase austenitic structure to be obtained in alloys which otherwise would contain ferrite. However, in 18/8 steel, the nickel plays another role in preventing corrosion (p. 340); attempts to ecionomize by replacing nickel with cheap elements which can act as austenite stablizers (e.g. manganese or even nitrogen) have led to alloys less satisfactory in regard to corrosion-resistance. The carbon also plays a part in stabilizing the austenitic phase, and, when (in order to prevent intergranular corrosion) the carbon content is reduced to a very low level, the composition may require adjustment if a purely austenitic structure is desired. Certain additions to the 18/8 type of steel, like molybdenum or silicon, which increase corrosion-resistance to certain reagents, cause the structure to become partly ferritic. Another phase, the so-called *sigma phase* (which occurs also in steels containing iron, chromium and no other metal at sufficiently high chromium-contents) is very liable to be present in the stainless steels containing molybdenum.

Mild steels are often classified into *killed* and *rimming*; an intermediate state is described as *balanced* steel. Killed steel is obtained if a sufficient supply of some element with a high affinity for oxygen is added to the molten steel before it is cast into ingots for rolling, so that at the moment of casting there is no effervescence. Rimming steel ingots are made from steel finished under an oxidizing slag, so that it enters into the ingot mould in an effervescent condition, probably due to the dissolved oxygen (or iron oxide) acting upon the carbon (or carbide) present and producing carbon monoxide. The effervescence keeps the steel well stirred, so that a rim of relatively pure iron is found around the outside of the ingot, and this survives when the metal is rolled into sheet. Rimming steel produces smooth sheet when rolled out, the two faces consisting of material purer and softer than the interior; such sheet is useful for tinplate and motor-car bodies, but it is considered that rimming steel should not be used for ships' rivets. The reason for the deposition of the rim of relatively pure iron is seen on turning back to fig. 195 (p. 980) which shows that the solid first deposited contains less carbon than the bulk of the liquid; it also contains less of the other minor constituents. Provided that the liquid is kept well stirred by the effervescence, columnar crystals are not likely to develop, and solidification, starting at the walls of the mould, should produce a pure rim. Under stagnant conditions, which may be expected with killed steel, columnar crystals will grow out from walls, with high-carbon liquid accumulating in between them; in due course this entrapped liquid will solidify so that in the end the outer part of the ingot may have much the same composition as the interior.

Iron-carbon alloys containing about 2 to 4% carbon are mainly suitable for castings and are then known as *cast iron*; when cast into bars or *pigs* of a type suitable for charging into a steel furnace with a view to conversion into steel by removal of the carbon, the material is known as *pig iron*. (In modern practice, the molten iron from the blast furnace is usually conveyed direct to the steel furnace, and is not allowed to solidify into pigs.)

Cast iron, if quickly cooled from the liquid state may consist largely of a eutectic of cementite and pearlite, with some pearlite representing primary austenite; it is then *white cast iron*. If the liquid is more slowly cooled, the cementite (which is an unstable phase) largely decomposes into graphite and iron, the product being then *grey cast iron*, which is less hard than the white variety. The composition of the iron as well as the cooling-rate is important in deciding between a grey or white structure. Silicon favours the formation of graphite, so that a silicon-rich iron will be grey after cooling at a rate where low-silicon iron would be white; in contrast, sulphur helps to stabilize cementite and causes the iron to be whiter, harder but more brittle. Even in grey iron, however, the presence of flaky graphite is the cause of planes of mechanical weakness in the material, and considerable success has been achieved at producing cast iron free from graphite in flakes. The graphite can be obtained in spherulitic form by adding magnesium to the liquid metal, and the material is then relatively ductile and is believed to be less liable to pitting corrosion. Pearlitic iron, with much of the carbon present as pearlite, can be obtained by adding small amounts of nickel (0·5 to 2%) and often chromium (up to 0·8%). Heat-resistance can be improved by adding aluminium and considerable amounts of silicon (about 6%).

Large amounts of silicon (14 to 17%) added to cast iron confers resistance to acids, but the material is very fragile (p. 347).

Carbon and Organic Compounds

Chain and Ring Compounds. Carbon stands in the Periodic Table half-way between two inert gases; either the loss or gain of four electrons should confer stability. The question as to whether carbon shows a positive or negative valency would seem likely to be decided by the other element with which it combines; certainly in carbon dioxide (CO_2) it is the oxygen, as usual, which captures the electrons. When one carbon atom joins to another they tend to share electrons. Diamond is built up or carbon atoms forming a structure such that if any given carbon atom is considered to be at the centre of a tetrahedron, the lines joining it to its four nearest neighbours are the lines pointing in the direction of the corners of the tetrahedron; each of the four bonds attaching it to its neighbours are considered to represent a *pair* of electrons, one contributed by each of the two carbon atoms joined by the band in question. Thus there are sufficient electrons around each carbon to bring it to a stable structure similar to that of an inert gas. Such an arrangement of carbon atoms can be continued indefinitely, and the hardness of diamond is a sign that the bonds produced by the pairing of the electrons can be very strong.

Sometimes the linkage between carbon atoms is represented thus $:\!\ddot{\underset{..}{C}}\!:$ the electron pairs being shown, but it is simpler to show carbon as having four bonds available to join it to other carbon atoms, $\begin{array}{c} | \\ -C- \\ | \end{array}$, where each line represents a pair of shared electrons. The links, however, do not really lie in the plane of the paper, but point as already explained, towards the corners of a tetrahedron.

Whilst the diamond structure represents a pattern which can be continued for an indefinite distance in three dimensions, graphite (another form of carbon) owes its properties to the arrangement of carbon atoms in infinite sheets, such that the carbon atoms form the corners of a continuous pattern of hexagons.

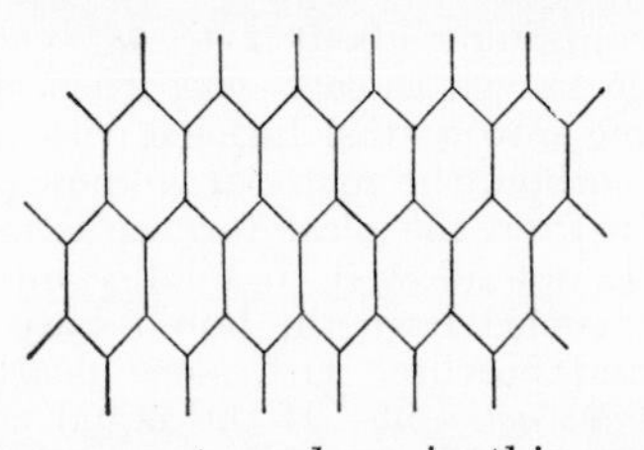

The whole of the linkages are not used up in this way, since each carbon is joined to only three other carbons. If there are a number of parallel sheets, forces are available to hold them together, but insufficient to prevent one sheet from sliding on another. It becomes easy to see why graphite is a good lubricant.*

Carbon compounds show certain characteristic features which deserve notice. One is the power of the atoms to join by sharing electrons (instead of an electron being captured by one atom from one to another) so that the linkages are largely non-polar, and possess no positive or negative ends. The other is the tendency to form long chains or rings of atoms (the rings being usually hexagonal or pentagonal, containing 6 or 5 atoms). Neither of these characteristics are peculiar to carbon compounds. The sharing of electrons occurs in the diatomic gases H_2, N_2, O_2, whilst the power to form chains or rings has been met with in the compounds of silicon or phosphorus—elements which stand close to carbon in the Periodic Table; in both those cases, however, the chains or rings contain oxygen atoms—as already explained.

Chain Hydrocarbons. Since each carbon atom throws out four bonds, a single carbon atom can combine with four hydrogen atoms, giving CH_4. If, however, the hydride of carbon (or hydrocarbon) is to contain two carbon atoms in the molecule, then one of the bonds from each carbon is needed to join them together, and the formula becomes $CH_3.CH_3$ or C_2H_6; similarly with three carbons we have $CH_3.CH_2.CH_3$ or C_3H_8 with four carbons $CH_3.CH_2.CH_2.CH_3$ and so on.

$$\begin{array}{ccccc} & & H & & \\ & & | & & \\ H & - & C & - & H \\ & & | & & \\ & & H & & \end{array} \qquad \begin{array}{ccccccc} & & H & & H & & \\ & & | & & | & & \\ H & - & C & - & C & - & H \\ & & | & & | & & \\ & & H & & H & & \end{array} \qquad \begin{array}{ccccccccc} & & H & & H & & H & & \\ & & | & & | & & | & & \\ H & - & C & - & C & - & C & - & H \\ & & | & & | & & | & & \\ & & H & & H & & H & & \end{array}$$

* Other compounds possessing lubricating properties show an arrangement in layers, including molybdenum sulphide, MoS_2, where the layers display a similar hexagonal honeycomb pattern, with the molybdenum atoms at alternate corners of each hexagonal, and the pairs of sulphur atoms at the other corners.

Thus we arrive at a " homologous series ", conforming to the general formula C_nH_{2n+2}, of which the first six members bear the names:—

Methane, CH_4; ethane, C_2H_6; propane, C_3H_8; butane, C_4H_{10} pentane, C_5H_{12}; hexane, C_6H_{14}.

It should be noted that two butanes exist, both having the composition C_4H_{10} but differing in structure, thus:—

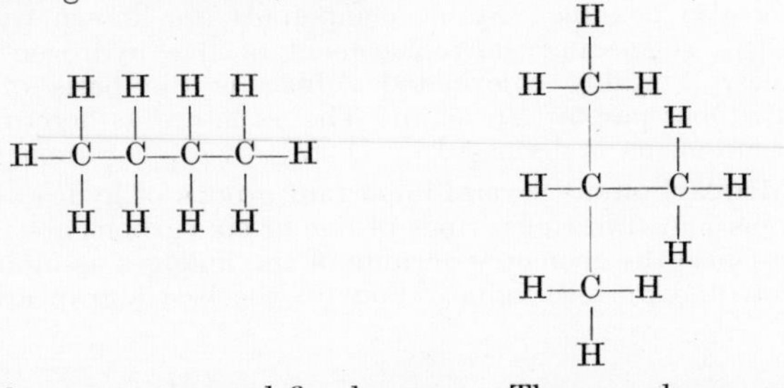

There can be three pentanes and five hexanes. These are known as *isomerides*; two isomerides, having the same molecular formula but different structural formulae usually differ slightly in physical properties (they have different boiling-points), but more strongly in chemical properties. Other examples of isomerism will be met with later.

The C_nH_{2n+2} series of hydrocarbons are known as the *Paraffins*. In such compounds all the bonds available are utilized, and it is possible for the bonds to extend outwards at their natural angles, as though pointing to the corners of a tetrahedron. Both facts contribute to the stability of paraffins, which form no addition compounds with bromine, although it is possible, under suitable circumstances, to form replacement compounds. On the other hand, there exists another homologous series, known as *Olefines*, with the general formula C_nH_{2n}, in which one pair of carbons are joined by two bonds. This double bonding does *not* increase stability, as might perhaps be expected, but decreases it, since the bonds are distorted; they cannot point outwards in their natural directions and the fact that the combining power of carbon atoms is not fully used up indicates the possibility of *adding on* atoms. Ethylene, $CH_2{=}CH_2$ or C_2H_4, the first of the olefines, *quickly* yields $CH_2Br.CH_2Br$ when brought into contact with bromine (many of the reactions involving carbon compounds proceed with extreme *slowness*). The general equations for the two types of reactions are

$$\text{(Replacement)} \quad C_nH_{2n+2} + Br_2 = C_nH_{2n+1}Br + HBr$$

$$\text{(Addition)} \quad C_nH_{2n} + Br_2 = C_nH_{2n}Br_2$$

This power of combining readily with bromine to give additive compounds is the typical feature of a double bond, and it was largely to explain this unusual reactivity that the concept of the double bond was introduced.

The series, C_2H_{2n-2}, which may contain a triple bond (or alternatively two double bonds), shows still greater instability and reactivity; the first and best known, acetylene (C_2H_2), was formerly used as an illuminant and is still used in welding.

As emphasized above, saturated hydrocarbons (paraffins) cannot add on chlorine or bromine, but it is possible to substitute Cl or Br atoms for one or more of the H atoms. Thus we have the series methane, CH_4; methyl chloride, CH_3Cl; methylene dichloride, CH_2Cl_2; chloroform $CHCl_3$; carbon tetrachloride, CCl_4—the first two being gaseous and the last three liquid at room temperature. Substitution may also occur in unsaturated hydrocarbons; examples include

Trichlorethylene: $CHCl{=}CCl_2$

which is useful in the degreasing of metallic surfaces, and

Monochlorethylene or vinyl chloride: $CH_2{=}CHCl$

—a gas which easily polymerizes (p. 1002) giving the well-known resin, polyvinyl chloride (P.V.C.); vinyl alcohol and vinyl acetate, analogous bodies with OH or acetate groups instead of Cl, also polymerize to form resins.

The hydrocarbons present in rubber have the empirical formula C_5H_8 but are really polymers; they can take up HCl when a solution in chloroform or benzene is treated with dry hydrogen chloride, but substitution of hydrogen by chlorine is also possible. Stable compounds are known with a composition $(C_{10}H_{11}Cl_7)_n$, suggesting the replacement of five hydrogen atoms in two C_5H_8 molecules. Suitably chlorinated rubber is the basis of the valuable " chlorinated rubber paints " (p. 575). The structure as determined by infra-red light measurements is discussed by D. L. Davies, *Verfkroniek* 1959, **32**, 82.

Ring Hydrocarbons. Several important groups of hydrocarbons are built of carbon atoms joined in rings; rings of five or six carbons are the most stable —probably because the geometry permits of the linkages assuming angles close to the natural (undistorted) value. Four of the best-known are

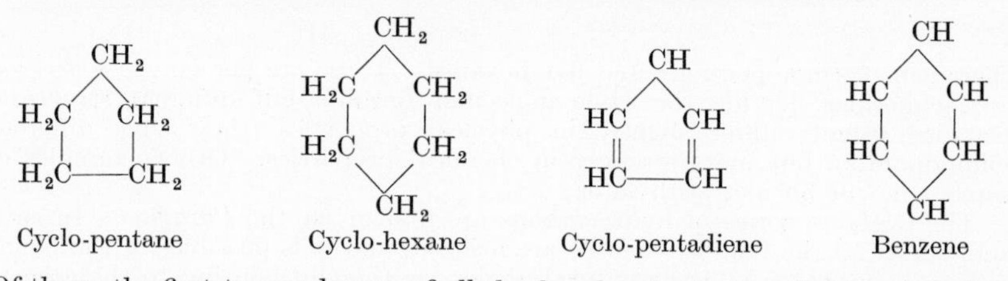

Of these, the first two make use of all the four bonds starting from each carbon, and their stability requires no further explanation. The comparative stability of benzene may seem surprising, since each of the carbons is using only three of its four linkages. Various proposals have been made for the disposal of the fourth bond. It was once thought that alternate carbons are joined by double bonds, as in I, but since there is no reason to choose I in preference to II, it is generally thought that the combining power represented by the extra bond of each atom is distributed among its two neighbours—a conception sometimes expressed by formula III.

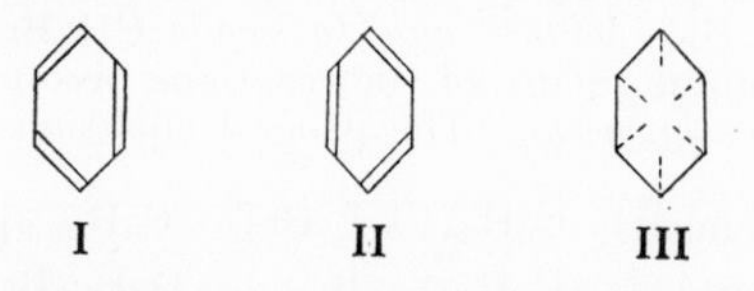

Chemists use in such cases the word *resonance*, which seems to imply that the bonds are *vibrating* between I and II as end-positions and III as mean positions; what is really meant is that chance of the molecule behaving as though it were I is equal to that of it behaving as though it were II.*

In diamond, each carbon atom is joined to its four neighbours and the three-dimensional pattern thus formed is capable of infinite continuation in all directions; thus diamond is a solid, and remains solid even when strongly heated (provided that oxygen, which would burn it to CO_2, is excluded). In contrast, each carbon atom of methane, being joined to four hydrogen atoms, has used up all its combining power, and there is no " primary " valency force left to cause the CH_4 molecules to join together further. However, any particle in

* X-ray analysis and electron diffraction show that in diamond and single-linked aliphatic compounds the carbon-to-carbon distances are 1.54 Å, whilst in doubly-linked and trebly-linked aliphatic compounds they are 1·33 Å and 1·20 Å respectively. In compounds containing benzene rings and also in graphite (which consists of sheets of hexagons) they are 1·42 Å—suggesting a linkage intermediate between the single and the double bond.

which positive and negative electricity occurs at different points has a slight stray field, and modern physics teaches that this will often be stronger than simple electrostatic ideas would predict. At all events, even a complete and stable molecule like CH_4 can join itself to others by means of " secondary " or " van der Waals " forces, but the pair will only remain in union if thermal motion is very slight. Thus methane is a gas at ordinary temperatures and will only condense to a liquid if it is cooled to a very low temperature. The secondary forces become more important as the chains become larger, and thus the condensation-point of the gases (or boiling-point of the liquid) becomes higher; in pentane (C_5H_{12}), we meet with a substance which is liquid at ordinary temperatures. The highest members of the hydrocarbon series are solid. Paraffin wax is a mixture of higher hydrocarbons, whilst the various petroleums found below the ground in oil fields are essentially liquid mixtures of hydrocarbons (often paraffin, olefine and ring compounds are found together); natural gas is a similar mixture of the earlier members of the series.

In writing the formulae for benzene derivations, a hexagon is taken to represent a ring of 6 carbon atoms with a hydrogen joined to each, *except* where some other atom or group is printed, which is then understood to have *replaced* the hydrogen at the point; thus

(phenol and toluene) are abbreviations for

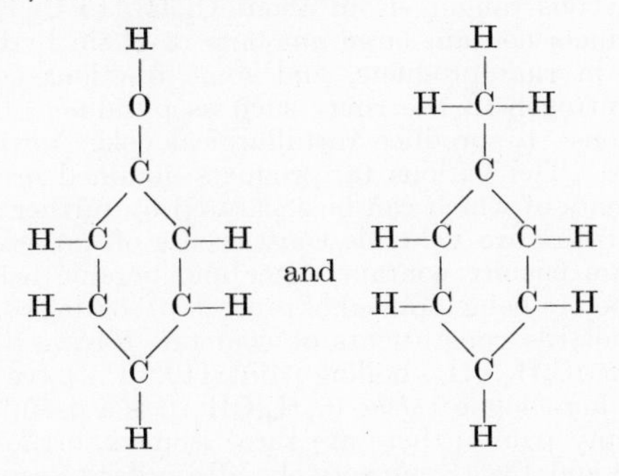

However, in heterocyclic groups containing nitrogen or sulphur, the N atom or S atom replaces CH in the ring as in Pyridine C_5H_5N, expressed as [hexagon with N] and Thiophene C_4H_4S [pentagon with S]. Hydrocarbons containing " fused hexagons " include Naphthalene $C_{10}H_8$ [two fused hexagons], Anthracence $C_{14}H_{10}$ [three fused hexagons] and Phenanthrene also $C_{14}H_{10}$ [three angularly fused hexagons, two positions marked *]. Replacement of carbon by nitrogen at the two positions asterisked gives ortho-phenanthroline, a useful reagent for iron. With ferrous iron it gives a red cation $[Fe(C_{12}H_8N_2)_3]^{++}$; the ferric cation is pale blue.

The fusion of the hexagons can be carried much further, so that the ratio of C to H may become very high; the products thus obtained are black, and compounds of that character occur in pitch, the black body left in the retort when petroleum products or coal are distilled. Graphite itself, consisting of large sheets of fused hexagons, can be regarded as the limiting case of ring-fusion.

Many of the substances used in producing anti-corrosive preparations are obtained during the distillation of oil or coal. The black pitch left in the retort is largely a mixture of non-volatile hydrocarbons containing much carbon and little hydrogen; many pitches are glassy or semi-glassy in character and resemble glass (and certain plastics) in showing a sharp transition of physical properties (density, specific heat, thermal conductivity and dielectric constant) at a certain temperature, often situated 60 or 70° below the point at which they soften perceptibly; below the transition temperature the molecules lose their translatory and rotational freedom, and become virtually locked in a disordered position.

Pitches can be dissolved in a solvent (possibly obtained during the same distillation process as has left the pitch in the retort) to yield protective paints drying by evaporation; or they can be fluxed with other products of the distillation process to give mixtures which are liquid when hot but solid when cold; these can be applied hot to metal to give relatively thick protective coatings (p. 280).

Of the products which pass over during the distillation of oil, *white spirit* is the fraction boiling between about 140 and 230°C.—that is, in the range higher than motor fuel (petrol, gasoline) and lower than the oils used for general fuel and lubrication purposes; it somewhat resembles petrol, possesses a characteristic but not unpleasant odour, and is much used as a solvent and thinner in the paint industry. Among the later fractions, *paraffin wax* is a mixture of mainly saturated hydrocarbons ranging from about $C_{23}H_{48}$ to $C_{30}H_{62}$.

Whilst oil products contain large amounts of chain hydrocarbons, coal-tar products are rich in ring products, and some fractions contain heterocyclic compounds with nitrogen in the ring—such as pyridine. Coal is distilled for a number of reasons—to produce metallurgical coke, town gas, and several industrial gas-fuels. The various tar-products obtained are complicated mixtures, the components of which can be separated by further distillation. Some of the black liquid tars are valuable constituents of anti-corrosive paints, but since the crude tars usually contain water and organic acids, they are often treated with lime before being applied as protective coatings on a metallic surface.

Of the more volatile constituents of coal tar, *benzene* (C_6H_6, boiling-point 80·2°C.) and *toluene* ($C_6H_5.CH_3$, boiling-point 110.6°C.) have already been mentioned. The next homologue *xylene* ($C_6H_4(CH_3)_2$) is a useful solvent and serves as a thinner in many paints; there are three isomers, ortho-, meta- and para-, boiling at 144, 139 and 138°C. respectively. In order to separate these various compounds in the pure state, several distillations are needed; a single distillation gives fractions which are a mixture, each boiling over a range of temperatures. For many purposes, the separation of the individual compounds is unnecessary, and indeed some of the most useful products are mixtures boiling over a wide range of temperatures. The fraction of coal tar boiling over the range 160 to 190°C. is a valuable solvent and thinner and is generally called *Solvent Naphtha.*

Many of the terms are used loosely. The name *Bitumen,* although defined in one dictionary as meaning the " solvent-extractable constituents of coal ", is more generally used to denote black solid mixtures of hydrocarbons as obtained from oil-fields or oil-refineries; it is rich in chain-hydrocarbons as opposed to the ring-compounds present in coal-tar products. Some of the black solid hydrocarbons occurring in the ground are useful components of protective compositions. *Gilsonite,* mined mainly in Utah, provides one example; *asphalt* (impure bitumen) coming from the famous Pitch Lake of Trinidad or elsewhere is another.

Wood tar, and other wood products, are also of interest to the corrosionist. *Oil of turpentine,* obtained by distilling the gummy exudation of pine trees, is still used as thinner for oil paints—although for many purposes it has been displaced by white spirit. It contains several hydrocarbons, notably α- and β-pinene, both of them two-ringed hydrocarbons of formula $C_{10}H_{16}$. The resinous substance left in the still is called *rosin* or *colophony*, and consists largely of abietic acid ($C_{20}H_{29}.COOH$), a complicated compound with three fused hexagon rings.

Emulsions. Although oils and waters are proverbially immiscible and separate into two layers if stirred together and allowed to stand, it is possible, by adding a suitable " emulsifying agent ", to obtain an emulsion consisting of very small globules of oil surrounded by water as the continuous phase; alternatively, by choosing a different type of emulsifying agent, we can obtain an emulsion of water particles with oil as the continuous phase. In the oil-in-water emulsions, the particles often carry negative charges which help to prevent them joining together; this tends to render the emulsion stable. The emulsifying agent is often a substance the anionic portion of which can be attached to the oil particles, thus contributing their negative charges; this type of emulsion is often cloudy, as the particles are large enough to scatter light. The water-in-oil emulsions usually contain substances which stabilize the water–oil interface, so that at room temperatures the water particles do not spontaneously unite, causing " demulsification " (separation into two phases); the nature of the substances introduced by individual manufacturers to stabilize the emulsion have not always been disclosed, but it is understood that long-chain amines along with cresols are often useful constituents. The water-in-oil emulsions are often clear, the particles being too small to scatter light, but when added to a large volume of water, they " invert " to give a cloudy oil-in-water emulsion, in which the oil particles are relatively large. In some cases, anti-corrosive properties are conferred by such additions.

Alcohols, aldehydes, acids. Besides hydrocarbons, numerous other homologous series of organic compounds are known; they can be regarded either as hydrocarbons with one or more hydrogen atoms replaced by some group such as —OH (in the alcohols), —CHO (in the aldehydes), or —COOH (in the acids). Alternatively they can be described as combinations of one of these groups with such radicals as methyl (CH_3), ethyl (C_2H_5), propyl (C_3H_7), butyl (C_4H_0), amyl (C_5H_{11}) and so on; these radicals, derived from paraffin by loss of one H atom, bear the generic name of *alkyl* radicals, and can be expressed by the general formula (C_nH_{2n+1}); the radicals derived from aromatic (ring) hydrocarbons, are called *aryl* radicals, examples being phenyl (C_6H_5) and benzyl ($C_6H_5.CH_2$). Most of the lower acids of the *aliphatic* or fatty series (chain-like without rings) are liquid at ordinary temperatures, and possess characteristic odours; some of the *aromatics* (ring-bodies acids) are colourless solids—for instance, benzoic acid $C_6H_5.COOH$.

The *alcohols* are not easily produced from the hydrocarbons, but methyl alcohol (CH_3OH) is the chief constituent of wood spirit, obtained by the dry distillation of wood; in modern practice, it is more largely made from a mixture of carbon monoxide and hydrogen by the reaction $CO + 2H_2 = CH_3OH$. Ethyl alcohol (C_2H_5OH) is obtained on the fermentation of sugar, and is the intoxicating constituent of many fermented drinks; the presence of methyl alcohol causes a liquid based on ethyl alcohol to be unsuitable for drinking, and methyl alcohol is accordingly added to duty-free alcohol intended solely for industrial purposes. Some drinks contain small amounts of " fusel oil ", which is a mixture of propyl alcohol (C_3H_7OH) and higher alcohols. The various constituents of fermented liquors can be separated by distillation.

By oxidation of alcohols, two hydrogen atoms are removed, leaving *aldehydes*; formaldehyde (H.CHO) can be obtained from methyl alcohol (CH_3OH) by passing its vapour mixed with air over a heated platinum spiral, whilst acetaldehyde ($CH_3.CHO$) is obtained from ethyl alcohol ($CH_3.CH_2OH$) by oxidizing it with chromic acid, CrO_3 (or a mixture of potassium dichromate with sulphuric acid, which is equivalent to CrO_3).* Further oxidation of the two aldehydes yields formic acid (H.COOH) and acetic acid (CH_3COOH) respectively, by up-take of an additional oxygen atom. This second oxidation-step proceeds more readily than the first—so that the aldehydes, although representing a higher state of oxidation than the alcohols, are more effective as reducing agents.

The structure of the aldehydes is easily understood if it is recollected that two linkages are necessary to bind an oxygen atom but only one to bind hydrogen. Thus ethyl alcohol can be written out in full as I and acetaldehyde as II; in old books acetic acid is represented as III but it is now considered that the position of the bonds joining hydrogen to oxygen " resonates " between the two oxygen atoms—as suggested in IV

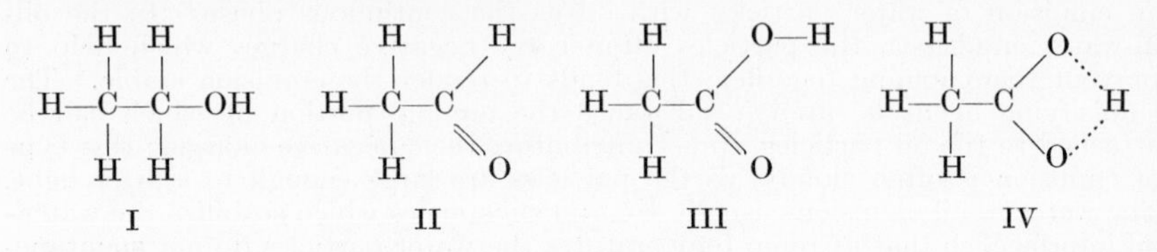

The compositions of some of the more important acids are shown in Table XL, which includes aliphatic (chain-) and aromatic (ring-) compounds, as well as some dibasic acids possessing *two* —COOH groups.

TABLE XL

Some Important Organic Acids

		M.P.	B.P.
Formic	H.COOH	8·3°C.	101°
Acetic	$CH_3.COOH$	16·6	118
Propionic	$C_2H_5.COOH$	− 22	141
Butyric	$C_3H_7.COOH$	− 7·9	162
Valeric	$C_4H_9.COOH$	− 58·5	186
Lauric	$C_{11}H_{23}.COOH$	44	225
Palmitic	$C_{15}H_{31}.COOH$	62·6	—
Stearic	$C_{17}H_{35}.COOH$	69·3	—
Oleic	$C_{17}H_{33}.COOH$	14	286
Linoleic	$C_{17}H_{31}.COOH$	− 5	230
Linolenic	$C_{17}H_{29}.COOH$	− 11	230 (?)
Oxalic	$COOH.COOH, 2H_2O$	189 (anhydrous)	—
Malonic	$COOH.CH_2.COOH$	132	—
Succinic	$COOH.(CH_2)_2.COOH$	182	—
Glutaric	$COOH.(CH_2)_3.COOH$	98	—
Adipic	$COOH.(CH_2)_4.COOH$	153	—
Pimelic	$COOH(CH_2)_5.COOH$	105·5	—
Suberic	$COOH.(CH_2)_6.COOH$	141	—
Azelaic	$COOH.(CH_2)_7.COOH$	106·5	—
Sebacic	$COOH.(CH_2)_8.COOH$	134·5	—
Benzoic	$C_6H_5.COOH$	122	249
Phthalic	$COOH.C_6H_4.COOH$	231	—
Salicylic	$OH.C_6H_4.COOH$	159	—
Glycollic	$CH_2OH.COOH$	80	—
Thio-glycollic	$CH_2SH.COOH$	− 16·5	—

* Such a method is mainly used in the laboratory. On a large scale acetaldehyde is manufactured by passing acetylene into sulphuric acid in presence of a catalyst, or by dehydrogenating ethyl alcohol by passing the vapour over heated copper.

The lowest di-basic acid, oxalic acid, although less strong than hydrochloric or nitric acid, is stronger than most organic acids, many of which are extremely weak. Unlike the lower monobasic acids, which are liquids, oxalic acid is a crystalline solid carrying two molecules of water, and generally written $(COOH)_2.2H_2O$, although the water may well be water of constitution (p. 246).

Salts of azelaic acid, $COOH.[CH_2]_7.COOH$

and suberic acid $COOH.[CH_2]_6.COOH$

occur, with other similar bodies, in the degradation products of oil-paints and appear to be largely responsible for their anti-corrosive properties. The sodium salt of benzoic acid $C_6H_5.COOH$ is a useful inhibitor.

Many organic acids occur in nature or in food-stuffs. Formic acid is found in nettles and ants, and is injected by bees when they sting other living creatures. Acetic acid is the chief constituent of vinegar. Butyric acid is contained in butter, and appears in larger quantities when the butter becomes rancid owing to the decomposition of the glyceride (see below) into glycerine and butyric acid. Stearic acid occurs in animal fats; its calcium and aluminium salts are of interest to the paint industry.

Salts of octoic acid are used as " driers " in paints, but these are apparently not derived from normal octoic acid, $C_7H_{15}.COOH$, but from an isomeride with two branches of carbon atoms. Naphthenic acids, the salts of which are also used as driers, is really a mixture of acids formed from rings containing five carbon atoms, with COOH groups attached such as

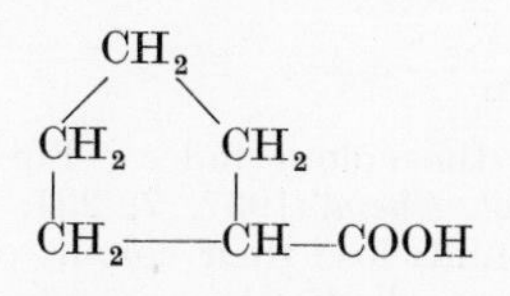

The mixture is a by-product of the petroleum industry. If left in oil, these acids render it very corrosive and are commonly removed or neutralized by lime or sodium hydroxide. For details see W. A. Derungs, *Corrosion* 1956, **12,** 617t.

Great importance attaches to alcohols carrying more than one OH group. Ethylene glycol $(CH_2OH.CH_2OH)$ is used in admixture with water as a non-freezing mixture in the radiators of cars. Glycerine

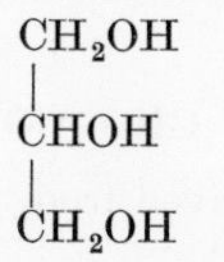

is important in connection with the glycerides.

By introducing OH groups into ring products (as substituents for H atoms), we obtain *phenols*. The most important to the corrosionist are phenol (carbolic acid), hydroquinone, pyrogallol, α-naphthol and

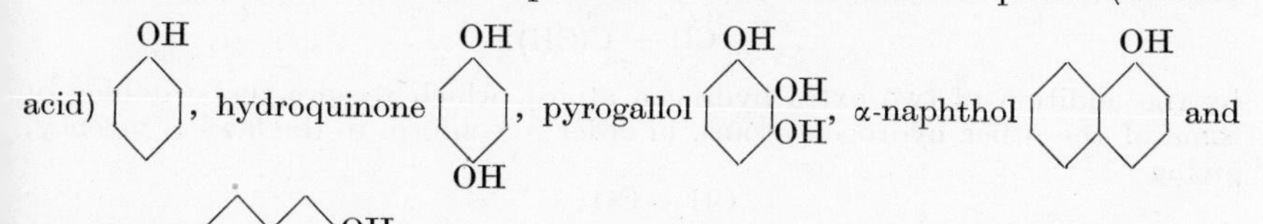

β-naphthol; α- and β-naphthols are used as addition agents in electro-plating. Hydroquinone, combined with its oxidation-product, quinone

O

O

is used in the quinhydrone electrode. The three cresols (*o*-, *m*- and *p*-) occur in tar; they are

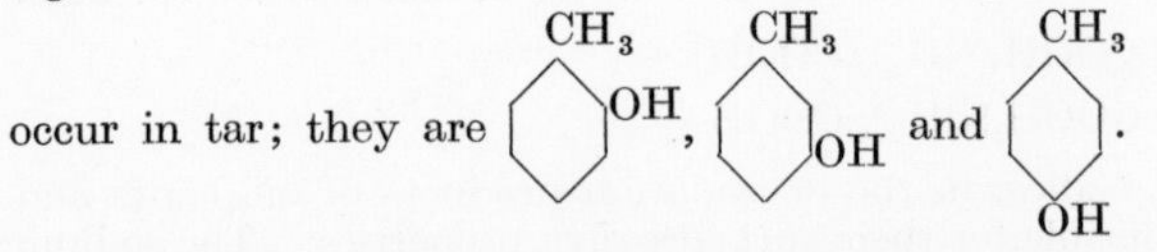

Some much more complicated ring-compounds containing hydroxyl-groups are important to corrosionists because they form the colouring matter of fruits, and sometimes play a part in the corrosion-process of tin cans. Red plums and raspberries owe their red colour mainly to chrysanthemin, which is really a chloride with a very large cation, and is capable of absorbing hydrogen to give chromene, and can thus provide a cathodic reaction for the corrosion process, even though oxygen is absent from the can

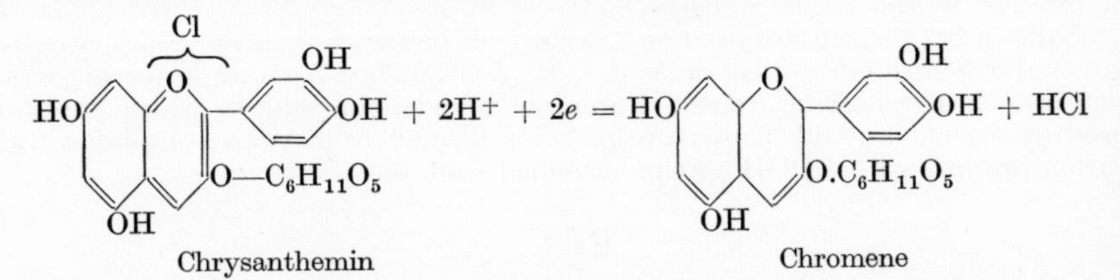

Chrysanthemin Chromene

The tin is corroded and the colour fades. For details, see F. W. Salt and J. G. N. Thomas, *J. appl. Chem.* 1957, **7**, 231.

Many coloured compounds lose their colour, or change it, when reduced to "leuco-bases"—as they are called. An example important for corrosionists is provided by indigo-carmine; the leuco-base can here take up dissolved oxygen, so that the reverse colour-change occurs, and this fact is utilized in the determination of small concentrations of oxygen in boiler-water (p. 450).

A compound of interest to corrosionists as a sensitive reagent for iron is iso-nitroso-dimethyl-dihydro-resorcinol,

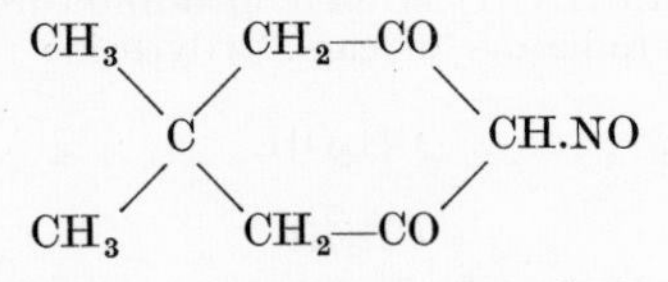

This can be regarded as being derived from the relatively simple phenolic body resorcinol (p. 1001)

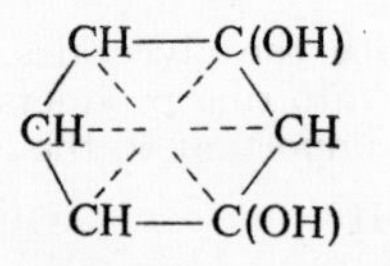

by the addition of two extra hydrogen atoms, which require the switching of some of the other hydrogen atoms, in order to conform to the laws of valency, giving

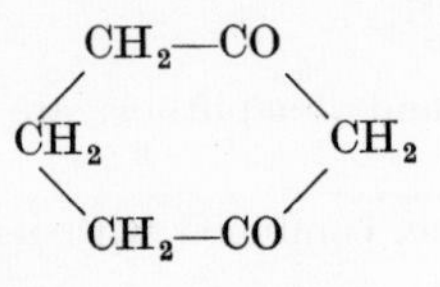

followed by the substitution of two CH_3 groups and one NO group for three hydrogen atoms.

Ethers, ketones, esters. On replacing the hydrogen of the OH group of an alcohol with an alkyl group, an ether is formed; similarly on replacing the hydrogen of the —CHO group of an aldehyde by an alkyl group, a ketone is formed. The general formula of the ethers is $R_1.O.R_2$, where R_1 and R_2 represent two alkyl groups, and that of the ketones is $R_1.CO.R_2$. In some important compounds R_1 and R_2 represent the same group. Common ether, $C_2H_5.O.C_2H_5$, is obtained by heating ethyl alcohol with sulphuric acid (which, in effect, abstracts water from two molecules $C_2H_5O\,\boxed{H + OH}\,C_2H_5 = C_2H_5.O.C_2H_5 + H_2O$, although the reaction really occurs in stages); acetone, $CH_3.CO.CH_3$, is obtained by the dry distillation of calcium acetate. A ketone which is important in the paint industry is MEK (methyl-ethyl-ketone), $C_2H_5.CO.CH_3$, a colourless liquid boiling at 78·6°C.

Esters are obtained by the removal of water from mixtures of alcohols and acids. Thus on heating ethyl alcohol with acetic acid, ethyl acetate, $CH_3.COOC_2H_5$, is slowly produced

$$CH_3.COOH + OH.C_2H_5 = CH_3.COOC_2H_5 + H_2O$$

An analogy is often drawn with the formation of inorganic salts

$$HCl + NaOH = NaCl + H_2O$$

but there are certain differences; the inorganic acid and alkali are largely ionized and the reaction is essentially

$$H^+ + OH^- = H_2O$$

In the case of alcohols and organic acids, the OH and H must be pulled off from the molecules and the reaction proceeds slowly and incompletely unless some substance is present having an affinity for water. In some "esterification" reactions, excess of alcohol serves; in others, sulphuric acid or hydrogen chloride is found to accelerate the change.

In contrast to the pungent and often unpleasant odours of many of the compounds hitherto mentioned, the odours of the esters recall fruits; iso-amyl isovalerate smells of apples, ethyl butyrate of pineapples and iso-amyl acetate of pears. The latter is used as a solvent for cellulose. Dibutyl phthalate, the ester of butyl alcohol C_4H_9OH and phthalic acid $(C_6H_4(COOH)_2)$, can be written $C_6H_4(COOC_4H_9)_2$; it is used as a plasticizer in paints. Phenyl salicylate (salol) $C_6H_4\langle^{OH}_{COOC_6H_5}$, once much used in pharmacy, has proved useful in experiments on growth of crystal faces (p. 383). Butyl titanate $((C_4H_9)_4TiO_4)$ is a constituent of heat-resisting paints.

Glycerides. The esters of glycerine with fatty acids are of special importance. The glycerine molecule possesses three OH groups,

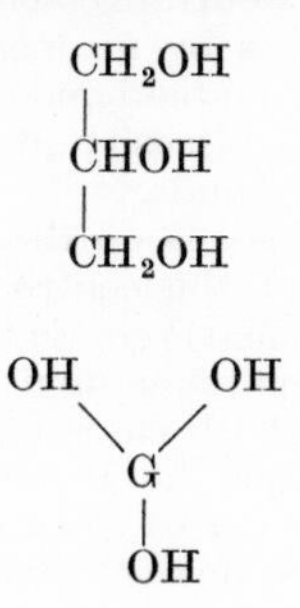

and may be written

where G represents C_3H_5. By interaction with an acid HX, we obtain a glyceride

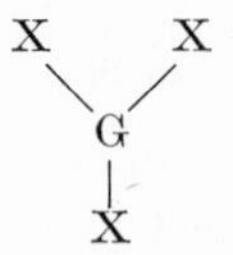

with the three X-chains pointed outwards in different directions. Of these, butyl glyceride occurs in butter, as already stated; palmitic and stearic glycerides, derived from palmitic acid ($C_{15}H_{31}.COOH$) and stearic acid ($C_{17}H_{33}COOH$) which have much longer chains, occur in most animal and vegetable fats.

Other important acids containing chains of seventeen carbon atoms (additional to the carbon of the —COOH group) contain less hydrogen, owing to the presence of one or more double bonds. Since the double bonds can be introduced in different positions in the chain, it is possible to have two compounds with the same " molecular formula " but differing in detailed structure and therefore in properties; these are known as " isomerides ", and further examples will be provided later. Again two double bonds can be replaced by one triple bond without change of molecular formula. Thus we have stearic acid $C_{17}H_{35}.COOH$, oleic acid $C_{17}H_{33}COOH$, inoleic acid $C_{17}H_{31}.COOH$; linolenic, α-elaeostearic and β-elaeostearic acids are isomerides, all possessing the formula $C_{17}H_{29}.COOH$. Of these, beef fat contains the glyceride of stearic acid, olive oil that of oleic acid, linseed oil those olf linoleic and linolenic acid, and tung oil those of α- and β-elaeostearic acids along with the " saturated " stearic acid (the word " saturated " implies that it has the maximum amount of hydrogen, owing to the absence of double bonds).

The glycerides of unsaturated acids possess industrial importance, since, when spread out as a thin film, they absorb oxygen from the air; after the absorption of the oxygen the various molecules are able to link up together at numerous points forming a three-dimensional net-work, so that translatory motion of the individual molecules becomes impossible, although vibration can continue; in other words, a viscous liquid film changes into a solid film. It is for that reason that linseed oil and tung oil are used in paints and varnishes. Double bonds are needed for this transformation; those fats and oils which only contain glycerides of saturated acids, such as stearic glyceride, are useless, whilst olive oil is unsatisfactory, since there is only one double bond in oleic acid. Tung oil " dries " (becomes rigid) more quickly than linseed oil, because it has two double bonds in the " conjugated " position —CH═CH−CH═CH—. Apparently this conjugation loosens the structure, and makes it easier for the molecule to lose a hydrogen atom, giving a free radicle which then combines with oxygen—as suggested later; the attachment of oxygen does not normally occur at the double bond itself—as was once thought. Linseed oil has no such conjugated set of bonds and some authorities think that the first stage in the drying of linseed oil is a re-arrangement of the double bonds, so as to produce conjugation. This view is by no means universally accepted, and it seems possible that even isolated double bonds can take up oxygen and initiate reactions which lead to increased viscosity and ultimately to rigidity, but do so more slowly than conjugated bonds.

An early paper by Blom laid stress on the fact that glycerides are 3-rayed stars with the glycerine group at the centre and the unsaturated groupings (where one star can link on to the next) remote from that centre; linkage between two stars can occur through oxygen taken up near the double bonds, or by direct union between carbon atoms; in both cases the double bonds will disappear as a result of the union of the " stars ". If raw linseed oil is heated in a pot closed to exclude air, it is converted to *stand-oil* or *litho-oil*, and here there is presumably a direct carbon-carbon linkage between the " stars "; if air is blown through the oil, we get *blown oil* in which OH groups are present, conferring polarity and improving the wetting power for pigments; here union through oxygen seems probable; *boiled oil* is obtained by heating linseed oil with " driers " (see below). All these methods of treating linseed oil produce some linking up of the molecules with consequent increase of viscosity, but the changes are not carried on so far that the oil ceases to be a liquid.

If a drying oil (with or without pigment) is spread out as a thin film on metal, glass or wood and exposed to air, the film ultimately becomes rigid; presumably the stars link up to form a three-dimensional net-work and lose their power of translatory motion. This process of drying, which in a pure oil film is extremely slow, is greatly accelerated by the presence of a *drier*, generally a compound of some metal of variable valency—such as lead, manganese or cobalt; iron can function as a drier. Zinc also possesses drying properties, despite the fact that in typical compounds of zinc, the valency is fixed at 2; however, zinc stands in the same group of the Periodic Table as mercury, a metal of variable valency, and the fact that zinc oxide becomes yellowish when heated suggests a tendency to variability of valency. Apart from zinc, all the metals capable of accelerating drying in notable degree are capable of existing in more than one valency state. This is surely no accident. The passage from one valency to another, usually accompanied by change of colour, has been studied by Krumbhaar; manganese compounds are light-coloured in the divalent and dark brown in the trivalent state; cobalt compounds are blue when divalent and green when trivalent. The blue cobaltous solution in an oil phase quickly becomes green if exposed to air as a film, and it seems reasonable to suppose that the cobaltic compound can then oxidize the drying oil—so that the cobalt acts as an oxygen-carrier (W. Krumbhaar, " Chemistry of Synthetic Surface Coatings " (Reinhold); A. V. Blom, *Kolloid-Zeitschrift* 1936, **75,** 223).

Mechanism of the drying process. The drying of oils is a complicated subject and even today divergent views seem to be held about the mechanism. Most authorities, however, believe that it consists of chain-reactions in which peroxides, hydroperoxides and free radicles take part; doubtless during the drying of a paint-film many reactions are proceeding in parallel, and whilst some of these cause linkage of the molecules, building a rigid net-work, others cause breakage of the molecules, giving volatile products, particularly aldehydes, which escape into the air, as well as fatty acids of low molecular weight. An account of the mechanism favoured today by the oil-chemist is provided by G. H. Hutchinson, *J. Oil Col. Chem. Assoc.* 1958, **41,** 474. See also L. A. O'Neill *Chem. and Ind.* (*Lond.*) 1954, p. 384; H. P. Kaufmann, *Verfkroniek* 1957, **30,** 317; M. Giesen, *ibid.* 1958, **31,** 316.

As already stated, it was formerly supposed that oxygen attached itself at the double bonds, so that different molecules became linked up through oxygen atoms, forming a net-work, and changing the liquid to a solid. It is now believed that by the action of ultra-violet light, or of traces of organic peroxide present, or by the action of a metallic catalyst (see below), a hydrogen atom is split off, leaving a " free radicle " to which an oxygen molecule attaches itself; the complex is then believed to react with other molecules of the oil, regenerating the free radicle, so that a chain reaction is set up, which might be expected to continue indefinitely (actually the " chain " sometimes becomes broken—if, for instance, two free radicles link up, forming an inactive molecule).

The formation of the first free radicle ($\dot{R}$) is generally written

$$RH \longrightarrow \dot{R} + H$$

whilst the chain is expressed

$$\dot{R} + O_2 \longrightarrow \dot{R}O_2$$

followed by

$$\dot{R}O_2 + RH \longrightarrow ROOH + \dot{R}$$

regenerating the free radicle. The ROOH is considered to decompose to some extent to give $\dot{R}O$ and $\dot{O}H$.

The part played by the metallic catalyst, or " drier ", is written (in the case of cobalt, which alternates between the divalent and trivalent state) thus:—

$$Co^{++} + ROOH \longrightarrow Co^{+++} + \dot{R}O + OH^-$$

or

$$Co^{++} + ROOH \longrightarrow Co^{+++} + RO^- + \dot{O}H$$

followed by

$$Co^{+++} + OH^- \longrightarrow Co^{++} + \dot{O}H$$

or

$$Co^{+++} + RO^- \longrightarrow Co^{++} + \dot{R}O$$

so that the Co^{++} is regenerated.

The various free radicles are considered to link up either to give carbon-linked dimerides

$$\dot{R} + \dot{R} \longrightarrow R—R$$

or ether-linked dimerides

$$\dot{R} + \dot{R}O \longrightarrow R—O—R$$

and these may undergo further oxidation or attack by free radicles, so as to give the cross-linking needed for rigidity.

At one time the metals which are useful as driers were added to the oil as oxides, which presumably reacted with the acids present in the oil to give oil-soluble soaps. It is clearly better to add ready-made oil-soluble substances, and later the linoleates of lead, manganese and cobalt were introduced as driers. Today the naphthenates are widely employed, and a more recent introduction is the so-called " octoate " group of driers (p. 991). Different metals confer different drying properties, some causing drying on the external surface of the film and others in the interior, so that it may be well to introduce compounds of all three metals, lead, cobalt and manganese, into a single paint. The exact manner in which oxygen, or an oxidizing agent, reaches the interior is not known, but there is evidence of convection cells of the hexagonal type in a drying oil-film; perhaps the metal is oxidized to high valency state on the external surface, and then moving inwards reacts with the oil, and is reoxidized on its return. This would provide a relatively rapid means of bringing an oxidizing agent into the interior; the oxidized metallic compound would be present in higher concentration than molecular oxygen, and would also probably react more quickly. Evidence of the convection cells is provided in an early paper by F. E. Bartell and M. van Loo, *Industr. engng. Chem.* 1925, **17**, 925.

Thus the ordinary oil-paint has four essential constituents: (1) a drying oil such as linseed or tung oil, (2) a drier, consisting of linoleates or naphthenates of lead, manganese and/or cobalt (sometimes all three metals are present), (3) a pigment or fine powder chosen to give colour, opacity and in some cases hardness, (4) a thinner, such as white spirit or turpentine. The thinner is added to obtain a mixture of consistency suited for being spread out on the metal as a thin, uniform coat. The application may be made by brushes, spraying, dipping or flowing. After a thin coat has been obtained, the thinner soon evaporates, and the dryer takes up oxygen, passes it on to the oil which " dries ", so that we obtain a solid coat consisting of pigment particles fairly closely packed in a matrix of oxidized oil. The drying process has features in common with the setting of plastics. Paints are discussed in greater detail in Chapter XIV.

Various other compounds are introduced into paints to improve the physical properties. These include resins, both natural and artificial (discussed later) and the soaps of calcium or aluminium. The couramone resins, obtained from coal tar, are important constituents of paints and varnishes; they contain polymers of benzfuran, a compound in which a benzene ring (six carbons) is fused on to a furane ring (four carbons and one oxygen).

Soaps. The two soaps just mentioned are salts of the higher fatty acids, such as stearic or palmitic. The soap used as a domestic detergent contains the sodium salts of these acids. Its detergent action is connected with the fact it possesses a " polar " portion (thc—COONa group) and a non-polar portion (the hydrocarbon chain). The former is stable in contact with water, itself a polar substance, and the latter in contact with grease or oil (non-polar). Other bodies possessing polar and non-polar ends have detergent properties and those made synthetically are called " Syndets ". If a metallic surface is greasy, it would involve an increase of interfacial energy for pure water to creep in between the grease and the metal, and thus loosen it; if the water contains soap or a syndet, this " creepage " can occur spontaneously (i.e. with decrease of free energy) since one end of the soap or syndet molecule will make easy contact with the grease and the other with the metal; thus the grease comes away without difficulty. Similar bodies with polar and non-polar portions are used as wetting agents, allowing water to enter crevices spontaneously where otherwise expenditure of energy would be required to force the water in.

The introduction of $—HSO_3$ groups into hydrocarbons produces the same valuable combination of polar and non-polar portions. Sulphonated petroleum products are valuable in connection with anti-corrosive oils and greases. Sulphonic acids, obtained by replacing H by the HSO_3 group in ring-hydrocarbons, (especially naphthol-sulphonic acids and their salts) are useful in electroplating.

Some natural oils and greases possess a polar end and require no sulphonation when used for providing protective covering. Lanoline which consists of the natural fat present in wool, provides better protection than the ordinary petroleum products. An important constituent is cholesterol $C_{27}H_{45}OH$, which is built up from four fused rings and a long side chain; the OH group is attached to the carbon of one of the rings.

Carbohydrates. The sugars are a group of compounds containing large numbers of hydroxyl groups and in general consist mainly of—CHOH—groups linked together; a CH_2OH group is often present, whilst two of the remote carbons may be joined through oxygen, thus producing a ring structure. The sugars are, however, a large group and details of their structure must be sought in a text-book on Organic Chemistry: for instance, L. F. Fieser and M. Fieser, " Organic Chemistry " 1956 edition, Chapter 14 (Reinhold).

Important sugars include glucose (grape sugar), $C_6H_{12}O_6$, and sucrose (beet or cane sugar), $C_{12}H_{22}O_{11}$. In the principal sugars hydrogen and oxygen occur in the ratio 2 : 1; this fortuitous circumstance has led to the group being called carbohydrates—wrongly suggesting carbon united with water. Other carbohydrates include cellulose, the chief constituent of wood; cotton, linen and paper are almost pure cellulose—apart from weighting or stiffening substances. By the action of a mixture of nitric and sulphuric acid, we obtain nitrocellulose —used in explosives and in paints.

Amines and amides. It is possible to replace one or more hydrogen atoms of ammonia, NH_3, by alkyl or aryl groups (R) giving *amines*, which may be primary NH_2R_1, secondary (NHR_1R_2) or tertiary ($NR_1R_2R_3$). Examples are ethylamine, $NH_2{\cdot}C_2H_5$, aniline (phenylamine), $C_6H_5NH_2$ and toluidine, $C_6H_4\langle{}^{CH_3}_{NH_2}$, which has three isomers (*o*, *m* and *p*); like ammonia, these are bases, forming salts with acids, such as aniline hydrochloride $C_6H_5{\cdot}NH_2{\cdot}HCl$— a crystalline solid; aniline itself is an oil. Aniline has given its name to a large class of dye-stuffs (not all made from aniline); one of the most famous (historically), rosaniline or fuschine, is made by the oxidation of a mixture of aniline, *o*- and *p*-toluidine and has the formula

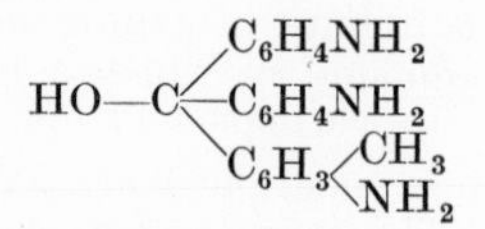

It has been used in bright plating-baths.

Certain amines have proved valuable as inhibitors of corrosion, especially in steam systems, where their volatility enables them to reach places inaccessible to ordinary inhibitors. They act, either by neutralizing acid substances or because they become absorbed on the metallic surface at points where corrosion would normally start. Another point of interest to corrosion workers is their use in the "curing" of epoxy-resin paints. Important amines include hexadecylamine (*cetylamine*) $C_{16}H_{33}NH_2$, octadecylamine (*stearylamine*), $C_{18}H_{37}NH_2$, cyclohexylamine

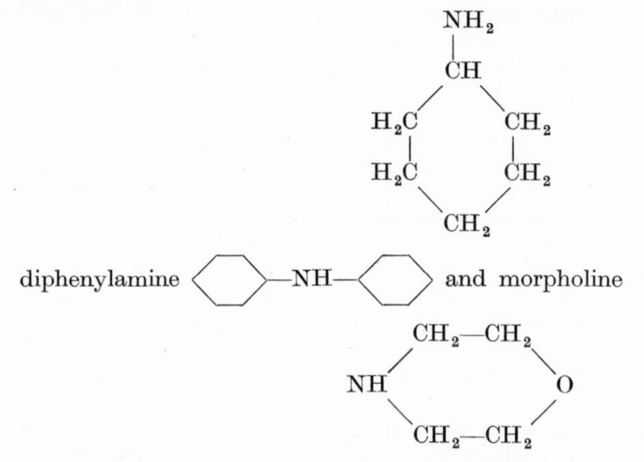

If the group used to replace the first hydrogen of ammonia is an "acid group" (obtained by removing OH from a molecule of an organic acid), we obtain *amides*. *Acetamide*, the amide of acetic acid, is $CH_3 \cdot CO \cdot NH_2$, whilst *urea*, $CO\langle^{NH_2}_{NH_2}$ can be regarded as the amide of carbonic acid $\left(H_2CO_3 \text{ or } CO\langle^{OH}_{OH}\right)$. More important to corrosion workers is its sulphur analogue, *thio-urea*, $CS\langle^{NH_2}_{NH_2}$ which can be regarded as urea in which sulphur plays the part of the oxygen—both being elements of the same group of the periodic table. Many of the "substituted" thio-ureas are restrainers of corrosion by acids, the most important being di-ortho-tolyl-thio-urea

$$\overset{CH_3}{C_6H_4}-NH-\underset{\underset{S}{\|}}{C}-NH-\overset{CH_3}{C_6H_4}$$

(The significance of the term "ortho" is explained below.)

Nitrogen can also replace —CH in benzene rings giving pyridine C_5H_5N and

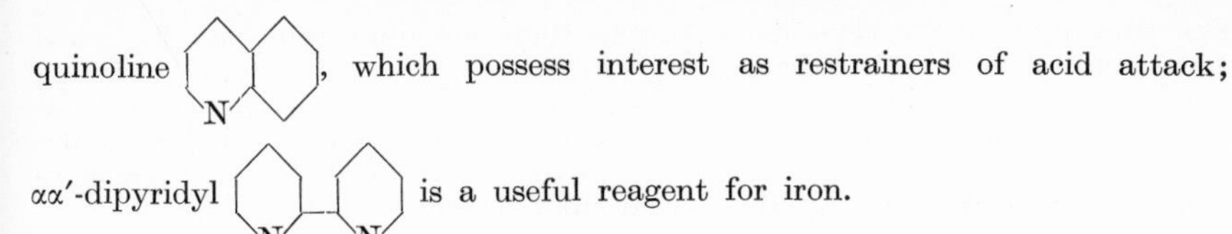

quinoline, which possess interest as restrainers of acid attack; αα′-dipyridyl is a useful reagent for iron.

Organic Compounds containing Sulphur. Mercaptans differ from alcohols in having a —SH group instead of an —OH group attached to a carbon atom. They are liquids with unpleasant odours, almost immiscible with water, but miscible with many organic liquids. Their boiling-points are lower than those of the corresponding alcohols; thus CH_3SH boils at 6° and CH_3OH at 66°C. The presence of mercaptan in certain mineral oils is responsible for much of their power to attack metal, although free sulphur, where present, is probably more dangerous; hydrogen sulphide also contributes to the corrosive properties of some oils.

Although the simple mercaptans are corrosive, some of their complicated derivations are inhibitors, such as the sodium salt of mercaptobenzothiazole

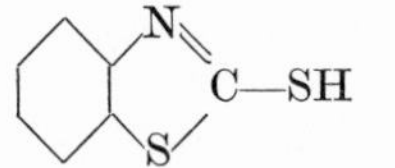

Cystine, R—S—S—R where R stands for $—CH_2—CH(NH_2)·COOH$, is important to the corrosionist, since it occurs in polluted sea-water and causes trouble to copper alloys; its copper derivative forms a film which is protective if continuous but gives rise to severe localized corrosion at any gaps; probably the film of copper derivative acts as an efficient cathode, and the dissolved cystine functions in effect as oxygen-carrier, being reduced to cystein R.SH, which is then oxidized once more by the oxygen present, regenerating the cystine.

Sulphur can also take part in ring-formation; thiophene

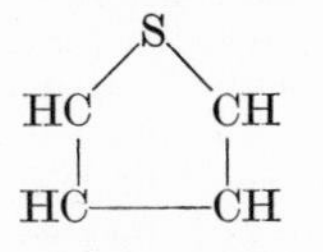

is a common impurity in coal-gas and is less easy to remove than H_2S or CS_2.

The *amino-acids*, formed by replacing one H atom in an organic acid by an—NH_2 group, require notice. Simple examples include amino-acetic acid (*glycine*), $NH_2.CH_2.COOH$, and amino-propionic acid (*alanine*)—both used in plating baths, and glutamic acid $COOH—CH_2—CH_2—CH(NH_2)—COOH$ used as a flavouring compound in food. The importance of the amino-acids lies in the power of their molecules to form long chains (through the elimination of water from NH_2 and COOH) thus giving a structure of the type

—NH—RH—CO—NH—RH—CO—NH—RH—CO—.

Other far more complicated forms are possible, and are found in the structure of the proteins, important constituents of foodstuffs. One protein of interest to corrosionists is *gelatine*, which occurs as an anhydride in bones and is extracted when they are boiled with dilute acid; it is an important constituent of glue, and possesses corrosion-inhibitive properties.

Sequestration. The ammines (p. 967) formed by union of ammonia with metallic salts are relatively unstable. Often the ammonia can be removed by heating the solution. Organic amines can unite with metallic salts giving complexes in which the metal is attached at two different points of the same molecule, providing a claw-like grip, and these are often more stable. Such compounds belong to a large group of organo-metallic compounds possessing abnormal stability; they are known as *chelate compounds*, from the Greek word *chele*, a crab's claw. An early case of chelating was provided by beryllium acetylacetonate, which contains two chelate rings, the co-ordinate valency-bonds being represented by the dotted lines in the formula

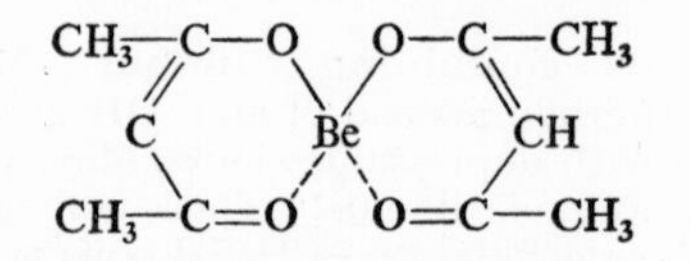

In general, these compounds fail to show the reaction of the metal in question, which is firmly linked with the organic portion, and they differ in several respects from typical organic compounds. For instance, a series of organic compounds built up from chains of carbon atoms have melting-points which rise as the length of the chains increases, but in a chelate series the reverse may be true. In the compound

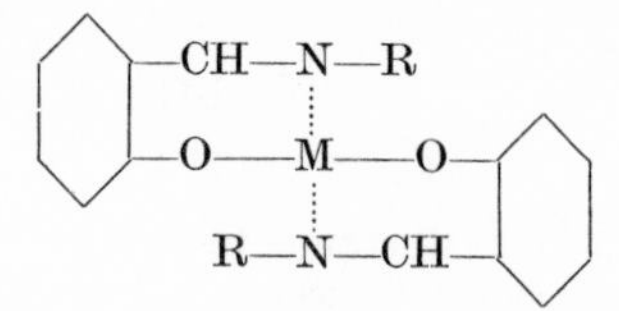

where M represents either Ni or Cu and R represents any alkyl group between CH_3 and $C_{14}H_{29}$, the melting-point declines (apart from some slight irregularity at C_4 and C_{10}) as the number of carbon atoms rises (R. G. Charles, *J. org. Chem.* 1957, **22**, 677). Compounds which provide opportunity for metal cations to join up so as to form stable chelate rings have the property of *sequestration*—although this is not confined to cases where chelate rings are formed. Sequestration has been defined as the power to form soluble complexes with metallic ions under conditions where normally a stable precipitate would be expected. In presence of a sequestering agent, metallic ions fail to show their usual analytical behaviour; unless the analyst is vigilant, a metal may escape detection.

The phenomenon of sequestration is met with in many classes of compounds. The condensed phosphates (p. 969) can act as sequestering agents for calcium, and prevent the separation of calcium carbonate under conditions where normally it would be thrown down as a sludge or scale. Lignin, which occurs in wood combined with cellulose or other carbohydrates, contains phenolic groupings and has been considered as a sequestering agent for Ca^{++} and Mg^{++} ions. Perhaps the most remarkable sequestering agent known is *E.D.T.A.* (ethylene-diamino-tetracetic acid)

$$\begin{matrix} COOH—CH_2 \\ COOH—CH_2 \end{matrix}\!\!>\!N—CH_2—CH_2—N\!<\!\!\begin{matrix} CH_2—COOH \\ CH_2—COOH \end{matrix}$$

which contains a variety of positions where a metallic cation can attach itself in chelate manner and thus lose its identity as a cation. It is much used in analysis. A survey of chelating agents is provided by J. K. Aiken, " Anti-corrosion Manual " 1958, pp. 158–170 (Corrosion, Prevention and Control).

Nomenclature of Isomers in ring compounds. It has been explained that two compounds can have the same formula, but may still differ in the arrangement of attached groups. There are, for instance, two propyl alcohols, $CH_3.CH_2.CH_2.OH$ and $CH_3.CHOH.CH_3$ having slightly different boiling-points (97° and 81°C.); there are four butyl alcohols and eight amyl alcohols. Specially important are the classes of isomers produced by attaching two groups to a benzene ring; these are known as ortho-, meta- and para-compounds (abbreviated to *o*-, *m*, and *p*-), according to the relative positions of the two groups

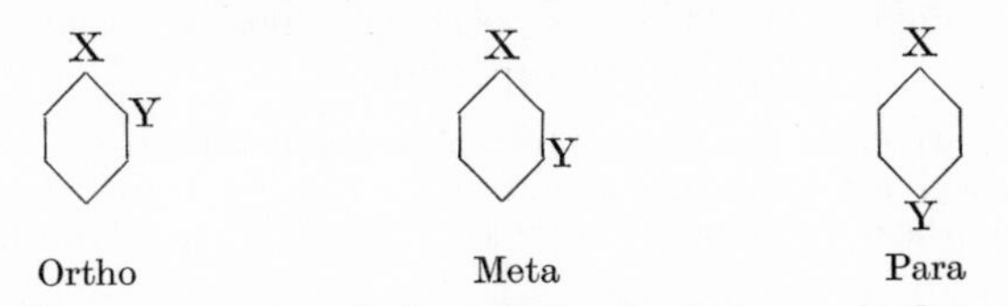

Ortho Meta Para

There is only one compound formed by replacing a hydrogen of benzene by OH, namely phenol C_6H_5OH (or carbolic acid), but three compounds containing two OH groups

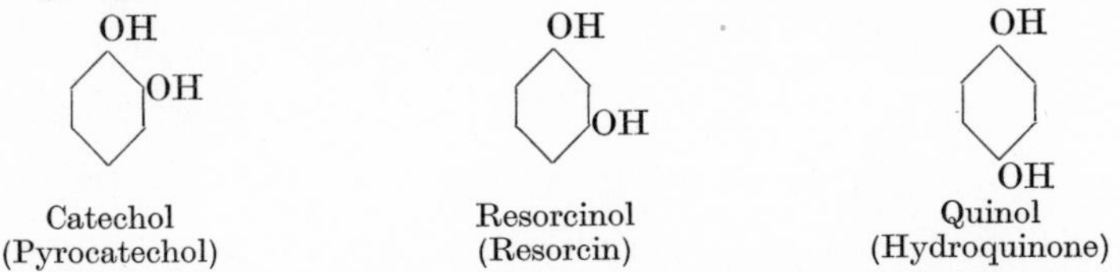

Catechol (Pyrocatechol) Resorcinol (Resorcin) Quinol (Hydroquinone)

Quinol is important in connection with the quin-hydrone electrode (p. **1034**); catechol is related to an important group of tannins, which are inhibitors of corrosion.

Plastics. The term "plastics" is today used, somewhat loosely, to cover many diverse substances of high molecular weight, which easily pass from a liquid form to a rigid form, thus permitting articles to be produced readily in any desired shape. Some are derived from natural products, but many are "synthesized". The cellulose compounds are sometimes regarded as plastics. Cellulose is a carbohydrate of much higher molecular weight than the sugars, mentioned above, and on "nitration" it yields nitric acid esters, wrongly termed nitrocellulose*.

Nitrocellulose varies with the conditions of formation and the number of nitrate groups introduced; some are explosive ("guncotton"), others are soluble in alcohol-ether mixtures and were at one time much used in photography ("collodion"). In paint manufacture, a type of nitro-cellulose soluble in such solvents as iso-amyl acetate has been used. If such a solution is spread out as a thin film, it will "dry" through simple evaporation of the solvent, leaving a coat of transparent lacquer, or, if a pigment has been added before the application, a coating of "paint". However, such coatings are liable to be brittle—possibly because the escape of the solvent leaves cavities, which act as "stress-raisers", so that under slight bending the coating cracks instead of being deformed. Consequently most lacquers and paints which dry by evaporation of the solvent (and some others) have a non-volatile *plasticizer* added which prevent this brittleness; according to one view, a plasticizer possesses molecules of suitable form to fill up the cavities which would otherwise act as stress-raisers, or according to another view it acts as an internal lubricant. The plasticizers are a large class, and knowledge about their relative suitability for any particular purpose is largely empirical; many of them are esters, for instance, dibutyl

* A distinction must be drawn between the true nitrates or nitric acid esters, formed when nitric acid acts on an alcohol

$$C_2H_5OH + HNO_3 = C_2H_5.O.NO_2 + H_2O$$

and the nitro-compounds formed when a mixture of nitric acid and sulphuric acid act on a ring hydrocarbon

$$C_6H_6 + HNO_3 = C_6H_5NO_2 + H_2O$$

The sulphuric acid is necessary to absorb the water formed and allow the action to proceed. For the corrosionist the removal of the water is important because it allows nitration-processes to be carried out in iron vessels (p. **331**).

phthalate $C_6H_4(COOC_4H_9)_2$. Another compound, mentioned in Chapter XIV, is dioctyl phthalate $C_6H_4(COOC_8H_{17})_2$.

One of the simplest of plastics is *polystyrene*, derived from styrene

$$C_6H_5.CH{=}CH_2,$$

a liquid boiling at 146°C., which polymerizes on heating and yields a glassy resin; in practice, catalysts are added to bring about the change. During polymerization, the simple molecules $CH{=}CH_2$ join up to form long chains

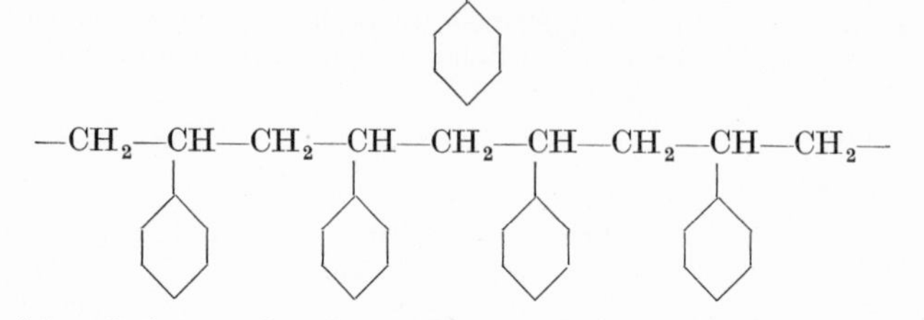

Although these chains can be of great length, the whole of the primary valencies are taken up, and there are no bonds left over to " cross-link " the molecules. Thus, although through the operation of secondary forces which are strong enough to prevent translatory motion, polystyrene is solid and fairly hard at room temperature, it softens when heated, and is soluble in various liquids—notably xylene. A solution of polystyrene containing a suitable plasticizer, spread out on metal produces, on evaporation of the solvent, a film of transparent lacquer. Alternatively, pigment may be added, giving a polystyrene paint. The paint pigmented with metallic zinc powder possesses valuable properties; it is possible to introduce enough zinc to obtain a coat which, after evaporation of the solvent, contains about 95% of metal; the metallic particles are in electrical contact with one another; this is impossible with metallic zinc in linseed oil, since the mixture containing the required content of zinc dust is too stiff to be applied—except with a palette knife (p. 615).

Styrene is a member of an important group known as the *vinyl compounds*, conforming to the general formula $CH_2{=}CHX$; they can be regarded as ethylene with one hydrogen atom replaced by X—, which may be C_6H_5— (as in styrene), but may equally well be Cl— or CH_3OOO— as in vinyl chloride and vinyl acetate; these two compounds also polymerize, giving *polyvinyl chloride* (" PVC ") and *polyvinyl acetate* respectively. It is possible to obtain a " co-polymer " in which some of the X groups of the chain

$$—CHX—CH_2—CHX—CH_2—CHX—CH_2—CHX—CH_2—$$

are chlorine whilst others are acetate groups. Another important vinyl resin, used in etch primers, is *polyvinyl butyral*

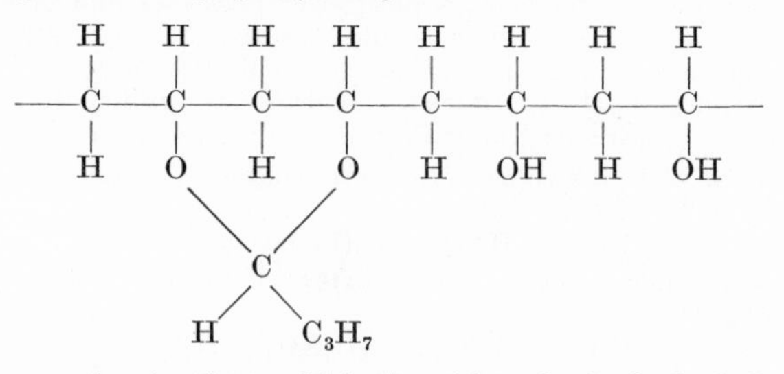

it is made by condensing butyraldehyde with polyvinyl alcohol. Some vinyl compounds undergo polymerization spontaneously on exposure to light, but in most cases the addition of a suitable accelerator—often a peroxide—is desirable.

Ethylene itself can be made to polymerize, under rather special manufacturing conditions, some of which involve high pressures, giving poly-ethylene (*polythene*). Other vinyl compounds of importance in the manufacture of resins include *acrylic acid*, $CH_2=CH—COOH$, *acrylo-nitrile*, $CH_2=CH.CN$ and *vinylidene chloride*, $CH_2=CCl_2$.

In most of the cases mentioned, the union of the small molecules to form large ones does not involve any second product; this is called *polymerization by addition*. In other cases the union involves the splitting off of water, hydrogen chloride or ammonia molecules, and such cases are properly called *condensation*. An example is the union of phenol and formaldehyde; when heated together they unite with expulsion of water, and produce chains which are cross-linked together, so that translatory motion becomes impossible even at high temperatures, and the mass becomes rigid. The resins thus produced are known as the P.F. (Phenol-Formaldehyde) Type; bakelite resins are familiar examples. Most of these are " thermo-setting " in contrast to the polystyrenes which are " thermo-softening "; the difference arises from the structure, since a three-dimensional net-work exists in the first group and is absent in the second. Any polymer which is not cross-linked to form a three-dimensional net-work gradually acquires the properties of a liquid if the temperature is raised sufficiently high, provided that there is no decomposition. However, the liquids produced on " melting " polymers differ somewhat from those produced by melting a compound composed of relatively small molecules. Viscosity measurement suggests that a polymer molecule does not move as a whole, but in sections. Its movement has been compared to that of " a rope lying on the ground in a randomly kinked and coiled shape; it can be moved bodily into a new position by taking a short length at a time, most of the rope being stationary whilst one length is moved " (G. Gee, *Proc. chem. Soc.* April 1957, p. 111, esp. p. 113).

The possibility of cross-linking (and thus the decision between thermo-setting and thermo-softening) is better understood by applying the useful concept of " functionality ", which may be described as the number of reactive groups present in the molecule. Glycol and phthalic anhydride both possess a functionality of 2, and when a molecule of glycol unites with one of phthalic anhydride, the combining power is exactly used up and there is no group left over to link up the molecules into a net-work, so that the product has no thermo-setting properties; it is used as a plasticizer in lacquers. By using glycerine, with functionality 3, in the place of glycol, cross-linking become possible, but the reaction is slow. Urea has a functionality of 4 and can react with formaldehyde to form a cross-linked polymer, and phenol (perhaps unexpectedly) has a functionality of 3, because the three H atoms in the ortho and para positions relatively to the OH, can interact with formaldehyde to give compounds such as

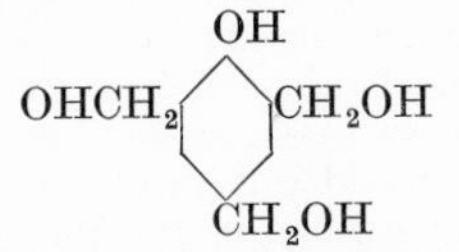

Where cross-linking occurs, relative movement of molecules is prevented, and the material hardens instead of softening on heating. The exact character of the reactions involved in the manufacture of P.F. resins cannot here be discussed in detail, but if we consider the formaldehyde to react as $OH.CH_2.OH$ (a hydrate $CH_2O.H_2O$ does exist), it is easy to see how two benzene rings might become joined. For instance

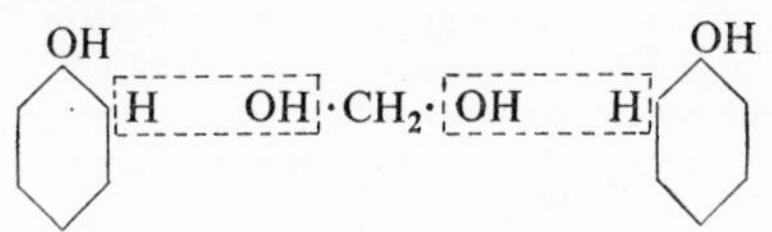

would give

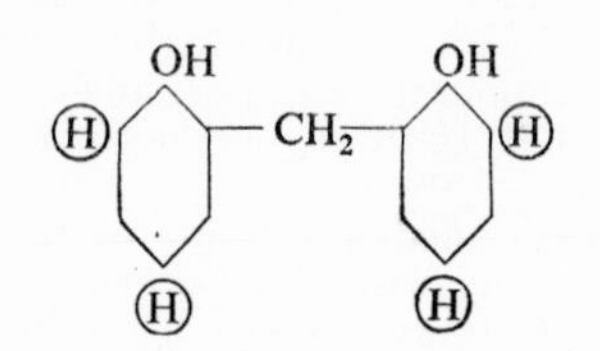

and since then the other four hydrogens indicated by being placed in circles can react with other formaldehyde molecules, we soon arrive at an infinite cross-linked net-work such as

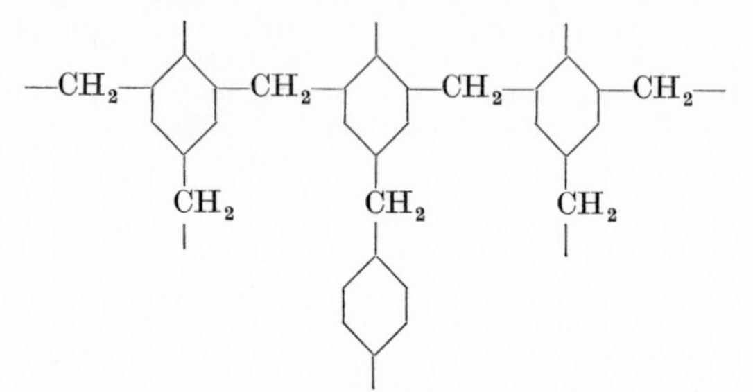

Probably the true facts are more complicated and the resulting structure less regular.

Synthetic Resins for Paints. There are many plastics analogous to the P.F. resins, since other aldehydes and other phenolic bodies join up in the same way. Their application in moulded articles of everyday life is well known. Before they can be used in paints or stoving lacquers, however, they require " modification ".

Natural resins such as rosin (p. 989), have long been used in " air-drying " paints and stoving varnishes. The first attempts to use synthetic resins were not successful, since most of the resins were found to be insoluble in the oils; however, by " modifying " the resin (e.g. attaching linoleate chains to the rings) it becomes soluble in linseed oils. Other methods for obtaining oil-soluble synthetic resins are based on heating with rosin or the use of a substituted phenol—such as butyl phenol.

The resins thus formed dry by the same mechanism as linseed oil. The drying can take place slowly by exposure to air at ordinary temperatures, or more rapidly at elevated temperatures. The so-called stoving lacquers or stoving enamels are usually sprayed on to the article to be coated, which is then placed in an oven. The coat produced on cooling is extremely serviceable. Instead of conducting the heating in an oven, which involves the body of the metal becoming heated, it can be carried out more speedily by exposure to radiant heat which is absorbed at the surface of the metal—just the place where it is needed. The use of radiant heat is particularly welcome for certain aluminium alloys which would deteriorate in mechanical properties—and sometimes in corrosion resistance—if the body of the metal became hot.

A few of the resins used in baking varnishes can produce coatings which harden in an oven without the up-take of oxygen, joining up by the condensation mechanism which is different from that involved in the resins mentioned above. These include the butylated amino resins—the " curing " of which is described by H. R. Touchin, *J. Oil Col. Chem. Assoc.* 1956, **39**, 653.

There are numerous other types of plastics, but only a few examples can be quoted. The urea-formaldehyde group—already mentioned—are much used in admixture in many paints and varnishes. Particularly important is the *alkyd* or *glyptal* group, produced when phthalic acid $C_6H_4\,(COOH)_2$, or similar acid, condenses with glycerine or other alcohol containing several hydroxyl groups. A glycerine molecule can bind together three phthalic acid molecules, and since each of the three carries another —COOH group which can interact with other glycerine molecules, there is every opportunity for producing a complex but rigid three-dimensional net-work, although this may not always be welcome for purposes of paint-manufacture.

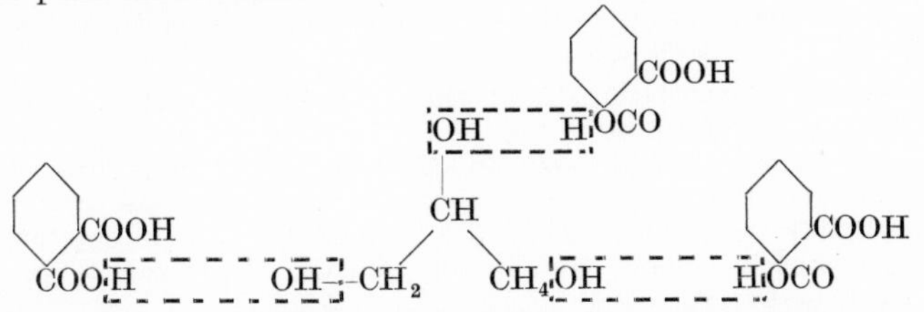

By using compounds with fewer hydroxyl groups than glycerine, thermo-softening resins can be obtained. Alkyd resins are much used in paints and varnishes, but these are generally " modified " compounds with acid groups corresponding to drying-oil glycerides as side-chains; they dry by a mechanism similar to that responsible for the drying of oils (p. 995).

The name *polyester resin* has come to be applied to those alkyds made from unsaturated acids or their anhydrides. Strictly speaking all alkyds are polyesters. The unsaturated compounds used include maleic and fumaric acids, both having the formula $COOH—CH{=}CH—COOH$, but with the different directional disposition of the two COOH groups to one another. The simplest polyester is made by heating glycerine, phthalic and maleic anhydrides. There are many other and more complicated systems. The " curing " or solidification of the film requires the presence of a catalyst and also an accelerator. In service, the user is supplied with (*a*) a solution of a polycondensation product in styrene or similar liquid, (*b*) a cobalt compound such as cobalt naphthenate and (*c*) an organic peroxide, such as methyl-ethyl-ketone peroxide. Of these (*a*) and (*b*) may be mixed at any convenient moment without loss of keeping power, but (*c*) must be added at the last moment, since as soon as admixture has been made, the hardening starts. This is something of an inconvenience. Another objection to the use of these resins as vehicles in paints is that air interferes with the curing; the reaction involves the formation of free radicles, some of which will combine with oxygen to give relatively inactive bodies; this trouble has been overcome to some extent by applying a wax covering.

The *epoxy-* or *ethoxylene* group of resins deserves special mention, since, although their use in paints and varnishes is a relatively recent development, there is already hope that useful products may be obtained, including metal-pigmented paints of great hardness. The starting-point is epichlorhydrin

$$\overset{O}{\overbrace{CH_2—CH}}—CH_2Cl,$$

which is usually condensed with the sodium salt of a dihydric phenol such as diphenylol-propane, $C_6H_4OH.C(CH_3)_2.C_6H_4OH$, with the elimination of NaCl. The solution of the polymer thus obtained in MEK and/or other solvent is mixed by the user with a curing agent just before being spread on the metallic surface; the agent may be an amine, an amide or some compound containing—NH_2 groups; sometimes the polymer is mixed with a poly-amide resin. Cross-linking, and consequent hardening, occurs rapidly, owing to the power of the oxygen to unhook itself from one of the two carbon atoms, providing, in effect, bonds for attachment to other molecules, thus $\overset{|}{C}H_2—\overset{|}{C}(OH)—CH_2—$.

The special importance of polyesters and epoxy-resins is their power to give cross-linking at room temperature; at slightly elevated temperatures (below

100°C.) the hardening occurs very rapidly. The epoxy-resin films obtained after curing are remarkably resistant to many chemicals. Another advantage of epoxy- and polyester resins as constituents of paint and varnish films is that during the curing no water, ammonia or volatile organic matter is evolved in the condensation reaction—which would seem to improve the chance of obtaining an impervious film. This refers only to the solidification of the material after it has been spread out by the user on the metal surface. In the reactions involved in the manufacture of the resin, water or other molecules are sometimes expelled.

Silicones. It has been stated that whereas in carbon compounds the atoms join directly forming —C—C—C—C— chains, the linkage in silicon-compounds occurs through oxygen so that the chains are of the type —Si—O—Si—O—Si—O—Si. The silicone resins are a class of compounds in which both linkages occur. They are formed through interaction of silicon chloride and various organic compounds, and the generalized type of structure can be expressed thus

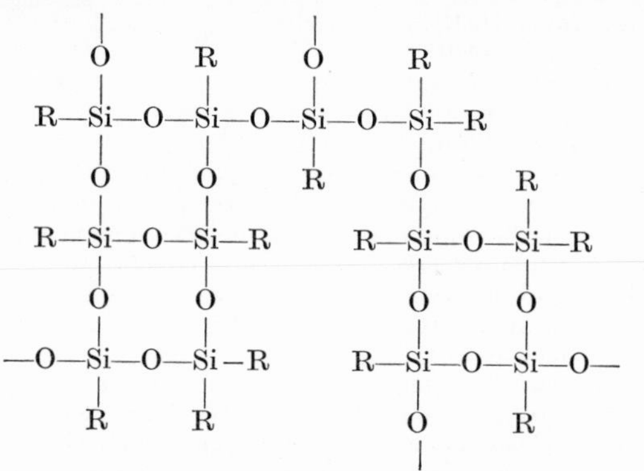

The —O—Si—O— linkage provides good heat-stability, and the silicones are therefore used in heat-resisting paints. They are resistant to many chemical reagents and often possess water-repellent properties.

APPENDIX II

PHYSICAL CHEMISTRY AND ELECTROCHEMISTRY

Chemical Kinetics

Possible and Impossible Reactions. If a truck starts at the top of a long hill, it will run down of its own accord—no work or energy being needed to push it. Indeed, the descent could be used to generate energy—if, for instance, the truck carried a small dynamo geared to its wheels and used for charging an accumulator. The accumulator would then gain electrical energy, as the truck was losing potential energy by passing from the high to the low level. Once arrived at the bottom, the truck could not move up the hill spontaneously, expenditure of energy being needed for the truck to ascend; but, since the accumulator is now charged, the dynamo, working as a motor, might in favourable circumstances accomplish this (although in practice the charge attained in the descent would not serve for the complete ascent, some energy having been changed to heat as a result of friction).

It would, however, be wrong to say that *any* journey which ends at a lower level than it starts can occur spontaneously. Supposing that a road commences by going up-hill, but finally descends to a lower level than the start, the journey will only occur if a certain amount of energy is first expended in getting the truck up the initial incline.

Chemical changes will only occur spontaneously if they result in a decrease of *free energy*; a decrease of free energy can, in favourable circumstances, be used to perform work, and the maximum work obtainable from a reaction is a convenient measure of the free-energy decrease accompanying that reaction. When a mixture of oxygen and hydrogen combines to form water vapour, energy is liberated and a gas-engine fed with such a mixture can be used to drive a dynamo and thus charge an accumulator. However, the water, once formed, will not spontaneously change back into oxygen and hydrogen; such a change can only occur if energy is supplied, for instance, by passing from the charged accumulator an electric current between two platinum wires dipped into the water. Combination of oxygen and hydrogen is a " possible " reaction, which can occur spontaneously; the reverse change is an " impossible " one, which cannot occur unless energy is supplied from an external source.

However, a possible reaction does not necessarily occur at any appreciable rate. At ordinary temperatures, a mixture of oxygen and hydrogen can be kept indefinitely in a glass vessel without appreciable formation of water. Only when a spark is passed through it, heating the mixture locally, does the combination start, leading to a violent explosion. The combination can, however, occur slowly, without any external addition of heat, if the glass vessel contains a piece of platinum wire covered with platinum black. On the surface of the platinum, the oxygen and hydrogen are adsorbed, and brought together in suitable proximity and in the right energy condition for combination, so that a slow formation of water occurs; much energy is evolved as heat (in the end the mixture may explode), but the combination does start at a low temperature —at which, in the absence of platinum, no combination would occur. Platinum is said to be a *catalyst* for such a reaction. Some other materials possess the same " catalytic " powers, but not all; zinc, for instance, fails to " catalyse " the union of oxygen and hydrogen. Although a catalyst can facilitate a " possible " reaction under conditions where normally it would take place slowly or not at all, it cannot bring about an " impossible " reaction; it is no use putting blackened platinum in water in the hope that the spontaneous decomposition to oxygen and hydrogen will occur.

The platinum black used to bring about the combination of oxygen and hydrogen is said to be a *surface catalyst* or *heterogenous catalyst*. It is probable that the combination proceeds preferably not over the whole surface but at certain favourable spots. It is noteworthy that many of the reactions important in corrosion processes (e.g. the evolution of hydrogen or the reduction of oxygen) appear to take place mainly at a limited number of points, probably where the structure of metal is such that the attachment of reacting molecules can be encouraged whilst unduly hindering the liberation of the product when formed.

Examples are also known of *homogeneous catalysis*, where the catalyst is dispersed as molecules in solution. It seems that the addition of a lead, manganese or cobalt salt to drying oil to assist drying may provide an example. The oxygen of the air converts the metal to a higher state of oxidation (e.g. divalent cobalt becomes trivalent) and the oxygen is passed on to the oil, which then becomes oxidized and finally sets (see p. 542). Sometimes the product of a reaction is itself a catalyst, and the reaction then becomes faster and faster as time proceeds, provided that the reaction-products are not swept away. Such a case is known as *autocatalysis*. An example is provided by the attack of nitric acid on copper, discussed in Chapter IX. Here the reaction proceeds most rapidly in corners where the liquid is stagnant and—in contrast to the attack on the baser metals—it occurs most slowly if the liquid is stirred.

Activation Energy. The reason why many " possible " reactions do not in fact occur at relatively low temperatures is that, although the final result of the reaction represents a lower energy condition than the initial state, it is necessary to pass through an intermediate stage which is richer in energy than either. This can easily be seen, if we consider the change in an organic compound from one atomic arrangement (X) to another arrangement (Y) which is more stable (lower in energy) than X. Clearly in the course of the re-arrangement we must expect to pass through a stage where the atoms, having been pulled away from their original, fairly stable, positions occupied in X, but not having assumed the very stable positions represented in Y, are in an exceedingly energy-rich state. Only those molecules of X which chance to possess much energy can pass through this intermediate condition, and at low temperature molecules rich in energy are rare; at high temperatures, energy-rich molecules become more common, and the change from X to Y may reach an appreciable velocity. If only molecules carrying energy equal to W_A (the *activation energy*) can pass through the intermediate condition, it becomes very important to know what proportion of the molecules at a given temperature do carry the requisite amount of energy.

It is possible to show, by considering the kinetics of gaseous molecules, that the chance of a *single* molecule possessing energy in excess of the value w is $e^{-w/kT}$ where k is Boltzmann's constant (the gas constant for a single molecule, in contrast to R, the gas constant for one gram-molecule). If the volume of a perfect gas contains n molecules, the number possessing energy exceeding w is $ne^{-w/kT}$ or $ne^{-W/RT}$ where W is the corresponding energy per gram-molecule. Such relationships are in fact not confined to gases—as is shown by the fact that in many chemical and physical reactions the logarithm of the time t_Q required to produce a standard quantity Q of product gives a straight line when plotted against $1/T$. This result is easily understood, since the quantity produced in time-element dt will be proportional to $e^{-W/RT}\,dt$. In simple cases, the gradient of the curve will enable us to calculate the activation energy W. An example is provided by the decomposition of hydrogen iodide (HI), a colourless gas which turns violet when heated owing to decomposition to iodine and hydrogen. This reaction—slow in the cold—becomes fast at high temperatures, log k being a rectilinear function of $1/T$, where k is the velocity constant defining the reaction rate.*

* If the equation is $2HI = H_2 + I_2$ and the concentration is c at time t, then

$$-\frac{dc}{dt} = kc^2$$

In certain types of chemical reactions, particularly reactions of gases, one step of the reaction liberates single atoms or portions of a molecule (*free radicals*) possessing high energy and capable of interacting with an ordinary molecule in such a way as to liberate another high-energy atom or radical which can react with another molecule and so on. These *chain reactions* are complicated and have at times been subjects of controversy. One view of the combination of oxygen and hydrogen—which over a certain range of temperatures takes place on the walls of a silica vessel, but not easily at points remote from the walls—is that hydroxyl radicals (OH) are formed which combine with hydrogen, and then later are regenerated

$$OH + H_2 = H_2O + H$$
$$H + O_2 + H_2 = H_2O + OH$$

A rather different example of a chain reaction is provided by a mixture of hydrogen and chlorine, which combine very rapidly to form hydrogen chloride when once a few single chlorine atoms have been produced, e.g. by the introduction of a trace of sodium vapour

$$Na + Cl_2 = NaCl + Cl$$

The chain consists of

$$Cl + H_2 = HCl + H$$
$$H + Cl_2 = HCl + Cl \quad \text{and so on.}$$

In this case, however, silica is unfavourable to the reaction, possibly by destroying one of the free atomic species, since it is found that the introduction of powdered glass or silica into the reaction vessel decreases the rate of production of hydrogen chloride.

An easier way to start the chain is by means of light; a mixture of hydrogen and chlorine can be preserved in the dark but explodes if exposed to sunshine. We have here an example of a *photochemical change*, but it differs from the simplest kind of photochemical change in which each light quantum* decomposes one molecule; in the case under consideration a light quantum serves to dissociate a molecule into its atoms

$$Cl_2 + h\nu = 2Cl,$$

where h is Planck's constant and ν the frequency, and the chain-reaction is then started, so that a single light quantum suffices to produce vast numbers of molecules of HCl.

Balanced Reactions. Although chemical reactions frequently proceed to completion—or nearly to completion—there are cases where a balance between two opposing chances is set up, so that equilibrium comes to exist between the two substances or two sets of substances. Consider the change $X \longrightarrow Y$. If both X and Y are solids, immiscible with one another, then the change from X to Y will complete itself. Suppose, for instance, we take a mixture of solid X and solid Y along with a small amount of liquid having a solvent action for both, but necessarily a greater power to dissolve X than Y, since the solubility of the less stable material *must* be greater than that of the more stable material. Evidently X will be dissolved, producing a saturated solution, which will in due course produce a super-saturated solution of Y, and thus deposit solid on Y; the original liquid, now unsaturated with X will dissolve more X, and this will continue until the whole of the solid X has been converted to solid Y—apart from a small amount which remains dissolved in the liquid.

Suppose, however, X and Y are themselves mutually soluble liquids, a

* Like all radiant energy light is divided into quanta each equal to $h\nu$ where ν is the true frequency (number of vibrations per second) and h is Planck's constant ($6{\cdot}6 \times 10^{-27}$ erg. secs.); evidently the size of the quantum increases as the frequency becomes larger, i.e. the wavelength shorter, so that short-wave light can produce reactions where long-wave light fails to do so—a fact well known to photographers.

balance is possible. Starting with pure X, we may get molecules of Y produced, but the reaction may slow down as the proportion of X in the liquid diminishes. Conversely, starting with pure Y, we may find that X is produced, at a rate which slows down as Y becomes exhausted. The proportions of X and Y present in the final state should be the same—whether we start from X or Y; the equilibrium mixture represents that in which the reaction $X \longrightarrow Y$ is proceeding at the same rate as $Y \longrightarrow X$ and we can write the state of balance $X \rightleftharpoons Y$. If Y is the more stable state, so that $W_Y > W_X$, where W_Y and W_X are the activation energies of the reactions starting from Y and X respectively, the equilibrium mixture will in general contain more Y than X. For if X and Y were present in concentrations C_X and C_Y, the amounts changing in the two directions would be respectively $P_X C_X e^{-W_X/RT}$ and $P_Y C_Y e^{-W_Y/RT}$ where P_X and P_Y are constants. If P_X and P_Y are of the same order of magnitude, the two opposing reactions will only come into balance when the proportion of Y in the mixture has come to exceed that of X.

Doubtless, if we could measure separately the time needed to produce quantity Q of Y from X, and the time needed to produce quantity Q of X from Y, we should find that in either case log t_Q plotted against $1/T$ would give a straight line; but since the gradient is given by the activation energy, it follows that (provided that P_X and P_Y are almost independent of temperature) the rate of change $Y \longrightarrow X$ will increase with rising temperature more rapidly than that of the change $X \longrightarrow Y$, so that equilibrium will occur at a higher ratio of X to Y. In other words, the rise of temperature favours the proportion of the unstable (energy-rich) constituent present in the equilibrium mixture. It is now possible, therefore, to explain the fact (of some industrial importance) that by blowing a mixture of oxygen and nitrogen through an electric arc where a very high temperature is reached, a sensible quantity of nitric oxide is produced, but that it is necessary to cool the gas very quickly, since the nitric oxide will largely break up again into oxygen and nitrogen if the cooling is slow. The equilibrium concentration of nitric oxide in air heated to 2400°C. is 2·33%, but only 0·377% at 1540°C.

Dissolution Processes

Three Types of Dissolution. The passage of metal into an aqueous solution —which is the essential feature of many corrosion processes—differs from other types of dissolution in that the outer layer of the metal enters the solution as *cations* (i.e. atoms which have lost electrons), *the electrons being left behind* in the surviving part of the metal; it is for this reason that corrosion is generally an electrochemical process. In other cases where a solid dissolves in water, it enters the liquid phase either as *uncharged molecules*, or as *ions carrying opposite charges* in such proportions that no charge is left on the undissolved part of the solid.

Before considering the dissolution of metals, it will be convenient to consider the two cases which are not complicated by the separation of electrons.

Dissolution as Molecules. Consider the case of glucose placed in water. Here a single number suffices to define the solubility, although it alters with temperature. At 17·5°C., when 100 c.c. of water have taken up 83 grams of glucose, the number of the glucose molecules leaving the crystal during an element of time is exactly balanced by the number being redeposited. In alcohol, the equilibrium solubility is only 1·94 per 100 c.c., whilst in ether and many organic solvents, glucose is almost insoluble. In contrast, the hydrocarbon, naphthalene, dissolves freely in ether, chloroform and many organic solvents, appreciably in alcohol but only very slightly in water (0·003 g. in 100 c.c.). It seems that " like things dissolve like "; the polar substances with plenty of OH groups are dissolved freely in water, but not in hydrocarbon solvents; the non-polar hydrocarbons are dissolved freely by hydrocarbon liquids but not by water. In this respect, dissolution presents a contrast to chemical combination where it is the unlike substances (e.g. sodium and chlorine) which combine with the greatest violence.

Dissolution as Pairs of Ions. When we pass to consider a salt dissolving in water, the situation is slightly less simple. If a crystal of sodium chloride is placed in water, sodium and chlorine ions pass into the liquid in equal numbers; the reverse process of deposition requires that sodium and chlorine ions must arrive at the right place in the right energy states. The chance that both may be present at a given place within a given time is proportional to the product of the two concentrations [Na][Cl], and when deposition is proceeding at the same rate as dissolution, producing equilibrium, the solution is said to be "saturated". Provided that nothing is present except sodium chloride and water we can again define the position by a single "solubility" value, which depends on temperature. However, it is possible to increase [Cl] without adding fresh sodium chloride. If to a saturated solution of sodium chloride which is already in equilibrium with the solid salt, we add hydrochloric acid, the rate at which Na^+ and Cl^- ions come together is increased, whilst the rate at which sodium chloride passes from the solid phase into the liquid remains almost unaltered. Thus the solution, previously just saturated, becomes supersaturated. Concentrated hydrochloric acid—added to a saturated solution of sodium chloride—throws down the salt as a solid; if the salt solutions contain impurities, they remain in solution, so that we have a convenient way of purifying salt. This is an example of *the effect of the common ion* in reducing solubility, the common ion being in this case an anion (Cl^-); a cation can also act in the same way, as when sodium chloride is added to a solution of a sodium soap in order to precipitate the soap as a solid—the process being known as *salting out.*

The effect of a common ion in depressing the solubility of substances which are in any case sparingly soluble is capable of precise mathematical prediction —since certain disturbing factors which come into play at high concentrations are unimportant at low concentrations. Silver chloride is sparingly soluble in water. The equilibrium

$$AgCl \rightleftharpoons Ag^+ + Cl^-$$

will be attained at that concentration at which the dissolution (left-to-right reaction) balanced deposition (right-to-left reaction). The latter will be proportional to the product of the concentrations $[Ag^+][Cl^-]$; there is some value of this product (the *Solubility Product*) at which the left-to-right reaction is exactly balanced. If now we add a little potassium chloride to the water, we shall increase $[Cl^-]$ and thus to obtain the same product at a lower value of $[Ag^+]$. If we increase [Cl] ten times, we must reduce $[Ag^+]$ to one-tenth of its former value; in other words we reduce the effective solubility of silver chloride to one-tenth. Evidently for such problems, we require to know the value of the solubility product, which in this case is $[Ag^+][Cl^-] = 1{\cdot}7 \times 10^{-10}$; having obtained it, we can easily calculate the solubility of silver chloride in a potassium chloride solution of any known strength. The solubility of calomel (mercurous chloride) in a potassium chloride solution depends on the same principle, which is important in connection with the construction of silver/silver-chloride and calomel electrodes—a subject discussed later.

Since the Na^+ and Cl^- ions making up a crystal of sodium chloride carry charges, it is not surprising that they retain these charges when the crystal dissolves in water. The cations and anions move in opposite directions to one another when an E.M.F. is applied to the solution; on this account an aqueous salt solution has a far better conductivity than pure water, although it is greatly inferior to the conductivity of a metal; the "ionic conductivity" of a salt solution differs from the "electronic conductivity" of a metal by the fact that it depends on two sets of ions carrying opposite charges but moving in opposite directions, so that in fact they both serve to move electricity in the same direction; the conductivity of a metal is due to the movement of electrons in a single direction. It should be noted that, in general, the ions as they exist in an aqueous solution are hydrated, but there is still some difference of opinion as to whether each ion has a definite number of water molecules attached to it, or whether the water molecules tend to orient themselves around the charged ions and are therefore dragged along with them as they move in the direction enforced by the potential gradient. This matter need not be discussed here, but the affinity of ions for water must never be forgotten, since it plays a part in deciding the energy change involved in the movement of metallic ions into the liquid phase during a corrosion process.

The mobility of different species of ions—as measured by their velocity when a unit potential gradient is applied to a solution—differs considerably, but there are two ions which display an abnormally high mobility, difficult to explain simply on the basis of size and weight. These are the *hydrogen ions* and *hydroxyl ions*, and they are believed to move in a manner different from other ions. The hydrogen ion is commonly written H^+, which, if taken literally, would denote a hydrogen atom which has lost its only electron, leaving a proton; it is generally believed that the proton is not free but is joined to a water molecule, so that the hydrogen ion is really H_3O^+, a water molecule with one extra proton, just as the hydroxyl ion OH^- is a water molecule which is short of one proton. If so, the apparent movement of the hydrogen ion is merely a switch of the surplus proton from one water molecule to another; thus when a potential gradient is applied to a chain of water molecules of which one possesses an extra proton, the following change can occur:—

$$HOH \quad HO^{+H}_{H} \quad HOH \quad HOH \quad HOH \quad HOH \quad HOH$$

becomes $$HOH \quad HOH \quad HO^{+H}_{H} \quad HOH \quad HOH \quad HOH \quad HOH$$

next $$HOH \quad HOH \quad HOH \quad HO^{+H}_{H} \quad HOH \quad HOH \quad HOH$$

then $$HOH \quad HOH \quad HOH \quad HOH \quad HO^{+H}_{H} \quad HOH \quad HOH$$ and so on.

Likewise the apparent movement of the hydroxyl ion in the opposite direction can be regarded as a proton switch, thus:—

$$HOH \quad HOH \quad HOH \quad HOH \quad HOH \quad \bar{O}H \quad HOH$$

becomes $$HOH \quad HOH \quad HOH \quad HOH \quad H\bar{O} \quad HOH \quad HOH$$

next $$HOH \quad HOH \quad HOH \quad H\bar{O} \quad HOH \quad HOH \quad HOH$$

then $$HOH \quad HOH \quad H\bar{O} \quad HOH \quad HOH \quad HOH \quad HOH$$ and so on.

This switch-mechanism involves no migration of particles in the usual sense of the word, and it is easy to see why the apparent movement of H^+ and OH^- proceeds much more rapidly than that of other ions, which must " thread their ways " between the crowd of particles.

Dissolution of Metals as Ions with Electrons left behind. The case of a metal standing in equilibrium with a solution introduces a new factor, since the metal dissolves, not as atoms, but as ions, thus leaving a negative charge on the undissolved part of the metal. Two factors come into play which tend to oppose further dissolution; first (as in the case of the dissolution of glucose or of salt) the metal ions in solution may be redeposited at an increasing rate as their concentration in the liquid increases; secondly (and more important) the electric potential difference produced by the electrons left behind will oppose passage of further ions from metal to liquid. In the case of silver standing in a solution of (say) silver nitrate, we can write the equilibrium

$$\underset{\text{(Metal)}}{Ag} \rightleftharpoons \underset{\text{(in hydrated form in the liquid)}}{Ag^+} + \underset{\text{(left in the metal)}}{e}$$

For any given concentration of silver ions in the liquid, a certain potential difference between metal and liquid is needed for equilibrium. If we increase the silver ion concentration, we clearly increase the right-to-left reaction, and equilibrium can only be restored if we alter the potential difference in a manner unfavourable to the right-to-left (deposition) reaction and favourable to the left-to-right (dissolution) reaction.

Electrochemical Equilibria

Measurement of an E.M.F. Before the potential drop at a single electrode such as Ag/Ag^+ can be discussed in greater detail, it is necessary to consider the measurement of the E.M.F. of a cell formed by combining two such electrodes. The well-known Daniell cell

$$Cu \mid CuSO_4 \mid ZnSO_4 \mid Zn$$

is merely a single example of a type of reversible cell expressed by the general formula

$$\text{Metal } M_1 \mid M_1 \text{ ions} \mid M_2 \text{ ions} \mid \text{Metal } M_2$$

The E.M.F. of such a cell is measured by balancing it on a potentiometer. The simplest form is shown in fig. 197. An accumulator is joined to the two ends of a resistance wire, and a sliding contact (x) can be made to move along

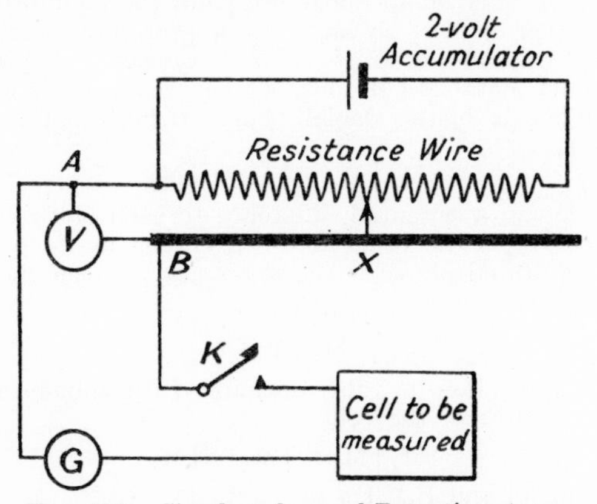

FIG. 197.—Simplest form of Potentiometer.

the wire, so that the potential difference between A and B, indicated on the voltmeter V, can be made anything between zero and 2 volts, according to the position of the slide. The reversible cell to be measured is placed so that its E.M.F. is in opposition to that provided from the potentiometer, and the slide is moved until the central-zero galvanometer G shows that no current is passing in either direction; when the E.M.F. of the cell slightly exceeds that provided, current will flow through G in one direction; when it is slightly deficient, current will flow in the other direction. Zero deflection indicates an exact balance, and the reading of the voltmeter represents the E.M.F. of the cell. By using the key K, it is possible to avoid current flowing for more than an instant, even when the set-up is out of balance.

The current passing through the voltmeter is provided by the accumulator, and in no way affects the accuracy of the measurement, since the cell under study is producing no current when balance has been reached. However, before the point of balance has been fixed, the cell is producing current—first in one direction and then in the other, and this may leave an after-effect, affecting slightly the accuracy of the determination; valve potentiometers have been introduced which allow much less current to flow when the system is out of balance, and, properly used, they achieve greater accuracy than the simple potentiometer of fig. 197. However, the fact that the simple potentiometer allows current to flow through the cell when it is out of balance, is not entirely a disadvantage, since small gas charges in the electrodes are likely to be used up during these periods of current flow, and there is less risk that the E.M.F. measured will represent a balance between reactions other than those which the experimenter has set out to examine.

The E.M.F. of the Daniell Cell

$$\mathrm{Cu} \mid \mathrm{CuSO_4} \mid \mathrm{ZnSO_4} \mid \mathrm{Zn}$$

is about 1·08 volts—the exact value depending on the two concentrations. The reactions which occur when current is flowing in one direction are equal and opposite to those which occur when it is flowing in the other direction. Thus if the balancing potential is a little below the equilibrium value, so that the current is flowing forwards (i.e. in the direction in which it would flow if the cell were being used to generate a current for some practical purpose), zinc is dissolved and metallic copper deposited; if the balancing potential slightly exceeds the equilibrium value, then zinc should be deposited and copper dissolved. A cell is only regarded as reversible if the reactions produced when " under-balanced " are exactly opposite to those produced when it is " over-balanced ", and if a very small departure from the equilibrium value, in one direction or the other, suffices to cause the appropriate reaction to take place.

By combining a number of *single electrode systems* or *half-cells*, each having the general formula Metal M | M Ions, we can make an indefinite number of complete reversible cells of the Daniell type. If both metals are divalent, the general formula is

$$M_1 \mid M_1^{++} \mid M_2^{++} \mid M_2$$

It is possible to assign a potential difference to each half-cell which it retains unchanged—whatever is the other half-cell chosen as its partner. Thus the E.M.F. of the cell stated above is the *difference* between the values of

$$M_1 \mid M_1^{++}$$

and

$$M_2 \mid M_2^{++}$$

Each half-cell will retain its value unchanged if combined with a third half-cell, so that the E.M.F. of the cell

$$M_1 \mid M_1^{++} \mid M_3^{++} \mid M_3$$

is roughly the sum of the E.M.F.s of

$$M_1 \mid M_1^{++} \mid M_2^{++} \mid M_2$$

and

$$M_2 \mid M_2^{++} \mid M_3^{++} \mid M_3$$

being given by

$$[M_1 \mid M_1^{++} - M_2 \mid M_2^{++}] + [M_2 \mid M_2^{++} - M_3 \mid M_3^{++}]$$

This statement is exactly true only if there is no appreciable potential difference at the liquid junctions (M_1^{++}/M_2^{++}, etc.). Such potentials may occur at junctions between liquids containing cations and anions of different mobility, so that one ion may tend to pass across the junction in greater numbers than the others; electrical equilibrium only becomes possible when a potential difference exists of such a sign and magnitude as to hold back the faster one and accelerate the slower ion. It is found that liquid junction potentials can be largely avoided if the two liquids are connected by a *bridge* containing a solution of concentrated potassium chloride or concentrated ammonium nitrate; the effectiveness of these two liquids in making the unwanted junction potentials very low is generally ascribed to the fact that the mobility of K^+ is almost the same as that of Cl^-, whilst that of NH_4^- is almost the same as that of NO_3^-; the high concentrations of the two liquids also help by making the (low) junction potentials remaining independent of the nature of the other solution.

It is clearly possible to adopt some particular half-cell as the *arbitrary zero* of a potential scale, and after combining it with a second half-cell to obtain the potential of the second half-cell on that scale—by measuring on a potentiometer the E.M.F. of the combination. It would be possible to choose for the arbitrary zero a single electrode or half-cell such as

$$\mathrm{Cu} \mid \mathrm{CuSO_4} \quad \text{or} \quad \mathrm{Ag} \mid \mathrm{AgNO_3}$$

—the concentration of the solution being clearly defined. Today, however, the half-cell commonly chosen for the arbitrary zero-point is the *Normal Hydrogen Electrode,* which consists of a platinum electrode (covered with black platinum to increase the effective surface area and promote the smooth working of the electrodic reactions), surrounded by hydrochloric acid of 1·2N concentration saturated with hydrogen gas at 1 atmosphere pressure (the reason for adopting 1·2N instead of N will become evident later). This electrode has no special theoretical claims to be regarded as a zero-point, nor is it particularly convenient to use,* but the " hydrogen scale " has, for some reason, become generally adopted.

It would be possible to obtain the value on the hydrogen scale directly. If we wished, for instance, to know the value of the half-cell

Cadmium | Cadmium chloride solution (of some stated concentration)

we could join it through a suitable liquid bridge to the normal hydrogen electrode, obtaining the combination,

$$\mathrm{Cd} \left| \begin{array}{c} \mathrm{CdCl_2} \\ \text{solution} \end{array} \right| \begin{array}{c} \text{Suitable} \\ \text{Bridge} \end{array} \left| \begin{array}{c} 1{\cdot}2\mathrm{N\ HCl,\ saturated} \\ \text{with } \mathrm{H_2} \end{array} \right| \mathrm{Pt}$$

and measure its E.M.F. by balancing it on the potentiometer. The value thus obtained will be the potential (on the hydrogen scale) of the half-cell Cd | $CdCl_2$ —assuming that the " suitable bridge " has really eliminated the junction potentials; it is doubtful whether saturated potassium chloride would do this. In practice, however, a less direct but more convenient and accurate method is generally adopted. If the potential of some new half-cell is required, it is usually combined with one of the practically convenient half-cells described later, such as a calomel half-cell or a silver/silver-chloride half-cell; the E.M.F. of the combination is then measured. It is possible subsequently to convert the value to the " hydrogen scale " by a suitable addition or subtraction, since the values of the various calomel or silver/silver-chloride half-cells relative to the Normal Hydrogen Electrode have been accurately established (p. 1019).

The manner of employment of the various electrodes has been suggested in figs. 64 and 142 (pp. 285, 801). In the calomel electrode, the mercury surface is covered with potassium chloride solution of the chosen concentration with a paste of solid calomel powder covering the mercury surface; silver may be " chloridized " (plated with silver chloride) by anodic treatment and placed in potassium chloride solution of appropriate strength. In laboratory work, the tube will lead to a vessel containing a connecting liquid so as to minimize liquid junction potentials. If it is desired to measure the potential at one point on a corroding metal surface, the tubulus—drawn out to a fine tip—may be brought

* Of the hydrogen electrode, Agar writes, " As a matter of interest, it is a very reliable electrode if one takes standard precautions in setting it up; I think distinctly better than the popular Ag | AgCl system " (J. N. Agar, Priv. Comm., May 23, 1958).

close to the metal at the point in which we are interested; the mercury and the corroding metal are connected by wires to a potentiometer (see fig. 150, p. 864).

Effect of Concentration. The value of the potential of a metal against a solution of one of its salts must depend on the concentration of that solution, since, if the balance is

$$M \rightleftharpoons M^{++} + 2e$$

an increased ion concentration will increase the right-to-left reaction whilst leaving the left-to-right reaction unchanged. Although ions can exist in a reasonably stable state either in the metal or in the liquid, there will be an intermediate position at the interface where an ion would possess high energy; only a few of the ions will possess sufficient energy to surmount the *energy hump* at the metal–liquid interface; in general, the activation energy needed to cross the barrier from the metal side (W_1) will not be equal to that needed to cross it from the liquid side (W_2). In the absence of any electrical potential difference the numbers per unit area crossing in each direction will be

$$K_1 e^{-W_1/kT} \quad \text{and} \quad CK_2 e^{-W_2/kT}$$

where C is taken for the moment to denote the concentration of ions (later, it will be shown that this idea may require some slight modifications), whilst K_1 and K_2 are constants independent of C, and k is Boltzmann's constant (see below). In general, these terms will not be equal, and a state of balance will only be obtained if there exists at the interface a potential difference V, distributed in such a way that V_1 is available to *help* the passage on the difficult side, and V_2 available to *hinder* the passage on the easy side, $(V_1 + V_2)$ being equal to V. The energy W_1 which a particle must possess to surmount the hump from the difficult side, is *decreased* to $(W_1 - n\varepsilon V_1)$ where n is the valency and ε the electronic charge, whilst that needed for surmounting it from the easy side is *increased* to $(W_2 + n\varepsilon V_2)$. At equilibrium

$$K_1 e^{-(W_1 - n\varepsilon V_1)/kT} = CK_2 e^{-(W_2 + n\varepsilon V_2)/kT}$$

or

$$C = \frac{K_1}{K_2} e^{-(W_1 - W_2 - n\varepsilon V)/kT}$$

(since $V_1 + V_2 = V$), which is

$$\log_e C = \log_e \frac{K_1}{K_2} - \frac{W_1 - W_2}{kT} + \frac{n\varepsilon V}{kT}$$

or

$$V = K + \frac{kT}{n\varepsilon} \log_e C$$

where K is a new constant independent of C. When C is unity, $V = K$. Therefore if ΔV is the alteration of potential when C departs from unity

$$\Delta V = \frac{kT}{n\varepsilon} \log_e C$$

Since Boltzmann's constant, the gas constant for a single molecule, is connected with R, the gas constant for one gram-molecule, by $R/k = F/\varepsilon$, where F is Faraday's Number (96,500 coulombs), the charge associated with one gram-molecule of a monovalent ion,* this may be written

$$\Delta V = \frac{RT}{nF} \log_e C = 2{\cdot}3 \frac{RT}{nF} \log_{10} C$$

Now at ordinary temperature (18°C., i.e. $T = 291$), $2{\cdot}3 \dfrac{RT}{F} = 0{\cdot}058$ volt,

* The value quoted by H. S. Harned and H. Owen, "Physical chemistry of Electrolytic Solutions" 3rd edition (1958) is 96,493·1 coulombs/gram-equiv.; J. N. Agar considers that the uncertainty is about ± 10 coulombs, so that 96,500 is good enough for our present purposes.

which means that every ten-fold shift of concentration will shift the potential by about 58 mv. for a monovalent metal and about 29 mv. for a divalent metal.*

On the whole, measurement confirms this expected relationship between concentration and potential. Nernst in 1891 found the E.M.F. of the " concentration cell "

$$Ag \mid N/100\ AgNO_3 \mid N/10\ AgNO_3 \mid Ag$$

to be 55 mv., which compares fairly well with the predicted 59 mv.; the boundary potential was not eliminated, but the error involved must have been small. Later workers have examined the matter, and find fairly good agreement between calculated and observed values of potential at ranges of low concentration, but less good agreement at higher concentrations. The reason appears to be that the argument developed above assumes not only that the metallic salt (say $AgNO_3$) is completely dissociated into its ions (Ag^+ and NO_3^-), but that these particles can move about without mutual hindrance—which may not be true, since the unlike charges may restrict movement in that direction which would increase separation. The effect of this is that, although N $AgNO_3$ may contain normal concentration of Ag^+ ions, the rate of delivery of such ions at a metal/salt solution interface is that characteristic of a lower concentration, and it is necessary to multiply the ionic concentration C by an *activity coefficient*, f, in calculating the potential needed to produce a balance; the product fC or a, is called the *Activity*.†

Although the variation of potential with *concentration* will be only approximately equal to $\frac{RT}{nF} \log_e C$, the variation with *activity* should be exactly $\frac{RT}{nF} \log_e a$. It is evident that as a basis for comparing the potentials of different metals, normal activities should be used rather than normal concentrations. In very dilute solutions, concentration and activity are practically equal, and if the potential at a point in this dilute range is measured, the potential at normal activity can be obtained by applying the correction

$$\frac{RT}{nF} \log_e a$$

The belief that the concentration of ions which is effective in determining the electrode potential is different from (and usually less than) the concentration which would be expected on the basis of the weight of salt introduced into the solution receives confirmation from a study of freezing points of salt solutions. Water and ice are in equilibrium at 0°C.—provided the water is pure; at that temperature molecules pass from ice to water as quickly as molecules pass from water to ice. If a little sugar or salt be dissolved in the water, some of the molecules striking the ice surface are not water, so that, whilst the passage of ice to water remains unchanged, that from water to ice becomes reduced, and the balance is upset; equilibrium is only restored if the temperature is reduced, which will hinder the passage of ice to water. Thus the presence of sugar in the water depresses the freezing-point. It can be shown that the depression is proportional to the number of independently moving particles in unit volume of solution—which provides a method of measuring the number of molecules. If the experiment is repeated with sodium chloride, the number of independent particles is found in very dilute solution to be double the expected number of " NaCl molecules "—indicating that the sodium chloride is completely dissociated into Na^+ and Cl^-; but at higher concentration the number is less than.

* The numbers 59·1 and 29·5 mv., found elsewhere in this book, refer to 25°C

† J. N. Agar prefers to put the matter thus: each positive ion tends to collect negative ions around it and vice versa so that the potential energy of each Ag^+ ion is lower than would be the case if the NO_3^- ions were removed far away—as in a very dilute solution; thus the energy barrier W_2 which must be crossed when an ion is discharged, is increased.

double the NaCl concentration, suggesting that the Na^+ and Cl^- ions are not moving independently. Here again it is necessary to introduce an activity coefficient.

In some cases it is possible to obtain the activity coefficient from measurements (1) of electric potential, (2) of freezing-point. Generally, the agreement is reasonably good, and such difference as exists between the two measurements which may be due to experimental error; some authorities, however, regard it as real, since theoretical arguments suggest that the activity coefficient operative in maintaining a state of balance at an electrode surface should be slightly different from that involved in maintaining a state of balance at an ice surface.

In addition to the departure at high concentration from the regular movement of potential with concentration (59·1 and 29·5 mv. for ions of valency 1 and 2 respectively), there is also a departure at very low concentrations. This partly arises from the fact that the argument developed above is statistical, and assumes that a very large number of particles are available to cross the interface in each direction—which at low concentrations will not be the case.* It also arises from the fact that the argument developed above rests on the assumption that the anodic passage of ions from the metal into the liquid is balanced solely by the opposite cathodic reaction—namely, the passage of ions from the liquid on to the metal. If the cathodic reaction consists (for instance) partly of the passage of hydrogen ions from liquid to give hydrogen in or on the metal, then the argument fails; the more dilute the solution, the more probable it becomes that the deposition of metallic ions will not be the only reaction proceeding. At very great dilutions the potential may be only slightly dependent on the metal ion concentration in the solution. Thus the potential of a cadmium electrode in an air-free solution containing potassium and cadmium chlorides becomes almost independent of the cadmium ion concentration when this is below $10^{-4\cdot4}$M, and indeed is much the same as that of cadmium placed in a solution of potassium chloride containing at the outset no cadmium ions (A. L. McAulay and E C. R. Spooner, *Proc. roy. Soc.* (*A*) 1932, **138,** 492; cf. O. Gatty and E. C. R. Spooner, " Electrode potential behaviour of corroding metals in aqueous solutions " 1938, p. 375 (Clarendon Press), where a slightly different interpretation of the matter is suggested).

For a metal like aluminium which cannot be deposited from aqueous solution, the potential of aluminium measured in an aluminium chloride solution has little significance; it cannot represent the balance

$$Al \rightleftharpoons Al^{+++} + 3e$$

but clearly represents a balance between two reactions which are not the opposites of one another; for instance, the anodic reaction might be $Al = Al^{+++} + 3e$ whilst the cathodic reaction, whether aluminium ions are present or absent from the solution, might be the evolution of hydrogen or the reduction of oxygen.

Table of Normal Potentials. The potentials of the commoner metals as measured in solutions containing normal *activity* of their salts are printed in Table XII (p. 312); the order of the metals would not be very different if solutions of normal *concentration* had been used, but the numbers would mostly be shifted appreciably in the negative direction. We can obtain the numbers for any activity value other than normal by applying the concentration correction $\frac{RT}{nF} \log_e a$ (which means, at room temperature, a 59·1 mv. shift in the *negative* direction upon each ten-fold dilution for a monovalent metal and a 29·5 mv. shift for a divalent metal); if we are dealing with solutions made up to definite concentrations, a knowledge of the activity coefficient is needed for exact calculation.

Table XII shows the noble (corrosion-resistant) metals standing at the top,

* J. N. Agar considers the statistical factor is unimportant and that the second effect (an alternative cathodic reaction) is the main cause of the disturbance.

whilst base (easily corroded) metals are placed lower down. There are, however, apparent exceptions to this rule; aluminium for instance is placed below iron and zinc, although it generally resists corrosion better than either—owing to the presence of a highly protective film; the numbers in the table refer, of course, to film-free metal.

The scale adopted in Table XII (the Normal Hydrogen Scale) is based on the Normal Hydrogen Electrode as Arbitrary Zero. As already explained, most of these numbers have not been obtained by direct coupling of a hydrogen electrode to the half-cell formed by the metal in its salt solution, since such a procedure would be inconvenient. Generally the half-cell has been coupled with a calomel electrode or silver/silver-chloride electrode, and the measured potential converted to the hydrogen scale, by " adding " one of the numbers shown in Table XLI. It should be noticed that the more concentrated the potassium chloride solution used in a silver/silver chloride electrode, the *lower* is the potential, since the $[Ag]^+$ activity is thereby diminished.

TABLE XLI

SOME PRACTICAL ELECTRODES AND THEIR POTENTIALS AT 25°C.

(from " Corrosion Handbook ": edited by H. H. Uhlig: published by Wiley)

Electrode	Potential
Hg | Solid Hg_2Cl, Solid KCl	+ 0·2415 volt
Hg | Solid Hg_2Cl_2, N KCl	+ 0·2800
Hg | Solid Hg_2Cl_2, N/10 KCl	+ 0·3337
Ag | Solid AgCl, N/10 KCl	+ 0·2881
Cu | Saturated $CuSO_4.5H_2O$	+ 0·316

The last-named electrode is much used in cathodic protection work (p. 285). Another electrode, used for the same purpose but in marine environments, is Ag | Solid AgCl | Sea-water; it is commonly regarded as having the same electrode potential as the saturated copper sulphate electrode, but " sea-water " varies greatly in composition from place to place in the sea.

Potential of a Metal forming two types of Ions. When a metal displays a variable valency, different numbers can be written down for the potential reached when it is in equilibrium with different ions, each at normal activity; the number representing equilibrium with the ion most stable at concentrations close to normal is generally quoted, but for some calculations regarding corrosion processes the others assume importance. Thus in the case of copper there exist, on paper, three equilibria

$$Cu \rightleftharpoons Cu^{++} + 2e \quad +0{\cdot}345 \text{ volt}$$
$$Cu \rightleftharpoons Cu^{+} + e \quad +0{\cdot}522$$
$$Cu^{+} \rightleftharpoons Cu^{++} + e \quad +0{\cdot}167$$

All the equilibria quoted refer to solutions containing the ions at normal activities, but the second equilibrium cannot in practice be realized since, if a solution containing Cu^+ at normal activity were somehow to be obtained and placed in contact with metallic copper, it would at once deposit metallic copper until the Cu^+ activity had sunk so low that the equilibrium

$$Cu \rightleftharpoons Cu^{+} + e$$

occurred at the same potential as

$$Cu \rightleftharpoons Cu^{++} + 2e$$

Conversely, if copper is placed in a liquid containing normal activity of Cu^{++}, it will react until sufficient Cu^+ has been formed to establish equilibrium at the potential at which Cu^{++} could also stand in equilibrium with the metal. In this particular case, the triangular equilibrium

$$\begin{array}{ccc} Cu & \rightleftharpoons & Cu^{++} + 2e \\ & \searrow\!\!\nwarrow \quad \nearrow\!\!\swarrow & \\ & Cu^{+} + e & \end{array}$$

is set up at a position where $[Cu^{++}]$ greatly exceeds $[Cu^{+}]$—provided that we are dealing with high *total* concentrations of copper ions. When the total amount of dissolved copper is small, the reverse is true, since the equilibrium potential of Cu^{++} shifts by only 29·5 mv. for each ten-fold dilution whereas that of Cu^{+} shifts by 59·1 mv. For this reason, the value of the equilibrium $Cu \rightleftharpoons Cu^{+} + e$ (+0·522 volt)—despite the fact that it cannot be realized at normal activities—must be taken as the starting-point for calculations in certain corrosion problems involving very small copper concentrations (p. 191).

The case of iron is the opposite to that of copper. Of the three equilibria

$$\begin{array}{llll} Fe & \rightleftharpoons Fe^{+++} + 3e & -0{\cdot}036 \text{ volt} \\ Fe & \rightleftharpoons Fe^{++} + 2e & -0{\cdot}440 \\ Fe^{++} & \rightleftharpoons Fe^{+++} + e & +0{\cdot}771 \end{array}$$

the first cannot be realized. Iron brought into contact with a liquid containing ferric ions at roughly normal activity interacts to give ferrous ions, and the triple equilibrium is only realized when $[Fe^{++}]$ is in great excess of $[Fe^{+++}]$.*

Potential of a highly reactive metal. Consider a metal like sodium which cannot be deposited from aqueous solution. Here a direct measurement of potential on the potentiometer would clearly be difficult and would certainly not provide the balance point of the equilibrium

$$Na \rightleftharpoons Na^{+} + e$$

In such a case an indirect method may be used. Thus although metallic sodium is rapidly attacked by water, a dilute amalgam made by dissolving about 0·2% of sodium in mercury will react only very slowly with a sodium chloride solution, owing to the high value of the hydrogen overpotential on a mercury surface. Allmand and Polack measured the E.M.F. of the cell

$$0{\cdot}206\% \text{ Na (in Hg)} \mid \underset{\text{(in water)}}{1{\cdot}022\text{M NaCl}} \mid \underset{\text{(solid)}}{\text{Calomel}} \mid \text{Hg}$$

and then tested pure sodium against the amalgam by using a solution of sodium iodide in anhydrous ethylamine, which has no action on sodium, in the cell

$$Na \mid NaI \text{ in } C_2H_5NH_2 \mid 0{\cdot}206\% \text{ Na (in Hg)}$$

Taking 0·65 as the activity coefficient of NaCl at 1·022M concentration, they arrived at —2·7149 volts as the normal electrode potential of sodium, which is close to the value (—2·72) obtained from considerations of free energy (A. J. Allmand and W. G. Polack, *J. chem. Soc.* 1919, **115,** 1020).

In other cases where, from data of a non-electrochemical character, the free energy of a change represented in the discharge of a cell is known, the balance E.M.F. (E) can be calculated, since the free energy change is equal to nEF. This free-energy change depends largely on the work needed to separate a cation existing in the surface layer of the metal from its neighbours, and transfer it—despite the opposition of the electric forces between the cation and the surplus electrons left behind in the metal—into the liquid; the transfer is helped by the affinity between the cation and the water-molecules, which probably orient themselves around it with their negative ends next to the positive cations, producing a stable arrangement.

Since this affinity of cations for water complicates the situation, studies have

* The various equilibria which can be set up at an iron electrode surface are discussed by K. Nagel, " Passivierende Filme und Deckschichten " p. 92 (edited by H. Fischer, K. Hauffe and W. Wiederholt; published by Springer).

been made of the potentials of metals in a contact with a compound in the fused state (generally the chloride). Moreover, theoretical calculations, based on free energy, have been made of the balance E.M.F. of the cell

Metal | Solid chloride | Chlorine

The order of the metals, arranged according to the magnitude of this E.M.F., does not differ greatly from that of the cell

Metal | Chloride solution (normal cation activity) | Chlorine

at the same temperature (25°C.). The cells containing *solid* chloride show the order

K, Na, Ca, Mg, Mn, Al, Cd, Pb, Sn, Fe (divalent), Co, Ni, Ag, Hg, Cu (divalent), Pt, Au,

whilst those containing *dissolved* chloride show

K, Ca, Na, Mg, Al, Mn, Zn, Fe (divalent), Cd, Co, Ni, Sn, Pb, Cu (divalent), Hg, Ag, Pt, Au.

The potentials of other metals not mentioned above, along with data at higher temperatures, will be found in the original paper by W. J. Hamer, M. S. Malmberg and B. Rubin, *J. electrochem. Soc.* 1956, **103**, 8.

In solutions containing cyanide, the simple cations of the heavy metal hardly exist, and a noble metal behaves electrochemically like a base metal. The negative value of the equilibrium

$$2CN^- + Ag \rightleftharpoons Ag(CN)^-_2 + e \quad (—0{\cdot}29 \text{ volt})$$

stands in strong contrast with the positive value of

$$Ag \rightleftharpoons Ag^+ + e \quad (+\ 0{\cdot}7991 \text{ volt})$$

High concentrations of Cl^- may also in some cases cause the metal to exist largely in the form of complex anions. Even in the absence of such complex-forming anions, cations may fail to be formed in certain cases. Thus Latimer writes the equilibrium between a platinum electrode and acid water as

$$Pt + 2H_2O \rightleftharpoons Pt(OH)_2 + 2H^+ + 2e \quad (+\ 0{\cdot}987 \text{ volt})$$

The equilibrium

$$Pt \rightleftharpoons Pt^{++} + 2e \quad (\text{about } +\ 1{\cdot}2 \text{ volts})$$

possesses only theoretical interest. Several equilibria between metal and oxide possess importance in corrosion work; an example is

$$3Fe + 8OH^- \rightleftharpoons Fe_3O_4 + 4H_2O + 8e \quad (-\ 0{\cdot}91 \text{ volt})$$

Electrochemical Kinetics

Distribution of Potential Drop in a Symmetrical Cell. Consider the cell

Ag | $AgNO_3$ solution | Ag

Since it is symmetrical it should balance exactly zero volts (if the establishment of balance on the potentiometer P is found to require a small E.M.F.—positive or negative—it is a sign that the two silver electrodes are not quite the same, differing perhaps in surface condition). If current is forced across the cell from the external battery B (fig. 198), in series with the key K, variable resistance, R, and ammeter, A, metallic silver will be deposited on the cathode, the electrode joined to the negative pole of the external battery, and a similar amount dissolved at the other electrode, the anode. The potentiometer will no longer register zero; the lower the external resistance is made, the higher becomes the current, and the greater the potential drop across the cell is found to be. This potential drop is needed for various purposes; it may have to provide for (1) an ohmic drop over the liquid within the cell, (2) concentration polarization,

(3) activation polarization, (4) approach-resistance polarization. These must be considered separately:—

(1) *Ohmic Drop.* If the current equal to I amperes is to pass through a cell of internal resistance R ohms, there must be a drop of IR volts across the cell; clearly this requirement can be diminished by bringing the two silver electrodes closer, or by adding, say, sodium nitrate or nitric acid to the liquid, since either of these changes will reduce R.

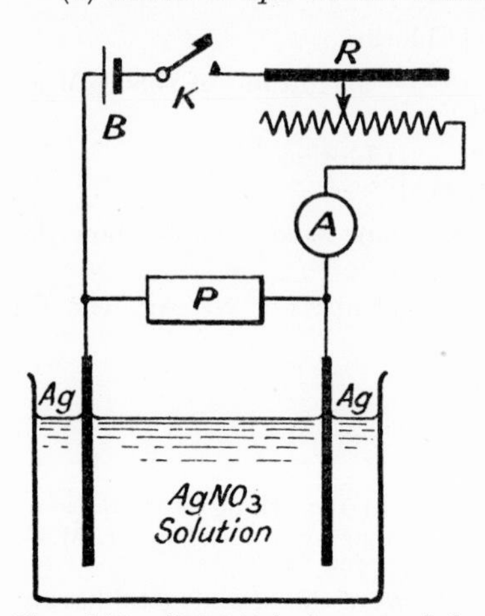

FIG. 198.—Potential drop needed to force current through a reversible cell.

(2) *Concentration Polarization.* As current flows, the concentration of silver nitrate near the anode increases whilst that near the cathode diminishes; thus the potential drop which would exist, quite apart from current, will become more positive at the anode and more negative at the cathode.

Both changes tend to oppose the flow of current. If current is switched off, by interrupting the circuit at the key K, the potential at each electrode will remain different from that at the outset, and the potentiometer will now balance the cell at an E.M.F. which is appreciably removed from zero. The fact that polarization remains after current has ceased to flow distinguishes concentration polarization from other sorts of polarization. Another distinction is that concentration polarization can be almost prevented by vigorous stirring. In stagnant solution, the amount of the polarization will depend largely on the geometry. If the electrodes are horizontal, with the anode below (fig. 199(I)), the heavy silver nitrate will accumulate on the anode surface,* whilst

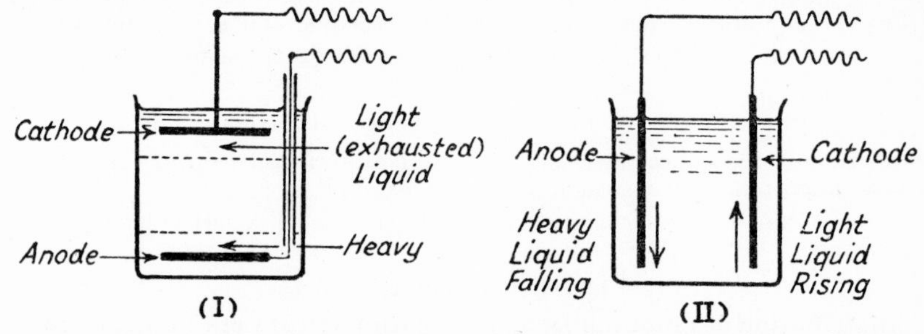

FIG. 199.—Effect of geometrical arrangement of electrodes on concentration polarization.

the light exhausted solution will remain above around the cathode; the polarization will thus increase steadily with time, the value ultimately reached being largely determined by freedom from mechanical disturbance. If the cathode is placed below, and the anode above, convection currents set up by gravity will prevent serious concentration polarization. Convection tends to reduce concentration polarization if the electrodes are vertical, with streams of light exhausted solution

* The " heaviness " of the liquid formed at the anode results from the fact that anions are moving continuously towards the anode but are not consumed when they arrive there, so that the concentration of NO_3^- (and consequently that of Ag^+ if electrical neutrality is to be preserved) becomes greater than elsewhere; the cause is analogous to that responsible for the concentration of chlorine ions at the anodic areas in stray-current corrosion exceeding that in the body of the ground (p. 265).

rising up the cathode surface, and streams of heavy concentrated solution descending on the anode (fig. 199(II)).

(3) *Activation Polarization.* So long as no current is flowing between the silver and the solution, the potential must be that at which the left-to-right reaction exactly balances the right-to-left reaction

$$Ag \rightleftharpoons Ag^+ + e$$

If current is to flow, then at the anode the left-to-right reaction must exceed the right-to-left reaction, and this needs a displacement of the potential in the positive direction; at the cathode the right-to-left reaction must exceed the left-to-right reaction, needing a displacement in the negative direction. Such displacements can be demonstrated by tubuli opening close to the two electrode surfaces and leading (through suitable bridges containing liquids chosen to eliminate liquid junction potential differences) to calomel or silver/silver-chloride electrodes (fig. 141, p. 800); the potentiometer reading, in the absence of current-flow, should be nearly the same for both electrodes, but when current is forced through it the two potentials will shift in reverse directions. At a silver electrode, this " activation " potential is relatively small, but for the " abnormal " metals, like iron and nickel, it may become large. It can be extremely large for the cathodic evolution of hydrogen on certain surfaces, and in such cases the polarization is generally known as *overvoltage*.

The placing of the tubuli requires care. If they press against the electrode surface, they will locally screen it from current; if they are placed at an appreciable distance from it, an *IR* drop, corresponding to the fall of potential in the liquid between the metallic surface and the tubulus mouth, will cause an error. One method of overcoming the difficulty consists in placing the tubulus tip at a small distance from the metal (which distance must be precisely measured) and then applying a correction to the measurements made so as to allow for the known *IR* drop. Other and more elegant methods of avoiding the difficulty —some based on tubuli of special design—have been introduced by Piontelli, and are described on p. 799.

In general, at any electrode, two currents may be regarded as flowing in opposite directions; the anodic (left-to-right) current density is given by $K_1 e^{-(W_1 - n\varepsilon\alpha(V+\Delta V))/kT}$ where $\alpha = V_1/V$ and ΔV is the departure of potential from the equilibrium value V; the cathodic (right-to-left) current density is given by $K_2 C e^{-(W_2 + n\varepsilon(1-\alpha)(V+\Delta V))/kT}$. These are shown by the thin lines on fig. 200, whilst the net current density, which is the difference between the opposing anodic and cathodic components, is shown by the thick lines. Two cases are presented; on normal metals, a small departure from the equilibrium value causes an appreciable flow of current, whereas on abnormal metals, a considerable departure is necessary. In drawing this diagram, it has been assumed that $\alpha = (1 - \alpha) = 1/2$, so that the curves are symmetrical about the origin.

The current flowing in each direction under equilibrium conditions ($\Delta V = 0$), where the anodic and cathodic currents are in balance, is known as the *exchange current*, I_0; the corresponding current density, I_0/A (where A is the electrode area), is written i_0. The curve relating the net current density i with the departure ΔV from the equilibrium potential is the difference between two exponentials; in the neighbourhood of the equilibrium point V it is approximately straight. At a distance from V, one exponential becomes small compared to the other. If, for instance, the potential is shifted well away from equilibrium in an anodic direction, the cathodic component can be neglected, and the net current density i is given with sufficient accuracy by the single term $K_1 e^{-(W_1 - n\varepsilon\alpha(V+\Delta V))/kT}$.

Since
$$i_0 = K_1 e^{-(W_1 - n\varepsilon\alpha V)/kT}$$
$$i = i_0 e^{n\varepsilon\alpha\Delta V/kT}$$

It follows that
$$\log_e (i/i_0) = n\varepsilon\alpha\Delta V/kT$$

or
$$\Delta V = 2{\cdot}303 \frac{kT}{n\varepsilon\alpha} \log_{10} \frac{i}{i_0} = b \log_{10} \frac{i}{i_0}$$

The equation is often written

$$\Delta V = a + b \log_{10} i$$

where a represents $- b \log_{10} i_0$; a and b are both constants independent of ΔV. This is known as *Tafel's equation* and b is called the *Tafel slope*.

It should be stated at once that for many electrode systems, the situation is far less simple than that represented by fig. 200. In some cases, it is likely that a definite gap should be left on each side of the equilibrium potential,

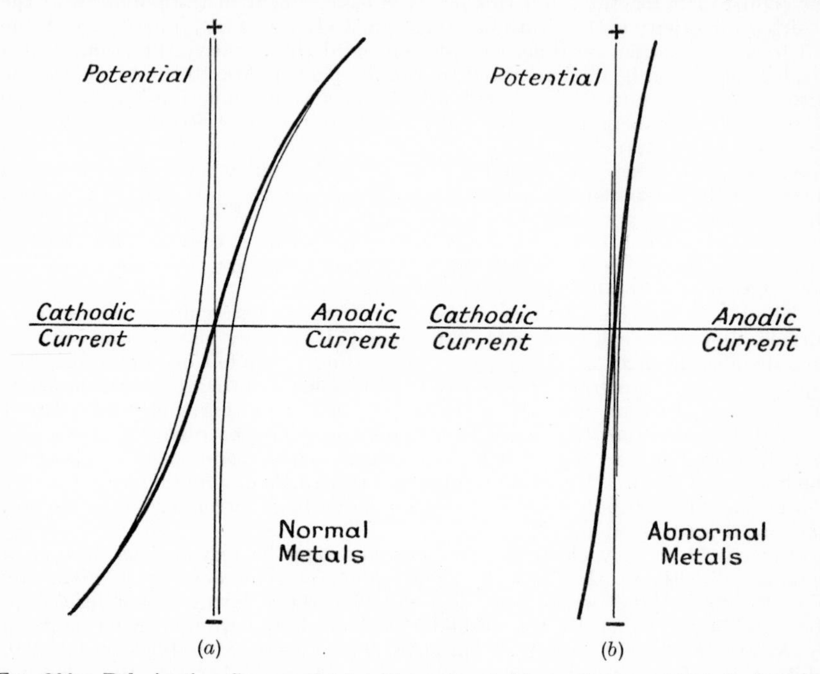

FIG. 200.—Polarization Curves of (a) a Normal and (b) an Abnormal Metal, obtained by compounding their anodic and cathodic components (resultant curves shown as thicker lines).

representing a range within which no current, either anodic or cathodic, can flow. In a theoretical paper, Vermilyea argues that, for a " perfect " surface, no (cathodic) current can flow until the potential has been moved a certain distance away from the theoretical balance-point; nor is this surprising, in view of the considerations presented on p. 1035. Of course, metal as we know it never possesses a perfect surface, but even imperfect surfaces of different characters behave differently; indeed, at potentials close to the theoretical balance-point, the surface structure is all-important in deciding the amount of current which can pass. The relation between i and V on an atomical rough surface is different from that on a surface which is smooth except for the steps present where screw dislocations emerge on to the surface (D. A. Vermilyea, *J. chem. Phys.* 1956, **25**, 1254).

It is sometimes thought that Tafel's rectilinear relation between log i and V is valid for the cathodic evolution (or anodic consumption) of hydrogen, but even this will only be true if complicating factors, such as concentration polarization, are absent; several other complicating factors could be named, including the fact that the passage between H^+ and H_2 proceeds by two main steps (essentially $H^+ \longrightarrow H$ and $H \longrightarrow H_2$)—a fact which can only be disregarded if one step proceeds very smoothly in comparison to the other. The value of ΔV needed to allow a given value of i depends on the metal forming the cathode on which hydrogen is being formed. Some values of the hydrogen overvoltage are presented in Table XLII. If the Tafel relation really represented the experimental facts, then the difference between the values in the columns headed 10^{-4} and 10^{-3} should be equal to that between the columns 10^{-3} and 10^{-2}; a glance will show that this is not the case.*

TABLE XLII

OVERVOLTAGE OF HYDROGEN EVOLUTION FROM N SULPHURIC ACID ON DIFFERENT CATHODE MATERIALS (numbers collected by W. M. Latimer, mostly from International Critical Tables and the *Handbuch der Physik*)

Current Density (amps/cm.^{2})	10^{-4}	10^{-3}	10^{-2}
Pt black	0·003	0·01	0·03
Pt bright	—	0·02	0·07
Au	0·12	0·24	0·39
Fe	0·22	0·40	0·56
Ni	0·20	0·56	0·74
C	0·29	0·55	0·70
Ag	0·30	0·48	0·76
Cu	0·35	0·48	0·58
Pb	0·45	0·52	1·09
Al	0·50	0·57	0·83
Zn	0·55	0·72	0·75
Hg	0·60	0·78	0·93

Although Table XLII sets forth the values of over-potential in convenient form, the figures may not represent the latest determinations. The data of Table XLIII extracted by J. N. Agar from the Russian work " Kinetics of Electrode Processes " by A. N. Frumkin, V. S. Bagotsky, Z. A. Iofa and B. N. Kabanov, p. 126 (Moscow University Press) are probably more accurate; this table gives the values of a a and b at 20°C. in the Tafel equation

$$\eta \text{ (volts)} = a + b \log_{10} i$$

where i is the current density in amps/cm.^{2}.

TABLE XLIII

OVERPOTENTIAL CONSTANTS (A. N. Frumkin)

Metal	Solution	a (volts)	b (volts)
Pb	N H_2SO_4	1·56	0·110
Hg	N H_2SO_4	1·415	0·113
	N HCl	1·406	0·116
	N KOH	1·51	0·105
Cd	1·3N H_2SO_4	1·40	0·120
Zn	N H_2SO_4	1·24	0·118
Sn	N HCl	1·24	0·116
Cu	N H_2SO_4	0·80	0·115
Ag	N HCl	0·95	0·116
Fe	N HCl	0·70	0·125
	2N NaOH	0·76	0·112
Ni	0·11N NaOH	0·64	0·100
Co	N HCl	0·62	0·140
Pd	1·1N KOH	0·53	0·130
Pt	N HCl	0·10	0·13

* J. N. Agar remarks that he has no great faith in the accuracy of the figures of Table XLII; he adds that for platinum black—perhaps for bright platinum also—the conditions $\overrightarrow{i} \gg \overleftarrow{i}$ is not fulfilled.

In comparing the hydrogen overvoltage of different materials, it should be remarked that the fact that black platinum gives a lower number than bright platinum at the same nominal current density is largely due to the fact that the effective area of the black platinum is much greater, so that the true current density is much lower than the nominal value. Conversely, the high overvoltage on liquid mercury may be partly (but not wholly) due to its flat surface, and the apparently high overvoltage on aluminium to the fact that, since much of the surface is covered with an oxide-film possessing a poor electronic conductivity, the cathodic reaction can take place only on exceptional points (where the film is thin or interrupted by gaps filled with other metals), so that the true current density is much higher than that tabulated. Table XLII shows that both on aluminium and mercury the increase of overvoltage on passing from 10^{-4} to 10^{-2} amps/cm.^{2} is less than on gold or nickel which, at first sight, seem to be metals of much lower overvoltage. In contrast, on lead, the overvoltage doubles itself on passing from 10^{-4} to 10^{-2} and here the high value appears to be a property of the metal and not simply due to the small area available. It will be recollected that lead is a catalytic poison.

Interesting papers developing other views on hydrogen overpotential come from P. van Rysselberghe, *J. chem. Phys.* 1949, **17**, 1226; 1952, **20**, 1522. P. J. Hillson, *Trans. Faraday Soc.* 1952, **48**, 462. J. F. Chittum, *Nature* 1959, **183**, 589.

(4) *Approach-resistance Polarization.* A fourth cause of potential drop should here be mentioned, since it appears to be important in connection with many corrosion processes.

Let us imagine that the silver deposit on the cathode has grown out—as indeed frequently occurs—forming a limited number of large crystals, and that deposition is occurring exclusively at the points forming the tips of a few of these crystals; in that case, the resistance of the bottle-neck approach to each favoured point may be high, and special *IR* drops occur in the parts of the liquid close to the point. Strictly speaking, this is part of the Ohmic Drop, but when we measure polarization with the tubulus in the ordinary way, it will be included in the value of the polarization recorded. It will, however, differ from activation and concentration polarization in being directly proportional to the current flowing. Similarly, it may happen that the attack on the anode is directed on small points—perhaps places where the atomic structure is highly disarrayed, in contrast with the rest of the surface, where, owing to the perfect structure, the activation energy of the passage from metal to liquid is much higher: in this case, also, the effect is to add to the polarization, as commonly measured, a term proportional to the current strength.

It is often found, in measuring the polarization at the anodic and cathodic areas of corrosion specimens (see p. 865) that the curves relating potential to current are almost straight. This may be due to the fact that, at any given moment, the anodic and cathodic reactions are proceeding at a few small points only; the points in question may change from time to time, so that in the end the anodic corrosion may be found to have affected large areas. The straightness of the polarization curve is difficult to explain except on the view that what is being measured is an " Approach-resistance Polarization " and that this is large enough to overwhelm the Activation Polarization or Concentration Polarization.

An interesting discussion of the division of over-potential into its component parts is provided by K. J. Vetter, *Z. Elektrochem.* 1952, **56**, 931. See also K. Nagel, *Z. Naturforschung* 1946, **1**, 433.

Electrical Balance and Chemical Balance. At the single electrode

$$Ag \mid AgNO_3$$

measurements on the potentiometer will indicate equilibrium at that value of the potential which will cause the " anodic " reaction

$$Ag = Ag^+ + e$$

to proceed at exactly the same rate as the "cathodic" reaction

$$Ag^+ + e = Ag$$

so that the electrode becomes neither anodic nor cathodic. Here we have not only an *electrical balance* (electrons passing in the two opposing directions at the same rate), but also a *chemical balance* (silver ions passing in the two opposing directions at the same rate).

Suppose, however, we have a zinc electrode partly immersed in potassium chloride solution and measure its potential by means of a tubulus, the mouth of which is placed at a distance from the zinc; if this is connected through a calomel electrode to a potentiometer, we can use the intrument to find the potential at which there is an electrical balance. However, the potential in question does not represent a chemical balance, since the cathodic reaction is not the reverse of the anodic reaction. Although the potentiometer procedure has been that designed for a study of an equilibrium system, it is now being applied to a system which is not in equilibrium. Near the water-line, oxygen is being reduced, perhaps to OH^- ions, by some such cathodic reaction as

$$O_2 + 2H_2O + 4e = 4OH^-$$

whilst further down, the zinc is passing into the liquid by the anodic reaction

$$2Zn = 2Zn^{++} + 4e$$

so that electrons produced on the lower part of the surface are used up on the upper part. The potential measured by a tubulus opening at a distance is merely the potential at which the cathodic reaction is proceeding at the same rate as the anodic reaction, but (in contrast to the Ag | $AgNO_3$ electrode) these two reactions are not equal and opposite to one another. The measurement is still not devoid of meaning, provided that we are quite sure what the two opposing reactions are; but if an undetermined amount of the oxygen is being reduced to hydrogen peroxide instead of to water, the potential measurement loses all its significance. Yet meaningless measurements of this kind are today frequently made by persons who can use a potentiometer without understanding what the values obtained signify, and unjustifiable conclusions are drawn from the results in such cases. The reader should be on his guard against placing reliance on arguments based on that sort of procedure.

If we bring the tubulus tip close to the zinc, we shall in general obtain a different potential near the cathodic area at the water-line from that read at the anodic area lower down. Such measurements possess considerable interest; the matter is discussed on p. 862.

Faraday's Law. Since the cathodic discharge of a metallic ion involves a number of electrons equal to the valency, there must clearly be a relation between the current passing and the amount of metal deposited at a cathode or, alternatively, dissolved at an anode. Faraday's Law states that the number of coulombs needed to deposit, or dissolve, one gram-equivalent, is the same for all metals; the number is, in fact, about 96,500 coulombs, or 26·8 ampere-hours—which means that 1 ampere-hour will produce, at an anode, the corrosion of 1·04 grams of pure iron (provided that it enters the solution as ferrous ions), 1·22 grams of zinc or 3·87 grams of lead; it will deposit these amounts at a cathode. This assumes that the current is wholly devoted to dissolution or deposition and that no alternative reactions are proceeding. If the amount of metal dissolved or deposited is found to differ from what is expected, the actual amount expressed as a percentage of the calculated amount is called the *Current Efficiency*.

Measurements of current efficiency may exceed 100%; for instance, at an anode, in addition to the current dissolved by the action of the current passing round the main circuit, there may also be dissolution through local action, which may include corrosion by currents flowing between different regions on the anode and not registered on the ammeter in the main circuit. The efficiency may fall far below 100% if the anode is partially passive.

Special Electrode Systems

Hydrogen Electrode. The normal hydrogen electrode has been mentioned as the arbitrary zero of the Hydrogen Scale. The potential of an inert electrode (usually platinum) and hydrogen-saturated acid can be altered not only by altering the hydrogen-ion concentration in the liquid, but also by altering the pressure of the gaseous hydrogen with which the liquid is in equilibrium. Hydrogen at high pressure will be more active (more like zinc and less like silver) than at low pressure, since clearly the increased pressure favours the left-to-right change of the equilibrium

$$H_2 \rightleftharpoons 2H^+ + 2e$$

Since two electrons per molecule are involved in this reaction, it follows that at ordinary temperature a ten-fold increase in hydrogen pressure shifts the potential (in the negative direction) by only 29·5 mv., although a ten-fold increase in hydrogen-ion concentration in the liquid shifts it (in the positive direction) by 59·1 mv., since here one electron is concerned in the discharge of one ion. In both cases, the relation is only approximately true if the gaseous *pressure* or ionic *concentration* is the variable which is being increased 10 times. For purposes of exact calculation, we must replace pressure by *fugacity* and concentration by *activity*; fugacity is the pressure corrected for the departure from ideal conditions—just as activity is concentration similarly corrected.

pH Scale. It has been mentioned that pure water consists mainly of H_2O molecules, but there is a very small amount of ionization to hydroxyl ions (OH^-) and hydrogen ions, generally written H^+ but regarded by many physical chemists as H_3O^+; according to this view, the ionization is not a breaking up of H_2O into OH^- and H^+ but is rather due to the fact that in the course of molecular movements, the proton belonging to one H_2O molecule attaches itself momentarily to a neighbour, thus

$$H_2O + H_2O = H_3O^+ + OH^-$$

This is a rare and fleeting occurrence; soon the proton flies back, but the same transfer then occurs between some other pair. At any moment, the concentration both of H^+ and OH^- in pure neutral water is about 10^{-7}N at ordinary temperature, although higher at elevated temperature; it is nearly 10^{-6}N at 100°C. If acid is added to the water, the $[H^+]$ increases and $[OH^-]$ diminishes; if alkali is added, $[OH^-]$ increases and $[H^+]$ diminishes.

In all cases, the equilibrium

$$H^+ + OH^- \rightleftharpoons H_2O$$

involves a constant value for $\dfrac{[H^+][OH^-]}{[H_2O]}$ at any fixed temperature, and here $[H_2O]$, being greatly in excess of $[H^+]$ and $[OH^-]$, can be regarded as constant. Hence the product $[H^+][OH^-]$ is constant, the value at room temperature being 10^{-14} (gram-ions/litre)^2.

The potential of blackened platinum in a liquid saturated with hydrogen at 1 atmosphere pressure affords a convenient way of expressing the hydrogen-ion concentration (or, strictly, hydrogen-ion activity) on a logarithmic scale. The values are known as *pH values*,* and the statement that a liquid possesses the pH value n, means that the hydrogen-ion activity is 10^{-n}. Similarly

* J. N. Agar adds the comment that there is much complication in defining the pH-scale since we do not know the activity of H^+ itself in HCl but only the mean value of the two ions H^+ and Cl^-; the single values are not necessarily equal to one another.

hydroxyl activity can be expressed on a pOH scale; pOH rises as pH falls, the sum (pOH + pH) being 14 at ordinary temperature. If concentration and activity were the same, normal acid would have a pH value equal to zero (pOH = 14), whilst normal alkali would have pH 14 (pOH = 0). However, the activity coefficient of normal hydrochloric acid is not unity but 0·809 and if a liquid of pH 0 is needed, we must choose 1·2N hydrochloric acid, not 1·0N; this explains the use of acid slightly in excess of normal in making up the normal hydrogen electrode. Since the potential moves through about 59 mv. for every ten-fold alteration of hydrogen activity, it follows that 59 mv. represents a shift of about one pH unit. The relation between potentials, pH and pOH values are shown in Table XLIV.

TABLE XLIV

POTENTIAL OF HYDROGEN AT ATMOSPHERIC PRESSURE IN DIFFERENT LIQUIDS

Hydrogen-ion Activity	Example of Liquid used	Potential of Platinum in the hydrogen-saturated liquid	pH	p(OH)
Over 1·0N	HCl above 1·2N	Positive	Negative	>14
1·000N	1·2N HCl	0·000 volt (arbitrary zero)	0	14
10^{-1}N	0·12N HCl	− 0·059	1	13
10^{-2}N	0·011N HCl	− 0·118	2	12
10^{-3}N	10^{-3}N HCl	− 0·177	3	11
10^{-4}N	10^{-4}N HCl	− 0·237	4	10
10^{-5}N	10^{-5}N HCl	− 0·296	5	9
10^{-6}N		− 0·355	6	8
10^{-7}N	Pure water	− 0·414	7	7
10^{-8}N		− 0·473	8	6
10^{-9}N	10^{-5}N KOH	− 0·532	9	5
10^{-10}N	10^{-4}N KOH	− 0·591	10	4
10^{-11}N	10^{-3}N KOH	− 0·650	11	3
10^{-12}N	0·011N KOH	− 0·710	12	2
10^{-13}N	0·12N KOH	− 0·769	13	1
10^{-14}N	1·2N KOH	− 0·828	14	0

It is possible, and at one time was customary, to use a hydrogen electrode in determining the hydrogen-ion activity of an unknown liquid. One method was to measure the E.M.F. of the combination

Pt	Unknown Liquid saturated with H_2 at 1 atm.	Connecting Liquid to avoid Junction Potentials	KCl	Calomel Electrode

If now the potential of the calomel electrode used against a *normal* hydrogen electrode is known, any departure (ΔV) from this value (0·28 volts if N KCl is used in the calomel electrode) is due to the fact that the hydrogen-ion activity of unknown liquid is not normal, but 10^{-n}N, where $n = 0{\cdot}059 \times \Delta V$. Thus the pH value (which is equal to n) can be found.

However, hydrogen-saturated electrodes are somewhat temperamental, partly through the effect of traces of impurity in poisoning the surface and partly from difficulty in keeping the liquid exactly saturated with hydrogen. It is today common practice to use a *glass electrode* consisting of a thin-walled bulb of special glass possessing appreciable electrical conductivity, filled with a *buffer* solution of known pH value containing a little quinhydrone (see below); a platinum wire is placed in it and the potential between that wire and the external liquid is a measure of the pH of the latter, moving by about 59 mv. for each pH unit; the exact number varies, however, from one glass electrode to another and it is usual to standardize each electrode in a number of solutions representing known hydrogen-ion concentrations. The mechanism of the glass electrode is still a subject of discussion, but a good account both of its theoretical and practical aspects is provided by M. Dole, " The Glass Electrode " (Wiley: Chapman & Hall, 1941 edition). See also R. G. Bates, " Electrometric pH determination " (Wiley: Chapman & Hall).

Comparison of Hydrogen and Oxygen Electrodes. The Normal Hydrogen Electrode makes use of platinum foil on which platinum has been deposited in a black, velvety, surface-rich form, so that the current density (current for unit area) corresponding to an appreciable current is extremely low; since it is the current density which determines activation polarization, the advantage of the blackened surface is evident. A further probable advantage is that such a surface possesses catalytic points on which the reactions

$$H_2 \rightleftharpoons 2H^+ + 2e$$

proceed smoothly in either direction; care must be taken not to poison these points by the adsorption of some firmly held impurity. If carefully prepared and preserved, the hydrogen electrode is approximately reversible. It balances the potential predicted by theory, and if the potential is moved even slightly away from the balance point, a current will pass in the appropriate direction; the electrochemical change indicated above occurs either in the left-to-right or right-to-left direction according as the potential has been moved in the positive or negative direction. In acid of normal activity and 1 atmosphere hydrogen pressure (pH = 0), the balance potential is by definition, *zero*, but as the pH rises the potential ought to shift at the rate of 59 mv. per pH unit —as shown in Table XLIV (p. 1029). On blackened platinum, the potentials indicated in the table can, in general, be realized experimentally, but on the surface of another metal, they will not usually be obtained; this may be partly due to lack of suitable catalytic spots, but partly because the metal itself tends to pass into solution, thus providing an alternative anodic reaction. In such a case, the potential measured by balance on the potentiometer is that value at which the sum of all the anodic reactions occurring balance the cathodic reaction or reactions; unless we know what all these reactions are, the measurement possesses no significance.

However, whilst the theoretical potential of the *hydrogen electrode* can be realized by observing reasonable precautions, the case is different with the *oxygen electrode.* Knowledge of the free energy of the reaction

$$\underset{\text{(gas)}}{2H_2} + \underset{\text{(gas)}}{O_2} = 2H_2O \text{ (liquid)}$$

predicts that cell

$$H_2 \left| \begin{array}{c} \text{Water} \\ \text{(acid, neutral or alkaline)} \end{array} \right| O_2$$

ought to balance an E.M.F. of 1·23 volts, irrespective of the pH value of the liquid. If the E.M.F. opposing the cell were raised slightly above 1·23 volts, then, supposing the cell were reversible, current should flow, and water should be decomposed to hydrogen and oxygen; if the E.M.F. were diminished by the smallest amount, then current should flow in the opposite direction, hydrogen and oxygen combining to form fresh water. In practice, if we try to obtain current from an hydrogen | oxygen cell, the E.M.F. obtained is far below 1·23 volts, whilst if we want to decompose water, we must use an E.M.F. far exceeding 1·23 volts—even though the electrodes may consist of blackened platinum. Thus, in practice we fail to realize the equilibrium

$$O_2 + 2H_2O + 4e \rightleftharpoons 4OH^-$$

at the theoretical potential, which should always be 1·23 volt positive to that of

$$2H^+ + 2e \rightleftharpoons H_2$$

in the same liquid.

The reason for the irreversible behaviour of the oxygen electrode is the high activation energy. Consider the cathodic reaction; this proceeds in steps; first presumably the oxygen molecules are adsorbed, which are reduced first to hydrogen peroxide and then to water (or perhaps are dissociated into atoms and reduced direct to water).* Comparatively seldom will a particle attain the correct energy condition to undergo the change unless it is helped by a shift of the potential away from the theoretical equilibrium value. The exchange current, in consequence, is very low. This should not in itself affect the potential at which there is a balance between the slow changes proceeding in each direction, provided that the only possible cathodic change is $O_2 \longrightarrow OH^-$ and the only possible anodic change $OH^- \longrightarrow O_2$. If, however, there is even the faintest trace of some oxidizable impurity in the liquid, an alternative anodic reaction become possible, and the potentiometer will indicate an electrical balance at that potential at which the cathodic reaction (reduction of oxygen) uses up exactly as many electrons as are supplied by the two anodic reactions. A trace (10^{-5} mol./litre) of sulphur dioxide, such as is likely to be present in any sulphuric acid distilled in air, might easily prevent the equilibrium potential from being obtained. Until 1956 attempts to establish the theoretical value experimentally failed. In that year, however, Bockris and Huq, after careful removal of the last trace of impurity by preliminary electrolysis (between auxiliary electrodes, *not* the electrode to be used for the ultimate potenial measurement) obtained a value of 1·24 volt—very close to the theoretical value (1·23 volt); in order to achieve this, it was necessary to reduce the impurity content below 10^{-11} mol./litre (J. O'M Bockris and A. K. M. Shamshul Huq, *Proc. roy. Soc.* (*A*) 1956, **237,** 277; cf. W. F. K. Wynne-Jones, *Trans. Faraday Soc.* 1957, **53,** 1527).

Alternatively anodic attack on the platinum, if capable of proceeding at a pace comparable to the very slow exchange reaction, would also be capable of preventing the attainment of equilibrium at the theoretical value. Oxide-films on a metallic surface usually become leaky intermittently by the crack-heal effect, and the passage of cations outwards may be comparable to the exchange reaction. Early work by Hoar suggested that this may be an important factor in explaining the repeated failures to obtain the reversible value of the oxygen electrode on a platinum electrode surrounded by liquid saturated with oxygen (T. P. Hoar, *Proc roy. Soc.* (*A*) 1933, **142,** 628).

Doubtless anodic attack on the metal explains why the less noble metals, even though they carry invisible oxide-films, fail to reach the reversible oxygen potential when placed in water containing oxygen, but a potential not very different from that of oxide-free metal, although slightly ennobled; this slight ennoblement, the amount of which depends on the degree of perfection of the oxide-film, can be regarded as the source of differential aeration currents (p. 128).

It is clear that the value of the exchange current is all-important in deciding the chance of obtaining the reversible value in measuring any potential in a liquid containing a trace of impurity; it also decides whether a slight departure from the equilibrium potential will serve to furnish a large flow of current or whether a large departure is needed. Thus the abnormal metals, iron and nickel, require a much greater shift from equilibrium if appreciable current is to flow (whether for cathodic deposition or anodic dissolution) than the normal metals, zinc and copper. Table XLV, giving some approximate values for the Exchange Currents, deserves study; the exact values doubtless depend on surface

* J. N. Agar points out that if the reduction proceeds by way of H_2O_2 in which the O—O bond persists, dissociation into atoms is not needed; this is why the reduction by way of H_2O_2 proceeds with relative smoothness; he adds that H_2O_2 is not formed in appreciable quantity at the potential corresponding to the equilibrium

$$O_2 + 2H_2O + 4e \rightleftharpoons 4OH^-.$$

conditions and adsorption of impurities. The high value for the redox reaction

$$Fe^{+++} + e \rightleftharpoons Fe^{++}$$

should be noted. The numbers are quoted from G. M. Willis, "Corrosion: a symposium" (Melbourne University) 1955–56, p. 51, esp. p. 57.

TABLE XLV

SOME VALUES FOR EXCHANGE CURRENTS

			amps/cm.^{2}
Hydrogen electrode	$(2H^+ + 2e \rightleftharpoons H_2)$	(on smooth Pt) .	10^{-3}
		(on mercury) . .	5×10^{-13}
Oxygen electrode	$(O_2 + 2H^+ + 2e \rightleftharpoons H_2O)$	(on bright Pt, in acid)	10^{-11}
Copper electrode	$(Cu^{++} + 2e \rightleftharpoons Cu)$	(normal metal) .	2×10^{-5}
Zinc electrode	$(Zn^{++} + 2e \rightleftharpoons Zn)$	(normal metal) .	2×10^{-5}
Iron electrode	$(Fe^{++} + 2e \rightleftharpoons Fe)$	(abnormal metal)	10^{-8}
Nickel electrode	$(Ni^{++} + 2e \rightleftharpoons Ni)$	(abnormal metal)	2×10^{-9}
Redox system	$(Fe^{+++} + e \rightleftharpoons te^{++})$	(on platinum, in acid)	10^{-3} to 10^{-2}

Potential of a Metal immersed in a liquid initially free from its ions. If zinc or copper is immersed in, say, potassium chloride solution, containing oxygen, the potential recorded varies greatly with motion in the liquid; stirring usually makes the potential of zinc more positive and that of copper more negative (p. 130). This can easily be explained.

If, at the outset, the solution is free from zinc and copper ions the potential is determined by the principle expressed by fig. 156 (p. 869), and will be given by the ordinate of the intersection point of a cathodic curve representing the reduction of oxygen with an anodic curve representing $Zn \rightarrow Zn^{++} + 2e$ or $Cu \rightarrow Cu^+ + e$. The current will be given by the abscissa of that point, and will be considerable in the case of zinc, so that the rate of oxygen-consumption may be rapid. In the steady state, oxygen may become largely exhausted, unless the liquid is stirred rapidly, which will keep the oxygen-concentration near to the saturation value; thus under conditions of rapid stirring, the cathodic curve descends less steeply than under stagnant conditions, and the potential shown by the intersection point is higher. This explains the differential aeration current set up between two zinc electrodes placed in two compartments of a divided cell both containing the same liquid, one compartment being stirred and the other stagnant; the zinc in the compartment with stirring is the *cathode*.

With copper, the anodic curve will occur at a much higher level and the current represented by the abscissa of the intersection point will be smaller, so that even under nearly stagnant conditions there is no appreciable exhaustion of oxygen. However, copper ions may accumulate if the conditions are nearly stagnant, and at these potential levels a second cathodic reaction $Cu^+ + e \rightarrow Cu$ has to be taken into account. The total cathodic current flowing at any given potential is obtained by adding the currents corresponding to the reduction of oxygen and redeposition of copper, and the effective cathodic curve is obtained by plotting the sums of the currents against potential. Stirring will cause the Cu^+ concentration in the steady state to become lower and thus shifts the curve downwards, so that the potential represented by the intersection point will be made more negative. This explains the moto-electric current set up between two copper electrodes placed in two compartments of a divided cell both containing the same liquid, one compartment being stirred and the other stagnant; the copper in the compartment with stirring is the *anode*.

An interesting factor is introduced if the solution contains an ion forming a sparingly soluble salt with the liquid in question (e.g. SO_4^{--} in the case of lead).

If, for reasons of nucleation, the sparingly soluble compound is precipitated out of contact with the metal, the removal of metallic ions will keep the potential lower than would otherwise be the case, and the rate of corrosion may actually be increased. If the compound forms a protective film in physical contact with the metal, and covering the whole surface, the corrosion-rate may be greatly reduced. The presence of minor constituents in the water will influence the position and character of the precipitation, and cases are on record where the presence of SO_4^{--} in a water has increased the corrosion-rate of lead, and others where it has diminished that rate. The effect on potential does not seem to have been investigated in detail, but it would be expected that if a loose precipitate is formed, the reduction of Pb^{++} concentration by SO_4^{--} would shift the potential in the negative direction, whilst the formation of an adherent film containing discontinuities might shift it in the positive direction if that film possessed sufficient electronic conductivity to allow the cathodic reduction of oxygen to occur on its outer surfaces. These matters deserve experimental study.

Redox Potential Systems. If a platinum electrode is placed in a solution containing two bodies in different states of oxidation, such as ferric and ferrous chlorides, it will take up a potential determined by the relative concentrations of the reduced and oxidized components. The potential represents the equilibrium

$$Fe^{++} \rightleftharpoons Fe^{+++} + e$$

and it can be shown that it is determined approximately by the formula

$$\frac{RT}{nF} \log_e \frac{C_0}{C_R}$$

and accurately by $$\frac{RT}{nF} \log_e \frac{A_0}{A_R}$$

where C_0, C_R are the concentrations, and A_0, A_R the activities, of the oxidized and reduced members respectively. The potential of this and some other *redox* systems (including mixtures of ferrocyanides and ferricyanides where the complex anions represent the oxidized and reduced members) are plotted in fig. 201 against the ratio in which the two members are present. It will be noticed that when the concentrations are nearly equal the exact value of the ratio has little effect. A compound formed by union of hydroquinone and quinone in equivalent amounts provides a redox system of

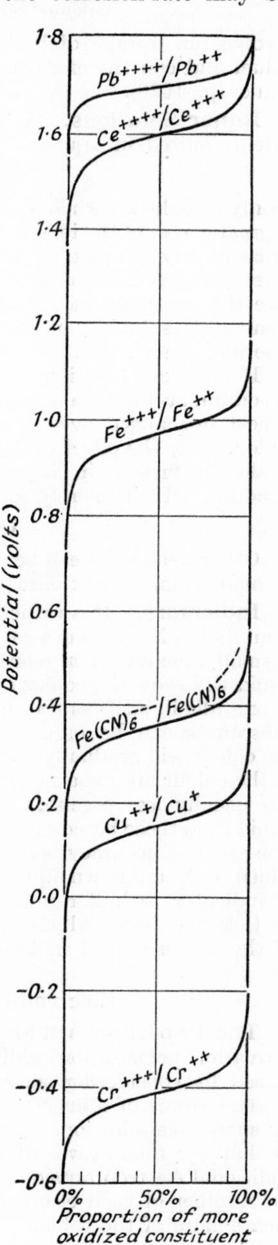

FIG. 201.—Some Redox Potentials as influenced by the concentration of the two constituents; the Pb^{++++}/Pb^{++} curve involving a two-electron transfer is less steep than the others, each of which involves only one electron.

considerable utility in electrochemical measurement; it is called *Quinhydrone* (p. 992). Since the equilibrium established when a platinum electrode is immersed in a liquid containing quinhydrone

$$\underset{\text{(Quinone)}}{C_6H_4O_2} + 2H^+ + 2e \rightleftharpoons \underset{\text{(Hydroquinone)}}{C_6H_4(OH)_2}$$

involves an uptake of hydrogen ions, the potential must depend on the pH value, and can therefore be used for estimating the pH value of an unknown liquid (p. 1028).

Buffer Solutions. A " weak " acid is one which is only to a very small extent ionized in aqueous solution, so that equilibrium of the type

$$H^+ + A^- \rightleftharpoons HA$$

is only established when the molecules of the undissociated acid, HA, are present in great excess of the ions H^+ and A^-. The sodium salts of such acids are usually ionized fairly completely into Na^+ and A^-, but the A^- thus produced is to some extent removed by joining with the small amount of H^+ present in water to give HA, so that aqueous solutions of sodium salts of weak acids react alkaline (conversely solutions of salts of weak bases with strong acids, like ammonium chloride, react acid).

If we take a solution containing, say, acetic acid, along with a large amount of sodium acetate, we obtain a *buffer mixture*—which will maintain its pH value almost unchanged even though small amounts of an acid or of an alkali are added. For the effect of adding acid (H^+) is merely to convert some A^- (which is already present in large quantity) into undissociated HA, whilst the effect of adding OH^- is merely to produce fresh A^- by the reaction

$$OH^- + HA = H_2O + A^-$$

Other weak acids can be used in preparing buffers. A buffer mixture can also be made from a weak base along with its salt.

Indicators. If a rather weak organic acid HX possesses a different colour from its ion X^-, it is clear that at low pH values (obtained by acidification with a small amount of a relatively strong acid, such as hydrochloric acid), the liquid will show the colour of HX, whilst at high pH values (obtained by adding strong alkali, such as sodium hydroxide) it will show the colour of X^-. Such substances can be used as " indicators ". Within a certain range of pH, the colour will gradually pass from that of HX to that of X^-, and the character of the colour affords a notion of the pH value. This provides a quicker, easier, but usually less precise, method of measuring the pH value of an unknown liquid than those based on a hydrogen electrode, a glass electrode or quinhydrone. The greatest accuracy is obtained if the colour produced by the indicator when added to the unknown liquid is compared with that produced by the same addition of indicator to buffer mixtures of accurately known pH value.

Coloured bases which produce ions possessing colours different from those of the undissociated molecules can also serve as indicators.

Electrodic Behaviour of Various Metals

The Two Classes of Metals. It has been stated that silver shows very little activation polarization whilst iron shows much. This difference may be discussed by considering their positions in the Periodic Table (p. 954).

The atoms of a succession of elements may be regarded as being built up by successive additions of protons (and usually neutrons also) to the nucleus, so that the atomic weight increases; each proton addition will necessitate an additional electron outside the nucleus, if the atom as a whole is to be neutral; the number of such electrons is indicated on by the atomic number shown in each square of the table.

The electrons are arranged in shells, and an atom in which all the shells are complete, so that there is no tendency to capture or lose electrons, displays complete stability; the inert gases of Group O are believed to owe their complete lack of chemical activity to the fact that they have a complete set of electrons in their outer shell.

In general, each new electron is added to an uncompleted shell and a fresh shell is not started until the existing one has been completed. In certain cases, however, a more stable arrangement is obtained if a new shell is started before an inner one is completed. This occurs in the part of the table enclosed by the thick line. The elements comprised therein are today called by Physicists and many Chemists the "Transition Elements" but, since Chemists at one time confined that name to elements of Groups VIIIA, IXA and Xa, it seems better to call them "Abnormal Elements"; they are often described as elements possessing an incomplete D-band. The elements outside the thick line will be called the "Normal Elements". Piontelli, who has performed a service by emphasizing the contrast between the electro-chemical behaviours of the two groups of metals, uses the word "inert" for the abnormal group. Since iron, which causes perhaps more corrosion trouble than any other metal, falls into this group, the terms "abnormal" and "normal" appear preferable.

The normal metals have large atomic volumes, comparatively low melting-points and comparatively low heats of sublimation; there is a big difference between the atomic and ionic radii. In contrast, the abnormal elements have close structures, the inter-atomic distance being small and the amplitude of thermal vibration small. It is more difficult to tear apart the atoms from one another in the case of a abnormal than in the case of a normal element, and this may be connected with the fact that the cohesion of the abnormal metal makes use of the electrons of the incomplete D-band—a view favoured by Pauling. If now it requires an abnormal amount of energy to separate the atoms of an abnormal metal during the evaporation of the metal, it would not be surprising to find a similar state of affairs if the atoms are to be separated by the passage of an atom (as a hydrated ion) into a solution; in other words, we may reasonably expect the activation energy of the anodic processes to be higher for abnormal than for normal metals. A strong field is needed to force the atoms (or ions) of an abnormal metal across the metal/liquid interface, so that both for anodic dissolution and cathodic dissolution a much larger shift of potential in one direction or the other is needed to produce a given current density, if the metal is an abnormal one than if it is a normal one.* Small ions like Cl^- which are strongly deformable in the electrostatic sense may help the movement in either the anodic or the cathodic direction. These views are largely due to R. Piontelli, *J. Chim phys.* 1948, **45**, 115, esp. p. 121, making use of some of the ideas of L. Pauling, *J. Amer. chem. Soc.* 1947, **69.** 542; *Phys. Rev.* 1938, **54,** 899.

If we consider a cell consisting of two electrodes of the same metal in a solution of its salt, and apply a constant current, the result may be different according as the metal belongs to the normal or abnormal class. The desired constancy of current can be obtained in different ways. The simplest arrangement is to use a rather high E.M.F. (say 100 volts), applied over a high resistance

* This refers to cases where an atom or ion has to cross the interface in one direction or the other—according as the change is anodic or cathodic. If the electrochemical reaction is a redox change such as between ferrous and ferric oxalate (complex) anions, involving only the transfer of an electron, the rate of reaction depends mainly on the cleanness of the surface. Thus at a dropping mercury electrode which provides an uncontaminated surface on each fresh drop, the exchange current may be higher than on a platinum electrode, although if the solution contains gelatine which will contaminate the surface of the drops as they form, the exchange current is greatly reduced (i.e. the activation energy increased). It is curious to find platinum acting as a less favourable surface than mercury, in view of the fact that the changes of the type $2H^+ + 2e \rightleftharpoons H_2$ occur so much more easily on platinum than on mercury. The exchange currents of redox reactions is studied by J. E. B. Randles and K. W. Somerton, *Trans. Faraday Soc.* 1952, **48,** 937.

(say 10,000 ohms, if 10 m.a. is required); since any alteration of resistance or potential distribution in the cell will be small compared to the external resistance or applied E.M.F., the current flowing must remain roughly constant—whatever the reactions involved. However, if there is only a restricted supply of atoms or ions in an energy state suitable to undergo the change

$$M = M^{n(+)} + ne$$

(and this may be the case if the metal is " abnormal "), then some alternative reaction must occur; such an alternative reaction usually demands a different potential, and a sharp jump of potential may sometimes be observed at the moment when the alternative reaction sets in.

Alternative Reactions at the Cathode. In the cell $Ag \mid AgNO_3 \mid Ag$ practically the whole current passing is devoted to deposition of silver at the cathode and its dissolution at the anode in accordance with Faraday's Law; indeed, in the instrument known as the " silver coulometer ", the increase of weight of the cathode is used as a quantitative indication of the total number of coulombs (ampere-seconds) which have passed. Moreover, since the electrochemical reaction occurs smoothly, the fresh silver will be deposited at the points which are crystallographically most suitable, such as the tips of the growing crystals; for less energy is needed to continue an unfinished crystal-face than to start a new one (p. 368). Even if the actual discharge-process occurs elsewhere, the new atom is likely to move along the surface to a " suitable " point, and the result is that deposition of silver from a nitrate solution results in a limited number of large crystals instead of a vast number of small ones. To obtain a smooth continuous coating of a normal metal (such as is desired in " plating "), it is necessary to add some substance which is easily adsorbed, so that, either by occupying the sites where deposition would otherwise take place most easily or by obstructing the movement of discharged atoms along the surface, the addition prevents the continued growth of existing crystals, and favours the initiation of new ones (p. 608).

With abnormal metals, the state of affairs is different. The potential moves a considerable distance from the equilibrium value, and energy is available for deposition of metal at any point on the surface, not merely on the tips of growing crystals. Moreover, unless the applied current is very low it cannot be entirely utilized in depositing metal; consequently an alternative reaction must occur. In general this must be the discharge of hydrogen ions, producing hydrogen atoms, which may sink into the cathode and perhaps render it appreciably brittle. Furthermore, even though the body of the bath may be appreciably acid, the consumption of hydrogen ions may raise the pH at the cathode surface, so that oxide or hydroxide is precipitated at the surface and thus incorporated in the deposited metal; the oxide or hydroxide may obstruct the increased growth of existing metal crystals and encourage the formation of fresh ones. Thus, without addition agents designed to improve the deposit, we may obtain a continuous plating covering the entire surface without serious discontinuities. In consequence the deposition of abnormal elements such as iron, nickel and chromium exhibits five characteristics; (1) high polarization, (2) currents efficiencies distinctly lower than 100%, (3) production of hydrogen on or in the metal, (4) co-deposition of oxide, hydroxide or basic salt, (5) continuous deposits rather than large isolated crystals, even in the absence of a special " addition agent ".

Alternative Reactions at the Anode. A silver anode will corrode at a rate not very different from that predicted by Faraday's Law, even when the current density is fairly high. However, an iron or nickel anode, unless the current density is very low, may fail to dissolve at the required rate, and the current must then be expended in some other way—such as the discharge of OH^- ions or other ions providing oxygen. This may first lead to the production of chemi-sorbed oxygen—that is oxygen attached to the metal by primary valencies, as opposed to physically adsorbed oxygen attached by Van der Waals forces; we can talk of a two-dimensional oxide-film, consisting of the layer of chemi-sorbed oxygen and the outer-most layer of metal atoms to which the oxygen is attached, but the two-dimensional oxide should not be identified with any oxide known in the massive, three-dimensional state. Only a limited amount of electricity can be used in producing chemi-sorbed oxygen, and thereafter the current must be expended either in building a three-dimensional oxide-film or in liberating oxygen as a gas. The first can thicken at a limited rate—which will become even slower as the oxide-film becomes more than a few atoms thick; thus the evolution of oxygen bubbles soon becomes inevitable, and the anodic potential must rise to the value which will suffice to provide energy for the liberation of oxygen. This explains the sudden potential-jump which accompanies the setting-in of passivity. Since the metal is now " plated " with oxide, the abnormally noble potential may remain after flow of current has been interrupted. Often the potential will ultimately collapse to a value close to that at which it started, but the collapse will usually take place less rapidly if the current has been flowing for a long time; for the prolonged passage of current will have enabled the three-dimensional film to thicken and will also produce a charge of oxygen upon the metal surface or dissolved in the liquid close to the surface, capable of repairing any discontinuities in the film which may spontaneously appear.

Thus it can be understood why the abnormal metals like iron, cobalt and nickel become passive more easily than the normal ones; the case of chromium, which has a soluble higher oxide CrO_3, is more complicated (p. 232).

Chemical Potentials

Analogy with Electrical Potential. It is commonly stated in elementary text-books that the electrical potential at a point is the energy needed to bring up unit quantity of electricity to that point from an infinite distance (or, better, from a standard position where the potential is arbitrarily taken to be zero). Such a definition is unsatisfactory, since in general the arrival of electricity in quantity far less than one " unit " would perceptibly alter the potential. It is better to consider the energy dW needed to bring up from the standard position an infinitesimal quantity of electricity dQ—too small to affect the potential—and define the electrical potential at the point in which we are interested as dW/dQ.

No chemical reaction can take place unless it involves a fall in free energy. Equilibrium represents a state in which an infinitesimal amount of reaction taking place in either direction would involve no change in free energy. It is convenient to assign to every substance a chemical potential, analogous to the electrical potential, and defined by the statement that if the free-energy increase attending the introduction of a very small additional quantity dQ of the substance is dW, the chemical potential is dW/dQ, expressed in suitable units.

Relation between Chemical Potential and Concentration, Activity or Fugacity. The chemical potential μ of a substance present in solution depends on the concentration, C (or more exactly on the activity, a); in the case of gas, it depends on the pressure, p (or more exactly, the fugacity, ϕ). At 25°C., where $RT = 592{\cdot}5$ calories per mole, the relationship in the four cases (expressed in calories per mole) is

$$\mu = \mu_0 + RT \log_e C = \mu_0 + 1363 \log_{10} C \quad \text{(approximate)}$$
$$\mu = \mu_0 + RT \log_e a = \mu_0 + 1363 \log_{10} a \quad \text{(exact)}$$
$$\mu = \mu_0 + RT \log_e p = \mu_0 + 1363 \log_{10} p \quad \text{(approximate)}$$
$$\mu = \mu_0 + RT \log_e \phi = \mu_0 + 1363 \log_{10} \phi \quad \text{(exact)}$$

where μ_0 is the value of μ under standard conditions (1 gram-molecule per litre for solutes, or 1 atmosphere for gases), and 1363 is 592·5 × 2·3. In a general reaction such as

$$aA + bB + cC + \ldots = a'A' + b'B' + c'C' + \ldots$$

which can be written

$$aA + bB + cC + \ldots - a'A' - b'B' - c'C' - \ldots = 0$$

the concentrations of the reactants A, B, C . . . and resultants A', B', C', . . . adjust themselves until, at equilibrium

$$\sum_y V_y \mu_y = 0$$

where μ_y represents the values of μ for the individual constituents A, B, C . . . A', B', C' . . . and V_y the number of molecules (a, b, c . . . a', b', c' . . .) taking part in the reaction. If we possess a table of the standard values μ^o for different substances, conditions of equilibrium can be calculated. For substances which commonly take part in corrosion reactions, such a table is conveniently provided as an appendix to M. Pourbaix's " Thermodynamics of Dilute Aqueous solutions ", translated by J. N. Agar, 1949 (Arnold); Pourbaix also indicates the manner in which chemical potential can be used in the calculation of equilibrium conditions. For more complete information, see *Cebelcor, Rapport Technique* No. **28** (1955) entitled " Enthalpies libres de formation standards, à 25°C."*

When the relation is an electrochemical one, involving transfer of electrons, we must introduce a term to cover this transfer, and Pourbaix, quoting Burgers, writes

$$-\sum_y V_y \mu_y + 23{,}060nE = 0$$

(Here μ_y is in calories and E in volts, requiring the conversion factor

$$\frac{96{,}540}{4{\cdot}18} = 23{,}060.)$$

As an example he takes the electrochemical reaction

$$MnO_4^- + 8H^+ + 5e = Mn^{++} + 4H_2O$$

which can be written

$$MnO_4 - Mn^{++} - 4H_2O + 8H^+ + 5e = 0$$

so that

$$-\mu_{MnO_4^-} + \mu_{Mn^{++}} + 4\mu_{H_2O} - 8\mu_{H^+} - 115{,}300E = 0$$

Under standard conditions

$$23{,}060nE^o = \sum_y V_y \mu_y^o$$

and since the values of μ^o can be obtained from the table, the value of the standard potential (referring to normal activities etc. and corresponding to the normal potential of such reactions as $M = M^{++} + 2e$) can be calculated.

The same principle is used to calculate equilibria involving no electron transfers, such as the solubilities of gas, the solubility products for solids standing in contact with a solution containing excess of one of its ions, or the ionization constants of weak acids or bases. Table XLVI shows the value of μ^o at 25°C., quoted from Pourbaix's writings, for 14 substances or ions which are important in corrosion.

* C.I.T.C.E. has adopted the term " Free Enthalpy ", the meaning of which is explained thus by M. Pourbaix (Priv. Comm., Aug. 1, 1957):—

" Réservant la notion d'*énergie libre* a l'énergie libre de Helmholtz, definie par la relation ($E - TS$) où E est l'énergie internelle, le C.I.T.C.E. appelle " *enthalpie libre* " la function ($H - TS$) où H est l'enthalpie. Cette enthalpie libre s'identifie avec l'énergie libre de Lewis et avec la function de Gibbs."

TABLE XLVI

VALUES OF μ_0 FOR VARIOUS SUBSTANCES IMPORTANT IN CORROSION
(M. Pourbaix)

		Calories
Solids	$Fe(OH)_2$	− 115,200
	$CaCO_3$ (calcite)	− 270,390
Liquid	H_2O	− 56,560
Ions in solutions	H^+ (aq.)	0
	OH^- (aq.)	− 37,455
	H_2CO_3 (aq.)	− 148,810
	HCO_3^- (aq.)	− 140,000
	CO_3^{--}(aq.)	− 125,760
	Fe^{++} (aq.)	− 20,310
	Ca^{++} (aq.)	− 133,600
Gases	H_2O	− 54,507
	H_2	0
	O_2	0
	CO_2	− 94,260

Any equilibrium constant, K, is defined by $\log_{10}K = -\frac{\Sigma V\mu^o}{1362}$, so that for the first and second ionization steps of carbonic acid

$$H_2CO_3 \rightleftharpoons HCO_3^- + H^+$$

and

$$HCO_3^- \rightleftharpoons CO_3^{--} + H^+$$

$$\log_{10} K_1 = \frac{\mu^o_{HCO_3^-} + \mu^o_{H^+} - \mu^o_{H_2CO_3}}{1363} = -\frac{-140{,}000 + 0 + 148{,}810}{1363} = -6{\cdot}46$$

$$\log_{10} K_2 = \frac{\mu^o_{CO_3^{--}} + \mu^o_{H^+} - \mu^o_{HCO_3^-}}{1363} = -\frac{-125{,}760 + 0 + 140{,}000}{1363} = -10{\cdot}42$$

Thus

$$\frac{[HCO_3^-][H^+]}{[H_2CO_3]} = 10^{-6{\cdot}46}$$

and

$$\frac{[CO_3^{--}][H^+]}{[HCO_3^-]} = 10^{-10{\cdot}42}$$

Similarly for the dissolution of ferrous hydroxide

$$\log_{10} K = \frac{\mu^o_{Fe^{++}} + 2\mu^o_{OH^-} - \mu^o_{Fe(OH)_2}}{1363} = -\frac{-20{,}310 - 74{,}910 + 115{,}200}{1363} = -14{\cdot}6$$

so that according to this calculation the solubility product $[Fe^{++}][OH^-]^2$ is $10^{14{\cdot}6}$ (cf., however, p. 438).

Movement of Material under a Chemical Potential Gradient. One of the places where the conception of chemical potential occurs in this book concerns the thickening of films (p. 821); material is stated to pass through the film either under an electrical potential gradient or under a chemical potential gradient, or under both. In elementary text-books "diffusion" is stated to take place under a concentration gradient; its rate of mass-transfer is generally described as being proportional to the concentration gradient dc/dy, where C is the concentration at distance y along the path representing the direction along which concentration varies most rapidly. In more advanced discussions the movement is associated to expressions such as $RT \log_e \frac{C_1}{C_2}$ or $RT \log_e \frac{a_1}{a_2}$ where C_1 and C_2 are the concentrations and a_1 and a_2 the activities at the two ends of the path. Clearly this expression for the driving mechanism in diffusion is closely analogous to the expression for the E.M.F. of a concentration cell $\frac{RT}{nF} \log_e \frac{C_1}{C_2}$ or $\frac{RT}{nF} \log_e \frac{a_1}{a_2}$, which represents the driving mechanism in ionic migration. But it is not immediately obvious how it is to be reconciled to the proportionality of the mass transfer rate to dc/dy. To help those who feel this difficulty, the connection between the two ways of regarding the subject will be indicated.

If we imagine a long cylinder with horizontal axis, containing particles of solute X dissolved in a solvent, and if the concentration of X is higher at the left end than at the right end, then, in absence of further information, considerations of probability predict that, across any vertical section, more particles of X will be passing from left to right than from right to left. A general left-to-right passage can occur spontaneously, that is with diminution of free energy, whereas a general right-to-left passage is only possible if energy is supplied. In other words the chemical potential is higher at the left than on the right, and the driving force tending to cause the X particles to move, on the whole, from left to right can be associated with the chemical potential gradient, $d\mu/dy$, just as the driving force for charged particles in a field where the electrical potential V varies is the electrical potential gradient dV/dy.

If we choose at random, say, 100 particles of X, more of them will be situated on the left than on the right, and consequently there will be more left-to-right than right-to-left motion. For each of the 100 particles randomly selected, the left-to-right movement will be $K\, d\mu/dy$ where K is a constant; but the *total* number of particles moving in *all* directions within a given volume will be proportional to the concentration and we can write for the effective left-to-right movement of material $K'C\, d\mu/dy$ where K' is a new constant.

Now, for ideal solutions obeying laws analogous to the gas laws

$$\mu = \mu^0 + RT \log_e C$$

which gives by differentiation

$$\frac{d\mu}{dC} = \frac{RT}{C}$$

Thus the effective rate of mass-transference (left-to-right) can be written

$$K'C\frac{d\mu}{dy} = K'C\frac{d\mu}{dC}\cdot\frac{dC}{dy} = K'RT\frac{dC}{dy}$$

It becomes clear, therefore, that the idea of the rate of mass-transfer being proportional to dC/dy is consistent with the idea of the driving mechanism being $RT \log_e \frac{C_1}{C_2}$; indeed they are two ways of stating the same thing.

Other References

An admirable elementary account of Electrochemistry, with special reference to corrosion reactions, is provided by E. C. Potter, " Electrochemistry: principles and applications " 1956 (Cleaver-Hume). Other books dealing with various parts of the subject include the early—but still useful—work of A. J. Allmand and H. J. T. Ellingham, " The Principles of Applied Electrochemistry " (Arnold); S. Glasstone, " Electrochemistry of Solutions " (Methuen). O. Gatty and E. C. R. Spooner, " Electrode Potential Behaviour of Corroding Metals in Aqueous Solutions " (Clarendon Press). Help will be derived from certain chapters of S. Glasstone, " Text-book of Physical Chemistry " (Macmillan) and S. Glasstone, K. J. Laidler and H. Eyring, " Theory of Rate Processes " (McGraw-Hill). The fundamental theory of electrodes and galvanic cells is authoritatively presented by E. Lange and P. van Rysselberghe, *J. electrochem. Soc.* 1958, **105**, 420; see also E. Lange, *Z. Elektrochem.* 1951, **55**, 76; 1952, **56**, 94; P. Rüetschi, *J. electrochem. Soc.* 1957, **104**, 176; cf. W. R. Harper, *Proc. roy. Soc.* (*A*) 1951, **205**, 83; 1953, **218**, 111; 1955, **231**, 388; *Phil. Mag.* (supplement) 1957, **6**, 365. The papers of R. Piontelli, quoted on p. 906, deserve careful study.

Information about electrode potentials, with related values for energy and enthalpy changes, will be found in the " International Critical Tables "; W. M. Latimer, " Oxidation state of the Elements and their Potentials in Aqueous Solution " (Prentice-Hall); G. N. Lewis and M. Randall, " Thermodynamics " (McGraw-Hill); also *Cebelcor, Rapport Technique* No. **28** (1955), " Enthalpies libres de formation standards à 25°C.".

ADDENDUM

INHIBITION AND PASSIVITY

Object. The purpose of this addendum is to correlate information given in Chapters V and VII regarding the passivity or inhibition produced respectively by simple immersion and by application of an external E.M.F., showing why certain passivating agents require an external E.M.F. whilst others do not. It also provides a simple example of the formation of an oxide-film from ions containing oxygen (such as NO_3^-, OH^-, SO_4^{--}, CrO_4^{--}), which some readers may find easier to follow than the cases considered in Chapters V and VII.

Iron in Copper Salt Solutions. Early work showed that pure iron, abraded and then exposed to air, behaved differently towards a copper salt solution according to the anion present. Copper chloride produced immediate deposition of copper everywhere. Copper sulphate produced deposition starting at isolated points, mostly situated at scratch-lines or near cut edges, and often spreading out over the surface within a second. Copper nitrate produced no visible change, provided that the solution was sufficiently concentrated and not too acid, and that the air-exposure had been sufficiently long; the air-exposure needed for passivity in copper nitrate was only a few minutes after fine abrasion but some hours after coarse abrasion; specimens rendered passive in copper nitrate solution have been kept in that solution for a year without visible change (U. R. Evans, *J. chem. Soc.* 1927, p. 1020, esp. p. 1030; 1929, p. 92, esp. p. 99; *Trans. Amer. Inst. Min. Met. Eng.* (*Inst. Met. Div.*) 1929, p. 7, esp. p. 10; *J. Iron Steel Inst.* 1940, **141,** 219P, esp. p. 227P).

The phenomena are easily explained by the mechanism suggested on pp. 140, 245. In copper chloride, around each discontinuity in the air-formed film, metallic copper will be deposited by the cathodic reaction, whilst at the discontinuity itself iron cations will enter the liquid by the anodic reaction. In *dilute* copper nitrate, the same change will occur. In a *concentrated* solution, however, a large number of iron cations would have to enter the liquid if the anodic attack was to keep up with the maximal possible rate of cathodic copper deposition, and these would have to penetrate the phalanx formed where the NO^-_3 ions have been driven up against the anodic surface by the electric-potential gradient, presumably oriented so that they are joined to the metal by (negative) oxygen ions.

Now a positive metallic ion, starting at (M), can easily reach (X), a point level with the negative oxygen ions, but before it can reach the body of the liquid, it must pass through a point such as (Y) where the energy is very high, owing to the positive groupings at this level; only an ion which starts from (M) with high energy can be expected to pass over the energy

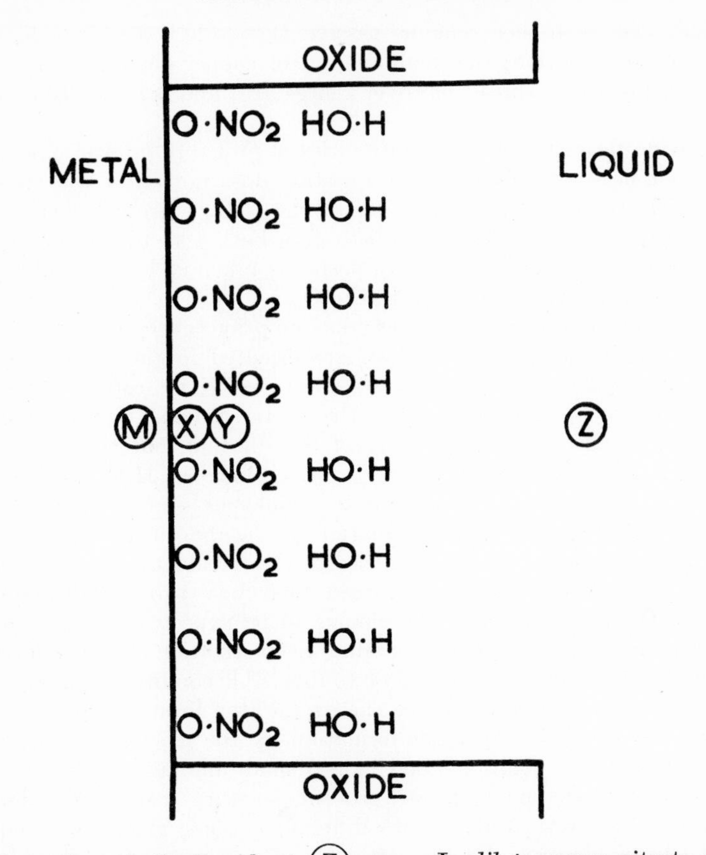

barrier and reach the liquid, at (Z), say. In *dilute* copper nitrate solution only a *slow* passage of cations into the liquid is needed to keep up with the cathodic copper deposition, and the necessary supply of high-energy cations will be forthcoming. At *high concentrations*, a *rapid* passage is needed, and the necessary number of high-energy cations will not be available, so that some will remain stuck at (X) and in analogous positions; this will liberate the appropriate number of cationic groups $(NO_2)^+$ to join the HO^- groupings of near-by water molecules, generating HNO_3; acidity will be produced in two ways (1) by the H^+ ions liberated from those water molecules which provided the HO^- groupings and (2) by ionization of the HNO_3. Meanwhile the metal cations, remaining at (X) along with the oxygen ions deserted by their $(NO_2{}^+)$ groups will form a layer of oxide. When once a layer of oxide has been formed, the passage of metallic cations into the solution will become more difficult still, and further layers of oxide will be deposited. The film will thus thicken by a mechanism similar to that discussed in Chapter II in connection with direct oxidation on air—until the potential gradient becomes too small to draw further ions through the oxide. The iron then remains passive, the oxide being invisible, as in the case of air-oxidation; the small amount of copper deposited during the passivation process is spread out over a large area and is normally invisible also.

The oxide formed at the discontinuities is probably γ-ferric oxide, but this has not been established. The question does not affect the principle involved; clearly the average number of the $(NO_2)^+$ groups liberated by each iron cation would have to be 1·0 to give FeO, 1·33 to give Fe_3O_4 and 1·5 to give Fe_2O_3, whilst the relative positions taken up by iron and oxygen ions would be different for γ- and α-Fe_2O_3.

The mechanism proposed involves no assumptions except that the anions driven up against the anodic surface are oriented in the position which electrical principles would lead us to expect. If NO_2^+ cations take part in the change as " proxies " for Fe^{++}, the electric transfer will be the same as if the Fe^{++} ions entered the body of the solution, but the mechanism involves no surmounting of an energy barrier, since the H_2O is already on the spot. It is, in fact, a mere switch of linkages similar to the proton switch generally accepted as an explanation for the abnormally high mobility of hydrogen and hydroxyl ions in electrolytic conduction—as compared with other ions which have to " thread their way through the crowd " (p. 1012). Oxide formation is the change to be expected at high current densities where only a reaction demanding low activation energy can occur, although at low current densities (close to reversible conditions) the reaction demanding the greater drop of free energy (i.e. the formation of a soluble salt) may be expected. No oxide-formation by the mechanism suggested can occur in chloride solutions, and indeed small amounts of Cl^- ions, by interrupting the NO_3^- phalanx, will prevent passivity. Moreover different anions containing oxygen along with different cationic groups will require different current densities for the setting in of passivity—as is the case.

Comparison with other cases of passivity. Passivation in copper nitrate shows a partial analogy to passivation in sodium hydroxide, where the metal cations move to position Ⓧ liberating the appropriate number of H^+ cations (portions of near-by OH^- anions), so that these can join other OH^- ions to form H_2O.

There is, however, an important difference between the two cases. In copper nitrate solution, a position in the liquid [Ⓩ, say] would represent the position of lowest energy, which will ultimately be reached—provided that the change is proceeding sufficiently slowly to render available high-energy ions able to surmount the energy barrier at Ⓨ. In dilute sodium hydroxide solution, when once it has become saturated with metallic hydroxide, Ⓧ will represent a situation where the energy is lower than Ⓩ, and cations will normally remain at Ⓧ. There is no need to demand a high anodic current density to obtain passivity, and immersion even in dilute alkali solution with only a small supply of oxygen to maintain the

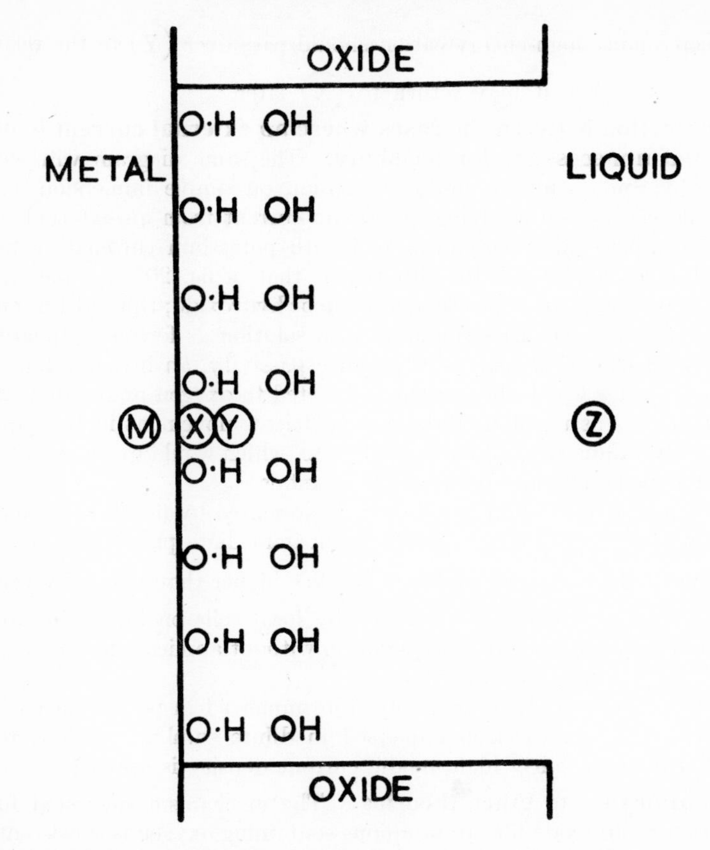

cathodic reaction on the film-covered portions will cause the iron to become passive by forming oxide at gaps.

Graphic representation. Fig. 202 serves to show the distinction between the two cases, potential energy being plotted against distance from the metallic surface. The situation is frankly over-simplified, since in practice there may be more than one energy hump and the fact that the electrical potential gradient may itself help the ions to cross the humps is not brought out; however, if it is borne in mind that certain factors are neglected, the diagram may assist understanding.

Fig. 202(A) represents the situation in copper nitrate, where low-energy cations stick at (X) and only high-energy cations pass over the energy-hump (Y) to reach the state of lower energy at (Z); thus low anodic current densities (as at gaps in the film in dilute copper nitrate) produce corrosion and high anodic current densities (as in concentrated copper nitrate) confer passivity. Fig. 202(B) represents the situation in sodium hydroxide, where both low-energy and high-energy cations will normally remain at (X); if an exceptionally high-energy cation should pass over (Y) to the relatively high position (Z), it may return to (X) later.

Distinction between the cases where an external current is necessary and unnecessary for passivity. The same diagram will serve to show why some solutions inhibit corrosion on simple immersion whereas others do so only on applying an anodic current from an external source. Let us compare dilute sulphuric acid with potassium chromate solution.

In the sulphuric acid, the situation is that of fig. 202(A), since ferrous ions can exist without deposition of oxide or hydroxide; thus, at low current density, the iron will pass smoothly into solution as ferrous sulphate, and only a high current density will produce passivity, an invisible film being produced—after which the current is devoted to oxygen production; in this argument it is assumed that there is sufficient movement in the liquid to avoid a crystalline crust of ferrous sulphate, which would cause an enhanced current density at gaps between the crystals.

In contrast, potassium chromate is represented by fig. 202(B), since the addition of Fe^{++} to CrO_4^{--} produces an immediate precipitate, as shown by Hoar (p. 152), suggesting that (Z) lies higher than (X). A very low current density—such as is produced by local cells on immersing an iron electrode in potassium chromate solution without application of an external E.M.F.—will suffice for passivity.

It becomes clear, therefore, why iron immersed in potassium chromate is unattacked, whereas iron immersed in dilute sulphuric acid is readily corroded unless a fairly high anodic current density is applied.

Relationship to other theories. The mechanism suggested for the formation of an oxide-film from anions containing oxygen is consistent with various views of inhibition and passivation expressed in Chapters V and VII,

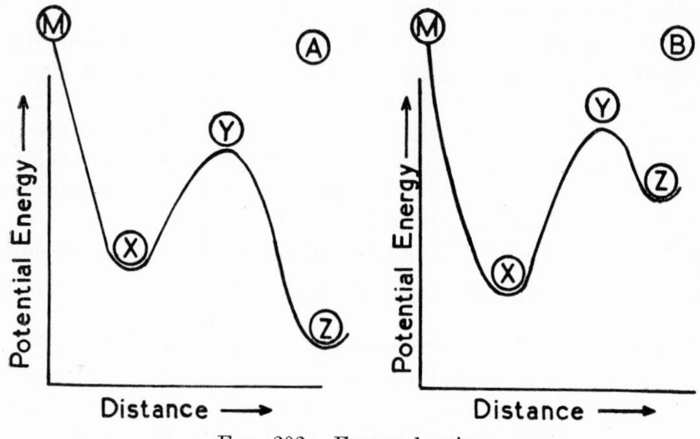

FIG. 202.—Energy barriers
(A) in cases where an external E.M.F. is needed for passivity
(B) in cases where simple immersion is sufficient.

and particularly with Hancock and Mayne's emphasis on the importance of conductivity (p. 147). A salt which increases conductivity increases the area available for cathodic oxygen reduction and thus increases the current flowing. When the anodic current density becomes too great for the supply of metal ions possessing sufficient energy to enter the liquid, an alternative reaction involving other cations or cationic groups must play its part, producing an oxide-film on the surface; the potential will, of course, rise to the level needed for this alternative reaction.

AUTHOR INDEX

[NOTE: Where an Author is quoted on numerous pages, it has seemed sufficient to indicate the Chapters in which his work can be studied]

SUBJECT INDEX